ENGINEERING CHEMISTRY

PAYAL B. JOSHI

Assistant Professor
Dept of Basic Science & Humanities, SVKM's NMIMS
Mukesh Patel School of Technology Management and Engineering, Mumbai

SHASHANK DEEP

Professor
Department of Chemistry
Indian Institute of Technology Delhi

OXFORD

UNIVERSITY PRESS

Preface

One of the most common questions on the minds of all first-year engineering students is 'Why do we need to study Chemistry to become an engineer? A convenient answer to this question is, 'Chemistry is everywhere: from the air we breath, to the food we eat.' Moreover, engineering is a profession that requires knowledge of materials, mathematics, and fundamental sciences.

Chemical engineers find direct correlation to chemistry as they are primarily involved in process design, mass–heat transfer reactions, optimizing chemical reactions in chemical industry, food engineering, etc. A closer relation of chemistry is found in civil and environmental engineering, where engineers work with various materials (cement, glass, concrete) and also study about environmental protection and pollution control. Environmental engineers need to understand the chemical reactions and mechanisms taking place in air, soil, and water. The year 2017 saw the introduction of artificial intelligence in chemistry, when IBM researchers developed an artificial neural network algorithm that could map the synthesis of molecules as well as predict bond energies and bond angles with utmost precision. Electronics engineers need to have knowledge of diodes, liquid crystals, semiconductors, etc. Thus, a deep understanding of chemicals, their properties and reaction mechanisms, will be an added advantage in today's competitive world.

ABOUT THE BOOK

This book *Engineering Chemistry* is primarily written for first-year engineering students keeping in mind the new AICTE curriculum. It will help them venture into the fascinating field of applications in chemistry for their chosen engineering field. It will also serve as a preliminary text students who have taken chemistry as a diploma course at the undergraduate level.

The contents of this book are such that students can gradually move from one topic to another to obtain comprehensive knowledge of the subject. The text is written in a simple language and supported with numerous examples, figures, and tables. Moreover, the rich pedagogy enables quick assessment. Students preparing for competitive examinations will also benefit from this book.

KEY FEATURES

- Provides comprehensive coverage of all important topics as per AICTE model curriculum and syllabi of various reputed universities
- Includes numerous self-explanatory figures, tables, and reactions that aid in the understanding of important topics
- Provides a large number of multiple-choice questions, review questions, and activity-based questions
- Includes simple as well as advanced solved numerical problems and check your progress questions interspersed in the text
- Includes summary and a list of key terms at the end of each chapter to enable recapitulation

CONTENTS AND COVERAGE

The contents of the book are arranged in 22 chapters and divided in to two parts: Part I: Basic Chemistry and Part II: Applied Chemistry.

Part I Basic Chemistry

Chapter 1, *Atomic and Molecular Structure*, elucidates in detail the structure of atom and includes de-Broglie equation, Schrödinger equation, particle-in-a-box model, atomic orbitals, molecular orbital theory, band theory of solids and Hückel's theory of aromaticity.

Chapter 2, *Periodic Properties and Chemical Bonding*, accounts for the periodic trends observed in Modern Periodic Table. The chapter includes a brief discussion on chemical bonding as well as a detailed account of various molecular interactions and hybridization with examples.

Chapter 3, *Thermodynamics and Chemical Equilibrium*, outlines the three simple laws known as laws of thermodynamics and their potential in explaining each and every process at equilibrium. The chapter also details how to make a spontaneous process non-spontaneous and vice versa and how to shift the position of equilibrium to the product/reactant side.

Chapter 4, *Phase Rule*, details Gibbs phase rule applied to one-, two- and multi-component systems. Iron–carbon phase diagram is discussed and illustrated with phase diagrams of all-component systems and congruent and incongruent systems.

Chapter 5, *Electrochemistry*, discusses cell potentials, EMF series, concentration cells, reference electrodes, pH determination using glass, hydrogen, and quinhydrone electrodes. Nernst equation is derived in a simple manner for easy understanding. A short account on potentiometric titrations and their various graphical representations are explained. Battery technologies ranging from acid-storage, alkali-storage to fuel cells are discussed.

Chapter 6, *Chemical Kinetics*, describes the details of rates of reactions, their dependence on concentration, temperature, and other factors. Potential energy surface and transition state theory are introduced to explain the molecular picture of the rate of reaction.

Chapter 7, *Surface Chemistry*, deals with mechanism of adsorption, catalysis, emulsions, colloids, detergents and surfactants. The chapter also introduces the concept of friccohesity of surfactants.

Chapter 8, *Solid State Chemistry*, explains the laws of crystallography, lattice planes, Miller indices, structure of different crystal structures, X-ray diffraction studies.

Chapter 9, *Coordination Chemistry and Organometallic Compounds*, discusses the nomenclature, Werner, valence bond and crystal field theory and stability of coordination compounds. The chapter also includes a discussion on HSAB principle, EAN rule, organometallic compounds and their use as catalyst in isomerization, polymerization, hydrogenation, and hydroformylation.

Chapter 10, *Organic Reactions and Synthesis of Drug Molecules*, discusses nucleophilic, addition, elimination, oxidation, reduction, and pericyclic reactions with examples. It also details the preparation, properties, and uses of drug molecules.

Chapter 11, *Stereochemistry*, deals with the representation of three-dimensional structures, concepts of chirality, isomers, and optical activity of organic compounds. It provides a comprehensive understanding of the relative and absolute configuration of organic molecules and conformational analysis of simple alkanes.

Chapter 12, *Instrumental Methods of Analysis*, explains a variety of techniques used to separate atoms/molecules, determine their structure, and characterize them qualitatively and quantitatively. These techniques include spectroscopy, microscopy, electrochemical and thermal analyses, and chromatography.

Part II Applied Chemistry

Chapter 13, *Water Chemistry*, details the sources of impurities in water, boiler problems, water softening methods, and desalination methods. It also discusses the significance of dissolved oxygen in water.

Chapter 14, *Corrosion*, explains the different forms of corrosion, their mechanisms and factors influencing them. The chapter also discusses various measures to control corrosion.

Chapter 15, *Metals and Alloys*, highlights the importance of metals and alloys and includes description of powder metallurgy, production of steel, metal ceramic powders and shape memory alloys.

Chapter 16, *Polymers*, explains the classification of polymers, methods of polymerization, preparation, properties and applications of commercially important polymers, compounding, plastic fabrication, and vulcanization. This chapter also includes the significance of specialty polymers in various applications.

Chapter 17, *Important Engineering Materials*, discusses the properties and applications of various types of materials such as cement, concrete, refractories, abrasives, adhesives, ceramics, glass, nanomaterials, liquid crystals, and composites. It also includes the manufacture of nanomaterials.

Chapter 18, *Lubricants*, discusses the types and properties of lubricants. The mechanism of lubrication is explained along with a note on selection of lubricants.

Chapter 19, *Energy Resources*, discusses the various renewable sources of energy, such as solar, tidal, wind, hydro, oceanic, biomass, nuclear, geothermal energy sources along with their advantages and limitations. It also includes a comprehensive account of the mechanism of nuclear fission and working of reactors.

Chapter 20, *Fuels and Combustion*, discusses different types of fuels, calorific values and their determination, coal analsysis, cracking of oils, and refining processes.. It also includes a short account of explosives and propellants.

Chapter 21, *Pollution and its Control*, discusses the causes and adverse effects of various types of pollution and their remedial measures.

Chapter 22, *Green Chemistry*, elucidates the basic principles of green chemistry along with examples. The chapter discusses the synthesis of adipic acid, indigo, ibuprofen, carbaryl, and acrylamide.

Appendix –*Laboratory Experiments*– includes the principle and procedure of a few laboratory experiments as prescribed by the AICTE syllabus.

Online Resources

The online resources centre provides resources for faculty and students using this text:

For Faculty
- Solutions Manual
- PowerPoint Slides

For Students
- Quizzes
- Extra Reading Material

ACKNOWLEDGEMENTS

I thank SVKMs NMIMS, Mukesh Patel School of Technology Management & Engineering (Mumbai Campus) for providing me all the necessary resources for this project. I would like to express deep gratitude to my parents, Dr Bhaskar Joshi and Smt. Shefali Joshi, for their constant encouragement to write. Their discussions over coffee enlightened me towards this fascinating journey of writing.

I thank my students for their constructive feedback about the course which made me realize so many ways of active learning. I would like to make a particular mention of my dear student S. Sarvanna Kumaran for his efforts in making the periodic table of elements.

I express my heartfelt gratitude to Oxford University Press for giving me the opportunity to publish this book through their esteemed publishing house. I am particularly thankful to the members of the editorial team for their guidance and cooperation in the successful publication of this book within the scheduled time.

I express my gratitude to Dr Subrata Sen Gupta, Former Reader in Chemistry, RPM College, West Bengal, for permitting the use of material on stereochemistry, and Prof. D.K. Bhattacharya, SSP Laboratory, New Delhi and Prof. Poonam Tandon, Associate Professor, Maharaja Agrasen Institute of Technology, New Delhi, for permitting the use of material on solid state chemistry.

Suggestions and feedback are welcome and can be sent to me at: payal.joshi@nmims.edu or payalchem@gmail.com

Payal B. Joshi

I have the opportunity to teach thermodynamics, kinetics, and spectroscopy for undergraduate and graduate level students at Indian Institute of Technology Delhi. At the outset, I would like to thank my students for their amazing questions and feedback that helped me to develop my own style of teaching and encouraged me to contribute to this book. I thank Oxford University Press for giving me the opportunity to contribute on these three topics.

I am grateful to my colleagues, especially Prof. Pramit K. Chowdhury, Prof. N.G. Ramesh, and Prof. Hemant K. Kashyap for their valuable time in going through the chapters. Their suggestions helped me in improving the chapters.

I would like to thank my Ph.D. scholar Dr Komal Singh Khatri for helping me in the writing of a few topics and preparing the questions. I would also like to thank Shilpa Sharma and Shubham Goyal for going through the manuscript and helping me in the correction of typos.

I would like to express my sincere thanks to my wife, Anshu Bhushan, for being extremely patient and for being there by my side with her prayers, unconditional love, and perpetual support. I am grateful to my parents who are the constant source of encouragement, provided me guidance throughout my life and kept faith in me. Last but not least, my two amazing children, Priyank and Priyal, keep me smiling even during tough times and helped me in my writing.

Shashank Deep

The Publisher and the authors would like to thank the following reviewers for their valuable feedback.
Pradip Kr Dutta, Motilal Nehru National Institute of Technology, Allahabad
Purvi B. Shukla, L.D. College of Engineering, Ahmedabad,
Rabi Narayan Panda, Dept of Chemistry, Birla Institute of Technology and Science, Pilani, Goa Campus
R. K. Mohapatra, Govt. College of Engineering, Keonjhar
Monideepa Chakrabortty, Assam Engineering College, Guwahati
P. Nagaraj, Anna University, Chennai
Ravindra Vadde, Kakatiya University, Warangal
M. V. Satyanarayana, P.V.P. Siddhartha Institute of Technology, Vijayawada
Jyotsna Kaushal, Chitkara University, Punjab
V. Shanmukha Kumar, KL University, Guntur
J. Chandra Rao, Sri Vasavi Engineering College, Pedatadepalli, AP
P. Sreenivasa Rao, Sri Vasavi Engineering College, Pedatadepalli, AP
Rekha Rani Dutta, The Assam Kaziranga University, Jorhat, Assam
Tapan Kumar Bastia, School of Applied Sciences, KIIT Bhubaneswar
T.V. Rajendran, SRM Institute of Science & Technology, Chennai
Ashish Kumar, Lovely Professional University, Jalandhar

Contents

Atomic and Molecular Structure

LEARNING OBJECTIVES

After reading this chapter, you will be able to:

- explain wave–particle duality of matter.
- deduce de Broglie relation and Schrödinger wave equation.
- understand Heisenberg's Uncertainty Principle and Born interpretation of Schrödinger wave function.
- apply particle-in-a-box model in conjugated molecules and nanoparticles.
- sketch the atomic orbitals and radial plots of hydrogen atom.
- discuss molecular orbital theory for diatomic (homonuclear and heteronuclear) molecules.
- illustrate band theory of metals, semiconductors, and insulators.
- introduce the concept of aromaticity in benzene and cyclobutadiene.

1.1 STRUCTURE OF ATOM — AN OVERVIEW

Atoms and molecules are the fundamental building blocks of matter. We all have learnt about atoms in the beginning of secondary school science classes; despite this, our understanding of the structure of atom is surprisingly low. In the 19th century, scientists were facing a major challenge to reveal the structure of atoms and explain their behaviour and properties. This led to a series of postulates and experiments validating them. The earliest investigations revealed that atoms are not indivisible. Various experiments have proved that the atom consists of charged particles. An atom is composed of protons and electrons, mutually balancing their charges. Protons are in the interior of an atom surrounded by electrons. J. Dalton, J.J. Thompson, E. Rutherford, and Niels Bohr successfully postulated atomic models and described the properties of the atom. (Fig. 1.1).

Postulates of Bohr's model

(a) There is a small, positively charged nucleus surrounded by electrons that travel in circular orbits around the nucleus.
(b) There is a presence of electrostatic forces between the electrons and the nucleus.
(c) Electrons move in circular orbits of fixed sizes called stationary orbits (or energy levels) K, L, M, and N and the energy of electrons is quantized.
(d) Atoms emit radiation: electrons jump from one orbit (allowed) to another and either absorb or emit light as electromagnetic radiation with a frequency as per Planck's relation, $\Delta E = E_2 - E_1 = h\nu$, where h is Planck's constant.

The limitations of Bohr's model are:

(a) The assumption of structured 'stationary fixed orbit' seems unjustified.
(b) It can only explain spectral lines of hydrogen atom, but after the first 20 elements in the periodic table, Bohr's model becomes difficult to predict the spectral details of complex atoms.

(c) It cannot explain chemical bonding of atoms.

(d) There is no explanation of the distribution of electrons within an atom.

Sommerfeld attempted to improvize Bohr's theory by postulating that electrons revolved around the nucleus in elliptical orbits and also introduced additional quantum numbers.

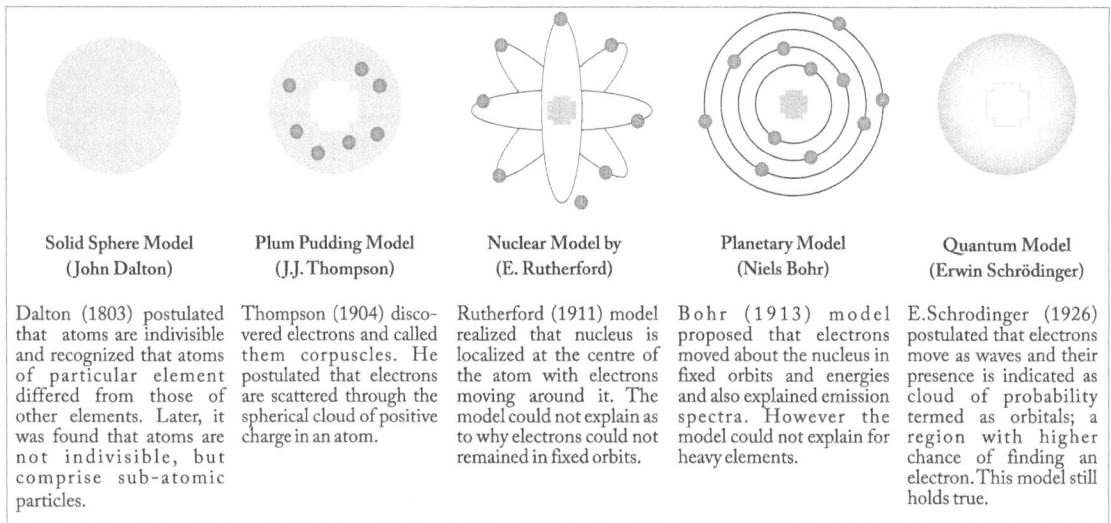

Solid Sphere Model (John Dalton)	Plum Pudding Model (J.J. Thompson)	Nuclear Model by (E. Rutherford)	Planetary Model (Niels Bohr)	Quantum Model (Erwin Schrödinger)
Dalton (1803) postulated that atoms are indivisible and recognized that atoms of particular element differed from those of other elements. Later, it was found that atoms are not indivisible, but comprise sub-atomic particles.	Thompson (1904) discovered electrons and called them corpuscles. He postulated that electrons are scattered through the spherical cloud of positive charge in an atom.	Rutherford (1911) model realized that nucleus is localized at the centre of the atom with electrons moving around it. The model could not explain as to why electrons could not remained in fixed orbits.	Bohr (1913) model proposed that electrons moved about the nucleus in fixed orbits and energies and also explained emission spectra. However the model could not explain for heavy elements.	E.Schrodinger (1926) postulated that electrons move as waves and their presence is indicated as cloud of probability termed as orbitals; a region with higher chance of finding an electron. This model still holds true.

Fig. 1.1 Various models describing the structure of atom

Further, in Bohr's theory, an assumption was made that the position and momentum of an electron were precisely known. A highly advanced theory, called 'wave mechanics' put forth by Erwin Schrödinger explained the spectra of one-electron system and even multi-electron systems. It also gave a detailed interpretation of chemical bond vibrations and other chemical phenomena.

1.2 DUAL NATURE OF MATTER (WAVE–PARTICLE DUALISM)

Bohr's theory was a giant step forward in understanding the atomic world; yet its limitations had to be broken down with the aid of quantum mechanics, which emerged very soon in the form of the dual nature concept. In 1905, Einstein put forth the photoelectric effect that described light as a photon. Scientists

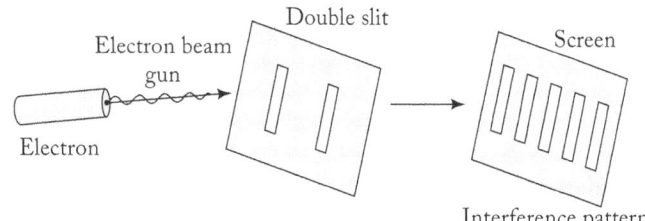

Fig. 1.2 Young's double slit experiment

were yet debating the dual nature of light and also reluctant to accept it. Einstein further introduced the concept of light as a continuous field of waves in his paper on special relativity—a complete contradiction of light considered as a stream of particles. Experimental evidence was given by Thomas Young's double-slit experiment.

As per this experiment (Fig.1.2), light travels away from a source as an electromagnetic wave. When it passes through the slits, it gets divided into two wavefronts. These wavefronts overlap and fall on to the screen and the entire wave field disappears and a photon appears.

1.2.1 Davisson and Germer Experimental Evidence of Electron Waves

The presence of matter waves was experimentally verified by C.J. Davisson and L.H. Germer at the Bell Telephone Laboratories. They showed that the beam of electrons reflected from a metal crystal produced a diffraction pattern. The wavelengths of electrons calculated from the experiments were found to be in agreement with de Broglie equation. G.P. Thompson demonstrated that an accelerated beam of electrons when passed through a thin gold film (~ 10^{-8} m) strikes on to a photographic plate, a diffraction pattern is obtained.

Figure 1.3 shows the experimental arrangement used by Davisson and Germer. It consisted of an electron gun comprising a tungsten filament (F), coated with barium oxide and heated with a low-voltage power supply. The electrons emitted by the tungsten filament were

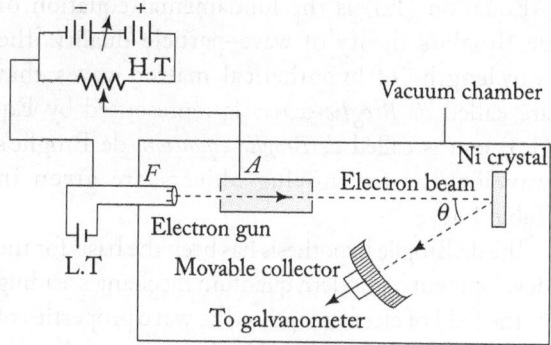

Fig. 1.3 Davisson–Germer electron diffraction arrangement

accelerated to a desired velocity by applying accurate voltage. The experiment was performed by varying the accelerating voltage from 44 V to 68 V. The entire apparatus (Fig. 1.3) was placed in an evacuated chamber. Electron beams were passed through a cylinder with fine holes along its axis that produced a fine collimated beam striking on to a nickel crystal. These electrons got scattered in all directions by atoms present in the solid crystal. The intensity of the scattered electron beam in a given direction was measured by a movable electron detector and galvanometer.

The deflection of the galvanometer was found to be proportional to the intensity of the electron beam entering the collector. By moving the detector on the circular scale to different positions, the intensity of the scattered electron beam was measured for different values of angle of scattering θ, that is, the angle between the incident and the scattered electron beams. The variation of intensity (I) of scattered electrons with an angle of scattering θ was obtained for different accelerating voltages.

Davisson–Germer experiment, thus, strikingly confirms the wave nature of electrons and the de Broglie relation. In 1989, the wave nature of a beam of electrons was experimentally demonstrated in a double-slit experiment, similar to that used for the wave nature of light. Moreover, in 1994, interference fringes were obtained with beams of iodine molecules that are around million times more massive than the electrons.

1.2.2 The de Broglie Equation and Derivation

In 1924, Louis de Broglie described the existence of matter waves. Already at that time, electromagnetic and sound waves were known. de Broglie suggested that wave–particle nature may exist even in material particles and electrons. He also derived an equation for the wavelength of photons (or particles) of light.

According to Planck's quantum theory, energy of a photon is given by $E = h\nu$ (h is Planck's constant, 6.626×10^{-34} Js, E = energy and ν = frequency of light, s^{-1}) and Einstein's equation for mass–energy equivalence is $E = mc^2$, where c is the velocity of light.

$$\therefore \quad h\nu = mc^2 \tag{1.1}$$

where m is the mass equivalent of photon. Further, it follows that

$$\frac{hc}{\lambda} = mc^2 \qquad \left(\because \nu = \frac{c}{\lambda} \right)$$

$$\lambda = \frac{h}{mc} \tag{1.2}$$

The product mc is the momentum of photon. de Broglie assumed that an equation of this type is also applicable to material particles. If a particle of mass, say m, travels with a velocity v, then

$$\lambda = \frac{h}{mv} \tag{1.3}$$

Equation (1.3) is the fundamental equation of de Broglie's theory of wave–particle duality. The wavelength of hypothetical matter waves that are called *de Broglie waves* is represented by Eq. (1.3) and is called *de Broglie equation*. de Broglie's wavelengths for moving objects are given in Table 1.1.

The de Broglie hypothesis has been the basis for the development of modern quantum mechanics leading to the field of electron optics. The wave properties of electrons have been utilized in the design of electron microscope used today.

Table 1.1 de Broglie wavelengths for moving objects

Object (moving)	Mass (g)	Wavelength (Å)
1 volt electron	9.11×10^{-28}	12.3
100 volt electron	9.11×10^{-28}	1.23
Helium atom (298 K)	6.65×10^{-24}	0.73
α-particle from radium	6.65×10^{-24}	6.6×10^{-5}
Dust particle	$\approx 10^{-6}$	6.6×10^{-13}
Driven golf ball	45	4.9×10^{-24}
Chemistry professor (walking)!	8×10^{4}	8.3×10^{-26}

1.2.3 Bohr's Theory Versus de Broglie Equation

Bohr (1913) postulated the atomic model in which nucleus of an atom is surrounded by particles known as electrons that revolve in defined shells or orbits. As per Bohr's planetary model, angular momentum is an integral multiple of $h/2\pi$. de Broglie gave a valid explanation supporting Bohr's model shown in Fig. 1.4.

de Broglie put forth that if one uses the wavelength associated with an electron and assume that an integral number of wavelengths must fit in the circumference of an orbit, one can deduce the same quantized orbital angular momentum postulated by Bohr's planetary model.

Let us say, an electron behaves as a standing wave that goes around the nucleus in a circular orbit. If one condition that the circumference of electron orbit should be equal to the integral number of wavelength of an electron (de Broglie wavelength, λ) is fulfilled, the electron will undergo constructive interference. If this condition is not satisfied, the electron may suffer destructive interference. As per this argument, if r is the radius of the circular orbit, then $2\pi r = n\lambda$.

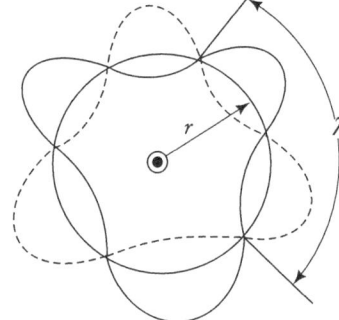

Fig. 1.4 Bohr's model of atom

We know that de Broglie equation, $\lambda = \dfrac{h}{mv}$.

$$\therefore\ 2\pi r = \frac{nh}{mv} \text{ or, } mvr = \frac{nh}{2\pi}$$

where $n = 1, 2, 3,$ and so on.

As mvr is the angular momentum of an electron, one can easily deduce that wave mechanical nature leads to Bohr's postulate, that is, angular momentum is an integral multiple of $h/2\pi$ and is quantized. Hence, it is clear that de Broglie concept supports Bohr's planetary model.

1.3 HEISENBERG'S UNCERTAINTY PRINCIPLE

In 1927, Werner Heisenberg put forth the Principle of Uncertainty, according to which, 'the simultaneous exact determination of position and momentum or any property related to momentum such as velocity is impossible'. If Δx is the uncertainty regarding position and Δp is the uncertainty about the momentum, then

$$\Delta x \times \Delta p = h; \text{ where, } h \text{ is Planck's constant.} \tag{1.4}$$

According to Uncertainty Principle, if the position of a particle such as an electron is known precisely, then there will be uncertainty about its momentum. If an electron with an exact known momentum strikes a fluorescent screen, a flash of light is emitted so that its position at that instant is known. However, continuous collisions of electron with the screen results in the loss of certain amount of energy and eventually the momentum of the electron will change. In an attempt to establish the precise position of the electron, an uncertainty is introduced regarding its momentum. Thus, the statements about the precise position and momentum will have no validity and shall be replaced by statements of probability that the electron has a given momentum and position. Heisenberg's Uncertainty Principle brings out the fact that nature only imposes a limit to accuracy with which the position and momentum of a particle are determinable experimentally and mathematically, stated by the equation as,

$$\Delta x \times \Delta p \geq \frac{h}{4\pi} \tag{1.5}$$

Hence, it can be concluded that the product of uncertainties cannot be less than $\dfrac{h}{4\pi}$.

1.4 Schrödinger Wave Equation

Erwin Schrödinger (1924) proposed and deduced the wave equation that forms the basis of the wave-mechanical behaviour of matter. It describes the particle motion and also its association to de Broglie wave. Schrödinger derived an equation for comparing the path taken by the particle with that of a ray of light and associated the wave with electromagnetic waves.

Let us consider the following equation,

$$y = f(x)\, g(t) \tag{1.6}$$

where, $f(x)$ is a function of coordinate x and $g(t)$ is a function of the time coordinate t.

For a stationary wave,

$$g(t) = A \sin(2\pi vt) \tag{1.7}$$

On substituting Eq. (1.7) in Eq. (1.6), we get,

$$y = f(x)\, A\, sin\,(2\pi vt) \tag{1.8}$$

$$\frac{\partial^2 y}{\partial t^2} = \text{f}(x) 4\pi^2 v^2\, A \sin(2\pi vt) = -4\pi^2 v^2 f(x) g(t) \tag{1.9}$$

Further, the one-dimensional classical wave equation is given as,

$$\frac{\partial^2 y}{\partial x^2} = \frac{1}{u^2} \frac{\partial^2 y}{\partial t^2}$$

Similarly, it follows Eq. (1.8) as,

$$\frac{\partial^2 y}{\partial x^2} = \frac{\partial^2 f(x)}{\partial x^2} g(t) \tag{1.10}$$

$$\frac{\partial^2 f(x)}{\partial x^2} = \frac{-4\pi^2 v^2}{u^2} f(x) \tag{1.11}$$

We know, velocity u can be expressed as, $u = v\lambda$

$$\frac{\partial^2 f(x)}{\partial x^2} = \frac{-4\pi^2}{\lambda^2} f(x) \tag{1.12}$$

Check Your Progress

1. How is Heisenberg principle different from Bohr's postulates about electrons?
2. What important information is obtained from Davisson–Germer experiment?
3. Justify the statement, 'de-Broglie relation supports Bohr's model of stationary orbit.'
4. State Heisenberg Uncertainty principle. Write its expression.

Equation (1.12) for wave motion in three directions represented by the co-ordinates x, y, and z is given as,

$$\frac{\delta^2 \Psi}{\delta x^2} + \frac{\delta^2 \Psi}{\delta y^2} + \frac{\delta^2 \Psi}{\delta z^2} = -\frac{4\pi^2}{\lambda^2}\Psi \tag{1.13}$$

where, Ψ is the amplitude function of the three co-ordinates. For simplicity, ∇ is written for x, y, and z co-ordinates. Equation (1.13) can then be written as follows,

$$\nabla^2 \Psi = -\frac{4\pi^2}{\lambda^2}\Psi \tag{1.14}$$

where, ∇ is Laplacian or differential operator given by,

$$\nabla^2 = \frac{\delta^2}{\delta x^2} + \frac{\delta^2}{\delta y^2} + \frac{\delta^2}{\delta z^2}$$

The fundamental assumption of wave mechanics is that Eq. (1.14) is applicable to all microscopic particles such as electrons, protons, and atoms. On substitution of λ in Eq. (1.14), de Broglie equation can be written as,

$$\nabla^2 \Psi = -\frac{4\pi^2 m^2 v^2}{h^2}\Psi \tag{1.15}$$

The kinetic energy of a particle is equal to $mv^2/2$ and this is equal to the difference between total energy E and potential energy U.

Hence, $E - U = \dfrac{mv^2}{2}$ and substitution of $mv^2/2$ in Eq. (1.15) gives,

$$\nabla^2 \Psi = \frac{8\pi^2 m}{h^2}(E - U)\Psi$$

or, $$\frac{\delta^2 \Psi}{\delta x^2} + \frac{\delta^2 \Psi}{\delta y^2} + \frac{\delta^2 \Psi}{\delta z^2} + \frac{8\pi^2 m}{h^2}(E - U)\Psi = 0 \tag{1.16}$$

Equation (1.16) is called *Schrödinger wave equation* (time-independent). As per this equation, if a particle of mass m moving with a velocity v has total energy E and potential energy U, then the particle has an associated wave, whose amplitude is wave function Ψ. It is a second degree differential equation with several solutions, of which only some are valid. The functions are satisfactory solutions of wave equation only for certain values of energy E and such values are called *eigen values*. The corresponding functions that are satisfactory solutions of Eq. (1.16) are called *eigen functions*. Eigen functions will be single value, finite, and continuous for all possible values of the three co-ordinates, that is, x, y, and z, including infinity (∞).

On further solving Eq. (1.16) and introducing $\hbar = \dfrac{h}{2\pi}$ we get,

$$-\frac{\hbar^2}{2m}\left(\frac{\partial^2}{\partial x^2} + \frac{\partial^2}{\partial y^2} + \frac{\partial^2}{\partial z^2}\right) + V = E\psi \tag{1.17}$$

We know that Laplacian operator is, $\nabla^2 = \dfrac{\delta^2}{\delta x^2} + \dfrac{\delta^2}{\delta y^2} + \dfrac{\delta^2}{\delta z^2}$

$\therefore$ $$\left[-\frac{\hbar^2}{2m}\nabla^2 + V\right]\psi = E\psi$$

Further, the Hamiltonian operator can be written as,

$$\widehat{H} = \left[-\frac{\hbar^2}{2m}\nabla^2 + V\right] \tag{1.18}$$

On comparing Eqs (1.17) and (1.18) we get,

$$\widehat{H}\,\psi = E\psi \tag{1.19}$$

Equation (1.19) is a compact form of Schrödinger equation.

1.4.1 Physical Significance of Wave Function

The function ψ is a mathematical function and is associated with moving particles and is not an observable quantity with any physical meaning. However, ψ^2 has significance and can be evaluated. Max Born (1926) proposed the statistical interpretation of wave function of electrons, called *Born interpretation*. As per Born interpretation, the electron is considered as a particle, and the square of the wave function ψ at any point in space represents the probability of finding an electron at that point at a given instant. In simpler terms, if ψ is large, the probability of finding an electron is also high. Born interpretation is in agreement with the Uncertainty Principle. The function ψ^2 is considered as a wave mechanical equivalent of the electron orbit of Bohr's theory and hence, the wave function is referred to as an orbital. An *orbital* represents a definite region in three-dimensional space around the nucleus where there is high probability of finding an electron of a definite energy.

1.4.2 Quantum Mechanical Model of Hydrogen Atom

For the hydrogen atom, Schrödinger wave equation is written as follows,

$$\nabla^2\Psi + \frac{8\pi^2 m}{h^2}\left(E - \frac{Ze^2}{r}\right)\Psi = 0 \tag{1.20}$$

where, U (potential energy) is replaced by $-\dfrac{Ze^2}{r}$.

The solution of Schrödinger equation is a complicated one. It is sufficient to know that solution of the wave equation for an electron in a hydrogen atom involves certain integers that determine the energy and momentum of an electron. These integers correspond to quantum numbers of Bohr–Sommerfield theory. On solving the wave equation, energy E of an electron is,

$$E = -\frac{2\pi^2 z^2 m e^4}{n^2 h^2} \tag{1.21}$$

Equation(1.21) is identical to the Bohr equation. The calculations of the values of wave functions corresponding to different values of quantum numbers have given probability distributions of an electron. These probability distributions have maxima and minima that signify that electron orbits have no significance.

For hydrogen atom, the maximum probability of finding an electron in the ground state is at a distance of 0.529 Å from the nucleus (Fig. 1.5 (a)). This is in accordance with Bohr's theory as the distance is similar to the radius of the first orbit. Figure 1.5 (b) shows the probability of finding an electron called electron cloud (see shaded portion). The density of electron cloud is proportional to the probability of finding an electron at that point in a given instant.

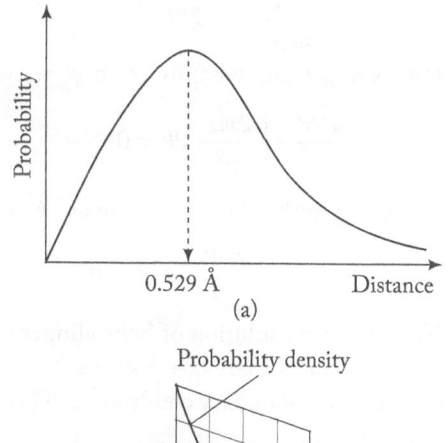

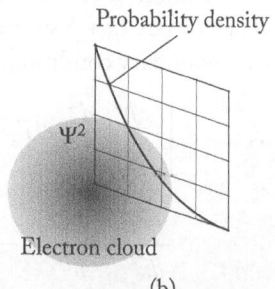

Fig. 1.5 Probability distribution of electron in hydrogen atom

1.4.3 Particle in a One-dimensional Box

Let us consider a particle of mass m which is allowed to freely move in a one-dimensional box of length l as shown in Fig. 1.6. The particle can only move parallel to the x-axis without friction, that is, interval of $x = 0$ to $x = l$. This interval is called *one-dimesional box* or *potential well*.

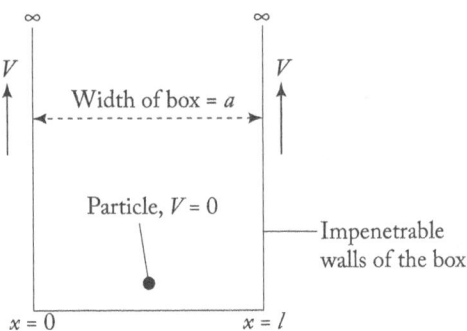

The potential energy, V of an electron at the bottom of the box is constant and taken as zero. Hence, inside the box $V = 0$. Let the width of the box be a. Also, potential energy V becomes infinity at the walls of the box. So, let the potential energy be infinite for $x < 0$ and $x > l$.

Fig. 1.6 Particle in a 1D box model

The assumptions for particle-in-a-box are as follows:

(a) It is assumed that the walls of the box possess infinite potential energy ensuring that the particle has zero probability of being at the walls or outside the box, called the *boundary conditions*.

(b) Further, the function is considered zero at $x = 0$ and for all negative values of x, as the particle is not allowed over the walls of the box.

(c) The function must necessarily be zero for all values of $x > l$. The boundary condition is hence set in such a way that the particle is strictly confined inside the box and cannot exist outside.

Inside the box, Schrödinger equation is,

$$\left[-\frac{\hbar^2}{2m}\frac{d^2}{dx^2}+V_x\right]\Psi(x)= E\Psi(x) \tag{1.22}$$

As $V_x = 0$, Eq. (1.22) becomes,

$$-\frac{\hbar^2}{2m}\frac{d^2}{dx^2} = E\Psi(x) \tag{1.23}$$

For solving E and wave function ψ_x we will mathematically rewrite Eq. (1.23) as,

$$\frac{d^2\Psi}{dx^2}+\left(\frac{2mE}{\hbar^2}\right)\Psi = 0$$

On reducing the above equation as $k^2 = \dfrac{2mE}{\hbar^2}$, we get,

$$\frac{d^2\Psi}{dx^2}+k^2\Psi = 0 \tag{1.24}$$

Now, a general solution of Schrodinger wave equation is,

$$\Psi(x) = a\cos kx + b\sin kx \tag{1.25}$$

Considering boundary conditions, $\Psi(x) = 0$ at $x = 0$ or $\Psi(0) = 0$. Outside the box, $V_x = \infty$.

$$\therefore \qquad \left[-\frac{\hbar^2}{2m}\frac{d^2}{dx^2}+V_x\right]\Psi_x = E\Psi(x) \tag{1.26}$$

Further,

$$\frac{d^2\Psi}{dx^2}+\frac{2m}{\hbar^2}(E-\infty)\Psi = 0 \tag{1.27}$$

When $\Psi = 0$ (outside the box), the particle cannot be found outside the box. Hence, $\Psi = 0$ is considered at the walls of the box and thus $x = 0$ and $x = l$. Figure 1.7 shows the wave functions of a one-dimensional particle in a box.

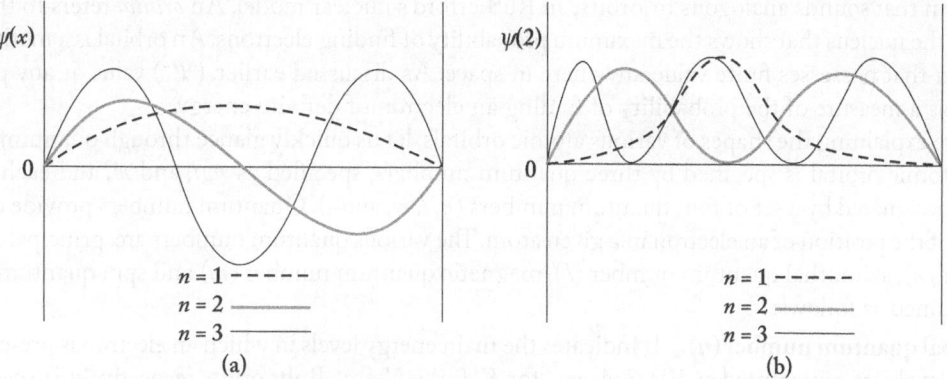

Fig. 1.7 Wave functions for one-dimensional particle in a box

From Eq. (1.25), we can say that if $a = 0$,
$$\Psi(x) = a \sin kx + b \cos kx \tag{1.28}$$
As we know $\Psi = 0$ at $x = 0$ and $x = l$.

Hence, one can solve the equations as follows: $b \sin kl = 0$

On rearranging, $\sin kl = 0$ and also, $kl = n\pi$ and $k = \dfrac{n\pi}{l}$

we can consider $k = \dfrac{n\pi}{l}$, where, $n = 0, 1, 2, 3$ $\tag{1.29}$

Hence, $\Psi = \Psi_n = b \sin\left(\dfrac{n\pi}{a}\right)$, where, $n = 0, 1, 2, 3$ $\tag{1.30}$

The term Ψ_n is called the *eigen function*. On considering Eqs (1.29) and (1.30) we can express,
$$k^2 = \frac{2mE}{\hbar^2}$$

Hence, $\dfrac{n^2\pi^2}{l^2} = \dfrac{2mE}{\hbar^2}$

$$E = \frac{n^2\pi^2\hbar^2}{l^2 2m} \tag{1.31}$$

On solving the above expressions and we know that

Total energy of an electron, E – Potential energy (U) = Kinetic energy of the electron

$$\therefore \quad E_n = \frac{n^2 h^2}{8ml^2}, \text{ where, } n = 0, 1, 2, 3 \dots \infty \tag{1.32}$$

Equation (1.32) clearly depicts that the particle in a box consists of discrete sets of energy values (energy is quantized). Some of the energy levels, say E_1, E_2 and E_3 can be written as follows:

$$E_1 = \frac{h^2}{8ml^2}; \quad E_2 = \frac{4h^2}{8ml^2}; \quad \text{and} \quad E_3 = \frac{9h^2}{8ml^2} \tag{1.33}$$

Hence, it is proven by the above equation that a bound particle possesses quantized energy, whereas a free particle has no quantized energy.

1.5 SHAPES OF ATOMIC ORBITALS AND PROBABILITY DISTRIBUTION

The solution of wave function (Ψ) of hydrogen atom led to three different types of quantum numbers that explain the spatial orientation of an electron relative to the nucleus. These solutions are called orbitals —

— a term that sounds analogous to 'orbits,' in Rutherford's nuclear model. An *orbital* refers to the region around the nucleus that shows the maximum probability of finding electrons. An orbital is a mathematical function that possesses finite value anywhere in space. As discussed earlier, (Ψ^2) value at any place and instant is a measure of the probability of finding an electron of definite energy.

Before explaining the shapes of various atomic orbitals, let us quickly glance through quantum number. Each atomic orbital is specified by three quantum numbers, specified as n, l, and m, and each electron can be designated by a set of four quantum numbers (n, l, m, and s). Quantum numbers provide complete details of the position of an electron in a given atom. The various quantum numbers are: principal quantum number (n), azimuthal quantum number (l), magnetic quantum number (m) and spin quantum number (s) explained as follows:

Principal quantum number (n)   It indicates the main energy levels in which an electron is present. These energy levels are represented as 1, 2, 3, 4, etc., for K, L, M, N, etc., Bohr orbits respectively. It specifies the energy of an electron in the given level and can be given by, $En = -1312/n^2$ kJ/mol. Hence, it is clear that energy of an electron is inversely proportional to square of the principal quantum number (i.e., energy of an electron increases with increasing (n). Further, maximum number of electrons that can be added in an energy level is $2n^2$.

Azimuthal quantum number (l)   It is also known as angular quantum number and was proposed by Sommerfield. It signifies the number of subshells to which the electron belongs and also the shape of the subshells. Further, it can express energies of all subshells, that is, s < p < d < f and value of l is always (n – 1). The values of l depend directly on n value, and for a given value of n, l can assume values as follows,

Value of l	0	1	2	3
Subshell	s	p	d	f
Shapes	Spherical	Dumbbell	Double dumbbell	Complex

> The origin of these letters designated for subshells is from the language used to describe the lines seen in earlier studies of atomic spectra: *s* was 'sharp,' *p* was 'principal,' *d* was 'diffuse,' and *f* was 'fundamental.' After *f*, an alphabetical order follows for designating subshells.

Magnetic quantum number (m)   It was proposed by Zeeman and denotes the number of permitted orientation of various subshells and also signifies the behaviour of electrons in a magnetic field. The values of m can vary from negative to positive through zero and can be calculated from l as per formula, m = +1, ..., 0, ..., 1. Hence, if l = 0, m = 0, if l = 1, m will be + 1, 0, 1 and if l = 2, m will be +2, +1, 0, 1, 2, and so on.

Spin quantum number (s)   Quantum mechanics necessitates a fourth quantum number so as to uniquely designate an electron and is termed as spin quantum number. The spin quantum number was proposed by S. Goudsmit and G. Uhlenbeck. A spin quantum number can have only two values, +1/2 and –1/2. Pauli Exclusion Principle (1945) clearly expresses these theoretical restrictions and states that 'only two electrons can be accommodated by a given atomic orbital.' Further, the two electrons assigned to a specific atomic orbital must be of opposite spin quantum number, that is, their spins must be paired. This led to the development of the electronic configuration of atoms in the periodic table.

Electronic Configuration of Atoms

An electronic configuration is defined as the distribution of electrons among the orbitals and subshells. Electrons are assigned to a specific atomic orbital one at a time so as to fill the orbitals of one energy level, before proceeding to the next higher energy level. The electrons in an atom fill the principal energy levels in order of increasing energy (the electrons get farther from the nucleus) and the order of levels filled can be depicted as:

1s, 2s, 2p, 3s, 3p, 4s, 3d, 4p, 5s, 4d, 5p, 6s, 4f, 5d, 6p, 7s, 5f, 6d, 7p

We apply **Aufbau principle** to fill up the energy levels according to which a maximum of two electrons are put into orbitals in the order of their increasing orbital energy. Further, we also consider **Hund's rule** which states that when electrons go into degenerate orbitals (i.e., orbitals of same energy), they occupy them singly before pairing begins.

An electron is commonly depicted by an upward ($\uparrow$) and downward ($\downarrow$) arrow thereby showing the two possible spin states. The distribution of electrons in various quantum levels can be depicted as shown in Table 1. 2.

Table 1.2 Electron distribution in orbitals

n	l	m	Atomic orbital	Orbitals in subshell
1	0	0	1s	1
2	0	0	2s	1
2	1	1, 0, +1	2p	3
3	0	0	3s	1
3	1	1, 0, +1	3p	3
3	2	1, 0, +1, +2	3d	5
4	0	0	4s	1
4	1	1, 0, +1	4p	3
4	2	1, 0, +1, +2	4d	5
4	3	1, 0, +1, +2, +3	4f	7

1.5.1 Forms of Hydrogen Atom and Wave Functions

The solution of wave function of a hydrogen atom with its electron in the lowest quantum energy level (principal quantum number = 1) depicts a spherical region as shown in Fig. 1.8 (a) of electron probability called 1s atomic orbital. The 1s atomic orbital has more than 95 per cent probability of finding an electron within a distance of 1.7 Å (170 pm) of the nucleus. The solution of wave equation for an electron in the next higher energy level

with principal quantum number 2, depicts two spherical regions of high electron probability called 2s atomic orbital. In the 2s orbital, one electron is nearer to the nucleus, similar to 1s atomic orbital, whereas the other electron is farther away from the nucleus. Similarly, the solution of wave equation depicting three spherical regions of high electron probability is called the 3s atomic orbital.

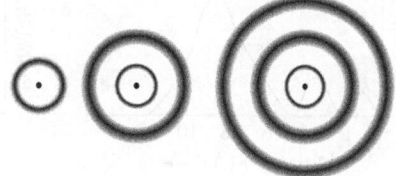

Fig. 1.8 (a) 1s, 2s, and 3s atomic orbitals

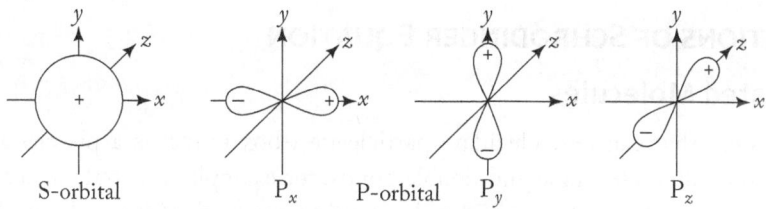

S-orbital   P$_x$ P-orbital   P$_y$ P$_z$

Fig. 1.8 (b) Shapes of s and p atomic orbitals

The solution of wave equation for the second quantum energy level of hydrogen atom described three additional atomic orbitals. These orbitals are known to be symmetrical in shape about three mutually perpendicular axes with higher electron probability in regions called *lobes* present on either side of the nucleus, as shown in Fig. 1.8 (b). One should bear in mind that Ψ^2 = 0 at the nucleus clearly represents that an electron cannot be present within the nucleus at any instant. When three orbitals of equal energy (but slightly higher than 2s) are oriented at right angles (90°) to each other, they are called p levels

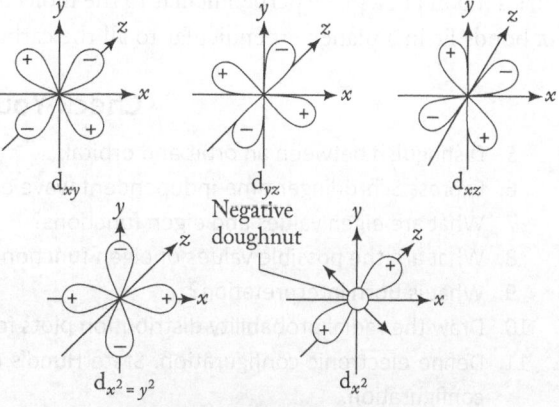

Fig. 1.9 Shapes of d orbitals

(see Fig. 1.8(b)). They are designated as $2p_x$, $2p_y$, and $2p_z$, at the orbitals with *x*, *y*, and *z* representing the Cartesian co-ordinates in three-dimensional space. The p orbital can accommodate six electrons and have dumb-bell shape along the three axes.

The five d orbitals that can accommodate 10 electrons as shown in Fig. 1.9 are designated as five orbitals namely $d_{(xy)}, d_{(yz)}, d_{(xz)}, d_{x^2-y^2}$ and d_{z^2}. As shown in Fig. 1.10, the probable distances of an electron are given by radial probability distribution plots. Hence, a plot of electron probability against *r* (distance of electron from the nucleus) for hydrogen atom is given for 1s, 2s, 2p, 3p, and 3d orbitals in this figure.

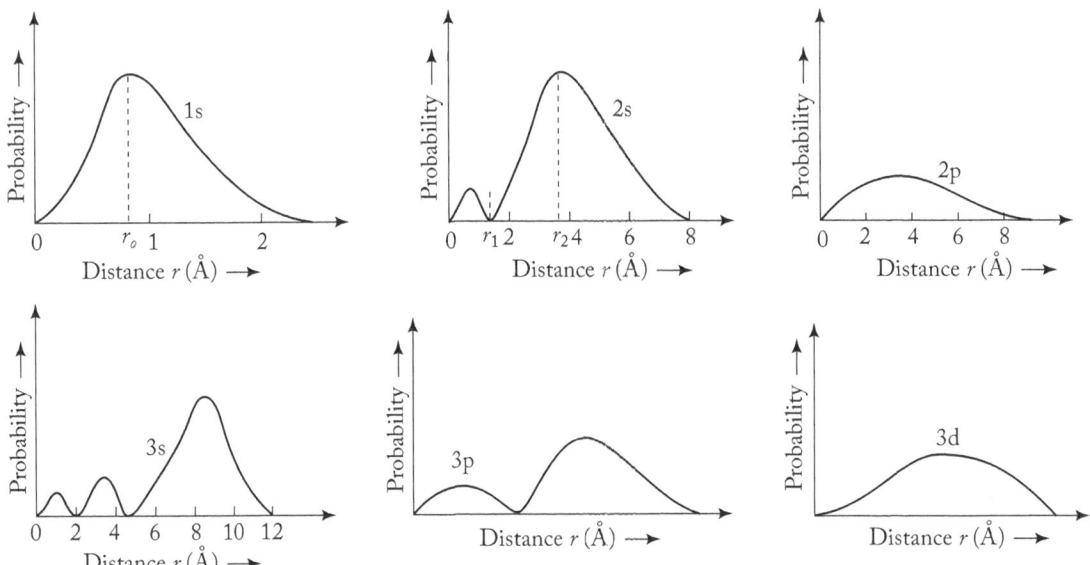

Fig. 1.10 Radial probability distribution plots for hydrogen atom

1.6 APPLICATIONS OF SCHRÖDINGER EQUATION

1.6.1 Conjugated Molecules

The chemical system that can best elucidate particle-in-a-box model is a pi- electron moving in a conjugated system of alternate single and double bonds; for example, 1, 3-butadiene. For simplicity, π bonding excluding sigma (σ) bonds is considered as a rigid framework of the molecule. Ethene molecule has a π bond in a plane perpendicular to the molecular plane, whereas in 1, 3-butadiene, the σ bonds and π bonds lie in a plane perpendicular to all the carbon and hydrogen atoms (see Fig 1.11).

Check Your Progress

5. Distinguish between an orbit and orbital.
6. Express Schrödinger time-independent wave equation.
7. What are eigen values and eigen functions?
8. What are the possible values of eigen function?
9. What is Born interpretation?
10. Draw the radial probability distribution plots for hydrogen atom.
11. Define electronic configuration. State Hund's rules and Aufbau principle for writing the electronic configuration.

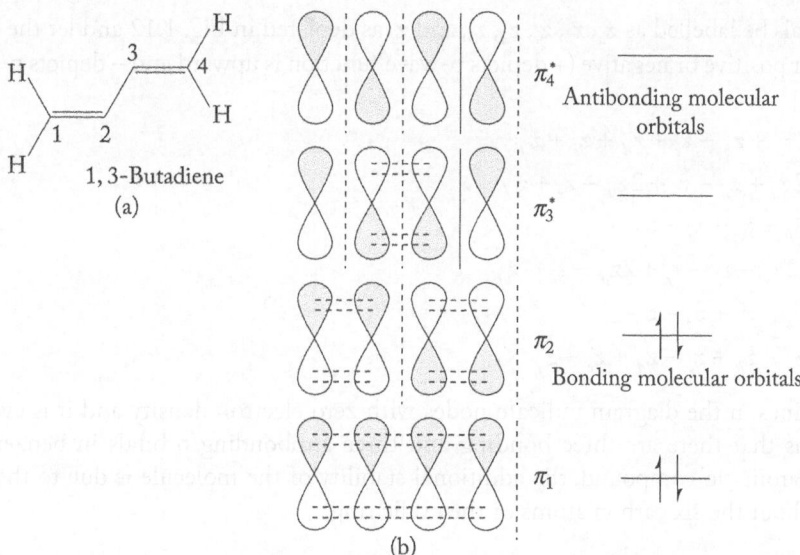

Fig. 1.11 (a) Structure of 1,3-butadiene (b) pi-molecular orbitals of 1,3-butadiene (shaded portion represents + sign on the lobes)

The four π electrons move freely over the four-carbon atom framework of single bonds. One can neglect the zig-zag C – C bonds and assume a one-dimensional box. We will overlook that π electrons have a node in the molecular plane. Since the electron wave function extends beyond the terminal carbons, one can add approximately one-half bond length at each end. This will give a bond of length equal to the number of carbon atoms times the C – C bond length. Thus, for butadiene the length will be 4×1.40 Å (1 Å $= 10^{-10}$ m). In the lowest energy state of butadiene, four delocalized electrons will fill the two lowest molecular orbitals and the total π-electron density is given (as shown in Fig. 1.11) by, $\rho = 2\Psi_1^2 + 2\Psi_2^2$. Further, equations for the four π orbitals can be written as follows.

$$\pi_1 = 0.37\ \Psi_1 + 0.60\ \Psi_2 + 0.60\ \Psi_3 + 0.37\ \Psi_4$$
$$\pi_2 = 0.60\ \Psi_1 + 0.37\ \Psi_2 + 0.37\ \Psi_3 + 0.60\ \Psi_4$$
$$\pi_3^* = 0.60\ \Psi_1 + 0.37\ \Psi_2 + 0.37\ \Psi_3 + 0.60\ \Psi_4$$
$$\pi_4^* = 0.37\ \Psi_1 + 0.60\ \Psi_2 + 0.60\ \Psi_3 + 0.60\ \Psi_4$$

The π-electron density is concentrated between carbon atoms 1 and 2, and between 3 and 4; the predominant structure of butadiene has double bonds between C_1, C_2 and C_3, C_4. Each double bond consists of a π bond, in addition to the underlying σ bond. Overall, butadiene can be described as a resonance hybrid with the contributing structures: major $\underset{1}{CH_2} = \underset{2}{CH} — \underset{3}{CH} = \underset{4}{CH_2}$ and,

minor $\circ CH_2—CH=CH—CH_2\circ$.

In the similar manner, one can understand benzene; a cyclic ring structure with six electrons each of which is present on carbon atoms in π orbitals perpendicular to the molecular plane. Benzene has six p orbitals and hence it has 6π orbitals, named a to f, as depicted in Fig. 1.12.

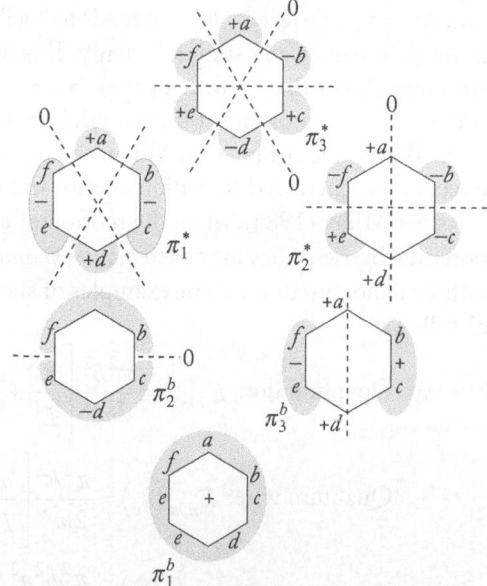

Fig. 1.12 pi molecular orbitals of benzene

Let the orbitals be labelled as z_a, z_b, z_c, z_d, z_e, and z_f as depicted in Fig. 1.12 and let the sign for each z orbital be either positive or negative (+ depicts p-wave function is upward and − depicts p- wave function is downward).

$$\pi_1^b = z_a + z_b + z_c + z_d + z_e + z_f$$
$$\pi_2^b = 2z_a + z_b - z_c + 2z_d - z_e + z_f$$
$$\pi_3^b = z_b + z_c - z_e - z_f$$
$$\pi_1^* = 2z_a - z_b - z_c + 2z_d - z_e - z_f$$
$$\pi_2^* = z_b - z_c + z_e - z_f$$
$$\pi_3^* = z_a - z_b + z_c - z_d + z_e - z_f$$

The dashed lines in the diagram indicate nodes with zero electron density and it is evident from the above equations that there are three bonding and three antibonding orbitals in benzene. Benzene is considered an 'aromatic' compound, the additional stability of the molecule is due to the presence of π orbitals throughout the six carbon atoms of the cyclic ring.

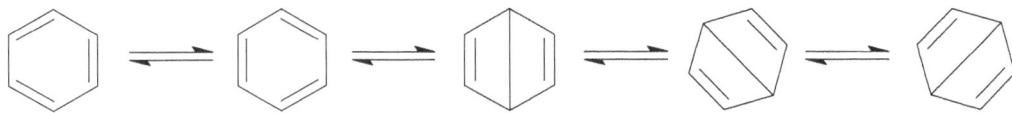

Fig. 1.13 Canonical structures of benzene

The stabilization energy of benzene is about 36 kcal and can be represented as resonance structures as shown in Fig. 1.13. Benzene represents a combination of all the above structures with the first two contributing largely.

1.6.2 Quantum Confinement in Nanoparticles

One of the major outcomes of size reduction of bulk materials to nanoscale levels is quantum confinement. Quantum confinement effect is a popular term in the nano world where the particle size ranges from 1–25 nm. At nanoscale levels, electron tends to 'feel' the presence of particle boundaries and respond to changes in particle size by adjusting its energy. This leads to discrete energy levels depending on the size of the structure. According to Yoffe (1993), Bohr radius of a particle can be written as, $a_B = \varepsilon m / m^* \, a_o$; where ε is the dielectric constant of the material, m^* is the particle mass, m is the rest mass of an electron, and a_o is the Bohr radius of H atom. When the particle size approaches Bohr radius, the quantum confinement effect causes increased transition energy and blue shift in the absorption spectra.

As per Miller (1984), when the motion of electrons and holes is confined in one or more directions by potential barriers, they are called *quantum confined structures*. Quantum well, quantum wire, and quantum dots or nanocrystals are some examples of such structures and their Schrödinger equation can be written as follows:

$$\text{Quantum dot: } E_{n,m,l} = \frac{\pi^2 \hbar^2}{2m^*}\left[\frac{n^2}{L_z^2} + \frac{m^2}{L_y^2} + \frac{l^2}{L_x^2}\right], \Psi = \phi(z)\phi(y)\phi(x)$$

$$\text{Quantum wire: } E_{n,m}(k_x) = \frac{\pi^2 \hbar^2}{2m^*}\left[\frac{n^2}{L_z^2} + \frac{m^2}{L_y^2}\right] + \frac{\hbar^2 k_x^2}{2m^*}, \Psi = \phi(z)\phi(y)exp(ik_x x)$$

$$\text{Quantum well: } E_n(k_x k_y) = \frac{\pi^2 \hbar^2 n^2}{2m^* L_z^2} + \frac{\hbar^2}{2m}\left(k_x^2 + k_y^2\right), \Psi = \phi(z)exp(ik_x x + ik_y y)$$

where n, m, l = 1, 2, ... quantum confinement numbers, L_x, L_y and L_z, are the confining dimensions, $exp(ik_x x + ik_y y)$ is called *wave function* that describes the electronic motion in x and y directions, same as electron wave functions.

1.7 MOLECULAR ORBITAL (MO) THEORY

F. Hund and R.S. Mulliken (1932) postulated the molecular orbital theory and its salient features are as follows:

(a) The electrons in a molecule are present in various molecular orbitals just like the electrons of atoms are present in various atomic orbitals.

(b) The atomic orbitals of similar energies and symmetry combine to form molecular orbitals.

(c) In an atomic orbital, an electron is influenced by one nucleus; it is not so in the case of molecular orbitals. As many atoms combine to form a molecule, electrons in a molecular orbital are under the influence of two or more nuclei depending on the number of combining atoms. Hence, an atomic orbital is monocentric, whereas a molecular orbital is polycentric in nature.

(d) When two atomic orbitals combine, two molecular orbitals are formed, namely bonding molecular orbital and antibonding molecular orbital.

(e) The bonding molecular orbital possess lower energy with greater stability than the corresponding antibonding molecular orbital. Electrons fill up the molecular orbitals following Pauli's, Aufbau, and Hund's rules just like atomic orbitals.

1.7.1 Molecular Orbitals in Homonuclear Diatomic Molecules

Generally, there are two types of diatomic molecules: homonuclear and heteronuclear molecules. If a molecule consists of two or more atoms belonging to the same element, they are called *homonuclear diatomic molecules*, for example, H_2, He_2, Li_2. The molecular orbital (MO) theory explains the formation of homonuclear diatomic molecules.

As per molecular orbital theory, when two atoms combine to form a molecule, the two nuclei are positioned at an equilibrium distance and their atomic orbitals lose their identity to form molecular orbitals. The electrons are added to these molecular orbitals which are quite similar to atomic orbitals. The s, p, d, f orbitals in atoms are determined by various sets of quantum numbers, whereas in molecules, there are σ, π, δ molecular orbitals determined by quantum numbers. In a molecule, an electron can move in a field of more than one nucleus, hence molecular orbitals are polycentric in nature and follows Aufbau principle, Pauli Exclusion principle, and Hund's rules. Just like an atomic orbital, a molecular orbital contains a maximum of two electrons with opposite spin.

An approximate quantum mechanical picture of electrons in a chemical bond can be derived by combining hydrogen-like wave functions, namely Ψ_1 and Ψ_2 for two atoms. The new wave function is called the *linear combination of atomic orbitals* (also called LCAO method). As per LCAO, molecular orbitals are formed by combination of the atomic orbitals of the combining atoms. Similar to ripples formed at the water surface, the electronic wave function can interact in a constructive or destructive manner to form molecular orbitals. If there is a constructive combination of atomic orbitals, an increase in electron probability occurs between the nuclei of approaching atoms leading to the formation of energetically favourable bonding molecular orbitals denoted as Ψ_B. In destructive combination of atomic orbitals, there is a zero-electron probability between the nuclei of approaching atoms leading to the formation of energetically unfavourable antibonding molecular orbital denoted as Ψ_A. The bonding molecular orbitals are formed by adding wave functions of electrons in the two atomic orbitals, whereas antibonding molecular orbitals are formed by subtracting their wave functions as,

$$\Psi_b = \Psi_A + \Psi_B \ldots (1.34) \text{ and } \Psi_a = \Psi_A - \Psi_B$$

$$(1.35)$$

As explained above, two atomic orbitals combine to form two molecular orbitals; hence it means that the number of molecular orbitals must always be equal to the number of atomic orbitals that are combined. The electron distribution in a given molecular orbital is obtained by squaring their wave functions, thus on squaring Eqs (1.34) and (1.35), we get,

$$\Psi_b^2 = \Psi_A^2 + 2\Psi_A \Psi_B + \Psi_B^2 \tag{1.36}$$

$$\text{and; } \Psi_a^2 = \Psi_A^2 - 2\Psi_A \Psi_B + \Psi_B^2 \tag{1.37}$$

Equations (1.36) and (1.37) depict the probability functions of bonding and antibonding molecular orbitals. The two equations differ by cross term $2\Psi_A \Psi_B$ and integral $\int \Psi A \Psi B \, d\tau$ is the overlap integral, S, which is infinitesimally small and hence neglected. For bonding S > 0; antibonding S < 0 and for non-bonding S = 0.

1.7.2 Shapes of Molecular Orbitals

Molecular orbitals can be sigma or pi depending on the mode of overlap of atomic orbitals. If a head-on collision occurs between atomic orbitals, sigma molecular orbitals will be formed. When atomic orbitals overlap laterally, pi-molecular orbitals are formed. Let us take the example of hydrogen molecule. If there is a favourable interaction between 1s atomic orbitals of two hydrogen atoms, it produces a molecular orbital cylindrically symmetrical along the inter-nuclear axis. The bond formed when two electrons occupy such a molecular orbital is called sigma (σ) bond and its associated antibonding orbital is called sigma star (σ^*). The electrons in such bonding orbitals are located nearer the inter-nuclear axis as shown in Fig. 1.14.

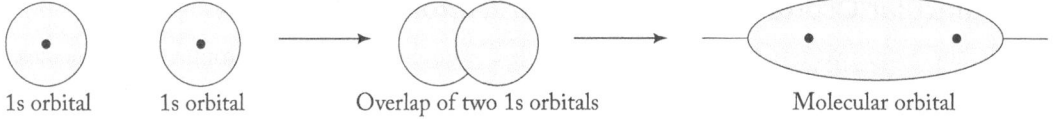

1s orbital   1s orbital   Overlap of two 1s orbitals   Molecular orbital

Fig. 1.14 1s orbitals leading to molecular orbital

Two different types of atomic orbitals can also result in the formation of molecular orbital of a sigma bond. Combining 1s and 2p atomic orbitals leads to the formation of molecular orbital of somewhat different shapes as shown in Fig. 1.15.

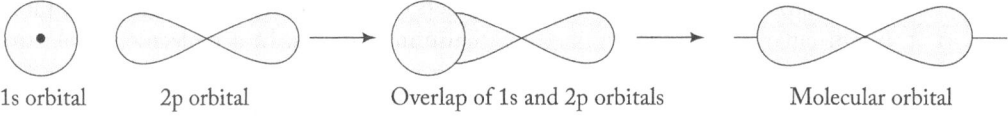

1s orbital   2p orbital   Overlap of 1s and 2p orbitals   Molecular orbital

Fig. 1.15 Atomic orbital overlap leading to molecular orbital

Another type of bonding that is generally seen in organic molecules is called the pi (π) bond that forms due to the interaction of parallel p orbitals located on adjacent atoms. Side-to-side interactions of p orbitals produce bonding pi (π) molecular orbital and an associated antibonding pi star (π^*) molecular orbital. In case of such bonding orbitals, electrons usually have the greatest probability of being located above and below the inter-nuclear axis as shown in Fig. 1.16.

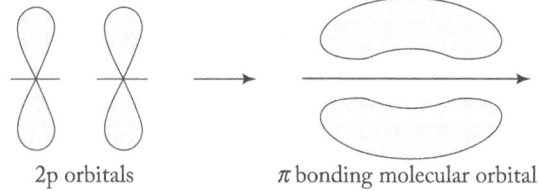

2p orbitals π bonding molecular orbital

Fig. 1.16 Formation of pi bonding molecular orbital

The following conditions must be met for effective atomic orbital overlap:

(a) Atomic orbitals involved in linear combination must possess similar energies. Hence, no combination is possible between 1s and 2s orbitals in a homonuclear diatomic molecule.

(b) There must be a considerable overlap between two atomic orbitals so as to form a molecular orbital.

(c) Atomic orbitals must have same symmetry about the molecular axes, that is, a $2p_z$ orbital will not combine with an atomic orbital due to varying symmetries, but a $2p_x$ orbital will combine with an s orbital to form a sigma molecular orbital. Further, a p_z orbital of one atom will not combine with a p_x or a p_y orbital of another atom.

(d) When p_x orbitals combine, bonding and antibonding molecular orbitals that are symmetrical about the inter-nuclear axis are denoted as σp and $\sigma^* p$, respectively. The combining p_y orbitals produces molecular orbitals of different shapes and do not remain symmetrical along the internuclear axis. They are usually denoted as $\pi 2p_y$ and $\pi^* 2p_y$ for bonding and antibonding molecular orbitals, respectively. Similarly, when p_z atomic orbitals combine, $\pi 2p_z$ and $\pi^* 2p_z$ molecular orbitals are formed.

(e) The wave functions that refer to two or more orbitals of same energy are called *degenerate*. So, $\pi 2p$ orbitals are *doubly degenerate* as there are two orbitals of equal energy; $\pi 2p_y = \pi 2p_z$ and their antibonding molecular orbitals are also doubly degenerate; $\pi^* 2p_y = \pi^* 2p_z$.

The sequence of energy levels in the increasing order of energy that helps in predicting the electronic structure of simple molecules is as follows:

$$\sigma 1s < \sigma^* 1s < \sigma 2s < \sigma^* 2s < \sigma 2p_x < \pi 2p_y = \pi 2p_z < \pi^* 2p_y = \pi^* 2p_z < \sigma^* 2p_x$$

1.7.3 Bond Order

The difference between the number of bonding and antibonding electrons that is divided by 2 is called bond order.

$$\text{Bond order} = \frac{\text{Number of bonding electrons} - \text{Number of antibonding electrons}}{2}$$

The reason for dividing the total number of electrons by 2 is because we always assume bonds as a pair of electrons. Hence, for a simple molecule such as hydrogen that has two electrons, its bond order will be, $2 - 0/2 = 1$.

This indicates that H_2 molecule has one bond. However, it is not necessary that bond order will always be a whole number.

Dihydrogen (H_2) The simplest homonuclear diatomic molecule is formed when atomic orbitals of two hydrogen atoms combine. The electrons occupy the molecular orbital of the lowest energy, the $\sigma 1s$ bonding orbital. A molecular orbital can hold two electrons, so both electrons in the dihydrogen molecule are in σ_{1s} bonding orbital and the electron configuration is $(\sigma 1s)^2$.

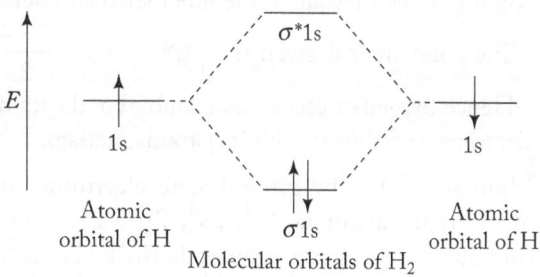

Fig. 1.17 Molecular orbital energy level diagram of dihydrogen H_2 molecule

$$\text{Bond order of dihydrogen molecule} = \frac{1}{2}[N_b - N_a] = \frac{2-0}{2} = 1$$

Nitrogen (N_2) The ground state electronic configuration of nitrogen atom is $1s^2, 2s^2, 2p^3$ and the electronic configuration of nitrogen molecule is,

$$2N\,(1s^2, 2s^2, 2p^3) = N_2\,[KK\,(\sigma 2s)^2\,(\sigma^* 2s)^2\,(\sigma 2p_x)^2\,(\pi 2p_y = \pi 2p_z)^4]$$

The 1s electrons from both the nitrogen atoms are referred to as K shell electrons (closed shell electrons); they do not participate in bonding as they are in the inner shell and denoted as KK in the electronic configuration.

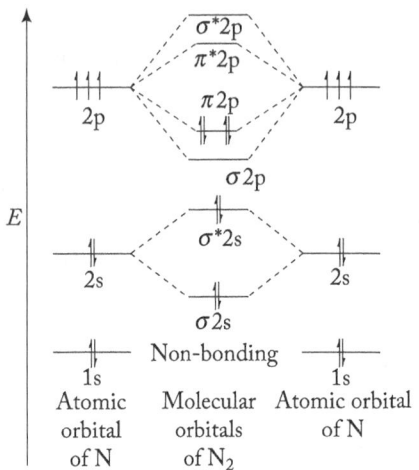

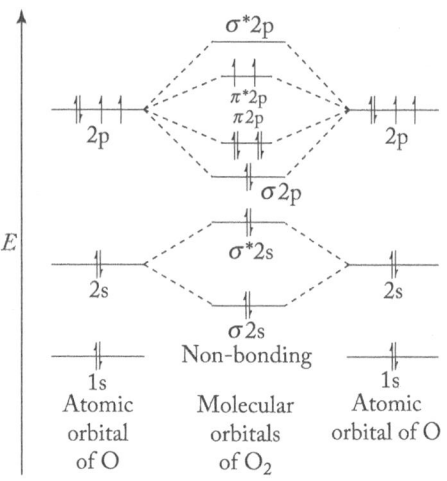

Fig. 1.18 Molecular orbital energy level diagram of N_2 molecule

Fig. 1.19 Molecular orbital energy level diagram of O_2

The bond order (i.e., number of covalent bonds) is given as: $\dfrac{1}{2}[N_b - N_a] = \dfrac{8-2}{2} = 3$

Hence, nitrogen is a triple bond molecule (N≡N) with one sigma and two pi bonds with diamagnetic properties.

Oxygen (O_2) The ground state electronic configuration of oxygen atom is $1s^2, 2s^2, 2p^4$ and electronic configuration of oxygen molecule is,

$$2O\ (1s^2, 2s^2, 2p^4) = O_2\ [KK\ (\sigma 2s)^2\ (\sigma^* 2s)^2\ (\sigma 2p_x)^2\ (\pi 2p_y = \pi 2p_z)^4\ (\pi^* 2p_y = \pi^* 2p_z)^2]$$

The 1s electrons from both oxygen atoms are referred to as K shell electrons as they do not take part in bonding since they are in the inner shell and denoted as KK in the electronic configuration.

The bond order is given as: $\dfrac{1}{2}[N_b - N_a] = \dfrac{8-4}{2} = 2$

Hence, oxygen molecule has a double bond with two unpaired electrons, and thus it exhibits paramagnetism.

Fluorine (F_2) The ground state electronic configuration of flourine atom is $1s^2, 2s^2, 2p^5$, and the electronic configuration of fluorine molecule is, $2F\ (1s^2, 2s^2, 2p^5)$

$$= F_2\ [KK\ (\sigma 2s)^2\ (\sigma^* 2s)^2\ (\sigma 2p_x)^2\ (\pi 2p_y = \pi 2p_z)^4\ (\pi^* 2p_y = \pi^* 2p_z)^4]$$

The 1s electrons from both flourine atoms are referred to as K shell electrons since they do not take part in bonding as they are in the inner shell and denoted as KK in the electronic configuration.

The bond order is given as: $1/2[N_b - N_a] = 8 - 6/2 = 1$

Thus, fluorine molecule has a single bond with no unpaired electrons, and thereby exhibits diamagnetism.

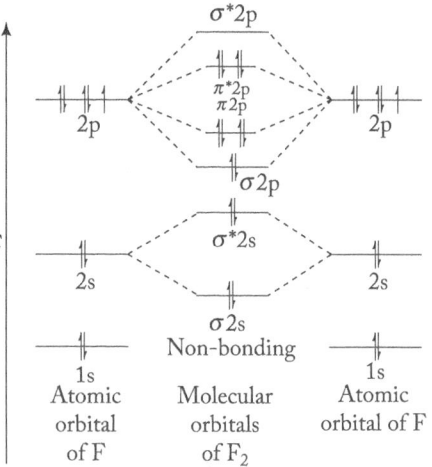

Fig. 1.20 Molecular orbital energy level diagram of F_2

1.7.4 Molecular Orbitals in Heteronuclear Diatomic Molecules

If two bonded atoms in a molecule are of different elements, they are called *heteronuclear diatomic molecules*, for example, CO, HCl, NO. The principles of chemical bonding in heteronuclear diatomic molecule are the same as those of the homonuclear diatomic molecules studied in the earlier section. However, some differences naturally appear in heteronuclear diatomic molecules, such as: (a) *loss of symmetry* and (b) *unequal electron cloud* due to different participating nuclei (or elements). As seen earlier, in homonuclear diatomic molecules, only the combination of atomic orbitals of equal energy and like-symmetry can form molecular orbitals. But such a limitation is not observed in heteronuclear diatomic molecules. When atomic orbitals of different elements combine, the following two factors affect the formation of molecular orbital.

Differing electronegativities   The two atomic orbitals of the combining elements are at different energies due to differing electronegativities between atoms. When a more electronegative atom approaches a strongly electropositive atom, electron density in such molecules is significantly polarized towards the more electronegative atom.

Let us consider carbon monoxide molecule, where C and O atoms (on Pauling scale, electronegativity = 2.6 and 3.5 respectively) combine, the atomic orbitals of oxygen will be lower in energy. When such atomic orbitals overlap, the resulting bonding molecular orbitals will resemble more like atomic orbitals of oxygen, whereas the antibonding molecular orbitals will resemble the atomic orbitals of carbon. Due to differing electronegativities, the electron cloud in the molecule will be drawn towards the atom with higher electronegativity and hence the heteronuclear diatomic molecule (CO) has an unsymmetrical electron distribution.

Further, the combining atomic orbitals in a heteronuclear diatomic molecule do not contribute equally to the bonding and antibonding molecular orbitals. Say, if a heteronuclear molecule AB has a more electronegative atom B, the atomic orbitals of atom B will be lower in energy than those of atom A. Thus, bonding molecular orbitals will be closer to atomic orbitals of atom B, whereas atomic orbitals will be contributing more to antibonding molecular orbitals. Hence, molecular orbitals of heteronuclear diatomic molecules can be written as:

$$\Psi_b = x\Psi_A + y\Psi_B; \text{ and } \Psi_a = y\Psi_A - x\Psi_B$$

where x and y are coefficients of atomic orbitals and $y > x$.

Reduced covalent bond energy   In a heteronuclear diatomic molecule, bonds formed from atomic orbitals of differing energies have reduced covalent bond energy. As the bonding MO will have lower energy than the atomic orbitals from which it is formed, the difference is called *exchange energy* (ΔE).

As shown in Fig. 1.21(c), the exchange energy in a heteronuclear molecule is reduced as the atomic orbitals do not match. It is evident from Fig. 1.21(a), that there is weakening in covalent bonding, but this is not true. Whenever there is loss of covalent character, it is compensated by an increase in the ionic character of bonds. If one adds up the ionic and covalent bonding, it results in a much stronger bonding as in Fig. 1.21(b).

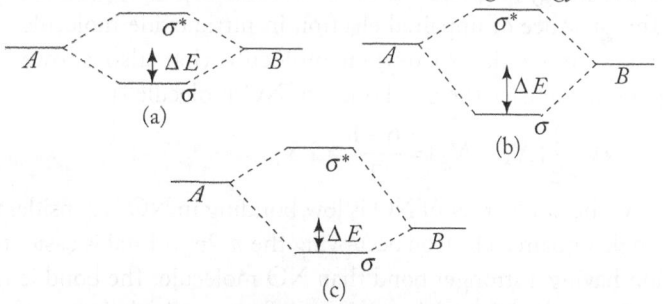

Fig. 1.21 (a) Covalent energy in a homonuclear diatomic molecule, (b) covalent energy In a heteronuclear diatomic molecule and (c) heteronuclear diatomic molecule with higher electronegativity difference

Carbon monoxide (CO)   The electronic configuration of carbon and oxygen atoms are: $_6C = 1s^2, 2s^2, 2p^2$ and $_8O = 1s^2, 2s^2, 2p^4$. The number of electrons available for bonding from carbon and oxygen are 4 and 6, respectively; thereby ten electrons need to be accommodated in the molecular energy levels. Carbon monoxide can be considered isoelectronic with nitrogen molecule and the electronic configuration of CO molecule can be expressed as, CO [KK $(\sigma 2s)^2 (\sigma^* 2s)^2 (\sigma 2p_x)^2$ $(\pi 2p_y = \pi 2p_z)^4]$.

As shown in Fig. 1.22, the bonding resulting from $\sigma 2s^2$ is effectively cancelled by antibonding $\sigma^* 2s^2$. This leaves $\sigma 2p_x^2$ to provide the bonding in CO molecule. As all the six electrons are present in the bonding molecular orbitals and none of it is in the antibonding molecular orbitals, hence just like nitrogen molecule, CO molecule also shows the bond order as,

$$\frac{1}{2}[N_b - N_a] = \frac{8-2}{2} = 3$$

Thus, in a CO molecule, there is a triple bond $C \equiv O$ with one σ and two π bonds with diamagnetic properties. With a high bond order of 3, CO is a stable molecule. All the electrons are paired and hence CO is a diamagnetic molecule.

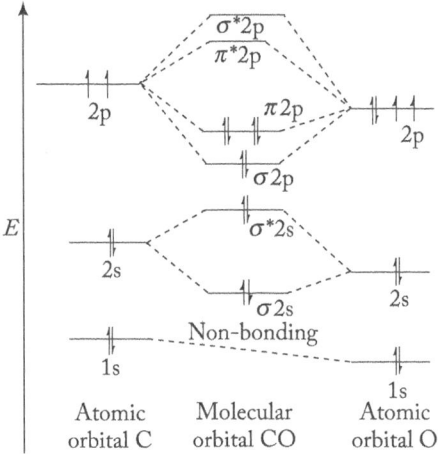

Fig. 1.22 Molecular orbital energy diagram of CO molecule

Nitric oxide (NO)   The electronic configuration of nitrogen and oxygen atoms are, $_7N = 1s^2, 2s^2, 2p^3$ and $_8O = 1s^2, 2s^2, 2p^4$. The molecular energy level diagram of nitric oxide will be quite similar to nitrogen molecule (refer to Fig. 1.23). In nitric oxide, there are 11 electrons to be filled in molecular orbitals. Thus, the configuration of NO molecule can be written as NO [KK $(\sigma 2s)^2 (\sigma^* 2s)^2 (\sigma 2p_x)^2 (\pi 2p_y)^2 \pi 2p_z)^2, \pi^* 2p_y^1, \pi^* 2p_z^0]$.

The four electrons of the two 2s orbitals fill up the bonding molecular orbitals, $\sigma 2s^2$ and antibonding molecular orbital, $\sigma^* 2s^2$. Out of the remaining seven electrons: three 2p electrons of nitrogen and four 2p electrons of oxygen, only six electrons will fill up the remaining higher molecular orbitals.

The only remaining electron will occupy the antibonding orbital, $\pi^* 2p_y^1$ and due to the presence of this single electron in the energy diagram, NO molecule exhibits paramagnetism. The presence of unpaired electron in nitric oxide molecule makes it similar to oxygen molecule that also shows paramagnetism. The bond order of NO molecule is

$$\frac{1}{2}[N_b - N_a] = \frac{6-1}{2} = 2.5$$

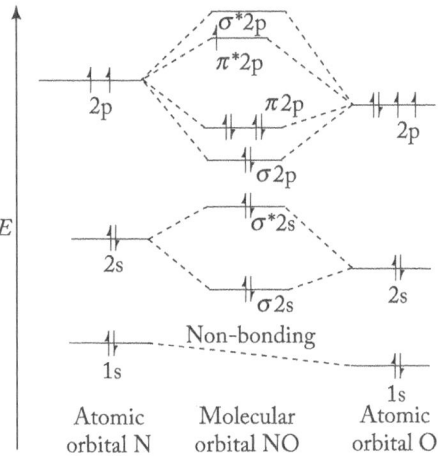

Fig. 1.23 Molecular energy level diagram of NO molecule

As the bond order of NO is low, bonding in NO is considerably weaker than in nitrogen molecule. The single unpaired electron occupying the $\pi^* 2p_y$ orbital is easier to be removed forming NO^+ (nitrosonium) ion having a stronger bond than NO molecule. The bond length in nitric oxide is greater than nitrogen molecule. In spite of the presence of an unpaired electron, nitric oxide molecule shows stability as this electron is well distributed over both nitrogen and oxygen atoms. The fact that nitrosonium ion can be easily obtained from nitric oxide clearly proves that NO^+ ion can exist as a stable species such as $NO^+HSO_4^-$ and $NO^+BF_4^-$.

Hydrogen chloride (HCl)   The electronic configuration of hydrogen and chlorine atoms are: $_1H = 1s^1$ and $_{17}Cl = 1s^2, 2s^2, 2p^6, 3s^2, 3p^5$. During the formation of hydrogen chloride, only three electrons of chlorine atom can combine with $1s^1$ electron of hydrogen. As the $3p_y$ and $3p_z$ orbitals of chlorine atom have no matching symmetry with 1s hydrogen orbital, there cannot be any overlap of these orbitals. The molecular energy level diagram is depicted in Fig. 1.24 considering no hybridization.

The electronic configuration of hydrogen chloride can be written as: HCl [KK, $2s^2$, $2p^6$, $3s^2$, $3p_y^2$, $3p_y^2$].

The shape of hydrogen chloride molecule clearly indicates the presence of a polar bond with a bond order of $\frac{1}{2}[N_b - N_a] = \frac{2-0}{2} = 1$. Since both the electrons are paired, hydrogen chloride is a diamagnetic molecule.

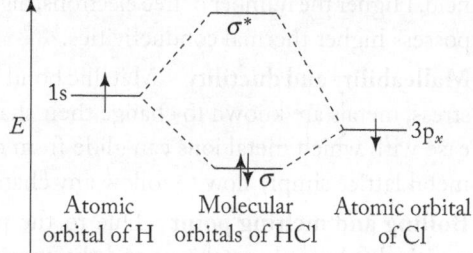

Fig. 1.24 Molecular energy level diagram of HCl molecule

1.8 METALLIC BOND

Metals exhibit crystalline properties possessing either body-centred cubic, face-centred cubic, or close-packed hexagonal lattices (Fig. 1.25). Each atom in the crystal lattice exhibits a high coordination number. Hence, bonding in such metallic crystalline structures cannot be explained using simple theories of bonding due to insufficient number of electrons.

It is observed that the metal atoms are closely packed in a crystal structure, which represents extensive overlap of electron orbitals such that the valence electrons are no longer associated with a particular nucleus; rather they are completely delocalized over all atoms in the crystal structure. An electrostatic attraction between metal atoms and valence electrons within its sphere of influence is called *metallic bond*. Metals are arrangements of positive ions as spheres of identical radii packed so as to completely fill the space. The theories put forth to explain bonding in metals are discussed here.

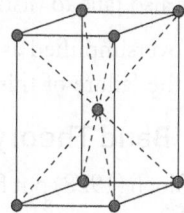

(a) Body-centred cubic

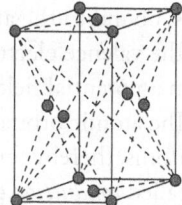

(b) Face-centred cubic

1.8.1 Free Electron Theory

Paul Drude (1900) put forth the free electron theory in which he considered metals as a lattice with electrons moving through it just similar to the movement of gaseous molecules. The theory was further improvized by Lorentz (1923) who stated that as metals have lower ionization potential, they easily lose valence electrons and hence are made of only a lattice of rigid spheres of positive ions and electrons delocalized in the lattice. Hence, one can model that metal behaves as an assembly of positive ions immersed in a sea of mobile, delocalized electrons as shown in Fig. 1.26.

As valence electrons in a metallic bond are spread over the crystal lattice, metallic bond is non-directional in nature. Free electron theory can explain the following properties of metals.

High strength The metallic bonds are very strong; hence metals can maintain a regular crystal structure.

Electrical and thermal conductivity The high electrical conductivity of metals can be attributed to the presence of free valence electrons as they can easily move under the influence of an electric

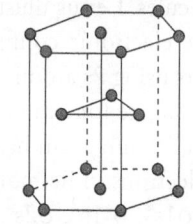

(c) Close packed hexagonal

Fig. 1.25 Crystal structures

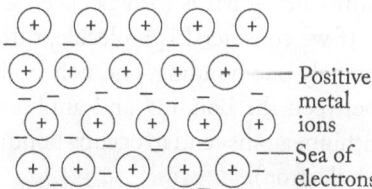

Fig. 1.26 Metal lattice showing delocalized electrons floating among positive ions

field. Higher the number of free electrons, higher will be its electric conductivity. In a similar way, metals also possess higher thermal conductivities.

Malleability and ductility Metallic bond is non-directional in character; hence on application of shear stress, metals are known to change their shape, this property of metals is called malleability. Further, the ease with which metal ions can glide from one lattice site to another is called ductility. Electrons in the metal lattice simply flow to follow any change in shape of the metallic crystal lattice.

Boiling and melting point Due to the presence of strong electrostatic attractive force between the positively charged metal ions and the surrounding valence electrons, metals exhibit higher boiling and melting points.

Demerits of free electron theory
 (a) It fails to explain specific heat of metals, marginally lower molar heat capacity of metals as compared to non-metals.
 (b) It also fails to distinguish between metals, insulators, and semiconductors.

  The oversimplified assumption that electron is free to move anywhere within the metal crystal lattice led to the failure of this theory. In order to explain all these characteristics, band theory was postulated.

1.8.2 Band Theory

Felix Bloch (1928) put forth a quantum mechanical model theory to explain metal bonding. The following are the assumptions of this theory.
 (a) All electrons present in completely filled energy levels of atoms are considered to be localized, that is, bound to the atoms with which they are associated.
 (b) The valence electrons in the outermost energy level of atoms are free to move; however, they move in a potential field that extends over all the atoms present in the crystal lattice.
 (c) The atomic orbitals of these free electrons can overlap with the atomic orbitals of electrons in other atoms, thereby forming delocalized molecular orbitals. Such molecular orbitals of free electrons are called *conduction orbitals* of a metal.

  It is obvious that band theory is merely an extension of molecular orbital concept applicable to diatomic molecules. Let us illustrate the above concept with the example of lithium (Li) metal.

  The electronic configuration of lithium atom is $1s^2$, $2s^1$ and if Li_2 molecule is considered, bonding occurs using 2s atomic orbitals. There are three vacant 2p atomic orbitals in the valence shell and this is a prerequisite for exhibiting metallic properties. MO theory can elucidate the formation of Li_2 molecule. Each lithium atom has two electrons in its inner shell, and one in its outermost shell, making a total of six electrons in its molecule. Hence, the electronic configuration of lithium molecule can be written as: Li_2: $\sigma 1s^2$, $\sigma^* 1s^2$, $\sigma 2s^2$ and bonding will occur, as the $\sigma 2s$ bonding molecular orbital is full and its corresponding antibonding orbital is vacant. If one ignores the innermost electrons, the 2s atomic orbitals from each lithium atom can combine to give two molecular orbitals, one bonding and the other antibonding MOs with valence electrons occupying the bonding orbitals.

  If we consider Li_3 molecule, three 2s atomic orbitals will combine forming three molecular orbitals, namely one bonding, one non-bonding, and one antibonding. The energy of non-bonding MO lies in between the bonding and antibonding molecular orbitals. Hence, three valence electrons from three lithium atoms tend to occupy bonding molecular orbital (2 electrons) and non-bonding molecular orbital (1 electron).

  When four lithium atoms combine to form Li_4, four 2s atomic orbitals with one electron each overlap, forming four molecular orbitals; two bonding and two antibonding orbitals. The presence of two non-bonding molecular orbitals between the bonding and anti-bonding molecular orbitals tends to reduce the energy band gap between these orbitals. Hence, the four valence electrons will occupy the two lowest energy bonding molecular orbitals.

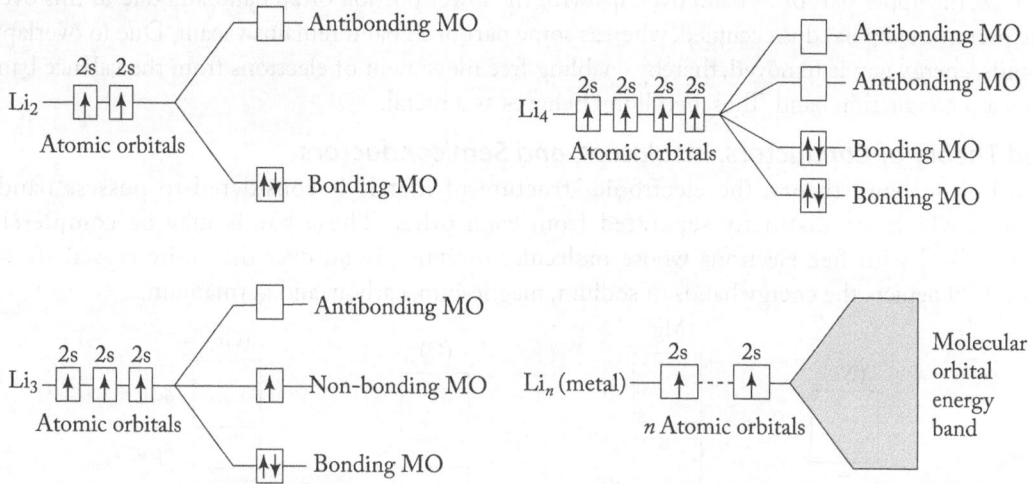

Fig. 1.27 Development of molecular orbitals into bands in metals

If n number of lithium atoms combine forming Li_n, there will be n number of 2s atomic orbitals with one electron each that will overlap forming n MOs; out of which half of them will be bonding and the remaining half will be antibonding. The electrons in n orbitals will only be enough to fill the $n/2$ number of bonding molecular orbitals, whereas antibonding molecular orbitals will remain vacant. Hence, as the number of lithium atoms increases, the spacing between the energy levels of molecular orbitals decreases, such that it virtually forms a band as in Fig. 1.27 (d). The band so formed is called the *molecular orbital energy band.*

Explanation of Electrical and Thermal Conduction (Band Theory Concept)

Metals contain either half-filled or partially-filled valence molecular orbital energy band because of the overlap with unoccupied molecular energy band. As there is only one valence electron per atom of lithium and a molecular orbital can hold up to two electrons, it follows that only half of the molecular orbitals in the 2s valence band are occupied, namely bonding molecular orbitals (Fig. 1.28 (a). Hence, it requires only an infinitesimal amount of energy to displace an electron to an unoccupied molecular orbital. This clearly elucidates that metals exhibit high thermal and electrical conductivities.

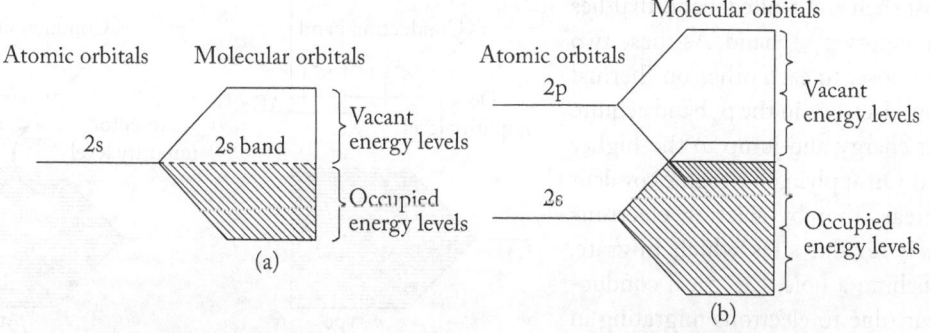

Fig. 1.28 (a) Metallic molecular orbitals of lithium showing half-filled band and (b) metallic molecular orbitals of beryllium showing overlapping bands

Beryllium has an electronic configuration $1s^2, 2s^2$ with two valence electrons that can fill the 2s band of molecular orbitals. Similarly, 2p atomic orbitals form a 2p band of molecular orbitals. As shown in Fig. 1.28, the upper part of 2s band overlaps with the lower portion of 2p band and due to this overlap, some part of the 2p band is occupied, whereas some part of 2s band remains vacant. Due to overlapping of bands, energy gap is removed, thereby enabling free movement of electrons from the valence band to the vacant conduction band. Thus, beryllium behaves as a metal.

Band Theory of Conductors, Insulators, and Semiconductors

According to band theory, the electronic structure of metals is considered to possess bands of electrons which are distinctly separated from each other. These bands may be completely or partially filled with free electrons whose molecular orbitals extend over the entire crystal structure. Figure 1.29 depicts the energy bands of sodium, magnesium, carbon, and germanium.

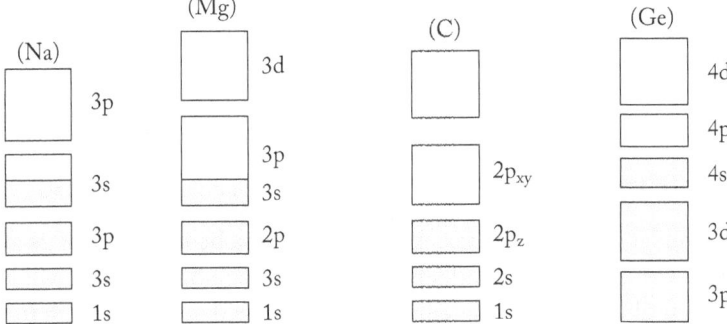

Fig. 1.29 Band models of (a) conductors (Na and Mg), (b) insulator (C), and (c) semiconductor (Ge)

Metals (conductors) In metals, electrical conductivities depend on the movement of electrons throughout the crystal structure under the influence of applied potential. This is possible only if electrons can be energized and jump to higher vacant band levels. As both the valence and conduction bands in metals are very close to each other, they exhibit excellent conductivity.

Non-metals (insulators) In this case, valence bands are fully occupied by electrons and there is a large energy gap between the valence and conduction bands. Hence, it is very difficult to excite an electron and a large amount of energy needs to be supplied for conductivity. Hence, non-metals are insulating materials.

Semiconductors These are materials that behave as insulators at lower temperatures and act as conductors at normal or higher temperatures. Silicon and germanium are classic examples of semiconductor materials. They have four electrons in their outermost shell and a filled band that lies below an empty p_{x-y} band. As these two bands are closer to each other, on thermal activation, electrons in the p_z band acquire sufficient energy and jump to the higher p_{x-y} band. On applying heat some covalent bonds break, thereby ejecting electrons from their regular sites which migrate, leaving behind a hole. Electrical conduction occurs due to electrons migrating in one direction and positive holes in the opposite direction; this is called *intrinsic*

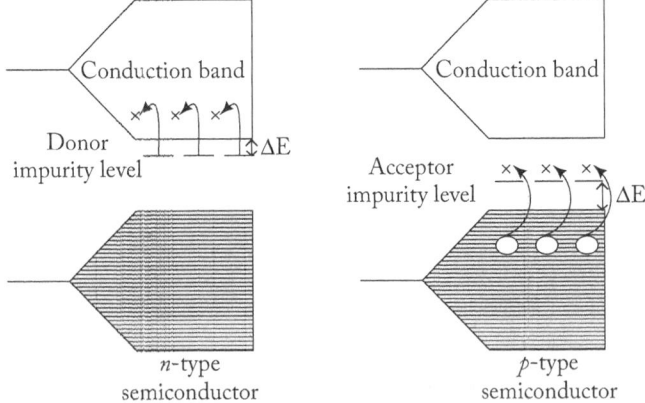

Fig. 1.30 Energy levels in n-type and p-type semiconductors

conduction. When trace impurities are added to such materials to further enhance the conductivity it is called *extrinsic conductivity*.

When arsenic possessing five valence electrons is doped with silicon or germanium, the four electrons of arsenic atom form a bond with four electrons of silicon, and the fifth electron is free to move. This extra electron occupies the donor impurity level just below the empty conduction band of silicon crystal. On applying thermal energy, the free electron can easily jump to the conduction band, thereby exhibiting conductivity and is termed as *n–type semiconductor*.

When indium or gallium having three valence electrons is added as an impurity, only three electrons of silicon are covalently bonded to the atoms of the dopant. Certain sites occupied by electrons are vacant called *positive holes* and occupy acceptor impurity level that lies closer to the filled valence band of silicon. On applying thermal energy, electrons get excited and jump from filled valence band to empty acceptor impurity conduction band consisting of positive holes. If a potential is applied, electrons from an adjacent atom jump and occupy the hole and in turn is replaced by an electron from another atom. It seems the positive holes are migrating and such materials are called *p–type semiconductors*.

1.9 CONCEPT OF AROMATICITY

Benzene (1825) was first isolated by Michael Faraday who extracted the compound from liquid residue obtained after heating whale oil under pressure. Eilhard Mitscherlich (1834) provided the molecular formula of benzene as C_6H_6 and called it 'benzin' due to its relationship to benzoic acid, but later was renamed as benzene. Alchemists called such compounds aromatic, because of their pleasing odour. However, today the term 'aromatic compound' signifies some chemical structures that fulfil certain criteria.

Benzene is a planar, cyclic compound with a cyclic cloud of delocalized electrons above and below the plane of the ring (Fig. 1.31). As π electrons are delocalized, all the C – C bonds have the same length. Further, it is also known that benzene is quite a stable compound with large resonance energy of 36 kcal/mol. The criteria to be fulfilled for a compound to be classified as aromatic are the following.

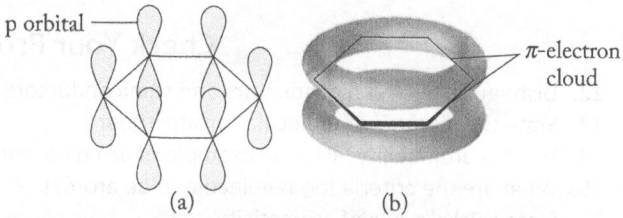

Fig. 1.31 Structure of benzene: (a) p orbitals on carbon atoms (b) π-electron cloud above and below benzene ring

(a) It should have an uninterrupted cyclic cloud of π electrons (also called π cloud) above and below the plane of the molecule.

(b) For a π-electron cloud to be cyclic and remain uninterrupted, the molecule must also be cyclic with every atom in the ring possessing a p orbital.

(c) To form an uninterrupted π-electron cloud, each p orbital must overlap with the p orbitals on either side of it, thus the molecule must essentially be planar with π-electron cloud containing an odd number of pair of electrons.

Erich Hückel (1931) was the first to recognize that an aromatic compound must possess an odd number of π electrons; this came to be called Hückel's rule or the $4n + 2$ rule. Hückel's rule is a mathematical way of expressing that an aromatic compound should have an odd number of pairs of π electrons. According to the Rule, for a planar, cyclic compound to be aromatic, its uninterrupted π cloud must contain $(4n + 2)$ π electrons, where n is any whole number. An aromatic compound must have $2(n = 0)$, $6(n = 1)$, $10(n = 2)$, $14(n = 3)$, and so on number of π electrons. As there are two electrons in a pair, Hückel's rule necessitates that an aromatic compound have 1, 3, 5, 7, etc. as pairs of π electrons.

Antiaromatic compounds   Some compounds fulfil the first criterion of Hückel's rule (listed above), but fail to satisfy the second criterion, that is, they possess an even number of pairs of π electrons. Cyclobutadiene is a planar molecule with two pairs of π electrons. Such compounds are called *antiaromatic compounds*. They are quite unstable and difficult to isolate. Figure 1.32 depicts the distribution of electrons in benzene and cyclobutadiene.

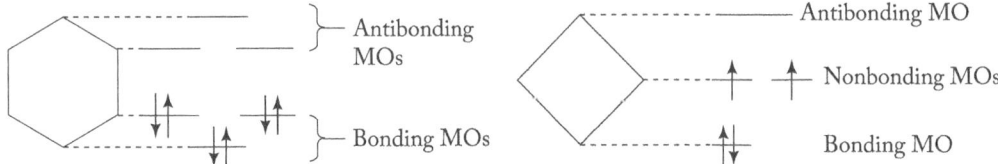

Fig. 1.32 Frost diagrams of (a) benzene and (b) cyclobutadiene

Arthur Frost proposed a simpler method to depict the distribution of electrons in aromatic systems called *Frost diagram*. In a Frost diagram, one needs to first draw the cyclic compound with one of its vertices pointed down. The molecular orbitals below the midpoint of the cyclic compound will be bonding molecular orbitals, whereas those above the midpoint are considered antibonding molecular orbitals. The midpoint of the cyclic structure in the Frost diagram will be considered as nonbonding molecular orbitals. The electrons are filled in the molecular orbitals as per Pauli Exclusion Principle and Hund's Rule, which states that if electrons are left over after filling up the bonding orbitals, they occupy non-bonding orbitals. It is evident from the diagram that in aromatic compounds such as benzene, all the bonding molecular orbitals are completely filled, whereas in a non-aromatic compound like cyclobutadiene, the presence of unpaired electrons explains its instability.

Check Your Progress

12. Distinguish between conductors and semiconductors.
13. State the features of molecular orbital theory.
14. What is aromaticity? Give an example of aromatic compound.
15. What are the criteria for a molecule to be aromatic?
16. State Hückel's rule of aromaticity.
17. List the assumptions of MO theory to explain metallic bond.
18. What are Frost diagrams? Illustrate Frost diagram of benzene and cyclobutadiene.
19. Draw the molecular energy level diagrams of
    (a) HCl   (b) NO   (c) O_2 (d) N_2 (e) H_2 (f) CO   (g) F_2
20. List the merits and demerits of free electron theory put forth to explain metallic bond.

SOLVED EXAMPLES

1. Calculate the wavelength of (a) a ball weighing 250 g and (b) an electron moving with a velocity of 50 m/s (Given: electron rest mass, m_e = 9.109 × 10^{-31} kg).

Solution: (a) According to de Broglie equation, $\lambda = \dfrac{h}{mv}$

$$\lambda = \frac{6.626 \times 10^{-34}\ \text{J.s}}{2.50 \times 10^{-1} \times 50\ \text{m.s}^{-1}} \quad (\text{as 1 kg = 1000 g})$$

Hence, λ = 5.3 × 10^{-35} m or 5.3 × 10^{-25} Å.

It is known that radius of an atom is in the order of 10^{-11} m; the above value is very small and is difficult to determine by any device.

(b) For electron,

$$\lambda = \frac{6.626 \times 10^{-34} \text{ J.s}}{9.109 \times 10^{-31} \times 50 \text{ m.s}^{-1}} = 1.46 \times 10^{-5} \text{ m } (1.46 \times 10^5 \text{ Å})$$

The wavelength obtained as above falls in the infrared region of EMR spectrum.

2. Calculate the kinetic energy of a moving electron of wavelength of 5.3 pm. (Given: mass of an electron = 9.11×10^{-31} kg and $h = 6.6 \times 10^{-34}$ J.s).

Solution: The velocity of an electron can be expressed as (on rearranging de Broglie relation),

$$v = \frac{h}{m\lambda} = \frac{6.6 \times 10^{-34} \text{ J.s}}{9.11 \times 10^{-31} \text{ kg} \times 5.3 \times 10^{-12} \text{ m}} \quad \text{(as 1 pm} = 10^{-12} \text{ m)}$$

$$= 1.3682 \times 10^8 \text{ ms}^{-1}$$

As, K.E of an electron is $\frac{1}{2} mv^2$, thus,

$$= \frac{9.11 \times 10^{-31} \text{ kg} \times (1.3682 \times 10^8 \text{ ms}^{-1})^2}{2} = 8.524 \times 10^{-15} \text{ kg.m}^2\text{s}^{-2}$$

3. If an electron moves with a velocity of 3.3×10^7 m/s, calculate the smallest possible uncertainty in its position. (Given: mass of an electron = 9.11×10^{-31} kg and $h = 1.05 \times 10^{-34}$ J.s).

Solution: As per Heisenberg's Uncertainty Principle, $\Delta x \times \Delta p = h$

Hence, $\Delta x = h/mv = \dfrac{1.05 \times 10^{-34} \text{ J.s}}{9.11 \times 10^{-31} \times 3.3 \times 10^7 \text{ m.s}^{-1}} = 3.492 \times 10^{-12}$ m (or, 0.0349 Å)

4. What is the wavelength of an electron moving at 5.31×10^6 m/s? (Given: mass of electron = 9.11×10^{-31} kg and $h = 6.626 \times 10^{-34}$ J·s)

Solution: According to de Broglie's equation,

$$\lambda = \frac{h}{mv} = \frac{6.626 \times 10^{-34} \text{ J.s}}{9.11 \times 10^{-31} \text{ kg} \times 5.31 \times 10^6 \text{ m/s}}$$

$$= \frac{6.626 \times 10^{-34} \text{ J.s}}{4.84 \times 10^{-24} \text{ kg. m/s}} = 1.37 \times 10^{-10} \text{ m or 1.37 Å}$$

5. Calculate the kinetic energy and de Broglie wavelength (nm) of C_{60} molecule moving at a speed of 100 m/s. (Given: atomic weight of C = 12.011 g, Avogadro's number = 6.022×10^{23} molecules/mol.

Solution: Molar mass of one C_{60} molecule = $60 \times 12.011 = 720.66$ g/mol

Mass of one molecule will be = $\dfrac{720.66 \text{ g/mol}}{6.022 \times 10^{23} \text{ molecules/mol}} = 1.1967 \times 10^{-21}$ g/mol

$$= \frac{1.1967 \times 10^{-21} \text{ g/mol}}{1000} = 1.1967 \times 10^{-24} \text{ kg}$$

Kinetic energy, $E = \dfrac{1}{2} mv^2 = \dfrac{1}{2} \times 1.1967 \times 10^{-24} \times (100)^2 = 5.9835 \times 10^{-21}$ J

According to de Broglie equation, $\lambda = \dfrac{h}{mv}$

$$= \frac{6.626 \times 10^{-34} \text{ Js}}{1.1967 \times 10^{-24} \times 100} = 5.5369 \times 10^{-12} \text{ m or } 5.537 \times 10^{-3} \text{ nm} \quad \text{(as 1 m} = 10^9 \text{ nm)}$$

6. Determine the minimum uncertainty in the velocity of a particle having a mass 1.1×10^{-27} kg if uncertainty in its position is 3×10^{-10} cm. (Given: $h = 6.6 \times 10^{-34}$ Js)

Solution: According to Heisenberg's Uncertainty principle,

$$\Delta x \times \Delta p \geq \frac{h}{4\pi} \quad \text{or,} \quad \Delta x \times m\Delta v = \frac{h}{4\pi}$$

On rearranging the above expression we get,

$$\Delta v = \frac{h}{4\pi m \Delta x} = \frac{6.6 \times 10^{-34}}{4 \times 3.143 \times 1.1 \times 10^{-27} \times \left(3 \times 10^{-10} \, cm \times \dfrac{1 \, m}{10^2 \, cm}\right)}$$

$$= 1.59 \times 10^4$$

Thus, the uncertainty in velocity of the particle $= 1.59 \times 10^4 \, ms^{-1}$

7. Calculate the energy of an electron in ground state confined to a box of 3Å in width and moving in one-dimension (*x*-axis only).

Solution: According to particle-in-a-box model,

$$E = \frac{n^2 h^2}{8ml^2} = \frac{1^2 \times (6.6 \times 10^{-34})^2}{8 \times 9.1 \times 10^{-31} \times (3 \times 10^{-10})^2}$$

On solving the above we get,

$\therefore \quad E_{ground \, state} = 6.648 \times 10^{-19} \, J$

SUMMARY

- Atomic and molecular structure forms the basis of chemistry. Learning about various postulates put forth by various atomic models helps us understand the structure of atoms.

- All sub-atomic particles have a wave-like nature called matter waves or de Broglie waves. The Davisson–Germer experiment practically demonstrated the wave nature of particles.

- According to Heisenberg's Uncertainty Principle, it is impossible to simultaneously determine the position and momentum of an electron.

- Schrödinger derived an equation for comparing the path taken by the particle with that of a ray of light and the associated de Broglie wave with electromagnetic waves.

- The quantum mechanical model of hydrogen atom and particle-in-a-box are deduced using Schrödinger equation.

- The solution of Ψ led to three different types of quantum numbers. As per Pauli Exclusion Principle, only two electrons can be accommodated by a given atomic orbital.

- Schrödinger equation is well studied for hydrogen atom. It can be applied to study conjugated systems and nanoparticles.

- According to the MO theory, atomic orbitals of similar energies and symmetry combine to form molecular orbitals, one of which is bonding and the other antibonding. Also, in an atomic orbital, an electron is influenced by one nucleus; whereas in a molecular orbital, electrons are under the influence of two or more nuclei depending on the number of combining atoms.

- MO theory finds application in understanding the structures of homo and heteronuclear diatomic molecules.

- Hydrogen, nitrogen, and oxygen are examples of elements forming homonuclear molecules, while carbon monoxide, nitric oxide, and hydrogen chloride are heteronuclear molecules.

- Theories put forward to explain metallic bonds include the free electron theory and the molecular orbital or band theory.

- Though the free electron theory could account for most of the properties of metals such as their high strength, electrical and thermal conductivity, and malleability and ductility, it failed to explain the specific heat of metals and distinguish between metals, insulators, and semiconductors.

- Band theory of solids proposed by Felix Bloch addressed the shortcomings of the free electron theory.

- Hückel's rule is a mathematical way of expressing that an aromatic compound should have an odd number of pairs of π electrons. According to the Rule, for a planar, cyclic compound to be aromatic, its uninterrupted π cloud must contain $(4n + 2)$ π electrons, where *n* is any whole number.

GLOSSARY

Antibonding molecular orbital: A molecular orbital whose occupation by electrons decreases the total energy of a molecule. Energy level of an antibonding MO lies higher than the average of the valence atomic orbitals of the atoms in a molecule.

Aufbau principle: A maximum of two electrons are put into orbitals in the order of increasing orbital energy.

Bond order: The number of chemical bonds in a molecule.

Crystal lattice: The 3D arrangement of atoms, ions, or molecules in a crystalline solid.

Hückel's rule: The mathematical expression denoting that an aromatic compound should have an odd number of pairs of π electrons.

Hund's rule: Rule for building up the electronic configuration of atoms and molecules. It states that when electrons go into degenerate orbitals, they occupy them singly before pairing begins.

Orbital (atomic or molecular): A wave function that depends on the spatial coordinates of a single electron.

Pauli Exclusion principle: A maximum of two electrons can occupy an orbital and their spins must be paired or opposed to each other.

KEY FORMULAE

- de Broglie relation: $\lambda = \dfrac{h}{mv}$

- Heisenberg Uncertainty relation: $\Delta x \times \Delta p = h$

- Schrödinger equation: $\dfrac{\delta^2 \Psi}{\delta x^2} + \dfrac{\delta^2 \Psi}{\delta y^2} + \dfrac{\delta^2 \Psi}{\delta z^2} = -\dfrac{4\pi^2}{\lambda^2}\Psi$

- Hückel's rule: $4n + 2\pi$ electrons, where n is any whole number.

- Schrödinger equation (for hydrogen atom):

$$\nabla^2 \Psi + \frac{8\pi^2 m}{h^2}\left(E - \frac{Ze^2}{r}\right)\Psi = 0$$

- Bond order: $\dfrac{1}{2}[N_b - N_a]$, where b and a are bonding and antibonding molecular orbitals.

EXERCISES

Multiple Choice Questions

1. Bohr's model of atom is supported by
   - (a) Dalton's theory
   - (b) de Broglie equation
   - (c) Uncertainty principle
   - (d) None of these

2. Bohr's model of atom is contradicted by
   - (a) Planck's quantum theory
   - (b) Pauli Exclusion principle
   - (c) Heisenberg Uncertainty principle
   - (d) All of the above

3. Uncertainty principle was stated by
   - (a) de Broglie
   - (b) Heisenberg
   - (c) Einstein
   - (d) Schrödinger

4. The region around the nucleus where the probability of finding an electron is maximum is
   - (a) orbit
   - (b) energy level
   - (c) shell
   - (d) orbital

5. Which orbital has dumb–bell shape?
   - (a) s orbital
   - (b) p orbital
   - (c) d orbital
   - (d) f orbital

6. The mass of an electron (me) is
   - (a) 9.109×10^{-32} g
   - (b) 8.1×10^{-31} kg
   - (c) 9.1×10^{-31} kg
   - (d) 9.1×10^{-31} mg

7. The atomic orbitals that possess same energy are.
   - (a) degenerate orbitals
   - (b) hybrid orbitals
   - (c) valence orbitals
   - (d) molecular orbitals

8. The size of the nucleus is approximately
   - (a) $1/100^{th}$ of the atom
   - (b) $1/1000^{th}$ of the atom
   - (c) $1/10000^{th}$ of the atom
   - (d) $1/100000^{th}$ of the atom

9. Eigen values correspond to
   (a) definite wave function values
   (b) quantum numbers
   (c) definite values of total energy
   (d) definite angular momentum of electrons
10. Which of the following statements is NOT correct about wave functions?
   (a) It is infinite is most cases.
   (b) It is single valued.
   (c) It is continuous.
   (d) It has a continuous slope.
11. In Schrödinger wave equation Ψ represents
   (a) orbit   (b) wave function
   (c) wave   (d) radial probability
12. Uncertainty Principle is applicable to
   (a) measuring radii of particles
   (b) all moving particles
   (c) only stationary particles
   (d) all small and fast moving particles
13. In the ground state of an atom, the electron is present
   (a) in the nucleus
   (b) in the second shell
   (c) nearest to the nucleus
   (d) farthest from the nucleus
14. The radial nodes present in 3s and 2p orbitals are
   (a) 0, 2   (b) 2, 0
   (c) 2, 1   (d) 1, 2
15. Quantum number denoted by symbol '*m*' is
   (a) magnetic quantum
   (b) principal quantum
   (c) spin quantum
   (d) azimuthal quantum
16. A spinning electron creates
   (a) electric field   (b) quantum field
   (c) magnetic field   (d) atom structure
17. The quantum number that describes the shape of an electron in an atom is:
   (a) principal quantum
   (b) azimuthal quantum
   (c) magnetic quantum
   (d) spin quantum
18. The value of Planck's constant 'h' is
   (a) 6.625×10^{-34} J s
   (b) 6.625×10^{-34} cal
   (c) 6.625×10^{-34} kJ
   (d) 6.625×10^{-34} k cal
19. Stabilization energy of benzene is
   (a) 35 kcal   (b) 36 kcal
   (c) 37 kcal   (d) 38 kcal
20. The region where there is probability of finding an electron is
   (a) node
   (b) particle-in-a-box model
   (c) electron cloud
   (d) orbit
21. The bond order of carbon monoxide molecule is
   (a) 2   (b) 2.5
   (c) 1.5   (d) 3
22. Antibonding molecular orbitals are formed by
   (a) destructive overlap of atomic orbitals
   (b) constructive overlap of atomic orbitals
   (c) overlap of excess negative ions
   (d) none of these
23. Band theory of solids can satisfactorily explain
   (a) nature of insulators
   (b) semiconducting behaviour
   (c) conduction in metals
   (d) All of these
24. A vacant or partially filled band is termed as
   (a) valence band   (b) conduction band
   (c) forbidden band   (d) molecular band
25. The highest energy band gap is exhibited by
   (a) semiconductor   (b) conductor
   (c) insulator   (d) metals
26. On increasing the temperature, conductivity of an intrinsic conductor
   (a) increases
   (b) decreases
   (c) remains constant
   (d) initially decreases and then increases
27. 'No two electrons in an atom can have the same set of quantum numbers.' This statement is called
   (a) Bohr's theory
   (b) Pauli Exclusion principle
   (c) Hückel's rule
   (d) Hund's rule
28. Wave nature of electrons was first experimentally verified by
   (a) Davisson–Germer   (b) Planck
   (c) de Broglie   (d) Pauli
29. The qantum number that determines the shape of the subshell is
   (a) magnetic   (b) principal
   (c) spin   (d) azimuthal
30. The number of orientations of each subshell is given by
   (a) magnetic quantum number
   (b) principal quantum number
   (c) azimuthal quantum number
   (d) spin quantum number

31. de Broglie equation has significance in explaining
    (a) subatomic particles   (b) molecules
    (c) only electrons   (d) electron pairing
32. The bond order of HCl molecule is
    (a) 3   (b) 2
    (c) 1   (d) 0.5
33. The bond order and magnetism of dinitrogen molecule are
    (a) 3 and paramagnetic, respectively
    (b) 3 and diamagnetic, respectively
    (c) 2 and paramagnetic, respectively
    (d) 2.5 and diamagnetic, respectively
34. An example of antiaromatic compound is
    (a) benzene   (b) naphthalene
    (c) cyclobutadiene   (d) none of these
35. Bond order of NO is
    (a) 2.5   (b) 2
    (c) 1.5   (d) 0.5

Review Questions

1. What is wave–particle dualism?
2. State and derive de Broglie equation.
3. State and explain Heisenberg's Uncertainty principle.
4. What is Bohr's frequency rule? State the difference between an orbit and an orbital.
5. How does Pauli Exclusion principle help in understanding the electronic configuration of atoms?
6. Discuss Heisenberg Uncertainty principle and Born approximation.
7. Describe Davisson–Germer experiment demonstrating the wave nature of electrons.
8. Explain the significance of Ψ and Ψ^2.
9. What are atomic orbitals? Draw the s, p, d orbitals with clear descriptions.
10. Deduce Schrödinger time-independent wave equation. Explain the terms involved in the expression and state its significance.
11. Explain Schrödinger equation for quantum model of hydrogen atom.
12. Describe the physical significance of Schrödinger wave functions.
13. Justify the statement, 'It is impossible to measure simultaneously the position and velocity of a fast moving body like an electron.'
14. Apply Schrödinger wave equation for a particle-in-a-box illustrating quantization of energy. Draw the radial plots for hydrogen atom.
15. Express Schrödinger wave equation for 1,3-butadiene and benzene using particle-in-a-box model.
16. Discuss the application of particle-in-a-box solution to conjugated butadiene and benzene systems and write the wave equations.
17. Write a short note on 'applications of particle-box model to nanoparticles.'
18. What is a metallic bond? Describe free electron theory to describe metal bonding. List their merits and demerits.
19. Discuss band theory to explain bonding in metals citing suitable examples.
20. With a neat labelled MO diagram, explain the bonding in F_2 molecule.
21. Explain the electrical conductivities of conductors, insulators, and semiconductors.
22. With a neat labelled MO diagram, explain the bonding in CO molecule. State the various features of CO molecule.
23. Draw a neat labelled MO diagram of dinitrogen and explain the bonding citing its electronic configuration.
24. What is bond order of a molecule? How is it calculated? Explain bonding in HCl molecule with an MO diagram and state its characteristics.
25. What is a Frost diagram? Draw Frost diagrams for benzene and cyclobutane and show their molecular orbital configurations.
26. Draw the molecular orbital diagram for oxygen molecule and explain its paramagnetic behaviour.
27. With a neat labelled MO diagram, explain the bonding in NO molecule.
28. What is aromaticity? Explain aromaticity of benzene.
29. Discuss aromaticity of compounds. Explain the criteria for a compound to be considered as aromatic.
30. Write a short note on Hückel's rule for aromaticity and Frost diagrams.

NUMERICAL PROBLEMS

1. Calculate the wavelength (in metres) of a proton travelling at a velocity of 2.55×10^8 m, assuming the proton mass as 1.673×10^{-27} kg.   (**Ans:** 1.533×10^{-15} m)

2. Determine the wavelength (in metres) of a wave associated with a 1 kg object moving at a speed of 1 km/h.   (**Ans:** 2.38×10^{-33} m)

3. What will be the wavelength (in pm) associated with an electron having a mass of 9.11×10^{-31} kg and travelling at a speed of 4.19×10^{-6} ms-1   (**Ans:** 174 pm)

4. Calculate the kinetic energy and de Broglie wavelength (in nm) of C_{60} molecule moving at a speed of 200 ms^{-1}. (Given: atomic weight of carbon = 12.011 g, Avogadro's number = 6.022×10^{-23} molecules/mol.   (**Ans:** KE = 2.393×10^{-20}J, 2.768×10^{-12} m)

5. What is the wavelength (in angstrom) of an electron moving at 5.31×10^6 m/s?   (**Ans:** 1.37 Å)

6. Determine the uncertainty in position of a dust particle of mass 1 mg if uncertainty in its velocity is 5.5×10^{-20} m/s. (assume $h = 6.623 \times 10^{-34}$ Js)   (**Ans:** 9.58×10^{-10} m)

7. Calculate de Broglie wavelength (in m) of dinitrogen molecule moving at a speed of 2800 ms^{-1}. (assume $h = 6.626 \times 10^{-34}$ Js)   (**Ans:** 5×10^{-12} m)

8. If uncertainties in position and velocity of a particle are 10^{-10} m and 5.27×10^{-24} ms^{-1} respectively, what is the mass of the particle? (assume $h = 6.625 \times 10^{-34}$ Js)   (**Ans:** 0.1 kg or 100 g)

9. If an electron is bound in a one-dimensional box of size 4×10^{-10} m, what will be its minimum energy? (assume $h = 6.6 \times 10^{-34}$ Js)   (**Ans:** $E = 3.739 \times 10^{-19}$ J)

10. If an electron is bound in a one-dimensional box of size 8×10^{-10} m, what will be its minimum energy? (assume $h = 6.6 \times 10^{-34}$ Js)   (**Ans:** $E = 9.349 \times 10^{-20}$ J)

FURTHER READING

1. Altmann, S., *Band Theory of Metals*, Pergamon Press, 2013.
2. de Broglie, L., (1924) XXXV. A Tentative Theory of Light Quanta', *Philosophical Magazine*, 47, 446-458. https://www. tandfonline com/doi/abs/10.1080/14786442408634378
3. Griffith, D.J. *Introduction to Quantum Mechanics*, Benjamin Cummings, 2004.
4. Kragh, H. *Niels Bohr and the Quantum Atom: The Bohr Model of Atomic Structure* 1913–1925, Oxford University Press, 2012.
5. Liboff, R. *Introductory Quantum Mechanics*, Addison–Wesley, 2002.
6. *Scientific Reports*, https://www.nature.com/srep/ Nature Publishing.
7. Vollhardt, K., C. Peter, and Neil E. Schore. *Organic Chemistry: Structure and Function*. W.H. Freeman and Company, 2007.

ANSWERS

Check Your Progress

1. According to Bohr's postulate, an electron travels a definite orbit around the nucleus, that is, the position and velocity of an electron in an atom is always known. The contradicting point by Heisenberg Uncertainty principle is that it is impossible to simultaneously determine both the position and velocity of an electron.

2. Davisson–Germer practically demonstrated that particles (i.e., electrons) possess wave nature.

3. According to de Broglie equation, $\lambda = h/mv$. We know that for a stationary orbit, its circumference must be an integer multiple of λ, such that $2\pi r = n\lambda$ or $\lambda = 2\pi r/n$. Thus, $h/mv = \dfrac{2\pi r}{n}$ or, mr = $n\,h/2\pi$. This is in accordance with Bohr's postulate, that is, J = $n\,h/2\pi$ and hence the statement is justified.

4. The simultaneous exact determination of position and momentum or any property related to momentum such as velocity is impossible, $\Delta x \times \Delta p \geq h/4\pi$

Orbit	Orbital
(a) They are definite circular paths present at definite distances from the nucleus where electrons revolve.	(b) They are regions around the nucleus that show the probability of finding electrons is maximum.
(b) Shape of orbit is circular.	(b) Shape of an orbital can be spherical (s orbital), dumb-bell (p orbital), or double dumb-bell (d orbital).
(c) It represents a 2-dimensional model with electrons moving in circular motion in one plane around the nucleus.	(c) It represents a 3-dimensional model with spherical movement of electrons around the nucleus.
(d) It can have $2n^2$ number of electrons, where n is the number of the orbits.	(d) It can accommodate a maximum of two electrons with opposite spins.

6. $\dfrac{\delta^2 \Psi}{\delta x^2} + \dfrac{\delta^2 \Psi}{\delta y^2} + \dfrac{\delta^2 \Psi}{\delta z^2} + \dfrac{8\pi^2 m}{h^2}\left(E - U\right)\Psi = 0$

7. The functions are satisfactory solutions of Schrödinger time-independent wave equation only for certain values of energy E. Such values are called *eigen values*. The corresponding functions that are satisfactory solutions of Schrödinger equation are called *eigen functions*.

8. Eigen function is single value, finite, and continuous for all possible values of the three co-ordinates, that is, x, y and z, including infinity ∞.

9. An electron is considered as a particle and the square of the wave function at any point in space represents the probability of finding an electron at that point at a given instant.

10. Refer to Section 1.5.1; Fig. 1.10

11. Refer to Section 1.5

12. The differences between conductors and semiconductors are as follows:

Conductor	Semiconductor
(a) No energy gap between valence and conduction band.	(a) Small energy gap between valence and conduction band.
(b) Valence band is either half-filled or partially-filled.	(b) Valence band is completely filled.
(c) Electrical conductivity decreases with increasing temperatures.	(c) Electrical conductivity increases with increasing temperatures.
(d) Impurities decrease electrical conductivity.	(d) Doping impurities enhance electrical conductivity.

13. Refer to Section 1.7

14. A compound possessing additional stability due to the presence of planar cyclic ring with uninterrupted $(4n + 2)$ π electrons is called aromaticity. Benzene is an example of aromatic compound.

15. Refer to Section 1.9

16. For a planar, cyclic compound to be aromatic, its uninterrupted π–cloud must contain $(4n + 2)$ π electrons, where n is any whole number.

17. Refer to Section 1.8.2

18. Frost diagrams are used to illustrate the distribution of electrons in aromatic systems. For illustration refer to Section 1.9
19. Refer to Figs 1.14 (H_2), 1.15 (N_2), 1.16 (O_2), 1.19 (CO), 1.20 (NO), 1.21 (HCl), 1.17 (F_2)
20. Refer to Section 1.7

Multiple Choice Questions

1. (b)	2. (c)	3. (b)	4. (d)	5. (b)	6. (c)	7. (a)
8. (d)	9. (d)	10. (c)	11. (b)	12. (d)	13. (c)	14. (b)
15. (a)	16. (c)	17. (b)	18. (a)	19. (b)	20. (c)	21. (d)
22. (a)	23. (d)	24. (b)	25. (c)	26. (a)	27. (b)	28. (a)
29. (b)	30. (a)	31. (a)	32. (c)	33. (b)	34. (c)	35. (a)

CHAPTER 2

Periodic Properties and Chemical Bonding

LEARNING OBJECTIVES

After reading this chapter, you will be able to:

- review the Modern Periodic Table of elements.
- understand the periodic trends of atomic properties in the Modern Periodic Table.
- explain the intermolecular forces of attractions in various molecules such as van der Waals forces, dipole–dipole forces, and hydrogen bonding.
- illustrate the various types of chemical bonding in molecules.
- discuss the types of hydridization with examples.

2.1 INTRODUCTION

The year 2019 was earmarked by IUPAC (International Union of Pure and Applied Chemistry) as the *Year of the Periodic Table* to mark the sesquicentenary of its formulation by Mendeleev 150 years ago. Since its development, scientists have been adding new elements and the number has grown to 118 after the announcement by IUPAC in November 2016 of four new elements—Nihonium (Nh), Moscovium (Mc), Tennessine (Ts), and Organesson (Og), of atomic numbers 113, 115, 117, and 118, respectively. If element 119 is confirmed in the coming years, an eighth period will have to be introduced in the periodic table. In June 2018 the article titled, 'The Limits of Nuclear Mass and Charge,' by W. Nazarewicz in *Nature Physics* tried to address the issue.

The earliest recorded attempt to sort elements was by J.W. Dobereiner in 1829, when elements were grouped in three on the basis of their similar properties. These groups were called *triads* (Table 2.1). He pointed out that *if elements in a triad are arranged in the order of increasing atomic weights, then the atomic weight of the middle element was the mean of the other two elements.* This is clearly reflected in the triad of lithium, sodium, and potassium whose atomic weights are 6.9, 23.0, and 39.0 amu, respectively.

Table 2.1 Dobereiner's triad

First triad	Atomic weight (amu)	Second triad	Atomic weight (amu)	Third triad	Atomic weight (amu)
Li	6.9	Ca	40.0	Cl	35.5
Na	23.0	Sr	87.6	Br	80.0
K	39.0	Ba	137	I	127.0
Mean	6.9 + 39.0/2 = 22.95	Mean	40 + 137/2 = 88.5	Mean	35.5 + 127.0/2 = 81.25

In 1865, J.A.R. Newlands observed that *if elements are arranged in the order of increasing atomic weights, every eighth element had properties similar to the first element.* As shown in Table 2.2, lithium and sodium will have similar properties, as sodium is placed at the eighth position with respect to lithium.

Table 2.2 Newland's octaves

Musical notes sa (do)	re (re)	ga (mi)	ma (fa)	pa (so)	dha (la)	ni (ti)
H	Li	Be	B	C	N	O
F	Na	Mg	Al	Si	P	S
Cl	K	Ca	Cr	Ti	Mn	Fe
Co, Ni	Cu	Zn	Y	In	As	Se
Br	Rb	Sr	Ce, La	Zr		

This came to be known as the *law of octaves*. Interestingly, law of octaves is same as the musical notes. However, Dobereiner's triad and law of octaves could not account for the properties of the numerous new elements which were being discovered at the time.

The Mendeleev Periodic Table was an attempt to comprehensively provide a scheme so as to explain the chemistry of known elements. In this case, elements were arranged in the order of increasing atomic weights so as to exhibit varying trends in their physical and chemical behaviour. In the Modern Periodic Table, atomic number is considered as a fundamental characteristic of an element, that is, the physical and chemical behaviour of elements vary with respect to their atomic numbers.

The Modern Periodic Table comprises 18 groups and 7 periods. Bury (1921) presented the arrangement of elements as per the electron distribution in Bohr atom model.

To understand chemical reactivity of elements, it is essential to establish a connection between the periodic table and electronic configuration of elements. On examining the electronic configuration of atoms, it is observed that a particular configuration repeats itself regularly when elements are arranged in the increasing order of their atomic numbers. This repetition at regular intervals is called *periodicity*.

2.2 GENERAL FEATURES OF MODERN PERIODIC TABLE

In the modern periodic table, elements are arranged in the order of increasing atomic numbers. The table comprises 18 vertical columns called *groups* or *families* and 7 horizontal rows. Each horizontal row is called *period* and the last vertical column is called the *zero group*. The remaining 17 columns form 8 groups. These groups are further subdivided into subgroups A and B. The subgroups lying to the right of group 8 are labelled as *subgroup A* and those lying on to the left are labelled as *subgroup B*. Group 8 is undivided and the seven periods contain 2, 8, 8, 18, 18, 32, and 18 elements, respectively and the 7th period is incomplete.

Recently, RSC (Royal Society of Chemistry) released an interactive Modern Periodic Table featuring the history, alchemy, podcasts, videos, and data trends across the periodic table. An RSC Periodic Table app is also available. For more info visit: http://www.rsc.org/periodic-table

2.2.1 Classification of Elements Based on Electron Configuration

On the basis of electron configuration, there are four types of elements.

Noble gases (or inert gases)   Elements in which all the available s, p, and d subshells of atoms are completely filled are called *noble* or *inert gases*. They do not have the tendency to gain or lose electrons and hence exhibit chemical inertness and are physically gaseous in nature.

s- and p-block elements   Elements where s and p subshells are progressively filled are called *representative elements*, namely s- and p-block elements.

Main transition elements   Elements in which d atomic orbitals are progressively filled are termed main transition or d-block elements.

Inner transition elements   Elements in which f atomic orbitals are progressively filled are called inner transition elements. Generally, there are two series of such elements in the periodic table, namely *lanthanide series* (4f orbitals are filled progressively) and *actinide* series (5f orbitals are filled progressively).

2.3 PERIODIC TRENDS IN PROPERTIES OF ELEMENTS

When elements are arranged according to their increasing atomic numbers, there is a recurrence of similar properties of elements at regular intervals. This is called *periodicity of elements*. Periodicity of elements occurs due to repetition of similar electronic configuration at regular intervals in the modern periodic table. There are two ways to explore periodicity in the modern periodic table, that is, within a group and within a period. The properties of elements and their periodicities are discussed in detail in the following sections.

2.3.1 Atomic Size

The distance between the centre of the nucleus and the outer shell of an atom is called its *atomic radius*. The atomic size is determined by the shell in which the valence electrons are found and the interaction between the nucleus and the valence electrons. As one goes across the period, the atomic radii decreases steadily with increasing atomic number. This is due to increasing nuclear charge across the period called the *effective nuclear charge*. It is known that s, p, d orbitals in a given shell possess slightly different energies. These energy differences

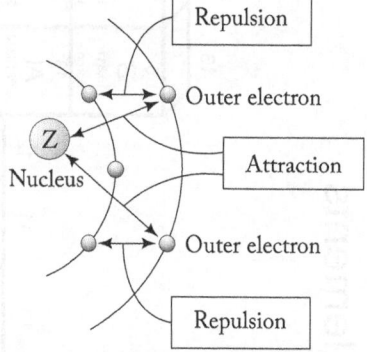

Fig. 2.1 Effective nuclear charge

between subshells cause electron–electron repulsion, which shields the outer electron from the nucleus. This net nuclear charge experienced by an electron is termed as *effective nuclear charge* (Z_{eff}).

For all elements except hydrogen, Z_{eff} is always less than the actual nuclear charge due to the presence of shielding effects. Greater the effective nuclear charge, more strongly are the outermost electrons attracted to the nucleus and consequently smaller the atomic radius. The effective nuclear charge increases toward the nucleus as per the trend: $ns > np > nd$. Figure 2.2 is a plot of atomic radius (in picometres) against atomic numbers (1–71). The decrease in atom size from left to right across each period is depicted here.

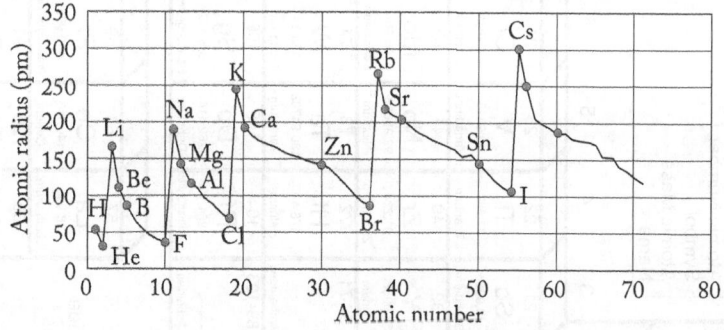

Fig. 2.2 Periodic trends in atomic radii

Further, as one goes down the group in the periodic table, an increase in atomic size is observed. This is because the number of energy levels increases from one element to the next and thus atomic size increases.

Periodic Table of Elements

Legend:
Atomic Number
Symbol
Atomic Mass
Name

Group	1	2	3	4	5	6	7	8	9	10	11	12	13	14	15	16	17	18
	1 H 1.008 Hydrogen																	2 He 8.002602 Helium
	3 Li 6.941 Lithium	4 Be 9.0121 Beryllium											5 B 10.811 Boron	6 C 12.0107 Carbon	7 N 14.0067 Nitrogen	8 O 15.9994 Oxygen	9 F 18.9984032 Fluorine	10 Ne 20.1797 Neon
	11 Na 22.990 Sodium	12 Mg 24.305 Magnesium											13 Al 26.9815386 Aluminum	14 Si 28.0855 Silicon	15 P 30.973762 Phosphorus	16 S 32.065 Sulfur	17 Cl 35.453 Chlorine	18 Ar 39.948 Argon
	19 K 39.098 Potassium	20 Ca 40.078 Calcium	21 Sc 44.996 Scandium	22 Ti 47.867 Titanium	23 V 50.9415 Vanadium	24 Cr 51.9961 Chromium	25 Mn 54.938045 Manganese	26 Fe 55.845 Iron	27 Co 58.933195 Cobalt	28 Ni 58.6934 Nickel	29 Cu 63.546 Copper	30 Zn 65.38 Zinc	31 Ga 69.723 Gallium	32 Ge 72.64 Germanium	33 As 74.92160 Arsenic	34 Se 78.96 Selenium	35 Br 79.904 Bromine	36 Kr 83.798 Krypton
	37 Rb 85.4688 Rubidium	38 Sr 87.62 Strontium	39 Y 88.90585 Yttrium	40 Zr 91.224 Zirconium	41 Nb 92.90638 Niobium	42 Mo 95.96 Molybdenum	43 Tc [98] Technetium	44 Ru 101.07 Ruthenium	45 Rh 102.90550 Rhodium	46 Pd 106.42 Palladium	47 Ag 107.8682 Silver	48 Cd 112.411 Cadmium	49 In 114.818 Indium	50 Sn 118.710 Tin	51 Sb 121.760 Antimony	52 Te 127.60 Tellurium	53 I 126.90447 Iodine	54 Xe 131.293 Xenon
	55 Cs 132.9054 Cesium	56 Ba 137.327 Barium	57-71 Lanthanides	72 Hf 178.49 Hafnium	73 Ta 180.94788 Tantalum	74 W 183.84 Tungsten	75 Re 186.207 Rhenium	76 Os 190.23 Osmium	77 Ir 192.217 Iridium	78 Pt 195.084 Platinum	79 Au 196.966569 Gold	80 Hg 200.59 Mercury	81 Tl 204.3833 Thallium	82 Pb 207.2 Lead	83 Bi 208.98040 Bismuth	84 Po [209] Polonium	85 At [210] Astatine	86 Rn [222] Radon
	87 Fr 223.020 Francium	88 Ra 226.025 Radium	89-103 Actinides	104 Rf 261.109 Rutherfordium	105 Db 262.114 Dubnium	106 Sg 266.122 Seaborgium	107 Bh 264.120 Bohrium	108 Hs 277 Hassium	109 Mt 268.139 Meitnerium	110 Ds 281 Darmstadtium	111 Rg 280 Roentgenium	112 Cn 285 Copernicium	113 Nh 286 Nihonium	114 Fl 289 Flerovium	115 Mc 288 Moscovium	116 Lv 293 Livermorium	117 Ts 294 Tennessine	118 Og 294 Oganesson

Lanthanide Series

57 La 138.905 Lanthanum	58 Ce 140.116 Cerium	59 Pr 140.908 Praseodymium	60 Nd 144.242 Neodymium	61 Pm 145 Promethium	62 Sm 150.36 Samarium	63 Eu 151.964 Europium	64 Gd 157.25 Gadolinium	65 Tb 158.925 Terbium	66 Dy 162.500 Dysprosium	67 Ho 164.930 Holmium	68 Er 167.259 Erbium	69 Tm 168.934 Thulium	70 Yb 173.054 Ytterbium	71 Lu 174.967 Lutetium

Actinide Series

89 Ac 227 Actinium	90 Th 232.038 Thorium	91 Pa 231.035 Protactinium	92 U 238.028 Uranium	93 Np 237 Neptunium	94 Pu 244 Plutonium	95 Am 243 Americium	96 Cm 247 Curium	97 Bk 247 Berkelium	98 Cf 251 Californium	99 Es 252 Einsteinium	100 Fm 257 Fermium	101 Md 258 Mendelevium	102 No 259 Nobelium	103 Lr 262 Lawrencium

Penetration of Orbitals

Penetration of orbitals refers to the extent with which the orbital of a shell interacts with lower quantum shell orbitals. This concept is relevant to polyelectronic atoms. We know that electrons get screened from the nuclear charge due to repulsion of electrons. Unlike hydrogen atom, a polyelectronic atom possesses many electrons in different shells or orbitals. These orbitals vary in energies and are designated with a specific principal quantum number. The variation in energies of different orbitals is given as, $E_{ns} < E_{np} < E_{nd} < E_{nf}$ where n is different values of shell n.

Energy of an electron for each shell and subshell follows the order:

1s < 2s < 2p < 3s < 3p < 4s < 3d < 4p < 5s < 4d < 5p < 6s < 4f < 5d < 6p < 7s <5f < 6d

Understanding the variation of orbital energies is a complicated matter; however it can be qualitatively explained by taking a simple example of 2s and 2p orbitals. We know that 2s orbital has lower energy than 2p orbital in a polyelectronic atom. The radial probability distribution plot of these two orbitals shows that 2p is actually preferable (due to lower energy) compared to 2s orbital. However, on looking closely, there is a small hump of electron density for 2s orbital very close to the nucleus. This means that an electron in 2s spends a little more time near the nucleus than an electron in 2p orbital. Hence, a 2s electron penetrates the nucleus more than a 2p orbital that causes the 2s electron to be attracted more towards the nucleus. Hence, the 2s orbital is lower in energy than the 2p orbital in a polyelectronic atom.

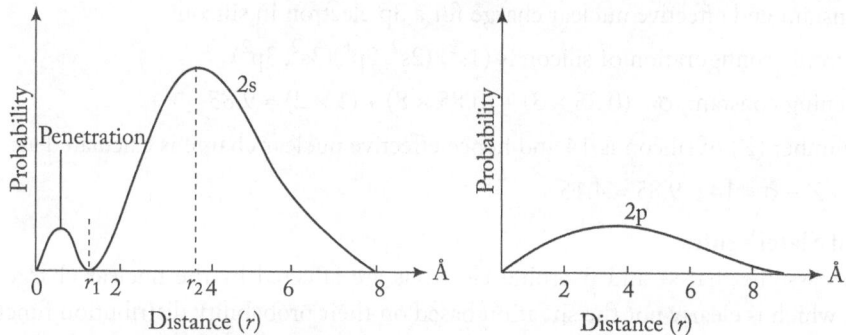

Fig. 2.3 Radial probability distribution of 2s and 2p orbitals

The penetration increases as the shell number increases. The penetration of orbitals of a given shell is in the order s > p > d > f.

Hence as per above order of energies, the 5s orbital penetrates both 4f and 4d orbitals. Also, 6s orbital penetrates 4f, 5d, and 5f orbitals. As the penetration of orbitals follow the above order, shielding of orbitals also follows the same order. The order of penetration of orbitals can hence be given as:

1s > 2s > 2p > 3s > 3p > 4s > 3d > 4p > 5s > 4d > 5p > 6s > 4f > 5d > 6p > 7s >5f > 6d

Slater's Rule

J.C. Slater gave a set of empirical generalizations to determine the atomic screening constants (σ). These rules predict the actual charge felt by an electron in a given shell. Slater's rule helps to estimate the effective nuclear charge Z_{eff} experienced by an electron.

(i) First write the electron configuration of the given atom in the below form:

(1s) (2s, 2p) (3s, 3p) (3d) (4s, 4p) (4d) (4f) (5s, 5p) . . .

(ii) Identify the electron of interest and ignore all electrons in higher groups (i.e., those on their right) of the order as shown in the above step. These groups are ignored as they do not shield electrons in lower groups.

(iii) Slater's rules are different based on the presence of an electron in a given shell and hence divided into two categories; (a) the shielding experienced by s and p electrons and (b) shielding experienced by d and f electrons.

Shielding experienced by ns and np electrons If the electrons are within the same group it will shield 0.35, except the 1s electron which shields 0.30. The electrons within the n−1 group shield 0.85, whereas electrons within the n−2 or lower groups shield 1.00.

Shielding experienced by nd or nf valence electrons The electrons within same group will shield 0.35, whereas electrons within the lower groups shield 1.00.

Let us take an example and apply Slater's rule to calculate screening constant and effective nuclear charge in nitrogen for 2p electron.

As per Slater's rule (i), the electronic configuration for nitrogen is written as: $(1s^2) (2s^2, 2p^3)$

There are three electrons in 2p shell and one of the electrons is under consideration (or question). We need to determine the nuclear charge felt by this electron. Hence, there are four electrons in group $(2s^2, 2p^2)$. Going by Slater's rule,

Screening constant, $\sigma = (0.35 \times 4) + (0.85 \times 2) = 3.10$

The atomic number (Z) of nitrogen is 7 and hence effective nuclear charge is calculated as:

Effective nuclear charge, $Z_{eff} = Z - \sigma = 7 - 3.10 = 3.90$

Screening constant and effective nuclear charge for a 3p electron in silicon:

Electronic configuration of silicon is $(1s^2) (2s^2, 2p^6)(3s^2, 3p^2)$.

Screening constant, $\sigma = (0.35 \times 3) + (0.85 \times 8) + (1 \times 2) = 9.85$

The atomic number (Z) of silicon is 14 and hence effective nuclear charge is calculated as:

$Z_{eff} = Z - \sigma = 14 - 9.85 = 4.15$

Drawbacks of Slater's rules

(a) The rules assume that s- and p-orbital electrons are affected by the nuclear charge in the same manner, which is clearly not the situation based on their probability distribution functions.

(b) All electrons in s, p, d, or f orbitals are taken to shield higher shell electrons with equal competence which is not true as the orbital energies are widely different.

(c) Penetration of higher orbitals in to inner core orbitals (particularly s orbitals) is completely ignored and hence the values so obtained are arbitrary.

2.3.2 Ionization Energy (Ionization Potential)

The energy required to pull off or remove an electron from a free or gaseous atom of an element is called its ionization energy or ionization potential.

$$X_{(g)} \rightarrow X^+_{(g)} + e^-$$

Generally, ionization energy is determined from atomic spectra and measured in electron volts (eV). The energy required for the above reaction is called *first ionization energy*. Further, when second and third electrons are removed from the atom, they are referred to as second and third ionization energies, respectively.

The factors that affect ionization energy of elements are as follows:

Atomic radius Greater the distance of an electron from the nucleus, smaller is the ionization energy, as the electron is less strongly held by the nucleus. Thus, ionization energy decreases with increase in atomic size.

Nuclear charge Greater the charge on the nucleus, greater is its hold over the electrons and hence higher will be its ionization energy.

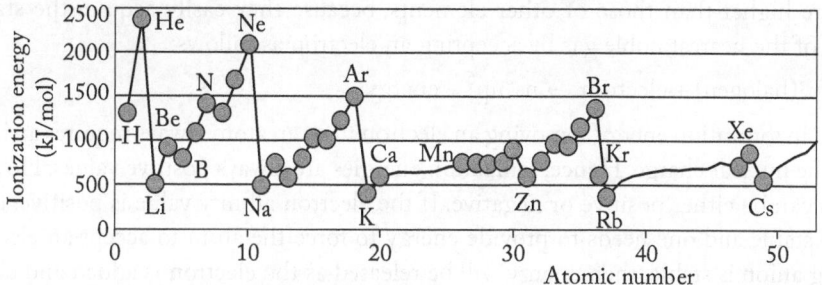

Fig. 2.4 First ionization energy (in kJ/mol) against atomic number for the first 50 elements in the periodic table

After removal of the first electron from an atom, there is an increase in the effective nuclear charge. The size of an ion decreases and hence the nuclear hold over the remaining electrons increases. The removal of the second electron becomes difficult and consequently, *second ionization energy is always greater than the first ionization energy.*

The periodic trend of ionization energy as a function of atomic number is depicted in Fig. 2.4. *In a given period, ionization energy increases from left to right in the periodic table.* This is due to increase in nuclear charge along the period. In any vertical column of the modern periodic table, *ionization energy steadily decreases from top to bottom, due to increasing atomic size.*

Shielding effect of other electrons The attractive force exerted by the nucleus on the most loosely bound electron is partially counterbalanced by repulsive forces exerted by inner electrons, resulting in decrease of ionization energies. More the number of electrons, highly effective will be its shielding and consequently, ionization energy decreases.

Type of electrons removed As seen in Chapter 1, the s orbital has electrons closer to the nucleus compared to p, d, f orbitals. Hence, an electron in s orbital is tightly held by the nucleus in comparison to d or f orbital electrons. Thus, ionization energy decreases in the order: s > p > d.

Exceptions Looking closer at Fig. 2.4, it is observed that there is a slight decrease in ionization energy from nitrogen to oxygen rather than the expected increase as per periodicity. This can be explained by considering electron pairing of both elements. Nitrogen has half-filled p subshell and one electron in each 2p orbital. In oxygen, to fulfil its p^4 configuration, it must pair two electrons in one of the 2p orbitals. Now, removing one electron will leave only one unpaired electron in all three 2p orbitals. This leads to lower electron–electron repulsion making it obvious that ionization energy of oxygen is lower than that of nitrogen. Similar exceptions in trend are also observed in sulphur and selenium.

We observe that the first ionization energy of copper is higher than that of potassium though in both these elements, outer electron is to be removed from 4s orbital. The electronic configurations of copper and potassium are $[Ar] 3d^{10}, 4s^1$ and $[Ar]3p^6, 4s^1$ respectively. The 3d subshell in copper is less effective in shielding the outer 4s electron as compared to p subshell in potassium. Hence, the 4s electron of copper is tightly held by the nuclear charge than in potassium. Thus, the first ionization energy of copper (745 kJ/mol) is more than that of potassium (418 kJ/mol). However, the second and third ionization energies of copper are lower than potassium due to loss of electrons from diffused d orbitals.

2.3.3 Electron Affinity

Electron affinity is the energy released when a free or a gaseous atom accepts an electron to form a free gaseous ion as per the following general reaction, $X_{(g)}$ + electron $\rightarrow X^-_{(g)}$ + energy.

Similar to ionization energy, electron affinity is also expressed as electron volts (eV) or kilojoules per mole (kJ/mol). Larger the energy released, larger will be the electron affinity. It is seen that electron affinities of halogens are higher than those of other elements, because they easily acquire the stable electronic configuration of the nearest noble gas by accepting an electron as follows:

$$\text{ns}^2\text{np}^5 \text{ (halogen)} + \text{electron} \rightarrow \text{ns}^2\text{np}^6 + \text{energy}$$

As discussed in ionization energy, removing an electron from an atom always requires an input of energy to overcome the nuclear charge. Hence, ionization energies are always positive values. However, electron affinity values can be either positive or negative. If the electron affinity value is positive, then the anion formed is not stable and one needs to provide energy to force the atom to accept an electron; whereas, if the resulting anion is stable, then energy will be released as the electron is added and electron affinity will be a negative value.

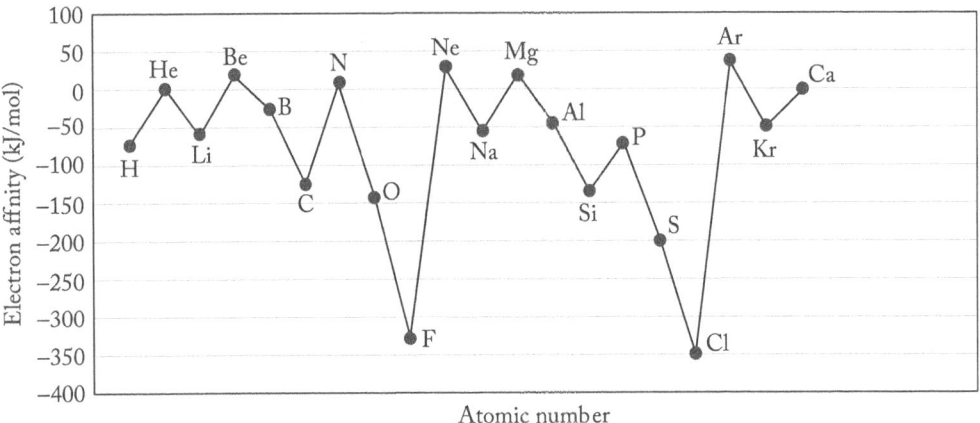

Fig. 2.5 Electron affinity (kJ/mol) as a function of atomic number for elements 1–20 (H to Ca)

As can be seen in Fig. 2.5, electron affinity values vary more erratically as compared to ionization energy. As one moves *from top to bottom in a group, electron affinity decreases (becomes less negative)*. When one moves from *left to right in the periodic table, electron affinity increases (become more negative)*. Also, more negative the electron affinity becomes, the more stable will be its anion. On a closer look at the trend, exceptions are noticed in the case of beryllium, magnesium, nitrogen, and the noble gases. The exceptions are similar to electron pairing as in the case of ionization energy.

Exceptions (a) Beryllium and magnesium have completely filled s subshell and an additional electron tends to go to the p subshell of considerably higher energy that is otherwise left empty in a neutral atom. This occupancy of a higher energy subshell is energetically unfavourable and accounts for the positive values of electron affinity in beryllium (8 kJ/mol) and magnesium (21 kJ/mol).

(b) Nitrogen and phosphorus have half-filled p orbital and thus possess extra stability. Adding an electron to the stable p orbital leads to increased electron–electron repulsion, which is energetically unfavourable. Thus, nitrogen has a positive electron affinity value (7 kJ/mol), whereas phosphorus has a negative value (–72 kJ/mol).

(c) Noble gases have completely filled shells that strongly discourage addition of an electron and hence they have positive electron affinity values.

2.3.4 Electronegativity

A relative measure of the tendency of a bonded atom to attract an electron towards itself is called electronegativity. Being a qualitative property, there is no standardized method for calculating electronegativity. Pauling (1932) derived electronegativity values by calculating the energies required to break chemical bonds in molecules. According to Pauling, electronegativity difference between two atoms is $0.18\sqrt{\Delta}$, where Δ is resonance energy (kJ/mol). Let us assume a hypothetical diatomic molecule A–B such that 100% covalent bond energy equals an arithmetic mean of covalent energies of A–A and B–B molecules; hence, $E_{100\% \text{ covalent A–B}} = 1/2 \left(E_{A–A} + E_{B–B} \right)$

where, $E_{A–A}$ and $E_{B–B}$ are bond energies in diatomic species A_2 and B_2, respectively.

To find the value for the additional stability of a molecule, Pauling suggested geometric mean instead of the above arithmetic mean as, $E_{100\% \text{ covalent A–B}} = \sqrt{E_{A–A} \cdot E_{B–B}}$

Hence, resonance energy D = actual bond energy $- \sqrt{E_{A–A} \cdot E_{B–B}}$

The above equation accounts for the electronegativity difference between atoms and additional stability of bonds in polar molecules.

However, relating electronegativity values on bond energies between atoms is not the only approach to understand the ability of atoms in a molecule to attract electrons, that is, electron affinity. One can also relate ionization potential and electron affinity of an atom to express its electronegativity. Mulliken (1934) proposed electronegativity (χ) expression of an atom (say A) as, $\chi_A = 1/2 \left[I_A + E_A \right]$, where I_A is ionization potential and E_A is electron affinity of atom A and expresses the average of these two properties to calculate the electronegativity of an atom. Mulliken successfully calculated electronegativity values for H, Li, B, C, N, O, F, Cl, Br, and I and these were found to be in agreement with Pauling scale.

Electronegativity increases from left to right across the periodic table It is known that elements follow the octet rule (see Section 2.4.1 Kössel–Lewis approach to Chemical Bonding), that is, tendency of atoms to have eight electrons in the valence or outer shell. As elements on the left side of the periodic table have less than a half-full valence shell, the energy required to gain electrons is significantly higher compared to the energy required to lose electrons. However, if valence shell is more than half-full, it is easier to attract an electron in to its shell. As the effective nuclear charge increases from left to right in a row, electronegativity increases in the same direction. As atomic number increases down a group, the atomic size increases with eventual decrease in effective nuclear charge. Thus, if one moves *from top to bottom along a group, electronegativity decreases.* According to these general trends, fluorine is the most electronegative element with 3.98 Pauling units.

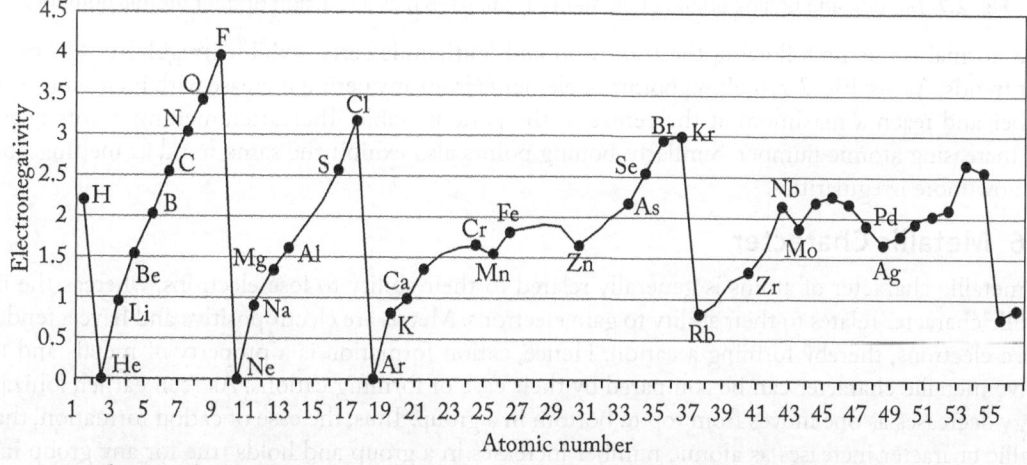

Fig. 2.6 Electronegativity as a function of atomic number of elements 1–56 (H to Ba)

Exceptions   Some exceptions to the electronegativity trend are observed in noble gases and lanthanide and actinide series. As noble gases have completely filled valence shells, they do not attract electrons. Elements in the lanthanide and actinide series have a complicated chemistry and do not exhibit the general trend. Hence, noble gases, lanthanides, and actinides do not have standard electronegativity values as other elements.

2.3.5 Melting and Boiling Points

To put it simply, melting point is the amount of energy required to break the bonds to change a substance from its solid phase into its liquid form. Stronger the bond between the atoms of an element, higher the energy required to break the bonds. As temperature is directly proportional to energy, a high bond dissociation energy correlates to a higher temperature. Melting and boiling points vary randomly as shown in Fig. 2.7; however, some of the pertinent conclusions drawn from the graph are as follows.

(a) Metals generally possess a high melting point.

(b) Most non-metals possess low melting points; but carbon (non-metal) possesses the highest boiling point of all the elements. Boron also has a high melting point.

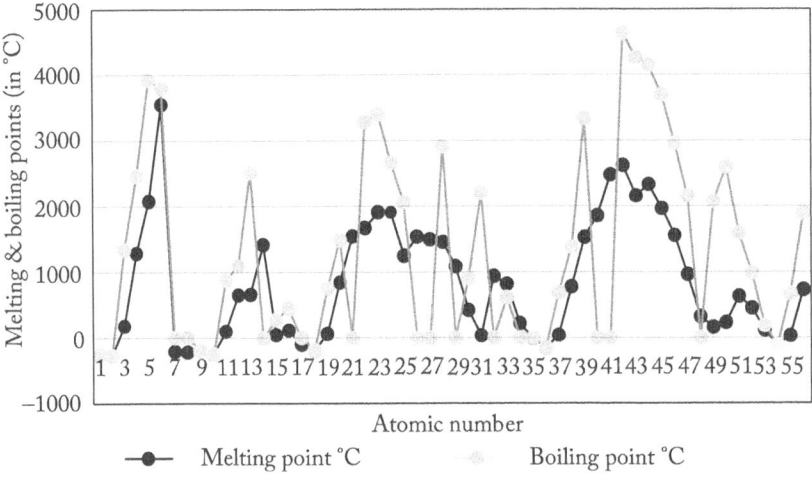

Fig. 2.7 Melting and boiling points of elements 1–56, (H–Ba) as a function of their atomic numbers

The normal elements following the transition and lanthanide series exhibit irregularities in melting point trends. As per Fig. 2.7, melting points of elements from any period increase with increasing atomic number and reach a maximum at the centre of the periodic table. Thereafter, melting points decrease with increasing atomic number. Similarly, boiling points also exhibit the same trend as melting points, but show more irregularities.

2.3.6 Metallic Character

The metallic character of atoms is generally related to their ability to lose electrons, whereas the non-metallic character relates to their ability to gain electrons. Metals are electropositive and have a tendency to lose electrons, thereby forming a cation. Hence, cation formation is a property of metals and their relative metallic character can be compared by their ease of forming cations. As seen earlier, ionization energy decreases as one moves from top to bottom in a group. Thus, the ease of cation formation, that is, metallic character increases as atomic number increases in a group and holds true for any group in the periodic table. Further, metallic character decreases from left to right across a period due to decrease in atomic radii that allows the outer electrons to ionize readily.

2.3.7 Polarizability

The extent to which electrons get disturbed from their bonding arrangement due to another approaching molecule is called polarizability. As electrons are charged particles, they respond to electric fields. When an electric field is applied to a molecule, electrons exhibit mobility to varying degrees and the electron cloud is said to be polarized in the applied electric field. This results in an induced dipole along with already existing permanent dipole in the molecule.

Polarizability decreases from left to right across the row of the periodic table because of increasing effective nuclear charge. Hence, atomic polarizabilities C > N > O > F and molecular polarizabilities CH_4 > NH_3 > H_2O are observed. It must be noted that atoms that hold onto their electrons tightly are not polarizable and smaller the particles, lesser will be their polarizability as their electrons are tightly held by the nucleus. *If one moves down the group in a periodic table, polarizability increases* because of increasing atomic sizes and a large electron cloud tends to get distorted more easily, that is, S > O; P > N (atomic polarizability); and H_2S > H_2O (molecular polarizability). While moving down the group, the greater polarizability stabilizes the negative charge as, F < Cl < Br < I, in which fluorine is the least polarizable and iodine most polarizable and hence the most stable.

2.3.8 Oxidation States

A number numerically equal to the charge and sign of the charge which the atom of an element appears to have acquired when in combination with atoms of other elements is called its oxidation state. Various possible oxidation states of a particular element can be determined by its electronic configuration.

The elements that have an incomplete outer shell tend to gain or lose electrons. Generally, metals have positive oxidation states, whereas non-metals have negative oxidation states. For s-block and p-block elements, the maximum oxidation state equals the group number. As for p-block elements, the lowest oxidation state equals to eight minus the group number and all transition (d-block) elements exhibit variable oxidation states as shown in Table 2.3.

Table 2.3 Oxidation states of transition elements forming a regular pyramid
(Underlined values are stable oxidation states)

Characteristic electrons	Sc d^1s^2	Ti d^2s^2	V d^3s^2	Cr d^5s^1	Mn d^5s^2	Fe d^6s^2	Co d^7s^2	Ni d^8s^2	Cu $d^{10}s^1$
Oxidation states				1					1
	2	2	2	2	2	<u>2</u>	<u>2</u>	<u>2</u>	<u>2</u>
	<u>3</u>	3	3	<u>3</u>	3	<u>3</u>	<u>3</u>	3	3
		<u>4</u>	4	<u>4</u>	4	4	4	4	4
			<u>5</u>	5	5	5	5		
				<u>6</u>	6	6			
					<u>7</u>				

Check Your Progress

1. What is Dobereiner's triad? State Newland's law of octaves.
2. Using only the periodic table, rank the following elements in order of decreasing atomic size: (a) Cr, Cs, F, Si, and Sr (b) Fe, K, Rb, S, and Se (c) Li, Be, B, C and N (d) C, Si, Sn, Pb and Ge (e) Ti, Cr, Mn, Fe and V.

2.4 CHEMICAL BONDING

When attractive forces hold various constituents (atoms, ions, etc.) together in different chemical species it is called a *chemical bond*. Let us enumerate the types of bonds that are commonly found in chemical compounds and various forces of attraction.

2.4.1 Kössel–Lewis Approach to Chemical Bonding

In 1916, Kössel and Lewis independently provided a successful theory about the formation of chemical bonds and explained valence based on noble gas inert behaviour. The theory of chemical combination between atoms put forth came to be known as *electronic theory of chemical bonding*. According to this, atoms can combine either by transfer of valence electrons from one atom to another (gain or lose electrons) or by sharing of valence electrons in order to have an octet in their valence shells. This is popularly known as the *octet rule*.

Kössel's Theory

In the periodic table, highly electronegative halogens and highly electropositive alkali metals are clearly separated by noble gases. Halogens form negatively charged ions by gaining an electron but alkali metals form positively charged ions by losing an electron. When positive and negative ions are formed, they attain a stable noble gas electronic configuration. The general electronic configuration of noble gases (except helium) is given by ns^2np^6.

The positive and negative ions are stabilized by electrostatic attraction. For example, in NaCl, the sodium ion (positive) and chlorine ion (negative) are held together by electrostatic attractive forces.

$$\underset{\text{[Ne]}\,3s^1}{\text{Na}} \rightarrow \underset{\text{[Ne]}}{\text{Na}^+} + e^- \qquad \underset{\text{[Ne]}\,3s^2 3p^5}{\text{Cl} + e^-} \rightarrow \underset{\text{[Ne]}\,3s^2\,3p^6 \text{ or [Ar]}}{\text{Cl}^-}$$

$$\text{Na}^+ + \text{Cl}^- \rightarrow \text{NaCl (or Na}^+\text{Cl}^-) \text{ Sodium chloride}$$

Similarly, formation of calcium fluoride can be depicted as:

$$\underset{\text{[Ar]}\,4s^2}{\text{Ca}} \rightarrow \underset{\text{[Ar]}}{\text{Ca}^{2+}} + 2e^- \qquad \underset{\text{[He]}\,2s^2 2p^5}{\text{F} + e^-} \rightarrow \underset{\text{[He]}\,2s^2\,2P^6 \text{ or [Ne]}}{\text{F}^-}$$

$$\text{Ca}^{2+} + 2\text{F}^- \rightarrow \text{CaF}_2 \text{ (or Ca}^{2+}(\text{F}^-)_2) \text{ Calcium fluoride}$$

Such type of chemical bonding that exists between positive and negative ions is called *electrovalent bond*.

Lewis Theory

(a) An atom can be considered as a positively charged 'kernel' (nucleus and inner electrons) with an outer shell which can accommodate not more than eight electrons.

(b) These eight electrons were assumed to occupy the corners of a regular cube that surround the kernel. Hence, the single shell electron of sodium will occupy one corner of a cube whereas all eight corners will be occupied by eight electrons in noble gases.

(c) The atoms having octet of electrons in the outermost shell represents a stable configuration. Lewis put forth the explanation that atoms achieve stable octet when linked by chemical bonds with other atoms.

(d) Such chemical bonds can be formed either by gaining or losing electrons such as sodium chloride and magnesium chloride. In some cases, chemical bond may arise due to sharing of a pair of electrons as in H_2, F_2, Cl_2 molecules. In the process, each atom attains a stable outer octet of electrons.

(e) During the formation of a molecule, only the outer shell electrons take part in chemical combination and are termed as *valence electrons*. The inner electrons are well protected and are not involved in bond formation.

Lewis introduced simple notations to represent valence electrons in an atom which are called Lewis symbols. Lewis symbols for the second period (lithium to neon) can be depicted as follows:

$$\dot{Li} \quad \dot{Be} \quad \cdot\dot{B}\cdot \quad \cdot\dot{C}\cdot \quad \cdot\dot{N}: \quad :\dot{O}: \quad :\dot{F}: \quad :\dot{Ne}:$$

In a Lewis symbol, the number of dots that surround the respective chemical symbol represents the number of valence electrons in that atom. The valency of an element is either equal to the number of dots in Lewis symbol or 8 minus the number of dots or valence electrons.

Lewis Representation of Simple Molecules

The Lewis dot structures provide a picture of bonding in molecules and polyatomic ions in terms of the shared pairs of electrons. The basic steps of writing Lewis dot structures are:

(i) Add valence electrons of the combining atoms as this gives the total number of electrons while writing the structure. If we consider methane (CH_4), there are eight valence electrons (four each from carbon and hydrogen).

(ii) In case of polyatomic ions, add one electron to the total number for each unit negative charge. Say, in case of carbonate (CO_3^{2-}) ion, the total number of electrons is given as:

Total electrons in CO_3^{2-} ion = 4 electrons from carbon + 3 (6 electrons from each oxygen) + 2 (1 electron for each negative charge) = 4 + 3 (6) + 2(1) = 24 electrons

Check Your Progress

3. Using only the periodic table, rank the following elements in ascending order of ionization energy: (a) He, Mg, N, Rb, and Si; (b) Li, P, F, Ca.
4. Justify the statement, 'Fluorine is the most electronegative amongst all elements.'
5. Write the equations of electronegativity proposed by Pauling and Mulliken.
6. List the factors affecting ionization potential of an atom.
7. Mark the locations for (a) third and fifth period noble gas (b) d-block element with one 4s electron and (c) p-block element that is a metal (d) element with the highest electronegativity (on Pauling scale) (e) s-block element with an effective nuclear charge of 1.3 (amu) in the blank modern periodic table given below:

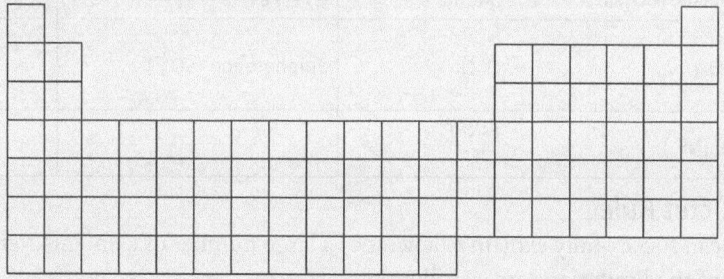

8. Define: (a) ionization potential, (b) electronegativity, and (c) electron affinity.
9. Justify: The first ionization energy of copper is higher than that of potassium though in both these elements, outer electron is to be removed from 4s orbital.
10. Why is the electron affinity of fluorine less than that of chlorine?
11. What is polarizability? State its periodicity in the modern periodic table.
12. Calculate the screening constant and effective nuclear charge of 3d electrons in zinc.
13. Define effective nuclear charge.

(iii) In case of a polyatomic cation, subtract one electron for each unit positive charge. For example, in ammonium ion (NH_4^+) one positive charge indicates loss of one electron.

(iv) Write the skeletal structure of the molecule or ions such that the least electronegative atom is at the centre and more electronegative atoms are written at terminal positions (an educated guess); for example, in CO_3^{2-} ion carbon will be the central atom and oxygen will occupy the terminal positions.

(v) Distribute the electrons properly as shared pairs that is in proportion to the total number of bonds. Once shared pairs of electrons are written accounting only for single bonds, any remaining electrons can be expressed for multiple bonding or indicative of lone pairs.

To understand the above rules, let us take an example of carbon monoxide.

(a) First count the total number of valence electrons of carbon and oxygen atoms. The valence shell electronic configurations of carbon and oxygen atoms are: $2s^2, 2p^2$ and $2s^2, 2p^4$ respectively. Hence, the valence electrons available for bonding are 4 + 6 =10 electrons.

(b) The skeletal structure of carbon monoxide is written as CO. Draw a single bond (i.e. one shared electron pair) between C and O and complete the octet on O, the remaining two electrons are the lone pair on C.

$$:C:\ddot{O}: \quad \text{or} \quad :C-\ddot{O}:$$

(c) This does not complete the octet on carbon and hence we use electrons to explain its multiple bonding (here it is a triple bond) between C and O atoms, satisfying the octet rule condition for both carbon and oxygen atoms. The final Lewis dot formula for carbon monoxide is:

$$:C::O: \quad \text{or} \quad :C\equiv O:$$

Table 2.4 Lewis dot symbols of some molecules and polyatomic ions

Hydrogen molecule (H_2)	H:H	Carbonate ion $(CO_3)^{2-}$	$\left[\begin{array}{c} :\ddot{O}: \\ :\ddot{O}:C:\ddot{O}: \end{array}\right]^{2-}$
Fluorine molecule (F_2)	$:\ddot{F}:\ddot{F}:$	Nitrite ion (NO_2^-)	$:\ddot{O}::N:\ddot{O}:^-$
Methane molecule (CH_4)	$\begin{array}{c}H\\H:\overset{}{\underset{}{C}}:H\\H\end{array}$	Nitric acid (HNO_3)	$\begin{array}{c}\ddot{O}::\overset{+}{N}\ :\ddot{O}:H\\ :\ddot{O}:\end{array}$
Carbon dioxide molecule (CO_2)	$\ddot{O}::C::\ddot{O}$	Hydrogen peroxide (H_2O_2)	$H:\ddot{O}:\ddot{O}:H$
Oxygen molecule (O_2)	$:\ddot{O}::\ddot{O}:$	Sulphate ion (SO_4^{2-})	$\begin{array}{c}:\ddot{O}:\\ ^-:\ddot{O}:\overset{}{\underset{}{S}}:\ddot{O}:^-\\ :\ddot{O}:\end{array}$
Nitrogen trifluoride (NF_3)	$\begin{array}{c}:\ddot{F}:N:\ddot{F}:\\ :\ddot{F}:\end{array}$		

Limitations of Octet Rule

Though octet rule can successfully explain valencies of a large number of elements, yet suffers from many limitations. Some of the limitations are as follows:

Incomplete octet of the central atom According to octet rule, elements of groups 1, 2, and 13 will not form covalent bonds as the central atoms have less than four electrons in their valence shells. These elements cannot achieve octet state by mutual sharing of electrons. However, it is found that some of the elements of these groups can form covalent compounds. Lithium chloride (LiCl), beryllium dihydride (BeH_2), and boron trichloride (BCl_3) are some examples of compounds showing covalency (Fig. 2.8). Li, Be, and B

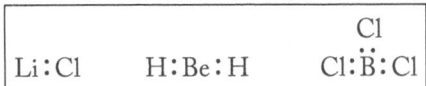

$$Li:Cl \qquad H:Be:H \qquad \begin{array}{c}Cl\\ :\ddot{}:\\ Cl:B:Cl\end{array}$$

Fig. 2.8 Lewis dot formulae for LiCl, BeH_2, and BCl_3

have only 1, 2, and 3 valence electrons. Some other such compounds are $AlCl_3$ and BF_3.

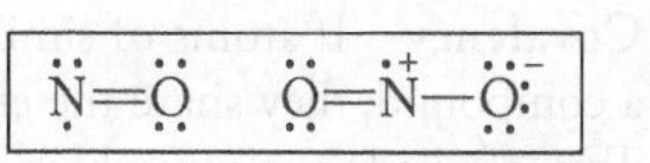

Odd-electron molecules When molecules possess odd number of electrons such as nitric oxide (NO) and nitrogen dioxide (NO_2), octet rule is not satisfied for all the atoms (Fig. 2.9).

$$\ddot{N}=\ddot{O} \qquad \ddot{O}=\overset{+}{\ddot{N}}-\ddot{O}^{-}$$

Fig. 2.9 Lewis dot formulae for NO and NO_2

Formation of super octet molecules Octet rule is violated in case of elements present in third period and beyond in the periodic table. This is because the central atom has also 3d orbitals besides 3s and 3p for bonding. Hence, in a number of compounds of these elements there are more than eight valence electrons around the central atom. This is referred to as the expanded octet. Phosphorus pentafluoride (PF_5), sulphur hexafluoride (SF_6), sulphuric acid (H_2SO_4), iodine heptafluoride (IF_7) are typical molecules, where the central atoms have more than eight valence electrons.

Xenon compounds As we know octet rule is primarily based on the fact that noble gases are inert. However, xenon and krypton being inert gases take part in bonding and form compounds with fluorine and oxygen, such as xenon fluorides (XeF_2, XeF_4), xenon oxyfluorides ($XeOF_2$, $XeOF_4$), and krypton fluoride (KrF_2). Thus, octet rule is invalid while explaining xenon compound formation.

(10 electrons around P atom)

(a) PF_5

(12 electrons around S atom)

(b) SF_6

(12 electrons around S atom)

(c) H_2SO_4

(14 electrons around I atom)

(d) IF_7

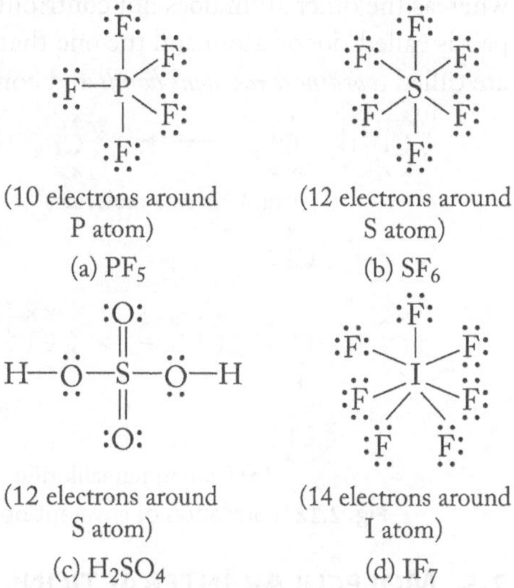

Fig. 2.10 Super octet molecules

Stability of molecules Octet theory does not explain the shape and geometry of molecules. Moreover, cannot explain the energies and stabilities of molecules.

2.4.2 Types of Chemical Bonds

Primarily, there are three types of chemical bonding—ionic or electrovalent, covalent, and coordinate covalent bonds.

Electrovalency or ionic valency When there is a combination of two atoms due to complete transfer of electrons from the outermost shell of one atom to another is called *electrovalency*.

Let us take the simple example of NaCl formation. The electronic structure of sodium is 2, 8, 1 and that of chlorine is 2, 8, 7. Thus, sodium atom can complete its octet by losing one electron, while chlorine can do so by gaining one electron.

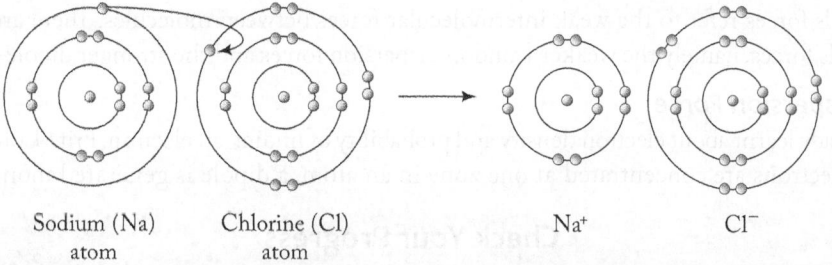

Sodium (Na) atom   Chlorine (Cl) atom Na^+ Cl^-

Fig. 2.11 Formation of sodium and chlorine ions

Such transfer of electrons from one atom to another atom results in the formation of sodium ions and chlorine ions as shown in Fig. 2.11. These ions are attracted towards each other due to electrostatic forces.

Covalency If atoms of similar or almost same electronegativity values combine to form molecules of a compound, they share the electrons in equal numbers to complete the octet in their outermost shells. This is called *covalency* which can be represented by dot-cross method as shown for chlorine and carbon tetrachloride (Fig. 2.12).

Coordinate covalency In coordinate bond formation, an electron pair is shared only by a single atom, whereas the other atom does not contribute towards bond formation. The atom that donates an electron pair is called 'donor' atom and the one that accepts the electron pair is called 'acceptor' atom. Such bonds are called *coordinate covalent bonds* and compounds so formed are called *coordination compounds*.

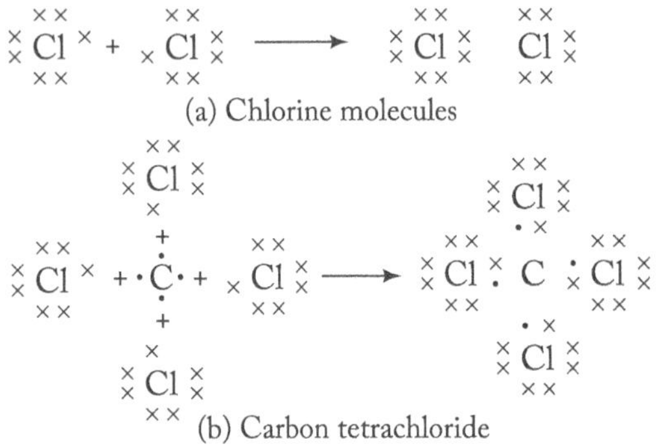

Fig. 2.12 Formation of covalent bonds

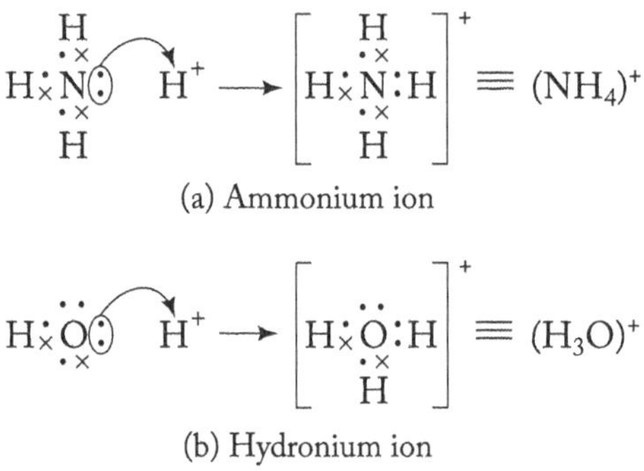

Fig. 2.13 Formation of coordinate bond

2.5 MOLECULAR INTERACTIONS

Even atoms and molecules with complete valency can interact with one another through various forms of interactions. The molecules as a whole can either attract or repel each other over several atomic diameters in distance.

Understanding such interactions is important while studying the correlation of complex molecular shapes such as proteins and enzymes and their physiological functions. Formation of complex molecules, polymeric aggregates, and condensation of gases to liquids, etc., involve hydrogen bonding, dispersion forces, van der Waals forces, and dipole–dipole interactions, which are pertinent examples of molecular interactions. Intermolecular forces of attraction determine the bulk properties of substances such as melting points, boiling points, viscosity, and so on.

The following is an overview of some of these molecular interactions.

2.5.1 van der Waals Forces

van der Waals forces refer to the weak intermolecular forces between molecules. There are two types of van der Waals forces, namely the weaker London dispersion forces and the stronger dipole–dipole forces.

London Dispersion Force
We have already learnt about electron density and probability of finding an electron. Fritz London explained that if the electrons are concentrated at one zone in an atom, a dipole is generated momentarily, even

Check Your Progress

14. State octet rule.
15. Write Lewis dot symbols for oxygen, carbonate ion, and nitric acid.
16. List the drawbacks of octet rule.
17. Write the significance and limitations of octect rule.

in a non-polar molecule. As the momentary electron gets dispersed at one end, the molecule becomes partially negatively charged at that end. This negative end makes the surrounding molecules produce an instantaneous dipole, thereby attracting the surrounding partial positive charge of the molecule. This process is known as the *London dispersion force of attrac*tion depicted in Fig. 2.9.

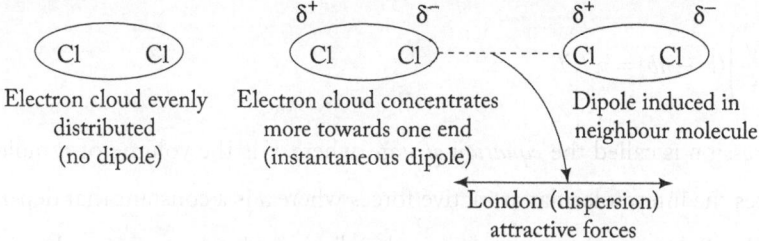

Fig. 2.14 London dispersion forces in chlorine molecule

Further, it can be generalized that more the number of electrons in a molecule, greater will be its ability to become polar and referred to as its polarizing power (or polarizability). London dispersion forces are stronger in those molecules that are loosely bound, but possess longer chains of elements. This is due to the fact that it is easier to displace electrons because the forces of attraction between electrons and protons in the nucleus of such molecules are weaker. London dispersion forces are responsible for a general trend of increasing boiling points with increasing molecular mass. Let us compare methane, ethane, propane, and n-butane (homologous series of alkanes). As the molecular weight increases, boiling points of alkanes also increase.

The strength of London dispersion forces also depends on the surface area of molecules, that is, greater the surface area of a molecule, higher will be its tendency for interacting with its neighbouring molecules. If we compare 2, 2-dimethyl pentane (neopentane) and n-pentane, the surface area of n-pentane is more than that of neopentane. Hence, neopentane is gaseous at room temperature, whereas n-pentane is a liquid.

Dipole–Dipole Forces

Quite similar to London dispersion forces, dipole–dipole forces are known to occur in molecules that are permanently polar. In this type of intermolecular interaction, a polar molecule such as water attracts the positive end of another polar molecule with the negative end of its dipole. The attraction between these two molecules is the dipole–dipole force; for example, in hydrogen chloride, dipole–dipole forces exist as shown in Fig. 2.15. In HCl molecule, there exists a strong covalent bond between hydrogen and chlorine atoms. Due to differences in electronegativities of both the atoms, they acquire charge separation and hence are polar in nature. Such polar molecules have a partial positive pole and a partial negative pole. As shown in the figure, the positive end of one HCl molecule will attract the negative end of the other HCl molecule. Such interactions are called *dipole–dipole interactions*.

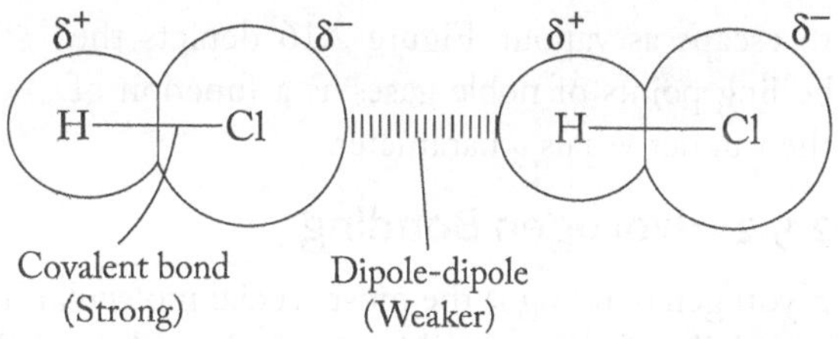

Fig. 2.15 Dipole–dipole interactions in HCl molecule

In 1837, J.D. van der Waals first proposed intermolecular forces to account for the properties of real gases. He recognized deficiencies in the ideal gas equation and proposed an alternate equation, particularly for real gases.

According to the ideal gas equation, for one mole of an ideal gas, $PV/RT = 1$, which is called the *compressibility factor*. This does not hold true for most real gases. There is a large deviation from the ideal value, especially at high pressures where the gas molecules are forced towards liquefaction. It is known that the molecules of the gas do not exist independently from each other because of forces of attraction between non-polar molecules. Dipole–dipole, dipole-induced dipole, and London forces are collectively termed as van der Waals forces, because all these types of forces result in deviations from the ideal gas behaviour. van der Waals equation is formulated for specific cases, such as non-ideal (real) gases, which is expressed as:

$$\left(P + \frac{n^2 a}{V^2}\right)(V - nb) = n RT$$

The above expression is called the *equation of state*, where V is the volume of n moles of real gas; the term, $\frac{n^2 a}{V^2}$ denotes the intermolecular attractive forces where a is a constant that depends on the nature of gas under study; P denotes the pressure; b is the eliminated volume per mole; R is a gas constant 0.08206 L atm mol^{-1} K^{-1}; and T denotes the temperature. Unlike most equations used for the calculation of real or ideal gases, van der Waals equation takes into account the volume of participating molecules and intermolecular forces of attraction.

The parameter a in the equation of state is predicted to show a correlation with other properties that are related to the forces between both organic and inorganic molecules.

Consider a simple case of interacting non-polar molecules (all of same type with existing London forces). In case of liquids, their boiling points provide a measure of strength of forces between the molecules, because these forces need to be overcome in order for the molecules to escape as vapour. Figure 2.16 depicts the boiling points of noble gases as a function of the van der Waals a parameter.

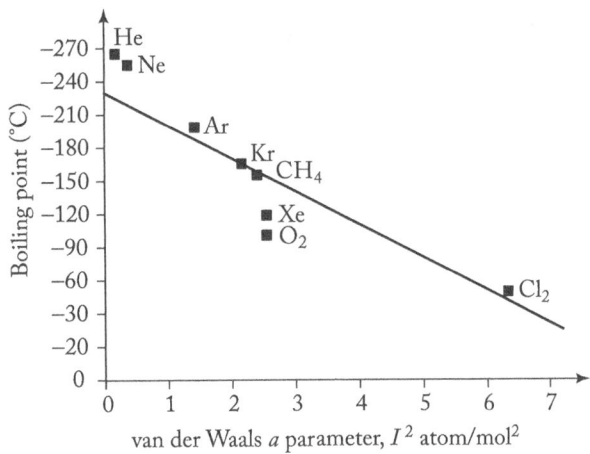

Fig. 2.16 Variation in boiling points of non-polar molecules with van der Waals a parameter

2.5.2 Hydrogen Bonding

Hydrogen bonding is the most crucial molecular interaction accounting for the characteristic properties especially of numerous biochemicals such as cellulose, proteins, starch, leather, and so on. When a hydrogen atom is bonded to a highly electronegative element such as F, O, N, Cl, or S, a covalent bond is formed. The electrons of the covalent bond shift towards the more electronegative atom and hydrogen carries a positive charge that allows it to be attracted to other atoms. Such attraction is called hydrogen bonding (hydrogen bridge).

Let us take the example of hydrogen fluoride that exhibits hydrogen bonding.

$$---\text{H}^{\delta+}\text{---F}^{\delta-}---\text{H}^{\delta+}\text{---F}^{\delta-}---\text{H}^{\delta+}\text{---F}^{\delta-}$$

Hydrogen bonding

Hydrogen bonding is represented as dotted lines (---) whereas covalent bonds are shown as solid lines (—). Hydrogen bond is defined as *the attractive force that binds the hydrogen atom of one molecule with the electronegative atom (F, O, N, Cl or S) of another molecule.* Hydrogen bonds are comparatively weaker than covalent bonds, but much stronger than dispersion and dipole–dipole forces. It is interesting to note that magnitude of hydrogen bonding depends on the physical state of the compound. In a solid, hydrogen bonding will be maximum, whereas minimum in the case of gaseous state. Hydrogen bonding have strong influence on the structural and physical properties of the compounds. Water is an example that shows extensive hydrogen bonding as shown in Fig. 2.17.

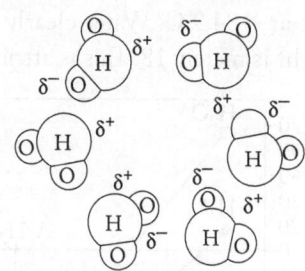

Fig. 2.17 Hydrogen bonding in ice

There are two types of hydrogen bonding encountered in most molecules: intermolecular and intramolecular hydrogen bonding. In *intermolecular hydrogen bonding*, bonding occurs between two different molecules of the same or different compounds. In *intramolecular hydrogen bonding*, the bond exists within two highly electronegative atoms present within the molecule. For example, ethanol forms intermolecular hydrogen bonding with trimethylamine; and 2-chlorophenol exhibits intramolecular hydrogen bonding as shown in Fig. 2.18.

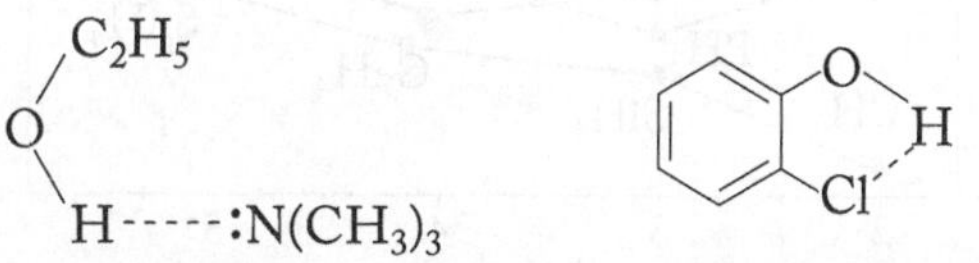

(a) Intermolecular hydrogen bonding between ethanol and trimethylamine

(b) Intramolecular hydrogen bonding in 2-chlorophenol

Fig. 2.18 Types of hydrogen bonding

Both the types of hydrogen bonds occur in liquids and also in solutions. There are some molecules that are associated in vapour phase due to hydrogen bonding. Hydrogen cyanide can remain associated in vapour phase by hydrogen bonding as shown here.

$$\text{---HCN---HCN---HCN---}$$

Alcohols also associate in liquid state as long chains due to the formation of several hydrogen bonds. In long chains of alcohol associated in liquid state, the $O—H---O$ bond distance is about 266 pm. Alcohol in its liquid state can also contain rings as shown in Fig. 2.19.

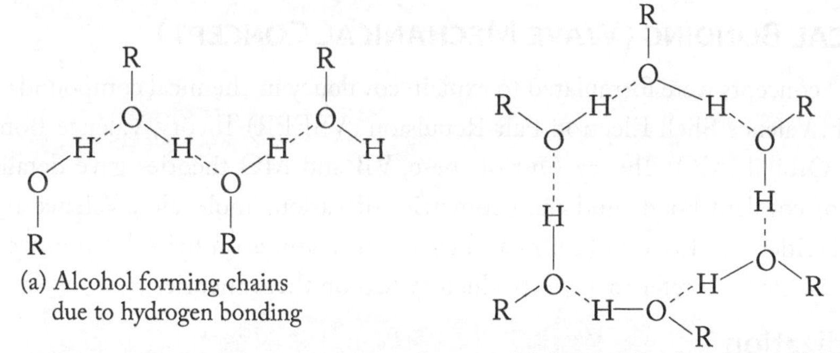

(a) Alcohol forming chains due to hydrogen bonding

(b) Alcohol forming rings due to hydrogen bonding

Fig. 2.19 Formation of H chains in alcohols

The extent of hydrogen bonding has an impact on the boiling points of compounds. Figure 2.20 shows the plot of boiling points of hydrogen compounds of elements in groups IVA to VIIA.

Methane (CH_4), silane (SiH_4), germanium tetrahydride (GeH_4), and tin tetrahydride (SnH_4) do not exhibit hydrogen bonding and thus it shows an expected increase in boiling points with increasing molecular weights of compounds. In group V elements, only ammonia (NH_3) shows significant hydrogen bonding with a boiling point of $-33.4°C$ and hence clearly out of line with other compounds such as phosphine (PH_3), arsine (AsH_3), and stibine (SbH_3). We have already seen the hydrogen bonding present in hydrogen fluoride (HF) which clearly has a higher boiling point of $19.4°C$ as compared to HCl which boils at $-84.9°C$. Water clearly stands out with highest boiling point of $100°C$, though its molecular weight is merely 18. This is attributed to the extensive hydrogen bonding in water molecule.

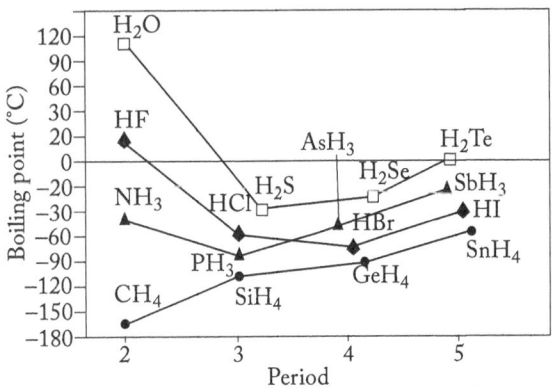

Fig. 2.20 Boiling points of hydrides of groups IVA, VA, VIA, and VIIA

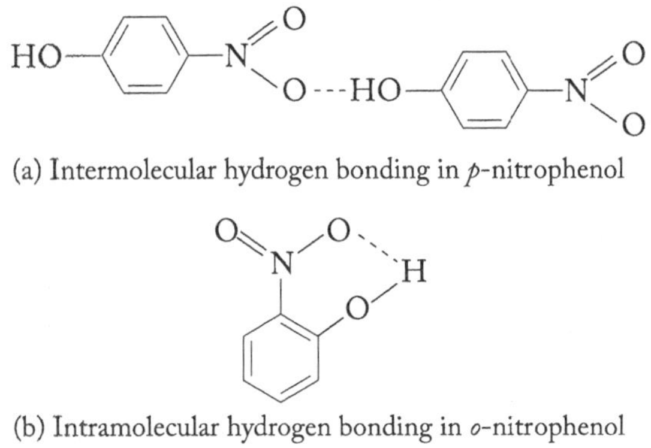

(a) Intermolecular hydrogen bonding in *p*-nitrophenol

(b) Intramolecular hydrogen bonding in *o*-nitrophenol

Fig. 2.21 Hydrogen bondings

Solubility of compounds also is dictated by the type of hydrogen bonding. The solubilities of *o* -, *m*- and *p*-nitrophenols ($NO_2C_6H_4OH$) are influenced by the presence of hydrogen bonding (Fig. 2.21).

o-Nitrophenol exhibits intramolecular hydrogen bonding and hence its —OH group is not available for bonding with water. Thus, *o*-nitrophenol is partially soluble in water. In *p*-nitrophenol, intermolecular hydrogen bonding is present and there are free —OH groups that can easily form hydrogen bonds with water molecule. Hence, *p*-nitrophenol is more soluble in water than *o*-nitrophenol. *o*-Nitrophenol will be completely soluble in benzene as —OH group is involved in intramolecular hydrogen bonding and hence interaction of benzene solvent with the ring of the *o*-nitrophenol is possible.

2.6 CHEMICAL BONDING (WAVE MECHANICAL CONCEPT)

Various classical concepts were formulated to explain covalency in chemical compounds such as Kössel–Lewis approach, Valence Shell Electron Pair Repulsion (VSEPR) Theory, Valence Bond (VB) Theory, and Molecular Orbital (MO) Theory. Out of these, VB and MO theories gave detailed accounts on the formation of covalent bonds and the geometries of various molecules. Valence bond theory was formulated by Heitler and London (1927) and its salient features on hybridization were explained for various molecules. One can refer to any introductory text on these theories.

2.6.1 Hybridization

As discussed in Chapter 1, the geometrical shapes of polyatomic molecules were put forth by Pauling which came to be known as hybridization. According to Pauling, atomic orbitals combine to form a new set of hybrid orbitals. An intermixing of atomic orbitals of slightly differing energies so as to redistribute their energies, resulting in the formation of a new set of orbitals called hybrid orbitals is called *hybridization*. These hybrid orbitals are involved in chemical bond formation and possess equivalent energy and distinct shapes.

Some characteristic features of hybridization are as follows.

(a) The number of hybrid orbitals is equal to the number of the atomic orbitals that get hybridized.

(b) The hybrid orbitals are always equivalent in energy and shape and are found to be effective in forming stable bonds than the pure atomic orbitals.

(c) These hybrid orbitals are directed in space in some preferred direction to have minimum repulsion between the electron pairs and thus a stable arrangement. Therefore, the type of hybridization indicates the geometry of the molecules.

Let us study some examples to understand the different types of hybridization that are possible.

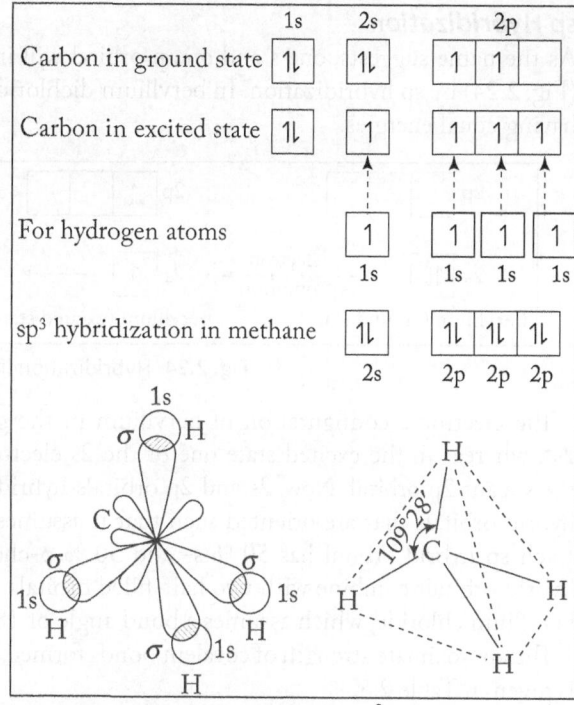

Fig. 2.22 Methane showing sp³ hybridization

sp³ Hybridization

When one s and three p orbitals intermix, four new hybrid orbitals are formed and the process is called sp³ hybridization. In methane, carbon atom forms four identical bonds with hydrogen of equal energies. The electronic configuration of carbon is $1s^2, 2s^2, 2p^2$ in the ground state, whereas in the excited state it is $1s^2, 2s^1, 2p_x^1, 2p_y^1, 2p_z^1$. Carbon forms four hybrid orbitals of equal energies out of 2s and 2p orbitals. These hybrid orbitals then overlap with s orbitals of hydrogen atom such that they are directed towards the vertices of a tetrahedron.

The geometry of the molecule is tetrahedral (Fig. 2.22). There is 25 % s-character and 75% p-character in each sp³ hybrid orbital and the bond angle between sp³ hybrid orbital is 109°28′.

sp² Hybridization

When one s orbital and two p orbitals mix up to hybridize, three new hybrid orbitals are formed and the process is called sp² hybridization.

In boron trichloride (Fig. 2.23), boron bonds with three chlorine atoms having equal energies. The ground state configuration of boron atom is $1s^2, 2s^2, 2p^1$, whereas in the excited state it is $1s^2, 2s^1, 2p_x^1, 2p_y^1$. One 2s orbital of boron intermixes with two 2p orbitals of excited boron atom to form three sp² hybrid orbitals. Each sp² hybrid orbital has 33.33 % s-character and 66.66 % p-character. The three sp² hybrid orbitals overlap and combine with half-filled p orbitals from three chlorine atoms and assume a trigonal planar geometry with a bond angle of 120°.

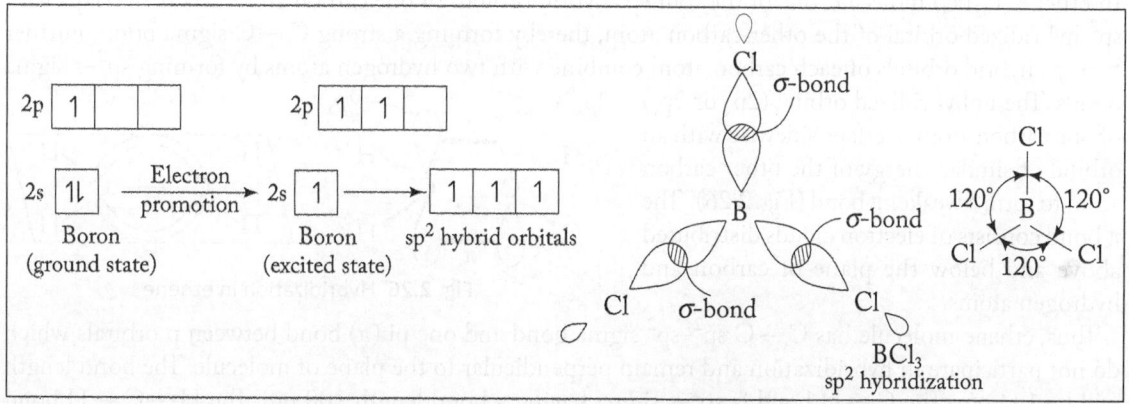

Fig. 2.23 Boron trichloride showing sp² hybridization

sp Hybridization

As the name suggests, one s and one p orbital will intermix to form two equivalent sp hybrid orbitals (Fig. 2.24) by sp hybridization. In beryllium dichloride, beryllium atom bonds with two chlorine atoms having equal energies.

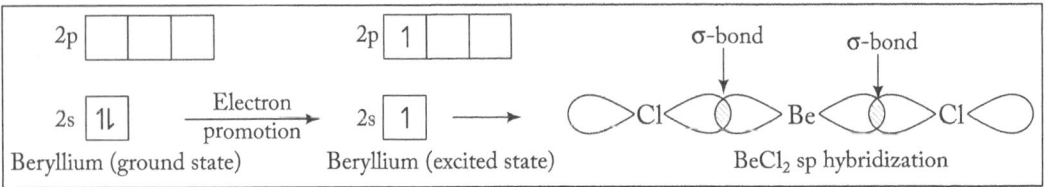

Fig. 2.24 Hybridization in beryllium dichloride

The electronic configuration of beryllium in the ground state is $1s^2$, $2s^2$, whereas in the excited state one of the 2s electrons is promoted to the vacant 2p orbital. Now, 2s and 2p orbitals hybridize to form two sp hybrid orbitals that are oriented such that it assumes a linear geometry. Each sp hybrid orbital has 50 % s- and 50 % p-character. The two sp hybrid orbitals combine with two half-filled orbitals of chlorine to form beryllium chloride, which assumes a bond angle of 180°.

The approximate strength of covalent bonds formed by different orbitals is given in Table 2.5.

Table 2.5 Bond strengths of different orbitals

Orbital	Relative bond strength
s	1.00
p	1.73
sp	1.93
sp^2	1.99
sp^3	2.00

2.6.2 Other Examples of sp^3, sp^2, and sp Hybridization

Ethane

The two carbon atoms in ethane (C_2H_6) molecule undergo sp^3 hybridization. One of the four sp^3 hybrid orbitals of carbon atom axially overlaps with orbitals of similar energy of the other carbon atom to form sp^3–sp^3 sigma bond. The remaining three hybrid orbitals of each carbon atom overlap with six hydrogen atoms forming six sp^3–s sigma bonds similar to that observed in methane (Fig. 2.25). The C—C and C—H bond lengths in ethane are 154 pm and 109 pm, respectively.

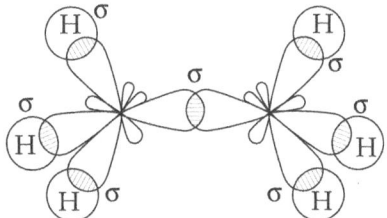

Fig. 2.25 Hybridization in ethane

Ethene

In ethene (C_2H_4) molecule, one of the four sp^2 hybrid orbitals of one carbon atom axially overlaps with sp^2 hybridized orbital of the other carbon atom, thereby forming a strong C—C sigma bond. Further two sp^2 hybrid orbitals of each carbon atom combine with two hydrogen atoms by forming sp^2–s sigma bonds. The unhybridized orbital ($2p_x$ or $2p_y$) of one carbon atom overlaps sidewise with an orbital of similar energy of the other carbon atom to form a weaker π bond (Fig. 2.26). The π bond consists of electron clouds distributed above and below the plane of carbon and hydrogen atoms.

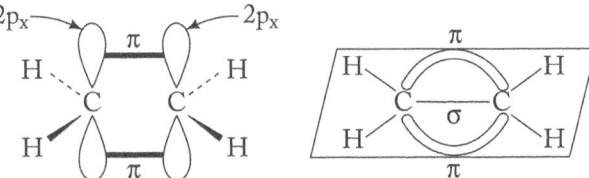

Fig. 2.26 Hybridization in ethene

Thus, ethene molecule has C—C sp^2–sp^2 sigma bond and one pi (π) bond between p orbitals which do not participate in hybridization and remain perpendicular to the plane of molecule. The bond length will be 134 pm. The C—H bond is sp^2–s sigma having a bond length 108 pm. The H—C—H bond angle is 117.6° while the H—C—C angle is 121°.

Ethyne

In the formation of ethyne molecule (C_2H_2), both the carbon atoms undergo sp hybridization having two unhybridized orbitals, that is, $2p_y$ and $2p_x$. One sp hybrid orbital of one carbon atom overlaps axially with sp hybrid orbital of the other carbon atom to form C—C sigma bond, while the other hybridized orbital of each carbon atom overlaps axially with the half-filled s orbital of hydrogen atoms forming σ bonds.

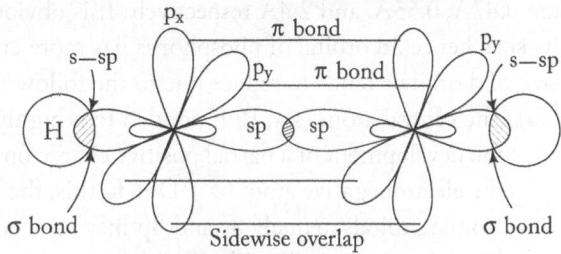

Each of the two unhybridized p orbitals of both the carbon atoms overlaps sidewise to form two π bonds between the carbon atoms. So, the triple bond between the two carbon atoms is made up of one sigma and two pi bonds as shown in Fig. 2.27.

Fig. 2.27 Hybridization in ethyne

2.6.3 Valence Shell Electron Pair Repulsion Theory (VSEPR)

The molecular hybridization of some molecules such as water and ammonia follow VSEPR (pronounced as *ves-pur* or *vuh-seh-per*) theory, that is, valence shell electron pair repulsion theory. Sidgwick and Powell (1940) put forth VSEPR theory to explain electron pair repulsions in the valence shell of atoms and it was further refined by Gillespie and Nyholm (1957). According to this theory, the valence electron pairs surrounding an atom are likely to repel each other and adopt an arrangement that minimizes this repulsion, consequently determining the various geometries of molecules. The repulsive interaction of electron pairs decreases in the order:

Lone pair (lp) – Lone pair (lp) > Lone pair (lp) – Bond pair (bp) > Bond pair (bp) – Bond pair (bp)

Thus, according to the above order lone pairs and bonding pairs of electrons behave differently. It is observed that the lone pairs are localized on the central atom, whereas each bonded pair is shared between two atoms. Hence, lone pair electrons in a molecule will tend to occupy more space as compared to the bonding electrons, and so on. This results in greater repulsion between lone pairs of electrons as compared to bonding pairs resulting in different molecular shapes and their bond angles.

Ammonia (NH_3) It contains a lone pair and a bonding pair of electrons (Fig. 2.28) and shows a distorted tetrahedron geometry. Here, sp^3 hybridization is seen as per VB theory, where one hybrid orbital possesses a lone electron pair. Due to lp–bp repulsion > bp–bp repulsion, the bond angle is 107° and if one ignores the forth corner of the distorted tetrahedron, ammonia molecule is pyramidal in shape.

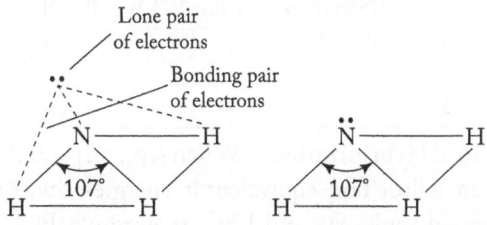

Fig. 2.28 Geometry of ammonia molecule

Water (H_2O) Oxygen atom is sp^3 hybridized with two hydrogen atoms lying at two corners of a tetrahedron. The remaining two corners are occupied by two sp^3 hybrid orbitals possessing unshaired electron pairs (Fig. 2.29).

If we ignore these corners of a tetrahedron, water molecule will be 'V' shaped with bond angle 105° due to the stronger repulsion of two unshared electron pairs due to which bonding pairs come closer to each other than in ammonia molecule.

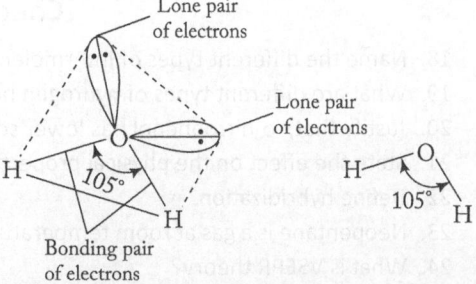

Fig. 2.29 Geometry of water molecule

2.7 HYBRIDIZATION INVOLVING d ORBITALS

Generally, compared to s and p orbitals, d orbitals are considerably large and high in energy when considered for hybridization. Say for example the sizes of 3s, 3p, and 3d atomic orbitals of phosphorus are 0.47Å, 0.55Å, and 2.4Å respectively. It is obvious that the energy of an orbital is directly related to its size; hence 3d orbital of phosphorus has more energy than 3s and 3p orbitals. It is observed that the size of d orbitals tends to reduce due to the following factors.

(a) The central atom (say, P) if bonded to a highly electronegative element (F, O, or Cl) will result in the development of a partial positive charge on the central atom (δ^+) and partial negative charge on the electronegative atom (δ^-). Due to this, the 3d orbital shrinks in size such that energies of other atomic orbitals, namely 3s and 3p may come closer to the 3d orbital so as to allow hybridization. Phosphorus pentachloride (PCl_5) and sulphur hexafluoride (SF_6) are two examples that exhibit hybridization with participating d orbitals.

(b) The size of d orbitals is also influenced by the number of electrons present in the atomic orbitals. The size of 3d orbital of sulphur is 2.46 Å when only one of it is occupied. However, when two such 3d orbitals are filled up, its size shrinks to 1.60 Å.

Hence, we can conclude that d orbitals can participate in hybridization when their contractions take place. Let us explain the various types of hybridization in simple compounds that involve the d orbitals.

dsp^2 Hybridization   When $d_{x^2-y^2}$, s, p_x, and p_y orbitals combine to form equivalent hybrid orbitals with lobes directed to the corners of a square in an xy plane, dsp^2 hybridization takes place. $[Ni(CN)_4]^{2-}$ and $[Cu(NH_3)_4]^{2+}$ are square planar complexes that exhibit dsp^2 hybridization with a bond angle of 90°.

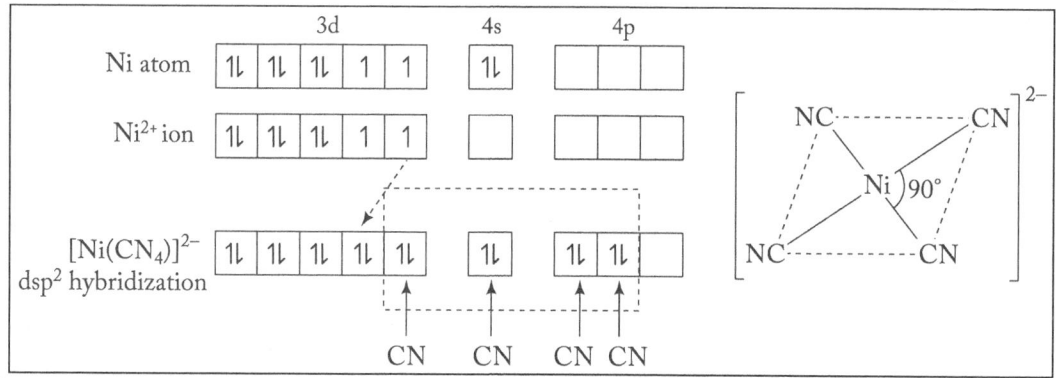

sp^{3}d Hybridization   When s, p_x, p_y, p_z, and d_{z^2} orbitals combine to form sp^{3}d hybridized orbitals, which are a little non-equivalent in energies, they orient their lobes at the corners of a trigonal bipyramid with bond angles 90° and 120°; for example, PCl_5.

Check Your Progress

18. Name the different types of intermolecular forces.
19. What are different types of hydrogen bonding?
20. Justify that, 'o-nitrophenol has lower solubility in water than p-nitrophenol.
21. State the effect on the physical properties of molecules due to hydrogen bonding.
22. Define hybridization.
23. Neopentane is a gas at room temperature whereas n-pentane is a liquid at room temperature. Why?
24. What is VSEPR theory?

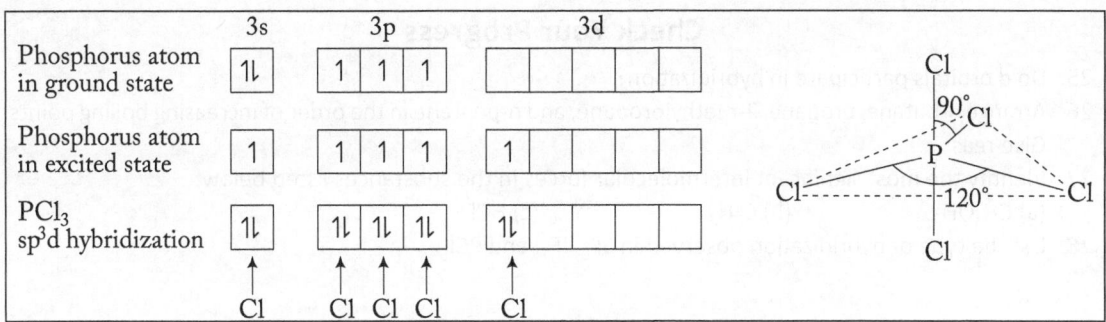

sp³d² Hybridization   In this case, $d_{x^2-y^2}$, and d_{z^2} orbitals combine with s, p_x, p_y, and p_z orbitals forming six hybridized orbitals oriented towards the corners of an octahedron; for example, SF_6.

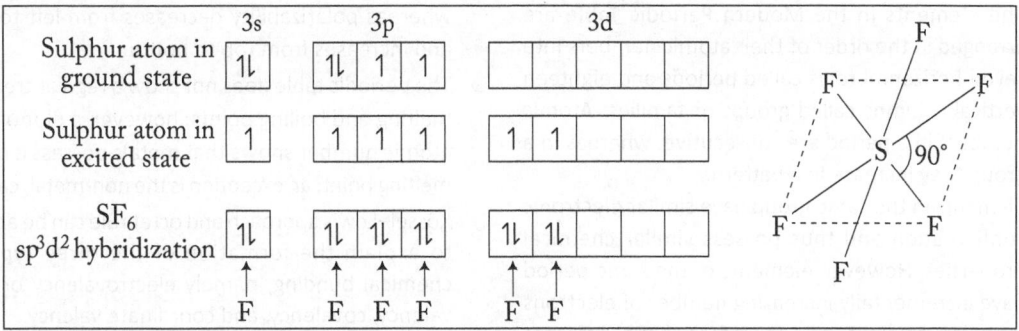

sp³d³ Hybridization   Here, $d_{x^2-y^2}$, d_{z^2}, and d_{xy} orbitals combine with s, p_x, p_y, and p_z orbitals forming seven hybrid orbitals oriented towards the corners of a pentagonal bipyramid; for example, iodine heptafluoride (IF_7).

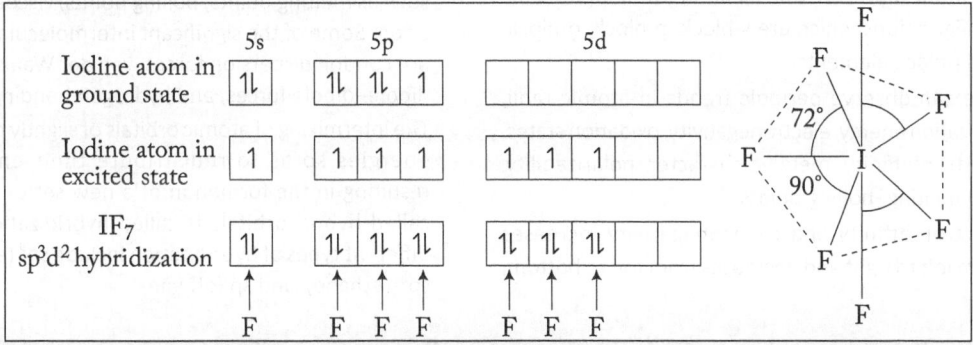

Table 2.6 gives a list of some types of hybrid orbitals and the corresponding shapes of molecules.

Table 2.6 Common types of hybrid orbitals and shapes of molecules

Hybridization	Orbitals involved	Number of bonds	Shape of molecule	Bond angle	Examples
sp	s, p_x	2	Linear or diagonal	180°	$BeCl_2$, C_2H_2
sp²	s, p_x, p_y	3	Trigonal planar	120°	BCl_3, C_2H_4
sp³	s, p_x, p_y, p_z	4	Tetrahedral	109°.28′	CH_4, $SiCl_4$, $SnCl_4$, NH_3, H_2O
dsp²	$d_{x^2-y^2}$, s, p_x, p_y	4	Square planar	90°	$[Ni(CN)_4]^{2-}$ $[Cu(NH_3)_4]^{2+}$
sp³d	s, p_x, p_y, p_z d_{z^2}	5	Trigonal bipyramidal	90° and 120°	PCl_5, $Fe(CO)_5$
sp³d²	$d_{x^2-y^2}$, d_{z^2}, s, p_x, p_y, p_z	6	Octahedral	90°	SF_6, SeF_6
sp³d³	$d_{x^2-y^2}$, d_{z^2}, d_{xy}, s, p_x, p_y, p_z	7	Pentagonal bipyramidal	72° and 90°	IF_7

Check Your Progress

25. Do d orbitals participate in hybridization?
26. Arrange n-butane, propane, 2-methylpropane, and n-pentane in the order of increasing boiling points. Give reason.
27. Identify the most significant intermolecular forces in the substances listed below:
    (a) CH_3OH (b) C_3H_8 (c) HCl
28. List the type of hybridization observed in SF_6, IF_7, and PCl_5.

SUMMARY

- The elements in the Modern Periodic Table are arranged in the order of their atomic numbers into seven horizontal rows called periods and eighteen vertical columns called groups or families. Atomic numbers in a period are consecutive, whereas in a group they increase in a pattern.

- Elements in the same group have similar electronic configuration and thus possess similar chemical properties. However, elements of the same period have incrementally increasing number of electrons from left to right, and therefore, have different valencies.

- Four types of elements can be recognized in the periodic table on the basis of their electronic configurations which are s-block, p-block, d-block, and f-block elements.

- One can observe periodic trends in atomic radii, ionization energy, electronegativity, oxidation states, electron affinity, metallic character, polarizability, and melting–boiling points.

- Electron affinity and electronegativity increases from left to right and decreases from top to bottom, whereas polarizability decreases from left to right and increases from top to bottom.

- The periodic table does not show a regular trend for melting and boiling points; however, a plot of their atomic number shows that metals possess a higher melting point; an exception is the non-metal, carbon.

- Kössel–Lewis approach and octet rule can be applied to explain the formation of the three types of chemical bonding, namely electrovalency or ionic valency, covalency, and coordinate valency.

- Intermolecular forces of attraction help us understand the different states of matter. They are responsible for bulk properties of various substances such as melting points, boiling points, viscosity, and so on. Some of the significant intermolecular forces are London dispersion forces, van der Waals forces, dipole–dipole forces, and hydrogen bonding.

- The intermixing of atomic orbitals of slightly differing energies so as to redistribute their energies, resulting in the formation of a new set of orbitals called hybrid orbitals is called hybridization. The different types of hybridization include sp^3 (ethane), sp^2 (ethane), and sp (ethyne).

GLOSSARY

Atomic radius: The distance between the centre of the nucleus and the outer shell of an atom.

Effective nuclear charge: The net nuclear charge experienced by an electron.

Electron affinity: The energy released when a free or gaseous atom accepts an electron to form a free gaseous ion.

Electronegativity: The relative measure of the tendency of a bonded atom to attract an electron towards itself.

Hybridization: The process of intermixing of atomic orbitals of slightly differing energies so as to redistribute their energies, resulting in the formation of a new set of orbitals called hybrid orbitals.

Hybrid orbital: The new set of orbitals formed due to intermixing of atomic orbitals of slightly differing energies.

Ionization energy: The energy required to pull off an electron from a free or gaseous atom of an element.

Metallic character: The tendency of atom to lose electrons.

Oxidation state: The numerical equivalent of the charge and sign of the charge which the atom of an element appears to have acquired when in combination with atoms of other elements.

Periodicity: A pattern in the Modern Periodic Table, where a particular configuration repeats itself regularly when elements are arranged with respect to their atomic numbers.

Polarizability: The extent to which electrons get perturbed from their bonding arrangement due to another approaching molecule.

KEY FORMULAE

- Effective nuclear charge, $Z_{eff} = Z - \sigma$ (where Z = atomic number, σ = screening constant).
- Pauling scale: $E_{100\% \text{ covalent A}-B} = \sqrt{E_{A-A} \cdot E_{B-B}}$ and Mulliken scale: $\chi_A = \dfrac{1}{2}\,[I_A + E_A]$
- van der Waals equation of state: $\left(P + \dfrac{n^2 a}{V^2}\right)(V - nb) = nRT$

EXERCISES

Multiple Choice Questions

1. The increasing order of electron affinity values of O, S, and Se is
   - (a) O < S < Se
   - (b) S < O < Se
   - (c) O < S > Se
   - (d) Se < O > S

2. Electron affinity is least in
   - (a) Krypton
   - (b) Oxygen
   - (c) Nitrogen
   - (d) Boron

3. The electronic configuration of an element is $1s^2$, $2s^2$, $2p^6$, $3s^2$, $3p^4$. The atomic number of the element present just below this element in the periodic table is
   - (a) 36
   - (b) 34
   - (c) 33
   - (d) 32

4. The correct order of ionic radius is
   - (a) Ce > Sm > Tb > Lu
   - (b) Lu > Tb > Sm > Ce
   - (c) Tb > Lu > Sm > Ce
   - (d) Sm > Tb > Lu > Ce

5. According to the Periodic Law of elements, the variation in properties of elements is related to their
   - (a) nuclear masses
   - (b) atomic numbers
   - (c) nuclear neutron–proton number ratio
   - (d) atomic masses

6. Non-metals belong to
   - (a) s-block elements
   - (b) p-block elements
   - (c) d-block elements
   - (d) f-block elements

7. Which of the following groups contain metals, non-metals, and metalloids?
   - (a) Group 1
   - (b) Group 17
   - (c) Group 14
   - (d) Group 2

8. Consider the following statements:
   I. The radius of an anion is larger than that of the parent atom.
   II. The I.E. increases from left to right in a period.
   III. The electronegativity of an element is the tendency of an isolated atom to attract an electron.
   The correct statements are:
   - (a) I
   - (b) II
   - (c) I and II
   - (d) II and III

9. Ethyne has _____ hybridization.
   - (a) sp
   - (b) sp^2
   - (c) sp^3
   - (d) dsp^2

10. Sulphur hexafluoride has _____ hybridization.
   - (a) sp
   - (b) sp^3d^2
   - (c) dsp^2
   - (d) sp^2

11. For a given value of principal quantum number, the order of increasing energy for different subshells is
   - (a) s < p < d < f
   - (b) p < d < f < s
   - (c) d < f < p < s
   - (d) f < d < p < s

12. As one moves from left to right in a period, electronegativity
    - (a) decreases
    - (b) increases
    - (c) remains same
    - (d) reaches zero

13. As one moves from top to bottom in a group, electron affinity
    - (a) decreases
    - (b) increases
    - (c) remains same
    - (d) becomes less negative

14. Ionization energy of boron is lesser than that of beryllium because
    - (a) Be has incomplete 2 orbital.
    - (b) Be has 2 pairs of electrons.
    - (c) 2p orbital is already higher in energy than 2s orbital.
    - (d) None of the above.

15. On moving from left to right in a period, electron affinity
    - (a) remains the same
    - (b) becomes more negative
    - (c) increases
    - (d) becomes zero

16. The determination of screening constant and effective nuclear charge is done by using
    - (a) VSEPR theory
    - (b) hybridization
    - (c) Slater's rule
    - (d) Lewis approach

17. Amongst H_2O, H_2S, H_2Se, H_2Te, the substance with the highest boiling point is
    - (a) H_2O
    - (b) H_2S
    - (c) H_2Se
    - (d) H_2Te

18. Coordinate covalent bond is formed by
    - (a) complete transfer of electrons
    - (b) sharing of electrons contributed by both atoms
    - (c) sharing of electrons by only one atom
    - (d) combining anions and cations

19. Of the following, intramolecular hydrogen bonding exists in

    - (a) water
    - (b) o-nitrophenol
    - (c) p-nitrophenol
    - (d) hydrogen sulphide

20. van der Waals equation of state is given by
    - (a) $\left(P + \dfrac{n^2 a}{V^2}\right)(V - nb) = nRT$
    - (b) $\left(P + \dfrac{na}{V}\right)(V - n - b) = nRT$
    - (c) $\left(V + \dfrac{n^2 b}{P^2}\right)(V - na) = nRT$
    - (d) $\left(PV + \dfrac{n^2 a}{V^2}\right)(V - nb) = n$

21. The bond angle(s) of trigonal bipyramid is (are)
    - (a) $90°$
    - (b) $90°, 120°$
    - (c) $120°$
    - (d) $109.5°$

22. The molecule that does not obey octet rule is
    - (a) water
    - (b) ammonia
    - (c) carbon tetrachloride
    - (d) phosphorus pentachloride

23. $[Ni(CN)_4]^{2-}$ has _____ hybridization.
    - (a) sp
    - (b) $sp^3 d^2$
    - (c) dsp^2
    - (d) sp^2

24. Valence bond theory was put forth by
    - (a) Slater
    - (b) Heitler and London
    - (c) Mulliken
    - (d) Pauli

25. In H_2O, NH_3, and CH_4 molecules, O, N, and C are
    - (a) sp^3 hybridized
    - (b) sp^2 hybridized
    - (c) sp^3, sp^2, sp hybridized
    - (d) sp, sp^2, sp^3 hybridized

Review Questions

1. Define the terms: (a) Ionization potential (b) Electron affinity (c) Polarizability (d) Electronegativity.
2. Define periodicity of elements and state its significance.
3. Justify the following statements with suitable examples:

    - (a) The properties of elements are a periodic function of their atomic numbers.
    - (b) Atomic radius increases in a group whereas it decreases along a period in the modern periodic table.
    - (c) Ionization energies of elements are always positive.

(d) Metallic character decreases across a period in the modern periodic table.

(e) Beryllium and magnesium tend to exhibit positive electron affinity values.

(f) Fluorine is the most electronegative element in the periodic table.

(g) Noble gases exhibit zero electronegativity value.

4. Explain the various factors that affect ionization potential of elements.

5. Write Slater's rule for estimating screening constant and effective nuclear charge. Mention the drawbacks of Slater's rules.

6. Discuss the relationship between melting and boiling points with respect to atomic numbers.

7. What is hybridization? State the salient features of hybridization. Explain hybridization of ethane molecule.

8. Explain hybridization involving d orbitals with phosphorus pentachloride as example.

9. Discuss the geometries of ammonia and water molecules.

10. Explain VSESPR theory. How is it helpful in predicting the geometries of various molecules?

11. Discuss the postulates of Kössel–Lewis approach to chemical bonding.

12. Explain octet rule and its limitations. Give examples of compounds that do not satisfy the octet rule.

13. Define electrovalency, covalency, and coordinate covalency of molecules. Explain each with suitable examples.

14. Why do different orbitals have variations in energies? Explain penetration of orbitals with a suitable example.

15. List the various elements found in the periodic table and discuss their peculiar features.

16. Graphically plot the periodic trends for electron affinity, electronegativity, atomic radius, and ionization energy.

17. What are various molecular interactions? Discuss any one in detail.

18. Define hydrogen bonding. Explain hydrogen bonding in *o*- and *p*-nitrophenol.

19. Explain London dispersion forces and dipole–dipole forces. Provide suitable examples for each.

20. With a suitable plot, explain the relation of hydrogen bonding to boiling points of various hydride compounds giving examples.

21. Write a short note on hydrogen bonding.

22. Discuss hybridization of $[Ni(CN_4)]^{2-}$ and its magnetic property.

23. Describe the hybridization of PCl_5 and draw the shape of the molecule.

24. Explain hybridization of iodine heptafluoride and state its geometry.

25. Write a short note on hybridization and geometries of various molecules.

NUMERICAL PROBLEMS

1. Calculate the screening constant and effective nuclear charge for 3d electrons in bromine (Z for bromine = 35)
   (Ans: $\sigma = 21.15$ and $Z_{eff} = 13.85$)

2. Calculate the screening constant and effective nuclear charge on one 6s electron in tungsten. (Z for tungsten = 74).
   (Ans: $\sigma = 70.55$ and $Z_{eff} = 3.45$).

3. Calculate the screening constant and effective nuclear charge for 4s electron in zinc (Z for zinc = 30)
   (Ans: $\sigma = 25.65$ and $Z_{eff} = 4.35$)

FURTHER READING

1. Bruce, P. Y. and K. J. R. Prasad, *Essential Organic Chemistry*, Pearson Education, New Delhi, 2008.

2. Carey, Francis A., *Advanced Organic Chemistry* Springer, 2001.

3. Loudon, G. M., *Organic Chemistry*, Oxford University Press, USA, 2001.

4. Pauling, L. 'The Nature of the Chemical Bond', *Journal of the American Chemical Society*, 1931, 53(4), 1367–1400 doi:10.1021/ja01355a027.

5. Nazarewicz, W., 'The Limits of Nuclear Mass and Charge,' *Nature Physics* (14), pp 537–541, 2018, URL: https://www.nature.com/articles/s41567-018-0163-3.

6. Olmsted, John and Gregory M. Williams *Chemistry: The Molecular Science*, Jones & Bartlett Publishers, USA, 1996.

7. Petrucci, Ralph H., et al. *General Chemistry: Principles and Modern Applications*. Upper Saddle River, NJ: Prentice Hall, 2007.

8. Slater, J.C., 'Atomic Shielding Constants', *Physical Review*, 36, p 57–64, 1930. https://journals.aps.org/pr/abstract/10.1103/PhysRev.36.57).

9. Wade Jr, L. G. *Organic Chemistry*, Pearson, 2012.

ANSWERS

Check Your Progress

1. Refer to Section 2.1

2. As per the modern periodic table, atomic size as a function of atomic number will be:
   (a) Cs (298) > Sr (219) > Cr (166) > Si (111) > F (42)
   (b) Rb (265) > K (243) > Fe (156) > Se (103) > S (88)
   (c) Li (167) > Be (112) > B (87) > C (67) > N (56)
   (d) Pb (154) > Sn (145) > Ge (125) > Si (111) > C (67)
   (e) Ti (176) > V (171) > Cr (166) > Mn (161) > Fe (156)

3. As per the modern periodic table, atomic size as a function of atomic number will be:
   (a) Rb (403) < Mg (738) < Si (786) < N (1400) < He (2372.5)
   (b) Li (520.3) < Ca (590) < P (1012) < F (1680)

4. Pauling scale: $E_{100\% \text{ covalent A-B}} = \sqrt{E_{A-A} \cdot E_{B-B}}$ and Mulliken scale: $\chi_A = \frac{1}{2}\left[I_A + E_A\right]$

5. The electronic configuration of fluorine is $1s^2, 2s^2, 2p^5$. It is observed that there are 5 electrons in the 2p shell of fluorine. Ideally, the optimum electronic configuration of a 2p orbital is 6 electrons; hence fluorine seems closer to this ideal configuration. Due to this, the electrons are held tightly to the nucleus that also explains the small atomic radius of fluorine.

6. Atomic size, nuclear charge, type of electrons removed, and shielding effect of electrons are factors that affect the ionization potential.

7.
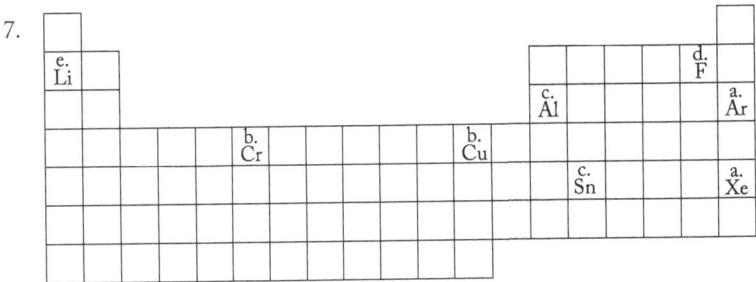

8. (a) The energy required to pull off or remove an electron from a free or gaseous atom of an element is called its ionization energy or ionization potential.
   (b) A relative measure of the tendency of a bonded atom to attract an electron towards itself is called electronegativity.
   (c) The energy released when a free or a gaseous atom accepts an electron to form a free gaseous ion is called electron affinity.

9. Copper has $[Ar]\, 3d^{10}, 4s^1$ configuration and potassium has $[Ar]3p^6, 4s^1$ configuration. 3d subshell in copper is less effective in shielding the outer 4s electron as compared to p subshell in potassium. Hence, the 4s electron of copper is tightly held by the nuclear charge than in potassium. Thus, the first ionization energy of copper

(745 kJ/mol) is more than that of potassium (418 kJ/mol). However, second and third ionization energies of copper are lower than potassium due to loss of electrons from diffused d-orbitals.

10. As per the periodic trend, the first element of a group is the smallest and should exhibit higher electron affinity. Amongst the halogens, fluorine is the smallest atom and more energy must be released on adding electrons, but this is not the case here. Fluorine has a small size and thus addition of any extra electron causes electron–electron repulsions. Hence, addition of an electron is less favourable in fluorine and thus its electron affinity value is lower than chlorine.

11. The extent to which electrons get disturbed from their bonding arrangement due to another approaching molecule is called polarizability. Polarizability decreases from left to right across the row of the periodic table because of increasing effective nuclear charge. On moving down the group in a periodic table, polarizability increases.

12. The electronic configuration of zinc is $(1s^2) (2s^2, 2p^6)(3s^2, 3p^6)(3d^{10})(4s^2)$.

Screening constant, $\sigma = (0.35 \times 9) + (1 \times 18) = 21.15$

Effective nuclear charge, $Z_{eff} = Z - \sigma = 30 - 21.15 = 8.85$

13. Net nuclear charge experienced by an electron is termed as *effective nuclear charge* (Z_{eff}).

14. Atoms can combine either by transfer of valence electrons from one atom to another (gaining or losing) or by sharing of valence electrons in order to have an octet in their valence shells. This is known as octet rule.

15. Refer to Table 2.4.

16. Refer to Section 2.4.1.

17. Refer to Section 2.4.1.

18. London forces, dipole–dipole, hydrogen bonding are types of intermolecular forces.

19. Intramolecular and intermolecular are the two types of hydrogen bondings.

20. The presence of intramolecular hydrogen bonding in ortho isomer makes it unavailable for bonding with water molecules.

21. If there is an extensive hydrogen bonding, boiling points of molecules tend to increase.

22. An intermixing of atomic orbitals of slightly differing energies so as to redistribute their energies, resulting in the formation of a new set of orbitals called hybrid orbitals is called *hybridization*.

23. If we compare 2, 2-dimethyl pentane (neopentane) and n-pentane, the surface area of n-pentane is more than neopentane. Hence, neopentane is a gas at room temperature whereas n-pentane is a liquid.

24. Refer to Section 2.6.3.

25. Yes, d orbitals participate in hybridization.

26. As the four compounds listed are alkanes and are non-polar, London dispersion forces will be predominant in this case that dictates boiling points of these alkanes. London forces are stronger for molecules with higher molecular weights. Hence n-pentane will have the highest boiling point. 2-Methyl propane is more compact than n-butane, hence larger surface area will result in higher boiling point of n-butane. Hence the order will be: propane (–42.1°C) < 2-methylpropane (–11.7°C) < n-butane (–0.5°C) < n-pentane (36.1°C).

27. CH_3OH (hydrogen bonding); C_3H_8 (alkane, non-polar molecule, London dispersion forces); HCl (dipole–dipole forces).

28. $SF_6 : sp^3d^2$, $IF_7 : sp^3d^3$, $PCl_5 : sp^3d$

Multiple Choice Questions

1. (c)	2. (a)	3. (b)	4. (a)	5. (b)	6. (b)	7. (c)
8. (c)	9. (a)	10. (b)	11. (a)	12. (b)	13. (d)	14. (c)
15. (b)	16. (c)	17. (a)	18. (c)	19. (b)	20. (a)	21. (b)
22. (d)	23. (c)	24. (b)	25. (a)			

Thermodynamics and Chemical Equilibrium

After reading this chapter, you will be able to:

- define a system, surroundings and walls, variables, path functions and state functions.
- discuss the first law of thermodynamics and apply it to calculate work done by/on a system or heat evolved/absorbed by the system in different processes involving an ideal gas.
- analyse the limitations of first law of thermodynamics.
- define Carnot's engine and Clausius inequality theorem.
- discuss the second law of thermodynamics and calculate the change in entropy of an ideal gas and in processes involving physical changes.
- discuss the third law of thermodynamics and analyse the criteria for spontaneous change.
- define Helmholtz and Gibbs Free energy change.
- discuss equilibrium and its types.

3.1 INTRODUCTION

Thermodynamics is the branch of physical chemistry that explains the changes in macroscopic equilibrium properties of a system during a process (chemical or physical), involving the flow of different forms of energy between the system and its surroundings. It help us to know whether or not a particular process will be spontaneous under a given set of conditions. The concepts from thermodynamics can be used to know the conditions under which yield in a particular process/reaction will be maximum at equilibrium.

On one hand, thermodynamics can tell us about what needs to be done if we want to carry out a particular process which is non-spontaneous. On the other hand, it helps us to know the conditions under which we can change a spontaneous process to non-spontaneous; for example, water boils at 100 °C at 1 atm and on further heating goes into gas form. But if we want to raise the temperature of water to 110 °C (retaining it in the liquid state)—which is needed to kill bacteria during sterilization of medical equipment—we need to make the process of vaporization non-spontaneous at 110 °C. This can be done by increasing the pressure as done in an autoclave.

3.2 IMPORTANT TERMS RELATED TO THERMODYNAMICS

3.2.1 System

A thermodynamic system is that part of the universe which is under investigation and needs to be studied. In chemistry, the most commonly used system to understand the concept of thermodynamics is gas in a cylinder. A system can be of the following three types.

Open system In an open system there is a possibility of exchange of mass and energy with its surroundings; for example, chemical reaction in a conical flask or a test tube without a stopper.

Closed system In a closed system there is the possibility of exchange of energy with the surroundings, but not that of mass; for example, chemical reaction in a conical flask/test tube with stopper. Herein heat energy is exchanged across the walls of the container.

Isolated system In an isolated system there is exchange of neither mass nor energy with the surroundings; for example, chemical reaction in an insulated flask.

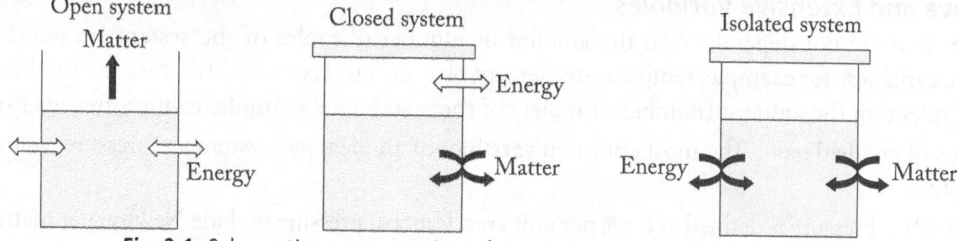

Fig. 3.1 Schematic representation of open, closed, and isolated systems

3.2.2 Ideal Gas System

For an understanding of chemical thermodynamics, the most commonly used system is a gaseous system. In this chapter, our focus will initially be on an ideal gas, a hypothetical gas in which the volume of molecules is assumed to be negligible and there is no interaction among these molecules. Finally, the concepts derived from thermodynamic treatment of ideal gases will be utilized to understand the direction of chemical reactions and physical processes.

3.2.3 Surroundings

The surroundings are a part of the universe which can affect the properties of the system and is separated from the system by a boundary.

Boundary Boundary or wall is the partition which separates a system from its surroundings.

Thermally conducting (diathermic) wall/boundary It allows exchange of heat, but not mass between the system and its surroundings.

Adiabatic wall/boundary It does not allow exchange of heat between the system and its surroundings.

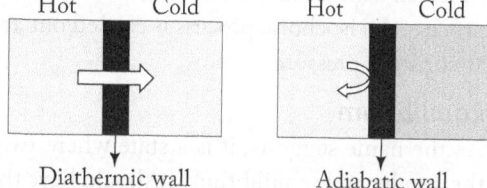

Fig. 3 2 Schematic representation of walls separating a system and its surroundings

3.2.4 Thermodynamic Properties/State of a System

The state of a system is defined as *the complete set of all of its properties which can change during various specified processes.*

State Function and Path Function

State function It is the property of the system. It is a function that changes by the same amount between two states, regardless of which path the system travels between the states. State functions depend only on initial and final states. Pressure, volume, and temperature are examples of state function.

Mathematically, a function z is a state function when its differential is an exact differential.

$$\int_{z_a}^{z_b} dz = z_b - z_a$$

According to Euler's criteria, a differential is called an *exact differential* when its mixed derivatives are equal. So, if z is a function of x and y, then z will be a state function only when

$$\frac{\partial}{\partial y}\left[\left(\frac{\partial z}{\partial x}\right)_y\right]_x = \frac{\partial}{\partial x}\left[\left(\frac{\partial z}{\partial y}\right)_x\right]_y$$

Path function   Path functions are variables which depend on the path taken between two states. Differential of a path function is called *inexact differential*. Heat and work are examples of path function.

At equilibrium, the state of a system is defined entirely by its state variables, and not by the path taken to achieve that state.

Intensive and Extensive Variables

Variables that are not dependent on the amount or number of moles of the system are referred to as intensive variables; for example: temperature, density, pressure etc. Extensive variables are variables which are dependent on the amount (number of moles) of the system; for example: volume, free energy etc.

Variables of an ideal gas   The most common variables of an ideal gas system are pressure, volume, and temperature.

Pressure (P)   Pressure is defined as force per unit area. Units of pressure include Newton per metre square (Nm^{-2}), atmosphere (atm), bar, and Torr. Nm^{-2} is SI unit of pressure and is also known as Pascal (Pa).

$$1 \text{ bar} = 0.98692 \text{ atm} = 10^5 \text{ pascals}$$
$$1 \text{ atm} = 101.325 \text{ kPa} = 1.01 \text{ bar} = 760 \text{ Torr}$$

Volume (V)   It is the amount of space which a system occupies. Units of volume are m^3, cm^3, dm^3 or L.

$$1 \text{ m}^3 = 10^6 \text{ ml} = 1000 \text{ L}$$

Process

A process is carried out to effect a change in the state of the system. Processes are given names based on the way in which they are carried out. A process which is carried out under constant temperature is called an isothermal process; whereas a process carried out without loss/gain of heat is called an adiabatic process. An isochoric process is carried out at constant volume; whereas an isobaric process is carried out at constant pressure.

Equilibrium

As the name suggests, it is a state where two opposite forces balance each other. For a system to be in thermodynamic equilibrium, it should have three different types of equilibrium: mechanical, material, and thermal equilibrium. A system is in **mechanical equilibrium** when there is no net force on the system. A system is in **material equilibrium** when no net chemical reaction is happening (chemical equilibrium) or the concentration of any of physical state (phase equilibrium) is not changing with time. A system is in **thermal equilibrium** with the surrounding when there is no net flow of heat between the system and its surroundings.

3.2.5 Laws of Thermodynamics

There are four laws of thermodynamics. The first and second laws deal with thermodynamic changes in a process, whereas the zeroth and third laws give the definition of temperature and entropy, respectively. The laws are empirical laws based on experimental observations.

Temperature and Zeroth Law of Thermodynamics

Temperature is an important state property. There is a need to define temperature since two different substances feel differently at the same temperature; for example, a metal table feels much colder than a wooden table on a cold winter day, although both of them are at the same temperature. In this context, the zeroth law was proposed. Zeroth law of thermodynamics states that *if two systems (A and C) are separately in thermal equilibrium with a third (B), then they must also be in thermal equilibrium with each other*. In simple terms, if $T_A = T_B$ and $T_B = T_C$, then $T_A = T_C$. This law allows us to define temperature;

temperature is the property which is equal for two systems in contact when there is no heat flow between them. If the two systems are at two different temperatures, there will be flow of heat between them when they are brought into contact. Heat will flow from high temperature to low till temperature becomes equal.

The law also helps in designing a tool to measure temperature, called thermometer. Suppose A is the system whose temperature need to be measured, and B is the glass of the thermometer, and C is some liquid in contact with glass. If B is in thermal equilibrium with system A and system C, then A and C are at same temperature. A property which is proportional to the temperature of the liquid can be used to measure temperature; for example: volume, resistivity, pressure, or colour. The most common example is mercury thermometer in which volume of mercury is used to measure temperature. The volume of mercury (V) increases linearly with increase in its temperature (T).

$T = a \times V + b$; where a and b are constants.

$T = a \times \text{Area}(A) \times \text{height} (h) + b$

$T = c \times h + b$; where $c = a \times A$

In celcius thermometer, temperature at which ice is in equilibrium with liquid water at 1 atm is taken as 0 degree celcius.

$0 = c \times h_{mp} + b$; h_{mp} is the height of mercury at 0 degree celsius.

The temperature at which liquid water is in equilibrium with water vapour at 1 atm is taken as 100 degree celcius.

$100 = c \times h_{bp} + b$; h_{bp} is the height of mercury at 100 degree celsius.

Thus, $c = \dfrac{h_{bp} - h_{mp}}{100}$

The height between h_{bp} and h_{mp} is divided into 100 equal interval parts and thus a scale to measure temperature is created.

The Zeroth law is so called because it came into existence about 100 years after the first and the second laws of thermodynamics.

Changing the State of a System

The state of a system can be influenced by either *doing work* (e.g. compression or expansion) on/by the system, or by *transferring heat* (e.g. heat with a flame) from/to the system.

Work It is defined as energy in transit between the system and its surroundings due to a force acting through a distance. The movement of piston in a cylinder containing gas against external pressure leads to work being done by/on the system. The system does work when there is expansion, whereas work is done by the surrounding on the system when there is contraction. SI unit of work is Joule.

During compression/expansion of gas, differential of the work done which causes the piston to move the distance dx against force F is given by

$\mathrm{d}w = -F\mathrm{d}x$

Check Your Progress

1. How does an intensive variable differ from an extensive variable?
2. Consider a derivative of two functions (z and l)(a) dz = xdy + ydx and (b) dl = mdn; Are z and l state functions?
3. What is the difference between an open and an isolated system?
4. A gas in a closed container undergoes a process, during which its pressure changes from 1 atm to 2 atm. Is the process isochoric, adiabatic, or isothermal?

A negative value of *Fdx* denotes that compression, for which d*x* is negative, is associated with positive work.

$$dw = -p_{ex} A dx$$

A is the area of the cylinder and *A*d*x* has the dimension of volume.

$$dw = -p_{ex} dV$$

$$w = -\int_{V_i}^{V_f} P_{ex} dV$$

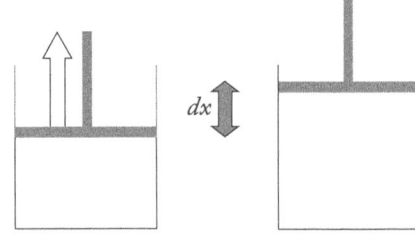

Fig. 3.3 Work done by gas during expansion against an external pressure

Work done by different processes:

(i) Free expansion: During free expansion, p_{ex} is zero. So, w is zero.

(ii) At constant external pressure

$$w = -p_{ex}\int_{V_i}^{V_f} dV = -p_{ex}(V_f - V_i)$$

Organized motion is associated with work. Orderly motions are stimulated in the system when work is done on it. Orderly motions are stimulated in the surrounding when work is done by the system.

Heat   Heat is the quantity of energy exchanged between a system and its surrounding due to difference in temperature. Heat is also transitory in nature. It is the transfer of energy which makes use of unorganized molecular motion. When energy is transferred to the system as heat, the transfer stimulates disordered motion of the atoms in the system. SI unit of heat is Joule.

3.3 HEAT CAPACITY OF A SYSTEM

Heat capacity   The thermal response of the system to heat flow is described by heat capacity. *It is the quantity of heat required to increase the temperature of a system by one K or °C.* Heat capacity C can be defined as

$$C = \frac{dq}{dT}$$

Since heat is an inexact differential and it depends on the path taken, there can be a large number of heat capacities depending on the paths taken such as isobaric, isochoric, isothermal paths, and so on.

Most reactions/processes take place at constant pressure or constant volume. Thus, of the various possible heat capacities, only two heat capacities are important: heat capacity at constant volume C_V and heat capacity at constant pressure C_p. SI unit of heat capacity is JK^{-1}.

Joule's Experiment   In this experiment, a mass of weight mg was dropped through a height z as shown in Fig. 3.4 thereby moving the side pulley and centre pulley. The centre pully in turn rotates the paddle wheel which does work on water kept in an adiabatic container. Thus, work (w) equivalent to mgz was done on water under adiabatic condition leading to increase in the temperature. Joule found that the increase in temperature was dependent only on the work (specifically on the z) but independent of how the work was performed (e.g. quick or slow dropping of mass, large or small paddle wheel).

Temperature rise means there is a change in the state. The property of the system which changed must have the unit of energy and thus given the name internal energy (U).

$$dU = dw \text{ (under adiabatic conditions)}$$

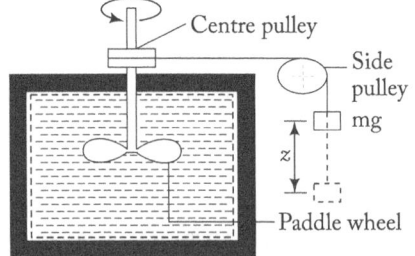

Fig. 3.4 Joule's experiment to show that the increase in temperature of water depends on work done by the system (mgz)

The change in state (temperature rise) can also be achieved by allowing heat (q) to flow in the system even if no work is done on/by the system. The temperature rise was shown to be independent of how heat has been allowed to flow in the system.

$dU = dq$ (no work involved)

Mathematical form of first law of thermodynamics   As discussed, the state of the system can be affected by either heat or work or both. This is the basis of the first law of thermodynamics.

According to the law, for an infinitesimal change of state, $dU = dq + dw$.

It should be noted that although heat and work are path functions, internal energy of the system is a state function. Thus, sum of two path functions can be a state function.

Sign of heat and work   If the exchange of heat or work between the system and its surroundings increases the energy of the system, they are assigned positive sign. Since heat absorption by the system leads to an increase in the energy of the system, a positive sign is assigned to heat absorption. Similarly, if work is done on the system, the energy of the system increases and thus work is taken as positive.

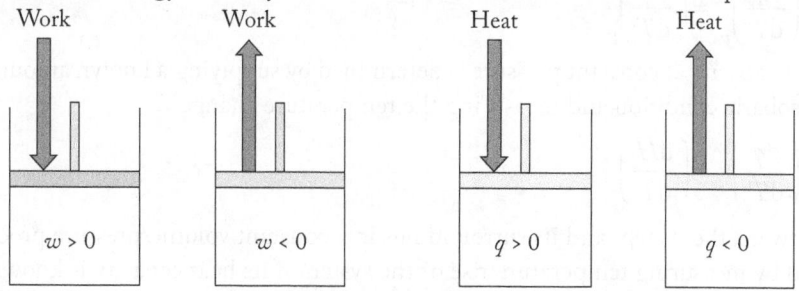

Fig. 3.5 The sign of work and heat will depend on whether the exchange of work/heat increases or decreases the energy of the system.

3.3.1 Internal Energy

It is the sum of the translation, rotational, and vibrational energies of the molecule. It also includes potential energy due to intermolecular forces. However, the internal energy of the system (U) is defined relative to a coordinate system fixed on the system. Thus, translational or rotational kinetic energy of a system as a whole is not included. It also does not include potential energy due to an induced electric or magnetic dipole moment, and energy due to relative mass. Any potential energy due to the location of the system in the external gravitational or electrostatic field is also not included. Internal energy is an extensive property, whereas 'internal energy per mole' is an intensive property.

Change of state at constant volume: Measurement of change in internal energy

From the first law of thermodynamics, $dU = dq + dw$

$$dU = dq - p_{ex}\, dV$$

At constant volume, $dU = dq_V$

On integration

$$\Delta U = q_V$$

Change in internal energy (ΔU) can be measured by measuring heat flow between the system and its surroundings at constant volume.

Change of state at constant pressure: Concept of enthalpy

Chemical and physical processes are generally carried out under constant pressure rather than under constant-volume condition. Thus, it is useful to define a state function which describes the energy of the system under constant-pressure condition.

$$dU = dq + dw$$

On integrating, $\Delta U = q_p - p\Delta V$

$$U_f - U_i = q_p - p(V_f - V_i)$$
$$q_p = (U_f + pV_f) - (U_i + pV_i)$$
$$q_p = (H_f) - (H_i) = \Delta H$$

The term '$U + pV$' is a state function and given the name enthalpy (H). Change in enthalpy (ΔH) can be measured by measuring the heat flow between the system and its surroundings in a constant-pressure process.

3.3.2 Measurement of Heat Capacity

Heat capacity at constant volume is determined by supplying a known amount of heat to the system under isochoric conditions and measuring the temperature change. If the measured temperature change is small, the heat capacity can be obtained as

$$C_V = \left(\frac{dq}{dT}\right)_V = \left(\frac{dU}{dT}\right)_V$$

Similarly, heat capacity at constant pressure is determined by supplying a known amount of heat to the system under isobaric condition and measuring the temperature change.

$$C_p = \left(\frac{dq}{dT}\right)_p = \left(\frac{dH}{dT}\right)_p$$

Heat flow between the system and its surroundings in a constant volume/pressure process (q_V and q_p) can be obtained by measuring temperature rise of the system if its heat capacity is known.

C_p of a gas is always greater than C_V. Under isochoric condition, heat supplied is totally used to raise the temperature (work is zero). However, under isobaric condition, heat supplied is partly used for doing work and the rest is used for raising temperature. Thus, heat required for unit change in temperature is more under constant-pressure condition than at constant volume.

Value of molar heat capacity at constant volume ($C_{V, m}$) for an ideal gas For each degree of freedom, there is contribution of 1/2 RT to internal energy for gases at temperature T, where R is a gas constant. Hence, each degree of freedom contributes 1/2 R to molar heat capacity at constant volume ($C_{V,m}$). Since a monoatomic gas has three translational degrees of freedom, the $C_{V,m}$ for a monoatomic gas is 3/2 R. A linear diatomic molecule has additional two rotational degrees of freedom and hence the $C_{V,m}$ for a linear diatomic gas is 5/2 R. A non-linear diatomic molecule has three rotational degrees of freedom and hence its $C_{V,m}$ is 3 R.

Value of $C_{p,m}$ for an ideal gas For an ideal gas, $C_{p,m} - C_{V,m} = R$. Thus, $C_{p,m}$ for a monoatomic, a linear diatomic, and a non-linear diatomic ideal gas are 5/2 R, 7/2 R, and 4 R, respectively.

3.3.3 Reversible Process

Reversibility is important in thermodynamics since we can get maximum work from the system when we carry out the process reversibly. A process that can be reversed by making an infinitesimal change at any point during the process is called a reversible process. Throughout an entire reversible process, the system is in thermodynamic equilibrium with surroundings.

A gas does more work if the same expansion is carried out in more number of steps rather than in one step; for example, Fig. 3.6 (a) shows expansion (V_1 to V_2) carried out in one step (left panel) and the same expansion in two steps (Fig. 3.6 b), first from initial state (p_1, V_1) to an intermediate state (p_{in}, V_{in}) and then from intermediate state (p_{in}, V_{in}) to final state (p_2, V_2). The work done by the system in one step and that done in two steps is equal to the shaded areas in Fig. 3.6 (left panel) and (right panel), respectively. A comparison of shaded areas suggests that work carried out in two steps is more than the work carried out in one step for expansion from V_1 to V_2.

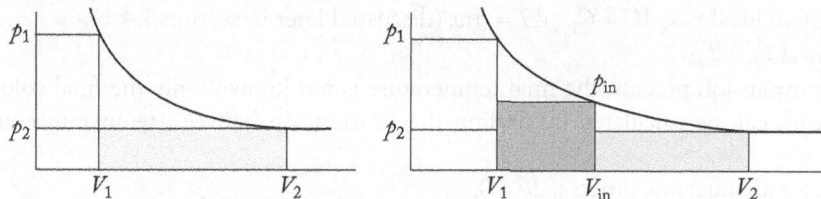

Fig. 3.6 Figure shows that the work done by the system is more if work is done in two steps rather than one.

Similarly, it can be shown that maximum work is obtained when work is carried out in infinite number of steps or in other words, reversibly. Work can be carried out reversibly if external force is varied by a very small amount.

Reversible work p-V work is reversible if expansion or compression is carried out in infinite number of steps. A reversible expansion or compression requires a balancing of internal and external pressure. Thus, in the expansion/compression of gas, reversibility is achieved when $p_{ex} \sim p_{gas}$.

3.4 APPLICATIONS OF FIRST LAW OF THERMODYNAMICS

(a) Work done in a reversible isothermal expansion/compression of an ideal gas

Suppose a gas undergoes reversible isothermal expansion from V_i to V_f.

From the definition of work, $w = -\int_{V_i}^{V_f} p_{ex} dV$.

For a reversible process, $p_{ex} = p$ of the gas; hence $w = -\int_{V_i}^{V_f} p\, dV$.

For an ideal gas, $w = -\int_{V_i}^{V_f} \dfrac{nRT}{V} dV$

For an isothermal process, $w = -nRT\int_{V_i}^{V_f} \dfrac{dV}{V}$ (T is constant, hence it can be taken out of integral.)

$$w = nRT \ln\frac{V_i}{V_f} = nRT \ln\frac{p_f}{p_i}$$

(b) Work done in a reversible adiabatic expansion/compression of an ideal gas

As shown previously, $w = -\int_{V_i}^{V_f} \dfrac{nRT}{V} dV$.

Temperature T is not constant so cannot be taken out of the integral. Hence the integral cannot be solved. However, the first law of thermodynamics can be used to obtain work done in this case.

Check Your Progress

5. What is the work done in an isochoric process involving an ideal gas? Is there a change when the same process is carried out reversibly?
6. What will be the relationship between work done and heat absorbed by the system in a cyclic process?
7. Prove that heat is not a state function using the first law of thermodynamics.
8. Using first law of thermodynamics, show that heat flow between a system and its surroundings at constant volume is equal to change in the internal energy.
9. Heat capacity at constant pressure of a gas is higher than heat capacity at constant volume. Why?
10. Explain why the reversibility in gas expansion is achieved when $p_{ex} \sim p_{gas}$?

From the first law of thermodynamics, $dU = dq + dw$

For an adiabatic process $dq = 0$.

$\Rightarrow \qquad dU = dw$

For one mole of an ideal gas, $dU = C_{V,m}dT = dw$. (discussed later in section 3.4.1)

$$w = C_{V,m}(T_f - T_i)$$

However, for expansion process, the final temperature is not known; only the final volume is known. In this case, work can be calculated by finding the relationship between temperature and volume as described below.

As we know, for an adiabatic process, $dq = 0$.

$\Rightarrow \qquad dU = dw$

For an adiabatic reversible process, $dU = -p\,dV$.

For an ideal gas, U is a function of only T (discussed below) and $pV = nRT$.

$$C_V dT = -\frac{nRT}{V}dV \text{ or, } C_V \frac{dT}{T} = -\frac{nR}{V}dV$$

$$C_V \ln T = -nR \ln V + \text{const or, } \ln T + \frac{R}{C_{V,m}}\ln V = \text{const}$$

$$\ln V^{R/C_{V,m}} T = \text{const}$$

$$V^{R/C_{V,m}} T = \text{const}$$

Thus, $\left(\dfrac{T_f}{T_i}\right) = \left(\dfrac{V_i}{V_f}\right)^{\gamma-1}$ where $\gamma = \dfrac{C_p}{C_V}$

Thus, if we know the final volume, work can be calculated by using the following equation.

$$w = C_{V,m}T_i\left(\left(\frac{V_i}{V_f}\right)^{\gamma-1} - 1\right)$$

Also, $\left(\dfrac{T_f}{T_i}\right) = \left(\dfrac{V_i}{V_f}\right)^{\gamma-1}$

For an ideal gas,

$$\left(\frac{V_i}{V_f}\right) = \left(\frac{T_i p_f}{T_f p_i}\right)$$

$\Rightarrow \qquad \left(\dfrac{T_f}{T_i}\right) = \left(\dfrac{p_f}{p_i}\right)^{(\gamma-1)/\gamma}$

Thus, if we know the final pressure, work can be calculated by using the following equation.

$$w = C_{V,m}T_i\left(\left(\frac{p_f}{p_i}\right)^{(\gamma-1)/\gamma} - 1\right)$$

(c) Reversible isochoric process for an ideal gas

Since volume does not change in an isochoric process, work done is zero.

(d) Reversible isobaric process for an ideal gas

Work done on a reversible isobaric process for an ideal gas,

$$w = \int_{V_i}^{V_f} -p\,dV = p(V_i - V_f) = nR(T_i - T_f)$$

3.4.1 Internal Energy Change in a Process

Internal energy change in a process can be calculated using the properties of state function and first law of thermodynamics.

In a closed system, $U = f(V, T)$.

$$dU = \left(\frac{\partial U}{\partial T}\right)_V dT + \left(\frac{\partial U}{\partial V}\right)_T dV$$

$$dU = C_{V,m} dT + \pi \, dV \quad \text{(for one mole)}$$

The quantity π has units of pressure and is known as *internal pressure*. It tells about how internal energy changes with change in volume at constant temperature. Since temperature is constant, only change in potential energy part of internal energy contributes to internal pressure. In case of a gas, the intermolecular forces between the gas molecules are responsible for the change in potential energy. However, the intermolecular force between ideal gas is assumed to be zero; so the internal pressure of an ideal gas is zero.

$$dU = C_{V,m} \, dT \text{ (For one mole of an ideal gas)}$$

Thus, internal energy of an ideal gas in a closed system is a function of only temperature. Internal energy is a state function; so change in internal energy between two states will be the same whether or not the process is carried out reversibly.

(a) Internal energy change in an isothermal process for an ideal gas:
Internal energy will not change because as mentioned above it is a function of only temperature.

(b) Internal change in an adiabatic process for an ideal gas:
Exchange of energy occurs as work done between the system and surroundings in an adiabatic process will change the temperature of the gas. If the process leads to temperature change from T_i to $T_{f,\text{adiabatic}}$, then

$$\Delta U = C_{V,m} \, (T_{f,\text{adiabatic}} - T_i) \text{ (per mole of an ideal gas)}$$

Internal change in an adiabatic process can also be calculated from the calculation of work done. For an adiabatic process, $dq = 0$.

$$\Rightarrow \quad dU = dw$$

(c) Internal change in an isochoric process for an ideal gas:
Exchange of heat between a system and its surroundings in an isochoric process leads to change in temperature. If heat exchange leads to temperature change from T_i to $T_{f,\text{isochoric}}$, then

$$\Delta U = C_{V,m} \, (T_{f,\text{isochoric}} - T_i) \text{ (per mole of an ideal gas)}$$

(d) Internal change in an isobaric process for an ideal gas:
Similarly, exchange of energy as heat/work between the system and its surroundings in an isobaric process also leads to change in temperature. If the process leads to temperature change from T_i to $T_{f,\text{isobaric}}$, then

$$\Delta U = C_{V,m} \, (T_{f,\text{isobaric}} - T_i) \text{ (per mole of an ideal gas)}$$

3.4.2 Enthalpy Change in a Process

Enthalpy change in a process can be calculated using the properties of state function and the first law of thermodynamics.

In a closed system, $H = f(p, T)$.

$$dH = \left(\frac{\partial H}{\partial T}\right)_p dT + \left(\frac{\partial H}{\partial p}\right)_T dp$$

$$dH = C_{p,m} \, dT + \mu_T \, dp$$

The term μ_T has unit of volume and is known as isothermal Joule–Thomson coefficient. It tells about how enthalpy changes with change in pressure at constant temperature. Since temperature is constant, only change in potential energy part of enthalpy contributes to Joule–Thomson coefficient. In case of a gas, it is the intermolecular forces between the gas molecules that are responsible for change in potential energy. However, the intermolecular force between ideal gas molecules is assumed to be zero; so μ_T of an ideal gas is zero.

$$dH = C_{p,m}\, dT$$

Thus, the enthalpy of an ideal gas in a closed system is a function of only temperature.

(a) Enthalpy change in an isothermal process for an ideal gas:
Enthalpy will not change because as mentioned above it is a function of only temperature.

(b) Enthalpy change in an adiabatic process for an ideal gas:
Exchange of energy as work between the system and its surroundings in an adiabatic process will change the temperature of the gas. If the process leads to temperature change from T_i to $T_{f,\text{adiabatic}}$, then

$$\Delta H = C_{p,m}\,(T_{f,\text{adiabatic}} - T_i) \text{ (per mole of an ideal gas)}$$

(c) Enthalpy change in an isochoric process for an ideal gas:
Exchange of heat between the system and surroundings in an isochoric process leads to change in temperature. If heat exchange leads to temperature change from T_i to $T_{f,\text{isochoric}}$, then

$$\Delta H = C_{p,m}\,(T_{f,\text{isochoric}} - T_i) \text{ (per mole of an ideal gas)}$$

(d) Enthalpy change in an isobaric process for an ideal gas:
Similarly, exchange of energy as heat/work between the system and surroundings in an isobaric process also leads to change in temperature. If the process leads to temperature change from T_i to $T_{f,\text{isobaric}}$, then

$$\Delta H = C_{p,m}\,(T_{f,\text{isobaric}} - T_i) \text{ (per mole of an ideal gas)}$$

Joule–Thomson Experiment

James Prescott Joule and William Thomson studied the change in temperature when pressure is suddenly changed under the condition of constant enthalpy. Constant enthalpy condition can be created by taking an insulated tube consisting of two chambers with a movable piston as in Fig. 3.7. The left chamber has gas at p_i, V_i, T_i whereas the right chamber is kept empty. Gas from the left chamber is pushed through a porous plug until all the gas is transferred to the region to the right of the porous plug. The gas is now at p_f, V_f, T_f.

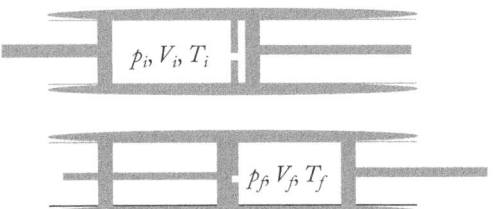

Fig. 3.7 Joule–Thomson experiment showing expansion of gas from one isolated chamber to another through a porous plug

Since the process is adiabatic, $q = 0$ or $\Delta U = w$.

Work done in this process is $w = p_iV_i - p_fV_f$. Substituting in the above equation, we get,

$$\Delta U = p_iV_i - p_fV_f$$

We know that change in energy is the difference between initial energy and final energy; substituting for ΔU and rearranging, we get

$$U_f - U_i = p_iV_i - p_fV_f$$
$$U_f + p_fV_f = U_i + p_iV_i$$
$$H_f = H_i$$

Thus, this process is *isoenthalpic* in nature.

H is a state function

$$H = f(p, T)$$

From Euler's cyclic rule

$$\left(\frac{\partial H}{\partial p}\right)_T \left(\frac{\partial p}{\partial T}\right)_H \left(\frac{\partial T}{\partial H}\right)_p = -1$$

$$\left(\frac{\partial H}{\partial p}\right)_T = -\left(\frac{\partial T}{\partial p}\right)_H C_p$$

Thus, by measuring the change in temperature with pressure, the change of enthalpy with pressure can be obtained. $\left(\dfrac{\partial H}{\partial p}\right)_T$ is known as isothermal Joule–Thomson coefficient, whereas $\left(\dfrac{\partial T}{\partial p}\right)_H$ is known as Joule–Thomson coefficient (μ).

Ideal gases have zero Joule–Thomson coefficient since isothermal Joule–Thomson coefficient is zero for an ideal gas. Real gases have non-zero Joule–Thomson coefficient. Since Δp is always negative in a Joule–Thomson experiment, a positive value of μ corresponds to cooling on expansion, and a negative μ value indicates warming.

3.5 HEAT IN A REACTION

A chemical reaction or a physical process is always accompanied by absorption or release of heat by the system. The enthalpy change in a chemical reaction or a physical process can be measured by measuring heat at constant pressure ($q_p = \Delta H$).

Endothermic and Exothermic Processes

Exothermic process is a process in which heat is released by the system leading to transfer of thermal energy to the surroundings. If an exothermic reaction is carried out in a vessel, the reaction vessel will feel warm.

Endothermic process is a process in which the system absorbs heat from the surroundings. If an endothermic reaction is carried out in a vessel, the reaction vessel will feel cold.

The *change in enthalpy*, ΔH, is the difference between the sum of enthalpy of the products (ΣH_P) and the sum of enthalpy of the reactants (ΣH_R).

$\Delta H = \Sigma H_P - \Sigma H_R$

For endothermic reaction: $\Sigma H_P > \Sigma H_R, \Delta H > 0$

For exothermic reaction: $\Sigma H_P < \Sigma H_R, \Delta H < 0$

Examples of exothermic processes Combustion reactions, rusting of iron (used in chemical hot pack) are examples of exothermic chemical reaction. Dissolving laundry detergent in water, crystallization of sodium acetate or 'hot ice' and freezing water into ice are also examples of exothermic physical processes.

Examples of endothermic processes Dissolving ammonium chloride in water, mixing water and ammonium nitrate (used in cold pack) are examples of endothermic chemical reactions. Melting of ice cubes, melting of solid salts, and evaporation of water vapour are some other examples of endothermic processes.

Thermochemical reaction:

Some important points about thermochemical reactions are:

Check Your Progress

11. The internal energy of an ideal gas depends only on the temperature. Why?
12. In the Joule–Thomson experiment, enthalpy is kept constant. How?
13. Why does hydrogen gas get warmer on expansion whereas most other gases on doing so become colder?
14. Why is pressure of an ideal gas not inversely proportional to its volume in an adiabatic process?
15. Enthalpy of an ideal gas does not depend on pressure under isothermal condition. Give reason.

- All reactants and products in thermochemical equations should be written along with symbol of their physical states.

$$C(\text{graphite}) + O_2(g) \rightarrow CO_2(g)$$

- For the reverse reaction, the sign of ΔH changes.

$$A(g) + B(g) \rightarrow C(g) + D(g), \Delta H = x \, J \, mol^{-1}$$

Then the ΔH for the reverse reaction will be

$$C(g) + D(g) \rightarrow A(g) + B(g), \text{ is } -x \, J \, mol^{-1}.$$

- The enthalpy change for n moles of reactant is n times the enthalpy change for one mole of reactant.

$$A(g) + B(g) \rightarrow C(g) + D(g), \Delta H = x \, J \, mol^{-1}$$
$$2\,A(g) + 2\,B(g) \rightarrow 2\,C(g) + 2\,D(g), \Delta H = 2x \, J$$

Enthalpy (Heat) of Formation

It is the enthalpy change when one mole of the substance in its standard state is formed from the corresponding constituent atoms, each atom being in its standard state.

The standard enthalpy of formation of any element in its most stable form is zero.

Standard state conditions

- Temperature: 25 °C or 298 K
- Pressure: 1.00 bar
- Element is in its stable state; for example, carbon in graphite form.

Example

$$C(\text{graphite}) + O_2(g) \rightarrow CO_2(g), \qquad \Delta_f H° = -394 \, kJ \, mol^{-1}$$

In this example, one mole of CO_2 is formed from its atoms in their standard states and thus, enthalpy in this reaction, that is $-394 \, kJ \, mol^{-1}$, is the enthalpy of formation of CO_2.

Hess's law   Enthalpy being a state function is the basis of Hess's law. Thus, a reaction taking place in one step or more than one step has the same enthalpy of reaction.

Application of Hess's law   It helps us to calculate the heat of a particular reaction if the reaction can be expressed in terms of different reactions (by adding or subtracting) for which change in the enthalpy is known. For example, enthalpies of reactions can be expressed in terms of heats of formation of reactants and products. If any two of enthalpies, enthalpy of sublimation, enthalpy of fusion, and enthalpy of vaporization are known, the third one can be calculated.   Enthalpies of formation of most of the compounds have been tabulated. Thus, heat of reactions can be obtained by tabulated heats of formation.

Enthalpy of Combustion

It is the enthalpy change when one mole of the substance is completely burnt in an excess of air.

$$C(\text{graphite}) + O_2(g) \rightarrow CO_2(g), \qquad \Delta H° = -394 \, kJ \, mol^{-1}$$

Binding energy   If we want to break the bond between two atoms, we need to supply energy to overcome the attractive force. Bond breaking is an endothermic process. On the other hand, heat is released when a new bond is formed. Thus, bond making is an exothermic process. The enthalpy change required to break a particular bond in one mole of gaseous molecule is the *bond energy*.

Lattice energy   Lattice energy is used to know the approximate bond strength in ionic compounds. It is defined as *the energy released when gaseous ions come together to form an ionic compound*.

$$M^+(g) + X^-(g) \rightarrow MX(s)$$

It is exothermic in nature and the value will be negative.

Lattice energy mainly depends upon two factors: charge and size of the ions. As charge on the ions increases, more energy is required to break the bonds and thereby the lattice energy increases. On the other hand, as the size of the ions increases, their interaction decreases and therefore lesser energy is required to break the lattice.

Born–Haber cycle As it is difficult to isolate gaseous ions, lattice energy cannot be determined experimentally. Born–Haber cycle is used to calculate the value of the lattice energy for an ionic crystal. There are a number of processes involved in Born–Haber cycle and Hess's law is used for addition and subtraction of energy associated with each step to give the overall lattice energy. The various steps involved are as follows.

(a) **Electron affinity** (EA): It is the energy released when an electron is added to a neutral atom in gaseous state.
$$X(g) + e^- \rightarrow X^-(g)$$

(b) **Ionization energy** (IE): It is the energy required to remove an electron from a neutral atom in gaseous state. It always has a positive value.
$$M(g) \rightarrow M^+(g) + e^-$$

(c) **Sublimation energy** ($\Delta_{sub}H$): It is the energy required for transition of reactants from solid phase to gaseous phase. Its value is positive.

(d) **Dissociation energy** ($\Delta_{diss}H$): It is the energy required to break a chemical bond. This is an endothermic process, therefore it has positive value.
$$1/2 \ X_2(g) \rightarrow X(g)$$

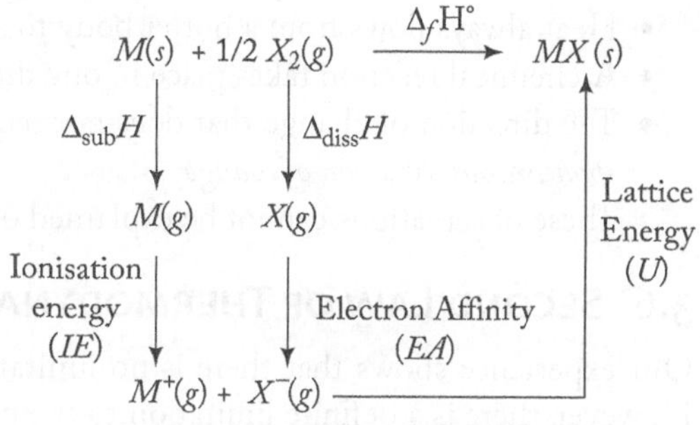

Fig. 3.8 Schematic representation of Born–Haber cycle

The general representation of a typical Born–Haber cycle is shown in Fig. 3.8.

The overall formation of a molecule MX from its constituents in gaseous state, that is, lattice energy is calculated using Hess's law as:
$$\Delta_f H° = \Delta_{sub}H + \Delta_{diss}H + IE + EA + U$$

Kirchoff's Law: The temperature Dependence of Heat of Reaction

For a reaction $X \rightarrow Y$,
$$\Delta H = H_Y - H_X$$

$$\frac{d}{dT}(\Delta H)_p = \left(\frac{\partial H_Y}{dT}\right)_p - \left(\frac{\partial H_X}{dT}\right)_p$$

$$\frac{d}{dT}(\Delta H)_p = C_{p,Y} - C_{p,X} = \Delta C_p$$

This is the well-known Kirchoff equation and can be used to calculate ΔH at a temperature, if ΔH at some other temperature is known.

$$\frac{d}{dT}(\Delta H)_p = \Delta C_p \ \text{ or, } d(\Delta H)_p = \Delta C_p dT$$

Integrating the above equation between two temperatures
$$\Delta H_2 - \Delta H_1 = \Delta C_p(T_2 - T_1)$$

Check Your Progress

16. If enthalpy of a reaction is known at some temperature, and heat capacities of reactants and products are known, the enthalpy of that reaction can be calculated at other temperatures. Explain.

17. The enthalpy of reaction can be calculated from enthalpy of formation of reactants and products. Explain.

18. Calculate the bond energy of 3 mol of CO_2 if average bond energy of CO bond is 347 kJ mol^{-1}.

19. Which of the given molecules has the highest lattice energy: MgO, $MgCl_2$, and NaCl? Explain.

20. Arrange the given molecules in increasing order of lattice energy: NaCl, KCl, and $AlCl_3$. Give reason.

Limitations of the first law of thermodynamics

(a) According to the first law, any energy-conserving process is permissible. It does not predict which of the several possible energy-conserving processes is possible.

(b) The first law places no limitation on the possibility of transforming one type of energy to another; for example, heat into work.

Some observations (not explained by the first law of thermodynamics)

- Heat always flows from a hotter body to a colder body.
- A chemical reaction takes place in one direction rather than another under a particular condition.
- The direction of change that does not require work to be done to bring about the change is called *spontaneous direction of change*.

These observations cannot be explained on the basis of the first law of thermodynamics.

3.6 SECOND LAW OF THERMODYNAMICS

Our experience shows that there is no limitation on the process of transformation of work into heat. However, there is a definite limitation to the possibility of transforming heat into work.

According to Kelvin: 'A transformation whose only final result is to transform into work the heat extracted from a source which is at the same temperature throughout is impossible'.

Carnot's cycle In 1824, a French engineer Sadi Carnot came up with a cycle, famously known as Carnot's cycle, to understand the limitation in converting heat into work. Carnot's cycle consists of four steps/processes, two isothermal and two adiabatic processes, in alternate arrangement. One of the two adiabatic/isothermal processes is an expansion process, whereas the other is compression (See Fig. 3.9).

Heat engine at constant temperature:

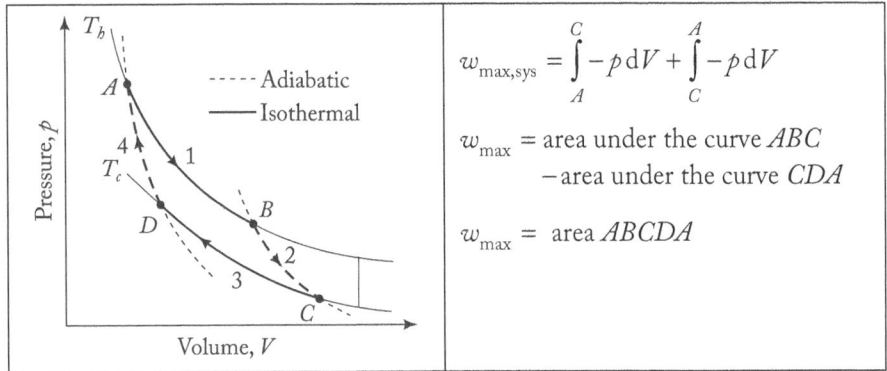

$$w_{max,sys} = \int_{A}^{C} -p\,dV + \int_{C}^{A} -p\,dV$$

w_{max} = area under the curve ABC
    $-$ area under the curve CDA

w_{max} = area $ABCDA$

Fig. 3.9 Carnot's cycle consisting of two isothermal steps (1 & 3) and two adiabatic steps (2 & 4)

The work done by the ideal gas, shown in right hand panel of Fig. 3.9, is given by area *ABCDA*. According to Carnot's cycle when T_c approaches T_h, the area defined by *ABCDA* approaches zero, and the maximum work approaches zero. In other words, no heat engine can convert heat to work at constant temperature. Thus, a heat engine must exchange heat between a source and a sink that are at different temperatures to do mechanical work, absorbing heat from the source and discarding heat to the sink.

Another important conclusion from this experiment can be understood from these equations.

From the first law of thermodynamics, $dq + dw = dU$.

If expressed in terms of work done by the system,

$$dq - dw_{sys} = dU$$
$$dw_{sys} = dq - dU$$

Thus work done by the system will be maximum if $dU = 0$.

$$dq = dw_{max} \text{ (if } dU = 0)$$

For a cyclic process, $dU = 0$. Hence,

$$dq = dw_{max}$$

In Carnot's cycle, $dw_{max} = dq$

$$w_{max} = q_{AB} - q_{CD}$$

This equation tells us that only part of the heat absorbed at higher temperature (q_{AB}) can be converted to work by the Carnot's engine. (Since q_{CD} is not equal to zero). This statement holds true for any other cycle which can be thought of.

Kelvin–Planck formulation 'It is impossible for a system to undergo a cyclic process whose sole effects are the flow of heat into the system from a hot reservoir and performance of an equal amount of work by the system on the surrounding.' Thus, there is a limitation in converting heat completely into work.

Entropy Conversion of heat to work is a spontaneous process whereas the opposite is not true. Application of heat and work stimulates the disorderly and orderly motion of the particles in the system, respectively. It means that isolated systems tend to go to that direction in which there is an increase in the disorder/randomness.

There was a need for measurement of disorder/randomness. Randomness can be related to heat; however, it is a path function. If we consider an ideal gas and apply the first law of thermodynamics, we have

$$dq_{rev} = C_V dT + \frac{nRT}{V} dV$$

dq is not a perfect differential since the term $dq_{rev} = \frac{nRT}{V} dV$ cannot be integrated.

However, $\dfrac{dq_{rev}}{T} = C_V \dfrac{dT}{T} + \dfrac{nR}{V} dV$

The above equation can be integrated. Thus, $\dfrac{dq_{rev}}{T}$ is a perfect differential.

$\dfrac{dq_{rev}}{T}$ *is a perfect differential: Proof from Carnot's cycle.*

Consider the Carnot's cycle in Fig. 3.9.

The heat absorbed by the engine during process AB is given by

$$q_{AB} = -w_{AB} = nRT_h \ln \frac{V_B}{V_A}$$

The heat released to the sink during process CD is given by

$$q_{CD} = -w_{CD} = nRT_c \ln \frac{V_C}{V_D}$$

For the adiabatic process BC,

$$V_B T_h^{\gamma-1} = V_C T_c^{\gamma-1}$$

For the adiabatic process DA

$$V_D T_c^{\gamma-1} = V_A T_h^{\gamma-1}$$

Multiplying the above two equations and removing the common factors from LHS and RHS gives

$$V_B V_D = V_C V_A$$

Thus, $q_{CD} = -w_{CD} = nRT_c \ln \dfrac{V_C}{V_D} = nRT_c \ln \dfrac{V_B}{V_A}$

$$\frac{T_c}{T_h} = \frac{q_{CD}}{q_{AB}}$$

$$\frac{q_{AB}}{T_h} - \frac{q_{CD}}{T_c} = 0$$

$$\oint \frac{dq_{rev}}{T} = 0$$

Thus, $\dfrac{dq_{rev}}{T}$ is the derivative of a state function called entropy and denoted by S.

$$dS = \frac{dq_{rev}}{T}$$

Thus, change in the entropy of the system can be obtained by integrating the above equation.

Clausius Inequality Theorem

Suppose a process from one state to another (say I to F) is carried out irreversibly;

From the first law of thermodynamics

$$dU = dq + dw = dq - p_{ex}dV$$

If the same process (I to F) is carried out reversibly, then

$$dU = dq_{rev} - pdV$$

Since initial (I) and final (F) states are same for both the processes and U is a state function

$$dU(\text{reversible}) = dU(\text{irreversible})$$

$$dq_{rev} - pdV = dq - p_{ex}dV$$

$$dq_{rev} - dq = (p - p_{ex})dV$$

If $p > p_{ex}$, expansion will take place.

$$\Delta V > 0$$

$$dq_{rev} - dq > 0$$

$dq_{rev} \geq dq$. This is also true for compression process.

Thus dq_{rev} is always greater than dq.

$$dq_{rev} - dq \geq 0$$

$$\frac{dq_{rev}}{T} - \frac{dq}{T} \geq 0$$

$$dS \geq \frac{dq}{T}$$

This is the famous Clausius Inequality theorem.

Entropy change during a spontaneous process   If we consider a cyclic process in which an isolated system changes its state from 1 to 2 irreversibly and then brought back to 1 reversibly.

According to Clausius Inequality theorem

$$\Delta S \geq \int_{1}^{2} \frac{dq_{12}}{T} + \int_{2}^{1} \frac{dq_{21}}{T}$$

$\Delta S = 0$ (S being a state function)

From the cyclic process

$$0 \geq \int_{1}^{2} \frac{dq_{12}}{T} + \int_{2}^{1} \frac{dq_{21}}{T}$$

$$dq_{12} = 0$$

$$0 \geq \int_{2}^{1} \frac{dq_{21}}{T}$$

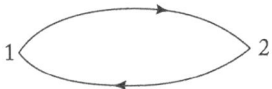

Fig. 3.10 A cyclic process in which an isolated system undergoes change from state 1 to state 2 irreversibly and back to state 1 reversibly

$$0 \geq S_1 - S_2$$

$$S_2 - S_1 \geq 0$$

$$\Delta S_{\text{(isolated system)}} \geq 0$$

Thus, according to the second law of thermodynamics, entropy of an isolated system always increases during a spontaneous process.

The universe itself can be thought of as one big isolated system.

$$\Delta S_{\text{universe}} \geq 0$$

In other words, entropy of the universe will increase during a spontaneous process.

Entropy of the system and its surroundings:

$$dS_{\text{sys}} = \frac{dq_{\text{sys, rev}}}{T_{\text{sys}}}$$

The thermodynamic parameters of a system are usually written without the subscript.

$$dS = \frac{dq_{\text{rev}}}{T}$$

The thermodynamic parameters of surroundings are written with the subscript 'surr'.

$$dS_{\text{surr}} = \frac{dq_{\text{surr, rev}}}{T_{\text{surr}}} = -\frac{dq_{\text{sys}}}{T_{\text{sys}}}$$

Thus, to know the entropy change in an irreversible process, we must devise a reversible process which undergoes the same change of state.

Interpretation of entropy: Entropy is measure of randomness/chaosness of a system.

$$dS \geq \frac{dq}{T}$$

$$dS - \frac{dq}{T} \geq 0$$

$$dS + dS_{\text{surr}} \geq 0$$

Total entropy of the system and its surroundings increases in the course of a spontaneous change.

Entropy change in a process for an ideal gas:

Suppose an ideal gas undergoes a process in which it goes from state 1 (T_1, V_1) to state 2 (T_2, V_2), then the entropy change of gas can be calculated as follows:

$$dq_{\text{rev}} = dU - dw_{\text{rev}}$$

For an ideal gas,

$$dq_{\text{rev}} = C_V dT + p \, dV$$

$$dq_{\text{rev}} = C_V dT + \frac{nRT}{V} dV$$

$$\frac{dq_{\text{rev}}}{T} = \frac{C_V dT}{T} + \frac{nR}{V} dV$$

Check Your Progress

21. The heat change in a reversible expansion/compression of a gas is always greater than that in an irreversible process. Why?
22. Prove that entropy of the system (S) is a state function.
23. Only a part of the heat absorbed at higher temperature (q_{AB}) can be converted to work by the Carnot's engine. Justify.
24. No heat engine can convert heat to work at constant temperature. Explain using Carnot's cycle.

Integrating,

$$\Delta S = C_v \ln \frac{T_2}{T_1} + nR \ln \frac{V_2}{V_1}$$

Thus, entropy increases with increasing temperature at constant volume and increases with increasing volume at constant temperature.

Change in entropy of a system undergoing phase change   Phase change takes place at constant temperature and pressure. Thus,

$$dS_{sys} = \frac{dq_{rev,\,sys}}{T} = \frac{dq_p}{T} = \frac{dH}{T}$$

$$\Delta S_{sys} = \frac{\Delta H}{T}$$

$$dS_{surr} = \frac{dq_{surr}}{T} = \frac{-dq_{sys}}{T} = \frac{-dH}{T}$$

$$\Delta S_{surr} = \frac{-\Delta H}{T}$$

$$\Delta S_{Total} = \Delta S_{sys} + \Delta S_{surr} = 0$$

General equation for change in entropy:

$$dq = dU - dw$$

For a closed system, $U = f(V,\ T)$

Then,

$$dU = \left(\frac{\partial U}{\partial T}\right)_V dT + \left(\frac{\partial U}{\partial V}\right)_T dV$$

So,

$$dq_{rev} = \left(\frac{\partial U}{\partial T}\right)_V dT + \left(\frac{\partial U}{\partial V}\right)_T dV + pdV$$

$$dq_{rev} = \left(\frac{\partial U}{\partial T}\right)_V dT + \left(\left(\frac{\partial U}{\partial V}\right)_T + p\right)dV$$

$$\frac{dq_{rev}}{T} = \frac{1}{T}\left(\frac{\partial U}{\partial T}\right)_V dT + \frac{1}{T}\left(\left(\frac{\partial U}{\partial V}\right)_T + p\right)dV$$

$$\frac{dq_{rev}}{T} = \frac{1}{T}\left(\frac{\partial U}{\partial T}\right)_V dT + \frac{1}{T}\left(\left(\frac{\partial U}{\partial V}\right)_T + p\right)dV$$

Here, we use thermodynamic equation of state which will be proved in Section 3.7. According to this,

$$T\left(\frac{\partial p}{\partial T}\right)_V = \left(\frac{\partial U}{\partial V}\right)_T + p$$

$$\Rightarrow \qquad dS = \frac{1}{T}\left(\frac{\partial U}{\partial T}\right)_V dT + \left(\frac{\partial p}{\partial T}\right)_V dV$$

$$= \frac{C_{V,m}dT}{T} + \left(\frac{\partial p}{\partial T}\right)_V dV$$

$$dS_{sys} = \frac{C_{V,m}dT}{T} + \left(\frac{\partial p}{\partial T}\right)_V dV$$

Similarly, it can be derived using the equation for enthalpy that

$$dS_{sys} = \frac{C_{p,m}dT}{T} - \left(\frac{\partial V}{\partial T}\right)_V dp$$

Specific cases:

(i) Processes taking place at constant volume

At constant volume,

$$dS_{sys} = \frac{C_{V,m}dT}{T}$$

$$\Delta S_{sys} = C_{V,m} \ln \frac{T_{f,\,isochoric}}{T_i}$$

(ii) Processes taking place at constant pressure

Similarly, it can be proved that under constant pressure,

$$\Delta S_{sys} = C_{p,m} \ln \frac{T_{f,\,isobaric}}{T_i}$$

(iii) Entropy change of an ideal gas in an isothermal process

At constant temperature,

$$dS_{sys} = \left(\frac{\partial p}{\partial T}\right)_V dV$$

For one mole of an ideal gas,

$$dS_{sys} = \frac{R}{V}dV$$

$$\Delta S_{sys} = R \ln \frac{V_{f,\,isothermal}}{V_i}$$

$$dS_{sys} = -\frac{R}{p}dp$$

$$\Delta S_{sys} = R \ln \frac{p_i}{p_{f,\,isothermal}}$$

Entropy change of the surroundings:

Entropy change of the system will not depend on whether the change is carried out reversibly or irreversibly. However, the entropy change of the surroundings will depend on the way the change is carried out.

For a reversible process, $\Delta S_{sys} + \Delta S_{surr} = 0$

For an isothermal process carried out reversibly,

$$\Delta S_{surr} = -\Delta S_{sys} = -RT \ln \frac{V_f}{V_i}$$

$$\Delta S_{surr} = -\Delta S_{sys} = -RT \ln \frac{p_i}{p_f}$$

For an isothermal process carried out irreversibly,

$$dS_{surr} = \frac{dq_{surr}}{T} = \frac{-dq_{sys}}{T} = \frac{dw}{T}$$

$$\Delta S_{surr} = \frac{-p_{ext}(V_2 - V_1)}{T}$$

Sign of various thermodynamic terms for a spontaneous change:

$dS_{sys} + dS_{surr} \geq 0$

$dS_{sys} \geq -dS_{surr}$

$dS_{sys} \geq dq/T$

$dq - TdS \leq 0$

Thus, for a spontaneous change, $dq - TdS \leq 0$

At constant volume, $dq_V - TdS \leq 0$

$dU - TdS \leq 0$

$dS_{U,V} \geq 0$

Thus, a spontaneous process will always be accompanied by increase in entropy of an isolated system (system with constant U and V).

However, most reactions take place under the conditions of (i) constant volume and temperature (ii) constant pressure and temperature.

(i) At constant volume and temperature

For a spontaneous change, $dq - TdS \leq 0$

At constant volume, $dq_V - TdS \leq 0$ or, $dU - TdS \leq 0$

At constant volume and temperature, $dU - TdS = dU - d(TS)$.

Thus, $d(U - TS)_{V,T} \leq 0$

$d(A)_{V,T} \leq 0$

Thus, at constant temperature and volume, the function $(U - TS)$ decreases during a spontaneous process and will achieve its minimum value at equilibrium. $U - TS$ has units of energy and is called Helmholtz free energy (A).

(ii) At constant pressure and temperature

On the other hand, at constant temperature and pressure,

$dq_p - TdS \leq 0$

$dH - TdS \leq 0$

$d(H - TS)_{p,T} \leq 0$

Function $(H - TS)$ decreases during a spontaneous process.

$d(G)_{p,T} \leq 0$

Thus, at constant temperature and pressure, the function $(H-TS)$ decreases during a spontaneous process and will achieve its minimum value at equilibrium. $H-TS$ has the unit of energy and is called the Gibb's free energy (G).

Combining the first and second laws of thermodynamics: Calculation of changes in Helmholtz and Gibbs energies

Change in internal energy can be directly calculated from the first law of thermodynamics.

Check Your Progress

25. The change in entropy between two states will be the same whether the change is carried out reversibly or irreversibly. Why?

26. Entropy change of an ideal gas in an adiabatic process is not zero. Explain.

27. How is the entropy change in a process expressed as a function of temperature and volume?

28. How is change in the entropy of surroundings calculated for an ideal gas undergoing an isothermal process?

29. What will be the entropy change when 1 mol of water goes to vapour at 100 °C?

An infinitesimal change in internal energy of a reversible process (from initial state I to final state F) is given by

$$dU = dq - p\,dV$$
$$dU = T\,dS - p\,dV \tag{i}$$

Although calculated for a reversible process from state I to F, the above equation is also equally valid for an irreversible process involving the same initial (I) and final states (F), since U is a state function.

Infinitesimal change in enthalpy can be expressed as

$$dH = d(U + PV)$$
$$dH = dU + p\,dV + V\,dp$$
$$dH = T\,dS - p\,dV + p\,dV + V\,dp$$
$$dH = T\,dS + V\,dp \tag{ii}$$

Infinitesimal change in Helmholtz energy can be expressed as

$$dA = d(U - TS)$$
$$dA = dU - T\,dS - S\,dT$$
$$dA = T\,dS - p\,dV - T\,dS - S\,dT$$
$$dA = -S\,dT - p\,dV \tag{iii}$$

Infinitesimal change in Gibbs free energy can be expressed as

$$dG = dH - T\,dS - S\,dT$$
$$dG = T\,dS + V\,dp - T\,dS - S\,dT$$
$$dG = V\,dp - S\,dT \tag{iv}$$

These four equations (i–iv) are fundamental equations of thermodynamics.

3.7 MAXWELL'S RELATIONS

There are a number of derivatives which cannot be determined experimentally. Thus, they need to be expressed in terms of derivatives which can be determined experimentally. For example change in entropy as a function of volume (at constant temperature) cannot be determined directly. However, it can be expressed as another derivative with the help of Maxwell's relations.

Maxwell's relations utilize the fact that mixed derivatives are equal for a state function.

Suppose x is a function of y and z.

$$dx = m\,dy + n\,dz$$

$$\left(\frac{\partial x}{\partial y}\right)_z = m$$

$$\left(\frac{\partial x}{\partial z}\right)_y = n$$

If x is a state function, its mixed derivatives $\dfrac{\partial}{\partial z}\left(\left(\dfrac{\partial x}{\partial y}\right)_z\right)_y$ and $\dfrac{\partial}{\partial y}\left(\left(\dfrac{\partial x}{\partial z}\right)_y\right)_z$ will be equal.

Thus, $\left(\dfrac{\partial m}{\partial z}\right)_y = \left(\dfrac{\partial n}{\partial y}\right)_x$

Thus, on equating the mixed derivatives of A from equation (iii), we get the relation

$$\left(\frac{\partial S}{\partial V}\right)_T = \left(\frac{\partial p}{\partial T}\right)_V$$

Change in pressure with changing temperature at constant volume can be easily measured in the laboratory, which in turn lets us know how entropy changes with volume at constant temperature.

Similarly, equating mixed derivatives of *G* from equation (iv) gives

$$\left(\frac{\partial V}{\partial T}\right)_p = -\left(\frac{\partial S}{\partial p}\right)_T$$

Equating mixed derivatives of *H* from equation (ii) gives

$$\left(\frac{\partial T}{\partial p}\right)_S = \left(\frac{\partial V}{\partial S}\right)_p$$

Equating mixed derivatives of *U* from equation (i) gives

$$\left(\frac{\partial T}{\partial V}\right)_S = -\left(\frac{\partial p}{\partial S}\right)_V$$

These four relations are the well-known Maxwell's relations.

Thermodynamic equation of state:

$$dU = TdS - pdV$$

$$\left(\frac{\partial U}{\partial V}\right)_T = T\left(\frac{\partial S}{\partial V}\right)_T - p$$

From Maxwell's relations,

$$\left(\frac{\partial S}{\partial V}\right)_T = \left(\frac{\partial p}{\partial T}\right)_V$$

$$\left(\frac{\partial U}{\partial V}\right)_T = T\left(\frac{\partial p}{\partial T}\right)_V - p$$

This is the first thermodynamic equation of state. This equation helps to obtain internal pressure $\left(\frac{\partial U}{\partial V}\right)_T$ in terms of parameters which can be measured experimentally.

$$dH = TdS + Vdp$$

$$\left(\frac{\partial H}{\partial p}\right)_T = T\left(\frac{\partial S}{\partial p}\right)_T + V$$

From Maxwell's relation,

$$\left(\frac{\partial V}{\partial T}\right)_p = -\left(\frac{\partial S}{\partial p}\right)_T$$

$$\left(\frac{\partial H}{\partial p}\right)_T = -T\left(\frac{\partial V}{\partial T}\right)_p + V$$

This is the second thermodynamic equation of state. This equation helps to obtain the isothermal Joule–Thomson coefficient $\left(\frac{\partial H}{\partial p}\right)_T$ in terms of parameters which can be measured experimentally.

Third law of thermodynamics   *The entropy of a perfectly crystalline substance at 0 K is zero.* The significance of this statement is that it helps in predicting the absolute value of entropy of a substance under specific condition.

Thus, entropy of a gas can be calculated as

$$S = S(0 \text{ K}) + \int_0^{T_m} C_{p,s}\frac{dT}{T} + \frac{\Delta H_m}{T_m} + \int_{T_m}^{T_b} C_{p,l}\frac{dT}{T} + \frac{\Delta H_v}{T_b} + \int_{T_b}^{T} C_{p,v}\frac{dT}{T}$$

where subscripts s, l, and v refer to solid, liquid, and vapour states, and m and b refer to melting and boiling points, respectively.

Another consequence of the third law of thermodynamics is that the temperature of a substance cannot be decreased to zero in finite number of steps. Thus, it is impossible to attain temperature of 0 K.

Significance of Helmholtz Free Energy

For a spontaneous process, $dq - TdS \le 0$.

For a reversible process, $dq - TdS = 0$

$dU - dw - TdS = 0$

Since the system does maximum work in a reversible process

$$dw_{max} = dU - TdS = dA \quad \text{(At constant temperature)}$$

Maximum work done (including expansion/compression also known as pV work) can be calculated by measuring the change in A.

Significance of Gibb's Free Energy

In the presence of works other than pV work (for example, electrical work, surface work),

$$dG = -SdT + Vdp + dw_{add, Rev}$$

At constant temperature and pressure

$$dG = dw_{add, Rev}$$

Thus, the measurement of change in free energy can be used to measure non-expansion reversible work (work other than pV work) or vice versa.

Thus, emf work obtained in an electrolytic cell at constant temperature and pressure is equal to ΔG for cell reaction.

$$dG = -dw_{EMF, Rev}$$

$$dG = -nFdE$$

Changing the free energy of atoms/molecules/species: To convert a non-spontaneous process to spontaneous or vice versa, the free energies of different components involved in the process need to be changed.

From the equation $dG = Vdp - SdT$

Change in molar free energy is given by

$$dG_m = V_m dp - S_m dT$$

It is evident that free energy of a component can be changed by changing the pressure and temperature.

Change of free energy with pressure:

At constant temperature, $dG = Vdp$ or, $dG_m = V_m dp$

(a) **For liquids and solids**

$$\Delta G_m = V_m(p_f - p_i)$$

(b) **For gases**

For an ideal gas,

$$\Delta G_m = \int_{p_i}^{p_f} \frac{RT}{p} dp$$

$$\Delta G_m = RT \ln \frac{p_f}{p_i}$$

Molar Gibb's Free Energy of a gas at pressure p is generally expressed with reference to its molar free energy at standard pressure (G_m°).

$$G_m - G_m^\circ = RT \ln \frac{p}{p^0}$$

$$G_m = G_m^\circ + RT \ln \frac{p}{p^0}$$

Thus, free energy of a liquid/solid/gas increases with increase in the pressure.

For a real gas,

For a real gas, pressure p is replaced by fugacity f.

$$G_m = G_m^0 + RT \ln \frac{f}{p^0}$$

$$\lim_{p \to 0} \frac{f}{p} = 1$$

As the pressure approaches zero, the gas behaves as an ideal gas and fugacity becomes equal to pressure.

Change of free energy with temperature (Gibb's–Helmholtz equation): The free energy of a species can be changed with temperature.

For an atom or a molecule, dG is given by

$$dG = V dp - S dT$$

At constant pressure, $dG = -S dT$

$$\left(\frac{\partial G}{\partial T} \right)_p = -S$$

Since S is positive, free energy of solid/liquid/gas decreases with increase in temperature.

$$\left(\frac{\partial G}{\partial T} \right)_p = \frac{G - H}{T}$$

$$\left(\frac{\partial G}{\partial T} \right)_p - \frac{G}{T} = -\frac{H}{T}$$

$$\frac{1}{T} \left(\frac{\partial G}{\partial T} \right)_p - \frac{G}{T^2} = -\frac{H}{T^2}$$

$$\left(\frac{\partial (G/T)}{\partial T} \right)_p = -\frac{H}{T^2}$$

Gibb's–Helmholtz equation helps us to determine the free energy of a species (atom/molecule) at a given temperature, if Gibb's energy at some other temperature is known.

For a process (chemical or physical)

$$\left(\frac{\partial (\Delta G/T)}{\partial T} \right)_p = -\frac{\Delta H}{T^2}$$

It also helps to determine the enthalpy of a process if the free energy change of the process can be measured at two different temperatures.

Check Your Progress

30. What is the physical significance of change in free energy?
31. Express the criteria for spontaneous change at constant volume and temperature.
32. Maxwell's relations can be used to find entropy change as a function of volume. Explain.
33. Mention whether free energy of an ideal gas will increase or decrease with change in temperature.
34. Measurement of change in free energy for a process at two temperatures can help us to know the enthalpy change in the process. Explain.

3.8 MATERIAL EQUILIBRIUM

When a reaction/process starts, the concentration of reactants decreases whereas that of products increases with time.

$$d(G)_{p,T} \leq 0$$

$d(G)_{p,T}$ will start from negative value and approaches zero. There will be a time when $d(G)_{p,T}$ will be zero, and the concentration of either reactants or products will not change. This is the time when material equilibrium is established. Even though reactants and products are changing into each other, there will be no net change in the concentration of reactants or products, since the rate of forward reaction is equal to the rate of backward reaction. Thus, chemical equilibrium is a dynamic equilibrium.

Reason for material equilibrium As discussed previously, reactant (R) gets converted to product (P) if free energy of the product (P) is less than that of the reactant. However, the reactions do not go to completion, that is, some amount of the reactant will not be used up and remain in the reaction mixture. Additionally, if we start with pure product (P), some amount of the product will get converted to reactant even if the free energy of product is less than the free energy of reactant. This is due to entropy of mixing which is positive. This makes the overall free energy of reaction (product to reactant) more negative leading to conversion of some product to reactant.

$$R \rightarrow P$$
$$\Delta_R G = n_P G_P - n_R G_R + \Delta G_{mix}$$

For an ideal mixture

$$\Delta G_{mix} = -T \Delta S_{mix}$$
$$\Delta_R G = n_P G_P - n_R G_R - T \Delta S_{mix} = n_P G_P - (n_R G_R + T \Delta S_{mix})$$

Free energy of mixing Consider that x moles of gas A and y moles of gas B are taken in a container separated by a partition, pressure being p for each gas.

For gas A,

since, $\Delta G_m = RT \ln \dfrac{p_f}{p_i}$

(G of a gas can be written with respect to standard state condition)

$$G_{m,A} = G^0_{m,A} + RT \ln \dfrac{p}{p^0}$$

$G_{m,A}$ is standard molar Gibb's Free energy of A and p^0 is standard pressure.
For gas B,

$$G_{m,B} = G^0_{m,B} + RT \ln \dfrac{p}{p^0}$$

$G_{m,B}$ is standard molar Gibb's energy of B.
Total free energy G will be equal to

$$G = x \times \left(G^0_{m,A} + RT \ln \dfrac{p}{p^0} \right) + y \times \left(G^0_{m,B} + RT \ln \dfrac{p}{p^0} \right)$$

$$G = x \times G^0_{m,A} + y \times G^0_{m,B} + (x + y) \left(RT \ln \dfrac{p}{p^0} \right)$$

Now, suppose the partition is removed allowing both gases to mix, and the total pressure remains the same. Then,

$$G_{m,A} = G^0_{m,A} + RT \ln \dfrac{f_A p}{p^0} \quad \text{and} \quad G_{m,B} = G^0_{m,B} + RT \ln \dfrac{f_B p}{p^0}$$

In the above equations, f_A and f_B are the fractions of gases A and B in the container. Therefore, total free energy after mixing (G') is

$$G' = x \times \left(G^0_{m,A} + RT \ln \frac{f_A p}{p^0} \right) + y \times \left(G^0_{m,B} + RT \ln \frac{f_B p}{p^0} \right)$$

$$G' = x \times G^0_{m,A} + y \times G^0_{m,B} + (x + y)\left(RT \ln \frac{p}{p^0} \right) + xRT \ln f_A + yRT \ln f_B$$

Change in free energy on mixing

$$G' - G = xRT \ln f_A + yRT \ln f_B$$
$$\Delta G_{\text{mix}} = xRT \ln f_A + yRT \ln f_B$$

Since f_A and f_B are less than 1, ΔG_{mix} will be negative.
Thus, mixing of two gases is a spontaneous process.

3.8.1 Types of Material Equilibrium

Material equilibrium can be of two types: (a) phase equilibrium and (b) chemical equilibrium.

Phase Equilibrium

Melting of ice, boiling of liquid, condensation of liquid vapour, solubility of salt, and change in phases are examples of physical processes, where a system changes from one physical state (phase) to another. An equilibrium dealing with such processes is called phase equilibrium. In this state, the amount of physical state of the system does not change with time; for example, if an ice cube is kept at 0 °C, ice will melt. After some time, a state will be achieved where the amount of ice and water will not change with time. This state is known as phase equilibrium.

Types of phase equilibrium:
(i)   Solid–liquid equilibrium: e.g., $H_2O(s) \rightleftharpoons H_2O(l)$.
(ii) Liquid–vapour equilibrium: $N_2(l) \rightleftharpoons N_2(g)$
(iii) Solid–vapour equilibrium: $CO_2(s) \rightleftharpoons CO_2(g)$
(iv) Solubility of sparingly soluble salt: $AgCl(s) \rightleftharpoons AgCl(aq)$
(v)   Dissolution of gases in liquid: $CO_2(g) \rightleftharpoons CO_2(aq)$
(vi) Solid–liquid–vapour equilibrium: $H_2O(s) \rightleftharpoons H_2O(l) \rightleftharpoons H_2O(g)$

Two phases can co-exist at only one temperature at a particular pressure. The temperature can change with pressure. However, three phases can co-exist only at a particular temperature and pressure.

Change of free energy for a physical process with pressure   Consider the conversion of carbon in graphite form to carbon in diamond form. If we change the pressure keeping the temperature constant, free energy of both diamond and graphite will increase.

$$\Delta G_{m,gr} = V_{m,gr}(p_2 - p_1)$$
$$\Delta G_{m,D} = V_{m,D}(p_2 - p_1)$$
$$\Delta \Delta G_m = (\Delta G_m)_{p2} - (\Delta G_m)_{p1} = (G_{m,D,p2} - G_{m,gr,p2}) - (G_{m,D,p1} - G_{m,gr,p1})$$
$$\Delta \Delta G_m = (G_{m,D,p2} - G_{m,D,p1}) - (G_{m,gr,p2} - G_{m,gr,p1}) = \Delta G_{m,D} - \Delta G_{m,gr} = (V_{m,D} - V_{m,gr})(p_2 - p_1)$$

(The subscripts m, gr, D, p_1, p_2 stand for molar, graphite, diamond, initial pressure, and final pressure, respectively.)

The free energy change ΔG_m of the process (conversion of graphite to diamond) is positive at 1 bar, making the process non-spontaneous at 1 bar. The molar volume of graphite is more than that of diamond. Thus, ΔG_m of the process decreases with increase in pressure and will become negative at very high pressure.

Change of transition temperature with pressure   The transition temperature between two phases can be changed by changing the pressure. This is utilized in the working of pressure cooker and autoclave.

Consider that the transition temperature (such as melting temperature) between two phases is T_1 at pressure p_1. This can be changed by applying pressure. Suppose the transition temperature is T_2 at p_2.

At equilibrium (T_1 and p_1), the molar Gibbs free energy of the two physical states (say α and β) are same.

$$G_{m,\alpha} = G_{m,\beta}$$

To change transition temperature to T_2, pressure needs to be changed. Suppose pressure is changed to p_2 to establish equilibrium at T_2, the molar Gibbs free energy of both state α and state β will change. Now suppose the molar Gibbs free energy of α and β are $G'_{m,\alpha}$ and $G'_{m,\beta}$ at T_2 and p_2.

Since T_2 is the new transition temperature at p_2,

$$G'_{m,\alpha} = G'_{m,\beta}$$

Substracting the two equations

$$G'_{m,\alpha} - G_{m,\alpha} = G'_{m,\beta} - G_{m,\beta}$$

For infinitesimal change,

$$dG_{m,\alpha} = dG_{m,\beta}$$

$$V_{m,\alpha}\,dp - S_{m,\alpha}\,dT = V_{m,\beta}\,dp - S_{m,\beta}\,dT$$

$$S_{m,\beta}\,dT - S_{m,\alpha}\,dT = V_{m,\beta}\,dp - V_{m,\alpha}\,dp$$

$$(S_{m,\beta} - S_{m,\alpha})dT = (V_{m,\beta} - V_{m,\alpha})dp$$

$$\Delta S_m\,dT = \Delta V_m\,dp$$

$$\frac{dp}{dT} = \frac{\Delta S_m}{\Delta V_m}$$

$$\Delta S_m = \frac{\Delta H_m}{T}$$

$$\frac{dp}{dT} = \frac{\Delta H_m}{T\Delta V_m}$$

This is the well-known Clayperon equation. This equation provides information about how transition temperature can be changed by applying pressure or vice versa.

Clausius–Clayperon equation   This is an extension of the Clayperon equation. The transition between a condensed phase (solid/liquid) and a gas is governed by this equation.

$$\frac{dp}{dT} = \frac{\Delta S_m}{\Delta V_m}$$

At the transition temperature

$$\frac{dp}{dT} = \frac{\Delta H_m}{T(V_{m,g} - V_{m,c})}$$

Since the volume of condensed phase is negligible in comparison to the volume of gas phase

$$\frac{dp}{dT} \cong \frac{\Delta H_m}{TV_{m,g}}$$

If we assume that gas behaves as ideal gas, $V_{m,g} = \dfrac{RT}{p}$

$$\frac{dp}{dT} = \frac{\Delta H_m\, p}{RT^2}$$

$$\frac{dp}{p} = \frac{\Delta H_m}{RT^2}\,dT$$

Integrating

$$\ln\frac{p_2}{p_1} = \frac{\Delta H_m}{R}\left(\frac{1}{T_1} - \frac{1}{T_2}\right) = \frac{\Delta H_m}{R}\left(\frac{\Delta T}{T_1 T_2}\right)$$

Vaporization/sublimation process is an endothermic process. Thus, boiling point of a liquid can be increased by increasing the pressure.

Free Energy Change in a Chemical Process and at Equilibrium

Any chemical process is accompanied by free energy change and a chemical process is spontaneous if the change in free energy is negative.

For a reaction, $aA + bB \rightleftharpoons cC + dD$

$$\Delta G = cG_{m,C} + dG_{m,D} - aG_{m,A} - bG_{m,B}$$

Suppose all the components are in gaseous phase, G of a gas can be written with respect to standard state condition ($p^0 = 1$).

$$G_m = G_m^0 + RT\ln\frac{p}{p^0} = G_m^0 + RT\ln\frac{p}{1}$$

$$\Delta G - \Delta G^0 = cRT\ln\left(\frac{p_C}{1}\right) + d\,RT\ln\left(\frac{p_D}{1}\right) - aRT\ln\left(\frac{p_A}{1}\right) - bRT\ln\left(\frac{p_B}{1}\right)$$

where $\Delta G^0 = cG_{m,C}^0 + dG_{m,D}^0 - aG_{m,A}^0 - bG_{m,B}^0$

$$\Delta G - \Delta G^0 = RT\ln\left(\frac{p_C}{1}\right)^c + RT\ln\left(\frac{p_D}{1}\right)^d - RT\ln\left(\frac{p_A}{1}\right)^a - RT\ln\left(\frac{p_B}{1}\right)^b$$

$$\Delta G - \Delta G^0 = RT\ln\frac{\left(\dfrac{p_C}{1}\right)^c\left(\dfrac{p_D}{1}\right)^d}{\left(\dfrac{p_A}{1}\right)^a\left(\dfrac{p_B}{1}\right)^b}$$

$$\Delta G = \Delta G^0 + RT\ln\frac{\left(\dfrac{p_C}{1}\right)^c\left(\dfrac{p_D}{1}\right)^d}{\left(\dfrac{p_A}{1}\right)^a\left(\dfrac{p_B}{1}\right)^b}$$

Thus, ΔG is related to pressure of gaseous reactants and products at different time.

At equilibrium, $\Delta G = 0$. Hence the above equation becomes,

$$0 = \Delta G^0 + RT\ln\frac{\left(\dfrac{p_C}{1}\right)^c_{eq}\left(\dfrac{p_D}{1}\right)^d_{eq}}{\left(\dfrac{p_A}{1}\right)^a_{eq}\left(\dfrac{p_B}{1}\right)^b_{eq}}$$

Thus, ΔG^0 is related to pressure of gaseous reactants and products at equilibrium.

$$\Delta G^0 = -RT\ln\frac{\left(\dfrac{p_C}{1}\right)^c_{eq}\left(\dfrac{p_D}{1}\right)^d_{eq}}{\left(\dfrac{p_A}{1}\right)^a_{eq}\left(\dfrac{p_B}{1}\right)^b_{eq}} = -RT\ln K_p$$

$$K_p = \frac{\left(\frac{p_C}{1}\right)^c_{eq} \left(\frac{p_D}{1}\right)^d_{eq}}{\left(\frac{p_A}{1}\right)^a_{eq} \left(\frac{p_B}{1}\right)^b_{eq}}$$

$$\Delta G = \Delta G^0 + RT \ln \frac{\left(\frac{p_C}{1}\right)^c \left(\frac{p_D}{1}\right)^d}{\left(\frac{p_A}{1}\right)^a \left(\frac{p_B}{1}\right)^b}$$

$$\Delta G = -RT \ln K_p + RT \ln \frac{\left(\frac{p_C}{1}\right)^c \left(\frac{p_D}{1}\right)^d}{\left(\frac{p_A}{1}\right)^a \left(\frac{p_B}{1}\right)^b}$$

$$\Delta G = -RT \ln Kp + RT \ln Q_P$$

$$Q_p = \frac{\left(\frac{p_C}{1}\right)^c \left(\frac{p_D}{1}\right)^d}{\left(\frac{p_A}{1}\right)^a \left(\frac{p_B}{1}\right)^b}$$

For a generalized equation for reaction involving real substances in gas, liquid, and solid state, G. N. Lewis introduced the term activity,

$$G_m = G_m^0 + RT \ln a_i$$

For a real gas, $a_i = \dfrac{f}{p_0}$

Thus, ΔG can be written as

$$\Delta G = -RT \ln K + RT \ln Q$$

$$K = \frac{(a_C)^c_{eq} (a_D)^d_{eq}}{(a_A)^a_{eq} (a_B)^b_{eq}}$$

$$Q = \frac{(a_C)^c (a_D)^d}{(a_A)^a (a_B)^b}$$

Here, a_i is the activity of i th component in the reaction mixture; Q is called reaction quotient. Thus, equilibrium constant K is related to the activity of reactants and products at equilibrium, whereas reaction quotient is related to the activity of reactants and products at any time. The direction of reaction will be decided by the value of K and Q.

$$\Delta G = RT \ln \frac{Q}{K}$$

Q decides the direction of reaction.

Condition 1: $Q = K$, $\Delta G = RT \ln \dfrac{Q}{K} = 0$ (no free energy change, hence the reaction will be at equilibrium)

Condition 2: $Q < K$, $\Delta G = RT \ln \dfrac{Q}{K} < 0$ (Direction of reaction will be towards products).

Condition 3: $Q > K$, $\Delta G = RT \ln \dfrac{Q}{K} > 0$ (Direction of reaction will be towards reactants).

Types of Equilibrium Constant for a Chemical Process

There are three types of equilibrium constants, in which activity is expressed in terms of concentration (c), partial pressure (p), and mole fraction (x) of reactants and products.

For a chemical reaction, $aA + bB \rightleftharpoons cC + dD$,

$$K_c = \frac{[C]_{eq}^c [D]_{eq}^d}{[A]_{eq}^a [B]_{eq}^b}$$

$$K_p = \frac{[p_C]_{eq}^c [p_D]_{eq}^d}{[p_A]_{eq}^a [p_B]_{eq}^b}$$

$$K_x = \frac{[x_C]_{eq}^c [x_D]_{eq}^d}{[x_A]_{eq}^a [x_B]_{eq}^b}$$

Equilibrium constant and the law of mass action

According to the law of mass action, the rate of a reaction is proportional to the product of effective concentration of reactants, each raised to a power equal to their stoichiometry in a balanced chemical reaction.

The rate of forward reaction $= k_f [A]_{eq}^a [B]_{eq}^b$

The rate of reverse reaction $= k_b [C]_{eq}^c [D]_{eq}^d$

$$K_c = \frac{k_f}{k_b} = \frac{[C]_{eq}^c [D]_{eq}^d}{[A]_{eq}^a [B]_{eq}^b}$$

$$k_f [A]_{eq}^a [B]_{eq}^b = k_b [C]_{eq}^c [D]_{eq}^d$$

Thus, at equilibrium, the rate of forward reaction is equal to the rate of reverse reaction.

Types of Chemical Reactions and expression for equilibrium constant

Homogeneous reactions In these reactions all the components are in the same physical state. All of them are either in gaseous state or in liquid state or in solid state; for example, in the formation /dissociation of ammonia, the reactants and product are all in the same gaseous phase.

Example: Formation of ammonia

$$NH_3(g) \rightleftharpoons \frac{1}{2}N_2(g) + \frac{3}{2}H_2(g)$$

$$K_p = \frac{\left[p_{N_2}\right]_{eq}^{1/2} \left[p_{H_2}\right]_{eq}^{3/2}}{\left[p_{NH_3}\right]_{eq}}$$

Heterogeneous reactions In these reactions the components are in more than one physical state.

Example: Dissociation of calcium carbonate

$$CaCO_3(s) \rightleftharpoons CaO(s) + CO_2(g)$$

$$K_p = p_{CO_2}$$

Relationship between K_C and K_p Assuming that all reactants and products behave as ideal gas

$$pV = nRT$$

$$p = \frac{n}{V}RT = cRT$$

$$K_p = \frac{[p_C]_{eq}^c [p_D]_{eq}^d}{[p_A]_{eq}^a [p_B]_{eq}^b}$$

$$K_p = \frac{([C]RT)_{eq}^c ([D]RT)_{eq}^d}{([A]RT)_{eq}^a ([B]RT)_{eq}^b}$$

$$K_p = \frac{[C]_{eq}^c [D]_{eq}^d}{[A]_{eq}^a [B]_{eq}^b}(RT)^{c+d-a-b}$$

$$K_p = \frac{[C]_{eq}^c [D]_{eq}^d}{[A]_{eq}^a [B]_{eq}^b}(RT)^{\Delta v} = K_c (RT)^{\Delta v}$$

Relationship between K_p and K_x

$$K_p = \frac{[p_C]_{eq}^c [p_D]_{eq}^d}{[p_A]_{eq}^a [p_B]_{eq}^b}$$

$$K_p = \frac{[x_C p]_{eq}^c [x_D p]_{eq}^d}{[x_A p]_{eq}^a [x_B p]_{eq}^b}$$

$$K_p = \frac{[x_C]_{eq}^c [x_D]_{eq}^d}{[x_A]_{eq}^a [x_B]_{eq}^b}(p)^{c+d-a-b}$$

$$K_p = \frac{(x_C)^c (x_D)^d}{(x_A)^a (x_B)^b}(p)^{\Delta v}$$

$$K_p = K_x (p)^{\Delta v}$$

3.9 LE-CHATELIER'S PRINCIPLE

If we alter the conditions (p, T, V), we can disturb the equilibrium of the chemical reaction. It will shift the course of reaction. The equilibrium will shift in that direction which tends to minimize the change.

The conditions which can be changed are (i) concentration, (ii) pressure, (iii) volume, and (iv) temperature.

Concentration of reactant or product   The addition of reactant or product in the reaction mixture will shift the reaction towards the direction in which the added reactant/product will be utilized. On the other hand, the removal of reactant or product will shift the reaction in the direction in which the removed reactant/product will be formed.

$$PCl_5(g) \rightleftharpoons PCl_3(g) + Cl_2(g)$$

Addition of reactant PCl_5 or removal of product (either PCl_3 or Cl_2) will shift the reaction in the forward direction, that is, more products will be formed.

Check Your Progress

35. The free energy of mixing is negative for ideal gases. Why?
36. Boiling point of water can be increased by increasing the pressure. Explain.
37. What is the difference between equilibrium constant (K) and reaction quotient (Q)?
38. Derive the relationship between K_p and K_c.
39. What is the relationship between ΔG, Q, and K?

Addition of reactant PCl_5 or removal of product (either PCl_3 or Cl_2) will disturb the equilibrium

$$Q = \frac{p_{PCl_3}\, p_{Cl_2}}{p_{PCl_5}}$$

$$K = \frac{p_{PCl_3}^{eq}\, p_{Cl_2}^{eq}}{p_{PCl_5}^{eq}}$$

In the case of addition of PCl_3, Q value will be greater than K and the reverse reaction will take place. Removal of PCl_3 will lead to Q value smaller than K and hence the forward reaction will take place.

Effect of pressure change   Pressure can be changed either by (a) increasing/decreasing the volume or (b) by adding an inert gas.

(a) By increasing/decreasing the volume

$$K_p = K_x (p)^{\Delta v}$$

$$K_p = \frac{(x_C)^c (x_D)^d}{(x_A)^a (x_B)^b} (p)^{\Delta v}$$

$$K_p = \frac{\left(\dfrac{n_C}{n_T}\right)^c \left(\dfrac{n_D}{n_T}\right)^d}{\left(\dfrac{n_A}{n_T}\right)^a \left(\dfrac{n_B}{n_T}\right)^b} (p)^{\Delta v}$$

On increasing the pressure or decreasing the volume at constant temperature, $(p)^{\Delta v}$ will increase if Δv is positive. Thus to keep K_p constant, $\dfrac{(n_C/n_T)^c (n_D/n_T)^d}{(n_A/n_T)^a (n_B/n_T)^b}$ will decrease. It means that the backward reaction will be favoured. On the other hand, if Δv is negative, the increase in pressure will shift the reaction in the forward direction. In summary, the increase in pressure will shift the reaction to that direction in which there is less number of molecules in a balanced chemical reaction.

(b) By adding an inert gas at constant volume

$$K_p = \frac{\left(\dfrac{n_C}{n_T}\right)^c \left(\dfrac{n_D}{n_T}\right)^d}{\left(\dfrac{n_A}{n_T}\right)^a \left(\dfrac{n_B}{n_T}\right)^b} (p)^{\Delta v}$$

$$K_p = \frac{(n_C)^c (n_D)^d}{(n_A)^a (n_B)^b} \left(\frac{p}{n_T}\right)^{\Delta v}$$

At constant volume, $\left(\dfrac{p}{n_T}\right)$ remains constant, even if pressure is increased by adding inert gas. Thus, there is no shift in the equilibrium.

Effect of addition of inert gas at constant pressure (volume changes)

$$K_p = \frac{(n_C)^c (n_D)^d}{(n_A)^a (n_B)^b} \left(\frac{p}{n_T}\right)^{\Delta v}$$

Pressure is constant but n_T increases and is positive, thus $\left(\dfrac{p}{n_T}\right)^{\Delta v}$ decreases. It means $\dfrac{(n_C)^c (n_D)^d}{(n_A)^a (n_B)^b}$ must increase. Thus, the addition of inert gas favours the product formation with increase in pressure.

Application of Le-Chatelier's Principle to Physical Equilibrium

(a) **Effect of temperature and pressure on the solubility of a salt:** Solubility of a salt will increase or decrease depending upon whether the dissolution happens with absorption of heat or with release of heat, respectively.

Dissolution of gas in liquid results in decrease of volume; so dissolution increases with increase in the pressure.

(b) **Effect of temperature and pressure on melting of ice:** Melting of ice is an endothermic process and volume decreases in the process. Thus, increase in temperature and pressure favours melting.

Application of Le-Chatelier's Principle in Chemical Equilibrium: Industrial Applications

The different aspects of thermodynamics and chemical equilibrium are very useful at industrial level. The concept of Le-Chatelier's principle has been widely utilized at industrial level to increase the yield of the product. Some of these applications are explained below.

Production of ammonia (Haber Process) Ammonia is primarily used as nitrogen source in the manufacturing of fertilizers. On industrial scale, ammonia gas is prepared by reacting nitrogen and hydrogen gas.

$$N_2(g) + 3H_2(g) \rightleftharpoons 2NH_3(l), \qquad \Delta H = 92 \text{ kJ mol}^{-1}$$

(a) The number of moles decreases in the reaction and hence, according to Le-Chatelier's principle, by increasing the pressure, the position of the equilibrium can be shifted to the right, thereby increasing the overall yield. The pressure used in manufacturing of ammonia is 200–250 atm.

(b) The reaction is an endothermic process, therefore with increase in the temperature, the position of equilibrium is supposed to shift to the right. Thus, the reaction is carried out at 400–450 °C.

(c) If ammonia is removed from the reaction mixture at regular intervals of time, then the position of equilibrium can be shifted to the right and the yield will increase.

(d) Also, if the reaction is performed in the presence of a catalyst such as iron and potassium hydroxide as a promoter, the yield can be increased.

Manufacture of sulphuric acid (Contact process) Sulphuric acid is one of the most important chemicals utilized in industry. It is used in the manufacture of fertilizer, paints, pigments and surfactants, and batteries. Sulphuric acid is prepared in several steps which are as follows.

(a) In the first step, sulphur is oxidized to sulphur dioxide which is further oxidized to sulphur trioxide. SO_3 is the important reactant in the preparation of sulphuric acid.

$$2SO_2(g) + O_2(g) \rightleftharpoons 2SO_3(g), \qquad \Delta H = -196 \text{ kJ/mol}$$

According to Le-Chatelier principle, the equilibrium shifts to the right by increasing the pressure and use of a catalyst like V_2O_5.

Similarly, a decrease in the temperature should shift the position of equilibrium to the right, increasing the overall yield. However, the rate of reaction decreases with decrease in temperature. Despite being an exothermic process, the reaction is performed at temperatures above 400 °C. The value of K_p is high enough to give the very high yield of ~99 %.

(b) In the second step, the SO_3 is reacted with conc. sulphuric acid to form fuming sulphuric acid known as *oleum*.

$$SO_3(g) + \text{conc. } H_2SO_4(l) \rightleftharpoons H_2S_2O_7(l)$$

In this step, overloading of SO_3 can shift the equilibrium to the right, thereby increasing the concentration of oleum.

(c) Oleum produced in the above step, is hydrolysed in excess of water to form sulphuric acid (~96% yield).

$$H_2S_2O_7(l) + H_2O(l) \rightleftharpoons 2H_2SO_4(l)$$

This is a highly exothermic process.

Lime production from limestone   Limestone (calcium carbonate) decomposes to form lime (calcium oxide) and carbon dioxide.

$$CaCO_3(s) \rightleftharpoons CaO(s) + CO_2(g), \qquad \Delta H = +178 \text{ kJ mol}^{-1}$$

In this process, as there is no gaseous reactant, the position of equilibrium shifts to the right on decrease in pressure and formation of product is favoured.

The reaction is endothermic; therefore according to Le-Chatelier's principle an increase in temperature will shift the position of the equilibrium to the right.

Production of methanol   Methanol is an important fuel used in a number of reactions. It is prepared by synthesis gas ($CO + H_2$).

$$CO(g) + H_2(g) \rightleftharpoons CH_3OH (g), \qquad \Delta H = -90 \text{ kJ mol}^{-1}$$

This is an exothermic gaseous reaction; therefore, the position of the equilibrium can be shifted to the right by decreasing the temperature and increasing the pressure. This can be achieved by using a moderately high temperature with a catalyst like Cu–Zn–Al_2O_3.

Effect of temperature on the equilibrium   It will depend on whether the reaction is exothermic or endothermic. An increase in the temperature will shift the reaction in the direction where heat is absorbed. The increase in temperature will favour the reaction in forward direction if the reaction is endothermic. Exothermic reactions are favoured at low temperature whereas endothermic reactions are favoured at high temperature.

K_p does not change with pressure; however, it depends on the temperature. The equation which governs the effect of temperature on equilibrium constant is called van't Hoff equation.

From Gibbs Hemholtz equation,

$$\left(\frac{\partial(\Delta G^\circ/T)}{\partial T} \right)_p = -\frac{\Delta H^\circ}{T^2}$$

Substituting for ΔG°, we get,

$$\left(\frac{\partial\left(-RT \ln K_p /T\right)}{\partial T} \right)_p = -\frac{\Delta H^\circ}{T^2}$$

$$\left(\frac{\partial(\ln K_p)}{\partial T} \right)_p = \frac{\Delta H^0}{RT^2}$$

$$d \ln K_p = \frac{\Delta H^0}{RT^2} dT$$

On integration

$$\ln \frac{K_{p,2}}{K_{p,1}} = \frac{\Delta H^0}{R} \left(\frac{1}{T_1} - \frac{1}{T_2} \right)$$

So, if the reaction is endothermic, K_p increases with increase in temperature.

Check Your Progress

40. Decomposition of limestone is favoured at low pressure and high temperature. Explain.
41. For the reaction $N_2(g) + O_2(g) \rightleftharpoons 2NO(g)$, what is the effect of increasing pressure by decreasing the volume of the container?
42. For the reaction, $CO(g) + H_2O(g) \rightleftharpoons CO_2(g) + H_2(g)$, the position of equilibrium shifts to the left on increasing the temperature. Is the reaction exothermic or endothermic?
43. How does K_p change with temperature? Derive the relationship.
44. Dissolution of KNO_3 in water is an endothermic reaction. What will be the effect of temperature on the solubility of KNO_3?

SOLVED EXAMPLES

1. Prove that the pressure of an ideal gas is a state function.

Solution: Pressure of one mole of an ideal gas is given by

$$p = \frac{RT}{V}$$

Thus, pressure p of an ideal gas is a function of volume and temperature.

$$p = f(T, V)$$

$$dp = \left(\frac{\partial p}{\partial T}\right)_V dT + \left(\frac{\partial p}{\partial V}\right)_T dV$$

$$\left(\frac{\partial p}{\partial T}\right)_V = \frac{R}{V}$$

$$\frac{\partial}{\partial V}\left[\left(\frac{\partial p}{\partial T}\right)_V\right]_T = \frac{-R}{V^2}$$

$$\left(\frac{\partial p}{\partial V}\right)_T = \frac{-RT}{V^2}$$

$$\frac{\partial}{\partial T}\left[\left(\frac{\partial p}{\partial V}\right)_T\right]_p = \frac{-R}{V^2}$$

Thus,

$$\frac{\partial}{\partial V}\left[\left(\frac{\partial p}{\partial T}\right)_V\right]_T = \frac{\partial}{\partial T}\left[\left(\frac{\partial p}{\partial V}\right)_T\right]_p$$

Mixed derivatives of p are equal and hence p is a state function.

2. One mole of an ideal gas undergoes an isothermal expansion from 2 m^3 to 6 m^3 at 300 K. Calculate the work done for expansion against a constant external pressure of 1 Nm^{-2}.

Solution: $w = -\int_{V_i}^{V_f} p_{ex} dV$

At constant external pressure

$$w = -p_{ex}\int_{V_i}^{V_f} dV$$

$$w = -1\ (6 - 2) = -4\ J$$

3. One mole of an ideal gas undergoes an isothermal reversible expansion from 2 m^3 to 6 m^3 at 300 K. Calculate the work done in this expansion.

Solution: For a reversible isothermal process, $w = -nRT \int_{V_i}^{V_f} \dfrac{dV}{V}$

$$w = nRT \ln \frac{V_i}{V_f} = -1 \times 8.314 \times 300 \times \ln \frac{6}{2} = -2.74 \text{ kJ}$$

4. A mole of hydrogen gas is expanded adiabatically and reversibly from 4 to 2 Nm^{-2} at 300 K. Calculate the final temperature and w.

$$\left(\frac{T_f}{T_i} \right) = \left(\frac{V_i}{V_f} \right)^{\gamma - 1}$$

$$w = C_{V,m} \, T_i \left(\left(\frac{V_i}{V_f} \right)^{\gamma - 1} - 1 \right)$$

$$= \frac{5R}{2} \times 300 \times (2^{5/3 - 1} - 1) = \frac{5R}{2} \times 300 \times (2^{2/3} - 1) = 1500 \, R$$

5. One mole of an ideal gas undergoes isothermal expansion from 2 m^3 to 8 m^3 at 300 K. Calculate the energy transferred as heat (q) and change in internal energy (ΔU) (a) for a reversible expansion and (b) for an expansion against a constant external pressure of 1 Nm^{-2}.

Solution: In an isothermal process, for an ideal gas, $\Delta U = 0$.

(a) For a reversible isothermal expansion

$$q = w = -nRT \ln \frac{V_f}{V_i} = nRT \ln \frac{V_i}{V_f} = -1 \times 8.314 \times 300 \times \ln \frac{8}{2} = -3.46 \text{ kJ}$$

for an isothermal expansion against a constant external pressure of 1 Nm^{-2}.

$$q = w = -p_{\text{ext}} \int_{V_i}^{V_f} dV = -1 \times (8 - 2) = -6 \text{ J}$$

6. A mole of hydrogen gas is compressed adiabatically and reversibly from 4 to 2 m^3 at 300 K. Calculate the ΔU.

Solution: In an adiabatic process, $q = 0$; or $\Delta U = w$.

$$\left(\frac{T_2}{T_1} \right) = \left(\frac{V_1}{V_2} \right)^{\gamma - 1}$$

$$w = C_{V,m} T_1 \left(\left(\frac{V_1}{V_2} \right)^{\gamma - 1} - 1 \right) = \frac{5R}{2} \left(2^{\frac{5}{3} - 1} - 1 \right) = \frac{5R}{2} \left(2^{\frac{2}{3}} - 1 \right)$$

7. Prove that $C_P - C_V = nR$

Solution: $C_P - C_V = \left(\dfrac{\partial H}{\partial T} \right)_P - \left(\dfrac{\partial U}{\partial T} \right)_V$

$$= \left(\frac{\partial (U + pV)}{\partial T} \right)_P - \left(\frac{\partial U}{\partial T} \right)_V$$

$$= \left(\frac{\partial U}{\partial T} \right)_P + p \left(\frac{\partial V}{\partial T} \right)_P - \left(\frac{\partial U}{\partial T} \right)_V \qquad \text{(i)}$$

$$U = f(V, T)$$

$$dU = \left(\frac{\partial U}{\partial T} \right)_V dT + \left(\frac{\partial U}{\partial V} \right)_T dV$$

$$\left(\frac{\partial U}{\partial T} \right)_P = \left(\frac{\partial U}{\partial T} \right)_V + \left(\frac{\partial U}{\partial V} \right)_T \left(\frac{\partial V}{\partial T} \right)_P \qquad \text{(ii)}$$

For an ideal gas, $\left(\dfrac{\partial U}{\partial V}\right)_T = 0$

Thus, from Eq. (ii)

$$\left(\frac{\partial U}{\partial T}\right)_P = \left(\frac{\partial U}{\partial T}\right)_V \qquad \text{(iii)}$$

Combining Eqs (i) and (iii), we get

$$C_P - C_V = \left(\frac{\partial U}{\partial T}\right)_V + p\left(\frac{\partial V}{\partial T}\right)_P - \left(\frac{\partial U}{\partial T}\right)_V$$

$$C_P - C_V = p\left(\frac{\partial V}{\partial T}\right)_P \qquad \text{(iv)}$$

For an ideal gas, $V = \dfrac{nRT}{p}$

$$\left(\frac{\partial V}{\partial T}\right)_P = \frac{nR}{p} \qquad \text{(v)}$$

From Eqs (iv) and (v),

$$C_P - C_V = p \times \frac{nR}{p} = nR$$

Hence proved.

8. The enthalpy change for the combustion of benzene to $H_2O(l)$ and $CO_2(g)$ is -3271 kJ mol^{-1}. Heat of formations of $H_2O(l)$ and $CO_2(g)$ are -286 and -394 kJ mol^{-1}, respectively. Calculate the enthalpy of formation of benzene.

Solution: $C_6H_6 + 15/2\ O_2 \rightarrow 6\ CO_2(g) + 3\ H_2O(l)$, $\qquad \Delta H = -3271$ kJ mol^{-1}

$\qquad\qquad$ $6\ CO_2(g) + 3\ H_2O(l) \rightarrow C_6H_6 + 15/2\ O_2$, $\qquad \Delta H = 3271$ kJ mol^{-1} $\qquad$ (i)

$\qquad\qquad$ $6\ C(gr) + 6\ O_2 \rightarrow 6\ CO_2(g)$, $\qquad \Delta H = -394 \times 6$ kJ mol^{-1} $\qquad$ (ii)

$\qquad\qquad$ $3\ H_2(g) + 3/2\ O_2 \rightarrow 3\ H_2O(l)$, $\qquad \Delta H = -286 \times 3$ kJ mol^{-1} $\qquad$ (iii)

Adding (i), (ii) and (iii)

$\qquad\qquad$ $6\ C(gr) + 3\ H_2(g) \rightarrow C_6H_6$ $\qquad\qquad \Delta H = 49$ kJ mol^{-1}

9. Calculate the entropy change when one mole of an ideal gas undergoes isothermal reversible expansion from 3 m^3 to 6 m^3.

Solution: $\Delta S = C_v \ln\dfrac{T_2}{T_1} + nR \ln\dfrac{V_2}{V_1}$

Under isothermal condition,

$$\Delta S = nR \ln\frac{V_2}{V_1} = 1 \times 8.314 \times \ln\frac{6}{3} = 5.76 \text{ JK}^{-1}$$

10. Calculate the entropy change when 1 mole of liquid water is converted to steam at 373 K and 1 atm. Given $\Delta H_{vap} = 40.668$ kJ mol^{-1}

Solution: $\Delta S_{sys} = \dfrac{\Delta H}{T} = \dfrac{40668}{373} = 109$ kJ K^{-1}mol^{-1}

11. Calculate the entropy change when 1 mole of water is heated from 298 K to 348 K at 1 atm. (Heat capacity of water = 75.3 J K^{-1} mol^{-1})

Solution: $\Delta S_{sys} = nC_V \ln\dfrac{T_f}{T_i} = 75.3 \times \ln\dfrac{348}{298} = 11.8 \text{ JK}^{-1}$

12. Assuming O_2 to be an ideal gas, calculate the molar entropy change for the following process.

$$O_2(g, 300 \text{ K}, 1 \text{ bar}) \rightarrow O_2(g, 400 \text{ K}, 2 \text{ bar})$$

Solution: $\Delta S = C_p \ln \dfrac{T_2}{T_1} + nR \ln \dfrac{p_1}{p_2}$

For a diatomic gas, $C_v = 5/2\ R$, $C_p = 7/2\ R$

$$\Delta S = \frac{7}{2} R \times \ln \frac{400}{300} + R \times \ln \frac{1}{2}$$

$$\Delta S = \frac{7}{2} R \times \ln \frac{400}{300} + R \times \ln \frac{1}{2} = 64.1 \text{ J K}^{-1}$$

13. In methanol–O_2 fuel cell, the combustion of liquid methanol takes place as

$$CH_3OH\ (l) + 3/2\ O_2(g) \rightarrow CO_2(g) + 2\ H_2O(l)$$

The standard free energy change for the formation of liquid methanol, carbon dioxide gas, and water are -183 kJ mol^{-1}, -354.9 kJ mol^{-1} and -293.7 kJ mol^{-1}, respectively. Determine the standard free energy change that can be converted into electrical work. Also calculate the efficiency of fuel cell, if standard enthalpy of combustion of methanol is -786 kJ mol^{-1}.

Solution: $\Delta_r G^\circ = 2\ \Delta_f G^\circ\ H_2O_{(l)} + \Delta_f G^\circ\ CO_{2(g)} - 3/2\ \Delta_f G^\circ\ O_2(g) - \Delta_f G^\circ\ CH_2OH_{(l)}$

$\qquad\qquad = 2 \times (-293.7) + (-354.9) - 3/2 \times (0) - (-183)$

$\qquad\qquad = -759.3$ kJ mol^{-1}

$\quad$% Efficiency $= (\Delta_r G^\circ / \Delta_r H^\circ) \times 100$

$\qquad\qquad = ((-759.3)/(-786)) \times 100$

$\qquad\qquad = 96.6$ %

14. For a reaction, $X \rightleftharpoons Y$ ($\Delta H^\circ = 38.4$ kJ mol^{-1}), the equilibrium constant is 2.7×10^{-6} at 298 K. Determine the standard free energy change and standard entropy change for the reaction.

Solution: We have, $\Delta G^\circ = -2.303\ RT \log K_{eq}$

$\qquad\qquad = -2.303 \times 8.314 \times 298 \times \log (2.7 \times 10^{-6})$

$\qquad\qquad = 31{,}773.79$ J mol^{-1} $= 31.77$ kJ mol^{-1}

Also,$\qquad\qquad \Delta G^\circ = \Delta H^\circ - T\Delta S^\circ$

$\qquad\qquad \Delta S^\circ = (\Delta H^\circ - \Delta G^\circ)/T = (38.4 - 31.77)/298$

$\qquad\qquad = 0.0222$ kJ K^{-1} mol^{-1} $= 22.2$ J K^{-1} mol^{-1}

15. K_p for the dissociation of sulphur trioxide gas into sulphur dioxide and oxygen is found to be 2.65×10^{-3} kPa at 650 K. Calculate K_c value in moles per litre for this reaction at the same temperature.

Solution: We have, $2\ SO_3(g) \rightleftharpoons 2\ SO_2(g) + O_2(g)$

Also,$\qquad\qquad K_p = K_c(RT)^{\Delta n}$

$\qquad\qquad \Delta n = 3 - 2 = 1$, $T = 650$ K, $R = 8.31$ kPa L mol^{-1} K^{-1}, $K_p = 2.65 \times 10^{-3}$ kPa

$\qquad\qquad 2.65 \times 10^{-3}$ kPa $= K_c(8.31$ kPa L mol^{-1} K$^{-1} \times 650$ K$)^1$

$$K_c = \frac{2.65 \times 10^{-3}}{8.31 \times 650} = 4.91 \times 10^{-7} \text{ mol L}^{-1}$$

15. The decomposition of 0.2 mole of PCl_5 takes place in a 10 L flask at 550 K. If pressure of 1 bar is attained at the equilibrium, calculate K_p and K_c for the reaction.

Solution:$\qquad\qquad PCl_5 \quad \rightleftharpoons \quad PCl_3 \quad + \quad Cl_2$

Initial moles$\qquad\quad$ 0.2 $\qquad\qquad\quad$ 0 $\qquad\qquad\quad$ 0

At equilibrium$\quad$ 0.2 $- x$ $\qquad\qquad x$ $\qquad\qquad x$

Total no. of moles at equilibrium $= (0.2 - x) + x + x = 0.2 + x$

Also, total no. of moles, $n = pV/RT = \dfrac{1 \text{ bar} \times 10 \text{ L}}{0.083 \text{ bar L mol}^{-1} \text{ K}^{-1} \times 550 \text{ K}} = 0.219$

$$0.2 + x = 0.219$$
$$x = 0.019$$

$$[PCl_5] = \frac{0.2 - 0.019}{10} M = 1.81 \times 10^{-2} M$$

$$[PCl_3] = [Cl_2] = \frac{0.019}{10} M = 1.9 \times 10^{-3} M$$

$$K_c = \frac{[PCl_3][Cl_2]}{[PCl_5]} = \frac{(1.9 \times 10^{-3})^2}{1.81 \times 10^{-2}} = 1.99 \times 10^{-4}$$

$$K_p = K_c(RT)^{\Delta n} = 1.99 \times 10^{-4} \times 0.083 \times 550 = 9.1 \times 10^{-3} \text{ bar}$$

16. The equilibrium constant for the formation of SO_3 from oxidation of SO_2 is found to be 3.4×10^{12} at 250 K. If the standard enthalpy change for the reaction is -1.8×10^5 J mol^{-1}, calculate the equilibrium constant at 400 K.

Solution: We have, $\ln \dfrac{K_2}{K_1} = -\dfrac{\Delta H^\circ}{R}\left(\dfrac{1}{T_2} - \dfrac{1}{T_1}\right)$

$$\ln \frac{K_2}{3.4 \times 10^{12}} = -\frac{-1.8 \times 10^5}{8.314}\left(\frac{1}{400} - \frac{1}{250}\right) = -32.47$$

$$K = 2.67 \times 10^{-2}$$

Equilibrium constant, $K = 2.67 \times 10^{-2}$

SUMMARY

○ The study of transformation of one form of energy to other is known as thermodynamics.

○ An open system allows exchange of matter and energy with the surroundings; a closed system allows only exchange of energy. In an isolated system, neither matter nor energy can be exchanged with its surroundings.

○ Extensive variables are the mass dependent properties of a system. On the other hand, intensive variables are the properties of a system which do not depend upon the mass of a substance.

○ State function is a property of a system which depends only upon the initial and final state of the system and not on the process involved during the change of state.

○ Zeroth law of Thermodynamics explains the concept of temperature. It states that if two systems are separately in thermal equilibrium with a third, then they must also be in thermal equilibrium with each other.

○ The major forms of energy which are exchanged between system and surroundings are work and heat.

○ Heat capacity of an ideal gas at constant volume (C_v) and at constant pressure (C_p) are interrelated as $C_p - C_v = R$.

○ The first law of thermodynamics states that for an isolated system, the total energy is constant. $\Delta U = q + w$

○ Hess's law states that the enthalpy change of the reaction remains the same whether the reaction takes place in one step or multiple steps. It is used to measure the ΔH for those reactions which are either very slow or whose ΔH cannot be determined experimentally.

○ Kirchoff's law defines the temperature dependence of the enthalpy change of a reaction.

○ Kelvin statement of Second law of Thermodynamics: A transformation whose only final result is to transform heat extracted from a source which is at the same temperature throughout is impossible.

○ Carnot's cycle consists of a sequence of isothermal and adiabatic expansion and compression processes. It helps to understand the limitation in converting heat into work.

- Second law of thermodynamics also states that the entropy of the universe increases during a spontaneous process ($\Delta S_{sys} + \Delta S_{surr} = \Delta S_{Total} > 0$)
- *Clausius Inequality* theorem states $\Delta S \geq dq/T$.
- The Helmholtz energy (A) is given by $A = U - TS$ and Gibb's free energy (G) is given by $G = H - TS$.
- Maxwell equations represent the interrelation between differential forms of different thermodynamic functions.
- Third law of Thermodynamics states that the entropy of all perfectly crystalline solids is zero at zero Kelvin temperature.

- Gibb's Helmholtz equation defines the temperature dependence of the Gibb's free energy: $(\mathrm{d}(G/T)\mathrm{d}T)_P = -H/T^2$.
- The *equilibrium constant* (K) in terms of Gibb's free energy is expressed as $\Delta G^o = -RT \ln K$.
- *Le Chatelier's Principle* states that when any system at equilibrium is subjected to change in temperature, concentration, pressure, or volume, a new equilibrium will be established to counteract the effect of the applied change.
- The temperature dependence of the equilibrium constant is explained by van't Hoff equation: $d(\ln K)/dT = \Delta_r H^\circ/RT^2$.

GLOSSARY

System: The part of universe under investigation

Surroundings: Everything else in the universe .other than the system.

Zeroth law of Thermodynamics: It states that if two systems are separately in thermal equilibrium with a third, then they must also be in thermal equilibrium with each other.

Work: The transfer of energy between a system and its surroundings due to force acting through a distance.

Heat: The quantity of energy in transit between the system and surroundings due to temperature difference.

Internal energy: The sum of all the forms of energies that a system can possess such as translational, rotational, vibrational, bond energies, energies due to intermolecular forces of attraction and electrostatic interactions.

Heat capacity: The amount of heat required to raise the temperature of a substance by one degree Celsius.

Enthalpy: The total heat content of the system. Mathematically, it is defined as the sum of the internal energy and pressure–volume.

Exothermic process: The process in which there is release of heat by the system to the surroundings ($\Delta H < 0$).

Endothermic process: The processes in which heat is absorbed by the system ($\Delta H > 0$).

Entropy: A thermodynamic property which measures the randomness or disorder of a system, ($dS = dq_{rev}/T$)

Helmholtz free energy: The change in Helmholtz Free Energy during a process is equal to maximum work done (including expansion/compression work, also known as PV work).

Gibb's Free Energy: The change in free energy during a process is equal to non-expansion reversible work (the work other than $p-V$ work).

Equilibrium: A state where the rate of forward and reverse reaction become equal in a physical and chemical process. The criterion for an equilibrium at constant temperature and pressure is $dG_{p,T} = 0$

KEY FORMULAE

- Work in a P-V process $dw = -p_{ex}dV$
- First Law of Thermodynamics $dU = dq + dw$
- Internal energy change in a isochoric process $\Delta U = q_V$
- Enthalpy $H = U + PV$
- Enthalpy change in a isobaric process $\Delta H = q_p$

- Heat capacity at constant volume
$$C_V = \left(\frac{dq}{dT}\right)_V = \left(\frac{dU}{dT}\right)_V$$
- Heat capacity at constant pressure
$$C_p = \left(\frac{dq}{dT}\right)_p = \left(\frac{dU}{dT}\right)_p$$

- Difference between $C_{p,m}$ and $C_{V,m}$ for an ideal gas
$$C_{p,m} - C_{V,m} = R$$

- P-V work for an ideal gas in a reversible isothermal process $w = nRT \ln \dfrac{V_i}{V_f} = nRT \ln \dfrac{p_f}{p_i}$

- P-V work for an ideal gas in a reversible adiabatic process $w = C_{V,m}(T_f - T_i) = C_{V,m} T_i \left(\left(\dfrac{V_i}{V_f} \right)^{\gamma-1} - 1 \right)$

- P-V work for an ideal gas in a reversible Isochoric process $w = p(V_i - V_f) = nR(T_i - T_f)$

- Internal energy change for an ideal gas $\Delta U = C_{V,m}\Delta T$

- ΔU for an ideal gas in an isothermal process $\Delta U = 0$

- ΔU for an ideal gas in a reversible adiabatic process $\Delta U = w$

- ΔU for an ideal gas in a reversible isochoric process $\Delta U = C_{V,m}\Delta T$

- Enthalpy change for an ideal gas $\Delta H = C_{p,m}\Delta T$

- ΔH for an ideal gas in an isothermal process $\Delta H = 0$

- ΔH for an ideal gas in a reversible adiabatic process $\Delta H = w + nR\Delta T$

- ΔH for an ideal gas in a reversible isochoric process $\Delta H = C_{p,m}\Delta T$

- Infinitesimal change in internal energy in a process $dU = C_{V,m}dT + \pi dV$

- infinitesimal change in enthalpy in a process $dH = C_{p,m}dT + \mu_T dp$

- Isothermal Joule—Thomson coefficient
$$\left(\frac{\partial H}{\partial p} \right)_T = -\left(\frac{\partial T}{\partial P} \right)_H C_p$$

- Kirchoff's Law (temperature dependence of enthalpy of reactions) $\dfrac{d}{dT}(\Delta H)_p = \Delta C_p$

- Infinitesimal change in entropy of the system
$$dS_{sys} = \frac{dq_{sys,\,rev}}{T_{sys}}$$

- Infinitesimal change in the entropy of surroundings
$$dS_{surr} = \frac{dq_{surr,\,rev}}{T_{surr}} = -\frac{dq_{sys}}{T_{sys}}$$

- Change in the entropy of one mole of an ideal gas
$$\Delta S = C_{V,m} \ln \frac{T_2}{T_1} + R \ln \frac{V_2}{V_1}; \quad \Delta S = C_{p,m} \ln \frac{T_2}{T_1} - R \ln \frac{p_2}{p_1}$$

- Change of entropy of one mole of an ideal gas in an isothermal process $\Delta S = R \ln \dfrac{V_2}{V_1}; \Delta S = R \ln \dfrac{p_2}{p_1}$

- Change of entropy of one mole of an ideal gas in an adiabatic process $\Delta S = 0$

- Change of entropy of one mole of an ideal gas in a isochoric process $\Delta S = C_{V,m} \ln \dfrac{T_2}{T_1}$

- Change of entropy of one mole of an ideal gas in a isobaric process $\Delta S = C_{p,m} \ln \dfrac{T_2}{T_1}$

- Internal pressure $\left(\dfrac{\partial U}{\partial V} \right)_T = T \left(\dfrac{\partial p}{\partial T} \right)_V - p$

- Isothermal Joule-Thomson coefficient
$$\left(\frac{\partial H}{\partial p} \right)_T = -T \left(\frac{\partial V}{\partial T} \right)_p + V$$

- Absolute entropy of a gas $S = S(0K) +$
$$\int_0^{Tm} C_{p,s} \frac{dT}{T} + \frac{\Delta H_m}{T_m} + \int_{T_m}^{T_b} C_{p,l} \frac{dT}{T} + \frac{\Delta H_v}{T_b} + \int_{T_b}^{T} C_{p,v} \frac{dT}{T}$$

- Clausius inequality Theorem $dS \geq \dfrac{dq}{T}$

- Free energy change for liquid/solid on change in pressure $\Delta G_m = V_m(p_2 - p_1)$

- Free energy change for a gas on change in pressure
$$\Delta G_m = RT \ln \frac{p_f}{p_i}$$

- Infinitesimal change in Helmholtz Free energy $dA = dw_{max}$

- Infinitesimal change in Gibb's Free Energy $dG = dW_{add,\,Rev}$

- Temperature dependence of free energy
$$\left(\frac{\partial (G/T)}{\partial T} \right)_p = -\frac{H}{T^2}$$

- ΔG_{mix} of mixing at same pressure and temperature
$$\Delta G_{mix} = xRT \ln f_A + yRT \ln f_b$$

- Clapeyron equation $\dfrac{dp}{dT} = \dfrac{\Delta H_m}{T \Delta V_m}$

- Clausius–Clapeyron equation
$$\ln \frac{p_2}{p_1} = \frac{\Delta H_m}{R} \left(\frac{1}{T_1} - \frac{1}{T_2} \right) = \frac{\Delta H_m}{R} \left(\frac{\Delta T}{T_1 T_2} \right)$$

- Free-energy of a chemical reaction $\Delta G = -RT \ln K + RT \ln Q$

- Equilibrium constant $K_c = \dfrac{[C]_{eq}^c [D]_{eq}^d}{[A]_{eq}^a [B]_{eq}^b}$

- Relationship between equilibrium constant $K_p = K_c (RT)^{\Delta v} = K_x (p)^{\Delta v}$

- van't Hoff equation $\ln \dfrac{K_{p,2}}{K_{p,1}} = \dfrac{\Delta H^0}{R} \left(\dfrac{1}{T_1} - \dfrac{1}{T_2} \right)$

EXERCISES

Multiple Choice Questions

1. The parameter, which is not a state function, is
   (a) internal energy   (b) work done
   (c) enthalpy   (d) entropy

2. The thermodynamic parameter, which is not a state function, is
   (a) q at constant pressure
   (b) q at constant volume
   (c) W under adiabatic condition
   (d) W under isothermal condition

3. The thermodynamic parameter, which is not an extensive property, is
   (a) enthalpy   (b) entropy
   (c) specific heat   (d) volume

4. For adiabatic process, the correct statement is
   (a) $\Delta T = 0$ (b) $\Delta U = 0$
   (c) $q = 0$ (d) $\Delta H = 0$

5. The temperature of the system decreases in an
   (a) adiabatic expansion
   (b) isobaric expansion
   (c) Joule-Thomson compression
   (d) adiabatic compression

6. In an isothermal expansion of an ideal gas,
   (a) $\Delta U = 0$ (b) $q = 0$
   (c) $\Delta G = 0$ (d) $\Delta S = 0$

7. Sublimation of dry ice is
   (a) an exothermic process
   (b) an endothermic process
   (c) an isoenthalpic process
   (d) a process accompanied by decrease in entropy

8. Joule-Thomson expansion for an ideal gas is an
   (a) isochoric process
   (b) isobaric process
   (c) isothermic process
   (d) isoenthalpic process

9. The reaction in which ΔH is equal to ΔU, is
   (a) $3 O_2(g) \rightarrow 2 O_3(g)$
   (b) $I_2(g) + H_2(g) \rightarrow 2 HI(g)$
   (c) $2 NO_2(g) \rightarrow 2 NO(g) + O_2(g)$
   (d) $2 SO_2(g) + O_2(g) \rightarrow 2 SO_3(g)$

10. For which of the following processes, is ΔS negative?
    (a) $H_2(g) \rightarrow 2H(g)$
    (b) $H_2(g, 298\ K) \rightarrow H_2 (g, 310\ K)$
    (c) $H_2(g, 1\ atm) \rightarrow H_2(g, 10\ atm)$
    (d) $H_2O(l, 373K, 1\ atm) \rightarrow H_2(g, 10\ 373K, 1\ atm)$

11. $\left(\dfrac{\partial H}{\partial P}\right)_T$ has the dimension of
    (a) pressure   (b) volume
    (c) energy   (d) heat Capacity

12. For an ideal gas at 300 K
    (a) $\left(\dfrac{\partial U}{\partial V}\right)_T = 0$ (b) $\left(\dfrac{\partial U}{\partial V}\right)_p = 0$
    (c) $\left(\dfrac{\partial H}{\partial T}\right)_p = 0$ (d) $\left(\dfrac{\partial H}{\partial V}\right)_p = 0$

13. For a reaction at equilibrium, $\Delta S = 100\ J\ K^{-1}$ at 300 K; ΔH of the reaction is
    (a) 0 kJ   (b) 30 kJ
    (c) 300 kJ   (d) 3 kJ

14. The parameter which always decreases during a spontaneous process at constant S and V, is
    (a) U (b) H
    (c) A (d) G

15. Given
    $2\ Fe_2O_3(s) + 6CO\ (g) \rightarrow 4Fe(s) + 3O_2(g);$
    $\Delta_R G° = +1487.0\ kJ\ mol^{-1}$ (i)
    $2\ CO(g) + O_2(g) \rightarrow 2\ CO_2(g);$
    $\Delta_R G° = -514.4\ kJ\ mol^{-1}$ (ii)
    Free energy change $\Delta_R G^0$ (in kJ mol^{-1}) for the reaction
    $Fe_2O_3\ (s) + CO_2\ (g) \rightarrow 2Fe\ (s) + CO\ (g) + 2\ O_2\ (g)$
    will be
    (a) 1000.7   (b) 2001.4
    (c) 972.6   (d) 486.3

16. 1.0 mole of an ideal gas undergoes reversible isothermal expansion from 1 m^3 to 5 m^3 at 298 K. ΔG for this process is
    (a) 3.98 kJ mol^{-1} (b) −3.98 kJ mol^{-1}
    (c) 4.56 kJ mol^{-1} (d) −4.56 kJ mol^{-1}

17. Partial derivative $\left(\dfrac{\partial S}{\partial V}\right)_T$ is equal to
    (a) $\left(\dfrac{\partial V}{\partial T}\right)_p$ (b) $\left(\dfrac{\partial p}{\partial T}\right)_V$
    (c) $-\left(\dfrac{\partial V}{\partial T}\right)_p$ (d) $-\left(\dfrac{\partial p}{\partial T}\right)_V$

18. The equilibrium constant for a reaction $A + B \rightleftharpoons C + D$ is 64. If the initial concentration of A and B is 1 M each, the equilibrium concentration of A and B is
    (a) 0.11 M   (b) 0.22 M
    (c) 0.44 M   (d) 0.88 M

19. The variation of the equilibrium constant (K) of a reaction with temperature is given by

    $$\ln K = 2 + \frac{2980}{T}$$

    The ΔH° and ΔS° of the reaction is given by
    (a) 2.4 kJ mol^{-1} and 2 J K^{-1} mol^{-1}
    (b) 3 kJ mol^{-1} and 16.6 J K^{-1} mol^{-1}
    (c) 24 kJ mol^{-1} and 16.6 J K^{-1} mol^{-1}
    (d) 24 kJ mol^{-1} and 2 J K^{-1} mol^{-1}

20. For a reaction $2AB(g) \rightleftharpoons A_2(g) + B_2(g)$, the degree of dissociation is z and is $\ll 1$. The relationship between z and equilibrium constant K_p is
    (a) $K_p \propto z$ (b) $K_p \propto z^2$
    (c) $K_p \propto 1/z$ (d) $K_p \propto 1/z^2$

Review Questions

1. Define system and surroundings.
2. How does a closed system differ from an isolated system?
3. Differentiate between a state and path function.
4. Choose state functions among the given variable: volume, temperature, heat, work, internal energy, enthalpy, heat capacity
5. Prove that volume is a state function using ideal gas equation.
6. What is an intensive function? How does it differ from an extensive function?
7. Choose intensive variables from among the given variables: volume, temperature, pressure, density, internal energy, enthalpy
8. What is a Carnot's cycle?
9. Can a heat engine convert heat to work at constant temperature?
10. How is an isothermal process different from an adiabatic process?
11. Define a reversible process.
12. How is work obtained from a system maximum when the process is carried out reversibly?
13. State Hess's law and explain it with an example.
14. What is Joule–Thomson experiment and what is its significance?
15. State the second law of thermodynamics in terms of entropy.
16. State Clausius inequality theorem.
17. Under what condition does change in internal energy become equal to heat exchanged between a system and its surroundings?
18. Under what condition does change in enthalpy become equal to heat exchanged between the system and its surroundings?
19. Under what condition does change in entropy of the system become a criterion for spontaneous change?
20. Under what condition does change in Helmholtz free energy of the system become a criterion for spontaneous change?
21. State the third law of thermodynamics and explain its importance.
22. What is Gibb's free energy? Explain its significance.
23. What is Helmshotlz's free energy? Explain its significance.
24. Prove that $dq/T = (R/p)dp - (R/T)dT$ is an exact differential.

NUMERICAL PROBLEMS

1. A balloon bag of capacity 4 litres is filled with 10 moles of hydrogen at 25 °C. Due to a sudden hole in the bag, all the gas escaped into the atmosphere. What will be the work done by the gas?   (**Ans:** −24.35 kJ)

2. For a reversible expansion from 12 litres to 15 litres, what will be the work done by 5 moles of an ideal gas? ($T = 298$ K)   (**Ans:** 2.76 kJ)

3. Determine the difference between enthalpy change and internal energy at constant volume for the combustion of three moles of naphthalene at 298 K.   (**Ans:** −14.86 kJ)

4. (a) If specific heat capacity of gold is 0.13 J K^{-1} g^{-1}, calculate the energy required to raise the temperature of 12 g of metal from 303 K to 573 K. (b) Determine the amount of iron that can be heated from 303 K to 523 K when the same amount of energy is supplied. (Specific heat capacity of iron is 0.45 J K^{-1} g^{-1})   (**Ans:** 421.2 J, 4.25 g)

5. 75.54 J of heat is required to raise the temperature of 10 dm^3 of an unknown gas by 8 °C at constant volume. Under STP conditions, calculate C_v, C_p, and atomicity of the gas. **(Ans:** Diatomic)

6. 2.56 g of an octane (C_8H_{18}) sample is burnt in a bomb calorimeter in excess of oxygen, which raises the temperature of calorimeter from 298 K to 307.93 K. Calculate the amount of heat transferred if the heat capacity of calorimeter is 9.83 kJ K^{-1}. Also determine ΔU and ΔH for the reaction at 298 K.
(Ans: 97.6 kJ, -4346.25 kJ mol^{-1}, 4357.4 kJ mol^{-1})

7. Given that the standard enthalpies of combustion of carbon and hydrogen gas are -349.6 kJ and -296 kJ, respectively. If the standard enthalpy of combustion of liquid ethanol is -1376.9 kJ, calculate the heat of its formation at constant volume under STP conditions. **(Ans:** -201.6 kJ mol^{-1})

8. If on burning, 27 g of benzene liberates 1573.9 kJ of heat, what is the standard enthalpy of combustion for benzene? **(Ans:** 4546.8 kJ mol^{-1})

9. The standard enthalpy for conversion of Mg to its oxide is -1202 kJ. If ΔS°_{sys} for the reaction is -217 J.K^{-1}, comment on the spontaneity of the reaction. **(Ans:** Reaction is spontaneous)

10. If 2 moles of a diatomic ideal gas is heated from 25 °C to 50 °C at constant pressure of 1 atm, what will be ΔU and ΔH for the process? **(Ans:** 519.63 J mol^{-1}, 727.48 J mol^{-1})

11. 1.00 mole of a monoatomic ideal gas at 298 K is expanded from 10 m^3 to 20 m^3 isothermally and reversibly. Calculate ΔS and ΔS_{surr} and ΔH in the process. **(Ans:** 5.76 J K^{-1} mol^{-1}, -5.76 J K^{-1} mol^{-1}, 0)

12. 1.00 mole of a monoatomic ideal gas at 298 K is expanded from 10 m^3 to 20 m^3 isothermally against a constant external pressure of 10^5 Pa. Calculate ΔS and ΔS_{surr} and ΔH in the process.
(Ans: 5.76 J K^{-1} mol^{-1}, -3355.7 J K^{-1} mol^{-1}, 0)

13. 1.00 mole of a monoatomic ideal gas at 298 K is expanded from 10 m^3 to 20 m^3 adiabatically and reversibly. Calculate ΔS and ΔS_{surr}, ΔH and ΔT in the process. **(Ans:** 0.058 J K^{-1} mol^{-1}, 0, -2.27 J mol^{-1}, -109.4 K)

14. In the transition from phase α to β, ΔG_m (298) = -500 J and ΔV_m = 0.05 m^3. Calculate the pressure at which a phase becomes more stable form at 298 K. **(Ans:** 10^4 Pa)

15. What will be ΔG of vaporization for 3 moles of a liquid at 60 °C (its normal boiling point) and 0.8 bar? Assume that the liquid vapour is an ideal gas. **(Ans:** -617.78 J mol^{-1})

16. The decomposition of 0.2 mole of PCl$_5$ is carried out in a 10 L flask at 550 K. If pressure of 1 bar is attained at the equilibrium, calculate K_p and K_c of the reaction. **(Ans:** 9.1×10^{-3}, 1.99×10^{-4})

17. An equilibrium is established when a mixture of dinitrogen and dioxygen gas is heated at 2650 K in a reactor, as:

$$N_2(g) + O_2(g) \rightleftharpoons 2NO(g)$$

The equilibrium constant, (K_c) for the reaction is 1.99×10^{-3}. If the equilibrium mole percentage of the product (NO) is 1.6, calculate the initial composition of N$_2$ and O$_2$ in terms of mole fraction. **(Ans:** 0.8595, 0.1405)

18. Thermodynamics parameters, K_p and standard free energy change for the dissociation reaction of dinitrogen tetraoxide into nitrogen dioxide at 303 K are 0.17 and 6.23 kJ mol^{-1}, respectively. Assuming initial pressures of NO$_2$ and N$_2$O$_4$ are 0.213 atm and 0.814 atm, respectively. Calculate ΔG for the reaction at these pressures. Also predict the direction of the reaction. **(Ans:** -1.04 kJ mol^{-1}, reaction is spontaneous)

19. Decarboxylation of silver carbonate takes place at 300 K to form silver oxide. The equilibrium constant for the reaction at 300 K is found to be 4.5×10^2. If the standard enthalpy change for the reaction is -1.69×10^5 J mol^{-1}, calculate the temperature at which the equilibrium constant will be 6.93×10^6. **(Ans:** 262.6 K)

20. For a reaction, $A + B \rightarrow C$, the equilibrium constant is 86.3 M^{-1} at 400 K. The standard enthalpy change for the reaction is found to be 92.5 kJ mol^{-1}. The equilibrium concentrations of the different reaction species, A, B, and C are 0.0333 M, 0.0333 M, and 0.1067 M, respectively. If the reaction mixture is cooled to room temperature, 298 K), calculate the new equilibrium constant. What will be the new equilibrium concentrations of A, B, and C? **(Ans:** 0.0029 M, 0.0029 M, 0.1368 M)

FURTHER READING

1. Atkins, Peter, and Julio de Paula, *Atkin's Physical Chemistry,* Oxford University Press, UK, 2017.
2. Deep, Shashank, *Lectures for EDUSAT.*
3. Levine, Ira N. *Physical Chemistry,* McGraw-Hill Companies, 2001.
4. Silbey , Robert J., Robert A. Alberty, and Moungi G. Bawendi, *Physical Chemistry,* John Wiley & Sons, 2004.

ANSWERS

Check Your Progress

1. Intensive variables (like temperature, density, viscosity, specific heat capacity, etc.) do not depend upon the quantity of the matter present in the system whereas extensive variables (mass, volume, energy, heat capacity etc.) do.

2. z is a state function, since $xdy + ydx$ is a perfect differential. Function l is not a state function.

3. An open system can exchange matter as well as energy with its surroundings (example: tea in a cup.) while an isolated system can neither exchange matter nor energy (example: tea in a thermos flask) with its surroundings.

4. The process is isochoric.

5. For an ideal gas, the work done in an isochoric process is zero. The work done is zero even if the process (isochoric) is carried out reversibly.

6. For a cyclic process, change in the internal energy (ΔU) is zero. Therefore, according to first law of thermodynamics, the net work done by the system is equal to the total heat added to the system.

7. To prove that heat is not a state function using first law of thermodynamics:

$$\mathrm{d}q = \mathrm{d}U - \mathrm{d}w$$

$$\mathrm{d}U = \left(\frac{\partial U}{\partial T}\right)_V \mathrm{d}T + \left(\frac{\partial U}{\partial V}\right)_T \mathrm{d}V$$

$$\mathrm{d}q = \left(\frac{\partial U}{\partial T}\right)_V \mathrm{d}T + \left(\frac{\partial U}{\partial V}\right)_T \mathrm{d}V + p\mathrm{d}V$$

$$\left(\frac{\partial q}{\partial T}\right)_V = \left(\frac{\partial U}{\partial T}\right)_V$$

$$\frac{\partial}{\partial V}\left(\left(\frac{\partial q}{\partial T}\right)_V\right)_T = \frac{\partial}{\partial V}\left(\left(\frac{\partial U}{\partial T}\right)_V\right)_T$$

$$\left(\frac{\partial q}{\partial V}\right)_T = \left(\frac{\partial U}{\partial V}\right)_T + p$$

$$\frac{\partial}{\partial T}\left(\left(\frac{\partial q}{\partial V}\right)_T\right)_V = \frac{\partial}{\partial T}\left(\left(\frac{\partial U}{\partial V}\right)_T\right)_V + \left(\frac{\partial p}{\partial T}\right)_V$$

Since U is a state function,

$$\frac{\partial}{\partial V}\left(\left(\frac{\partial U}{\partial T}\right)_V\right)_T = \frac{\partial}{\partial T}\left(\left(\frac{\partial U}{\partial V}\right)_T\right)_V$$

Thus,

$$\frac{\partial}{\partial T}\left(\left(\frac{\partial q}{\partial V}\right)_T\right)_V - \frac{\partial}{\partial V}\left(\left(\frac{\partial q}{\partial T}\right)_V\right)_T = \left(\frac{\partial p}{\partial T}\right)_V$$

$\left(\dfrac{\partial p}{\partial T}\right)_V$ is not zero, hence $\dfrac{\partial}{\partial T}\left(\left(\dfrac{\partial q}{\partial V}\right)_T\right)_V - \dfrac{\partial}{\partial V}\left(\left(\dfrac{\partial q}{\partial T}\right)_V\right)_T$ is not zero. Thus $\dfrac{\partial}{\partial T}\left(\left(\dfrac{\partial q}{\partial V}\right)_T\right)_V$ is not equal to

$\dfrac{\partial}{\partial V}\left(\left(\dfrac{\partial q}{\partial T}\right)_V\right)_T$ and q is not a state function.

8. From the First law of thermodynamics,

$$dU = dq + dw \; ; \;\; dU = dq - p_{ex}dV$$

At constant volume,

$$dU = dq_V$$

$$\Delta U = q_V$$

9. C_p of a gas is always greater than C_V. Under isochoric condition, work is zero and thus heat supplied is totally used to raise the temperature. However, under isobaric condition, heat supplied is partly used for doing work and the rest is used for raising the temperature. Thus, heat required to increase the temperature by 1 K is greater under isobaric condition than at isochoric condition.

10. A reversible process is one where the change in reaction can be reversed by an infinitesimal change in any of the variables. Thus, p_{ex} needs to be equal to p_{gas}.

11. $U = f(V, T)$

$$dU = \left(\dfrac{dU}{dV}\right)_T dV + \left(\dfrac{dU}{dT}\right)_V dT = \left(\dfrac{dU}{dV}\right)_T dV + c_V dT \qquad\qquad ...(1)$$

At constant temperature, K.E. part of internal energy is constant and thus, only potential energy part of internal energy can change on change in the volume. Since there is no attractive or repulsive forces between molecules of an ideal gas, potential energy of an ideal gas does not change with volume. Thus, $\left(\dfrac{dU}{dV}\right)_T$ is zero and hence U of an ideal gas is a function of only temperature.

12. In Joule–Thomson experiment, the process proceeds under adiabatic condition, i.e., $q = 0$. As per first law of thermodynamics, $\Delta U = w$ (total work).

The arrangement in Joule–Thomson experiment (Fig. 3.7) is such that

$$U_f - U_i = w_f + w_i = -p_f V_f + p_i V_i$$

$$U_f + p_f V_f = U_i + p_i V_i$$

$\Delta H = H_f - H_i = 0$, i.e., the process is isenthalpic expansion.

13. All gases show cooling effect upon expansion because of positive Joule–Thomson coefficient. On the contrary, hydrogen gas has negative Joule–Thomson coefficient, i.e., its inversion temperature (193 K) is lower than the ambient temperature. Therefore, at ambient temperature hydrogen gas will heat up upon expansion.

14. The pressure of an ideal gas is inversely proportional to the volume of the gas under isothermal condition. Under adiabatic condition,

$$p \propto \dfrac{1}{V^\gamma}$$

15. Under isothermal condition, K.E. part of enthalpy is constant. For an ideal gas, P.E. part is also constant since there is no intermolecular forces of attraction or repulsion. Thus, enthalpy of an ideal gas does not change on change in either pressure or volume.

16. We have, $d\Delta H = \Delta C_p dT$;

$\Delta C_p = C_p$ of product $- C_p$ of reactant

On integrating with respect to temperature, we have

$$\Delta H_{T_2} = \Delta H_{T_1} + \Delta C_p(T_2 - T_1)$$

17. Enthalpy of formation is the amount of energy required to form one mole of substance from its constituent elements in its standard state. In a particular reaction, if we know the enthalpy of formation of each of the reactants and products, then the enthalpy of reaction can be calculated using simple mathematics. For example: consider the reaction

$C_2H_5OH + 3O_2 \rightarrow 2CO_2 + 3H_2O$

Here, we know the enthalpies of formation for the following reactions:

$C_2H_5OH \rightarrow 2C + 3H_2 + 0.5O_2 = 228$ kJ mol^{-1}

$2C + 2O_2 \rightarrow 2CO_2 = -394 \times 2 = -788$ kJ mol^{-1}

$3H_2 + 1.5O_2 \rightarrow 3H_2O = -286 \times 3 = -858$ kJ mol^{-1}

Adding all these three equations, we will get the resultant reaction: $C_2H_5OH + 3O_2 \rightarrow 2CO_2 + 3H_2O$.

Therefore, the enthalpy of the reaction is obtained by simple addition of the enthalpies of formation for all three processes, i.e.,

$228 + -788 + -858 = -1418$ kJ/mol.

18. Bond energy $= (347 \times 2) \times 3$ kJ mol^{-1} = 2082 kJ mol^{-1}

19. Lattice energy is proportional to $\left(\dfrac{q_+ q_-}{r_+ + r_-} \right)$, therefore the lattice energy trend will be MgO > MgCl$_2$ > NaCl.

20. Lattice energy is proportional to $\left(\dfrac{q_+ q_-}{r_+ + r_-} \right)$; therefore the lattice energy trend will be AlCl$_3$ > NaCl > KCl.

21. The reversible expansion does the maximum amount of work, since a reversible expansion requires expansion in infinite number of steps. A system does more work if expansion is carried out between two states in more number of steps.

22. $dq_{rev} = C_V dT + \dfrac{nRT}{V} dV$

    dq is not a perfect differential since the term $dq_{rev} = \dfrac{nRT}{V} dV$ cannot be integrated. However,

    $dS = \dfrac{dq_{rev}}{T} = C_V \dfrac{dT}{T} + \dfrac{nR}{V} dV$

    The above equation can be integrated. Thus, dS is a perfect differential. Thus, S is a state function.

23. In Carnot's Engine

    $dw_{max} = dq$

    $w_{max} = q_{AB} - q_{CD}$

    This equation tells us that only part of the heat absorbed at higher temperature (q_{AB}) can be converted to work by the Carnot's engine. (Since q_{CD} is not equal to zero).

24. According to Carnot's cycle when T_c approaches T_h, the area defined by ABCDA (See Fig. 3.9) approaches zero, and the maximum work approaches zero. In other words, no heat engine can convert heat to work at constant temperature.

25. Being a state function, the entropy change will remain the same whether the process occurred reversibly or irreversibly.

26. Suppose the change in state is from (T_1, V_1, p_1) to (T_2, V_2, p_2) in adiabatic process, $(dq = 0)$. If carried out the same change reversibly, q will not be zero $(dq_{rev} > dq)$ and hence entropy change of an ideal gas in an adiabatic process is not zero.

27. $S = f(T, V)$

    $dS = \left(\dfrac{\partial S}{\partial T} \right)_V dT + \left(\dfrac{\partial S}{\partial V} \right)_T dV$

$$T \mathrm{d}S = T \left(\frac{\partial S}{\partial T} \right)_V \mathrm{d}T + T \left(\frac{\partial S}{\partial V} \right)_T \mathrm{d}V \qquad \qquad \ldots(1)$$

$$T \left(\frac{\partial S}{\partial T} \right)_V = \left(\frac{T \partial S}{\partial T} \right)_V = \left(\frac{\partial q_{\mathrm{rev}}}{\delta T} \right)_V = c_V \text{ and } \left(\frac{\partial S}{\partial V} \right)_T = \left(\frac{\partial p}{\partial T} \right)_V = \frac{nR}{V} \qquad \ldots(2)$$

Substituting Eq. (2) in Eq. (1), we have

$$\mathrm{d}S = \frac{c_V}{T} \mathrm{d}T + \frac{nR}{V} \mathrm{d}V$$

On integration, we have

$$\Delta S = C_V \ln \frac{T_1}{T_2} + nR \ln \frac{V_1}{V_2}$$

28. $\Delta S_{\mathrm{surr}} = -\dfrac{q}{T}$

For an isothermal process, $\Delta U = 0$.

$$w = -q$$

$$\Delta S_{\mathrm{surr}} = \frac{w}{T} = \frac{-p_{\mathrm{ext}}(V_2 - V_1)}{T}$$

29. $\Delta S = \dfrac{\Delta H_{\mathrm{vap}}}{T} = \dfrac{40.65 \times 1000 \text{ J mol}^{-1}}{373 \text{ K}} = 108.98 \text{ J K}^{-1}\text{mol}^{-1}$

30. It is the measure of the total non-expansion work done by the system.

31. At constant volume and temperature, the process will be spontaneous if Helmholtz free energy change is negative, i.e., $\Delta A < 0$.

32. $\left(\dfrac{\partial S}{\partial V} \right)_T = \left(\dfrac{\partial p}{\partial T} \right)_V$

33. Free energy of an ideal gas will decrease with increase in temperature. As we know, $\left(\dfrac{\partial G}{\partial T} \right)_P = -S$ and S is a positive quantity.

34. Measurement of change in free energy for a process at two different temperatures can be used to calculate the enthalpy change for the process using Gibb's Helmholtz equation.

$$\left(\frac{\partial (\Delta G / T)}{\partial T} \right)_p = -\frac{\Delta H}{T^2}$$

35. For an ideal gas mixture or ideal solution, $\Delta_{\mathrm{mix}} H$ is zero, therefore the Gibbs free energy change is given as $\Delta_{\mathrm{mix}} G = -T \Delta_{\mathrm{mix}} S$. Entropy increases on mixing of two gases.

This implies that for ideal conditions, free energy change of mixing is always negative.

36. The relationship between boiling point *(T)* and pressure *(p)* is given by

$$\ln \frac{p_2}{p_1} = \frac{\Delta H_m}{R} \left(\frac{1}{T_1} - \frac{1}{T_2} \right) = \frac{\Delta H_m}{R} \left(\frac{\Delta T}{T_1 T_2} \right)$$

Vaporization/ sublimation process is an endothermic process. Thus, boiling point of a liquid can be increased by increasing the pressure.

37. The equilibrium constant *(K)* is the ratio of the concentration of the products and reactants at equilibrium. The reaction quotient *(Q)* is the ratio of the concentrations of products and reactants at state other than equilibrium state.

38. For a reaction: $aA + bB \rightarrow cC + dD$

$$K_p = \frac{p_C^c \, p_D^d}{p_A^a \, p_B^b}$$

For ideal gas, $P = \dfrac{nRT}{V} = cRT$, $c = \text{concentration} = \dfrac{n}{V}$

Substituting the pressure term in K_p equation, we have

$$K_p = \frac{[c_C RT]^c [c_D RT]^d}{[c_A RT]^a [c_B RT]^b} = \frac{[C]^c [D]^d}{[A]^a [B]^b} (RT)^{(c+d)-(a+b)}$$

$$K_p = K_c (RT)^{\Delta n_g}$$

39. $\Delta_r G = -RT \ln K + RT \ln Q = RT \ln \dfrac{Q}{K}$

40. In this process, $\Delta n_g > 0$ and $\Delta H > 0$ and thus decomposition of limestone is favoured at low pressure and high temperature.

41. No change in the direction of reaction.

42. The reaction is endothermic.

43. If the reaction is endothermic, K_p increases with an increase in temperature.

44. Increase in temperature will increase the solubility of KNO_3.

Multiple Choice Questions

1. (b)	2. (d)	3. (c)	4. (c)	5. (d)	6. (a)	7. (b)
8. (c)	9. (b)	10. (c)	11. (b)	12. (a)	13. (b)	14. (a)
15. (a)	16. (b)	17. (b)	18. (a)	19. (c)	20. (d)	

Phase Rule

After reading this chapter, you will be able to:

- understand and derive Gibbs phase rule.
- define various terms such as phases, components, and degrees of freedom.
- derive the phase rule equation.
- explain condensed phase rule for lead–silver system.
- explain congruent and incongruent phase diagrams.
- describe the multicomponent system of Bi–Pb–Sn using ternary phase diagram.
- understand the phase diagram of iron–carbon system and various phases in them.

4.1 INTRODUCTION

Phase rule was generalized by Josiah Willard Gibbs in 1874 and was later developed by H.W.B. Rooseboom in 1884. Phase rule is based on the principles of thermodynamics. While dealing with chemical systems, we come across two or more phases that are present in equilibrium, referred to as a *heterogeneous system*. In this chapter we will discuss Gibbs phase rule and its applications in such heterogeneous systems. The concept of phase rule finds its relevance in material science, metallurgy, and geology. It includes characterizing the chemical states of minerals, melts, solids, liquids, and vapours, and predicting their phases at equilibrium as a function of temperature and pressure. We will also list the applications of phase rule in electrical components and their preservation.

4.2 GIBBS PHASE RULE

A heterogeneous system consists of two or more different phases having different physical states or possibly different chemical compositions, or both. These phases are separated from one another by surface of separation or interfaces and there exists a dynamic equilibrium between them. The equilibrium between these phases is called *poly-phase equilibrium*. Such systems are studied by applying Gibbs phase rule.

According to Gibbs phase rule, *If the equilibrium in a heterogeneous system is not influenced by gravity, electrical or magnetic forces, then the number of degrees of freedom (F) of the system is related to the number of components (C) and the number of phases (P) existing at equilibrium with one another by the equation.*

$$F = C - P + 2$$

Here, F is the number of degrees of freedom, P is the number of phases, and C is the number of components.

Josiah Willard Gibbs devoted his working on chemical thermodynamics at Yale University. He was known to spend years on his findings before publishing original concepts and theories. His papers were published in now an obscure journal *The Transactions of the Connecticut Academy of Arts and Science*. Willard Gibbs is regarded as the first American theoretical scientist.

4.3 Important Terms in Gibbs Phase Rule

4.3.1 Phase

Phase, denoted as P, is defined as the homogeneous, physically distinct, and mechanically separable part of a heterogeneous system. Generally, each phase is separated in a given system by interfaces.

To understand the concept of phases, let us take some simple examples. Instructors can demonstrate the following to the students in the classroom.

Consider the simple example of the water system that has ice, water, and vapour as three phases. Each phase in itself is homogeneous, but the system as a whole is heterogeneous. All phases exist in dynamic equilibrium with one another along the interfaces in the system. If one considers a gas phase system, it is known that all gases are completely miscible; hence the system comprising different gases will be a *single phase system*, that is, $P = 1$. Now, let us imagine that a beaker is partially filled with water. This means that water (liquid) is in dynamic equilibrium with its vapour (gas). Hence, $P = 2$ and will be considered as a *two-phase system*. Preparing a mixture of solids intimately will consist of as many phases as the number of substances added, such as, if we take solid pepper and salt together, there are two distinct phases. If there are two or more liquids, the number of phases will depend on its miscibility. If we intimately mix water with benzene, two distinct liquid phases are obtained along with a vapour phase. As $P = 3$, it will be a *three-phase system*.

The classic example of water and salt solution will demonstrate the case even further. Take some water in a beaker and keep adding a known quantity of salt. After some time, the composition of water will change, but the number of phases remains the same, that is, two (liquid solution and water vapour). Continue adding salt until some portion remains undissolved in water. Now, the system has three phases, that is, water (liquid), vapour (gas), and salt (solid).

4.3.2 Component

Component, denoted as C, is defined as the minimum number of independently variable constituents by means of which composition of all the phases can be expressed either directly or by means of a chemical equation. In this we are considering the 'smallest number' of components of a system.

In a water system, comprising ice, water, and vapour, the phase composition can be expressed in terms of one chemical constituent, that is, H_2O and hence is a one-component system. In a sulphur system, the various forms of sulphur such as rhombic, monoclinic, liquid, and vapour can be represented by a single chemical constituent, sulphur; hence this is also a one-component system. A mixture of ethanol and water is an example of a two-component system. One needs to indicate both ethanol and water to express its composition. Let us now take the example of thermal decomposition of solid $CaCO_3$. There are three phases, namely (i) solid $CaCO_3$ (ii) solid CaO and (iii) gaseous CO_2.

$$CaCO_{3\,(s)} \overset{\Delta}{\rightleftharpoons} CaO_{\,(s)} + CO_{2\,(g)}$$

Though there are three species, the number of components is only two, because of the equilibrium. Any two of the three constituents can be chosen as the components. If CaO and CO_2 are considered, then the composition of the phase $CaCO_3$ is expressed as one mole of component CO_2 plus one mole of component CaO. If $CaCO_3$ and CO_2 are considered, then the composition of the phase CaO can be expressed as one mole of $CaCO_3$ minus one mole of CO_2.

Now, during the decomposition of ammonium chloride, as

$$NH_4Cl_{(s)} \xrightarrow{\Delta} NH_{3(g)} + HCl_{(g)}$$

there are three constituents and two phases (solid and gas) and the equilibrium that exists in the reaction can be depicted as:

$$NH_4Cl_{(s)} \rightleftharpoons NH_4Cl_{(g)} \rightleftharpoons NH_{3(g)} \rightleftharpoons HCl_{(g)}$$

If the above reaction is carried out in a closed container, then the number of components will be one, since the gaseous phase of both NH_3 and HCl are formed in fixed stoichiometric proportions and is represented as $NH_4Cl_{(g)}$.

If an excess of either NH_3 or HCl is added, the composition of gaseous phase cannot remain the same and thus cannot be represented as only NH_4Cl, but one more component is needed. Either, an excess of NH_3 or HCl would have invoked as a second component. Thus, it is a two-component system.

In the sodium sulphate and water system, the following are observed.

Phases	Components
Na_2SO_4	$Na_2SO_4 + 0H_2O$
$Na_2SO_4.7H_2O$	$Na_2SO_4 + 7H_2O$
$Na_2SO_4.10H_2O$	$Na_2SO_4 + 10H_2O$
$Na_2SO_{4\,(aq)}$	$Na_2SO_4 + xH_2O$
$H_2O_{(s)}, H_2O_{(l)}, H_2O_{(g)}$	$0\,Na_2SO_4 + H_2O$

As shown in the table, the different phases are: Na_2SO_4, $Na_2SO_4.7H_2O$, $Na_2SO_4.10H_2O$, solution of Na_2SO_4 in water, ice, water and vapour. In this case, the composition of certain phases is expressed as either H_2O or Na_2SO_4. In other cases, two independent components, viz. Na_2SO_4 and H_2O, are required.

4.3.3 Degrees of Freedom

Degrees of freedom, denoted as F, is defined as the number of variables such as temperature, pressure, and composition that must be specified in order to define a system at equilibrium. It is also referred to as *variance*.

If we consider water system, at a specific temperature and pressure, the three phases of water—ice, liquid, and vapour, are in equilibrium; hence the degrees of freedom $F = 0$. When the degree of freedom of a given system is zero, it is referred to as *nonvariant or invariant system*. Similarly, equilibrium of liquid water ⟺ vapour has two phases and hence, $F = 1$. When the degree of freedom of a given system is one, it is referred to as a *univariant system*. In the same way, if the degree of freedom of a given system is 2, it is referred to as a *bivariant system*. Consider a system existing in one single phase of pure nitrogen gas whose composition is expressed as N_2. Two variables that are required to be specified are temperature and pressure, and hence the system is bivariant. A system of a gaseous mixture of 70 % N_2 and 30 % O_2 at 22 °C at 1 atm pressure is defined in terms of composition, temperature, and pressure. Such a system of a mixture of gases has $F = 3$ and it is referred to as a *trivariant system*.

4.4 DERIVATION OF PHASE RULE EQUATION

Phase rule allows the prediction of the number of degrees of freedom F for heterogeneous systems when the number of components C and number of phases are known. Let us consider a heterogeneous system at equilibrium with components C existing in P phases having temperature, pressure, and concentration as variables. If a system is at equilibrium, there can be one temperature and one pressure variable, but the

Check Your Progress

1. What will happen if one thermally decomposes calcium carbonate? State the number of phases.
2. Consider the reaction, $NH_4Cl_{(s)} \rightarrow NH_{3\,(g)} + HCl_{(g)}$. State the number of phases.
3. Consider sodium sulphate in water system. Comment on the number of components and phases.
4. State Gibbs phase rule.

concentration will be more. If temperature and pressure are same for all phases, one should consider one variable each. The concentration for one phase having C components is expressed as $(C-1)$. Hence, concentration variable for P phases will be $P(C-1)$.

Hence, total number of variables is $P(C-1) + 1 + 1$, i.e., $P(C-1) + 2$ (4.1)

(concentration) (temp) (press)

On the basis of the principles of thermodynamics, when a heterogeneous system is in equilibrium at constant temperature and pressure, the chemical potential (μ) of a given component must be same in every phase. Hence, for one component in three phases (say 1, 2, and 3) we get,

$$\mu_1 = \mu_2 \text{ and } \mu_1 = \mu_3 \qquad (4.2)$$

Thus, for every component at equilibrium in three phases, two equations are known. For C components in P phases, the number of equations will be, $C(P-1)$. Let F be the difference between the number of variables and the number of equations connecting them.

So, number of degrees of freedom (F) is given as,

$$F = [P(C-1) + 2] - [C(P-1)] \qquad (4.3)$$

On further simplifying Eq. (4.3), we get,

$$F = CP - P + 2 - CP + C$$

Hence, $F = C - P + 2$ (4.4)

Equation (4.4) is known as Gibbs phase rule equation, sometimes alternatively written as,

$$P + F = C + 2$$

A phase diagram is are generally employed to study the phase equilibria of various systems. It is a graphical plot that depicts the different phases in a heterogeneous system at equilibrium.

To understand phase rule, let us apply the above equation to various polyphase equilibria and deduce the number of degrees of freedom for all curves, boundaries, and areas in a phase diagram.

4.5 ONE-COMPONENT WATER SYSTEM

As shown in Fig. 4.1, the three phases of water, namely solid, liquid, and vapour, exist in dynamic equilibrium. The various equilibria observed are: liquid $\rightleftharpoons$ vapour (evaporation), solid $\rightleftharpoons$ vapour (sublimation) and solid $\rightleftharpoons$ liquid (fusion). Each of these equilibria involves two phases and the composition of each phase can be expressed in terms of the chemical constituent, water. Hence, it is a one-component system.

Curves and Points

Curve OA in Fig. 4.1 is the vapour pressure curve of water. As temperature increases, vapour pressure (VP) of water increases. At the normal boiling point of 100 °C, vapour pressure becomes equal to 1 atm. Along this curve, water and vapour exist in equilibrium.

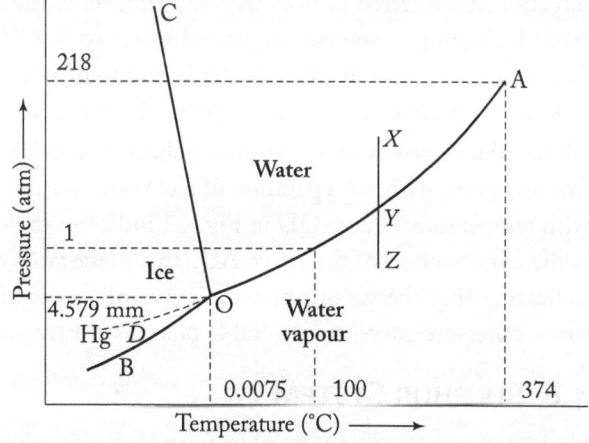

Fig. 4.1 Water system

Check Your Progress

5. Define (a) phase (b) component (c) degrees of freedom.
6. In a system of benzene and water, how many phases will be present?

On applying phase rule, $F = C - P + 2$

In this case, the number of phases, $P = 2$; so, $F = 1 - 2 + 2 = 1$ (univariant)

Curve OA is vapour pressure of liquid, ending at point A, beyond which vapour pressure of liquid does not increase. The critical pressure of 218 atm. is attained at critical temperature, that is, 374°C. Above point A, liquid and vapour phases are indistinguishable and the two phases become identical. No liquid can exist beyond the critical temperature point.

Curve OB is called the *sublimation curve of ice* (or vapour pressure curve of ice) depicting the vapour pressure of ice at different temperatures. Here, ice is in equilibrium with the vapour phase. Curve OC is called the *fusion curve of ice*, where again $F = 1$, as there is dynamic equilibrium between ice and water.

Along all these curves, two phases exist in equilibrium. Applying phase rule, $P = 2$ and $C = 1$. Hence, $F = C - P + 2$, so $F = 1$ (univariant).

All the curves OA, OB, and OC meet at point O. At triple point O (4.579 mm Hg, 0.0075°C), all three phases co-exist in dynamic equilibrium $P = 3$, and hence, $F = 0$ (nonvariant). The system has no degree of freedom which means that one cannot change pressure or temperature variables without changing the number of phases.

Regions or Areas

These indicate the existence of single phase; area *BOC* depicts solid phase; area *AOC* depicts liquid phase; and area *AOB* depicts the vapour phase. On applying phase rule,

$F = C - P + 2$ or $P = 1, C = 1$ or $F = 2$ (bivariant)

Similarly, for other areas, $F = 2$. Thus, the water system is a bivariant system.

Metastable State

When the same state of a system can be attained under a given set of conditions from either direction using any procedure, a *true equilibrium* is established. Let us take the example of the water system, in which ice and water attain true equilibrium at 1 atm. pressure and 0°C. This can be attained by either partial freezing of water or partial melting of ice; ice $\rightleftharpoons$ water.

However, if a state of the system can be attained by a careful change of system conditions in only one direction, it is referred to as *metastable equilibrium*. The state of the system so obtained is called *metastable state*; for example, one can supercool water to − 2°C or lower temperatures very slowly without the formation of ice crystals. Thus, water is said to be in metastable state at − 2°C. The mere addition of one tiny crystal of ice can freeze the supercooled water by raising the temperature to 0°C.

Metastable state of water   From the above discussion, we know that it is possible to cool water below its freezing point without separation of ice. Water is said to be supercooled whose vapour pressure changes with temperature. Curve OD in Fig. 4.1 indicates the vapour pressure curve of supercooled water which is the continuation of the curve AO. This metastable curve lies above the vapour pressure curve of ice, indicating that the vapour pressure curve of supercooled water is greater than that of ice. Hence, the vapour pressure curve of metastable phase is always greater than that of its corresponding stable phase.

4.6 SULPHUR SYSTEM

Under ordinary conditions, sulphur exists in the rhombic form, but on heating it exhibits peculiar behaviour which can be depicted in a phase diagram. On rapid heating, sulphur melts at 114.85°C whereas if heated slowly, it transforms to the monoclinic form melting at 119 °C.

$$\text{Rhombic sulphur} \xrightleftharpoons{95.45°C} \text{Monoclinic sulphur}$$

The above phase reaction taking place at 95.45°C is called the *transition temperature* above which monoclinic sulphur is stable. At the transition temperature both forms of sulphur can co-exist together and are stable. Thus, sulphur is known to exist in four possible phases, viz., rhombic sulphur (S_R), monoclinic sulphur (S_M), liquid sulphur (S_L), and vapour sulphur (S_V), but all these phases cannot co-exist together. Figure 4.2 depicts the phase diagram of sulphur.

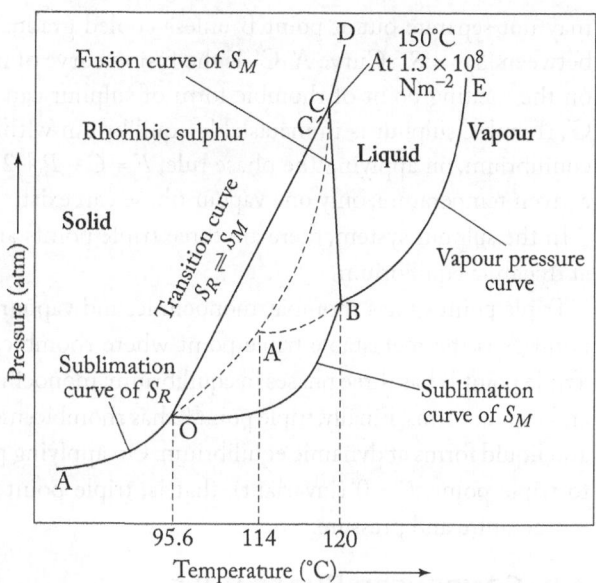

Fig. 4.2 Phase diagram of sulphur system

Areas

On observing the phase diagram, there are four areas: (i) AOCD indicates stable existence of rhombic sulphur; (ii) BOC is triangle-shaped showing monoclinic sulphur; (iii) DCBE represents liquid sulphur; and (iv) AOBE is the vapour phase of sulphur. On considering phase rule on all the areas of sulphur system, sulphur exists in a single phase in all the areas, that is, $P = 1$. The varying forms of sulphur can be explained by one chemical formula (S) and thus is a one-component system. On applying phase rule, we get

$$F = C - P + 2 = 1 - 1 + 2 = 2$$

Hence, to represent any point in an area, both temperature and pressure are necessary, and thus the sulphur system is a bivariant system.

Boundary Lines or Curves

Curve OA in Fig. 4.2 is the sublimation curve of rhombic sulphur and two phases are in equilibrium along this curve ($S_R \rightleftharpoons S_V$). Curve OB is the sublimation curve of monoclinic sulphur where there is equilibrium between two phases ($S_M \rightleftharpoons S_V$).

Curve OC is the transition curve of rhombic and monoclinic sulphur where both forms exist at equilibrium. The curve is inclined slightly towards the right side with a positive slope. This happens as there is an equilibrium along the transition curve of rhombic and monoclinic sulphur. The densities of rhombic sulphur and monoclinic sulphur are 2.07 and 1.95 (g/cm^3), respectively. Hence, the conversion from rhombic to monoclinic form is accompanied with an increase in volume and pressure which causes a rise in transition temperature.

Curve BC is the fusion curve of monoclinic sulphur; the effect of pressure on the melting point of monoclinic sulphur can be inferred from this curve. Curve CD is the fusion curve of rhombic sulphur where equilibrium exists between rhombic sulphur and liquid sulphur ($S_R \rightleftharpoons S_L$).

Curve BE is the vapour pressure curve of liquid sulphur that is in equilibrium with its vapour ($S_L \rightleftharpoons S_V$).

Metastable state   In Fig 4.2, the dotted lines represent metastable equilibria existing in the sulphur system. Curve OA′ is an extension of curve AO depicting the metastable equilibrium between $S_R \rightleftharpoons S_V$ and if temperature of the rhombic form is allowed to raise rapidly, it can remain in equilibrium with its vapour phase without transforming to the monoclinic form. Curve BA′ is an extension of curve EB. When liquid sulphur is allowed to cool along this curve, solid monoclinic form may not separate out at point B unless cooled gradually. Curve BA′ represents metastable equilibrium between $S_L \rightleftharpoons S_V$. Curve A′ C′ is the fusion curve of metastable rhombic sulphur; the effect of pressure on the melting point of rhombic form of sulphur can be inferred from this curve. Along the curve A′ C′, rhombic sulphur is in metastable equilibrium with its liquid form, $S_R \rightleftharpoons S_L$. As two phases exist at equilibrium, on applying the phase rule, $F = C - P + 2 = 1 - 2 + 2 = 1$ (univariant). This denotes that at a given temperature, only one vapour phase can exist.

In the sulphur system, there are three triple points and at any given triple point, three phases co-exist at dynamic equilibrium.

Triple point O has rhombic, monoclinic, and vapour forms of sulphur in dynamic equilibrium. Triple point A′ is the metastable triple point where rhombic, liquid, and its vapour forms are in equilibrium. Triple point B has three phases in equilibrium; monoclinic, liquid, and vapour forms. Finally, triple point C has rhombic, monoclinic, and liquid forms at dynamic equilibrium. On applying phase rule to triple point, $F = 0$ (invariant), that is, triple point has fixed temperature and pressure.

Triple point	Temperature (°C)	Pressure (Nm^{-2})
O	95.6	0.8
A′	114	4.0
B	120	5.333
C	150	1.3×10^8

4.7 CONDENSED PHASE RULE

When two independent components are present in a heterogeneous system, it is called a *two-component system*. According to phase rule, a two-component system with one phase will be,

$$F = C - P + 2 = 2 - 1 + 2 = 3$$

This implies that a two-component system with one phase will have three degrees of freedom or rather needs to be defined by three variables, namely temperature, composition, and pressure. In such circumstances, a three-dimensional phase diagram needs to be constructed which is complex. If one can eliminate one variable which has negligible effect, such systems are called *condensed systems* and the rule is called *condensed phase rule*, expressed as,

$$F = C - P + 1 \tag{4.5}$$

In a two-component system, the effect of pressure is very small and hence neglected; instead an arbitrary constant pressure of 1 atm is considered.

A simple temperature–composition diagram is drawn to explain two-component systems. Some of the important two-component systems are: lead–silver system; zinc–magnesium system; and sodium chloride–water system.

Check Your Progress

7. Give an example of invariant system.
8. How many triple points are present in water system?
9. What is a metastable state? Give an example.
10. Justify the statement, 'Transition curve of rhombic sulphur has a positive slope.'
11. Draw the phase diagram of a one-component system that has more than one solid phase.
12. Give the number of components in the following systems: (a) aqueous acetic acid; (b) magnesium carbonate in equilibrium with its decomposition products.

4.7.1 Lead–Silver Eutectic System

An *eutectic system* refers to a homogeneous mixture of substances that melt or solidify at a temperature that is lower than the melting points of either of its constituents. Consider two components, lead and silver; in the solid state; these combine to neither produce a compound nor form a solid solution by dissolving in each other (immiscible and remain separate). As there are two independent components representing the system, Pb–Ag system is a two-component system. But, in liquid state, they are completely miscible with one another. On solidification it gives rise to an eutectic mixture.

The temperature–composition phase diagram of Pb–Ag system shown in Fig. 4.3, which shows the phases of the system existing under different conditions of temperature and composition.

Curves and Points

Point A is the melting point of lead (Pb); Point B is the melting point of silver (Ag). Pure lead melts at 327 °C. Addition of silver lowers its melting point along AC. Curve AC is called freezing point curve of Pb and curve BC is the freezing point curve of silver.

Areas and Boundaries

In area above ACB, the components form a single homogeneous liquid phase. Applying condensed phase rule, we get $F = 3 - 1 = 2$; that is, two degrees of freedom; hence the system is bivariant, that is, two variables temperature (T) and composition of liquid phase can be varied independently.

Along curves AC and BC, there are two phases in equilibrium. Applying condensed phase rule on curves AC and BC, we get, $F = 3 - 2 = 1$.

As the degree of freedom is one, the system is monovariant and its composition varies with temperature.

The curves intersect at C, where Pb and Ag are in equilibrium with the liquid phase. Hence, at point C, there are three phases in equilibrium. On applying phase rule, we get $F = 3 - P = 0$.

As the degree of freedom is zero, the system is invariant which can exist at definite temperature and composition of liquid. The temperature of the eutectic is 303°C and the composition of the solution phase is 2.6 % Ag.

Point C represents the lowest temperature at which Pb and Ag can exist at equilibrium with the liquid. Point C is the eutectic point. An eutectic point is a point in the phase diagram that corresponds to definite composition and melting temperature of an eutectic alloy. The corresponding temperature is called the *eutectic temperature* (303 °C) and the corresponding composition is the *eutectic composition* (97.4 % lead and 2.6 % silver).

If a liquid mixture of composition represented by point a is cooled, then the temperature drops without any change in composition. This happens till point b, corresponding to temperature t_1. As cooling is continued, the composition changes along curve bc. When eutectic temperature (303°C) is reached, both Pb and Ag crystallize out in fixed ratio c, which gives the eutectic composition. This principle is

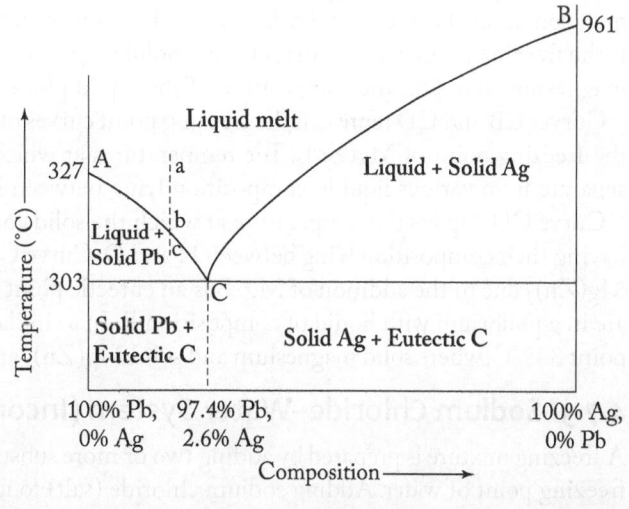

Fig. 4.3 Pb–Ag system

used in desilverization of lead by Pattinson's process. In this process, the argentiferous Pb containing small percentage of Ag is first heated to above 327°C. It is then allowed to cool till the temperature falls along the line ab. When point b is reached Pb will crystallize out and the solution will be richer in terms of Ag. Cooling is continued further, when Pb separates out and is removed. The liquid melt becomes richer in Ag till eutectic composition of 2.6 % Ag is reached. Thus, enrichment of silver from very low concentration of less than 0.1 % to about 2.6 % is possible by this process.

4.7.2 Zinc–Magnesium System (Congruent Melting Points)

Figure 4.4 elucidates the phase diagram of zinc–magnesium (Zn–Mg) system which is a two-component

system. Zinc and magnesium form alloys and also are known to be completely miscible in their liquid state. The melting points of Zn and Mg are 419°C and 650°C respectively. These metals form inter-metallic alloy compound, $Mg(Zn)_2$ with a melting point of 590°C. The compound $Mg(Zn)_2$ remains as a solid upto its melting point, beyond which it fuses sharply forming a liquid melt with the same composition as the solid along with constant temperature. Such melting points, wherein both liquid melt and solid possess same composition co-existing together at constant temperature is called *congruent melting point.*

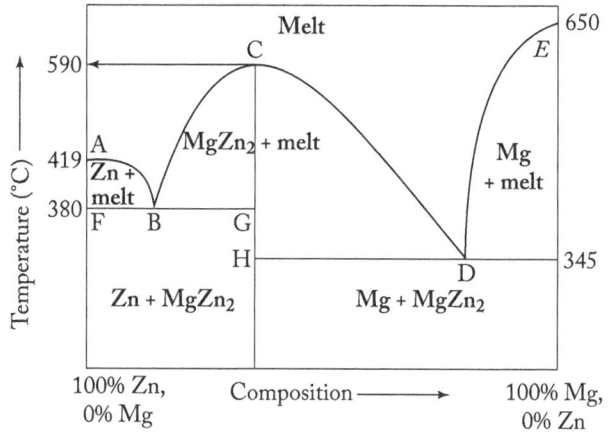

Fig. 4.4 Zn–Mg congruent system

Curves and Points

The melting points of pure forms of Zn and Mg are depicted as points A and E respectively in the phase diagram. Points B and D depict the eutectic points, wherein B is the least eutectic point of Zn–$MgZn_2$ system. At point B, solid zinc, solid $MgZn_2$, and liquid $MgZn_2$ co-exist together at dynamic equilibrium. Applying reduced phase rule at point B, $P = 3$ and $C = 2$; thus

$F = C - P + 1 = 2 - 3 + 1 = 0$ (non-variant)

Curve AB is the freezing point curve of zinc. In the area ABFA, solid zinc is in equilibrium with liquid melt containing both zinc and magnesium, the composition of which is given by the curve AB. Curve ED is the freezing point curve of magnesium. Solid magnesium is in equilibrium with liquid melt containing magnesium and zinc, the composition of the liquid phase along the curve DE.

Curves CB and CD represent the freezing point curves of the alloy $Mg(Zn)_2$. Addition of zinc depresses the freezing point of $Mg(Zn)_2$. The temperatures at which the solid compound $Mg(Zn)_2$ will begin to separate from various liquids, composition lying between B and G, fall on the curve CB.

Curve CD depicts the temperature at which the solid compound $Mg(Zn)_2$ starts freezing from liquids having their composition lying between H and D. Curve CD gives the depression in the freezing point of $Mg(Zn)_2$ due to the addition of Mg. B is an eutectic point 380 °C at which solid zinc and solid $Mg(Zn)_2$ are in equilibrium with liquid of composition B. In a similar manner, point D represents another eutectic point 345 °C, where solid magnesium and solid $Mg(Zn)_2$ are in equilibrium with liquid of composition D.

4.7.3 Sodium Chloride–Water System (Incongruent Melting Points)

A freezing mixture is prepared by adding two or more substances that allow obtaining temperatures below freezing point of water. Adding sodium chloride (salt) to ice is one such freezing mixture. Such mixtures are commonly employed to cool suspensions to form crystals of organic compounds in the laboratory.

Figure 4.5 depicts the phase transition of sodium chloride and water system. It is a solid–liquid system in which two substances form a solid unstable compound, sodium hydrate (NaCl.2H$_2$O) that does not have a true melting point and decomposes before reaching it. The temperature at which decomposition of NaCl hydrate occurs is called the *incongruent melting point* or *peritectic* or *transition temperature*. The compound NaCl.2H$_2$O can be seen as a salt crystal with dissolved impurity, that is, two moles of H$_2$O occupying one unit in NaCl. The dihydrate of sodium chloride is considered to undergo transition as follows.

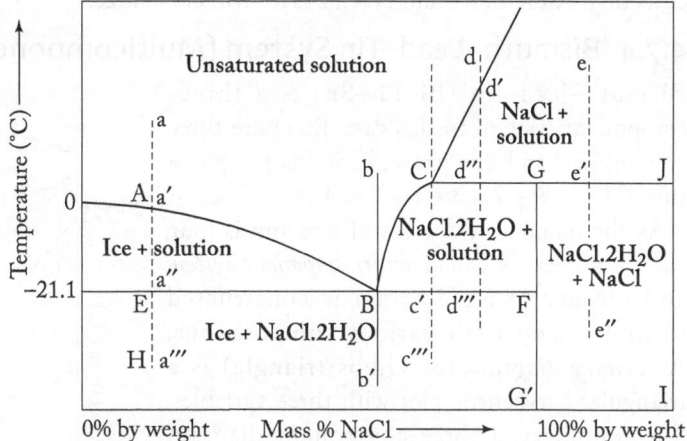

Fig. 4.5 Phase diagram of sodium chloride–water system

$$NaCl.2H_2O \rightarrow NaCl \text{ (anhydrous)} + solution$$

At the point when such a transition takes place, $F = C - P + 1 = 2 - 3 + 1 = 0$ (invariant). The peritectic reaction takes place at a definite temperature with a definite composition.

The freezing point of water is labelled as A in the phase diagram. Now, let us imagine that pure crystals of sodium chloride are added to ice slowly and continuously. Continuous addition of salt will lower the freezing point of water along the curve AB. This phenomenon is called the *freezing point depression*. Curve AB is called the freezing curve of water. The lowering of freezing point is continued till it reaches point B, the eutectic point of the system. Once sufficient amount of sodium chloride is added, the eutectic temperature obtained is –21.1°C with eutectic composition of 23.3 % by weight of NaCl in water. At point B, ice, solid sodium chloride, and the solution (melt) coexist at equilibrium, and hence the degree of freedom, F of this system is 0 (invariant).

Curve BC represents the solubility curve of compound NaCl.2H$_2$O and curve CD represents the freezing point curve of anhydrous sodium chloride.

The point G' corresponds to the composition of NaCl.2H$_2$O. On heating, this compound will remain as a solid until it reaches point F, that is, peritectic temperature of – 21.1°C. Hence, point G' is labelled as the *incongruent melting point of the system*, where NaCl.2H$_2$O is in dynamic equilibrium with anhydrous sodium chloride and solution ($F = 0$, invariant). If heating at G is continued, the compound will decompose to form NaCl and further the temperature will remain constant until the reaction is complete. If heating is further carried out, there will be an increase in temperature of the system and sodium chloride will be in equilibrium with its solution ($F = 1$, univariant).

The dashed lines depict various phase transitions possible in the system. Let the solution of NaCl in water (labelled as a) be cooled; then at a', ice will start separating out till it reaches a'. At a", NaCl.2H$_2$O will also start to separate out from the solution and if cooling is continued further, say at point a''', it becomes invariant ($F = 0$). Similarly, if a system at b which is an unsaturated solution of NaCl in water is cooled, the temperature of the solution will decrease along the line bB and at B, ice and NaCl.2H$_2$O will separate out. A concentrated solution of NaCl in water is represented by c' cools till it reaches point C, when NaCl.2H$_2$O separates out. If cooling is continued, more and more amount of NaCl.2H$_2$O separates out and will be in equilibrium with the solution whose composition is given along curve BC. If the system cools from d to d', anhydrous NaCl starts to separate out till d". Beyond d" peritectic reaction will start and continue till it meets point d'''. At d''', the solution reaches the eutectic composition and a eutectic mixture of ice and NaCl.2H$_2$O will crystallize.

4.7.4 Bismuth–Lead–Tin System (Multicomponent System)

Bismuth–lead–tin (Bi–Pb–Sn) is a three-component system. In this case, there are three components and one phase; hence as per phase rule $F = C - P + 2 = 3 - 1 + 2 = 4$

As the number of degrees of freedom is four, such a system is called *multicomponent system* and ternary phase diagram is constructed to understand their various compositions. A *ternary diagram* (or Gibbs triangle) is a triangular barycentric plot with three variables. For simplicity, a three-component diagram will have two variables that are fixed so as to obtain a two-dimensional phase diagram. The phase diagram of bismuth–lead–tin system is elucidated in Fig. 4.6 which is a temperature–composition phase diagram at constant pressure in three-dimensional space. (Note: We are considering a solid–liquid phase diagram for simplicity). All these three metals dissolve appreciably into each other in solid phase, but

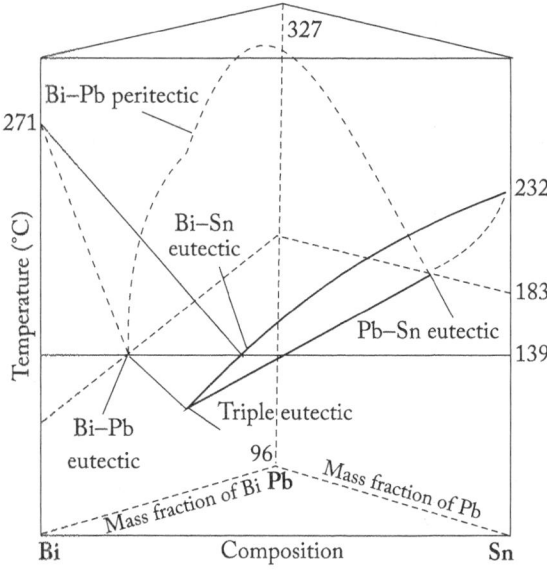

Fig. 4.6 Three-dimensional temperature–composition phase diagram of bismuth–lead–tin system

solid solution regions are considered as zero so as to simplify its diagrammatic representation in a ternary phase diagram.

Each face of the triangular prism is a two-component temperature–composition. The inner portions of the triangular prism depict compositions in which all three components are present. The triple eutectic point occurs at 96°C as shown in Fig. 4.6. The surface shown in the phase diagram is the lower boundary of the three-dimensional one-phase liquid region and further surface has three grooves that end up towards the triple eutectic thereby forming three two-component eutectics.

In Fig. 4.7, the Bi–Pb–Sn system is represented by a composition–composition phase diagram. The ternary phase diagram is obtained by passing a plane through a prism slightly above the triple eutectic point. The central triangular portion represents a one-phase liquid region in which all the points represent the composition of single liquid phase. Every tie-line drawn from the edge of the central region connects to the corners of the triangle. Tie lines connect compositions of liquid and vapour phases in equilibrium. It is evident that only one solid phase freezes out from a composition corresponding to a point on edge of the region. At a composition corresponding to the corners of a region, two components will freeze out. The triangular regions along the triangular sides depict three-phase regions. The two ends of a

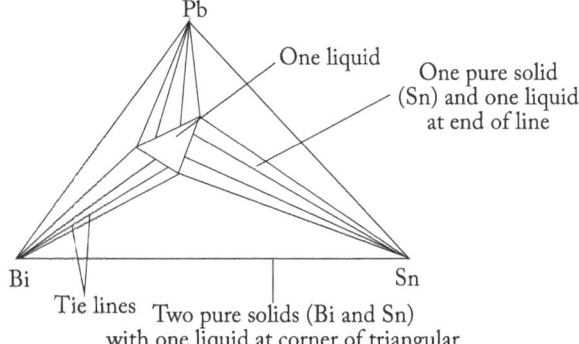

Fig. 4.7 Solid–liquid composition phase diagram of Bi–Sn–Pb at 100 °C and 1 atm

tie-line denote possible composition of two phases and corners of these regions provide details on the composition of three existing phases. At triple eutectic temperature, the area of the liquid region diminishes to zero from which three tie-lines originate and extend towards each corner that indicates that at triple eutectic three solid phases are in equilibrium with eutectic liquid.

4.7.5 Iron–Carbon System

Understanding the iron–carbon system (Fe–C system) forms the basis of ferrous metallurgy and the developments of various microstructures. Iron–carbon system is not a true equilibrium phase diagram, since Fe_3C (iron carbide) is in metastable phase. The phase diagram (Fig. 4.8) depicts the type of alloys formed under slow cooling and the effect of heat treatment on the properties of cast iron and commercial steels. Ideally, phase diagram should extend from 100 % iron to 100 % carbon, but we will study up to 6.67 % carbon, since cast iron contains not more than 5 % carbon content.

The various phases of the diagram are explained as below:

α- **Ferrite** It is an interstitial solid solution of carbon in α-Fe existing at room temperature with a body centred cubic structure. It is a stable form of iron below 912°C. The maximum solubility of carbon in α-ferrite is 0.022 % at 727 °C that decreases with lowering temperatures. α-Ferrite is the softest amongst all phases, fairly ductile, paramagnetic material that transforms to γ-austenite at 912 °C.

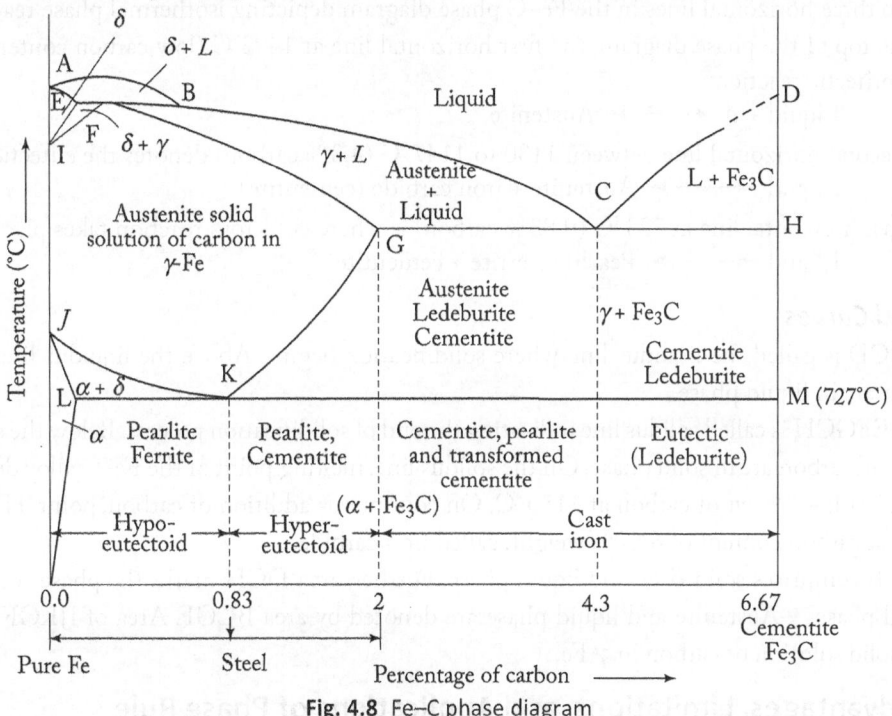

Fig. 4.8 Fe–C phase diagram

Check Your Progress

13. Differentiate between (a) peritectic and eutectic points; (b) triple and critical points.
14. Justify the statement, 'The eutectic solution has definite composition and sharp melting point but is not a separate compound.'
15. State the condensed phase rule.
16. What are the eutectic composition and eutectic temperature of lead–silver system?
17. Is it true that an invariant system has no degree of freedom?
18. What is a freezing mixture? Mention its use.

γ-**Austenite**   It is an interstitial solid solution of carbon in *γ*-Fe with face-centred cubic structure. The maximum solubility of carbon is 2.14% at 1147 °C and decreases to about 0.77% carbon at 727 °C. *γ*-Austenite is a soft, ductile, malleable, non-magnetic material that transforms to *γ*-ferrite at 1395 °C.

δ-**Ferrite**   It is an interstitial solid solution of carbon in *δ*-Fe with body-centred cubic structure. It is stable at only higher temperatures above 1394 °C. The maximum carbon solubility is about 0.09% at 1495 °C and exhibits paramagnetic behaviour.

Cementite (iron carbide, Fe_3C)   It is an interstitial intermetallic compound having 6.67% carbon content. It possesses a complex orthorhombic structure and exists as a compound form at room temperature for a long time. Of the all phases described above, cementite is the hardest microstructure and is brittle with low tensile strength. Cementite tends to decompose to *α*-ferrite and carbon (graphite) at temperatures in the range of 650–700 °C.

Pearlite   An eutectoid mixture of ferrite (88% wt) and iron carbide (12% wt), pearlite consists of about 0.8% carbon content and has a lamellar microstructure.

Ledeburite   An eutectic mixture of austenite and iron carbide, ledeburite contains 4.35% carbon having a melting point of 1147 °C.

There are three horizontal lines in the Fe–C phase diagram depicting isothermal phase reactions:

(a) At the top of the phase diagram, the first horizontal line at 1493 °C (low carbon content) denotes the peritectic reaction:

$$\text{Liquid} + \delta \longleftrightarrow \text{Austenite}$$

(b) The second horizontal line between 1130 to 1147 °C (4.3% carbon) denotes the eutectic reaction:

$$\text{Liquid} \longleftrightarrow \text{Austenite} + \text{iron carbide (cementite)}$$

(c) The last horizontal line at 723 °C (0.83% carbon) is where eutectoid reaction takes place:

$$\text{Liquid} \longleftrightarrow \text{Pearlite (ferrite} + \text{cementite)}$$

Areas and Curves

Curve ABCD is called the liquidus line where solidification begins. Above the liquidus line both iron and carbon are in liquid phase.

Curve AEFGCH is called solidus line indicating the end of solidification process. Below the solidus line both iron and carbon are in solid phase. On the solidus line, melting point of the Fe–C alloy decreases up to point 'C' with 4.3% wt of carbon at 1150 °C. On continuous addition of carbon, point 'H' is reached with highest carbon content of 6.67% weight, called iron carbide.

Area AEB comprises solid *δ*-Fe and liquid phase. Further, area DCH marks the phases of cementite with liquid phase. *γ*-Austenite and liquid phase are denoted by area BCGF. Area of IJKGF comprises austenite solid solution of carbon in *γ*-Fe.

4.7.6 Advantages, Limitations, and Applications of Phase Rule

Advantages

(a) Gibbs phase rule is a simple and convenient method of classifying the states of equilibrium. The systems having same number of degrees of freedom will behave in same manner under the influence of variations in temperature, pressure, and concentration.

(b) Phase rule is widely applicable to macroscopic systems and thus it is not necessary to have information about the molecular structure.

(c) It is also easily applicable to physical and chemical phase reactions.

(d) There is no need to formulate any assumptions with respect to the constituents of matter, as it only considers the phases present in dynamic equilibrium.

(e) It helps in deciding that under a certain set of conditions, the number of substances, if placed together will remain in equilibrium.

Limitations

(a) Phase rule is applicable only to heterogeneous systems and only when all the phases are in equilibrium.

(b) All the phases of the system must be present under the same pressure, temperature, and gravitational forces.

(c) It considers only the number of phases present for the given substance and does not relate to the concentration of substance present in each phase at equilibrium.

(d) While applying the phase equation to various systems, the variables chosen to study are temperature, pressure, and composition. Electric, magnetic, gravity, kinetic effects, surface tension, and other influences are completely excluded.

Applications

Electronic industries, alloy industries, medical science, pharmacy are some of the fields where phase rule finds its applications.

Safety plugs or fuses   These are made of an alloy of bismuth, lead, tin, and cadmium (50:25:12.5:12.5%) that has a melting temperature of 65°C. As it is a low melting alloy system, it can ensure safe and hazard-free working as a safety fuse in buildings.

Solder   It is an alloy whose melting point is lower than that of the corresponding metal pieces and is employed to manufacture electric wires. Due to high content of tin, it is not employed in many types of electrical equipment.

Freezing mixture   Often, routine organic laboratory syntheses require preparation of freezing mixtures, particularly during the crystallization step. When ice is mixed with salts, the freezing point of ice drops and can be utilized as freezing mixtures. An ideal freezing mixture is one that causes depression in freezing point and salts should have greater solubility with ice. Calcium chloride ($CaCl_2.6H_2O$) added in ice is an excellent freezing mixture with a cryohydric point of −55.9°C. Sodium chloride ($NaCl.2H_2O$) in ice is a cheaper alternative as a freezing mixture having a cryohydric point of −22.0°C.

Lyophilization   Freeze drying technique is commonly employed to preserve food products, microbial cultures, botanical samples, agricultural products, and pharmaceutical substances. It involves the removal of water from the sample to be preserved under vacuum conditions. The process involves deep-freezing all the samples that are to be preserved so that water content solidifies. Further, ice is evaporated through sublimation (ice to vapour) under vacuum of 1 mm of Hg without harming the integrity of the sample. Water eliminated from the process is recycled in condensers and reused. Hence, it is evident that knowledge of water phase diagram comes in handy in such cases.

Check Your Progress

19. State the application of lead–silver system (from phase diagram).
20. State the applications of phase rule.
21. Mention any two advantages of phase rule.

SOLVED EXAMPLES

1. If an alloy of bismuth (Bi) and cadmium (Cd) contains 25% Cd, determine the mass of eutectic in 1 kg of the alloy, if eutectic contains 40% Cd.

Solution: As alloy contains 25% Cd, hence remaining will be Bi; i.e., 75%.

In 1 kg alloy, Cd is found to be 40% ; i.e., 400 g.

Hence, mass of Bi in the eutectic is $\dfrac{750 \times 40}{60} = 500$ g

So, mass of eutectic in 1 kg alloy = 750 + 500 = 1250 g

Result: Mass of eutectic = 1250 g.

2. 100 kg molten alloy of Pb–Ag contains 0.3% silver that is cooled to form an eutectic mixture with 2.6% silver. Calculate the amount of silver that can be recovered from the eutectic mass and mass of lead that separates out.

Solution: Let us say, mass of Ag in 100 kg alloy is 0.3 kg.

Mass of Ag in the eutectic will be $\dfrac{0.3}{2.6} \times 100 = 11.538$ kg

Mass of Pb that will separate out = 100 − 11.538 = 88.462 kg

Result: Mass of Ag = 11.538 kg; mass of Pb = 88.462 kg

SUMMARY

○ Gibbs phase rule provides the fundamental concept to explain various heterogeneous systems in dynamic equilibrium. A heterogeneous system consists of two or more phases having different physical states or possibly different chemical compositions, or both.

○ Condensed phase rule can be used to study solid–liquid equilibria, as the pressure variable is ignored.

○ The phase diagrams of water, lead–silver, zinc–magnesium, sodium chloride–water, Fe–C and Bi–Pb–Sn systems have been discussed.

○ Once the concept of phase rule is understood, alloy systems can be studied for their characteristic properties and applications.

○ Various applications of phase rule include lyophilization, preparation of solders and safety plugs for electrical components. However, there are some limitations of phase rule as electric, magnetic, gravity, kinetic effects, surface tension, and other influence are completely excluded.

GLOSSARY

Alloy: An engineering material obtained by the solidification of a solution of two or more elements, of which one is essentially a metal.

Austenite: A metallic, non-magnetic form of iron that is softer than martensite.

Chemical potential: A form of energy that can be released or absorbed during phase transition.

Component: The minimum number of independently variable constituents by means of which composition of all the phases can be expressed either directly or by means of a chemical equation.

Congruent melting point: A point wherein both liquid melt and solid possess same composition co-existing together at constant temperature.

Degree of freedom: The number of variables such as temperature, pressure, and composition that must be specified in order to define a system at equilibrium.

Incongruent melting point: A point where solid does not melt uniformly.

Metastable water: Water that is supercooled to below 0°C in a water phase system.

Phase diagram: A graphical plot that depicts the different phases in a heterogeneous system at equilibrium.

Interstitial solid solution: Solute atoms occupying interstitial lattice spaces of solvent.

Phase: The homogeneous, physically distinct, and mechanically separable part of a heterogeneous system.

Ternary diagram: (or Gibbs triangle) is a triangular barycentric plot with three variables.

KEY FORMULAE

- Gibbs phase equation: $F = C - P + 2$

- Condensed phase rule equation: $F = C - P + 1$

EXERCISES

Multiple Choice Questions

1. In a one-component condensed system, if the degree of freedom is zero, the maximum number of phases that can co-exist will be
   - (a) 0
   - (b) 1
   - (c) 2
   - (d) 3

2. The number of phases in a heterogeneous system can be deduced using
   - (a) phase rule
   - (b) incongruent phase diagram
   - (c) both A and B
   - (d) none of them

3. The composition of an eutectic mixture of lead–silver system is
   - (a) 2.6 % Pb and 97.4 % Ag
   - (b) 2.6 % Ag and 97.4 % Pb
   - (c) 26 % Ag and 74 % Pb
   - (d) 2.5 % Ag and 96.5 % Pb

4. In water system, at triple point, F is found to be
   - (a) invariant
   - (b) bivariant
   - (c) trivariant
   - (d) none of these

5. The number of phases when ammonium chloride is heated in a closed container will be
   - (a) 2
   - (b) 1
   - (c) 3
   - (d) 4

6. The congruent melting point of Zn–Mg system is
   - (a) 590 °C
   - (b) 419 °C
   - (c) 650 °C
   - (d) –21.1 °C

7. How many phases will be present if two immiscible phases like benzene and water are mixed together?
   - (a) 1
   - (b) 3
   - (c) 4
   - (d) 5

8. The eutectic temperature of NaCl and water system is
   - (a) –21.1 °C
   - (b) –21.5 °C
   - (c) –21.05 °C
   - (d) +21 °C

9. In lead–silver system, the eutectic temperature is
   - (a) 273 °C
   - (b) 300 °C
   - (c) 310 °C
   - (d) 303 °C

10. The number of components present in $CaCO_{3(s)} \rightarrow CaO_{(s)} + CO_{2(g)}$ is
    - (a) 2
    - (b) 3
    - (c) 1
    - (d) 4

11. The process of preserving food products, microbial cultures, plants, and medicines is called
    - (a) sublimation
    - (b) melting
    - (c) freezing
    - (d) lyophilization

12. For water ⇔ vapour, the degrees of freedom will be
    - (a) 0
    - (b) 2
    - (c) 1
    - (d) 3

13. When pressure increases, melting point of ice will
    - (a) decrease
    - (b) increase
    - (c) remain unchanged
    - (d) reach equilibrium

14. The hardest structure that appears on Fe–C equilibrium is
    - (a) ferrite
    - (b) pearlite
    - (c) cementite
    - (d) ledeburite

15. The degree of freedom for Bi–Pb–Sn system is
    - (a) 3
    - (b) 4
    - (c) 2
    - (d) 1

16. The eutectic point for Bi–Pb–Sn system is
    - (a) 95 °C
    - (b) 96 °C
    - (c) 96.5 °C
    - (d) 95.5 °C

17. The carbon content in ledeburite is
    - (a) 4.3 %
    - (b) 4.5 %
    - (c) 4.35 %
    - (d) 4 %

18. Calcium chloride ($CaCl_2.6H_2O$)–ice freezing mixture has a cryohydric point of
    - (a) –55.9 °C
    - (b) – 22.0 °C
    - (c) –22 °C
    - (d) –55.6 °C

19. Zn–Mg system is an example of
    (a) incongruent melting point
    (b) congruent melting point
    (c) freezing mixture
    (d) none of these

20. In the sulphur system, a triple point has rhombic, monoclinic, and vapour forms of sulphur in dynamic equilibrium at
    (a) 95.6 °C
    (b) 114 °C
    (c) 120 °C
    (d) 150 °C

21. The structure of α-ferrite at room temperature is
    (a) hcc
    (b) bcc
    (c) fcc
    (d) linear

22. Gibbs phase rule is
    (a) $F = C - P + 3$
    (b) $F = P + C$
    (c) $F = C - P - 2$
    (d) $F = C - P + 2$

23. Freeze-drying is based on the principle of
    (a) condensation
    (b) sublimation
    (c) evaporation
    (d) fusion

24. A binary system that consists of two substances but are miscible in all proportions in liquid phase without chemically reacting is called
    (a) univariant system
    (b) invariant system
    (c) eutectic system
    (d) metastable system

25. Sulphur system is a
    (a) one-component system
    (b) two-component system
    (c) multi-component system
    (d) none of these

Review Questions

1. What is phase rule? Derive the phase rule equation.
2. State Gibbs phase rule. State the mathematical expression and explain the terms involved in it.
3. Explain the terms: (a) phase (b) degrees of freedom (c) components.
4. State the advantages and limitations of Gibbs phase rule.
5. Predict the number of phases in the following systems:
    (a) KCl dissolved in water
    (b) $Fe_{(s)} + H_2O_{(g)} \Leftrightarrow FeO_{(s)} + 4H_2O_{(g)}$
    (c) Solid I_2 in a closed container
6. Determine the number of components for the following systems: (a) Ammonia gas (b) Table salt in water (c) Solid ice melting in a closed jar
7. State the number of degrees of freedom for the following systems: (a) Nitrogen gas (b) Partially miscible liquids with vapours (c) Hydrated copper sulphate
8. What is triple point in water system? Explain its significance. Can four phases co-exist at equilibrium in a one-component system? Support your answer with reasons.
9. What is metastable equilibrium? Explain its significance.

10. With a suitable diagram, discuss the lead–silver system.
11. What is congruent melting point? With a suitable diagram, explain *any one* system with a congruent melting point.
12. Describe the characteristics of Pb–Ag system and explain Pattinson's process.
13. What is incongruent melting point? With a suitable diagram, explain *any one* system with an incongruent melting point.
14. What is condensed phase rule? Derive the condensed phase rule equation.
15. Explain iron–carbon phase diagram mentioning the phases observed in the system. Add a note on its significance.
16. What is a one-component system? Explain sulphur system with a suitable phase diagram.
17. What are multicomponent systems? Explain any one in detail.
18. Discuss Bi–Pb–Sn system using a ternary phase diagram.
19. Write a short note on 'congruent and incongruent systems'.
20. Write a short note on 'multicomponent system' with an example.

Numerical Problems

1. If an alloy of lead (Pb) and tin (Sn) contains 73% Sn, determine the mass of eutectic in 1 kg of the alloy, if eutectic contains 64% Sn. **(Ans.** 750 g)

2. 100 kg molten alloy of Pb–Ag contains 0.28% Ag that is cooled to form an eutectic mixture with 2.6% Ag. Calculate the amount of silver that can be recovered from the eutectic mass and mass of lead (Pb) that separates out. **(Ans.** mass of Ag = 10.76 kg and mass of Pb = 89.24 kg).

FURTHER READING

1. Alper, J.S., 'The Gibbs Phase Rule Revisited: Interrelationships between Components and Phases', J. *Chem. Educ,* 76 (11), p 1567, 1999.

2. Bauccio, M. L. (ed.) *ASM Metals Reference Book.* ASM International, 1993.

3. Mortimer, Robert, *Physical Chemistry,* Academic Press, 2000.

4. Predel, Bruno, Michael J. R. Hoch, Monte Pool, *Phase Diagrams and Heterogeneous Equilibria: A Practical Introduction.* Springer, 2013.

5. Reisman, Arnold, *Phase Equilibria: Basic Principles, Applications, Experimental Techniques,* Academic Press, 1970.

ANSWERS

Check Your Progress

1. $CaCO_3$ on decomposition will form solid calcium oxide and release carbon dioxide.

   $CaCO_3 \text{ (s)} \rightleftharpoons CaO \text{ (s)} + CO_2 \text{ (g)}$

   The number of phases is 3–2 solid phases of $CaCO_3$ and CaO and 1 CO_2 vapour phase.

2. The reaction $NH_4Cl_{(s)} \rightleftharpoons NH_{3 \text{ (g)}} + HCl_{\text{ (g)}}$ depicts the decomposition of ammonium chloride, forming ammonia and hydrogen chloride gases. So, there is one solid phase of NH_4Cl and two gaseous phases of ammonia and HCl; thus P = 2.

3. Refer to Section 4.3.2

4. Refer to Section 4.2

5. Refer to Section 4.3

6. In a system of benzene and water, there are two phases, as benzene and water are immiscible.

7. An example of an invariant system is Water system: solid $\rightleftharpoons$ liquid $\rightleftharpoons$ vapour which is invariant.

8. The water system has one triple point.

9. Refer to Section 4.5

10. There is an equilibrium along the transition curve of rhombic and monoclinic sulphur. The densities of rhombic sulphur and monoclinic sulphur are 2.07 and 1.95 (g/cm^3) respectively. Hence, the conversion from rhombic to monoclinic form is accompanied with increase in volume and pressure which causes rise in transition temperature. Thus, it is justified that transition curve is inclined towards the right side with a positive slope.

11. Sulphur system is an example of a one-component system that has more than one solid phase. Refer to Fig. 4.2.

12. The number of components in both aq. acetic acid and magnesium carbonate in equilibrium with its decomposition products is 2.

13.

Eutectic point	Peritectic point
It is a point in the phase diagram that corresponds to definite composition and melting temperature of an eutectic alloy. The reaction can be depicted as, $$\text{Liquid} \underset{\text{cooling}}{\overset{\Delta}{\rightleftharpoons}} \text{Solid (1)} + \text{Solid (2)}$$	It is a point in the phase diagram at which incongruent melting compound decomposes below its melting point to form a new stable solid compound. $$\text{Solid (1)} \overset{\Delta}{\longrightarrow} \text{Solid (2)} + \text{melt}$$ where, solid (1) is unstable and solid (2) is stable form.
Triple point	Critical point
All the phases coexist in dynamic equilibrium with F = 0.	This is the point beyond which liquid and vapour phase merge into a single phase with F = 2.

14. When an eutectic mixture reaches the eutectic temperature, it undergoes a reversible eutectic reaction as,

$$\text{Liquid} \underset{\text{cooling}}{\overset{\Delta}{\rightleftharpoons}} \text{Solid (1)} + \text{Solid (2)}$$

Hence, the above reaction occurs at a definite temperature and exhibits a sharp melting point.

15. According to the condensed phase rule, in a two-component system, the effect of pressure is very small and is disregarded and arbitrary constant pressure at 1 atm is considered. In this case, $F = C - P + 1$.

16. The eutectic composition and eutectic temperature of lead–silver system is 97.4% Pb and 2.6% silver at 303 °C.

17. Yes, an invariant system has no degree of freedom.

18. Freezing mixture is prepared by adding two or more substances that allow obtaining temperatures below freezing point of water. Salt–ice is an example. Such mixtures are commonly employed to cool suspensions to form crystals of organic compounds in the laboratory.

19. Desilverization of lead by Pattinson's process to obtain lead with 2.6% silver.

20. The phase rule is used to understand the composition of safety plugs, solders, and preserve foodstuff and medicines by lyophilization.

21. Refer to Section 4.7.6

Multiple Choice Questions

1. (d)	2. (a)	3. (b)	4. (a)	5. (a)	6. (a)	7. (b)
8. (a)	9. (d)	10. (a)	11. (d)	12. (c)	13. (a),	14. (c)
15. (b)	16. (b)	17. (c)	18. (a)	19. (b)	20. (a)	21. (b)
22. (d)	23. (b)	24. (c)	25. (a)			

Electrochemistry

After reading this chapter, you will be able to:

- define electrode potential and standard electrode potential.
- explain the construction and working of various electrodes, such as calomel, quinhydrone, and glass electrodes.
- derive Nernst equation and determine pH using glass and quinhydrone electrodes.
- understand the electrochemical series and its applications in emf calculations.
- understand the principle of potentiometric titrations and their various graphical representations.
- elucidate battery technologies of lithium cells, lead–acid battery, and lithium–ion cells.
- explain the concepts of electroplating and electroless plating of nickel.

5.1 INTRODUCTION

In the late 17th century, Sir Humphry Davy studied the electrolysis of salts in water and laid the foundation of electrochemistry. L.A. Galvani (1800) demonstrated the effects of electricity on various life forms and pioneered his work in bioelectromagnetics. In 1800, Alessandro Volta invented the electrochemical battery comprising zinc and copper electrodes in sulphuric acid solution. Michael Faraday (1832) postulated the laws of electrolysis. Svante Arrhenius (1884) studied the conductivities of various electrolytes that earned him Nobel Prize (1903) in chemistry. Walther Nernst (1887) correlated the principles of electrochemistry and thermodynamics and derived the famous 'Nernst equation.' Today, electrochemistry has led to numerous developments, especially in battery technology.

Understanding electrochemical cell design helps in understanding the classical concepts of corrosion theory. Energy storage devices such as batteries are inconceivable without understanding the electrochemical concepts. This topic finds direct correlation to electronic engineering and its allied fields. Electrochemistry involves the study of reactions that involve ionic or electron transfer that is associated with oxidation and reduction reactions. The measurement of electromotive force provides a quantitative measure of these electron-transfer reactions occurring near the electrodes. An electrochemical cell usually consists of a set-up comprising two electrodes placed (or dipped) in its own solution. In redox reactions, free energy is released due to the movement of electrons that gives rise to potential difference called the electromotive force (emf) of the cell.

Michael Faraday (1791–1867) was a British physicist and chemist and a renowned experimental scientist. The son of a blacksmith, Faraday was self-educated and his experiments on electricity and magnetism led to the invention of electric motor and generator. Faraday discovered the phenomenon of electrolysis and laid the foundation of electrochemistry. The specialized terms such as electrode, anode, cathode, and others were coined by him. He is also credited with the discovery of benzene.

5.2 ELECTROCHEMICAL CELL

A device that produces electric current from a chemical reaction (redox reaction) is called an *electrochemical cell*. In an electrochemical cell, chemical energy formed during a redox reaction is converted into electrical energy. Generally, electrochemical cells comprise two dissimilar electrodes immersed in a conducting ionic solution (Fig. 5.1). The electrodes are further connected by an external metallic conductor and each electrode/electrolyte system is termed as a *half cell*.

Electrochemical cells are of four types: galvanic cell, electrolytic cell, concentration cell, and fuel cell.

Galvanic or voltaic cells These are typical electrochemical cells in which dissimilar electrodes are immersed in one type of electrolyte and the chemical reactions result in electrical energy. Dry cells, Ni–Cd battery, lead–acid battery are some examples of galvanic cells.

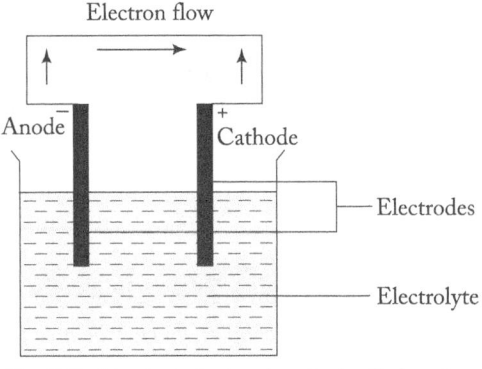

Fig. 5.1 A typical electrochemical cell showing two electrodes in an electrolyte

As shown in Fig. 5.2, a typical Daniell cell is a Zn–Cu couple. In a Daniell cell, zinc (anode) electrode is placed in 1M zinc sulphate solution. Copper electrode is immersed in 1M copper sulphate solution. As per electrochemical convention, zinc half cell is placed on the left and copper half cell is placed on the right side. Both zinc and copper electrodes are connected to a voltmeter or a potentiometer. The two half cells are interconnected using a salt bridge which contains potassium nitrate–agar gel. On setting up the cell, the following reactions take place:

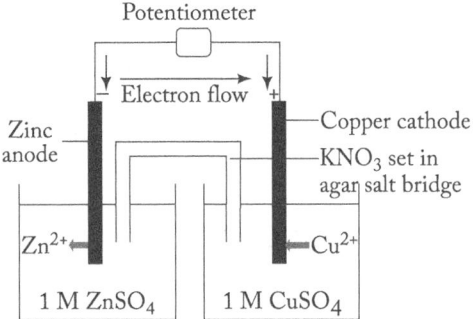

Fig. 5.2 Daniell cell with 1M $ZnSO_4$ and 1M $CuSO_4$ connected with a KNO_3 agar salt bridge

Anode (Zn half cell): $Zn \rightarrow Zn^{2+} + 2e^-$

Cathode (Cu half cell): $Cu^{2+} + 2e^- \rightarrow Cu$

Overall cell reaction: $Zn + Cu^{2+} \rightarrow Zn^{2+} + Cu$

During the working of the cell, electrons flow from the zinc electrode towards the copper electrode. There is solvation of zinc electrode, that is, zinc dissolves into the zinc sulphate solution and forms Zn^{2+} ions. The electrons are taken up by Cu^{2+} ions and form copper atoms at the surface of the cathode. Further, sulphate ions from the cathode half cell move towards the anode and zinc ions move towards the cathode half cell through the salt bridge. The movement of ions from each half cell completes the electrical circuit and thus there is continuous supply of electric current. Over a period of time, zinc electrode and copper ions get completely exhausted and the cell will cease to function.

Galvanic cells can be of chemical type or concentration type. A *chemical cell* produces electricity due to the overall chemical reactions taking place in it. Chemical cells could be primary or secondary. In *primary cells*, the chemical reactions are not reversible and hence cannot be recharged. Dry cell is an example of a primary cell. *Secondary cells* can be recharged due to reversibility of their chemical reactions. Lead–acid battery is an example of such cells.

Concentration cell is a typical electrochemical cell, where electrical energy is generated when electrodes of the same metal are immersed in a solution of its ions at different concentrations (or activity). In these cells, electrode reactions generate electrical energy.

Contrary to a chemical cell, concentration cells produce electricity due to decrease in free energy that occurs due to transfer of matter from one half cell to the other. This means that the nature of the two electrodes will remain the same, but the concentration of the solutions in the two half cells will differ, or vice versa. Concentration cells are of two types, namely *concentration cells without transference* (indirect transfer of material) and *concentration cells with transfer* (direct transfer of material).

Electrolytic cell   It is a device that uses electrical energy from an external source to bring about chemical reactions. It forms the basis of electroplating process discussed later in the chapter.

Fuel cell   It is an electrochemical device that efficiently converts chemical energy of fuels directly in to electricity that can be utilized for power generation. Today, researchers focus on fuel cell technology to provide environmentally sustainable power generation in place on fossil fuels.

5.2.1 Reversible and Irreversible Cells

The reversibility of an electrochemical cell depends on the electrodes involved in the reaction. When an external emf applied to a cell is more than its emf (E_{cell}), the cell reactions get reversed and current flows in the opposite direction. In a Daniell cell, having a theoretical potential of 1.1 V, if one applied an external emf of the same value, the cell stops working. However, if an emf higher than 1.1 V is applied, the cell reactions in the Daniell cell will be reversed. Such a type of cell is called a reversible cell.

In irreversible electrochemical cells, reactions cannot be reversed by providing a higher external emf than the actual emf. Dry cell is an example of irreversible cell that is discussed later in the chapter.

5.2.2 Electrode Potential

Electrode potential (E) is defined as the potential developed at the interface of an electrode and the solution of its own salt. In simpler terms, it depicts the measure of tendency of an electrode to gain or lose electrons when brought in contact with a solution of its own salt. Experimentally, it is not possible to determine the potential of a single electrode and hence, two electrodes with slight differences in their electrode potentials are taken as, $E°_{anode} < E°_{cathode}$. A single electrode potential ($E°$) develops at the interface between the electrode and the conducting electrolyte solution. However,

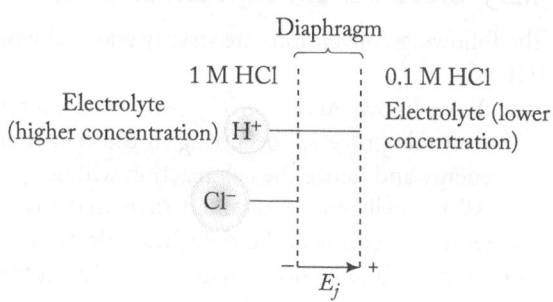

Fig. 5.3 Illustration of liquid junction potential. Ions move at different rates across the diaphragm resulting in the separation of charges and creating a potential difference.

the situation changes drastically when two dissimilar electrodes are brought in contact by immersing in an electrolyte; the electrode with higher reduction potential (i.e., tendency of gaining electrons) acts as the cathode and the other acts as the anode. The electrons are forced to move from anode to cathode and a dynamic equilibrium is established at both the half cells at the electrode interfaces. There is a constant interaction of opposite charges at both the electrodes, resulting in the formation of an *electrical double layer*. Hence, there is always a certain potential between the interface of the electrode and its ionic solution at a given temperature. This is called the electrode potential. *Electrode potential is defined as the tendency of an electrode to gain or lose electrons when placed in a solution containing its own ions.* If an electrode tends to lose electrons it is called *oxidation potential*, whereas if an electrode tends to gain electrons, it is called *reduction potential*.

Consider an electrochemical cell with two different electrolytes in contact with each other (just like in Daniell cell); in this case, an additional potential difference is present across the electrolytic interface.

This is called *liquid junction potential* (denoted as E_j). For example, in two differing concentrations of hydrochloric acid in an electrochemical cell, the highly mobile hydrogen ions will diffuse in to the dilute solution, followed by the bulkier chlorine ions until equilibrium is attained. This causes junction potential to set at the interface giving a potential value of about 2 mV. In other words, differing mobilities of ions across the electrolyte cause junction potential as shown in Fig. 5.3. A salt bridge is placed to eliminate or reduce the junction potential across the electrolyte interface.

Standard electrode potential ($E°$) for half cells can be measured using a reference electrode such as standard hydrogen electrode (SHE) or standard calomel electrode (SCE). *Standard electrode potential* is defined as *the potential of an electrode measured in 1 M metal ion concentration at 298 K* and is expressed as, $E°_{M_n^+/M}$ for two half cells.

Let us relate electromotive force of an electrochemical cell with some thermodynamic parameters.

The maximum amount of work that is produced by an electrochemical cell (w_{max}) is related to cell potential (E_{cell}) and the total charge transferred during the reaction (nF) as follows:

$$w_{max} = nFE_{cell}$$

As work is performed by the electrochemical cell, it is expressed as a negative value. Further, the free energy change (ΔG) is also a measure of maximum amount of work performed in a chemical reaction; hence, $\Delta G = w_{max}$.

Hence, there exists a relation between electrode potential and free energy change as, $\Delta G = -nFE_{cell}$.

Under standard conditions, the above equation becomes $\Delta G° = -nFE°_{cell}$.

This equation represents a spontaneous redox reaction with a negative free energy value and positive value for cell potential.

5.2.3 Electrochemical Conventions

The following conventions are strictly adopted while studying electrochemical cells as recommended by IUPAC.

(a) As we know, $\Delta G = -nFE_{cell}$, electrical energy is produced due to decrease in free energy of redox electrode processes occurring in the cell. If emf of the cell is positive, it indicates decrease in free energy and hence the cell reaction will be spontaneous ($E°_{cathode} > E°_{anode}$).

(b) All the cells are represented such that oxidation takes place at the left hand electrode, whereas reduction occurs at the right hand electrode.

(c) In case of a reduction reaction occurring at the cathode, the sign of its electrode potential is denoted as positive; whereas if reduction reaction occurs at the anode, electrode potential will be denoted as negative.

(d) The net cell reaction is an algebraic sum of the chemical reactions occurring in the two half cells (or electrodes).

(e) While representing a cell, if oxidation occurs at the anode, it is written as $M|M^{n+}_{(C1)}$; for example, $Zn|Zn^{2+}_{(C1)}$. The reduction reaction occurring at the cathode is written as $M^{n+}_{(C1)}|Cu$ that is, $Cu^{2+}_{(C2)}|Cu$. A single vertical line indicates interface between metal electrode and its ions. Double vertical lines indicate a salt bridge: $Zn| Zn^{2+}_{(C1)}||Cu^{2+}_{(C2)}|Cu$.

(f) The standard emf of a cell is the algebraic difference between reduction potentials of the cathode and the anode.

5.3 ELECTRODES OR HALF CELLS

While constructing an electrochemical cell, choosing the correct electrode material or half cell is an essential requirement. Depending upon the purpose of electrochemical work, various types of electrodes are employed. Some of the important electrodes discussed here are hydrogen, calomel, quinhydrone, and glass electrodes.

5.3.1 Hydrogen Electrode

Normally employed as a primary reference electrode, a standard hydrogen electrode (SHE) has potentials on a hydrogen scale. The standard electrode potential of any electrode is obtained by coupling it with SHE and measuring the emf of the cell.

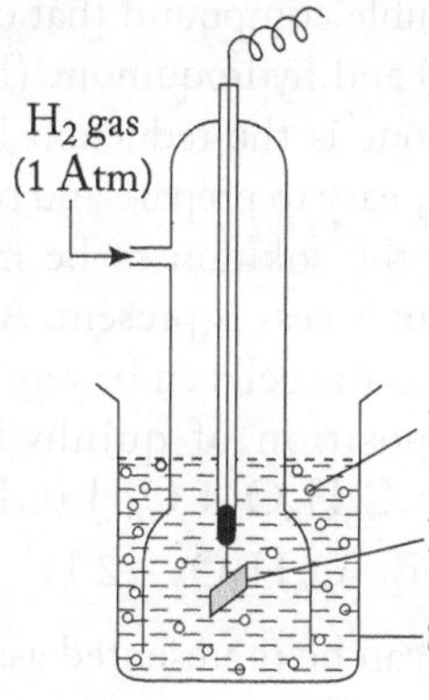

Construction   As shown in Fig 5.4, pure hydrogen gas at 1 Atm is bubbled through a solution of hydrogen ions at 1M concentration. A platinum electrode coated with platinum black is immersed in the hydrogen ion solution. A platinum wire attached to the platinum electrode completes the electrical connections.

Working   A standard hydrogen electrode is represented as, $Pt\,|\,H_{2(g)}\,1\,Atm, H^+(C = 1\,M)$. On coupling SHE with another electrode whose electrode potential needs to be determined can be represented as follows,

$$Pt\,|\,H_{2(g)}\,1\,Atm, H^+(C = 1\,M)\,\|\,M^{n+}(C = 1M)\,|\,M$$

The electrode potential of SHE is zero and the equilibrium reaction at hydrogen electrode is,

$$H^+ + e^- \rightleftharpoons 1/2H_{2(g)}$$

The equilibrium reaction can be depicted as:

$$E = E^\circ + 0.0592 \log_{10}[H^+]$$

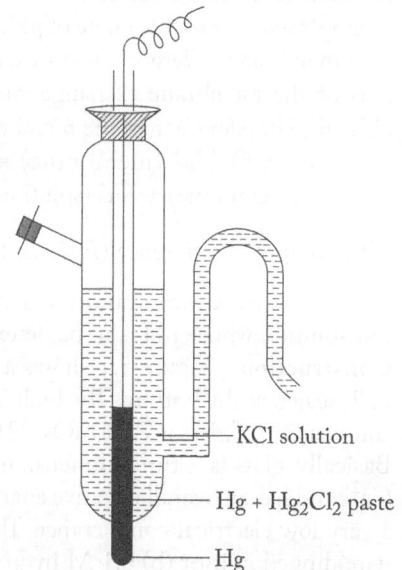

Fig. 5.4 A typical standard hydrogen electrode

5.3.2 Calomel Electrode

For routine emf measurements, SHE is not used as platinum wire tends to easily get poisoned in the presence of trace impurities in hydrogen gas. Also, maintaining hydrogen pressures at 1 Atm makes it difficult to handle. Hence, a secondary reference calomel electrode is usually employed for the purpose.

Construction   A typical calomel electrode is a narrow glass tube containing mercury in contact with a solution of potassium chloride saturated with a paste of mercurous chloride and mercury as shown in Fig. 5.5.

A clean platinum wire is dipped in the mercury layer for electrical contact. The calomel electrode can be represented as, $Hg\,|\,Hg_2Cl_2\,|\,Cl^-_{(sat)}$ and its electrode potential is 0.2422 V. The reduction reaction is, $Hg_2Cl_2 + 2e^- \rightarrow 2Hg + 2Cl^-$. The reduction potential can be written as:

$$E_{SCE} = E^\circ_{SCE} - 0.0592\,\log_{10}[Cl^-]$$

Calomel electrode cannot be used above 50 °C as mercurous chloride decomposes and results in erratic emf readings. Due to toxicity of mercury, it is hazardous to handle the electrodes if there is a leakage.

Fig. 5.5 A typical saturated calomel electrode

5.3.3 Quinhydrone Electrode

It is a secondary reference electrode which consists of an equimolar mixture of quinone and hydroquinone (Fig. 5.6). Quinhydrone (QH) is a slightly soluble compound that consists of equimolar amounts of quinone (Q) and hydroquinone (H$_2$Q). Quinone is the oxidant and hydroquinone is the reductant in this reaction. Quinhydrone electrode is very easy to prepare and handle. Pure solid quinhydrone is dissolved in the solution to be measured until the solution is saturated and an excess is present. A platinum wire is dipped into this solution and the cell can be represented as: Pt | H$_2$Q, Q

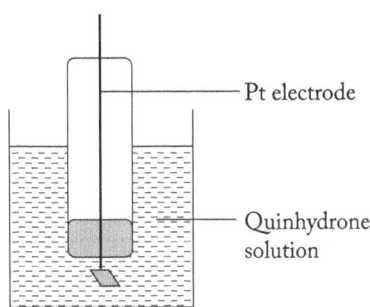

Fig. 5.6 Quinhydrone electrode

The decomposition of quinhydrone is given as: C$_6$H$_4$O$_2$.
$$C_6H_4(OH)_2 \rightarrow C_6H_4O_2 + C_6H_4(OH)_2$$
The reaction is: $C_6H_4O_2 + 2\,H^+ + 2e^- \rightarrow C_6H_4(OH)_2$

The half cell can be represented as: $Pt(s)|H_2Q,Q,H^+(aq)$ and its reaction will be,
$$Q + 2\,H^+ + 2e^- \rightarrow H_2Q$$

5.3.4 Glass Electrode

Principle   Glass electrode is reversible to hydrogen ions. The pH of aqueous solution depends on hydrogen ion concentration and is measured using a glass electrode. When a glass electrode is immersed in a solution (unknown pH), a potential difference develops across its ion-selective membrane which is proportional to the hydrogen ion concentration. The measured potential is highly accurate since glass electrode remains unaffected by oxidizing or reducing agents and can be employed over a wide range of pH. During pH determination, the glass membrane undergoes an ion-exchange reaction in which sodium ions of the membrane exchange with hydrogen ions. A potential difference develops across the membrane and can be denoted as,

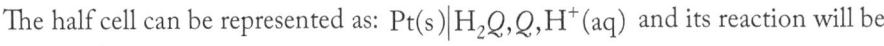

Fig. 5.7 A typical glass membrane electrode

--SiO-Na$^+$ (membrane) + H$^+$ (solution) → SiO-H$^+$ (membrane) + Na$^+$ (solution)

The potential difference (E_G) can be written as, $E_G = \dfrac{RT}{nF} \ln \dfrac{C_2}{C_1}$

Here, C_1 is the concentration of acid solution present in the glass bulb and C_2 is the concentration of test solution whose pH is to be determined.

Construction   Figure 5.7 shows a typical glass electrode that consists of a glass tube that ends in a bulb sealed at the bottom. The bulb is also called the glass membrane, which is as thin as ~0.05 mm. The composition of glass is 72 % SiO$_2$, 22 % Na$_2$O, and 6 % CaO and it possesses higher electrical conductance. Basically, glass is a three-dimensional network of Si–O and certain cations such as Li$^+$, Na$^+$, K+, and Ca^{2+} and has an overall negative charge. The movement of these cations inside the Si–O network imparts a very low electrical conductance. The bulb contains either (a) buffer solution of pH ~4 with platinum wire dipped in it or (b) 0.1 M hydrochloric acid with Ag wire coated with AgCl dipped in it. Keeping potentials of the glass electrode, dipped wires, and the glass membrane constant, when such an electrode is dipped in a test solution, only hydrogen ion activity will affect the potential of the glass electrode.

5.4 NERNST EQUATION

In an electrochemical cell, the electrode potential depends on the nature of the electrode, concentration of the electrolyte, and temperature. Nernst derived a fundamental equation relating the free energy change

Walther Hermann Nernst (1864–1941) was a German chemist who developed Nernst equation in 1887. The Nernst equation forms the basis of electrochemistry. He has immensely contributed to the fields of thermodynamics, electrochemistry, solid state, and photochemistry. Nernst formulated a heat theorem called 'Nernst Heat Theorem' that led to the third law of thermodynamics for which he was conferred with Nobel Prize in 1920.

of any electrode processes to the potential developed at the electrode along with the concentration of ions in the solution and its temperature.

Let us consider a reduction reaction occurring in an electrochemical cell,

$$M^{n+} + ne^- \rightleftharpoons M$$

As electrochemical cell is doing work, there will be decrease in free energy for the reaction. The decrease in free energy change ($-\Delta G$) for the above reaction can be expressed as,

$$\Delta G = \Delta G^\circ + 2.303 \, RT \log_{10} \frac{[M]}{[M^{n+}]} \tag{5.1}$$

We already know that under standard conditions, $\Delta G^\circ = -nFE^\circ$, where E° denotes standard electrode potential (in volts), F denotes the quantity of electricity produced (in Coulombs), and n is the total number of electrons liberated at the single electrode.

Hence, Eq.(5.1) can be written as,

$$-nFE = -nFE^\circ + 2.303RT \log_{10} \left[\frac{1}{M^{n+}} \right] \tag{5.2}$$

Dividing Eq. (5.2) by $-nF$, we get

$$E = E^\circ - \frac{2.303 \, RT}{nF} \log_{10} \left[\frac{1}{M^{n+}} \right] \tag{5.3}$$

On rearranging Eq (5.3), we get

$$E = E^\circ + \frac{2.303 \, RT}{nF} \log_{10} [M^{n+}] \tag{5.4}$$

Equation (5.4) represents the Nernst equation for single electrode potential
On solving further, we get

$$E = E^\circ + \frac{0.0591}{n} \log_{10} [M^{n+}] \dots \text{(at cathode)} \tag{5.5}$$

$$E = E^\circ - \frac{0.0591}{n} \log_{10} [M^{n+}] \dots \text{(at anode)} \tag{5.6}$$

Equations (5.5) and (5.6) are generally used for practical purposes.

5.4.1 EMF of a Cell

Let us consider a cell that is represented as, $M_1 | M_2^{n+}(C_1) M^{n+}(C_2) | M_2$, where C_1 and C_2 are the concentration of the two half cells and $E^\circ_{\text{cathode}} > E^\circ_{\text{anode}}$.

The net reaction can be expressed as, $M_1 + M_2^{n+} \rightleftharpoons M_2 + M_1^{n+}$; where M_1^{n+} denotes oxidized state of metal M_1 and M_2^{n+} denotes oxidized state of metal M_2.

Hence, emf of the cell will be, $E_{\text{cell}} = E^\circ_{\text{cell}} - \frac{2.303 \, RT}{nF} \log_{10} \left[\frac{M_1^{n+}}{M_2^{n+}} \right]$, where

$$E^\circ_{\text{cell}} = E^\circ_{M_2^{n+}/M_2} - E^\circ_{M_1^{n+}/M_2} = E^\circ_{\text{cathode}} - E^\circ_{\text{anode}}$$

5.4.2 Determination of pH Using Hydrogen Electrode

Though rarely used in pH determination, hydrogen electrode is considered as the standard reference for most electrochemical analysis. A cell is set up by combining hydrogen electrode with SCE and can be represented as: $Pt|H_2$ (g, 1 atm)|test solution|SCE.

The emf of the above cell will be calculated as: $E_{cell} = E_{SCE} - E_{H_2}$

Further, we know that $H^+ + e- \Leftrightarrow \frac{1}{2} H_2$

$$E_{H_2} = E^\circ_{H_2} + 0.0592 \log_{10} a_{H^+} = -0.0592 pH$$

Hence, $E_{cell} = E_{SCE} + 0.0592$ pH

On rearranging the above expression, we get, $pH = \dfrac{E_{cell} - E_{SCE}}{0.0592}$.

5.4.3 Determination of pH Using Quinhydrone Electrode

E. Billmann (1921) introduced quinhydrone electrode to determine the pH of various solutions.

The potential of quinhydrone electrode is obtained using Nernst equation as follows:

$$E_{QH} = E^\circ_{QH} - \frac{0.0592}{2} \log_{10} \left[\frac{a_{H_2Q}}{a_Q \times a^2_{H^+}} \right] = E^\circ_{QH} - \frac{0.0592}{2} \log_{10} \left[\frac{a_{H_2Q}}{a_Q} \right] + \frac{0.0592}{2} \log_{10} a^2_{H^+}$$

We know that quinone and hydroquinone are present in equimolar amounts; hence $a_Q = a_{H_2Q}$.

$$E_{QH} = E^\circ_{QH} + 0.0592 \log_{10} a_{H^+} \tag{5.7}$$

The standard electrode potential of quinhydrone is $+ 0.6994$ V; hence, Eq. (5.7) becomes,

$$E_{QH} = + 0.6994 + 0.0592 \log_{10} a_{H^+} \tag{5.8}$$

Equation (5.8) clearly shows that potential of quinhydrone electrode changes with hydrogen ion activity. Thus, when pH of the solution is to be determined, a cell is set up by combining quinhydrone electrode with saturated calomel electrode and can be represented as:

$$(-)SCE \parallel \text{test solution saturated with } QH|Pt(+)$$

The emf of the above cell can be calculated as follows:

$$E_{cell} = E_{QH} - E_{SCE}$$

$$E_{SCE} = + 0.242 \text{ V and } E_{QH} = \left(0.6994 + 0.0592 \log_{10} a_{H^+} \right)$$

$$E_{cell} = 0.6994 + 0.0592 \log_{10} a_{H^+} - 0.242$$

We know that, $pH = - \log_{10} a_{H^+}$

$\therefore \qquad E_{cell} = 0.4574 + 0.0592 \log_{10} a_{H^+} = 0.4574 - 0.0592$ pH

Finally, one can calculate pH of the test solution by rearranging the above equation,

$$pH = \frac{0.4574 - E_{cell}}{0.0592} \tag{5.9}$$

5.4.4 Determination of pH Using Glass Electrode

The most commonly employed electrode for pH determination is the glass electrode since it is not affected by oxidizing or reducing agents, works well over a wide range of pH (0 – 9), and is simple to operate. We know that the glass membrane on contact with test solution will give rise to a potential difference that is dependent on pH and can be expressed using Nernst equation as,

$$E_G = E^\circ_G + 0.0592 \log_{10} a_{H^+} \tag{5.10}$$

For determining the standard electrode potential of glass electrode, one can measure the emf of known pH (usually buffers such as potassium hydrogen phthalate) coupled with a reference electrode and using Eq. (5.10) can be reframed as

$$E_G = E_G^\circ - 0.0592\, pH \tag{5.11}$$

During pH determination, if the glass electrode is combined with SCE, the cell set-up can be represented as: $Hg|Hg_2Cl_2|Cl^-||$ test solution $H^+|$glass electrode

It is observed that the electrode potential of glass electrode is higher than that of the reference calomel electrode ($E_G > E_{SCE}$). If E_G° and E_{SCE} are known, pH using glass electrode can be calculated as:

$$E_{cell} = E_{cathode} - E_{anode} = E_G - E_{SCE}$$

$$= [E_G^\circ - 0.0592\, pH] - E_{SCE}$$

Here, $E_G^\circ = [K + E_{Ag/AgCl} + E_{as}]$ and E_{as} = small potential from glass electrode.

$$\therefore\ 0.0592\, pH = E_G^\circ - E_{SCE} - E_{cell}$$

On rearranging the above expression, we get, $pH = \dfrac{E_G^\circ - E_{SCE} - E_{cell}}{0.0592}$.

Alternatively, glass electrodes can also be coupled with $Ag/AgCl/Cl^-$ dipped in 0.1 M HCl solution as an electrode system and the cell can be depicted as:

$Ag(s)|AgCl(s)|Cl^-(aq)||\ H+\ inside||\ H+\ outside|Cl^-(aq)|AgCl(s)|Ag(s).$

The expression for calculating pH will be $pH = \dfrac{E_G^\circ - E_{Ag/AgCl} - E_{cell}}{0.0592}$.

5.5 CONCENTRATION CELLS

In a concentration cell, electrical energy is generated when electrodes of the same metal are immersed in a solution of its ions at different concentrations (or activities) (Fig. 5.8). It is observed that there is a decrease in the free energy accompanying the transfer of materials from one electrode to the other electrode due to differing concentrations. Electrode concentration cells and electrolyte concentration cells are two common types of cells encountered which may be reversible either to cations or anions.

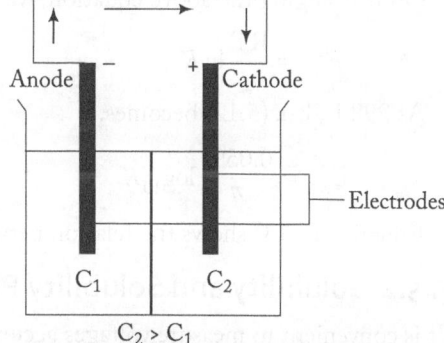

Fig. 5.8 A typical concentration cell with differing electrolyte concentrations

Consider the following electrolyte concentration cell

$$Ag\ |\ AgNO_3\ (C_1)\ ||\ AgNO_3\ (C_2)\ |\ Ag$$

When current is passed through the cell, the following cell reactions take place:

$$Ag - e^- \rightleftharpoons Ag^+(C_1)\ and\ Ag^+(C_2) + e^- \rightleftharpoons Ag^\circ$$

In the above case, concentration of Ag^+ in cell 1 is less than that in cell 2 and the number of electrons (n) is unity. The emf of the cell depends on the relative concentrations of the electrolyte and hence can be given as, $E = \dfrac{RT}{F}\ln\dfrac{C_2}{C_1}$

Consider an electrode concentration cell which is reversible to cations, such as

$$Pt\ |\ H_{2\,(g)}\ (P_1\ Atm)\ |\ HCl\ (aq)\ |\ H_2\ (P_2\ Atm)\ |\ Pt$$

In the above cell, P_1 is greater than P_2 which depicts unequal pressures of hydrogen gas; the cell reactions can be represented as follows:

$1/2\ H_2(P_1) \rightleftharpoons H^+ + e^-$ and $H^+ + e^- \rightleftharpoons 1/2\ H_2(P_2)$. Thus, the overall reaction of the cell will be,

$$1/2\ H_2(P_1) \rightleftharpoons 1/2\ H_2(P_2)$$

The above cell is called *gas concentration cell* and the cell reaction during the passage of 1 faraday current transfers 0.5 mole of hydrogen gas from pressure P_1 to P_2 and emf of the cell is written as,

$$E = \frac{RT}{2F} \ln \frac{P_1}{P_2}$$

However, in case electrode concentration cell is reversible to anions with spontaneous cell reaction where $P_2 > P_1$, the emf of the cell will be represented as,

$$E = \frac{RT}{2F} \ln \frac{P_2}{P_1}$$

If P_2 is equal to P_1, the concentration cell does not generate electrical energy, if $\ln \frac{P_2}{P_1}$ is positive, E_{cell} will be positive and higher the ratio of $\frac{P_2}{P_1}$, higher will be its E_{cell} value.

5.5.1 Equilibrium Constant for Cell Reactions

It is essential to determine the equilibrium constants of a cell reaction. Using emf measurements and thermodynamics, it is possible to calculate equilibrium constants for any cell reaction.

We know that equilibrium constant is related to standard free energy change of a cell reaction as:

$$\Delta G^\circ = -RT \ln K$$

and, $\Delta G^\circ = -nFE^\circ_{cell}$

On rearranging the above equation, we get

$$\therefore \qquad E^\circ_{cell} = \frac{RT}{nF} \ln K \qquad\qquad\qquad (5.12)$$

At 298 K, Eq. (5.12) becomes,

$$E^\circ_{cell} = \frac{0.0591}{n} \log_{10} K \qquad\qquad\qquad (5.13)$$

Equation (5.13) shows the relation between emf of a cell reaction with equilibrium constant.

5.5.2 Solubility and Solubility Product

It is convenient to measure voltages accurately using a voltmeter; thus electrochemical methods provide a convenient way to determine the solubility products (K_{sp}) of sparingly soluble salts.

Solubility is defined as the maximum amount of solute that dissolves in a solvent at equilibrium. Here, equilibrium refers to a state when concentration of reactants and products remains the same after the chemical reaction has taken place.

Solubility product is the mathematical product of the dissolved ion concentrations raised to the power of its stoichiometric coefficients. Let us take the example of barium sulphate precipitate which is a sparingly soluble salt that dissolves and forms ions as:

$$BaSO_4(s) \rightleftharpoons Ba^{2+}(aq) + SO_4^{2-}(aq)$$

The solubility product (K_{sp}) for the above reaction can be written as:

$$K_{sp} = [Ba^{2+}][SO_4^{2-}]$$

The solubility and solubility product of sparingly soluble salts such as AgCl, $PbSO_4$, and $BaSO_4$, can be determined by measuring the emf of a concentration cell comprising the salt under study. Let us take the example of solubility of silver chloride.

$$AgCl\ (s) \rightleftharpoons Ag^+ + Cl^-$$

Solubility product is given as the product of activities of the two ions. As solubility product (K_{sp}) is an equilibrium constant it can be related to standard emf (E°_{cell}) of the cell where the reaction takes place.

On applying the law of mass action,

$$K = \frac{a_{Ag^+} \times a_{Cl^-}}{a_{AgCl}}, \text{ where } a \text{ is the acitivity of ions}.$$

$$= a_{Ag^+} \times a_{Cl^-} \ (\because a_{AgCl} = 1) = K_{sp}$$

We know that, $E^\circ_{cell} = \dfrac{RT}{nF} \ln K$

Hence, at 298K, $E^\circ_{cell} = \dfrac{0.0591}{n} \log_{10} K_{sp}$ \hfill (5.14)

Suppose one needs to determine the solubility of AgCl; the emf of the cell can be measured for,

$$Ag|0.01\ N\ AgNO_3|Glass\|Saturated\ AgCl|Ag$$

The emf of the above cell can be obtained using the relation, $E = \dfrac{0.0591}{n} \log_{10} \dfrac{C_2}{C_1}$.

The valence of Ag ions is 1 and its concentration in 0.01 N $AgNO_3$ is 0.01 gram of silver ions per litre.

Thus, $E = \dfrac{0.0591}{n} \log_{10} \dfrac{0.01}{C_1}$

One can easily calculate the concentration of AgCl as grams of silver ions per litre. On multiplying this value with the molecular weight of silver chloride (143.5 g), one can obtain the solubility of silver chloride in grams per litre.

5.6 ELECTROCHEMICAL SERIES AND ITS APPLICATIONS

When electrodes are arranged in the order of their increasing standard reduction potentials on a hydrogen scale, it is called electrochemical series. Table 5.1 lists some of the common electrodes and their standard reduction potentials. These values are on hydrogen scale since they are measured with standard hydrogen electrode (SHE) as the reference electrode whose potential is considered as zero.

Applications of Electrochemical Series

Calculation of emf One can predict the feasibility of the redox reaction using the electrochemical series. The standard emf of a cell can be calculated as the standard electrode potentials are known as,

$$E^\circ_{cell} = E^\circ_{cathode} - E^\circ_{anode}$$

If E_{cell} is positive, the reaction is feasible whereas if it is negative, the reaction is not feasible. Let us take an example,

$$2\,Ag(s) + Zn^{2+}\ (aq) \rightarrow Ag+\ (aq) + Zn(s)$$

Table 5.1 Standard electrode potential of important electrodes at 298 K

Electrode	E° (Volts)
Li^+, Li	−3.05
Mg^{2+}, Mg	−2.370
Al^{3+}, Al	−1.660
Zn^{2+}, Zn	−0.763
Fe^{2+}, Fe	−0.441
Cd^{2+}, Cd	−0.403
Ni^{2+}, Ni	−0.236
Sn^{2+}, Sn	−0.136
$2H^+$, H	0.00
Cu^{2+}, Cu	+0.34
Ag^+, Ag	+0.80
F_2, $2F^-$	+2.87

For the above, half-cell reactions are:

Anode: $2\,Ag_{(s)} \rightarrow 2Ag^+_{(aq)} + 2e^-;\; E^\circ = 0.80$ V

Cathode: $Zn^{2+}_{(aq)} + 2e^- \rightarrow Zn_{(s)};\; E^\circ = 0.763$ V

$E^\circ_{cell} = E^\circ_{cathode} - E^\circ_{anode}$

$E^\circ_{cell} = -0.763 - 0.80 = -1.563$ V

As E°_{cell} is negative, the reaction is not feasible.

Now, consider the feasibility of the reaction: $2\,Al_{(s)} + 2\,Sn^{4+}_{(aq)} \rightarrow 2\,Al^{3+} + 3\,Sn^{2+}_{(aq)}$

For the above, half-cell reactions are:

Anode: $2\,Al_{(s)} \rightarrow 2\,Al^{3+} + 6e^-;\; E^\circ = -1.66$ V

Cathode: $3\,Sn^{4+}_{(aq)} + 6e^- \rightarrow 3\,Sn^{2+}_{(aq)};\; E^\circ = +0.15$ V

$E^\circ_{cell} = E^\circ_{cathode} - E^\circ_{anode} = 0.15 - (-1.66) = +1.81$ V

As E°_{cell} is positive, the reaction is feasible.

Reactivity of metals The activity of metals can be decided by the electrochemical series. The tendency of metals to go in to solution can be predicted from the magnitude of standard reduction potential. Those metals which are on the top of the series easily ionize and go into solution. Hence, metals higher up in the series are active than those at the lower end of the series.

Predicting displacement reactions One can easily predict displacement reactions using the electrochemical series. The metal with a negative reduction potential will displace hydrogen from an acid solution; for example, zinc can displace hydrogen as,

$$Zn + H_2SO_4 \rightarrow ZnSO_4 + H_2; \quad E^\circ_{Zn} = -0.76 \text{ V}$$

whereas metals with positive reduction potential will not displace hydrogen from acid solution; say, silver cannot displace hydrogen as,

$$Ag + H_2SO_4 \rightarrow \text{no reaction}; \quad E^\circ_{Ag} = +0.80 \text{ V}$$

Further, metals at the higher end of the series will easily displace the metals which are at the bottom of the series. Zinc can easily displace copper from a solution.

Check Your Progress

1. What is an electrochemical cell?
2. What is electrode potential?
3. What are reversible and irreversible cells?
4. Define oxidation and reduction potentials.
5. What is liquid junction potential? How can it be eliminated?
6. Define reference electrode. Name any three reference electrodes.
7. Why is standard hydrogen electrode not routinely employed as a reference electrode?
8. Can we store silver nitrate solution in a copper vessel?
9. What will happen if a nickel rod is used to stir copper sulphate solution? Give the reason.
10. Is it feasible to determine single electrode potential? Comment on your answer.
11. Can 1 M $CuSO_4$ solution be stored in a nickel container? Justify your answer.
12. What is electrochemical series?
13. Is the cell reaction, $Ni_{(s)} | Ni^{2+}_{(aq)} \| Au^{3+}_{(aq)}\, Au_{(s)}$ possible?
14. It is preferable to employ glass electrode over quinhydrone electrode for pH measurements. Why?

Corrosion Understanding the concepts of electrochemistry enables us to apply them in corrosion science. Reactive metals are prone to be anodic and hence liable to corrosion. The cathodic or noble metals (lower end in the series) will be resistant to corrosion. However, galvanic series is a better predictor for determining the relative corrosivity of metals.

Dental fillings and electrochemistry Silver amalgam is a commonly used dental filling. If a person with silver amalgam dental filling accidentally bites into a piece of aluminum, it generates current in the mouth. This happens due to the formation of Ag – Al galvanic cell in the mouth where saliva acts as the electrolyte. The reactions occurring in such a case will be:

$Ag^+(aq) + e^- \rightarrow Ag(s); E° = + 0.80$ V and $Al^{3+}(aq) + 3e^- \rightarrow Al(s); E° = -1.66$ V

As per the electrode potential values, Al will reduce Ag, since $E°_{cell} = + 0.80 - (-1.66) = + 2.46$ V

$E°_{cell} = + 0.80 - (-1.66) = + 2.46$ V

5.7 ELECTROPLATING

The electro-deposition of metals on metals, non-metals, and alloys is called electroplating. It involves depositing a thin film of noble metal or an alloy over an active base metal by passing direct current through an electrolytic solution containing the soluble salt of the coating metal. Nickel, chromium, gold, and tin plating methods are some common examples of electroplating. The metal object to be coated is thoroughly cleaned and made cathodic. Further, the coating metal block acts as the anode. These electrodes are suitably suspended in an electrolytic bath

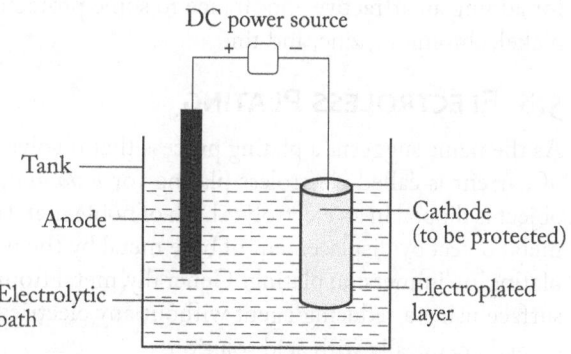

Fig. 5.9 Electroplating process

containing salt solution of the metal to be coated (Fig. 5.9). The metal salt in aqueous solution undergoes ionization and a potential difference is applied to the salt solution through the electrodes. Due to the applied potential difference, metal ions migrate towards the cathode and get deposited.

The electroplating process can be best represented by the reaction: $M \rightarrow M^{n+} + ne^-$. In the reaction, metal (M) is to be plated and n is the number of moles of electrons in the reaction, and the coating occurs due to the reduction reaction at the cathode as: $M^{n+} + ne^- \rightarrow M$

Faraday's law of electrolysis forms the basis of electroplating process. It is stated as: *The amount of substance that gets deposited at an electrode is proportional to the quantity of electricity passed through an electrolytic solution.* Further, the amount of different substances liberated by a given quantity of electricity is inversely proportional to their equivalent weights.

If a current of I ampere passes through an electrolyte for t seconds, carried out in an electrolytic cell, metal ions reduce to metal at the cathode. The total quantity of charge (Q) is given by the equation, $Q = It$. If W g of a metal gets deposited by Q coulombs of electricity, then $W \propto Q$. Now if N_A is Avogadro's number and e is the electrical charge per electron, then the total quantity of electricity, Q can be shown as

$$Q = W \times n \times N_A \times e,$$

where, $N_A \times e$ is called Faraday's constant (F) and has the value 96500 coulombs/mol. Thus, the number of moles of metal deposited at cathode by Q will be, $W = \dfrac{Q}{nF}$.

Further, we can write a general expression, $W = \dfrac{I \times t \times A}{nF}$

where W is the weight of metal that gets plated on the surface in grams, I is the current in amperes, t is the time in seconds, A is the atomic weight of metal in g/mol, and F is Faraday's constant. The quantity A/nF is called electrochemical equivalent (ECE) of a metal.

The factors that govern the electroplating process are listed as follows:

(a) If anode is of same metal of which salt is in solution, the salt is reformed by anode material passing into the solution in the form of ions.

(b) The conducting power, solubility, and metal content of the electrolyte should be high. It should not undergo hydrolysis, oxidation, or other chemical or environmental changes. It should possess sufficient covering power and the deposit obtained should be compact and adherent, thereby giving a uniform deposit on the base metal.

Applications of electroplating include the following:

(a) It improves the appearance of base metal, thereby also making it resistant to corrosion.

(b) Engineering coatings (also called functional coatings) are used for enhancing specific properties of the surface, such as solderability, wear resistance, reflectivity, and conductivity. Metals for engineering purpose include gold (Au), silver (Ag), and lead (Pb). Decorative protective coatings are primarily used for adding an attractive appearance to some protective qualities. Metals in this category include copper, nickel, chromium, zinc, and tin.

5.8 ELECTROLESS PLATING

As the name suggests, a plating process that involves deposition of a metal layer without the application of current is called electroless plating (or *autocatalytic plating*) process. In this process, the base metal object is dipped in an electrolyte bath of noble metal salt. The noble metal layer gets deposited on the base metal object by displacement of base metal by the noble metal. Hence, it is also termed as 'displacement plating' or 'immersion plating.' Generally, metal from its salt solution is deposited on a catalytically active surface using a reducing agent without any electrical energy. To make a surface catalytically active, it is etched chemically with acid solution.

Overall, it can be best depicted by the following general reaction as:

$$\text{Metal ions + Reducing agent} \xrightarrow{\text{catalytically active}} \text{Metal deposit + Oxidized product}$$

If nickel coating is desired, the base metal is dipped in nickel sulphate bath and sodium hypophosphite at 100 °C maintained at pH of 4.5 to 5.0. Nickel ion from the electrolyte solution reduces to nickel and a nickel phosphide adherent film is formed on the base metal object.

The various steps of performing electroless plating of nickel are as follows:

(i) Substrate surface preparation: The metal surface on which the desired coating is to be plated is degreased using organic solvents or alkali treatment followed by pickling.

(ii) Plating process: The plating bath solution consists of nickel chloride (20 g/dm^3), hypophosphite (reducing agent) (20g/dm^3) solutions, acetate buffer (pH = 4 – 5), and sodium succinate as a complexing reagent. The electroless plating is carried out at 346 K and the chemical reactions are:

$$H_3PO_2^- \text{(hypophosphite)} + H_2O \rightarrow H_3PO_3^- \text{(Phosphite)} + 2H^+ + 2e^-$$

$$Ni^{2+} + 2e^- \rightarrow Ni$$

Hence the net cell reaction will be,

$$H_3PO_2^- + H_2O + Ni^{2+} \rightarrow H_3PO_3^- + 2H^+ + Ni$$

In the net reaction, it can be concluded that $H_3PO_2^-$ is oxidized to $H_3PO_3^-$, whereas Ni^{2+} is reduced to nickel and forms an adherent plate on the surface. Both the above oxidation and reduction reactions occur on the surface of the substrate along with the elimination of H^+ ions, thereby decreasing the pH of the buffered solution in the presence of acetate salts.

5.9 BATTERIES

A battery is a portable power device that converts chemical energy to electrical energy. It is prepared by connecting electrochemical cells in series or parallel to provide necessary voltage or current, or both. These compact batteries are widely used in flashlights, watches, cameras, toys, calculators, television remotes, laptops, computers, tape recorders, cordless devices, telephones, medical devices, electric bells, space vehicles, and many other appliances.

Advantages of batteries

(a) The batteries need not be connected to an external electrical system.

(c) They act as a compact and portable power device which can be easily replaced.

Disadvantages of batteries

(a) Batteries can be used only for a limited period of time. It requires recharging using electricity.

(b) These devices are prone to explosion and fire hazards, if not maintained properly.

Characteristics of an ideal battery

An ideal battery should be portable, lightweight with long life cycle, temperature stable with no voltage drop while in use. It should have lower cell resistance with high power output. The battery case should be leak-proof and resistant to thermal shocks. It should be easily rechargeable (numerous charge–discharge cycles) and withstand overcharging or overdischarge.

Commercial batteries can be classified into: (a) primary cells, (b) secondary cells and, (c) reserve batteries.

Primary cells The chemical energy stored in these batteries gets converted to electrical energy only once, until the active materials are exhausted, after which the cell is of no use. They cannot be recharged or reused. Dry cells, lithium cells, mercury cells are common primary cells.

Secondary cells These batteries can be recharged by passing current through them and reused numerous times. Nickel–cadmium battery and lead–acid battery are common examples.

Reserve batteries In these, the main component (usually electrolyte) is kept separate from the rest of the battery components before activation. These batteries are used only in emergency situations in weaponry, missiles, bomb fuses, projectiles. During use, the battery is activated by adding electrolytes such as water, gases, or sometimes the solid electrolyte is heated. Ag–Zn battery, Li-thionyl chloride, and thermal battery are some examples of reserve batteries. As the main component is kept separate until activation, these batteries do not undergo self-discharge and have longer lives.

5.9.1 Zn–MnO$_2$ Dry Cell

A typical Zn–MnO$_2$ shown in Fig. 5.10 is a primary battery also called a dry cell. A dry cell consists of a graphite rod at the centre with a metal cap. The graphite rod is surrounded by a moist paste of graphite and manganese dioxide (MnO$_2$). The graphite rod with graphite–MnO$_2$ paste acts as the cathode. A paste of ammonium chloride and zinc chloride is present as the electrolyte layer. All these are placed in a zinc cylinder that acts as the anode.

Further, the zinc cylinder is wrapped with polypropylene to avoid leakage of the battery material. The dry cell can be represented as

$$Zn \,|\, Zn^{2+} \,||\, NH_4^+ \,|\, MnO_2 \,|\, C$$

The electrode reactions of the dry cell are:

Anode: $Zn \rightarrow Zn^{2+} + 2e^-$

Cathode: $2\,MnO_2 + 2\,H_2O + 2e^- \rightleftharpoons 2\,MnO(OH) + 2\,OH^-$

Net cell reaction: $Zn + 2\,MnO_2 + 2\,H_2O \rightarrow 2\,MnO(OH) + Zn^{2+} + 2\,OH^-$

During working of the dry cell, MnO_2 reduces to Mn_2O_3. Secondary reactions also take place in the dry cell, but do not contribute significantly to the working of the dry cell. The secondary reactions are as follows:

$$2NH_4Cl + 2OH^- \rightarrow 2NH_3 + 2H_2O + 2Cl^-$$
$$Zn^{2+} + 2NH_3 + 2Cl^- \rightarrow Zn(NH_3)_2Cl_2$$

As shown in the above reactions, OH ions formed during the electrode reactions react with ammonium chloride to form NH_3 which combines with $ZnCl_2$ to form $Zn(NH_3)_2Cl_2$ complex.

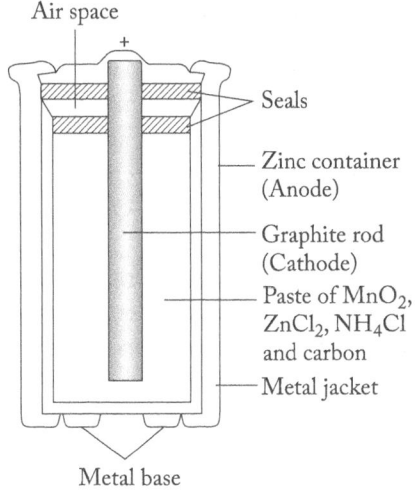

Fig. 5.10 Zn–MnO₂ dry cell

The **advantages** of dry cell are:

(a) It is a portable source to power small electric and electronic gadgets, such as cameras, lighting devices, watches, and calculators.
(b) Dry cell is a low-cost and easily available battery.
(c) It can generate stable voltages for small discharge of current, though for shorter period of time.

Some of the **limitations** of dry cell are:

(a) Dry cells cannot be recharged once all the cell materials are exhausted.
(b) Battery capacity is quite low and thus not useful for advanced applications such as electric vehicles.
(c) They cannot be stored for a longer period of time and works well only within the temperature range of $20 - 40\,°C$.

5.9.2 Lead Storage Cell

A lead storage cell consists of two lead electrodes immersed in 20 % sulphuric acid solution which acts as the electrolyte. The specific gravity of the acid electrolyte is about 1.2 at 298 K. Lead electrodes are in grid forms as shown in Fig. 5.11. The grid of one electrode is filled with spongy lead acting as the anode and the other electrode grid is filled with a paste of lead dioxide that acts as the cathode. The entire battery is encased in a secure plastic container and the cell can be represented as follows:

$$- Pb\,|\,PbSO_{4(s)}H_2SO_{4(aq)}\,\|\,PbSO_{4(s)},\,PbO_{2(s)}\,|\,Pb +$$

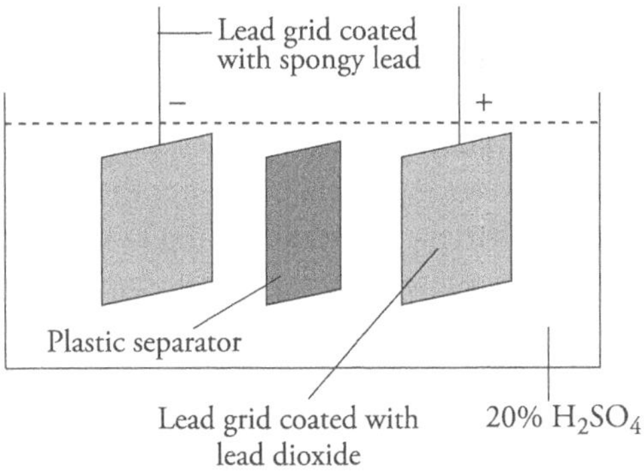

Fig. 5.11 Electrode–electrolyte system in an acid storage cell

The reactions taking place in the cell during discharge are:

At anode: $Pb_{(s)} \rightleftharpoons Pb^{2+} + 2e^-$

$$Pb^{2+} + SO_4^{2-} \rightleftharpoons PbSO_{4(s)}$$

Hence, the net reaction will be: $Pb^{2+} + SO_4^{2-} \rightleftharpoons PbSO_{4(s)} + 2e^-$

At cathode: $PbO_{2(s)} + 4H^+ + 2e^- \rightleftharpoons Pb^{2+} + 2H_2O$

$$Pb^{2+} + SO_4^{2-} \rightleftharpoons PbSO_{4(s)}$$

Hence, the net reaction will be: $PbO_2 + 4H^+ + SO_4^{2-} + 2e^- \rightleftharpoons PbSO_{4(s)} + 2H_2O$

The overall reaction while discharging of the cell will be the sum of the two net reactions and can be represented as follows:

$$Pb_{(s)} + PbO_{2(s)} + 2H_2SO_{4(aq)} \rightleftharpoons 2PbSO_{4(s)} + 2H_2O$$

During discharge, lead sulphate is formed at both the electrodes and sulphuric acid is consumed in the reaction. While charging, lead sulphate gets stripped from the electrode surface and sulphuric acid is reformed. This mechanism called *double sulfation theory* was put forth by Gladstone and Tribe. In fully charged mode, the density of sulphuric acid is 1.2 with a potential of about 2 V. To obtain higher potentials, numerous electrode pairs are connected in series.

Applications   Lead storage battery is commercially used in automotive engines, rail-trains, invertors (power), laboratories, etc. Certain **limitations** of lead batteries include (a) decrease in potential on decreasing the electrolyte concentration and (b) overcharging damages the electrodes thereby reducing their shelf life.

5.9.3 Nickel–Cadmium Battery

Nickel–cadmium (Ni–Cd) battery is a secondary rechargeable alkaline storage battery. Figure 5.12 shows a Ni–Cd battery and can be represented as:

$$Cd \,|\, CdO \,|\, KOH(6M) \,\|\, NiOOH, Ni(OH)_2 \,|\, Ni$$

In a typical Ni–Cd battery, cadmium electrode acts as the anode and cathode uses nickel oxyhydroxide as an active material. The two electrodes are separated using an alkaline electrolyte of 6M potassium hydroxide solution. Ni–Cd batteries have a potential of about 1.4 V. During discharge, electrode reactions of Ni–Cd are as follows:

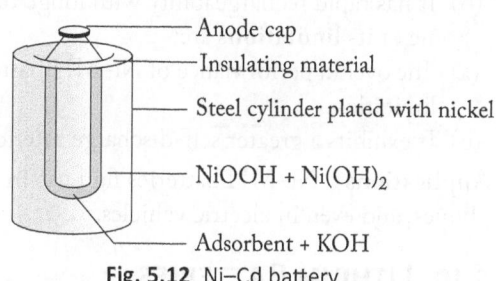

Anode cap
Insulating material
Steel cylinder plated with nickel
$NiOOH + Ni(OH)_2$
Adsorbent + KOH

Fig. 5.12 Ni–Cd battery

$$\text{Anode: } Cd + 2\,OH^- \rightarrow Cd(OH)_2 + 2\,e^-$$
$$\text{Cathode: } 2\,NiO(OH) + 2\,H_2O + 2\,e^- \rightarrow 2\,Ni(OH)_2 + 2\,OH^-$$

Overall cell reaction: $Cd + 2\,NiO(OH) + 2\,H_2O \rightarrow Cd(OH)_2 + 2\,Ni(OH)_2$

During recharging of Ni–Cd battery, the above cell reactions go from right to left (i.e. cell reactions are reversible). During working of the cell, electrolyte is not consumed in the reaction which makes these batteries rechargeable.

The **advantages** of Ni–Cd battery are:

(a) They have long life and can be operated at higher discharge rates and wide temperature ranges.

(b) They can be stored for longer periods of time without any deterioration.

Some of the **disadvantages** are self-discharge, environmentally toxic due to the presence of heavy metals and hence harder to recycle.

Applications   Ni–Cd batteries are used to power flashlights, radios, cordless phones, pacemakers, hearing aids, cameras, and so on.

5.9.4 Nickel–Metal Hydride Battery

As shown in Fig. 5.13, a nickel–metal hydride battery consists of nickel oxyhydroxide (NiOOH) as the cathode and hydrogen in metal hydride form (usually hydrogen is adsorbed on a metal alloy) as the anode. The metal alloy on which hydrogen is adsorbed could be made of lanthanum–nickel or titanium–zirconium alloy metals. Both the electrodes are immersed in an aqueous solution of potassium hydroxide electrolyte and separated by a synthetic insulating

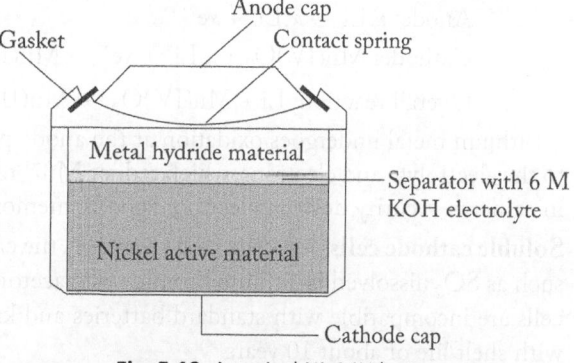

Gasket
Anode cap
Contact spring
Metal hydride material
Separator with 6 M KOH electrolyte
Nickel active material
Cathode cap

Fig. 5.13 A typical Ni–MH battery

material. The cathode is a highly porous substrate on which nickel oxyhydroxide is coated. The anode consists of a nickel wire grid on which hydrogen is bonded.

The electrode reactions can be depicted as follows:

Anode: $MH + OH^- \rightarrow M + H_2O + e^-$

Cathode: $NiOOH + H_2O + e^- \rightarrow Ni(OH)_2 + OH^-$

Hence the overall cell reaction will be: $MH + NiOOH \rightarrow M + Ni(OH)_2$ (emf = 1.25 – 1.35 V)

Some of the **advantages** are:

(a) Ni–MH rechargeable batteries are environment friendly, possess high capacity, and requires negligible maintenance.

(b) It has rapid rechargeability with longer cycle lives.

Some of its **limitations** are:

(a) The overall performance of Ni–MH batteries is poor, but the presence of safer materials offsets this limitation.

(b) It exhibits a greater self-discharge rate to about 50 % thereby leading to limited service life.

Applications Ni–MH batteries find use in consumer electronics, such as computers, cameras, cellular phones, and even in electric vehicles.

5.10 LITHIUM BATTERIES

In lithium cells, lithium metal acts as the anode due to its lightweight, lower electrode potential, and good conductivity. Lithium cells are safer, cheaper, and non-toxic that exhibit high performance. It can produce voltages ranging between 1.5 V and 3.7 V that are quite high as compared to ordinary batteries. Based on the type of electrolyte, cathode, etc., primary lithium cells can be of three types: (a) soluble cathode cells, (b) solid cathode cells, and (c) solid electrolyte cells.

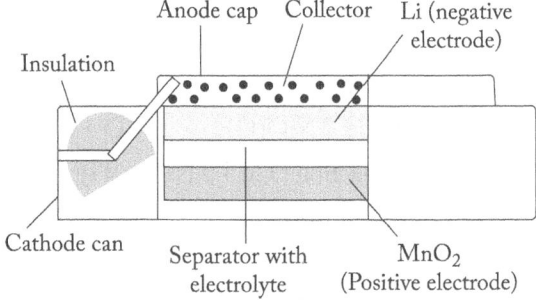

Fig. 5.14 Li/MnO$_2$ battery

Li cells with solid cathode Li/MnO$_2$ is an example of a solid cathode cell, wherein manganese dioxide acts as the solid cathode. As shown in Fig. 5.14, lithium anode and heat-treated MnO$_2$ cathode are placed in an electrolyte containing lithium salts mixed in an organic solvent of propylene carbonate and 1, 2-dimethoxy ethane. Li/MnO$_2$ cell gives a high voltage of about 3.2 V and is known to perform over a wider temperature range with a shelf life of about 7 years.

The cell reactions can be shown as:

Anode: $x.Li \rightarrow x.Li^+ + xe^-$

Cathode: $Mn(IV)O_2 + x.Li^+ + xe^- \rightarrow Mn(III)O_2\,Li^+$

Overall reaction: $Li + Mn(IV)O_2 \rightarrow Mn(III)O_2Li^+$

Lithium metal undergoes oxidation at the anode producing positively charged Li$^+$ ions that diffuse in to the electrolyte and electrons, which reduce Mn^{4+} to Mn^{3+} at the cathode. Li/MnO$_2$ is widely employed in cameras, security devices, electronic goods, memory back-ups, etc.

Soluble cathode cells As the name suggests, the cathode material could be liquid or gaseous in nature, such as SO$_2$ dissolved in lithium bromide and acetonitrile, or SOCl$_2$ solvent with LiAlCl$_4$ solute. These cells are incompatible with standard batteries and known to give voltages ranging from 2.8 V to 3.5 V with shelf life of about 10 years.

In Li/SOCl$_2$ cell, lithium anode, carbon cathode, and electrolyte of lithium aluminium chloride in liquid thionyl chloride are used. Hence the cell reactions can be shown as:

Anode: $2\,Li \rightarrow 2\,Li^+ + 2\,e^-$

Cathode: $4\,Li + 4\,e^- + 2\,SOCl_2 \rightarrow 4\,LiCl + SO_2 + S$

Overall reaction: $4\,Li + 2\,SOCl_2 \rightarrow 4\,LiCl + S\,O_2 + S$

The carbon cathode accepts electrons, whereas thionyl chloride reacts with lithium producing LiCl, S, and SO$_2$. LiCl gets precipitated on the cathode and SO$_2$ and S dissolve in the electrolyte. Li/SOCl$_2$ cells are employed in military radios, automatic meter reading systems, etc.

Solid electrolyte cells These have long life with lower discharge rate. Memory backups, heart pacemakers, and devices requiring low current employ solid electrolyte cells. Lithium iodide is a common solid electrolyte used in such batteries. The cell usually comprises lithium–metal anode, polymeric cathode, and solid lithium iodide as an electrolyte providing a voltage of 2.8 V due to the reaction:

$2\,Li + p2Vp + n\,I_2 \rightarrow 2\,LiI + p2Vp + (n-1)I_2$, where p2Vp is poly-2-vinyl pyridine.

Lithium-ion cells These are secondary rechargeable batteries where lithium acts as the cathode and graphite acts as the anode. As shown in Fig. 5.15, there is migration of only lithium ions between the two electrodes through a lithium-based salt as an electrolyte (LiPF$_6$ or LiBF$_4$ dissolved in ethylene carbonate). During discharge, lithium ions move from anode to cathode through a non-aqueous electrolyte. For charging the lithium cell, an external power source is connected and current flows in the opposite direction.

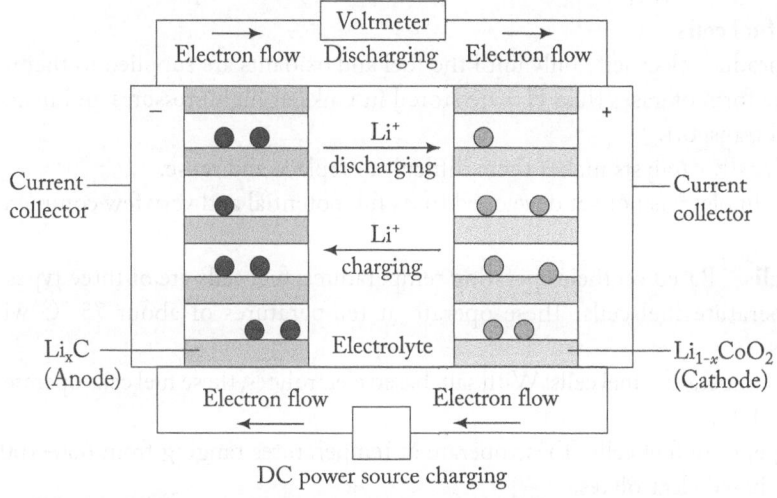

Fig. 5.15 Lithium-ion cell

The cell reactions are:

Charging (electrolytic): $LiCoO_2$ (lithium cobalt oxide) $\rightarrow Li_{1-x}CoO_2 + x\,Li^+ + x\,e^-$ (cathode)

$x\,Li^+ + x\,e^- + x\,C_6 \rightarrow x\,LiC_6$ (anode)

Discharging (galvanic): reverse of above reactions.

Lithium-ion batteries are lightweight, undergo minimal structural deformation, have large surface area along with small mobile (Li) ions moving across the electrolyte. Portable electronic gadgets, power backups, alarm systems are some examples where lithium-ion cells are used.

5.11 FUEL CELLS

Fuel cell is an electrochemical device that efficiently converts chemical energy of fuels directly in to electricity that can be utilized for power generation. The major attraction of fuel cells is that it does not require recharging as the fuels constantly get replenished at the electrodes. An ideal fuel cell consists of two electrodes and an electrolyte. During operation, a fuel (such as hydrogen, methanol) is continuously sent to the anode, whereas oxidizers are sent to the cathode. Due to redox reactions occurring at the electrodes, electrical energy is generated without any evolution of harmful products.

Table 5.2 Differences between conventional and fuel cells

Conventional cells (or batteries)	Fuel cells
The reactants are placed inside the cell or battery compartment.	The reactants are to be supplied from outside. In this case, fuel and oxidants are added from outside the cell.
Secondary batteries can be recharged.	Fuel cells cannot be recharged.
The cell reaction products are highly toxic and hence proper disposal is a controversial issue.	The cell reactions result in non-toxic products, hence they are environment-friendly while in use or when disposed.
The use of costly catalysts is low, but heavy metals are used in batteries.	Fuel cells are used on a lower scale due to the use of costly catalysts.

Advantages of fuel cells
(a) Fuel cells do not require recharging.
(b) They provide high power efficiency and produce direct current for long periods of time.
(c) These cells are eco-friendly as the products of cell reactions are non-toxic.
(d) They are silent during operation; hence lesser maintenance is required.

Limitations of fuel cells
(a) Fuel cells produce electricity only until the fuel and oxidants are supplied to them.
(b) Fuels in the form of gases (like H_2) are stored in tanks at high pressures and hence are difficult to handle and transport.
(c) The use of costly catalysts makes them difficult to replace and reuse.
(d) Fuel cell technology is not yet developed to its full potential and very few commercial products are available.

Types of fuel cells   Based on their operating temperatures, fuel cells are of three types.
(a) Low-temperature fuel cells: These operate at temperatures of about 75 °C with water-based electrolytes.
(b) Moderate-temperature fuel cells: With salt-based electrolytes, these fuel cells operate at temperatures of about 600 °C.
(c) High-temperature fuel cells: These operate at temperatures ranging from 600–1000 °C with solid or ceramic-based electrolytes.

5.11.1 Hydrogen–Oxygen Fuel Cell

As shown in Fig. 5.16, hydrogen is supplied to one electrode (anode) and oxygen (oxidant) is supplied to the cathode. Electrodes are hollow tubes of porous compressed carbon with finely divided platinum impregnated in the anode and silver oxide (Ag_2O) incorporated in the cathode. The electrolyte is potassium hydroxide solution maintained at 200 °C at 20–40 Atm. Both platinum and silver oxide act as catalysts for the cell reactions.

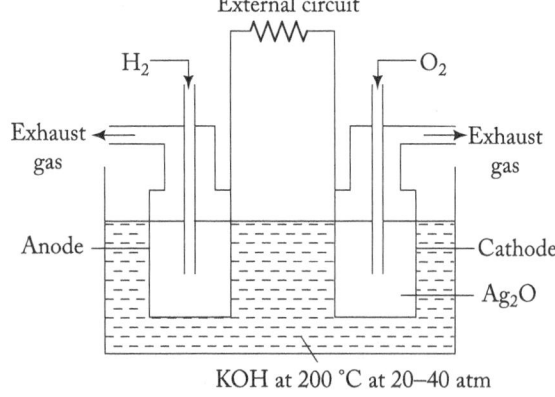

Fig. 5.16 Hydrogen–oxygen fuel cell

The electrode reactions taking place in a H_2–O_2 fuel cell are:

At anode: $2H_2 \rightarrow 4H^+ + 4e^-$ (oxidation)

At cathode: $O_2 + 2H_2O + 4e^- \rightarrow 4OH^-$ (reduction)

Hence the overall reaction is, $2H_{2(g)} + O_{2(g)}^- \rightarrow 2H_2O_{(l)}$

Applications The voltage of H_2–O_2 fuel cell is 1.15 V based on the continued supply of hydrogen and oxygen. The only limitation is the large-scale use of hydrogen and its tendency to explode in the presence of platinum catalyst. These fuel cells are employed in space vehicles to provide water.

5.11.2 Methanol–Oxygen Fuel Cell

In methanol–oxygen fuel cell, methanol is used as fuel, oxygen is used as oxidant to provide electrical energy, and H_2SO_4 acid acts as the electrolyte (Fig. 5.17).

Construction Methanol–oxygen fuel cell consists of two electrodes, where the anode is a porous nickel electrode impregnated with Pt/Ru catalyst and the cathode is porous nickel electrode coated with silver catalyst.

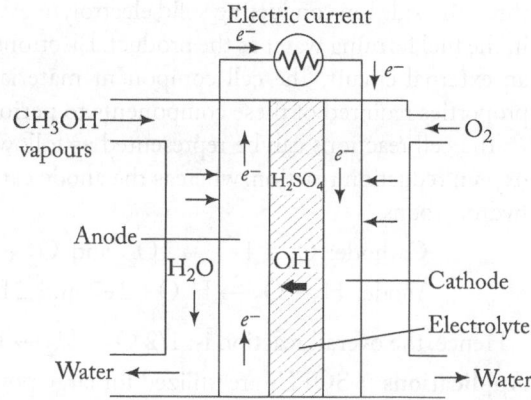

Methanol vapour mixed with sulphuric acid (care is taken while mixing them) is supplied through the anode chamber. Clean, dry, and ultrapure oxygen is passed through the cathode chamber. Sulphuric acid electrolyte is placed in the central chamber of the fuel

Fig. 5.17 A typical direct methanol–oxygen fuel cell

cell. Continuous fuel–oxidant supply needs to be maintained in the cell for uninterrupted electrical supply. Further, the cell is also provided with a proton conducting membrane near the cathode so as to avoid diffusion of anode reactants (methanol) to enter the cathode chamber. The membrane allows migration of only protons towards the cathode. After the reactions, products are carbon dioxide and water which are non-toxic in nature.

Working The cell reactions occurring in the methanol–oxygen fuel cell are as follows:

At anode: $CH_3OH + 6OH^- \rightarrow CO_2 + 5H_2O + 6e^-$

At cathode: $3/2\,O_2 + 3H_2O + 6e^- \rightarrow 6OH^-$

Net reaction: $CH_3OH + 3/2\,O_2 \rightarrow CO_2 + 2H_2O$

Applications Methanol–oxygen fuel cells are used for military purposes and for large-scale power production. They are also used in fuel cell vehicles and space shuttles.

5.11.3 Solid Oxide Fuel Cells

High-temperature solid oxide fuel cell (SOFC) is an electrochemical device that generates electrical energy directly from oxidizing a fuel. The main feature is its electrolyte material which is either a solid oxide or a ceramic electrolyte (Fig. 5.18). Yttrium-stabilized zirconium-based solid electrolyte is commonly employed for generating electricity. These fuel cells operate at high temperatures in the range of 1000–1100 °C. SOFCs exhibit higher efficiencies of over 60 % with greater fuel

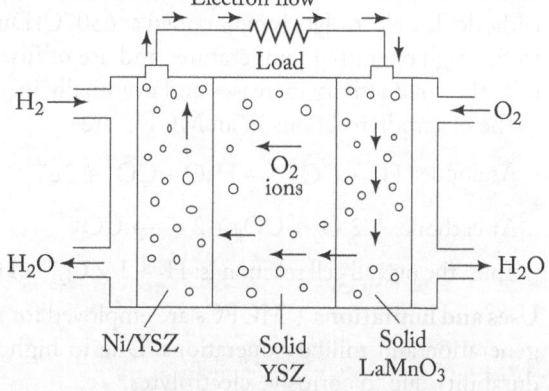

Fig. 5.18 A typical solid oxide fuel cell

adaptability and reliability for electrical energy generation. These fuel cells are eco-friendly due to lower levels of NO_x, SO_x emissions during their operations.

Construction SOFC comprises two porous electrodes that are separated by a dense, oxygen-ion conducting electrolyte. The anode is made of highly conducting Ni/ytrria-stabilized zirconia cermet (i.e., Ni/YSZ) and the cathode is a mixed conducting material of perovskite, that is, lanthanum manganate ($LaMnO_3$). For obtaining higher voltages, SOFCs are connected in layers interconnected by lanthanum manganate that joins anodes and cathodes that are adjacently placed.

Working Clean, dry, and ultrapure oxygen is supplied to the cathode (air/oxidant electrode) which reacts with incoming electrons to form oxide ions. These oxide ions then migrate to the anode (fuel electrode) through oxide ion conducting solid electrolyte. At the anode, oxygen ions combine with hydrogen present in the fuel forming water as the product. Electrons are released that travel from anode to cathode through an external circuit. The cell component materials are chosen based on suitable electrical conducting properties required of these components to perform their intended functions.

The cell reactions can be represented as follows: The cathode is typically an oxide that catalyses the oxygen reduction reaction, whereas the anode catalyses the oxidation of fuel, with hydrogen or reformed hydrocarbons.

$$\text{Cathode: } O_2 + 4\,e^- \rightarrow 2\,O_2^- \text{ and } O_2 + 4\,e^- + 4\,H^+ \rightarrow 2\,H_2O$$
$$\text{Anode: } H_2 + O_2 \rightarrow H_2O + 2\,e^- \text{ and } 2\,H_2 \rightarrow 4\,H^+ + 4\,e^-$$

Hence, the overall reaction is: $1/2\ O_2 + H_2 \rightarrow H_2O$

Applications SOFCs are utilized for large power generation plants but yet its full potential is subject to pilot research. These fuel cells also find use in the transportation sector to fabricate on-board auxiliary power units (APUs).

5.11.4 Molten Carbonate Fuel Cell

As the name suggests, the molten carbonate fuel cell (MCFC) uses molten or fused carbonate salts as electrolyte (Fig. 5.19). Fused alkaline salt is the preferred electrolyte to overcome the limitation of aqueous electrolytes at high operating temperatures (loss of water on evaporation). Usually, a eutectic mixture of lithium carbonate and sodium carbonate is used as the electrolyte. The electrode system has porous nickel powder alloyed with copper as the anode and porous NiO doped with lithium as the cathode. The electrolyte is maintained at 650 °C. Due to the high operating temperatures and use of fused salts the conductivity increases and eventually increases the electrode reactions.

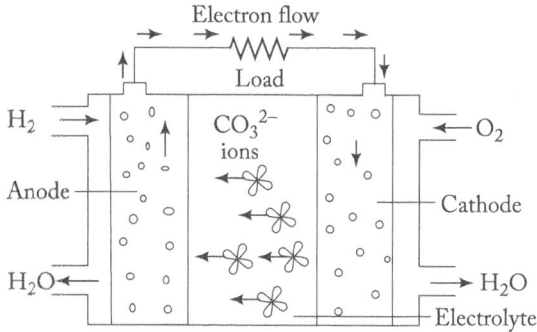

Fig. 5.19 Molten carbonate fuel cell

The electrode reactions in an MCFC are:

At anode: $H_2 + CO_3^{2-} \rightarrow H_2O + CO_2 + 2\,e^-$

At cathode: $1/2\ O_2 + CO_2 + 2\,e^- \rightarrow CO_3^{2-}$

Thus, the overall cell reaction is: $H_2 + 1/2\ O_2 \rightarrow H_2O$

Uses and limitations MCFCs are employed for natural gas- and coal-based power plants for electricity generation and military operations. Due to high operating temperatures, these cells may have limited durability due to corrosive electrolytes.

5.11.5 Phosphoric Acid Fuel Cell

A commonly employed fuel cell, phosphoric acid fuel cell (PAFC) comprises an anode and a cathode made of a finely dispersed platinum catalyst on carbon and a silicon carbide structure that holds the phosphoric acid electrolyte.

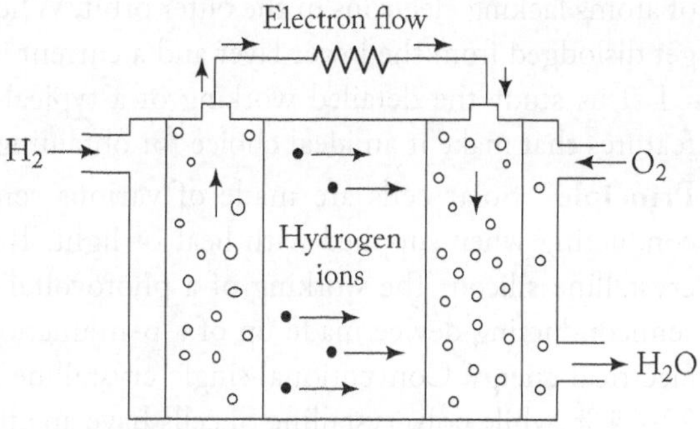

Figure 5.20 depicts a typical phosphoric acid fuel cell where liquid phosphoric acid contained in a Teflon-bonded silicon carbide matrix is the electrolyte. The porous carbon electrodes contain a platinum or rhodium catalyst.

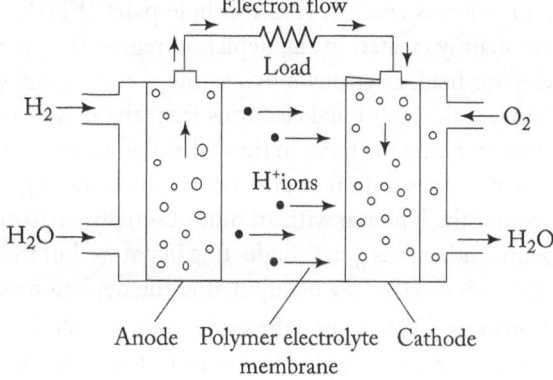

Fig. 5.20 Phosphoric acid fuel cell

The cell reactions in a PAFC are:

At anode: $H_2 + 2OH^- \rightarrow 2H_2O + 2e^-$

At cathode: $1/2 O_2 + H_2O + 2e^- \rightarrow 2OH^-$

Thus, the overall cell reaction is, $H_2 + 1/2 O_2 \rightarrow 2H_2O$

Usually the operating temperature is maintained at 200°C to obtain a higher efficiency of 85 % for simultaneous electricity and heat generation. During the cell reaction, protons move through the electrolyte to the cathode to combine with oxygen and electrons, producing water and heat.

5.11.6 Polymer Electrolyte Membrane Fuel Cell

Polymer electrolyte membrane fuel cell (PEMFC) uses an ion-exchange polymeric membrane as an electrolyte along with platinum electrodes (Fig. 5.21). These cells operate at relatively low temperatures of about 80 °C delivering an efficiency of about 60% for electric current generation.

The cell reactions occurring in PEMFC are:

At anode: $H_2 \rightarrow 2H^+ + 2e^-$

At cathode: $1/2 O_2 + 2H^+ + 2e^- \rightarrow H_2O$

Hence, the overall cell reaction is: $H_2 + 1/2 O_2 \rightarrow H_2O$

Unlike the earlier discussed fuel cells, PEMFC has a solid polymeric membrane electrolyte that

Fig. 5.21 Polymer electrolyte membrane fuel cell

allows protons to move through towards the cathode and combine with oxygen and electrons, producing water and heat. These fuel cells are used in electric vehicles and portable electronic devices.

5.12 Photovoltaic Cell

Alexandre-Edmond Becquerel (1839) discovered the photovoltaic effect in the junction formed between a platinum electrode and silver chloride electrolyte. Russell Ohl (1939) built the first photovoltaic device that comprised silicon p-n junction. A solar cell, popularly called a photovoltaic cell, can directly convert solar radiations into electric current. Generally used to power calculators and satellites, a photovoltaic cell is made of a semiconducting material and looks just like a flat wafer-like material. When sunlight is incident on the solar cell, it gives out an electric current equivalent to a battery-run flashlight in just one step.

Many such solar cells are connected together to harness more amount of power. Two thin layers of semiconducting materials (normally doped silicon) separated by a junction-layer form a solar cell. The lower layer consists of atoms with a single electron in their outer orbit, whereas the upper layer consists of atoms lacking electrons in the outer orbit. When light photons strike the solar cell material, electrons get dislodged from the lower layer and a current is generated that flows back to the upper layer.

Let us study the detailed working of a typical photovoltaic cell, its preparation, and various salient features that make it an ideal choice for obtaining cleaner energy.

Principle Solar cells are made of various semiconducting materials which are made electrically conducting when supplied with heat or light. The major element of such semiconducting materials is crystalline silicon. The working of a photovoltaic cell is based on photoelectric effect. It consists of a semiconducting device made up of a p-n junction diode, which in the presence of sunlight, generates electrical energy. Conventional single crystalline layer Si-based solar cells have an efficiency of around 22–24 %, while polycrystalline Si cells have an efficiency of 18 %.

Construction A silicon-based photovoltaic cell is made of a thin wafer made of ultra-thin layer of phosphorus doped (n-type) silicon on top of a thicker layer of boron-doped (p-type) silicon. An electric field is created at the point of contact of these layers and when light strikes the surface, current is generated (Fig. 5.22).

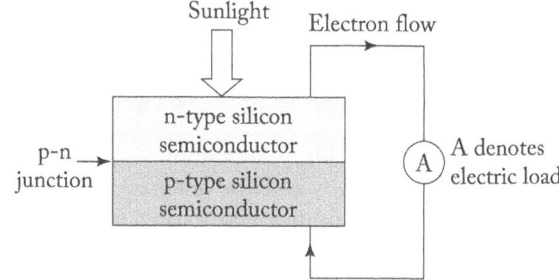

Fig. 5.22 A p-n junction in solar cell

p-n junction of a solar cell When one combines p-type and n-type semiconductors in close contact, a junction is created. Electron hole pairs (EHPs) are mainly created in the depletion region (i.e., p-type layer) and because of the built-in potential and electric field, electrons move to the n region and the holes to the p region. When an external load is applied, the additional electrons from the n-type layer migrate towards the p-type layer depletion zone, thereby filling the holes in them. The depletion zone is devoid of any mobile charges and helps in keeping other charges from p- and n-type layers from migrating across it. A region depleted of carriers is present around the junction with an infinitesimally small electrical imbalance of about 0.7 V. A p- and n-doped semiconductor is quite conducting in nature, but their junction is non-conducting which is called *depletion zone p-n junction*. By manipulating the depletion zone, such diodes can be put to a myriad of uses.

Working When sunlight strikes a photovoltaic cell, the energy-rich photons are absorbed by the cell. The absorption of a photon results in the dislocation of an electron from the silicon atom and a positive 'hole' is created. The free electron and positive hole together are neutral and hence need to be separated so as to generate electricity. Thus, when photon hits the cell, free electrons tend to combine with positive holes present on the p-layer.

Due to the presence of p-n junction, it allows electrons to move only in one direction. When an electrical contact is created on the front and rear of the cell connected through an external circuit (see Fig. 5.23), free electrons can return to positive holes only by flowing through this external circuit, thereby generating current. The electric power generated from a photovoltaic cell is directly proportional to its area and intensity of the sun's rays that strike the cell, usually measured in watts (W).

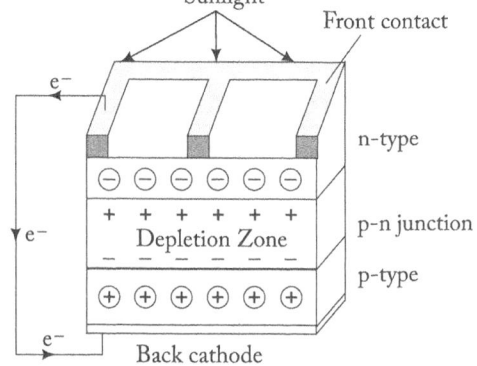

Fig. 5.23 Working of photovoltaic cell

Preparation of Solar Grade Silicon by Union Carbide process

Before discussing the production of solar-grade silicon, let us look into the salient physical and chemical properties of silicon relevant to photovoltaics.

Physical properties

(a) Silicon is an electropositive element found abundantly in the earth's crust.

(b) A metalloid material found in group IV of the periodic table, silicon has good metallic luster and is considerably brittle.

(c) It acts as a semiconductor with a band gap of 1.2 eV at 25 °C.

(d) At the atmospheric pressure, silicon crystallizes to a diamond cube-like structure.

(e) Silicon possess absorption and transmission properties in the wavelength range of 0.4 – 0.5 μm which is a crucial parameter for the performance of photovoltaic cells.

Chemical properties

(a) Silicon is quite stable and is tetravalent having strong affinity for oxygen, thereby forming stable oxides and silicates.

(b) It can also form strong bonds with carbon with resulting Si–C network.

(c) It also forms hydrides and silanes (SiH_4), which is a key chemical compound for production of amorphous silicon and purification of silicon to semiconductor grade quality.

(d) With chlorine, silicon forms trichlorosilanes and tetrachlorosilanes which are intermediate and by-products of the purification process in the conversion of metallurgical grade silicon to semiconductor grade quality.

In the production of solar grade silicon by Union Carbide method, two major steps are involved: (a) Manufacture of metallurgical grade silicon and (b) its conversion to solar grade silicon.

Metallurgical grade silicon contains high levels of impurities while semiconductor grade silicon possesses very low levels of impurities (about ppb level). Hence, the objective is to obtain silicon which is intermediate to metallurgical and semiconductor grade quality. Such type of silicon is called solar grade silicon.

Manufacture of metallurgical grade silicon

(i) Quartz (SiO_2) is taken in an electric arc furnace and heated with carbon to high temperatures in range of 1500–2000 °C. Liquid silicon is obtained from the bottom of the arc furnace which is collected,

$$SiO_2 \text{ (quartz)} + 2\,C \rightarrow Si_{(i)} + 2\,CO \uparrow$$

(ii) Next, liquid silicon is cooled and sent for refining process. If silicon contains aluminium, magnesium, or calcium, they are precipitated as alumina, magnesium oxide, and calcium oxide, respectively as shown in the following reactions.

$$4\,Al + 3\,SiO_2 \rightarrow 3\,Si_{(1)} + 2\,Al_2O_3$$

$$2\,Mg + SiO_2 \rightarrow Si_{(1)} + 2\,MgO$$

$$2\,Ca + SiO_2 \rightarrow Si_{(1)} + 2\,CaO$$

(iii) The silicon obtained after the refining process is called metallurgical grade silicon (denoted as MG–Si) and is about 98 % pure.

Conversion of metallurgical grade silicon to solar grade silicon

(i) Synthesis of silane: Metallurgical grade silicon obtained from the earlier step is treated with HCl at 300 °C to form tetrachlorosilane, $Si_{(1)} + 4\,HCl \rightarrow SiCl_4 + 2\,H_2$

Tetrachlorosilane undergoes hydrogenation to form trichlorosilane which is passed into an ion-exchange resin column where silane (SiH_4) will be obtained.

$$SiCl_4 + 2\,H_2 + Si \rightarrow 4\,HSiCl_3 \rightarrow SiH_4$$

(ii) Purification of silane: Silane so obtained is purified by distillation process where impurities will be left behind and silane will vaporize.

(iii) Decomposition of silane: In the last step, silane undergoes decomposition to form solar grade silicon with ppb levels of impurities, $SiH_4 \rightarrow Si\,(solar\ grade) + 2\,H_2$

This step involves pyrolysis where covalent bonds are broken at very high temperatures to yield silicon with the evolution of hydrogen.

Advantages of photovoltaic cell

(a) There are no emissions of toxic combustion products or radioactive residues during operation; hence it is eco-friendly.

(b) It is low on operating costs as no fuels are involved in the operation.

(c) Photovoltaic cells can be integrated in various civil structures, such as buildings with greater reliability.

Disadvantages of photovoltaic cell

(a) As sunlight is diffuse in nature, it provides very low energy density.

(b) The installation costs are quite high and energy can be generated only during the day time.

Applications of photovoltaic cells are as follows:

(a) Photovoltaic technology is used to power a variety of commercially available consumer-based products such as toys, watches, calculators, radio, televisions, flashlights, and fans.

(b) Using arrays of photovoltaic cells integrated as a roof-top or building integrated systems provide several kilowatts of electricity. It has shown great promise in electrifying rural areas.

(c) Photovoltaic cells have been used for electric power in space. Solar-powered electric vehicles have also been devised to achieve eco-friendly transportation.

(d) Various communication signals such as mobiles, radio, and television require amplification; and photovoltaics that run low-power transmitters are used in hilly locations.

(e) In scientific and climatic studies such as seismic activity monitoring, highway conditions, meteorological information, and other research activities can be powered by photovoltaic cell systems. Even portable traffic lights can be powered by these systems.

5.13 POTENTIOMETRIC TITRATIONS

Potentiometry is an electroanalytical technique that involves measuring potentials in an electrochemical cell, typically under the conditions of no current flow. It is known that potential develops in an electrochemical cell due to free energy changes occurring during chemical reactions until equilibrium is attained. The measured potential is proportional to the concentration of the analyte (compound under study) and is utilized in a variety of reactions, such as acid–base, redox, and precipitation. A potentiometric set-up comprises an electrochemical cell with two electrodes; the *reference electrode* whose potential is kept constant and the other is called the *indicator electrode* where analyte potential is measured during the titration. Potentiometric titration involves end-point detection by measuring emf of a cell comprising an indicator and reference electrodes.

As shown in Fig. 5.24, a saturated calomel electrode (SCE) is coupled with a platinum indicator electrode. Nernst equation is employed to determine the potential, which depends on the concentration of the ionic species in the solution,

given as $E = E^\circ - \dfrac{0.0592}{n} \log_{10} \dfrac{[M_1^{n+}]}{[M_2^{n+}]}$, where n is

the change in oxidation number or the number of

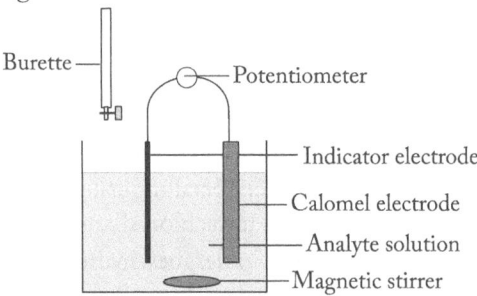

Fig. 5.24 A typical potentiometric titration cell

moles of electrons involved in the reaction. During potentiometric titration, the concentration of the analyte decreases and consequently, the electrode potential of the indicator electrode also decreases. This serves as an indication of the equivalence point of the titration which is accurately measured in the presence of the reference electrode.

The **advantages** of potentiometric titrations are:

(a) It can be carried out for coloured solutions where indicators cannot be used.

(b) One does not need prior information about relative strength of titrants before titration.

Fig. 5.25 elucidates typical potentiometric titration curves from which one can derive quantitative data.

The plot of E versus volume of titrant is also called the *direct plot* where at first when the titrant is added the change in potential is slow, but at equivalence point the potential rises sharply giving a typical S-shaped curve. Due to flatter slope nearing equivalence point, (Fig. 5.25(a)), it becomes harder to interpret. The first derivative curve is a plot of slope of the curve (i.e, $\Delta E/\Delta V$-ratio of change in emf and change in volume added) against volume of the titrant added. Here, equivalence point occurs at the titration volume corresponding to the maxima of the curve. In the second derivative curve, values from the first derivative curve are taken and plotted, that is, $\Delta^2 E/\Delta V$ against volume of titrant added. The point on volume axis where the curve cuts through zero on the ordinate gives the equivalence point.

The various important potentiometric titrations are: (a) acid–base, (b) oxidation–reduction (redox) and (c) precipitation.

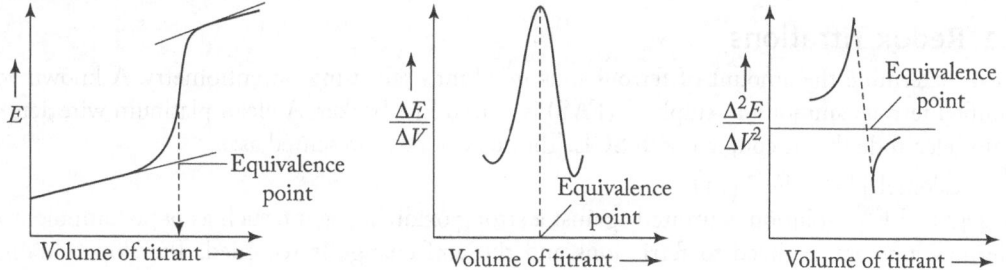

Fig. 5.25 Potentiometric titration curves: (a) potential vs volume of titrant, (b) first differential curve, and (c) second differential plot

Check Your Progress

15. What is electroplating?
16. What is electroless plating?
17. How is electroless plating carried out?
18. State Faraday's law of electrolysis.
19. Lead–acid battery can be easily recharged. Give the reason.
20. State the differences between conventional cells and fuel cells.
21. Name the components of a Ni–Cd battery.
22. Define the terms: (a) primary cells (b) reserve battery.
23. State the characteristics of an ideal battery.
24. State the application of hydrogen–oxygen fuel cell.
25. What are the components of a lead storage battery?
26. Write the major applications of lithium-ion cells.
27. State the role of MnO_2 and $ZnCl_2$ in a dry cell.
28. Name the electrolyte used in SOFC.
29. What is molten carbonate fuel cell? Why is it preferred over AFC? Give electrode reactions in MCFC.
30. State the working of PEMFC.

5.13.1 Acid–Base Titrations

In acid–base titrations, generally quinhydrone electrode is employed as an indicator electrode which is coupled with SCE. An indicator electrode which will be reversible to H^+ ions is selected, for example, hydrogen electrode, glass electrode, or quinhydrone electrode. The cell can be represented as:

$$(-) \, Pt \mid Hg, Hg_2Cl_2 \mid KCl \, (sat) \parallel H^+(test \, solution) \mid Q, QH_2 \mid Pt \, (+)$$

A known volume of the acid is taken in the beaker and the stoichiometric amount of quihydrone powder is added. Further, a clean platinum electrode is dipped in the acid solution. A strong base such as NaOH which acts as the titrant is filled in a burette. During titration, as NaOH titrant is added to the acid solution, H^+ ion concentration in the half cell containing quinhydrone will decrease. However, there will be a sharp rise in the potential and the steepest portion of the curve indicates the equivalent point of the titration. It is known that E_{cell} is a function of pH; hence, the cell reactions can be shown as:

At anode: $2 \, Hg + 2 \, Cl^- \rightarrow Hg_2Cl_2 + 2 \, e^-$

At cathode: $Q + 2 \, H^+ + 2 \, e^- \rightarrow QH_2$

Thus, $E_{cell} = E_{cathode} - E_{anode} = 0.6996 - 0.0592 \, pH$

The quinhydrone electrode cannot be used in solutions that would react with quinone or hydroquinone. Hydroquinone being a weak acid, the electrode cannot be used above pH 8.5 when the dissociation of hydroquinone becomes appreciable.

5.13.2 Redox Titrations

One can determine the amount of ferrous ions in Mohr's salt using potentiometry. A known volume of acidified ferrous ammonium sulphate (FAS) is placed in a beaker. A clean platinum wire acts as the indicator electrode that is coupled with SCE. The cell can be represented as:

$$calomel \parallel Fe^{2+}, Fe^{3+} \, (Pt)$$

The acidified FAS solution is titrated against a strong oxidizing agent, such as potassium dichromate. The ferrous ions get oxidized to ferric ions and the emf change is recorded. The slope obtained on plotting the potential values against the volume of titrant added, will give the equivalence point and can be calculated as, $E = E^{\circ} + \dfrac{0.0592}{n} \log_{10} \dfrac{[Fe^{3+}]}{[Fe^{2+}]}$ at $25 \, ^{\circ}C$

5.13.3 Precipitation Titrations

While determining chloride ions in solution, it is usually titrated with silver nitrate and emf changes are measured. In the potentiometric set-up, a saturated KCl solution is placed in a beaker equipped with a silver wire as the electrode. When silver nitrate is added from the burette, sparingly soluble precipitate of AgCl is formed and a reversible electrode of $Ag|AgCl_{(s)}|Cl^-$ is obtained which is coupled with SCE. The cell can be represented as: $Ag|AgCl_{(s)}|Cl^- \parallel SCE$

As precipitation proceeds, concentration of silver ions decreases due to the formation of sparingly soluble precipitate of silver chloride, varying the emf of the cell slightly. At equivalence point, a small addition of $AgNO_3$ causes a large change in the emf. On plotting a graph of emf values against the volume of titrant added, the steepest portion of the slope will give the equivalence point of the titration. One can calculate the emf as per the equation given as, $E = Constant - 0.0592 \log C_{Ag^+}$.

Check Your Progress

31. What is potentiometry?
32. What is a potentiometric titration? Mention its advantages.
33. Draw the different potentiometric titration curves.

SOLVED EXAMPLES

1. Calculate the standard electrode potential of $Cu|Cu^{2+}$ if the electrode potential at 25 °C is 0.296 V when the concentration of copper ions is 0.015 M.

Solution: As per Nernst equation,

$$E = E° + \frac{0.0592}{2} \log_{10}[Cu^{2+}]$$

$$E° = E - 0.0296 \log_{10}[Cu^{2+}] = 0.296 - 0.0296 \log_{10}[0.015] = 0.350 \text{ V}$$

Result: Standard electrode potential $E° = 0.350$ V

2. Calculate the emf of a Daniell cell at 25 °C with standard potential of 1.1 V, if concentrations of $ZnSO_4$ and $CuSO_4$ are 0.002 M and 0.2 M, respectively.

Solution: As per Nernst equation,

$$E = E° + \frac{0.0591}{2} \log_{10} \frac{[Cu^{2+}]}{[Zn^{2+}]}$$

$$= 1.1 + \frac{0.0592}{2} \log_{10} \frac{[0.2]}{[0.002]} = 1.1592 \text{ V}$$

Result: $E = 1.1592$ V

3. Calculate the standard emf of the cell: $Cd \,|\, Cd^{2+}(a=1) \,|\, Cu^{2+}(a=1) \,|\, Cu$.

Solution: $E°_{cell} = E°_{right} - E°_{left}$

$$E°_{cell} = E°_{Cu^{2+}/Cu} - E°_{(Cd/Cd^{2+})} = 0.3370 - (-0.4003) = 0.737 \text{ V}$$

Result: $E° = 0.737$

4. Calculate the standard free energy change for the following cell at 25 °C:

$$Ni \,|\, Ni^{2+}(a=1) \,|\, Ag^{+}(a=1) \,|\, Ag$$

Solution: At 25 °C, $E°_{Ni^{2+}/Ni} = -0.24$ V and $E°_{Ag^{+}/Ag} = 0.80$ V

$E°_{cell} = E°_{right} - E°_{left} = 0.80 - (-0.24) = 1.04$ V

We know, $\Delta G° = -nFE° = -2 \times 96500 \times 1.04 = -200720$

Result: $\Delta G° = -200720$ J/mol

5. The emf of a cell is represented as:

$$(-)SCE \,\|\, \text{test solution saturated with QH} \,|\, Pt(+); \ 0.2334 \text{ V}$$

at 298 K. Determine the pH of the test solution.

Solution: As per Eq. (5.9), $pH = \dfrac{0.4574 - E_{cell}}{0.0592}$

On substituting the values, we get,

$$pH = \frac{0.4574 - 0.2334}{0.0592} = 3.78$$

Result: The pH of the test solution = 3.78

6. The emf of a cell made by coupling glass electrode with SCE at 298 K was found to be 0.1153 V in the presence of a buffered solution at pH 4.2. When a test solution was analysed, its emf was found to be 0.1981 V. Determine the pH of the test solution.

Solution: We know that, $E_{cell} = E_{SCE} - E_G$ and $E_G = E°_G + 0.0592 \text{ pH}$

Hence, $E_{cell} = E_{SCE} - E°_G + 0.0592 \text{ pH}$

$$0.1153 = E_{SCE} - E°_G + 0.0592 \times 4.2$$

Further, for test solution, $0.1981 = E_{SCE} - E_G^\circ + 0.0592 \text{ pH}$

Equating the above expressions, we get,

$$\text{pH} = \frac{0.1981 - 0.1153}{0.0592} + 4.2 = 5.6$$

Result: The pH of the test solution = 5.6

7. Calculate the reduction potential of copper electrode if it is in contact with 0.6 M $CuSO_4$ solution at 298 K (Given E° of Cu = 0.34 V).

Solution: As per Nernst equation,

$$E = E^\circ + \frac{0.0592}{2} \log_{10}[Cu^{2+}]$$

$$E = 0.34 + \frac{0.0592}{2} \log_{10}[0.6] = 0.346 \text{ V}$$

Result: The reduction potential of copper electrode = 0.346 V at 298 K

8. Calculate the voltage of the below cell, if E° is 1.98 V.

$$Mg \mid Mg^{2+}(c = 1 \text{ M}) \parallel Cd^{2+}(c = 7 \times 10^{-11} \text{ M}) \mid Cd$$

Solution: At anode: $Mg \rightarrow Mg^{2+} + 2e^-$ and at cathode: $Cd^{2+} + 2e^- \rightarrow Cd$

$$E = E^\circ - \frac{0.0592}{2} \log_{10} \frac{[Mg^{2+}]}{[Cd^{2+}]} = 1.98 - \frac{0.0592}{2} \log_{10} \frac{[1]}{[7 \times 10^{-11}]} = 1.68 \text{ V}$$

Result: Voltage of the cell = 1.68 V

9. Write the electrode reactions and determine the emf of the following cell at 298 K if E° is 0.47 V.

$$Cu \mid Cu^{2+}(c = 1 \times 10^{-2}) \parallel Ag^+(c = 1 \times 10^{-1} M) \mid Ag$$

Solution: At anode: $Cu \rightarrow Cu^{2+} + 2e^-$ and at cathode: $2Ag^+ + 2e^- \rightarrow 2Ag$

Net cell reaction: $Cu + 2Ag^+ \rightarrow Cu^{2+} + 2Ag$

$$E = 0.47 - \frac{0.0592}{2} \log \frac{[1 \times 10^{-2}]}{[1 \times 10^{-1}]^2} = 0.44 \text{ V}$$

Result: emf of the cell = 0.44 V

10. Calculate the emf of the following concentration cell at 25 °C. $Cu \mid Cu^{2+}(0.05 \text{ M}) \parallel Cu^{2+}(5 \text{ M}) \mid Cu$

Solution: The cathode in the concentration cell has higher Cu^{2+} concentration. According to Nernst equation,

$$E_{cell} = \frac{0.0592}{n} \log_{10} \frac{[M_2]}{[M_1]} = \frac{0.0592}{2} \log_{10} \frac{[5]}{[0.05]} = 0.0592 \text{ V}$$

Result: emf of the cell = 0.0592 V

11. A cell is represented as $Hg \mid HgNO_3 (0.001M) \parallel HgNO_3 (0.01M) \mid Hg$. If the emf at 291 K is 0.029V, find the valence of Hg cell.

Solution: As per Nernst equation,

$$E_{conc.} = \frac{2.303RT}{n} \log_{10} \frac{[0.01]}{[0.001]}$$

$$0.029 = \frac{2.303 \times 8.314 \times 291}{n} \log_{10} \frac{[0.01]}{[0.001]} = \frac{5571.8 \times 1}{n \times 96500}$$

$$0.029 \times n \times 96500 = 5571.8$$

Thus, on rearranging the above equation and solving we get,

$$2798.5n = 5571.8$$

$$n = 1.99 \approx 2$$

Result: The valence of Hg cell = 2

12. What is the standard emf of an electrochemical cell made of Cd electrode in a 1.0 M $Cd(NO_3)_2$ solution and Cr electrode in 1.0 M $Cr(NO_3)_3$ solution?

Solution: The cell can be represented as: $Cr_{(s)}|Cr^{3+}_{(aq)}||Cd^{2+}_{(aq)}|Cd_{(s)}$

Anode (oxidation): $Cr_{(s)} \rightleftarrows Cr^{3+}$ (1 M) $+ 3e^-$

Cathode (reduction): $2e^- + Cd^{2+}$ (1 M) $\rightleftarrows Cd_{(s)}$

Net cell reaction: $2Cr_{(s)} + 3Cd^{2+}$ (1 M) $\rightleftarrows 3Cd_{(s)} + 2Cr^{3+}$ (1 M)

$E^{\circ}_{cell} = E_{cathode} - E_{anode} = -0.40 - (-0.74) = 0.34$

Result: Standard emf $E^{\circ} = 0.34$ V

13. Calculate the emf of the concentration cell: $Ag|AgNO_3(0.00486$ M$)||AgNO_3(0.048$ M$)|Ag$.

Solution: The electrode with higher electrolyte concentration is the cathode, whereas the other electrode will be the anode.

Hence the cell reactions will be as follows:

At anode: $Ag_{(s)} \rightarrow Ag^+ + e^-$ and at cathode: $Ag^+ + e^- \rightarrow Ag_{(s)}$

The emf of the cell is

$$E = \frac{0.0592}{1} \log_{10} \frac{[0.048]}{[0.00486]} = 0.0588 \text{ V}$$

Result: The emf of the cell = 0.0588 V

14. Calculate the emf of the following concentration cell at 25 °C: $Ni | Ni^{2+}(0.01$ M$) || Ni^{2+}(0.1$ M$) | Ni$

Solution: According to Nernst equation,

$$E_{cell} = \frac{0.0592}{n} \log_{10} \frac{[M_2]}{[M_1]} = \frac{0.0592}{2} \log_{10} \frac{[0.1]}{[0.01]} = 0.0296 \text{ V}$$

Result: The emf of the cell = 0.0296 V

15. If emf of the cell: $Pt|H_2|$test solution$|SCE$ is 0.392 V at 298K, what will be the pH of the test solution?

Solution: We know,

$$\text{pH} = \frac{E_{cell} - E_{SCE}}{0.0592} = \frac{0.392 - 0.242}{0.0592} = 2.53$$

Result: pH of the solution = 2.53

16. Consider that a current of 3 A is allowed to pass through a solution of copper sulphate for 20 minutes. The amount of copper deposited at the cathode can be calculated. (We know that atomic weight of Cu = 63).

Solution: According to Faraday's law, $W = \dfrac{I \times t \times A}{nF}$

Hence, weight of copper deposited is $W = \dfrac{3 \times 1200 \times 63}{2 \times 96500} = 1.175$ g

Result: Weight of copper deposited is 1.175 g.

17. The emf of the following cell is 1.3279 V at 298 K: $Pt|Sn^{2+}(0.1$M$), Sn^{4+}(0.1$M$)||Cl^-(0.01$M$), Cl_2(1$ Atm$)|Pt$

Calculate the standard potential of chlorine electrode, if standard potential of $Pt | Sn^{2+}, Sn^{4+}$ electrode is 0.15 V.

Solution: Cell reactions are:

$Sn^{4+} + 2e^- \rightarrow Sn^{2+}$

$2Cl^- \rightarrow Cl_2 + 2e^-$

Net cell reaction: $Cl_2 + Sn^{2+} \rightleftarrows Sn^{4+} + 2Cl^-$

$$E = E^{\circ}_{cell} - \frac{0.0592}{2} \log_{10} \frac{[Cl^-]^2 \times [Sn^{4+}]}{[Sn^{2+}][Cl_2]}$$

$$1.3279 = E^{\circ}_{cell} - \frac{0.0592}{2} \log_{10} \frac{[0.01]^2 \times [0.1]}{[0.1][1]}$$

$$= E^{\circ}_{cell} - 0.059 \times -2$$

$$= E^{\circ}_{cell} + 0.1184$$

∴ $E^{\circ} = 1.3279 - 0.1184 = 1.2095$ V

∴ $E^{\circ}_{cell} = E^{\circ}_{R} - E^{\circ}_{L}$

$1.2095 = E^{\circ}_{Cl_2} - 0.15$

∴ $E^{\circ}_{Cl_2} = 1.3595$ V

Result: The standard potential of chlorine electrode = 1.3595 V

18. Calculate the equilibrium constant for the reaction in the cell: $^{-}Zn \mid Zn^{2+}(1 \text{ M}) \parallel Sn^{2+}(1 \text{ M}) \mid Sn^{+}$

Solution: Reduction reaction: $Sn^{2+} + 2e^{-} \rightarrow Sn$

Oxidation reaction: $Zn \rightarrow Zn^{2+} + 2e^{-}$

$$E^{\circ}_{cell} = \frac{0.0591}{n} \log_{10} K$$

$$E^{\circ}_{cell} = E^{\circ}_{R} - E^{\circ}_{L} = -0.140 - (-0.763) = 0.623 \text{ V}$$

∴ $$0.623 = \frac{0.0591}{2} \log_{10} K$$

On solving,

$\log_{10} K = 21.08$

Thus, $K = 1.202 \times 10^{21}$

Result: The equilibrium constant $K = 1.202 \times 10^{21}$. The high value of equilibrium constant indicates that the cell reaction is complete.

19. Calculate the equilibrium constant for the reaction between silver nitrate and metallic zinc.

Solution: The cell reaction is, $2Ag^{+} + Zn \rightleftharpoons Zn^{2+} + 2Ag$

We know, $E^{\circ}_{cell} = \frac{0.0591}{n} \log_{10} K$

$E^{\circ}_{cell} = 1.56$ V

On solving,

$$\log_{10} K = \frac{n \, E^{\circ}_{cell}}{0.0591} = \frac{2 \times 1.56}{0.0591} = \frac{3.12}{0.0519} = 52.791$$

Result: Thus, equilibrium constant $K = 6.18 \times 10^{52}$

20. Consider the following cell: $Ag \mid AgNO_3(0.0093 \text{ N}) \parallel AgNO_3(x \text{ N} = ?) \mid Ag$. The emf of the above cell is 0.086 V at 25 °C. Determine the concentration of x.

Solution: $E = \frac{0.0591}{n} \log \frac{C_2}{C_1}$ $0.086 = \frac{0.0591}{1} \log \frac{x}{0.0093}$

Antilog $\frac{0.086}{0.0591} = \frac{x}{0.0093}$

$28.520 = \frac{x}{0.0093}$ or $x = 0.265$ M

Result: Hence, the concentraiton of x is 0.265 M

21. The E° values of Li/Li^{+}, Zn/Zn^{2+}, Cu/Cu^{2+}, and Ag/Ag^{+} are -3.0 V, -0.77 V, $+0.33$ V and $+0.80$ V, respectively. Which combination of electrodes will you use to construct a cell of higher emf if ionic concentrations are 0.1 M, 1.0 M, 10 M, and 0.01 M in same order?

Solution: As per Nernst equation, the potential of half cell is:

$$E = E° + \frac{0.0591}{n} \log_{10} [M^{n+}]$$

(a) For Li/Li$^+$, electrode potential will be,

$$E = -3.0 + \frac{0.0591}{1} \log_{10} [0.1] = -3.0 - 0.0591 = -3.0591 \text{ V}$$

(b) For Zn/Zn^{2+}, electrode potential will be,

$$E = -0.77 + \frac{0.0591}{2} \log_{10} [1.0] = -0.77 \text{ V}$$

(c) For Cu/Cu^{2+}, electrode potential will be,

$$E = 0.34 + \frac{0.0591}{2} \log_{10} [10] = 0.34 + 0.0295 = 0.3694 \text{ V}$$

(d) For Ag/Ag$^+$, electrode potential will be,

$$E = 0.80 + \frac{0.0591}{1} \log_{10} [0.010] = 0.80 - 0.1182 = 0.6818 \text{ V}$$

Result: From the above values, the highest value of cell emf can be obtained by coupling Ag/Ag$^+$ (cathode) and Li/Li$^+$ (anode) half cells.

$$E_{cell} = E_{cathode} - E_{anode} = 0.6818 - (-3.0591) = 3.7409 \text{ V}$$

22. Consider the cell: $Mg_{(s)} | Mg^{2+} (aq) \| Ag^+_{(aq)} | Ag_{(s)}$. Calculate $E°_{cell}$ and the maximum work that can be obtained by operating the cell. $\left(\text{Given: } E°_{Mg/Mg^{2+}} = -2.37 \text{ V and } E°_{Ag/Ag^+} = 0.80 \text{ V} \right)$.

Solution: $E°_{cell} = E°_R - E°_l = 0.8 - (-2.37) = 3.17 \text{ V}$

$w_{max} = -nFE°_{cell} = 2 \times 96500 \times 3.17 = 611810 \text{ J} = 611.810 \text{ kJ}$

Result: $E°_{cell} = 3.17 \text{ V}; w_{max} = 611.810 \text{ kJ}$

23. Calculate the standard free energy change at 25 °C for the reaction: $Zn_{(s)} + 2 Ag^+_{(aq)} \rightarrow Zn^{2+}_{(aq)} + 2 Ag_{(s)}$

Solution: For the above reaction, the free energy change can be calculated as:

$$\Delta G = -nFE = -2 \times 96500 \times 1.56 = -301080 \text{ J} = -301.08 \text{ kJ (coulombs} \times \text{volts} = \text{joules)}$$

Result: Free energy change = -301.08 kJ

24. If the concentration of Ca^{2+} in a saturated solution of CaF$_2$ (fluorite) is 2.1×10^{-4} M, what is the solubility product of fluorite?

Solution: The solubility of fluorite can be shown as:

$$CaF_{2(s)} \rightarrow Ca^{2+}_{(aq)} + 2F^-_{(aq)}$$

As per the above expression, Ksp = $[Ca^{2+}] [F^-]^2$

Thus, Ksp $= (2.1 \times 10^{-4}) (4.2 \times 10^{-4})^2 = 3.7 \times 10^{-11}$

Result: Solubility = 3.7×10^{-11}

25. Calculate solubility and solubility product of silver iodide in pure water at 298 K. (Given: $E°$ of Ag | Ag$^+$ = +0.7991 V and Ag | AgI, I$^-$ = −0.152 V ; Atomic weight of I = 126.92).

Solution: The solubility of silver iodide can be expressed as,

$$AgI \rightleftharpoons Ag^+ + I^-$$

The above reaction takes place in a typical cell as:

$$Ag | Ag^+ \| I^-, AgI(s) | Ag$$

Reduction: $AgI + e^- \rightarrow Ag + I^-$
Oxidation: $Ag \rightarrow Ag^+ + e^-$

Net reaction $AgI \rightleftharpoons Ag^+ + I^-$

The emf of the cell is equal to the standard emf, i.e., E°_{cell}.

$$E^\circ_{cell} = -0.152 - 0.7991 = -0.9511 \text{ V}$$

We know that, $E^\circ_{cell} = \dfrac{0.0591}{n} \log_{10} K_{sp}$

$\therefore$ $-0.9511 = \dfrac{0.0591}{1} \log_{10} K_{sp}$

Thus, $K_{sp} = 8.072 \times 10^{-17}$

Solubility (s) in pure water = $\sqrt{K_{sp}}$

$$s = \sqrt{8.072 \times 10^{-17}} = 8.983 \times 10^{-9} \text{ mol/dm}^3$$

Solubility in g/dm^3 = $8.983 \times 10^{-9} \times 235 = 2.111 \times 10^6$ (235 = mwt of AgI)

Result: Thus solubility of AgI = 2.111×10^{-6} g/dm^3; Solubility product = 8.072×10^{-17}

SUMMARY

○ Electrode potential is a measure of the driving force of an electrochemical reaction. All E° values are independent of stoichiometric coefficients for the half-cell reaction. If E° cell is positive, the reaction will occur spontaneously under standard conditions. However, if it is negative, then the reaction is non-spontaneous under standard conditions, but will proceed spontaneously in the opposite direction.

○ Nernst equation allows the determination of cell potential under non-standard conditions. It also relates cell potentials to the reaction quotients and provides accurate determination of equilibrium constants.

○ Reference electrodes are of different types, such as hydrogen and calomel electrodes. Quinhydrone, hydrogen, and glass electrodes are employed in routine pH determination.

○ Electroplating involves deposition of a noble metal layer on an active base metal to provide aesthetics

and reduce the chance of corrosion. It follows the principles of Faraday's law of electrolysis.

○ Electroless plating or autocatalytic plating occurs without the application of current. In electroless plating of nickel, the base metal is dipped in nickel sulphate bath and sodium hypophosphite, when Ni ions from the electrolyte forms a nickel phosphide adherent film on the metal object.

○ Batteries are portable power devices and could be primary, secondary, or reserve batteries. Examples include dry-cell batteries, lead storage battery, Ni–Cd, Ni–MH, lithium-based cells, fuel cells, and photovoltaic cell.

○ Potentiometric titration is an important electroanalytical technique which involves measuring emf of cell comprising indicator and reference electrodes. These titrations can be of three types: acid–base, redox, and precipitation titrations.

KEY FORMULAE

- Nernst equation:

$$E_{cell} = E^\circ_{cell} - \frac{2.303RT}{nF} \log_{10} \left[\frac{M_1^{n+}}{M_2^{n+}} \right]$$

$E^\circ_{cell} = E^\circ_{cathode} - E^\circ_{anode}$

$\Delta G^\circ = -nFE^\circ_{cell}$

- For pH calculation using hydrogen electrode:

$$pH = \frac{E_{cell} - E_{SCE}}{0.0592}$$

- For pH calculation using quinhydrone electrode:

$$pH = \frac{0.4574 - E_{cell}}{0.0592}$$

- For pH calculation using glass electrode:

$$pH = \frac{E^\circ_G - E_{SCE} - E_{cell}}{0.0592}$$

- Concentration cell: $E = \dfrac{0.0591}{n} \log_{10} \dfrac{C_2}{C_1}$

- emf and equilibrium constant:

$$E^{\circ}_{cell} = \frac{0.0591}{n}\log_{10}K$$

- Solubility of sparingly soluble salt: Solubility (s) in pure water = $\sqrt{K_{sp}}$

- Solubility product: $E^{\circ}_{cell} = \frac{0.0591}{n}\log_{10}K_{sp}$

- Faraday's law, $W = \frac{I \times t \times A}{nF}$

GLOSSARY

Anode: A half cell in an electrochemical cell, where oxidation occurs.

Battery: A portable power device that involves conversion of chemical energy to electrical energy.

Cathode: A negative half cell in an electrochemical cell where reduction occurs.

Cell potential: The overall (or net) electrode potential of an electrochemical cell.

Electrochemistry: The study of the relationship between chemical reactions and electricity.

Electrode (half cell): Usually a conducting metal immersed in an electrolyte.

Electrode potential: The potential developed at the interface of an electrode and the solution of its own salt.

Electroless plating: Also called displacement plating, here the noble metal layer gets deposited on the base metal object by displacing the base metal.

Electroplating: The electro-deposition of metals on metals, non-metals, and alloys to prevent corrosion.

Glass electrode: An ion-selective electrode, made of silver wire coated with silver chloride dipped in dilute HCl solution; sensitive towards H^+ ions.

Liquid junction potential: The potential difference developed at the interface of two electrolytic solutions of differing concentrations.

Potentiometry: An electroanalytical technique that involves measuring potentials in an electrochemical cell due to free energy changes occurring during chemical reactions until equilibrium is attained.

Redox reaction: The gain and loss of electrons (oxidation–reduction) in a reaction.

Salt bridge: A U-shaped tube comprising KCl or KNO_3 salt set in agar or gelatin that allows migration of ions between two half cells along with maintaining electrical neutrality.

Saturated calomel electrode: A subsidiary reference electrode consisting of mercury in contact with KCl solution saturated with calomel (Hg_2Cl_2).

EXERCISES

Multiple Choice Questions

1. During electroless plating process, the base metal object is dipped in
   - (a) highly active metal salt
   - (b) more noble metal salt
   - (c) in a mixture of active–passive metal salts
   - (d) none of these

2. While charging of lead–acid battery, the concentration of sulphuric acid
   - (a) increases
   - (b) decreases
   - (c) remains unchanged
   - (d) first decreases and then increases

3. The glass membrane present in glass electrode undergoes exchange of sodium ions with
   - (a) Ca^{2+}
   - (b) H^+
   - (c) Mg^{2+}
   - (d) NH_4^+

4. The emf of a cell at equilibrium is
   - (a) 100 V
   - (b) > 100 V
   - (c) 0 V
   - (d) < 0 V

5. The graph plotted in a typical potentiometric titration is
   - (a) electrode potential vs pressure of titrant
   - (b) electrode potential vs intensity
   - (c) electrode potential vs concentration of titrant
   - (d) electrode potential vs titrant volume

6. Leclanché cell is an example of
   - (a) secondary cell
   - (b) fuel cell
   - (c) primary cell
   - (d) none of these

7. Fuel cells convert chemical energy to
   - (a) potential energy
   - (b) electrical energy
   - (c) heat
   - (d) Pressure

8. The electrode potential of standard calomel electrode
   - (a) 0.2400 V
   - (b) 0.2300 V
   - (c) 0.2422 V
   - (d) 0.1 V

9. The electrode used to determine pH is
   - (a) glass electrode
   - (b) calomel electrode
   - (c) redox electrode
   - (d) silver electrode

10. Calomel electrode comprises calomel in a solution of
    - (a) sat. NaCl
    - (b) sat. AgCl
    - (c) Ag–AgCl
    - (d) sat. KCl

11. An electrochemical series is arranged in
    - (a) increasing order of standard reduction potential
    - (b) decreasing order of standard reduction potential
    - (c) increasing order of oxidation potentials
    - (d) decreasing order of equivalent weights

12. The potential of a single electrode in a half cell is called
    - (a) reduction potential
    - (b) single electrode potential
    - (c) cell potential
    - (d) half-wave potential

13. Which of the following constitutes a Daniell cell?
    - (a) Cu–Mg
    - (b) Zn–Ag
    - (c) Zn–Cu
    - (d) Ag–AgCl

14. The cathode material in nickel–metal hydride battery is made of
    - (a) NiOOH
    - (b) NiO_2
    - (c) H_2 adsorbed on metal alloy
    - (d) $Ni(OH)_2$

15. The standard emf of a Daniell cell is
    - (a) 1.1 V
    - (b) 0.1 V
    - (c) 1.12 V
    - (d) 1.4 V

16. The emf of a cell in terms of reduction potential of its left and right electrodes is calculated as
    - (a) $E = E_{left} + E_{right}$
    - (b) $E = E_{left} - E_{right}$
    - (c) $E = E_{right} - E_{left}$
    - (d) none of these

17. If standard emf of Daniell cell is 1.1 V, the free energy change will be
    - (a) 212300 J/mol
    - (b) - 212300 J/mol
    - (c) 200 J/ mol
    - (d) 210 J/mol

18. Solar grade silicon is prepared by
    - (a) pyrolysis
    - (b) junction method
    - (c) electroplating

    - (d) Union Carbide process

19. Which of the following is NOT a fuel cell?
    - (a) solid oxide based
    - (b) molten carbonate based
    - (c) solar cell
    - (d) phosphoric-acid based

20. Electroplating process is based on
    - (a) corrosion
    - (b) fuel cells
    - (c) solar power
    - (d) electrolysis

21. Dry cell is an example of
    - (a) reversible cell
    - (b) irreversible cell
    - (c) solar cell
    - (d) fuel cell

22. Hydrogen–oxygen fuel cell is used in space vehicles to supply
    - (a) O_2
    - (b) H_2
    - (c) H_2O
    - (d) NaCl

23. During discharge of lead storage battery
    - (a) H_2SO_4 is used.
    - (b) Pb is deposited.
    - (c) $PbSO_4$ is consumed.
    - (d) The battery stops working.

24. The role of salt bridge in a galvanic cell is to
    - (a) eliminate liquid junction
    - (b) transfer electrolyte
    - (c) dissolve ions
    - (d) none of these

25. In lithium-ion cell, graphite acts as the
    - (a) cathode
    - (b) electrolyte
    - (c) barrier
    - (d) anode

26. The cathode in Ni–Cd battery is
    - (a) NiOH
    - (b) $Ni(OH)_2$
    - (c) NiO(OH)
    - (d) NiO_2

27. Which of the following statements associated with batteries is incorrect?
    - (a) In a dry cell, the reaction $Zn \rightarrow Zn^{2+}$ continues to occur even when the battery is not being used.
    - (b) Secondary batteries are rechargeable.
    - (c) Cell reaction in a primary battery is not reversible.
    - (d) Electrodes with larger surface area generate more electrode potential.

28. Which of the following is an advantage of fuel cells?
    - (a) They can be recharged by adding more material to be oxidized and reduced.

(b) They produce little or no harmful pollutants while working.

(c) They can be made to run very quietly.

(d) All of the above.

29. The salt bridge in the electrochemical cell serves to

(a) increase the rate at which equilibrium is attained

(b) increase the voltage of the cell

(c) maintain electrical neutrality

(d) increase the oxidation/reduction rate

30. Standard cell potential is measured

(a) at 25 °C

(b) when ion concentrations of aqueous reactants are 1.00 M

(c) under the conditions of 1.00 Atm for gaseous reactants

(d) all of the above

Review Questions

1. Explain concentration cell with a suitable example and derive an expression for emf of the concentration cell.

2. Discuss the construction and working of a standard hydrogen electrode.

3. What is electrochemical series? State its applications.

4. Write a note on reference electrodes.

5. Explain electroplating and state the various factors affecting the metal deposition.

6. Explain electroless plating of nickel.

7. Explain liquid junction potential. State the role of salt bridge.

8. Derive Nernst equation and state its importance in studying electrochemical cells.

9. Derive the relation between equilibrium constant and emf of the cell.

10. Explain the determination of solubility and solubility product using emf measurements.

11. List the various electrochemical conventions one employs while depicting electrochemical cells and their reactions.

12. Explain electrode and electrolyte concentration cells with suitable examples.

13. Explain the construction and working of a calomel electrode.

14. Distinguish between calomel and hydrogen electrodes.

15. What is standard hydrogen electrode? How are single electrode potentials determined?

16. Explain the construction and working of quinhydrone electrode.

17. What is pH? How does one determine the pH of a solution using glass electrode.

18. Describe pH determination using quinhydrone electrode.

19. Describe the construction of a glass electrode. How does one determine the pH of a solution using glass electrode? State the equation and explain the terms involved in it.

20. Write short notes on: (a) saturated calomel electrode; (b) standard hydrogen electrode; (c) glass electrode electrode; (d) quinhydrone electrode.

21. Describe the principle, construction, and working of hydrogen–oxygen fuel cell.

22. Outline the electrochemical processes occurring in: (a) Ni–MH battery; (b) lead–acid battery; (c) lithium cells.

23. List the advantages and disadvantages of fuel cells.

24. With neat labelled diagrams, explain the working of solid oxide fuel cell and methanol-oxygen fuel cell.

25. Differentiate between conventional batteries and fuel cells.

26. Discuss the construction and working of lead–acid battery.

27. What are fuel cells? Explain any one type of fuel cell in detail.

28. Explain the construction and working of methanol–oxygen fuel cell.

29. Explain the following w.r.t. construction, working, and applications: (a) Molten carbonate fuel cell; (b) solid oxide fuel cell; (c) phosphoric acid fuel cell; (d) polymer electrolyte membrane fuel cell; (e) alkaline fuel cell.

30. Explain the principle of photovoltaic cell.

31. Discuss the construction and working of a solar cell.

32. Discuss the manufacture of solar grade silicon by Union Carbide process.

33. Explain redox titration using potentiometry.

34. State the principle of potentiometric titrations. Explain the various potentiometric curves.

35. What are potentiometric curves? Explain first and second differential plots of a potentiometric titration.

36. Write a short note on potentiometric titrations.

NUMERICAL PROBLEMS

1. Calculate emf of the cell: $Al \mid Al^{3+}(c = 0.1\ M) \parallel Ni^{2+}(c = 0.01\ M) \mid Ni$
   Given that: $E^{\circ}_{Al^{3+}/Al} = -1.66\ V$ and $E^{\circ}_{Ni/Ni^{2+}} = -0.23V$ **(Ans: 1.43V)**

2. Calculate E°_{cell} and w_{max} for:
   $Al \mid Al^{3+}(c = 0.01M) \parallel Fe^{2+}(c = 0.2M) \mid Fe$ if $E^{\circ}_{Al^{3+}/Al} = -1.66\ V$ and $E^{\circ}_{Fe/Fe^{2+}} = -0.44\ V$ **(Ans: $E^{\circ}_{cell} = 1.22\ V$ and w_{max} = 706.38 kJ)**

3. Calculate the reduction potential of copper electrode if it is in contact with 0.5 M $CuSO_4$ solution at 298 K (Given E° of Cu = 0.34 V). **(Ans: 0.331V)**

4. The emf of a cell represented as: $(-)\ SCE \parallel$ test solution saturated with QH $\mid$ Pt (+); 0.234 V at 298 K. Determine the pH of the test solution. **(Ans: pH = 3.773)**

5. The emf of a cell is made by coupling glass electrode with SCE at 298 K and was found to be 0.135 V in the presence of a buffered solution at pH 4.2. When the test solution was analysed, its emf was found to be 0.191 V. Determine the pH of the test solution. **(Ans: pH = 5.14)**

6. Calculate the standard free energy change for Cu–Zn cell at 25 °C if the cell develops a potential of 1.1 V. **(Ans: ΔG = – 212.3 kJ/mol**

7. Calculate the emf of a Daniell cell at 25 °C with standard potential of 1.1 V, if concentrations of $ZnSO_4$ and $CuSO_4$ are 0.0015 and 0.15 M respectively. **(Ans: emf = 1.1592 V)**

8. Calculate the voltage of the cell, if E° is 1.998 V, $Mg \mid Mg^{2+}(c = 1\ M) \parallel Cd^{2+}(c = 6 \times 10^{-11}M) \mid Cd$ **(Ans: Voltage: 1.695 V)**

9. Determine the electrode potential of Pt $\mid$ Sn^{2+} (c=0.03), Sn^{4+} (c=0.02) if E° is 0.15 V. **(Ans: 0.1448 V)**

10. Determine the equilibrium constant for the following reaction:
    $Cu(s) + 2\ Ag^+ \rightarrow Cu^{2+}(aq) + 2\ Ag(s)$, if $E^{\circ}_{cell} = 0.46\ V$ **(Ans: K = 3.69 × 10^{15})**

11. If in an electroplating process, a current of 3.2 A is allowed to pass through a solution of $CuSO_4$ for 25 min, calculate the amount of copper deposited at the cathode. (Given: at. wt of Cu = 63). **(Ans: Wt of Cu = 1.566 g)**

12. Determine the emf of the following concentration cell: $Cu \mid Cu^{2+}(c = 0.072\ M) \parallel Cu^{2+}(c = 0.32\ M) \mid Cu$
    Also, comment on its spontaneity. **(Ans: 0.0191 V and cell emf is positive, spontaneous).**

13. Consider the following cell: $Cd(s) \mid Cd^{2+}(c = 0.0409\ M = ?) \parallel Ni^{2+}(c = x\ M) \mid Ni(s)$
    The emf of the cell is 0.20 V. Calculate the molar concentration of nickel ions in solution. **(Ans: 0.193 M)**

14. If the concentration of Ca^{2+} in a saturated solution of CaF_2 is 3.1 × 10⁻⁴ M, what is the solubility product of fluorite? **(Ans: K_{sp} = 1.191 × 10⁻¹⁰)**

15. Calculate the solubility and solubility product of silver iodide in pure water at 298 K. (Given: E° of Ag $\mid$ Ag^+ = +0.7991 V and $Ag \mid AgI$, I^- = –0.152 V; At. wt of I = 126.92). **(Ans: s = 2.11 × 10⁻⁶ g/dm³ and K_{sp} = 8.072 × 10⁻¹⁷)**

FURTHER READING

1. Bard, Allen J. and Faulkner R. Larry, *Electrochemical Methods—Fundamentals and Applications*, USA, 2006.
2. Ives, David J. G. and George J. Janz, *Reference Electrodes – Theory and Practice*, Academic Press, New York, 1961.
3. Koppel, Tom, *Powering the Future: The Ballard Fuel Cell and the Race to Change the World*. John Wiley & Sons, Incorporated, New York, 1999.
4. Oldham, Keith B., Jan C. Myland, and Alan M. Bond, *Electrochemical Science and Technology: Fundamentals and Applications*. John Wiley and Sons, UK, 2012.
5. O'Hayre, Ryan et.al., *Fuel Cell Fundamentals*, John Wiley and Sons, UK, 2006.

ANSWERS

Check Your Progress

1. A device that produces electric current from a chemical reaction (redox reaction) is an electrochemical cell.

2. Electrode potential is the tendency of an electrode to gain or lose electrons when placed in a solution of its own ions.

3. Refer to Section 5.2.1

4. If an electrode tends to lose electrons it is called oxidation potential, whereas if an electrode tends to gain electrons, it is called reduction potential.

5. Refer to Section 5.2.2

6. Reference electrode is an electrode whose standard potential is known and is used to measure the electrode potentials of other electrodes. SHE, SCE, quinhydrone electrode are examples.

7. SHE is not employed as a reference electrode since platinum wire tends to easily get poisoned in the presence of trace impurities in hydrogen gas. Also, maintaining hydrogen pressures at 1 Atm makes it difficult to handle. Hence, a calomel electrode is commonly employed for the purpose.

8. Yes, we can store $AgNO_3$ solution in a copper vessel as the reduction potential of Ag^+/Ag electrode (0.80 V) is higher than that of Cu^{2+}/Cu (0.34 V).

9. Since the reduction potential of Ni^{2+}/Ni electrode (0.25 V) is lower than Cu^{2+}/Cu (0.34 V), nickel will displace copper from its solution as per the chemical reaction, $Ni_{(s)} + Cu^{2+} \rightarrow Ni^{2+}_{(aq)} + Cu_{(s)}$

10. No, it is practically not possible to determine single electrode potential. There is a need of reference electrode whose potential is known and remains stable. Only the difference between these two electrodes can be measured practically.

11. If copper sulphate is stored in a nickel container, the following reaction will take place, $Ni + CuSO_4 \rightarrow NiSO_4 + Cu$ and the cell is represented as:

    $$Ni_{(s)}|Ni^{2+}_{(aq)}||\ Cu^{2+}_{(aq)}|Cu(s)$$

    $$E°_{cell} = E°_{cathode} - E°_{anode} = 0.34 - (-0.25) = 0.59\ V$$

    As the emf of the cell is positive, copper sulphate will react with nickel container and thus cannot be stored in it.

12. The arrangement of electrodes in the order of their increasing standard reduction potentials on a hydrogen scale is called electrochemical series.

13. $Ni\ (s)\ |\ Ni^{2+}\ (aq)\ |\ |\ Au^{3+}\ (aq)\ Au(s)$

    $$E°_{cell} = E°_{cathode} - E°_{anode} = 1.5 - (-0.25) = 1.75\ V$$

    As the value of $E°_{cell}$ is positive, the cell reaction is feasible.

14. Glass electrode is simple, stable, resistant to oxidation, and attains rapid equilibrium. It can be used up to pH 10. However, quinhydrone electrode cannot be used in redox reactions beyond pH 8.

15. Refer to Section 5.7

16. Refer to Section 5.8

17. Refer to Section 5.8

18. Refer to Section 5.7

19. In a lead–acid battery, lead sulphate gets deposited on the electrodes and hence electrode reactions can be reversed during recharge of the battery.

20. Refer to Table 5. 2

21. The components of a Ni–Cd battery include cadmium anode, nickel oxyhydroxide cathode, and 6 M KOH electrolyte.

22. Refer to Section 5.9

23. Refer to Section 5.9

24. Hydrogen–oxygen fuel cell is used in space vehicles to provide water.

25. The components of a lead storage battery include lead electrodes, positive electrode deposited with lead oxide, which is dipped in sulphuric acid.
26. Lithium-ion cells are used in cell phones and laptops.
27. MnO_2 is an oxidant in a dry cell. $ZnCl_2$ is added in a dry cell to avoid cracks in batteries. $ZnCl_2$ reacts with ammonia forming a complex salt $[Zn(NH_3)_2Cl_2]$. Due to this, the pressure of ammonia gas is lowered and thus cracks do not appear in a dry cell thereby preventing leakage.
28. The electrolyte used in SOFC is solid yttria-stabilized zirconium.
29. Refer to Section 5.11.4
30. PEMFC has a solid polymeric membrane electrolyte that allows protons to move through towards the cathode and combine with oxygen and electrons, producing water and heat.
31. Refer to Section 5.13
32. Refer to Section 5.13
33. Refer to Fig. 5.25

Multiple Choice Questions

1. (a)	2. (a)	3. (b)	4. (c)	5. (d)	6. (c)	7. (b)
8. (c)	9. (a)	10. (d)	11. (a)	12. (b)	13. (c)	14. (c)
15. (a)	16. (c)	17. (b)	18. (d)	19. (c)	20. (d)	21. (b)
22. (c)	23. (a)	24. (a)	25. (d)	26. (c)	27. (d)	28. (d)
29. (c)	30. (d)					

Chemical Kinetics

LEARNING OBJECTIVES

After reading this chapter, you will be able to:

- understand the difference between thermodynamics and kinetics.
- know the utility of kinetics.
- calculate the rate of a reaction.
- understand the concept of order and molecularity of a reaction.
- derive the integrated rate laws and calculate the half-life of a reaction.
- know the dependence of the rate of a reaction on concentration and temperature.
- understand the concept of pre-exponential factor and activation energy.
- determine the pre-exponential factor using transition state theory.
- understand the concept of potential energy surface.

6.1 INTRODUCTION

Kinetics involves the study of progress of a process from one state to another during a physical and/or chemical change. It starts with the measurement of rates of reactions, followed by the determination of the rate laws at the macroscopic level. Finally, the rate law is used to understand the mechanism of a reaction at the microscopic level.

Thermodynamics is concerned with only the initial and final states, whereas kinetics deals with what happens to a particular process at any time between two states. Thermodynamics helps us to find out the amount of product at equilibrium or on completion of a reaction. On the other hand, kinetics helps us to find the yield at any given time.

Some reactions are slow whereas others are fast. Some reactions happen only at high temperature. The knowledge of this can be used to our advantage. The degradation of paint and plastic should be slow, whereas an air–fuel mixture should burn rapidly at ignition. Burning of cooking gas should be fast whereas rusting of iron needs to be slowed down. The rate of oxidation of a fuel should be negligible at room temperature, but it should be very fast at the high temperature at which an engine operates.

Kinetic studies can be used to know the following.

- Amount of product or reactant: The rate law and its integrated form can be used to know the amount of product or reactant at any time. It also helps in finding the time required to obtain the maximum amount of product.
- Stability and rate of drug metabolism: A drug's stability is measured over time at typical room temperature and humidity conditions. This helps in deciding its expiry date. Moreover, the knowledge of its metabolic rate is essential in prescribing the interval of drug intake.
- Degradation/decomposition rate: Knowledge of the rate of degradation is important not only in drug industry, but also in bioremediation, degradation of environmental pollutants, deterioration of food, etc.

- Rate of product formation at different temperatures: Temperature can have dramatic effects on the rate of a reaction. Kinetics can help us quantify the effect of temperature on the reaction rate.
- Mechanism of reaction, microscopic nature of a reaction: One of the most important uses of kinetics is to know the mechanism of reaction. It helps us to devise alternative strategies for obtaining products in higher yield and/or shorter time.
- Effect of solvent on the reaction rate: Understanding the mechanism of a reaction helps us to know how the polarity and nature of solvent (protic or aprotic) can affect the rate of reaction.
- Slow and fast reactions: Some reactions are instantaneous and some are slow. Kinetics enables us to know the rate of reaction.
- Structure and nature of transition state: Knowledge of potential energy surface of reactants can help us to know the structure and nature of transition state.
- Role of catalysis: One of the most important applications of kinetics is in the field of enzyme catalysis. Kinetics helps us to understand the mechanism of enzyme action.
- Radiometric dating: The age of ancient relics, such as paintings, dinosaur fossils, and remains from ancient civilizations can be obtained by kinetic studies involving radiometric dating.

In order to understand and predict the behaviour of a physical or chemical process, knowledge of both thermodynamics and kinetics is required.

6.2 IMPORTANT CONCEPTS IN KINETICS

Stoichiometric coefficient of a reactant/product Stoichiometry represents the quantitative relationship between the number of moles of reactants and products based on the law of conservation of mass. It is negative for reactant(s) and positive for product(s).

For the reaction:

$$N_2(g) + 3H_2(g) \rightleftharpoons 2\,NH_3(g)$$

the balanced equation suggests that one mole of $N_2(g)$ reacts with three moles of $H_2(g)$ to give two moles of $NH_3(g)$. Thus, in this reaction, the stoichiometric coefficients of reactants/products $N_2(g)$, $H_2(g)$, and $NH_3(g)$ are −1, −3, and 2, respectively.

Kinetic reaction profile This is obtained by measuring the concentration of one of the reactants or product as a function of time during the course of a reaction.

Rate of reaction Every chemical or physical process takes place at its own speed. The speed of a chemical reaction is quantitatively measured in terms of the rate of the reaction. Rate of the reaction (ϑ) is defined as the change in the extent of reaction (ξ) with time (t) per unit volume (V).

$$\vartheta = \frac{1}{V}\frac{d\xi}{dt} \tag{6.1}$$

The infinitesimal change in number of moles of 'i' th reactant or product (dn_i) is given by

$$dn_i = v_i\,d\xi$$

where v_i is the stoichiometric coefficient of 'i' th species (reactants or products). It is negative for reactants and positive for products.

$$\vartheta = \frac{1}{V}\frac{1}{v_i}\frac{dn_i}{dt} = \frac{1}{v_i}\frac{dC_i}{dt} \tag{6.2}$$

where $C_i\,(= n_i/V)$ is the concentration of 'i' th reactant or product.

For a reaction

$$A + 2B \rightarrow 3C + D$$

$$\vartheta = \frac{d[D]}{dt} = \frac{1}{3}\frac{d[C]}{dt} = -\frac{d[A]}{dt} = -\frac{1}{2}\frac{d[B]}{dt}$$

$$\frac{d[D]}{dt} \Big/ \frac{d[C]}{dt} = 1/3$$

Thus, the rate at which any reactant/product is consumed or formed is proportional to its stoichiometric coefficient. The rate of a reaction is zero at equilibrium. It is important to optimize the rate of a reaction to get maximum yield in less time.

6.3 LAW OF MASS ACTION (RATE LAW)

The rate of a reaction (ϑ) depends on the concentration of the reactants (A, B, ...). This is known as the law of mass action.

$$\vartheta = f([A], [B], \ldots)$$

$$\vartheta = k[A]^a [B]^b \tag{6.3}$$

Here, k is the rate constant and is independent of concentration. $[A]$ and $[B]$ are the concentrations of reactants A and B, respectively; and a and b are numbers obtained from the experimental methods. It should not be determined by simply looking at the stoichiometry of the balanced equation.

Order of a reaction   It is obtained experimentally by measuring the dependence of the rate of a reaction on the concentration of reactants. The overall order of the reaction is equal to the sum of exponents of the concentration terms in the rate law.

$$\vartheta = k[A]^a [B]^b$$

Order of the reaction = $a + b$
Order of the reaction with respect to $A = a$
Order of the reaction with respect to $B = b$
The unit of k will depend on the order of reaction.

Elementary and composite reactions   Some of the reactions take place in one step, whereas others proceed via multiple steps. A reaction taking place in one step is called an elementary reaction, whereas a reaction which involves more than one elementary step is known as composite reaction.

Molecularity of a reaction   Molecularity is defined only for an elementary reaction. It is the number of reactant molecules that takes part in an elementary reaction. Thus, for an elementary reaction, $A \rightarrow P$, the molecularity is one. For an elementary reaction, $A + B \rightarrow P$, molecularity is two. Molecularity is same as the overall order of a reaction for an elementary reaction.

For a composite reaction, the molecularity is the total number of reactants taking part in the rate determining step of a reaction. Molecularity is always an integer, whereas order can be fraction or zero.

6.3.1 Determination of Rate of a Reaction

The rate of a reaction can be obtained by looking at the rate of change of concentrations of one of the reactants or product with time t. The slope of the plot of concentration versus time at any given instant will give the rate of change of concentration at that time.

Factors Affecting the Rate of a Reaction

Concentration of reactants   The concentration of reactants, in general, increases the rate of reaction. However, it may also not affect the rate of a reaction; for example, concentration of iodine does not affect the rate of reaction of iodination of acetone in the presence of H^+. The dependence of reaction rate on concentration of the reactant is given by the rate law. A type of reactant, called inhibitor, decreases the rate of a reaction.

Nature of reactants The nature of reactants greatly influences the rate of reaction; for example, the rate of reaction of a nucleophilic substitution at aliphatic halide (R–X) will differ for different R and X. The rate of a nucleophilic substitution reaction follows the order: R–Cl < R–Br < R–I since the bond gets weaker from R–Cl to R–I.

Stability of intermediates The stability of intermediates may also affect the rate of reaction. In S_N1 nucleophilic reaction, the rate of reaction increases with increase in the stability of the intermediate.

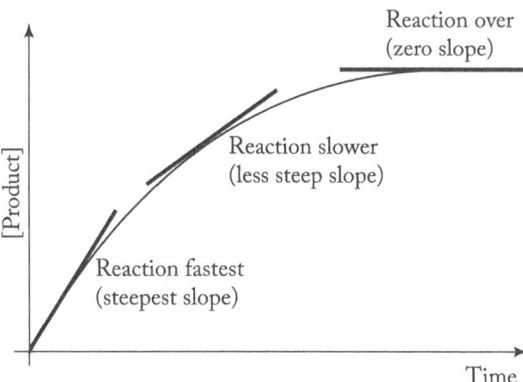

Fig. 6.1 Concentration of product as a function of time can be used to calculate the rate of reaction

Nature of solvent The solvent can stabilize the intermediates and can affect the rate of a reaction. A polar solvent stabilizes the carbocation formed and hence increases the rate of a reaction where carbocation is involved (in S_N1 substitution and E1 elimination).

Temperature In most cases, rate of a reaction increases with increase in temperature. The increase in the reaction rate can be explained by Arrhenius hypothesis (explained in Section 6.5). In most cases, the rate gets doubled with every 10 K increase in temperature.

Surface area of reactant Increase in surface area of reactant leads to increase in the rate of a reaction. Thus, the rate of reaction is much faster with a finely divided reactant than with its larger pieces.

Catalysts They increase the rate of a reaction without themselves undergoing any change. They do so by decreasing the activation energy.

Photochemical radiations The rate of some reactions, called photochemical reactions, increases when the reactants are irradiated by electromagnetic radiation of certain wavelength.

6.3.2 Determination of Rate Law

Reactions Involving Single Reactant
The following methods can be used to find the order of reactions involving a single reactant.

Differential Method It can be used to determine the order with respect to a particular reactant.
   If the reaction is of nth order with respect to a reactant A, then

$$\vartheta = k[A]^n \tag{6.4}$$

$$\frac{\vartheta}{[A]^n} = k \tag{6.5}$$

Check Your Progress

1. Reactions with high molecularity are rare. Give the reason.
2. Define the following: (a) order (b) molecularity of a reaction
3. Can order and molecularity be same for a reaction? Explain.
4. Define rate of reaction. What are the factors which can influence the rate of the reaction?
5. The rate of formation of AC in a gaseous hypothetical reaction,
   $4 A_2B + 10 C \rightarrow 8 AC + 2 B_2C$ is 4.87×10^{-3} mol.L^{-1}.s^{-1}.
   Find the rate of disappearance of C.

The order of the reaction with respect to A will be the value of n at which $\dfrac{\vartheta}{[A]^n}$ is constant. A more simplified approach can be obtained as follows:

Taking log on both sides of Eq. (6.4), we get

$$\log \vartheta = \log (k[A]^n)$$

$$\log \vartheta = \log k + n \log [A] \tag{6.6}$$

Thus, the plot of $\log \vartheta$ versus $\log [A]$ will be a straight line and the slope of the plot will be equal to the order of reaction.

Integrated Rate Laws   Rate laws are differential equations. They can be integrated to obtain the integrated rate law. These laws can be used to determine the order of a reaction. They can be derived for reactions of different order as follows:

Integrated rate law for a zero order reaction   A reaction, $A \rightarrow P$, is of zero order if the rate of reaction is independent of the concentration of reactant A as shown below.

$$-\frac{d[A]}{dt} = k$$

The unit of k for a zero order reaction is same as the rate of the reaction, that is, mol $L^{-1}s^{-1}$.

$$-d[A] = kdt$$

Integrating between time $t = 0$ to $t = t$, we get the integrated rate law for zero order.

$$-\int_{[A]_0}^{[A]} d[A] = \int_{t=0}^{t=t} k\, dt$$

$$[A]_0 - [A] = kt$$

$[A]_0$ is the concentration of A at $t = 0$, $[A]$ is the concentration of A at time t.

Examples of zero order reaction   A zero order reaction is generally observed in reactions involving a surface or a catalyst saturated with the reactant.

(i) Decomposition of N_2O on hot platinum surface: In this reaction, platinum surface is saturated with N_2O and the rate of reaction does not depend on the concentration of N_2O.

(ii) Iodination of acetone in the presence of H^+: The rate of reaction is zero order with respect to iodine.

Integrated rate law for first order reaction   A reaction is of first order if the rate of the reaction is proportional to the concentration of the reactant.

$$-\frac{d[A]}{dt} = k[A]$$

$$-\frac{d[A]}{A} = kdt$$

Integrating between time $t = 0$ to $t = t$, we get integrated rate law for a first order reaction.

$$-\int_{[A]_0}^{[A]} \frac{d[A]}{A} = \int_{t-0}^{t=t} k\, dt$$

$$\ln\frac{[A]_0}{[A]} = kt$$

$$\ln[A] = \ln[A]_0 - kt$$

For a first-order reaction, a plot of $\ln [A]$ vs t will yield a straight line with a slope of $-k$. Unit of k is s^{-1}.

Examples of first order reactions

(i) Decomposition of H_2O_2: The rate of reaction depends on concentration of H_2O_2.

(ii) Inversion of cane sugar: The rate of reaction depends on concentration of sugar.

Integrated rate law for second order reaction A reaction is of second order if the rate of the reaction is proportional to the square of the concentration of reactant.

$$-\frac{d[A]}{dt} = k[A]^2$$

$$-\frac{d[A]}{[A]^2} = kdt$$

Integrating between time $t = 0$ to $t = t$

$$-\int_{[A]_0}^{[A]} \frac{d[A]}{[A]^2} = \int_{t=0}^{t=t} kdt$$

$$\frac{1}{[A]} - \frac{1}{[A]_0} = kt$$

So if a process is second-order in A, a plot of $1/[A]$ vs t will yield a straight line with a slope of k.

Examples of second order reactions

(i) Decomposition of NO_2 to NO

(ii) Decomposition of N_2O to N_2

(iii) Saponification of ester

Integrated rate law for nth order reaction

$$-\frac{d[A]}{dt} = k[A]^n$$

$$-\frac{d[A]}{[A]^n} = kdt$$

Integrating between time $t = 0$ to $t = t$

$$-\int_{[A]_0}^{[A]} \frac{d[A]}{[A]^n} = \int_{t=0}^{t=t} kdt$$

$$\frac{1}{[A]^{n-1}} - \frac{1}{[A]_0^{n-1}} = (n-1)kt$$

Table 6.1 Integrated rate equation and linearized plot for different orders of reaction

Order of reaction	Integrated rate equation	Linear plot
Zero order	$[A]_0 - [A] = kt$	$[A]$ vs t
First order	$\ln\frac{[A]_0}{[A]} = kt$	$-\ln [A]$ vs t
Second order	$\frac{1}{[A]} - \frac{1}{[A]_0} = kt$	$\frac{1}{[A]}$ vs t
nth order	$\frac{1}{[A]^{n-1}} - \frac{1}{[A]_0^{n-1}} = (n-1)kt$	$\frac{1}{[A]^{n-1}}$ vs t

Table 6.1 shows the integrated rate equation and linearized plot for different orders of reaction.

Check Your Progress

6. Derive the integrated rate law for the first order reaction $R \rightarrow P$. If the rate constant is 3.06×10^{-3} s^{-1}, then calculate the % of the reactant left after 200 s.

7. In the reaction, $X + Y \rightarrow Z$; the rate of the reaction depends only upon the concentration of X and the plot of $1/[X]$ vs time is a straight line. What is the possible rate law for the reaction?

8. For a first order reaction $A \rightarrow B$, the rate constant is found to be $2\ s^{-1}$. What will be the concentration after 1 s, if the initial concentration is 1 M?

9. For reactions of which order, the units of rate of the reaction and rate constant are same?

10. State the condition when a bimolecular reaction follows first order kinetics. What are such reactions called?

	Concentration vs Time	Straight line plot
Zero order		
First order		
Second order		

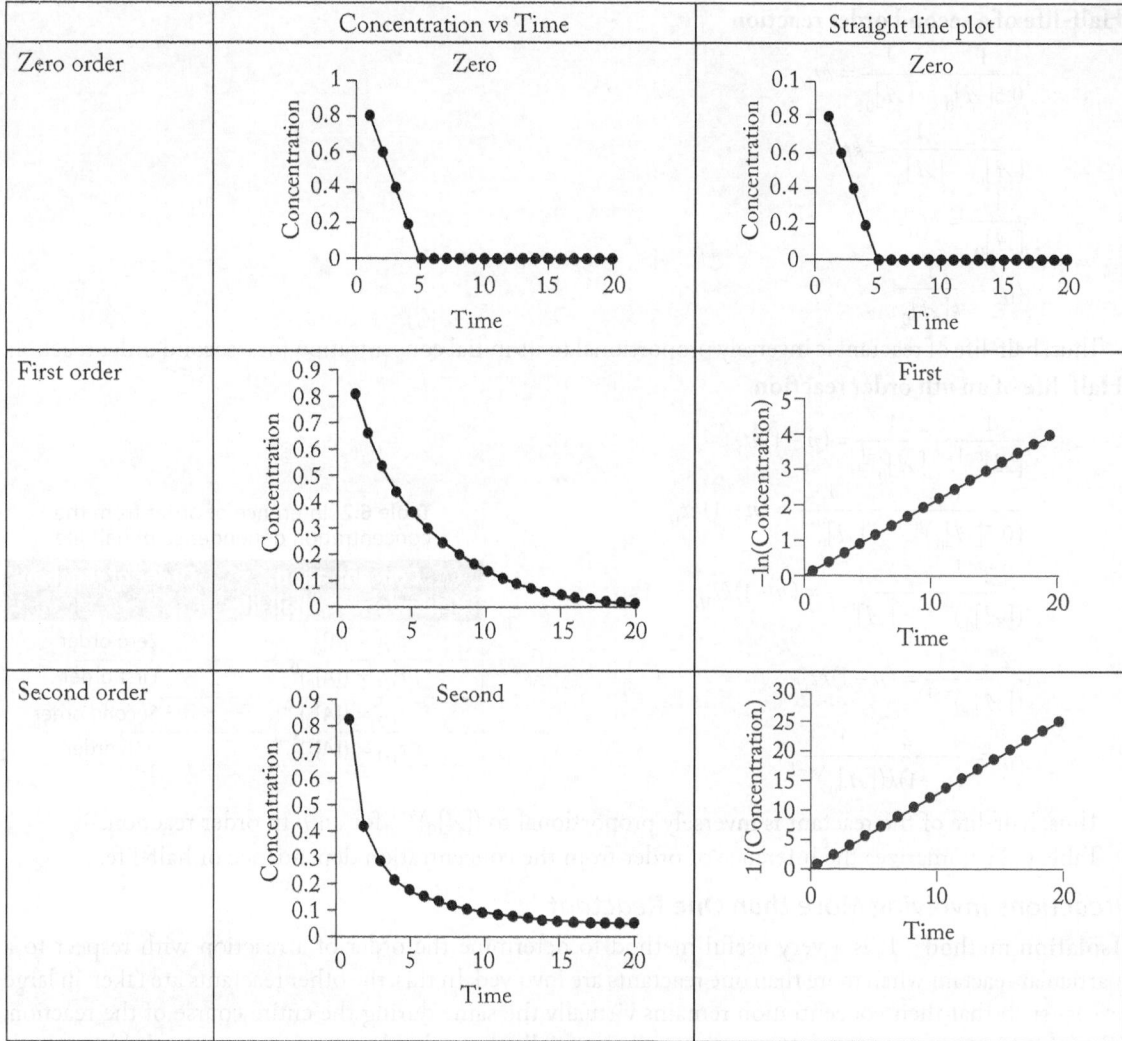

Fig. 6.2 Concentration of reactant versus time (left panel) and linearized plot for different orders of reaction (right panel)

Half-Life Method Half-life ($t_{1/2}$) is defined as the time required for one-half of a reactant to react. Because [A] at $t_{1/2}$ is one-half of the original [A], [A] = 0.5 [A]$_0$. The dependence of half-life on reactant concentration is different for reactions of different order. Thus, this fact can be exploited to find the order of the reaction.

Half-life of a first order reaction

$$\ln \frac{[A]_0}{0.5[A]_0} = kt_{1/2}$$

$$\ln 2 = kt_{1/2}$$

$$t_{1/2} = \frac{0.693}{k}$$

Thus, half-life of the reactant does not depend on its concentration for a first order reaction.

Half-life of a second order reaction

$$\frac{1}{0.5[A]_0} - \frac{1}{[A]_0} = kt_{1/2}$$

$$\frac{2}{[A]_0} - \frac{1}{[A]_0} = kt_{1/2}$$

$$\frac{1}{[A]_0} = kt_{1/2}$$

$$t_{1/2} = \frac{1}{k[A]_0}$$

Thus, half-life of reactant is inversely proportional to its initial concentration for a second order reaction.

Half-life of an *n*th order reaction

$$\frac{1}{[A]^{n-1}} - \frac{1}{[A]_0^{n-1}} = (n-1)kt_{1/2}$$

$$\frac{1}{(0.5[A]_0)^{n-1}} - \frac{1}{[A]_0^{n-1}} = (n-1)kt_{1/2}$$

$$\frac{2^{n-1}}{([A]_0)^{n-1}} - \frac{1}{[A]_0^{n-1}} = (n-1)kt_{1/2}$$

$$\frac{2^{n-1}-1}{([A]_0)^{n-1}} = (n-1)kt_{1/2}$$

$$t_{1/2} = \frac{2^{n-1}-1}{(n-1)k([A]_0)^{n-1}}$$

Table 6.2 Inference of order from the concentration dependence of half-life

Concentration dependence of half-life	Order of reaction
$t_{1/2} \propto [A]_0$	Zero order
$t_{1/2} \propto ([A]_0)^0$	First order
$t_{1/2} \propto ([A]_0)^{-1}$	Second order
$t_{1/2} \propto ([A]_0)^{1-n}$	*n* th order

Thus, half-life of the reactant is inversely proportional to $([A]_0)^{n-1}$ for an *n* th order reaction.
Table 6.2 summarizes the inference of order from the concentration dependence of half-life.

Reactions Involving More than One Reactant

Isolation method   It is a very useful method to determine the order of a reaction with respect to a particular reactant when more than one reactants are involved. In this, the other reactants are taken in large excess such that their concentration remains virtually the same during the entire course of the reaction. Thus, in this case, the rate of the reaction will essentially depend only on one reactant and the methods used for determining the rate order for one-reactant reactions can be applied.

Suppose a reaction involves two reactants, *A* and *B*.

$$\vartheta = k[A]^\alpha [B]^\beta$$

To determine the order of reaction with respect to *A*, *B* is taken in large excess such that concentration of *B* remains virtually the same throughout the reaction.

$$[B] \simeq [B]_0$$

Check Your Progress

11. Define half-life. If a first order decomposition reaction takes 30 min for 25 % completion, calculate the half-life of the reaction.
12. If half-life of a first order reaction is 3.9×10^{12} s, calculate the time required for 100 % completion of the reaction.
13. For a reaction, $X \rightarrow Y$; the initial concentration of *X* and the initial rate of the reaction are found to be 0.01 mol L^{-1} and 0.00154 M min^{-1}. If the reaction follows first order kinetics, calculate the half-life.
14. A first order reaction takes 60 minutes for 45% completion. Calculate its rate constant and half-life.

Here, $[B]_0$ and $[B]$ are initial concentration of B and concentration of B at a given time, respectively. Thus

$$\vartheta = k[A]^\alpha [B]_0^\beta$$
$$\vartheta = k'[A]^\alpha$$

It is evident that the rate of the reaction is dependent on only the concentration of A on addition of large excess of B and thus the order with respect to A can be measured as measured in case of reactions involving single reactant. Similarly, the order of reaction with respect to reactant B can be measured by taking A in large excess.

Method of initial rate   This method is based on the measurement of rate at the initial stages of the reaction for different concentrations of only one reactant while keeping the concentration of others constant.

Suppose a reaction involves two reactants A and B.

$$\vartheta = k[A]^\alpha [B]^\beta$$

The rate of reaction is determined at initial stage of the reaction where $[A] \sim [A]_0$ and $[B] \sim [B]_0$.

$$\vartheta = k[A]_0^\alpha [B]_0^\beta$$

Now, to get the order of reaction with respect to A, change the initial concentration of A keeping the initial concentration of B constant. Suppose that initial concentration of A is changed to $[A]_0'$.

$$\vartheta' = k[A]_0'^\alpha [B]_0^\beta$$

$$\frac{\vartheta'}{\vartheta} = \left(\frac{[A]_0'}{[A]_0} \right)^\alpha$$

If the reaction is of first order with respect to A

$$\frac{\vartheta'}{\vartheta} = \frac{[A]_0'}{[A]_0}$$

If the reaction is of second order with respect to A

$$\frac{\vartheta'}{\vartheta} = \left(\frac{[A]_0'}{[A]_0} \right)^2$$

Thus, the order can be obtained by the measurement of initial rates at two different concentrations of A. Similarly, the order of reaction with respect to B can be obtained by changing $[B]_0$ and keeping $[A]_0$ constant.

In this method, there is no need to take the other reactants in large excess. However, this method may not work for fast reactions since A will be consumed fast and $[A]$ will not be the same as $[A]_0$.

6.4 INTEGRATED RATE LAW IN TERMS OF PHYSICAL PARAMETER

Monitoring of concentration with time is difficult and in some cases not possible. In these cases monitoring of a physical property that is proportional to concentration can help us in obtaining the order of reaction and thereby the rate constant. For a gaseous reaction in a closed container, the pressure of gaseous reactants and products at a given temperature T is proportional to their number of moles. Thus, measurement of the total pressure during the course of reaction can help in obtaining the integrated rate law in terms of a physical parameter.

For a reaction, $2\,NO_2(g) \rightarrow N_2O_4(g)$

Total pressure will be given by

$$p_T = \frac{n_{NO_2} RT}{V} + \frac{n_{N_2O_4} RT}{V} = C n_{NO_2} + C n_{N_2O_4}$$

C is the proportionality constant

$$p_T \propto (n_{NO_2} + n_{N_2O_4})$$

and $C = RT/V$ will be same for reactant or product, since both are at same temperature and have same volume.

Let us consider the decomposition of ammonia in a closed container at a certain temperature. Ammonia breaks into nitrogen gas and hydrogen gas. Let P_0, P_t and P_∞ be the initial pressure, pressure at time t, and final pressure of gas in the container, respectively. Pressure is proportional to the number of moles at constant volume and temperature. C_1 is proportionality constant.

$$NH_3(g) \ \rightarrow \ \frac{1}{2}\,N_2(g) \ + \ \frac{3}{2}\,H_2(g)$$

@ time $t = 0$	a	0	0	$p_0 = C_1 a$
@ time $t = t$	$(a - x)$	$x/2$	$3x/2$	$p_t = C_1(a + x)$
@ time $t = \infty$	0	$a/2$	$3a/2$	$p_\infty = C_1 2a$

$$p_t - p_\infty = -C_1(a - x)$$
$$p_0 - p_\infty = -C_1 a$$
$$\frac{p_t - p_\infty}{p_0 - p_\infty} = \frac{(a - x)}{a}$$

Any other physical parameter can also be used to get the ratio of $(a - x)/a$. Thus, optical rotation can be used to obtain this ratio if the reaction involves chiral reactant(s) and or product(s). Ionic conductivity can be used if we are dealing with ionic reactions. However, these parameters are a bit different than pressure. In case of pressure, one mole of any of the reactant or product gases exerts the same pressure in a closed container and thus proportionality constant between pressure and number of moles is same for any of the reactants or products. On the other hand, optical rotation and ionic conductivity of reactant/product is proportional to its concentration, the proportionality constant for each reactant and product will be different.

For the reaction,

Sucrose $\rightarrow$ Glucose + Fructose

Total optical rotation = $C_{\text{sucrose}}\, n_{\text{sucrose}} + C_{\text{glucose}}\, n_{\text{glucose}} + C_{\text{Fructose}}\, n_{\text{Fructose}}$

Here, the value of C will be different for sucrose, glucose, and fructose, since optical rotation per mole of sucrose, glucose, and fructose is not the same.

Consider a reaction, $A \rightarrow P$. Let Y_0, Y_t and Y_∞ be the values of a certain physical parameter at time 0, at time t, and at the completion of reaction, respectively. The physical parameter of a species is proportional to the number of moles of that species and C_1 and C_2 are the proportionality constant for reactant and product, respectively.

	A	$\rightarrow$ P	
@ time $t = 0$	a	0	$Y_0 = C_1 a$
@ time $t = t$	$(a - x)$	x	$Y_t = C_1(a - x) + C_2 x$
@ time $t = \infty$	0	a	$Y_\infty = c_2 a$

$$Y_t - Y_\infty = C_1(a - x) - C_2(a - x) = (C_1 - C_2)(a - x)$$
$$Y_0 - Y_\infty = (C_1 - C_2)a$$
$$\frac{Y_t - Y_\infty}{Y_0 - Y_\infty} = \frac{(a - x)}{a}$$

In fact $a/(a - x)$ is related to any of the physical parameter in the similar way, independent of the order of the reaction.

For another reaction,

$$A \quad + \quad 3B \rightarrow 2C \quad + \quad D$$

@ time $t = 0$	a	b	0	0	$Y_0 = C_1 a + C_2 b$
@ time $t = t$	$(a - x)$	$(b - 3x)$	$2x$	x	$Y_t = C_1(a - x) + C_2(b - 3x) + C_3(2x) + C_4(x)$
@ time $t = \infty$	0	$(b - 3a)$	$2a$	a	$Y_\infty = C_2(b - 3a) + C_3(2a) + C_4 a$

$$Y_t - Y_\infty = -(a - x)(2C_3 + C_4 - C_1 - 3C_2)$$
$$Y_0 - Y_\infty = -a(2C_3 + C_4 - C_1 - 3C_2)$$
$$\frac{Y_t - Y_\infty}{Y_0 - Y_\infty} = \frac{(a - x)}{a}$$

This relation can be used to obtain the integrated rate law for reactions of different order.

For a first order reaction

$$\ln\frac{[A]_0}{[A]} = \ln\frac{a}{(a - x)} = kt$$

$$\ln\frac{[A]_0}{[A]} = \ln\frac{Y_0 - Y_\infty}{Y_t - Y_\infty} = kt$$

Thus, a plot of $\ln\dfrac{Y_0 - Y_\infty}{Y_t - Y_\infty}$ versus t will be a straight line for a first order reaction and the slope will give the value of first order rate constant.

For a second order reaction

$$\frac{1}{[A]} - \frac{1}{[A]_0} = kt$$

Multiplying the equation by $[A]_0$

$$\frac{[A]_0}{[A]} - 1 = kt[A]_0$$

$$\frac{[A]_0}{[A]} = kt[A]_0 + 1$$

$$\frac{Y_0 - Y_\infty}{Y_t - Y_\infty} = kt[A]_0 + 1$$

Thus, a plot of $\dfrac{Y_0 - Y_\infty}{Y_t - Y_\infty}$ versus t will be a straight line for a second order reaction and the slope/$[A]_0$ will give the value of second order rate constant.

Check Your Progress

15. Consider a reaction between A and B. In the presence of excess of A, the rate of reaction gets doubled if the concentration of B is doubled. In the presence of excess of B, the rate of reaction gets quadrupled if the concentration of A is doubled. What is the order of the reaction?

16. Consider the reaction, $2A \rightarrow P$. Let Y_0, Y_t and Y_∞ be the values of a certain physical parameter (proportional to the concentration) at time zero, at time t, and at the completion of reaction, respectively. Derive the relationship between Y_0, Y_∞, Y_t, A_0, t, and rate constant k for the given reaction, where reaction is of order 2 with respect to A.

17. Does the method of initial rate work for a fast reaction? Explain.

6.5 TEMPERATURE DEPENDENCE OF RATE CONSTANT

Arrhenius equation   The rate of a reaction depends strongly on temperature. This is due to dependence of the rate constants on the temperature. The temperature dependence of a large number of chemical reactions can be explained by Arrhenius equation.

Arrhenius suggested that the reactants must get additional energy to get activated and form the products. This additional energy is called 'activation energy'.

The rate constant of a reaction is proportional to the concentration of the activated molecules. Activated molecules are those which have energies greater than activation energy. The concentration of activated molecule is given by $\exp\left(-\dfrac{E_a}{RT}\right)$. Thus, the rate constant is given by

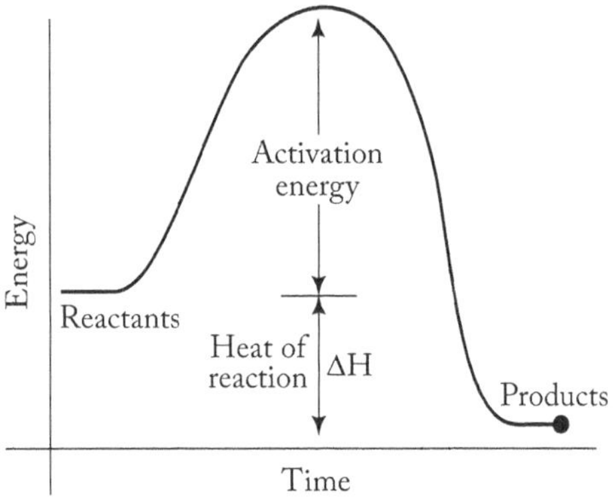

Fig. 6.3 The energy versus time profile for a reaction, showing the activation energy and heat of reaction

$$k = A \exp\left(-\frac{E_a}{RT}\right)$$

where k = rate constant

  A = Pre-exponential factor

  E_a = Activation energy

$$\ln k = \ln A - \frac{E_a}{RT}$$

Thus a plot between $\ln k$ and $1/T$ will result in a straight line and the slope and intercept will be equal to $\ln A$ and $-\dfrac{E_a}{R}$. Such a plot is known as *Arrhenius plot*.

Experimental activation energy is defined by

$$\left(\frac{\partial \ln k}{\partial T}\right)_p = \frac{E_a}{RT^2}$$

$$\ln \frac{k_2}{k_1} = \frac{E_a}{R}\left[\frac{1}{T_2} - \frac{1}{T_1}\right]$$

For reactions taking place in aqueous solution, activation energies typically lie between 40 kJ mol^{-1} and 120 kJ mol^{-1}. The rate of a reaction is thus, generally increased by two to three times for each 10 K rise in temperature.

Check Your Progress

18. How does temperature affect the rate constant?
19. What will be the activation energy if the rate of the reaction increases three times on 10 K increase in temperature from 300 K?
20. A first order gaseous reaction takes 30 minutes for 75 % completion at 273 °C, while it takes 8 minutes at 293 °C. Calculate the activation energy for the reaction.
21. For a chemical reaction, a large number of reactant molecules possess energy more than threshold energy, despite of this the reaction rate is very slow. Give the possible reason.

6.6 CONCEPT OF ACTIVATION ENERGY

The concept of activation energy can be obtained from potential energy surface. Potential energy surface is a plot of potential energy as a function of its conformation/configuration/orientation and locations of atoms relative to each other.

6.6.1 1D Potential Energy Surface (Potential Energy of Two-atom System)

The potential energy of a two-atom system is the energy resulting due to change in the distance between atoms. Potential energy changes as a consequence of change in attraction/repulsion between two atoms as the distance between two atoms changes. Potential energy of a two-atom system is a function of distance between two atoms (r). Potential energy (U) is related to force (F) as

$$F = -\frac{dU}{dr}$$

Thus, if the force between two atoms tries to decrease/increase the r, potential energy will change in a way to negate its effect on r. It means that if there is attractive force between two atoms, the potential energy will increase with increase in distance between two atoms.

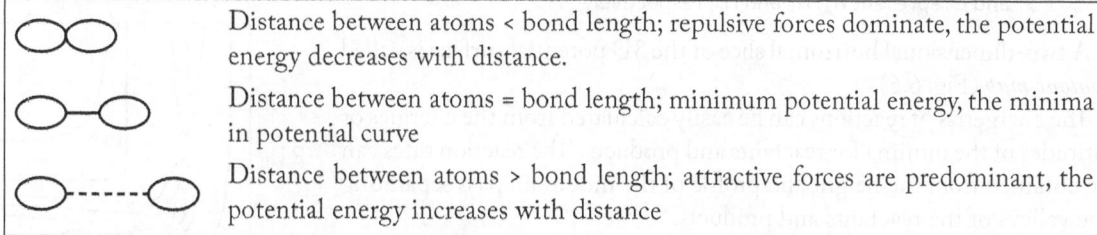

	Distance between atoms < bond length; repulsive forces dominate, the potential energy decreases with distance.
	Distance between atoms = bond length; minimum potential energy, the minima in potential curve
	Distance between atoms > bond length; attractive forces are predominant, the potential energy increases with distance

Fig. 6.4 Variation of potential energy as a function of distance

6.6.2 Potential Energy Surface of Multi-atom Systems

Potential energy of a multi-atom system depends on more than one factor. The dimensionality of the potential energy surface will depend on vibrational degrees of freedom, that is, 3N–5 for a linear system and 3N–6 for a non-linear system.

(a) Potential energy surface of H_3 system: Since there are three atoms in the H_3 system, dimensionality of PES is 3. However, if we restrict to a linear H_3 molecule, the dimensionality can be decreased to 2. Consider the reaction

$$H_A - H_B + H_C \rightarrow H_A \text{ --- } H_B \text{ --- } H_C \rightarrow H_A + H_B - H_C$$

A typical energy 2D potential energy surface of H_3 system is shown in Fig. 6.5.

(i) When H_C is far away from $H_A - H_B$: the potential energy surface is similar to 1D potential energy surface for $H_A - H_B$ (front face of the 3D surface).

(ii) When H_A is far away from $H_B - H_C$: the potential energy surface is similar to 1D potential energy surface for $H_B - H_C$ (side-face of 3D surface).

The PES is analogous to a hilly landscape, where

(i) reactants and products are represented by valleys on front and side face, respectively.

(ii) As H_C approaches $H_A - H_B$, the bond between H_A and H_B starts to stretch and potential energy starts to increase. As stretching increases, potential energy reaches maximum and starts to decrease as the distance between H_B and H_C decreases. The reaction pathway is the lowest energy route along mountain passes. The maximum point on this path is called *saddle point*. Thus, a reaction starts from a valley and ends in another valley through the mountain pass.

(iii) The highest point on the pass represents the saddle point. Saddle point is the minimum between mountain cliffs but maximum between valleys.
(iv) A transition state is first order saddle point in potential energy surface; energy increases in all directions but one.

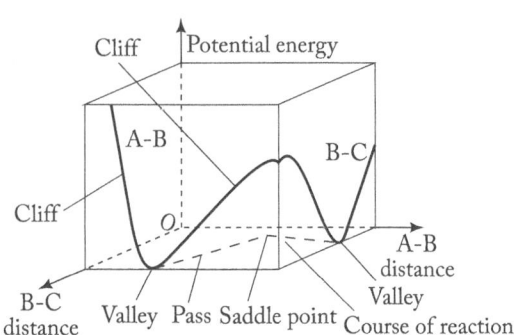

Fig. 6.5 Potential energy surface of $H_A - H_B + H_C$; A, B, and C represent H_A, H_B and H_C, respectively

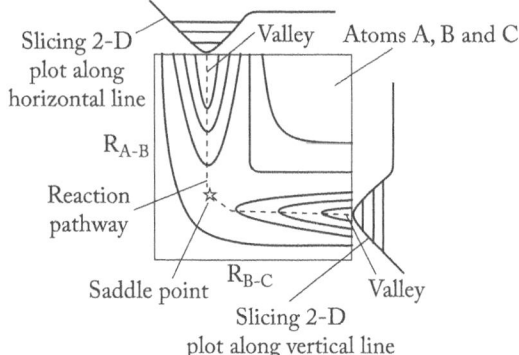

Fig. 6.6 Contour map of a three-atom system

A two-dimensional horizontal slice of the 3D potential surface is called *contour map* (Fig. 6.6).

The energetics of reactions can be easily calculated from the energies or altitudes of the minima for reactants and products. The reaction rates can be obtained from the height and profile of the mountain pass separating the valleys of the reactants and products.

Attractive surfaces for exothermic reaction   This type of surface is typical of the reactions of the type A + BC, in which saddle point is quite early in the reaction path, that is, activated complex is formed much before a larger lengthening of B–C bond (Fig. 6.7). Example: $F + H_2 \rightarrow FH + H$. Reactions with attractive surfaces take place more efficiently if the reactants are collided together with a lot of translational kinetic energy. A considerable portion of released energy after passing the activation energy goes into vibration of A–B bond.

Repulsive surfaces for exothermic reaction   This type of surface is characteristic of reactions in which saddle point, that is, activated complex is formed much later at a larger lengthening of B–C bond (in a reaction A + BC) (Fig. 6.8). Example: Reactions with repulsive surfaces take place more efficiently if the energies are in vibration of reactants. In this case, released energy after passing the activation energy does not go into vibration of A–bond. Example,

$$H + Cl_2 \longrightarrow HCl + Cl$$

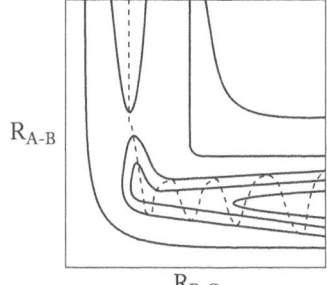

Fig. 6.7 Contour map for attractive surfaces of a three-atom system

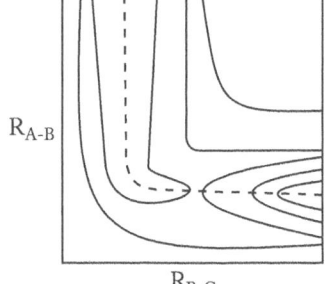

Fig. 6.8 Contour map for repulsive surfaces of a three-atom system

Check Your Progress

22. What is potential energy surface? Draw a typical 2D potential energy surface of H_3 system.
23. The reaction pathway is the lowest energy route along mountain passes. Why?
24. Which point on the potential energy surface and contour map represents transition state?
25. How does contour map for attractive surface differ from that of repulsive force?

6.7 ACTIVATED COMPLEX THEORY (TRANSITION STATE THEORY)

Activated (transition) complex theory proposed by Henry Erying provided the most elegant conceptual framework for understanding the rate of a reaction. The main features of the activated complex theory are the following.

(i) Reactant molecules combine to form an intermediate before changing to product.

$$AB + C \rightleftharpoons ABC^{\neq} \xrightarrow{k_2} P$$

(ii) The intermediate ($ABC^{\neq}$) is called *transition state* and it is first order saddle point on the potential energy surface (Fig. 6.9).

(iii)

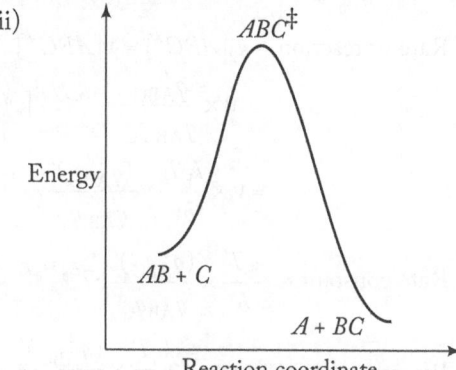

Fig. 6.9 Energy profile of reaction $AB + C \rightleftharpoons ABC^{\neq} \xrightarrow{k_2} P$, shown as potential energy versus reaction coordiantes plot. Reaction coordinate represents the path taken by the system during the reaction. The transition state ($ABC^{\neq}$) corresponds to the structure having energy maximum

(iv) Reactants and transition state are in pseudo-equilibrium and the rate of decay of transition state is of first order.

$$K^{\neq} = \frac{[ABC^{\neq}]}{[AB][B]}$$

Rate of reaction = $k_2[ABC^{\neq}]$

(v) The concentration of activated complex can be calculated using statistical thermodynamics. Statistical properties of a molecule in thermodynamic equilibrium are described by partition function. Molecular partition function (q) gives an indication of the number of states that are thermally accessible to a molecule at the temperature of the system.

$$q = \sum e^{-(E/k_b T)}$$

E is the energy of state and is the sum of electronic (E_{el}), vibrational (E_{vib}), rotational (E_{rot}), and translational energy (E_{trans}), k_b is Boltzmann constant, and T is the temperature.

$$q = \sum e^{-((Eel + Evib + Erot + Etrans)/k_b T)} = q_{el} \, q_{vib} \, q_{rot} \, q_{trans}$$

Equilibrium constant can be expressed in terms of molecular partition function of reactants and products (q_{AB}, q_C, and $q_{ABC}^{\neq}$ as partition functions of AB, C, $ABC^{\neq}$).

$$K^{\neq} = \frac{[ABC^{\neq}]}{[AB][C]} = \frac{q_{ABC}^{\neq}}{q_{AB} q_C} e^{-E_0/k_b T}$$

$$[ABC^{\neq}] = \frac{q_{ABC}^{\neq}}{q_{AB} \, q_C} e^{-E_0/k_b T} [AB][C]$$

(vi) The saddle point has positive curvature in all degrees of freedom except the one which corresponds to crossing the barrier which is negative.

$$q_{ABC}^{\neq} = q_{vib} (q_{ABC}^{\neq})'$$

Here, q_{vib} is the degree of freedom along the reaction pathway.

The expression $(q_{ABC}^{\neq})'$ represents partition function of all other (normal) thermodynamic degrees of freedom of the transition state structure.

Since bond along the reaction pathway is only partly formed, vibration along a reaction pathway is a loose vibration.

$$q_{vib} = \lim_{v \to \infty} \frac{1}{1 - \exp\left(-\dfrac{hv}{k_b T}\right)}$$

where k_b is Boltzmann constant.

$$q_{vib} = \cfrac{1}{1 - \left(1 - \cfrac{hv}{k_b T}\right)}$$

$$q_{vib} = \frac{k_b T}{hv}$$

Rate of reaction = $k_2[ABC^{\neq}] = v[ABC^{\neq}]$

$$= v \times \frac{q_{ABC}^{\neq}}{q_{AB}q_C} e^{E_0/kT}[AB][C] = v \times \frac{q_{vib}(q_{ABC}^{\neq})'}{q_{AB}q_C} e^{-E_0/kT}[AB][C]$$

$$= v \times \frac{k_b T}{hv} \times \frac{(q_{ABC}^{\neq})'}{q_{AB}q_C} e^{-E_0/k_b T}[AB][C] = \frac{k_b T}{h} \times \frac{(q_{ABC}^{\neq})'}{q_{AB}q_C} e^{-E_0/k_b T}[AB][C]$$

Rate constant = $\dfrac{k_b T}{h} \times \dfrac{(q_{ABC}^{\neq})'}{q_{AB}q_C} e^{-E_0/k_b T}$

Pre-exponential factor = $\dfrac{k_b T}{h} \times \dfrac{(q_{ABC}^{\neq})'}{q_{AB}q_C}$

6.7.1 Energetics of Formation of Transition State

The rate constant and equilibrium constant of the reaction $AB + C \rightleftharpoons ABC^{\neq}$ is related by the equation,

$$\text{Rate constant } (k) = \frac{k_b T}{h} \times \frac{(q_{ABC}^{\neq})'}{q_{AB}q_C} e^{-E_0/kT} = \frac{k_b T}{h} \times K^{\neq}$$

Taking log on both sides,

$$\ln k = \ln \frac{k_b}{h} + \ln T + \ln K^{\#}$$

$$\frac{d(\ln k)}{dT} = \frac{1}{T} + \frac{d(\ln K^{\#})}{dT}$$

$$\frac{d(\ln k)}{dT} = \frac{1}{T} + \frac{\Delta^{\#}U^0}{RT^2}$$

$$\frac{d(\ln k)}{dT} = \frac{\Delta^{\#}U^0 + RT}{RT^2}$$

From Arrhenius equation,

$$\frac{d(\ln k)}{dT} = \frac{E_a}{RT^2}$$

So,

$$\frac{E_a}{RT^2} = \frac{\Delta^{\#}U^0 + RT}{RT^2}$$

$$E_a = \Delta^{\#}U^0 + RT$$

$$\Delta^{\#}U^0 = E_a - RT$$

For a gaseous reaction,

$$\Delta^{\#}H^0 = \Delta^{\#}U^0 + \Delta n_g RT$$

$$E_a = \Delta^{\#}H^0 + RT(1 - \Delta n_g)$$

For a bimolecular reaction

$$\Delta n_g = -1$$

$$E_a = \Delta^{\#} H^0 + 2RT$$

Thus, enthalpy and internal energy of transition can be calculated if energy of activation is known.

Rate constant $(k) = \dfrac{k_b T}{h} \times K^{\#}$

$$k = \frac{k_b T}{h} exp\left(-\frac{\Delta^{\#} H^0}{RT}\right) exp\left(\frac{\Delta^{\#} S^0}{R}\right)$$

$$k = \frac{k_b T}{h} exp\left(-\frac{(E_a - RT(1 - \Delta n_g))}{RT}\right) exp\left(\frac{\Delta^{\#} S^0}{R}\right)$$

$$k = \frac{k_b T}{h} exp\left(-\frac{E_a}{RT}\right) exp\left(\frac{\Delta^{\#} S^0}{R}\right) exp(1 - \Delta n_g)$$

Pre-exponential factor (A) $= \dfrac{k_b T}{h} exp\left(\dfrac{\Delta^{\#} S^0}{R}\right) exp(1 - \Delta n_g)$

Entropy of transition can be obtained from pre-exponential factor.

SOLVED EXAMPLES

1. Consider a chemical reaction in which peroxydisulphate ions are used to oxidize iodide ions to triiodide. If rate of disappearance of peroxydisulphate ions is 1.37×10^{-3} mol L^{-1} s^{-1}, calculate the rate with which iodide ions will disappear. Also calculate the rate of formation of sulphate ions.

Solution: We have, $3\,I^- + S_2 O_8^{2-} \rightarrow I_3^- + 2\,SO_4^{2-}$

$$\text{Rate of reaction} = -\frac{1}{3}\frac{d[I^-]}{dt} = -\frac{d[S_2 O_8^{2-}]}{dt} = \frac{d[I_3^-]}{dt} = \frac{1}{2}\frac{d[SO_4^{2-}]}{dt}$$

Given, $-\dfrac{d[S_2 O_8^{2-}]}{dt} = 1.37 \times 10^{-3}\,\text{mol}\,L^{-1}\,s^{-1}$

Rate of disappearance of $I^- = -\dfrac{d[I^-]}{dt} = -3 \times \dfrac{d[S_2 O_8^{2-}]}{dt} = 3 \times 1.37 \times 10^{-3}\,\text{mol}\,L^{-1}\,s^{-1}$

Rate of disappearance of $I^- = 4.11 \times 10^{-3}\,\text{mol}\,L^{-1}\,s^{-1}$

Rate of formation of $SO_4^{2-} = \dfrac{d[SO_4^{2-}]}{dt} = -2 \times \dfrac{d[S_2 O_8^{2-}]}{dt} = 2.74 \times 10^{-3}\,\text{mol}\,L^{-1}\,s^{-1}$

2. The rate constant of a zero order reaction, $A \rightarrow$ Products, is 2.1×10^{-2} mol L^{-1} s^{-1}. If the initial concentration of A is 21 moles per litre, calculate its concentration after 15 minutes.

Solution: The integrated rate expression for a zero order reaction is

$$[A]_0 - [A] = kt$$

$$[A] = [A]_0 - kt = 21 - (2.1 \times 10^{-2} \times 15 \times 60) = 21 - 18.9 = 2.1\,\text{mol}\,L^{-1}$$

3. The initial concentration of $N_2 O_5$ in the following first order reaction is 2.37×10^{-3} mol.L^{-1} at 298 K.

$$N_2 O_5 \rightarrow 2 NO_2 + 1/2\,O_2$$

Calculate the rate constant for the reaction if the concentration of $N_2 O_5$ is reduced to 0.38×10^{-3} mol.L^{-1} after 50 min.

Solution: For a first order reaction, $\log \dfrac{[N_2O_5]_0}{[N_2O_5]_t} = \dfrac{kt}{2.303}$

$$k = \log \frac{[N_2O_5]_{t^0}}{[N_2O_5]_t} \times \frac{2.303}{(t-t^0)} = \log \frac{2.37 \times 10^{-3}}{0.38 \times 10^{-3}} \times \frac{2.303}{(60-0)} = 0.031 \text{ min}^{-1}$$

4. The rate constant for a first order reaction is $1.75 \times 10^{-2} \text{ s}^{-1}$ at 298 K. Calculate the time required to reduce the initial amount of reactant from 10 g to 4 g.

Solution: Initial amount = 10 g, Final amount = 4 g.

$$k = 1.75 \times 10^{-2} \text{s}^{-1}$$

$$t = \log \frac{[A]_o}{[A]} \times \frac{2.303}{k} = \log \frac{10}{4} \times \frac{2.303}{1.75 \times 10^{-2}} = 52.37 \text{ s}$$

5. In a second order dimerization reaction, the rate of reaction is $8.76 \times 10^{-4} \text{ mol L}^{-1} \text{ s}^{-1}$. If concentration of the monomer at the time of observation is 0.05 mol L^{-1}, calculate the rate constant.

Solution: $2A \rightarrow A_2$

As it is a second order reaction, the rate expression will be

$$\text{Rate} = k[A]^2$$

$$8.76 \times 10^{-4} = k \times (0.05)^2$$

$$k = (8.76 \times 10^{-4})/(0.05)^2 = 3.504 \text{ L mol}^{-1}\text{s}^{-1}$$

6. For a first order chemical reaction,

$$NH_2NO_2 \ (aq) \rightarrow N_2O \ (g) + H_2O \ (l)$$

99 % decomposition of 6 g of NH_2NO_2 happened in 13 hours at 283 K. Determine the half-life of the reaction.

Solution: For a first order reaction,

$$k = \frac{2.303}{t} \log \frac{a}{(a-x)}$$

$$k = \frac{2.303}{t_{99\%}} \log \frac{a}{(a-0.99a)} = \frac{2.303}{13} \log 10^2 = 0.354 \, h^{-1}$$

Half-life of the reaction will be

$$t_{1/2} = \frac{0.693}{k} = \frac{0.693}{0.354} = 1.96 \, h$$

7. For a chemical reaction: $A + B \rightarrow$ Products, an experiment was performed with three different conditions. The observations for the same are tabulated as:

Exp.	Initial [A] (mol L^{-1})	Initial [B] (mol L^{-1})	Init. Rate of Formation of products (M s^{-1})
1.	0.05	0.05	7.8×10^{-4}
2.	0.10	0.05	1.56×10^{-3}
3.	0.10	0.025	7.8×10^{-4}

Determine the rate law for this reaction and the rate constant. What will be the initial rate of reaction when $[A]_o = 0.132$ M and $[B]_o = 0.012$ M?

Solution: In experiments 1 and 2, the concentration of *B* is same. On doubling the concentration of *A*, the rate of reaction also gets doubled. Therefore, the reaction is first order in *A*.

In experiments 2 and 3, the concentration of *A* remains same whereas the concentration of *B* gets halved which results in halving of the rate of the reaction. Therefore, the reaction is first order in *B*.

Hence, the rate law for this reaction is:

rate = $k\,[A]\,[B]$

Calculation of the rate constant:

7.8×10^{-4} M s^{-1} = k(0.05 M) (0.05 M)

$k = 0.312$ M^{-1} s^{-1}

When $[A]_o = 0.132$ M and $[B]_o = 0.012$ M

Rate of reaction = (0.312 M^{-1} s^{-1}) (0.132 M) (0.012 M)

Rate of reaction= 4.94×10^{-4} M s^{-1}

8. In a flask containing inert gas, PH_3 gas is introduced at 500 °C. The decomposition of PH_3 gas to P_4 (g) and H_2 (g) happens with time. The total pressure is as a function of time and is given in the table below.

Time (s)	0	60	120	∞
p (mm)	246.4	263.9	267.2	268.1

What is the order and rate constant of the reaction?

Solution: If the reaction is of first order,

$$k = \frac{2.303}{t} \log \frac{a}{(a-x)} = \frac{2.303}{t} \log \frac{(p_\infty - p_o)}{(p_\infty - p_t)}$$

at $t = 60$ s,

$$k = \frac{2.303}{60} \log \frac{(268.1 - 246.4)}{(268.1 - 263.9)} = 0.0274 \text{ s}^{-1}$$

at $t = 120$ s,

$$k = \frac{2.303}{120} \log \frac{(268.1 - 246.4)}{(268.1 - 267.2)} = 0.0265 \text{ s}^{-1}$$

As k comes out to be nearly constant, it is a first order reaction.

$$\text{Mean value of } k = \frac{0.0274 + 0.0265}{2} = 0.027 \text{ s}^{-1}$$

9. For the first order decomposition of hydrogen peroxide, the half-life is 486 min at 373 K. Assuming the activation energy of the reaction to be 36 kJ mol^{-1}, calculate the time required for 60 % completion of the reaction at 480 K.

Solution: We have, at 373 K, $t_{1/2}$ = 486 min

$$k_{373\,K} = \frac{0.693}{t_{1/2}} = \frac{0.693}{486} = 1.43 \times 10^{-3} \text{ min}^{-1}$$

We have, $E_a = 36$ kJ mol^{-1} and k_{480} = ?

According to Arrhenius equation,

$$\log \frac{k_2}{k_1} = \frac{E_a}{2.303\,R} \left(\frac{T_2 - T_1}{T_1 T_2} \right)$$

$$\log \frac{k_{480}}{k_{373}} = \log \frac{k_{480}}{1.43 \times 10^{-3}} = \frac{36 \times 10^3}{2.303 \times 8.314} \left(\frac{480 - 373}{480 \times 373} \right) = 1.12$$

$$k_{480} = 1.43 \times 10^{-3} \times 13.29 = 1.9 \times 10^{-2} \text{ min}^{-1}$$

Time required for 60 % completion of the reaction at 480 K,

$$t_{60\%} = \frac{2.303}{1.9 \times 10^{-2}} \log \frac{a}{(a - 0.6a)} = \frac{2.303}{1.9 \times 10^{-2}} \log 2.5 = 48.2 \text{ min}$$

10. The rate of a hydrogenation reaction carried at 450 K is equal to the rate of the same reaction but in the presence of a catalyst at 300 K. What will be the activation energy of the reaction if it drops by 10 kJ mol^{-1} in the presence of a catalyst?

Solution: Let E_a be the activation energy in the absence of catalyst and E_b be the activation energy in the presence of catalyst. Then we have,

$$k = Ae^{-E_a/450\,R} = Ae^{-E_b/300\,R}$$

Hence, $\dfrac{E_a}{450\,R} = \dfrac{E_b}{300\,R}$ or $E_b = \dfrac{3}{4.5}E_a$

Also, $E_a - E_b = 10$ or $E_a - \dfrac{3}{4.5}E_a = 10$ or $E_a = 30$ kJ mol^{-1}

11. The rate constant for a gaseous first order dissociation reaction $AB \rightarrow A + B$, is represented as $\log k\,(s^{-1}) = 13.97 - \dfrac{1.73 \times 10^3\,K}{T}$. What will be the activation energy for the reaction? Also determine the temperature at which half-life of the reaction is 300 min.

Solution: According to Arrhenius equation,

$$\log k = \log A - \dfrac{E_a}{2.303\,RT}$$

On comparing this with the given equation, we have

$$\dfrac{E_a}{2.303\,R} = 1.73 \times 10^3 \text{ or } E_a = 33.12 \text{ kJ mol}^{-1}$$

$t_{1/2} = 300$ min $= 300 \times 60$ s

$$k = \dfrac{0.693}{300 \times 60\,s} = 3.85 \times 10^{-5}\,s^{-1}$$

As per the given rate constant equation, we get

$$\log(3.85 \times 10^{-5}) = 13.97 - \dfrac{1.73 \times 10^3\,K}{T}$$

The temperature for half-life to be 300 min is

$$T = -\dfrac{1.73 \times 10^3}{\log(3.85 \times 10^{-5}) - 13.97} = 94.1\,K$$

12. In a gaseous chemical reaction $2A \rightarrow A_2$, dimerization took place at 450 K. As per transition state theory, the rate constant is expressed as: $k_2 = 10^3 \exp\left(-\dfrac{75.89}{RT}\right)$. Calculate the thermodynamic parameters $\Delta H^{\ddagger}$, $\Delta S^{\circ\ddagger}$, $\Delta U^{\circ\ddagger}$ and $\Delta G^{\circ\ddagger}$ of the reaction.

Solution: According to Arrhenius equation, the rate constant for the reaction is given as:

$$k_2 = 10^3 \exp\left(-\dfrac{75.89}{RT}\right)$$

where, $A = 1 \times 10^3$ and $E_a = 75.89$ kJ mol^{-1}

$\Delta^{\#}U^0 = E_a - RT = 72.148$ kJ mol^{-1}

$\Delta^{\#}H^0 = \Delta^{\#}U^0 + \Delta n_g\,RT = \Delta^{\#}U^0 - RT = 68.407$ kJ mol^{-1}

Pre-exponential Factor $(A) = \dfrac{k_bT}{h}\exp\left(\dfrac{\Delta^{\#}S^0}{R}\right)\exp(1 - \Delta n_g)$

$\exp\left(\dfrac{\Delta^{\#}S^0}{R}\right) = \dfrac{hA}{k_bT}\exp(-2) = 1.438 \times 10^{-11}$

$\Delta^{\#}S^0 = -207.5$ J K^{-1} mol^{-1}

$\Delta^{\#}G^0 = \Delta^{\#}H^0 - T\Delta^{\#}S^0 = 68.407 - (450 \times 207.5 \times 10^{-3}) = -24.9$ kJ mol^{-1}

SUMMARY

○ Kinetics involves the study of progress of a process from one state to another during a physical and/ or chemical change; Thermodynamics helps us to find out the amount of product at equilibrium or on completion of a reaction.

○ Kinetic studies can be used to know the amount of product or reactant; stability and rate of drug metabolism, mechanism of reaction, and the nature of a reaction.

○ The rate of a reaction can be obtained by studying the rate of change of concentration of one of the reactants or product with time t. Some of the factors that affect the rate of reaction are: concentration, nature of reactants, temperature of reaction medium, and stability of intermediates.

○ Integrated rate law is the integrated form of the rate law in which the differential rate equations are integrated to give a relationship between rate constant and concentrations of reactants at different times.

○ Zero order reactions are the reactions in which the rate of reaction does not change with the concentration of reactants. $\vartheta = k[A]^\circ = k$ or $kt = [A]_o - [A]$. The unit of the zero order rate constant is M/s or $mol.L^{-1}.s^{-1}$.

○ First order reactions are the reactions in which the rate of reaction is directly proportional to the concentration of the reacting species. $\vartheta = k[A]$ or $kt = 2.303 \log [A]_o/[A]$. The unit of first order reaction rate constant is inverse of time i.e. s^{-1}.

○ Rate law expression for the nth order of the reaction is given as $kt = \dfrac{1}{[A]^{n-1}} - \dfrac{1}{[A]_o^{n-1}}$. The unit of rate constant for n^{th} order reaction is $(mol.L^{-1})^{1-n}.s^{-1}$.

○ Half-life period for zero order reactions is expressed as $t_{1/2} = \dfrac{[A]_0}{2k}$. For first order reactions, it is given as $t_{1/2} = \dfrac{0.693}{k}$. The general expression for n^{th} order reaction is $t^{1/2} \propto 1/[A]_0^{n-1}$.

○ Differential method is used to determine the order of a reaction involving one reactant; isolation method

and method of initial rate are employed when there is more than one reactant.

○ In isolation method, the concentrations of all reactants except one are taken in excess so that the rate depends upon the concentration of the reactant which is not in excess.

○ The method of initial rate is another method to determine the rate of reaction, which involves the initial rates of the reactants. In this method, several experiments are performed with different sets of initial concentration of reactants. The comparison of the observations is then the deciding factor for the determination of the rate of the reaction.

○ Arrhenius equation describes the relationship between temperature and the rate constant. It is given as $k = A.e^{-E_a/RT}$, where A represents the pre-exponential factor while the exponential term $(e^{-E_a/RT})$ is known as Boltzmann factor.

○ Potential energy surfaces are the mapping of the potential energy as a function of the relative position of the atoms actively involved in the reaction.

○ *Saddle point* represents the maxima between the two minima of the energy pathway. For exothermic reactions, the saddle point is observed quite early in the reaction path while it appears quite late for repulsive surfaces. It also represents transition-state structure in the reaction.

○ Activated complex is the highly energetic arrangement of the atoms formed during the reaction and corresponds to the peak of the energy profile of the reaction progress.

○ In transition state theory, the activated complex is supposed to be in equilibrium with the reactant species. The rate of formation of products depends on the rate at which the activated complex passes through the transition state. This theory is described by the Eyring equation which defines the rate constant as $k = \dfrac{k_b T}{h} K^{\neq}$, where $K^{\neq}$ is the equilibrium constant expressed in terms of partition function of the reactants and products.

GLOSSARY

Rate of reaction: The rate of change of concentration of any of the reactant or product with time at any particular instant of time. The unit of rate of the reaction is M s^{-1} or mol L^{-1} s^{-1}.

Order of a reaction: The sum of the exponents of the concentration terms in the experimental rate law expression. It can be zero, 1, 2, 3, or any fractional value.

Molecularity of a reaction: The number of reacting species which collide simultaneously to bring about the chemical change. It is a theoretical concept. It cannot be zero and also cannot be more than 3.

Rate law: The mathematical expression between rate of reaction and concentrations of reacting species based on the experimental observations.

Half-life period of the reaction ($t_{1/2}$): The time at which the concentration of the reactants is reduced to half of their initial concentration.

Activation energy (E_a): The additional energy required by the reacting species over and above the average potential energy to enable it to overcome the energy barrier between the reactants and the products.

Chemical kinetics: The branch of chemistry which deals with the study of reaction rates and their mechanism.

Stoichiometric coefficients of a reactant or product: The number of moles of reactants or products participating in a balanced chemical equation.

EXERCISES

Multiple Choice Questions

1. For a zero order reaction, the time required for 100 % completion of the reaction is:

   (*a* and *k* are initial concentration of reactant and rate constant of reaction)
   - (a) ak
   - (b) a^2k
   - (c) a/k
   - (d) k/a

2. What would be the ratio of the rate constant and Arrhenius factor at temperature, if the activation energy is $2.303RT$ J mol^{-1}?
   - (a) 10
   - (b) 100
   - (c) 1/10
   - (d) 1/100

3. A radioactive atom (atomic mass is 22 u) took 10 years to reduce its radioactivity to 90 %. Its half-life is
   - (a) 2 years
   - (b) 3 years
   - (c) 5 years
   - (d) 10 years

4. The hydrolysis of an organic chloride is carried out as RCl + H$_2$O → ROH + HCl. If the reaction takes place in excess of water, then which of the following statements is correct?
   - (a) Both molecularity and order of the reaction is 1.
   - (b) Both molecularity and order of the reaction is 2.
   - (c) Molecularity is 1, but order of the reaction is 2.
   - (d) Molecularity is 2, but order of the reaction is 1.

5. For every 10 °C rise in temperature, the rate of reaction doubles. What will be the increase in rate of the reaction if temperature is increased by 40 °C?
   - (a) 4 times
   - (b) 8 times
   - (c) 16 times
   - (d) 32 times

6. For a first order decomposition reaction 2 A(g) → B(g) + C(s), the total pressure after 10 min is found to be 300 pascal. If after 100 % completion, total pressure is 200 pascal, calculate the rate constant in min^{-1}.
   - (a) 0.0693
   - (b) 0.693
   - (c) 6.93
   - (d) 69.3

7. For a chemical reaction $A + B \rightarrow$ products, the initial rate of the reaction doubles on doubling the concentration of B. Also the initial rate of reaction increases by a factor of 8 if the concentration of both A and B doubles. The rate law expression for the reaction is
   - (a) Rate = k[A][B]
   - (b) Rate = k[A]^2[B]
   - (c) Rate = k[A][B]^2
   - (d) Rate = k[A]^2[B]^2

8. What will be the specific rate constant for a first order chemical reaction if its half-life is 1386 s?
   - (a) 0.5×10^{-1} s^{-1}
   - (b) 0.5×10^{-2} s^{-1}
   - (c) 0.5×10^{-3} s^{-1}
   - (d) 0.5×10^{-4} s^{-1}

9. For a first order reaction, the half-life is found to be 6.93 min. The time required for 99 % completion of the reaction is
   - (a) 12 min
   - (b) 24 min
   - (c) 32 min
   - (d) 46 min

10. For two different reactions, the first order rate constants k_1 and k_2 are found to be $10^{16} \cdot e^{(-2000/T)}$ and $10^{15} \cdot e^{(-1000/T)}$, respectively. The temperature at which both rate constants are equal is
    - (a) 1000 K
    - (b) 1000/2.303 K
    - (c) 2000 K
    - (d) 2000/2.303 K

11. The correct expression for the reaction: $1/2\ A \to 2B$ is
    (a) $-d[A]/dt = d[B]/dt$
    (b) $-2d[A]/dt = d[B]/dt$
    (c) $-d[A]/dt = 1/2\ d[B]/dt$
    (d) $-d[A]/dt = 1/4\ d[B]/dt$

12. For a first order reaction $A \to B$, k represents the rate constant for the reaction. If the initial concentration of reactant is 0.5 M, the half-life for the reaction is
    (a) $\log 2/k$
    (b) $1/k$
    (c) $\ln 2/k$
    (d) $0.693/5k$

13. For a chemical reaction $2A + B \to$ products, the half-life of the reaction doubles on doubling the concentration of A while it remains unchanged if concentration of B gets doubled. The unit of rate constant for the given reaction is
    (a) Unitless
    (b) $\text{mol L}^{-1}\text{s}^{-1}$
    (c) s^{-1}
    (d) $\text{L mol}^{-1}\text{s}^{-1}$

14. The expression which is not correct for the reaction $2A + B \to 3C + D$ is
    (a) Rate $= -d[B]/dt$
    (b) Rate $= -1/3\ d[C]/dt$
    (c) Rate $= -1/2\ d[A]/dt$
    (d) Rate $= d[D]/dt$

15. For a chemical reaction, $A + B \to C + D$, the reaction is found to be of second order with respect to B, what will be the change in initial rate of reaction if the concentration of B doubles keeping the rest unchanged?
    (a) Rate will get doubled.
    (b) Rate will increase by 3 times.
    (c) Rate will increase by a factor of 4.
    (d) Rate will remain unchanged.

16. In Arrhenius equation, $k = Ae^{-E/RT}$, what will be the term E represent?
    (a) The total energy of the reacting molecules at temperature T
    (b) Fraction of the molecules having energy greater than activation energy
    (c) The energy below which colliding molecules will not react
    (d) The energy above which colliding molecules will not react

17. The reaction between nitrous oxide and bromine to form NOBr follows a two-step mechanism.
    Step 1: $NO + Br_2 \to NOBr_2$
    Step 2: $NOBr_2 + NO \to NOBr$.
    The second step is the rate determining step. Calculate the order of reaction with respect to NO.
    (a) 0
    (b) 1
    (c) 2
    (d) 3

18. The factor which does not affect the rate of the reaction is
    (a) Molecularity of the reaction
    (b) Temperature of the reaction
    (c) Concentration of the reactants
    (d) Nature of the reactants

19. The statement that is not true is
    (a) Order of the reaction cannot be fractional.
    (b) Order of the reaction is the sum of the coefficient of the concentration terms in the rate law.
    (c) Order of the reaction can be determined experimentally.
    (d) Order of the reaction is not influenced by the stoichiometric coefficient of the reactants.

Review Questions

1. Define stoichiometry of a reaction.
2. Can order and molecularity be fractional? Explain.
3. How is the half-life of a reaction dependent on concentration for a zero order reaction?
4. How does a solvent affect the rate of a reaction?
5. How can the kinetics help in determining the age of an old material?
6. For which order of reaction, is half-life independent of concentration?
7. For which order of reaction, does rate constant has same unit as rate of reaction?
8. What is the unit of rate constant for a third order reaction?
9. What is the integrated rate law for a second order reaction?
10. What is the integrated rate law for a zero order reaction in terms of a physical parameter (Y) for a zero order reaction?
11. Explain how kinetics studies can be useful in the discovery of a drug?
12. Which kind of reactant can decrease the rate of a reaction?

13. Explain how surface area of a reactant can affect the rate of reaction?

14. Explain how a catalyst increases the rate of a reaction?

15. Give two examples of first and second order conditions?

16. Explain how isolation method helps in determining the order of a reaction when more than two reactants are involved.

17. What is the order of reaction if $\dfrac{1}{[A]}$ vs t is linear?

18. If the slope of plot of versus is 1, what is the order of reaction?

19. For a gaseous reaction, how is $a/(a{-}x)$ related to total pressure at different time $(P_0, P_t$ and $P_\infty)$?

20. What is the difference between activation energy and enthalpy of a reaction?

21. Draw the potential energy surface of reaction between H_2 and H. Explain.

22. How is a transition state represented on potential energy surface?

23. What is partition function and how is it related to equilibrium constant?

24. Explain the different features of activated complex theory.

25. What is the relationship between enthalpy of transition and activation energy?

NUMERICAL PROBLEMS

1. For the chemical reaction, $x + 2y \rightarrow z$, calculate the rate of disappearance of y if the rate of appearance of z is found to be 1.38 mol L^{-1} h^{-1}. (**Ans:** 2.76 mol L^{-1} hr^{-1})

2. In a gaseous reaction, nitrogen dioxide is formed from the reduction of nitrogen pentoxide. If dioxygen is formed at the rate of 68 g mm^{-1}, calculate the rate of formation of NO_2. Also determine the rate of disappearance of N_2O_5. (**Ans:** 8.5 mol mm^{-1}, 4.25 mol mm^{-1})

3. For the chemical reaction $A \rightarrow B$, on doubling the concentration of A, the rate of reaction increases 4 times. If the rate of reaction is expressed as $k[A]^n$, determine the order of the reaction. (**Ans:** 2)

4. The liberation of gases like methane, dihydrogen, and carbon monoxide takes place in decomposition of dimethyl ether. In terms of partial pressure of dimethyl ether, the rate of reaction is expressed as $k\left(p_{CH_3OCH_3}\right)^{3/2}$. Assuming that pressure is measured in bar and time in minutes, determine the units of rate and rate constant. (**Ans:** mol L^{-1} min^{-1}, bar$^{-1/2}$ min^{-1})

5. For a hypothetical reaction $A \rightarrow P$, the initial concentration of A(0.6 mole per litre) is reduced to 0.3 moles per litre within 10 minutes. In the next 10 minutes, it further reduces to 0.15 moles per litre. Calculate the order and rate constant of the reaction. (**Ans:** First oder, 0.0693 min^{-1})

6. In a first order decomposition reaction of a compound, 20 % of the initial amount is decomposed within 20 minutes. Calculate the rate constant. What will the time required if only 10 % of initial amount remains unreacted? (**Ans:** 1.12×10^{-2} min^{-1}, 205.6 min)

7. Thermal decomposition of a gas X, is represented at 473 K as

   $5X \rightarrow 3Y + 2Z$

   The reaction is second order with respect to X. If second order rate constant for the reaction is 5.96×10^{-4} M^{-1} s^{-1} and initial concentration of X is 0.15 M, calculate the initial rate of reaction.
   (**Ans:** 1.34×10^{-5} mol L^{-1} s^{-1})

8. A second order reaction has the rate equation $R = k[A][B]$, and the rate constant of the reaction is found to be 4.78 $M^{-1}s^{-1}$. Assuming concentrations of $[A]$ and $[B]$ are 2.3 M and 99 M, determine the rate constant of its pseudo-1st-order reaction. (**Ans:** 4.8×10^{-2} s^{-1})

9. For a chemical reaction $A \rightarrow B + C$, the half-life of the reaction is directly proportional to reactant concentration. Calculate the increase in rate of the reaction if the concentration of the reactant is increased by 10 times.

   (**Ans:** No change)

10. Disproportionation reaction of ClO_2 solution in alkaline medium is represented as:

$$2\,ClO_2\,(aq) + 2\,OH^-\,(aq) \rightarrow ClO_3^-\,(aq) + ClO_2^-\,(aq) + H_2O\,(l)$$

The initial rate of formation of ClO_3^- was observed at different initial concentrations of ClO_2 and OH^-.

Obs.	$[ClO_2]_o$ (M)	$[OH^-]_o$ (M)	Init. Rate of Formation of ClO_3^- (M s^{-1})
1	1.37×10^{-3}	1.82×10^{-4}	3.07×10^{-5}
2	2.74×10^{-3}	1.82×10^{-4}	1.31×10^{-4}
3	2.74×10^{-3}	3.64×10^{-4}	2.59×10^{-4}

Determine the rate equation for the chemical reaction and thereafter calculate the rate constant, k for the reaction. Also calculate the initial reaction rate for the reaction if $[ClO_2]_o = 9.87 \times 10^{-4}$ M and $[OH^-]_o = 1.79 \times 10^{-3}$ M. **(Ans: rate = $k[ClO_2]^2\,[OH^-]$, 1.57×10^{-4} mol L^{-1} s^{-1})**

11. Using the given data, determine the rate law for the decomposition reaction of ethanal:

$$CH_3CHO\,(g) \rightarrow CH_4\,(g) + CO\,(g)$$

Exp.	$[CH_3CHO]$ (M)	$[CO]$ (M)	Rate (M s^{-1})
1	0.63	0.40	0.76
2	0.21	0.55	0.086
3	0.21	0.40	0.086

What is the rate constant for the reaction? **(Ans: 1.206 mol L^{-1} s^{-1})**

12. For a second-order reaction, the rate constant at 5 °C was found to be 10.67×10^{-4} L mol^{-1} and 9.8×10^{-2} L/mol at 55 °C. What is the activation energy of this reaction? **(Ans: 68.55 kJ mol^{-1})**

13. The rate constants of a chemical reaction at 400 K and 700 K are found to be 0.035 s^{-1} and 0.075 s^{-1}, respectively. Calculate E_a and A. **(Ans: 5.915 kJ mol^{-1} and 0.2072)**

14. Two similar order reactions possess same activation energies but their entropy of activation differs by 38.3 J K^{-1}mol^{-1}. Determine the ratio of rate constants of two reactions at room temperature. **(Ans: 100)**

FURTHER READING

1. Atkins, P. and J.D. Paula, *Physical Chemistry*, Oxford University Press.
2. Laidler, K.J. *Chemical Kinetics*, Pearson Education.
3. Mortimer, M. and P. Taylor, *Chemical Kinetics and Mechanism*, Royal Society of Chemistry, Thomas Graham house, Science Park, Milton.

ANSWERS

Check Your Progress

1. Molecularity cannot be more than 3 as the probability of simultaneous collision of all reacting species is very low. Therefore, such reactions proceed via a sequence of steps, and in each step, only two or three species can take part.
2. (a) Order of reaction is the sum of exponents of the concentration terms in the rate law. (b) Molecularity of reaction is defined as the number of reactant molecules that takes part in an elementary reaction.
3. Yes, the order and molecularity of a reaction can be same for the reactions where one or more reactants reacts with each other directly to form the product in a single step. Also only single transition state must exist.
4. Rate of a reaction (υ) is defined as the change in the extent of reaction (ξ) with time (t) per unit volume (V). The factors that influence the rate of reaction are concentration, surface area, and nature of reactants, nature of solvent, temperature and catalyst.
5. The rate of reaction for the given gaseous reaction

$$= -\frac{1}{10}\frac{d[C]}{dt} = +\frac{1}{8}\frac{d[AC]}{dt}$$

Rate of disappearance of $C = \dfrac{10}{8}$ (Rate of appearance of AC)

$$= \dfrac{10}{8} \times 4.87 \times 10^{-3} = 6.08 \times 10^{-3} \text{ mol L}^{-1}\text{s}^{-1}$$

6. The integrated rate law for the first order reaction will be

$$[A] = [A]_0\, e^{-kt}$$

The fraction remaining after 200 s will be

$$\dfrac{[A]}{[A]_0} = e^{-3.06 \times 10^{-3} \times 200} = 0.542 = 54.2\,\%$$

7. For the given reaction, the rate law will follow second order kinetics and can be expressed as *rate* $= k[X]^2$.

8. $[A] = [A]_0\, e^{-kt} = e^{-2} = 0.135$ M

9. Zero order reactions

10. One of the reactants is in large excess. Therefore, its concentration does not change with time appreciably and remains almost constant and thus concentration term can be taken as a constant. The reaction under such condition will then follow first order kinetics. Such reactions are called pseudo unimolecular or first order reaction.

11. The half-life of a reaction is the time in which the concentration of a reactant is reduced to one half of its initial concentration. It is denoted as $t_{1/2}$.

$$k = \dfrac{2.303}{t} \log \dfrac{a}{(a-x)} = \dfrac{2.303}{30} \log(4/3) = 9.59 \times 10^{-3}\,\text{min}^{-1}$$

For first order reaction, half-life will be

$$t_{1/2} = \dfrac{0.693}{k} = \dfrac{0.693}{9.59 \times 10^{-3}} = 72.3 \text{ min}$$

12. The reaction would be 100 % complete only after infinite time, thus practically impossible.

13. For a first order reaction: $\dfrac{dx}{dt} = k[X]$

$$0.00154 = k \times 0.01$$

$$k = 0.154 \text{ min}^{-1}$$

Half-life will be

$$t_{1/2} = \dfrac{0.693}{k} = \dfrac{0.693}{0.154} = 4.5 \text{ min}$$

14. The rate constant is given as

$$k = \dfrac{2.303}{t} \log \dfrac{a}{(a-x)} = \dfrac{2.303}{60} \log(100/55) = 9.97 \times 10^{-3}\,\text{min}^{-1}$$

And the half-life will be

$$t_{1/2} = \dfrac{0.693}{k} = \dfrac{0.693}{9.97 \times 10^{-3}} = 69.5 \text{ min}$$

15. Overall order of the reaction = 3.

16. $\dfrac{Y_0 - Y_\infty}{Y_t - Y_\infty} = kt[A]_0 + 1$, Refer to section 6.4.

17. No, since the concentration of A will change fast and will not be equal to initial concentration.

18. Arrehenius equation: $k = Ae^{-E_a/RT}$. Thus, the rate constant of a reaction increases with increase in temperature.

19. $\log \dfrac{k_2}{k_1} = \dfrac{E_a}{2.303\,R} \left(\dfrac{T_2 - T_1}{T_1 T_2} \right)$

$k_2 = 2k_1,\ T_1 = 300$ K, $T_2 = 310$ K

therefore, $\log 2 = \dfrac{E_a}{2.303 \times 8.314} \left(\dfrac{310 - 300}{310 \times 300} \right)$

$E_a = \dfrac{\log 2 \times 2.303 \times 8.314 \times 310 \times 300}{10} = 53.6 \text{ kJ mol}^{-1}$

20. First case: 75 % completion at 273 °C i.e., 546 K in 30 minutes

$k_1 = \dfrac{2.303}{t_1} \log \dfrac{a}{(a-x)} = \dfrac{2.303}{30} \log \dfrac{4}{1} = 4.62 \times 10^{-2} \text{ min}^{-1}$

Second case: 75 % completion at 293 °C i.e., 566 K in 8 minutes

$k_2 = \dfrac{2.303}{t_2} \log \dfrac{a}{(a-x)} = \dfrac{2.303}{8} \log \dfrac{4}{1} = 17.3 \times 10^{-2} \text{ min}^{-1}$

Applying $\log \dfrac{k_2}{k_1} = \dfrac{E_a}{2.303R} \left(\dfrac{T_2 - T_1}{T_1 T_2} \right)$

We get $E_a = 169.5 \text{ kJ mol}^{-1}$

21. The possible reason of very slow reaction rate is the orientation of the reacting molecules. Although the reacting molecules may possess higher energy than the threshold energy, they do not have proper orientation for effective collisions.

22. Potential energy surface is a plot of potential energy as a function of its conformation/configuration/orientation and locations of atoms relative to each other. For 2D contour map, please refer to Fig. 6.6.

23. Reaction takes the lowest energy path, among the multitude of possible paths, between two valleys (reactant and product) of low energy.

24. Transition state represents the saddle point.

25. Attractive surface is typical of the reactions of the type A + BC, in which saddle point, i.e., activated complex, is quite early in the reaction path. Repulsive surface is characteristic of reactions in which saddle point is formed much later.

Multiple Choice Questions

1. (c)	2. (c)	3. (b)	4. (d)	5. (c)	6. (a)	7. (b)
8. (c)	2. (d)	10. (b)	11. (d)	12. (c)	13. (d)	14. (b)
15. (c)	16. (c)	17. (c)	18. (a)	19. (a)		

Surface Chemistry

After reading this chapter, you will be able to:

- define adsorption and classify its types into physical and chemical adsorption.
- explain adsorption on the basis of various adsorption isotherms.
- appreciate the role of catalysts in industry.
- discuss the types of eznzyme-catalysed reactions and the mechanism of action of enzymes.
- describe the types and mechanism of action of colloids, emulsions, surfactants, and detergents.
- understand the concept of friccohesity.

7.1 INTRODUCTION

Surface chemistry is a branch of molecular physical chemistry that deals with the understanding of surface characteristics of various materials, adsorption phenomena, action of catalysts, and colloidal chemistry. Engineering students who are particularly interested in material science and molecular catalysis will require a thorough understanding of molecular surface chemistry, which will enable them to apply the concepts in innovative technologies that rely on metals, semiconductors, superconductors, and polymeric surfaces. The predominant application of surface chemistry is catalysis. Solar energy conversions, selective catalysts in chemical industries, phase transfer reactions, and nano-catalysts are some current trending fields that are expected to grow manifold in the near future. In this chapter, we will learn about surface chemistry under three categories — adsorption, catalysis, and colloids.

7.2 ADSORPTION

Adsorption refers to a surface phenomenon in which a substance collects or concentrates on the surface of another substance. The surface of a solid is under strain because molecules in the solid interior are surrounded from all sides by other molecules. Hence, they are attracted equally from all directions, whereas the molecules on the surface are attracted inwards by other molecules of the solid. Hence, surface particles experience an inward pull due to sudden changes in the intermolecular forces on the surface. When a new solid surface is created, some intermolecular forces are broken and valency of some surface atoms is left unoccupied. Therefore, unbalanced residual forces existing on the surface results in higher surface energy that tends to attract and retain other molecules of other substances, so as to reduce surface strain. The substance held on to the solid surface does not penetrate in to the bulk of the solid and hence the phenomenon is termed as adsorption.

7.2.1 Mechanism of Adsorption

Adsorption can be defined as a mass transfer process that involves accumulation of a substance at the interface of two phases, such as liquid–liquid, liquid–solid, gas–liquid, or gas–solid. The substance accumulated or collected on the surface is called *adsorbate*, whereas the substance that adsorbs or collects the adsorbate on its surface is the *adsorbent*. Charcoal, silica gel, clay, alumina, colloids, etc., are classic examples of adsorbents. The process of removing adsorbed substances from an adsorbent is called *desorption*.

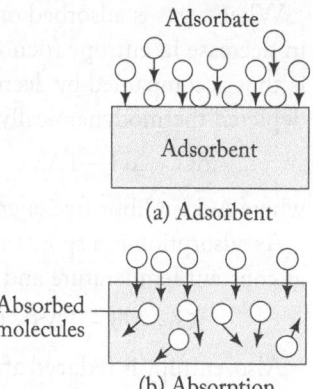

(a) Adsorbent

Adsorption should not be confused with absorption (Fig. 7.1). The term *absorption* means that a substance is not only retained at the surface but gets uniformly distributed throughout the solid or liquid bulk. We can say that water gets absorbed by anhydrous calcium chloride, but adsorbed by silica gel. Ammonia gas is absorbed in water, but gets adsorbed on activated coal. In some processes it is difficult to differentiate these processes and hence McBain suggested a general term 'sorption,' in which both adsorption and absorption take place simultaneously.

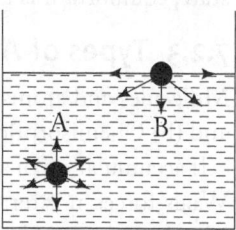

(b) Absorption

Fig. 7.1 Sorption processes

Adsorption in Liquids

In Fig. 7.2, molecule 'A' in the interior of the liquid is surrounded by other molecules and hence attracted from all sides equally.

However, molecule 'B' present on the liquid surface experiences an inward attraction. This is due to the fact that the number of molecules per unit volume is greater in the liquid bulk than in the vapour, thereby giving rise to surface tension. Hence, particles present at the liquid surface and inside give rise to difference in the free energies, both at the surface and in the bulk. Thus, surface tension exists at the surface of liquids.

Fig. 7.2 Adsorption in liquids

Adsorption in Solids

If a solid crystal is cleaved into smaller units, the surface area increases. Due to cleavage of solid crystal, attractive residual forces will be formed on the solid surface, thereby resulting in adsorption of other molecular species.

Various gases can be easily adsorbed on to solids with larger surface area. Silica gel and charcoal have porous structures and can take up appreciable volumes of gases. Ideally, charcoal is activated by heating at temperatures beyond 300 °C in vacuum and acts as an efficient adsorbent. One can determine the amount of gas adsorbed per unit mass of the solid using volumetric or gravimetric analysis. In gravimetry, the amount of adsorbed gas is determined by comparing the weights of solid sample before and after the adsorption process. In volumetric analysis, the gas is filled in a vessel of known volume at a given temperature. The pressure of the gas is measured by a manometer and a known weight of solid is introduced in to it. As adsorption proceeds, pressure of the gas drops quickly. The amount of gas adsorbed is calculated from the net drop in pressure of gas, assuming Boyle's law holds true in ideal conditions.

Solids are excellent in adsorbing dissolved solutes from aqueous solutions. Activated charcoal can adsorb various dyes, acids, bases, toxins, heavy metals from aqueous solutions at a given temperature. Adsorption at the solid–liquid interface occurs as a monolayer adsorption as solvating power of solvent disallows multilayer adsorption. Today, this concept is widely applied in waste water treatment.

7.2.2 Thermodynamics of Adsorption

The rate and extent of adsorption increase with the increase of adsorbent surface area per unit volume at a given temperature and pressure. During adsorption, the residual forces decrease, that is, decrease in surface energy that appears as heat (denoted as H). Hence, adsorption is an exothermic process; so, ΔH is always negative.

When a gas is adsorbed on to a solid surface, the movement of its molecules is restricted. This results in decrease in entropy (denoted as S) of gas after adsorption; hence, ΔS is always negative. Adsorption is thus accompanied by decrease in enthalpy and decrease in entropy of the system. Any process can be depicted thermodynamically at constant temperature and pressure by Gibbs equation,

$$\Delta G = \Delta H - T\Delta S \tag{7.1}$$

where ΔG is Gibbs free energy, ΔH is heat of enthalpy, and ΔS is entropy of the system.

As adsorption is a spontaneous process, there is a decrease in the free energy of the system and hence, at constant temperature and pressure, Eq. (7.1) can be depicted as,

$$\Delta G = \Delta H - T\Delta S < 0 \tag{7.2}$$

Also, entropy is reduced after adsorption; hence, $\Delta S < 0$ or negative. Hence, for a spontaneous reaction, ΔG is less than 0 (or negative), and exothermic reaction, $\Delta H < 0$ (or negative). As adsorption proceeds, ΔH becomes less and less negative; ultimately ΔH becomes equal to $T\Delta S$ and ΔG becomes zero. At this state, equilibrium is attained.

7.2.3 Types of Adsorption

Molecules and atoms can attach to a surface in two ways, by: (a) Physisorption (or physical adsorption) and (b) Chemisorption (or chemical adsorption).

If accumulation of adsorbate atoms or molecules on the surface of adsorbent occurs due to weak van der Waals' forces, it is termed as *physisorption*. Physisorption takes place with the formation of a multilayer of adsorbate on the adsorbent surface. It has low enthalpy of adsorption, that is, ΔH adsorption is about 20 – 40 kJ mol^{-1} due to weaker forces of attraction along with lower activation energy. Hence, physisorption is practically a reversible process.

When adsorbate atoms or molecules are held to the adsorbent by chemical bonds, it is termed as *chemisorption*. These chemical bonds can be ionic or covalent in nature. The enthalpy values are higher, ~80 – 240 kJ mol^{-1} as it involves formation of chemical bonds along with higher activation energy. Hence, it is also called *activated adsorption* and is found to be practically irreversible. At times, both physisorption and chemisorption occur simultaneously and hence it is difficult to differentiate the processes. A physical adsorption at lower temperatures could easily transverse to chemisorption at higher temperatures; for example, nickel catalyst can easily adsorb hydrogen by weak van der Waals forces. Further, hydrogen molecules undergo dissociation forming hydrogen atoms that are held on the surface by chemisorption.

7.2.4 Characteristics of Physisorption

Lack of specificity   As van der Waals forces are universal, adsorbent surfaces do not exhibit any preference for any particular adsorbate, that is, it is not specific to the adsorbent.

ΔH of adsorption   Physisorption is an exothermic process with low enthalpy of adsorption, ~20–40 kJ mol^{-1}. Lower ΔH values are attributed to weak van der Waals forces between adsorbate and the adsorbent surface.

Effect of pressure   If one considers gas–solid adsorption, the extent of adsorption increases with increase in pressure as the volume of the gases decreases during adsorption (as per Le Chatelier's principle).

Effect of temperature   As physical adsorption is an exothermic process, it occurs readily at lower temperatures and decreases on increasing the temperature in accordance with Le Chatelier's principle).

Reversible in nature   Adsorption of gases on solids is generally reversible.

$$\text{Solid} + \text{Gas} \rightleftharpoons \text{Gas} / \text{Solid} + \text{heat}$$

Nature of adsorbate   The extent of adsorption depends on the type of adsorbate. Easily liquefiable gases with higher critical temperatures are readily adsorbed as the van der Waals forces are stronger near their critical temperatures.

Surface area of the adsorbent   The extent of adsorption increases with increase in surface area of the adsorbent. Hence, finely powdered metals and substances with large surface areas are excellent adsorbents and functional catalysts.

7.2.5 Characteristics of Chemisorption

High specificity   Chemisorption occurs only when there is any possibility of chemical reaction between the adsorbate and adsorbent. Hence, chemisorption is a highly specific process.

ΔH of adsorption   Chemisorption is an exothermic process with high enthalpy of adsorption, that is, 80–240 kJ mol^{-1}. Higher ΔH values are due to the formation of strong chemical bonds between adsorbent and the adsorbate.

Effect of pressure   Chemisorption is not appreciably affected by negligible changes in pressure, but higher pressures are favourable for chemisorption.

Effect of temperature   The extent of chemisorption increases with increase in temperature up to a certain limit and after which it decreases.

Irreversibility   Chemisorption involves compound formation on the adsorbent surface and is usually irreversible in nature.

Surface area of the adsorbent   Both physisorption and chemisorption increase with increase of adsorbent surface area.

7.3   ADSORPTION ISOTHERMS

Usually, adsorption processes are studied using graphs called adsorption isotherms. When variation in the amount of gas adsorbed by an adsorbent is plotted against equilibrium pressure at constant temperature, an adsorption isotherm curve is obtained (Fig. 7.3). Let us have a look at various adsorption isotherms and their characteristics.

7.3.1 Freundlich Adsorption Isotherm

Freundlich adsorption isotherm (1909) provides the relationship between the amount of gas adsorbed per unit mass of solid adsorbent and pressure at a given temperature.

$$\frac{x}{m} = k.p^{1/n} \text{ where } n > 1 \tag{7.3}$$

In the above expression, x is the mass of adsorbed gas on m grams of adsorbent, p is the pressure of gas in the space above the adsorbent, and k and n are constants that depend on the type of adsorbent and gas at a particular temperature.

For the purpose of calculation, taking logarithms on both sides of Eq. (7.3) gives

$$\log \frac{x}{m} = \log k + \frac{1}{n} \log p \tag{7.4}$$

As shown in Fig. 7.3, the validity of Freundlich isotherm can be verified by plotting $\log x/m$ on the y-axis (ordinate) against $\log p$ on the x-axis (abscissa).

A straight line as in Fig. 7.4 indicates that Freundlich isotherm is valid, otherwise not. The slope of the straight line provides the value of $1/n$ and the intercept on the y-axis gives the value of $\log k$.

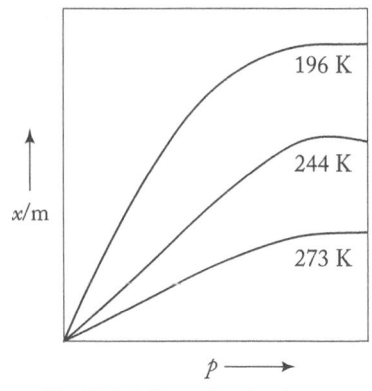

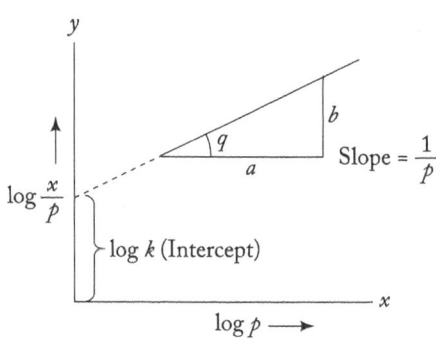

Fig. 7.3 Adsorption isotherm   **Fig. 7.4** Freundlich isotherm

The factor $1/n$ can have values ranging between 0 and 1; hence Eq. (7.4) holds true over a limited range of pressure.

If $1/n = 0$, x/m = constant, adsorption is independent of pressure; and if $1/n = 1$, $x/m = kp$, then adsorption varies proportionally with pressure.

7.3.2 Langmuir Adsorption Isotherm

As Freundlich adsorption isotherm is an approximation, Langmuir (1916) proposed the theory of adsorption of gases on solids, called Langmuir's unimolecular adsorption isotherm. The following are the assumptions of this isotherm.

During chemisorption, a chemical bond is formed between the molecules of the adsorbent and the adsorbate. As the intramolecular forces tend to fall off rapidly with distance, adsorbed layers are not more than one molecular layer thick (i.e., monolayer or unimolecular layer). The rate of adsorption is proportional to the unoccupied surface area of the adsorbent and pressure of the gas.

There are three stages in the adsorption process:

(i) Molecules get adsorbed on to the surface (adsorption).

(ii) Adsorbed molecules leave the surface (desorption).

(iii) When the rate of adsorption is equal to that of desorption, the process attains adsorption equilibrium.

Hence, as per above assumptions, Langmuir derived an equation shown as follows.

Let us consider the adsorption equilibrium, $G + S \rightleftharpoons G - S$, where G is the adsorbed gas, S is the adsorption site on the solid surface, and $G - S$ is the chemisorption complex formed on the surface. Let v be the volume of gas adsorbed per unit mass of the adsorbent at pressure p and v_m be the volume of gas needed to cover unit mass of the adsorbent with a complete, uniform monolayer.

Then $\dfrac{v}{v_m} = \theta$ (7.5)

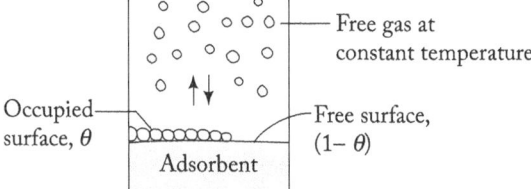

where θ is the fraction of surface occupied by gas molecules. Hence the free surface for fresh molecules to get adsorbed can be considered as $(1 - \theta)$ shown schematically in Fig 7.5.

Fig. 7.5 Schematic representation of Langmuir adsorption

Hence, as per law of mass action,

Rate of adsorption = $k_a[G][S]$ and

Rate of desorption = $k_d[G - S]$

Now, [G] can be replaced by p, (pressure of G at equilibrium) and [S], (concentration of vacant sites) be replaced by $(1 - \theta)$. Also, [G − S] can be written as θ.

On equating the rates of adsorption and desorption, we get,

$$k_a.p(1-\theta) = k_d\theta \tag{7.6}$$

$$\therefore \frac{\theta}{1-\theta} = \frac{k_a.p}{k_d} \tag{7.7}$$

$$\theta = \frac{k_a.p}{k_{d+k_a.}p} = \frac{bp}{1+bp} \tag{7.8}$$

$$\therefore v = \frac{v_m.bp}{1+bp} \tag{7.9}$$

Here, b is k_a/k_d, that is, equilibrium constant termed as adsorption coefficient of G on solid surface being utilized. Equation (7.7) represents Langmuir adsorption isotherm.

Equation (7.7) on rearranging gives,

$$\frac{p}{v} = \frac{p}{v_m} + \frac{1}{b.v_m} \tag{7.10}$$

If a plot of $\frac{p}{v}$ against p is drawn, a straight line with slope $\frac{1}{v_m}$ and intercept of $\frac{1}{b.v_m}$ on the $\frac{p}{v}$ axis will be obtained as shown in Fig. 7.6. This graph is a straight line and hence verifies Langmuir adsorption isotherm.

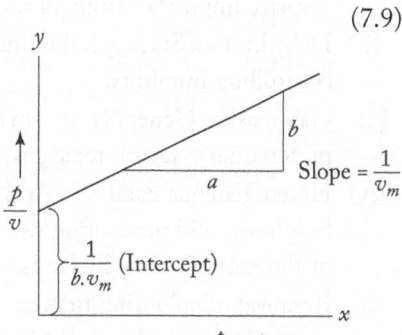

Fig. 7.6 Verification of Langmuir adsorption isotherm

BET Equation for Multimolecular Adsorption

Extending the Langmuir adsorption concept to explain multilayer adsorption on non-porous solid surfaces, Brunauer, Emmett, and Teller postulated that equilibrium exists between the solid adsorbent and the first layer of gas molecules adsorbed on it. Further, they also proposed that equilibrium can exist between the first layer of gas molecules and those in the second adsorbed layer, and so on. The BET equation is derived by balancing the rates of evaporation and condensation for various adsorbed molecular layers. BET equation follows the assumption that heat of adsorption, ΔH_{ad} is applicable only to the first adsorbed monolayer, whereas heat of liquefaction, ΔH_l of vapour is applicable to second and subsequent adsorbed molecular layers. The BET equation is expressed as,

$$\frac{p}{v(p_o - p)} = \frac{1}{v_m c} + \frac{(c-1)}{v_m c} \times \frac{p}{p_o} \tag{7.11}$$

where v is equilibrium volume of gas adsorbed per unit mass of adsorbent at pressure p, and p_o is saturated vapour pressure at the given temperature, v_m is the volume of gas needed to cover unit mass of adsorbent with a complete monolayer, and c is a constant for the given adsorbent such that

$$c = e^{\frac{(\Delta H_l - \Delta H_{ad})}{RT}}$$

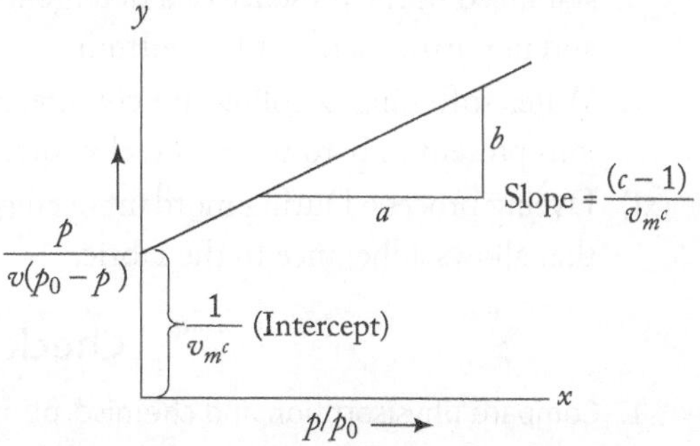

Fig. 7.7 BET plot for multilayer adsorption

If a plot of $\frac{p}{v(p_o - p)}$ against $\frac{p}{p_o}$ is linear, BET equation holds true. The intercept $\frac{1}{v_m c}$ and slope $\frac{(c-1)}{v_m c}$ can be obtained from the linear BET plot as shown in Fig. 7.7.

In 1938, Stephen Brunauer, Paul Hugh Emmett, and Edward Teller published an iconic article about BET theory in *Journal of the Americal Chemical Society*.
For more info visit the link: https://pubs.acs.org/doi/abs/10.1021/ja01269a023

7.4 APPLICATIONS OF ADSORPTION

Some of the important applications of adsorption are listed here.

(i) Surface active agents: Surface active agents in lubricants, cleaning solutions, detergents, paints, etc., displace impurities from all the surfaces by wetting phenomenon.

(ii) Dehydration: Silica gel, alumina, etc., are known to act as adsorbents for removing moisture, thereby controlling humidity.

(iii) Gas masks: Generally, gas masks comprise activated charcoal or a mixture of adsorbents that preferentially adsorb toxic gases from air and are used for breathing in coal mines, etc.

(iv) Heterogeneous catalysis: Adsorption theory explains gaseous reactants adsorbing on to solids. Sulphuric acid production using nickel catalyst, ammonia production with iron catalyst are some of the examples in which adsorption of reactants increases the rate of the reaction.

(v) Removal of colouring matter: Activated charcoal is an excellent adsorbent that is used to remove or decolourize industrial effluents.

(vi) Curing diseases: Many drugs are known to kill germs by getting adsorbed on them. To treat arsenic poisoning, colloidal $Fe(OH)_3$ acts as an antidote that adsorbs arsenic and eliminates it from the human body.

(vii) Adsorption indicators: The surfaces of some precipitates have the property to adsorb certain dyes. For example, in precipitation titrations, Eosin indicator is used when KBr is titrated against $AgNO_3$.

(viii) Chromatographic analyses: Column chromatography involves the use of a solid adsorbent that has selective adsorption capacity for certain substances present in aqueous solutions. Common adsorbents include silica gel and zeolites to separate isomers, lipids, sugars, and proteins.

(ix) Froth floatation process: Finely divided sulphide ores (ZnS, PbS) are mixed with pine oil and agitated with detergent water (frothing agent). Air is bubbled through the mixture and these bubbles get stabilized in the presence of a detergent that can adsorb mineral particles and float on the surface and impurities settle at the bottom.

(x) Water softening: Zeolites are commercial adsorbents that preferentially adsorb Ca^{2+}, and Mg^{2+} ions present in hard water, thereby softening it.

(xi) Dyeing process: During mordant dyeing, various metals such as Fe^{3+}, Al^{3+} and Cr^{3+} adsorb the dye that allows adherance to the fabric.

Check Your Progress

1. Compare physisorption and chemisorption.
2. List some examples of hydrophilic and hydrophobic adsorbents.
3. What is desorption?
4. What is an adsorption isotherm?
5. Potash alum is employed as an adsorbent to remove impurities from water. Why?
6. Write the equations for Langmuir and Freundlich adsorption isotherms.

7.5 CATALYSIS AND ENZYMOLOGY

Berzelius (1835) observed that the presence of certain substances increased the rate of the reaction, but the substances themselves remain unchanged at the end of the process. He suggested the term 'catalysts' for such substances. *Compounds or substances that alter the rate of a chemical reaction and themselves remain chemically and quantitatively unchanged after the reaction are called catalysts.* The phenomenon is termed as *catalysis.* Primarily, there are two types of catalysts, *positive catalysts* enhance the reaction rate and *negative catalysts* retard the reaction rate.

> Jöns Jacob Berzelius was a Swedish chemist renowned for developing the principles of stoichiometry, classical analytical techniques, atomic weights of elements, isomerism, and catalysis. Berzelius Day is celebrated on 20 August every year to honour his remarkable contribution to modern chemistry.

Some of the classic examples include:
1. On heating, potassium chlorate undergoes slow decomposition at 653–873 K giving oxygen.

$$2\,KClO_3 \rightarrow 2\,KCl + 3\,O_2$$

When a trace amount of manganese dioxide (MnO_2) is added to the above reaction, decomposition takes place at a rapid rate at considerably lower temperatures (473–633 K). Since MnO_2 remains unchanged after the reaction with respect to its mass and composition, it can be considered as a positive catalyst.
2. The decreased rate of decomposition of hydrogen peroxide in the presence of glycerol is a classic example of negative catalysis.
3. Conversion of starch to sugar in the presence of an acid catalyst is positive catalysis.

7.5.1 Characteristics of Catalytic Reactions (Criteria of Catalysis)

Catalyst remains unchanged chemically and quantitatively at the end of the reaction The mass and composition of catalysts should remain unaltered at the end of the reaction. However, they may undergo certain physical changes in its molecular structure and design; for example, granular manganese dioxide catalysing $KClO_3$ decomposition changes to a fine powdery material at the end of the reaction. As catalysis is a surface phenomenon, any physical changes caused to the catalyst reflect its involvement in the chemical reaction.

A trace amount of catalyst is sufficient to considerably affect the rate of chemical reaction As the catalyst is not consumed in the reaction and also is regenerated at the end of the process, even a trace amount of it can result in large amounts of reactants to combine. It is observed that low concentrations of cupric ions, i.e, 1g ion in 10^6 litres catalyses the oxidation of sodium sulphite. Generally, finely divided materials provide more surface area, thereby enhancing the reaction rate to a greater extent.

Catalyst does not affect the equilibrium position in a reversible reaction It is known that catalysts remain unchanged at the end of the chemical reaction; hence they do not contribute energy to the system. According to second law of thermodynamics, same equilibrium position must be attained, even in the presence of a catalyst. Further, catalysts affect forward and backward reactions in a reversible process to an equal extent. Hence, catalyst assists in attaining quicker equilibrium, but the position remains unaffected and can be depicted as shown in Fig. 7.6.

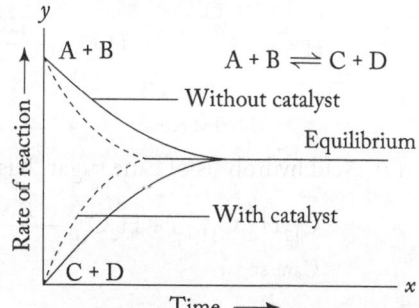

Fig. 7.8 Effect of catalyst on the equilibrium attainment

Catalysts cannot initiate a chemical reaction   The primary function of a catalyst is to only alter the reaction rates of a given chemical reaction. It can accelerate the desired reaction and also decelerate unwanted side reactions.

Catalysts are specific in their action   A catalyst effective in accelerating a particular reaction may not be as effective for another reaction. It is also observed that the same reactant will give different products on changing the catalyst; for example, in the presence of hot copper catalyst, ethanol converts to ethanal, while it converts to ethene in the presence of hot alumina catalyst.

$$C_2H_5OH \xrightarrow{\text{Hot Cu}} CH_3OH \text{ (Ethanal)}$$
$$\text{Ethanol} \xrightarrow{\text{Hot Al}_2O_3} CH_2{=}CH_2 + H_2O \text{ (Ethene)}$$

The physical state of the catalyst determines its efficiency   In heterogeneous catalysis, it is found that finely powdered materials act as efficient catalysts than course grained ones, for example, finely divided platinum is more efficient than its undivided form.

Change in temperature alters the rate of a catalytic reaction   The overall efficiency of a catalyst is altered by changing the temperature and the temperature at which the reaction rate is maximum is called its optimum temperature.

Catalytic activity is altered in the presence of a foreign substance   A foreign substance that enhances the activity of a catalyst is called *promoter*, whereas the one that inhibits or destroys the catalytic activity is called *poison* or *anti-catalyst*. Molybdenum acts as an efficient promoter for iron catalyst in ammonia manufacture by Haber's process.

7.5.2 Types of Catalysis

According to the phases of reactants and catalysts, catalysis can be categorized into homogeneous and heterogeneous catalysis.

Homogeneous catalysis   This involves reactants and catalysts of same phase.

(a) Oxidation of sulphur dioxide in to sulphur trioxide in the presence of oxides of nitrogen as a catalyst in the lead chamber process of manufacture of sulphuric acid. The reactants sulphur dioxide and oxygen, and catalyst nitric oxide are all in the same gaseous phase.

$$2SO_{2(g)} \xrightarrow[\text{NO(g)}]{O_2(g)} 2SO_{3(g)}$$

(b) In the hydrolysis of methyl acetate in the presence of acid catalyst, both the reactant, methyl acetate and the catalyst are in aqueous phase.

(c) Acid hydrolysis of cane sugar has both reactants and catalyst in the aqueous phase.

$$C_{12}H_{22}O_{11(aq)} + H_2O_{(l)} \xrightarrow{H_2SO_4 \text{ (l)}} C_6H_{12}O_{6(aq)} + C_6H_{12}O_{6(aq)}$$
Cane sugar   Glucose   Fructose

Heterogeneous catalysis   This involves reactants and catalysts in different phases. Some examples are as follows:

(a) Oxidation of sulphur dioxide to sulphur trioxide in the presence of platinum catalyst involves gaseous reactants and solid metal catalyst.

$$2SO_{2(g)} \xrightarrow[\text{Pt(s)}]{O_2(g)} 2SO_{3(g)}$$

(b) In Haber's process, nitrogen and hydrogen combine to form ammonia in the presence of solid iron catalyst.

$$N_{2(g)} + 3H_{2(g)} \xrightarrow{\text{Fe(s)}} 2NH_{3(g)}$$

(c) The vegetable oil hydrogenation using finely divided nickel involves one reactant in the liquid state and hydrogen in the gaseous state in the presence of solid catalyst.

$$\text{Vegetable oil}_{(l)} + H_{2(g)} \xrightarrow{\text{Ni(s)}} \text{Vegetable ghee}_{(s)}$$

Catalytic poisons Any substance that inhibits or destroys catalytic activity, thereby retarding the overall reaction is called catalytic poison or inhibitor. Most of the heterogeneous catalysts are rendered ineffective due to the presence of trace quantities of impurities that inhibit or retard the catalytic activity.

(a) Activity of platinum catalyst in a reaction between hydrogen and oxygen is poisoned due to hydrogen sulphide, carbon disulphide, and carbon monoxide.

$$2H_2 + O_2 \xrightarrow[\text{inhibited by CO}]{\text{Pt(s)}} 2H_2O$$

(b) In the reaction between ethene and hydrogen forming ethane, copper catalyst gets poisoned due to the presence of mercury or carbon monoxide.

$$CH_2 = CH_2 + H_{2(g)} \xrightarrow[\text{inhibited by Hg or CO}]{\text{Cu(s)}} CH_3 - CH_3$$

Autocatalysis It is observed that in certain reactions, products themselves can catalyse the reaction. Such a phenomenon is called autocatalysis and the substance that effects the catalysis is called *autocatalyst*. Say, if we consider a reaction A → P, where P denotes products, the rate law can be written as, $v = k[A][P]$.

Here the reaction rate will be proportional to the concentration of product P. The first studied BZ reaction (Belousov–Zhabotinsky reaction) is exhibited in a mixture of potassium bromate, malonic acid, and cerium (IV) salt in an acid medium. The two steps of its mechanism show that the product $HBrO_2$ acts as an autocatalyst.

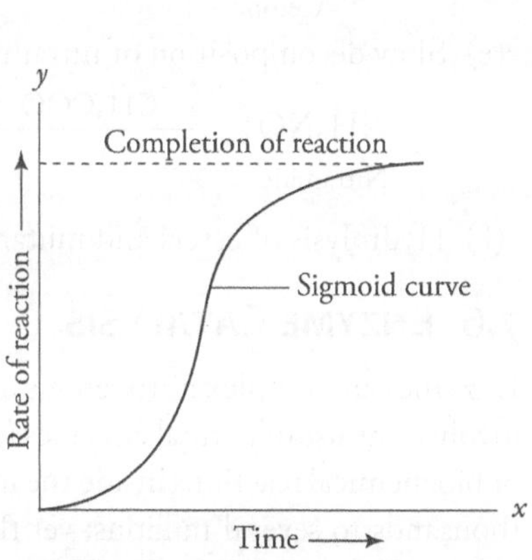

Fig. 7.9 Progress of an autocatalytic reaction with time (in arbitrary units)

(i) $BrO_3^- + HBrO_2 + H_3O^+ \rightarrow 2BrO_2. + 2H_2O$

(ii) $2\,BrO_2. + 2Ce(III) + 2\,H_3O^+ \rightarrow 2\,HBrO_2 + 2\,Ce(IV) + 2\,H_2O$

From the above reactions, it is clear that $HBrO_2$ is a reactant in the first step and further effects the increase in the rate of formation of $HBrO_2$.

For an autocatalytic reaction, the reaction rate increases with time since the concentration of catalytic products keeps on increasing; this is depicted as a plot in Fig. 7.9 showing a sigmoid curve reaching maximum on completion of the reaction.

7.5.3 Acid–Base Catalysis

When reactions are catalysed by an acid or a base or by both, they are termed as acid–base catalysis. Usually a homogeneous reaction, acid–base catalysis involves reactants, catalysts, and products in the same phase.

If reactions are catalysed by H^+ ions, they are termed as *acid-catalysed reactions*, whereas if reactions are catalysed by OH^- ions, they are termed as *base-catalysed reactions*.

Various Brønsted acids (proton donors) such as H^+, acetic acid, cations of weak bases (NH_4^+), hydronium ions (H_3O^+) are called *general acid catalysts*. Most Brønsted bases such as OH^- and acetate ions are called *general base catalysts*. If reactions are catalysed by both general acids and bases, they are called *general acid–base catalysed reactions*. Some of the classic examples of acid–base catalysis are listed below.

(a) Inversion of cane sugar is a type of specific acid catalysis.

$$C_{12}H_{22}O_{11(aq)} + H_2O_{(l)} \xrightarrow[H^+]{\Delta H_2SO_4\,(l)} C_6H_{12}O_{6(aq)} + C_6H_{12}O_{6(aq)}$$
$$\text{Cane sugar} \qquad\qquad\qquad\qquad \text{Glucose} \qquad \text{Fructose}$$

(b) Formic acid decomposition catalysed by hydrogen ions is also a type of specific acid catalysis.

$$\text{HCOOH} \xrightarrow{\Delta, H^+} CO + H_2O$$
$$\text{Formic acid} \qquad \text{Carbon monoxide}$$

(c) Iodination of propanone is catalysed by acetic acid and follows general acid catalysis.

(d) Acetone conversion to diacetone alcohol catalysed by hydroxide ions is a type of specific base catalysis.

(e) Slow decomposition of nitramide is catalysed by acetate ions and is a type of general base catalysis.

$$NH_2NO_{2(aq)} \xrightarrow{CH_3COO^-} N_2O_{(g)} + H_2O_{(l)}$$
$$\text{Nitramide} \qquad\qquad \text{Nitrous oxide}$$

(f) Hydrolysis of esters and mutarotation of glucose are catalysed by either acids or bases.

7.6 ENZYME CATALYSIS

Enzymes are complex nitrogenous organic compounds produced by plants and animals. Enzyme catalysis involves the use of natural enzymes (proteins) to perform chemical transformations on organic compounds or biochemical reactions inside the living organisms. The molecular weights of enzymes range from several thousands to several millions; yet they can efficiently catalyse transformations of molecules as small as carbon dioxide and nitrogen. The catalysis of biochemical reactions in the cell is vital due to the very low reaction rates of the uncatalysed reactions.

7.6.1 Characteristics of Enzyme Catalysis

(a) Almost all biochemical reactions are effectively catalysed by enzymes.
(b) It is known that all enzymes are proteins; hence their ability to catalyse chemical reactions is attributed to their primary, secondary, tertiary, and quaternary structures.
(c) Enzymes possess high degree of specificity for their substrates and are known to accelerate chemical reactions tremendously.
(d) The rate of an enzyme-catalysed reaction is maximum at a definite pH called *optimum pH* generally in the range of $5 - 7$ on the pH scale.

(e) The presence of certain substances called *co-enzymes* tends to enhance the rate of the reaction (promoters). Various metal ions like Na^{2+}, Co^{2+}, and Cu^{2+}, tend to weakly form bonds with the enzyme, thereby enhancing the overall rate of the chemical reaction.

(f) Just like ordinary catalysts, biocatalysts too get poisoned in the presence of foreign substances. The mechanism of action of most drugs is related to their enzyme inhibition ability in the human body.

(g) Enzymes can effectively function in aqueous solution under mild conditions contrary to organic chemical reactions that require extreme conditions such as high temperature and pressure.

(h) They are effective in trace concentrations as they are not consumed in the reactions where they are used as catalysts.

(i) They do not affect the direction of the reaction but enables the reaction to reach equilibrium sooner.

7.6.2 Mechanism of Enzyme Catalysis

All enzymes have numerous cavities called *active sites* on their surface that are characterized by definite shape and possess various functional groups such as –COOH (carboxylic acid), –NH$_2$ (amino), –SH (thiol), –OH (alcoholic), and so on. The reactant molecules (substrate) that have complementary shapes will fit in to the enzyme active sites just like a lock and its key (lock-and-key mechanism). Due to the presence of active functional groups within enzymatic cavities, the enzyme–substrate activated complex is formed that later decomposes yielding products.

A two-step enzyme-catalysed reaction can be shown as follows.

(i) Binding of enzyme to the reactant molecules (substrate) forming an activated complex $E + S \rightarrow ES^*$.

(ii) Decomposition of activated complex forming products, $ES^* \rightarrow E + P$

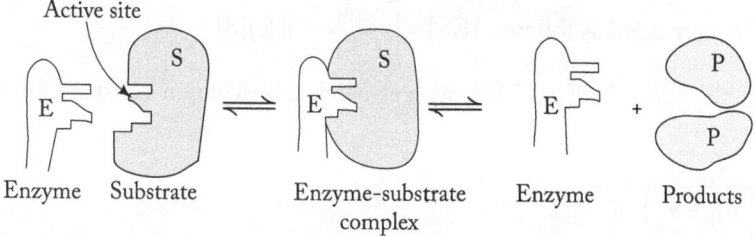

Fig. 7.10 Mechanism of enzyme-catalysed reaction

The progress of chemical reaction in the presence of enzyme is shown in Fig 7.10. The peak of the curves denotes the formation of activated complexes after which there is a decline in reactant concentration leading to product formation.

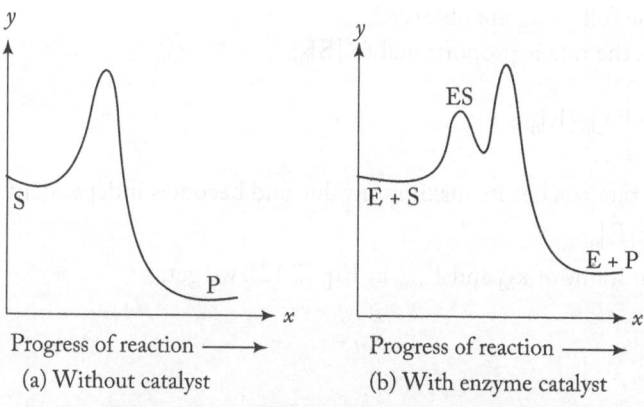

Fig. 7.11 Comparison of reaction progress

Michaelis–Menten Enzyme Kinetics

For understanding the mechanism of enzyme catalysis, it is essential to study the kinetic behaviour of reaction systems. Enzyme–substrate complex formation is a critical event in such reactions. Michaelis and Menten proposed enzyme–substrate binding to be reversible in nature and derived a kinetic model to understand the factors influencing the enzymatic reactions, such as temperature, concentration, and pH of the system.

The characteristics of an enzyme-catalysed reaction are listed below.

(a) For initial concentration of substrate $[S]_0$, the initial rate of product formation is directly proportional to the total concentration of enzyme, $[E]_0$.

(b) For a given $[E]_0$ and lower concentration of $[S]_0$, the rate of product formation is proportional to $[S]_0$.

(c) If for a given $[E]_0$ one gets higher values of $[S]_0$, the rate of product formation will be independent of $[S]_0$ further reaching a maximum velocity denoted as V_{max}.

Let us consider a simple single substrate reaction, in which a free enzyme, E binds with substrate S to form the complex ES. The ES complex results in the formation of product P.

$$E + S \underset{k_{-1}}{\overset{k_1}{\rightleftharpoons}} ES \xrightarrow{k_2} E + P$$

As per the above reaction, the rate of product formation will be, $V = k_2 [ES]$ (7.12)

On applying steady-state approximation, we get,

$$\frac{d[ES]}{dt} = k_1[E][S] - k_{-1}[ES] - k_2[ES] = 0$$

Further, it can be rearranged as follows, $[ES] = \left[\dfrac{k_1}{k_{-1} + k_2} \right][E][S]$

On substituting the above in Eq. (7.12), we get Michaelis–Menten constant for enzyme-catalysed reaction:

$$V = \frac{k_2[E]_0}{1 + \left[\dfrac{k_{-1} + k_2}{k_1} \right]\dfrac{1}{[S]_0}}$$ (7.13)

The constant $\left[\dfrac{k_{-1} + k_2}{k_1} \right]$ is characteristic for a given enzyme acting on a particular substrate and is termed as Michaelis constant denoted as k_M and can be further expressed as $k_M = [E][S]/[ES]$.

As per Eq. (7.13), the following are observed,

1. When $[S]_0 << k_M$, the rate is proportional to $[S]_0$,

$$V = \left[\frac{k_1 k_2}{k_{-1} + k_2} \right][S]_0 [E]_0$$

2. If $[S]_0 >> k_M$, the rate reaches its maximum value and becomes independent of $[S]_0$,

$$V = V_{max} = k_2 [E]_0$$

On substituting definitions of k_M and V_{max} in Eq. (7.12), we get,

$$V = \frac{V_{max}}{1 + \dfrac{k_M}{[S]_0}}$$

The above expression can be rearranged to allow data analysis using linear regression as,

$$\frac{1}{V} = \frac{1}{V_{max}} + \left[\frac{k_M}{V_{max}}\right]\frac{1}{[S]_0}$$

This is called the Lineweaver–Burk equation and a plot called Linewaever–Burk plot (or double reciprocal plot) is prepared which helps in determining V_{max}, k_M, and also classifying various mechanisms of enzyme catalysis (Fig. 7.12).

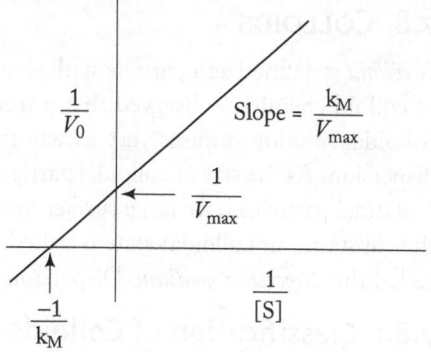

Fig. 7.12 Lineweaver–Burk plot

7.7 CATALYSTS IN INDUSTRY

Almost 90 % of industrial processes are realized only because of catalysts. Catalysts are employed in the manufacture of important chemical products such as hydrogen, ammonia, sulphuric acid, alcohols, esters, synthetic resins, plastics, fuels, dyes, and medicines. Some of the major sectors of world economy that rely completely on catalysis are: petroleum and energy production, chemical and polymer production, pharmaceuticals, food industry, and pollution prevention. Platinum, nickel, iron, cobalt, manganese, chromium, vanadium, aluminium are commonly utilized as metal catalysts. Some of the reactions catalysed by important metals are given in Table 7.1.

Table 7.1 Some industrially important reaction catalysts

Process	Typical catalysts	Reactions (or conversions)
Dehydration	Al_2O_3, ThO_2	Olefins and water vapour to alcohols
Hydrogenation	Ni	Oils to fats
	Fe	Ammonia synthesis
	Pt, Pb, Rh	Hydrogenation of double bonds
	ZnO, Cr_2O_3	Methanol synthesis
Dehydrogenation	Fe_2O_3, MoO_2, Cr_2O_3	High-temperature hydrogenation
Haber's ammonia manufacture	FeO, K_2O, Al_2O_3 mixture	Ammonia manufacture
Ostwald's process	Platinized asbestos, 573 K	Nitric acid manufacture
Contact process	Platinized asbestos, V_2O_5, 673 – 723 K	Sulphuric acid manufacture
Deacon's process	Metallic halides, $CuCl_2$	Halogenation reactions

Check Your Progress

7. What are catalysts?
8. List any four characterisitics of catalysts.
9. Why does the rate of chemical reactions increase due to enzymes?
10. How is specific acid catalysis different from general acid catalysis?
11. What are catalyst poisons?
12. Provide any two reactions which are catalysed by both acids and bases.
13. Write the two steps in Belousov–Zhabotinsky reaction.

7.8 COLLOIDS

A *colloid* is defined as a particle with size (diameter) between 10 to 100 nm. Thomas Graham, the father of colloid chemistry observed that a true solution can easily pass through a parchment paper, whereas colloidal solution diffuses very slowly through it. The dispersion of colloidal particles is called colloid dispersion. As the size of colloidal particles is very small, they behave quite differently from a true solution. Colloidal particles have large surface area per unit mass due to their small size. In a colloidal dispersion, the substance in colloidal state is called the *dispersed phase* and the medium in which it is suspended is called the *dispersion medium*. Dispersion medium is not limited to liquids; it can also be a solid or gas.

7.8.1 Classification of Colloids

The criteria for classifying colloids are:
  (i) Physical state of dispersed phase and dispersion medium.
 (ii) The nature of interaction between dispersed phase and dispersion medium.
  Based on the above criteria, Table 7.2 lists eight types of colloidal dispersion.

Table 7.2 Types of colloidal dispersion with examples

Dispersed phase	Dispersion medium	Name	Examples
Solid	Solid	Solid sol	Coloured glasses, ruby and gemstones, carbon in steel
Solid	Liquid	Sol	Toothpaste, cellular fluids, paints
Solid	Gas	Aerosol	Dust, smoke
Liquid	Solid	Gel	Butter, jelly, cheese, ice cream
Liquid	Liquid	Emulsion	Milk, hair cream, mayonnaise
Liquid	Gas	Liquid aerosol	Fog, cloud, mist, sprays
Gas	Solid	Solid sol	Insulating foams, pumice stone
Gas	Liquid	Foam	Whipped cream, fire extinguisher foam, soap lather, froth

Sols (solid in liquid), gels (liquid in solid), and emulsions (liquid in liquid) are common examples of colloids. Based on the nature of interaction between dispersed phase and dispersion medium, colloidal sols are broadly of two types, viz., lyophilic (solvent attracting) and lyophobic (solvent repelling). If the dispersion medium is water, the two types of colloidal solutions are hydrophilic and hydrophobic sols.

Lyophilic sols (solvent loving) These are colloidal sols formed by directly mixing gum, gelatin, starch, rubber, etc. with a suitable solvent. An important feature of these sols is that if dispersion medium is separated from its dispersed phase, it can be reconstituted by remixing with the same dispersion medium. Hence, they are also termed as *reversible* sols.

Lyophilic sols are quite stable and possess higher viscosities and lower surface tension than that of its dispersion medium. A dispersed phase in a sol can be easily precipitated by adding an electrolyte and this process is called *coagulation*. A large amount of electrolyte is needed to coagulate the dispersed phase in a lyophilic sol. The particles in a lyophilic sol are not easily detected under an ultramicroscope; they show poor electrophoretic effect. Examples include starch solutions, glue, gelatin.

Lyophobic sols (solvent hating) These are sols of metals, sulphur, and silver halides. As these sols have repelling dispersed phase and dispersion medium, lyophobic sols are comparatively unstable than lyophilic sols. These sols are non-viscous and possess viscosity and surface tension almost nearly same as that of its dispersion medium. As against lyophilic sols, a small amount of electrolyte is sufficient to coagulate the dispersed phase in lyophobic sol. Once precipitated, lyophobic sols cannot be brought back in to the colloidal state. Examples include gold sol, sulphur sol, metal sulphides, and Prussian Blue dye.

7.8.2 Preparation of Colloids

Lyophilic sols are prepared by warming the solid in a liquid dispersion medium. For example, starch solution is prepared by warming in water with continuous stirring. Lyophobic sols are prepared by special methods, namely aggregation and dispersion methods. Dispersion methods include Bredig's Arc method and peptization.

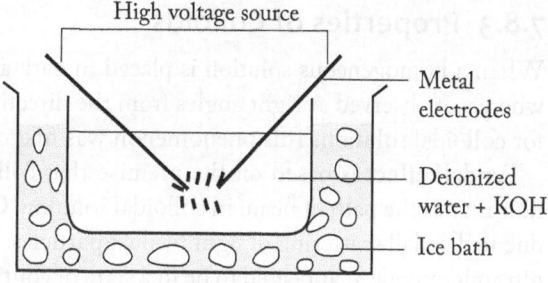

Fig. 7.13 Bredig's Arc Method

Bredig's Arc Method

In this method, colloidal sols of metals (such as gold and silver) are prepared by striking an electric arc between electrodes of the metal immersed in a suitable dispersion medium of deionized water and traces of potassium hydroxide (electrolyte) added to it. The dispersion medium is placed in an ice bath (Fig. 7.13).

When an electric arc is struck across the metal electrodes, the metal vaporizes and condenses in water which aggregate to form colloidal particles in water.

Peptization The dispersal of precipitate into a colloidal sol by the action of an electrolyte in solution is called peptization. It is the reverse of coagulation of sols. The electrolyte used for peptization is called peptizing agent. A freshly precipitated solid is dispersed in a colloidal solution in water by adding a trace quantity of

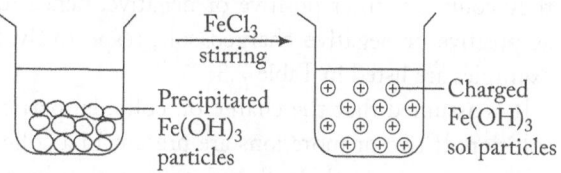

Fig. 7.14 Peptization

an electrolyte which contains a common ion. The dispersed precipitate will adsorb the common ions resulting in positive and negative charges on the precipitate. These electrically charged particles will split up as colloidal particles.

Example: A colloidal sol of ferric hydroxide is prepared by agitating a fresh precipitate of ferric hydroxide with a trace electrolyte such as ferric chloride as shown in Fig. 7.14.

Aggregation Method

In this method, colloidal sols are prepared by chemical reactions or using different solvents. The atoms or molecules of dispersed phase tend to aggregate as colloidal particles under suitable conditions such as temperature, concentration, and so on. Some of the chemical reactions are explained below.

Double decomposition Arsenic sulphide colloidal sol is prepared by passing a stream of hydrogen sulphide gas through a solution of arsenic oxide. A yellow coloured sol of arsenic sulphide is obtained.

$$As_2O_3 + 3H_2S \longrightarrow As_2S_3(sol) + 3H_2O$$

Reduction Sols of metals can be prepared by treating dilute solutions of metal chlorides with reducing agents such as tannic acid and formaldehyde. Gold sol is obtained by treating gold chloride with formaldehyde.

$$2AuCl_3 + 3HCHO + 3H_2O \longrightarrow 2Au(sol) + 3HCOOH + 6HCl$$

Oxidation When a stream of hydrogen sulphide gas is allowed to pass into a solution of sulphur dioxide, sulphur sol is obtained.

$$SO_2 + 2H_2S \longrightarrow 3S(sol) + 2H_2O$$

Hydrolysis A distinct red sol of ferric hydroxide is obtained by adding a few drops of 30% ferric chloride solution in excess of boiling water with continuous agitation.

$$FeCl_3 + 3H_2O \longrightarrow Fe(OH)_3(sol) + 3HCl$$

7.8.3 Properties of Colloids

When a homogeneous solution is placed in dark and is observed in the direction of light, it looks clear, whereas if observed at right angles from the direction of light beam, it appears dark. This holds true even for colloidal solution. This phenomenon was first observed by Tyndall and is called Tyndall effect.

Tyndall effect is based on the premise that colloidal particles scatter light in all directions and this illuminates the path of beam in colloidal solution. One can visibly see dust particles in a beam of sunlight due to Tyndall scattering of light by dust particles. When R. Brown observed colloidal solution under an ultramicroscope, it appeared to be in a state of continuous zig-zag motion; this is known as **Brownian movement**. Smaller colloidal particles moved rapidly than the heavier colloidal particles and bombard with other particles along the way. Due to Brownian movement, it is difficult to settle colloidal particles, thus rendering stability to colloidal sols. There is always an electric charge on colloidal particles and they could be either positive or negative, hence termed as positive or negative charged sols, respectively. Some examples are listed in Table 7.3.

Table 7.3 Types of colloidal sols

Positive charged sols	Negative charged sols
Titanium dioxide sol	Starch, gum, coal
Blood	Dyes such as eosin, Congo Red
Metallic oxides like $CrO_3.xH_2O$, $Fe_2O_3.xH_2O$	Copper, gold, Ag sols

It is assumed that the charge on colloidal particles arises due to *preferential adsorption* of ions from solution. If two or more ions are present in a colloidal dispersion, there will be preferential adsorption of ion common to the colloidal particle thereby rendering a charge on it. On adding silver nitrate to potassium iodide solution, silver iodide precipitate is formed that adsorbs iodide ions from the dispersion medium resulting in a negatively charged colloidal sol. But if potassium iodide solution is added to silver nitrate solution, there is adsorption of silver ions from dispersion medium and positively charged sol is formed. [AgI/I^- (negative charged) and AgI/Ag^+ (positively charged)]. Sometimes, a combination of multiple layers of opposite charges termed electrical double layer is formed around a colloidal particle. If an electric potential is applied across platinum electrodes immersed in a colloidal solution, colloidal particles tend to move towards one of the electrodes and is termed as *electrophoresis*. One can coagulate colloidal particles from its dispersion medium using electrophoresis.

Applications of Colloids

Colloids have found wider applications in industry. Some of the pertinent examples are as follows.

Drinking water purification   Natural waters are known to contain numerous suspended particles and impurities. Alum is a common coagulant added to water that removes all suspended particles and make water deem fit for drinking.

Pharmaceutical drugs and medicines   Most of the medicines are colloidal in nature. Colloidal gold is administered as an intramuscular injection. Milk of magnesia, $Mg(OH)_2$ suspension is administered orally to treat patients with stomach disorders such as constipation, acidity, heartburn, and so on. Colloidal silver is popularly used in wound dressing and antibiotic coatings on medical devices such as catheters.

Air purification   Smoke is a colloidal dispersion of solid particles of carbon, dust, particulate matter, etc., in air. Electrical precipitation of smoke takes place in a Cottrell precipitator in which smoke before leaving the exhaust is led through a chamber containing plates of charge opposite to those of smoke particles. When smoke particles come in contact with the charged plates, hazardous particles of smoke lose their charges and precipitate as ash which settles at the bottom of the chamber.

Commercial products   Inks, synthetic polymers, soap, rubbers, paints, varnishes, lubricants, cement are important commercial colloidal solutions that have varied uses in industry.

Leather tanning   Raw animal skin is a positively charged colloidal system made up of proteins and is soaked in tannins obtained from tree bark that are negatively charged. The two oppositely charged sols coagulate and form a hard adherent layer on the leather that resists decomposition. This process is called leather tanning. Chromium salts are also alternatively used in place of tannins.

Photography   Photographic films are prepared by coating an emulsion of light-sensitive silver bromide in gelatin over celluloid film.

Rubber industry   Latex is a colloidal solution of rubber particles that are negatively charged and raw rubber is coagulated from latex using acids.

7.9 EMULSIONS

A liquid–liquid colloidal system in which finely divided droplets are dispersed in another liquid is called an emulsion. An emulsion is obtained when two immiscible or partially miscible liquids are mixed together. Emulsions are of two types: (a) oil-in-water (O/W type) and (b) water-in-oil (W/O type).

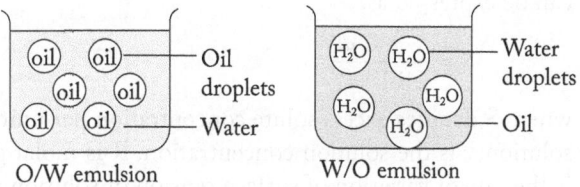

Fig. 7.15 Types of emulsion

In **O/W emulsion**, water acts as the dispersion medium with oil droplets as the dispersed phase. Milk, face creams are classic examples of O/W emulsion type. In **W/O emulsion,** oil is the dispersion medium. Butter, cream are common examples. Today, oil-in-oil (O/O type) emulsion involving a polar oil like propylene glycol in a non-polar paraffin oil is also studied.

Oil-in-water emulsions are unstable and at times separate out in to two distinct layers on standing. Hence, for stabilizing emulsions, a third component called *emulsifying agent* is added that forms an interfacial film between the suspended particles and the dispersion medium. The suitable choice of an emulsifier is crucial in emulsion formation and its long-term stability. Proteins, gums, soaps are emulsifiers for O/W emulsions, whereas long-chain alcohols and metal salts of fatty acids are used as emulsifiers for W/O emulsions. Adding a large amount of dispersion medium can dilute the emulsions with dispersed phase as a separate layer. The emulsion droplets are generally negatively charged and hence can be precipitated by various electrolytes. Some pertinent industrial applications of emulsions are:

(a) Food-based emulsions such as mayonnaise, beverages, desserts
(b) Cosmetics such as face creams, hair sprays, sunscreen lotions
(c) Pharmaceuticals like anesthetic O/W emulsion types, certain medicines
(d) Agrochemicals like crop-oil sprays

Emulsions exhibit colloidal properties such as Brownian movement and Tyndall effect. Heating, freezing, and centrifugation tend to destabilize the emulsions.

Check Your Progress

14. Define colloids and emulsion.
15. What is Tyndall effect and Brownian motion?
16. Colloidal particles move in zig zag fashion. Why?
17. Comment: 'Colloidal particles precipitate on addition of electrolytes.'
18. Provide one example: (a) solid sol (b) liquid aerosol (c) solid emulsion.
19. Name the different methods to prepare colloids.
20. Name any two methods to purify colloids.
21. Only colloidal systems exhibit Brownian and Tyndall effects. Why?
22. What is an electrical double layer?
23. What are different types of emulsions? Give examples.
24. List the applications of emulsions.

7.10 SURFACTANTS

Substances that decrease the surface tension at the interface and get concentrated at the interface are called *surfactants* or *surface-active agents*. Surfactants are amphiphilic in nature and consists of a hydrophilic (water soluble) head and a hydrophobic tail (water insoluble) as shown in Fig. 7.16.

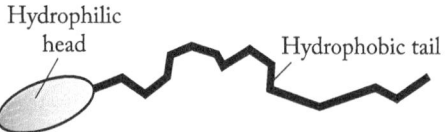

Fig. 7.16 A typical surfactant molecule

J.W. Gibbs (1878) observed that the concentration of the solute at liquid surface could be greater than in the bulk of the solution; the thermodynamic relation expressing this behaviour is called Gibbs adsorption equation. Hence, for dilute solutions, Gibbs equation can be expressed as,

$$S = -\frac{c}{RT} \cdot \frac{dr}{dc} \qquad (7.15)$$

where S denotes excess solute concentration per unit area of the surface as compared to the bulk of the solution, c is the solution concentration, R is molar gas constant, T is absolute temperature, and dr/dc is the rate of variation of surface tension of solution with solute concentration. As per Eq. (7.15) if any solute causes surface tension of solvent to decrease, dr/dc will be negative and hence S will be positive. However, if solute increases the surface tension, dr/dc will be positive and S will have a negative value. This depicts higher concentration of solute in the bulk and exhibits negative adsorption.

Surfactants are categorized into the following.

(a) *Cationic surfactants* such as cetyltrimethylammonium bromide $C_{16}H_{33}N(CH_3)_3^+Br^-$ (CTAB): the surface activity resides in the cation.

(b) *Anionic surfactants* like sodium cetyl sulphate $C_{16}H_{33}SO_4^-Na^+$; the surface activity resides in the anion.

(c) *Non-ionic surfactants* like cetyl polyglycol ether $C_{16}H_{33}(OCH_2CH_2)_nOH$ (where, n = 15–30).

(d) *Zwitterionic surfactants* such as lauryl amido propyl dimethyl betaine have two different charge groups and are hence also referred to as amphoteric surfactants. However, these charges are either permanent or dependent on specific pH range.

Lowering of Surface Tension

Water has a high surface tension (72 dyne/cm at 25 °C) and most solutes reduce its value when dissolved in it.

Consider a beaker filled with water. Water molecules present in the bulk are pulled by neighbouring molecules from all directions and thus the resultant pull on it may be negligible.

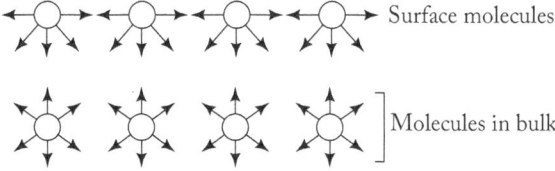

Fig. 7.17 Illustration of surface tension

However, the water molecules on the surface are pulled only by the molecules below them. Hence water molecules at the surface experiences a net pull from molecules directly below them. This causes the surface tension phenomenon at the surface of liquid.

When a surfactant is introduced in a beaker containing water, the surfactant molecule enters the air/water interface and replaces the water molecules. This reduces the net pull on the surface layer of water molecules, thereby reducing the surface tension of water.

Methods of Preparation

As discussed, surfactants can be anionic, cationic, or zwitterionic in nature. Let us have a look at the methods of their preparation.

Anionic surfactants These have carboxylate, phosphate, sulphate, or sulphonate polar head groups along with a counter ion such as Na^+, K^+, Ca^{2+}, Mg^{2+}, protonated alkyl amines, etc. These counter ions determine the solubility of the surfactants. Na^+, K^+ promote water solubility, whereas Ca^{2+}, Mg^{2+} promote oil solubility. Protonated alkyl amines tend to promote both oil and water solubilities.

Alkylbenzene sulphonate is prepared by sulphonation of alkylbenzene using sulphur trioxide.

$$R-\langle\bigcirc\rangle \quad + 2\,SO_3 \longrightarrow R-\langle\bigcirc\rangle-SO_2OSO_3H \xrightarrow[\text{Slow}]{R-\langle\bigcirc\rangle} R-\langle\bigcirc\rangle-SO_3H$$

Alkyl benzene Sulphur trioxide Pyrosulphonic acid

$$\downarrow NaOH$$

$$R-\langle\bigcirc\rangle-SO_3^-\,Na^+$$

Alkyl benzene sulphonate

A sulphonic acid derivative called pyrosulphonic acid is obtained which is neutralized using caustic soda. Alkylbenzene sulphonate salts are commonly used as detergents which possess linear alkyl chains; hence they are also referred to as LABS (linear alkylbenzene sulphonate).

Sodium dodecyl sulphate (SDS) is a common anionic surfactant. Dodecanol is treated with sulphuric acid to form dodecyl sulphate. On treating dodecyl sulphate with sodium hydroxide (caustic soda), sodium dodecyl sulphate is obtained.

$$CH_3(CH_2)_{10}CH_2OH + H_2SO_4 \rightarrow CH_3(CH_2)_{10}CH_2O\overset{O}{\underset{O}{-\overset{\|}{\underset{\|}{S}}-}}OH + H_2O \xrightarrow{NaOH} CH_3(CH_2)_{10}CH_2O\overset{O}{\underset{O}{-\overset{\|}{\underset{\|}{S}}-}}O^-Na^+$$

Dodecanol

Dodecylsulphate

Sodium dodecyl sulphate (SDS)

For large-scale production, sulphur trioxide is used to manufacture SDS surfactants. SDS is used as laundry detergents, toothpastes, shampoos, etc.

Non-ionic surfactants Liquid detergents are usually non-ionic type of surfactant that is formed by the reaction between stearic acid and polyethylene glycol.

$$CH_3(CH_2)_{16}COOH + HO(CH_2CH_2O)_nCH_2CH_2OH \xrightarrow{H_2O} CH_3(CH_2)_{16}COO(CH_2CH_2O)_nCH_2CH_2OH$$

Stearic acid   Polyethylene glycol   Non-ionic detergent

Cationic surfactants These are quaternary ammonium salts used as shampoos, conditioners, fabric softeners, and antistatic agents. Ester-containing quaternary ammonium surfactants are prepared by the esterification of fatty acids with an amino alcohol and N-alkylation.

$$2\,RCOOH + N(CH_2CH_2OH)_3 \xrightarrow{-H_2O} (RCOOCH_2CH_2)_2NCH_2CH_2OH$$

$$\downarrow (CH_3)_2SO_4 \quad \overset{CH_3}{|}$$

$$(RCOOCH_2CH_2)_2\overset{+}{N}-CH_2CH_2OH + CH_3SO_4^-$$

CTAB (cetyl trimethyl ammonium bromide) is a common cationic detergent prepared by the alkylation of 1-bromohexadecane and trimethyl amine.

$$CH_3(CH_2)_{15}-Br \ + \ \underset{CH_3}{\overset{CH_3}{N}}-CH_3 \longrightarrow \left[CH_3(CH_2)_{15}-\underset{CH_3}{\overset{CH_3}{\overset{+}{N}}}-CH_3 \right] Br^-$$

1-Bromohexadecane   Trimethyl amine   CTAB

Zwitterionic surfactants   These have two charged groups of different signs and are also called *amphoteric* surfactants. However, these charges are either permanent or dependent on specific pH range. CAPB (cocamidopropyl betaine) is a common zwitterionic surfactant.

$$HO-\overset{O}{\overset{\|}{C}}-CH_2(CH_2)_9CH_3 \ + \ N-NH_2 \longrightarrow N-N-CH_2(CH_2)_9CH_3$$

Lauric acid   DMAPA   LAPDMA

CAPB

When lauric acid reacts with dimethylaminopropylamine (DMAPA), lauramidopropyldimethylamine (LAPDMA) is obtained. LAPDMA is treated with sodium monochloroacetate to form cocamidopropyl betaine (CAPB).

Zwitterionic surfactants are used in shampoos, conditioners, shower gels, bubble baths, and other personal care products.

7.11 MICELLES AND REVERSE MICELLES

Micelles are associated colloids that behave as strong electrolytes at lower concentrations, whereas exhibit colloidal behaviour at higher concentration leading to the formation of aggregated particles. The process of formation of micelles is known as *micellization*. The number of surfactant molecules in a micelle is called *aggregation number*.

Miceller clusters have a wide variety of shapes, such as spherical and lamellar as shown in Fig. 7.18. The micelle structure is shown as an approximate spherical body with hydrophilic groups on the surface and hydrophobic groups directed towards the interior. Sodium stearate $(C_{17}H_{35}COO^-Na^+)$ is a common component of many soaps and when dissolved in water, it forms $C_{17}H_{35}COO^-$ and Na^+ ions. The COO^- group is the polar head and $C_{17}H_{35}$ is the hydrophobic tail.

Potassium oleate is an example of colloidal electrolyte or associated colloids. When potassium oleate is added in infinitesimally increasing amounts to water at 323 K, it dissolves in water forming potassium and oleate ions and surface tension of the solution continuously decreases from that of pure water. When a molar concentration of 10^{-3} M is reached, a break appears in surface tension–concentration curve and thereafter, surface tension remains constant. This break or discontinuity is due to association of oleate ions in clusters called micelles.

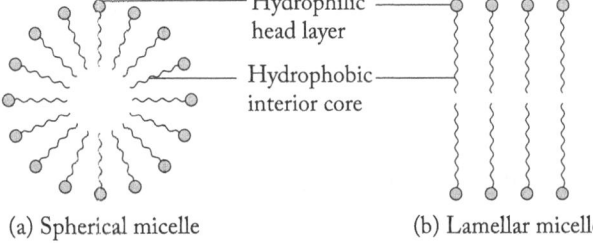

(a) Spherical micelle   (b) Lamellar micelle

Fig. 7.18 Structure of micelles

The threshold concentration at which micelles appear is called *critical micelle concentration* (CMC). Below CMC, oleate ions exist as simple ions in solution, but above CMC, they get associated as micelles of colloidal sizes. The change of ions to micelles is a reversible process and micelles can be destroyed by dilution of solutions. Soaps, higher alkyl sulphates, sulphonates, polyethylene oxides, amine salts are some common colloidal electrolytes.

7.11.1 Reverse Micelles

Reverse micelles are nanosized (1–10 nm) water droplets dispersed in a non-polar solvent due to the action of surfactants.

Surfactants naturally form aggregates in non-polar solvent. The aggregation number in reverse micelles is much smaller than the aggregation number in aqueous micelles. In a non-polar solvent, the polar head groups of surfactants are directed inward and the hydrophilic (or lipophilic) tail is outwards directed towards the non-polar solvent as shown in Fig 7.19.

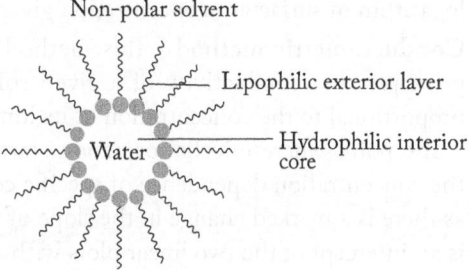

Fig. 7.19 Reverse micelles

Such micelles are called *reverse micelles* or *inverse micelles*. Thus, a reverse micelle will possess a polar inner core that can solubilize water molecules and the lipophilic chains will protect the inner core from the non-polar medium. The most widely studied reverse micelle is the sodium salt of bis-(2-ethylhexyl) sulphosuccinate referred to as Aerosol OT (AOT) (Fig. 7.20).

Chemically, Aerosol OT is an anionic surfactant and non-toxic in nature. It is a versatile surface-active agent which has many applications, some of which are as follows.

(a) Aerosol OT is used as a wetting agent in cosmetics, medical formulations, and food preparations.

(b) It is used as a surfactant in paints and dyes, tablet formulations, protein extraction.

(c) It acts as reaction vessels for many organic reactions. The interior hydrophilic core of AOT allows transport phenomenon of water in non-polar solvents.

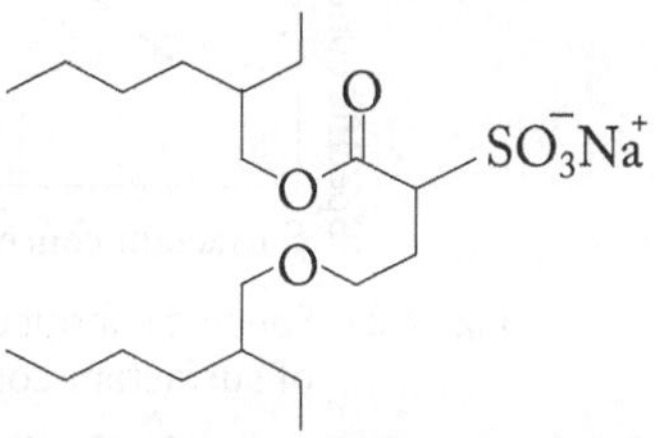

Fig. 7.20 Structure of Aerosol OT

Hydrophobic and Hydrophilic Interactions

Surfactants have a natural tendency to concentrate at the interface. If a surfactant is dissolved in water, its molecules tends to come to air/water interface in such a way that the hydrophobic chain is oriented towards the air and hydrophilic head remains in water.

As more surfactant molecules are added to water, it keeps moving towards the surface of water. The reduction in surface tension is directly proportional to the concentration of surfactant molecules in water. Surface tension continues to decrease with increasing surfactant concentration till the surface saturates. Above CMC, surface tension becomes constant as there is no further increase in concentration of surfactant at the interface.

7.11.2 Critical Micelle Concentration and Its Determination

The experimental evidence of micelle formation lies in the quantitative determination of critical micelle concentration. The different techniques to determine CMC are as follows.

Surface tension method   It is a classical and convenient method to determine CMC. It is observed that surface tension decreases at air/water interface when surfactant concentration increases. This is due to adsorption of surfactant molecules at the interface.

A sudden drop in surface tension occurs when micellization begins. The inflection point in a plot of surface tension versus logarithm of surfactant concentration gives the CMC.

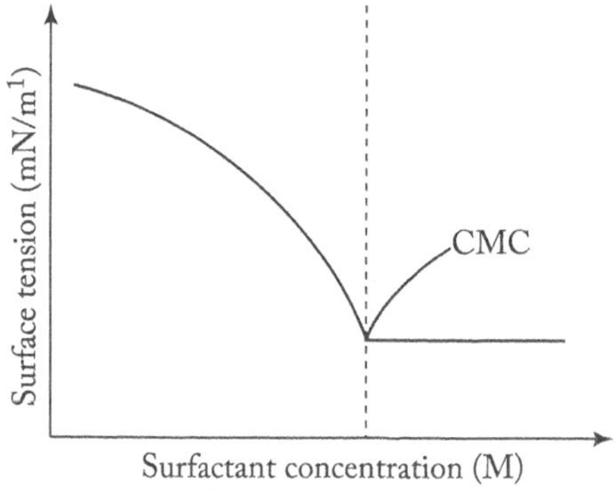

Conductometric method   This method is based on the principle that conductivity of a given solution is directly proportional to the concentration of its ions.

The point where micelle formation starts is indicated on the concentration dependence of specific conductivity (κ) as an inflection point, which is easy to locate as there is a marked change in the slope of a linear plot $[\kappa = f(c)]$ as shown in Fig 7.22. The CMC value is an intercept of the two linear plots with different slopes.

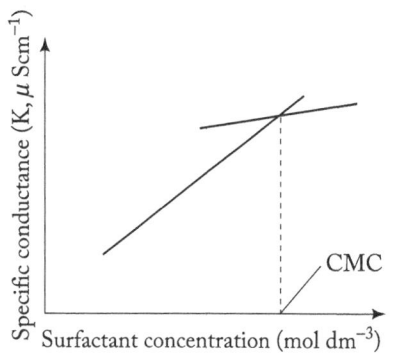

Fig. 7.21 Surface tension as a function of surfactant concentration

Fig. 7.22 Specific conductance as a function of surfactant concentration

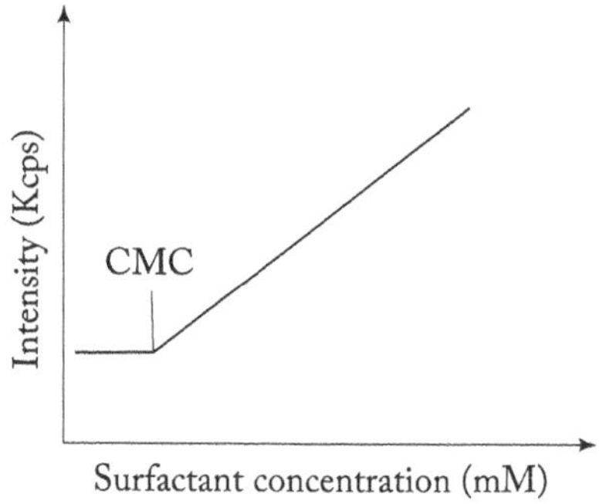

Fig. 7.23 Intensity of scattered light (in kilo counts/s Kcps) as a function of surfactant concentration

Light scattering method   Dynamic light scattering (DLS) method is employed to determine CMC of surfactant solution by studying samples with particle sizes in the sub-micron range. It is based on the principle that as the concentration of particles in a given solution increases, the intensity of scattered light increases proportionally.

It involves measuring the time-dependent fluctuations in the intensity of scattered light from a suspension of particles undergoing random movement. The intersecting lines provide the CMC value for the surfactant under study. CMC values of some common surfactants are listed in Table 7.4.

Table 7.4 CMC values of some common surfactants

Surfactant	Chemical formula	CMC (mole/m^3)
CTAB	$C_{16}H_{33}.N.(CH_3)_3^+ Br^-$	0.90
Sodium dodecyl sulphate	$C_{12}H_{25}SO_4^- Na^+$	8.10
Triton X-100	$C_{14}H_{22}O(C_2H_4O)_n$ (where n = 9 – 10)	0.20
Aerosol OT	$C_{20}H_{37}NaO_7S$	2.0×10^{-5}

7.12 DETERGENTS

A detergent (Latin word: *detergere* means to wipe or clean) is a surface active agent that 'wets' the dirt particles and fabric by water. The process of removal of dirt, stains, oily soil from fabrics is called *detergency*. A typical detergent consists of a mixture of surfactants and (hydrolytic chiefly lipolytic) enzymes.

Detergent formulation is known to be quite complex due to the diverse nature of fabrics and dirt.

Particulate soil matter contains clay and other minerals which are removed by wetting and dispersing processes. Highly charged anionic polyelectrolytes are used as surfactants in this case.

There are some resistant stains such as blood, tea, coffee, food colours, which cannot be removed by simple surfactant–enzyme combination. Thus, bleaching agents are added which oxidize the colour (chromophore) in to non-coloured products that are washed away using surfactants.

Some detergents contain sequestering agents such as zeolites for divalent ions which can be used in hard water. The sequestering agents prevent surfactant precipitation as calcium and magnesium salts in hard water. Apart from these, perfumes and fluorescing agents are also added in detergents for pleasant smell post-washing and brightening of fabric, respectively.

Mechanism of Cleaning

Various mechanisms have been put forth to explain cleaning using surfactants. Here, we will focus on fabric cleaning for oily soil removal.

Roll-up mechanism The interfacial tension between between oily soil and fabric, especially cotton is reduced. If the contact angle between soil and fabric is larger than 90° after addition of surfactant, it can easily lift the stain which can be removed by washing.

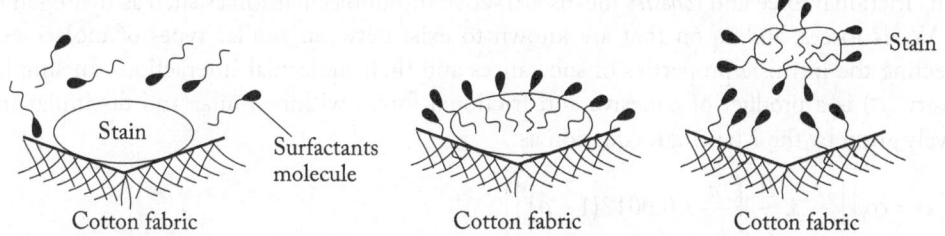

Fig. 7.24 Roll-up mechanism to remove oily soil from fabric

Emulsification The oily soil is emulsified by surfactant and trapped in foam, which is prevented from settling down on the fabric. This mechanism is independent of the fabric type.

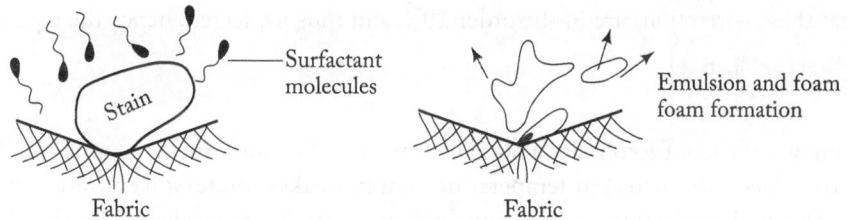

Fig. 7.25 Emulsification mechanism to remove oily soil from fabric

To achieve cleaning through emulsification, the interfacial tension between oil and surfactant must be low.

Solubilization In this mechanism, oily soil is solubilized in situ by forming a microemulsion. Microemulsions are macroscopic homogeneous mixture of oil, water, and surfactant.

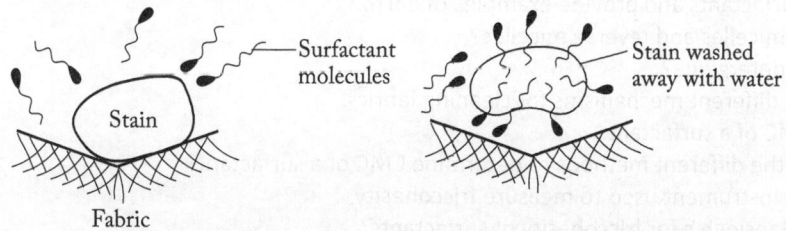

Fig. 7.26 Solubilization mechanism to remove oily soil from fabric

On a microscopic level, microemulsions are those with an amphiphilic surfactant present as a monolayer between oil/water interface. For effective solubilization, the surfactant should efficiently lower the interfacial tension between oil and water.

7.13 Friccohesity of Surfactants

There are numerous publications on the preparation of newer molecules and exploiting their potential for various applications. Especially, studies on structural properties of proteins are well reported, but lacks thermodynamic data as to how these newer protein molecules behave in a given solution, usually surfactants and ionic liquids. Protein–surfactant interactions form a fundamental premise for friccohesity. It is quite a complex concept which was simplified by Man Singh who studied the behaviour of simple amino acids in surfactants. These values can then be extrapolated to understand the conformational behaviour of polymers, especially proteins in solutions.

Surfactants are extensively utilized in pharmaceutical processes such as drug delivery, nanoemulsions, in-vivo studies, and in biotechnology such as wetting–foaming agents and emulsifier–demulsifier, making protein–surfactant interactions important.

Man Singh (2006) and his coworkers put forth a physicochemical concept called 'friccohesity' where *fric* means frictional force and *cohesity* means attractive intramolecular forces such as hydrogen bonding, van der Waals forces, and so on that are known to exist between similar types of molecules thereby also reflecting the intrinsic properties of substances and their molecular interactions. In simpler terms, friccohesity (σ) is a product of cohesive and frictional forces within similar and dissimilar molecules respectively given by the Mansingh equation as:

$$\sigma = \sigma_0 \left[\left(\frac{t}{t_0} \pm \frac{B}{t} \right) \left(\frac{n}{n_0} \pm 0.0012(1 - \rho) \right) \right]$$

where, t and t_0 are sample and reference flow times respectively, σ_0 is called reference friccohesity. The terms n_0 and n denote pendant drop number of reference and sample respectively. The fractional term B/t denotes kinetic energy correction and $\pm 0.0012(1 - \rho)$ is buoyancy correction. Experimentally, it was found that these corrections are of the order 10^{-7}, and thus neglected; hence the equation becomes,

$$\sigma = \sigma_0 \left[\left(\frac{t}{t_0} \right) \left(\frac{n}{n_0} \right) \right]$$

For determining surfactant friccohesity, a novel instrument called survismeter is employed. The decreased friccohesity (σ) values with increased temperature signify weaker solute–solvent interactions at higher temperatures. The weakened solute–solvent interactions occur due to weakened frictional and cohesive forces between the solute and solvent molecules with simultaneous lowering of intermolecular forces.

Check Your Progress

25. What are surfactants?
26. Classify surfactants and provide examples of each.
27. What are micelles and reverse micelles?
28. What are detergents?
29. Name the different mechanisms for cleaning fabrics.
30. Define CMC of a surfactant.
31. What are the different methods to determine CMC of a surfactant?
32. Name the instrument used to measure friccohesity.
33. Express Mansingh c for friccohesity of surfactants.

Activity-based Questions

A. Replacing conventional adsorbents for wastewater treatment.

Several attempts have been made to replace old, toxic conventional adsorbents with low-cost adsorbents. Primarily low-cost greener adsorbents can be categorized in to five groups, viz.,

1. agricultural waste, 2. industrial by-products 3. seawater minerals, 4. novel green adsorbents, and 5. ore materials. Investigations about their adsorbent capacities, non-toxicity, and applications in waste water treatment is an on-going research area attracting world-wide attention.

Present a case study about various adsorbents from the above categories, their sorption capacities (for dyes, pollutants, heavy metals), costs, reusability post-adsorption and their mechanisms. Also, identify the probable adsorbent candidates that can successfully replace conventional adsorbents in waste water treatments.

B. Phase-transfer catalysts (PTCs) in industrial processes.

Conventional transition metal catalysts are important in various industrial processes but face challenges in terms of product desorption and recycling of catalysts. Phase transfer catalysis is known since the early 1960s. PTC allows rapid migration of a reactant from one phase into another where chemical reaction occurs. Quaternary ammonium salts, crown ethers, surfactants, etc. are commonly employed to accelerate the chemical reaction without using toxic solvents.

Present a case study on the different types of phase transfer catalysts, catalytic mechanism involved, and their industrial applications.

SOLVED EXAMPLES

1. If 0.129 dm^3 of N$_2$ gas at 1 atm pressure and 273 K was needed to cover 1g of silica gel adsorbent, calculate the surface area of silica gel, if each nitrogen molecule occupies an area of 16.2×10^{20} m^2.

Solution: Number of moles of nitrogen in 0.129 dm^3 = $\dfrac{0.129}{22.4} = 5.758 \times 10^{-3}$ moles

Hence, total number of nitrogen molecules

$$= 5.758 \times 10^{-3} \times 6.023 \times 10^{23}$$
$$= 34.66 \times 10^{20}$$

Thus, total area of 1g silica gel

$$= 34.66 \times 10^{20} \times 16.2 \times 10^{20} = 561.492$$

Result: Surface are of silica gel = 561.492 m^2

2. When hydrogen is adsorbed on powdered copper, a monolayer is formed with 1.45×10^{-3} dm^3 of hydrogen measured at STP per gram of adsorbent. If density of liquid hydrogen is 0.07 kg/dm^3, determine the specific surface area of copper.

Solution: Number of hydrogen molecules in 1.45×10^{-3} dm^3 at STP is

$$= \frac{1.45 \times 10^{-3} \times 6.02 \times 10^{23}}{22.4} = 3.9 \times 10^{19}$$

If we assume that liquid hydrogen is closely packed, molar volume of hydrogen is N_v where N is Avogadro's number and v is the volume of one hydrogen molecule. If ρ is density of liquid, and M for H$_2$ is 2×10^{-3}, then

$$v = \frac{M}{N\rho} = \frac{2 \times 10^{-3}}{6.02 \times 10^{23} \times 0.07}$$

$$= 4.743 \times 10^{-26} \text{ dm}^3 = 4.743 \times 10^{-29} \text{ m}^3$$

Assuming H_2 molecules to be spherical, $\quad v = \dfrac{4}{3}\pi r^3$

Hence, $r = \sqrt[3]{\dfrac{3v}{4\pi}}$

$\therefore \qquad r = \sqrt[3]{\dfrac{3 \times 4.743 \times 10^{-29}}{4 \times 3.14}} = 2.246 \times 10^{-10}\,\text{m}$

The cross-sectional area of a hydrogen molecule = $\pi r^2 = 3.14 \times (2.246 \times 10^{-10})^2 = 15.84 \times 10^{-20}\,\text{m}^2$

Hence, specific surface area of Cu = Number of H_2 molecules × Cross-sectional area of H_2

$\qquad = 3.9 \times 10^{19} \times 15.84 \times 10^{-20} = 6.178\,\text{m}^2/\text{g}$

Result: Specific surface area of copper = 6.178 m²/g

SUMMARY

○ Adsorption is the surface phenomenon of attracting and retaining various substances on the surface of a solid resulting in a higher surface concentration than in the bulk. The substance that gets adsorbed is the adsorbent and the substance on which the process of adsorption takes place is the adsorabate.

○ Adsorption is of two types, physisorption and chemisorption. Physisorption is driven by weak van der Waals forces, whereas chemisorption involves strong covalent bond formation.

○ It is experimentally observed that almost all solids tend to absorb gases. The extent of adsorption of a gas on a solid depends upon the type of gas, surface area of solid, and temperature and pressure of the gas.

○ The graphical relationship showing the extent of adsorption (x/m) and pressure of the gas at constant temperature is called adsorption isotherm.

○ Catalyst is any substance added in trace quantities that alters the reaction rate without itself getting consumed in the reaction. It could be positive or negative and the phenomenon is known as catalysis.

○ Homogeneous catalysis is when both the substrate and catalyst are in the same phase, whereas in heterogenous catalysis they are of different phases.

○ Reactions catalysed by an acid or a base or by both, are termed as acid–base catalysis. Enzyme catalysis involves the catalysis of biochemical reactions by enzymes, which are proteins present in plants and animals.

○ Enzyme catalysis is facilitated due to the presence of active sites on the structure of the enzymes that are highly specific to the substrate.

○ Colloidal solutions have particles of size ranging from 10 to 100 nm and are thus intermediate between true solutions and suspensions. A colloidal system has two phases—dispersed phase and dispersion medium.

○ Colloidal systems exhibit interesting electrical, optical, and mechanical properties.

○ Emulsions are colloidal systems as a liquid–liquid phase and they can be oil-in-water type and water-in-oil type. The process of making emulsions is known as emulsification. An emulsifying agent or emulsifier is added during their preparation for stabilization. Soaps and detergents are most frequently used as emulsifiers.

○ Colloids encompass various fields of applications in industry and daily life. Surfactants or surface-active agents reduce the surface tension developed at interfaces and get concentrated there. Surfactants could be cationic, anionic, or neutral in nature.

○ Friccohesity (σ) can be defined as the product of cohesive and frictional forces within similar and dissimilar molecules respectively. It can be measured using survismeter.

GLOSSARY

Activated complex: The intermediate (unstable) structure at the maximum energy point in an energy profile diagram that forms along the reaction path.

Adsorbate: The substance accumulated or collected on to the surface.

Adsorbent: The substance that adsorbs or collects the adsorbate.

Adsorption: The mass transfer process that involves accumulation of substance at the interface of two phases, such as liquid–liquid, liquid–solid, gas–liquid, or gas–solid.

Adsorption isotherm: A graph or plot prepared to study adsorption processes.

Autocatalysis: A catalytic process, where the products catalyse the reaction.

Catalyst: A substance that alters the rate of a chemical reaction and itself remains chemically and quantitatively unchanged after the reaction.

Catalysis: The phenomenon exhibited by catalyst.

Chemisorption: The process where adsorbate atoms or molecules are held to the adsorbent by chemical bonds.

Coagulation: The process of changing colloidal particles in a sol into an insoluble precipitate by addition of an electrolyte.

Colligative: The properties of a colloidal solution that depend on the ratio of the number of colloidal particles to the number of dispersion medium molecules in a system.

Colloid: A particle with size (diameter) between 10 to 100 nm.

Detergent: A surface active agent that wets the dirt particles and fabric by water.

Emulsion: A liquid–liquid colloidal system in which finely divided droplets are dispersed in another liquid.

Friccohesity: A physicochemical parameter which shows solute–solvent interactions (usually protein–surfactant) and is a product of cohesive and frictional forces within similar and dissimilar molecules.

Micelle: An aggregate of molecules in a colloidal solution.

Mutarotation: A phenomenon of gradual change in specific rotation of an optically active compound with time in a solvent.

Physisorption: The accumulation of adsorbate atoms or molecules on the surface of adsorbent occurs due to weak van der Waals forces.

Promoters: Additive substances that enhance the catalytic activity, but are not true catalysts themselves.

Sorption: The process when both adsorption and absorption occur simultaneously.

Surface energy: The energy associated with the intermolecular forces at the interface between two phases.

Surfactant: A substance that decreases the surface tension of the liquid in which it is dissolved.

KEY FORMULAE

- Freundlich adsorption isotherm

$$log\frac{x}{m} = log\,k + \frac{1}{n}log\,p$$

- Langmuir adsorption isotherm

$$\frac{p}{v} = \frac{p}{v_m} + \frac{1}{b.v_m}$$

- BET equation

$$\frac{p}{v(p_o - p)} = \frac{1}{v_m c} + \frac{(c-1)}{v_m c} \times \frac{p}{p_o}$$

- Michaelis–Menten constant

$$v = \frac{k_2[E]_0}{1 + \left[\dfrac{k_{-1}+k_2}{k_1}\right]\dfrac{1}{[S]_0}}$$

- Gibbs adsorption equation

$$S = -\frac{c}{RT}\cdot\frac{dr}{dc}$$

- Mansingh equation (friccohesity)

$$\sigma = \sigma_0\left[\left(\frac{t}{t_0}\pm\frac{B}{t}\right)\left(\frac{n}{n_0}\pm 0.0012(1-\rho)\right)\right]$$

EXERCISES

Multiple Choice Questions

1. The phenomenon of concentration of gas or liquid molecules on a solid surface is called
   - (a) absorption
   - (b) sorption
   - (c) adsorption
   - (d) desorption

2. Which of the following is a good adsorbent?
   - (a) Silica gel
   - (b) Alumina
   - (c) Clay
   - (d) All of these

3. The relation between equilibrium pressure of a gas and its amount adsorbed onto a solid adsorbent at constant temperature is called
   - (a) adsorption isobar
   - (b) adsorption isotherm
   - (c) physisorption
   - (d) chemisorption

4. At which temperature does chemical adsorption occur?
   (a) at high temperature
   (b) at very low temperature
   (c) at temperatures favouring absorption
   (d) temperature does not affect

5. In autocatalysis
   (a) reactant act as catalyst.
   (b) heat evolved during reaction, acts as catalyst.
   (c) product act as catalyst.
   (d) solvent act as catalyst.

6. Tyndall effect is shown by
   (a) ideal solution
   (b) colloidal solution
   (c) saturated solution
   (d) true solution

7. Which of the following is an example of surface catalysis?
   (a) Leather tanning
   (b) Ammonia manufacture by Haber's process
   (c) Isomerism of complexes
   (d) Photographic film development

8. Adsorption of gas on solid adsorbent depends on
   (a) temperature
   (b) pressure of gas
   (c) type of adsorbent
   (d) all of them

9. Which of the following is oil/water emulsion?
   (a) Milk
   (b) Vanishing cream
   (c) Butter
   (d) All of these

10. The assumption which was **not** put forth as Langmuir adsorption isotherm is
    (a) layer of gas adsorbed onto a solid surface is one-molecule thick.
    (b) the absorbed layer is uniform.
    (c) there is no attraction between adjacent molecules.
    (d) attraction between the adsorbent molecules is very high.

11. Which of the following is **not true** about Freundlich adsorption isotherm?
    (a) This isotherm is applicable in certain limits of pressure.
    (b) Constant k and n change with temperature.
    (c) It shows deviation at low pressures.
    (d) Freundlich isotherm is an empirical equation.

12. If $1/n$ value turns zero in Freundich adsorption isotherm, then adsorption is independent of
    (a) pressure
    (b) temperature
    (c) quantity
    (d) Both (a) and (b)

13. Smoke is an example of .... type of colloidal system.
    (a) gas in solid
    (b) solid in gas
    (c) gas in gas
    (d) gas in liquid

14. Langmuir adsorption isotherms fails at
    (a) low temperature
    (b) high pressure
    (c) intermediate pressure
    (d) none of these

15. At CMC, the surface molecules
    (a) decompose
    (b) dissociate
    (c) associate
    (d) become completely soluble

16. When a beam of light passes through a colloidal solution, it is
    (a) reflected
    (b) scattered
    (c) deflected
    (d) absorbed

17. Langmuir adsorption isotherm relation is
    (a) $\dfrac{p}{v} = \dfrac{p}{v_m} + \dfrac{1}{b.v_m}$
    (b) $\dfrac{p}{v} = \dfrac{p}{v_m} + \dfrac{1}{v_m}$
    (c) $\dfrac{p}{v.b} = \dfrac{p}{v_m} + \dfrac{1}{b.v_m}$
    (d) $\dfrac{p}{v} = 0$

18. Multilayer adsorption can be explained by
    (a) absorption process
    (b) Langmuir adsorption
    (c) Freundlich adsorption
    (d) BET isotherm

19. The graphical representation of autocatalytic reaction is
    (a) sigmoid curve
    (b) straight line
    (c) bell curve
    (d) parabolic curve

20. Of the following which statement about enzyme catalysis is correct?
    (a) The rate of formation of transition state intermediate determines overall free energy change of the reaction.
    (b) The active site of an enzyme is perfectly complementary to the substrate in its ground state.
    (c) The rate of formation of the transition state intermediate determines the overall reaction rate.
    (d) Natural substrates bind to enzymes more tightly than transition state analogues.

21. Which of the following statements about Lineweaver–Burk plots are correct?
    (a) All enzyme-catalysed reactions give a linear Lineweaver–Burk plot.
    (b) The plot provides estimates of v_{max}, k_M for all types of enzyme catalysis.
    (c) For a classical enzyme, v_{max}, k_M can be determined by linear extrapolation.
    (d) All of the above statements are correct.

22. In peptization reaction
    (a) colloid converts in to precipitates.
    (b) precipitates convert in to colloids.
    (c) true solution is formed from suspension particles.
    (d) true solution is formed by dissolving precipitates.

23. Of the following, the one that forms a cationic micelle is
    (a) Sodium dodecyl sulphate
    (b) Cetyltrimethyl ammonium bromide
    (c) Urea
    (d) Sodium acetate

24. Which of the following is not a colloidal solution?
    (a) Brine solution   (b) Fog
    (c) Smoke   (d) Butter

25. An arsenic sulphide sol (AS_2S_3) is prepared by the reaction
    $$AS_2O_3 + H_2S \rightarrow AS_2S_3 \text{ (sol)} + 3H_2O$$
    This method of preparing colloidal solution is
    (a) reduction
    (b) oxidation

    (c) hydrolysis
    (d) double decomposition

26. Which of the following processes is an example of heterogenous catalysis?
    (a) Contact process
    (b) Haber's process
    (c) Hydrogenation of vegetable oils
    (d) All of these

27. Belousov–Zhabotinsky reaction is
    (a) autocatalytic   (b) acid–base type
    (c) hydrolysis type   (d) none of these

28. Catalytic poisoning
    (a) reduces the activity of the catalyst
    (b) reduces the reaction rate of the reaction
    (c) increases the activation energy of the reaction
    (d) increases the temperature of the reaction

29. The methods to determine CMC of a surfactant are
    (a) surface tension
    (b) dynamic light scattering
    (c) conductometric
    (d) all of them

30. The term catalyst was coined by
    (a) Chadwick   (b) J.J. Thomson
    (c) Berzelius   (d) Rutherford

31. The instrument used to determine friccohesity of surfactants is
    (a) survismeter   (b) conductometer
    (c) viscometer   (d) None of these

32. Aerosol OT surfactant is
    (a) cationic   (b) anionic
    (c) zwitterionic   (d) non-ionic

Review Questions

1. Define the terms: (a) adsorbate (b) adsorbent (c) adsorption (d) desorption.

2. Distinguish between physisorption and chemisorption.

3. Enumerate the characteristics of physisorption and chemisorption.

4. Discuss Langmuir's theory of adsorption. Deduce an expression for Langmuir's monomolecular adsorption isotherm.

5. Explain Langmuir's theory of adsorption. How is it verified?

6. State BET equation. Explain the terms involved. How can one graphically validate the BET equation?

7. Discuss the various applications of adsorption processes.

8. What is a catalyst? State the types of catalysts with suitable example for each type.

9. List the various characteristics of catalysis.

10. Justify the statement, 'Catalysts enhance the reaction rate of chemical processes.'

11. Explain homogeneous catalysis with suitable examples.

12. Distinguish between homogeneous and heterogeneous catalysis.

13. Describe the following terms with suitable examples: (a) Catalytic poisons (b) promoters (c) autocatalysis.

14. With a schematic diagram, describe adsorption theory of heterogeneous catalysis with a suitable example.

15. Explain acid–base catalysis giving examples.

16. How does the Michaelis–Menten equation explain why the rate of an enzyme-catalysed reaction is proportional to the amount of enzyme?

17. Derive Michaelis–Menten equation for enzyme catalysis.

18. Explain Lineweaver–Burk plot in enzyme-catalysed reaction. State its importance.

19. What is an adsorption isotherm? Describe Freundlich adsorption isotherm.

20. What do you understand by activation of adsorbent? How is it achieved?

21. What is Tyndall effect? Explain in detail.

22. What are lyophilic and lyophobic sols? Give one example of each type. Why are hydrophobic sols easily coagulated?

23. What are micelles? Give an example of a miceller system.

24. Explain reverse micelles with a suitable example.

25. Discuss the different techniques to determine CMC of surfactant solutions.

26. Discuss the preparation of various types of surfactants.

27. Explain the mechanism of cleaning using surfactants.

28. Explain detergency or cleaning mechanism for oily soil on fabrics.

29. Comment: 'Fincly divided substance is more effective as an adsorbent.'

30. Explain detergency and the mechanism of cleaning of fabrics.

31. What is Mansingh equation? State its significance.

32. Write short notes on (a) emulsions (b) catalysts in industry.

33. Write a note on 'friccohesity of surfactants.'

FURTHER READING

1. Chandra, A., V. Patidar, M. Singh, and R.K. Kale, 'Physicochemical and Friccohesity Study of Glycine, L-Alanine and L-Phenylalanine With Aqueous Methyltrioctylammonium and Cetylpyridinium Chloride from T = (293.15 to 308.15) K', *J. Chem. Thermodynamics* 65 (2013) 18–28.

2. Farrauto, R.J. and C.H. Bartholomew, *Fundamentals of Industrial Catalytic Processes*, Blackie Academic & Professional, London,1997.

3. Fogler, H.S. *Elements of Chemical Reaction Engineering*, Prentice Hall of India, 2002.

4. Somorjai, G.A. and Y. Li., *Introduction to Surface Chemistry and Catalysis*, John Wiley & Sons Inc, Hoboken, N.J., 2010.

5. Singh, M., 'Survismeter Type I and II for Surface Tension, Viscosity Measurements of Liquids for Academic and Research, and Development Studies'. *Journal of Biochemical Biophysical Methods*, 67 (2–3) 151–161, 2006.

6. Rothenberg, G., *Catalysis: Concepts and Green Applications*, John Wiley & Sons. Inc, 2017.

ANSWERS

Check Your Progress

1. Refer to Sections 7.2.4 and 7.2.5

2. Silica gel, zeolites are hydrophilic adsorbents that adsorb moisture from the reaction system. Polymeric substances like polystyrene divinylbenzene and coal are hydrophobic adsorbents.

3. When adsorbed species are removed from the adsorbent surface, it is called desorption.

4. A plot of gas adsorbed per unit mass of adsorbent (x/m) on a solid against pressure of gas at constant temperature is called adsorption isotherm.

5. Al^{3+} ions in potash alum tend to form precipitates with impurities present in the water and coagulate them. Due to the coagulating effect of alum, impurities quickly settle down and purified water is obtained.

6. Refer to Section 7.5.

7. Catalysts are compounds or substances that alter the rate of a chemical reaction and themselves remain chemically and quantitatively unchanged after the reaction.

8. Refer to Section 7.6.1.

9. Enzymes lower the activation energy of the chemical reaction and thus lesser energy is required for product formation.

10. In specific acid catalysis, H^+ ions are the only ones that catalyse the reactions, whereas in general acid catalysis, other acid catalysts including H^+ ions can catalyse the reactions.

11. Refer to Section 7.5.2.

12. The two examples of reactions that can be catalysed by acids as well as bases are: hydrolysis of methyl acetate and mutarotation of glucose.

13. Refer to Section 7.5.2 (Autocatalysis)

14. See Glossary

15. Refer to Section 7.8.3

16. Refer to Section 7.8.3

17. On adding electrolyte, the charge on the colloidal particles gets neutralized by oppositely charged ions provided by the electrolyte. The colloidal particles coagulate and become aggregated particles that easily settle at the bottom due to gravity.

18. (a) Solid sol: ruby glass; (b) Liquid aerosol: clouds (c) Solid emulsion: creams.

19. Some of the different methods to prepare colloids include Bredig's arc method, peptization, and chemical reactions (double decomposition, reduction, oxidation, hydrolysis).

20. Dialysis and ultrafiltration are two methods to purify colloids.

21. Colloidal particles cannot settle rather remain dispersed and are continuously bombarded with dispersion medium molecules leading to these effects.

22. A combination of multiple layers of opposite charges around a colloidal particle is called an electrical double layer.

23. Refer to Section 7.9

24. Refer to Section 7.9

25. Refer to Section 7.10

26. Refer to Section 7.10

27. Refer to Section 7.11

28. Refer to Section 7.12

29. Roll-up, emulsification, and solubilization are different mechanisms of cleaning fabrics.

30. Refer to Section 7.11.2.

31. Surface tension method, DLC method, and conductometry are some of the methods of determining CMC.

32. Survismeter is used to measure friccohesity.

33. Mansingh expression for friccohesity of surfactants is given as

$$\sigma = \sigma_0 \left[\left(\frac{t}{t_0} \pm \frac{B}{t} \right) \left(\frac{n}{n_0} \pm 0.0012 (1-\rho) \right) \right]$$

Multiple Choice Questions

1. (c)	2. (d)	3. (b)	4. (a)	5. (c)	6. (b)	7. (b)
8. (d)	9. (a)	10. (d)	11. (c)	12. (a)	13. (a)	14. (b)
15. (c)	16. (b)	17. (a)	18. (d)	19. (a)	20. (c)	21. (d)
22. (b)	23. (b)	24. (a)	25. (d)	26. (d)	27. (a)	28. (a)
29. (d)	30. (c)	31. (a)	32. (b)			

Solid State Chemistry

After reading this chapter, you will be able to:

- understand the concepts of lattice and unit cell.
- state the laws of crystallography.
- explain the various types of Bravais lattices.
- comprehend lattice planes and the method of determining the Miller indices.
- calculate the number of atoms per unit cell and the atomic radius.
- understand the concepts of coordination number, packing factor, and radius ratio rule.
- explain important crystal structures.

8.1 OVERVIEW OF SOLID STATE CHEMISTRY

The field of material science and engineering deals with synthesizing novel materials such as semiconductors, superconductors, nanomaterials, magnetic ferrites, photoconductors, and so on. To prepare solid materials one needs a fundamental understanding of the structure of solids and their properties.

A solid is defined as any matter that possesses definite volume and shape. Solids are incompressible, rigid, and mechanically strong when compared to liquids and gases. In a solid, the constituent atoms, ions, or molecules remain bound to each other arranged in a regular manner by cohesive forces and hence it possesses a definite crystal lattice.

In general, solids can be classified into two types depending upon the arrangement of the constituent atoms: (a) crystalline solids and (b) amorphous solids. *Crystalline solids* possess an orderly arrangement of atoms resulting in well-defined solid shapes in three-dimensional space. Such solids have sharp melting points and exhibit anisotropy (i.e., thermal, electrical, and magnetic properties are direction-dependent). Sodium chloride (NaCl), caesium chloride (CsCl), sugar, diamond, quartz are examples of crystalline solids.

Amorphous solids do not possess any orderly arrangement of atoms and thus do not have well-defined shapes in three-dimensional space. Amorphous solids do not have sharp melting points and are isotropic in nature. Glass, polymers, wax, etc., are examples of amorphous solids.

8.2 CRYSTALLOGRAPHY

When one studies various geometric forms of crystalline substances and their properties, it is collectively known as *crystallography* which is based on three fundamental laws: (a) law of constancy of interfacial angles, (b) law of rational indices, and (c) law of symmetry.

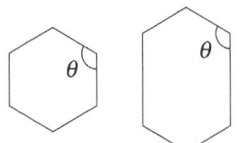

Fig. 8.1 Interfacial angle in a crystal

8.2.1 Law of Constancy of Interfacial Angles

The shape and size of a particular compound or an element may vary based on the conditions under which crystallization occurs. However, the angle between the faces will always be constant irrespective of the nature of crystallization. Thus, the term 'face' refers to the planar surface binding the crystals and the angle between the two intersecting faces is called interfacial angle (θ).

8.2.2 Law of Rational Indices

Hauy (1784) proposed the law of rational indices based on the fact that the external form of a crystal was a reflection of its inner regular arrangement of units. Thus, one can view a crystal as being made up of several smaller crystals. If a crystal is repeatedly divided, each of the resulting fragments will be the replica of the original one. Thus, one can deduce the smallest crystal which is the building block of the crystal and is called the *unit cell*.

While describing the geometry of a crystal, three mutually perpendicular axes are chosen so that all the crystal faces either intercept the axes at definite distances from the point of origin or remain parallel to some of the axes, where intercepts will be at infinity. The law of rational indices states that, 'It is possible to choose along the three coordinate axes unit distances (a, b, c) which may or may not be of same length, such that the ratio of three intercepts of any crystal plane is given as ($ha : kb : lc$) where h, k, and l are integral numbers. Hence, all faces cut at a given axis at distances from the origin bear a simple ratio to one another.

In Fig. 8.2, OX, OY, and OZ are three perpendicular axes and let us assume that a, b, and c are the intercepts along the axes for crystal plane ABC. If any other plane, say DEF, makes an intercept $2a : 2b : 3c$, then the ratio of intercepts can be represented as 2:2:3 and depicts the plane. The coefficients of a, b, and c are termed as *Weiss indices* of the crystal plane. If the plane is parallel to one axis, then the plane cuts at that axis at infinity. At such times, Weiss indices are difficult to represent and are replaced by Miller indices. *Miller indices* of a plane are obtained by taking the reciprocals of Weiss coefficients and multiplying by the lowest common multiple to determine the integral values. Hence, for Fig. 8.2, Miller indices for plane DEF will be 1/2, 1/2, 1/3 or simply 3:3:2.

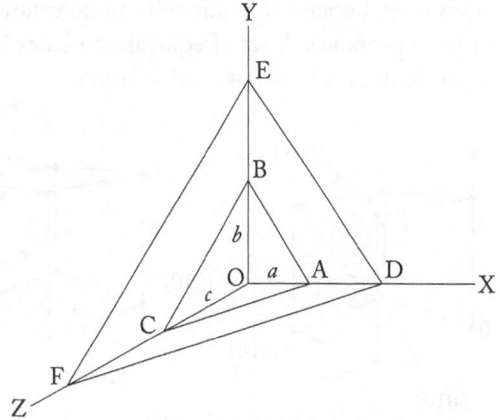

Fig. 8.2 Crystallographic axes

8.2.3 Miller Planes (Crystal Planes and Directions)

Single-crystal materials are used in a variety of applications. Many of these applications rely on properties that are orientation- or direction-dependent. To remove any ambiguity, a standard method of indicating planes and directions in a crystalline material is used. In this method, a set of three integers h, k, and l, placed within parentheses [e.g., (hkl)] is generally used to indicate a particular crystalline plane. The set h, k, l is called the *Miller indices*. Thus, Miller indices are the set of smallest three integers used to designate a group of parallel, equidistant planes in a crystal. These integers are determined using the following procedure:

Step 1. Determine the intercepts of the given plane on the crystal axes and express them as integral multiples of the respective basis vectors. A translation of the particular plane with respect to the origin is allowed as long as the basic direction of the plane is maintained.

Step 2. Evaluate the reciprocal of the three integers obtained in Step 1 and reduce these reciprocals to the smallest set of integers, while maintaining their relationship. Designate these integers as h, k, and l.

Step 3. Label the plane as (hkl). A few common lattice planes for a cubic crystal are shown in Fig. 8.3

(shaded regions) and also indicate the Miller indices for these planes. One must, however, remember that the letters *h*, *k*, and *l* define a set of parallel planes inside a lattice, not just a single plane.

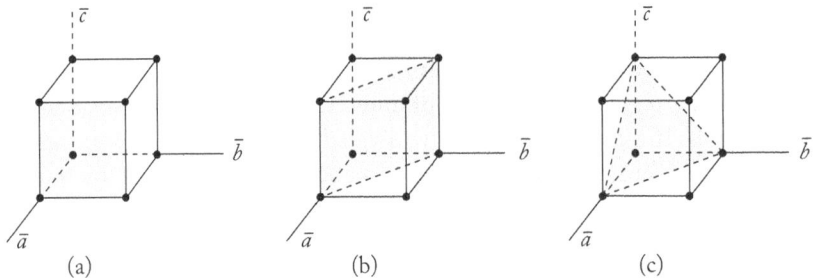

Fig. 8.3 Three-lattice planes of cubic lattice (a) (100) plane (b) (110) plane (c) (111) plane

Any lattice has many planes that are equivalent. Any given plane, characterized by a set of Miller indices can be shifted within the lattice to an equivalent plane position by suitably moving or rotating the unit cell. As an example, consider a cubic lattice. The faces of the cubic lattice are crystallographically equivalent, because the unit cell can be rotated in different directions, without affecting either its form or its appearance. A set of equivalent planes is represented by curly brackets {*hkl*}. Figure 8.4 shows six equivalent faces {100} of a cubic lattice.

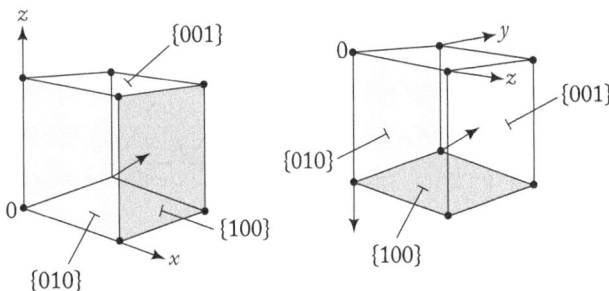

Fig. 8.4 Equivalent {100} planes

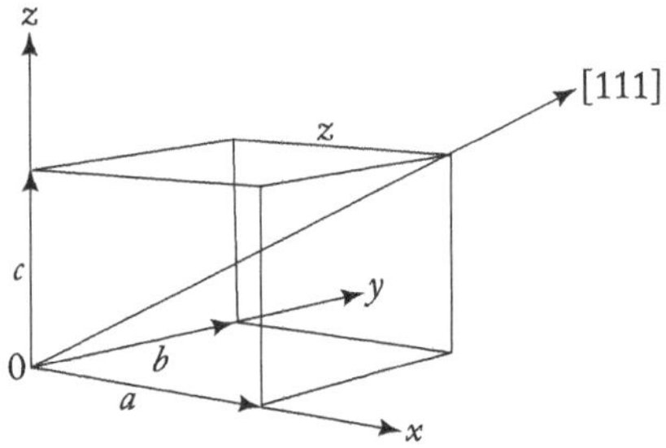

Fig. 8.5 Schematic diagram of [111] direction

The following procedure is used to determine the directions in a lattice:

(i) Choose the axis vectors with respect to a suitable origin.

(ii) The vector components of a particular direction are then expressed in multiples of the basis vectors.

(iii) The three integers are then reduced to their smallest values, while retaining the relationship between them.

(iv) The given direction is then expressed within square brackets; for example, [*abc*].

Figure 8.5 depicts the [111] direction, which is also the body diagonal of a cubic lattice.

Like planes, many directions in a lattice are equivalent. These depend on the choice of orientation of the axes. Equivalent directions are expressed within angular brackets (e.g., <*hkl*>).

A few equivalent <100> directions are shown schematically in Fig. 8.6.

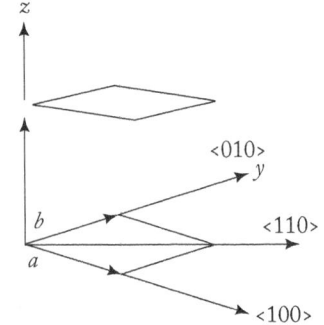

Fig. 8.6 Equivalent <100> directions

8.2.4 Law of Symmetry

An element of symmetry is the total number of planes, axes, and centre of symmetry possessed by a crystal. According to the law of symmetry, all crystals of the same substance possess the same elements of symmetry. Various types of elements of symmetry are possible, but we will discuss the three main types as follows.

Plane of symmetry   An imaginary plane by which if a crystal is divided into two parts, one will be the mirror image of the other imaginary half; this is called the plane of symmetry.

Axis of symmetry   An imaginary line if passed through the crystal and if the crystal is rotated about 360°, it will have the same appearance more than once in the course of its complete revolution; this is called the axis of symmetry. Based on the number of times a crystal presents the same appearance, it can possess two-fold, three-fold, four-fold, or six-fold symmetry.

Centre of symmetry   It is a point within the crystal such that any line drawn through it will intersect the crystal surface at equidistant ways in both the directions.

8.3 Lattice — Unit Cell

In a single crystal, a basic three-dimensional pattern, which may consist of a single atom or a group of atoms, is repeated at regular intervals. This periodic arrangement of atoms in a crystal is called a lattice. Due to this periodic arrangement of atoms, the environment around any point within a crystal appears to be the same. A lattice may thus be defined as an array of points in space such that the environment around each point is the same. The distance between the constituent atoms and the relative orientation of these atoms can have different magnitudes. The fundamental unit that undergoes periodic repetition to create the lattice is called a *unit cell*, which is the basic building block of the crystal.

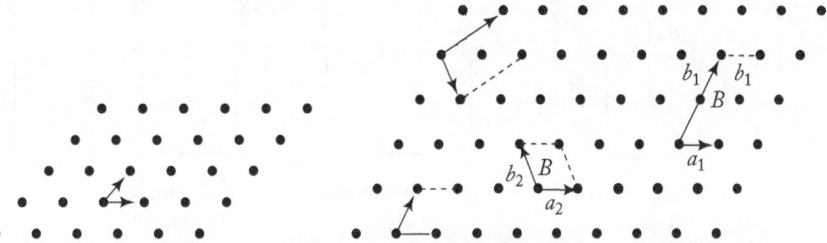

Fig. 8.7 (a) 2D Single crystal lattice (b) A 2D single crystal lattice showing various possible unit cells

Figure 8.7 is a schematic representation of an infinite two-dimensional array of lattice atoms. In this representation, an atom is denoted by a dot, referred to as the *lattice point*. Any lattice point may be translated through a distance *a* in one direction and by *b* in a second non-collinear direction, to generate a two-dimensional lattice. The two translational directions can have any angle between them, but the angle is fixed for a particular crystal. A three-dimensional lattice can similarly be created by translating a lattice point in a third non-collinear direction through a distance *c*. A general three-dimensional lattice can be obtained by carrying out a periodic repetition of the unit cell. A unit cell is not unique to a particular lattice and thus can possess several possible unit cells for a two-dimensional lattice.

Check Your Progress

1. State the properties of crystalline and amorphous solids. Give examples.
2. What is crystallography?
3. Name the laws of crystallography.
4. Define Miller indices.
5. Define (a) plane of symmetry (b) axis of symmetry (c) centre of symmetry.
6. Distinguish between crystalline and amorphous solids.
7. How many atoms are assigned to a unit cell of (a) face-centred and (b) body-centred cubic lattice?

A typical unit cell is shown in Fig. 8.8. Unit cells may vary in size; the smallest unit cell that can generate a crystal is called a *primitive cell*. It must, however, be remembered that a primitive cell is not always the best choice for a good representation of a lattice. For example, a unit cell based on orthogonal directions can lead to some simplifications, but it may not be a primitive cell. A unit cell can be characterized by a set of three vectors *a*, *b*, and *c*, which in general may or may not be orthogonal or equal in length.

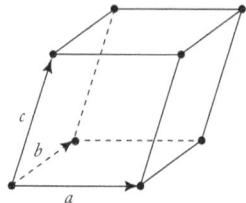

Fig. 8.8 A unit cell

8.3.1 Bravais Lattice

Any lattice point is indistinguishable from another lattice point as long as the displacement vector between the two lattice points can be represented by the following equation:

$$r = p\boldsymbol{a} + q\boldsymbol{b} + s\boldsymbol{c} \tag{8.1}$$

where p, q, and s are integers, and $\boldsymbol{a}$, $\boldsymbol{b}$, and $\boldsymbol{c}$ are called the *basis vectors*.

Seven unique lattice point arrangements can be used to fill up a three-dimensional space. These seven unique arrangements are called *crystal systems*, which include cubic, tetragonal, orthorhombic, rhombohedral (or trigonal), hexagonal, monoclinic, and triclinic systems. There are a total of 14 distinct arrangements of lattice points, and these are called *Bravais lattices*, named after Auguste Bravais (1811–1863), a French crystallographer. Figure 8.9 shows the schematic representations of these 14 types of Bravais lattices.

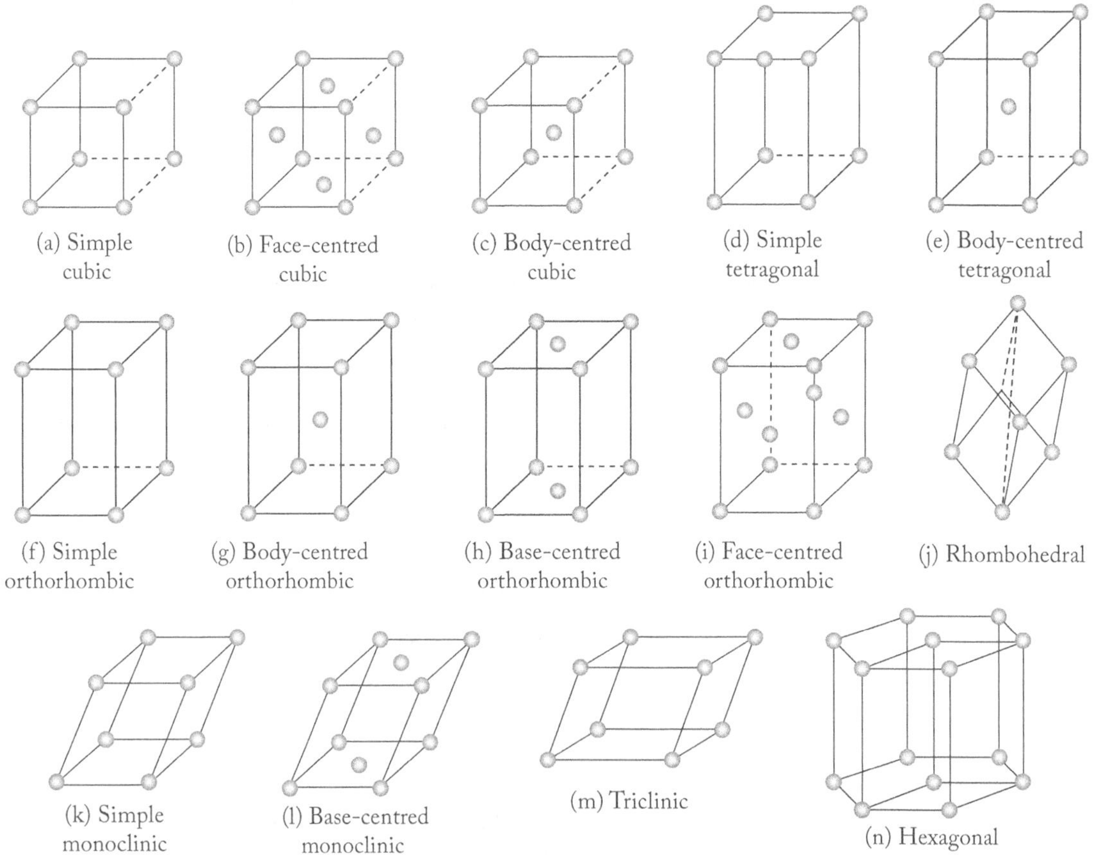

(a) Simple cubic	(b) Face-centred cubic	(c) Body-centred cubic	(d) Simple tetragonal	(e) Body-centred tetragonal

(f) Simple orthorhombic	(g) Body-centred orthorhombic	(h) Base-centred orthorhombic	(i) Face-centred orthorhombic	(j) Rhombohedral

(k) Simple monoclinic	(l) Base-centred monoclinic	(m) Triclinic	(n) Hexagonal

Fig. 8.9 Schematic representation of Bravais lattices

Table 8.1 highlights the characteristic features of the various crystal systems.

Table 8.1 Salient features of the seven crystal systems

Structure	Basis vectors	Angles between basis vectors	Volume of the unit cell
Cubic	$a = b = c$	All angles equal 90°	a^3
Tetragonal	$a = b \neq c$	All angles equal 90°	a^2c
Orthorhombic	$a \neq b \neq c$	All angles equal 90°	abc
Rhombohedral or trigonal	$a = b = c$	All angles are equal (α), but $\alpha \neq 90°$	$a^3\sqrt{3\cos^2\alpha + 2\cos^3\alpha}$
Monoclinic	$a \neq b \neq c$	Two angles equal 90° and one angle $\beta \neq 90°$	$abc\sqrt{1 - \cos^2\alpha - \cos^2\beta}$
Triclinic	$a \neq b \neq c$	All angles are different and none equals 90°	$abc\sqrt{\begin{array}{c}1 - \cos^2\alpha - \cos^2\beta - \cos^2\gamma \\ +2\cos\alpha\cos\beta\cos\gamma\end{array}}$
Hexagonal	$a = b \neq c$	Two angles equal 90° and one angle equals 120°	$0.866a^2c$

The simplest form of a lattice is the cubic lattice, because the unit cell of such a lattice has a cubic volume. In a simple cubic structure, an atom is located at each corner. For a cubic lattice, the vectors a, b, and c are equal in length and perpendicular to each other. A cubic system has three variants of Bravais lattices, namely simple cubic (SC), body-centred cubic (BCC), and face-centred cubic (FCC). As shown in Fig. 8.10, a body-centred cubic structure has an additional atom at the centre of the cube. The face-centred cubic structure has additional atoms on each face of the cube. On the same lines, a tetragonal system has two variants, namely simple tetragonal and body-centred tetragonal Bravais lattices. In general, a body-centred Bravais lattice system implies a lattice point located at the centre of the unit cell, as shown in Fig. 8.10(b).

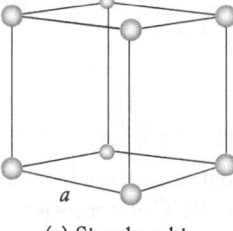 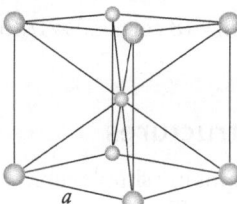 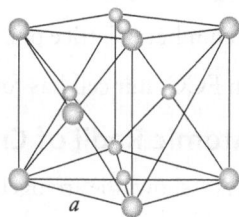

(a) Simple cubic   (b) Body-centred cubic   (c) Face-centred cubic

Fig. 8.10 Unit cells for cubic lattices

8.3.2 Interplanar Spacing in Cubic Lattices

For cubic lattices, the direction represented by [hkl] is perpendicular to the (hkl) plane. The adjacent parallel planes of atoms in a crystal have the same Miller indices, and the distance between two adjacent parallel planes is called the *interplanar spacing*, represented by d_{hkl}. Lattice parameters describe the size and shape of the unit cell. In a cubic system, the length of one of the sides of the cube completely describes the unit cell. This length is called the *lattice parameter* a_0. For cubic materials, the interplanar spacing is given by the equation, $d_{hkl} = \dfrac{a_0}{\sqrt{h^2 + k^2 + l^2}}$, where a_0 is the lattice parameter.

8.3.3 Number of Atoms Per Unit Cell

Each unit cell is defined by a specific number of lattice points. A given lattice point can, however, be shared by more than one unit cell.

From Fig. 8.11, we can see that a lattice point at a corner of one unit cell is shared by seven adjacent unit cells. Thus, one lattice point is shared by eight unit cells. This implies that one-eighth of each corner lattice point belongs to any one particular unit cell. Each unit cell in turn has eight corners. Thus, the number of lattice points contributed by all the corner positions in one unit cell is given by the following expression:

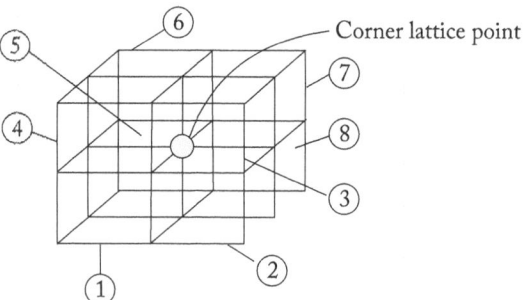

Fig. 8.11 Corner lattice point shared by eight unit cells

$$1/8(\text{lattice point/corner}) \times 8(\text{corner/unit cell})$$
$$= (1 \text{ lattice point/unit cell})$$

Finally, the number of atoms per unit cell is given by the product of the number of atoms per lattice point and the number of lattice points per unit cell. For most metals, each lattice point has one atom. For a simple cubic unit cell, lattice points are present only at the corners of the cube. Thus,

Number of lattice points/unit cell = (8 corners) × (1/8) = 1

Since one atom is located per lattice point, the number of atoms per unit cell for a simple cubic unit cell is 1. For a BCC unit cell, lattice points are present at each corner and at the centre of the cube. Thus, the number of lattice points per unit cell is as follows:

Number of lattice points/unit cell = (8 corners) × 1/8 + (1 centre) × 1 = 2

The number of atoms per unit cell for a BCC unit cell is thus 2. For an FCC unit cell, lattice points are present at all the corners and all the faces of the cube. Note that each face is shared by two unit cells. Hence, the number of lattice points per face is 1/2. The number of lattice points per unit cell is, therefore, given by the following expression:

Number of lattice points/unit cell = (8 corners) × 1/8 + (6 faces) × 1/2 = 4

Thus, an FCC unit cell has four atoms.

8.3.4 Atomic Radii of Crystal Structures

Figure 8.12(a) shows the arrangement of atoms in a simple cubic structure and it can be concluded that atomic radius, $r = a_0/2$, where, a_0 is the lattice parameter. The arrangement of atoms in a BCC structure is shown in Fig. 8.12(b). From the figure, it can be seen that $4r = \sqrt{3}\, a_0$ or $r = \sqrt{3}/4\, a_0$. Similarly, the corresponding schematic arrangement of atoms in an FCC structure is shown in Fig. 8.12(c) and has the atomic radius, $4r = \sqrt{2}\, a_0$ or $r = \sqrt{2}/4\, a_0$.

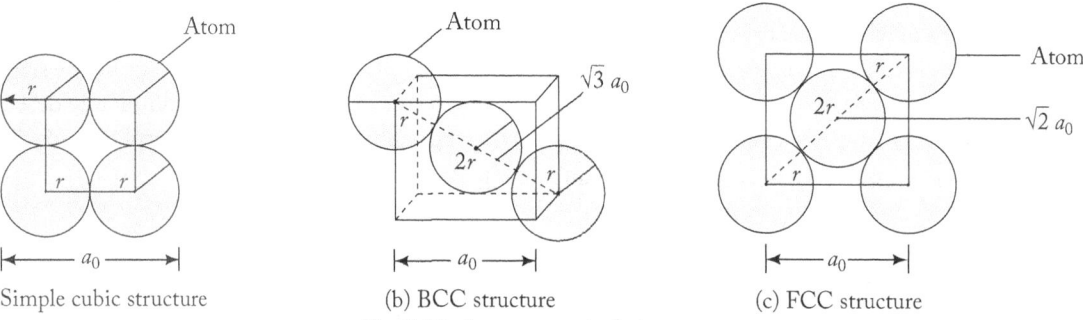

(a) Simple cubic structure   (b) BCC structure   (c) FCC structure

Fig. 8.12 Arrangement of atoms

8.3.5 Coordination Numbers and Packing Factor

The coordination number of an atom is the number of atoms sharing boundary with this atom, or in other words, the number of nearest neighbours of the atom. Thus, the coordination number indicates how tightly and efficiently atoms are packed together in a lattice. In ionic solids, the coordination number of anions is the number of nearest cations. Similarly, the coordination number of cations is the number of nearest anions. Figure 8.13(a) shows the nearest neighbours for a simple cubic lattice and is evident that any atom in a simple cubic lattice has six nearest neighbours. The coordination number of an atom in a simple cubic lattice is thus 6.

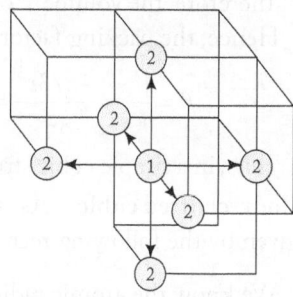

(a) of a simple cubic lattice

Figure 8.13(b) shows the nearest neighbour configuration of a BCC lattice. The coordination number of an atom in a BCC structure is 8. Similarly, the coordination number of an atom in an FCC structure is 12.

Figure 8.14 shows the schematic of the hexagonal close-packed structure (HCP). The coordination number of HCP is 12.

The diamond lattice is obtained by inserting one FCC lattice into another FCC lattice and displacing the same along the body diagonal by one-fourth of its length. Figure 8.15 shows the diamond unit cell and the tetrahedron formed within the diamond lattice.

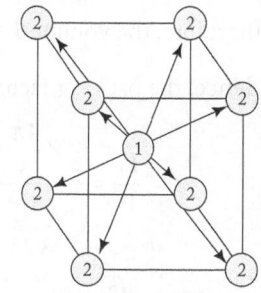

(b) of a BCC lattice

Fig. 8.13 Nearest neighbours

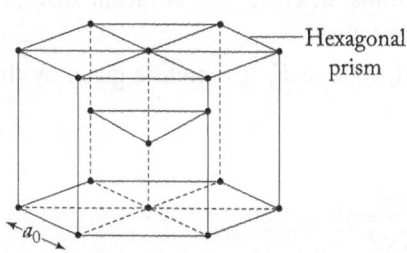

Fig. 8.14 Schematic diagram of HCP structure

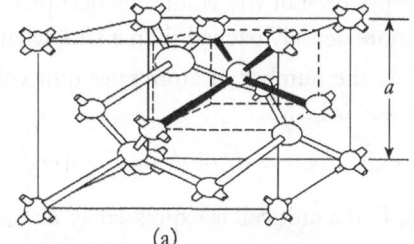

Fig. 8.15 (a) Diamond unit cell and (b) Tetrahedron within diamond unit cell

Since each atom is covalently bonded with four neighbouring atoms, the coordination number of each atom is four.

Atoms in a crystal can be visualized as hard spheres, whose sizes are such that they touch their closest neighbours. Vacant spaces would exist in such a model of a crystalline material. In such a crystal model, the fraction of space occupied by atoms is called the *packing factor*. Packing factor f is defined as follows:

$$\text{Packing factor, } f = \frac{\text{Number of atoms/unit cell} \times \text{Volume of each atom}}{\text{Volume of unit cell}} \quad \ldots(8.2)$$

The volume of each atom in this equation is calculated, assuming the atom to be spherical in shape. Packing factors for some important structures are evaluated in the following pages.

Simple cubic As the number of atoms per unit cell is 1, the atomic volume per unit cell V_1, is given by $V_1 = 4\pi/3 r^3$. Here, r represents the atomic radius and we know that atomic radius of a simple cubic structure is given as, $r = a_0/2$ or $a_0 = 2r$.

Therefore, the volume V of a unit cell is given as $V = a_0^3 = 8r^3$.

Hence, the packing factor is given by the following equation:

$$f = \frac{V_1}{V} = \frac{4\pi/3 r^3}{8r^3} = \frac{\pi}{6} \text{ or } f \cong 0.52 \qquad \text{...(8.3)}$$

Thus, in a simple cubic structure, 52% of the volume is occupied by atoms, whereas 48% is vacant space.

Body-centred cubic   As the number of atoms per unit cell is 2, the atomic volume per unit cell V_1, is given by the following relation: $V_1 = 2 \times 4/3 \, \pi r^3$.

We know the atomic radius for BCC is, $r = \sqrt{3}/4 \, a_0$ or $a_0 = 4r/\sqrt{3}$

Therefore, the volume V of a unit cell is expressed as $V = a_0^3 = \dfrac{64r^3}{3\sqrt{3}}$.

Hence, the packing factor is given by the following relation:

$$f = \frac{V_1}{V} = \frac{2 \times \dfrac{4\pi}{3} r^3}{\dfrac{64r^3}{3\sqrt{3}}} = \frac{8\pi}{3} \times \frac{3\sqrt{3}}{64}$$

or $\qquad f = \dfrac{\sqrt{3}}{8}\pi \approx 0.68 \qquad \text{....(8.4)}$

Thus, in a BCC structure 68% of the volume is occupied by atoms, whereas 32% is vacant space. A BCC structure is thus more densely packed than a simple cubic structure.

Face-centred cubic   As the number of atoms per unit cell is 4, volume V_1 is therefore given by the following relation: $V_1 = 4 \times 4/3 \, \pi r^3$.

We know the atomic radius as, $r = \sqrt{2}/4 \, a_0$ or $a_0 = 4r/\sqrt{2}$.

Therefore, the volume V of a unit cell is expressed as $V = a_0^3 = \dfrac{64}{2\sqrt{2}} r^3$.

Hence, the packing factor is given by the following relation:

$$f = \frac{V_1}{V} = \frac{16/3 \, \pi r^3}{\dfrac{64r^3}{2\sqrt{2}}} = \frac{16\pi}{3} \times \frac{2\sqrt{2}}{64}$$

$$f = \frac{\pi\sqrt{2}}{6} \cong 0.74 \qquad \text{...(8.5)}$$

Thus, in an FCC structure 74% of the volume is occupied by atoms, whereas 26% is vacant space. The FCC configuration of atoms thus represents the most densely-packed structure among the various cubic structures. Further, one can calculate the mass of one atom present in a unit cell which is given as follows:

$$\text{Mass of one atom} = \frac{\text{Molar mass of substance}}{\text{Avogadro's number}} \qquad \text{...(8.6)}$$

Radius ratio rule   Goldschmidt observed that crystal structures of many inorganic compounds such as NaCl, KCl, CsCl, ZnS, CaF$_2$, TiO$_2$, and SiO$_2$, are dependent on radius ratio and coordination numbers. The ions in an ionic compound are held together by electrostatic forces of attraction. If one takes the ratio of radius of a cation (say r_+) to that of the anion (say r_-), it is referred to as the radius ratio rule, which can be expressed as,

$$\text{Radius ratio } (R) = \frac{r_+}{r_-} \qquad\qquad\qquad ...(8.7)$$

The radius ratio rule assumes that all ions are hard solid spheres. Higher the radius ratio, larger is the size of the cation and thus greater will be the coordination number. Radius ratio rule helps in determining the coordination numbers and shapes of ionic crystals. Table 8.2 shows the various crystal shapes and their characteristic features.

However, radius ratio rule suffers from some limitations:

(a) The assumption that each ion is a solid hard sphere is not true in a practical sense. Particularly, anions are not hard and are polarizable in nature. These anions can form covalent bond with the cations.

(b) Further, the geometries predicted by radius ratio rule may be general and not exact information, hence it can only predict possible geometries of ionic crystal.

Table 8.2 Coordination numbers, radius ratios, and shapes of ionic crystals

Coordination no.	Radius ratio	Crystal shape
2	0 – 0.155	Linear
3	0.155 – 0.225	Trigonal planar
4	0.225 – 0.414	Tetrahedral
4	0.414 – 0.732	Square planar
6	0.414 – 0.732	Octahedral
8	0.732 – 1.00	Body-centred cubic
12	1.00	Close cubic packed or hexagonal cubic packed

8.4 LATTICE ENERGY AND BORN–HABER CYCLE

In an ionic crystal, positive and negative ions are held together by electrostatic forces and its bond energy is expressed by a term referred to as lattice energy. *Lattice energy* is defined as change in enthalpy that occurs when one mole of a solid crystalline substance is formed from its gaseous ions. Born–Haber cycle put forth by Max Born and Fritz Haber (1919) used Hess's law of constant heat summation to determine the lattice energy of a given crystal structure. According to Hess's law of heat summation, 'total change in enthalpy during the complete course of a chemical reaction is the same whether the reaction involves one or multiple steps.

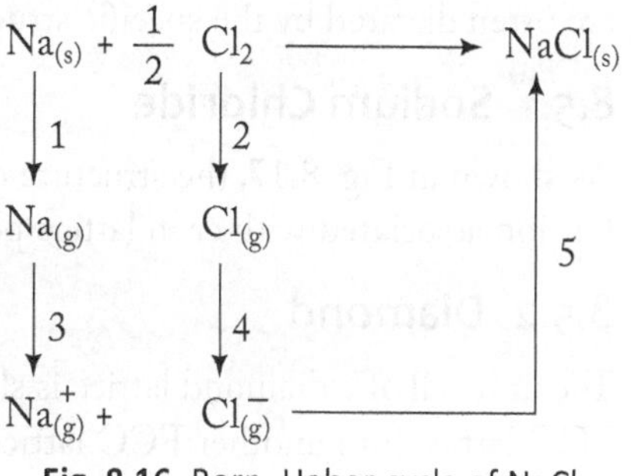

Fig. 8.16 Born–Haber cycle of NaCl crystal formation

Let us take the example of sodium chloride (NaCl) which undergoes change in enthalpy $\Delta H°$ when Na^+ and Cl^- ions in the gaseous phase form 1 mole of NaCl crystal.

Determination of Lattice Energy of NaCl

According to Hess's law, one can determine the lattice energy of ionic crystal (in this case, NaCl). The enthalpy change for direct formation of sodium chloride from sodium metal and chlorine is – 411 kJ.

$$Na(s) + \frac{1}{2} Cl_2 \rightarrow NaCl(s) \ (\Delta H° = -411 \text{ kJ})$$

The enthalpy change of NaCl can be calculated by a series of steps.

(i) The first step is *sublimation* where sodium metal is converted to its gaseous atoms.

$$Na(s) \rightarrow Na(g) \ (\Delta H_1° = +108 \text{ kJ})$$

(ii) In the second step, chlorine molecules dissociate to chlorine atoms.

$$\frac{1}{2}\,Cl_2 \to Cl(g)\;(\Delta H_2^\circ = +121\ kJ)$$

(iii) Further, there is conversion of gaseous sodium to sodium ions due to loss of electrons and the energy required to perform this reaction is called *ionization energy*.

$$Na(g) \to Na_{(g)}^+ + e^-\;(\Delta H_3^\circ = +495\ kJ)$$

(iv) Chlorine atoms gain an electron to form chloride ions and the energy so released during the reaction is called *electron affinity*.

$$Cl + e^- \to Cl_{(g)}^-\;(\Delta H_4^\circ = -348\ kJ)$$

(v) Lastly, sodium and chlorine ions come together and form a crystal lattice.

$$Na_{(g)}^+ + Cl_{(g)}^- \to NaCl_{(s)}\;(\Delta H_5^\circ = -lattice\ energy)$$

If we take all the enthalpies calculated for each of the above steps and equate with the enthalpy of formation of NaCl, we get,

$$\Delta H_1^\circ + \Delta H_2^\circ + \Delta H_3^\circ + \Delta H_4^\circ + \Delta H_5^\circ = 108 + 121 + 495 - 348 - lattice\ energy = -411\ kJ \quad ...(8.8)$$

On solving Eq. (8.8) we get, lattice energy = + 787 kJ/mol.

The cyclic changes shown above is called Born–Haber cycle and is useful in determining the lattice energies of crystalline substances.

8.5 CRYSTAL STRUCTURES

Crystalline materials can exist in a variety of structures. The characteristic properties of these materials are often dictated by the specific structure.

8.5.1 Sodium Chloride

As shown in Fig. 8.17, the structure of sodium chloride (NaCl) crystal consists of one Na^+ ion and one Cl^- ion associated with each lattice point in an FCC configuration.

8.5.2 Diamond

The unit cell of a diamond lattice is shown in Fig. 8.18. The diamond lattice is obtained by inserting one FCC lattice into another FCC lattice and displacing the same along the body diagonal by one-fourth of its length.

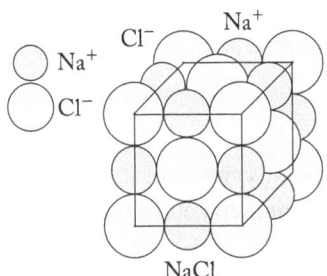

Fig. 8.17 Schematic diagram of NaCl lattice

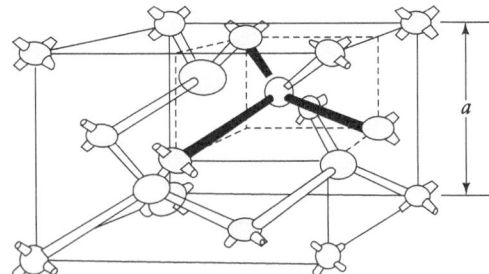

Fig. 8.18 Diamond crystal structure

This results in the formation of a tetrahedron within the diamond lattice. Each atom in the diamond lattice is surrounded by four nearest neighbours, which are located at the apexes of the tetrahedron having an edge of *a*/2. The basic lattice structure of many semiconductor materials used in the modern semiconductor industry, for example, Si and Ge, is a diamond lattice.

8.5.3 Zinc Blende (ZnS)

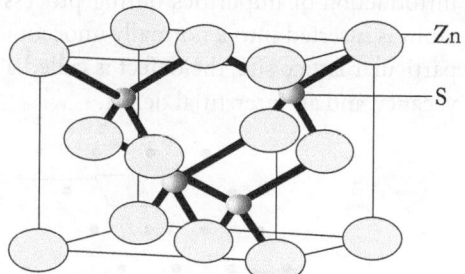

Zinc ions have a charge of +2 and S ions have a charge of –2. Some common compound semiconductors like GaAs have a zinc blende lattice (Fig. 8.19), which is closely related to the diamond lattice, with two different types of atoms in the lattice.

In a typical GaAs lattice, one sub-lattice is of Ga and the other is of As.

Fig. 8.19 Schematic diagram of zinc blende

8.5.4 Graphite

Graphite consists of carbon atoms arranged in regular hexagons in flat parallel layers. The bonding between the different parallel layers is not strong. The layers are, therefore, easily separable from each other.

A schematic representation of the graphite structure is shown in Fig. 8.20. Weak bonding between the layers lends softness and lubricating property to graphite. Carbon atoms in the hexagonal layers are joined together by covalent or metallic bonds. The good electrical conductivity of graphite can be attributed to these metallic bonds.

8.5.5 Titanium Dioxide

Titanium dioxide (Rutile) exhibits rutile-type structure. Each Ti^{4+} ion is surrounded by six oxide ions that are arranged octahedrally (Fig. 8.21). Further, each oxide ion is surrounded by three titanium ions in a trigonal planar manner.

Hence, the coordination number of Ti^{4+} and O^{2-} are 6 and 3, respectively. TiO_2 shows excellent catalytic activity and good electrical conductivities.

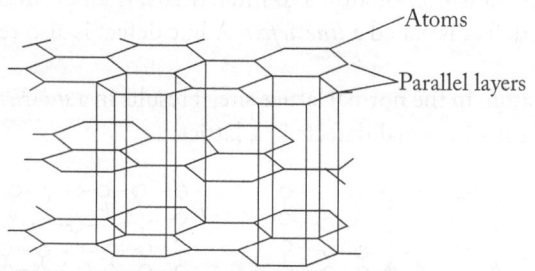

Fig. 8.20 Typical representation of layered structure of graphite

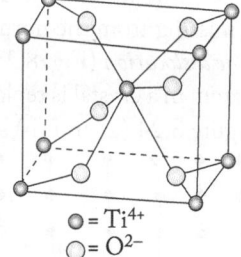

$\bullet = Ti^{4+}$
$\circ = O^{2-}$

Fig. 8.21 Structure of titanium dioxide

8.6 IMPERFECTIONS IN ATOMIC PACKINGS

Crystal imperfections are the defects in the regular geometrical arrangement of the atoms in a crystalline solid. These imperfections result from the deformation of the solid due to rapid cooling from high temperature, or high-energy radiation (X-rays or neutrons) striking the solid. These defects influence the mechanical, electrical, and optical behaviour of the solid at single points, along lines, or on its whole surfaces.

8.6.1 Point and Surface Defects

Localized disruptions in the otherwise perfect arrangement of atoms in a crystal structure are called point defects. A *point defect* in a crystal refers to a missing/misplaced atom/ion in the lattice. It is also known as a *zero-dimensional defect*. A localized disruption affects not only one atom at a particular location, but also several atoms in the region around it. Several processes are responsible for the creation of these defects. For example, atoms may gain energy due to heat and result in defects. Defects can also be created by the introduction of impurities during processing or doping. An *interstitial defect* is formed when an extra atom is inserted into a normally unoccupied position in a crystal structure. If an atom is missing from a particular lattice site, the defect is called a *vacancy*. Figure 8.22 shows the schematic representation of a vacancy and an interstitial defect.

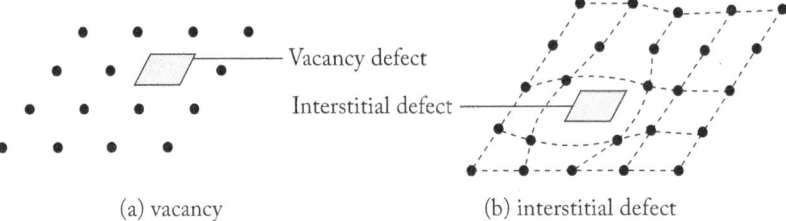

(a) vacancy   (b) interstitial defect

Fig. 8.22 Two-dimensional single-crystal lattice showing defects

Vacancies and interstitial defects can change the electrical properties of a material. Quite often, this change is due to the deviations produced in the nature of chemical bonding between atoms. Sometimes, a vacancy and an interstitial defect may occur in close proximity. One common way in which this may happen is when atoms move from their natural sites to interstitials, thereby creating a vacancy. This type of vacancy–interstitial defect is called a *Frenkel defect*. Frenkel defects produce effects that are characteristically different from those produced by simple vacancies or interstitials alone. Detailed studies show that equilibrium concentration of interstitial atoms at a given temperature is as follows:

$$n_I = ANe^{-E_{FI}/k_BT} \qquad\qquad \ldots(8.9)$$

where, E_{FI} is the formation energy of the interstitial (generally several eV), N is the number of sites in the given volume, and A is an integer (generally, ~1) indicating the number of identical interstitial positions per lattice atom.

More complex defects can also occur in crystals, such as the *Schottky defect*. In this type of defect, vacancies are not accompanied by a simultaneous transition of atoms to interstitials. If an entire row of atoms is missing from the normal lattice site, the defect is called a *line defect*. A line defect is also referred to as a *line dislocation* (Fig. 8.23).

If one atom in a crystal is replaced by a different atom in the normal lattice site, it results in a *substitutional defect*. Figure 8.24 (a) and (b) show two types of substitutional defects in a lattice.

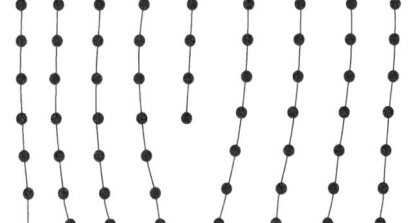

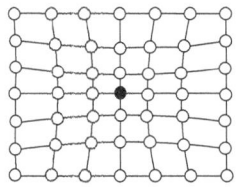

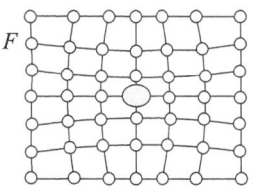

(a) Small substitutional atom   (b) Large substitutional atom

Fig. 8.23 Line dislocation in two-dimensional lattice   **Fig. 8.24** Two types of substitutional atoms

Table. 8.3 Common point defects in various crystals

Crystal	Structure of the crystal	Common intrinsic defect
Alkali halides (except caesium)	NaCl type	Schottky
Alkaline earth oxides	NaCl type	Schottky
Silver halides	NaCl type	Frenkel (cation type)
Alkaline earth fluorides, oxides of cerium, thorium	Fluorite (or CaF_2) type	Frenkel (anion type)

Boundaries or planes that separate a material into different regions are called *surface defects*. The individual regions have the same crystal structure, but differ in orientations. A material's surface shows abrupt disruption of the crystalline structure. Surface atoms suffer distortions in coordination number and atomic bonding. These properties often play an important role in the operation of microelectronic devices. Surfaces of materials can also exhibit defects such as roughness and notches, which make them more reactive than the bulk portion of the materials.

8.6.2 Line and Volume Defects

Line defects refer to missing of a partial plane of atoms/ions/molecules in a crystal. A line discontinuity in a crystalline structure is called a *dislocation*. There are three basic types of dislocations, namely edge dislocation, screw dislocation, and mixed dislocation. Insertion of extra half planes of unit cells leads to the development of an *edge dislocation*. The regions that surround the dislocation line are made up of perfect crystals. There is, however, a disruption in the crystal structure along the dislocation line. In a *screw dislocation*, the atomic planes do not exist separately from each other. The planes form a single surface like the threads of a screw and spiral from one end of the crystal to the other. In three-dimensional visualization, it appears like a helical structure and is not flat like a spiral. In *mixed dislocations*, both edge and screw dislocations are present together. The absence of a number of atoms within a crystal leads to the formation of volume defects or voids. *Voids* result in the formation of internal surfaces within the crystal and give rise to broken bonds at the surface. These broken bonds have properties similar to those of micro-cracks, which are areas where a solid is liable to fracture easily.

Role of Imperfections in Plastic Deformations

Metals and alloys generally undergo two types of deformation, namely elastic deformation and plastic deformation. *Elastic deformation* is a temporary change in shape that occurs due to an applied stress on the material. The elastic deformation however vanishes as soon as the applied stress is removed. Elastic deformation generally occurs due to stretching of interatomic bonds and no dislocation related motion occurs. Irreversible deformation or change in shape is called *plastic deformation*. Such a deformation or change in shape does not disappear when the applied stress is removed. This is because in plastic deformation, stress causes dislocation related motion. In fact, a plastic deformation results due to the cumulative effect of slip of numerous dislocations.

8.7 X-Ray Diffraction and Bragg's Law

One can study the structures of various crystalline materials using the non-destructive method of X-ray powder diffraction. Phase identification, quantitative analysis, pattern indexing, peak position unit cell parameter refinement, and structural imperfections are usually studied using an X-ray diffractometer.

A known amount of randomly oriented fine powder of sample (~10 µm) is allowed to be impinged with a coherent beam of monochromatic X-rays of a known wavelength. Usually, copper (Cu) is used as an X-ray tube which has a wavelength of 1.54 Å and a typical X-ray diffractometer is shown in Fig. 8.25.

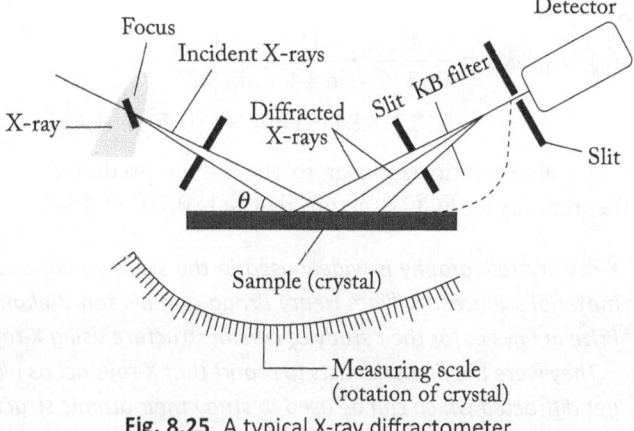

Fig. 8.25 A typical X-ray diffractometer

8.7.1 Bragg's Equation

W. H. Bragg and W. L. Bragg studied the optical properties of X-rays and deduced an equation to determine the interplanar distances in a crystal. We know that a crystal possesses space lattice arrangement of atoms or ions and thus are equally spaced with numerous parallel planes containing lattice points or atoms. Bragg's law states that,

When a beam of coherent monochromatic X-rays of same wavelength strikes a crystal, it gets diffracted in a manner such that it causes an interference pattern or may be reinforcement of diffracted beam from the crystal and the whole beam may behave as if it was reflected from crystal surface.

To understand the above statement, let us consider a beam of monochromatic X-rays that strikes on a set of parallel and equidistant planes shown as lines AA, BB, CC in Fig. 8.26.

These parallel planes are assumed to be regular arrangements of points indicating the position of atoms in a crystal. Let d be the distance between the successive planes with Miller indices as (100), (110) and (111) planes; d will be represented as, d_{100}, d_{110}, d_{111}. In Fig. 8.26, LMN is the wavefront of X-rays; LS and MT are a series of parallel monochromatic

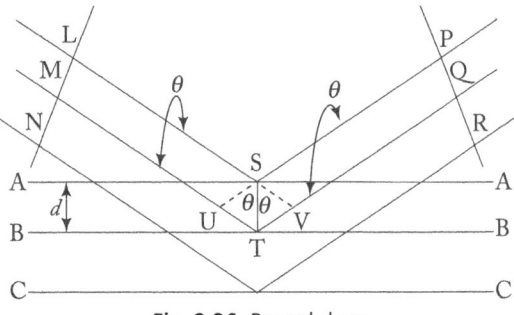

Fig. 8.26 Bragg's law

X-rays of wavelength λ that strikes the crystal face at an angle θ called glancing angle. LSP denotes the path of X-rays reflected from the first layer and MTQ denotes the path of X-rays reflected from the second layer. If the length of MTQ is greater than the length of LSP path by a whole number of wavelength, then the two beams from L and M will reinforce one another after reflection resulting in a stronger beam. Bragg's equation is represented as

$$n\lambda = 2d \sin \theta \qquad \qquad \text{...(8.11)}$$

where n is an integer having values 1, 2, 3, etc., and is called the *order of reflection*. The intensity of the reflected beam is sinusoidal with respect to the glancing angle and the X-ray spectrum obtained is shown in Fig. 8.27.

Sodium chloride has an FCC structure and its first order reflection from (100), (110), (111) faces were observed at 5.9°, 8.4°, and 5.2° respectively. Thus, the ratio of spacings of principal lattice planes of NaCl is given as,

$$d_{100}, d_{110}, d_{111} = \frac{1}{\sin 5.9°} : \frac{1}{\sin 8.4°} : \frac{1}{\sin 5.2°}$$

$$= 9.727 : 6.844 : 11.04 = 1 : 0.7037 : 1.135$$

The above ratio is closer to the values predicted theoretically for FCC structure, that is, 1 : 0.707 : 1.154.

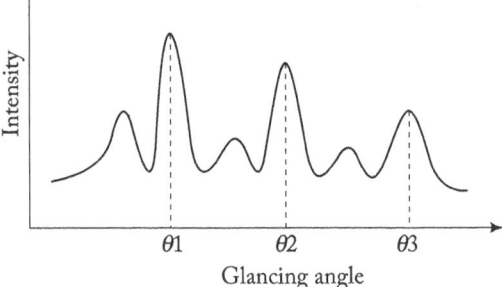

Fig. 8.27 X-ray spectrum

X-ray crystallography is widely used in the study of various crystal structures and forms a foundation in material sciences. William Henry Bragg and his son William Lawrence Bragg jointly won the 1915 Nobel Prize in Physics for their study of crystal structure using X-rays.

They were the first scientists to report that X-rays act as waves. When X-rays fall on crystal structure they get diffracted which can be used to study their atomic structures.

8.7.2 Determination of Avogadro's Number

Figure 8.28 shows the unit cell of sodium chloride which extends in all directions to form a bigger crystal. The edge of the cube is the distance between sodium ions or two chloride ions and this edge length is twice the d_{100}. One can calculate the interplanar distance from the dimensions of the crystal and the number of ions associated with each unit cell. In Fig. 8.28, 14 points denote sodium ions and 13 points denote chloride ions. If each sodium ion at the corner of the cube is shared equally by eight unit cubes, then only 1/8th of each of the eight corner sodium ions will be $1/8 \times 8 = 1$ sodium ion associated with the unit cell. Further, there are six sodium ions at the centres of each face of the unit cell, each of which is shared by four unit cubes; hence, $1/2 \times 6 = 3$ sodium ions are associated with the unit cell. Hence, there are 4 sodium ions in the unit cell. In the same way, 12 chlorine ions are at the centres of edges each of which is shared by four unit cubes; hence, $1/4 \times 12 = 3$ chloride ions are associated with the unit cell. One chlorine ion is at the centre of the unit cell and thus in all, 4 chlorine ions are present in the unit cell. It is thus proven that unit cell of NaCl has four sodium and four chlorine ions; that is, four NaCl molecules.

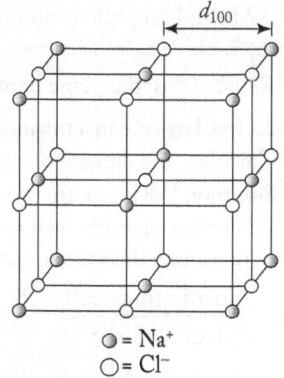

$= Na^+$
$= Cl^-$

Fig. 8.28 Unit cell of NaCl

If the molar volume of sodium chloride is M/ρ where, ρ is density of the crystal and M is the molecular weight, the volume of four NaCl molecules is,

$$V = \frac{\text{Mass}}{\text{Density}} = \frac{4M}{N_A} \times \frac{1}{\rho} \quad \left(\text{where } \frac{M}{N_A} \text{ is the mass of each molecule} \right)$$

If $2\,d_{100}$ is edge of the unit cell, then $V = [2d_{100}]^3$, which can be also expressed as,

$$[2\,d_{100}]^3 = \frac{4M}{N_A\,\rho} \qquad \qquad ...(8.11)$$

Equation (8.11) can be used to determine the Avogadro's number.

Check Your Progress

8. List the different Bravais lattices.
9. State and express Bragg's equation.
10. How many atoms are assigned to a unit cell of (a) face-centred and (b) body-centred cubic lattice?
11. Justify the statement: A unit cell of sodium chloride contains four molecules of NaCl.
12. What is a point defect and line defect? List the various types of line defects.
13. Consider a gold unit cell which has FCC structure. How many atoms will occupy gold unit cell? What will be the mass of gold atom present in the unit cell?
14. State the expression to determine Avogadro's number using X-ray diffraction of NaCl crystal.
15. Draw the crystal structures of NaCl, rutile TiO_2, ZnS, diamond, graphite.

SOLVED EXAMPLES

1. A plane in a crystal is depicted in the figure. Determine the corresponding Miller indices.
Solution: The plane shown in the figure has intercepts a, $2b$, and $3c$ along the three crystal axes. We know that the lattice points in a three-dimensional lattice are given by: $r = pa + qb + sc$.

From the figure and Eq. (8.1), it can be concluded that the intercepts are, $p = 1$, $q = 2$, and $s = 3$.

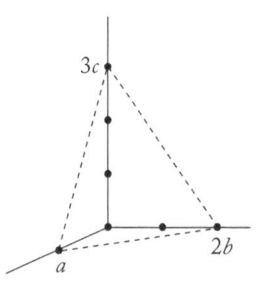

On taking the reciprocals, we get $\left[1, \dfrac{1}{2}, \dfrac{1}{3}\right]$.

Multiplying all the numbers by the lowest common denominator 6, we get $(6, 3, 2)$.

Result: Thus, the plane depicted in the figure has the Miller indices $(6, 3, 2)$.

2. The lattice constant a of Si is 5.43×10^{-8} cm and its atomic weight is 28.1. Calculate the density of Si.

Solution: We know that for Si,

Number of atoms/cell $= 8$, $a = 5.43 \times 10^{-8}$ cm

Therefore, the atomic concentration is,

$$\frac{\text{No. of atoms/cell}}{\text{Cell volume}} = \frac{8}{a^3} = \frac{8}{(5.43 \times 10^{-8})^3} \text{ atoms/cm}^3 = 5 \times 10^{22} \text{ atoms/cm}^3$$

Hence, the density of Si is calculated as follows: Density $= \dfrac{5 \times 10^{22} \times 28.1}{6.02 \times 10^{23}} = 2.33$

where 6.02×10^{23} is Avogadro's number.

Result: Density of Si $= 2.33$ g/cm^3

3. A plane in a crystal is shown in the figure. Calculate the corresponding Miller indices.

Solution: The plane shown in the figure has intercepts $2a$, $2b$, and $2c$ along the three crystal axes. The lattice points in a three-dimensional lattice are given by the following expression: $r = pa + qb + sc$

Thus, the intercepts of the plane are, $p = 2$, $q = 2$, and $s = 2$.

On taking the reciprocals, we get, $[1/2, 1/2, 1/2]$.

Multiplying all the numbers by the lowest common denominator 2, we get $(1, 1, 1)$.

Result: Thus, the plane depicted in the figure has the Miller indices $(1, 1, 1)$.

4. The lattice constant of a unit cell of aluminium is 4.049 Å. Calculate the spacing of (220) planes.

Solution: Let (hkl) be the Miller indices; then from Eq. (8.2),

$$d_{hkl} = \frac{a_0}{\sqrt{h^2 + k^2 + l^2}} \text{ with } a = 4.049 \text{ Å}$$

where $h = 2$, $k = 2$, $l = 0$; thus we get,

$$d_{hkl} = \frac{4.049}{\sqrt{4 + 4 + 0}} = 1.432 \text{Å}$$

Result: Spacing $= 1.432$ Å

5. The lattice constant of a crystalline material with cubic lattices is found to be 4.24 Å. Calculate the spacing of (110) planes.

Solution: Let (hkl) be the Miller indices,

$$d_{hkl} = \frac{a_0}{\sqrt{h^2 + k^2 + l^2}} \text{ with } a = 4.24 \text{ Å}$$

where $h = 1$, $k = 1$, and $l = 0$; thus we get,

$$d_{hkl} = \frac{4.24}{\sqrt{1 + 1 + 0}} = 2.998 \text{ Å}$$

Result: Spacing $= 2.998$ Å

6. The spacing between (220) planes is found to be 1.41 Å. Determine the lattice constant.

Solution: We know, $d_{hkl} = \dfrac{a_0}{\sqrt{h^2 + k^2 + l^2}}$ or, $a = d\sqrt{h^2 + k^2 + l^2}$

Substituting the given values in the above equation we get, $a = 1.41\sqrt{2 + 4 + 0} = 3.988$ Å

Result: Lattice constant = 3.988 Å

7. Sodium chloride has an FCC lattice and length of the cube edge is 5.64Å. Calculate $d_{100}, d_{110}, d_{111}$.

Solution: In a face-centred cubic lattice $d_{100} = \dfrac{a}{2}$.

As the length of the cube edge = 5.64 Å = 5.64×10^{-10} m
$$= 564 \text{ pm}$$
$$\therefore \qquad\qquad 2d_{100} = 564 \text{ pm}$$

Hence, $d_{100} = \dfrac{564}{2} = 282$ pm

Interplanar spacing in NaCl is, $d_{100} : d_{110} : d_{111} = 1 : \dfrac{1}{\sqrt{2}} : \dfrac{2}{\sqrt{3}}$

$\therefore \qquad$ If $d_{100} = 282$, then $d_{110} = \dfrac{1}{\sqrt{2}} \times 282 = 199.4$ pm

and, $\qquad d_{111} = \dfrac{2}{\sqrt{3}} \times 282 = 325.6$ pm

Result: $d_{100}, d_{110}, d_{111}$ are 282, 199.4, 325.6 pm

8. The structure of a crystalline material is identical to silicon. If the atomic concentration of the material is 1×10^{22} atoms/cm^3, calculate the lattice constant. If the density of the material is 1.55 g/cm^3, calculate the atomic weight of the material.

Solution: Number of atoms/cell = 8

$$\text{Atomic concentration} = \frac{\text{Number of atoms per cell}}{\text{Cell volume}} = \frac{8}{a^3}$$

where a is the lattice constant.

Thus, $\qquad a^3 = \dfrac{8}{1 \times 10^{22}} = 8 \times 10^{-22}$

or, $\qquad a = 9.28$ Å

The atomic weight is calculated as follows:

$$\text{Density} = \frac{\text{Atomic concentration} \times \text{Atomic weight}}{6.02 \times 10^{23}}$$

On rearranging the above equation, we get,

$$\text{Atomic weight} = \frac{\text{Density} \times 6.02 \times 10^{23}}{\text{Atomic concentration}} = \frac{1.55 \times 6.02 \times 10^{23}}{1 \times 10^{22}} = 93.3$$

Result: Atomic weight = 93.3 amu

9. Calculate the distance between the lattice planes that give first order reflection at an angle of 26.42° with molybdenum X-rays of wavelength 0.710 Å.

Solution: According to Bragg's law, $n\lambda = 2d \sin \theta$

or, $\qquad d = \dfrac{n\lambda}{2\sin\theta} = \dfrac{1\times 0.710}{2\times\sin 26.42°} = 0.799$

Result: Distance = 0.799 Å

10. The unit cell of CaO is a cube having an edge length 420 pm. Each cell contains an equivalent of 4 atoms of calcium and 4 atoms of oxygen. If molecular weight of Ca is 56 g/mol, calculate its density.

Solution: We know, $[2d_{100}]^3 = \dfrac{4M}{N_A\rho}$

$\therefore \qquad \rho = \dfrac{4M}{[2d_{100}]^3 N_A} = \dfrac{4\times 56}{[420\times 10^{-12}]^3 \times 6.022\times 10^{23}}$

$\qquad\qquad = \dfrac{224}{4.4615\times 10^{-5}} = 5.020\times 10^6 \text{ gm}^{-3}$

Result: Density = 5.020×10^6 gm^{-3}

11. The first order reflection of X-ray beam from (100) plane of NaCl occurs at 6.2°. Determine the wavelength of X-rays. What will be the angle of reflection if X-rays of wavelength 154 pm are used? (Given: d_{100} NaCl = 282 pm).

Solution: According to Bragg's equation, $n\lambda = 2d \sin\theta$

$\therefore \qquad \lambda = \dfrac{2d\sin\theta}{n} = \dfrac{2\times 282\times\sin 6.2°}{1} = 60.91 \text{ pm}$

Further, $\quad \sin\theta = \dfrac{n\lambda}{2d} = \dfrac{1\times 154}{2\times 282} = 0.2730$

Hence, $\theta = 15.84°$

Result: Angle of reflection = 15.84°

12. The first order reflection maxima from (100), (110), and (111) planes of a given cubic crystal occur at 7.2°, 10.2°, and 12.5° respectively. What type of cubic lattice does the crystal possess?

Solution: The ratio of distances between the planes to found using Bragg's equation, $n\lambda = 2d \sin\theta$

or, $\qquad d = \dfrac{n\lambda}{2\sin\theta}$

$\therefore \qquad d_{100} : d_{110} : d_{111} = \dfrac{n\lambda}{2\sin 7.2°} : \dfrac{n\lambda}{2\sin 10.2°} : \dfrac{n\lambda}{2\sin 12.5°}$

For all reflections, n and λ are constant,

$\therefore \qquad d_{100} : d_{110} : d_{111} = \dfrac{1}{\sin 7.2°} : \dfrac{1}{\sin 10.2°} : \dfrac{1}{\sin 12.5°}$

$\qquad\qquad = \dfrac{1}{0.1253} : \dfrac{1}{0.1771} : \dfrac{1}{0.2164} = 1 : 0.707 : 0.579$

Result: Hence, the crystal is a simple cubic lattice type.

13. The first order reflection maximum for (100) NaCl plane occurs at 5.9° using X-rays of wavelength 58 pm. If density and molecular weight of NaCl are 2.17×10^6 gm^{-3} and 58.5 g/mol respectively, determine the Avogadro's number.

Solution: According to Bragg's equation, $n\lambda = 2d \sin\theta$.

$\qquad d_{100} = \dfrac{n\lambda}{2\sin\theta_{100}} = \dfrac{1\times 58}{2\sin 5.9°} = 281.6 \text{ pm}$

Since, $\quad [2d_{100}]^3 = \dfrac{4M}{N_A \rho}$

$\therefore \qquad N_A = \dfrac{4M}{[2d_{100}]^3 \rho} = \dfrac{4 \times 58.5}{[2 \times 281.6 \times 10^{-12}]^3 \times 2.17 \times 10^6} = 6.02 \times 10^{23}\,\text{mol}^{-1}$

Result: Avogadro's number = $6.02 \times 10^{23}\,\text{mol}^{-1}$

SUMMARY

- Solids are of two types: crystalline and amorphous. The periodic arrangement of atoms in a crystal is called a lattice.
- A unit cell is the fundamental unit that undergoes regular repetition to create a lattice. The smallest unit cell is called a primitive cell.
- There are 14 Bravais lattices. Simple cubic, body-centred cubic, and face-centred cubic lattices are the three most common cubic lattices.
- Equivalent planes in a crystal are characterized by a set of Miller indices. The number of atoms sharing boundary with a particular atom gives its coordination number.

  The fraction of space occupied by atoms in a crystal is called the packing factor (*f*); it is 0.52 for simple cubic, 0.68 for BCC, and 0.74 for FCC structure. NaCl has an FCC configuration which is well studied using X-ray diffraction.
- A diamond lattice is obtained by inserting one FCC lattice into another FCC lattice. GaAs has a zinc blende lattice. Graphite contains carbon atoms arranged in regular hexagons in flat parallel layers. TiO_2 has a rutile-type crystal structure which finds application in catalysis.
- Born–Haber cycle is used to determine the lattice energies of ionic crystals. It applies the concept of Hess's law of constant summation to determine lattice energy by comparing it with standard enthalpy of the given ionic crystal.
- Various imperfections are possible in a crystal structure. Localized disruptions in a crystalline structure are called point defects. An extra atom inserted into a normally unoccupied position in a crystal structure results in an interstitial defect.
- A vacancy–interstitial defect is called a Frenkel defect. In a Schottky defect, vacancies are not accompanied by a simultaneous transition of atoms to interstitials.
- The absence of an entire row of atoms from normal lattice sites results in a line defect. Substitutional defect is due to the replacement of an atom in a crystal by a different atom in the normal lattice site.
- X-ray diffraction and Bragg's law help in structural characterization of various crystalline solids. One can determine Avogadro's number for atoms present in a crystalline unit cell using X-ray diffraction and Bragg's equation.

GLOSSARY

Bravais lattice: Any of 14 possible 3D configurations of points used to describe the ordered arrangement of atoms in a crystal.

Crystal face: The surface that is planar binding the crystals.

Interfacial angle: The angle formed between two intersecting faces of a crystal.

Lattice: An array of points in space such that the environment around each point is the same.

Lattice point: A two-dimensional representation of an atom denoted by a dot.

Miller indices: A set of three numbers used to designate a group of parallel, equidistant planes in a crystal.

Packing factor: The fraction of space occupied by atoms.

Primitive cell: The smallest unit cell that can generate a crystal.

Unit cell: The fundamental unit that undergoes a regular repetition to create the lattice, a basic building block of the entire crystal.

KEY FORMULAE

- Displacement between two lattice points:
  $r = pa + qb + sc$
- Packing factor (f):

$$f = \frac{\text{No. of atoms/unit cell} \times \text{Vol. of each atom}}{\text{Vol. of unit cell}}$$

- Miller indices: $d_{hkl} = \dfrac{a_0}{\sqrt{h^2 + k^2 + l^2}}$

- Equilibrium concentration of interstitial atoms at a given temperature:

$$n_I = ANe^{-E_{Fi}/K_B T}$$

- Bragg's equation: $n\lambda = 2d \sin \theta$
- Avogadro's number: $[2d_{100}]^3 = \dfrac{4M}{N_A \rho}$
- Radius ratio rule: $R = \dfrac{r_+}{r_-}$

EXERCISES

Multiple Choice Questions

1. Which of the following is an amorphous solid?
   - (a) Sugar
   - (b) Table salt
   - (c) Plastic
   - (d) Diamond

2. The fundamental unit that repeats to form a lattice is called a
   - (a) cell
   - (b) unit cell
   - (c) crystal
   - (d) defect

3. The smallest unit cell is called
   - (a) prime cell
   - (b) simple cell
   - (c) primitive cell
   - (d) fundamental cell

4. The number of Bravais lattices is
   - (a) 13
   - (b) 12
   - (c) 15
   - (d) 14

5. For a cubic structure
   - (a) $a = b = c$
   - (b) $a = b \neq c$
   - (c) $a \neq b \neq c$
   - (d) $a + b = c$

6. A set of equivalent planes is represented by
   - (a) $[hkl]$
   - (b) $\{hkl\}$
   - (c) (hkl)
   - (d) $<hkl>$

7. For a simple cubic lattice, d_{hkl} is given by
   - (a) $\dfrac{a_0}{(h^2 + k^2 + l^2)}$
   - (b) $\dfrac{a_0}{\sqrt{h^2 + k^2 + l^2}}$
   - (c) $\dfrac{a_0^2}{\sqrt{h^2 + k^2 + l^2}}$
   - (d) $\dfrac{a_0}{h + k + l}$

8. The number of atoms in a unit cell of FCC structure is:
   - (a) 2
   - (b) 1
   - (c) 3
   - (d) 4

9. For a simple cubic structure, f is
   - (a) 0.52
   - (b) 0.52
   - (c) 0.51
   - (d) 0.49

10. The semiconductor GaAs has
    - (a) an FCC lattice
    - (b) a zinc blende lattice
    - (c) a diamond-like lattice
    - (d) a hexagonal lattice

11. Zn ions in ZnS have a charge of
    - (a) +2
    - (b) −2
    - (c) +1
    - (d) −1

12. X-ray diffraction helps in
    - (a) studying phase changes
    - (b) identifying crystalline structures
    - (c) distinguishing compounds
    - (d) preparation of various isomers

13. The elements of symmetry are
    - (a) plane of symmetry
    - (b) centre of symmetry
    - (c) axis of symmetry
    - (d) all of these

14. NaCl and CsCl are examples of
    - (a) cubic crystal system
    - (b) orthorhombic crystal system
    - (c) tetragonal crystal system
    - (d) rhombhohedral crystal system

15. Bragg's equation for X-ray diffraction is
    - (a) $n\lambda = 2d \sin \theta$
    - (b) $\lambda = d \sin \theta$
    - (c) $n^2 = 2d \sin \theta$
    - (d) $n\lambda = 2d \tan \theta$

16. In an ionic crystal, a cation and an anion leave the lattice causing two vacant sites. Such a defect is called
    - (a) Frenkel defect
    - (b) Schottky defect
    - (c) interstitial defect
    - (d) none of these

17. The change in enthalpy that occurs when one mole of solid crystal is formed from gaseous ions is called
    - (a) enthalpy of formation
    - (b) sublimation energy
    - (c) lattice energy
    - (d) standard enthalpy

18. In a solid lattice, a cation has left a lattice site and is present in an interstitial position. Such a defect is called
    - (a) Frenkel defect
    - (b) Schottky defect
    - (c) interstitial defect
    - (d) none of these

19. If the radius ratio lies in the range of 0.732 – 1.00, the coordination number will be
    - (a) 3
    - (b) 8
    - (c) 4
    - (d) 6

20. The numbers of atoms in a simple cubic, FCC and BCC are
    - (a) 1, 2, 4
    - (b) 1, 4, 2
    - (c) 4, 2, 1
    - (d) 2, 4, 1

21. If a crystal plane has intercepts 3, 4 and 2 along x, y and z axes respectively, its Miller indices will be
    - (a) 4, 4, 6
    - (b) 4, 4, 8
    - (c) 1, 2, 3
    - (d) 4, 3, 6

22. If a particular solid (A) with mass of 30 crystallizes in FCC structure having density of 3 gcm-3, its unit cell length will be
    - (a) 4.00 Å
    - (b) 44.32 Å
    - (c) 4.049 Å
    - (d) 4.11 Å

23. For an orthorhombic crystal system, the incorrect expression is
    - (a) $a = b \neq c$
    - (b) $a \neq b \neq c$
    - (c) $\alpha = \beta = \gamma = 90°$
    - (d) none of these

24. The coordination number of BCC is
    - (a) 2
    - (b) 4
    - (c) 6
    - (d) 8

25. The permitted coordination number in an ionic crystal is 6; hence the arrangement of anions around the cation will be
    - (a) octahedral
    - (b) planar
    - (c) tetrahedral
    - (d) BCC

Review Questions

1. Define: (a) lattice (b) unit cell (c) Miller indices (d) Weiss indices (e) coordination number (f) packing factor.

2. Explain the procedure to obtain the Miller indices. How does one convert Weiss indices to Miller indices?

3. Explain the notations (a) [*hkl*] and (b) {*hkl*}.

4. Deduce the percentages of volume of vacant space in a simple cubic, FCC, and BCC structures.

5. Write a note on 'Bravais lattice.'

6. State and explain the laws of crystallography.

7. State the law of symmetry. Discuss the various elements of symmetry in a crystal structure.

8. Give the coordination numbers for (a) an BCC structure and (b) FCC structure.

9. Explain the relation between coordination number, packing fraction, and radius ratio of an ionic crystal.

10. Explain radius ratio rule and mention its limitations.

11. Deduce packing fractions for SCC, FCC, and BCC crystal structures.

12. What is lattice energy? Explain Born–Haber cycle elucidating one example.

13. Discuss X-ray diffractometer and explain its application in studying NaCl structure.

14. Discuss Bragg's equation. Explain the experimental set-up used to determine X-ray diffraction for NaCl structure.

15. How does one determine Avogadro's number using X-ray diffraction for NaCl structure?

16. Explain the crystal structure of (a) NaCl (b) diamond (c) ZnS (d) graphite.

17. Discuss the crystal structure of zinc blende.

18. Explain the crystal structure of diamond and graphite.

19. Explain Frenkel and Schottky defects in a crystal structure.

20. Write an informative note on 'imperfections in atomic packings.'

NUMERICAL PROBLEMS

1. Calculate Miller indices of the faces having the following intercepts with the three axes perpendicular to each other: (i) 2a, 3b, 4c (ii) 5/2 a, 5/2 b, 3c **(Ans: (i) (643); (ii) (665))**

2. Aluminium forms FCC structure and density of Al is 2.70 gcm^{-3}. Calculate the edge length of unit cell. (Atomic weight of Al = 27) **(Ans: 4.053×10^{-8} cm)**

3. Find the interplanar distance in a crystal in which the series of planes produces first order of reflection from a copper X-ray tube (λ = 1.539Å) at an angle of 22.5°. **(Ans: d = 2.01Å)**

4. The first order reflection maxima from (100), (110) and (111) planes of a given cubic crystal occur at 13°24′, 9°32′ and 23°50′ respectively. What type of cubic lattice does the crystal possess?

(Ans: Body centred cubic lattice)

5. The first order reflection of X-ray beam from (100) plane of NaCl occurs at 6.3°. Determine the wavelength of X-rays. What will be the angle of reflection if X-rays wavelength of 156 pm are used? (Given: d_{100} NaCl = 282 pm). **(Ans: λ = 61.89 pm, θ =16.05°)**

6. Sodium chloride has FCC structure. The distance between sodium and chloride ions using X-rays was found to be 282.1 pm. If its density is 2.175, calculate the Avogadro's number. **(Ans: = 5.99×10^{23} mol^{-1})**

7. CaF_2 crystallizes in FCC lattice with an edge length of 546 pm. Its molecular weight is 78g/mol and density is 3.18g/cm^3. Determine the Avogadro's number. **(Ans: 6.03×10^{23} mol^{-1})**

8. Polonium crystallizes as SCC and has an atomic mass of 209 and density of 91.5 kgm-3. What will be the edge length of unit cell? **(Ans: 15.597×10^{-8} cm)**

9. Calculate the distance between the lattice planes that give first order reflection at an angle of 26.42° with molybdenum X-rays of wavelength 0.710Å. **(Ans: 0.799 Å)**

10. The angles of reflection for first order from (100), (110) and (111) planes of a cubic crystal are 11°54′, 17° and 10°30′ respectively. What type of lattice does the crystal possess? **(Ans: FCC)**

FURTHER READING

1. DiSalvo, F. J. 'Solid State Chemistry – A Rediscovered Chemical Frontier', *Science*, 247, 649.
2. Rao, C.N.R and J.Gopalakrishnan, *New Directions in Solid State Chemistry*, Cambridge University Press.
3. Suryanarayana, C. M. and Grant Norton, *X-ray Diffraction – A Practical Approach*, Springer-Verlag, New York Inc, 2013.
4. West, Anthony, *Solid State Chemistry and its Applications*, John Wiley & Sons, 2014.

ANSWERS

Check Your Progress

1. Crystalline solids have sharp melting points and exhibit anisotropy whereas amorphous solids do not exhibit sharp melting points. NaCl, ice, salt, CsCl are crystalline solids whereas glass, wax, and polymers are amorphous solids.
2. Crystallography is the study of geometric forms of crystalline substances and their properties.
3. Laws of crystallography include the law of constancy of interfacial angles, law of rational indices, and law of symmetry.
4. A set of smallest three integers used to designate a group of parallel, equidistant planes in a crystal.
5. Refer to Section 8.2.
6. Crystalline vs Amorphous solids

Crystalline solids	Amorphous solids
Possess sharp melting points	Melt over a wide range of temperatures
Constituent atoms are arranged in a regular manner	Constituent atoms are not arranged in a regular manner
Anisotropic	Isotropic
Eg., NaCl, sugar, CsCl	Eg., glass, wax, polymers

7. (a) In an FCC unit cell, there are 8 atoms at each corner and 6 atoms at the centre of each face. Hence the number of atoms assigned to the unit cell is, $= (8 \times 1/8) + (6 \times 1/2) = 1 + 3 = 4$ atoms

   (b) In a BCC unit cell, there are 8 atoms at the corner and one atom at the centre of body-centred cubic lattice. Thus, the total number of atoms assigned to the unit cell is, $= (8 \times 1/8) + 1 = 1 + 1 = 2$ atoms

8. There are 14 Bravais lattices: (a) simple cubic (b) face-centred cubic (c) body-centred cubic (d) simple tetragonal (e) body-centred tetragonal (f) simple orthorhombic (g) body-centred orthorhombic (h) base-centred orthorhombic (i) face-centred orthorhombic (j) rhombohedral (k) simple monoclinic (l) base-centred monoclinic (m) triclinic (n) hexagonal.

9. Refer to Section 8.7.1

10. (a) In an FCC unit cell, there are 8 atoms at each corner and 6 atoms at the centre of each face. Hence the number of atoms assigned to the unit cell is, $(8 \times 1/8) + (6 \times 1/2) = 1 + 3 = 4$ atoms

    (b) In a BCC unit cell, there are 8 atoms at the corner and one atom at the centre of body-centred cubic lattice. Thus, the total number of atoms assigned to the unit cell is, $= (8 \times 1/8) + 1 = 1 + 1 = 2$ atoms

11. Refer to Section 8.7.2: Determination of Avogadro's number.

12. Localized disruptions in the otherwise perfect arrangement of atoms in a crystal structure are called point defects. Line defects refer to missing of a partial plane of atoms/ions/molecules in a crystal. Line defects are of three types viz, edge dislocation, screw dislocation, and mixed dislocation.

13. Solution: Eight corners at 1/8 atom each $= 1/8 \times 8 = 1$ atom

    Six faces at the centre of each face $= 1/2 \times 6 = 1$ atom

    Thus, 4 atoms occupy unit cell of gold.

    $$\text{Mass of gold atom} = \frac{\text{Molar mass}}{\text{Avogadro's number}} = \frac{197}{6.022 \times 10^{23}} \times \frac{1}{4} = 8.178 \times 10^{-23}\, gm$$

14. If the molar volume of sodium chloride is M/ρ where ρ is density of the crystal and M is the molecular weight, the volume of 4 NaCl molecules is,

    $$V = \frac{\text{Mass}}{\text{Density}} = \frac{4M}{N_A} \times \frac{1}{\rho}\left(\text{where, } \frac{M}{N_A}\text{ is mass of each molecule}\right)$$

    If 2 is edge of the unit cell, then $V = [2d_{100}]^3$ which can be also expressed as,

    $$\left([2d_{100}]^3 = \frac{4M}{N_A \rho}\right)$$

    The above equation (can be used to determine the Avogadro's number.

15. Refer to Section 8.5.

Multiple Choice Questions

1. (c) 2. (b) 3. (c) 4. (d) 5. (a) 6. (b) 7. (b)
8. (d) 9. (a) 10. (b) 11. (a) 12. (b) 13. (d) 14. (a)
15. (a) 16. (b) 17. (c) 18. (a) 19. (b) 20. (b) 21. (d)
22. (c) 23. (a) 24. (d) 25. (a)

Coordination Chemistry and Organometallic Compounds

LEARNING OBJECTIVES

After reading this chapter, you will be able to:

- understand the various terminologies used in coordination chemistry.
- explain the rules of nomenclature of coordination compounds.
- list the postulates of Werner and valence bond theories of coordination compounds.
- introduce crystal field theory of coordination compounds with suitable examples.
- explain Pearson's principle of hard and soft acids and bases (HSAB).
- account for the stability of coordination compounds and their importance and applications in various fields.
- classify various types of organometallic compounds and explain EAN rule.
- discuss isomerization, polymerization, hydrogenation, and hydroformylation reactions using organometallic catalysts.

9.1 INTRODUCTION

Coordination chemistry is a branch of inorganic chemistry that deals with the formation of coordination compounds, particularly the transition elements. Even before coordination chemistry came to be treated as a separate branch of chemistry, applications of coordination compounds were already known. Alizarin (bright red coloured complex) as a textile fabric dye was known as early as the 15th century. In nature, chlorophyll, vitamin B12, haemoglobin are coordination compounds of magnesium, cobalt, and iron respectively. Tetraamminecupric ion $[Cu(NH_3)_4]^{2+}$, Prussian blue, $Fe_4[Fe(CN)_6]_3 \cdot xH_2O$, potassium hexachloroplatinate, $K_2[PtCl_6]$ are some classic examples of early coordination compounds.

9.2 IMPORTANT TERMINOLOGIES IN COORDINATION CHEMISTRY

Coordination compound   Compounds containing one or more coordinate covalent bonds.

Coordinate covalent bond (dative bond)   A covalent bond in which both the bonding electrons are derived from the same atom. Metal ions and ligands usually bond through dative linkages.

Central ion (centre of coordination)   A cation attached or bonded to one or more neutral molecules, atoms, or ions.

Ligand   Any atom, ion, or neutral molecule that donates an electron pair and bonds with the central metal ion through secondary valency. Ligands possessing one, two, or more donor atoms are called *monodentate*, *bidentate*, and *polydentate*, respectively. This is called *dentate character*.

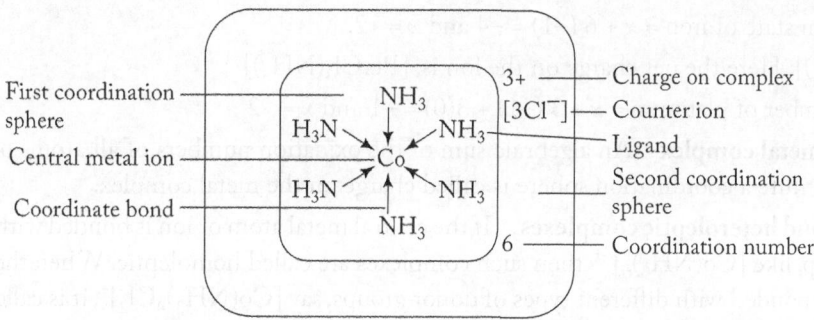

Fig. 9.1 Coordination compound terminologies (for [Co(NH$_3$)$_6$]Cl$_3$)

When a bi- or polydentate ligand uses its two or more donor atoms to form a bond with the central metal atom/ion, it is called a *chelate ligand*. Coordination compounds can form many rings with such ligands and the process is called *chelation* (means 'to claw').

The ligands that can coordinate through two different atoms are called *ambidentate* ligands. For example, NO$_2$ ion can form a coordinate bond either through nitrogen or oxygen; SCN$^-$ ion can bond through sulphur (thiocyanato) or nitrogen (isothiocyanato) with metals. The total number of chelating ligands is termed as *denticity of the ligand*.

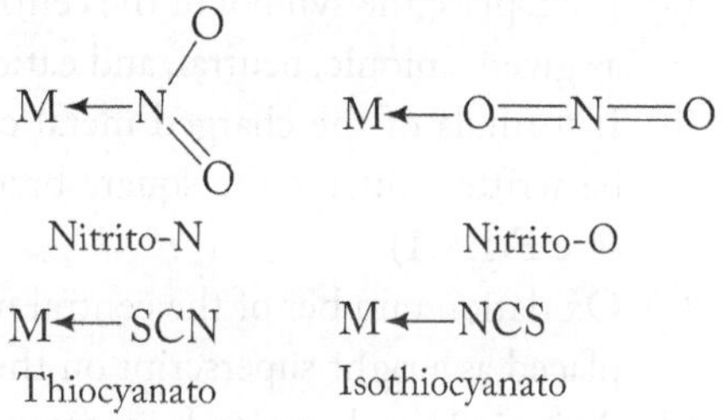

Table 9.1 Typical ligands showing dentate character

Monodentate	Sharing one electron pair	F$^-$, Cl$^-$, NH$_3$, H$_2$O, OH$^-$
Bidentate	Sharing two electron pairs	Ethylenediamine (H$_2$NCH$_2$CH$_2$NH$_2$), Glycinate ion (NH$_2$CH$_2$COO$^-$) Oxalate ion ($^-$OOC – COO$^-$)
Polydentate	Sharing many electron pairs	Ethylenediamminetetracetic acid (EDTA) – hexadentate (6 electron pairs)

Complex A molecule or an ion possessing a central metal atom or ion surrounded by a definite number of ligands held by coordinate covalent bonds.

Coordination sphere It refers to a collective term when the central metal atom is surrounded by ligands usually denoted as a square bracket.

Primary valency The charge on a central metal ion. Say, in Co(III) complex, +3 can be balanced with −3 forming a compound like CoCl$_3$. Hence, primary valence is satisfied in the second coordination sphere.

$$M \leftarrow N \underset{O}{\overset{O}{\diagup}} \qquad M \leftarrow O = N = O$$

Nitrito-N   Nitrito-O

$$M \leftarrow SCN \qquad M \leftarrow NCS$$

Thiocyanato   Isothiocyanato

Fig. 9.2 Ambidentate ligands

Secondary valency It refers to the number of ions of molecules that can coordinate to the metal atom or ion, for example, in [Co(NH$_3$)$_6$]Cl$_3$, cobalt (III) ion has six vacant valence orbitals and its secondary valence is six that is satisfied in the first coordination sphere of the metal ion (Fig. 9.1).

Coordination number (or ligancy) The total number of ligands surrounding a central metal ion in a coordination sphere.

Calculation of oxidation number of coordination compounds

[Ni(CO)$_4$] Say we consider it as, [Nix(CO)$_4$]^0:

Thus, oxidation number of nickel = $x + 4(0) = 0$ and hence it bears 0 oxidation state.

[Cu(NH$_3$)$_4$]SO$_4$: Let us consider it as, [Cu x (NH$_3$)$_4$]$^{2+}$ as net charge on the complex.

Thus, oxidation number of copper = $x + 4(0) = +2$ and so, $x = +2$.

K$_4$[Fe(CN)$_6$]: In this complex, the net charge on the ion is [Fe x (CN)$_6$]$^{4-}$.

Thus, oxidation state of iron = $x + 6\,(-1) = -4$ and $x = +2$.

$K[PtCl_3(NH_3)]$: Here the net charge on the ion is, $[PtxCl_3(NH_3)]^{-1}$

Oxidation number of platinum = $x + 3\,(-1) + 3(0) = -1$ and $x = +2$

Charge on a metal complex An algebraic sum of the oxidation numbers of all atoms or molecules or ions that constitute a coordination sphere is called charge on the metal complex.

Homoleptic and heteroleptic complexes If the central metal atom or ion is bonded with only one type of donor group, like $[Co(NH_3)_6]^{3+}$, then such complexes are called homoleptic. When the central metal atom or ion is bonded with different types of donor groups, say $[Co(NH_3)_4Cl_2]^+$, it is called heteroleptic complexes.

Complexes and double salts Double salts and coordination complexes are formed by the combination of two or more stable compounds in stoichiometric ratio. However, double salts dissociate in water in to simple ions completely. For example, Mohr's salt, $FeSO_4(NH_4)_2SO_4.6H_2O$ and potash alum, $KAl(SO_4)_2.12H_2O$ are examples of double salt. Coordination complex, on the other hand, does not dissociate in to simpler ions on dissolving in water. For example, the complex, potassium ferrocyanide, $K_4[Fe(CN)_6]$ does not dissociate into K, Fe^{2+}, and CN^- ions.

9.3 NOMENCLATURE OF COORDINATION COMPOUNDS

Coordination compounds are named according to the rules and recommendations of Nomenclature of Inorganic Chemistry [IUPAC (2005)]. The rules are given below.

9.3.1 Writing the Coordination Formulae

(a) First, place the symbol of the central metal atoms followed by the symbol of the ligand in the order as given: anionic, neutral, and cationic. Enclose the complex using a square bracket.

(b) If formula of the charged metal complex is written without any counter-ions, the charge should be written outside the square bracket as a right superscript with the number preceding the sign (see Fig. 9.1).

(c) Oxidation number of the central metal atom or ion may optionally be depicted as Roman numeral placed as a right superscript on the element symbol, for example, iron carbonyl, $[Fe^{II}(CO)_4]$.

(d) Anionic ligands are to be written first and in alphabetical order as per the first symbols of their formulae. Neutral and cationic ligands are then listed in the following order: water, ammonia, other inorganic ligands, and organic ligands in alphabetical order.

(e) Structural information can be depicted by prefixes such as, *cis -*, *trans -*, *mer -*, *fac-* etc.

(f) Ligand abbreviations and formulae of polyatomic ligands are placed in parentheses ().

9.3.2 Writing the Names of Complexes

(a) The cation is named first followed by the anion irrespective of whether the cation or anion is the complex species.

(b) In a metal complex, ligand names should be written in alphabetical order irrespective of their charge and there should be no space between words or between the name and the charge.

(c) The names of anionic complexes should end with *-ate* or *-ic* (for acids) while there is no such specification for cationic and neutral complexes.

(d) Naming ligands: Organic and inorganic ligands should end with *-o*. For ligands ending with '-ide', '-ite', or '-ate', 'e' is replaced by 'o', renaming them as, '-ido', '-ito', or '-ato', respectively. Thus, -azide becomes -azido, -nitrite becomes -nitrito, and -sulphate becomes -sulphato. Some anionic ligands are exception to this nomenclature (Table 9.2).

Table. 9.2 Names of anionic ligands based on IUPAC nomenclature

Ligand	Name	Ligand	Name	Ligand	Name
F^-	Fluoro	OH^-	Hydroxo	O_2^{2-}	Peroxo
Cl^-	Chloro	CN^-	Cyano	CH_3O^-	Methoxo
Br^-	Bromo	HS^-	Thiolo	$C_5H_5^-$	Cyclopentadienyl
I^-	Iodo	S_2^-	Thio	$C_6H_5^-$	Phenyl
H^-	Hydrido	O_2^-	Superoxo	O^{2-}	Oxo

(e) There is no change in naming neutral ligands and these are generally named as a molecule. Say, ethylenediamine (en) for $NH_2CH_2CH_2NH_2$, or say, diethylenetriamine for $NH_2CH_2CH_2NH$. $CH_2CH_2NH_2$. When water, ammonia (NH_3), nitrosyl (NO), and carbonyl (CO) are the ligands, the terms 'aqua', 'ammine', 'nitrosyl', and 'carbonyl' are used.

(f) Cationic ligands have specific ending like, 'ium', as in NH_2-NH_3^+, named as hydrazinium, and H_2N-CH_2-CH_2-NH_3^+ can be written as, 2-aminoethylammonium.

(g) The number of ligands is denoted by prefixes like, 'mono' for one ligand and usually omitted, 'di' for two bonded ligands, and so on without any space. For complicated ligands, prefixes such as *bis* for two ligands, *tris* for three, *tetrakis* for four ligands are used. The ligand names are usually denoted in the parenthesis ().

(h) The coordinating atom of a ligand to a central metal atom is indicated by the symbol 'κ' (Kappa) followed by element symbol after ligand name. So, for M-SCN- (where M denotes metal), it is written as thiocyanoto-κS and for M-NCS-, it is isothiocyanoto-κN.

(i) The bridging ligand (one that connects two metal atoms) is denoted by the symbol 'μ.'

As per IUPAC nomenclature, some examples are listed here.

Table 9.3 Formulae and names of selected coordination complexes

Formula	Name
$[Fe(CN)_6]^{4-}$	Hexacyanoferrate(II) ion
$[CoCl_3(NH_3)_4]$	Triamminetrichlorocobalt(III)
$[CuCl_2(CH_3NH_2)_2]$	Dichlorobis(methylamine)copper(II)
$H_2[PtCl_6]$	Hexachloroplatinic(IV) acid
$K[SbCl_5(C_6H_5)]$	Potassium pentachloro(phenyl) antimonite(V)
$[Cr(NH_3)_3(H_2O)_3]Cl_3$	Triamminetriaquachromium(III) chloride
$[Ag(NH_3)_2][Ag(CN)_2]$	Diamminesilver(I) dicyanidoargentate(I)

(j) Some coordination complexes have proven to be industrially important and named after scientists and the examples are as follows:
Zeise's salt: $K[Pt(C_2H_4)Cl_3]$; Magnus green salt: $[Pt(NH_3)_4][PtCl_4]$
Wilkinson catalyst: $[Rh(PPh_3)_3Cl]$; Erdmann's salt: $K[Co(NH_3)_2(NO_2)_4]$
Vaska's complex: $[Ir(CO)(PPh_3)_2Cl_2]$

9.4 THEORIES OF BONDING IN COORDINATION COMPOUNDS

Various theories have been put forth to explain bonding in coordination complexes. Some of the pertinent theories include Werner's theory, Valence bond theory, and Crystal field theory.

9.4.1 Werner's Theory

Alfred Werner (1893) examined various compounds of Co (III) chloride and ammonia. According to Werner, metal ions exhibit two types of linkages or valences, namely primary and secondary valences. The *primary valence* is equal to the oxidation number of the metal ion, whereas *secondary valence* is the number of atoms directly bonded to the metal called the *coordination number*. He also stated that metal ions will always try to satisfy both their valences. While naming coordination complexes, all molecules and ions are represented within the sphere in the brackets and the 'free' anions that dissociate from the complex ion on dissolution in water are shown outside the brackets.

To demonstrate coordination theory, Werner treated compounds containing cobalt chloride and ammonia with hydrochloric acid. It was found that ammonia was not completely removed from the cobalt compounds bringing about the conclusion that ammonia is tightly bound to the cobalt ion. When aqueous silver nitrate was added, one of the products formed was silver chloride. The amount of silver chloride

Table 9.4 Werner's coordination complexes

Initial compound	Colour	Final compound formed on adding silver nitrate
$CoCl_3.6NH_3$	Orange	$[Co(NH_3)_6]Cl_3$
$CoCl_3.5NH_3$	Purple	$[Co(NH_3)_5Cl]Cl_2$
$CoCl_3.4NH_3$	Green	*trans*-$[Co(NH_3)_4Cl_2]Cl$
$CoCl_3.4NH_3$	Violet	*cis*-$[Co(NH_3)_4Cl_2]Cl$

formed was equal to the number of ammonia molecules bonded to cobalt chloride. The observations made by Werner can be tabulated as in Table 9.4.

The result of Werner's observations was based on conductance measurements, reactions of complexes, and isomer analysis. In the inner coordination sphere, ligands are directly bound to the central metal. In the outer coordination sphere, other ions are attached to the complex ion. Werner's coordination theory gave the following postulates:

(a) Metal ions in coordination compounds exhibit two types of linkages (or valences), namely primary and secondary. Metal ions always try to satisfy both the valences. As the primary valences are ionizable, they are satisfied by negative ions. The secondary valence is non-ionizable and usually satisfied by neutral molecules or negative ions. Secondary valence is equal to the coordination number and is fixed for the central metal ion.

(b) The ions or groups linked by secondary bonds to the central metal possess characteristic spatial orientations corresponding to different coordination numbers referred to as *coordination polyhedral*.

Check Your Progress

1. Give the formulae for Zeise's salt and Wilkinson catalyst.
2. Define the term ligand. State their types.
3. Give two examples of different types of ligands.
4. Calculate the oxidation number of the central metal atom of $Ni(CO)_4$ and $[Cu(NH_3)_4]SO_4$.
5. Write the formulae for the following coordination compounds:
   (a) Tetraammineaquachloridocobalt (III) chloride (b) Potassium tetrahydroxidozincate (II) (c) Potassium trioxalatoaluminate (III) (d) Dichloridobis (ethane-1, 2-diamine) cobalt (III).
6. Write the IUPAC names of the following coordination compounds:
   (a) $[Pt(NH_3)_2Cl(NO_2)]$ (b) $K_3[Cr(C_2O_4)_3]$ (c) $[CoCl_2(en)_2]Cl$
   (d) $[Co(NH_3)_5(CO_3)]Cl$ (e) $Hg[Co(SCN)_4]$ (f) $K_2[CrCO(CN)_5]$

(c) The ions or species present within the square bracket are called *coordination entities* or *complexes* whereas ions present outside the square bracket are called *counter ions*.

(d) The common geometrical shapes in transition metal complexes are tetrahedral, square planar, and octahedral forms. For instance, various cobalt complexes such as $[Co(NH_3)_6]^{3+}$, $[CoCl(NH_3)_5]^{2+}$, and $[CoCl_2(NH_3)_4]^+$ are octahedral, while $[Ni(CO)_4]$ and $[PtCl_4]^{2-}$ are tetrahedral and square planar, respectively.

9.4.2 Valence Bond Theory

Linus Pauling and Slater (1935) proposed the valence bond (VB) theory to explain bonding in coordination compounds. The postulates of VB theory are as follows:

(a) The metal atom or ion possesses vacant orbitals, which are equivalent to its coordination number for covalent bond formation with filled ligand orbitals.

(b) The covalent bond is formed due to overlap of vacant metal and filled ligand orbitals giving rise to M–L σ bond.

(c) Apart from σ bond, π bonds are formed in certain coordination compounds due to sideways overlap of filled metal and vacant ligand orbitals.

(d) The metal atom or ion under the influence of ligands uses its $(n-1)$ d orbital, ns, np or ns, np, nd orbitals for hybridization to obtain a set of equivalent orbitals of similar energy. This leads to the formation of complexes with distinct geometries such as octahedral, tetrahedral, square planar, and many other fascinating shapes. These hybridized orbitals overlap with ligand orbitals (Lewis bases) that donate their electron pairs for bonding.

Table 9.5 Number of orbitals, their hybridization types, and geometries of molecules

Coordination number	Hybridization	Geometry (or shape)
2	sp	Linear
3	sp^2	Trigonal planar
4	sp^3	Tetrahedral
4	dsp^2	Square planar
5	sp^3d	Trigonal bipyramid
6	sp^3d^2	Octahedral
6	d^2sp^3	Octahedral

Let us look at some of the examples illustrating the case. VB theory also gave justification for magnetism displayed by various coordination complexes.

In hexamminecobalt(III) complex, $[Co(NH_3)_6]^{3+}$, cobalt ion has the electronic configuration $3d^6$. The hybridization scheme is as shown in Fig. 9.3.

Six ammonia molecules bond with cobalt ion by donating their six electron pairs. The complex assumes an octahedral geometry. As

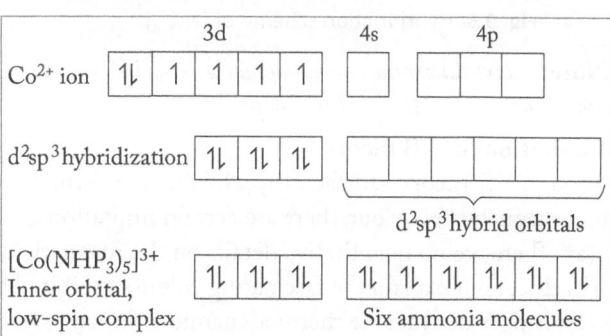

Fig. 9.3 Hybridization scheme of $[Co(NH_3)_6]^{3+}$

seen in the hybridization, all the electrons are paired and hence the complex is diamagnetic. As inner 3d metal orbital takes part in hybridization, $[Co(NH_3)_6]^{3+}$ is called *inner orbital* or *low-spin* (or spin-paired) *complex*.

Another cobalt complex is $[CoF_6]^{3-}$ in which the central cobalt ion undergoes sp^3d^2 hybridization utilizing its outer 4d orbital; hence the complex is called an *outer orbital* or *high-spin* (or spin-paired) complex. The hybridization scheme is as shown in Fig. 9.4.

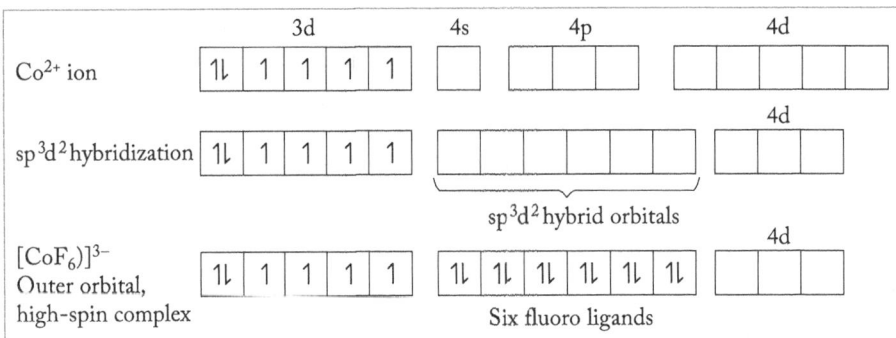

Fig. 9.4 Hybridization scheme of $[CoF_6]^{3-}$

In $[NiCl_4]^{2-}$, nickel has the electronic configuration of $3d^8$. Each chlorine ion donates a pair of electrons and the complex has tetrahedral geometry. The hybridization scheme (Fig. 9.5) shows that there are two unpaired electrons, thus exhibiting paramagnetism. On the other hand, nickel carbonyl, $[Ni(CO)_4]$ has tetrahedral geometry and is diamagnetic. Another example of nickel complex is of $[Ni(CN)_4]^{2-}$ (Fig. 9.6) that is a square planar complex with diamagnetic behaviour.

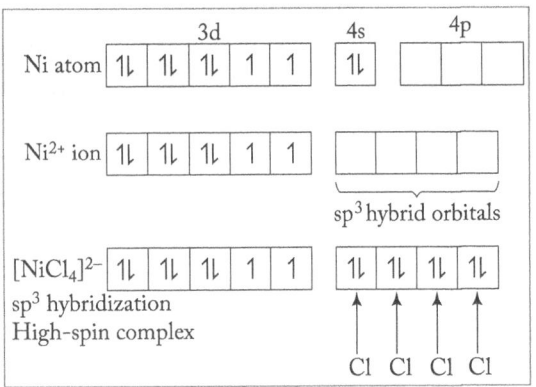

Fig. 9.5 Hybridization scheme of $[NiCl_4]^2$

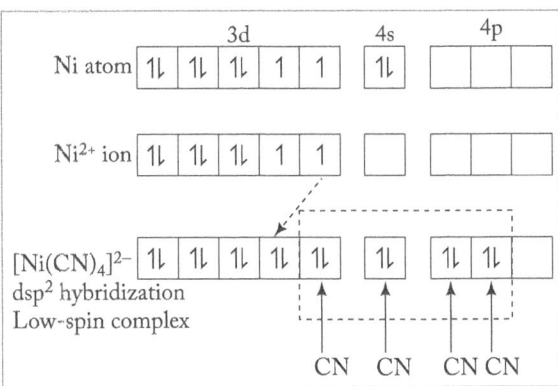

Fig. 9.6 Hybridization scheme of $[Ni(CN)_4]^{2-}$

Note: *Hybrid orbitals do not actually exist; it is a mere mathematical manipulation of wave equation for the atomic orbitals as discussed in Chapter 1.*

Limitations of VB theory

Though VB theory satisfactorily explains the formation of coordination compounds, their geometries and magnetic behaviour, there are certain limitations:

(a) There are no quantitative details on the magnetic data of coordination complexes.
(b) It cannot explain the fascinating colours exhibited by these complexes that lead to spectral analysis.
(c) It does not include thermodynamic or kinetic stabilities exhibited by coordination compounds.
(d) It fails to distinguish between different types of ligands such as weak or strong field ligands.
(e) It fails to explain complexes with coordination number 4, that is, tetrahedral and square planar structures.

9.4.3 Crystal Field Theory

H. Bethe and van Vleck (1935) developed the crystal field theory quite synonymous to the valence bond theory. The salient features of crystal field theory are as follows:

(a) It is based on the assumption that the metal ion and ligand act as point charges since their interaction is purely electrostatic.

(b) The ligands are either negatively charged or neutral molecules which donate electron pairs to the metal ion.

(c) The ligands tend to surround the metal ion and create an electrical field, thereby resulting in magnetism and spectral behaviour.

(d) Due to the interaction of d orbitals of a transition metal ion with ligand orbitals, crystal field effects are generated and hence the theory was named as 'crystal field theory.'

(e) When the ligands approach the metal ion, d orbitals of the metal ion possessing same energy (degeneracy) get destroyed and acquire different energies and a new complex is formed.

As discussed in Chapter 1 (Sec. 1.5.1, Fig. 1.9), the shapes of the five d orbitals are different and divided in to two sets depending on the nature of their orientation in space. The grouping of the five d orbitals of metals are as follows:

(a) The d_{xy}, d_{yz} and d_{xz} set of orbitals have their lobes lying between the co-ordinate axes and are termed as *non-axial orbitals* and designated as t_{2g} orbitals. The term t_{2g} implies a triply degenerate set.

(b) The second orbital set includes $d_{x^2-y^2}$ and d_{z^2} which have their lobes lying along the co-ordinate axes are designated as e_g orbitals. The term e_g orbitals refer to a doubly degenerate set.

In a free metal ion, all the five d orbitals are degenerate, but when ligands approach them, electron repulsion occurs causing the energy levels of d orbitals to rise. The degenerate metal orbitals are split into two groups of orbitals of differing energies under the influence of the approaching ligands. This phenomenon is called *crystal field splitting* and forms the basis of crystal field theory.

Crystal Field Splitting in an Octahedral Complex

As shown in Fig. 9.7, an octahedral complex with coordination number 6 has a central metal ion surrounded by six ligands at six corners of the octahedron.

The three axes x, y, and z pointing along the corners of the octahedron will have $d_{x^2-y^2}$ and d_{z^2} orbitals lie along these axes and point directly towards the ligands. These orbitals will experience greater electron repulsion than the remaining d orbitals, viz., d_{xy}, d_{yz}, d_{xz} that are directed in between any two coordinate axes. Hence, energy of

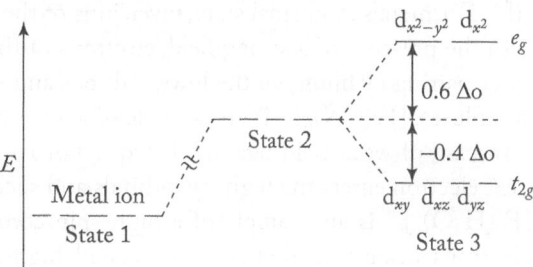

Fig. 9.7 Splitting of d orbitals in an octahedral complex

$d_{x^2-y^2}$ and d_{z^2} orbitals will be increased much more and acquires lesser stability. In the figure, state 1 depicts the degeneracy of all five d orbitals of the central metal ion. State 2 represents the hypothetical degeneracy of all the orbitals at higher energy levels, if all ligands approach the metal ion at an equal distance from each of the d orbitals. Finally, state 3 represents crystal field splitting and the energy difference between t_{2g} and e_g orbitals is denoted as Δ_o (subscript Δ_o denotes octahedral). The energy difference Δ_o is measured by a parameter called Dq and arbitrarily taken as 10 Dq. After splitting, e_g levels lie 6 Dq above and t_{2g} levels are 4 Dq below with respect to the unperturbed d orbitals.

Weak and Strong Field Splitting

The magnitude of crystal field splitting depends on three factors, namely the central metal ion, oxidation state of the metal ion, and the nature of ligands.

Greater the charge on the metal ion, closer will be the ligands to the central metal ion and hence greater will be the crystal field splitting. Therefore, on the basis of the extent of d orbital separation, ligands can be strong field or weak field ligands. Crystal field splitting ability of common ligands can be summarized in the following order of increasing field strength as:

$$I^- < Br^- < S^{2-} < SCN- < Cl^- < NO_3^- < F^- < OH^- < Ox^{2-} < H_2O < NCS^-$$
$$< CH_3CN < NH_3 < en < dipy < phen < NO_2^- < CN^- < CO$$

The above series is called the *spectrochemical series* which is based on the data obtained from spectra of coordination complexes.

The concept of formulating spectrochemical series is based on the fact that d orbital splitting and relative frequencies of visible absorption bands can be predicted for complexes having same metal ions, but different ligands. Say, if we have two cobalt complexes, namely $[CoF_6]^{3-}$, where cobalt has d^6 configuration. As per the series, it is evident that F^- is a weaker ligand than NH_3 and thus, the d-splitting ($10Dq$) will be smaller for $[CoF_6]^{3-}$ than $[Co(NH_3)_6]^{3+}$ as shown in Fig. 9.8.

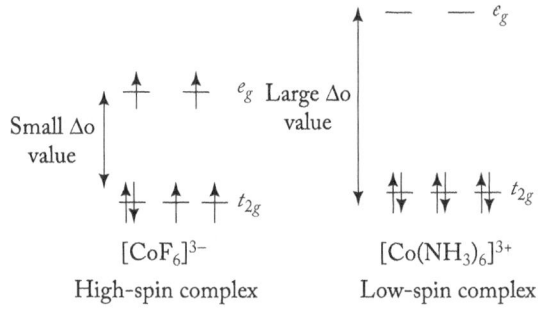

Fig. 9.8 Magnitude of splitting in $[CoF_6]^{3-}$ and $[Co(NH_3)_6]^{3+}$ complexes

However, this rule is not followed universally by all complexes or else life would be very easy for coordination chemistry indeed! Hence, certain issues need to be addressed while applying the spectrochemical series.

(a) The series is based on the spectral analysis of metal ions in common oxidation states, since the metal–ligand interactions of unusually high/low oxidation states of metal could be different from those found in metals in normal state.

(b) For metals in normal state, inversions of the order of the series are at times observed.

In the presence of a strong field, electrons of the ligands pair up, since $10Dq$ is large enough to allow force pairing by filling up the lower orbitals and hence such complexes will be diamagnetic or low-spin complexes. $[Fe(CN)_6]^{4-}$ is an example of a low-spin complex.

In case of weak field ligands, 10Dq is not large enough to allow force pairing of electrons and the next electron enters the higher e_g orbitals and such complexes are paramagnetic or high-spin complexes. $[Fe(H_2O)_6]^{2+}$ is an example of a high-spin complex. In short, complexes with maximum number of electrons in an unpaired state are called high-spin (paramagnetic) complexes, whereas those with minimum number of unpaired electrons are called low-spin (diamagnetic) complexes. Figure 9.9 shows the distribution of electrons in Fe^{3+} (d^5) complexes in different magnetic fields.

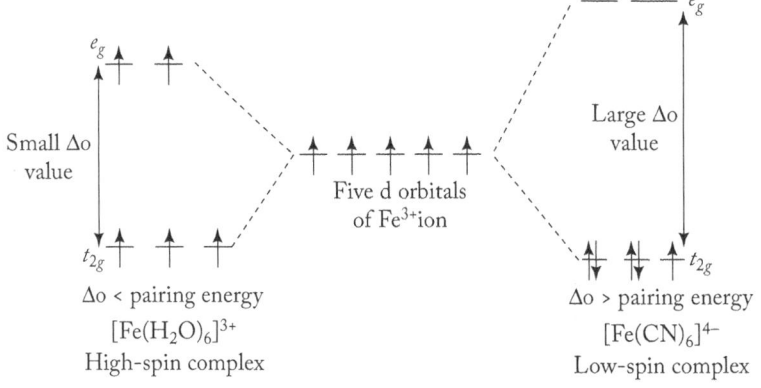

Fig. 9.9 Distribution of electrons in Fe^{3+} (d^5) complexes in different fields

High-spin and low-spin complexes are also known as *spin free* and *spin paired* complexes respectively. An increase in stability of a complex due to increased attraction of the metal ion and the ligand is termed as *crystal field stabilization energy* (CFSE). Just like most molecules, the electron distribution in d orbitals in octahedral complexes is based on the following rules.

(a) Electrons prefer occupying orbitals of the lowest energy.

(b) It obeys Hund's rule, that is, pairing of electrons will not occur unless all available orbitals in a given set have one electron each.

(c) If energy to promote an electron in e_g level is greater than the energy to pair electrons in the lower t_{2g} level, then electrons will preferentially first undergo pairing.

(d) If energy to promote an electron in e_g level is lower than the energy to pair electrons in the lower t_{2g} level, then electrons will preferentially occupy higher e_g orbitals.

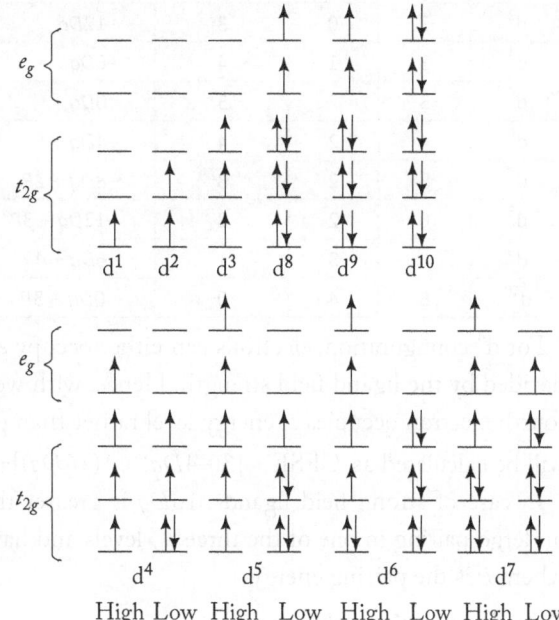

Fig. 9.10 Electron distribution of d^1 to d^{10} configurations in octahedral complexes

The magnitude of crystal field splitting is an important factor in predicting the magnetic properties of an octahedral complex. A lower value of 10Dq, that is, in configurations of d^4, d^5, d^6, d^7 species imparts paramagnetic behaviour. Figure 9.10 shows the electron distribution from d^1 to d^{10} octahedral complexes. A higher value of 10Dq results in diamagnetism. This is observed since these species have only two alternatives in their electron arrangement, which are:

(a) Δ_o has a greater value than the pairing energy and favours low-spin complexes.

(b) Δ_o is lower than the pairing energy and favours formation of high-spin complexes.

Calculation of CFSE

In an octahedral complex, it is observed that e_g level is raised by 6Dq and t_{2g} level is lowered by 4Dq. Hence, every t_{2g} electron will denote increasing stability corresponding to -4Dq and each electron in e_g level will denote decreasing stability corresponding to $+6$Dq. Now, if one considers that x number of electrons is present in both the energy levels, CFSE can be calculated by the formula,

$$CFSE = [x(-4Dq) + y(+6Dq)] \qquad \qquad ...(9.1)$$

Table 9.6 depicts the CFSE values of all configurations in octahedral complexes. As shown in Fig. 9.9, in octahedral complexes with d^1 configuration, electrons will occupy lower t_{2g} orbitals irrespective of the ligand field strength with CFSE of -4Dq. In d^2 configuration, two electrons will occupy the lower t_{2g} orbitals with CFSE of -12Dq. Hence, octahedral complexes with d^1, d^2, d^3 configurations will exhibit paramagnetism irrespective of ligand field strength.

Table 9.6 Electron distribution (d_1 to d_{10}) in octahedral complexes and their CFSE values

d^n	Weak field ligands				Strong field ligands			
	Distribution of electrons		No. of unpaired electrons	CFSE (P = pairing energy)	Distribution of electrons		No. of unpaired electrons	CFSE
	t_{2g}	e_g			t_{2g}	e_g		
d^1	1	0	1	$-4Dq$	1	0	1	$-4Dq$
d^2	2	0	2	$-8Dq$	2	0	2	$-8Dq$
d^3	3	0	3	$-12Dq$	3	0	3	$-12Dq$
d^4	3	1	4	$-6Dq$	4	0	2	$-16Dq + P$
d^5	3	2	5	$0Dq$	5	0	1	$-20Dq + 2P$
d^6	4	2	4	$-4Dq + P$	6	0	0	$-24Dq + 3P$
d^7	5	2	3	$-8Dq + 2P$	6	1	1	$-18Dq + 3P$
d^8	6	2	2	$-12Dq + 3P$	6	2	2	$-12Dq + 3P$
d^9	6	3	1	$-6Dq + 4P$	6	3	1	$-6Dq + 4P$
d^{10}	6	4	0	$0Dq + 5P$	6	4	0	$0Dq + 5P$

For d^4 configuration, electrons can either occupy e_g level or pair up in the lower t_{2g} level. This will be decided by the ligand field strength. Hence, with weak field ligands, where $10Dq$ < pairing energy, the fourth electron occupies e_g energy level rather than pairing up leading to destabilization and the CFSE will be calculated as, CFSE = $[3(-4Dq) + 1(+6Dq)] = -6Dq$.

In case of strong field ligands, $10Dq$ is greater than pairing energy, hence the fourth electron will undergo pairing in one of the three t_{2g} levels and have CFSE = $[4(-4Dq) + y(+6Dq)] + P = -16Dq + P$, where P is the pairing energy.

Tetrahedral Complexes

Figure 9.11 shows the crystal field splitting in a tetrahedral complex. There are four ligands bonded to the central metal ion in a tetrahedral complex and the coordination number is 4. In such complexes, there is a poor overlap between the metal and ligands, where ligand field interacts more with d_{xy}, d_{xz} and d_{yz} orbitals. Hence, the d orbitals are split in two different energy levels where the higher level will comprise d_{xy}, d_{xz} and d_{yz} orbitals and the lower energy level will comprise $d_{x^2-y^2}$ and d_{z^2} orbitals. Further, a tetrahedral complex does not have centre of symmetry, the term 'g' in t_{2g} and e_g does not hold true and simply written as t_2 and e, respectively.

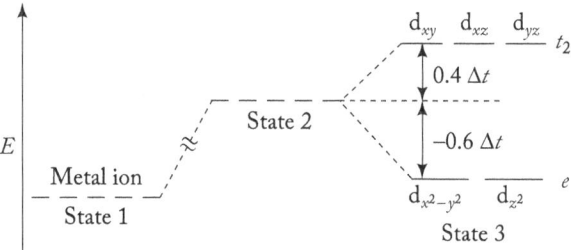

Fig. 9.11 Splitting of d orbitals in a tetrahedral complex

Consequently, the crystal field splitting (Δ_t) will be reverse to that in Δ_o and will be lower in value due to lesser number of ligands in a tetrahedral complex. Further, crystal field splitting will follow the relation $\Delta_t = 0.44 \Delta_o$. Tetrahedral complexes generally have more number of unpaired electrons and hence are high-spin complexes.

Square Planar Complexes

Just like in a tetrahedral complex, a square planar complex has four ligands bonded to the central metal ion. The only difference is that the ligand electrons are attracted to the xy plane and any orbital in the plane will have a higher energy level. The complexes with d^8 configuration usually attain a distorted octahedral complex or a square planar complex.

In d^8 complexes, electrons are completely paired and hence form low-spin complexes. In a square planar geometry, ligands are removed from z-axis and thus electrons in d_{z^2} orbital will become more stable than electrons in $d_{x^2-y^2}$ orbitals due to lowering of electrostatic repulsion. Further, there will be reduced electrostatic repulsion amongst electrons in d_{xz} and d_{yz} orbitals compared to those in d_{xy} orbitals. The crystal field splitting (Δ_{sp}) will be larger than Δ_o and can be related as $\Delta_{sp} = 1.74 \Delta_o$. The crystal field splitting of square planar complex is shown in Fig. 9.12.

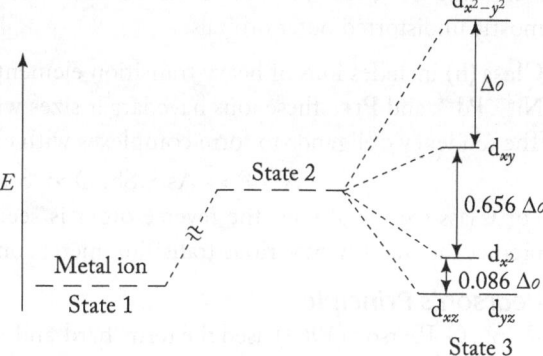

Fig. 9.12 Splitting of d orbitals in a square planar complex

Merits of crystal field theory

(a) It provides an explanation for most geometries of coordination complexes.
(b) It explains that complexes with coordination number 4 can be oriented as square-planar and tetrahedral geometries.
(c) One can easily explain magnetic properties by considering orbital contributions.
(d) CFT can successfully explain colour and spectral properties of transition metal complexes.

Limitations of crystal field theory

(a) Bonding strength and certain chemical properties cannot be explained, as CFT considers only the electrostatic interactions in complexes.
(b) The relative ligand strength does not fall in the scope of crystal field theory.

9.5 HARD AND SOFT ACIDS AND BASES (PEARSON ACID–BASE CONCEPT)

Certain ligands form quite stable complexes with metal ions such as Ag^+, Hg^{2+}, Pt^{2+} which have nearly-full d orbitals. Also, some ligands preferentially form stable complexes with metal ions with no d electrons such as Al^{3+}, Co^{3+}, and Ti^{4+}. On the basis of this preferential ligand bonding, S. Arhland, J. Chatt, and N. R. Davies (1958) classified metal ions and ligands into class (a) and class (b) types.

Check Your Progress

7. Write the various theories put forth for coordination compounds.
8. Write any two postulates of Werner's coordination theory.
9. $[Co(NH_3)_6]^{3+}$ is an inner orbital complex whereas $[CoF_6]^{3-}$ is an outer orbital complex. Justify.
10. $[Ni(CO)_4]$ is a tetrahedral complex, whereas $[Ni(CN)_4]^{2-}$ is a square planar complex. Why?
11. State any 2 limitations of VBT.
12. What is crystal field splitting?
13. What is 10 Dq?
14. What is spectrochemical series?
15. Justify the statement '$[Fe(CN)_6]^{4-}$ is a low-spin complex, whereas $[Fe(H_2O)_6]^{2+}$ is a high-spin complex.
16. Define CFSE. Calculate CFSE for d^4, d^5, d^6 and d^7 configurations for weak and strong field ligands in octahedral complexes.
17. State the merits and limitations of CFT.

Class (a) includes ions of alkali and alkaline earth metals, light transition elements with higher oxidation states like Cr^{3+}, Co^{3+}, Fe^{3+}, and Ti^{4+}. Due to their small size, they have greater polarizing power and mostly undistorted outer orbitals.

Class (b) includes ions of heavy transition elements with lower oxidation states such as Ag^+, Cu^+, Hg^{2+}, Ni^{2+}, Pd^{2+}, and Pt+; these ions have larger sizes with distorted outer orbitals.

The tendency of ligands to form complexes with class (a) metal ions is in the following order:

$$N > P \gg As > Sb; O \gg S \ggg Se > Te; F^- > Cl^- > Br^- > I^-$$

For Class (b) metal ions, the reverse order is seen. Such classification helps coordination chemists to predict the stability of various transition metal complexes.

Pearson's Principle

Ralph. G. Pearson (1963) used the term 'hard' and 'soft' to describe class (a) and (b) metal ions. Hard acids are those included as Class (a) metal ions and soft acids include those in Class (b). Hard bases could be ligands such as fluoride or ammonia, whereas soft bases may include ligands such as phosphine or iodide ion. A distinction between hard and soft acids and bases is directly linked to their polarizing power.

According to Pearson, hard acids and bases are small, low polarizable species, whereas soft acids and bases are larger species with greater polarizability. Hence the HSAB principle can be stated as, 'Hard acids prefer to bind to hard bases and soft acids prefer to bind to soft bases. Table 9.7 lists the metal ions and ligands classified as per Pearson's principle. Let us take an example of a simple equilibrium reaction, where B may be a hard or soft acid based on equilibrium behaviour.

$$BH^+ + CH_3Hg^+ \rightleftharpoons CH_3\,HgB^+ + H^+$$

In this reaction, one can sense the competition between hard acid H^+, and soft acid CH_3Hg^+, (methyl mercury ion). A hard base will allow the reaction to move towards the left and a soft base will allow the reaction to proceed towards the right. Methyl mercury ion is a typical soft acid with monovalency just like a proton. Hence, there is no stark distinction between hard and soft acids. Such ions are categorized as borderline acids and bases.

Table 9.7 Pearson's classification of acids and bases

Hard acids	Borderline acids	Soft acids
H^+, Li^+, Na^+, K^+	Fe^{2+}, Co^{2+}, Ni^{2+}, Cu^{2+}	Pt^{2+}, Pd^{2+}, Cd^{2+}, Cu^+
Be^{2+}, Mg^{2+}, Ca^{2+}, Sr^{2+}	Zn^{2+}, Pb^{2+}, Sn^{2+}, Sb^{2+}	Ag^+, Au^+, Hg_2^{2+}, $B(CH_3)_3$
Al^{3+}, BF_3, $Al(CH_3)_3$, $AlCl_3$	Bi^{3+}, Rh^{3+}, NO^+, SO_2	B_2H_6, $GaCl_3$, $[Fe(CO)_5]$, $[Co(CN)_5]^{3-}$
Sc^{3+}, Ti^{4+}, Zr^{4+}, Cr^{3+}		
Ce^{3+}, Lu^{3+}, CO_2, SO_3		

Hard bases	Borderline bases	Soft bases
NH_3, RNH_2, N_2H_4, H_2O, *ROH	Br^-, NO_2^-, SO_3^{2-}	H^-, CN^-, SCN^-, $S_2O_3^{2-}$
R_2O, OH^-, NO_3^-, ClO_4^-, CO_3^{2-}	C_6H_5N, $C_6H_5NH_2$	Br^-, Br^-, R_2S, CO
SO_4^{2-}, PO_4^{3-}, CH_3COO^-, F^-, Cl^-		C_2H_4, R_3P, $P(OR)_3$, S^{2-}

9.6 Magnetism and Colours in Coordination Compounds

The magnetic properties of coordination compounds are usually determined as a measure of their magnetic susceptibility. Coordination complexes possess unpaired electrons in their d orbitals and hence they are considered paramagnetic and are known to get attracted in the presence of a magnetic field due to free electron spin. This measure of magnetism is called *magnetic susceptibility*.

(a) Metal ions such as Ti^{3+} (d^1), V^{3+} (d^2), and Cr^{3+} (d^3), have three electrons in their d orbitals and two empty d orbitals that participate in octahedral hybridization with 4s and 4 p orbitals. The magnetic behaviour of these ions and their coordination species are quite similar. If more than 3 d electrons are present some not available for octahedral hybridization (due to Hund's rule, refer to Chapter 1). Hence, for configurations d^4 (Cr^{2+}, Mn^{3+}), d^5 (Mn^{2+}, Fe^{3+}), d^6 (Fe^{2+}, Co^{3+}) an empty pair of d orbitals results only by pairing of 3d electrons that leaves two, one, and zero unpaired electrons, respectively.

(b) For d^6 configurations, VB theory explains anomalies in magnetic moments satisfactorily by applying the case of inner and outer orbital coordination complexes. $[Mn(CN)_6]^{3-}$ and $[MnCl_6]^{3-}$ are paramagnetic but with two and four unpaired electrons respectively. Similarly, $[Fe(CN)_6]^{3-}$ and $[FeF_6]^{3-}$ are paramagnetic but with one and five unpaired electrons respectively.

(c) In cobalt complexes, we come across an interesting anomaly; $[Co(C_2O_4)_3]^{3-}$ and $[CoF_6]^{3-}$ are diamagnetic and paramagnetic respectively. VB theory states that $[Mn(CN)_6]^{3-}$, $[Fe(CN)_6]^{3-}$, and $[Co(C_2O_4)_3]^{3-}$ are inner orbital complexes that undergo d^2sp^3 hybridization, the former two complexes are paramagnetic and the latter is found to be diamagnetic. Further, $[MnCl_6]^{3-}$, $[FeF_6]^{3-}$, and $[CoF_6]^{3-}$ are outer orbital complexes that undergo sp^3d^2 hybridization and are paramagnetic, corresponding to four, five, and four unpaired electrons.

Coordination compounds are known to exhibit fascinating colours, which often is a fundamental premise for identification of various compounds. Crystal field theory successfully explained colours of transition metal complexes. Hexaaquatitanium(III) complex, $[Ti(H_2O)_6]^{3+}$ is a violet coloured octahedral complex in which a single electron in its 3d orbital is present in t_{2g} level (ground state).

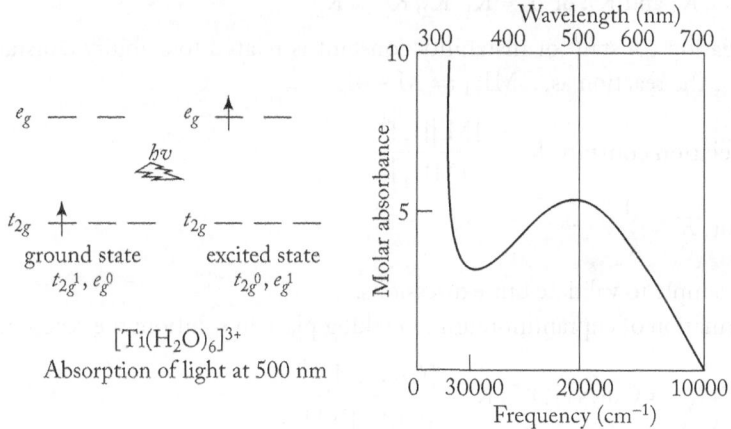

Fig. 9.13 (a) Transition of an electron (b) Absorption spectrum of $[Ti(H_2O)_6]^{3+}$

When light corresponding to the energy of the blue–green region is absorbed by titanium complex ($t_{2g}^1 e_g^0$), the sole electron quickly jumps to the higher e_g level, that is, ($t_{2g}^0 e_g^1$) and hence the colour of the compound is violet. CFT attributes the colour of the coordination compounds to d–d transition of an electron. In the absence of ligands, crystal field splitting will not occur and thus the substance will be colourless. If you heat $[Ti(H_2O)_6]Cl_3$ complex, it turns colourless. Anhydrous copper sulphate is white in colour whereas its hydrated form ($CuSO_4.5H_2O$) has a characteristic blue colour. Hence, there is a direct influence of ligand on the colour exhibited by coordination compounds.

9.7 STABILITY OF COORDINATION COMPOUNDS

In a coordination complex, stability refers to the degree of association between the two species involved at equilibrium state. The magnitude of (formation) equilibrium constant for the association, quantitatively expresses the stability and is termed as stability constant. Generally, there are two types of stability studies, viz., thermodynamic and kinetic. *Thermodynamic stability* refers to the strength of metal–ligand linkages. It indicates the measure of the extent to which the coordination complex will form or transform in to a different species under a set of given conditions at equilibrium. *Kinetic stability*, on the other hand, refers to the speed at which the formation or dissociation of a complex will occur, leading towards equilibrium.

Thus, if we have a reaction of the type: $M + nL = MLn$

where M is the metal and nL is the number of ligands bonded to it. Hence, quantitatively, greater the degree of association, greater is the stability of the resulting complex. We can express stabilities using equilibrium constant (K) for the above reaction. Equilibrium constant is also termed as *stability constant* (or even called *formation constant*). Let us write some general reactions and their equilibrium constants as:

$$M + L \rightleftharpoons ML; K_1 = \frac{[ML]}{[M][L]} \qquad\qquad ML_2 + L \rightleftharpoons ML_3; K_3 = \frac{[ML_3]}{[ML_2][L]}$$

$$ML + L \rightleftharpoons ML_2; K_2 = \frac{[ML_2]}{[ML][L]} \qquad\qquad ML_3 + L \rightleftharpoons ML_4; K_4 = \frac{[ML_4]}{[ML_3][L]}$$

$K_1, K_2, K_3,$ and K_4 are stepwise stability constants. Further, we can express the overall stability constant as:

$$M + 4L \rightleftharpoons ML_4; \beta_4 = \frac{[ML_4]}{[M][L]^4}$$

Thus, one can conclude that stepwise and overall stability constants are related as:

$\beta_4 = K_1, K_2, K_3$ and K_4; or $\beta_n = K_1, K_2, K_3 \dots K_n$

Similarly, dissociation constant or instability constant is related to stability constant as its reciprocal and is expressed for the reaction as,, $ML_4 \rightleftharpoons M + 4L$.

$$\text{So, dissociation constant,} K_a = \frac{[M][L]^4}{[ML_4]}$$

Hence, $K_a = \dfrac{1}{K}$ or, $K = \dfrac{1}{K_a}$

Let us take an example to validate our expressions.

Consider the formation of cuprammonium ion taking place in solution; the reaction will be as follows.

$$Cu^{2+} + NH_3 \rightleftharpoons Cu(NH_3)^{2+}; K_1 = \frac{[Cu(NH_3)^{2+}]}{[Cu^{2+}][NH_3]}$$

$$Cu(NH_3)^{2+} + NH_3 \rightleftharpoons Cu(NH_3)_2^{2+}; K_1 = \frac{[Cu(NH_3)^{2+}]}{[Cu(NH_3)^{2+}][NH_3]} \text{ and so on}$$

$K_1, K_2 \dots K_n$ are stepwise stability constants and overall stability constant.

$$\beta_4 = \frac{[Cu(NH_3)^{2+}]}{[Cu^{2+}][NH_3]^4}$$

The addition of the four amine groups to copper shows a pattern found for most formation constants, in that the successive stability constants decrease. In this case, the four constants are:

$\log K_1 = 4.0, \log K_2 = 3.2, \log K_3 = 2.7, \log K_4 = 2.0$ or $\log \beta_4 = 11.9$

An interesting relationship exists between free energy change, overall stability constant, entropy, and enthalpy.

We know, standard free energy $\Delta G° = 2.303\ RT \log \beta$

$$\Delta G° = \Delta H° - T\Delta S°$$

Thus, $2.303R \log \beta = \left(\Delta S° - \dfrac{\Delta H°}{T} \right)$

From the above expression, it is clear that formation of the complex is favoured by negative enthalpy changes but with positive entropy changes too. Thus, greater the evolution of heat in a reaction, higher will be the stability of reaction products. If there is greater entropy (disorder) in the formation of reaction product relative to the starting reactants, higher will be the stability of the complex. The size and charge on the metal atom/ion, crystal field effects, electropositivity of metal, nature of ligands, chelate effects, steric factors are various factors that influence the stability of complexes. It is beyond the scope of this book to explain all these factors.

9.8 CHELATES

Chelates (pronounced as 'key-late') are a special type of coordination compounds that contain one or more rings in their structure. The cyclic ring structures are formed due to linking of a metal ion with polydentate ligand. These cyclic ring structures are called chelates (derived from Greek word *chēlē* means 'claw'). The ligand that forms linkages with a metal ion is called the *chelating agent* and the process is referred to as *chelation*. The presence of cyclic rings in the structure provides stability to the coordination complexes.

Ethylenediamine is a bidentate ligand that can form coordination complex with copper ions as follows:

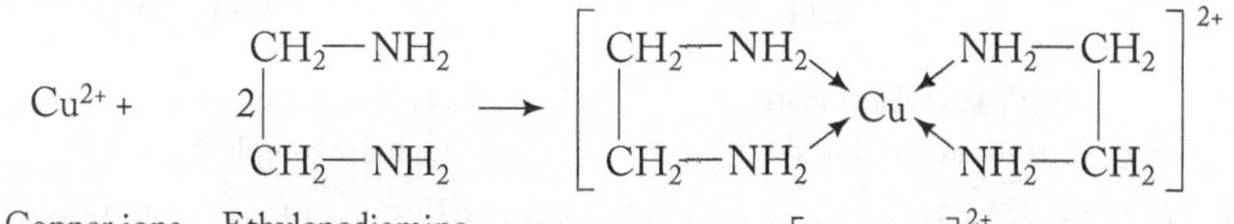

Copper ions   Ethylenediamine

$\left[Cu\,(en)_2 \right]^{2+}$

Copper (II) ethylenediamine

The complex copper(II)ethylenediamine contains a pair of five membered rings which is called a chelate. Another interesting example is the formation of nickel–DMG complex due to the reaction of Ni^{2+} with dimethyl glyoxime (DMG) in ammoniacal solution.

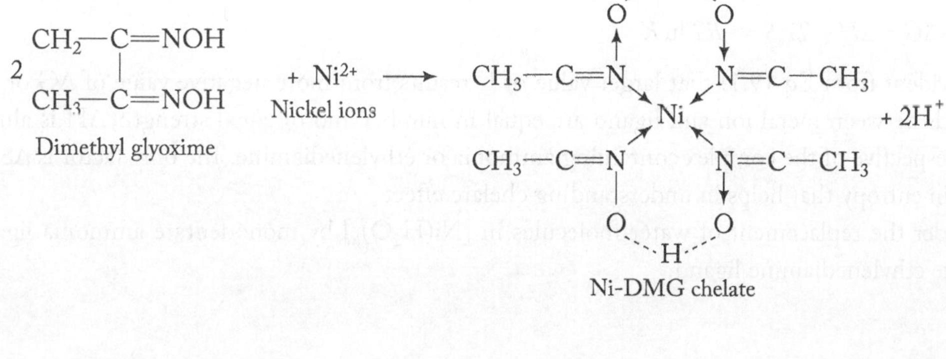

Dimethyl glyoxime

Ni-DMG chelate

Types of Chelates

Chelates are classified into bidentate, tridentate, quadridentate, and hexadentate based on the type of chelating ligand. Depending on the coordination number and charge on the complex, chelates are classified in to the below two types.

Inner metallic complexes of first order   These complexes are formed by coordinating with ligands that satisfy both the charge (primary valency) and the coordination number of the metal ion (secondary valency) simultaneously. Such ligands participate with both neutral and acidic groups that coordinate with the metal ion. The inner metallic complexes are non-electrolytes and are also called chelates of the first order; examples are copper–glycine complex and beryllium acetylacetonate:

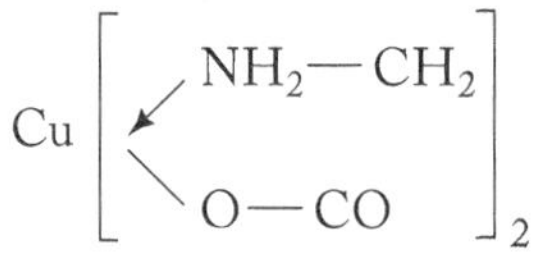

Copper–glycine complex 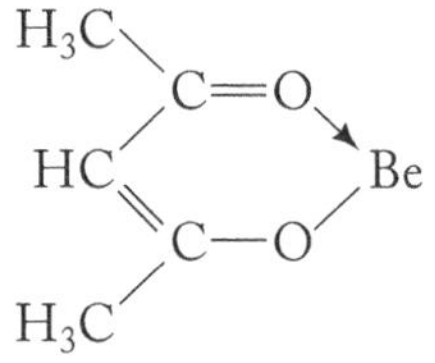   Beryllium acetylacetonate

Inner metallic complexes of second order   If the number of acido groups in an inner metallic complex is less or more than the charge on the central metal atom, the complex will be ionic in nature. Such complexes are called inner metallic complexes of second order. Acetylacetonates of Si (IV) and Co(II) are examples of inner metallic complexes of second order.

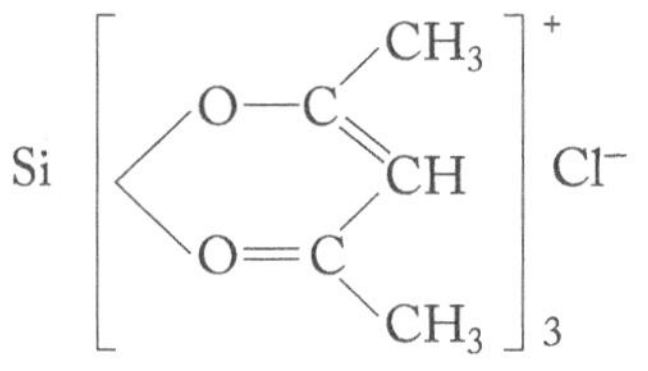

Si(IV)acetylacetonate   Co(II)acetylacetonate
(cationic complex)   (anionic complex)

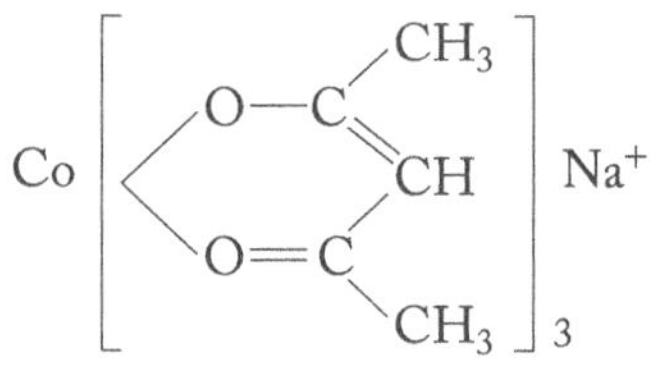

Chelate effect   It is interesting to note that the complexes formed with chelating agents such as ethylenediamine are more stable than the analogous complexes with monodentate ligands. $[Zn(en)_2]^{2+}$ complex is more stable than $[Zn(NH_3)_4]^{2+}$ though the basicity of ethylenediamine and ammonia are similar. Since ethylenediamine has a tendency to form chelate rings with the metal ion, they exhibit increased stability compared to ammonia-based metal complexes. This is called the *chelate effect*. In both ligands, nitrogen atoms donate the electron pairs; hence the stability differences are not related to the bond strength between the metal ion and nitrogen atoms. Consider the equilibrium constant and free energy of a reaction, related by the equation

$$\Delta G = \Delta H - T\Delta S = -RT\ln K \tag{9.1}$$

It is evident from Eq. (9.1) that larger value of K results from more negative value of ΔG or ΔH. As the bonds between metal ion and ligand are equal in number and of equal strength, ΔH is almost the same irrespective of the complex containing ammonia or ethylenediamine. The only factor is ΔS, that is, change in entropy that helps in understanding chelate effect.

Consider the replacement of water molecules in $[Ni(H_2O)_6]$ by monodentate ammonia ligand and bidentate ethylenediamine ligand.

When four ammonia molecules enter the coordination sphere of the central metal ion, four water molecules depart from the coordination sphere. Thus, there are equal numbers of free molecules before and after the complex formation. In this case, change in entropy is about zero. However, when two ethylenediamine molecules enter the coordination sphere of metal ion, four water molecules depart from the coordination sphere. Hence, there is an increase in the number of free molecules and the overall disorder of the system and entropy increases. A positive value of ΔS will result in more negative value of ΔG and consequently larger value of K.

$$[Ni(H_2O)_6] + 2NH_3 \rightarrow [Ni(NH_3)_2 (H_2O)_4] + 2H_2O$$
$$[Ni(H_2O)_6] + (en) \rightarrow [Ni(en)(H_2O)_4] + 2H_2O$$

From the above reactions, it is seen that replacing the first water molecule by ammonia or ethylenediamine is equally probable. However, replacing the second water molecule by the second amine group of ethylenediamine is more probable than ammonia molecule, since one of the amine groups is already attached to the central metal ion, and thus the other end of amine group is closer to water molecule than ammonia. Hence, $[Ni(en)(H_2O)_4]^{2+}$ complex is more probable to form than the less stable $[Ni(NH_3)_2 (H_2O)_4]^{2+}$ complex.

Apart from the presence of chelate rings, ring size also contributes to the stability of complexes. It is observed that chelate rings with five or six membered rings are more stable than those of other sizes. Consider a set of ligands with the general formula $H_2N(CH_2)_n NH_2$, where $n = 2, 3,$ or 4 that form complexes with same type of central metal ion. The most stable complex will be possible with ethylenediamine ($n = 2$) resulting in a five-membered chelate ring. If $n = 3$, the ligand 1,3-diaminopropane will form a six-membered chelate ring which will be less stable than ethylenediamine. A plausible explanation is that there is less strain in a five-membered chelate than a chelate ring of any other size.

9.9 Applications of Coordination Compounds

Besides their use in chemical laboratories and industries, coordination compounds have numerous applications in other fields.

(a) Quantitative analysis: The brilliant colour reactions exhibited by metal ions on bonding with various ligands form a fundamental premise for quick detection and estimation using classical and instrumental methods. Ethylenediaminetetraacetic acid (EDTA), Prussian blue, Cupron, dimethylglyoxime (DMG), and α–nitroso–β–naphthol, are some commonly employed reagents for determining the presence of various transition metals in a test solution.

(b) Hardness of water: This is determined using EDTA reagent on the principle of complex formation between the reagent and metal ions in water.

(c) Catalysis: Zeigler–Natta catalyst ($TiCl_4$ and trialkyl aluminium) is employed for preparing polyethene. Wilkinson catalyst $[RhCl(PPh_3)_3]$ is employed in the hydrogenation of alkenes. Monsanto acetic acid process utilizes $[RhCl(CO)(PPh_3)_3]$ or $[RhCl(CO)_2]_2$ as a catalyst in the presence of methyl iodide and iodine.

(d) Photography: When photographic film is developed, it is fixed using hypo solution that dissolves all undecomposed silver bromide and forms a complex ion $[Ag(S_2O_3)_2]^{3-}$.

(e) Chelate therapy: Toxic metals can be eliminated from the body by complexing them with various ligands. Say, copper and iron can form complexes with polydentate ligands such as D-penicillamine and desferrioxamine B respectively. Lead poisoning can be treated using ethylenediaminetetraacetic acid. *cis*-Platin and related platinum complexes are known to effectively inhibit tumour cells.

9.10 ORGANOMETALLIC COMPOUNDS

Coming under the branch of inorganic chemistry, organometallic chemistry deals with chemical compounds having at least one metal–carbon (M–C) bond, where C is the organic part and M could be any transition, inner-transition, or main group metal atom. Bio-inorganic chemistry deals with compounds of physiological importance such as the chlorophyll, haemoglobin, myoglobin, and so on.

Organolithium, organomagnesium, dimethylzinc, and metal carbonyls are commonly employed organometallic compounds in a variety of chemical reactions. Chauvin, Grubbs, and Schrock (2005) won Nobel Prize for their work on metal-catalysed alkene metathesis. In 2010 Richard Heck, Negishi and A. Suzuki won Nobel Prize for their work on palladium-catalysed cross coupling in organic synthesis. In this chapter we will focus our study on the application of organometallic compounds in isomerization, polymerization, hydroformylation, and hydrogenation reactions.

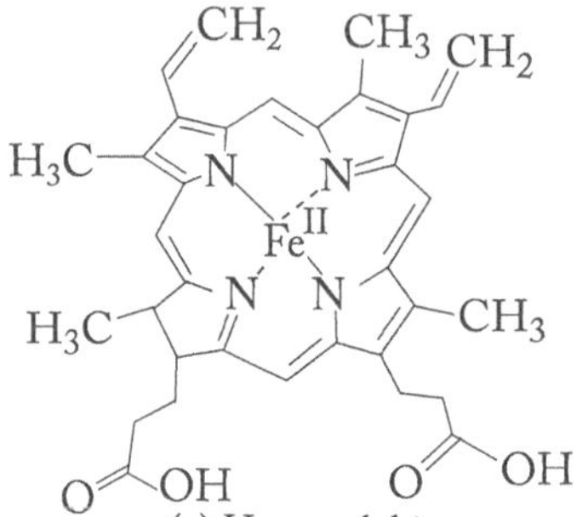

Fig. 9.14 Structure of organometallic compounds

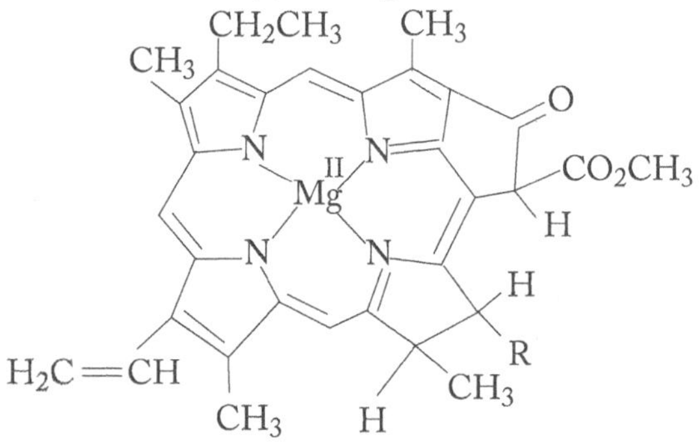

9.10.1 Classification of Organometallic Compounds

Organometallic compounds are broadly classified into the following three types: ionic organometallics, covalent organometallics, and metal carbonyls.

Ionic organometallic compounds These are usually alkali metal salts (except that of lithium). These are highly reactive and kinetically unstable with shorter lifetimes.

Covalent organometallic compounds These are characterized by the type of bonding between the metal and the organic portion. Grignard reagent (R-Mg-X, where R is the alkyl group and X is a halide), diethyl zinc ($(C_2H_5)_2Zn$) are examples of σ-bonded organometallic compounds. Zeise's salt ($K[Pt(C_2H_4)Cl_3]$), ferrocene, dibenzene are examples of π-bonded organometallic compounds.

Metal carbonyls These are an important class of organometallic compounds, in which carbon monoxide acts as the ligand. Such compounds possess both σ and π characters and follow the 18-electron rule. $Ni(CO)_4$ (nickel carbonyl) and $Fe(CO)_5$ (iron carbonyl) are examples of metal carbonyls.

Check Your Progress

18. State HSAB principle.
19. Calculate the overall dissociation equilibrium constant for $[Cu(NH_3)_4]^{2+}$ ion, if β_4 for this complex is 2.8×10^{13}.
20. What are chelates? Give examples.
21. What is chelate effect? Are chelates stable?
22. What are thermodynamic and kinetic stability of complexes?
23. Classify chelates on the basis of coordination number and charge on the complex; give examples.
24. State EAN rule.
25. Name any two compounds that disobey EAN rule.

9.10.2 18-Electron Rule

The stability of organometallic complexes, especially those involving the first row transition metals, can be predicted using the 18-electron rule. When a metal achieves an outer electronic configuration of ns^2 $(n-1)d^{10}np^6$, there will be 18 electrons in the valence orbitals resulting in a stable configuration. However, it should be noted that the second and third row transition metals tend to exhibit 16-electron complexes; thus the 18-electron rule may not be a precedent. According to the 18-electron rule, it is predicted that valence shells of transition metals consist of nine valence orbitals, that is, one s orbital, three p orbitals, and five d orbitals, which can accommodate 18 electrons as either bonding or non-bonding electron pairs. If a metal complex has 18 valence electrons, it has achieved the same electron configuration as the noble gas in the period and thus can be considered as a stable complex. It is important to note that there are exceptions to the rule.

To determine whether a complex follows an 18-electron rule, the steps are:

(a) Determine the oxidation state of the central transition metal of the coordination complex.

(b) Add up the electrons of the central metal and the ligands and predict the stability of the complex.

Let us now apply the rule to a few examples by electron counting:

$Cr(CO)_6$ Here, chromium has six outer shell electrons and 6 coordinating CO ligands which will contribute 12 electrons. Thus, chromium carbonyl follows the 18-electron rule.

$[TiF_6]^{2-}$ The oxidation state of titanium is +4 (4 electrons) and there are 6 coordinating fluorine ligands (12 electrons), thus the complex has 16 electrons and does not obey the 18-electron rule.

$Re(R_3P)_2CH_3(C_2H_2)(CO)_2$ Here, R_3P is phosphine group and R is alkyl group. There is no charge on the complex and one anionic ligand, that is, methyl group is present. Due to this, Re will have + 1 charge. Thus, the contributing electrons are as follows:

Central metal = 6; $2R_3P$ groups = 4, 2 CO ligands = 4, anionic ligands (CH_3) = 2; ethylene = 2

Hence there are a total of 18 electrons. Thus, the complex follows the 18-electron rule.

$Fe(CO)_2(NO)_2$ Here the contributing electrons are Fe = 8; 2 CO = 4, and 2 NO = 6.

As the number of electrons adds up to 18, it follows the 18-electron rule.

For most coordination compounds, electron count should add up to 18 electrons. However, there are many exceptions to the 18-electron rule.

The naming of organometallic compounds is similar to the naming of coordination compounds, but certain ligands are known to exhibit multiple modes of bonding, referred to as hapticity.

Hapticity refers to coordination of a ligand to the central metal through an uninterrupted series of atoms denoted as η (pronounced eta). Eta-x was developed to indicate the number of carbon atoms of a π system coordinated to the metal centre. For instance, η^5 – cyclopentadienyl ligand represents that all five carbon atoms of the cyclopentadiene ring are bonded to the transition metal centre.

9.10.3 Effective Atomic Number (EAN Rule)

N. Sidgwick (1927) extended the Lewis concept of electron pair bond formation to explain bonding in coordination compounds. According to Sidgwick, a metal ion accepts electron pairs from the ligands until it achieves the next noble gas configuration. Hence, ligands are called electron donors (Lewis base) and metal ions are electron acceptors (Lewis acid). Hence, the bond between donor and acceptor is called coordinate or dative bond. The rule came to be known as *effective atomic number rule*. Sidgwick put forth the EAN rule which states that the effective atomic number of a central metal ion is the total number of electrons around them and the electrons gained by central metal ion from coordination with ligands. If the EAN is equal to the atomic number of the nearest higher inert gas, the coordination compound satisfies the EAN rule. The general formula for calculating EAN is:

EAN = At. no. of central metal ion − No. of electrons lost during metal formation + Total no. of electrons gained from ligands

or EAN = Z − Oxidation no. + 2 × No. of ligands

Limitations of Sidgwick theory Many complexes are stable but do not follow EAN rule. Hence, mere obeying of EAN rule does not hold true in all cases.

Sidgwick theory does not predict the geometry of the complexes. It cannot predict the magnetic behaviour of coordination compounds. Table 9.8 shows EAN values calculated for various complexes with some exceptions to the rule.

Table 9.8 EAN values of various complexes

Complex	Central metal	Atomic no. of metal	Oxidation no. of metal	No. of electrons on central metal	No. of electrons donated to metal ion	EAN
$[K_4Fe(CN)_6]$	Fe	26	+2	24	6 x 2 = 12	24 + 12 = 36 (Kr)
$[Ni(CO)_4]$	Ni	28	0	28	4 x 2 = 8	28 + 8 = 36 (Kr)
$[Co(NH_3)_6]^{3+}$	Co	27	+3	24	6 x 2 = 12	24 + 12 = 36 (Kr)
$[Fe(CN)_6]^{4-}$	Fe	26	+2	24	6 x 2 = 12	24 + 12 = 36 (Kr)
$[Fe(CO)_5]$	Fe	26	0	26	5 x 2 = 10	26 + 10 = 36 (Kr)
$[Ag(NH_3)_4]^+$	Ag	47	+1	46	4 x 2 = 8	46 + 8 = 54 (Xe)
Exceptions to the EAN rule						
$[NiF_6]^{4-}$	Ni	28	+2	26	6 x 2 = 12	26 + 12 = 38
$[V(CO)_6]$	V	23	0	23	6 x 2 = 12	23 + 12 = 35
$[Mn(CO)_4]$	Mn	25	0	25	4 x 2 = 8	25 + 8 = 33
$[Ni(NH_3)_6]^{2+}$	Ni	28	+ 2	26	6 x 2 = 12	26 + 12 = 38
$[Co(CN)_6]^{4-}$	Co	27	−2	25	6 x 2 = 12	25 + 12 = 37
$[PdCl_4]^{2-}$	Pd	46	−2	44	4 x 2 = 8	44 + 8 = 52

9.11 REACTIONS OF ALKENES (ORGANOMETALLIC COMPOUNDS)

There are many examples where coordination compounds are employed as industrial catalysts. Catalysts are substances that alter the rate of the reaction without themselves undergoing any change. During the progress of the chemical reaction, metal catalyst in the complex may undergo changes in coordination number, bonding, geometry, etc.

The following reactions of alkenes using metal complexes are homogeneous in nature (same phase of reactants and catalyst) and carried out in near-ambient conditions in aqueous solutions, as opposed to severe conditions when the reactions are carried out in the gas phase.

9.11.1 Hydrogenation of Alkenes

During hydrogenation ethene (ethylene) reacts with hydrogen forming ethane; this is an exothermic reaction that is thermodynamically favoured, but does not occur under ambient temperature or pressure conditions.

$$H_2C = CH_2 + H_2 \xrightarrow{\text{Pd/C}} H_3C - CH_3$$

Ethene   Ethane

The use of transition metal catalysts such as Ni, Pt, Cu, or Pd allows hydrogenation to occur rapidly.

Wilkinson's catalyst is obtained by refluxing rhodium (III) chloride hydrate with an excess of triphenylphosphine in ethanol. Triphenylphosphine behaves as a two-electron reducing agent that oxidizes itself; three equivalents of triphenylphosphine act as ligands in the product, while the fourth reduces rhodium (III) to rhodium (I) species.

$$RhCl_3(H_2O)_3 + 4\,PPh_3 \rightarrow RhCl(PPh_3)_3 + O = PPh_3 + 2HCl + 2H_2O$$

Figure 9.15 depicts the mechanism of hydrogenation using Wilkinson's catalyst.

(i) The first step involves the dissociation of one triphenyl phosphine ligand giving a 14-electron complex.

(ii) This is followed by an oxidative addition of hydrogen to the rhodium metal resulting in an octahedral coordination complex. Usually, hydrogen atoms occupy the *cis* position in the octahedron complex.

(iii) Propene enters and forms a π complex with rhodium metal.

(iv) As hydrogen transfers (migratory insertion) to the alkene, triphenyl phosphine enters the coordination sphere of rhodium metal. Further, another hydrogen transfer takes place along with reductive elimination of alkane, resulting in the regeneration of the catalyst.

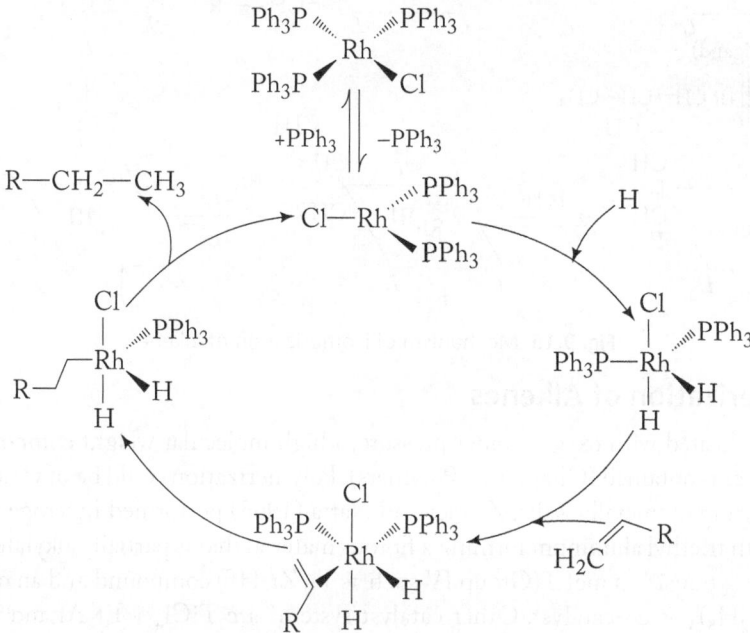

Fig. 9.15 Mechanism of hydrogenation of alkene

Sir Geoffrey Wilkinson who pioneered the work of hydrogenation of alkenes is also known for his work on the structure of ferrocene for which he received Nobel Prize in 1973.

9.11.2 Isomerization of Alkenes

An industrially important reaction, isomerization of alkenes is used to prepare specific isomers that are employed as monomers in polymerization. One step in the isomerization process involves a change in the bonding mode of an alkene; for example, isomerization of 1-alkene to produce 2-alkene occurs as the alkene changes from η^2 to η^3 in the transition state. The general isomerization reaction can be depicted as:

$$R-CH_2-CH=CH_2 \rightleftharpoons \underset{RCH}{\overset{H_2C}{HC}}-M \rightleftharpoons R-CH=CH-CH_3$$

$$\underset{M}{\qquad} \qquad\qquad \underset{M}{\qquad}$$

(M = metal catalyst)

The isomerization process begins with a substitution reaction where an alkene replaces a ligand (L) followed by addition of H^+ and a ligand forming a six-bonded coordination complex. A hydrogen transfer changes the alkene from η^2 to η^1, which is reversed as the alkyl group is converted into a 2-ene as a ligand enters the coordination sphere of rhodium.

Further, reductive elimination takes place by loss of one ligand and H^+, forming a square planar complex in which the alkene is present as the 2-ene. Finally, elimination of $RCH=CH_2CH_3$ occurs as an incoming ligand enters the coordination sphere of rhodium and liberates the product and reforms the catalyst. The changes in the oxidation state of rhodium metal are shown in Fig. 9.16.

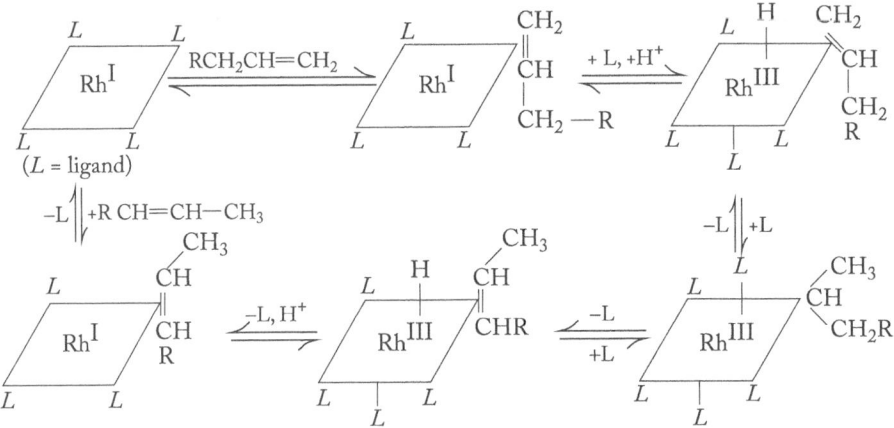

Fig. 9.16 Mechanism of isomerization of alkene

9.11.3 Polymerization of Alkenes

When ethylene is heated with oxygen under pressure, a high molecular weight compound, essentially a long-chain polymer is obtained (Chapter 16 Polymers). Polymerization could be of various types and can be catalysed using organometallic salts. Zeigler and Natta (1963) performed heterogeneous catalysis by treating $TiCl_4$ with triethyl aluminum forming a fibrous material that is partially alkylated. Zeigler–Natta catalyst comprises a transition metal (Group IV such as Ti, Zr, Hf) compound and an organoaluminium compound, $Al(C_2H_5)_3$ as co-catalyst. Other catalyst systems are $TiCl_4 + Et_3Al$ and $TiCl_3 + AlEt_2Cl$ (where Et is ethyl group).

The crystal structure of titanium chloride can be visualized as each Ti atom coordinated to six chlorine atoms. On the crystal surface, Ti atom is surrounded by five chlorine atoms with one vacant orbital. When an organoaluminium compound (co-catalyst) approaches, it donates its ethyl group to the titanium atom and the Al atom is coordinated to one of the chlorine atoms. Further, one chlorine atom from titanium is eliminated. Thus, the catalyst system still has an empty orbital and is activated by the coordination of the co-catalyst to the titanium atom (Fig. 9.17).

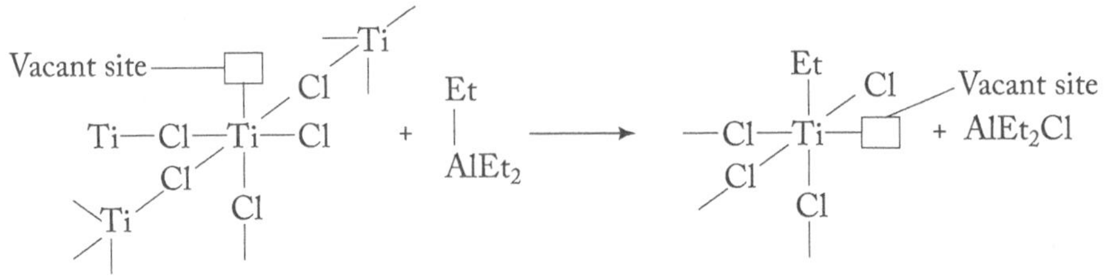

Fig. 9.17 Activation of catalyst system by coordination of $AlEt_3$ to Ti atom

Initiation When propylene approaches the active metal centre of the catalyst, it gets coordinated to titanium atom due to overlap of their orbitals. Alkene–metal complex so formed goes through electron shuffling. The pi electrons from carbon–carbon bond shifts, thereby forming a Ti–C bond, whereas the electron pair from the bond between Ti atom and ethyl group of co-catalyst shifts forming a bond between the ethyl group and methyl-substituted carbon as shown in Fig. 9.18. Again, titanium will be available with a vacant orbital ready for the propagation step.

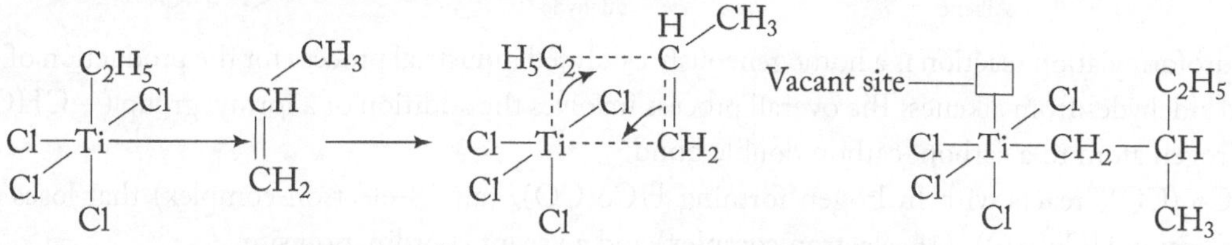

Fig. 9.18 Coordination of Ziegler-Natta Catalyst with alkene

Propagation As shown here, another propylene molecule approaches and the above process starts over, resulting in the formation of a linear polypropylene.

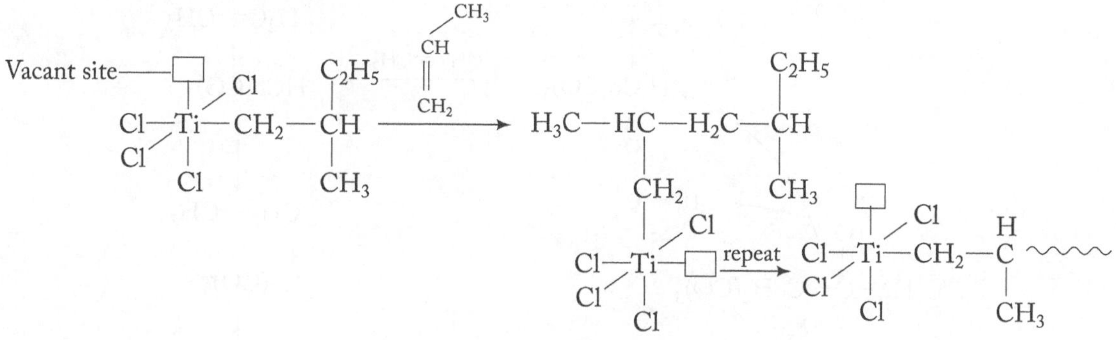

Fig. 9.19 Propagation of chain reaction

Termination The final step of polymerization, forming the desired polymers can take place as (a) β-elimination from the polymer chain forming metal hydride; (b) β-elimination with hydrogen transfer to the monomer, and (c) via hydrogenation.

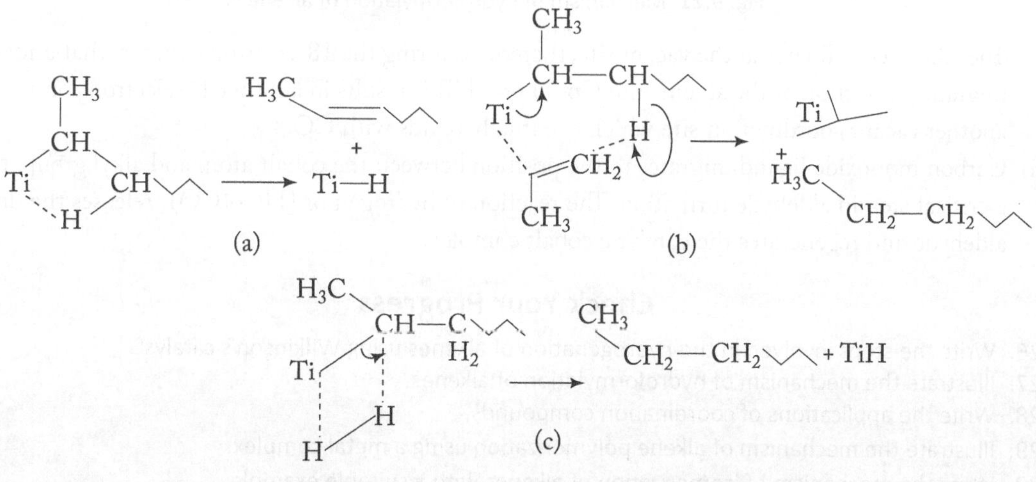

Fig. 9.20 Termination steps in polymerization

9.11.4 Hydroformylation of Alkenes

When an alkene reacts with CO and H_2 in the presence of a catalyst (Co or rhodium salt), an aldehyde is formed by hydroformylation (or oxo process). The general reaction can be shown as:

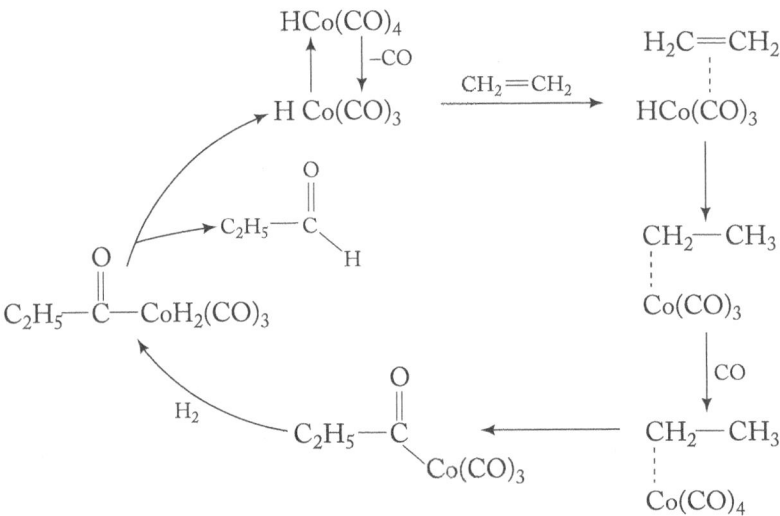

Hydroformylation reaction is a homogeneously catalysed industrial process for the production of value-added aldehydes from alkenes. The overall process involves the addition of a formyl group (—CHO) and a hydrogen atom to a carbon–carbon double bond.

(i) $Co_2(CO)_8$ reacts with hydrogen forming $HCo(CO)_4$ (an 18-electron complex) that loses cobalt forming $HCo(CO)_3$ (16-electron complex) and a vacant coordination site.

$$Co_2(CO)_8 + H_2 \longrightarrow 2\,HCo(CO)_4$$

Fig. 9.21 Mechanism of hydroformylation of alkene

(ii) The alkene coordinates at the vacant site, thereby restoring the 18-electron complex that undergoes migratory insertion of the alkene into Co–H bond. This results in another 16-electron complex and another vacant coordination site which eventually bonds with CO.

(iii) Carbon monoxide ligand migrates to the position between the cobalt atom and alkyl group; this is a critical step in aldehyde formation. The reaction of hydrogen or $[HCo(CO)_4$ releases the desired aldehyde and regenerates the catalytic cobalt complex.

Check Your Progress

26. Write the steps involved in the hydrogenation of alkenes using Wilkinson's catalyst.
27. Illustrate the mechanism of hydroformylation of alkenes.
28. Write the applications of coordination compounds.
29. Illustrate the mechanism of alkene polymerization using a metal complex.
30. Write the mechanism of isomerization of alkenes with a suitable example.

SUMMARY

- The pathbreaking and startling advances on the bonding and molecular structures of coordination compounds in the past seventy years have led to understanding their properties, resulting in their application in chelate therapy.
- Werner postulated two types of linkages called primary and secondary by a metal atom or ion in a coordination compound. Today, we are aware that these linkages are covalent in nature.
- Valence bond theory successfully explained the formation of complexes, magnetism, and even their fascinating geometries. However, it could not quantitatively interpret the magnetic data of these compounds. Further, it did not mention about their optical properties.
- Crystal field theory is based on observing the effects of various ligand fields and their influence on the degeneracy of metal d orbitals. Crystal field splitting, spectrochemical series, and crystal field stabilization energies are some of the concepts that provide quantitative information on orbital separation energies, magnetic moments, and even stability of various complexes. The mere assumption that ligands are point charges raises questions on theoretical difficulties.
- Magnetic behaviour of coordination complexes can be accounted for by the unpaired electrons in the d orbitals of the metal; whereas their colour can be attributed to the d–d electron transition when light falls on them.
- The stability of coordination compounds is measured in terms of stepwise stability (or formation) constant (K) or overall stability constant (β).
- The stabilization of coordination compound is due to formation of chelates and one can correlate to Gibbs energy, enthalpy, and entropy terms.
- A ligand that forms a bond with a metal ion is called chelating agent and the process is referred to as chelation.
- Coordination compounds have significant applications in catalysis and metallurgical processes. The applications of coordination chemistry encompass fields of analytical chemistry, industrial chemistry, photography, medicine, pharmaceutics and health care.
- Organometallic compounds form a group of physiologically important compounds such as chlorophyll and haemoglobin, which have at least one metal–carbon bond, where the metal can be any transition metal. The 18-electron rule represents the stability of various organometallic compounds. One needs a sound knowledge of coordination chemistry to understand organometallic bonding.
- Isomerization polymerization, hydrogenation, and hydroformylation reactions of alkenes are industrially important and are driven by a variety of organometallic compounds or catalysts.

GLOSSARY

Coordination compounds: Compounds containing one or more coordinate covalent bonds.

Coordination isomerism: A type of structural isomerism, where the ratio of ligands to the metal ion remains the same, but its bonding to the metal ion changes.

Coordination number (or ligancy): The total number of ligands surrounding a central metal ion in a coordination sphere.

Chelate: A coordination compound, where a ligand (usually polydentate) is bonded to a central metal atom forming various chelate rings.

Crystal field splitting: The destruction of degeneracy of d orbitals in metals under the influence of ligand field.

Crystal field stabilization energy (CFSE): An increase in stability of a complex due to increased attraction of metal ion and the ligand.

Diamagnetic: The absence of any unpaired electrons in a complex.

Dq (or 10Dq): An arbitrary number denoting energy differences between metal orbitals of approaching ligands.

Hard acid: A small highly charged cation or molecule with a high positive charge on the central atom and shows slight polarizability.

Heteroleptic: Refers to complexes where central metal atom or ion is bonded with different types of ligands.

Homoleptic: Refers to complexes where central metal atom/ion is bonded only with one type of ligands.

Hydrate isomerism: Here water is a solvent in the coordination compound.

Kinetic stability: The speed at which the formation or dissociation of a complex will occur leading towards equilibrium.

Organometallic compounds: Compounds having at least one metal–carbon bond, where C is organic portion and M could be any transition, inner-transition, or main group metal atom.

Paramagnetic: The presence of unpaired electrons in a complex that behaves like tiny magnets (weak attraction) under the influence of magnetic field.

Primary valency: The charge on a central metal ion.

Secondary valency: The number of ions of molecules that can coordinate to the metal atom or ion.

Spectrochemical series: The order of ligands based on ligand strength, which lists the various metal atoms/ions as per their oxidation number, group, and identity.

Thermodynamic stability: A measure of the extent to which the coordination complex will form or transform to different species under a set of given conditions at equilibrium.

Topology: The way in which ligands are interrelated or arranged around the central metal atom or ion.

EXERCISES

Multiple Choice Questions

1. The exchange of coordination group by a water molecule in a complex molecule can exhibit:
   - (a) ionization isomerism
   - (b) ligand isomerism
   - (c) hydrate isomerism
   - (d) geometrical isomerism

2. $[Co(NH_3)_6]^{3+}$ complex is
   - (a) paramagnetic
   - (b) diamagnetic
   - (c) square planar
   - (d) none

3. The type of isomerism shown by $[Co(en)_2(NCS)_2]$ Cl and $[Co(en)_2(NCS)Cl]NCS$ is:
   - (a) coordination
   - (b) ionization
   - (c) linkage
   - (d) all of these

4. Which among the following has square planar geometry?
   - (a) $[Cu(en)_2]^{2+}$
   - (b) $[Ag(NH_3)_2]^+$
   - (c) $[MnCl_4]^{2-}$
   - (d) $[Ni(CO)_4]$

5. Of the following, the strongest ligand is
   - (a) Br^-
   - (b) HO^-
   - (c) F^-
   - (d) CN^-

6. The complex ion which has no d electrons in the central metal atom is
   - (a) $[MnO_4]^-$
   - (b) $[Co(NH_3)_6]^{3+}$
   - (c) $[Fe(CN)_6]^{3-}$
   - (d) $[Cr(H_2O)_6]^{3+}$

7. Colour of tetramminecopper(II) sulphate is
   - (a) blue
   - (b) red
   - (c) violet
   - (d) green

8. In $K_4[Ni(CN)_4]$ complex, the oxidation state of nickel is
   - (a) zero
   - (b) –1
   - (c) +2
   - (d) +3

9. Potasssium ferrocyanide is an example of
   - (a) normal salt
   - (b) double salt
   - (c) complex
   - (d) mixed salt

10. All ligands are
    - (a) Lewis acids
    - (b) Lewis bases
    - (c) neutral
    - (d) reactive intermediates

11. An octahedral complex is formed when the central metal atom undergoes hybridization among the _____ orbitals
    - (a) sp^3
    - (b) dsp^2
    - (c) sp^3d
    - (d) sp^3d^2

12. The metals and ligands were classified into class (a) and class (b) by
    - (a) Alfred Werner
    - (b) S. Arhland, J. Chatt and N.R. Davies
    - (c) Ralph Pearson
    - (d) Linus Pauling

13. According to the HSAB principle, a hard acid:
    - (a) is slightly polarizable
    - (b) has a low charge density
    - (c) shows a preference for soft bases
    - (d) shows a preference for donor atoms of low electronegativity

14. HSAB concept was put forth by
    - (a) Alfred Werner
    - (b) Linus Pauling
    - (c) H. Bethe and van Vleck
    - (d) Ralph Pearson

15. According to crystal field theory, the metal–ligand bond is
    - (a) covalent
    - (b) electrostatic
    - (c) polarized
    - (d) metallic

16. Triply degenerate orbitals are _____ and denoted as _____.
    - (a) d_{xy}, d_{yz}, d_{xz} and e_g
    - (b) d_{xy}, d_{yz}, d_{xz} and t_{2g}
    - (c) $d_{x^2-y^2}, d_{z^2}$ and t_{2g}
    - (d) $d_{x^2-y^2}, d_{z^2}$ and e_g

17. $[Fe(CN)_6]^{4-}$ and $[Fe(H_2O)_6]^{2+}$ are _____.
    - (a) low- and high-spin complexes respectively
    - (b) high-and low-spin complexes respectively
    - (c) both low-spin complexes
    - (d) both high-spin complexes

18. $[Ti(H_2O)_6]^{3+}$ complex absorbs light at
    - (a) 100 nm
    - (b) 200 nm
    - (c) 350 nm
    - (d) 500 nm

19. For complexes with d^3 configuration, its CFSE will be
    - (a) $-4Dq$
    - (b) $-10Dq$
    - (c) $-12Dq$
    - (d) $-12Dq + 3P$

20. EDTA as a ligand is
    - (a) monodentate
    - (b) bidentate
    - (c) tridentate
    - (d) hexadentate

21. The overall dissociation equilibrium constant for $[Cu(NH_3)_4]^{2+}$, if β_4 for this complex is 1.8×10^{13} will be
    - (a) 5.55×10^{-14}
    - (b) 6.55×10^{-14}
    - (c) 3.24
    - (d) 2.558×10^{-7}

22. The formulae for Wilkinson catalyst and Zeise's salt are
    - (a) $[Rh(PPh_3)_3Cl]$ and $[Pt(NH_3)_4][PtCl_4]$
    - (b) $[Rh(PPh_3)_3Cl]$ and $K[Pt(C_2H_4)Cl_3]$
    - (c) $K[Pt(C_2H_4)Cl_2]$ and $K[C_6(NH_3)_2(NO_2)_4]$
    - (d) $K[Pt(C_2H_4)Cl_2]$ and anhydrous $CuSO_4$

23. For $[Ti(H_2O)_6]^{3+}$ in excited state, its electron configuration as per CFT will be
    - (a) $t_{2g}^1 e_g^1$
    - (b) $t_{2g}^1 e_g^0$
    - (c) $t_{2g}^0 e_g^1$
    - (d) $t_{2g}^1 e_g^2$

24. $Cr(CO)_6$ is known to
    - (a) follow 18-electron rule
    - (b) act as Wilkinson catalyst
    - (c) act as ligand
    - (d) act as multidentate ligand

25. Zeigler–Natta catalyst comprises
    - (a) transition metal compound with $Cr(CO)_6$
    - (b) transition metal compound with Al
    - (c) only transition metal compound
    - (d) transition metal compound with $Al(C_2H_5)_3$

26. When an alkene reacts with CO and H_2 in the presence of Co or Rh salts to form an aldehyde, the reaction is
    - (a) hydroformylation
    - (b) isomerization
    - (c) carbonation
    - (d) hydrogenation

27. The ammoniacal solution of nickel reacts with dimethylglyoxime to form
    - (a) Ni–DMG complex
    - (b) $NiCl_4$ precipitate
    - (c) Wilkinson catalyst
    - (d) Zeise's salt

28. The complex that does not follow EAN rule is
    - (a) $K_4Fe(CN)_6$
    - (b) $[Fe(CO)_5]$
    - (c) $[Ag(NH_3)_4]^+$
    - (d) $[Co(CN)_6]^{4-}$

Review Questions

1. Define the terms: dative bond, ligand, complex, coordination sphere, primary and secondary valency, dentate character, charge on complex.

2. Explain the following terms:
   - (a) Crystal field stabilization energy
   - (b) Crystal field splitting
   - (c) $10Dq$
   - (d) High-and low-spin complexes

3. State and explain Pearson's principle of acids and bases and give examples of each type.

4. Explain on the basis of valence bond theory that $[Ni(CN)_4]^{2-}$ ion with square planar structure is diamagnetic and $[NiCl_4]^{2-}$ ion with tetrahedral geometry is paramagnetic.

5. What is spectrochemical series? Explain the difference between a weak field ligand and a strong field ligand.

6. Write the distribution of d electrons in octahedral complex and their corresponding crystal field stabilization energies.

7. What are monodentate, bidentate, and ambidentate ligands? Give two examples of each.

8. How does one calculate crystal field stabilization energy? State its significance.

9. Justify the following:

    (a) $[NiCl_4]^{2-}$ is paramagnetic, but $[Ni(CO)_4]$ is diamagnetic though both are tetrahedral.

    (b) $[Fe(H_2O)_6]^{3+}$ is strongly paramagnetic whereas $[Fe(CN)_6]^{3-}$ is paramagnetic. $[Co(NH_3)_6]^{3+}$ is an inner orbital complex whereas $[Ni(NH_3)_6]^{2+}$ is an outer orbital complex.

    (c) Hexaquamanganese(II) ion contains five unpaired electrons, while the hexacyano ion contains only one unpaired electron.

10. What is crystal field splitting? Show the splitting of d orbitals in an octahedral crystal field with a figure.

11. What is the basic difference between a double salt and complex?

12. On the basis of HSAB concept, explain class (a) and (b) acids and bases.

13. Write the formulae for the following coordination compounds:

    (a) Tetraamminediaquacobalt(III) chloride

    (b) Potassium tetracyanidonickelate(II)

    (c) Tris(ethane–1,2–diamine) chromium (III) chloride

    (d) Amminebromidochloridonitrito-N-platinate(II)
    (**Ans:** (a) $[Co(NH_3)_4(H_2O)_2]Cl_3$ (b) $K_2[Ni(CN)_4]$ (c) $[Cr(en)_3]Cl_3$ (d) $[Pt(NH_3)BrCl(NO_2)]^-$)

14. Write the IUPAC names of the following coordination compounds:

    (a) $[Co(NH_3)_6]Cl_3$; (b) $[Co(NH_3)_5Cl]Cl_2$

    (c) $K_3[Fe(CN)_6]$; (d) $K_3[Fe(C_2O_4)_3]$; (e) $K_2[PdCl_4]$

    (f) $[Pt(NH_3)_2Cl(NH_2CH_3)]Cl$

    (**Ans:** (a) Hexaamminecobalt(III) chloride

    (b) Pentaamminechloridocobalt(III) chloride

    (c) Potassium hexacyanidoferrate(III) (iv) Potassium trioxalatoferrate(III)

    (d) Potassium tetrachloridopalladate(II) (vi) Diamminechlorido(methanamine)platinum(II) chloride)

15. Discuss the nature of bonding in the following coordination entities as per VBT:

    (a) $[Fe(CN)_6]^{4-}$ (b) $[FeF_6]^{3-}$
    (c) $[Co(C_2O_4)_3]^{3-}$ (d) $[CoF_6]^{3-}$

16. Calculate the overall dissociation equilibrium constant for complex, $[Cu(NH_3)_4]^{2+}$ ion, if its β_4 is 5.9×10^{13}. (**Ans:** $1/\beta_4 = 1.695 \times 10^{-14}$)

17. List the factors which govern stability of complexes.

18. How is stability of a complex related to thermodynamic parameters like enthalpy and entropy? State the expression and mention its significance.

19. List the various applications of coordination compounds.

20. Write a short note on 'magnetism and colour of coordination compounds.'

21. Discuss colour and absorption spectrum of $[Ti(H_2O)_6]^{3+}$ coordination complex.

22. Write a note on 'thermodynamic and kinetic stability of complexes.'

23. Explain chelate effect with a suitable example. Discuss the factors that attribute to the stability of chelates.

24. Write a note on 'chelation and chelate effect.'

25. What are borderline acids and bases? State Pearson's concept about borderline species.

26. What are organometallic compounds? Classify them.

27. What is 18-electron rule? Determine whether $Cr(CO)_6$ and $[TiF_6]^{2-}$ complexes obey 18-electron rule.

28. Explain EAN rule. Give the examples of EAN rule followed by complexes.

29. Explain the mechanism of hydroformylation of alkenes with a suitable example.

29. What is hydrogenation of alkenes? Describe the mechanism with a suitable example.

30. What is isomerization of alkene? Explain the mechanism using rhodium catalyst.

31. What is Zeigler–Natta catalyst? Explain polymerization of alkenes with an example.

32. Write a note on 'metal complex catalysts for alkene preparation.'

FURTHER READING

1. Bhatt, Vasishta. *Essentials of Coordination Chemistry: A Simplified Approach with 3D Visuals*, Academic Press, Elsevier Inc., 2016.

2. Calvin, M. and K. W. Wilson. 'Stability of Chelate Compounds, *Journal of American Chemical Society*, 67, 67 (11), pp. 2003–07, (1945).

3. Eisch, John J. *The Chemistry of Organometallic Compounds - The Main Group Elements.* The Macmillan Company, New York, 1967.

4. Kuffaman, G.B. 'Alfred Werner's Research on Structural Isomerism', *Coordination Chemistry Review*, (11), p. 161, (1973).

5. *Nomenclature of Inorganic Chemistry*, IUPAC *Recommendations 2005*, RSC Publishing.

6. Pfenning, Brian (2015). *Principles of Inorganic Chemistry.* Hoboken, New Jersey: John Wiley & Sons, Inc.

7. Powell, P. *Principles of Organometallic Chemistry*, Chapman and Hall, London, 1988.

8. Ralph G. Pearson, 'Hard and Soft Acids and Bases', *Journal of American Chemical Society*, 1963, 85 (22), pp 3533–39, URL: https://pubs.acs.org/doi/pdf/10.1021/ja00905a001

9. Tsutsui, M., M. N. Levy, A. Nakamura, M. Ichikawa, and K. Mori, *Introduction to Metal-π-Complex Chemistry*, Plenum Press, New York, 1970.

ANSWERS

Check Your Progress

1. Zeise's salt: $K[Pt(C_2H_4)Cl_3]$ and Wilkinson catalyst $[Rh(PPh_3)_3Cl]$

2. Refer to Section 9.2

3. (a) Monodentate – F, Cl^-, NH_3, H_2O, OH^-

    (b) Bidentate – Ethylenediamine ($H_2N–CH_2–CH_2–NH_2$), Glycinate ion ($H_2N–CH_2–COO^-$), Oxalate ion ($^-OOC – COO^-$)

    (c) Polydentate – Ethylenediamminetetraacetic acid (EDTA) – hexadentate (6 electron pairs)

4. Refer to P. 2

5. (a) Tetraammineaquachloridocobalt(III) chloride – $[Co(NH_3)_4(H_2O)Cl]Cl_2$

    (b) Potassium tetrahydroxidozincate(II) – $K_2[Zn(OH)_4]$

    (c) Potassium trioxalatoaluminate(III) – $K_3[Al(C_2O_4)_3]$

    (d) Dichloridobis(ethane-1,2-diamine)cobalt(III) – $[CoCl_2(en)_2]^+$

6. (a) $[Pt(NH_3)_2Cl(NO_2)]$ – Diamminechloridonitrito-N-platinum(II)

    (b) $K_3[Cr(C_2O_4)_3]$ – Potassium trioxalatochromate(III)

    (c) $[CoCl_2(en)_2]Cl$ – Dichloridobis(ethane-1,2-diamine)cobalt(III) chloride

    (d) $[Co(NH_3)_5(CO_3)]Cl$ – Pentaamminecarbonatocobalt(III) chloride

    (e) $Hg[Co(SCN)_4]$ – Mercury (I) tetrathiocyanatocobaltate(III)

    (f) $K_2[CrCO(CN)_5]$ – Potassium carbonylpentacyanochromium(III)

7. Werner's theory, Sidgwick theory, valence bond theory, crystal field theory.

8. (a) Metal ions in coordination compounds exhibit two types of linkages, namely primary and secondary. Metal ions always try to satisfy both the valences.

    (b) The ions or groups linked by secondary bonds to the central metal possess characteristic spatial orientations corresponding to different coordination numbers referred to as coordination polyhedral.

9. According to VBT, as inner 3d metal orbital takes part in hybridization, $[Co(NH_3)_6]^{3+}$ is called inner orbital or low-spin (or spin-paired) complex. $[CoF_6]^{3-}$ in which the central cobalt ion undergoes sp^3d^2 hybridization utilizing its outer 4d orbital; hence the complex is called an outer orbital or high-spin (or spin-paired) complex.

10. In $[NiCl_4]^{2-}$, nickel has the electronic configuration of $3d^8$. Each chlorine ion donates a pair of electrons and the complex has tetrahedral geometry. As shown in the hybridization scheme, there are two unpaired electrons, thus exhibiting paramagnetism. On the other hand, nickel carbonyl, $[Ni(CO)_4]$ has tetrahedral geometry and is diamagnetic. Another example of nickel complex is of $[Ni(CN)_4]^{2-}$ that is a square planar complex with diamagnetic behaviour.

11. (a) No quantitative details on the magnetic data of coordination complexes.

    (b) Cannot explain colours exhibited by these complexes that lead to spectral analysis.

12. According to CFT, in a free metal ion, all the five d orbitals are degenerate, but when ligands approach them, electron repulsion occurs causing the energy levels of d orbitals to rise. The degenerate metal orbitals are split into two groups of orbitals of differing energies under the influence of the approaching ligands. This phenomenon is called crystal field splitting.

13. The crystal field splitting and energy difference between t_{2g} and e_g orbitals is denoted as Δ_o (subscript o denotes octahedral). The energy difference Δ_o is measured by a parameter called Dq and arbitrarily taken as 10Dq.

14. Crystal field splitting ability of common ligands can be summarized in the following order of increasing field strength as:
$$I^- < Br^- < S^{2-} < SCN^- < Cl^- < NO_3^- < F^- < OH^- < O^{2-} < H_2O < NCS^- < CH_3 \, CN$$
$$< NH_3 < en < dipy < phen < NO_2^- < CN^- < CO$$
The above series is called the spectrochemical series.

15. In the presence of a strong field, electrons of the ligands pair up since 10Dq is large enough to allow force pairing by filling up the lower orbitals and hence such complexes will be diamagnetic or low-spin complexes. $[Fe(CN)_6]^{4-}$ is an example.

    In case of weak field ligands, 10Dq is not large enough to allow force pairing of electrons and the next electron enters the higher e_g orbitals and such complexes are paramagnetic or high-spin complexes. $[Fe(H_2O)_6]^{2+}$ is an example.

16. An increase in the stability of a complex due to increased attraction of metal ion and the ligand is termed as crystal field stabilization energy (CFSE). Refer to Table 9.6 for the values.

17. Refer to p. 12.

18. HSAB principle states that hard acids prefer to bind to hard bases and soft acids prefer to bind to soft bases.

19. Solution: Overall dissociation constant is the reciprocal of overall stability constant and will be calculated as: $1/\beta_4 = 3.571 \times 10^{-14}$

20. Chelates are a special type of coordination compounds that contain one or more rings in their structure; examples: Copper (II) ethylenediamine, Ni–DMG complex.

21. When polydentate ligands coordinate with a metal ion to form chelate, they impart greater stability than those complexes with monodentate ligands. This additional stability due to formation of chelates is called chelate effect. Yes, chelates are stable.

22. Refer to Section 9.7

23. Refer to Section 9.8

24. Refer to Section 9.10.3

25. Refer to Section 9.10.3

26. Refer to Section 9.11.1

27. Refer to Section 9.11.4

28. Refer to Section 9.9

29. Refer to Section 9.11.3

30. Refer to Section 9.11.2

Multiple Choice Questions

1. (c)	2. (b)	3. (b)	4. (a)	5. (d)	6. (a)	7. (a)
8. (a)	9. (c)	10. (b)	11. (d)	12. (b)	13. (a)	14. (d)
15. (b)	16. (b)	17. (a)	18. (d)	19. (c)	20. (d)	21. (a)
22. (b)	23. (c)	24. (a)	25. (d)	26. (a)	27. (a)	28. (d)

Organic Reactions and Synthesis of Drug Molecules

After reading this chapter, you will be able to:

- differentiate various organic reaction intermediates.
- illustrate the mechanism, energetics, and stereochemistry of nucleophilic and elimination reactions.
- understand the mechanism of various oxidation and reduction reactions.
- discuss the mechanism of hydroboration of alkenes.
- analyse some important rearrangement, cyclization, and ring opening reactions.
- explain the synthesis and applications of paracetamol and aspirin.

10.1 INTRODUCTION

Organic chemistry involves studying the reactivity of carbon compounds. Till date more than 20 million organic compounds have been synthesized, isolated, and characterized in literature. The mere understanding of the start and end of an organic reaction is not sufficient. Today, chemical engineers need to understand the types of organic reactions, their mechanisms, energetics, reagents, etc. The term 'reaction mechanism,' implies the stepwise illustration of bond formation and bond breaking in an organic compound resulting in the formation of product.

10.2 OVERVIEW OF ORGANIC REACTIONS

10.2.1 Bond Fission

An organic reaction can occur in any one of the following ways.
 (a) One-step concerted reaction: In this case, starting reactant molecules directly convert into products.
 (b) Multi-stepwise reaction: The reaction involves more than one step, initially forming an unstable reactive intermediate which then forms the product.

Homolytic Fission

If the bond breaks by equally dividing the electrons between the two atoms in the bond, the process is called homolysis, homolytic cleavage, or homolytic fission.

For example, if molecule AB undergoes homolysis, both atoms A and B will get one electron each (Fig. 10.1(a)).

Heterolytic Fission

If the bond breaks by unequal division of electrons between the two reacting atoms, it is called heterolysis, heterolytic cleavage, or heterolytic fission. Atoms of different electronegativities will result in electrons moving towards more electronegative atom during bond fission.

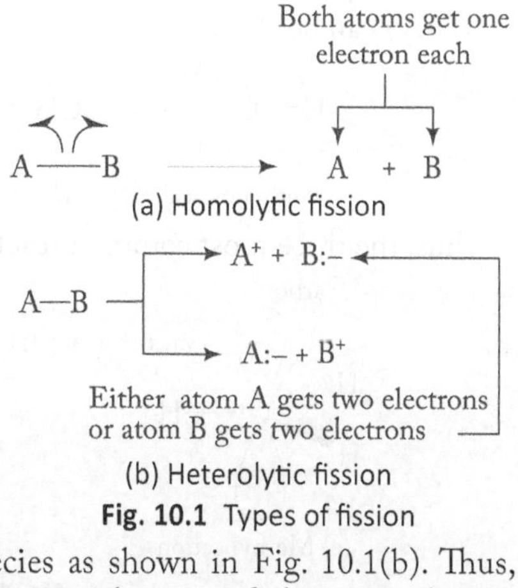

Here, molecule AB undergoes heterolysis giving different species as shown in Fig. 10.1(b). Thus, homolysis results in the generation of uncharged reactive intermediates with unpaired electrons, whereas

Both atoms get one electron each

A——B $\longrightarrow$ A + B

(a) Homolytic fission

A—B $\longrightarrow$ A^+ + $B{:}^-$

$\longrightarrow$ $A{:}^-$ + B^+

Either atom A gets two electrons or atom B gets two electrons

(b) Heterolytic fission

Fig. 10.1 Types of fission

heterolysis generates charged reactive intermediates. Reactive intermediates are transient, unstable, and reactive with short lifetimes. Carbocations, carbanions, radicals, and carbenes are reactive intermediates formed in a variety of organic reactions, though carbenes are less encountered in organic reactions.

10.2.2 Carbocations, Carbanions, Radicals, and Carbenes

When a covalent bond undergoes heterolytic fission, it results in the formation of unstable charged species as a pair of electrons and the charge is carried by the carbon atom, which is known as *carbocation*. Thus, carbocations are electron-deficient intermediates that possess a carbon atom surrounded by six electrons. Carbocations have a planar structure with positively charged three coordinated carbon atoms in sp^2 hybridization. It has three sp^2 hybridized orbitals in a plane with an empty p-orbital lying perpendicular to the plane. The relative stabilities of carbocation follow the order: tertiary > secondary > primary > methyl.

For example, methyl cation, ethyl carbocation (primary), isopropyl carbocation (secondary), and *tert*-butyl carbocation (tertiary) can be shown as:

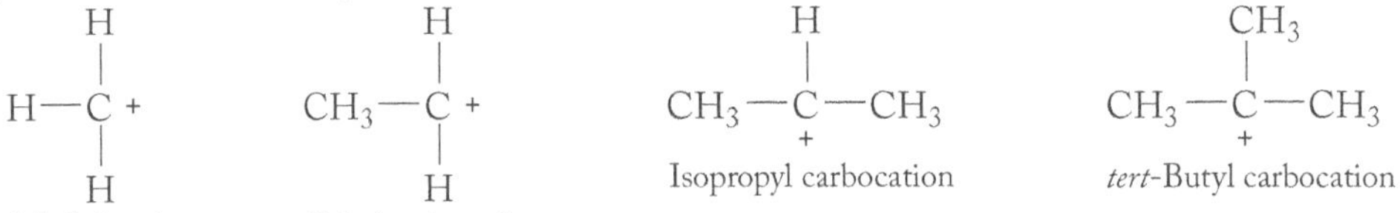

$$\underset{\text{Methyl cation}}{\overset{\overset{\textstyle H}{|}}{\underset{\underset{\textstyle H}{|}}{H\!-\!C\,+}}} \qquad \underset{\text{Ethyl carbocation}}{\overset{\overset{\textstyle H}{|}}{\underset{\underset{\textstyle H}{|}}{CH_3\!-\!C\,+}}} \qquad \underset{\text{Isopropyl carbocation}}{\overset{\overset{\textstyle H}{|}}{\underset{\underset{\textstyle +}{}}{CH_3\!-\!C\!-\!CH_3}}} \qquad \underset{\textit{tert}\text{-Butyl carbocation}}{\overset{\overset{\textstyle CH_3}{|}}{\underset{\underset{\textstyle +}{}}{CH_3\!-\!C\!-\!CH_3}}}$$

Carbanions are formed by heterolytic cleavage of a covalent bond where carbon is linked to a less electronegative atom. These intermediates possess a tetrahedral configuration with electron pairs occupying one of the sp^3 hybridized orbitals. The order of relative stabilities of carbanions is tertiary < secondary < primary < methyl.

For example, methyl anion, ethyl carbanion (primary), isopropyl carbanion (secondary), and *tert*-butyl carbanion (tertiary) can be shown as:

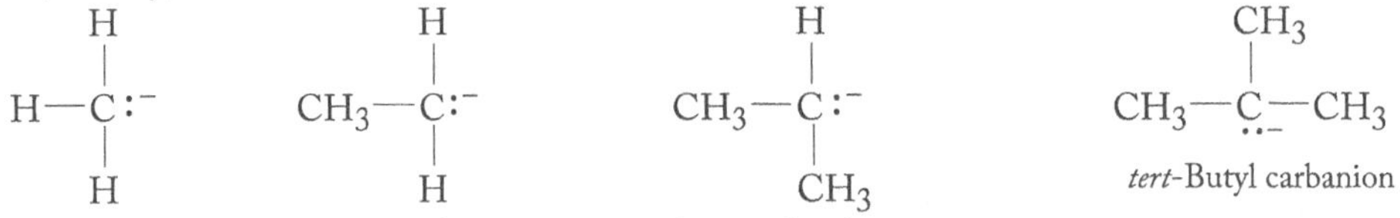

$$\underset{\text{Methyl anion}}{\overset{\overset{\textstyle H}{|}}{\underset{\underset{\textstyle H}{|}}{H\!-\!C\!:^-}}} \qquad \underset{\text{Ethyl carbanion}}{\overset{\overset{\textstyle H}{|}}{\underset{\underset{\textstyle H}{|}}{CH_3\!-\!C\!:^-}}} \qquad \underset{\text{Isopropyl carbanion}}{\overset{\overset{\textstyle H}{|}}{\underset{\underset{\textstyle CH_3}{|}}{CH_3\!-\!C\!:^-}}} \qquad \underset{\textit{tert}\text{-Butyl carbanion}}{\overset{\overset{\textstyle CH_3}{|}}{\underset{\underset{\textstyle \cdot\cdot}{}}{CH_3\!-\!C\!-\!CH_3}}}$$

If a covalent bond undergoes homolysis, it generates two uncharged products with unpaired electrons, known as *radicals*. Hence, radicals are unstable reactive intermediates possessing a single unpaired electron with no charge. The relative stabilities of free radicals follow the order tertiary > secondary > primary > methyl.

For example, methyl radical, ethyl radical (primary), isopropyl radical (secondary), and *tert*-butyl radical (tertiary) can be shown as:

$$\underset{\text{Methyl radical}}{H\!-\!\overset{\textstyle \cdot}{C}{\overset{\displaystyle /H}{\diagdown_H}}} \qquad \underset{\text{Ethyl radical}}{CH_3\!-\!\overset{\textstyle \cdot}{C}{\overset{\displaystyle /H}{\diagdown_H}}} \qquad \underset{\text{Isopropyl radical}}{CH_3\!-\!\overset{\textstyle \cdot}{C}{\overset{\displaystyle /H}{\diagdown_{CH_3}}}} \qquad \underset{\textit{tert}\text{-Butyl radical}}{CH_3\!-\!\overset{\textstyle \cdot}{C}{\overset{\displaystyle /CH_3}{\diagdown_{CH_3}}}}$$

Thus, the three most common reactive intermediates are carbocations, carbanions, and radicals.

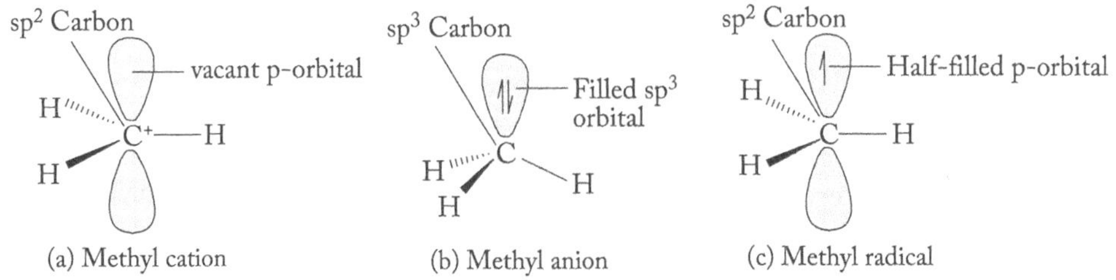

(a) Methyl cation   (b) Methyl anion   (c) Methyl radical

Fig. 10.2 Common reactive intermediates

Figure 10.2 shows (a) the simplest carbocation (methyl carbocation) with 6 valence electrons and has a trigonal planar structure with sp^2 hybridized carbon and a vacant p orbital; (b) the carbanion methyl anion that has a trigonal pyramidal structure with an sp^3 hybridized carbon with a lone pair of electrons and (c) methyl radical ($\cdot$CH$_3$) which is trigonal planar with sp^2 hybridized carbon and a single electron in its unhybridized p orbital.

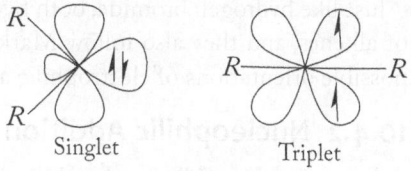

Singlet   Triplet

Fig. 10.3 Carbenes

A less common group of highly reactive divalent intermediates is called *carbenes*. The carbon atom in carbenes has six electrons with no charge (Fig. 10.3). It exists in two different configurations, namely singlet and triplet. *Singlet carbenes* have spin-paired electrons and are amphiphilic in nature; *triplet carbenes* have two singly occupied orbitals and are also called *diradicals*.

Some examples of carbenes are methylene carbene (:CH$_2$), phenylic carbene (:CH-Ph), dichlorocarbene (:CCl$_2$), and dimethoxycarbene [:C(OMe)$_2$].

10.3 REAGENTS IN ORGANIC REACTIONS

During any chemical reaction, the attack of reagent on substrate molecule initiates the transformation leading towards product formation. Either a fresh reagent is added or may be generated in situ during the reaction. The common reagents used in organic chemical reactions are the following.

Electrophilic reagents or electrophiles   These are Lewis acids, that is, they are capable of accepting electrons from the substrate. Electrophiles are either positively charged species or electron-deficient molecules; examples are H$^+$, Cl$^+$, Br$^+$ NO$_2^+$, AlCl$_3$, FeCl$_3$, SO$_2$, BF$_3$, RMgX.

Nucleophilic reagents or nucleophiles   These are Lewis bases and act as electron pair donors. Generally, nucleophiles are either electrically neutral or negatively charged species; for example, H$_2$O, NH$_3$, ROR, RNH$_2$, F$^-$, Br$^-$, Cl$^-$, OH$^-$, CN$^-$.

Radicals   These reactive intermediates have an odd number of electrons and are generated in situ during homolysis of a covalent bond. One of the most common examples is the photolysis of chlorine. The reaction is shown as follows: Cl—Cl → 2Cl$^\bullet$

Organic reactions can be of the following types: (a) addition, (b) substitution, (c) elimination, (d) rearrangement, and (e) cyclization and ring opening.

10.4 ADDITION REACTIONS

When an organic compound having multiple bonds (double or triple) combine with a reagent it leads to the formation of various saturated products. If there is a double bond in the organic compound that undergoes addition reaction, a saturated product is formed; whereas, if there is a triple bond, addition reaction results in the formation of lesser unsaturated or saturated products. Addition reactions can occur in the following three ways—electrophilic addition, nucleophilic addition, and free radical addition.

10.4.1 Electrophilic Addition

V. Markovnikov (1869) explained electrophilic addition of HBr to alkenes. When hydrogen halides react with alkenes, it follows Markownikov's rule. As per the rule, 'addition of hydrogen to an unsymmetrical olefin occurs at those carbon atoms with maximum number of hydrogen atoms. (i.e., carbon with least substitution). If there is an electronegative group in the reagent, it goes to the most substituted carbon atom. Such addition reactions are termed *regioselective* in nature and lead to the formation of stable carbocations.

Let us take the example of addition of hydrogen bromide to 1-propene (Fig. 10.4). The first step involves protonation of the pi bond of 1-propene that results in the formation of a carbocation. The carbocation, itself a strong electrophile, reacts with the halide ion resulting in the formation of products.

Just like hydrogen bromide, both hydrogen chloride and hydrogen iodide can add to the double bonds of alkenes, and they also follow Markovnikov's rule and exhibit regioselectivity, that is, one of the two possible orientations of electrophilic addition results preferentially over the other one.

10.4.2 Nucleophilic Addition

Alkenes are generally nucleophilic in nature and hence favour electrophilic addition reactions instantly. When an alkene is attached with electron-withdrawing groups, it withdraws the electron density from its pi bonds and allows for nucleophilic addition reactions. The presence of carbonyl groups in a compound allows both electrophilic and nucleophilic addition reactions.

[Positive charge on less substituted carbon; less stable and hence will not form]

Fig. 10.4 Addition of HBr to 1-propene

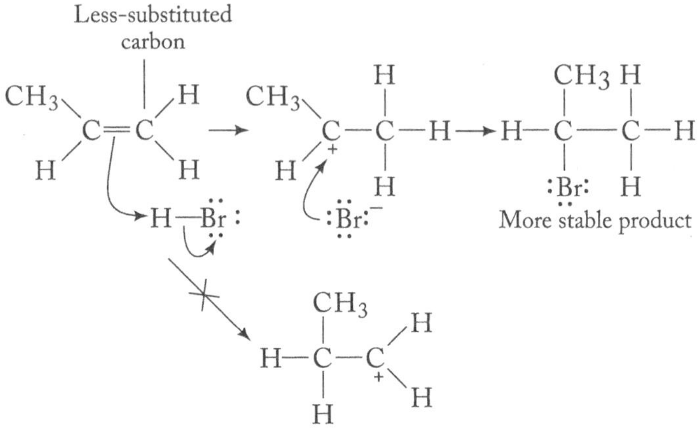

Cyanohydrin reaction is a classic example of nucleophilic addition that involves an unstable alkoxide intermediate (Fig. 10.5).

(a) The cyanide ion attacks the carbonyl group and leads to the formation of an unstable intermediate.
(b) The unstable intermediate abstracts a proton from the solvent (here it is water) and cyanohydrin is formed.

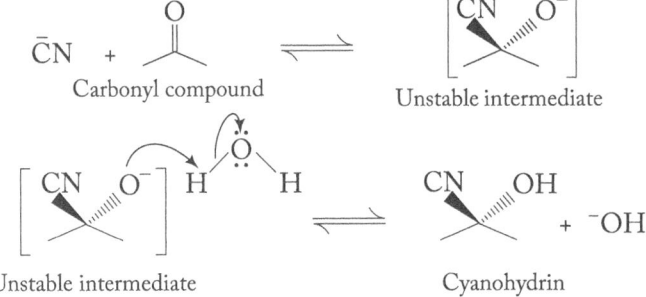

Fig. 10.5 Mechanism of cyanohydrin reaction

Conjugate addition, hydroamination, addition to Grignard reagent, addition to acids are some common examples of nucleophilic addition.

10.4.3 Free Radical Addition

A classic example of free radical addition reactions is addition of HBr to 1-propene to form 1-bromopropane.

$$\underset{\text{1-Propene}}{H_3C{-}CH{=}CH_2} + \underset{\text{Hydrogen bromide}}{HBr} \xrightarrow{\text{Peroxide}} \underset{\text{1-Bromopropene}}{CH_3{-}CH_2{-}CH_2{-}Br}$$

M. Kharasch and F. Mayo (1933) studied the addition of hydrogen bromide to alkenes in the presence of peroxides and found to form products very different from those expected as per Markownikov's rule. Hence if hydrogen halide adds to an unsymmetrical alkene in the presence of a peroxide, it follows free radical addition as per anti-Markownikov's rule. The peroxides result in the formation of free radicals that initiate addition and hence the overall reaction mechanism is led by free radicals. The O—O bond of peroxides is quite weak and hence undergoes fission to form two alkoxyl radicals. The alkoxyl radicals initiate addition reaction, such that hydrogen atom from the reagent becomes bonded to the carbon atom of alkene with the least number of hydrogen atoms (Fig. 10.6).

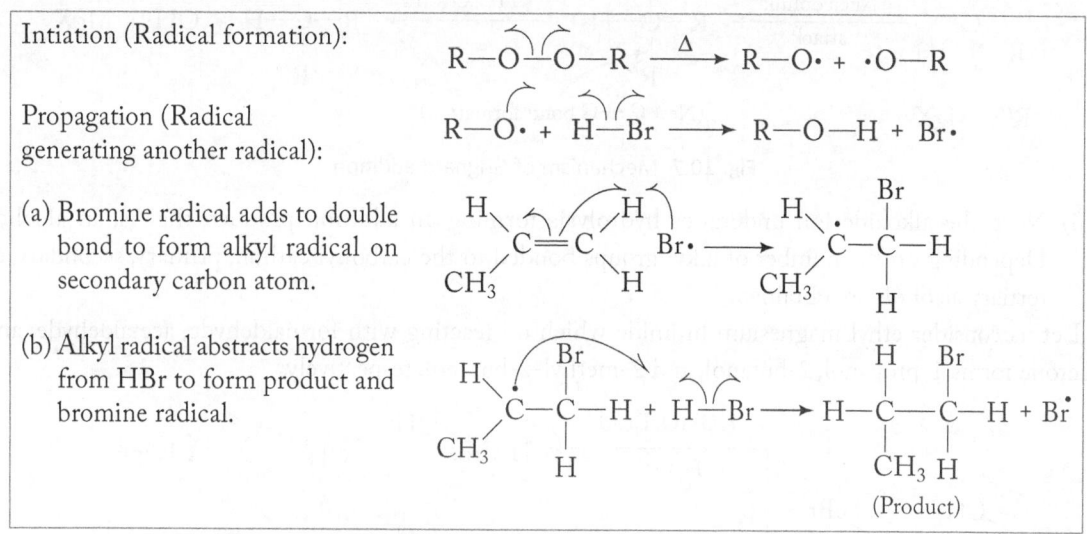

Fig. 10.6 Mechanism of free radical addition

Note that Markownikov's rule is not applicable in the presence of peroxides. This is because the peroxide-type reactions are generally rapid and even if peroxide is present in trace amounts, a mixture of Markownikov and anti-Markownikov products will be formed. Another interesting feature is that such peroxide effects are observed with only hydrogen bromide additions to alkene and not with hydrogen chloride. This is because the second step that involves the reaction of hydrogen chloride with alkyl radical is highly endothermic in nature. The same case is observed with hydrogen iodide, since the reaction of alkene with iodine atom is also endothermic.

10.4.4 Grignard Additions on Carbonyl Compounds

Grignard V. (1912) won Nobel Prize in Chemistry for preparing organomagnesium halides called Grignard reagent. The Grignard reaction is a popular and versatile method for C—C bond formation in organic compounds. In this section, we will understand the addition of Grignard reagent to various carbonyl compounds.

Grignard reagent is prepared by reacting an alkyl halide with the corresponding metal in the presence of the solvent, diethyl ether. The reaction is highly exothermic and results in the formation of an organohalogen (in this case, organomagnesium halide) or the Grignard reagent.

$$\underset{\text{Alkyl halide}}{\text{R—X}} \quad + \quad \underset{\substack{\text{Magnesium} \\ \text{metal}}}{\text{Mg}} \quad \xrightarrow[\text{Diethyl ether solvent}]{(CH_3CH_2)_2O} \quad \underset{\substack{\text{Organomagnesium} \\ \text{reagent or Grignard reagent}}}{\text{R—Mg—X}}$$

Grignard reagents are always prepared in situ and must be anhydrous because they react strongly with water. They react with carbonyl compounds to form various alcohols, hence is a reduction reaction.

Mechanism of reaction The mechanism of Grignard reaction is quite simple and illustrated in Fig. 10.7.

(i) The first step involves nucleophilic attack by the organic part of the reagent (R″) on the carbonyl carbon (electrophile) forming an alkoxide ion, initiating the formation of a new C — C bond.

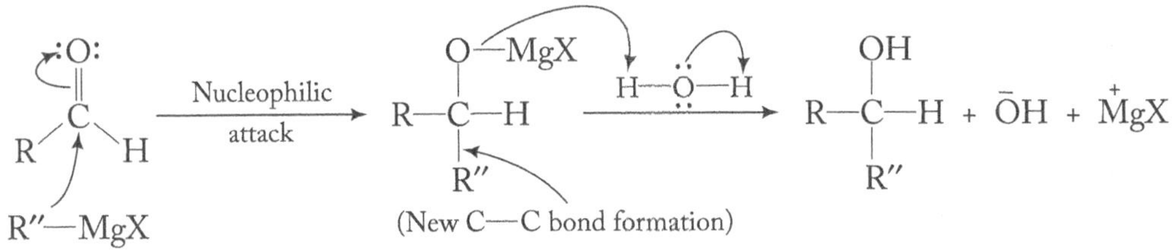

Fig. 10.7 Mechanism of Grignard addition

(ii) Next, the alkoxide ion undergoes hydrolysis forming an addition product, that is, an alcohol. Depending on the number of alkyl groups bonded to the carbonyl carbon, primary, secondary, or tertiary alcohols are obtained.

Let us consider ethyl magnesium bromide which on reacting with formaldehyde, acetaldehyde, and acetone forms 1-propanol, 2-butanol, and 2-methyl-2-butanol, respectively.

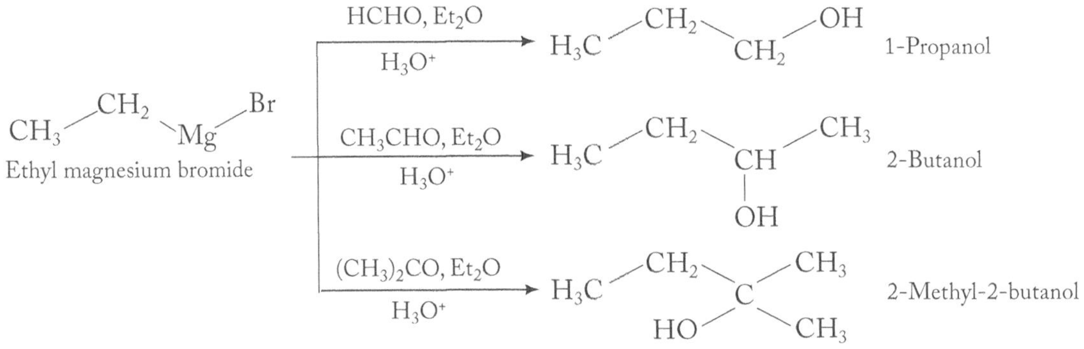

Some examples of Grignard reactions are shown here:

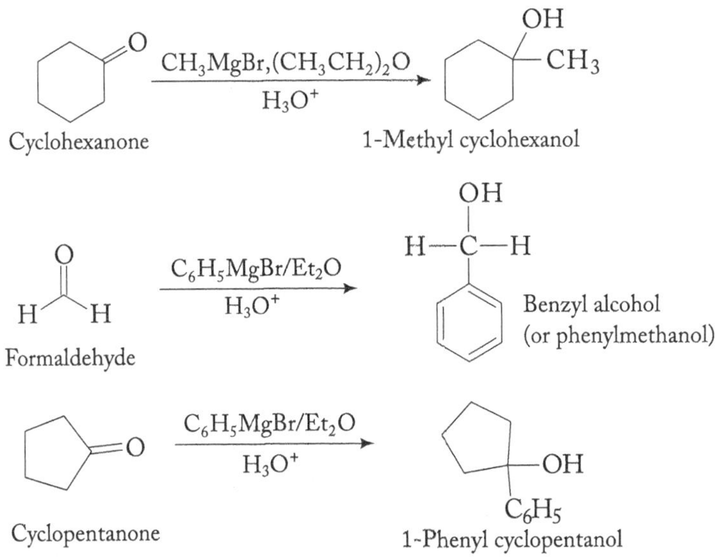

1-Methyl cyclohexanol

Benzyl alcohol
(or phenylmethanol)

1-Phenyl cyclopentanol

Stereochemically, Grignard reagent adds from both sides of the trigonal planar carbonyl carbon resulting in the formation of two alkoxides. On hydrolysis, it will result in the formation of a racemic mixture, that is, an equal amount of the two enantiomers.

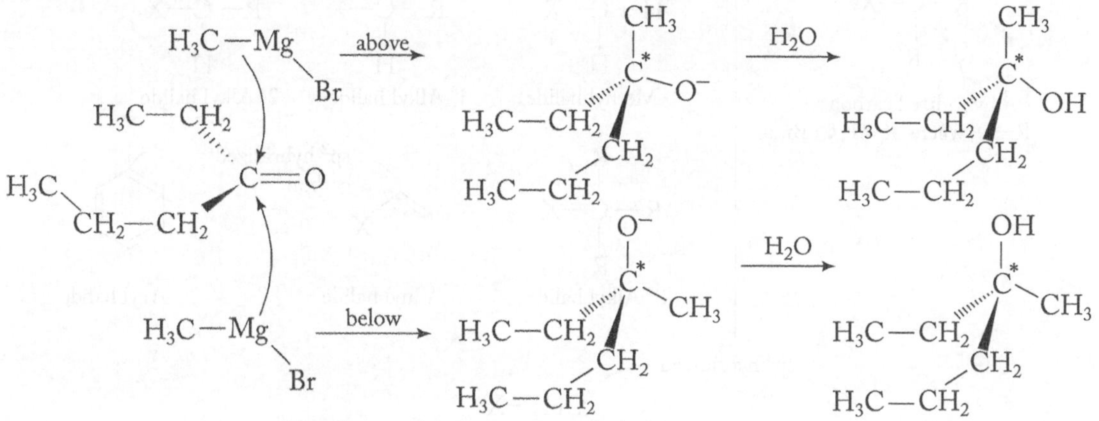

(Here *represents stereogenic centre.)

3-Hexanone on reacting with methyl magnesium bromide gives alcohol products with chiral (stereogenic) centres. Since, the methyl group can be added from either face of the carbonyl carbon equally, a racemic mixture is obtained and hence will be optically inactive.

10.5 SUBSTITUTION REACTIONS

Substitution reactions are a class of organic reactions in which an atom or a group of atoms is replaced by another atom or a group of atoms. Based on the mechanism of reaction involved, these are classified into the following three types—nucleophilic, electrophilic, and free radical substitution reactions.

10.5.1 Nucleophilic Substitution Reactions

When a nucleophile attacks the carbon atom in an alkyl halide, it is called nucleophilic substitution reaction. Alkyl halides have an sp^3 hybridized carbon atom bonded to a halogen atom. They are either primary (1°), secondary (2°), or tertiary (3°) based on the number of carbon atoms bonded to the C—X bond (where, X = halogen).

During nucleophilic substitution, the electron-rich nucleophile donates its electron pair to the carbon atom of the alkyl halide. The reaction occurs along with the heterolysis of the polar C—X bond, generally called the *leaving group*. Hence, nucleophilic substitution reactions are typically Lewis-acid base reactions.

Check Your Progress

1. Illustrate homolytic and heterolytic fission reactions.
2. Name the various organic reactions encountered in organic chemistry.
3. Name the various reagents used in organic reactions.
4. What is Markownikov and anti-Markownikov rule in radical reactions?
5. How does one prepare Grignard reagent?
6. Illustrate the reaction of Grignard reagent with carbonyl compound.

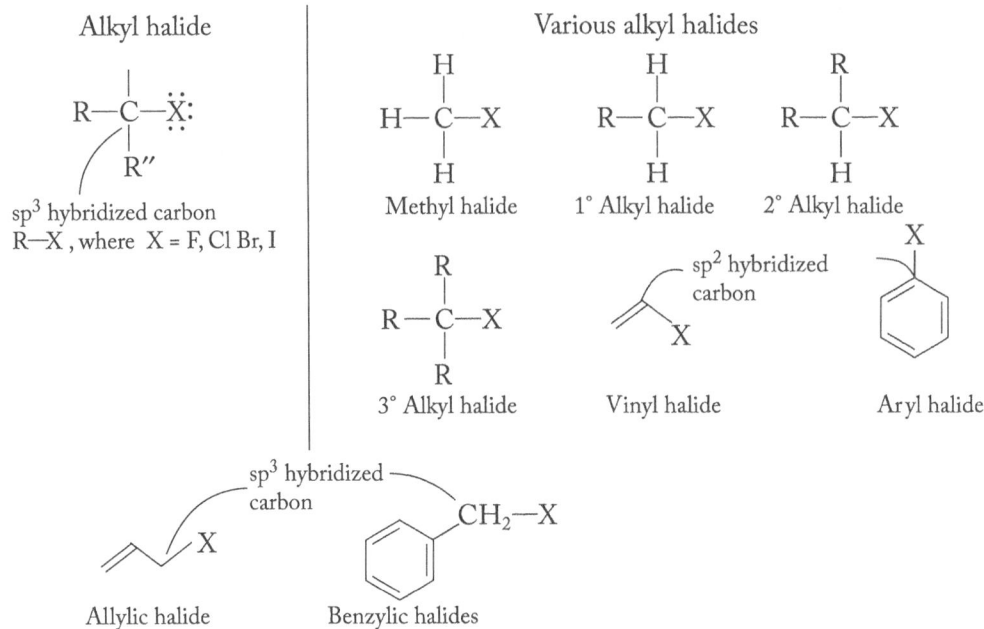

Fig. 10.8 Types of alkyl halides

During nucleophilic substitution of RX, the halogen group (X) is replaced by the incoming nucleophile. The C—X σ bond gets cleaved and a new C—Nu σ bond is formed. Here, X is the leaving group.

$$\underset{\text{Alkyl halide}}{R-X} \ + \ \underset{\text{Nucleophile}}{\overset{\bullet\bullet}{\underset{\bullet\bullet}{Nu}}}^{-} \ \xrightarrow{\quad\text{Substitution by Nu}\quad} \ \underset{\text{}}{R-Nu} \ + \ \underset{\text{Leaving group}}{X:-}$$

Vinyl halides have a halogen atom bonded to a C=C double bond and aryl halides have a halogen atom bonded to the benzene ring. These two organic halides possess an X atom bonded directly to sp^2 hybridized carbon atom and thus do not undergo nucleophilic substitution reactions. Allylic halides and benzylic halides possess halogen atoms bonded to sp^3 hybridized carbon atoms and can undergo nucleophilic substitution reactions.

On hydrolysis, 2-chloro-2-methylpropane forms 2-methyl-2-propanol which is a nucleophilic substitution product. Here, chloro group is substituted by hydroxyl group of water in the presence of acetone solvent.

2-Chloro-2-methylpropane $\xrightarrow[\text{Acetone}]{H_2O}$ 2-Methyl-2-propanol $+ \ HCl$

Some more examples of nucleophilic substitution are:

(a) $\underset{\text{Ethyl bromide}}{CH_3-CH_2-Br} + \underset{\text{Methoxide ion}}{:\overset{\bullet\bullet}{\underset{\bullet\bullet}{O}}CH_3}^{-} \longrightarrow \underset{\text{Ethyl methyl ether}}{CH_2-CH_2-\overset{\bullet\bullet}{\underset{\bullet\bullet}{O}}CH_3} + Br^{-}$

(b) $\underset{\text{Ethyl chloride}}{CH_3-CH_2-Cl} + \underset{\text{Hydroxide ion}}{:\overset{\bullet\bullet}{\underset{\bullet\bullet}{O}}H}^{-} \longrightarrow \underset{\text{Ethyl alcohol}}{CH_3-CH_2-\overset{\bullet\bullet}{\underset{\bullet\bullet}{O}}H} + Cl^{-}$

As shown in the above examples, the identity of cation is usually inconsequential and thus often ignored while writing chemical reactions.

(Carbocation) Nucleophile can attack from either sides

Based on molecularity of the reaction, nucleophilic substitution reactions are classified into two types, namely (a) substitution nucleophilic unimolecular (S_N1) and (b) substitution nucleophilic bimolecular (S_N2) reactions.

A nucleophilic substitution mechanism that undergoes a two-step process involving a carbocation intermediate is called S_N1 reaction. S_N1 is an abbreviation for 'Substitution Nucleophilic Unimolecular.' A nucleophilic substitution mechanism that involves a one-step concerted process is called S_N2 reaction. S_N2 is an abbreviation for 'Substitution Nucleophilic Bimolecular'. The numbers 1 and 2 in S_N1 and S_N2 denote the kinetic order of the reaction and **do not** indicate the number of steps in the mechanism.

Adrenaline Rush

Adrenaline (epinephrine) is a hormone secreted by our adrenal glands when we sense danger or stress and during strenuous workouts. It causes an increase in heart rate, pulse, and blood pressure. Adrenaline is produced instantly in our body by a simple nucleophilic substitution reaction of its precursor, norepinephrine.

S_N1 Reaction Mechanism

In an S_N1 reaction mechanism, there are two steps (Fig. 10. 9).
(a) first C—Cl bond cleaves and the organic substrate forms a planar carbocation.
(b) The nucleophile rapidly attacks the planar carbocation either from the front or the rear side to form products.

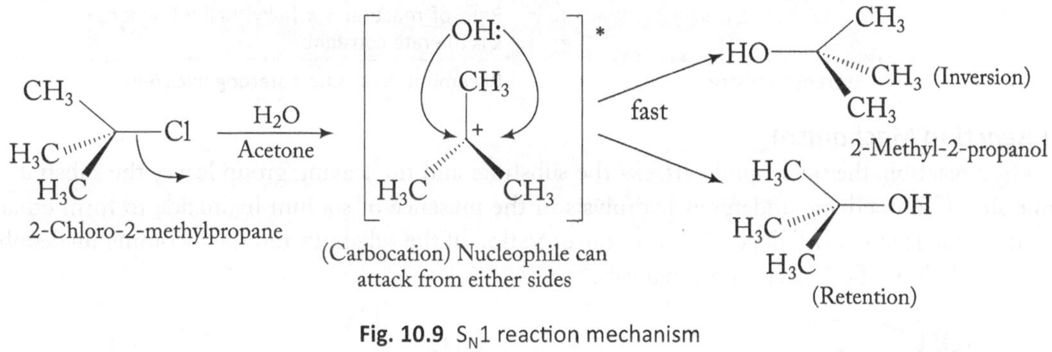

Fig. 10.9 S_N1 reaction mechanism

If the nucleophile attacks from the front side of the carbocation, the product so formed has the same configuration as the original substrate. This is called *retention of configuration*. If the nucleophile attacks from the rear side of the carbocation, it causes *inversion of configuration*. Hence, the final product is considered as a racemic mixture as both front and rear attacks happen. Generally, tertiary halides undergo S_N1 reaction because the carbocation intermediate is stable due to resonance.

The energy changes taking place during hydrolysis of 2-chloro-2-methylpropane can be shown by an energy profile diagram. In Fig. 10.10, the first energy barrier causes C—Cl bond fission leading to the formation of a carbocation. The second energy barrier depicts the partial bond formation with the incoming nucleophile. Following this, there is rapid product formation with the removal of proton.

Overall, the S_N1 reaction follows first order kinetics which depends only on the concentration of the alkyl halide. The type of nucleophile and its concentration have no effect on the rate of the reaction. Hence, doubling the concentration of nucleophile will not hasten the rate of S_N1 reaction.

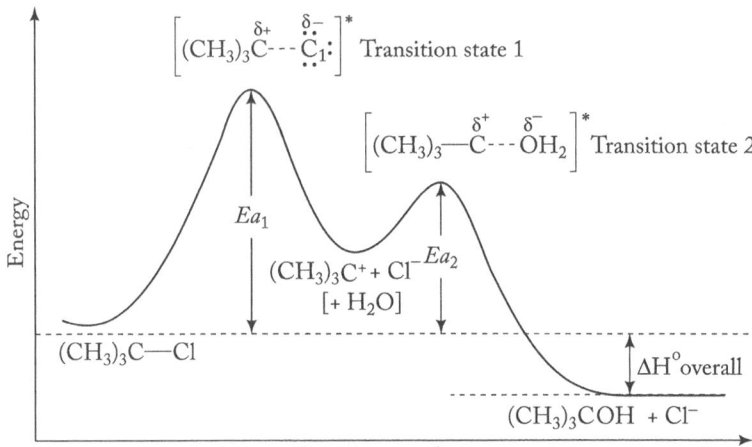

Fig. 10.10 Energy profile diagram of S_N1 reaction of 2-chloro-2-methylpropane with water, where Ea is energy of activation; $E_{a1} > E_{a2}$, since step 1 involves breaking of bond and step 2 involves bond formation, and $\Delta H°_{overall}$ refers to heat of product formation

Table 10.1 Features of S_N1 reaction mechanism

Features	Outcome
Mechanism	Two steps
Identity of alkyl group, i.e., R	Highly substituted halides react faster $R_3CX > R_2CHX > RCH_2X > CH_3X$
Energy profile	Two energy barriers, first-order kinetics Rate of reaction = k [alkyl halide], where k is the rate constant.
Stereochemistry	Racemization at single stereogenic center

S_N2 Reaction Mechanism

In an S_N2 reaction, the nucleophile attacks the substrate and the leaving group leaves the substrate, all in one step. Chloroethane undergoes hydrolysis in the presence of sodium hydroxide to form ethanol. The rate of the reaction will depend on the concentration of the substrate and the incoming nucleophile. Figure 10.11 shows S_N2 reaction mechanism.

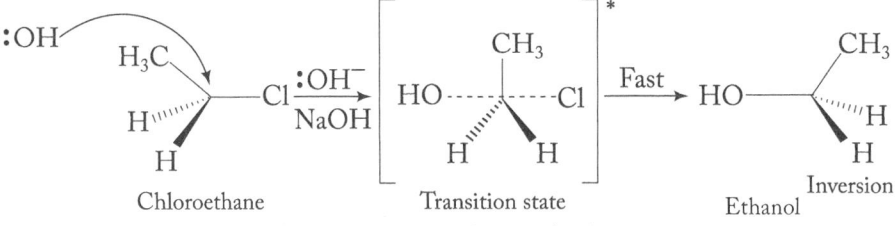

Fig. 10.11 S_N2 reaction mechanism

Where's Walden? The stereochemical outcome of an S_N2 reaction is the inversion of configuration first reported by Latvian chemist Dr. Paul Walden (1896) and is also popularly called Walden inversion.

Mechanism of reaction

(i) In the first step, the nucleophile attacks from the rear end of the substrate thereby forming an unstable transition state.

(ii) In the transition state, the carbon atom is sp^2 hybridized, as it is fully bonded to three substituents and partially bonded to the attacking nucleophile and leaving group.

(iii) Finally, the leaving group cleaves and the product is obtained.

As the nucleophile attacks from the rear end, the product exhibits *inversion of configuration*. Figure 10.12 depicts the energy changes during the hydrolysis of chloroethane following S_N2 mechanism.

Primary alkyl halides readily undergo S_N2 reaction as the rate of the reaction decreases with increasing bulkiness of the substituents on the carbon atom. Hence, one can say that S_N1 reactions are preferred at the tertiary carbon atom, whereas S_N2 reactions are favoured at the primary carbon atom.

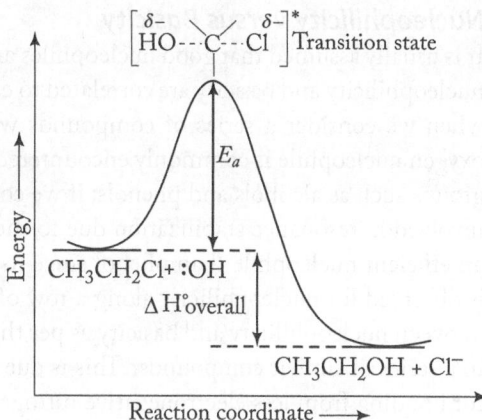

Fig. 10.12 Energy profile diagram for S_N2 reaction of 2-chloro-2-methylpropane with water, where E_a is the energy of activation and $\Delta H°_{overall}$ refers to heat of product formation

Overall, the S_N2 reaction follows second-order kinetics and the rate of the reaction depends on the concentration of alkyl halide and the incoming nucleophile. Thus, changing the concentration of either of the reactants affects the rate of reaction.

Table 10.2 Characteristics of S_N2 reaction mechanism

Features	Outcome
Mechanism	One step
Identity of alkyl group, i.e., R	Unhindered halides react faster $CH_3X > RCH_2X > R_2CHX > R_3CX$
Energy profile	One energy barrier, second-order kinetics Rate of reaction = k [alkyl halide] [nucleophile] where k is the rate constant.
Stereochemistry	Inversion of configuration at stereogenic centre.

The factors affecting nucleophilic substitution reactions are as follows:

Leaving group In both S_N1 and S_N2 reactions, leaving groups depart with electrons and hence a good leaving group is one which can accommodate original bonding electrons. It is found that good leaving groups are those who are conjugate bases of strong acids. Halogen groups like iodo, bromo, chloro, fluoro (in same order), carboxylates, cyano, amino, etc., are some good leaving groups. Hydroxy, alkyl, amino, alkoxy groups are poor leaving groups which on protonation turn into good leaving groups.

Nucleophile As nucleophilic attack takes place after carbocation formation, its presence does not influence the rate of S_N1 reaction. Contrary is the case in S_N2 reactions, where the attacking nucleophile pushes off the leaving group and hence is a critical parameter in predicting the reaction kinetics. Generally, good nucleophiles are good electron donors such as amino, alkoxy, hydroxy, etc.

Solvent polarity S_N1 reactions are favourable in polar solvents because the carbocation formation is the rate determining step. Water, methanol, ethanol, and formic acid are commonly used solvents. In S_N2 processes, the transition state is polar with charges dispersed all over it. If a polar solvent forms a solvent cage around the transition state, it will need additional energy to be broken. Hence, the rate of S_N2 reaction decreases with increasing solvent polarity.

Nucleophilicity versus Basicity

It is usually assumed that good nucleophiles are good electron donors, that is, good Lewis bases. However, nucleophilicity and basicity are correlated to each other in many cases. This comparison becomes apparent when we consider a series of compounds where the same atom is the nucleophile. Say, for example, oxygen nucleophile is commonly encountered in many reactions as the reactive component of functional groups such as alcohols and phenols; if we compare an alcohol and a phenol, the electrons in phenol are involved in resonance stabilization due to the presence of benzene ring. Hence, alcohol comes across as an efficient nucleophile than phenol since its electrons are easily available for donation. A similar trend is observed for nucleophilicity along a row of the periodic table. Tables 10.3 and 10.4 show the relation between nucleophilicity and basicity as per the periodic table. Nitrogen is more nucleophilic than oxygen in a series of similar compounds. This is due to the greater availability of electron pair in nitrogen atom for bonding from less electronegative nitrogen atom.

Table 10.3 Relation of nucleophilicity to basicity for atoms in the same row of periodic table

	N atom nucleophiles	O atom nucleophiles	N vs O nucleophiles	
Decreasing nucleophilicity	$H_2\ddot{N}:^-$	$C_2H_5\ddot{O}:^-$	$H_2\ddot{N}:^-$	**Decreasing basicity**
	$C_2H_5\ddot{N}H_2$	$H\ddot{O}:^-$	$H\ddot{O}:^-$	
	$H_3\ddot{N}$	$C_6H_5\ddot{O}:^-$	$H_3\ddot{N}:$	
	$C_6H_5\ddot{N}H_2$	$CH_3CO_2^-$	$H_2\ddot{O}:$	

The correlation does not hold true for atoms if we move down a family of the periodic table. It is seen that as the atomic number increases, nucleophilicity increases but basicity decreases.

Table 10.4 Relation of nucleophilicity to basicity for atoms in the same family of periodic table

	Group V nucleophiles	Group VI nucleophiles	Group VII nucleophiles	
Decreasing nucleophilicity	$R_3P:$	$R\ddot{S}:^-$	$\ddot{I}:^-$	**Decreasing basicity**
	$R_3N:$	$R\ddot{O}:^-$	$:\ddot{Br}:^-$	
			$:\ddot{Cl}:^-$	
			$:\ddot{F}:^-$	

The fundamental difference lies in the fact that basicity is an equilibrium phenomenon that measures the position of equilibrium of a reagent with proton (commonly water). Nucleophilicity, on the other hand, depicts the kinetics of a reaction involving carbon atom usually in non-aqueous solvents. The size and shape of a reagent is yet another feature associated with nucleophilicity; for example, methoxide and *tert*-butoxide ions have similar basicity.

Due to the bulky nature of *tert*-butoxide ion, there will be certain stereochemical constraints while attacking a carbon atom, thereby reducing its ability to function as an efficient nucleophile in comparison to a lighter methoxide ion.

Methoxide ion

tert-Butoxide ion

10.6 ELIMINATION REACTIONS

Alkyl halides also undergo elimination reactions wherein the halogen group from one carbon and hydrogen atom from another carbon is removed resulting in a double bond (pi bond). The carbon atom bonded to the halogen atom is called α-*carbon*. The carbon adjacent to α-carbon is the β-*carbon*. As the elimination begins with removal of hydrogen from β-carbon, such a reaction is called *1,2–elimination* or β-elimination. Hence, the product of elimination is generally an alkene or alkyne. Dehydration, dehydrogenation, dehalogenation, and dehydrohalogenation are common examples of elimination reactions.

The removal of hydrogen halide (HX) from an alkyl halide is called *dehydrohalogenation*. This is one of the most common methods to introduce a π bond and prepare an alkene. Dehydrohalogenation is an example of α-elimination since it involves the loss of elements from two adjacent atoms, that is, the α-carbon bonded to the leaving group X and the carbon atom adjacent to it. As shown in Fig. 10.13 the elimination mechanism has the following steps.

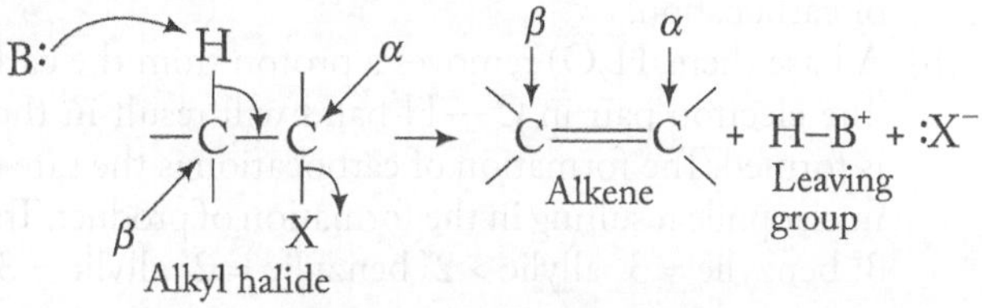

Fig. 10.13 Dehydrohalogenation of alkyl halide

(a) The base (B:) abstracts a proton from the β-carbon and forms H—B⁺.
(b) The electron pair in the β C—H bond forms a new π bond between the α- and β-carbon atoms.
(c) The electron pair in the C—X bond ends up on the halogen, forming the leaving group, : X⁻.

Some examples of elimination reactions are:

a.

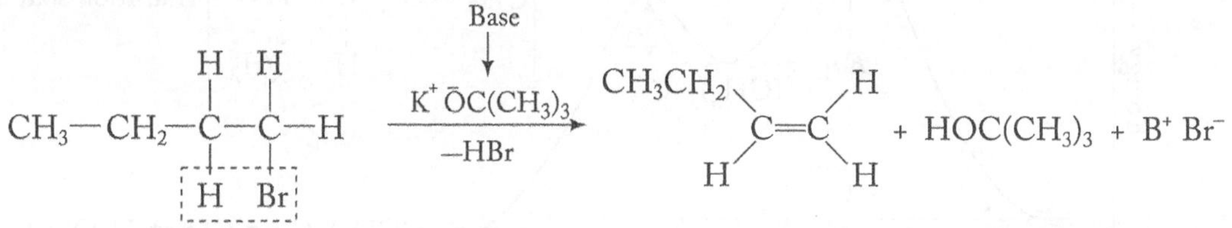

1-Bromobutane   1-Butene

b.

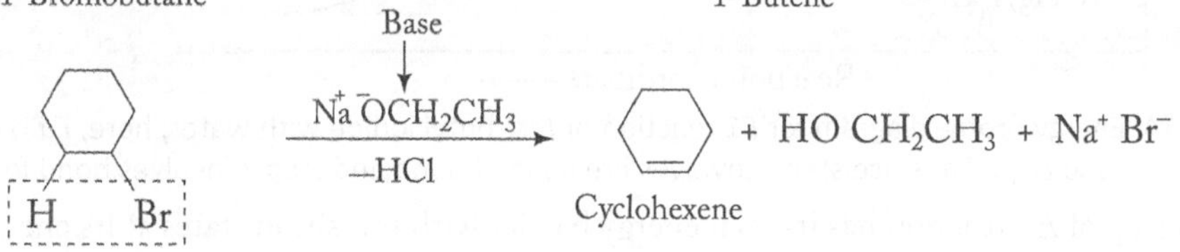

Cyclohexyl bromide   Cyclohexene

Examples (a) and (b) are some common dehydrohalogenation reactions. Sodium hydroxide, potassium hydroxide, sodium methoxide, sodium ethoxide, potassium *tert*-butoxide are commonly employed bases in elimination reactions.

Types of elimination reactions

Similar to nucleophilic substitution reactions, there are two types of elimination reactions, namely E_1 and E_2 reactions, where E denotes elimination and the number denotes the molecularity of the reaction. During elimination reaction, the nucleophile attacks the organic substrate molecule (alkyl halide).

E1 Mechanism

When *tert*-butyl iodide reacts with water it forms 2-methyl propene and follows E1 mechanism. The mechanism of E_1 reaction can be shown as:

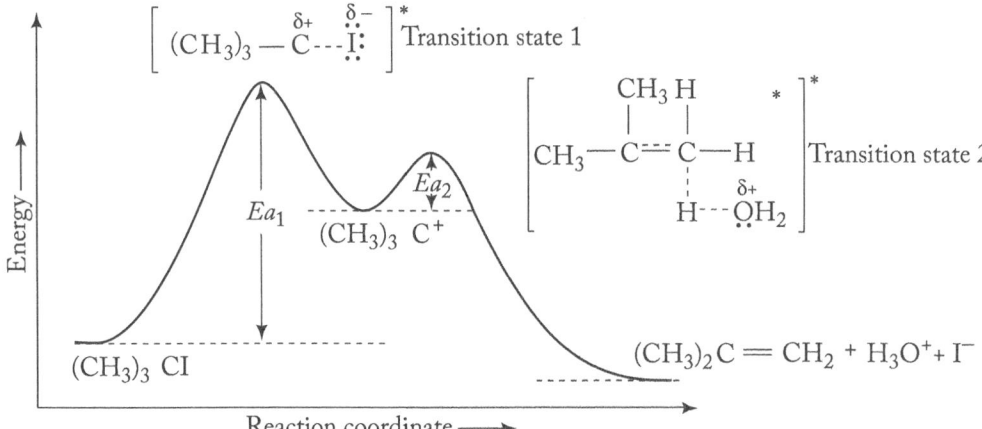

$E1$ mechanism is a two-step reaction

(a) The first step involves heterolytic dissociation of —C—I bond of alkyl halide leading to the formation of carbocation.

(b) A base (here, H_2O) removes a proton from the carbon adjacent to the carbocation (i.e., β carbon). The electron pair in C—H band will result in the formation of a new π bonds, that is, an alkine is formed. The formation of carbocation is the rate determining step followed by rapid attack of the nucleophile resulting in the formation of product. The reactivity of alkyl halides decreases in the order: $3°$ benzylic $\approx 3°$ allylic > $2°$ benzylic $\approx 2°$ allylic $\approx 3° > 1°$ benzylic $\approx 1°$ allylic $\approx 2° > 1° >$ vinyl

As $E1$ reaction involves the formation of carbocations, the carbon skeleton may undergo rearrangement before proton abstraction to form a more stable carbocation.

A typical energy profile diagram for E_1 reaction is shown in Fig. 10.14.

Fig. 10.14 Energy profile diagram for E1 reaction of *tert*-butyl iodide with water; here, *Ea* is energy of activation and $Ea_1 > Ea_2$ since step 1 involves breaking of bond and step 2 involves bond formation

Each step of $E1$ reaction has its own energy barrier with transition states at its energy maximum peak. $E1$ reaction follows first-order kinetics and the rate of the reaction depends on concentration of the alkyl halide.

Zaitsev rule Alexander Zaitsev formulated a rule for easier prediction of alkene product likely to be favoured in elimination reactions. Named as Zaitsev's rule (or Saytzeff rule), it states that, 'preferred alkene formed in greater amount is the one that corresponds to hydrogen removal from β-carbon having the fewest hydrogen substituents.' In the reaction of 2-chloropentane with water, there are two possibilities and as per the rule, 2-pentene is the major product.

Saytzeff rule is a mere generalization about the regiochemistry of elimination reactions and gives no idea about the stereochemistry of the alkene product.

E2 Mechanism

The reaction of hydroxide on tertiary butyl bromide forming 2-methyl propene is an example of an E2 reaction.

$$H_3C-\underset{\underset{\text{Br}}{|}}{\overset{\overset{\text{CH}_3}{|}}{C}}-CH_3 \xrightarrow[-Br, -H_2O]{OH^-} H_2C=\underset{}{\overset{\overset{\text{CH}_3}{|}}{C}}-CH_3$$

tert-Butyl bromide

2-Methyl propene

The mechanism of E2 reaction is,

$$HO: \quad \overset{H}{\underset{H_2C}{\frown}} \overset{CH_3}{\underset{|}{\overset{|}{C}}}-CH_3 \xrightarrow{-Br, -H_2O} H_2C=\overset{\overset{CH_3}{|}}{C}-CH_3$$

A proton is removed   Br
Bromide ion is eliminated

The mechanism of E2 reaction is a one-step reaction where the nucleophile attacks the *tert*-butyl bromide. Both the proton and bromide ions are eliminated in one step without any formation of an intermediate. The decreasing order of reactivities of alkyl halides in E2 reaction will follow the order: RI > RBr > RCl > RF.

A typical energy profile diagram for E2 mechanism is shown in Fig. 10.15.

Overall, E2 reactions follow second-order kinetics and the rate of reaction depends on the concentration of alkyl halide and the incoming nucleophile.

Table 10.5 Characteristics of E1 reaction mechanism

Features	Outcome
Mechanism	Two steps
Type of alkyl group, i.e., R	Highly substituted halides react faster $R_3CX > R_2CHX > RCH_2X > CH_3X$
Energy profile	Two energy barriers first –order kinetics Rate of reaction = k [alkyl halide] where k is rate constant.
Base	Favoured by weaker bases like H_2O, alcohols.
Solvent	Polar protic solvents; favour E1 reactions as they solvate ionic intermediates formed in the mechanism.

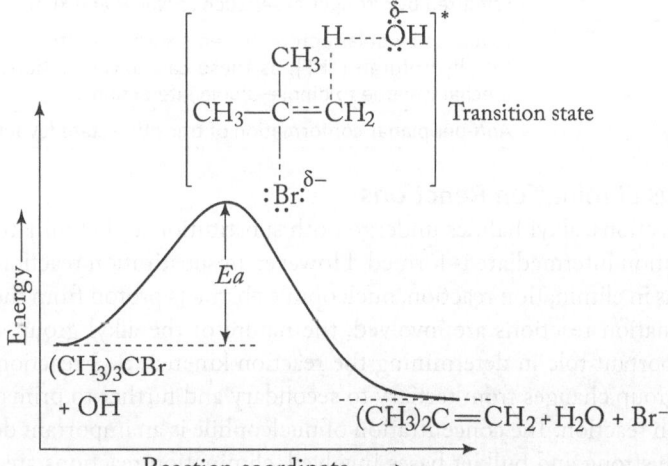

Fig. 10.15 Energy profile diagram for E2 reaction mechanism of *tert*-butyl bromide with sodium hydroxide

Stereochemistry of E2 reaction

The product formed in an E2 reaction does not possess tetrahedral stereogenic centre, but its transition state contains four atoms which react simultaneously that seems like working in a typical stereochemical manner.

There are four atoms in a transition state, namely two carbon atoms, one hydrogen, and one leaving group that can align in two ways, that is, *syn*-periplanar and *anti*-periplanar as shown in Fig. 10.16.

If H and X atoms are oriented on the same side of the molecule, it is called *syn*-periplanar, whereas if they are on opposite sides of the molecule, it is called *anti*-periplanar.

Since, *anti*-periplanar conformation resembles staggered conformation, it is clear that E2 elimination is favourable when it assumes *anti*-periplanar conformation (Fig. 10.17). In the same way, *syn*-periplanar conformation resembles a typical eclipsed conformation and is less preferred orientation in E2 reaction.

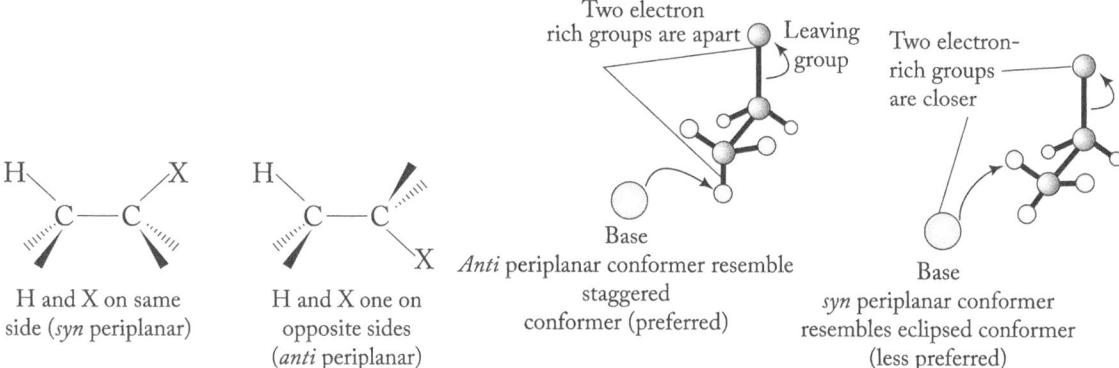

H and X on same side (*syn* periplanar)	H and X one on opposite sides (*anti* periplanar)	*Anti* periplanar conformer resemble staggered conformer (preferred)	*syn* periplanar conformer resembles eclipsed conformer (less preferred)

Fig. 10.16 Orientation of groups in E2 transition state   **Fig. 10.17** Possible orientations of E2 transition state

Table 10.6 Characteristics of E2 reaction mechanism

Features	Outcome
Mechanism	One step
Type of alkyl group, i.e., R	Highly substituted halides react faster $R_3CX > R_2CHX > RCH_2X > CH_3X$
Energy profile	One energy barrier, Second-order kinetics Rate of reaction = k [alkyl halide] [base] where k is rate constant.
Base	Favoured by stronger bases such as NaOH and KOH.
Solvent	Favoured by polar aprotic solvents such as acetone, acetonitrile, and tetrahydrofuran (THF), as these can solvate cations formed in the mechanism due to dipole–dipole interactions.
Stereochemistry	*Anti*-periplanar conformation of transition state favours products.

Substitution versus Elimination Reactions

As seen in previous sections, alkyl halides undergo both substitution and elimination reactions. In both the reactions, carbocation intermediate is formed. However, in substitution reaction, nucleophile attacks the α-carbon, whereas in elimination reaction, nucleophile abstracts proton from the β-carbon. Whether substitution or elimination reactions are involved, the nature of the alkyl group and concentration of the base plays an important role in determining the reaction kinetics. S_N2 reaction is favoured over E2 reaction if the alkyl group changes from tertiary to secondary and further to primary, whereas reverse is the case in elimination reaction. The concentration of nucleophile is an important decider of the reaction dynamics. If there are strong and bulkier bases involved, elimination reactions are favoured. Table 10.7 provides an overall comparison of these two reaction mechanism.

Table 10.7 Substitution versus elimination reactions

	Preferred Mechanism	
Alkyl halide	**S$_N$1 versus E1**	**S$_N$2 versus E2**
Primary	No reaction.	Mainly substitution. If steric hindrance in alkyl halide or nucleophile, elimination is favoured.
Secondary	Both are possible.	Both. Stronger nucleophiles prefer elimination.
Tertiary	Both are possible.	Only elimination.
Reactivity of alky halide		
S$_N$1: 3° > 2° > 1°	E1: 3° > 2° > 1°	
S$_N$2: : 1° > 2° > 3°	E2: 3° > 2° > 1°	

10.7 KINETIC VERSUS THERMODYNAMIC CONTROL OF REACTIONS

The study of chemical reactivity involves the concepts of stoichiometry, kinetics, and thermodynamics. *Stoichiometry* helps in classifying, balancing net reactions, and calculating other quantitative parameters. *Reaction kinetics* tells us about rate laws, molecularity, and reaction mechanism of organic reactions. Further, *thermodynamics* tells us about reaction efficiency, chemical equilibrium, and stabilities of products. The study of kinetic and thermodynamic control is necessary to understand competing reactions. Consider the example of electrophilic addition of hydrogen bromide to a conjugated diene, 1-3-butadiene, which yields 1,2- and 1,4- addition products, namely 3-bromo-1-butene and 1-bromo-2-butene.

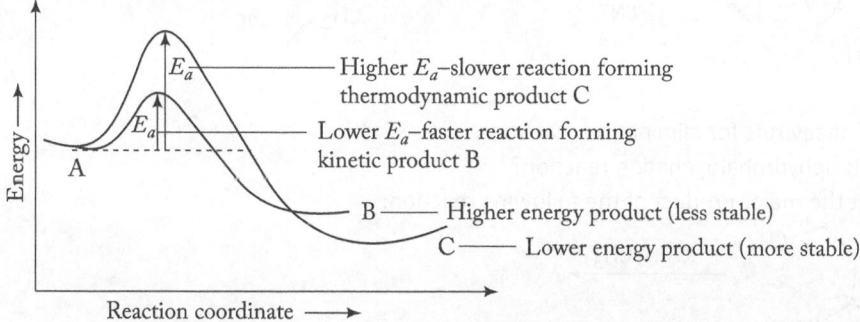

	1,3-Butadiene	3-Bromo-1-butene (1,2-product)	1-Bromo-2-butene (1,4-product)
% Yield	At low temperature (−80°C)	80 %	20 %
	At high temperature (40°C)	20 %	80 %

The 1, 2-addition product is preferentially formed quickly at lower temperatures and hence called *kinetic product*, whereas the 1, 4-product is formed slowly at higher temperatures and is called *thermodynamic product*. Most organic reaction kinetics show that products that form at a faster rate are generally stable, and in such cases, both kinetic and thermodynamic products are the same. This is not true in this case. Here, the stable 1, 4-product forms at a much slower rate than the 1,2-product. The energy profile diagram of this reaction is shown in Fig 10.18 and will help predict this behaviour.

Fig. 10.18 Kinetic vs thermodynamic products of reaction A → B + C

Energy of activation helps to determine the reaction rates, whereas the amount of product obtained at equilibrium is determined by its stability. If the starting reactant A forms a mixture of two products, B and C, the relative energy barriers will indicate how fast products B and C are formed. The conversion of reactant A to B is faster because the energy of activation is low, whereas its conversion to C will be slower due to higher energy of activation. This deviation from normal behaviour can be explained from the fact that the 1,4 product (1-bromo, 2-butene), has two alkyl groups attached to the carbon–carbon double bond, whereas there is only one alkyl group in the 1,2-product (3-bromo,1-butene). As we know, the more substituted the alkene, greater is its stability.

In case the temperature is kept sufficiently low, product C will be formed faster and effectively; this will be an irreversible reaction making C a major product. This is called *kinetic control* and the resulting product will be called *kinetic product*. Even if we increase the temperature, product C will still be formed but the reaction will be reversible and hence reactant A will be regenerated. However, now at higher temperature, free energy will be minimized resulting in chemical equilibrium thereby forming product B. This is called *thermodynamic control* and here B will be called *thermodynamic product*. The ratio of products formed in such reactions will entirely depend on the temperature conditions. Figure 10.19 shows the energy profile diagram of the two-step mechanism of electrophilic addition of hydrogen bromide to 1,3-butadiene:

Check Your Progress

7. Complete the following reactions:

(a) 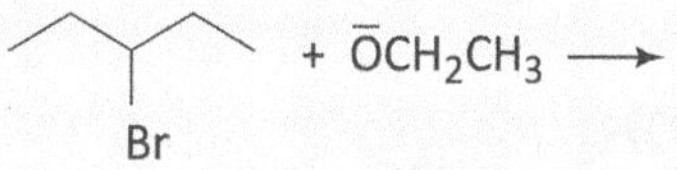 $+ \bar{O}CH_2CH_3 \longrightarrow$

(b) 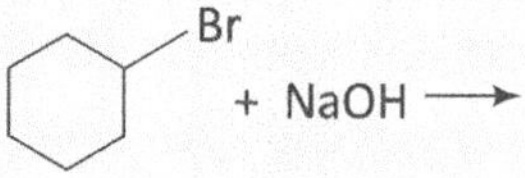 $+ NaOH \longrightarrow$

8. What type of solvents favour S_N1 and S_N2 reactions? Justify your answer.
9. Do the following groups be considered as good leaving group during nucleophilic substitution Cl^-, OH^-, NH_2, OH_2?
10. What will happen to the rate of S_N1 reaction under the following conditions?
    (a) [RX] is tripled and [: Nu^-] is kept the same.
    (b) [RX] is halved and [: Nu^-] is kept the same.
    (c) Both [RX] and [: Nu^-] are doubled.
    (d) [RX] is halved and [: Nu^-] is doubled.
11. What will happen to the rate of S_N2 reaction if concentration of both alkyl halide and nucleophile are doubled?
12. What are the stereochemical outcomes of S_N1 and S_N2 reactions?
13. Draw the product (including stereochemistry) of the following reactions:

(a) 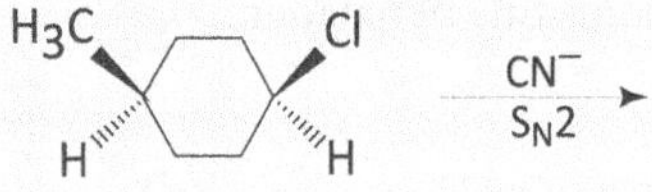 $\xrightarrow[S_N2]{CN^-}$

(b) 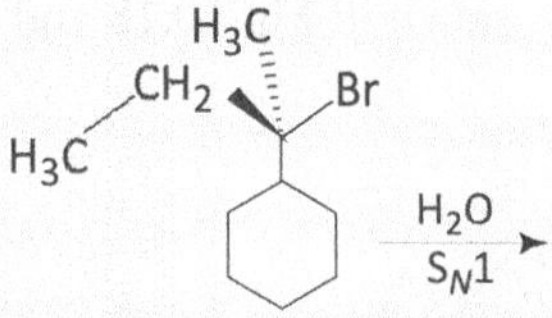 $\xrightarrow[S_N1]{H_2O}$

14. State Zaitsev rule for elimination reactions.
15. What is dehydrohalogenation reaction?
16. Predict the major product of the following reaction:

$\xrightarrow{- :OCH_2CH_3} ?$

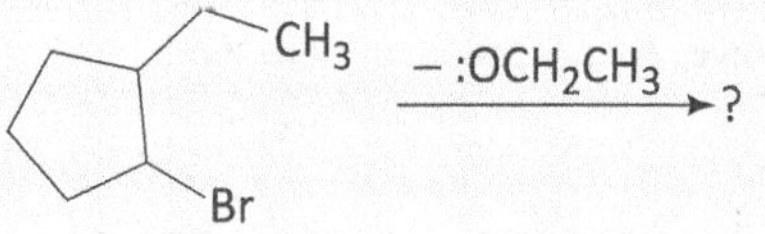

$$CH_2-CH=CH-CH_2 + H-Br \longrightarrow CH_3CH(Br)CH=CH_2 + CH_3CH=CHCH_2Br$$

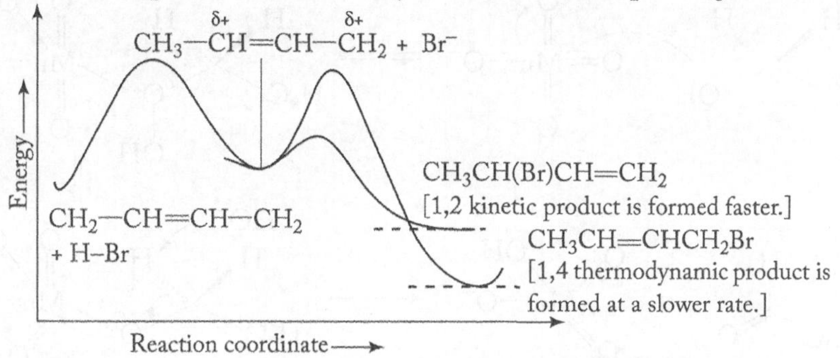

Fig. 10.19 Energy diagram for the two-step mechanism of the reaction

From the figure, it is evident that energy of activation (E_a) is a critical parameter at lower temperatures. This is due to the fact that most molecules do not have sufficient kinetic energy to overcome higher energy barrier at lower temperatures. Hence, they react via a faster pathway to form the kinetic product. At higher temperatures, the molecules possess sufficient kinetic energy to reach the transition state and form stable products.

10.8 OXIDATION REACTIONS: OXIDATION OF ALCOHOLS USING KMNO$_4$ AND CHROMIC ACID

Redox reactions refer to loss and gain of electrons, or an increase (oxidation) or a decrease in oxidation number (reduction). Oxidizing reagents generally possess O—O bond and some classic examples are oxygen, ozone, peroxides, hydroperoxides, and peroxyacids. However, it is not limited to these reagents. The reagents with metal–oxygen bonds are also known to act as oxidizing agents such as CrO_3, $K_2Cr_2O_7$, MnO_2, $KMnO_4$, OsO_4, Ag_2O, and are generally used in acidic media. Let us elucidate various functional group transformations using these oxidizing reagents. Generally, primary alcohols and aldehydes oxidize to carboxylic acids, and secondary alcohols to ketones. As tertiary alcohols do not have hydrogen atoms on carbon along with hydroxyl group, they do not exhibit oxidation.

10.8.1 Potassium Permanganate

In potassium permanganate, manganese has the highest oxidation state (+7) and hence it is a strong oxidizing agent. It oxidizes primary alcohols and aldehydes to the corresponding carboxylic acids; whereas secondary alcohols are oxidized to the corresponding ketones. The oxidation reaction is carried out in basic aqueous solution, from which MnO_2 precipitates out. After the oxidation is complete, the reaction mixture is filtered and acidified to obtain carboxylic acid.

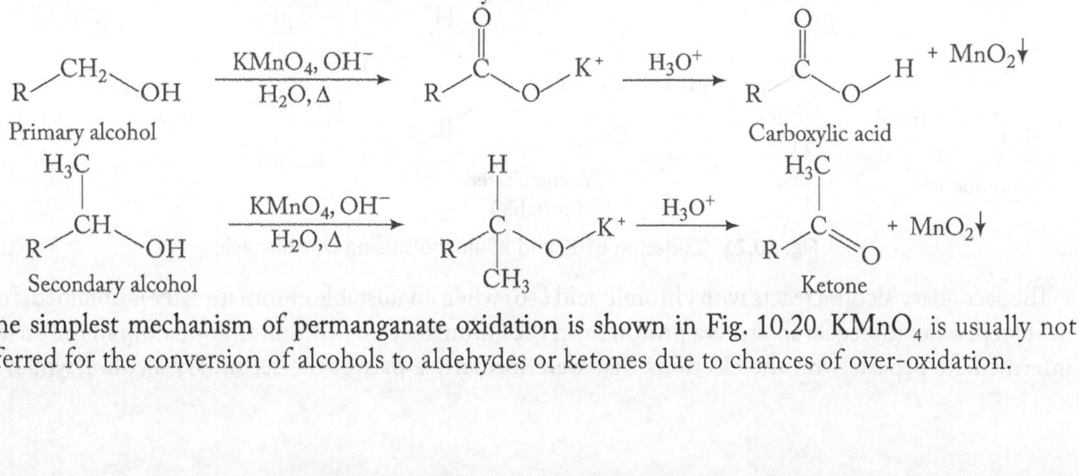

The simplest mechanism of permanganate oxidation is shown in Fig. 10.20. $KMnO_4$ is usually not preferred for the conversion of alcohols to aldehydes or ketones due to chances of over-oxidation.

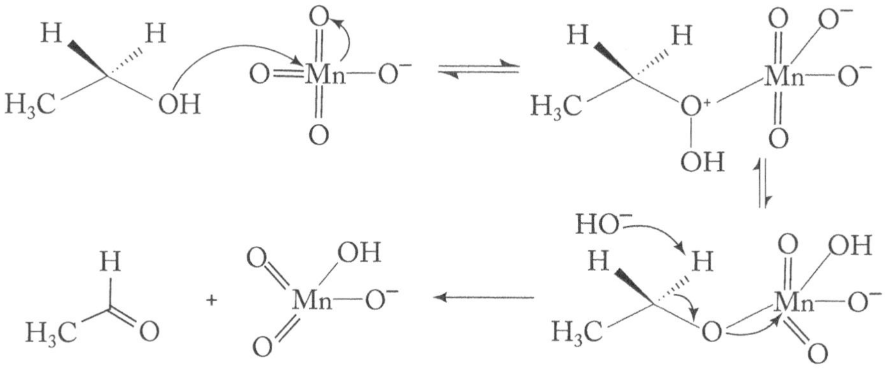

Fig. 10.20 Mechanism of permanganate oxidation

Some more examples of oxidation reaction using $KMnO_4$ as oxidizing agent are:

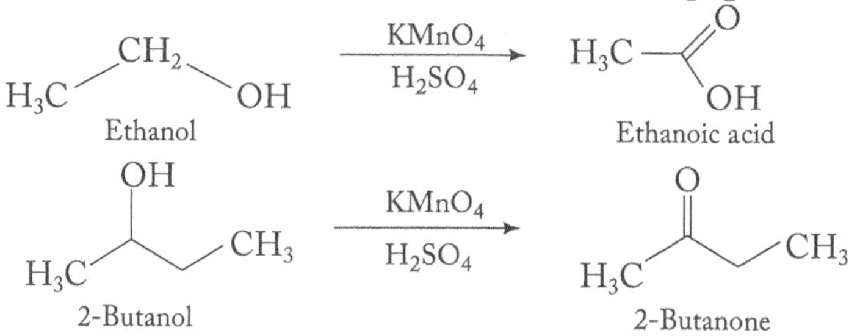

10.8.2 Chromic Acid

Chromium trioxide on reaction with dilute sulphuric acid (CrO_3/H_2SO_4) in situ forms chromic acid and is popularly called *Jones reagent*. It is used for the oxidation of primary and secondary alcohols to carboxylic acids and ketones, respectively. Chromic acid reacts with alcohol to form an intermediate chromate ester, which in the presence of a base (water in this case) forms a carbonyl compound. The mechanism of the reaction (Fig. 10.21) can be explained as follows.

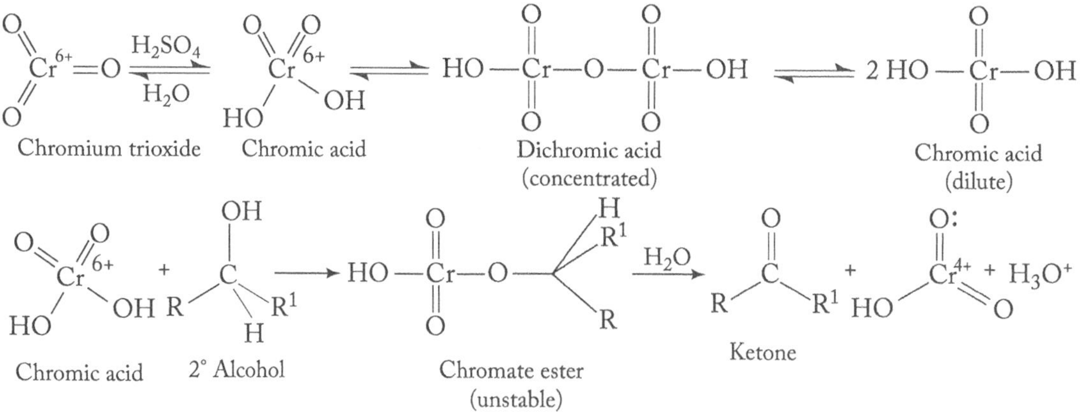

Fig. 10.21 Oxidation of 1° and 2° alochols using chromic acid

The secondary alcohol reacts with chromic acid (+6) when an unstable chromate ester is obtained. In the next step, water molecule abstracts a proton from the chromate ester. The chromium atom in the unstable intermediate departs with the electrons that were formerly a part of the secondary alcohol, which gets oxidized to form ketone. Chromium itself undergoes reduction and acquires +4 charge. Chromic acid oxidation is not useful for acid-sensitive organic substrates. Some examples of chromic acid oxidation are as follows:

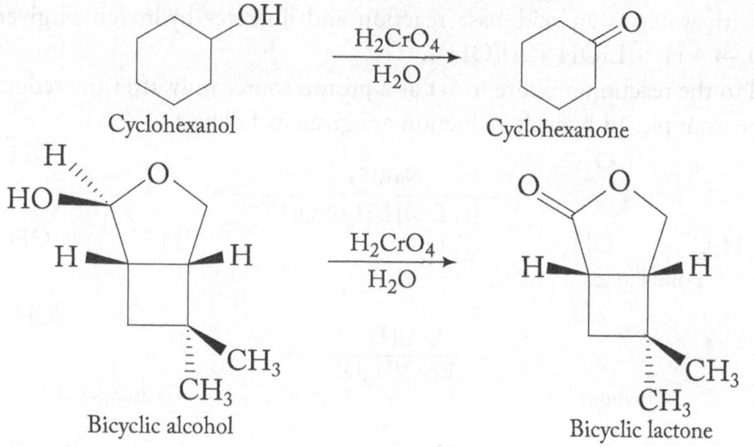

Cyclohexanol Cyclohexanone

Bicyclic alcohol Bicyclic lactone

10.9 REDUCTION REACTIONS

Complex metal hydrides are a class of reagents used as reducing agents for various carbonyl compounds. Lithium aluminium hydride ($LiAlH_4$) and sodium borohydride ($NaBH_4$) are two commonly used reagents to reduce aldehydes to primary alcohols and ketones to secondary alcohols. Both the reagents contain a polar metal–hydrogen bond that serves as a source of the nucleophile hydride, ($H:^-$). $LiAlH_4$ is a stronger reducing agent than $NaBH_4$, because the Al—H bond is much more polar than the B—H bond.

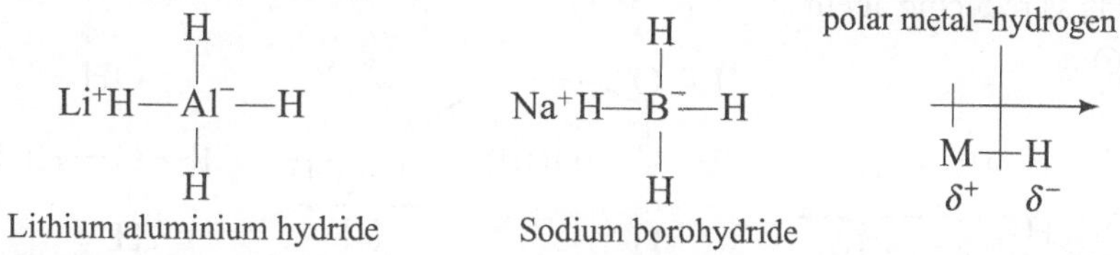

Lithium aluminium hydride Sodium borohydride

Lithium aluminium hydride is prepared by treating lithium hydride with an ethereal solution of aluminium chloride as per the reaction,

$$4\,LiH + AlCl_3 \xrightarrow{\text{ether}} LiAlH_4 + 3\,LiCl$$

Sodium borohydride is prepared by reacting sodium hydride with trimethyl borate at 250°C as per the reaction,

$$B(OCH_3)_3 + 4\,NaH \xrightarrow[\Delta]{250°C} NaBH_4 + 3\,NaOCH_3$$

The general reduction reaction using hydride reagents can be shown as:

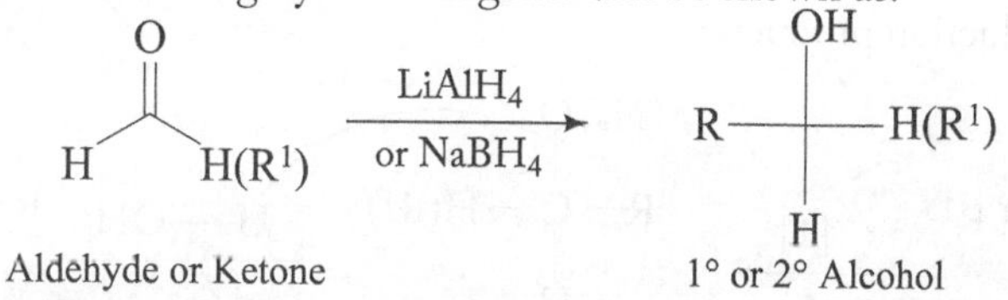

Aldehyde or Ketone 1° or 2° Alcohol

Check Your Progress

17. Name any two oxidizing and reducing agents studied in this chapter.
18. Illustrate the oxidation mechanism carried out in the presence of $KMnO_4$.
19. Illustrate the mechanism of oxidation using chromic acid.

These reduction reactions are usually carried out in an ether solvent such as diethyl ether or tetrahydrofuran. Reductions with $NaBH_4$ are typically carried out in methanol as solvent. Anhydrous conditions are maintained during the reduction reactions because lithium aluminium hydride reacts violently with water as an acid–base reaction and liberates hydrogen as given in the reaction, $LiAlH_4 + 4\ H_2O \rightarrow 4\ H_2 + LiOH + Al(OH)_3$.

Water is added to the reaction mixture to act as a proton source only after the reduction with $LiAlH_4$ is complete. Some examples of hydride reduction are given as follows:

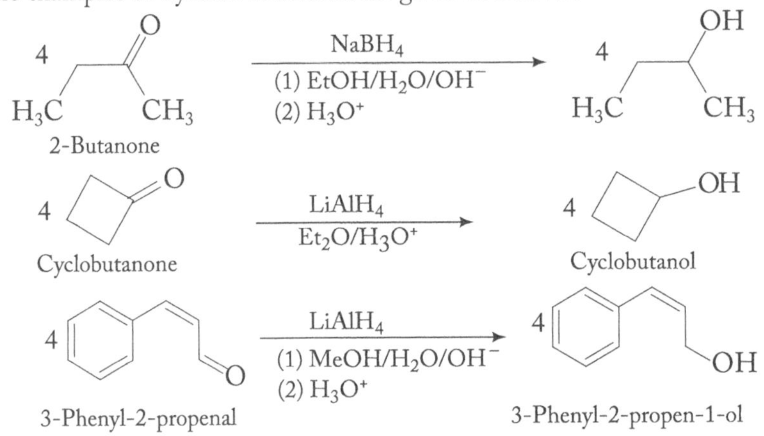

Sodium borohydride is a less reactive but more selective reducing agent than lithium aluminium hydride. Hydride reduction of aldehydes and ketones occurs by the general mechanism of nucleophilic addition, that is, nucleophilic attack followed by protonation. Figures 10.22(a) and (b) show the mechanism using $LiAlH_4$ and $NaBH_4$ as reducing agents.

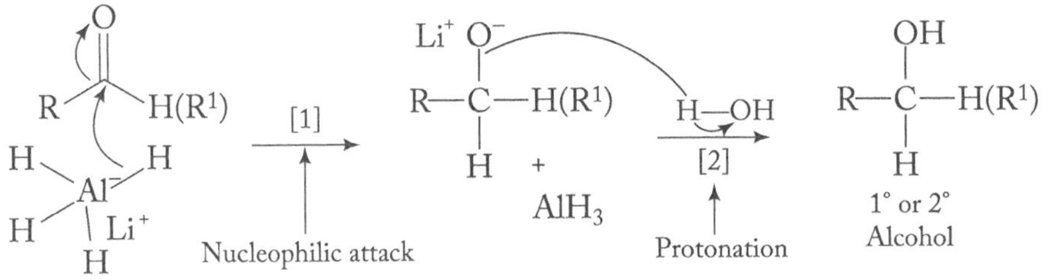

Fig. 10.22 (a) $LiAlH_4$ reduction of carbonyl compound

Step 1 involves donation of $H:^-$ from the nucleophile (AlH_4^-) to the carbonyl group of the aldehyde or ketone. This is followed by fission of the π bond of $C{=}O$ group and movement of an electron pair on to oxygen as shown in Fig. 10.22 (a). This results in the formation of a new $C{-}H$ bond. In step 2, the alkoxide is protonated by a solvent such as water or methanol (acid–base reaction) to form the final alcohol, which is the reduction product.

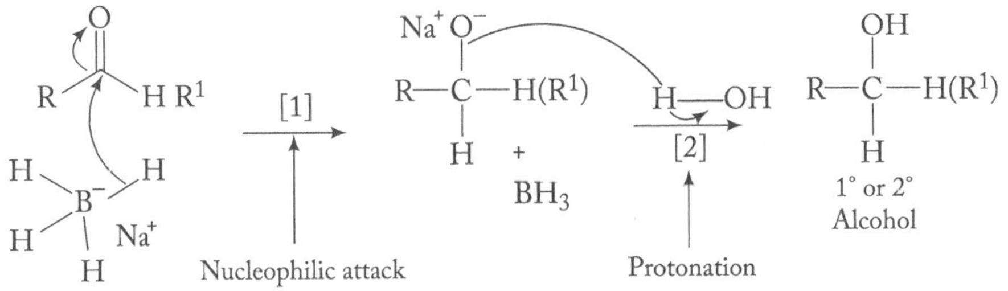

Fig. 10.22 (b) Mechanism of $NaBH_4$ reduction of carbonyl compound

Stereochemically, the approach of hydride normally takes place with equal probability from either side of the planar carbonyl group. If a chiral centre is formed in the conversion of carbon atom from trigonal to tetrahedral form, the alcohol product will be a racemic mixture, that is, optically inactive. When 2-pentanone is treated with sodium borohydride in anhydrous ethanolic solvent, R-2-pentanol and S-2-pentanol are obtained as a racemic product in equimolar quantities.

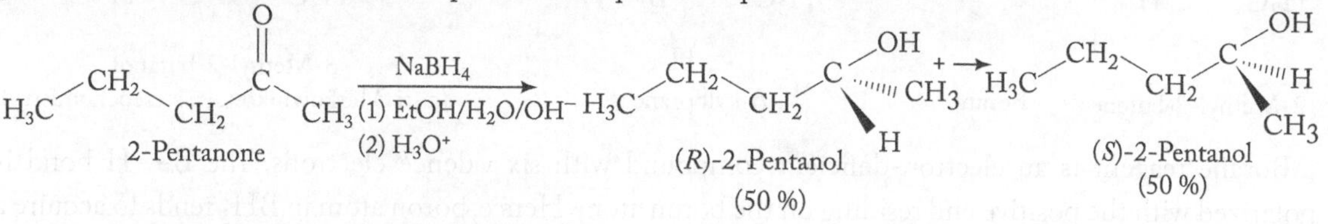

10.10 HYDROBORATION OF ALKENES

When alkenes (olefins) undergo hydration, alcohols are formed as per Markownikov's rule and such reactions are regioselective and stereospecific. Hydroboration reaction is a typical hydration method that transforms alkenes to anti-Markonikov alcohol. Studies by H.C. Brown (1979) showed that diborane (B_2H_6) adds to alkenes forming alkylboranes which oxidize to alcohols. Diborane (B_2H_6) reagent is a hydride source comprising two molecules of borane (BH_3). Diborane is extremely toxic and a flammable gaseous reagent used as a complex with tetrahydrofuran solvent ($BH_3 \cdot THF$). Diborane is obtained by reacting sodium borohydride with boron trifluoride in the presence of the solvent, diglyme.

$$3\,NaBH_4 + 4\,BF_3 \xrightarrow{\text{diglyme}} 2\,(BH_3)_2 + 3\,NaBF_4$$

Sodium borohydride   Boron trifluoride   Diborane

$$(BH_3)_2 \rightleftharpoons 2\,BH_3$$

Diborane   Borane

Borane adds to the double bond of an alkene to form alkylborane. Next, hydrogen peroxide oxidizes the alkylborane to an alcohol as per anti-Markownikov's rule. Another interesting feature is that oxidation results in the retention of configuration as the hydroxy group replaces the boron atom. Generally, three molecules of alkene react with one mole of borane to form tri-alkylboranes. Each B—H bond can add across the double bond of the alkene forming trialkylboranes. The formation of alkylboranes is highly spontaneous and tends to be inflammable, hence they are not isolated. The general stoichiometric hydroboration reaction can be depicted as:

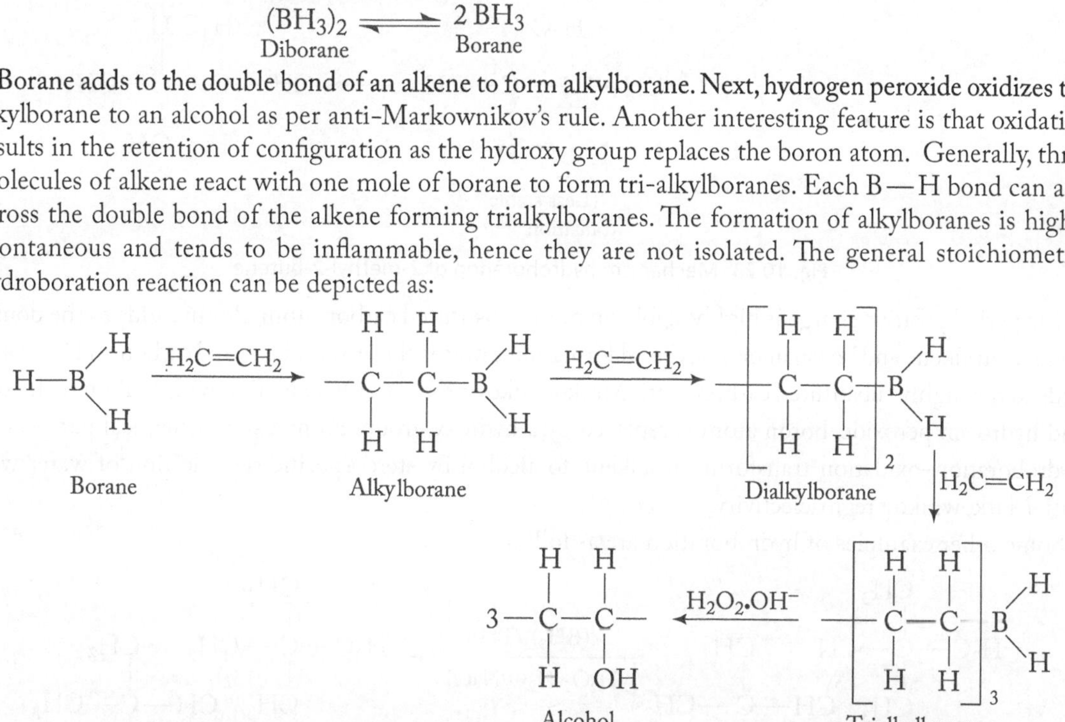

It is observed that these reactions are regioselective and boron atom tends to add to a lesser substituted carbon atom with hydrogen bonding to a more substituted carbon atom.

The mechanism of hydroboration can be explained by taking the example of borane addition to 2-methyl-2-butene forming 3-methyl-2-butanol.

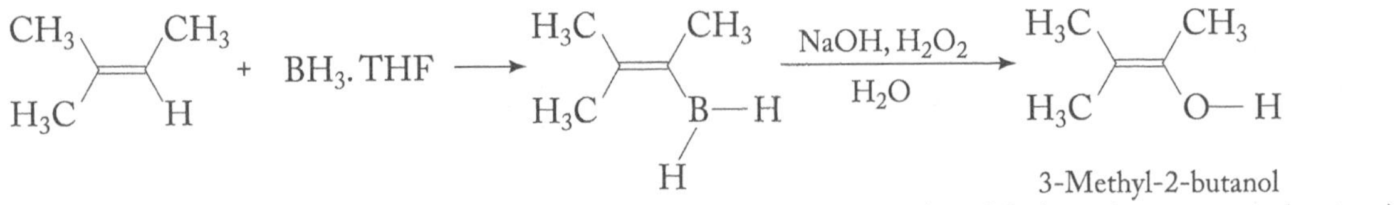

| 2-Methyl-2-butene | Borane | Alkylborane | 3-Methyl-2-butanol (anti-Markownikov *syn* stereochemistry) |

Borane reagent is an electron-deficient compound with six valence electrons. The B—H bond is polarized with the positive end residing on the boron atom. Hence, boron atom in BH_3 tends to acquire a complete octet that forms unusual bridge bonds with hydrogen. BH_3 is a strong electrophilic reagent and in the transition state, boron atom attracts electrons from the π bond resulting in partial positive charge on carbon (Fig. 10.23). The mechanism of hydroboration for the above reaction can elucidated as below:

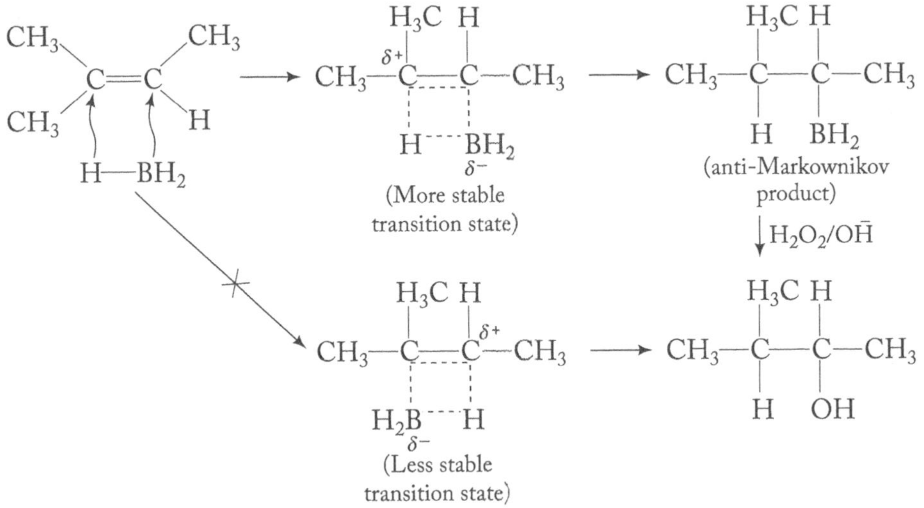

Fig. 10.23 Mechanism hydroboration of 2-methyl-2-butene

The partial positive charge is highly stable on more substituted carbon atom. Borane adds to the double bond of an alkene and boron undergoes bond formation towards the lesser substituted carbon and hydrogen adds to the highly substituted carbon (anti-Markownikov product). On oxidation with sodium hydroxide and hydrogen peroxide, boron atom is replaced by a hydroxy group. From a stereochemical perspective, hydroboration–oxidation transforms an alkene to alcohol by stereospecific *syn*-addition of water with anti-Markownikov regioselectivity.

Some other examples of hydroboration are as follows:

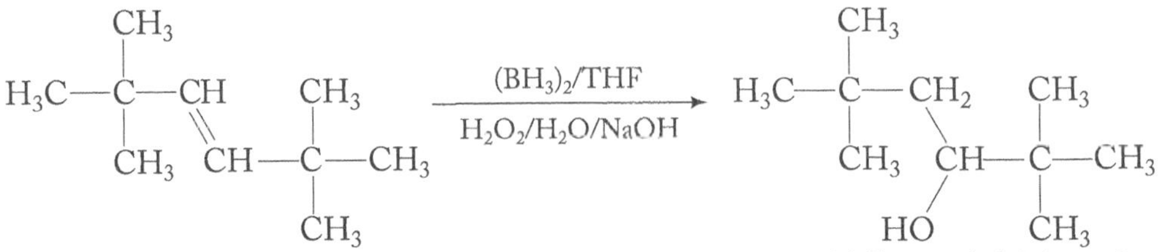

| 2,2,5,5-Tetramethyl-3-hexene | 2,2,5,5-Tetramethyl-3-hexanol |

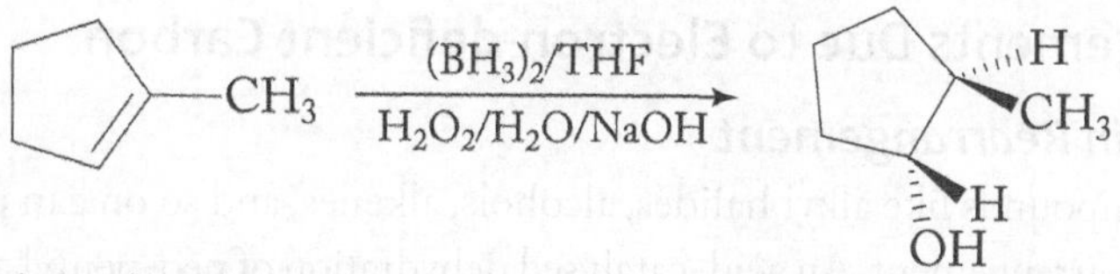

1-Methyl cyclopentene   *trans*-2-Methyl cyclopentanol

Hydroboration reactions also occur with carbon–carbon triple bonds, though not yet explored to a greater detail and are beyond the scope of this book.

10.11 REARRANGEMENT REACTIONS

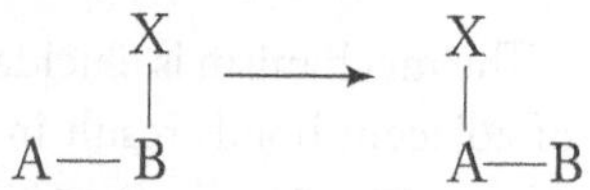

Rearrangement reactions involve the migration of an atom or a group from one centre (called migration origin) to another (called migration terminus) within the same organic molecule. The migrations depend on the reaction conditions, nature of rearrangement, etc. Say, we have a molecule ABX, where X is the migrating group attached to B which can rearrange and migrate from B to A as shown above:

Here, X is the migrating group, atom B is the migration origin, and atom A is the migration terminus. On the basis of the *nature* of the migrating atom or group, rearrangement reactions are classified into the following three types.

(a) Electrophilic: migrating group migrates without its electron pair.
(b) Nucleophilic: migrating group migrates with its electron pair.
(c) Free radical: migrating group migrates with only one electron.

Rearrangement reactions can be intermolecular or intramolecular.

Intermolecular rearrangement The migrating group (X,Y) is detached from its migration origin (A) from two different molecules. In this case, migration of a group/atom can take place to a different molecule.

$$X—A—B + Y—A—C \longrightarrow A—B—Y + A—C—X$$

Intramolecular rearrangement The migrating group (X) does not completely detach from the migration origin and occurs within the same organic molecule.

$$X—A—B \longrightarrow A—B—X$$

In this section we will focus on the simpler skeletal rearrangements commonly observed in organic synthesis taking place at electron-deficient carbon and nitrogen.

Check Your Progress

20. Predict the product:

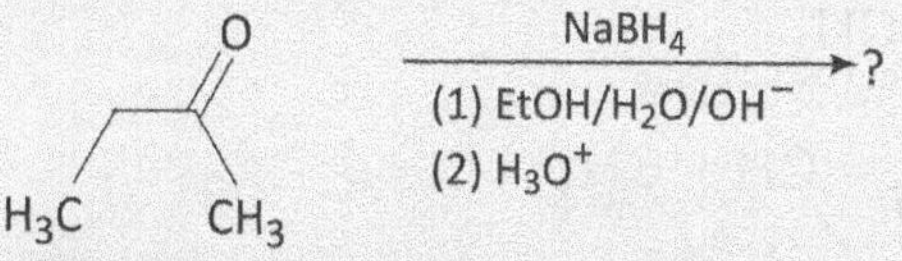

21. Illustrate the mechanism of lithium aluminum hydride reduction of carbonyl compound.
22. What is hydroboration reaction?
23. Identify the reaction and predict the products:

2-pheylpropene $\xrightarrow[\text{H}_2/\text{O}_2/\text{H}_2\text{O}/\text{NaOH}]{\text{(BH}_3\text{)}_2/\text{THF}}$

10.11.1 Rearrangements Due to Electron-deficient Carbon

Wagner–Meerwein Rearrangement

A wide variety of compounds like alkyl halides, alcohols, alkenes, and so on can undergo 1, 2-shift called Wagner–Meerwein rearrangement. An acid-catalysed dehydration of neo-pentyl alcohol to give 2-methyl-2-butene is an example depicting 1, 2-shift which can be shown as,

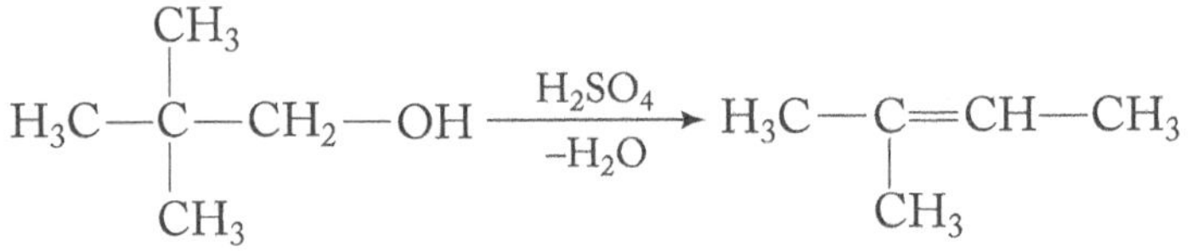

$$H_3C-\underset{\underset{CH_3}{|}}{\overset{\overset{CH_3}{|}}{C}}-CH_2-OH \xrightarrow[-H_2O]{H_2SO_4} H_3C-\underset{\underset{CH_3}{|}}{C}=CH-CH_3$$

The mechanism is elucidated (Fig. 10.24) wherein a rearrangement of carbocation using sigma electrons of adjacent bonds result in change in the carbon skeleton and can be explained in three steps.

Step 1: The first step is the generation of an electron-deficient centre or carbocation through protonation and loss of water molecule.

Step 2: The lesser stable primary carbocation undergoes rearrangement by migration of an adjacent alkyl group to an electron-deficient carbon atom giving rise to a stable tertiary carbocation. This 1, 2-alkyl shift to an electron-deficient carbon results in changes in the carbon skeleton. Alkyl group migrates with the electrons and thus is called an *internal nucleophile*.

Step 3: The rearranged carbocation either combines with a nucleophile or loses a proton forming products. Such reactions that involve methyl group migration are also called *Nametkin rearrangement*.

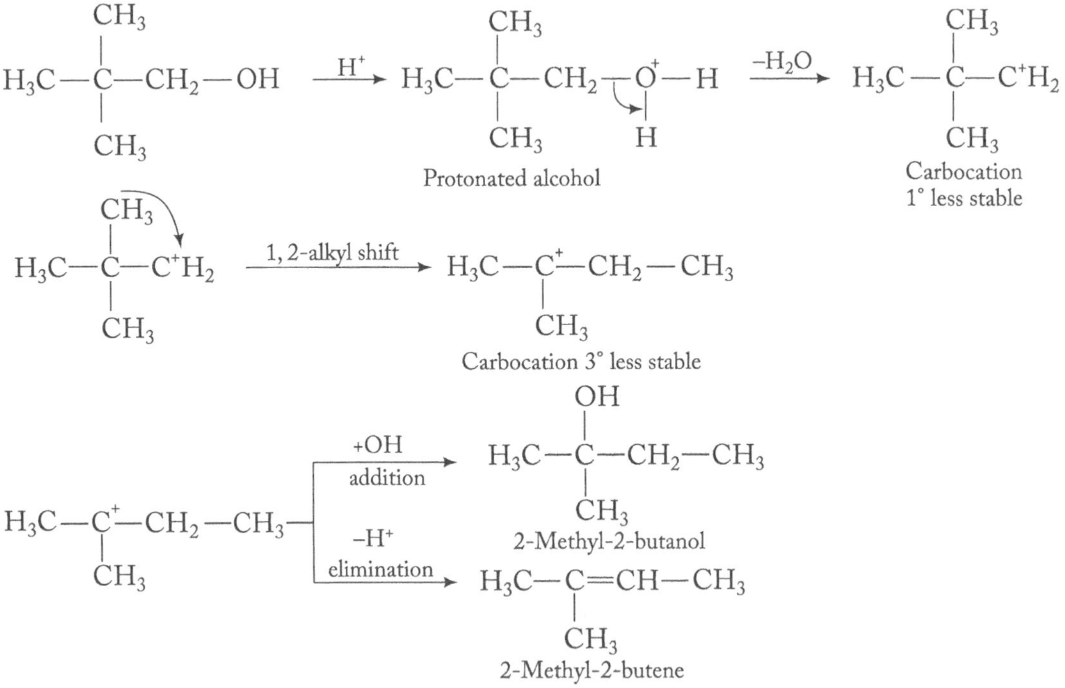

Fig. 10.24 Mechanism of Wagner–Meerwein rearrangement

A classic example of Wagner–Meerwein rearrangement reaction is the conversion of α-pinene to bornyl chloride in the presence of HCl at −20°C as per Markownikov's rule.

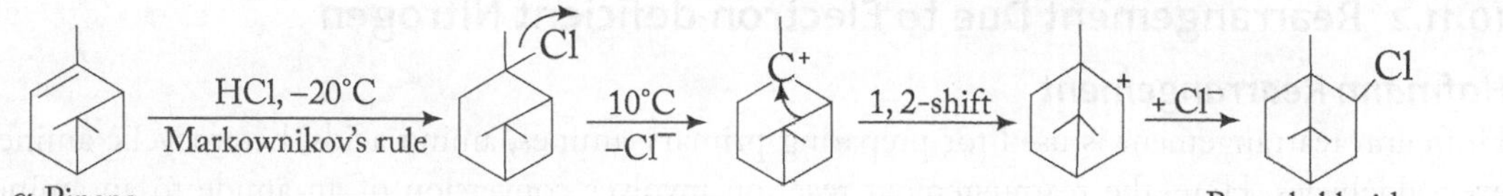

The reaction scheme at top (α-Pinene to Bornyl chloride):

$$\alpha\text{-Pinene} \xrightarrow[\text{Markownikov's rule}]{\text{HCl}, -20^\circ\text{C}} \xrightarrow[-\text{Cl}^-]{10^\circ\text{C}} \text{[carbocation]} \xrightarrow{1,2\text{-shift}} \text{[carbocation]} \xrightarrow{+\,\text{Cl}^-} \text{Bornyl chloride}$$

Pinacol–Pinacolone Rearrangement

When 1, 2-vicinal diol (also called pinacol) is heated with acids (sulphuric acid), it undergoes dehydration with molecular rearrangement to give a highly substituted ketone called pinacolone. Such a rearrangement reaction is called pinacol rearrangement.

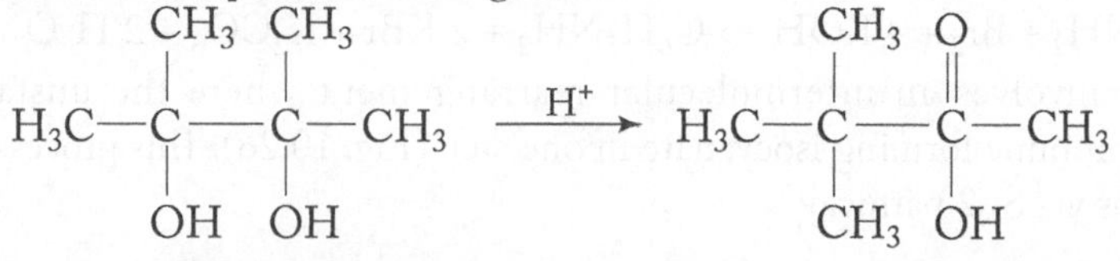

$$\underset{\text{Pinacol}}{\text{H}_3\text{C}-\overset{\overset{\text{CH}_3}{|}}{\underset{\underset{\text{OH}}{|}}{\text{C}}}-\overset{\overset{\text{CH}_3}{|}}{\underset{\underset{\text{OH}}{|}}{\text{C}}}-\text{CH}_3} \xrightarrow{\text{H}^+} \underset{\text{Pinacolone}}{\text{H}_3\text{C}-\overset{\overset{\text{CH}_3}{|}}{\underset{\underset{\text{CH}_3}{|}}{\text{C}}}-\overset{\overset{\text{O}}{\|}}{\underset{\underset{\text{OH}}{|}}{\text{C}}}-\text{CH}_3}$$

The mechanism follows 1, 2-alkyl shift as seen in Wagner–Meerwein reaction. In the first step, pinacol is protonated and loses a water molecule forming a carbocation. The adjacent alkyl group migrates with its bonding electrons to an electron-deficient carbon (1, 2-alkyl shift) to form a protonated ketone. Of the two groups, the one which is electron rich will act as the migrating group (Fig. 10.25).

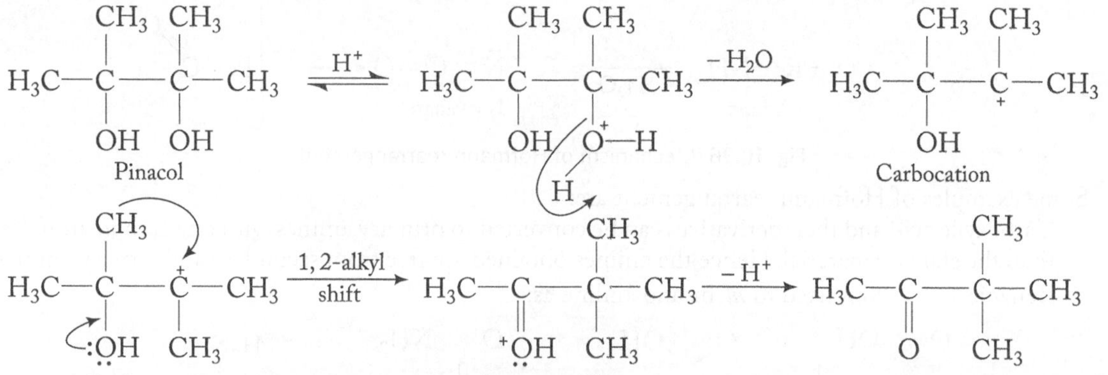

Fig. 10.25 Mechanism of pinacol–pinacolone rearrangement

If 1, 1, 2-triphenyl-1, 2-propane diol is heated with sulphuric acid, it undergoes rearrangement to form methyl triphenyl ketone. Here, diphenyl methyl carbocation (a) is preferred over benzyl carbocation (b) due to higher stability of the former and it is observed that phenyl group migrates in preference to methyl group due to its electron-rich nature.

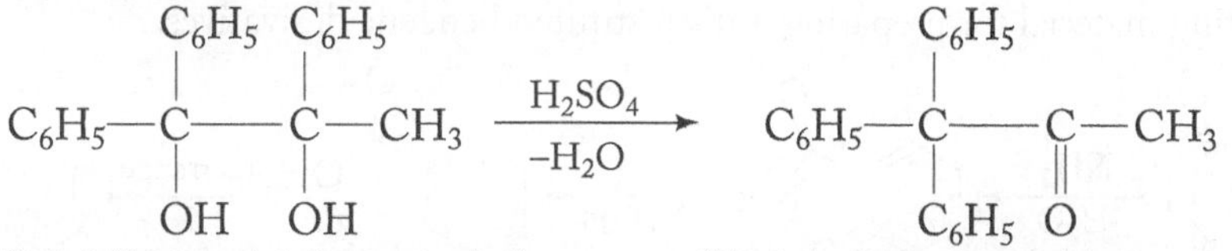

$$\underset{(1,1,2\text{-Triphenyl-1, 2-propanediol})}{\text{C}_6\text{H}_5-\overset{\overset{\text{C}_6\text{H}_5}{|}}{\underset{\underset{\text{OH}}{|}}{\text{C}}}-\overset{\overset{\text{C}_6\text{H}_5}{|}}{\underset{\underset{\text{OH}}{|}}{\text{C}}}-\text{CH}_3} \xrightarrow[-\text{H}_2\text{O}]{\text{H}_2\text{SO}_4} \underset{(\text{Methyl triphenyl methyl ketone}}{\text{C}_6\text{H}_5-\overset{\overset{\text{C}_6\text{H}_5}{|}}{\underset{\underset{\text{C}_6\text{H}_5}{|}}{\text{C}}}-\overset{\overset{}{}}{\underset{\underset{\text{O}}{\|}}{\text{C}}}-\text{CH}_3}$$

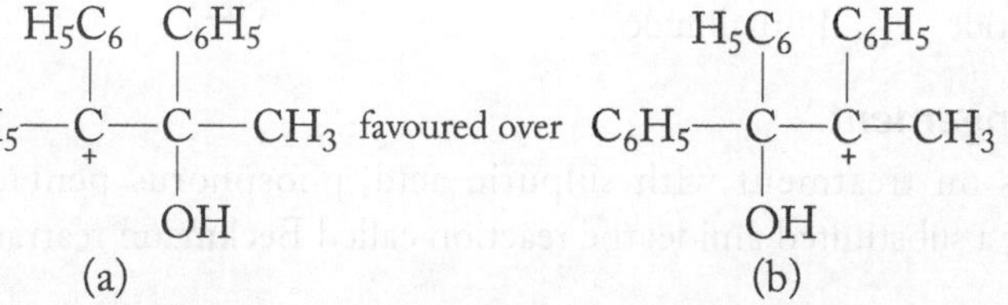

$$\underset{(a)}{\text{C}_6\text{H}_5-\overset{\overset{\text{H}_5\text{C}_6}{|}}{\underset{\underset{\text{OH}}{|}}{\text{C}}}-\overset{\overset{\text{C}_6\text{H}_5}{|}}{\underset{+}{\text{C}}}-\text{CH}_3} \quad \text{favoured over} \quad \underset{(b)}{\text{C}_6\text{H}_5-\overset{\overset{\text{H}_5\text{C}_6}{|}}{\underset{+}{\text{C}}}-\overset{\overset{\text{C}_6\text{H}_5}{|}}{\underset{\underset{\text{OH}}{|}}{\text{C}}}-\text{CH}_3}$$

10.11.2 Rearrangement Due to Electron-deficient Nitrogen

Hofmann Rearrangement

Hofmann rearrangement is used for preparing primary amines, amino acids, heterocyclic amines, and even aldehydes. Here, the rearrangement reaction involves conversion of an amide to an amine with one carbon atom less in the presence of alkaline hypohalite or bromine in alkaline solution. The overall reaction can be shown as:

$$RCONH_2 + Br_2 + 4\,KOH \rightarrow RNH_2 + 2\,KBr + K_2CO_3 + 2\,H_2O$$

Let us take an example of benzamide that rearranges to form aniline (a popular benzene derivative):

$$C_6H_5CONH_2 + Br_2 + 4\,KOH \rightarrow C_6H_5NH_2 + 2\,KBr + K_2CO_3 + 2\,H_2O$$

The mechanism involves an intermolecular rearrangement where the unstable N-bromamide intermediate loses bromine forming isocyanate in one step (Fig. 10.26). This process is quite a slow step and usually proceeds via S_N2 pathway.

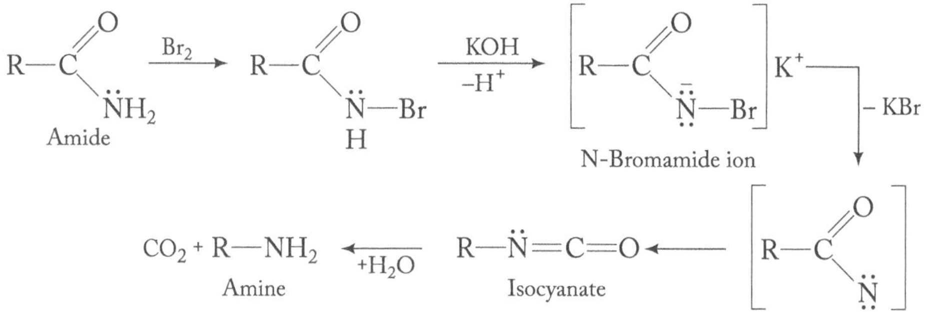

Fig. 10.26 Mechanism of Hofmann rearrangement

Some examples of Hofmann rearrangement are:

(a) Carboxylic acid and their derivatives can be converted to primary amines with one carbon atom less than the starting material. Hence the amines obtained are free from secondary and tertiary amines. Benzoic acid is converted to *m*-bromo aniline as:

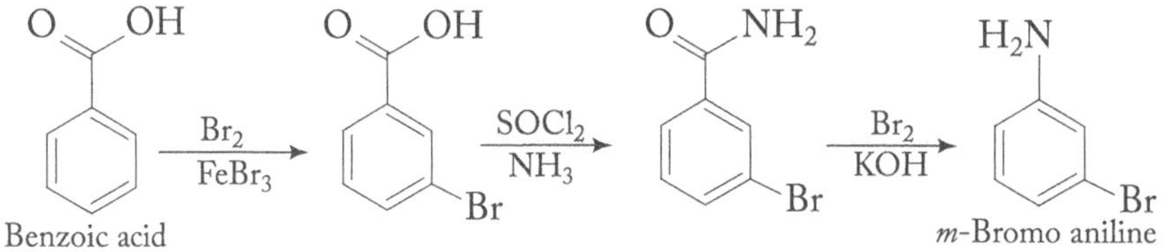

(b) An industrial application of Hofmann rearrangement is the preparation of anthranilic acid, which is the starting material for preparing *o*-disubstituted benzene derivatives.

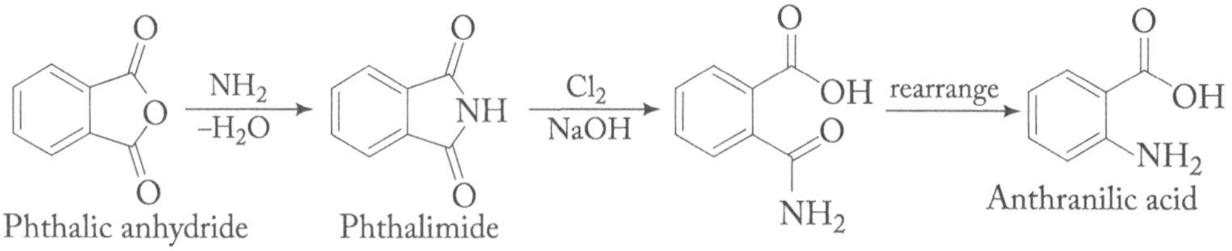

Beckmann Rearrangement

Aromatic ketoximes on treatment with sulpuric acid, phosphorus pentachloride, etc., undergo rearrangement giving a substituted amide; the reaction called Beckmann rearrangement is depicted as:

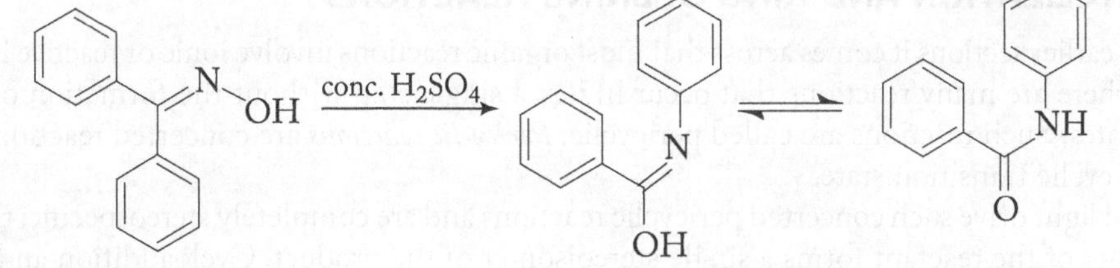

Benzophenone oxime   N-Phenyl benzamide (benzanilide)

The first step involves the reaction of an oxime with the reagent giving an intermediate that ionizes to form a carbocation. The carbocation, so formed reacts with water forming amide (enol form). It is an intramolecular rearrangement reaction where the alkyl or aryl group does not detach from the original molecule during migration. Beckmann rearrangement can be shown as per the reagent employed during the course of reaction as follows.

(a) With sulphuric acid:

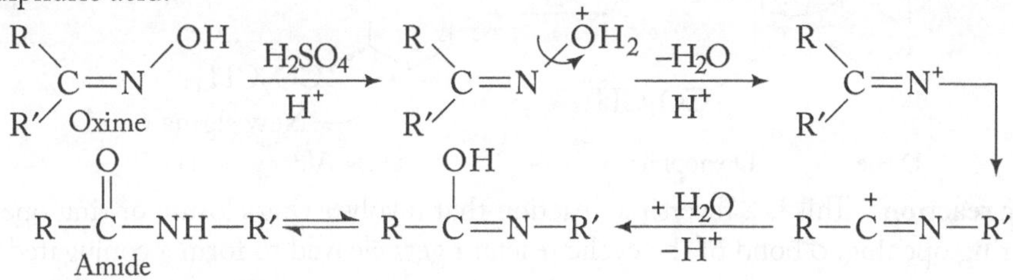

(b) With phosphorus pentachloride:

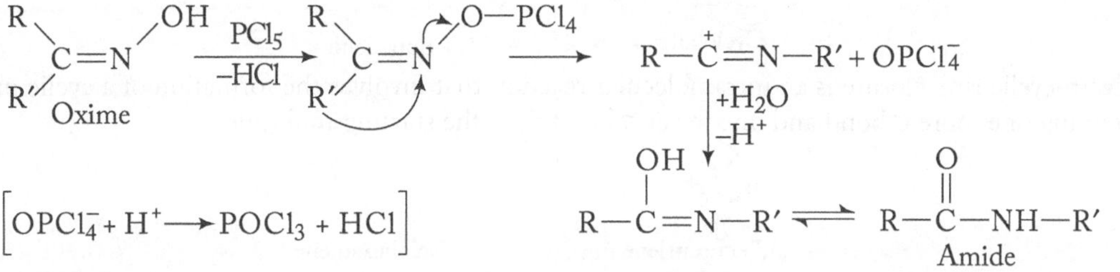

$$\left[OPCl_4^- + H^+ \longrightarrow POCl_3 + HCl \right]$$

Caprolactum, which is an important industrial raw material used to prepare nylon-6,6, is obtained by Beckmann rearrangement as:

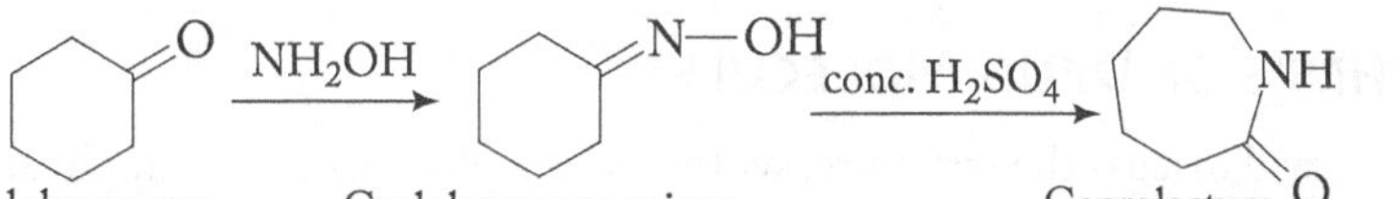

Cyclohexanone   Cyclohexanone oxime   Caprolactum

Beckmann rearrangements also enable to describe the configuration of ketoximes.

Check Your Progress

24. List the different types of rearrangement reactions.
25. Identify and complete the following reaction:

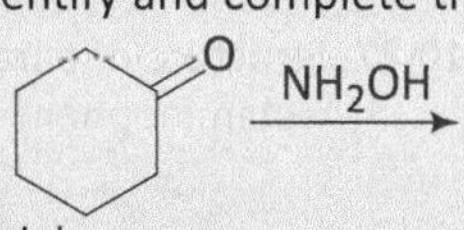

cyclohexanone

26. How can we prepare anthranilic acid using Hofmann rearrangement? Illustrate the reaction.
27. Illustrate the mechanism of Wagner–Meerwein rearrangement.

10.12 CYCLIZATION AND RING OPENING REACTIONS

As seen in earlier sections it comes across that most organic reactions involve ionic or reactive intermediates; however, there are many reactions that occur in just a single step without the formation of any reactive intermediates. Such reactions are called pericyclic. *Pericyclic reactions* are concerted reactions proceeding through a cyclic transition state.

Heat and light drive such concerted pericyclic reactions and are completely stereospecific; that is, a single stereoisomer of the reactant forms a single stereoisomer of the product. Cycloaddition and electrocyclic reactions are two types of pericyclic reactions.

Cycloaddition   It occurs between two compounds with π bonds to form a cyclic product with two new σ bonds initiated by heat or light. Diels–Alder is an example of cycloaddition reaction. In a Diels–Alder reaction, a diene and a dienophile react in the presence of heat as follows:

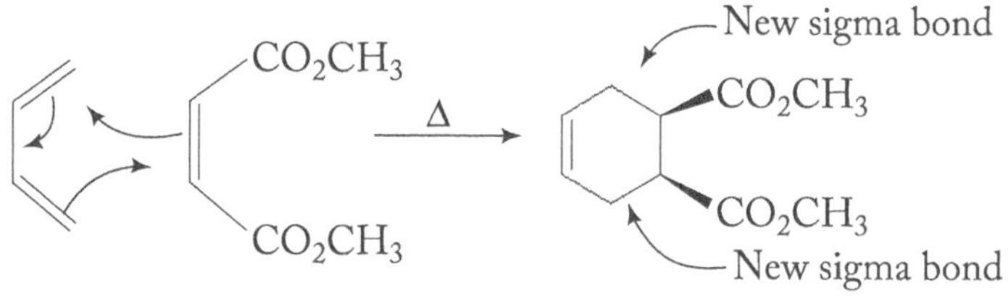

Diene   Dienophile   Diels-Alder product

Electrocyclic reaction   This is a reversible reaction that involves ring closure or ring opening. During electrocyclic ring opening, σ bond of the cyclic reactant gets cleaved to form a conjugated product with one more π bond.

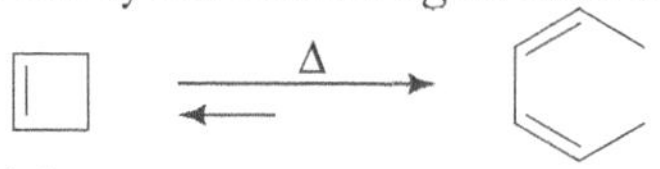

Cyclobutene   1,3-Butadiene

Electrocyclic ring closure is an intramolecular reaction that involves the formation of a cyclic product containing one more σ bond and one fewer π bond than the starting molecule.

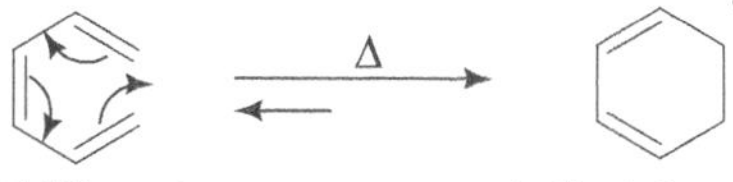

1,3,5-Hexatriene   1,3-Cyclohexadiene

To understand pericyclic reactions, one needs to understand π molecular orbitals which are beyond the scope of this book.

10.13 SYNTHESIS OF DRUG MOLECULES

In this section, we will discuss the structure, synthesis, and pharmaceutical applications of two important drug molecules, namely paracetamol and aspirin.

10.13.1 Paracetamol

Structure   The structure of paracetamol is shown in Fig. 10.27.

Commonly known as paracetamol, its IUPAC name is N-(4-hydroxyphenyl)acetamide. It is a white crystalline solid that melts at 169–171°C with solubility in hot water, (5 g/100 cm^3) and in ethanol (14 g/100 cm^3). It has a molecular weight of 155.189 g/mol.

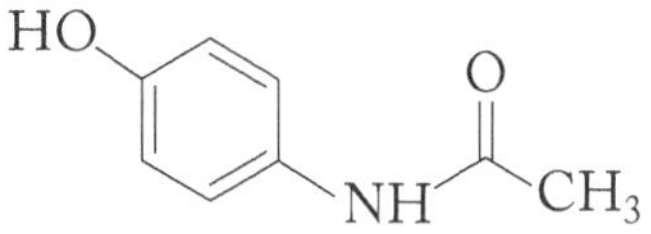

Fig. 10.27 Structure of paracetamol (Acetaminophen)

Check Your Progress

28. What are pericyclic reactions? Give one example.
29. Illustrate Diels–Alder reaction with a suitable example.

Synthesis   The synthesis of paracetamol is an excellent example for elucidating selective reactivities related to nucleophiles and leaving groups.

(a) When 4-aminophenol is treated with an excess of acetic anhydride, both amino and phenolic groups (nucleophilic groups) will undergo acetylation forming a diacetyl derivative.

(b) The diacetyl derivative so formed is hydrolysed using aqueous sodium hydroxide. The alkaline hydrolysis of diacetyl derivative results in slower hydrolysis of amide than that of the ester since $ArNH^-$ is a poor leaving group than ArO^- group. However, if 4-aminophenol is treated with 1 mole equivalent of acetic anhydride, paracetamol is obtained selectively. This is because the amino group is a better nucleophile than the phenolic group.

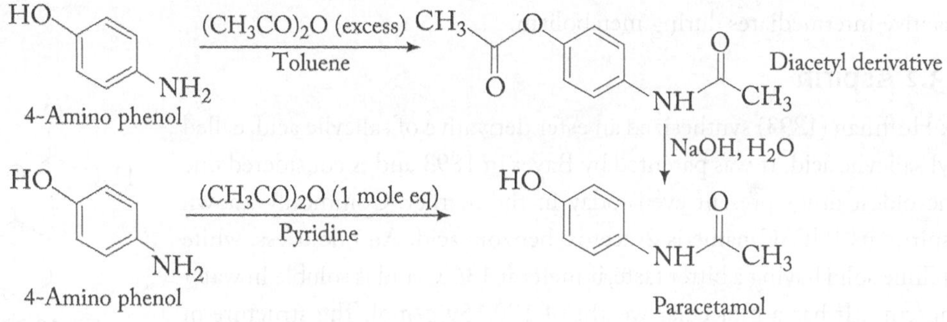

Mechanism of reaction

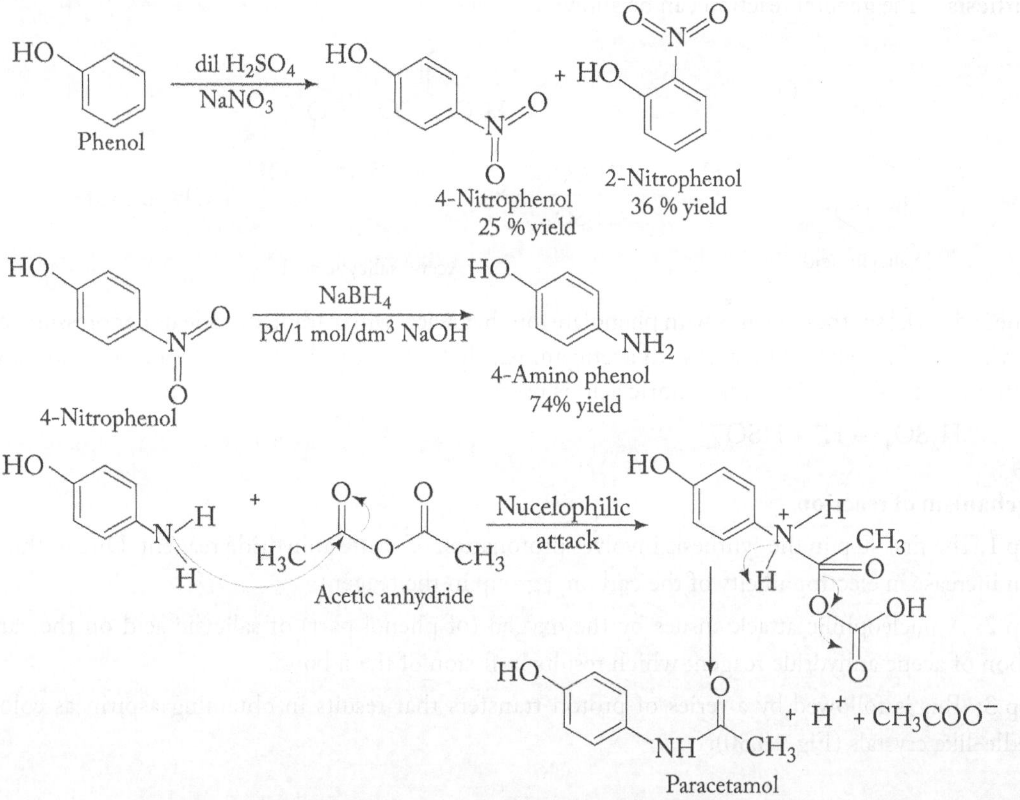

Fig. 10.28 Mechanism of paracetamol formation

Step 1: It involves the nucleophilic attack of nitrogen present in 4-amino phenol on the carbonyl carbon of acetic anhydride. This results in the formation of a tetrahedral carbon intermediate with positive charge on nitrogen atom and negative charge on one of the oxygen atoms making the acetate group to cleave.

Step 2: The acetate ion undergoes protonation and finally acetaminophen is formed along with acetic acid as the byproduct (Fig. 10.28).

Pharmaceutical applications Paracetamol is a widely popular medicine used as a mild painkiller. It is analgesic and antipyretic in action. It is also used to relieve mild arthritic pain, post-operative pain, and dental aches. Paracetamol is a relatively safer drug, but at higher doses could be toxic leading to formation of reactive intermediates during metabolism.

10.13.2 Aspirin

Felix Hoffman (1893) synthesized an ester derivative of salicylic acid, called acetyl salicylic acid. It was patented by Bayer in 1893 and is considered one of the oldest drugs present even today in the market. Commonly known as aspirin, its IUPAC name is 2-acetoxybenzoic acid. An odourless, white crystalline solid having a bitter taste, it melts at 136°C and is soluble in water (3 mg/cm$^{3)}$. It has a molecular weight of 180.159 g/mol. The structure of aspirin is shown in Fig. 10.29.

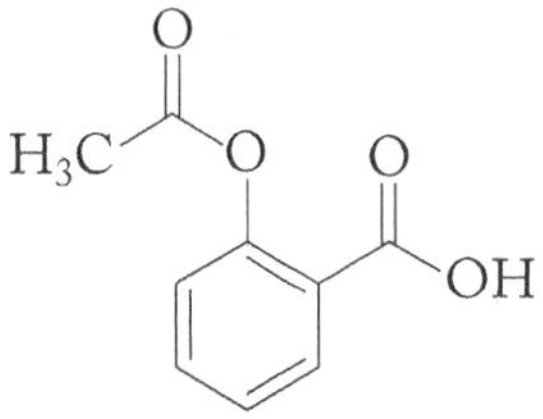

Fig. 10.29 Aspirin

Synthesis The general reaction can be shown as:

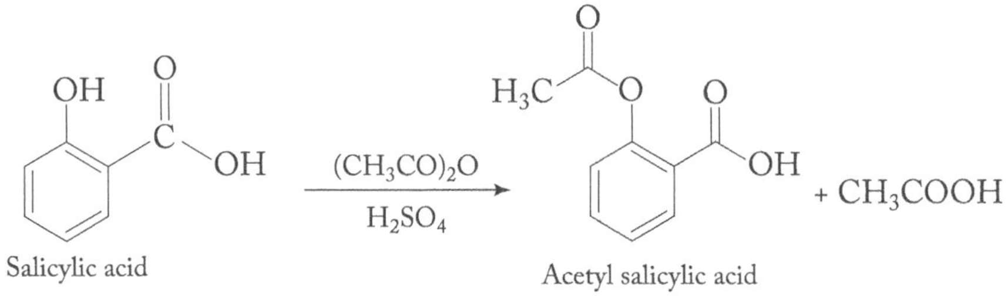

Salicylic acid   Acetyl salicylic acid

The industrial synthesis begins with phenolate ion that reacts with carbon dioxide under pressure to form salicylic acid. Salicylic acid reacts with acetic anhydride to give aspirin. It is generally an acid-catalysed reaction, where sulphuric or phosphoric acid is used.

$$H_2SO_4 \rightarrow H^+ + HSO_4^-$$

Mechanism of reaction

Step 1: The first step in the synthesis involves protonation of acetic anhydride reagent. Due to this, there is an increase in electrophilicity of the carbonyl group in the reagent.

Step 2: A nucleophilic attack ensues by the oxygen (of phenol part) of salicylic acid on the carbonyl carbon of acetic anhydride reagent which results in fission of the π bond.

Step 3: This is followed by a series of proton transfers that results in obtaining aspirin as colourless needle-like crystals (Fig. 10.30).

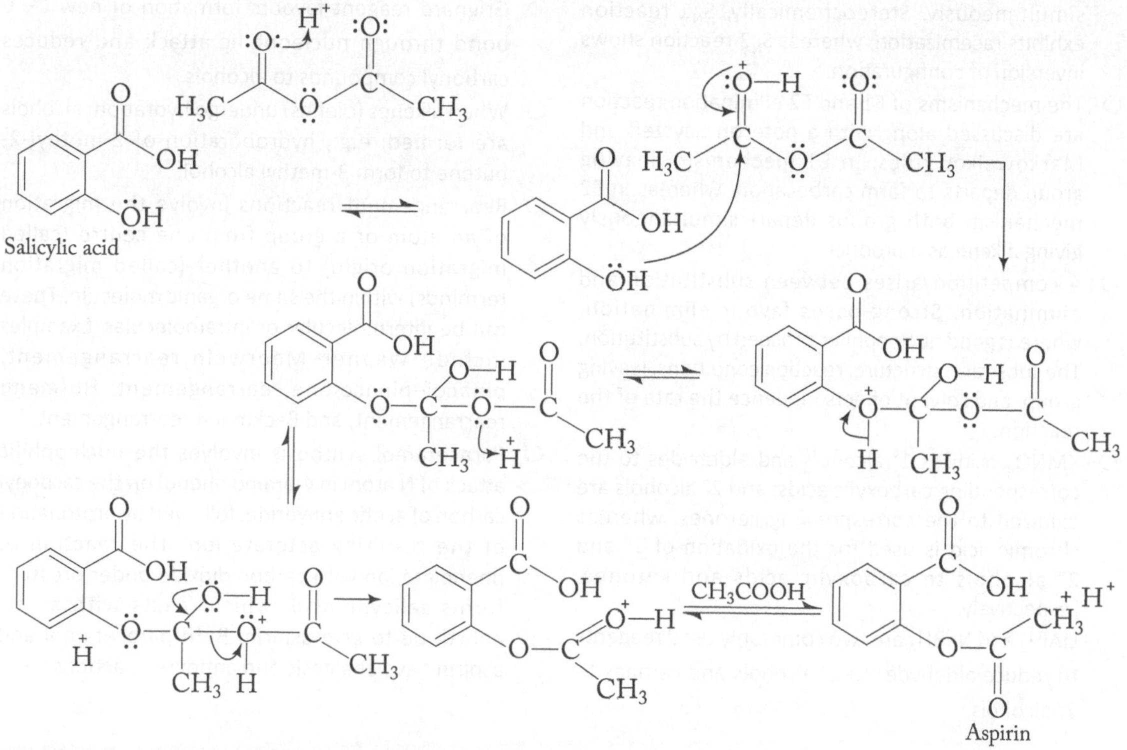

Fig. 10.30 Mechanism of formation of aspirin

Pharmaceutical applications Aspirin is frequently used as a non-steroidal anti-inflammatory drug. It is effective in relieving symptoms of pain (analgesic) due to headaches, injury, and mild arthritic pain. It is also antipyretic (lowering fever) and prevents blood clots.

Check Your Progress

30. List any two applications of aspirin and paracetamol.
31. Write the structures of paracetamol and aspirin.
32. Illustrate the mechanism for preparing aspirin.
33. Illustrate the mechanism for preparing paracetamol.
34. State the properties of paracetamol and aspirin.

SUMMARY

○ Reaction mechanism is a detailed stepwise description of the pathway by which reactants are converted to products. Typically, reactions take place either in one step or multiple steps, and their energetics can be represented using an energy profile diagram. Addition, substitution, and elimination are common reaction mechanisms studied in organic chemistry.

○ Nucleophilic substitution reactions proceed in two ways: S_N1 and S_N2. In an S_N1 reaction, a two-step mechanism occurs, where first a halide group departs and a carbocation is formed. The incoming nucleophile attacks the carbocation to form the product. An S_N2 reaction is a one-step concerted process, i.e., the leaving of halide group and the attack of incoming nucleophile occur simultaneously. Stereochemically, S_N1 reaction exhibits racemization, whereas S_N2 reaction shows inversion of configuration.

○ The mechanisms of E1 and E2 elimination reaction are discussed along with a note on Saytzeff and Markownikov rules. In E1 mechanism, leaving group departs to form carbocation, whereas in E2 mechanism both groups depart simultaneously giving alkene as a product.

○ A competition arises between substitution and elimination. Strong bases favour elimination, whereas good nucleophiles proceed by substitution. The substrate, structure, reaction conditions, leaving group, and solvent effects influence the rate of the reaction.

○ $KMnO_4$ oxidizes 1° alcohols and aldehydes to the corresponding carboxylic acids; and 2° alcohols are oxidized to the corresponding ketones; whereas chromic acid is used for the oxidation of 1° and 2° alcohols to carboxylic acids and ketones, respectively.

○ $LiAlH_4$ and $NaBH_4$ are two commonly used reagents to reduce aldehydes to 1° alcohols and ketones to 2° alcohols.

○ Grignard reagent favours formation of new C—C bond through nucleophilic attack and reduces carbonyl compounds to alcohols.

○ When alkenes (olefins) undergo hydration, alcohols are formed; e.g., hydroboration of 2-methyl-2-butene to form 3-methyl alcohol.

○ Rearrangement reactions involve the migration of an atom or a group from one centre (called migration origin) to another (called migration terminus) within the same organic molecule. These can be intermolecular or intramolecular. Examples include Wagner–Meerwein rearrangement, pinacol–pinacolone rearrangement, Hofmann rearrangement, and Beckmann rearrangement.

○ Paracetamol synthesis involves the nucleophilic attack of N atom in 4-amino phenol on the carbonyl carbon of acetic anhydride, followed by protonation of the resulting actetate ion. The reaction of phenolate ion with carbon dioxide under pressure forms salicylic acid, which reacts with acetic anhydride to give aspirin. Both paracetamol and aspirin have analgesic and antipyretic actions.

GLOSSARY

Carbanion: A reactive intermediate formed by heterolysis; has tetrahedral configuration with electron pairs occupying one of the sp^3 hybridized orbitals.

Carbocation: Electron-deficient intermediates that possess a carbon atom surrounded by six electrons.

Carbene: A divalent reactive intermediate with no charge.

Electrophile: A compound or molecule capable of accepting electrons from the substrate.

Nucleophile: A compound or molecule that acts as electron pair donor.

Zaitsev rule (in elimination reaction): The preferred alkene formed in greater amount is the one that corresponds to hydrogen removal from β-carbon having the fewest hydrogen substituents.

S_N1 and S_N2: Substitution nucleophilic unimolecular and substitution nucleophilic bimolecular.

EXERCISES

Multiple Choice Questions

1. What is the major product of the following reaction?

   (a) (*S*)-2-Butanol
   (b) (*R*)-2-Butanol
   (c) Racemic mixture of (*R*) and (*S*)- 2-butanol
   (d) No reaction

2. In S_N1 reaction occuring between *tert*-butyl chloride and iodide ion, if the concentration of iodide ion is doubled, the rate of formation of *tert*-butyl iodide will:

(a) double   (b) increase 4-fold

(c) decrease   (d) remain the same

3. Which of the following alkyl halides would undergo S_N2 reaction most rapidly?

   (a) CH_3CH_2-Br   (b) CH_3CH_2-I

   (c) CH_3CH_2-F   (d) CH_3CH_2- Cl

4. Which of the following alkyl halides would undergo S_N1 reaction most rapidly?

   (a) [I]

   (b) [II]

   (c) [III]

   (d) All will react at same rate

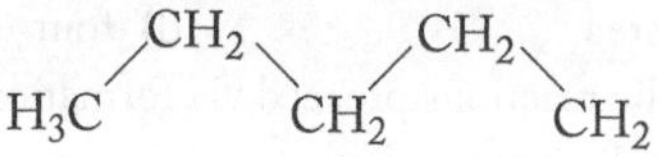

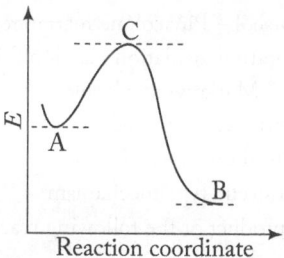

[I]

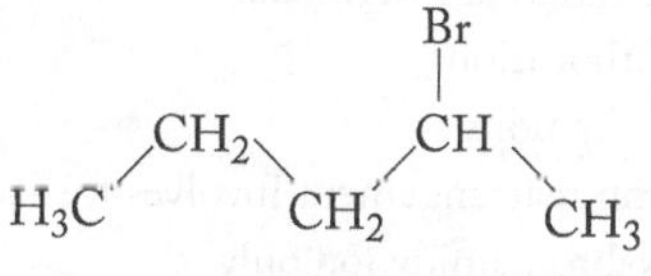

[II]

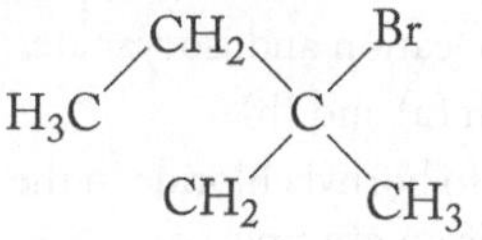

[III]

5. According to the following energy profile, the rate of reaction from A to B is determined by

(graph of E vs Reaction coordinate, showing A, C peak, B)

Reaction coordinate

   (a) energy of only A

   (b) energy of only B

   (c) energy difference between B and A

   (d) energy of only C

6. The product of the below reaction will be:

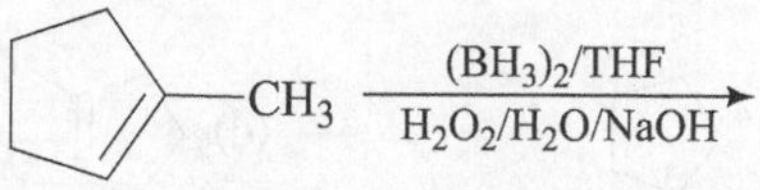

(cyclopentene-CH₃ $\xrightarrow[\text{H}_2\text{O}_2/\text{H}_2\text{O}/\text{NaOH}]{(\text{BH}_3)_2/\text{THF}}$)

(A., 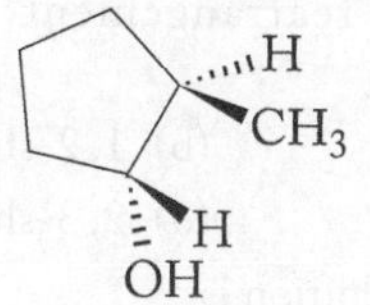 B., 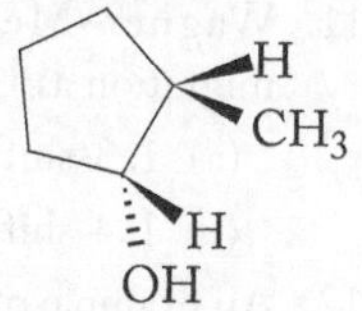 C. 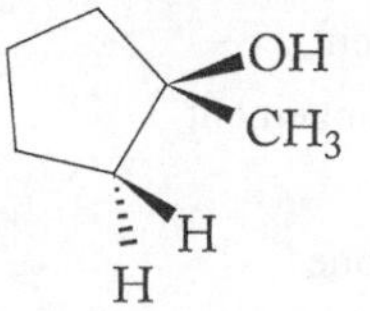 structures with cyclopentane rings; D. No reaction)

7. The following reaction using a Grignard reagent will yield:

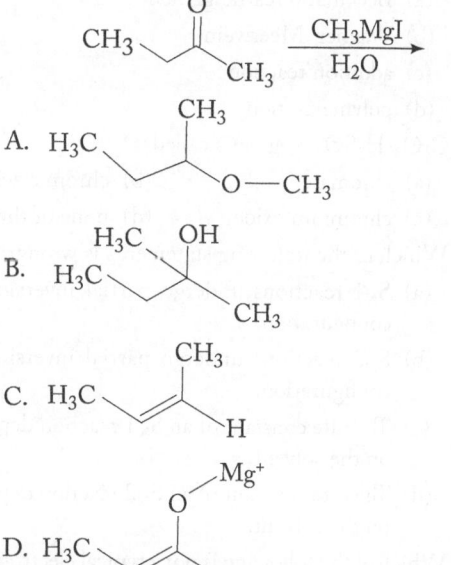

$$CH_3\text{-CO-}CH_3 \xrightarrow[\text{H}_2\text{O}]{\text{CH}_3\text{MgI}}$$

A. (structure)

B. (structure)

C. (structure)

D. (structure)

8. In permanganate oxidation, the oxidation number of Mn is

   (a) +6   (b) +5

   (c) +7   (d) zero

9. The alcohol product formed after hydride reduction is

   (a) optically active   (b) optically inactive

   (c) R- isomer   (d) planar structure

10. IUPAC names of aspirin and paracetamol are:

   (a) 2-Acetoxybenzoic acid and N-(4-hydroxyphenyl) acetamide respectively

   (b) 2-Acetoxycinnamic acid and N-(2-hydroxyphenyl) acetamide respectively

   (c) 2,4-Acetoxybenzoic acid and N-(4-hydroxyphenyl) acetamide respectively

   (d) 2-Acetoxycinnamic acid and N-4-phenyl acetamide respectively

11. Wagner–Meerwein rearrangement involves migration as
    - (a) 1, 3-shift
    - (b) 1, 2-shift
    - (c) 1, 4-shift
    - (d) 2, 3-shift

12. An example of cycloaddition is
    - (a) Wagner–Meerwein
    - (b) Beckmann rearrangement
    - (c) Diels–Alder
    - (d) Pinacol–pinacolone

13. The conversion of cyclohexanone to caprolactum is an example of
    - (a) Beckmann rearrangement
    - (b) Wagner–Meerwein
    - (c) addition reaction
    - (d) polymerization

14. CrO_3/H_2SO_4 reagent is called
    - (a) chromate
    - (b) chromic acid
    - (c) chromium oxide
    - (d) none of these

15. Which of the following statements is wrong?
    - (a) S_N1 reactions undergo partial inversion of configuration.
    - (b) S_N2 reactions undergo partial inversion of configuration.
    - (c) The rate constant of an S_N1 reaction depends on the solvent.
    - (d) The rate constant of an S_N2 reaction depends on the solvent.

16. Which of the following is not a typical electrophile?
    - (a) Cl_2
    - (b) HCl
    - (c) $(CH_3)_4N^+$
    - (d) Br_2

17. Which of the following statements regarding E1 mechanism is incorrect?
    - (a) E1 mechanism is a 1-step process.
    - (b) E1 mechanisms are unimolecular in the rate-determining step.
    - (c) E1 mechanisms are generally first order.
    - (d) E1 mechanism is a 2-step process.

18. Which of the following is not a carbene intermediate?
    - (a) $:CH_2$
    - (b) $:CCl_2$
    - (c) CCl_3
    - (d) $:CBr_2$

19. The number of transition state(s) in an S_N2 mechanism is/are
    - (a) 1
    - (b) 2
    - (c) 3
    - (d) 4

20. Choose the correct reaction–intermediates.
    - (a) Pinacol–pinacolone – carbocation
    - (b) Wagner–Meerwein – carbene
    - (c) Addition of HBr to propene – carbanion
    - (d) Hofmann rearrangement – free radical

21. The MOST preferred conformation in E2 reaction mechanism is
    - (a) *anti* periplanar
    - (b) *syn* periplanar
    - (c) tetrahedral
    - (d) *gauche* conformation

22. The number of transition state(s) is an S_N1 mechanism is/are
    - (a) one
    - (b) two
    - (c) three
    - (d) four

23. Pericyclic reactions proceed via formation of
    - (a) cyclic intermediate
    - (b) tetrahedral intermediate
    - (c) carbocations
    - (d) carbanions

24. Hofmann rearrangement involves formation of
    - (a) N-bromamide ion only
    - (b) isocyanate only
    - (c) carbocation and isocyanate
    - (d) Both (a) and (b)

25. α-pinene to bornyl chloride in the presence of HCl at −20°C is an example of
    - (a) Hofmann rearrangement
    - (b) Wagner – Meerwein rearrangement
    - (c) Beckmann rearrangement
    - (d) Pinacol – Pinacolone rearrangement

26. Hydroboration-oxidation reaction follows
    - (a) anti-Markownikov's rule
    - (b) Markownikov's rule
    - (c) radical type mechanism
    - (d) pericyclic type mechanism

27. The E2 product of the following reaction is

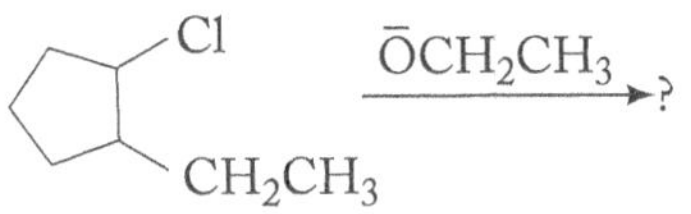

- (a) 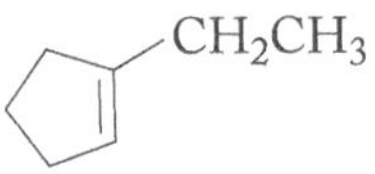
- (b) 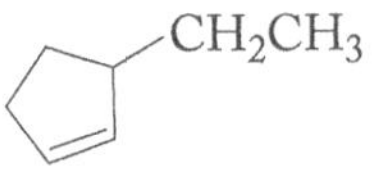
- (c) 
- (d)

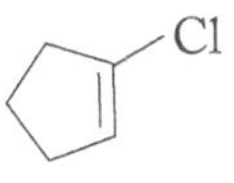

28. The mechanism of hydride reduction reactions using $LiAlH_4$ and $NaBH_4$ follows
    (a) nucleophilic substitution
    (b) nucleophilic addition
    (c) electrophilic addition
    (d) free radical type

29. Cyanohydrin reaction is an example of
    (a) nucleophilic substitution
    (b) pericyclic reaction

(c) free radical reaction
(d) nucleophilic addition

30. The CORRECT statement is:
    (a) B_2H_6 adds to alkenes forming alkylboranes that oxidizes to alcohols
    (b) Diel's Alder reaction involves free radical mechanism
    (c) Chromic acid reacts with alcohol to form an intermediate chromate ester
    (d) Only (a) and (b)

Review Questions

1. What are reactive intermediates? Illustrate and explain the structures of carbocation and carbanion.

2. Explain cyanohydrin reaction with suitable mechanism.

3. Discuss the salient features of S_N1 reaction mechanism with a suitable example.

4. Discuss the salient features of S_N2 reaction mechanism with a suitable example. Draw the energy profile diagram of the mechanism.

5. Name the various reagents commonly encountered in organic reactions. State their characteristics with suitable examples.

6. Explain the mechanism of nucleophilic addition reaction. Illustrate the mechanism of preparing cyanohydrin.

7. Explain S_N1 reaction mechanism with a suitable example. State the various factors affecting the rate of reaction. Draw the energy profile diagram to explain the energetics of the reaction.

8. With a suitable example, explain S_N2 reaction mechanism. State the various factors affecting the rate of the reaction.

9. Justify the statement: 'Alkyl halides undergo nucleophilic substitution reactions, whereas vinyl and aryl halides do not undergo nucleophilic substitution reactions.'

10. Discuss the stereochemistry of S_N1 and S_N2 reactions citing suitable examples.

11. Distinguish between E1 and E2 reaction mechanisms.

12. Explain E1 reaction mechanism. Draw the energy profile diagram of E1 reaction mechanism with a suitable example.

13. Explain E2 reaction mechanism. Draw the energy profile diagram of E1 reaction mechanism with a suitable example.

14. Discuss the stereochemical outcome of E2 reaction.

15. What are pericyclic reactions? Write Diels-Alder reaction and explain the features of the reaction.

16. Discuss the various ring closure reactions with examples. (Hint: pericyclic reactions)

17. Write a note on Hofmann rearrangement reaction.

18. Differentiate between the following : (a) homolytic and heterolytic cleavage; (b) carbanion and carbocation.

19. State Saytzeff rule for predicting elimination products with an example.

20. Explain addition reactions. State Markownikov's rule with a suitable example.

21. Predict the major products of the following reactions and propose mechanisms to support your predictions. (a) pent-1-ene + HCl (b) 2-methylpropene + HBr.

22. Explain hydroboration of alkenes. Illustrate the mechanism.

23. What is 1,2-alkyl shift observed in pinacol-pinacolone rearrangement? Explain with an example.

24. Discuss Hofmann rearrangement with a suitable example.

25. Explain the conversion of benzoic acid to m-bromo aniline by Hofmann rearrangement.

26. Explain the mechanism of Beckmann rearrangement citing a suitable example.

27. Discuss the mechanism of Wagner–Meerwein rearrangement reaction. Give one example.

28. Write the mechanism for conversion of cyclohexanone to caprolactum. (Hint: Beckmann rearrangement).

29. Illustrate the mechanism for the conversion of α-pinene to bornyl chloride. Discuss each step of the mechanism in detail. (Hint: Wagner–Meerwein rearrangement)

30. Write the mechanism for the conversion of 1,1, 2-triphenyl-1, 2-propanediol to methyl triphenyl methyl ketone. (Hint: Pinacol–pinacolone rearrangement).

31. Complete the following reactions:

(a) $\xrightarrow[\text{H}_2\text{O}_2/\text{NaOH}]{\text{BH}_3.\text{THF}}$?

(**Ans:**)

(b) $\xrightarrow[\text{H}_2\text{O}_2/\text{NaOH}]{\text{BH}_3.\text{THF}}$?

(**Ans:** OH)

(c) $\underset{\text{OH}}{\square} \xrightleftharpoons{\Delta}$ (**Ans:**)

(d) $\xrightarrow[\text{H}_2\text{O}]{\text{H}_2\text{CrO}_4}$

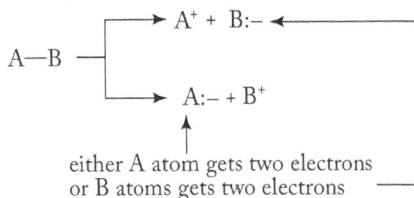

(**Ans:**)

32. Explain the synthesis and applications of paracetamol.

33. Describe the structure, synthesis and applications of aspirin.

FURTHER READING

1. Pine, S. H. *Organic Chemistry,* McGraw Hill, 1987.
2. Solomons, G., C. Fryhle, and S. Snyder, *Organic Chemistry,* Wiley, 2014.
3. Smith, Janice G. *Organic Chemistry,* McGraw Hill, 2016.
4. Royal Society of Chemistry. (2007). 'Aspirin - The Wonder Medicine'. Retrieved from http://www.rsc.org/learn-chemistry/resource/res00000287/aspirin
5. http://www.compoundchem.com/Students and teachers are encouraged to visit the site as it provides interesting graphical posterson everyday chemistry prepared by Andy Brunning.

ANSWERS

Check Your Progress

1. If we consider an organic molecule say, AB which undergoes fission and can be illustrated as:

$$\text{A—B} \begin{cases} \rightarrow \text{A}^+ + \text{B:}^- \leftarrow \\ \rightarrow \text{A:}^- + \text{B}^+ \end{cases}$$

either A atom gets two electrons
or B atoms gets two electrons

2. 1. Addition 2. Substitution 3. Elimination 4. Rearrangement 5. Cyclization and ring opening

3. Electrophilic, nucleophilic and radical-based reagents are commonly used in organic reactions.

4. Markownikov's rule: Addition of hydrogen to an unsymmetrical olefin occurs at those carbon atoms with maximum number of hydrogen atoms.

   Anti-Markownikov's rule: Addition of hydrogen to an unsymmetrical olefin occurs at those carbon atoms with least number of hydrogen atoms.

5. Refer to Section 10.4.4.

6. Grignard reagents react with carbonyl compounds to form various alcohols. For the reaction mechanism, refer to Section 10.4.4.

7. The reaction is nucleophilic substitution.

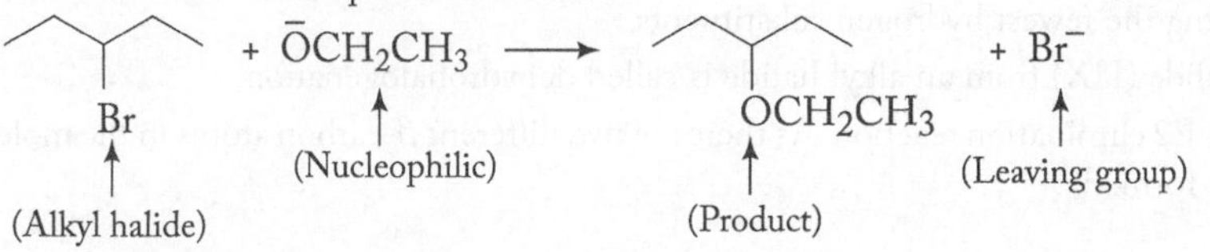

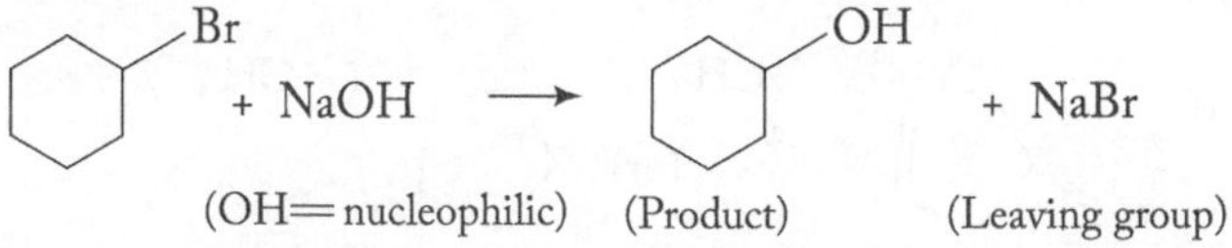

(Alkyl halide) — Br, (Nucleophilic) $\bar{O}CH_2CH_3$, (Product) OCH_2CH_3, (Leaving group) Br^-

(OH = nucleophilic)   (Product)   (Leaving group)

8. Polar, protic solvents favour S_N1 reactions, whereas polar, aprotic solvents favour S_N2 reactions. This is due to the fact that a carbocation intermediate and a halide ion are formed during S_N1 reaction. Polar solvents are known to stabilize these ions due to solvation effect. In S_N2 reactions, there are no intermediates, hence are not affected greatly by polar solvents.

9. Chloro group is a good leaving group as it is a conjugate base of strong acid, HCl. Hydroxy group is a poor leaving group as it is a conjugate base of water (a weak acid). Amino group is an extremely poor leaving group as it is a conjugate base of ammonia and a very strong base too. Water molecule is a good leaving group as it is a conjugate base of hydronium ion (a strong acid).

10. The rate of S_N1 reaction under the following conditions
   In case (a), as concentration of alkyl halide is tripled, the rate of reaction will also be tripled, i.e. Rate = k 3[RX] for SN1 reaction. In case (b), rate of the reaction is halved. In case (c), rate of reaction will be doubled. In case (d), rate of reaction is halved.

11. Rate of reaction will be doubled.

12. S_N1 reaction proceeds with the formation of carbocation that is attacked by nucleophile from either side resulting in racemization. In S_N2 reaction, nucleophile attacks from the rear end of the alkyl halide, resulting in inversion of configuration.

13. (a) For the above reaction, chloro is the leaving group and cyano group (CN^-) is the incoming nucleophile. As S_N2 reaction proceeds by inversion of configuration, the reaction can be shown as follows:

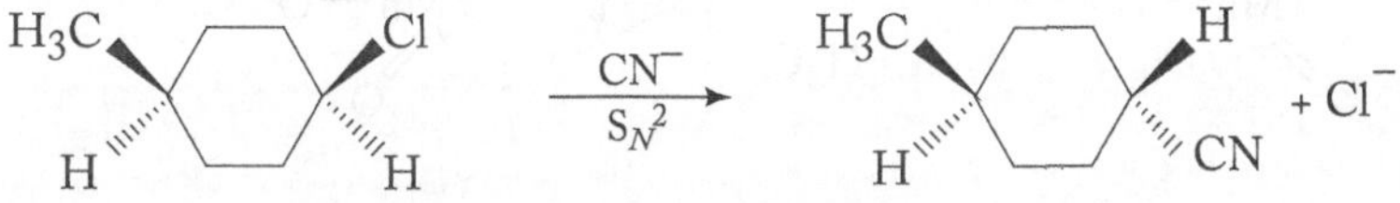

The inversion occurs at the C—Cl bond and product is obtained.

(b) Water molecule is the incoming nucleophile and bromo group is the leaving group. The reaction follows S_N1 mechanism as shown in figure:

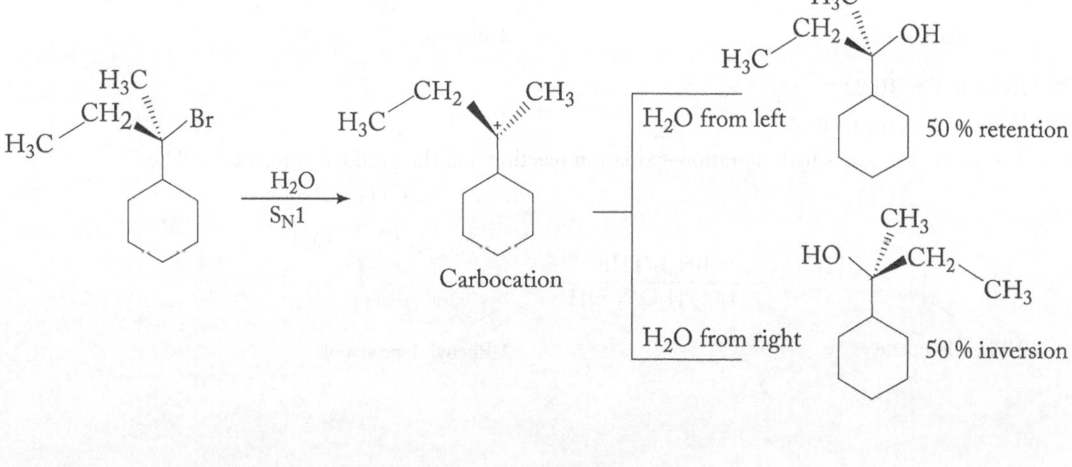

14. According to Zaitsev rule the preferred alkene formed in greater amount is the one that corresponds to hydrogen removal from β-carbon having the fewest hydrogen substituents.

15. The removal of hydrogen halide (HX) from an alkyl halide is called dehydrohalogenation.

16. The reaction is an example of E2 elimination reaction. As there are two different β-carbon atoms in the molecule, two different alkenes can be formed.

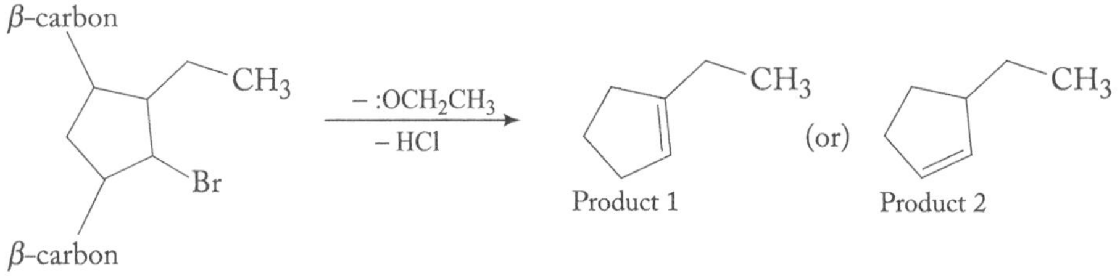

β-carbon

β-carbon

According to Zaitsev rule, the major product should be 1, as it has the more substituted double bond and thus it is predicted as the major product in the above reaction.

17. Oxidizing reagents: chromic acid (CrO_3/H_2SO_4), acidified potassium permanganate ($KMnO_4$).

    Reducing reagents: lithium aluminium hydride ($LiAlH_4$), sodium borohydride ($NaBH_4$).

18. Refer to Section 10.8.1

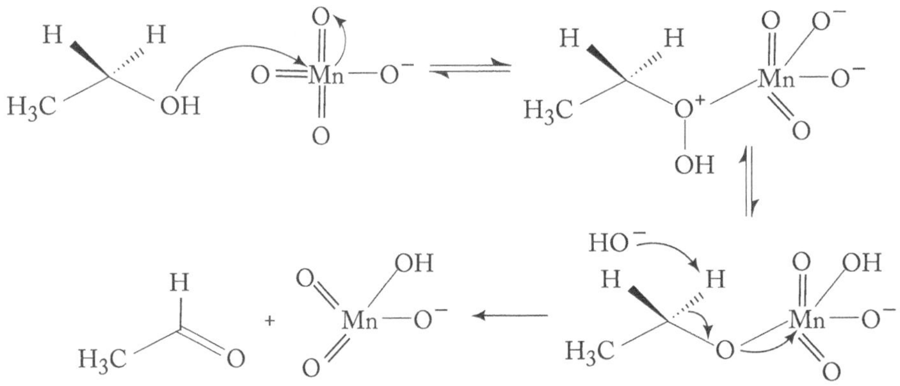

19. Refer to Section 10.8.2

20.

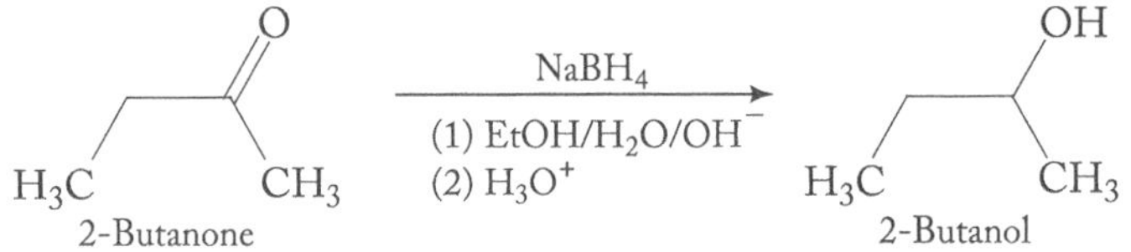

21. Refer to Fig. 10.22

22. Refer to Section 10.10

23. The above reaction is hydroboration–oxidation reaction and the product so formed will be:

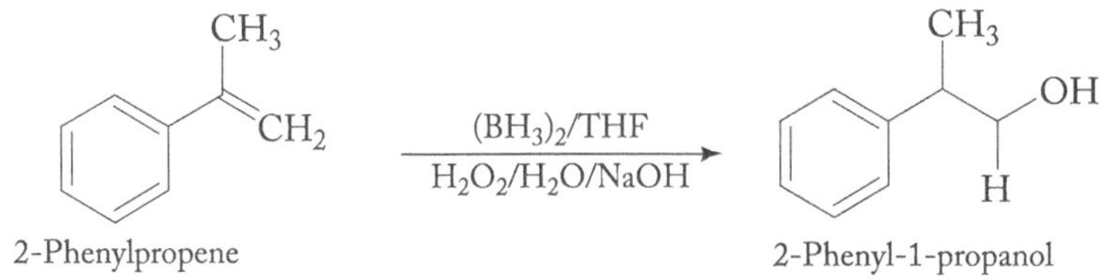

24. Refer to Section 10.11
25. Refer to Section 10.11.2
26. Refer to Section 10.11.2
27. Refer to Section 10.11.1
28. Pericyclic reactions are concerted reactions proceeding through a cyclic transition state; e.g., Diels–Alder reaction.
29. Refer to section 10.12
30. Aspirin and paracetamol are administered as analgesics and antipyretics.
31. Refer to Section 10.13
32. Refer to Section 10.13.2
33. Refer to Section 10.13.1
34. Refer to Section 10.13

Objective Type Questions

1. (c)	2. (d)	3. (b)	4. (c)	5. (d)	6. (a)	7. (b)
8. (c)	9. (b)	10. (a)	11. (b)	12. (c)	13. (a)	14. (b)
15. (b)	16. (c)	17. (a)	18. (c)	19. (a)	20. (a)	21. (a)
22. (b)	23. (a)	24. (d)	25. (b)	26. (a)	27. (a)	28. (b)
29. (d)	30. (d)					

Stereochemistry

CHAPTER 11

LEARNING OBJECTIVES

After reading this chapter, you will be able to:

- represent various molecules on paper in three dimensions using molecular models and projection formulae.
- differentiate various types of isomers—structural, stereoisomers, enantiomers, and diastereomers.
- understand chirality and optical activity of molecules.
- depict torsional curve to describe various conformations of ethane, propane, and n-butane.
- describe isomerism in transition metal compounds.

11.1 INTRODUCTION

The term 'stereochemistry' refers to the three-dimensional structure of molecules. Louis Pasteur (1848) is considered the father of stereochemistry as he first described dissymmetric grouping of atoms in a molecule. Further Van't Hoff and Le Bel (1874) laid a strong foundation of stereochemical studies and added a third dimension to the age-old two-dimensional approach of viewing chemical structures. Barton and Hassel proposed the concept of conformational analyses of various molecules. In this chapter, we will delve in to various fundamental concepts of chirality, symmetry, isomerism, configuration assignment, and also conformational analyses of a few simple molecules.

Establishing Molecular Chirality–Pasteur's Experiment

Louis Pasteur was a trained crystallographer. He prepared an aqueous solution of sodium ammonium tartrate racemate, which when crystallized by slow evaporation gave two types of crystals. When viewed under the microscope, one type of crystal was found identical to the earlier known tartaric acid salt (dextro form) that rotated plane polarized light to the right (dextrorotatory), while the other rotated the plane of polarized light to the left (laevorotatory) to the same extent. Both the forms were related to each other as an object and its non-superimposable mirror image, very much like our hands, and were termed enantiomorphic (*enantios* = opposite, *morph* = shape). It was for the first time that laevorotatory tartaric acid was obtained and a manual resolution of racemate was performed in the laboratory. All these developments led towards the establishment of molecular chirality. Zigzag structures (discussed later) of different forms of tartaric acid are shown in Fig. 11.1.

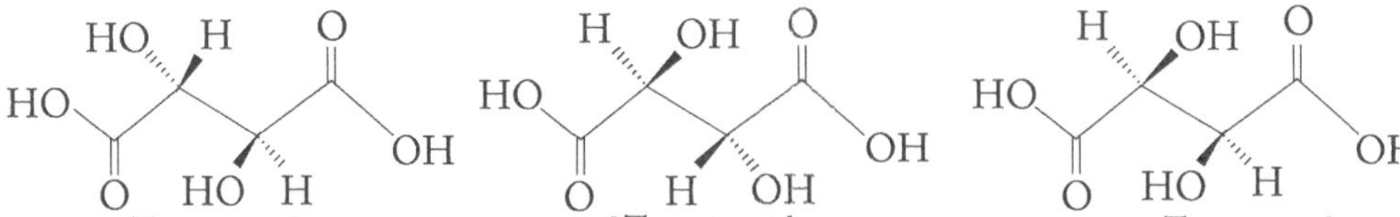

d-Tartaric acid *l*-Tartaric acid *meso*-Tartaric acid

Fig. 11.1 Different forms of tartaric acid

As there are two stereogenic centres in tartaric acid, four stereoisomers can be represented for the compound. The four forms of tartaric acid are (+)-tartaric acid, (−)-tartaric acid, (±)-tartaric acid and *meso*-tartaric acid. The first two acids are enantiomers exhibiting optical activity. The (±)− tartaric acid and *meso*-forms are optically inactive.

> **Tartaric Acid: Adam of Stereoisomer!**
> One can refer to tartaric acid as the 'Adam of stereoisomers,' since it was among the first ones to be studied to explain molecular chirality.

11.2 MOLECULAR MODELS: REPRESENTATION OF 3-DIMENSIONAL STRUCTURES

Chemists have devised various molecular models and a set of conventions for better picturization and understanding molecules in three dimensions. Basically, there are two types of models, viz., one that shows the framework (bonds and nuclei) and the other which shows the entire bulk of each atom in a molecule, called space-filling model. Ball-and-stick models are commonly used as an aid to study stereochemistry and can be utilized in classroom teaching. However, sometimes these models could be deceptive. They are also stiff to work with and hence it becomes difficult to explain bond angles and their rotations.

11.2.2 Two-dimensional Representation of 3D Structures

In-class teaching involves chalkboard or slideshows, and one can use computational models to show molecules in three dimensions. However, in order to present their answers, students should know how to view molecules in three dimensions.

The different 2D views of three-dimensional molecules[*] are discussed as follows.

Flying Wedge

The most common representation is the flying wedge formula, in which normal lines (—) depict the bonds that are in the plane of the paper. The bonds to any atoms above the plane of the drawing are shown as a solid wedge (◀), starting from an atom in the plane of a drawing at the narrow end of the wedge; and bonds to atoms below the plane are shown as a hashed wedge (⦙⦙⦙⦙). Note that the narrow end of the wedge begins at the atom in the plane of the drawing. For example, two enantiomorphic lactic acids can be shown by the flying wedge structures as in Fig. 11.2.

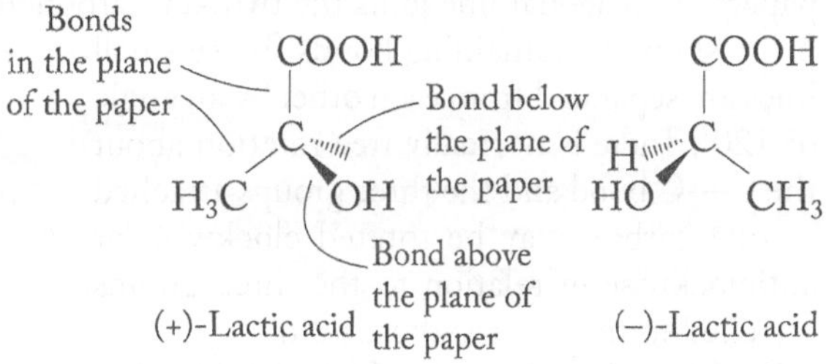

Fig. 11.2 Flying wedge formula of lactic acid

Flying wedge formula may be used for compounds having any number of chiral centres.

For the tetrahedral configuration, the following representations are recommended:

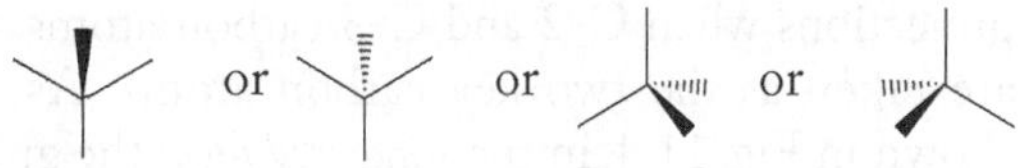

Tetrahedral configuration may also be shown as:

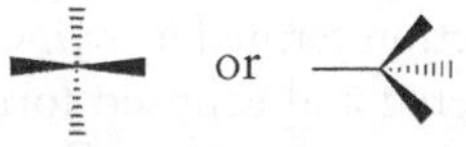

A wavy line (〜) can be used to indicate either that the configuration is unknown but only one form is present, or if shown in the text that both isomers are present and will be defined when required.

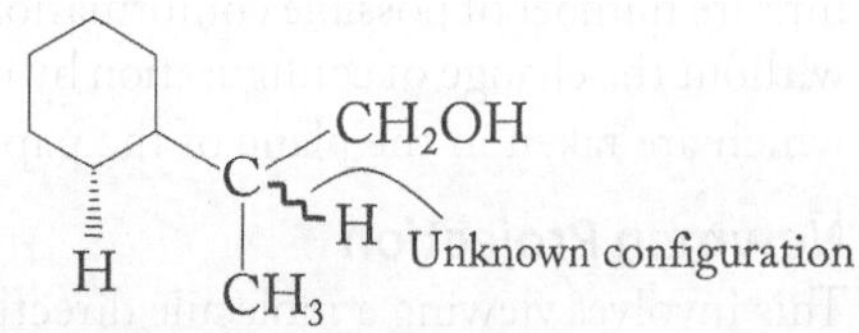

---

[*] ACD/ChemSketch software can be used to draw 3D diagrams of various organic molecules. Refer to https://www.acdlabs.com/resources/freeware/chemsketch/

When it is intended not to show any configuration, it is best to use only plain lines for all bonds. Double bonds should be shown as far as possible with accurate angle (ca. 120°), when configuration is implied. To show the absence of any configurational information, a linear representation should be used.

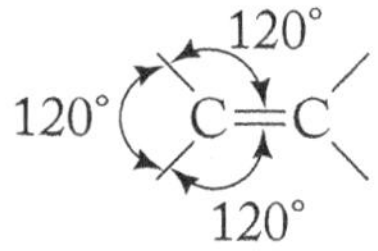

Zigzag Formulae

Zigzag formulae are normally used to represent compounds containing two or more chiral centres. It is always shown in staggered conformation and the whole carbon backbone is placed in the plane of the paper or blackboard. Bonds above or in front of the plane are shown by solid wedges and bonds below or in back of the plane are shown by hashed wedges or broken lines. The zigzag formula, although simpler, is used only for chiral centres bearing H atom as one of the substituents. The hydrogen atom is assumed to be in the structure but is not usually drawn in the projection formulas. Zigzag formulas of D-glucose and D-mannose are shown in Fig. 11.3. The chiral centres with front OH groups have H atoms in the back position, and vice versa.

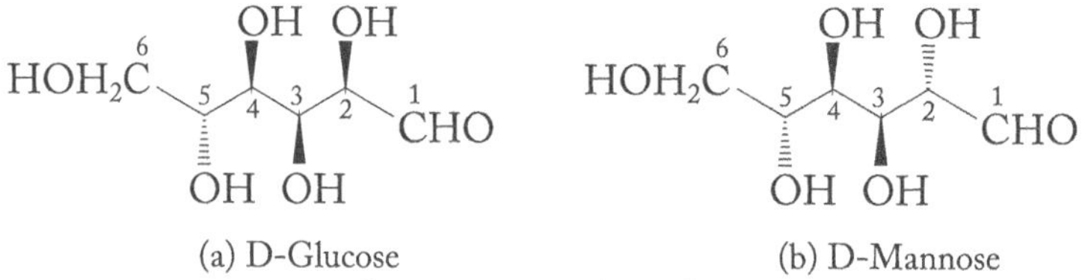

(a) D-Glucose   (b) D-Mannose

Fig. 11.3 Zigzag formulae

Sawhorse Formula

Sawhorse formula is yet another common way to represent three-dimensional molecules on paper. When it is necessary to know the spatial relationship between the substituents attached to two adjacent carbon atoms (may be chiral or achiral), sawhorse projections are found to be very convenient. Here, the C—C bond is viewed slightly from above and sideways and all the bonds are drawn as straight lines.

To draw sawhorse projection formula, a diagonal line is drawn and considered to be in the plane of the paper. The diagonal line joins the two key carbon atoms and small lines projected above and below the plane show the remaining bonds. These small lines are separated from each other by an angle of 120°. There is normally free rotation about the C—C bond and the three groups attached to one carbon may be rotated clockwise or anticlockwise in relation to the three groups attached to the other carbon atom.

For example, n-butane, $CH_3CH_2CH_2CH_3$, can be represented by the above sawhorse projections when C-2 and C-3 carbon atoms are taken as the two key carbon atoms. As

n-Butane (fully staggered) n-Butane (fully eclipsed)

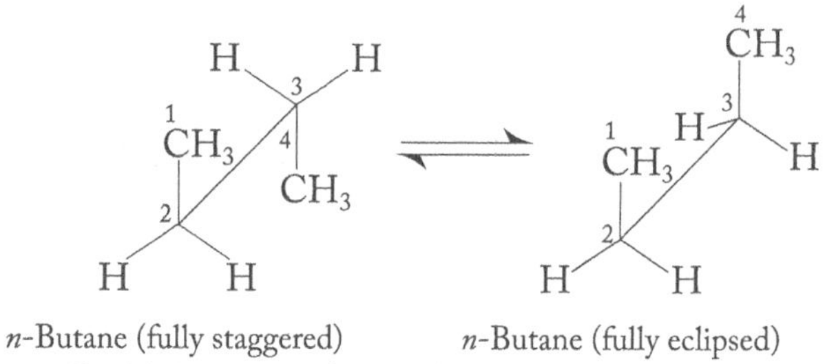

Fig. 11.4 Staggered and eclipsed conformations of n-butane

shown in Fig. 11.4, in the *staggered form* the groups are oriented as far away from each other as possible. The other sawhorse projection termed as *eclipsed form* have the groups on C-2 and C-3 present nearest to each other. Both staggered and eclipsed forms represent two conformations of n-butane among the infinite number of possible conformations. One sawhorse representation can be transformed into another without the change of configuration by rotating the molecule about the axis joining the two carbon atoms, which are taken in the plane of the paper. Sawhorse projection is also called *perspective formula*.

Newman Projection

This involves viewing a molecule directly down the C—C bond axis. In this projection, the front carbon is represented as a central point with bonds emerging from it, whereas the rear carbon is shown as a circle with the respective bonds emerging from it.

To draw a Newman projection, the molecule bearing the front and back carbon atoms is viewed along the bond joining the carbon atoms, and these atoms are represented as superimposed circles, only one circle being drawn. The centre of the circle represents the front carbon atom and the circumference represents the rear carbon atom. The remaining bonds on each carbon are shown by small straight lines at angles of 120° joined to the centre and to the circumference. In Newman projection, the line joining the two key carbon atoms is represented by circle and its centre is not visible.

For example, Newman projections of *n*-butane in Fig. 11.5 shows that the rear carbon atom of staggered form is rotated 60° anticlockwise to get an eclipsed conformation. These conformations are Newman projections of *erythro (pref)* and *threo (parf)* forms of *n*-butane.

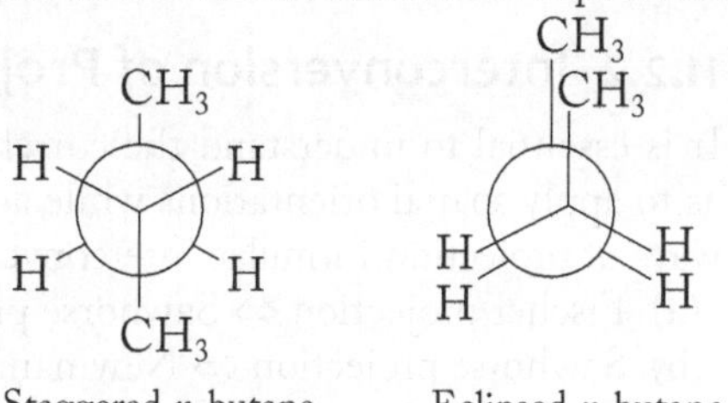

Staggered *n*-butane   Eclipsed *n*-butane

Fig. 11.5 Newman projection formulae of *n*-butane

Fischer Projection

Fischer formula is a convention for displaying the three-dimensional configurational relationship of molecules with chiral centres in a planar representation. In this representation, the molecule is so oriented that each chiral carbon is in the plane of the projection (paper or blackboard) and the four bonds are shown by two vertical and two horizontal lines. The point of intersection of these lines is the seat of the chiral centre. The chiral centre is not shown by any atomic symbol. Fischer projection formula is particularly useful to represent complex molecules such as carbohydrates, sugars, proteins, steroids, and drug molecules. For example, Fischer projection of one stereoisomer of lactic acid, $CH_3CHOHCO_2H$, is shown in three different ways in Fig. 11.6. Fischer projections are not normally used to represent compounds that do not have chiral centres.

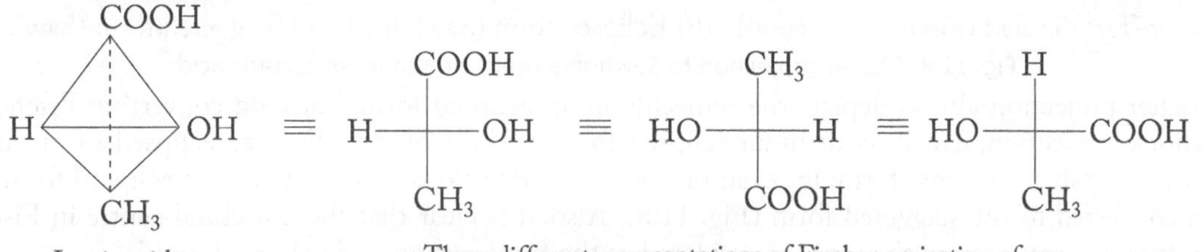

Lactic acid   Three different representations of Fischer projection of one enantiomer of lactic acid

Fig. 11.6 Fischer projection formula of lactic acid

In drawing the Fischer projection formula of compounds, certain conventions should be clearly understood to avoid misinterpretation of the stereochemical aspects of the structures.

(i) Although all bonds are represented by plain straight lines, it is to be kept in mind that vertical bonds project below the paper or blackboard and horizontal bonds project above the paper or blackboard.

(ii) For the purpose of comparison between two Fischer projections, one projection may be rotated 180° within the plane of the paper about an axis perpendicular to the paper. It is not permissible to rotate the projection formula within the plane of the paper either by 90° or 270°.

(iii) The projection formula, being two-dimensional, must never be lifted out of the projection plane and turned over. If this is done, then the vertical bonds (which were below the plane of the projection) will come above the plane, the horizontal bonds will become projected below the plane, and thus the conventions of Fischer projection will be violated. If the number of chiral carbons in a molecule is n, then n number of horizontal lines will cut the vertical line and each point of intersection presents a chiral centre in that projection.

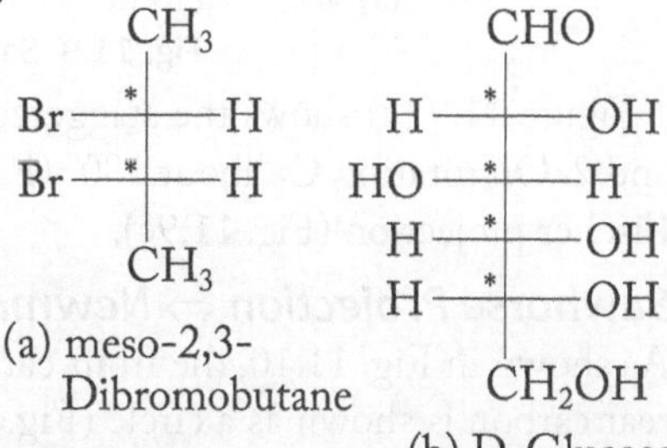

(a) meso-2,3-Dibromobutane

(b) D-Glucose

Fig. 11.7 Fischer projection formulae

For example, *meso*-2,3-dibromobutane has two chiral centres and naturally occurring glucose, has four chiral centres (asterisked) (Fig. 11.7). It should be noted that in Fischer projection, only chiral centres are shown by horizontal and vertical bonds.

11.2.2 Interconversion of Projection Formulae

It is essential to understand the correlation between various representations and molecular models so as to apply spatial orientations while assigning various configurations to molecules. Some of the simple ways of projection formulae interconversions are as follows:

(a) Fischer projection ⇔ Sawhorse projection
(b) Sawhorse projection ⇔ Newmann projection ⇔ Fischer projection
(c) Fischer projection ⇔ Flying wedge formula
(d) Zigzag projection ⇔ Fischer projection

While performing the above projection formulae interconversions, some rules must be followed.

Fischer Projection ⇔ Sawhorse Projection

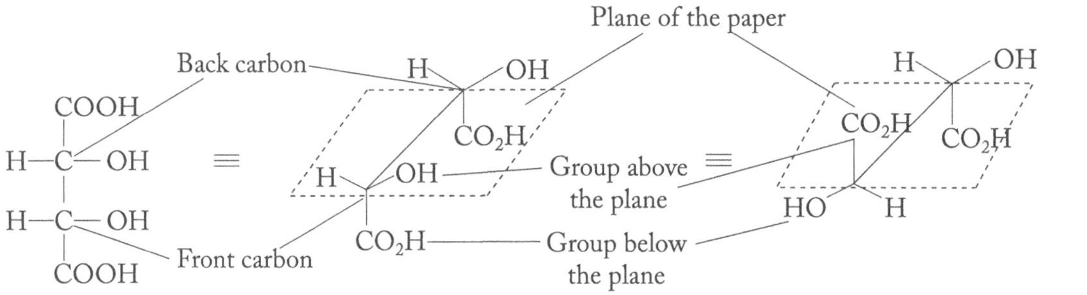

(a) *meso*-Tartaric acid (Fischer projection) (b) Eclipsed form (Sawhorse) (c) Staggered form (Sawhorse)

Fig. 11.8 Fischer projection to Sawhorse projection of *meso*-tartaric acid

Fischer projection always depicts the molecule in an eclipsed form. Thus, on converting Fischer to Sawhorse projection, it will result in an eclipsed form of the molecule. Further, eclipsed form can be drawn as a stable staggered form; for example, *meso*-tartaric acid is first written as an eclipsed form and then converted to the staggered form (Fig. 11.8). Also, it is clear that the last chiral centre in Fischer projection (counting from the top) is considered as the front carbon in Sawhorse formula.

Now, if we want to convert Sawhorse to Fischer projection, the molecule should be drawn as an eclipsed form before conversion as shown in Fig. 11.9.

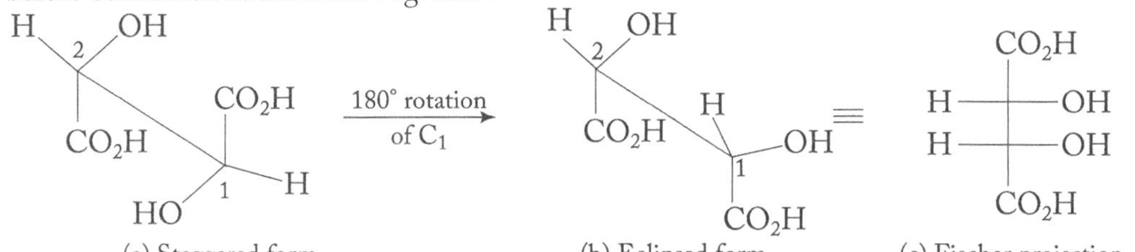

(a) Staggered form (b) Eclipsed form (c) Fischer projection

Fig. 11.9 Sawhorse to Fischer projection of tartaric acid

Figure 11.9 (a) shows the staggered form of tartaric acid, where the carbon atoms are numbered as 1 and 2. On rotating C_1 about 180° (Fig. 11.9b), we obtain the eclipsed form which can be written as the Fischer projection (Fig. 11.9c).

Sawhorse Projection ⇔ Newman Projection ⇔ Fischer Projection

As shown in Fig. 11.10, the front carbon of Sawhorse projection (Fig. 11.10a) is drawn as it is and the rear carbon is shown as a circle (Fig. 11.10b), thereby transforming into a Newman projection. For the conversion of Newman to Fischer projection, draw an eclipsed form of Newman projection. Note that in the eclipsed form of Newman projection the vertical bonds are arranged in such a way that they remain below the horizontal plane, as shown by dotted lines in Fig. 11.10 (c).

Further, the transformation of eclipsed Newman projection to Fischer projection is then carried out keeping the front chiral atom as the lowest chiral centre in the Fischer projection (counting from the top) (Fig. 11.10d).

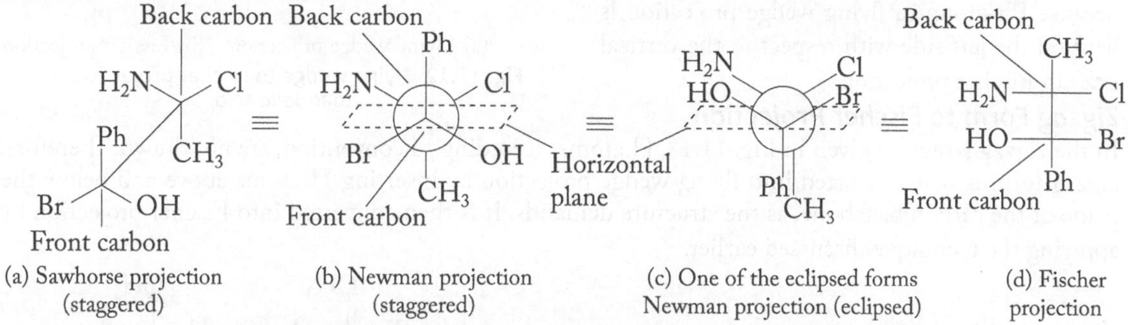

Fig. 11.10 Sawhorse to Newman and Fischer projection formulae

Fischer Projection ⇔ Flying Wedge

Conversion of Fischer projections to flying wedge and vice versa can be carried out as follows. In the following example (Fig. 11.11) Fischer projection of one of the stereoisomers of lactic acid is converted into flying wedge forms. In doing so, the vertical bonds in Fischer projection are considered to be in the plane of the paper and the horizontal bonds are above and below the plane.

Note that when the lower vertical bond linked to $-CH_3$ group in Fischer projection (a) is bent on the right side in Fig. (b) the group ($-OH$) on the right side in the horizontal bond in Fischer projection should be written above the plane of the paper (represented by solid wedge) the H atom, therefore should be placed below the plane (represented by hashed wedge). If the lower bond linked to $-CH_3$ group in Fischer projection is bent on the left side (as in Fig. (c) then the group (H) on the left side of the horizontal bond in Fischer projection is to be placed above (shown as solid wedge) the plane of the paper and the group on the right side ($-OH$) should be placed below (shown as hashed wedge) the plane. Thus, both Figs (b) and (c) represent the flying wedge forms of Fischer projection of the same isomer of (+)-lactic acid.

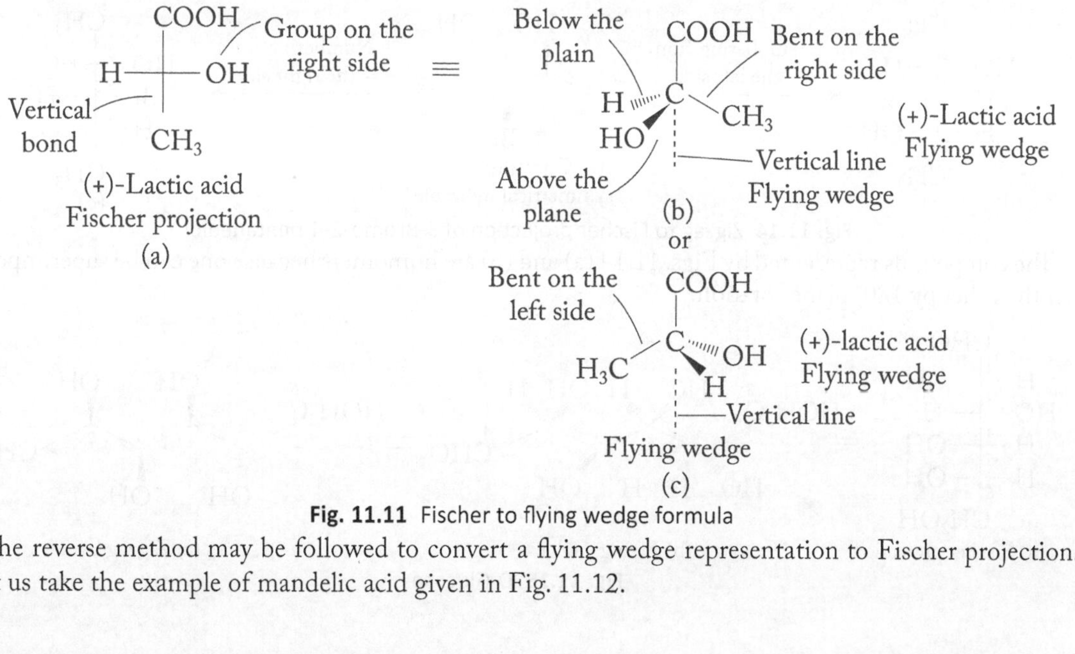

Fig. 11.11 Fischer to flying wedge formula

The reverse method may be followed to convert a flying wedge representation to Fischer projection. Let us take the example of mandelic acid given in Fig. 11.12.

In Fig 11.12 (a), CO_2H and Ph groups are in the plane of the paper in flying wedge representation. In transforming into Fischer projection, the OH group, which is above the plane in flying wedge structure, is put on the left of the horizontal bond, because Ph group in flying wedge projection is bent on the left side with respect to the vertical lines in Fischer projection.

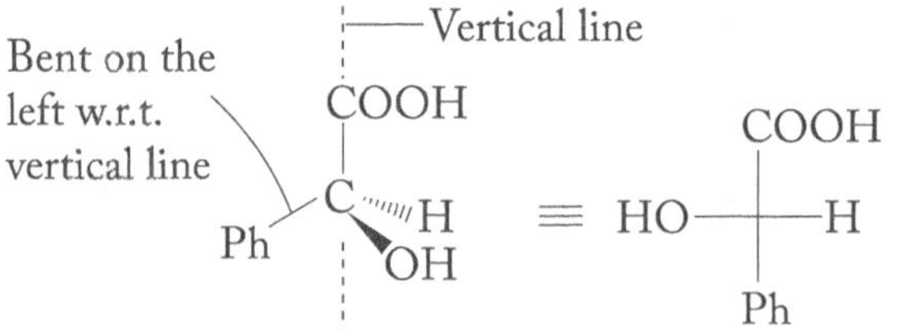

(a) Flying wedge projection (b) Fischer projection
Fig. 11.12 Flying wedge to Fischer projection of mandelic acid

Zigzag Form to Fischer Projection

In the zigzag structure given in Fig. 11.13, H atoms, according to convention, are not drawn. Therefore, zigzag form is first converted into flying wedge projection by inserting H atoms above and below the plane of the carbon backbone, as the structure demands. It is then converted into Fischer projection by applying the technique discussed earlier.

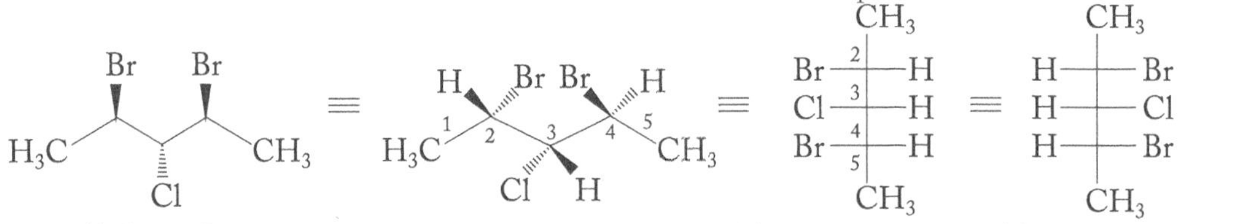

(a) Zigzag form (b) Flying wedge projection (c) Fischer projection (*meso*-compound) (d) Fischer projection (*meso*-compound)

Fig. 11.13 Zigzag to Fischer projection formulae

It is to be noted that in the above example, the carbon backbone is numbered from the left side and C-1 is placed at the top in the Fischer projection. Since the compound is symmetrical, the structure can also be numbered from the right side, when we get the mirror image of the previous Fischer projection of the same compound. However, in this particular case, both are same *meso*-compound (superimposable). In case of a symmetrical molecule, numbering can be done from either end of the carbon chain leading to two apparently different Fischer projections but they are homomers. In unambiguous case, the numbering should follow IUPAC rules.

Some of the examples of such interconversions are:

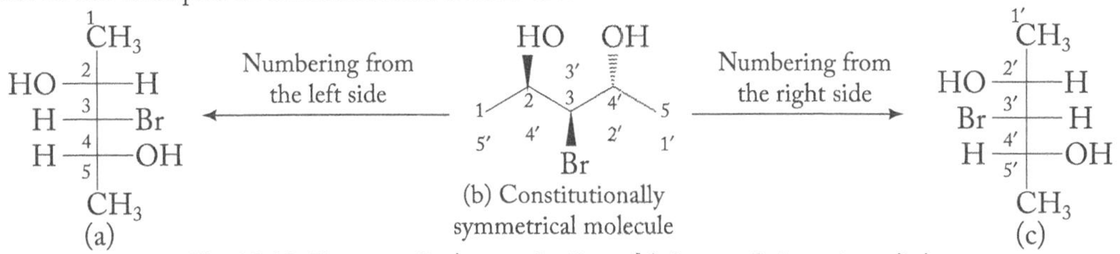

(a) (b) Constitutionally symmetrical molecule (c)

Fig. 11.14 Zigzag to Fischer projection of 3-Bromo-2-4-pentanediol

The compounds represented by Figs. 11.14 (a) and (b) are homomers because one can be superimposed on the other by 180° plane rotation.

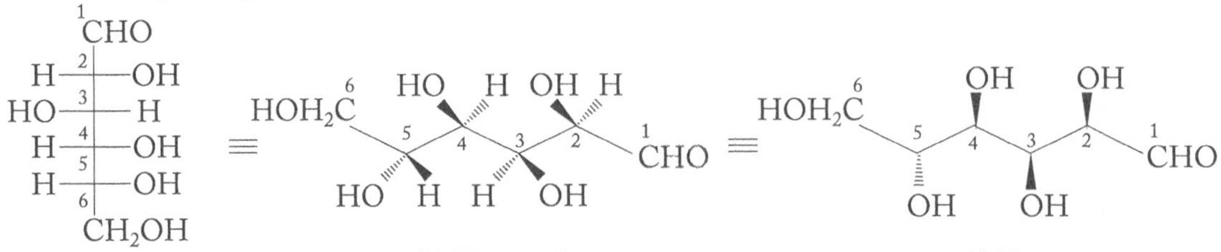

(a) Fischer projection (b) Flying wedge (c) Zigzag

Fig. 11.15 D-Glucose

In the above example (Fig. 11.15), the numbering of the carbon atoms has been done according to IUPAC nomenclature.

11.3 ISOMERISM

Understanding isomerism forms the basis of stereochemistry. The term isomerism is used to describe the phenomenon when two or more molecules have identical molecular formula involving the same number and type of atoms, but different arrangements of atoms in consideration to atom connectivity and their spatial arrangements. There are three major classes of isomers—structural, stereochemical, and conformational.

11.3.1 Structural Isomers

Molecules having the same molecular formula but different sequence of their atom-connectivity are called structural isomers. They are also known as *constitutional isomers*. This term is more relevant because the term structure may also include configurational and conformational aspects. The term constitution connotes the number, kind, and connectivity of the atoms in a molecule. The constitution of a molecule is usually represented by a two-dimensional graph, in which atoms linked to each other are connected by single, double, and triple bonds. These types of structures are called *Lewis dash structures*. Constitutional isomers are further classified into the following three types.

(a) Chain isomers: CH_3—CH_2—CH_2—CH_3 and CH_3—$\overset{\overset{\textstyle CH_3}{|}}{CH}$—$CH_3$

                                         n-Butane   Isobutane

(b) Position isomers: CH_2—CH_2—CH_2—OH   and CH_3—$CH(OH)$—CH_3

                                   Propan-1-o1   Propan-2-o1

(c) Functional group isomers: CH_3CH_2OH and CH_3—O—CH_3

                                  Ethanol   Methoxymethane

11.3.2 Stereoisomerism

Molecules having the same connectivity of atoms but differing in the relative three-dimensional arrangement of their atoms are known as *stereoisomers* and the phenomenon is called stereoisomerism. Hirschman and Hanson have classified compounds on the basis of constitutional isomers and stereoisomers. (Refer to Fig. 11.16)

11.3.3 Configurational Isomers

Stereoisomers differing in configurations are called *configurational isomers*. The term configuration means the fixed relative arrangement of atoms in space of a molecule of defined constitution. The term does not include those arrangements in space of a molecule that can be achieved by rotation of the part of the molecule about one or more single bonds.

Each stereoisomer of defined constitution has a unique configuration that can only be converted into a different configuration by chemical means involving breaking and making of bonds (Mosher, 1980). Since bond breaking is a high-energy process, configurational stereoisomers are isolable at ordinary room temperature.

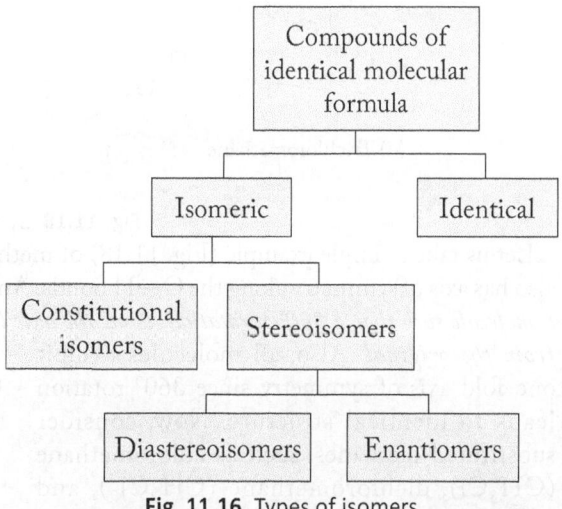

Fig. 11.16 Types of isomers

11.4 Symmetry and Chirality

There are various objects which we encounter in our daily life which are chiral or achiral in nature. The term 'chiral' is derived from the Greek word *'cheir'* that means 'hand.' Our hands are chiral in nature, that is, the right and left hands are mirror images of each other (non-superimposable) but not identical. For example, shoes we wear are chiral, but the socks are achiral. A chiral centre is defined as a tetrahedral carbon atom bonded to four different groups and is designated by an asterisk (*). If a molecule has more than one chiral centre, its isomers exist as only enantiomers as they are non-superimposable mirror images of each other. Now, if all the atoms in a molecule

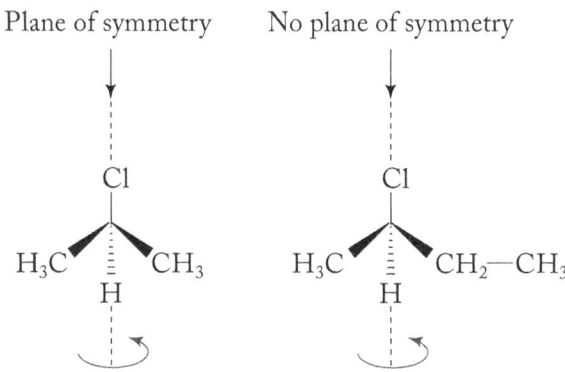

Fig. 11.17 Plane of symmetry in 2-chloropropane and its absence in 2-chlorobutane

have two or more similar groups, it does not have a chiral centre and the molecule will be superimposable on its mirror image and is called an *achiral molecule*.

The mere presence of chiral centre is not sufficient to predict chirality in a molecule. There are various symmetry elements which help in deciding the molecular chirality. The most common symmetry element is called *plane of symmetry*. A plane of symmetry (or mirror plane) is defined as *an imaginary plane that bisects the molecule (or object) in to two halves which are mirror images of each other*. The imaginary plane may pass through atoms, between atoms, or both; for example, 2-chloropropane has a plane of symmetry, whereas 2-chlorobutane does not (refer to Fig. 11.17).

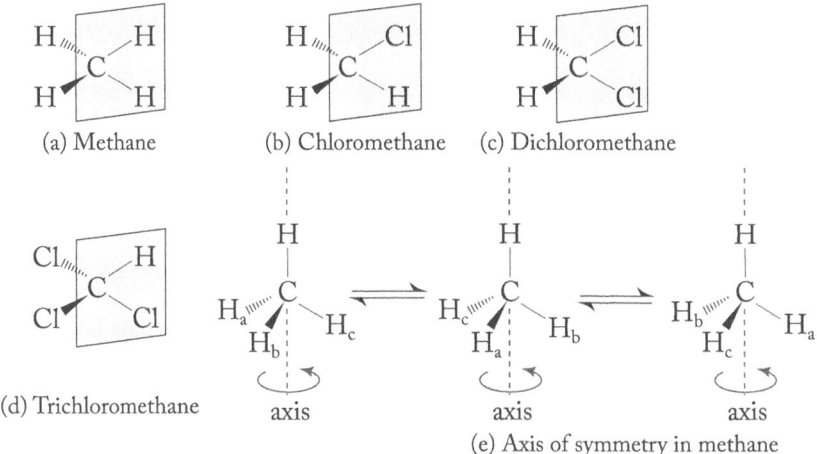

(a) Methane   (b) Chloromethane   (c) Dichloromethane

(d) Trichloromethane   (e) Axis of symmetry in methane

Fig. 11.18 Symmetry planes

Let us take a simple example (Fig. 11.18) of methane, a molecule that possesses plane of symmetry. It also has axis of symmetry along the C—H bonds. An *axis of symmetry* is defined as *a line that passes through a molecule such that a 360°/n rotation about the axis leads to a three-dimensional structure indistinguishable from the original*. Also, all molecules exhibit one-fold axis of symmetry since 360° rotation leads to identical structure. Now, consider substituted methanes, such as chloromethane (CH_3Cl), dichloromethane (CH_2Cl_2), and trichloromethane ($CHCl_3$); we observe that each

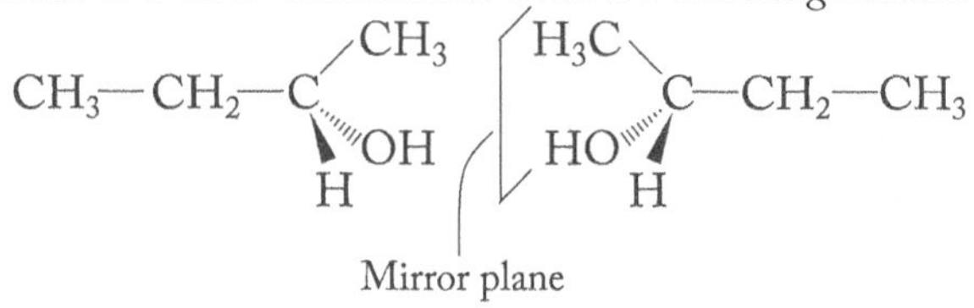

$$CH_3—CH_2—C\overset{CH_3}{\underset{H}{\overset{|}{\diagdown}}}OH \qquad \overset{H_3C}{\underset{HO}{\overset{\diagup}{\diagdown}}}C—CH_2—CH_3$$

Mirror plane

Fig. 11.19 Enantiomers of butan-2-ol

will have one plane of symmetry. The molecules which have plane of symmetry are essentially achiral. Let us take one more example of butane-2-ol. It will exist as non-superimposable mirror images of each other. Hence, butane-2-ol is chiral due to the presence of asymmetry in it (Fig. 11.19).

11.4.1 Enantiomers

A pair of stereoisomers having non-superimposable mirror images is called enantiomers. For example, mandelic acid (PhCHOHCOOH) has two stereoisomers (Fig. 11.20) that are mirror images of each other, but one is not superimposable on the other. Therefore, they represent a pair of enantiomers. A stereoisomer has only a pair of enantiomers because a molecule can have only one mirror image. On the other hand, a stereoisomer can have more than two diastereoisomers.

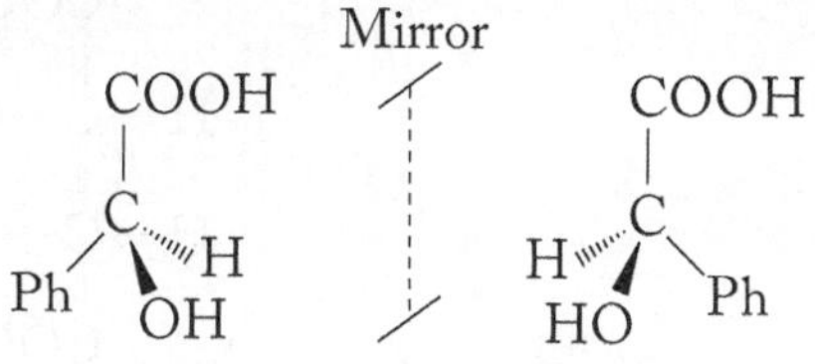

Fig. 11.20 Enantiomers of mandelic acid

By the term superimposable, we mean that two structures can be placed on each other in such a way that atoms with bonds of one structure coincide with the other. Enantiomers occur only among those compounds that are chiral. In other words, a chiral molecule and its mirror image are called a pair of enantiomers and the relationship between them is termed as enantiomeric.

11.4.2 Racemic Modifications

A racemic modification is an equimolecular mixture of a pair of enantiomers independent of whether it is crystalline, liquid, or gaseous. The stereochemical descriptor *rac* stands for racemic or racemate.

The racemic modification is optically inactive due to external compensation, that is, (+)-rotation of one enantiomer is compensated by the (−)-rotation of the other. Since racemic modification is a mixture, it can be separated into pure enantiomers (Fig. 11.21). The process is known as resolution. The racemic modification may exist in three different forms in the solid state.

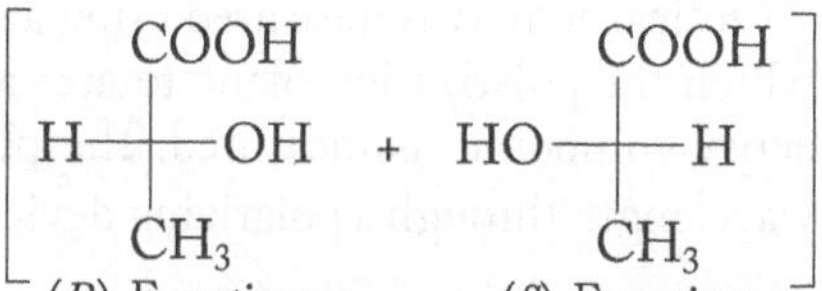

Fig. 11.21 *rac*-2-Hydroxypropanoic acid (Lactic acid)

Chiral Drugs

Many of the common drugs such as Brufen (ibuprofen) are chiral and exist as enantiomers. At times, one enantiomer of a drug can alleviate or cure a disease whereas its other enantiomer may be ineffective or even poisonous.

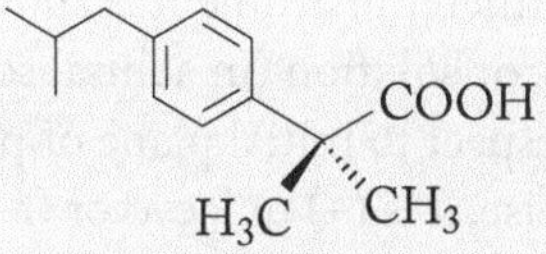

(*S*)-Fluoxetine
(Anti-inflammatory)

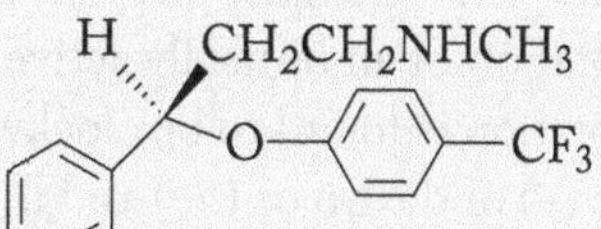

(*R*)-Ibuprofen
(Anti-depressant)

Ibuprofen and fluoxetine have stereogenic centres. These compounds exist as enantiomeric pairs of which only one of them will exhibit biochemical activity. (*S*)-Ibuprofen is an active component of the anti-inflammatory drug product Motrin. (*R*)-Ibuprofen has no anti-inflammatory activity itself, but is known to convert rather slowly to the (*S*)-enantiomer in vivo. (*R*)-Fluoxetine is the active component found in the anti-depressant Prozac. [For (*R*) and (*S*) refer to Section 11.6]

11.4.3 Diastereomers

A pair of stereoisomers having no mirror image relationship is known as diastereoisomers; for example, *meso*-tartaric acid and active tartaric acid (Fig. 11.22) represent a pair of diastereoisomers.

The structures shown in Fig. 11.22 (a) and (b) are stereoisomers but not mirror images of each other.

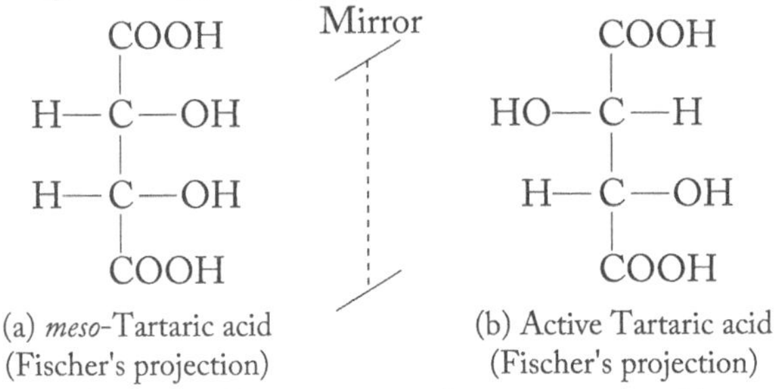

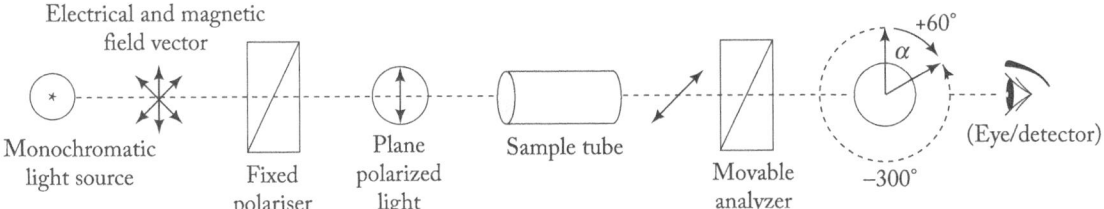

(a) *meso*-Tartaric acid
(Fischer's projection)

(b) Active Tartaric acid
(Fischer's projection)

Fig. 11.22 Diastereomers of *meso*-tartaric acid

11.4.4 Optical Activity

Stereoisomers that are capable of rotating the plane of a plane-polarized monochromatic light, when such light is passed through them, are said to be optically active and these stereoisomers are called optically active stereoisomers or simply optical isomers. The term is now obsolete. Optical isomers are now called *enantiomers*.

Optical activity is measured using a simple instrument known as polarimeter, in which the angle (α) by which the polarization plane rotates a plane polarized light when passed through a sample of optically active compound is measured. The plane polarized light is created by passing ordinary light of single wavelength through a polarizing device which may be a calcite crystal ($CaCO_3$ crystal) or polarizing film.

Fig. 11.23 Polarimeter

Such devices transmit selectively only that component of a light beam having electrical and magnetic field vectors oscillating in a single plane. This type of light wave is specifically called 'plane polarized monochromatic light.' The schematic diagram of the polarimeter instrument is shown in Fig. 11.23. The observed angle of rotation of an optically active liquid, gas, or solution (in achiral solvent) is denoted by 'α'. If the angle of rotation is found to be clockwise with respect to initial plane of polarization, then the rotation is said to be (+) or dextro or (*d*–) and if anticlockwise, it is (–) or laevo or (*l*–) with respect to the observer. The prefixes 'dextro' and 'laevo' come from the Latin 'dexter' and 'laevus', meaning right and left. The observed optical rotation (α) is proportional to the concentration of the optically active molecules and path length of the solution through which polarized light has traversed. This is known as Biot's law, $\alpha = [a] \times c \times l$, where c = concentration of the compound in g/ml, l = length of the tube in dm and $[a]$ = proportionality constant and is called *specific rotation*.

Specific rotation can be calculated using the formula:

$$[\alpha]_D^\theta = \frac{\text{Observed rotation in degrees}}{\text{Path length in dm} \times \text{Concentration in g/ml}} = \frac{\alpha}{l \times C}$$

11.5 *E–Z* NOTATIONAL SYSTEM

When the groups present on either end of a double bond (alkene) are the same or are structurally similar to each other, it is always easier to describe the configuration of the double bond as *cis* or *trans*. For example, *cis*-but-2-ene and *trans*-but-2-ene are stereoisomers, and the terms '*cis*' and '*trans*' denote the configuration of the double bond. However, *cis–trans* notation for alkenes becomes impossible when one of the doubly bonded carbon atoms possesses two identical substituents; for example, but-*1*-ene and 2-methylpropene do not have stereoisomers:

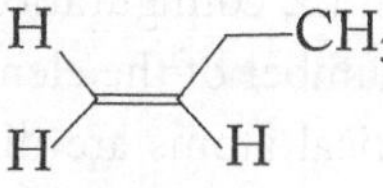

But-1-ene
(no stereoisomers possible)

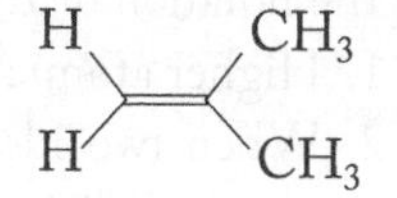

2-Methylpropene
(no stereoisomers possible)

The stereochemical designations, '*cis*' and '*trans*' seem ambiguous when it is not obvious which substituent on stereogenic carbon is similar or merely analogous to a reference substituent. This gave inception to a completely unambiguous system for specifying double bond stereochemistry based on ranking substituents present on carbon atoms as per atomic number criterion called *E–Z* nomenclature system. If atoms of higher atomic number are present on the same side of the double bond, it has '*Z* configuration' (*Z* comes from German word *zusammen* that literally means together). If atoms of higher atomic numbers are present on the opposite sides of a double bond, it is denoted as '*E* configuration' (*E* comes from German word *entgegen* that means opposite). The compounds shown here have *Z* and *E* configurations.

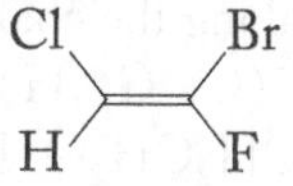

(*Z*)-configuration

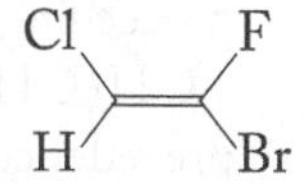

(*E*)-configuration

Check Your Progress

1. Do the following pair of structures represent identical molecules or a pair of enantiomers?

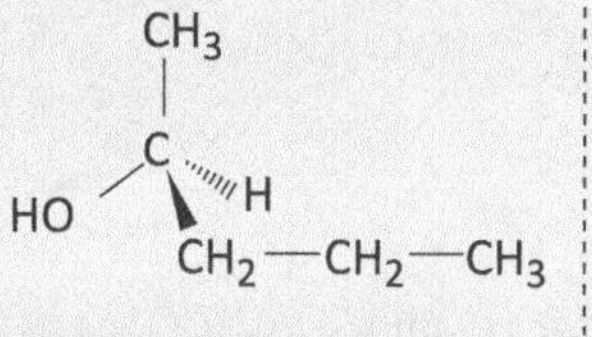

(I)

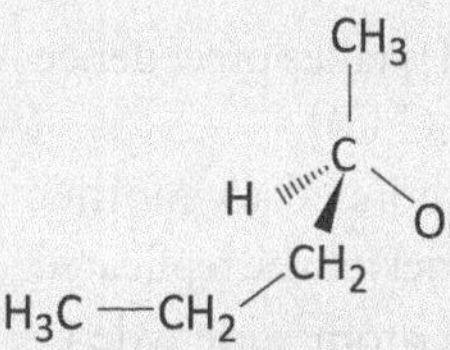

(II)

2. Define: (a) enantiomers, (b) diastereomers and (c) meso compounds.
3. Define constitutional isomer with a suitable example.
4. Glycerol, $CH_2OHCHOHCH_2OH$, is an important constituent in the biosynthesis of fats. Does glycerol have symmetry? Is it chiral?
5. Draw all possible stereoisomers of the below compound and then pair up enantiomers and diastereomers.

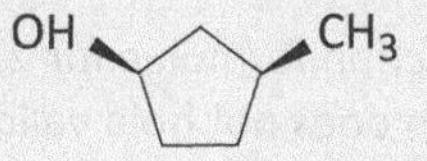

6. What are optically active compounds? How are they measured?
7. One of the enantiomers of 2-pentanol is placed in a polarimeter; the observed rotation is 4.05°counterclockwise. The solution was made by diluting 6 g of 2 pentanol to a total of 40 ml, and the solution was placed into a 200-mm polarimeter tube for the measurement. Determine the specific rotation for this enantiomer of 2-pentanol.
8. The observed rotation of 8.0 g of a compound (A) in 50 ml of solution in a polarimeter tube of 20 cm long is + 3.4° . What is the specific rotation of the compound?

The compound on the left hand side has Cl and Br as higher ranked substituents on the same side of the double bond, thus having *Z*-configuration; whereas Cl and Br higher ranked groups are present on opposite sides of the double bond of the compound on the right and thus it has *E*-configuration.

The notation of *E* and *Z* configurations follows the CIP rules, which are as follows:

1. Higher atomic number of the element will take precedence over lower atomic number.
2. When two identical atoms are directly attached to an olefin (double) bond, compare the atoms attached with these two with their atomic numbers. Precedence is then decided at the first point of difference; say, ethyl [−C(C, H, H)] will precede methyl [-C(H,H,H)], *tert*-butyl will take precedence over isopropyl, and isopropyl over ethyl, and so on.

$$-C(CH_3)_3 > -CH(CH_3)_2 > -CH_2CH_3$$
$$-C(C, C, C) > -C(C, C, H) > -C(C, H, H)$$

3. One needs to work outwards from the point of attachment and compare all the atoms attached to a particular atom before proceeding further along the chain: $-CH(CH_3)_2$ [±C(C, C, H)] takes precedence over $-CH_2CH_2OH$ [−C(C, H, H)]

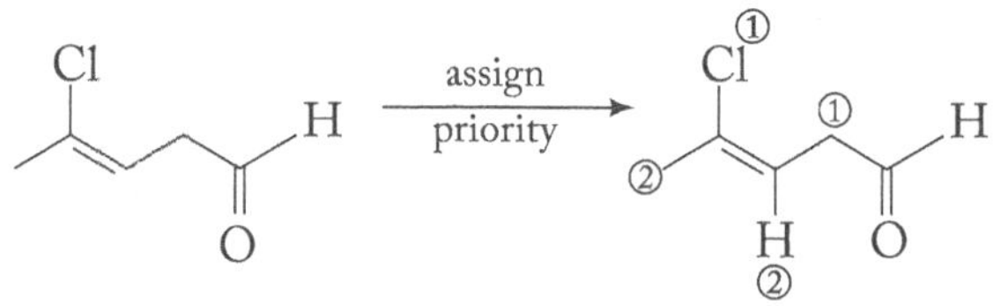

High priority groups on same side of double bond
(3Z)-4-Chloro-3-pentenal

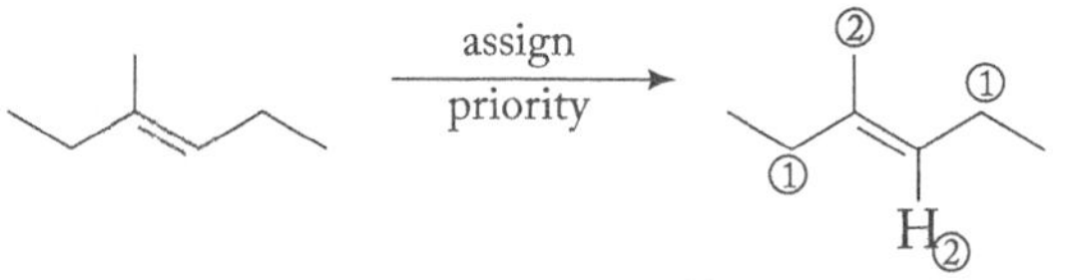

4. On working outwards from the point of attachment, check substituent atoms one by one and never assume it as a group. Oxygen has a higher atomic number than carbon; thus for instance, $-CH_2OH$ [−C(O,H,H)] takes precedence over $-C(CH_3)_3$ [−C(C,C,C)]

High priority groups on opposite side
(3E)-3-Methyl-3-hexene

5. If an atom is multiply bonded to another atom, then one can consider it to be replicated as a substituent on that atom: aldehyde (−CHO) will be considered as −C(O, O, H) and thus, the group ±CH=O [−C(O, O, H)] precedes over the alcoholic group ($-CH_2OH$) as [−C(O, H, H)].

Check Your Progress

9. Designate the following compounds as *E* or *Z*.

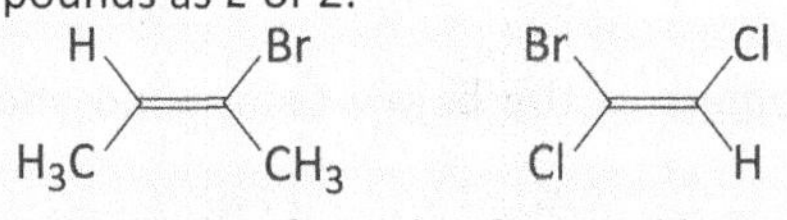

10. Draw any four possible Fischer projection formulae for 2,4-dibromohexane.
11. In which case does *cis–trans* notation does not hold valid for alkenes?
12. What is chirality? Is the molecule shown below chiral or achiral?

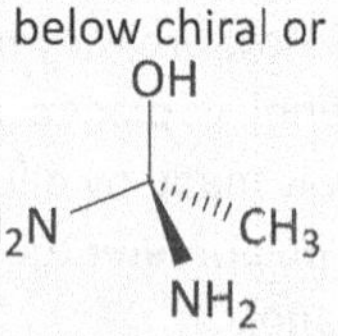

13. If 1.5 g of a sample x is dissolved in ethanol and made up to 50 ml of the solution showing an observed rotation of + 2.79° in a 10-cm polarimeter tube, determine its specific rotation at 20°C (sodium light λ = 589.3 nm D line).

11.6 ABSOLUTE CONFIGURATION

An absolute configuration describes the actual spatial arrangement of atoms in a chiral molecule (or group) and its stereochemical description is denoted as *R* or *S*, referring to *rectus* or *sinister* respectively.

Cahn, Ingold, and Prelog (CIP) rule is used to designate the configurations of stereoisomers having chiral centres. It is often called 'chirality rule'. The method has also been extended to compounds having chiral axes and chiral planes. This system is commonly called *R, S*-system and since this type of specifying configurations is independent of any reference compound, the system is often termed as absolute configuration assignment.

11.6.1 Method of Assigning R, S, Notations

The CIP system is based on rules and a strict hierarchical order of decisions until a single stereodescriptor can be used to describe a given configuration. For tetrahedral stereogenic atoms having four different atoms or groups, the chirality rule is based on the arrangement of these atoms or groups, including chain and rings, in an order of precedence, often referred to as an order of priority. For assigning *R, S*-nomenclature to any chiral centre of a molecule, certain procedures are to be followed.

1. Identify the number of chiral centres in the molecule.
2. Identify the four different atoms or groups attached to each of these chiral centres. Assign to each of the substituents on a chiral centre a priority symbol 1, 2, 3, 4 or *a, b, c, d* based on *sequence rules* (discussed latter), such that the decreasing order of priority is 1 > 2 > 3 > 4 or *a > b > c > d*, where '>' denotes 'has priority over'. The name 'ligand' should not be used for atoms or groups in case of organic molecules.
3. The molecule is then viewed from the position remotest from the lowest priority group and a hypothetical path is drawn moving from 1 to 2 and then to 3 (1 → 2 → 3) or *a → b → c*. If this path traces a clockwise path, then the stereocentre is said to have *R* configuration (*R* from *rectus*, Latin for *right*).

If the said path traces a counterclockwise path, then the stereocentre is said to have *S* configuration (*S* from *sinister*, Latin for *left*). This is illustrated in Fig. 11.24.

The designation (*R*)- or (*S*)- is written in italics within parentheses followed by a hyphen before the name of the compound, preceded when necessary by the appropriate locants for the chiral centres. It is to be noted that if a molecule contains a single chiral centre, then a pair of enantiomers is possible. One of them is *R*-isomer and its mirror image is *S*-isomer. For example, in case of lactic acid, $CH_3CHOHCOOH$, we can name them as (*R*)-lactic acid and (*S*)- lactic acid.

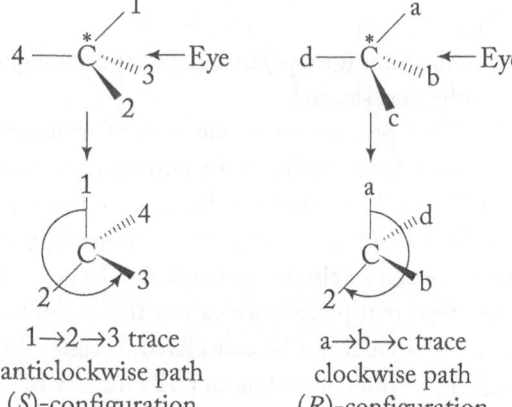

Fig. 11.24 Designating (*R*) and (*S*) configurations

11.6.2 Sequence Rules for Determination of Priority of Ligands

According to CIP chirality rules, developed by Cahn, Ingold, and Prelog, there are six (0 to 5) rules for the determination of priorities of atoms or groups attached to a chiral carbon centre. Below, the methods have been discussed pointwise and not according to the rules published in the original paper (Cahn, et. al., 1966).

1. Four different atoms or groups attached to a chiral centre get their priorities according to the atomic number of the atoms of each group directly attached to the chiral centre. For example, in 1-bromo-1-iodoethane, CH_3CHIBr, the chiral centre is attached to CH_3, I, Br, and H. The atomic numbers of atoms directly attached to the chiral centre show that priority order should be I > Br > methyl > hydrogen. If one of the four groups is replaced by a lone pair of electrons (i.e, carbanion, nitrogenous compounds, etc., having tetrahedral geometry) then the lone pair will have the lowest priority.

2. When preferences cannot be determined on the basis of the atomic number of the atoms directly attached to the chiral carbon, that is, when two or more atoms directly attached to the chiral centre are the same, then the atomic numbers of next sets of atoms in the unassigned groups are taken into consideration. This outward exploration is continued until a decision can be made. The priorities are then assigned at the first point of difference.

For example, in case of 2-bromobutane, $CH_3CHBrC_2H_5$, the four ligands attached to the chiral centres are CH_3, C_2H_5, Br, and H. Of these Br is the highest and H is the lowest priority group. To decide preference between CH_3— and CH_3CH_2—, decision cannot be made on the basis of first atom (C) of each group directly attached to the chiral centre. However, in case of —CH_3, other three atoms are hydrogen, that is, (H, H, H) but in case of —CH_2CH_3, two atoms are hydrogen and the third one is carbon, that is, H, H, C. Therefore, —CH_2CH_3 gets priority over —CH_3. (i) If a chiral centre contains branched groups containing several carbon atoms then exploration requires a well-defined hierarchy of paths.

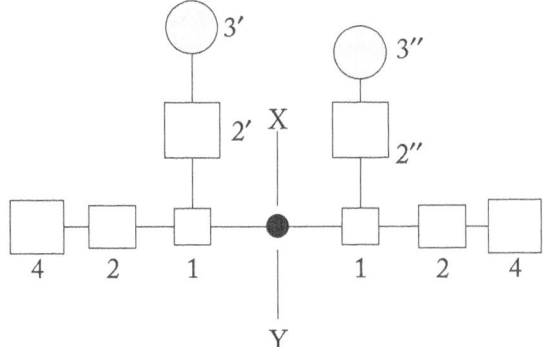

Fig. 11.25 Quick (*R*) and (*S*) designation tree

The most important principles are:
(a) All atoms (groups) in a given set of branches (Fig. 11.25) must be explored before one proceeds to the next set, and
(b) Once precedence of one path of exploration over another has been established in one set, that precedence carries to the next set of branches till the decision is conclusive.

In Fig. 11.25, • represents the chiral centre. X and Y are two ligands, which pose no ambiguity according to CIP rules. Other two ligands are shown by geometric figures. Boxes marked 1 in each of the two groups are same and so also boxes marked 2. In boxes marked 2′ and 2″, say, 2′ > 2″ in terms of sequence rule. Therefore, that precedence carries the exploration to two circles, marked 3′ and 3″. The boxes marked 4 and 4″ should not be considered in spite of the fact that one of them (4″) may have groups having higher priorities according to CIP chirality rules.

A list of names of ligands arranged in increasing order of sequence-rule preference is given below.

1. Hydrogen	16. *sec*-Butyl	31. Trityl	46. Diethylamino
2. Methyl	17. Cyclohexyl	32. *o*-Nitrophenyl	47. Trimethylamino
3. Ethyl	18. 1-Propenyl	33. Formyl	48. Nitroso
4. Propyl	19. *tert*-Butyl	34. Acetyl	49. Nitro
5. Butyl	20. Isopropenyl	35. Benzyl	50. Hydroxy

6. Pentyl	21. Ethynyl	36. Carboxy	51. Methoxy
7. Hexyl	22. Phenyl	37. Methoxycarbonyl	52. Ethoxy
8. Isopentyl	23. *p*-Tolyl	38. Benzyloxycarbonyl	53. Benzyloxy
9. Isohexyl	24. *p*-Nitrophenyl	39. *tert*-Butyloxycarbonyl	54. Phenoxy
10. Allyl	25. *m*-Tolyl	40. Amino	55. Formyloxy
11. Neopentyl	26. 3, 5-Xylyl	41. Methylamino	56. Fluoro
12. 2-Propynyl	27. *m*-Nitrophenyl	42. Phenylamino	57. Mercapto
13. Benzyl	28. 3,5-Dinitrophenyl	43. Acetylamino	58. Chloro
14. Isopropyl	29. 1-Propynyl	44. Benzoylamino	59. Bromo
15. Vinyl	30. *o*-Tolyl	45. Dimethylamino	60. Iodo

11.6.3 R, S-Nomenclature of Structures in Fischer Projections

When the lowest priority group is in any of the vertical bonds in Fischer projection then it is farthest from the observer, and therefore, its position should not be disturbed. In the below structure lowest priority group is in vertical upper bond and there is no need to reshuffle the groups.

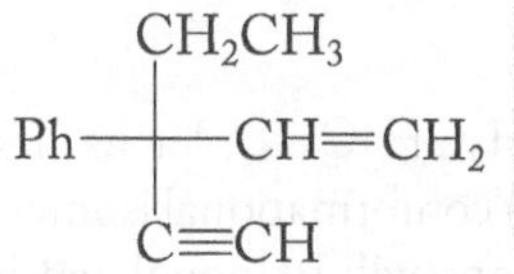

In this case, priority order of the groups, according to sequence rule is

$$Ph > -CH \equiv CH > -CH = CH_2 > CH_2CH_3$$

(a) (b) (c) (d)

The arrangement is

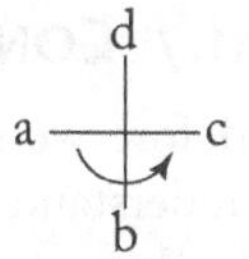

Now we will draw a path from a → b → c, without involving d. In the present case, the path is anticlockwise, therefore, the absolute configuration of the original compound is (*S*). The complete IUPAC name of the compound is (*S*)-3-ethyl-3-phenylpent-1-en-4-yne. If in a Fischer projection, the lowest priority atom or group is not in any vertical bonds, it should be brought into a vertical bond by changing the positions among any three atoms (or groups) sequentially on the chiral centre without disturbing the position of the fourth. Such changes of positions of atoms or groups on a chiral centre do not change the absolute configuration of the chiral centre. Exchanging two pairs of substituents simultaneously can convert one Fischer projection to another having the same configuration. A few examples are given here:

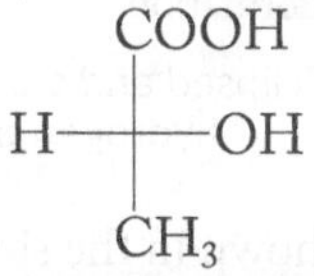

COOH

H——OH

CH₃

2-Hydroxypropanoic acid

Priority of groups:

OH > COOH > CH₃ > H

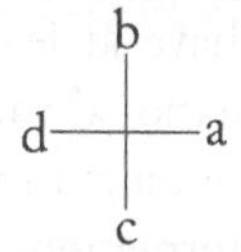

Here 'd', the lowest priority group is not on any vertical bond. But we can bring it into a vertical position, either above or below, by interchanging the positions among d, b, and c, without disturbing the position of a, we get the arrangement of groups as:

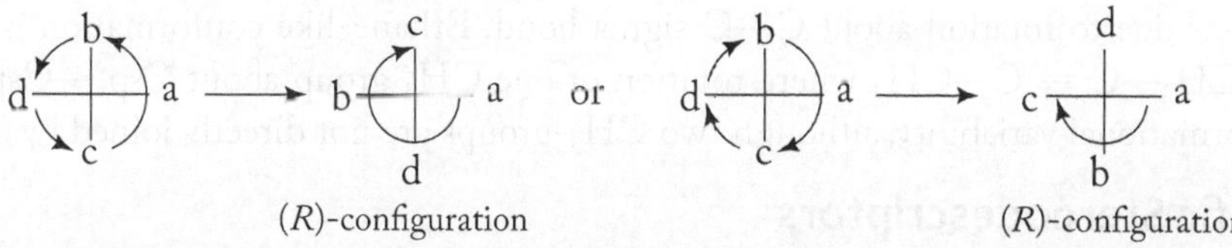

(*R*)-configuration   (*R*)-configuration

We trace a path from a → b → c, a clockwise path is observed. Therefore, the absolute configuration of the given acid is R, that is, the compound is (*R*)-2-hydroxypropanoic acid. The same result is obtained when the lowest priority group is brought to any of the vertical bonds by simultaneous changes of position of groups pairwise as:

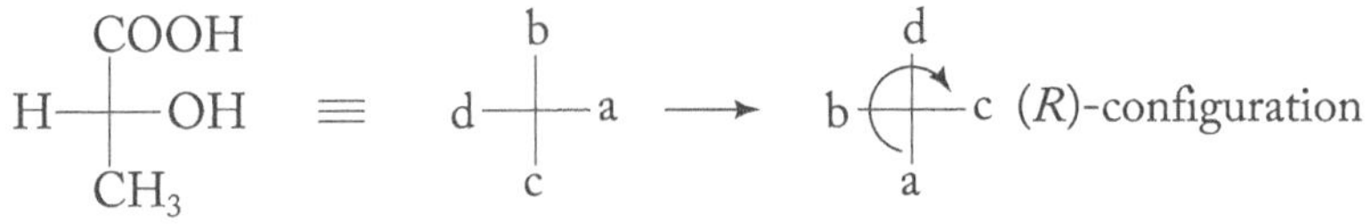

COOH
H——OH ≡
CH₃

2-Hydroxypropanoic acid
priority of groups:

b
d——┼——a
c

→

d
b——(⟳)——c (*R*)-configuration
a

Exchange of positions
between b, d and between a, c

OH > COOH > CH₃ > H
(a)   (b)   (c)   (d)

It is to be noted that interchange of positions between any two atoms or groups on a chiral centre changes the configuration from (*R*) to (*S*) and vice versa.

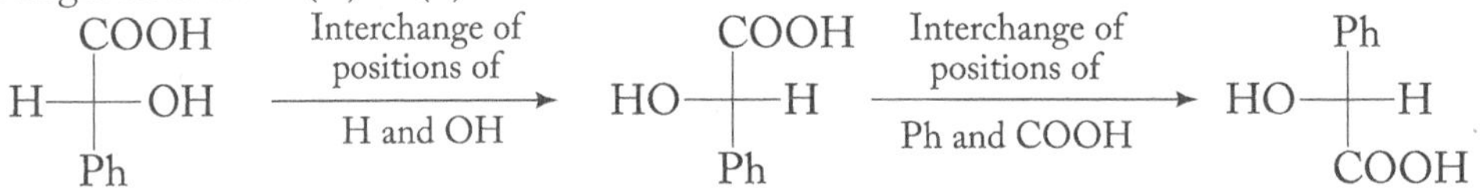

COOH
H——OH
Ph

Interchange of
positions of
———————→
H and OH

COOH
HO——H
Ph

Interchange of
positions of
———————→
Ph and COOH

Ph
HO——H
COOH

(*R*)-2-Hydroxy-2-phenylpropanoic acid
(Mandelic acid)

(*S*)-Isomer

(*R*)-Isomer

11.7 CONFORMATIONAL ANALYSIS OF BUTANE

Before we study about conformations of acyclic compound butane ($CH_3CH_2CH_2CH_3$), let us first understand about rotations about single bonds. We have already defined the term conformational isomers. Each conformation will have a different energy, the lower energy conformations will be populated in preference to those of higher energy. When a molecule has a number of conformations that correspond to energy minima, they are called *conformers*. The term conformational analysis means the analysis of the physical and chemical properties of a compound in relation to its geometry and population of its conformers.

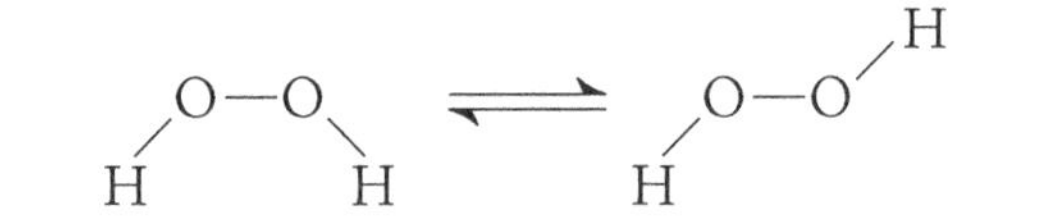

Different conformations of a molecule differ in torsion angles about one or more bonds. The concept of torsion angle has been given later in this chapter. Since it requires four atoms to define a torsion angle, a molecule must have at least four atomic centres to display conformational variability. Moreover, the four atoms must be linked in a row, A—B—C—D

O—O
H   H

⇌

O—O
H   H

Eclipsed conformation   Staggered conformation

Fig. 11.26 Eclipsed and staggered conformations of hydrogen peroxide

to display the conformational change. Bonding of four centres as shown in the side figure cannot exhibit conformational diversity. For example, pyramidal NH_3 or planar BF_3 displays no torsion angles and consequently they have no conformational isomers.

On the other hand, H—O—O—H represents a tetra-atomic molecule and exhibits conformational isomerism having infinite number of conformations (Fig. 11.26). H_3C—CH_3 can have infinite number of conformations due to rotation about C—C sigma bond. Ethane-like conformation is also possible in but-2-yne, CH_3—C ≡ C—CH_3 where rotation of one CH_3 group about Csp^3—Csp sigma bond gives the conformational variability, although two CH_3-groups are not directly joined by a single bond.

11.7.1 Specific Stereodescriptors

Conformations of acyclic compounds are sometimes referred to using specific stereodescriptors, namely eclipsed, staggered, and gauche (skew). Two atoms or groups attached to two adjacent centres are said to be 'eclipsed' if the torsion angle between them is zero. The conformation is said to be 'staggered' when the atoms or groups attached to adjacent centres are as far apart as possible compared to an eclipsed conformation. 'Gauche' or 'skew' are synonymous with synclinal conformation, that is, torsion angle is larger than 30° but less than 90°. The designation *syn* or eclipsed is often used when torsion angle ω is 0°. The designation 'gauche is frequently used for $\omega = 60°$ and the designation is *anti* when $\omega = 180°$. The stereodescriptors '*trans*' or '*anti*' are not recommended in place of anticlinal, nor '*cis*' or '*syn*' in place of synclinal. Usually, eclipsed and staggered stereodescriptors are used to denote the conformations when all atoms or groups are identical on adjacent centres (Fig. 11.27).

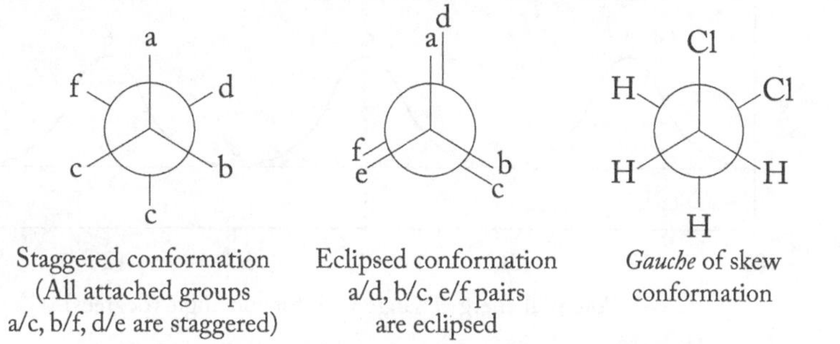

Staggered conformation
(All attached groups
a/c, b/f, d/e are staggered)

Eclipsed conformation
a/d, b/c, e/f pairs
are eclipsed

Gauche of skew
conformation

Fig. 11.27 Stereodescriptors of acyclic compounds

11.7.2 Torsional Curves of a Few Simple Acyclic Compounds

The curve showing the change of potential energy of a molecule with change in torsion angle (about some chosen C—C or other s-bond) is known as torsional curve.

Torsional Curve of Ethane (CH₃—CH₃)

In ethane, CH_3—CH_3, the attachment of two tetrahedral carbon atoms does not determine, unequivocally, the relative position of the three atoms of hydrogen situated on one carbon atom with reference to the hydrogen atoms on the other carbon atom. Work of Pitzer in 1936 shows that rotation of the molecule about C—C bond axis is not completely free. There exists an energy barrier to this free rotation but this hindrance does not prevent rotation, but it destabilizes certain arrangements. There are infinite possible positions for the hydrogen atoms on C-2 atom relative to those on C-1. Each arrangement is a conformation. The aim of the conformational analysis is to examine the various conformations adopted by molecules and to put emphasis on those, which are energetically favoured (conformers). The conformation of ethane changes as one carbon atom rotates relative to the other about the C—C σ-bond. Figure 11.28 shows the variation of potential energy with torsion angle (dihedral angle) for ethane. The two extremes, in terms of energy, are the staggered conformation, which has the minimum energy, and the eclipsed conformation, which has the maximum energy. Both may be conveniently represented as Newman projections as shown in the torsional curve of ethane.

Since there are three equivalent hydrogen atoms on C-1 and also on C-2, it is necessary to specify one hydrogen atom on C-1 and one on C-2 to determine the dihedral angle. The torsion angle (w), for the staggered conformation is 60° while that for eclipsed conformation is 0°. There is an energy barrier to rotation of 12.09–12.26 kJ mol^{-1} (2.89–2.93 kcal mol^{-1}). Torsional barrier in ethane is not due to steric interaction between the two eclipsed hydrogen atoms, which can be seen by a simple calculation, using van der Waals radii of hydrogen atoms. The separation of H atom on C-1 from its nearest neighbour on C-2 is 2.3 Å (0.230 nm) in the eclipsed and 2.5 Å (0.250 nm) in the staggered conformation. The explanation is more likely to lie in the eclipsing of C—H bonds.

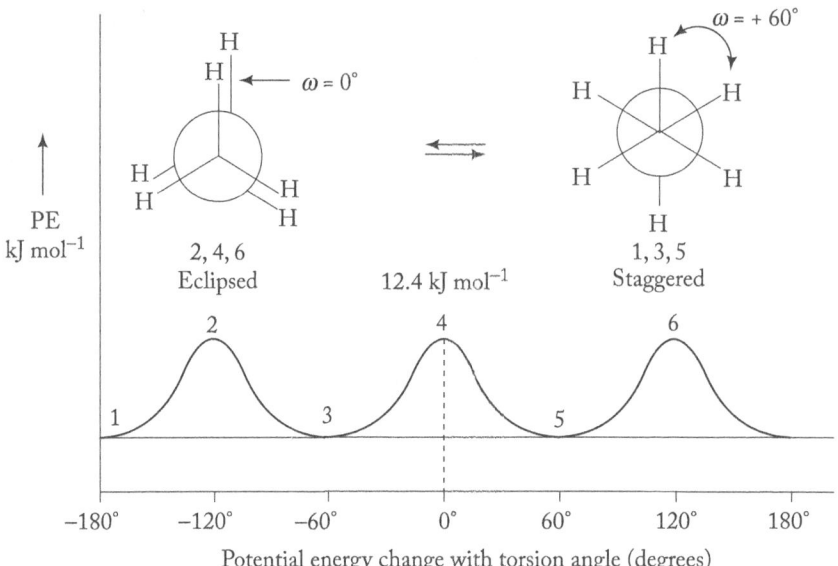

Fig. 11.28 Potential energy curve with torsion angle of ethane

Studies have shown that van der Waals repulsion and electrostatic interaction of weakly polarized C—H bonds are not of importance for the presence of energy barrier for staggered and eclipsed forms. Sovers et al. (*J. Chem. Phys*, 49, 2592, 1968) suggested that instability of eclipsed form of ethane is due to unfavourable overlap interaction between the bond orbitals (Pauli exclusion principle). Badar et al. (*J. Am. Chem. Soc.*, 112, 6530, 1990) concluded that stability of staggered conformation is due to favourable interaction between the bonding–antibonding orbitals.

Population of different conformations can be calculated using Boltzmann formula. At 25 °C, there is only one molecule of eclipsed ethane for each 160° of staggered ethane, that is, population of eclipsed ethane is negligible. Recent studies have shown that the energy barrier is around 2.89–2.93 kcal mol^{-1} (12.09– 12.26 kJ mol^{-1}). The torsional potential (also known as Pitzer potential) is approximately shown by the expression 5.6, where w is the torsion angle and E_0 is the energy barrier as,

$$E = 1/2\ E_0\ (1 + \cos 3w)$$

Three indistinguishable energy-minima conformations correspond to staggered conformation which appear at $w = +60°$, $(-60°)$, and $+180°(-180°)$. The energy maxima conformations are formed at $w = 0°$, $+120°$, and $-120°$.

Torsional Curve of Propane (CH$_3$CH$_2$CH$_3$)

Propane gives a torsional curve similar to that of ethane but energy barrier between the eclipsed and staggered forms is slightly higher; that is, 14.2 kJ mol^{-1}. The small difference between the rotational energy barrier of ethane and propane further shows that steric interaction has insignificant role (at least in this case) in the origin of energy barrier. The eclipsing of C—H with C—CH$_3$ in propane is hardly more unfavorable than the eclipsing of C—H with C—H in ethane, despite the greater steric hindrance of the methyl group. Potential energy curve of propane as a function of the torsion angle is given in Fig. 11.29.

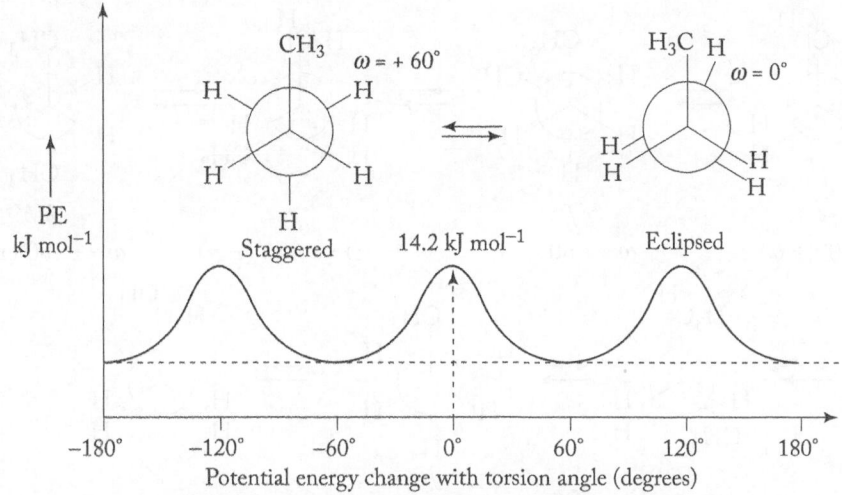

Fig. 11.29 PE curve with torsion angle of propane

In the case of *n*-butane the change of potential energy with torsion angle, which is the angle between the Me—C-2—C-3 and Me—C-3—C-2 planes, is to be plotted. In *n*-butane, there are three energy barriers to rotation about C(2)—C(3) bond and three minimum energy conformations (conformers) are distinguishable. The two energy-minima conformations are enantiomeric gauche forms and the third one is achiral anti-form. The anti-form is of lowest energy because it is free from the repulsive CH_3/CH_3 van der Waals interactions of the gauche forms. There are also three barriers for 360° rotation of conformations. Two have lower energy involving only CH_3/H eclipsing. These eclipsed conformations are enantiomeric and therefore, equi-energetic and the third one is high-energy one involving CH_3/CH_3 eclipsing. Potential energy diagram against torsion angle of *n*-butane, when rotated about C-2/C-3 σ-bond is shown in Fig. 11.30.

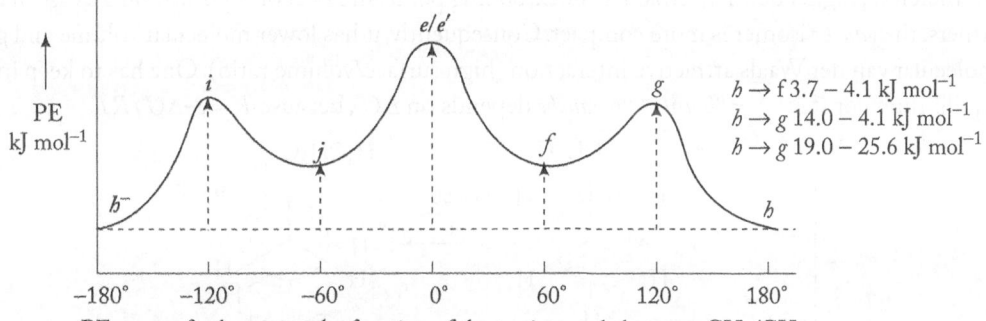

PE curve of *n*-butane as the function of the torsion angle between CH_3/CH_3

Fig. 11.30 Torsional curve of *n*-butane

The Newman projections of *e, f, g, h, i, j* are shown in Fig. 11.31.

e (*synperiplanar* or fully eclipsed) is the maximum-energy conformation. The higher (25 kJ mol^{-1}) energy barrier rotation arises from van der Waals net repulsive interactions, principally between the eclipsed methyl groups, and torsional strain owing to the eclipsing of three pairs of vicinal bonds.

f and *j* (*synclinal*) are *gauche* conformations. The average energy difference between the *anti* and *gauche* conformations in *n*-butane is about 3.3 kJ mol^{-1}. This amount of energy barrier, with respect to staggered form, results from the van der Waals repulsive interaction between the two methyl groups.

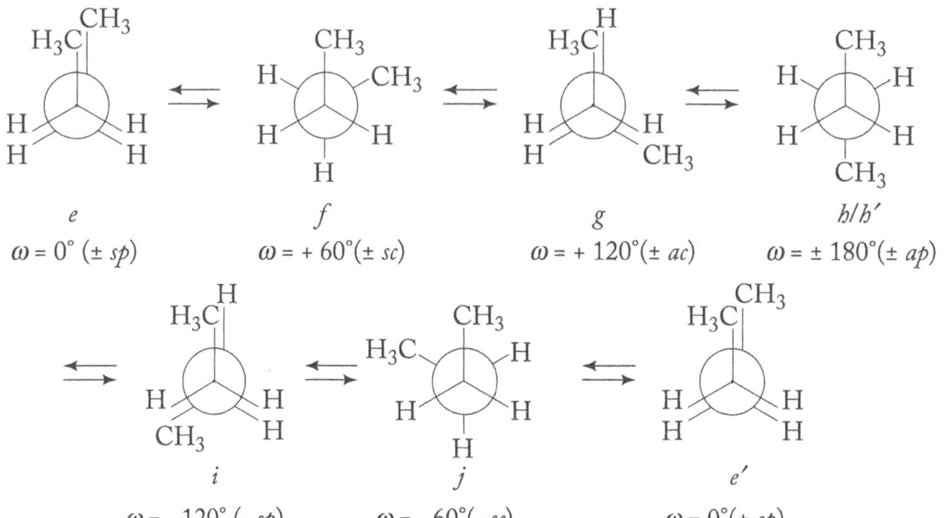

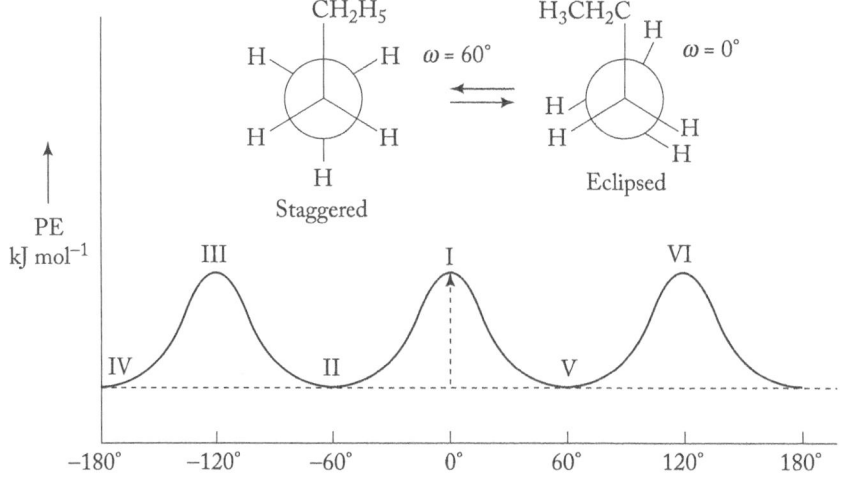

Fig. 11.31 Various conformations of *n*-butane

g and i (*anticlinal*) are partially eclipsed conformations. The energy barrier from staggered form in these cases is similar to the barrier in propane (14 kJ mol^{-1}), van der Waals repulsive interactions between H and CH$_3$ is considered to be very insignificant 1.26 kJ mol^{-1} for each methyl–hydrogen repulsion in this case.

Conformational Energy Curve of *n*-Butane about C-1/ C-2 Rotation

In *n*-butane there are three C—C bonds against which rotation is possible. The potential energy diagram of *n*-butane (CH$_3$CH$_2$CH$_2$CH$_3$) due to rotation about C-1—C-2 bond is shown in Fig. 11.32. I, III, VI, are eclipsed conformations and II, IV, V are staggered conformations. In fact, X-CH$_2$CH$_2$-Y (X π Y or X=Y) type molecules are capable of giving torsion curves similar to *n*-butane. In case of *n*-butane, X=Y=CH$_3$. Auwers Skita or conformational rule states that the isomer of high enthalpy has lower molecule volume, therefore, higher density, refractive index, boiling point, and heat of vaporization. In case of butane conformers, the *gauche* isomer is more compact. Consequently, it has lower molecular volume and greater intermolecular van der Waals attractive interaction (high surface/volume ratio). One has to keep in mind that equilibrium constant K = % *anti* / % *gauche* depends on $\Delta G°$, because $K = e{-}\Delta G°/RT$.

Fig. 11.32 Potential energy curve of *n*-butane about C-1/ C-2 rotation

Butane *Gauche* Interaction

The lower energy barrier of *n*-butane, passing through conformation *f* or *j* (Fig. 11.32) is equivalent to that of ethane. Essentially torsional strain constitutes this barrier. However, in the higher barrier (e) of *n*-butane, the eclipsing of two methyl groups introduces steric interactions, which superimpose on the torsional energy of ethane and thereby increases the height of the barrier. The comparison of the energies of conformations *h* (*anti*) and *f* (*gauche*) shows that the latter becomes destabilized by about 3.3 kJ/mol. This amount of energy represents the steric interaction energy of two methyl groups at a torsion angle ≡ dihedral angle of 60°, by comparison with the *anti*-position. This result may be simplified by stating that *butane–gauche* interaction is about 3.3 kJ/mol. This fundamental interaction (steric in origin) is encountered in many organic molecules and is specifically called *butane–gauche interaction*.

11.7 ISOMERISM IN TRANSITION METAL COMPOUNDS

J. Liebig (1823) demonstrated that silver salt of fulminic acid, silver fulminate ($Ag—O—N≡C$) and silver isocyanate ($Ag—N≡C≡O$) have the same composition, but varying properties. J. Berzelius (1830) coined the term 'isomerism' which was later documented by A. Butlerov after 30 years.

Transition metal compounds or coordination compounds exhibit various types of isomerism due to their differing geometries and coordination numbers. Figure 11.33 shows the general classification of coordination compound isomers. Further, it is important to note that most of these isomers interconvert rapidly and fast instrumental analysis are employed to study them.

Check Your Progress

14. Draw the Fischer projection of (*S*)-2-hydroxybutanoic acid [$CH_3CH_2CH(OH)COOH$].
15. Draw the structures of (*R*)-2-chlorobutane and (*S*)-2,3-dihydroxypropanal.
16. Draw the zigzag formula and Fischer formula of (i), (ii)

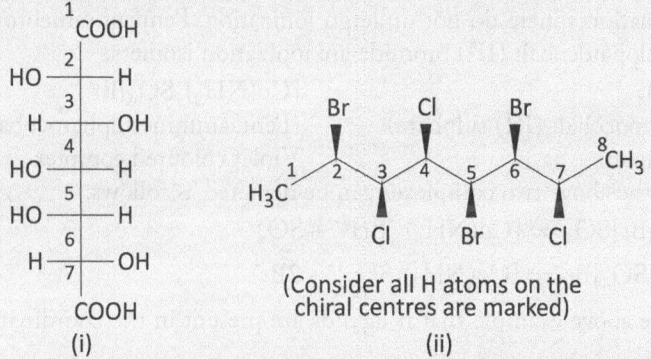

(i)   (ii)

(Consider all H atoms on the chiral centres are marked)

17. Draw the torsional curve of ethane and propane.
18. Draw the Fischer projection formula of (3*R*)-6-Bromohex-1-en-3ol.
19. Define conformers and conformational analysis.
20. Are all substances with chiral atoms optically active? Justify.
21. Assign (*R*) or (*S*) notation for the following molecules:

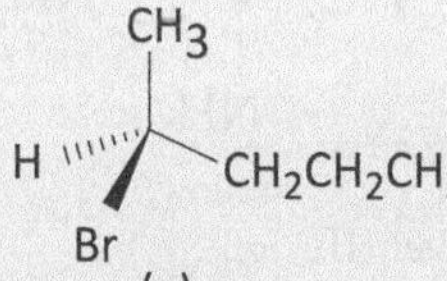

(a) 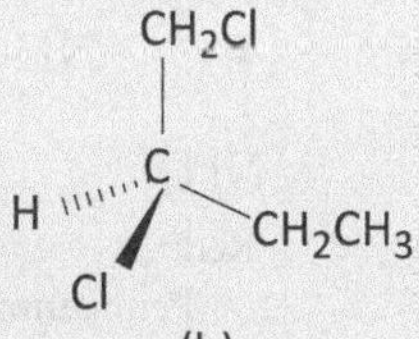   (b) 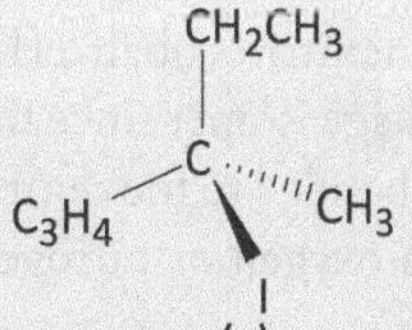   (c)

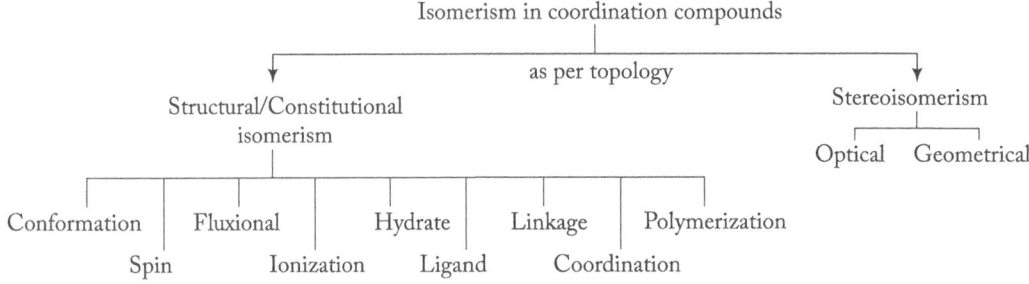

Fig. 11.33 Classification of isomerism in coordination compounds

Based on topology, isomerism in coordination (or metal complex) compounds are broadly classified in to the following types:

Structural Isomerism (or Constitutional Isomerism)

Here, all the types of isomerism exhibited by metal complexes are included except stereoisomers, that is, isomers have different topology. Structural isomerism in organometallic compounds is further classified into nine subclasses: (a) conformation, (b) spin, (c) fluxional, (d) ionization, (e) hydrate, (f) ligand, (g) linkage, (h) coordination, (i) polymerization. In this section, we will discuss ionization, linkage, coordination, and hydrate isomeric forms.

Stereoisomerism

In stereoisomerism, isomers have the same topology, but the arrangement of ligands in space is different. Geometrical isomerism and optical isomerism are subclasses of stereoisomerism.

Ionization isomerism   This type of isomerism is exhibited by coordination complexes that have the same empirical formula, but give different ions on ionization in solution. Such complexes are called ionization isomers. Such type of isomerism occurs due to differences in the position of ligands and ligands if present in coordination sphere do not undergo ionization. Pentaamminebromocobalt (III) sulphate and pentaamminesulphatocobalt (III) bromide are ionization isomers.

$[Co(NH_3)_5Br]SO_4$
(Pentaamminebromocobalt (III) sulphate)
Red coloured complex

$[Co(NH_3)_5SO_4]Br$
(Pentaamminesulphatocobalt (III) bromide)
Violet coloured complex

The ionization of the above two complexes can be depicted as follows:

$$[Co(NH_3)_5Br]SO_4 \rightleftharpoons [Co(NH_3)_5\,Br]^{2+} + SO_4^{2-}$$

$$[Co(NH_3)_5SO_4]Br \rightleftharpoons [Co(NH_3)_5SO_4]^+ + 2Br^-$$

It is clear from the above example that if ligands are present in the coordination sphere, they do not undergo ionization.

Linkage isomerism   Certain ligands possess two or more different atoms that can coordinate with the central metal ion in different ways. Such type of bonding results in linkage isomerism and the resulting complexes are called linkage isomers. Pentaamminenitritocobalt (III) ion can exist as linkage isomer since the nitro group can be bonded to the central metal ion through oxygen or nitrogen (Fig. 11.34).

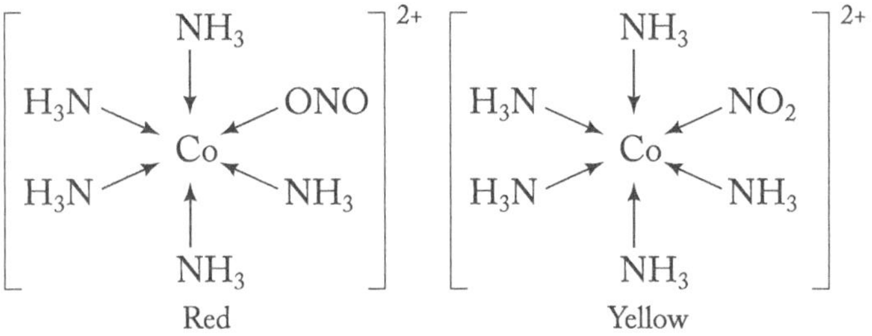

Red
Yellow
Pentaamminenitritocobalt (III) ion
Fig. 11.34 Linkage isomerism

Isocyanate (SCN^-) and isothiocyanate (NCS^-)) ligands also exhibit linkage isomerism when bonded to a central metal ion either through sulphur or nitrogen.

Coordination isomerism   In this type of isomerism, the composition of the complex ion tends to vary. The ratio of ligands to the metal ion remains the same but the ligand to the metal ion changes. Let us imagine we have a solution of $[Co(NH_3)_6]^{3+}$ and $[Cr(CN)_6]^{3-}$ and another solution of $[Cr(NH_3)_6]^{3+}$ and $[Co(CN)_6]^{3-}$; then both solutions are considered as coordination isomers. $[Co(NH_3)_5(SCN)]Cl_2$ and $[Co(NH_3)_5(NCS)]Cl_2$ are also coordination isomers.

Geometrical isomerism (or *cis–trans* isomerism)   This is observed in disubstituted metal complexes with a coordination number of four or six having square-planar and octahedral geometries, respectively. If similar ligand groups are adjacent to each other, they are called *cis*-isomers whereas those lying opposite to each other are called *trans* isomers. Geometrical isomerism is not exhibited by tetrahedral complexes, because the ligand bonding to the metal ion is the same with respect to each other. Diamminedichloroplatinum (II) square planar complex and dichlorotetraamminecobalt (III) octahedral complex exhibit geometrical isomerism (Fig. 11.35).

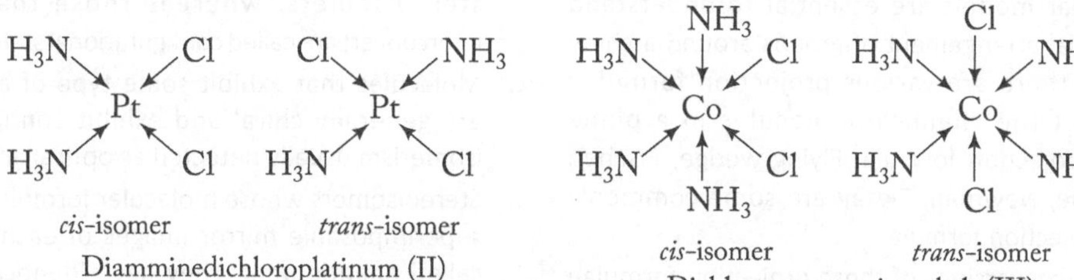

cis-isomer   *trans*-isomer
Diamminedichloroplatinum (II)

cis-isomer   *trans*-isomer
Dichlorotetraamminecobalt (III)

Fig. 11.35 Geometrical isomerism

Yet another interesting geometrical isomerism occurs in octahedral complexes such as $[Co(NH_3)_3(NO_2)_3]$. When the ligand occupies adjacent position at the corners of an octahedron, it is called *facial (fac)* isomer, whereas if they occupy around the meridian, it is called *meridional (mer) isomer* (Fig. 11.36).

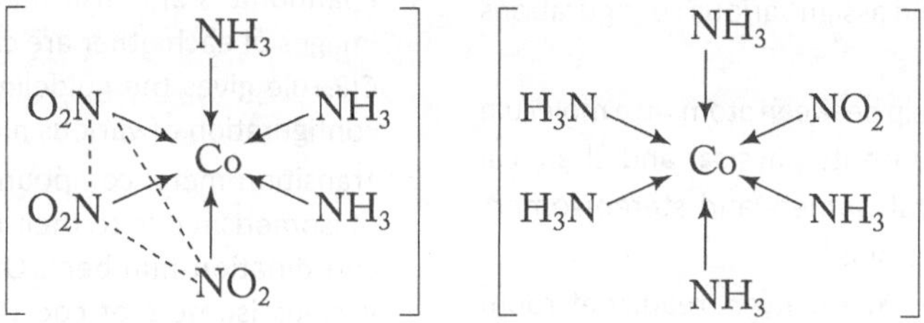

fac-isomer   *mer*-isomer

Fig. 11.36 Facial and meridional isomers of $[Co(NH_3)_3(NO_2)_3]$

Optical isomerism   This is seen in metal complexes that are mirror images of each other and are non-superimposable on each other. Such complexes lack symmetry and rotate the plane of polarized light either towards the right (dextro) or towards the left (laevo). The dextro and laevo forms of $[Co(en)_2Cl_2]^+$ are shown in Fig. 11.37.

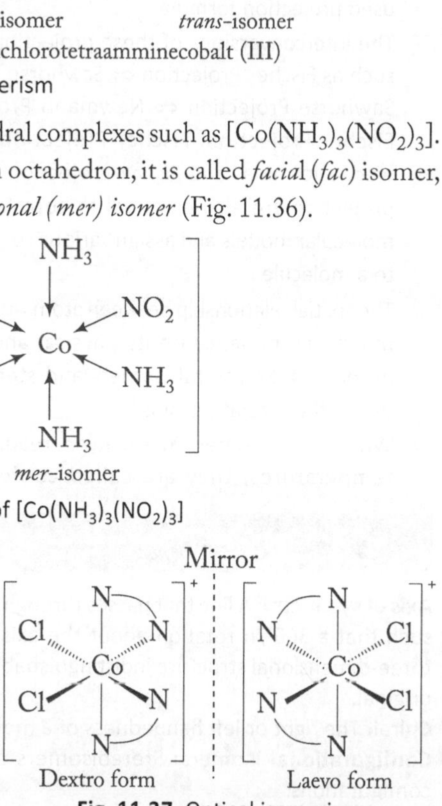

Solvate (or hydrate isomerism)   As the name suggests, here water is usually involved as a solvent in the coordination compound. Solvate isomers differ on the basis of whether a

Mirror

Dextro form   Laevo form

Fig. 11.37 Optical isomerism

solvent molecule is bonded to the metal or is merely present as a free form in the crystal lattice. An aqua complex, $[Cr(H_2O)_6]Cl_3$ and its solvate isomer $[Cr(H_2O)_5Cl]Cl_2.H_2O$ is one such example. Further, such isomers also exhibit various colours, like in this case aqua complex is violet and the solvate isomer is greyish-green in colour.

Check Your Progress

22. Why do tetrahedral complexes not exhibit geometrical isomerism in spite of having two different monodentate ligands coordinated with the central metal ion?
23. Draw the structures of geometrical isomers of $[Fe(NH_3)_2(CN)_4]^-$.
24. Indicate the types of isomerism exhibited by the following complexes:
    (a) $K[Cr(H_2O)_2(C_2O_4)_2]$; (b) $[Co(en)_3]Cl_3$; (c) $[Co(NH_3)_5(NO_2)](NO_3)_2]$; (d) $[Pt(NH_3)(H_2O)Cl_2]$

SUMMARY

◯ Molecular models are essential to understand the spatial arrangement of groups around a chiral centre. There are various projection formulae to depict the chemical molecules in a plane called projection formula. Flying wedge, Fischer, Sawhorse, Newman, Zigzag are some commonly used projection formulae.

◯ The interconversions of these projection formulae such as Fischer Projection ⇔ Sawhorse Projection, Sawhorse Projection ⇔ Newmann Projection ⇔ Fischer Projection, Fischer Projection ⇔ Flying wedge formula, and Zigzag projection ⇔ Fischer projection are also essential to correlate various molecular models and assign various configurations to a molecule.

◯ The spatial relationship between atoms in a molecule has direct influence on its physical and chemical properties. Structural isomers and stereoisomers depict these relationships.

◯ When stereoisomers interconvert readily at room temperatures, they are called conformational stereoisomers, whereas those that do not interconvert are called configurational stereoisomers.

◯ Molecules that exhibit some type of asymmetry are generally chiral and exhibit configurational isomerism usually detected as optical activity.

◯ Stereoisomers whose molecular formulae are non-superimposable mirror images of each other are called enantiomers. They have identical physical properties but rotate the plane polarized light in opposite direction in equal amounts.

◯ Compounds can be denoted as their *cis–trans* and *E* and *Z* isomers.

◯ Molecules with two or more chiral centres that are enantiomers and also non-superimposable mirror images of each other are called diastereomers. The CIP rule gives the guidelines for denoting *R* and *S* configuration of various molecules.

◯ Transition metal compounds exhibit various types of isomerism due to their differing geometries and coordination numbers. On the basis of topology various isomers of coordination compounds are possible.

GLOSSARY

Axis of symmetry: A line that passes through a molecule such that a 360°/n rotation about the axis leads to a three-dimensional structure indistinguishable from the original.

Chiral: The right or left handedness of a molecule.

Configurational isomers: Stereoisomers differing in configurations.

Conformational analysis: Analysis of the physical and chemical properties of a compound in relation to its geometry and population of its conformers.

Diastereomers: A pair of stereoisomers having no mirror images.

Enantiomers: A pair of stereoisomers having non-superimposable mirror images.

Isomerism: The phenomenon when two or more molecules that have identical molecular formula but different arrangements of their atoms in connectivity and spatial arrangements.

Optical activity: The phenomenon due to which stereoisomers can rotate plane-polarized light.

Optical isomers: Metal complexes which are mirror images of each other and are non-superimposable.

Plane of symmetry: The imaginary plane that bisects the molecule into two halves which are mirror images.

Polarimeter: The apparatus to measure optical activity of molecules.

Racemic modification: An equimolecular mixture of a pair of enantiomers.

Structural isomers: The molecules having same molecular formula but different sequence of their atom-connectivity.

Stereoisomers: The molecules having same connectivity of atoms but differing in the relative three-dimensional arrangement of their atoms.

Ionization isomerism: A type of structural isomerism, where complexes have same empirical formulae that form different ions in solution.

Linkage isomerism: When ligands with different atoms coordinate with a central metal ion in different ways.

EXERCISES

Multiple Choice Questions

1. Which of the following is the enantiomer of the following substance (S)?

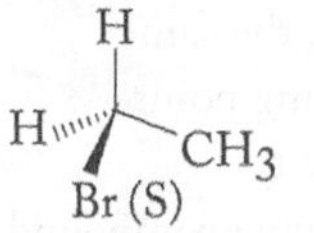

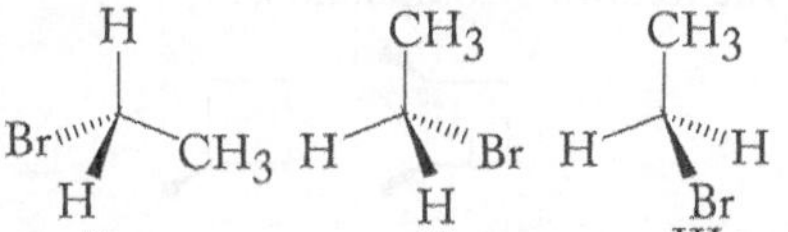

   (a) I
   (b) II
   (c) Both II and III
   (d) It does not have a non-superimposable enantiomer.

2. The molecules shown below are:

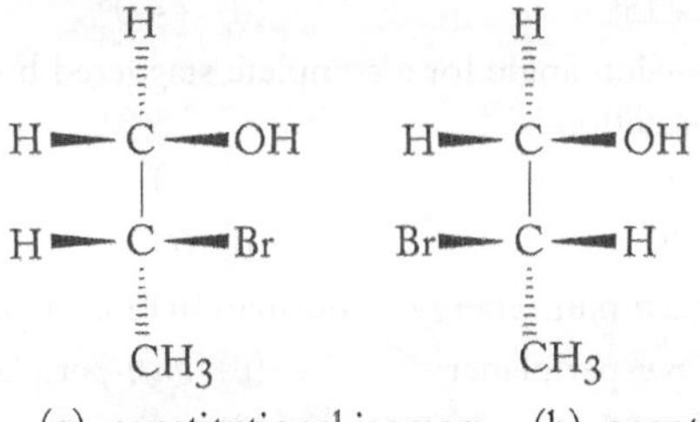

   (a) constitutional isomers   (b) enantiomers
   (c) diastereomers   (d) identical

3. Molecules I and II are:

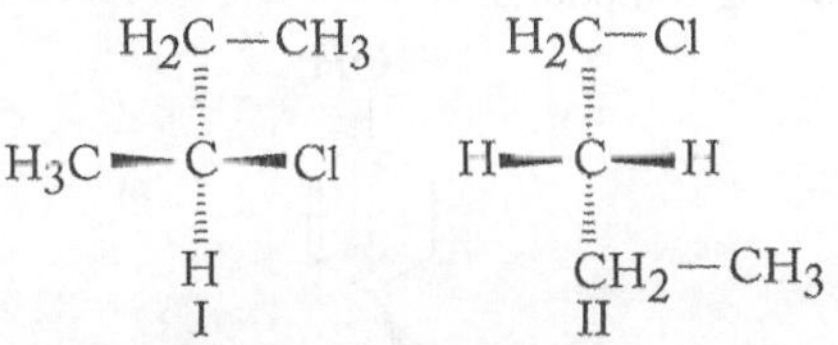

   (a) constitutional isomers
   (b) enantiomers

   (c) non-superposable mirror images
   (d) diastereomers

4. Hexane and 3-methylpentane are examples of
   (a) enantiomers
   (b) stereoisomers
   (c) diastereomers
   (d) constitutional isomers

5. Which structure represents (*S*)-2-bromobutane?

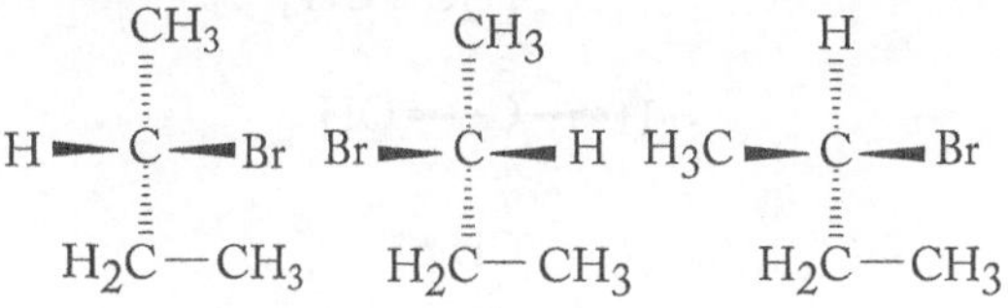

   (a) I   (b) II
   (c) III   (d) None of these

6. Which of the following groups has the highest priority according to the CIP sequence rules?
   (a) Methyl   (b) CH_2Cl
   (c) CH_2OH (d) CHO

7. Which of the following has (*R*) configuration?

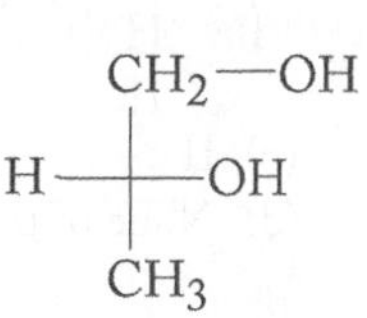

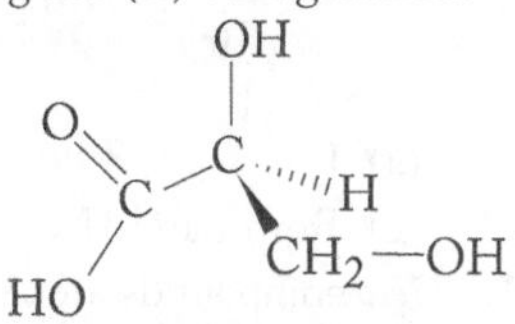

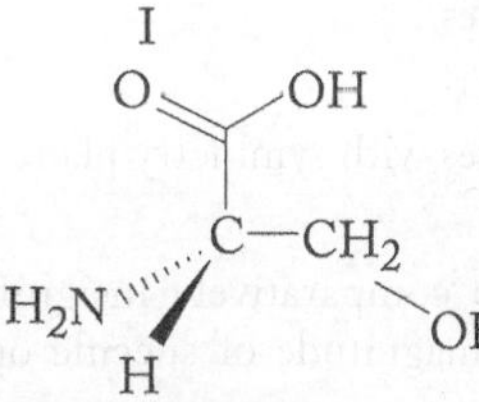

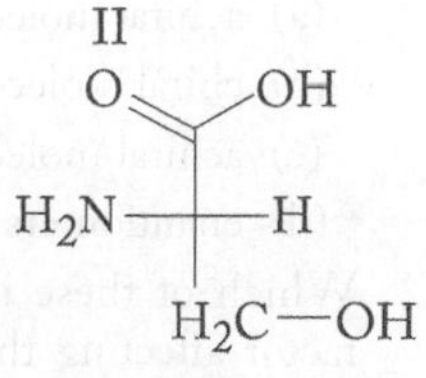

(a) I
(b) II
(c) III
(d) II and IV

8. Which of the following is true of any (*S*)-enantiomer?

   (a) It rotates plane-polarized light to the right.

   (b) It rotates plane-polarized light to the left.

   (c) It is a racemic form.

   (d) It is the mirror image of the corresponding (*R*)-enantiomer.

9. Which of the following represent (*R*)-2-butanol?

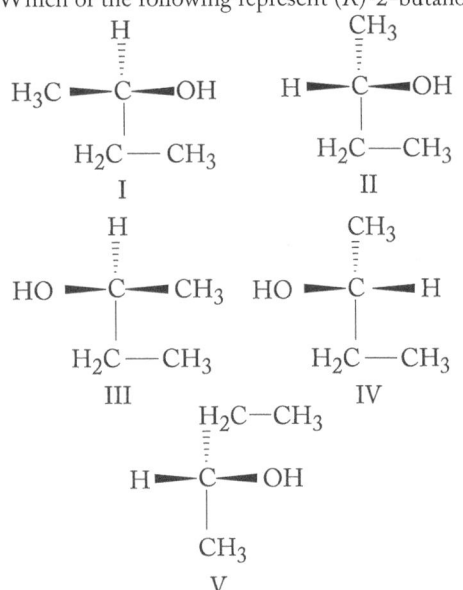

(a) III and V
(b) I, III, IV and V
(c) I, IV and V
(d) I and III

10. Which of the following structures will exhibit optical activity?

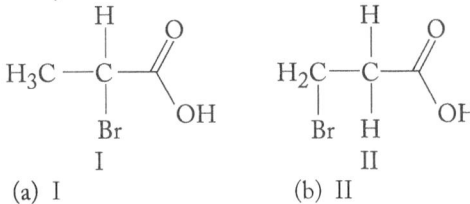

(a) I
(b) II
(c) Both I and II
(d) None of these

11. *Meso* compounds are
   (a) achiral molecules
   (b) chiral molecules
   (c) achiral molecules with symmetry plane
   (d) enantiomers

12. Which of these is a comparatively insignificant factor affecting the magnitude of specific optical rotation?

   (a) Concentration of the substance
   (b) Purity of substance
   (c) Temperature of measurement
   (d) Length of sample tube

13. Which statement is true of 1,3-dimethylcyclobutane?
   (a) Only one form of the compound is possible.
   (b) Two diastereomeric forms are possible.
   (c) Two sets of enantiomers are possible.
   (d) Two enantiomeric forms and one *meso* compound are possible.

14. What can be said with certainty if a compound has $[\alpha]_D^{25} = -9.25°$?
   (a) Compound has (S) configuration.
   (b) Compound has (R) configuration.
   (c) Compound is NOT a meso form.
   (d) It possesses many stereogenic centres.

15. Which of the following is NOT true of enantiomers? They have the same:
   (a) boiling point   (b) melting point
   (c) density   (d) specific rotation

16. The following compound is:

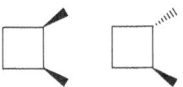

   (a) meso   (b) homomer
   (c) stereoisomers   (d) identical

17. If a solution of a compound (30.0 g/100 ml of solution) has a measured rotation of +15° in a 2 dm tube, the specific rotation is:
   (a) +50°   (b) +25°
   (c) +15°   (d) +4.0°

18. The torsion angle for a complete staggered butane conformation is
   (a) 0°   (b) 20°
   (c) 40°   (d) 60°

19. The maximum-energy conformation of *n*-butane is
   (a) *syn*-periplanar   (b) *anti*-periplanar
   (c) gauche   (d) eclipsed

20. How many asymmetric carbons are present in the following compound?

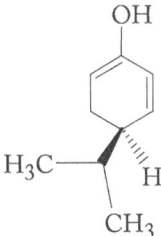

(a) 0   (b) 2

(c) 3   (d) 1

21. $[Co(NH_3)_5NO_2]Cl_2$ and $[Co(NH_3)_5(ONO)]Cl_2$ are

    (a) geometrical isomers   (b) linkage isomers

    (c) solvate isomers   (d) coordination

22. Which of the following exhibits coordination isomerism?

    (a) $[Cr(en)_2Cl_2]$

    (b) $[Co(en)_2Cl_2]$

    (c) $[Cr(NH_3)_6]Cl_3$

    (d) $[Cr(NH_3)_6][Co(CN)_6]$

Review Questions

1. Write a short note on representing 3D molecules using various formulae.

2. What is chirality? How does one predict molecular chirality? Give suitable examples.

3. What are structural isomers and stereoisomers? Give examples of each.

4. Elucidate the various ways of representing molecules in three dimensions.

5. Explain Fischer projection formula. Draw Fischer projection formula of lactic acid.

6. Draw zigzag formulae of D-glucose and D-mannose.

7. What is optical activity? How does one determine optical activity using polarimeter?

8. What is a racemic mixture? Explain with a suitable example.

9. Assign *E–Z* configuration to the following compounds.

(a)   (b)   (c)

(d)   (e)

(**Ans:** a. (*E*)- b. (*Z*)- c. (*E*), d.(*Z*), e. (*E*))

10. Explain the rules for assigning *E* and *Z* descriptors to organic molecules

11. How does one interconvert Fischer to a Sawhorse projection formula? Explain with a suitable example.

12. Explain the projection formulae for interconversion from Sawhorse to Newman to Fischer projection formulae.

13. Write a short note on interconversion of Flying wedge and zigzag formulae to Fischer projection formula.

14. What is absolute configuration of a molecule? How is it determined? Mention the steps involved while assigning *R/S* configuration to a molecule.

15. Assign *R* or *S* configuration to the following compounds:

(**Ans:** *S*, achiral, *R*, *R*, *S*)

16. With the help of a torsional curve, explain conformational analysis of ethane and propane.

17. Discuss conformational analysis of *n*-butane. Draw various conformers with their names and torsional curve.

18. What is isomerism in coordination compounds? List the various types of isomerism exhibited by coordination compounds with suitable examples.

19. Differentiate between geometrical and optical isomerism of coordination compounds.

FURTHER READING

1. Cahn, R.S., G.I. Ingold, and V. Prelog, *Angew. Chem., Int. Edn. En.* 5, 385 (1966).

2. *Compound Interest, http://www.compoundchem.com/

---

* Students and teachers are encouraged to visit the site as it provides interesting graphical posters on everyday chemistry prepared by Andy Brunning

3. Eliel, E.L., *Stereochemistry of Carbon Compounds*, Wiley, 1975.
4. McMurry, John and Simanek, Eric. *Fundamentals of Organic Chemistry*. Brooks Cole, 2006.
5. Mosher, H.S., *Glossary to Audio Course 'Stereochemistry'*, American Chemical Society, Washington DC, 1980.
6. Nasipuri, D. *Stereochemistry of Organic Compounds*, New Age International, 2005.
7. Sen Gupta, S. *Basic Stereochemistry of Organic Molecules*, Oxford University Press, 2014.
8. Schore and Vollhardt. *Organic Chemistry Structure and Function*, W.H. Freeman & Company, 2007.

ANSWERS

Check Your Progress

1. Structure (I) has (*S*) configuration, and Structure (II) has (*R*) configuration. As both these structures have opposite configurations, they represent as enantiomeric pair.

2. (a) Enantiomers are stereoisomers that are related as non-superimposible mirror images and the only property that differ is the direction (+ or –) of optical rotation.
   (b) Diastereomers are stereoisomers that are not mirror images and are different compounds with varying physical properties.
   (c) *Meso* compounds are achiral molecules with chiral centres, possessing a plane of symmetry.

3. Molecules having the same molecular formula but different sequence of their atom-connectivity are called constitutional isomers. 2-Methyl propane and *n*-butane are constitutional isomers.

4. Glycerol has a plane of symmetry. Using proper orientation of the molecule as shown in the figure we can visualize its plane of symmetry. Glycerol is achiral because it has a plane of symmetry.

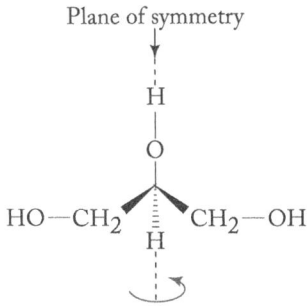

Plane of symmetry in glycerol

5. The stereoisomers can be drawn as follows:
   A and B are enantiomers, and C and D are enantiomers. A is a diastereomer of C and D. B is also a diastereomer of C and D.

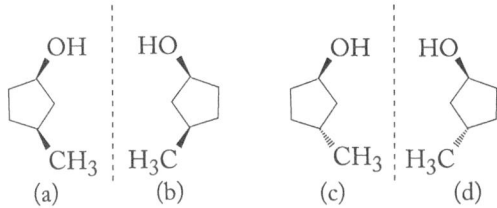

6. Stereoisomers that are capable of rotating plane-polarized monochromatic light, when such light is passed through them, are said to be optically active compounds and are measured using a polarimeter.

7. As the observed rotation is counterclockwise, it is laevorotatory, (–)-2-pentanol. The concentration is 6 g per 40 ml = 0.15 g/ml, and the path length is 200 mm = 2 dm. The specific rotation is, $[\alpha]_D^{25} = \dfrac{-4.05°}{(0.15)(2)} = (-) - 13.5°$

8. The concentration is 8 g/50 ml = 0.16 g/ml, and the path length is 20 cm = 2 dm. The given compound (A) is dextrorotatory and its specific rotation will be, $[\alpha]_D^{25} = \dfrac{+3.4°}{0.16 \times 2} = (+) - 10.6°$

9. (a) (*E*)-2-bromo-2-butene. (b) (*Z*)-1-bromo-1,2-dichloroethene
10. The four possible Fischer projection formulae of 2, 4-dibromohexane are:

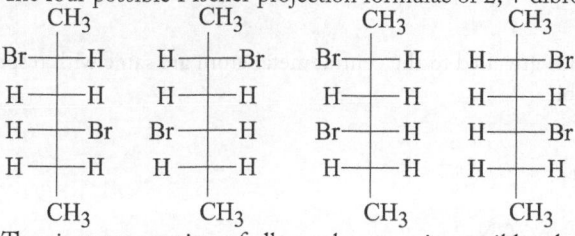

11. The *cis–trans* notation of alkenes becomes impossible when one of the doubly bonded carbon atoms possess two identical substituents.
12. Chirality describes the handedness of a molecule, i.e., non-superimposability of a molecule on its mirror image. The given molecule is achiral as there are two similar groups.
13. $[\alpha]_{589.3}^{20} = \dfrac{+2.79°}{1 \times 0.03} = +93°$ (in ethanol)
14. Fischer projection of (*S*)-2-hydroxybutanoic acid:

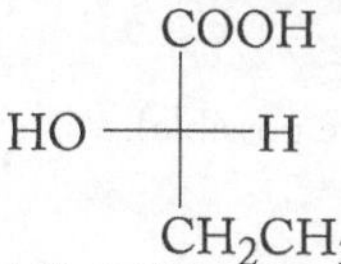

$$
\begin{array}{c}
\text{COOH} \\
\text{HO} \!-\!\!\!-\!\!\!-\! \text{H} \\
\text{CH}_2\text{CH}_3
\end{array}
$$

15. (*R*)-2 Chlorobutane   (*S*)-2, 3-Dihydroxypropanal

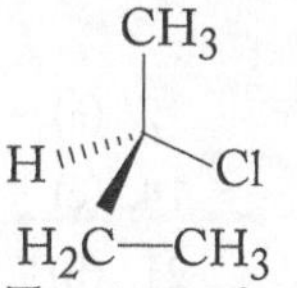

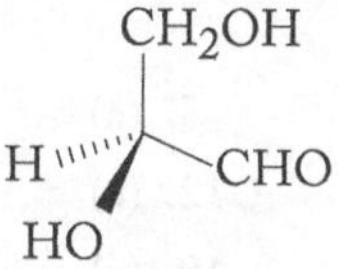

16. The zigzag formula and Fischer projection of the given molecules are shown below.

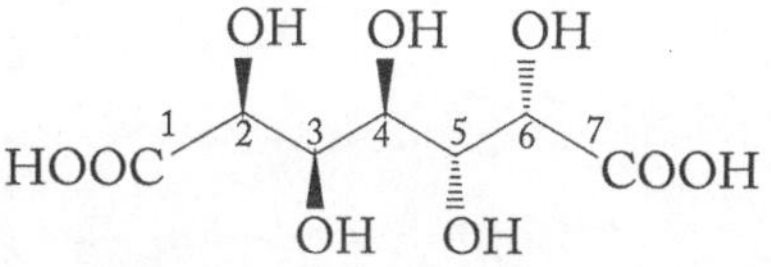

Zigzag formula
(i)

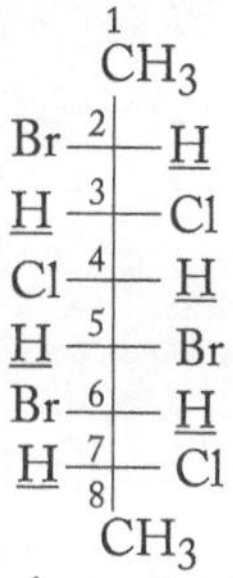

$$
\begin{array}{c}
\overset{1}{\text{CH}_3} \\
\text{Br} \overset{2}{-\!\!\!-} \text{H} \\
\text{H} \overset{3}{-\!\!\!-} \text{Cl} \\
\text{Cl} \overset{4}{-\!\!\!-} \text{H} \\
\text{H} \overset{5}{-\!\!\!-} \text{Br} \\
\text{Br} \overset{6}{-\!\!\!-} \text{H} \\
\text{H} \overset{7}{-\!\!\!-} \text{Cl} \\
\overset{8}{\text{CH}_3}
\end{array}
$$

Fischer projection
(ii)

17. Refer to Figs 11.28 and 11.29
18. Fischer projection formula of (3*R*)-6-bromo-1-hexen-3-ol.

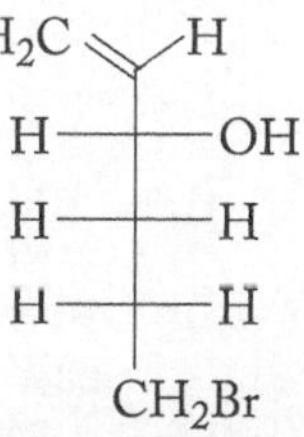

$$
\begin{array}{c}
\text{H}_2\text{C} \!=\! \text{H} \\
\text{H} -\!\!\!- \text{OH} \\
\text{H} -\!\!\!- \text{H} \\
\text{H} -\!\!\!- \text{H} \\
\text{CH}_2\text{Br}
\end{array}
$$

19. When a molecule has a number of conformations that correspond to energy minima, then they are called conformers. Conformational analysis is the analysis of the physical and chemical properties of a compound in relation to its geometry and population of its conformers.

20. No, the mere presence of a chiral centre does not make a compound optically active. For example, *meso*-tartaric acid has two chiral centres, yet is optically inactive due to plane of symmetry.

21. Ans: a. (S)- b. (R)- and c. (R)-.

22. As the relative positions of monodentate ligands attached to the central metal atom are same with respect to each other, geometrical isomerism is not possible.

23. Geometrical isomers of $[Fe(NH_3)_2(CN)_4]^-$

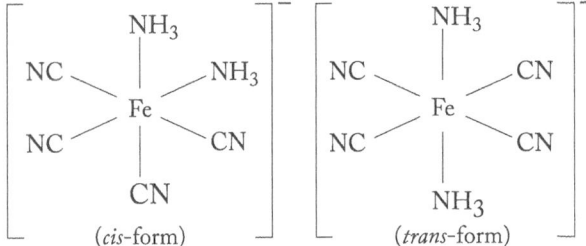

24. (a) Both geometrical (*cis-*, *trans-*) and optical isomers can exist.

(b) Two optical isomers can exist.

(c) There are 10 possible isomers. (Hint: geometrical, ionization, and linkage isomers are possible).

(d) Both geometrical (*cis-*, *trans-*) and optical isomers can exist.

Check Your Progress

1. (d)	2. (b)	3. (a)	4. (d)	5. (a)	6. (b)	7. (a)
8. (d)	9. (c)	10. (a)	11. (c)	12. (c)	13. (b)	14. (c)
15. (d)	16. (c)	17. (b)	18. (d)	19. (a)	20. (d)	21. (b)
22. (d)						

Instrumental Methods of Analysis

After reading the chapter, you will be able to understand:

- the principle of electromagnetic radiation and its interaction with matter.
- discuss the types of spectroscopy such as rotational, IR, UV/visible, and fluorescence spectroscopy and their applications.
- understand the principle of working of flame photomerty and its applications.
- discuss the types of electron microscopes and their working and applications.
- understand diffraction spectroscopy and its applications.
- understand the working of conductometer and its use in chemical analysis.
- discuss the principle of thermal methods (TGA, DTA) and their applications.
- discuss the principle, types, and applications of various chromatographic methods.

12.1 INTRODUCTION

Analytical methods are tools employed in chemistry for the separation, structural determination, qualitative and quantitative analysis of the elements/compounds. These methods are divided into a few classes of techniques based on the principles of analysis employed. These classes are

1. Spectroscopic methods
2. Microscopy
3. Electrochemical methods
4. Thermal methods
5. Chromatography

12.2 SPECTROSCOPIC METHODS

Spectroscopic methods constitute a set of methods in which interaction of electromagnetic radiation with matter/chemical molecules is observed and analysed to obtain information about their structure (e.g., bond lengths, bond angles, planarity, presence of functional groups) and other fundamental properties such as the energy levels of a molecule. In most forms of spectroscopy, the behaviour of electromagnetic radiation was observed after its passage through the sample.

12.2.1 Electromagnetic Radiation

Electromagnetic radiation (light) is the form of energy having dual property of wave and particle. Electromagnetic radiation is a radiation/wave propagated by varying electric and magnetic field oscillating at right angles to each other. In Fig. 12.1, electric field is oscillating in and out (in xy plane),

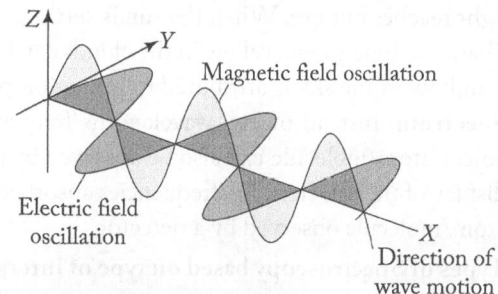

Fig. 12.1 Representation of electromagnetic radiation

whereas magnetic field is oscillating up and down (in xz plane) and wave is moving in x-direction. Electromagnetic radiation travels in a straight line with a velocity of 3×10^8 ms^{-1}.

Characteristics of a Wave

Wavelength   It is the distance between two successive maxima/crests or minima/troughs of the wave. It is generally expressed in units of nm or Å.

Frequency   It is measure of the number of waves passing through a given point in unit time. It is expressed in units of sec^{-1} (Hz).

Electromagnetic radiation as a particle   Electromagnetic radiation also behaves as discrete packets of particle called photon or quanta. The particle and wave properties are not mutually exclusive, they complement each other. Energy of photon is related to frequency/wavelength which is a wave attribute.

Dual behaviour of matter   Matter has dual behaviour. It also has wavelike property and travels in space like a wave, rather than along a definite path. The classical concept of particle trajectory is replaced by wave equation, and the spatial amplitude of wave is called a wavefunction, ψ(psi).

Interaction of Electromagnetic Radiation With Molecules

Matter consists of atoms, whereas atoms consist of electrons, protons, and nucleus. Both matter and light have particle and wave properties. Electrons are described by wave functions and photon is a packet of energy associated with electromagnetic radiation.

   Electromagnetic radiation can induce/undergo different processes on interaction with a molecule

(a) Absorption
(b) Transmission
(c) Emission
(d) Reflection
(e) Scattering
(f) Diffraction (specialized case of scattering)
(g) Optical rotation
(h) Refraction
(i) Interference

   Absorption and emission phenomena can be explained by taking into account particle nature of the electromagnetic radiation whereas reflection, refraction, interference and diffraction can be explained by its wave nature.

   When an electromagnetic radiation falls on an object, the radiations of some wave lengths are absorbed whereas others are get reflected or transmitted. The color of an object we see is due to the wavelengths that reach our eyes reflected, transmitted, emitted or scattered. A leaf looks green since it reflects green light. The light transmitted/reflected is complementary to the absorbed light. If an object absorbs radiations of all wavelengths, it appears black. If an object reflects radiation of all wavelengths, it appears white. The colour of a transparent object depends on what wavelength of light is transmitted though the object. A transparent bottle will look green if green light is transmitted through the bottle. Low-pressure sodium vapour lamps emit monochromatic bright yellow light on heating the sodium atom to its vapour. Scattering can change the direction of short wavelength light (blue and violet). The sky appears blue since scattered light reaches our eye. When the sun is setting, the unscattered light reaches our eye, and the appears red. Shades of blue, green, yellow in the clouds can be observed due to diffraction of light from water droplets. Rainbow in the sky is attributed to refraction phenomenon.

Spectrum: Instead of eye, wavelengths/frequencies absorbed, emitted, transmitted, or scattered by an object/atom/molecule can also be observed by placing a detector in appropriate direction. Spectrum is a display of the wavelengths/frequencies absorbed, reflected, scattered, transmitted or emitted by an object/atom/molecule observed by a detector.

Types of Spectroscopy based on type of interaction

Absorption spectroscopy: This is the kind of spectroscopy in which absorption of radiation is observed as a function of wavelength. Gases, placed between source and the observer, show pattern of dark spectral lines where light within a number of narrow frequency ranges has been removed. Different atoms and molecules show different absorption spectra.

Emission spectroscopy: This is the kind of spectroscopy in which emission of radiation is observed as a function of wavelength. Emission spectra is of three types: line spectrum, band spectrum and a continuum spectrum. Emission from excited atom leads to sharp, well defined peak in the line spectrum. Small molecule and radicals when heated leads to emission of overlapping lines resulting into a band spectrum. Continuous spectrum is a type of spectrum observed from sun or star or hot solids in which radiation is distributed over all frequencies, not just a few specific frequency ranges.

Reflectance spectroscopy: This is the kind of spectroscopy in which reflection of radiation is observed as a function of wavelength. Oxygenated haemoglobin, when illuminated with white light, absorbs the blue light and reflects back the red light, giving its characteristic colour.

Scattering spectroscopy: Spectroscopy based on scattering phenomenon is called scattering spectroscopy. There are two types of scattering spectroscopy: elastic and inelastic scattering. Scattering where wavelength of scattered light is same as that of incident light is called *elastic scattering* whereas scattering where wavelength of scattered light is different from that of incident light is called *inelastic scattering*. Elastic scattering is of two types: *Rayleigh scattering* where scattering intensity is independent of scattering angle and big particle scattering where scattering intensity is dependent on scattering angle. *Raman spectroscopy* is a spectroscopic technique based on inelastic scattering. Diffraction is a special case of scattering spectroscopy.

There are several other forms of spectroscopy based on the way of interaction between matter and electromagnetic radiation such as optical rotation dispersion spectroscopy, reflectometric interference spectroscopy, etc.

12.2.2 Blackbody Radiation and Quantization of Energy

A blackbody is a hypothetical object that absorbs all of the radiation falling on it and emits all of the absorbed radiation. Since no radiation is reflected, blackbody appears black.

Characteristics of Blackbody Radiation
(i) A blackbody emits radiation of all frequencies (or wavelengths) at temperature greater than zero.
(ii) A hot blackbody emits radiation of higher energy at all frequencies (or wavelengths) than a colder one.
(iii) Increase in the temperature of a blackbody leads to a shift of blackbody spectrum peak (maximum energy) to higher frequency (shorter wavelength). Thus, there is blue shift of peak on increase in temperature. The colour of blackbody changes from black to red, then to yellow and blue on increase in temperature. A yellow star, such as our own sun, is hotter than a red star. A blue star is hotter than a red star and yellow star.

Failure of Classical Physics in Explaining Blackbody Radiation
Rayleigh-Jeans law (classical physics) is used to explain hot body radiation. According to classical physics, spectral brightness (amplitude of radiation) should increase with increase in frequency. Thus, spectral brightness of even relatively cold objects should be high in the UV and visible regions. According to Rayleigh-Jeans law, the energy density per unit frequency interval at frequency v is given by

$$\rho = \frac{8\pi v^2}{c^3} <E> \text{ where } <E> = kT \text{ as per Classical physics, } k = \text{Boltzmann constant, } c = \text{velocity of light}$$

Thus, the spectral brightness is expected to increase with increase in v. Prediction by Rayleigh-Jeans law matches with experimental observations at low frequencies (v), but fails at high frequencies. This failure at high frequencies is known as the ultraviolet catastrophe. Experimentally, the spectral brightness always tapers off at the high frequency.

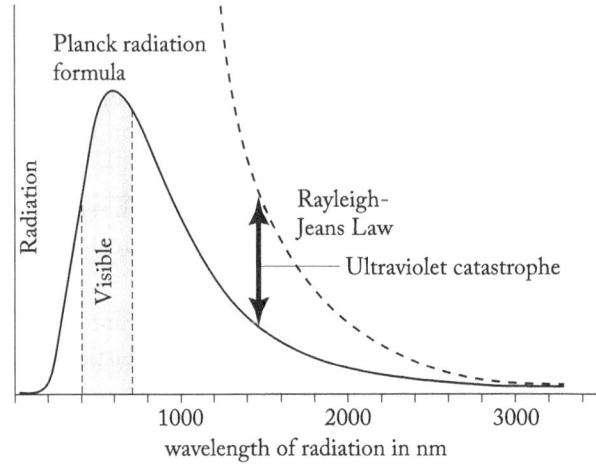

Fig. 12.2 Deviation between the plots obtained from Rayleigh-Jeans law and Planck's radiation formula

12.2.3 Quantization of Energy

Max Planck suggested that blackbody radiation is due to oscillations of electrons in the constituent particles of the body. He assumed that the energies of the oscillators are discrete and is proportional to integral multiples of hv. h is called Planck's constant.

$$<E> = \frac{hv}{e^{hv/kT} - 1}$$

At low frequency,

$$<E> = \frac{hv}{(1 + hv/kT) - 1} = kT$$

At high frequency,

$$<E> = \frac{hv}{e^{hv/kT}} = hve^{-hv/kT}$$

At high frequency, the term $e^{-hv/kT}$ dominates, and energy density decreases. The high frequency radiation at a particular temperature requires comparatively higher amount of energy and hence the chances of high-frequency transition is low. However, at high temperatures, the probability for high-frequency radiation increases.

The presence of several lines in the hydrogen spectrum, line spectra for atoms, and molecules supports the hypothesis of quantization of energy. Thus, atoms and molecules exist in certain discrete energy levels.

Transition between different energy levels due to interaction between atom/molecule and electromagnetic radiation Spectroscopic techniques based on transition between two different energy levels (absorption or emission spectra) are most interesting as these provide a wealth of information

Check Your Progress

1. What are the different ways in which electromagnetic radiation can interact with matter?
2. What property of electromagnetic radiation is responsible for absorption and reflection?
3. What are the characteristics of a black body?
4. Rayleigh-Jeans law is not able to account for black body radiation. Explain.

about the system under investigation. Spectroscopic techniques based on reflection, refraction, scattering, diffraction, and interference are used to study bulk properties of matter and are out of scope at this level.

12.2.4 Types of Molecular Energy Levels

A molecule in space can have different kinds of energy based on its motion.

Energies of different energy levels can be obtained by solving wave equation (Schrödinger equation). Schrödinger used wave equation to derive well-known Schrödinger equation. The solution of the equation can give wavefunctions and energy associated with different levels. (explained in Chapter 1)

$$-\left(\frac{h^2}{8\pi^2 m}\right)\frac{d^2\psi(x)}{dx^2} + V\psi(x) = E\psi(x)$$

In short, it is represented as

$$H\psi(x) = E\psi(x)$$

The Born–Oppenheimer Separation of Electronic and Nuclear Motions
Electrons are quite lighter than nucleus and hence movement of nucleus during electronic motion is negligible. Thus, according to the Born–Oppenheimer approximation, the electronic motions can be separated from the vibrational/rotational motions of nuclei.

$$\psi = \psi_n \psi_e$$
$$H_n\psi_n = E_n\psi_n$$
$$H_e\psi_e = E_e\psi_e$$
$$E = E_e + E_n$$

Subscripts e and n stand for electronic and nuclear motion, respectively.

Nuclei   Nuclei undergo three different types of motion with respect to the centre of mass of the molecule— vibration; rotation, and translation.

Quantization of nuclear degrees of freedom   All 3N nuclear degrees of freedom of molecule having N atoms, translational, rotational and vibrational degree of freedom of a molecule are quantized. Thus, translational, rotational, and vibrational energy levels of a molecule are quantized.

$$\psi_n = \psi_t \psi_v \psi_r$$
$$E_n = E_t + E_v + E_r$$

Subscripts t, v, and r stand for translation, vibration, and rotation, respectively.

Vibration   In a molecule, atoms are linked by bonds. Atoms and a bond between two atoms can be modelled as ball and spring, respectively.

(Quantum mechanics treats a vibrating molecule as a quantum mechanical harmonic oscillator. For such as an oscillator, the energy levels are quantized. The energies of these discrete vibrational energy levels are given by:

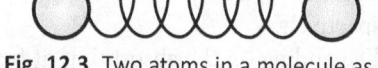

Fig. 12.3 Two atoms in a molecule as a ball and bond between them as spring

$$E_v = \left(v + \frac{1}{2}\right)h\nu$$

v is the vibrational quantum number, $v = 0, 1, 2\ldots$

Even in the lowest state, $E_{v=0} = \frac{1}{2}h\nu$. It is known as the *zero point* energy. Energy gap between two vibrational levels is of the order of $10^3 - 10^5$ J mol^{-1}. Thus a molecule vibrates on the absorption of infrared radiation.

Rotation Similar to vibrational energies, rotational energy is also quantized. The energy of rotational energy levels can be obtained by applying Schrödinger equation by assuming diatomic molecule as a rigid rotor. The energies of these level are given by

$$E_J = \frac{\hbar^2}{2I} J(J+1)$$

where J is the rotational quantum number and I is the moment of Inertia. Energy gap between two rotational levels is of the order of $10^1 - 10^3$ J mol^{-1}. Thus a molecule rotates on the absorption of microwave radiation.

Translation (motion of centre of mass) The energy of translational energy levels can be obtained by solving Schrödinger equation of model particle in a box.

$$E_n = \frac{n^2 h^2}{8\ ml^2}$$

n = quantum number; l = length of the box, m is the mass of electron and h is Planck's constant.

Transition between two energy levels and associated wavelength:

On absorption of light of energy ΔE, the species goes to a higher energy level.

$$\Delta E = h\nu = \frac{hc}{\lambda}$$

h = Planck's constant, ν = frequency, c = velocity of light, λ = wavelength of light

Types of Molecular Spectroscopy based on Transition between Different Energy Levels

The different groups of frequency comprising the electromagnetic radiation lead to transition between different kinds of energy levels (nuclear, electronic, rotational.....) and thus differ in the way they interact with molecules, giving rise to different kinds of spectroscopy.

Table 12.1 Types of spectroscopy and associated properties

Type	Nature of transition	Energy range	Frequency range	Wavelength range
γ-rays	Nuclear	10–12 GJ mol^{-1}	30 EHz–300 EHz	1000 to 10 pm
X-ray	Inner electronic	10–100 MJ mol^{-1}	30 PHz–30 EHz	10 pm-10 nm
UV	Outer electronic	300–1000 kJ mol^{-1}	800 THz–30 PHz	10 nm–380 nm
Visible	Outer electronic	100–300 kJ mol^{-1}	400 THz–800 THz	380 nm–750 nm
Infrared	Vibration, rotation	10^3–10^5 J mol^{-1}	300 GHz–400 Thz	750 nm–1 mm
Microwave	rotation	10^1–10^3 J mol^{-1}	300 MHz–300 GHz	1 mm–100 mm
Radiofrequency	Nuclear spin	10^{-3}–10^1 J mol^{-1}	1–1000 MHz	10 m – 1 cm

EHz, PHz, THz, GHz and MHz stand for exahertz, petahertz, terahertz, gigahertz and megahertz, respectively.

Depending on the absorption of frequency, absorption spectroscopy can be further subdivided to
(i) UV-Visible absorption spectroscopy: Absorption of electromagnetic radiation in UV-Vis region leads to electronic transition.
(ii) IR absorption spectroscopy: Absorption of infra-red radiation leads to transition between quantized vibrational energy levels.
(iii) Microwave absorption spectroscopy: Absorption of microwave radiation leads to transition between quantized rotational energy levels.

Similarly, emission spectroscopy can also be divided into different types based on absorption of light in different region (i) fluorescence: emission from excited electronic state (ii) IR emission spectroscopy (emission from excited vibrational level) (iii) Rotational emission spectroscopy (emission from excited rotational level).

Types of Spectroscopy Based on Mode of Collection of Signal

Spectroscopy can be of two types depending on the mode of collection.

Frequency domain spectroscopy   It measures the absorption/intensity of electromagnetic radiation as a function of wavelength/frequency.

Time domain spectroscopy   This is a mode of spectroscopy in which the amplitude of electromagnetic radiation is measured as a function of time. Time domain spectra need to be converted into frequency domain spectra through Fourier transformation. Data can be collected rapidly in this mode since the sample can be bombarded with electromagnetic radiation consisting of a wide range of frequencies at one go. This reduces the time required for a single scan and thus is a method of choice for collection of signal in low sensitive techniques such as NMR.

Fourier transformation (transformation of time domain data to frequency domain)   This is a way of processing all wavelengths/frequencies simultaneously.

Two key ideas
  (i) A time-domain spectrum is basically a combination of several cosine waves of different frequencies and amplitude.
 (ii) Cosine waves of different frequencies are orthogonal to each other, that is, integral, taken between $t = 0$ and $t = \infty$, the product of any two cosine waves of different frequency is zero. Thus, cosine wave of 1 Hz and 2 Hz are orthogonal to each other.

$$\int_{t=0}^{t=\infty} \cos(\omega_1 t) \times \cos(\omega_2 t) = 0$$

The Fourier transform of any time domain function is itself a function of frequency, whose magnitude represents the amount of that frequency present in the original time domain function.

12.2.6 Properties of Spectra

Spectral resolution   It is a measure of the ability to separate nearby features in the wavelength space.

$$R = \frac{\lambda}{\Delta \lambda}$$

where $\Delta \lambda$ is the minimum wavelength separation of two well-resolved features.

Line broadening   The absorption or emission peaks in spectra are not confined to a single wavelength/frequency/wave number. It is spread over a range of frequencies. This spreading is known as line broadening.

Line broadens due to the following
  (i) Heisenberg uncertainty principle (HUP): Line width due to HUP is known as natural line width
 (ii) Doppler effect
(iii) Pressure
 (iv) Power or saturation broadening
  (v) Electric and magnetic fields

Check Your Progress

  5. What type of transition happens when a molecule absorbs IR light?
  6. What happens when a molecule absorbs microwave radiation?
  7. What kind of radiation is required for electronic transition?
  8. What is the difference between time-domain spectroscopy and frequency-domain spectroscopy?
  9. How can one convert a time-domain data to frequency-domain data?
 10. What wavelength of light is needed for transition between two energy levels differing by 1000 kJ mol^{-1}.

(i) *Heisenberg uncertainty principle* According to Heisenberg uncertainty principle (HUP),

$$\Delta E \Delta t \geq \frac{h}{2\pi}$$

$$\Delta v \Delta t \geq \frac{1}{2\pi}$$

Δv is related to $\Delta \lambda$ by the equation

$$|\Delta \lambda| = \frac{\Delta v \lambda^2}{c}$$

Thus, the uncertainty in wavelength (line width) will depend on life time (Δt) of the excited state. A longer life time of the excited state leads to sharper line. The natural line width is Lorentzian in shape.

(ii) *Doppler Effect* Atoms move randomly. The wavelength of a spectral line is dependent on velocity of the source relative to the observer. This is known as Doppler effect.

$$\frac{\Delta \lambda}{\lambda_0} = \frac{v}{c}$$

$\Delta \lambda$ = Wavelength shift, λ_0 = Wavelength of source if it is not moving,

v = Velocity of source, c = Speed of light

Rotation of star/gas will produce a broadening of spectral lines. Photons emitted from side spinning toward the observer/detector, gets blueshifted. Photons emitted from side spinning away from us, gets redshifted. Doppler effect is not limited to only electromagnetic waves, but also holds true for sound waves. When a train approaches us, the sound wave that reaches us has shorter wavelength and higher frequency. Thus, we hear sound with higher pitch.

(iii) *Pressure broadening* When an excited state atom is hit with another high-energy atom, energy is transferred which changes the energy of the excited state, and hence, the energy of the photon emitted. This results in linewidth broadening. The broadening is **Lorentzian** in shape.

(iv) *Electric and magnetic fields* Interaction of an emitting particle with an electric field causes a shift in energy. This is known as Stark broadening.

Intensity of spectral lines Intensity of spectral lines depends on

(i) The *transition probability* between the two states (selection rules) Ψ_{fi} is given by

$$\mu_{fi} = \int \Psi_f \hat{\mu} \Psi_i \, d\tau$$

Ψ_i is the wavefunction of state from where transition starts whereas Ψ_f is the wavefunction of the final state involved in transition. $\hat{\mu}$ is the transition dipole moment operator. The transition is allowed, only if the above integral (also referred to as the Transition moment integral) is non-zero.

(ii) *Population of states*: The intensity of any spectroscopic signal increases with increase in the population difference between the two energy levels involved in transition.

$$\frac{N}{N_0} = e^{-\frac{E_x}{kT}}$$

N and N_0 are population of excited and ground level at temperature T, E_x is the difference between two energy levels, and k is Boltzmann's constant.

For transition between two levels, i to f, with population N_i and N_f and energy E_i and E_f, respectively

$$\frac{N_i}{N_f} = \frac{e^{-\frac{E_i}{kT}}}{e^{-\frac{E_f}{kT}}} = e^{\frac{\Delta E}{kT}}$$

$$\Delta E = E_f - E_i$$

$$\frac{\dfrac{N_i}{N_f}-1}{\dfrac{N_i}{N_f}+1}=\frac{\exp\left(\dfrac{\Delta E}{kT}\right)-1}{\exp\left(\dfrac{\Delta E}{kT}\right)+1}$$

$$\frac{N_i-N_f}{N_i+N_f}=\frac{\exp\left(\dfrac{\Delta E}{kT}\right)-1}{\exp\left(\dfrac{\Delta E}{kT}\right)+1}$$

$$=\frac{1+\dfrac{\Delta E}{kT}+\cdots-1}{1+\dfrac{\Delta E}{kT}+\cdots+1}=\frac{\Delta E}{2kT}$$

$$\frac{\Delta N}{N_t}=\frac{\Delta E}{2kT}$$

$$\Delta N=2kT\frac{\Delta E}{N_t}$$

Table 12.2 Various types of spectroscopy and population ratio between excited and ground state

Type	Nature of transition	Energy (Jmol^{-1})	N1/N0
UV-Visible	Electronic	120000	1×10^{-21}
Infrared	Vibration	12000	0.008
Microwave	Rotation	12	1
Radiofrequency	Nuclear spin	0.12	1

At 300 K, kT is equivalent to 208 cm^{-1} or 2.5 kJ mol^{-1}

Energy levels corresponding to UV/visible spectrum are far apart in energy, and according to the Boltzmann distribution, the population difference between two energy levels is large. As a result, UV/visible spectroscopy is an extremely sensitive technique. In NMR, the energy separation of the spin states is comparatively very small and considered to be an insensitive technique. Several scans are needed to get an appreciable signals in NMR.

(iii) *Concentration and path length of the sample* Concentration and path length affect the intensity of the electromagnetic radiation according to Beer–Lambert law.

$$\ln\frac{I}{I_0}=-\varepsilon cl$$

Here, I_0 and I are intensities of electromagnetic radiation before and after passing through the sample, ε is the molar extinction coefficient of the chromophore in the sample, c is the concentration, and l is the path-length of the sample. ε is high for allowed transition.

Check Your Progress

11. In a hydrogen type atom, the energy associated with the transition of electrons from ground state to higher state is found to be -1.9×10^{-21} J. Calculate the ratio of the population of the higher state to that at ground state at room temperature.

12. Life time of an excited state is 10 ns. Calculate the natural line width of transition from ground to excited state.

13. What is the Doppler-shifted wavelength of a red (680 nm) traffic light approached at 100 km h^{-1}

14. Estimate the life-time of a state that gives rise to a line of width 0.5 cm^{-1}.

15. What are the factors which affect intensity of spectroscopic signal?

12.2.7 Rotational Spectroscopy

Rotational spectroscopy is based on the transition between the quantized rotational levels. Transition takes place on absorption in the microwave region.

The solution of Schrödinger equation for a rigid rotor gives

$$E_J = \frac{\hbar^2}{2I} J(J+1)$$

Here, I is the moment of inertia of molecule and is equal to μr_0^2, where μ is reduced mass of the molecule and r_0 is equilibrium bond length.

For a diatomic molecule,

$$\mu = \frac{m_1 \times m_2}{m_1 + m_2}$$

The energy difference between two successive rotational energy levels is given by

$$\Delta E = E_{J+1} - E_J$$

$$\Delta E = \frac{\hbar^2}{2I}\left[(J+1)(J+2) - J(J+1)\right]$$

$$\Delta E = \frac{\hbar^2}{I}(J+1)$$

$$\Delta E = 2hcB(J+1)$$

$$h\nu = 2hcB(J+1)$$

$$\frac{1}{\lambda} = 2B(J+1)$$

$$B = \frac{h}{8\pi^2 \mu r_0^2 c}$$

B has a unit of wavenumber and is generally expressed in terms of cm^{-1}.

$$B = \frac{h}{8\pi^2 I c}$$

Spacing between two successive lines   If a line corresponds to transition from J to $J+1$ level,

$$\frac{1}{\lambda} = 2B(J+1)$$

The successive line will correspond to transition from $J+1$ to $J+2$

$$\frac{1}{\lambda'} = 2B(J+2)$$

Thus, the gap between two successive lines will be

$$\frac{1}{\lambda'} - \frac{1}{\lambda} = 2B(J+2) - 2B(J+1) = 2B$$

The gap between the successive rotational energy levels is proportional to B and inversely proportional to I. There are many molecules whose energy gap between rotational energy levels are less than kT and thus many of these rotational levels will be appreciably occupied at room temperature.

Important points about rotational energy level

Rotational levels are not equally spaced.

Rotational levels have degeneracy. $M_J = 0 \ldots \pm J$; $g(J) = (2J+1)$;

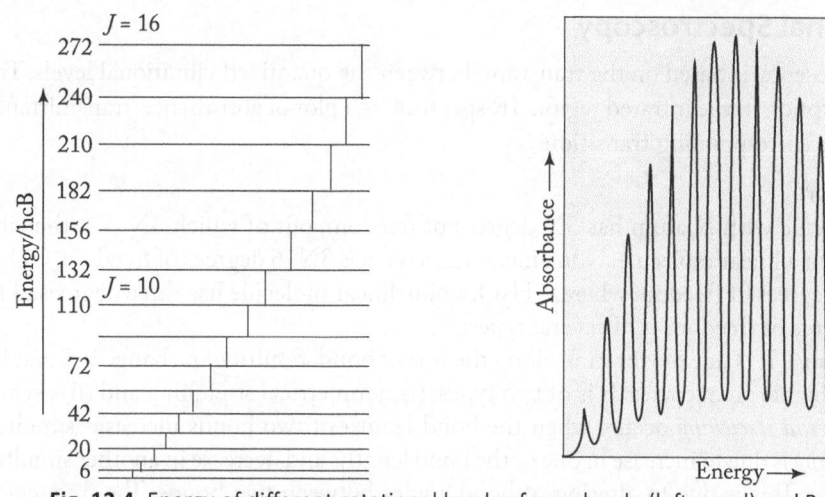

Fig. 12.4 Energy of different rotational levels of a molecule (left panel) and Peaks in a rotational spectrum (right panel)

Intensity of rotational lines Intensity of rotation depends on
(i) **Transition dipole moment:** For transition from rotational state $(Y_{J,i}^{m_l}(\theta,\phi))$ to state $(Y_{J,f}^{m_l}(\theta,\phi))$. The transition dipole moment is given by

$$\mu_{fi} = \int_0^{2\pi}\int_0^{\pi} Y_{J,f}^{m_l}(\theta,\phi)\hat{\mu}Y_{J,i}^{m_l}(\theta,\phi)\sin\theta d\theta d\phi$$

It must not be zero. For this,
  (a) The molecule must have permanent dipole moment. This is known as gross selection rule for the rotational spectrum. Thus, the homonuclear diatomic molecules, with no permanent dipole moment, do not show rotational spectra.
  (b) $\Delta J = \pm 1$. The transition following this criteria is known as allowed transition.

(ii) **Boltzmann distribution and degeneracy**
  If N_0 and N_J are population of ground level and j th level and ΔE is the energy difference between the two levels, then

$$\frac{N_J}{N_0} = (2J+1)\exp\left(-\frac{\Delta E}{kT}\right)$$

  $(2J+1)$ increases but $\exp\left(-\dfrac{\Delta E}{kT}\right)$ term decreases with increase in the J level from where transition is taking place. Thus, the intensity of lines first increases (multiplicity being more predominant) then decreases (exponential term becomes more predominant)

Applications of rotational spectroscopy:
  (i) Relative abundance of isotopes can be obtained through rotational spectroscopy.
     If the intensity of rotational line of molecule A*X (A* is isotope of A) is x times that of AX, then natural abundance of A*X (y) is given by

$$y = \frac{x}{(1+x)} \times 100\%$$

 (ii) Rotational spectra can be used to determine the bond length (See example 5).
(iii) Rotational spectroscopy can be used to differentiate between conformational isomers.
 (iv) Dipole moment of molecules can be obtained using rotational spectroscopy.
  (v) The presence of chemical species in interstellar bodies can be ascertained.
 (vi) Temperature changes the intensity of rotational spectrum and thus temperature of interstellar species can be obtained.

12.2.8 Vibrational Spectroscopy

This form of spectroscopy is based on the transition between the quantized vibrational levels. Transition takes place on absorption in the infrared region. IR spectrum is a plot of absorbance/transmittance versus wave number of the corresponding transition.

Types of Vibration

A polyatomic molecule with N atom has 3N degrees of freedom, out of which 3N–5 is the vibrational degree of freedom for a linear molecule. Non-linear molecule has 3N–6 degrees of freedom. CO_2, a linear molecule has four degrees of freedom whereas H_2O, a non-linear molecule has three degrees of freedom. The vibrational degree of freedom is of several types:

Stretching vibration   It is due to vibration along the line of bond, resulting in change in bond length. It generally occurs at higher frequencies. It is of two types: (i) symmetrical stretching and (ii) asymmetrical stretching. *Symmetrical stretching* occurs when the bond length of two bonds increases simultaneously. *Asymmetrical stretching* is due to increase in one of the bond lengths and decrease in another simultaneously.

Bending vibration   This is due to altering of bond angles between two bonds. The frequency of this vibration is low in comparison to stretching vibration.

In-plane bending: This is of two types (a) *Scissoring*: Two atoms approach each other; (b) *Rocking*: Two atoms move in the same direction.

Out-plane bending: (a) *Wagging*: Two atoms move to one side of the plane (b) *Twisting*: one atom moves above the plane and the other moves below the plane.

Modes of vibration of H_2O:
   Symmetric H_2O stretching: 3657 cm^{-1}; **asymmetric H_2O stretching:** 1595 cm^{-1}; **bending:** 3756 cm^{-1}

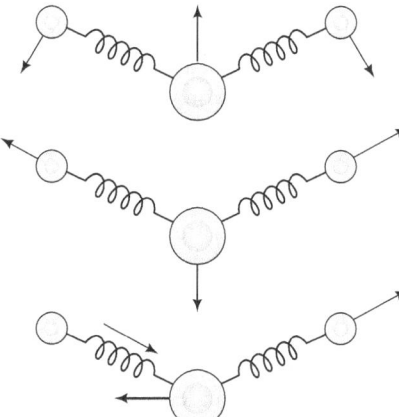

Frequency of vibration   The solution of Schröndinger equation of harmonic oscillator gives energy associated with vibrational level

$$E_v = (v + 1/2)\, hv$$

The frequency of the IR transition depends on:

$$\Delta E = hv = \frac{h}{2\pi}\sqrt{\frac{k}{\mu}}$$

v = frequency of transition
k = force constant of spring and is related to bond energy
μ = reduced mass of the molecule

Fig. 12.5 Vibrational modes of water

Check Your Progress

16.   Among the following, find out the molecules which are rotationally active.
   (i) HCl (ii) BCl_3 (iii) CH_3Cl (iv) SF_6 (v) C_2H_4 (vi) CH_4
17.   In a molecule of HCl, the masses of H and Cl are 1.673×10^{-27} kg and 5.68×10^{-26} kg, respectively. The equilibrium bond length of the molecule is 1.353 Å, calculate the moment of inertia of the molecule.
18.   What will be the rotational constant for a molecule of CS if the masses of the both the atoms mC and mS are 12.00 u and 31.97 u, respectively and the equilibrium bond length of the molecule is 1.392 Å?
19.   For a molecule *AB*, the observed rotational constant is 0.142 cm^{-1}. If the masses of atom A (m_A) and B (m_B) are 126.9 u and 34.97, respectively, calculate the equilibrium bond length of the molecule.
20.   If the rotational line of $^{12}C^{16}O$ is 124 times intense than that of $^{13}C^{16}O$, then what will be the natural abundance of $^{13}C^{16}O$ in CO?

Frequency of transition is directly proportional to the strength of the bonding between the two atoms ($v \propto k$) and inversely proportional to the reduced mass of the two atoms ($v \propto 1/\mu$).

(i) Frequency decreases with increasing atomic weight (higher μ). Thus, the stretching frequency of bond C–H is more than C–D, which in turn is more than C–C.

(ii) Frequency increases with increasing bond energy (higher k). Thus, the stretching frequency of bond C–C is lower than C=C, which in turn is lower than C $\equiv$ C.

Each functional group has a characteristic frequency. Thus, IR is one of the most important techniques used in the detection of functional groups.

Table 12.3 Vibrational frequencies of groups

Group	Nature of vibration	Type of compound	Wave number (cm^{-1})	Intensity
–C–H	Stretching Bending	Alkane	2980–2840 1480–1380	Strong
O–H	Stretching (H-bonded)	Primary, secondary alcohols	3600–3200	Strong, broad
N–H	Stretching Bending	Primary Amine	3400–3200 1600	Medium, (two bands)
C–O	Stretching	Esters, alkyl ketones	1150–1050	Strong
C–O	Stretching	Ether	1300–1000	Strong
C–F C–Cl C–Br C–I	Stretching Stretching Stretching Stretching	alkyl halide	1400–1000 820–580 680–520 500	Strong Strong Strong Strong
=C–H =C–H C=C	Stretching Bending Stretching	Alkene	3080–3020 900–680 1680–1630	Medium Strong Variable
-C$\equiv$C-	Stretching	Alkyne	2180–2100	Variable, not present in symmetrical alkynes
C–H C=C	Stretching Stretching	Aromatic hydrocarbon	3100–3000 1600–1400	Medium Medium weak, multiple bands
C=O	Stretching	Carbonyl compounds	1740–1700	Strong

Intensity of spectral lines in IR Intensity of spectral lines depends on the transition probability between the two states (selection rules). The probability for a vibrational transition to occur, that is, the intensity of the different lines in the IR spectrum, is given by the transition dipole moment μ_{fi} between an initial vibrational state v_i and a vibrational final state v_f:

$$\mu_{fi} = \int v_f \, \hat{\mu} \, v_i \, d\tau$$

$\hat{\mu}$ is dipole moment operator and is a function of internuclear distance (x). If the product $v_f \, \hat{\mu} v_i$ is odd, the transition is forbidden, since integral of odd function is zero.

Since $\hat{\mu}$ is a function of x, it can be expanded using Taylor series

$$\mu(x) = \mu_0 + \left(\frac{\partial \mu}{\partial x}\right)_0 x + \frac{1}{2}\left(\frac{\partial^2 \mu}{\partial x^2}\right)_0 x^2 + \cdots$$

$$\mu_{fi} = \int v_f \left(\mu_0 + \left(\frac{\partial \mu}{\partial x}\right)_0 x + \frac{1}{2}\left(\frac{\partial^2 \mu}{\partial x^2}\right)_0 x^2 + \cdots \right) v_i \, d\tau$$

$$\mu_{fi} = \mu_0 \int v_f v_i \, d\tau + \left(\frac{\partial \mu}{\partial x}\right)_0 \int v_f x v_i \, d\tau + \frac{1}{2}\left(\frac{\partial^2 \mu}{\partial x^2}\right)_0 \int v_f x^2 v_i \, d\tau + \cdots$$

The first term is zero since the two wavefunctions are orthogonal to each other.

Neglecting the higher order term,

$$\mu_{fi} = \left(\frac{\partial \mu}{\partial x}\right)_0 \int \upsilon_f x \upsilon_i d\tau$$

In order to have a vibrational transition visible in IR spectroscopy, the electric dipole moment (μ) of the molecule must change with vibration in the molecule. Such vibrations are 'infrared active'. Vibrations in homonuclear diatomic molecules do not create a variation of μ, thus they are infrared inactive and cannot be studied through IR spectroscopy. However, a molecule without a permanent dipole moment can be studied if there is variation of the dipole moment with nuclear displacement.

Substituting the solutions of the Schröndinger equation for an *harmonic* oscillator wherein one obtains the wavefunctions of the initial state υ_i and final state υ_f, the integral part of transition dipole moment will be non-zero only when $\Delta \upsilon = \pm 1$.

Applications of IR Spectroscopy

Identification of functional groups and structure elucidation Different functional groups absorb characteristic frequencies of IR radiation. Thus, the peak values in IR spectra help us to know the functional groups present in an unknown molecule.

Identification of molecules No two unique molecules have the same IR spectrum. Thus, the IR spectrum can be used to identify a particular molecule.

Studying the progress of a reaction Most chemical reactions are accompanied by change in the functional groups. Thus, the progress of a reaction can be monitored by either looking at the disappearance of the characteristic peak of reactant or by observing the appearance of the characteristic peak of product.

Presence of moisture in a sample Water has a characteristic absorption in the IR region. Thus, presence of moisture can be detected through IR spectroscopy.

Presence of water as lattice water or as ligand The presence of characteristic bands in the following regions, viz. (a) 3600–3200 cm^{-1} (ii) 1650 cm^{-1} and (iii) 300–600 cm^{-1} suggests the presence of lattice water.

Detection of impurities This can be done by comparing the spectrum of the given sample with the spectrum of a pure reference compound. The presence of any additional peak will suggest the presence of impurity in a sample. An impurity can also be detected by observing a distinct peak particular to that impurity.

Quantitative analysis The concentration of a particular analyte in a sample can be determined by comparison of the peak area characteristic to that analyte in a sample to a standard in which concentration of that molecule is known.

Differentiation between geometrical isomers *Trans*-isomers are more symmetrical than *cis*- and hence have lesser number of peaks in the spectrum.

Secondary structure of proteins The secondary structures of protein, α-helix, β-sheet, and random coil have characteristic peaks. These peaks can be used to know the secondary structure of the protein.

Analysis of food product IR spectroscopy is used in the analysis of milk and milk products. IR has been used to analyse conjugated linoleic acid which is associated with many important biological functions.

In forensic analysis Most of the evidence available at a crime spot is basically that of an organic molecule. IR spectra can be used to detect human body fluids, sweat, hair, various inflammables, and other toxic molecules.

12.3 UV/VISIBLE SPECTROSCOPY

UV/Visible spectroscopy deals with the absorption of light in the UV/visible part of the spectrum (210 – 900 nm).

12.3.1 Types of Electronic Transitions

Electronic transitions in organic molecules (Fig. 12.6)

(i) **$\sigma \to \sigma^*$ transition** It does not fall in the UV/visible region. Transition takes place on the irradiation of electromagnetic radiation in the 120–150 nm range.

(ii) **$n \to \sigma^*$ transition** Organic molecules containing carbonyl groups typically undergo this transition. Transition takes place on the irradiation of electromagnetic radiation in the 150–250 nm range.

(iii) **$n \to \pi^*$ transition** One of the most common transitions in organic molecules containing carbonyl groups. The transition takes place on the irradiation of electromagnetic radiation in the range of 400–700 nm. These transitions are forbidden by symmetry and hence have low molar absorptivity (around 10–100 L mol^{-1}cm^{-1}).

(iv) **$\pi \to \pi^*$ transition** One of the most common transitions in conjugated organic molecules. The transition takes place on the irradiation of electromagnetic radiation in the range of 200–400 nm. These transitions are high in symmetry and hence have low molar absorptivity (around 1000–10000 L mol^{-1}cm^{-1}).

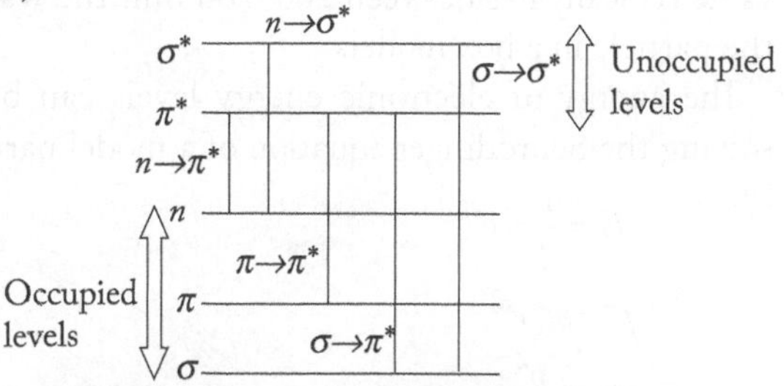

Fig. 12.6 Different types of electronic transitions

Table 12.4 Typical absorptions of simple chromophores

Groups	Transition	Wavelength max (nm)
R-OH	$n \to \sigma^*$	180
R-O-R	$n \to \sigma^*$	180
R-NH$_2$	$n \to \sigma^*$	190
R$_2$C=CR$_2$	$\pi \to \pi^*$	175
R-C≡C-R	$\pi \to \pi^*$	170
R-CHO	$\pi \to \pi^*$	190
R-CHO	$n \to \pi^*$	290
R$_2$CO	$\pi \to \pi^*$	180
R$_2$CO	$n \to \pi^*$	280
RCOOH	$n \to \pi^*$	205
R-COOR′	$n \to \pi^*$	205
R-CHO	$\pi \to \pi^*$	190

Electronic transition in inorganic transition metal complexes The five degenerate d orbitals split in the presence of ligand. The manner in which they split depends on the nature of coordination (octahedral/tetrahedral/square planar). The energy difference between levels will depend on the position of the transition metal in the periodic table, its oxidation state, and the nature of ligand bonded to it. Absorption by transition metal complexes leads to d–d transition. This transition is symmetry forbidden. However, some of the vibration can remove symmetry and thus d–d transitions are weakly allowed.

Charge transfer transition This transition takes place in complexes in which there is a donor and acceptor group of electrons. The most common examples of such cases are where a metal donates electrons to π^* orbital.

Transition and their molar absorptivity:

(i) Transition between states with different spin multiplicity is not allowed.

(ii) Parity (symmetry) must change during transition. In molecules with centre of symmetry, p–p or d–d transitions are forbidden.

(iii) The molecular orbitals of homonuclear diatomic molecule and other centrosymmetric molecules are called gerade (g) or ungerade (u) depending upon their change in phase on inversion operation, gerade orbitals show no change whereas ungerade one shows change. π orbital is ungerade whereas π^* is gerade. $\pi \rightarrow \pi^*$ transition is allowed since it involves changes in parity. σ orbitals are gerade whereas σ^* are ungerade.

Electronic transition in conjugated system The most significant transition is $\pi \rightarrow \pi^*$ transition. The value typically lies between 200–700 nm. The wavelength of such transitions can be calculated by using the particle in a box model.

The energy of electronic energy levels can be obtained by solving the Schrodinger equation of a model particle in a box.

$$E = \frac{1}{2}mV^2$$

$$p^2 = m^2 V^2$$

$$E = \frac{p^2}{2m}$$

$$E = \frac{h^2}{2\,m\lambda^2}$$

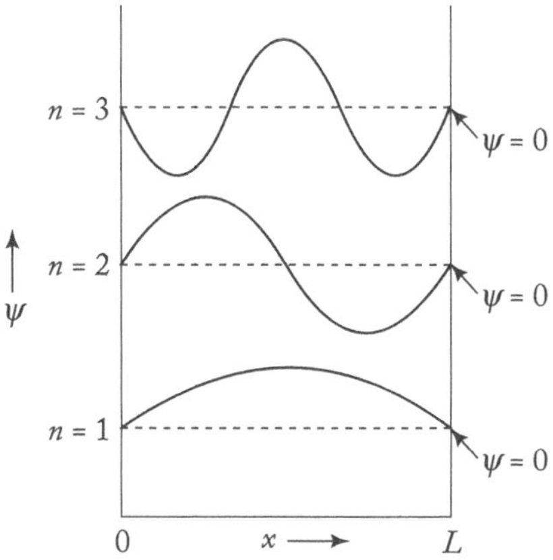

Fig. 12.7 Length of molecule should be integral multiple of $\lambda/2$

m, V are mass and velocity of electon, respectively. The length of molecule should be an integral multiple of $\lambda/2$ as shown in Fig. 12.7.

$$\lambda = 2L/n$$

$$E_n = \frac{n^2 h^2}{8\,ml^2}$$

Wavelength of transition in a conjugated system Molecules like butadiene can be treated as particle confined in a one-dimensional box. The energy of such a molecule can be given by

$$E_n = \frac{n^2 h^2}{8\,ml^2}$$

where n is the energy level. Transition of an electron takes place from the highest occupied molecular level (HOMO) to the lowest unoccupied molecular level (LUMO) level. HOMO level (i) and LUMO level (j) will depend on the number of conjugated electrons (p).

Check Your Progress

21. How many degrees of vibrational freedom are in (a) H_2O and (b) SO_2?

22. Which one out of the given bonds exhibits highest stretching wavenumber? (a) C-I and C-Cl and (b) C=O and C≡N?

23. In an IR spectrum of hex-1-yne, the strong absorption bands are observed at 3309 cm^{-1}, 2967 cm^{-1}, 2341 cm^{-1}, 2865 cm^{-1} and 2112 cm^{-1}. Identify the band which corresponds to stretching vibration of carbon-carbon triple bond and the terminal C-H bond.

24. If the force constant for fluorine molecule (F_2) is 450 N.m^{-1}, calculate its stretching vibrational wavenumber.

25. An IR spectrum of unknown compound with molecular formula $C_7H_6O_2$ shows following absorption bands: (a) 2996 cm^{-1}: a broad absorption band, (b) 2935 cm^{-1}: some sharp bands, (c) 1700 cm^{-1}: a srong absorption band, (d) 1495 cm^{-1}: a sharp absorption band and (e) 640-790 cm^{-1}: a series of sharp bands. Identify the compound.

$$\Delta E_{electron} = \frac{(j^2 - i^2)h^2}{8\ mL^2}$$

$$= \frac{\left[\left(\frac{p}{2}+1\right)^2 - \left(\frac{p}{2}\right)^2\right]h^2}{8m((p-1)l)^2} = \frac{(p+1)}{(p-1)^2}\frac{h^2}{8\ ml^2}$$

Since $E_{photon} = \Delta E_{electron}$,

$$\frac{hc}{\lambda_{photon}} = \frac{(p+1)}{(p-1)^2}\frac{h^2}{8\ ml^2}$$

Canceling an h on both sides gives us,

$$\frac{c}{\lambda_{photon}} = \frac{(p+1)}{(p-1)^2}\frac{h}{8\ ml^2}$$

Effect of substituents/solvents on the absorption spectra   Substituents and solvents can shift the peak of the spectrum towards both sides (higher or lower wavelength).

 (i) A shift towards longer wavelength is known as *Bathochromic shift* (red shift). The increase in the size of conjugated system from ethylene to butadiene to hexatriene results in bathochromic shift (175 to 217 to 258 nm). A polar solvent stabilizes the π^* orbital and decreases the energy of $\pi \rightarrow \pi^*$ transition leading to Bathochromic shift. Auxochrome with a lone pair also extends the conjugation resulting in bathochromic shift.

 (ii) A shift towards shorter wavelength is called hypsochromic shift (blue shift). This can happen with change in solvent. The polar solvent stabilizes n orbitals more than π^* and thus increases the energy of $n \rightarrow \pi^*$ transition leading to hypsochromic shift.

 (iii) Substituents can also affect the intensity of the peak. An increase in intensity is referred to as *hyperchromic* shift. Introduction of auxochrome can lead to hyperchromic shift. The presence of an amino group in aniline leads to hyperchromic shift in comparison to benzene.

 (iv) Decrease in intensity is referred as hypochromic shift. *cis*-Stilbene absorbs with weak intensity than *trans*-stilbene.

12.3.2 Applications of UV/Visible Spectroscopy

Quantitative determination of transition metal ion complexes   Several transition metal complexes are coloured and absorb in the visible region at a specific wavelength. The absorption coefficient at wavelength maxima for various complexes is tabulated. Measurement of absorption at this wavelength helps in knowing the concentration of the metal complexes in a given sample.

Qualitative and quantitative determination of conjugated double bond organic compounds   Organic molecules with conjugated double bond absorb in UV/visible region. Their wavelength maxima (wavelength at which absorption is maximum) changes with the length of conjugation. The peak at certain wavelengths not only gives us an idea about an organic molecule, the absorbance coefficient of the organic compound can also be used to calculate the concentration of the compound using Beer–Lambert law.

Quantitative determination of protein molecule   Protein molecule has three aromatic amino acids: tryptophan, tyrosine, and phenylalanine. Of these, tryptophan has high quantum yield and absorbs around 280 nm. The concentration of protein is generally calculated by measuring absorbance at 280 nm using Beer–Lambert law.

Quantitative determination of a drug   Drugs are sold in market in a particular formulation. The quantity of drug in such formulations can be calculated using UV-Vis spectrophotometer. Diazepem tablet is analysed by dissolving it in methanol and measuring the absorption at 284 nm.

Quantitative determination of drug release   The rate of drug release from pharmaceutical doses can be measured by measuring the absorbance of drug with time.

To identify the product of a reaction   The product of a reaction can be identified if it absorbs at a distinct wavelength. The conjugated and aromatic products can be identified using UV/visible spectroscopy.

To know the progress of a reaction   UV/visible spectrum can be used to look at the progress of a reaction, and analyse if the product absorbs a distinct wavelength than the reactants.

To measure the binding constant between a substrate and drug molecule   The absorbance of a drug at a particular wavelength is directly proportional to the concentration. This can be utilized to know the binding constant of a reaction.

To measure the equilibrium constant of a reaction   The equilibrium constant of any reaction can be calculated by measuring absorbance.

To identify impurity in the sample   The impurity which absorbs in UV/visible region can be easily detected.

Kinetics of a reaction   Absorbance measurements can be used to know the order and rate constant of a reaction.

Quantification of DNA/RNA   Proteins absorb at 280 nm. DNA and RNA absorb at 260 nm. Protein also absorbs at 260 nm. The ratio of absorbance at 260 nm and 280 nm gives an idea about DNA–protein ratio in the sample.

Food quality control   UV/visible spectroscopy is routinely employed to estimate carbohydrate contents in beer. Spectroscopy is used to know the amount of hydroxyproline which in turn gives the meat content of sausages.

Detection of pollutants   Quantification of heavy metal and organic pollutants in water is carried out using UV-Visible spectroscopy. Concentration of toxic metals can be obtained by converting them to a complex which absorbs in UV-Vis region. For example, concentration of lead can be obtained by complexing with dithizone to form lead dithizonate which absorbs at 750 nm.

As a detector in liquid chromatography   UV/Vis detectors are quite often used in molecular spectroscopy. Size exclusion chromatography in combination with UV/Vis detector is used to know the molecular size/weight of proteins.

12.4 FLUORESCENCE SPECTROSCOPY

When a molecule absorbs light, an electron is promoted to a higher excited state (generally a singlet state, but may also be a triplet state). The excited state can get depopulated in several ways.

Check Your Progress

26. What is the difference between hyperchromic shift and hypochromic shift?
27. The UV-VIS spectrum of an organic compound shows a band at 225 nm. If the absorbance of the compound (0.1 mM) in water is 1.25, calculate the molar absorptivity of the band at 225 nm. (cell length = 1 cm).
28. In acetone, what is the effect on the $n \rightarrow \pi^*$ transition if solvent changes from water to methanol?
29. Which one has longest wavelength for $n \rightarrow \pi^*$ transition: (a) Propanone: CH_3COCH_3 and (b) Butane-2,3-dione: $CH_3COCOCH_3$?
30. If the molar absorptivity of an absorption band of 0.2 mM organic compound is 1000 at 265 nm, calculate the intensity ratio.

The molecule can lose its energy non-radiatively by giving its energy to another absorbing species in its immediate vicinity (energy transfer) or by collisions with other species in the medium.

(i) **Inter system crossing:** Excited state singlet to triplet
(ii) **Phosphorescence:** Triplet to ground state singlet by emitting light
(iii) **Fluorescence:** Excited singlet to ground state singlet by emitting light

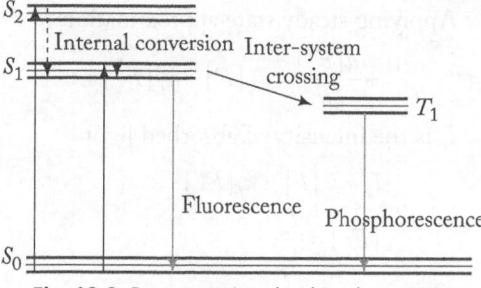

Fig. 12.8 Processes involved in absorption and emission

Stoke's shift Since fluorescence takes place from the lowest vibrational state of the excited singlet state, it occurs at lower energies or longer wavelength. Thus, it is red shifted in comparison to absorption spectra. This shift is known as Stoke's shift.

Emission spectrum The spectrum in which excitation wavelength is kept constant and fluorescence intensity is measured as a function of wavelength, that is, spectrum of emitted light is determined.

Excitation spectrum This spectrum is obtained in which fluorescence intensity at a particular fixed wavelength is measured as a function of the excitation wavelength.

Kasha's rule The same fluorescence emission spectrum is generally observed irrespective of the excitation wavelength. This happens since the time scale of the internal conversion is very fast in comparison to the time scale of fluorescence.

The emission spectrum and absorption spectrum of the S_0 to S_1 transition are typically mirror images of each other (Fig. 12.9).

This is due to:

(i) The spacing between vibrational levels of ground state is same as that of the excited state.
(ii) The fluorophore from ground state vibrational level of excited state can return to any of the ground state vibrational levels.

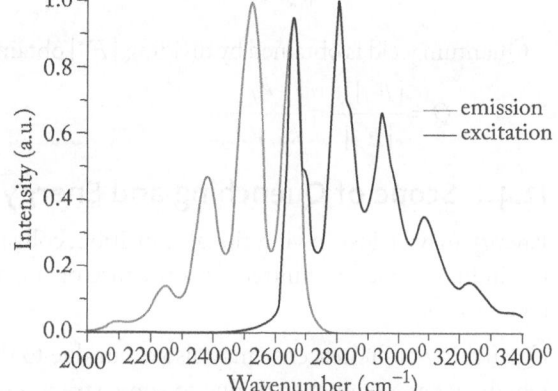

Fig. 12.9 Excitation and emission spectra of anthracene

Life time The life time of the excited state is the average time that a fluorophore spends in the excited state before returning to its ground state.

Absorption versus emission spectra The time scale of absorption is in femtoseconds to picoseconds. It is much shorter in comparison to the time scale of molecular motion that might take place during the absorption process. Thus absorption spectra are not sensitive to molecular motions. In contrast to absorption, emission occurs over a longer period of time, typically in nanoseconds. Thus, this longer life time enables fluorescent molecules to interact with other molecules in solution such as oxygen.

Quantum yield It is the ratio between the number of photons emitted and that absorbed.

$$Q = \frac{\text{number of photons emitted}}{\text{number of photons absorbed}}$$

Let us consider only radiative emission

$$F \xrightarrow{k_a} F^* \text{(absorption)}$$

$$F^* \xrightarrow{k_f} F \text{(Fluorescence)}$$

$$\frac{d[F^*]}{dt} = k_a[F] - k_f[F^*]$$

Applying steady state approximation to F^*

$$+\frac{d[F^*]}{dt} = k_a[F] - k_f[F^*] = 0$$

I_a is the intensity of absorbed light

$$I_a = k_a[F] = k_f[F^*]$$

$$[F^*] = \frac{I_a}{k_f}$$

If non-radiative transition is also present

$$F \xrightarrow{k_a} F^* \text{(absorption)}$$

$$F^* \xrightarrow{k_f} F \text{(Fluorescence)}$$

$$F^* \xrightarrow{k_{nr}} F \text{(non-radiative transfer)}$$

$$+\frac{d[F^*]_{nr}}{dt} = k_a[F] - k_f[F^*]_{nr} - k_f[F^*]_{nr} = 0$$

$$[F^*]_{nr} = \frac{I_a}{k_f + k_{nr}}$$

Quantum yield is obtained by dividing $[F^*]$ obtained in the presence and absence of non-radiative decay

$$Q = \frac{[F^*]_{nr}}{[F^*]} = \frac{k_f}{k_f + k_{nr}}$$

12.4.1 Scope of Quenching and Energy Loss During Fluorescence

Energy may be lost in vibrational transition, collision with the solvent, heat transfer, etc. Only a part of the light absorbed is emitted. It is because of this that the quantum yield in most practical cases is not equal to one.

Quenching of fluorescence may also occur due to the presence of some foreign molecules in the solution which act as a quencher, or due to some structural rearrangement in the molecule (say, protein), which drives the fluorophore to a conformation where it is in the vicinity of a quencher (any amino acid residue or disulphide bond).

Fluoresence intensity may also decrease due to the non-radiative transfer of the emitted energy to some other chromophore, which absorbs at that energy. This phenomenon is called *Förster resonance energy transfer* (FRET). However, FRET and quenching should not be treated synonymously.

Effect of solvent on the fluorescence spectra:
Typically, the fluorophore has a larger dipole moment in the excited state than the ground state. Polar solvents stabilize the excited state and shift the emission to lower energy. As the solvent polarity is increased, this effect becomes larger.

Fluorescence quenching It is a process that decreases fluorescence intensity of a fluorophore. A variety of molecular processes can result in quenching. These include collision with a quencher, binding with quencher, energy transfer, etc.

Collisional quenching This is a type of quenching in which a fluorophore transfers its energy to a colliding molecule and the transfer is non-radiative.

$$F^* + Q \rightarrow F + Q$$

Overall, the reaction will be given by

$$F \xrightarrow{k_a} F^* \text{(absorption)}$$

$$F^* \xrightarrow{k_f} F \text{(Fluorescence)}$$

$$F^* \xrightarrow{k_{nr}} F \text{(non-radiative transfer)}$$

$$F^* + Q \xrightarrow{k_Q} F + Q \text{ quenching}$$

In the absence of quencher

$$+\frac{d[F^*]}{dt} = k_a[F] - k_f[F^*] - k_{nr}[F^*] = 0$$

$$[F^*] = \frac{I_a}{k_f + k_{nr}}$$

In the presence of quencher

$$+\frac{d[F^*]_q}{dt} = k_a[F] - k_f[F^*]_q - k_{nr}[F^*]_q - k_q[F^*]_q[Q] = 0$$

$$[F^*]_q = \frac{I_a}{k_f + k_{nr} + k_q[Q]}$$

$$\frac{F_0}{F} = \frac{[F^*]}{[F^*]_q} = \frac{k_f + k_{nr} + k_q[Q]}{k_f + k_{nr}} = 1 + \frac{k_q[Q]}{k_f + k_{nr}} = 1 + k_q\tau_o[Q]$$

In case of a quenching only due to collision, the Stern Volmer equation holds true.

$$\frac{F_0}{F} = 1 + K_{SV}[Q]$$

F_0 and F are the fluorescence intensities of the fluorophore alone and in the presence of a quencher, respectively. $[Q]$ is the concentration of quencher and K_{SV} is the Stern–Volmers's constant. K_{SV} is equal to $k_q\tau_o$, where τ_o is the life time of fluorophore and k_q is the quencher rate coefficient.

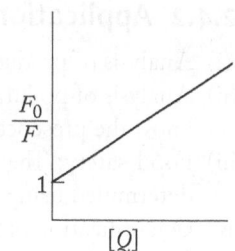

Fig. 12.10 Stern–Volmer plot

A plot between $\dfrac{F_0}{F}$ versus $[Q]$ is a straight line with slope equal to K_{SV}.

Static quenching In static quenching, quenching takes place due to quencher binding to the fluorophore in ground state.

$$F + Q \rightleftharpoons FQ$$

$$K_S = \frac{[F-Q]}{[F][Q]}$$

$$[F]_0 = [F] + [F-Q]$$

$$K_S = \frac{[F]_0 - [F]}{[F][Q]}$$

$$K_S[Q] = \frac{[F]_0}{[F]} - 1$$

$$\frac{[F]_0}{[F]} = 1 + K_S[Q]$$

Similar to collisional quenching, a plot of $\dfrac{F_0}{F}$ versus $[Q]$ is a straight line with slope equal to K_S. K_S is binding constant for binding between fluorophore and quencher.

Förster's Resonance Energy Transfer (FRET): Fluorescence emission from the donor (D) is absorbed by the acceptor (A).

- The emission spectrum of donor and the absorption spectrum of the acceptor must have a spectral overlap.
- FRET is a non-radiative process. The FRET efficiency (E) is dependent on the distance between the donor and acceptor. E is calculated from the fluorescence intensities of the donor in the presence (I_{DA}) and in the absence of an acceptor (I_D)

$$E = 1 - \frac{I_{DA}}{I_D}$$

$$R = \left(\frac{1}{E} - 1\right)^{1/6} R_0$$

where R_0 is known as Forster radius. It is the distance between the centre of donor and acceptor at which FRET efficiency is 50%. The Forster radius of many pairs of donor and acceptor is known. By measuring the FRET efficiency and using Forster radius, the distance between donor and acceptor can be calculated under a particular condition.

12.4.2 Applications of Fluorescence Spectroscopy

(i) Analysis of products: Products can be analysed for their purity.

(ii) Analysis of pollutants: Fluorescence markers for different pollutants are available. It can be used to know the presence of a particular pollutant and its concentration.

(iii) Food safety: The presence of toxic elements/compunds can be qualitatively and quantitatively determined using fluorescence markers for the particular toxic elements/ pollutants.

(iv) Quantification of DNA and RNA: There is an increase in fluorescence intensity of fluorescent dyes Hoechst 33258 and ethidium bromide upon binding to DNA. Excitation and emission wavelength of Hoechst 33258 is 355 nm and 460 nm, respectively. Ethidium bromide is excited at 530 nm and monitored at 620 nm. These dyes are used to quantify DNA and RNA.

(v) Analysis of disease markers in pathological labs: A fluorophore can be attached to glucose binding protein and can be used as sensor for determination of glucose (related to diabetic). FITC-Dextran and FITC-Dextran/RuBpy based pH and urea biosensors have shown great potential for monitoring kidney functions.

(vi) Calculation of binding constant: Binding constant can be measured if one of two reactants has a fluorophore or can be tagged with a fluorophore.

(vii) Tracking the progress of reactions: The progress of a reaction can be monitored if the reactant or product has fluorophore.

(viii) In–vivo monitoring: Any process can be monitored if there is change in fluorophore signal between reactant and product.

12.4.3 Fluorescence Microscopy

The resolution of an unaided human eye is of the order of 0.1 mm (at the optimum viewing distance of 25 cm). The limit of resolution is the minimum distance by which two nearby objects are separated and still appear as two distinct ones. Microscopy is a technique which is used to see objects of size much smaller than 0.1 mm. In general, there are two types of microscopy: (i) optical microscopy and (ii) electron microscopy.

Fluorescence microscopy is an optical microscopy based on the principle of fluorescence. Specimens are irradiated with specific band of wavelengths of high intensity and weak emitted light is separated from the intense excitation light with filters designed for that specific wavelength corresponding to emitted light allowing the viewer to see only that which is fluorescing. Two main sources of light are mercury vapour and xenon arc lamps. The most common fluorescence microscope is epifluorescence microscope. One problem with this microscope is that it collects out-of-focus light and thus images obtained are not sharp. In confocal microscopy, out-of-focus emitted light is blocked using a pinhole in front of the detector and thus a sharp three-dimensional image can be obtained.

Uses of fluorescence microscopy:
1. It is used in the study of living cells and tissues.
2. It can be used to view DNA and RNA within the cell.
3. It can be used for measurement of pH and Ca ion concentration inside a cell.
4. Viability studies on cell population can be carried out by fluorescence microscopy.
5. It is widely and routinely used for study of coals.
6. It can be used in the study of ceramics and semiconductors.

12.5 NMR SPECTROSCOPY

NMR deals with the interaction of electromagnetic radiation in radiofrequency range with nuclear spin of certain nuclei.

Principle of NMR:

Table 12.5 Spin for different types of nuclei

Number of neutron	Number of proton	Spin (I)
Even	Odd	1/2, 3/2, 5/2
Odd	Even	1/2, 3/2, 5/2
Even	Even	0
Odd	odd	1,2,3

- Neutrons and protons, composing any atomic nucleus, are associated with quantum property of spin (I). Nuclear spin is the sum of spin of protons and neutrons. I depends on the mass and atomic number of the nuclei and is calculated using the Nuclear Shell model. (Refer Table 12.5)
- Rotating object including spin possesses angular momentum $(\vec{I})$

$$\vec{I} = m\vec{v}\,\vec{r} = \hbar\sqrt{I(I+1)}$$

- If spinning object has charge, then a magnetic dipole will be generated. Then strength and direction of magnetic moment $(\vec{\mu})$ will be given by

$$\vec{\mu} = \frac{g_N \mu_N \vec{I}}{\hbar} = \gamma\vec{I}$$

g_N is known as nuclear g factor, μ_N is nuclear magneton and is equal to $\left(\dfrac{e\hbar}{2m_N}\right)$ where m_N is the mass of nucleus. γ is known as gyromagnetic ratio. Gyromagnetic ratio is a property of spin. (Refer Table 12.6)

The nuclei with I spin has $(2I + 1)$ degenerate levels. The degeneracy of the energy levels is lifted by the interaction of nuclear dipole moment with an intense external magnetic field (B_0).

Check Your Progress

31. What are the advantages of fluorescence over UV-Vis spectroscopy?
32. What is the effect on fluorescence of a compound if solvent is replaced by a highly polar one?
33. What is quantum yield of a fluorophore? What is typical life time of fluorescence?
34. If a compound fluoresced at 345 nm after excitation at 275 nm, what will be its Stoke's shift ?
35. If 3 mM of a quencher (kq = 3.58×1011 M^{-1}.s^{-1}) is required to decrease the fluorescence intensity of a compound to half of its unquenched value, calculate the life time of the compound.

Energy of each state is given by

$$E = \bar{\mu} B_0$$

$$E = \gamma B_0 \bar{I}$$

Since spins align with the magnetic field (in z direction)

$$E = \gamma B_0 \bar{I}_z$$

$$\overline{I_z} = m\hbar$$

where m is magnetic quantum number. $m = +I, I-1 \ldots -I$. Thus, nuclei with $+1/2$ spin has 2 energy levels with m $= +1/2$ and $-1/2$ whereas nuclei with $+1$ spin has 3 energy levels with m $= +1, 0, -1$.

$$E = m\hbar\gamma B_0$$

The nuclei with half spin has two energy levels (α and β) and their energies are given by

$$E_\alpha = \frac{1}{2}\hbar\gamma B_0$$

$$E_\beta = -\frac{1}{2}\hbar\gamma B_0$$

For nuclei with positive γ, E_β is lower in energy.

Table 12.6 Gyromagnetic ratio of different nuclei

Nucleus	$\gamma \times 10^7$ (rad T^{-1}s^{-1})	$\frac{\gamma}{2*\pi}$ (MHz T^{-1})
^{1}H	26.75	42.577
^{13}C	6.73	10.708
^{15}N	-2.7116	-4.316
^{19}F	25.1665	40.052
^{35}P	10.8289	17.235

Tipping away of magnetization from magnetic field direction (z-axis)

If magnetization is somehow tipped away from z-axis, the magnetization rotates around the z-axis making a cone of constant angle with z-axis. This rotational motion is called *precession*.

Negative γ = positive sense of precession

Positive γ = negative sense of precession

Excitation of transitions between energy levels (or tipping of magnetization away from z-axis) is stimulated using radio-frequency electromagnetic radiation.

The flip angle (in radians) is proportional to magnetic field of radio-frequency electromagnetic radiation (B_1), gyromagnetic ratio, and duration of pulse (t).

$$\theta = \omega_1 \times t = \gamma B_1 t$$

The absorbed photon in radiofrequency range promotes a nuclear spin from its ground state to its excited state.

$$\Delta E = \hbar\gamma B_0$$

Frequency is given by $\nu = \dfrac{1}{2\pi}\gamma B_0$

Angular frequency is given by $\omega = \gamma B_0$

This frequency is known as *Larmor Frequency*. Thus, Precession frequency = Larmor frequency. The allowed transition occurs at Larmor frequency.

The frequency of transition depends on the (i) gyromagnetic ratio and (ii) magnetic field. 1T = 42.5781 MHz (^{1}H), 10.71 MHz (^{13}C), 4.34 MHz (^{15}N).

Table 12.7 NMR nuclei frequency in different external magnetic fields

Magnetic field	^{1}H Frequency (MHz)	^{13}C Frequency (MHz)	^{15}N Frequency (MHz)
9.36 T	400	100.8	40.8
11.7 T	500	126	51
14.04 T	600	151.2	61.2

Chemical Shift scale The frequency of absorption for two protons will not be the same (refer to factors affecting chemical shift). It is important to note that frequency of absorption will be different for same spin at NMR instruments with different magnetic strengths since ΔE depends on B; for example, if frequencies of two proton lines are at 400.0004 and 400.0008 MHz for a sample in a 400 MHz machine. These two lines will appear at 600.0006 and 600.0012 in a 600-MHz NMR machine. Thus, separation between two lines is 400 Hz on 400 MHz machine and 600 Hz on 600 MHz machine, exactly proportional to magnetic field strength.

To keep the position of peaks same on two NMR machines, the position in NMR peak is not expressed in terms of frequency/ wavelength/wavenumber, but in terms of chemical shift. The chemical shift is given by

$$\delta(ppm) = \frac{\nu - \nu_{REF}}{\nu_{REF}} \times 10^6$$

Thus, if reference frequency is 400 MHz at 400-MHz NMR machine, it will be 600 MHz at 600-MHz NMR machine.

$$\text{Spin 1 position in 400-MHz NMR} = \frac{400.0004 - 400}{400} \times 10^6 = 1 \text{ ppm}$$

$$\text{Spin 1 position in 600-MHz NMR} = \frac{600.0006 - 600}{600} \times 10^6 = 1 \text{ ppm}$$

Shielding constant

Every nucleus is associated with circulating electrons, which create a local magnetic field (B_e) opposite to the direction of external magnetic field. Thus, a nucleus experiences a smaller field than external magnetic field. In other others, spins are shielded from the external magnetic field.

$$\nu = \frac{1}{2\pi}\gamma(B_0 - B_e)$$

Shielding is proportional to the external magnetic field.

$$B_e = \sigma B_0$$
$$\nu = \frac{1}{2\pi}\gamma(B_0 - \sigma B_0)$$

σ is called shielding constant and depends on the structure of the molecule.

Effect of chemical environment on shielding constant/ chemical shift:

Electronegativity The surrounding electron density of proton shields the nucleus from external magnetic field. Electron-withdrawing substituents, when attached to the same or an adjacent carbon, de-shield the protons, and resonance occurs at a lower field or higher chemical shift. (Table 12.8)

Anisotropic Effect

Aromatic compounds The circulation of electrons around benzene ring produces a ring current in the presence of an external magnetic field. This causes the protons in the molecular plane to be deshielded, whereas protons above and below the plane are shielded. (Table 12.9)

Double bonds Similar to aromatic ring, there is a deshielding region in the plane of double bond. Thus, vinylic and allylic protons are downfield shifted. (Table 12.10)

Anisotropy of acetylenes Electron circulation around the triple bond π system takes place in such a way that protons

Table 12.8 Chemical shift of different types of methyl protons

Type of methyl proton	Structure	Chemical shift
Fluorides	H_3CF	4.0–4.5
Chlorides	H_3CCl	3.0–4.0
Bromides	H_3CBr	2.5–4.0
Iodides	H_3CI	2.0–4.0
Alkane	H_3CR	0.9–1.5

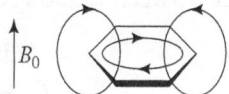

Fig. 12.11 Ring current around benzene ring in the presence of external magnetic field

Table 12.9 Chemical shift of aromatic compounds

Protons near to benzene ring		
Aromatic proton	Ar–H	6.0–8.5
Benzylic proton	Ar–C–H	2.2–3.0

attached to a triple bond experience a strong diamagnetic effect resulting in unusual upfield shift of C ≡ C–H signals. (Table 12.11)

Table 12.10 Chemical shift of protons near to carbon–carbon double bond

| Vinylic proton | C=C–H | 4.6–5.9 |
| Allylic proton | C≡C–CH3 | 1.7 |

Anisotropy of carbonyl groups Protons in the plane of carbonyl group are strongly deshielded due to the anisotropy of C=O. Thus, aldehydic protons and formyl protons of formate esters are deshielded and appear at high delta value. (Table 12.12)

Table 12.11 Chemical shift of protons near to carbon–carbon triple bond

| Acetylenic proton | Triple bond, C ≡C - H | 2.0 – 3.0 |

Hydrogen Bonding Effects on Chemical Shifts of –OH, NH and SH Protons Chemical shifts are affected by hydrogen bonding. There is a down-field shift of –OH, –NH, and –SH protons if they are hydrogen-bonded. The NH protons of carboxylic amides and sulfonamides are downfield shifted in comparison to their related amines due to their acidic nature. Acidic protons have higher tendency to form hydrogen bonding. Similarly, OH groups of phenols and carboxylic acids are downfield shifted in comparison to aliphatic alcohols. (Table 12.12)

Spin–Spin Interaction and Coupling The interaction between one spin with neighbouring spin results in the splitting of lines in the NMR spectrum into two or more components. The spacing between lines obtained due to spin–spin coupling is called coupling constant. It is independent of the applied field.

Table 12.12 Chemical Shift of protons attached to different functional groups

Group	Types of proton	Chemical Shift (in ppm)
Alcohols	RO(H)	3.5–3.8
Enolic	C = C – O(H)	15.0– 7.0
Phenolic	ArO(H)	4.2–6.8
Ethers	ROC(H)RR'	3.2–3.8
Acids	RCOO(H)	8.8–12.5
Acids	R'R(H)CCOOH	2.0–2.6
Esters	R'R(H)CCOOR	2.0–2.4
Carbonyl compounds	R'R(H)C C = O	2.0–2.7
Aldehydic	RC(H)O	9.0–10.0
Amino	RN(H$_2$)	1.0–5.0

For isolated spin, two states are possible, up (α) and down (β). Thus one transition is possible. For two isolated spins I and S, two transitions are possible, one for each spin. For two coupled spins (I and S), four different states (Both up ($\alpha\alpha$), both down ($\beta\beta$), first one up second one down ($\alpha\beta$), first one down and second one up ($\beta\alpha$)) are possible. Four different transitions are possible where $\Delta S = 1$. These are allowed transitions. There are two transitions where I goes from α to β (known as I_1 and I_2 transition) and there are two other transitions where S goes from α to β (known as S_1 and S_2 transition). The frequency difference between I_1 and I_2 (or S_1 and S_2) is called *coupling constant*.

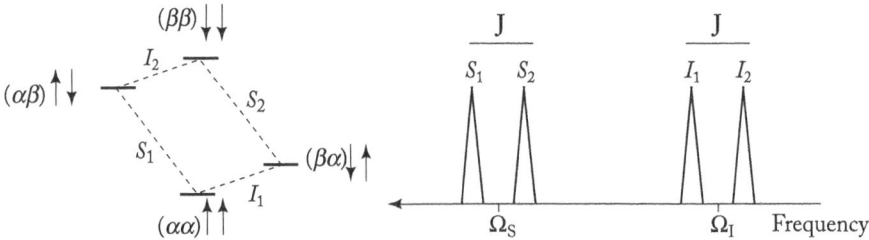

Fig. 12.12 Splitting of signals due to coupling between two spins

The multiplicity of the signal for a set of equivalent protons is dependent on the number of protons of the adjacent atoms. The multiplicity is given by $n + 1$, where n is the number of protons attached to adjacent atoms.

Coupling constants for different protons

(i) Germinal protons (protons on same C): 10–18 Hz
(ii) Vicinal protons (protons separated by 3 bonds): 0–12 Hz
(iii) *Cis* and *trans* protons in an alkene group: *trans*: 11–19 Hz; *cis*: 5–14 Hz
(iv) Aromatic protons: *ortho*: 7–10 Hz, *meta*: 2–3 Hz, *para*: 0–1 Hz

Difference from other spectroscopies

- The generation of the ground and excited NMR states requires the existence of an external magnetic field.
- NMR excited state has a life time that is of the order of 10^9 times longer than the life time of excited electronic states.
- According to Heisenberg Uncertainty principle, $\Delta E \, \Delta t \approx h/2\pi$. The longer an excited state exists, the narrower the line width. Thus, NMR is high resolution technique.
- Longer life time also facilitates multidimensional spectroscopy by allowing the resonance frequency information associated with one spin to be passed to another.
- The longer life time permits the measurement of molecular dynamics over a wide range of time scales.

Applications of NMR spectroscopy

Structure determination of organic molecule   There are several million organic compounds which are being synthesized and isolated, and therefore determination of the structure is often demanding. NMR is found to be most useful tool for the structural determination of various organic compounds.

Structure determination of proteins   In addition to the application of NMR in small molecules, the technique is well suited for determining secondary and tertiary structures of proteins. Various two-dimensional and three-dimensional pulse sequences are designed for the purpose.

Kinetics of the reaction   The potential of NMR in assigning the compounds are proven useful in studying kinetics of the reaction. The kinetic parameters are determined by observing the reactant's or product's NMR peak with time.

Differentiation between geometrical isomers   Chemical shift and coupling constants are two important parameters in solution state NMR. Of these, coupling constant is the strength of interaction between two spins via intervening electrons, and is dependent upon the geometry of the molecule. Hence, geometrical isomers, that is, *cis* and *trans* forms can be differentiated with the information of coupling constant.

Conformational analysis   The coupling constant is dependent upon the dihedral angle. Hence, the coupling constant plays a decisive role in determining different conformers of the molecules.

Analysis of polluted water   The characteristic signal of water is a singlet in ^{1}H-NMR. Impurity of any other compound, like oil and another pollutant, results in a complex ^{1}H-NMR spectrum. Hence, impurities can be determined quantitatively by analysing relative intensities of the pollutant peaks.

Food analysis   Food consists of different compounds, mainly primary and secondary metabolites. Primary metabolites contain organic acids, amino acids, and sugars and secondary metabolites are phenolic compounds, terpenes, and sterols. NMR, the potential candidate for structure determination of small and large molecules, is hence, useful in food analysis.

Drug–substrate and protein–drug reaction   Chemical shift of the nuclei is typically dependent upon its environment. When a drug is chemically bound to any substrate or protein then the change in environment can be witnessed through NMR in the form of change in the chemical shift. The non-bonding interactions do not change the chemical shift; however, the spatial interaction changes. The spatial interactions are determined with Nuclear Overhauser Spectroscopy (NOESY).

Magnetic Resonance Imaging (MRI): It is an imaging techniques based on the principles of NMR. It uses strong magnetic field, magnetic field gradient and radiofrequency waves to generate images. This technique is extensively used by medical professionals to get detailed images of internal organs and tissues inside the body. It is a non-invasive technique.

Uses of MRI:
  (a) It can be used for disease detection, diagnosis, and treatment monitoring.
  (b) It can be used for screening of tumor or cyst in various parts of the body.
  (c) It can be used for locating fracture and abnormalities in bones and joints.
  (d) It can be used for screening of anomalies in brain and spinal cord.
  (e) Functional MRI can be used to look at areas of activation within the brain.
  (f) Cardiac imaging can be used to look at the structure and function of heart.
  (g) It can be used in diagnosis of breast, prostate, and liver cancer.

12.6 MASS SPECTROMETRY (MASS SPEC OR MS)

In a simple mass spectrometer, the molecules (M) are bombarded with high-energy electrons. The molecules lose an electron forming a radical cation, a species with positive charge and one unpaired electron (M^+). The molecule or the radical cation generally breaks into fragments radical plus cation or radical cation plus small molecules like CO. Only cations are detected in MS, not the free radical. The cations formed are separated by magnetic deflection. The amount of deflection is proportional to the mass-to-charge ratio (m/z). The mass spectrum shows mass of each cation versus its relative abundance.

Molecular ion peak   The peak of radical cation corresponding to the mass of the original molecule is called molecular ion peak or parent ion peak. The molecular ion is usually the highest mass in the spectrum with some exception. Molecular ion peaks are generally absent in highly branched molecules.

Base peak   It is most intense peak in the spectrum. The base peak is not necessarily the same as the parent ion peak.

Instrumentation
  (i) Atomizer: It is of two types.
      Continuous atomizer: Samples are introduced continuously as in flame. Nebulizers are used to introduce sample into flame. Nebulizers introduce the sample in the form of aerosol.
      Discrete atomizer: Samples are introduced in time intervals using syringes or autosamplers.

Check Your Progress

36. How many peaks are observed in proton NMR spectrum of dihydrogen (H_2), ethane (C_2H_6) and benzene (C_6H_6)?

37. Define coupling constant? If a hydrogen nuclei say H_a is coupled with the other neighboring hydrogen nuclei say H_b, state the multiplicity and the relative intensity of lines in the signal of H_a.

38. If the gyromagnetic ratio of the ^1H is 9.68 MHz T^{-1}, what will be the ^1H resonance frequency at 3.1 T?

39. What will be the possible number of spin states of ^1H nuclei ($I = 1/2$) when placed in a magnetic field?

40. In a magnetic field (1.5 T), calculate the frequency of bare ^1H in MHz. Given is $g_N = 5.585$ and $\mu_N = 5.05 \times 10^{-27}$ J.T^{-1}.

41. In a ^1H NMR spectrum of an organic compound (molecular formula: $C_4H_9NO_2$), the signals are observed at delta 5.2 (broad, 1H), 4.16 (q, 2H), 2.73 (d, 3H) and 1.28 (t, 3H) ppm. Identify the compound.

Two most common used ionization sources are

Sources	Method of ionization
Electron Impact (EI)	High energy electrons
Chemical ionization (CI)	Gaseous ions

Mass analyzers for mass spectrometry: Two most common mass analyzers are

Table 12.13 Mass analysers for mass spectrometry

Type	Principle of analysis	Quantity measured
Magnetic sector	Generated ions are deflected in the magnetic field depending on *m/z* value	Momentum/charge
Double focussing	Ions are electrostatically focussed followed by deflection in the magnetic field depending on *m/z* value	Momentum/charge

Peaks with masses higher than molecular ion peak In nature, most elements exist as a mixture of isotopes. The presence of heavier isotopes leads to peaks that have masses that are higher than the parent ion peak. Most commonly observed high-molecular weight peaks are: (i) M+1 and (ii) M+2 peaks.

The separation and analysis of the fragments provides information about: (a) molecular weight and (b) structure of the molecule.

Some of the peaks in mass spectrum can be easily recognized.

(i) Odd molecular ion peak can indicate the presence of odd number of N.

(ii) Isotopes of C, N, and S contribute to M+1 peak. However, the contribution is significant for C. The number of carbon atoms in the molecule is calculated by dividing the relative abundance of M+1 peak by 1.1.

(iii) Equally intense M+ and M+2 indicates the presence of bromine.

(iv) If M+2 peak intensity is 1/3 rd of M peak intensity, it indicates the presence of chlorine.

(v) If M+2 peak is about 4 % of the M+ peak, it indicates the presence of sulphur.

(vi) A peak at 127 indicates the presence of iodine in the molecule.

Fragmentation Patterns

Fragmentation pattern in alkanes

(a) Loss of methyl: Peak will be observed at M-15

(b) Loss of ethyl: Peak will be observed at M-29

(c) Loss of propyl: Peak will be observed at M-43

(d) Loss of butyl: Peak will be observed at M-57

Thus, mass spectrum of pentane or 2-methyl butane shows peak at (i) 72 (molecular ion peak) (ii) 57 (M-15) (iii) 43 (M-29) and (iv) 29 (M-43)

Fragmentation pattern in alkenes Fragmentation typically forms resonance stabilized allylic carbocations. Peaks at 41, 55, 69 .....will be obtained.

Fragmentation pattern in aromatic compounds Peaks corresponding to 91 (benzylic cation) and 77 can indicate the presence of aromatic benzene ring in the molecule. Thus, mass spectrum of propyl benzene shows peaks at 120 (molecular ion peak), 91 and 77.

Fragmentation pattern in alcohol M-17 or M-18 can indicate the presence of alcohol group in the compound. Primary alcohol has a prominent peak at 31. 1-Propanol has molecular ion peak at 60 and base ion peak at 31. Small peaks at 42 and 43 are also observed.

Fragmentation pattern in amine The odd M+ peak points toward odd number of nitrogen in the molecule. Cleavage at α-position to form iminium ion is predominant. There will be a loss of alkyl group.

Fragmentation pattern in ether Fragmentation will result in oxonium ions. Molecular ion peaks with loss of methyl, ethyl, and propyl will be obtained.

Fragmentation pattern in aldehydes Peaks corresponding to M-1, M-29 may be obtained. Mass spectrum of 1-propanal shows peaks at 58 (molecular ion peak), 57 (M-1), and 29 (M-29).

Fragmentation pattern in aldehydes Fragmentation will lead to the loss of methyl, ethyl, and propyl.

McLafferty Rearrangement It is the rearrangement of the molecular and parent ion to form a radical ion and a neutral molecule (such as alkene). In this rearrangement, the transfer of the hydrogen from a part of the molecule to another takes place via a six-membered cyclic transition state.

An American chemist F. W. McLafferty observed that the carbonyl compounds with hydrogen at γ position have an intense peak quite far from molecular ion peak. This is due to the β-cleavage with transfer of hydrogen at γ position to C=O to give a neutral alkene molecule and a radical cation.

Rearrangement in ketones:

Rearrangement in esters

Applications of Mass Spectroscopy Mass spectroscopy possesses both qualitative as well as quantitative applications. Some of the important applications of mass spectroscopy are listed below.

(i) *Determination of precise atomic mass of an element*: An element is a mixture of its isotopes, with each isotope having different abundance. Thus, the precise mass of an element is in fraction. Mass spectroscopy helps in determination of precise mass of an element.

(ii) *Determination of relative abundance of isotope of an element*: The relative percentage abundance of an isotope can be obtained by comparing the peak height corresponding to isotope relative to maximum intensity peak of an element.

(iii) *Identification of molecules from fragmentation pattern*: Mass spectroscopy is particularly useful in identification of molecules having same formula with different structures. Thus, butane and isobutane can be differentiated based on peaks at 29 and 27, respectively.

(iv) *Determination of molecular formula of a compound*: Analysis of molecular ion peak and fragments provides information about molecular formula of the compound.

(v) *Determination of molecular weight of other molecules*: It is also used in the determination of molecular weight of other small bio-molecules like amino acids, sugars, alcohols, sugar phosphates, fatty acids, vitamins etc. Here, mass spectroscopy coupled with liquid chromatography gives better results as loss of bio-molecules is reduced.

(vi) Application in *proteomics*: Mass spectroscopy is widely used to identify and quantify protein in larger set of proteins expressed in cell.

(vii) *Application in food industry*: Food can be contaminated by various types of microbial pathogens, biotoxins, chemical pollutants, pesticides used for preservation of food and many more. Mass spectroscopy is a powerful tool in detecting these various types of contaminants in food stuffs.

(viii) *Application in clinical research*: Mass spectrometers are used as very powerful tool in clinical research such as (a) identifying microbial proteins, (b) drug screening and (c) analysis of steroid hormones.

(ix) *Environmental studies*: There are number of contaminants in environment which harm humans and animals like industrial wastes, landfills, pesticides and pharmaceutical wastes. Mass spectroscopy coupled with chromatography is a sensitive, strong, fast and cost-effective analytical technique to quantify these contaminants.

(x) Space *exploration*: Mass spectroscopy has a very important application in quantitative analysis of plasma during space missions.

12.7 FLAME PHOTOMETRY

Flame photometry is a branch of atomic absorption spectroscopy, developed by Bowling Barnes, David Richardson, John Berry, and Robert Hood in the 1980s. It is also known as *flame emission spectroscopy* as it is based on the measurements of the emitted photons produced when the sample is introduced into the flame. The wavelength of the emitted light helps in the identification of the element present in the sample, while the intensity of the flame determines the amount of the element present in the sample. Flame photometry has become a very important tool in analytical chemistry for the determination of concentration of the certain metal ions like sodium, potassium, caesium and so on.

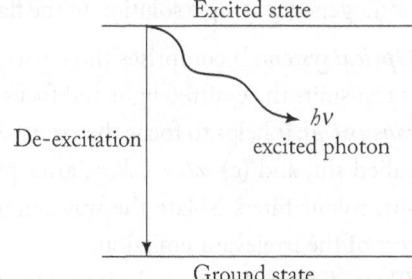

Fig. 12.13 Return of atoms from excited state to ground state

Principle In flame photometry, the solution of salts metals, mainly alkali and alkaline earth metals, are introduced to the flame, where they first vaporize or sublime into gaseous form and then dissociate to constituent atoms. Some of these atoms get excited to higher energy levels, but being unstable, they return back to the ground state along with the emission of photons (Fig. 12.13).

The wavelengths of these emitted photons generally lie in the visible region of the spectrum. Each metal (alkali and alkaline earth metal) has their characteristic wavelength tabulated in Table 12.14.

Within certain concentration ranges, the intensity of the emitted radiation is proportional to the number of the atoms coming back to the ground state, which in turn, is proportional to the concentration of the sample (Scheibe–Lomakin equation).

$$I = kC$$

where I is the intensity of the emitted light, C is the concentration of the sample, k is the proportionality constant.

Check Your Progress

42. State the principle of mass spectroscopy.
43. What is the use of analyser in a mass spectrometer?
44. In a mass spectrum of an unknown compound, a molecular ion peak at $m/z = 181$ is observed with a relative intensity of 100. If the relative intensities of $m + 1$ peak and $m + 2$ peak are 8.06 and 95, respectively, identify the molecular formula of the compound.
45. What is McLafferty rearrangement? Explain with an example.
46. What is difference between molecular ion peak and base peak?
47. What is the fragmentation pattern in aromatic compounds?

Instrumentation Flame photometer is a simple instrument (Fig. 12.14) consisting of the following basic components:

Flame source: In flame photometry, there is no source of light. A burner is used as the flame source to provide the energy for excitation of electrons. The flame in the burner is maintained at a constant temperature. The flame temperature is an important factor in flame photometry which depends on the proportion of oxidant and fuel. The oxidants used are mainly air, oxygen, or nitrous oxide. A number of flame burners are available depending upon the temperature required for excitation process (Table 12.15).

Nebulizer or mixing chamber: It helps the flow of homogeneous sample solution to the flame at a steady rate.

Optical system: It comprises three parts. (a) *convex mirror*: it transmits the emitted light and focuses it to the lens; (b) *convex lens*: it helps to focus the emitted light on to a point called slit; and (c) *colour filters*: after passing through the slit, colour filters isolate the wavelength of interest from rest of the irrelevant emission.

Photo-detector: It detects the intensity of emitted radiations which are then converted into electric signals.

Working of flame photometer: The working overview of the flame photometer is shown in Fig. 12.14. In the first step, the sample solution is dried. This is called desolvation, where sample particles are dehydrated using flame and the solvent is evaporated leaving behind solid particles

Table 12.14 Characteristic emission wavelength of some metals

Metal	Emitted wavelength (λ) nm
Antimony (Sb)	253
Magnesium (Mg)	285
Copper (Cu)	325
Nickel (Ni)	355
Iron (Fe)	372
Calcium (Ca)	423
Barium (Ba)	455
Sodium (Na)	586
Lithium (Li)	671
Potassium (K)	766

Table 12.15 Different flame burners with their characteristic temperature

Flame burner Fuel–Oxidant mixture	Temperature (°C)
Natural gas–Air	1700
Propane–Air	1800
Hydrogen–Air	2000
Acetylene–Air	2300
Hydrogen–Oxygen	2650
Acetylene–Nitrous oxide	2700
Acetylene–Oxygen	3200
Cyanogen–Oxygen	4800

(*vaporization*). These solid particles are then introduced in the flame, leading to the formation of gaseous molecules. They then undergo thermal dissociation and form gaseous atoms or ions (*atomization*). These atoms or ions at ground state absorb energy from the flame and are excited to higher energy state (*excitation*). Being unstable, they come back to the ground state with emission of photons of a characteristic wavelength (*emission*). The intensity of the emitted photon is directly proportional to the concentration of the sample solution.

Applications: Flame photometer exhibits both quantitative and qualitative applications which are as follows.

(i) Flame photometer detects the emitted radiations of characteristic wavelengths which help in determining the presence of alkali or alkaline earth metals in soil sample.

(ii) The flame test analysis of soil sample helps in choosing an appropriate fertilizer for the soil cultivation.

(iii) This method of analysis can also be used for the determination of the concentration of sodium and potassium ions in body fluids, muscles, and heart.

(iv) Flame photometry is also used for analysis of soft drinks, fruit juices, and alcoholic beverages.

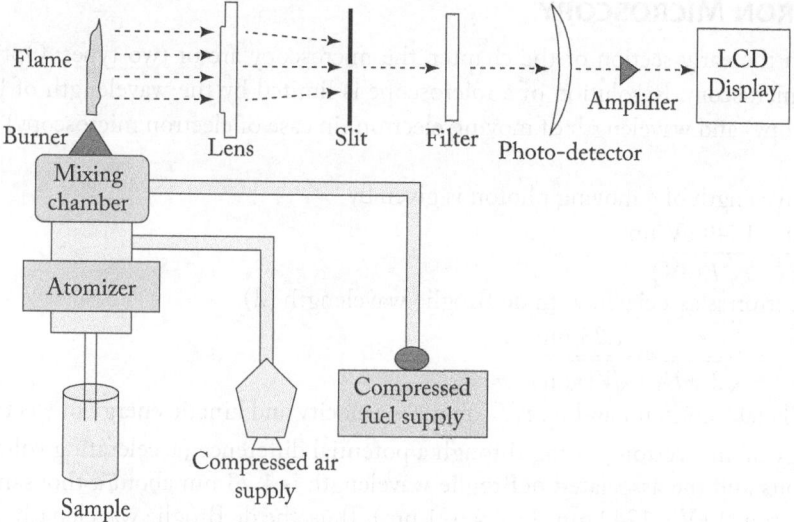

Fig. 12.14 Diagramatic representation of flame photometer

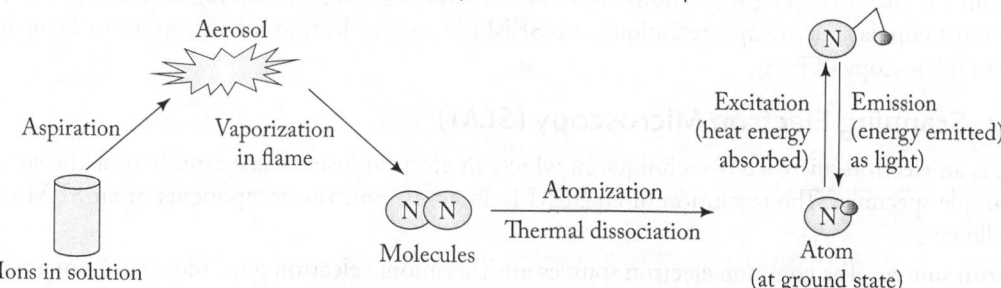

Fig. 12.15 Pictorial representation of working of flame photometry

Advantages

(i) Flame photometry is very simple and inexpensive analytical technique.

(ii) It is an easy, reliable, and convenient technique for the determination of alkali and alkaline earth metals.

(iii) It is sensitive to concentrations in the range of parts per million (ppm) to part per billion (ppb).

Disadvantages Flame photometry has a few disadvantages also.

(i) It cannot be used for measurement of metals in solution.

(ii) Solutions with higher concentrations cannot be analysed by this technique.

(iii) A standard solution with known concentration is required to analyse the emission results.

(iv) The detection of non-metals such as carbon and halides cannot be done using flame photometry due to their non-radiating nature.

Check Your Progress

48. What is the principle of atomic absorption spectroscopy? Is atomic absorption spectroscopy also known as atomic flame photometry?

49. What is function of flame in flame photometry?

50. What does an atomizer do in the emission system of the flame photometry?

51. Name some fuels used in Flame photometry.

52. What are the advantages and disadvantages of flame photometry?

12.8 ELECTRON MICROSCOPY

As described in the early section of the chapter, the microscopy are of two types: Optical microscopy and Electron microscopy. Resolution of a microscope is limited by the wavelength of light (in case of optical microscopy) and wavelength of moving electron (in case of electron microscopy) used to interact with sample.

de Broglie wavelength of a moving photon is given by

$$\lambda = \frac{hc}{E} = \frac{1240 \text{ eV.nm}}{E(\text{eV})}$$

A moving electron is associated with de Broglie wavelength (λ).

$$\lambda = \frac{h}{mv} = \frac{h}{\sqrt{2\,mE}} = \frac{1.23 \text{ nm}}{\sqrt{V(\text{volt})}}$$

Where h is Planck's constant and m, v, E are mass, velocity and kinetic energy of electron, respectively

Kinetic energy of an electron passing through a potential difference (accelerating voltage field) of 1 V is 1 electron volts and the associated deBroglie wavelength is 1.23 nm about a thousand times smaller than a 1 eV photon (1 eV = 1242 nm, 2 eV = 621 nm). Thus, the de Broglie wavelength of the electron is much smaller than that of light (photons) and thus is useful in getting a much higher resolution of image.

The most popular microscopic techniques are SEM (Scanning electron microscopy) and Transmission electron microscopy (TEM).

12.8.1 Scanning Electron Microscopy (SEM)

SEM is an electron microscopy technique in which an electron beam scans rapidly over the surface of the sample specimen. The resolution of an SEM is about 10 nm. The components of an SEM include the following.

Electron source: The common electron sources are Thermionic electron guns (emitted from a tungsten or lanthanum hexaboride (LaB_6) cathode) and cold field emission guns.

Condenser lens: The beam of electrons from the electron gun is focused into a small, thin, coherent beam (focal spot sized 1–5 nm) by the use of the condenser lens.

Condenser aperture: It is used to control the fraction of the beam which is allowed to hit the specimen. It therefore helps to control the depth of field.

Objective lens: The electron beam passes through scanning coils in the objective lens that deflect the beam in a raster fashion over a rectangular area of the sample surface.

As the primary electrons strike the surface they are inelastically scattered by atoms in the sample. With the scattering the primary electron beam effectively spreads a teardrop-shape that extends less than 100– 5000 nm into the surface.

The interaction of the primary electron beam with the atoms near the surface causes the emission of particles at each point in the raster as shown in Fig. 12.16.

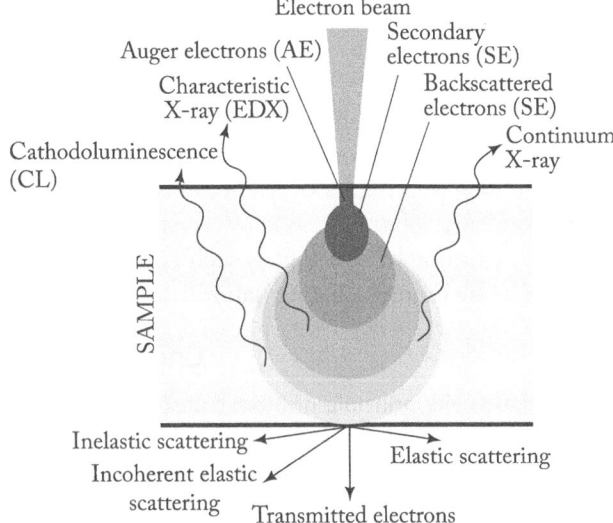

Fig. 12.16 Interaction of electrons with matter

(i) Secondary electrons give information about topography.
(ii) Back-scattered electrons give information about composition and topography. Atoms with higher atomic number backscatter more electrons than a lower atomic number atom.
(iii) Characteristic X-rays: The displacement of inner shell electron by collision with an incoming electron results in an ionized atom. The ionized atom returns to the ground state when an outer shell electron moves to inner shell by emission of X-ray. The characteristics of emitted X-ray gives information about the chemical composition of the sample.
(iv) Auger electron: Sometimes the emitted energy is transferred to the outermost electron, causing it to eject from the atom. Such an electron is called auger electron.

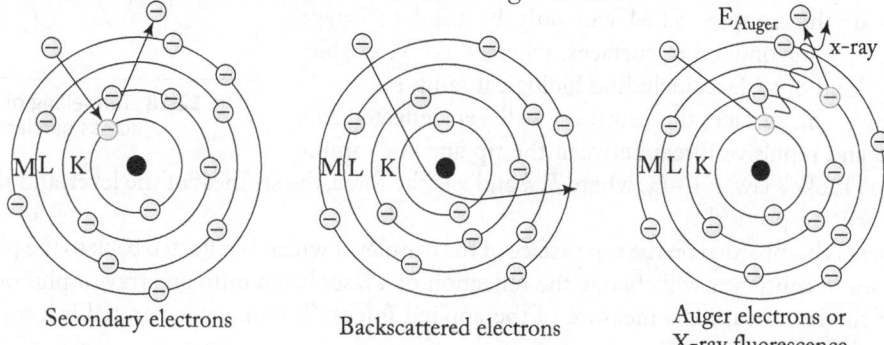

Secondary electrons   Backscattered electrons   Auger electrons or X-ray fluorescence

Fig. 12.17 (a) Secondary electrons, (b) backscattered electrons, and (c) characteristic X-rays

The relative number of emitted electrons and radiation are translated to brightness at each equivalent point on a cathode ray tube. The final picture is a magnified image of the specimen.

SEM gives characteristic information about topography, morphology, composition, and crystallographic information.

12.8.2 Transmission Electron Microscopy

Transmission electron microscopy (TEM) is an electron microscopy technique in which the wave-like property of a moving electron and its interaction with a sample is utilized. The transmitted electron after passing through a sample forms an image. The image is magnified and focused on an imaging device. The resolution of a TEM is about 0.2 nm. The components are similar to that of SEM.

Applications
(i) TEM can be used to get information about surface topology, morphology, composition and crystalline nature of materials, polymers, and biomaterials.
(ii) TEMs can be used in semiconductor analysis and in the production and manufacturing of computer and silicon chips.
(iii) Transmission electron microscopy (TEM) is one of the most powerful techniques to measure particle size, size distribution, sample homogeneity, morphology of the nanoparticles.
(iv) It is used to study the morphology of protein aggregates; they can be utilized to look at the arrangement of protein molecules in cell membrane.
(v) It can be used to visualize the organization of molecules in viruses.
(vi) It is used to image the interior of cells.

12.8.3 Scanning Probe Microscope

STM It is a type of scanning probe microscope which uses tunnelling property of electron to image the sample. Electrons tunnel to the forbidden region due to the wave-like nature of electrons.

In scanning tunneling microscope (STM), a very sharp metal wire tip scans over a surface. When the tip is very close to the surface of a conductor, tunneling of electrons through the air gap takes place. A piezoelectric rod is attached to the tip and voltage of rod is altered to maintain a constant distance between the tip and the surface. Measurement of the changes required in this voltage allows to obtain a three-dimensional picture of the material surface. Apart from topographical properties of materials, STM also provides information about electrical properties which is needed for understanding of the behaviour of microelectronic devices.

AFM: Atomic force microscopy (AFM) is a type of scanning probe microscope which makes use of a cantilever with a sharp probe to scan the sample. STM can only be used to image conducting or semiconducting surfaces, whereas AFM enables imaging any kind of surface, including biological samples.

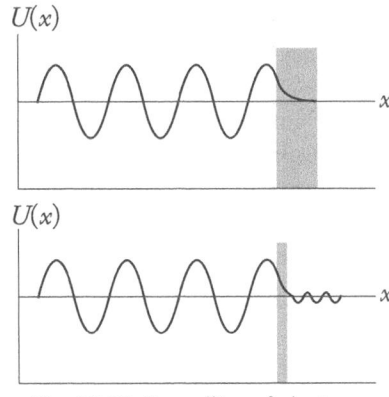

Fig. 12.18 Tunnelling of electrons across barrier

On scanning the surface, the cantiliver will get deflected due to attractive and repulsive forces between the tip and the sample according to Hooke's law, $F = -kx$, where F, k and x is the force, the stiffness of the lever, and the distance the lever is bent, respectively.

A laser beam is bombarded on the top surface of the cantilever which is reflected back to the photodiodes. The deflection in cantilever will change the reflection of a laser beam onto an array of photodiodes. The variation of the laser beam is a measure of the applied forces. The image generated is a topographical illustration of the sample surface.

12.9 X-RAY DIFFRACTION

X-ray diffraction is one of the best analytical techniques for the determination of the crystal structure of organic and biomolecules. X-rays are discovered by a German physicist W. Roentgen in 1895. X-rays are the electromagnetic radiations whose wavelengths lie in the range 0.02 Å – 100 Å. Since the wavelengths of X-rays lie at atomic levels and are smaller than that of visible rays, they possess more energy and higher penetration power. The penetration power depends upon the density of the matter; therefore, these are very good in exploring atomic structures. X-rays are bundles of separate waves which interact with each other either in a constructive way (waves with same phase) or a destructive way (waves are out of phase). First x-ray diffraction pattern of a crystal was reported by Knipping and von Laue in 1914.

Principle A crystalline substance consists of atoms arranged in parallel rows separated by a specific distance called spacing. When an intense beam of X-ray strikes the crystal, it diffracts it depending upon atomic arrangement in the crystal and orientation. X-ray diffraction patterns are produced when X-rays scattered from atoms in each set of lattice planes of a sample interfere constructively. For the exact structure of a crystal, the X-ray diffraction patterns need to be recorded at different orientations. The comparison of the entire patterns will give a true structure of a crystal.

Instrumentation and Working The set-up of a typical XRD instrument is shown in Fig. 12.19. In this technique, X-rays are generated by a cathode ray tube and passed through the primary or incident beam optics. The primary optics consists of a number of slits, where first the generated X-rays are converted into monochromatic rays and then collimated to concentrate (a set of parallel rays). When these

Check Your Progress

53. What is the advantage of electron microscope over light microscope?
54. State the principle of SEM.
55. What are the factors upon which the intensity of the secondary electrons depends?
56. How does TEM differ from SEM?

concentrated monochromatic X-rays fall on the crystal sample, it produces a constructive interference. Now, the diffracted beam is passed through the secondary optics (a set of slits and a single crystal monochromator) and made to reach the detector, where the diffracted X-rays are processed and counted.

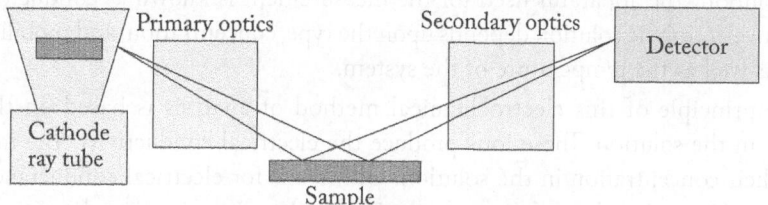

Fig. 12.19 Instrumental set-up of XRD

The theory behind the determination of crystal structure from its diffraction pattern is given by W. H. Bragg and his son W. L. Bragg. They proposed a law which correlates the incident angle with maximum diffraction intensity (θ) to the wavelength of the incident X-ray (λ) and the inner atomic spacing (d) of the crystal lattice (Fig. 12.20).

$$n\lambda = 2d \sin \theta$$

where n is an integer and known as the order of reflection.

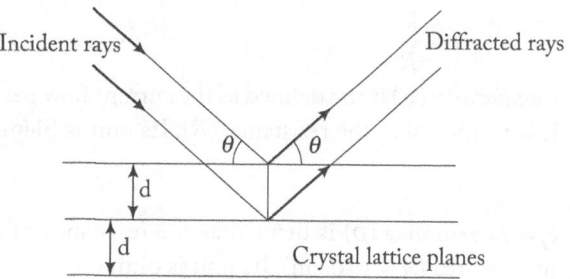

Fig. 12.20 Illustration of Bragg's law

The number of scans with a wide range of 2θ angles will help in extracting the true structure of the sample. Each mineral has a unique set of d-spacings; therefore, comparison of the resultant d-spacing (extracted from the diffraction pattern) with that of standard reference patterns makes the identification of the sample mineral easy.

Applications   It is the most widely used analytical technique for identification of the unknown crystalline samples. This technique is highly useful in studies of geology, material science, and environmental studies as they characterize the crystalline materials and identify the fine-grained minerals like clays. It is also used to measure sample purity.

This technique is also widely used in medicinal field. XRD has been extensively used to get the structure of proteins and their complexes. This is also used to study the rapid biological and chemical processes.

Limitations   In-spite of being a powerful and rapid technique for the identification of unknown sample with minimal sample requirement, XRD has also some limitations. Only homogeneous and single phase material can be identified properly by XRD. A mixture of minerals (heterogeneous sample) may not give accurate peak data. The sample must be powdered fine enough to get good results. Also, the analysis cannot be done at a fast rate, that is, degrees/minutes. For high angle reflections, the resultant peaks may overlay which disturbs the analysis. Non-crystalline materials like glass are not appropriate sample for XRD study; only crystalline material can be used.

Check Your Progress

57. State the factors that govern the X-ray techniques.
58. Which one shows the higher extent of intensity of scattering of x-ray from the ions, Na^+ or Li^+?
59. If the second order reflection from the crystal takes place with x-ray ($\lambda = 225$ pm) at an angle of 25.3°, what will be the inter layer distance, d?

12.10 CONDUCTOMETRY

Conductometry is an electrochemical technique of analysis used to measure the electrical conductance of an electrolyte solution. The apparatus used for the measurement is known as conductometer. Electrical conductivity of an electrolytic solution depends upon the type, concentration, and mobility of ions present in the solution as well as the temperature of the system.

Principle The principle of this electrochemical method of analysis is based on the movement of the ions present in the solution. These ions produce the electrical conductivity. The movement of ions depends upon their concentration in the solution. Ohm's law for electrical conductance states that the strength of current (I) passing through a conductor is directly proportional to the potential difference or EMF (V) and inversely proportional to the resistance of the conductor (R).

$$I = \frac{V}{R}$$

Conductance (C) is the defined as the current flow per unit area of the conductor per unit potential applied. It is reciprocal to the resistance (R). Its unit is Siemens (S) or ohm^{-1} or mho.

$$C = \frac{1}{R}$$

Specific resistance (ρ) is defined as the resistance of a substance per unit length ($l = 1$ cm) and per unit surface area ($A = 1$ sq.cm). Its unit is ohm.cm

$$\rho = \frac{AR}{l} \text{ or } R = \rho\frac{l}{A}$$

Specific conductivity (κ) is the conductivity of a substance per unit length and per unit surface area. Its unit is mhos.cm^{-1}.

$$\kappa = \frac{1}{\rho} = \frac{1}{R}\frac{l}{A} = C\frac{l}{A}$$

where, l/A is known as cell constant and cannot be easily determined accurately. Therefore, the indirect method is employed for its determination:

$$\text{Cell constant,} \frac{l}{A} = \frac{\text{Specific conductance}}{\text{Observed conductance}}$$

Equivalent conductivity (Λ) is defined as the specific conductance of a solution containing one equivalent of electrolyte per liter of the solution. If N is the normality of the solution,

$$\Lambda = \frac{1000 \ \kappa}{N}$$

Molar conductivity (Λ_m) is the specific conductance of the solution containing one mole of electrolyte per liter of the solution. If M is the molarity of the electrolyte

$$\Lambda_m = \frac{1000 \ \kappa}{M}$$

The conductance of the solution depends on various factors. It depends on the size, molecular weight and charge carried by the ions present in the solution. Conductance of the solution increases with increase in concentration of the ions. Also it increases with increase in temperature due to increase in the mobility of the ions. Since conductance is related to cell constant (i.e. l/A), it also depends upon the size of the electrodes.

Instrumentation The apparatus used to measure conductance of the solution is called conductometer. The diagrammatic representation of a typical conductometer is shown in Fig. 12.21. It consists mainly of three parts.

1. Current source: A high-frequency AC supply is employed in the circuit of conductometer. DC source is not suitable to use as it will polarize the electrode resulting in high resistance.
2. Conductivity cells: These are made up of pyrex or quartz and fitted with two platinum electrodes. To maintain constant temperature these are placed in a vessel filled with water. There are three different types of cells available. 1. Cell with wide mouth; 2. Cell designed for precipitation reactions and 3. Dip type cells.
3. Electrodes: Electrodes are platinum sheets of surface area of 1 cm^2 placed at a distance of 1 cm apart. The surface of the platinum electrodes is coated with platinum black to avoid the polarization and increase the surface area. This coating of platinum electrode was done by dipping the electrode in a solution of 2–3 % choloroplatinic acid and 0.02–0.03 % of lead acetate followed by the passage of current for about 20 minutes. If concentration of the solution is very low, then electrodes are largely and closely packed.

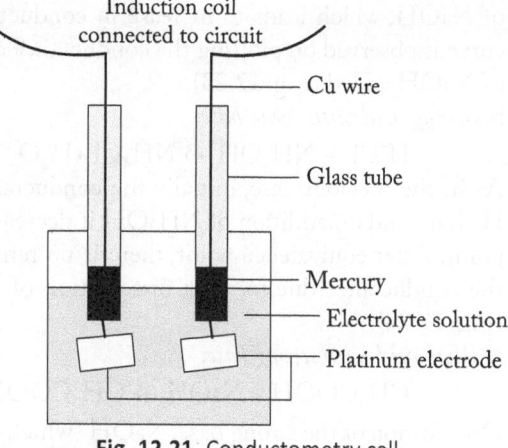

Fig. 12.21 Conductometry cell

Procedure The conductance cell is immersed in the sample solution whose conductance is to be determined. The circuit employed for the measurement is known as Wheatstone bridge (Fig. 12.22). The measurement is to be carried at constant temperature using a thermostat. On connection of cell with the resistance box, the current (AC) passes through the cell using induction coils. The conductivity of the solution is measured using the relation:

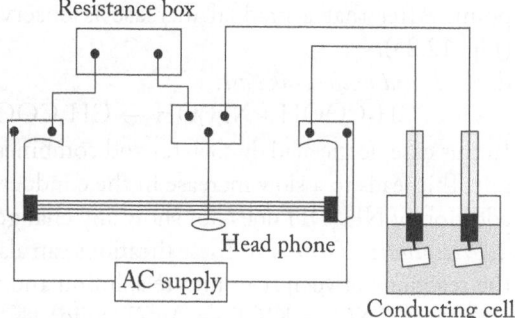

Fig. 12.22 Schematic representation of set up for measurement of conductance by conductivity cell

$$\text{Conductivity of solution} = \frac{1}{\text{Resistance of solution}}$$

Conductometric Titrations: Conductometric measurements are used for determination of the end points in various titrations. In this method, the observed conductance values are plotted against the volume of titrant added (linear function). Initially, on addition of titrant conductance decreases to a certain point and then starts increasing. This point of divergence is called equivalence point or end-point of the titration. The different types of conductometric titrations are as follows.

1. *Acid-base Titrations*: In this type of titration, the conductance of hydrogen ions and hydroxyl ions are compared with the conductivity of the solution. Again, there are different combinations of acid/base pair.

a. *Strong acid with strong base:* For example:

$$H^+Cl^- + Na^+OH^- \rightarrow H_2O + Na^+Cl^-$$

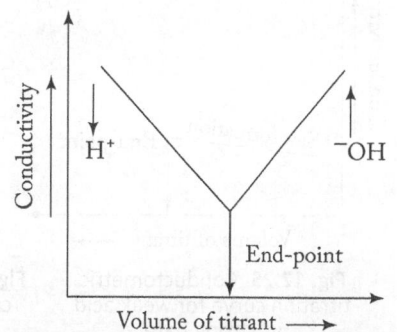

Fig. 12.23 Conductometric titration curve for strong acid vs strong base

Initially, the solution shows high conductance due to H^+ ions obtained due to complete dissociation of acid. On addition of NaOH, hydroxyl ions neutralize hydrogen ions to form water which reduces the conductance. After equivalence point, the concentration of free hydroxyl ions increases on addition of NaOH, which leads to increase in conductance. A V-shaped curve is observed on plotting the conductance against the volume of NaOH added (Fig. 12.23).

b. *Strong acid with weak base*:

$$H^+Cl^- + NH_4OH \rightarrow NH_4Cl + H_2O$$

As in the previous case, initially the conductance is high due to H^+ ions, and on addition of NH_4OH it decreases till equivalence point. After equivalence point, there is no remarkable change in the conductance due to weak dissociation of the base NH_4OH (Fig. 12.24).

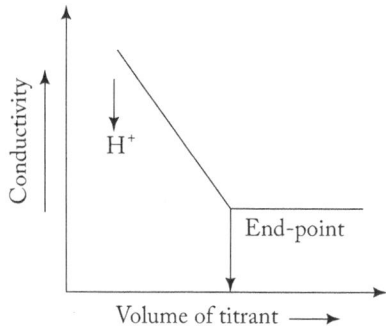

Fig. 12.24 Conductometric titration curve for strong acid vs weak base

c. *Weak acid with strong base*:

$$CH_3COOH + NaOH \rightarrow CH_3COONa + H_2O$$

On addition of the strong base, NaOH (which itself dissociated to give OH^- ions), acetic acid dissociates to produce H^+ ions. Here, on addition of titrant, the conductance increases with a slow rate till equivalence point. After that a gradual increase is observed in the conductance on further addition of NaOH (Fig. 12.25).

d. *Weak acid with weak base*:

$$CH_3COOH + NH_4OH \rightarrow CH_3COONH_4 + H_2O$$

In this case, acetic acid dissociates and combines with ammonium hydroxide to form ammonium acetate salt. This leads to a slow increase in the conductance. After equivalence point, being a weak base, further addition of NH_4OH does not show any change in conductance (Fig. 12.26).

2. *Precipitation Titrations*: These titrations can also be monitored by conductometeric method; for example, the reaction between siver nitrate solution and potassium chloride solution.

$$Ag^+NO_3^- + K^+Cl^- \rightarrow AgCl \text{ (solid)} + K^+NO_3^-$$

Initially, the conductance is high due higher conductance of silver ions as compared to that of potassium ions. On reaction of two solutions, the conductance decreases till equivalence point. After that, there is an increase in conductance due to increase in potassium and chloride ions on further addition of potassium chloride (Fig. 12.27).

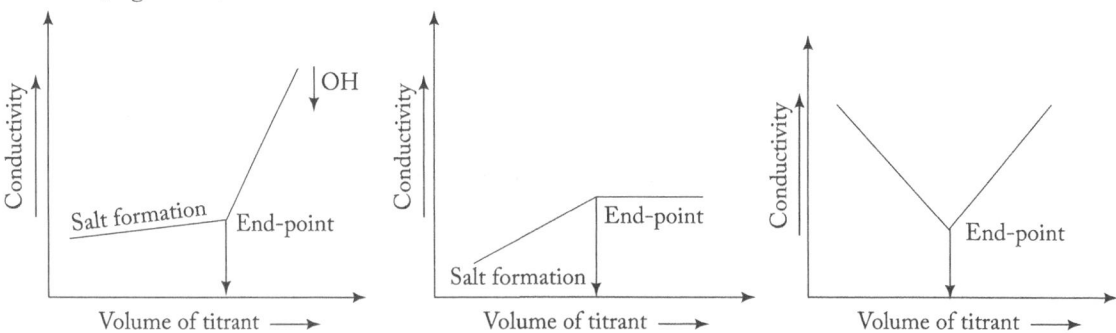

Fig. 12.25 Conductometric titration curve for weak acid vs strong base

Fig. 12.26 Conductometric titration curve for weak acid vs weak base

Fig. 12.27 Conductometric titration curve of $AgNO_3$ vs KCl

Applications Conductometry is an electrochemical analytical method which possesses a number of applications in chemistry and chemical industries.

1. This technique can be used for the determination of basicity of acids.
2. It is a simple analytical technique for the determination of solubility of sparingly soluble salts such as $BaSO_4$, $PbSO_4$, $AgCl$, and AgI. For the measurement, the specific conductance of a saturated solution of salt (κ_s) is measured. From the known value of specific conductance of water (κ_w), the specific conductance of the salt (κ_{salt}) can be determined as: $\kappa_{salt} = \kappa_s - \kappa_w$

   Let c be the equivalent concentration (equivalents per litre) or solubility of the salt, then
   $$\Lambda_m = \frac{1000 \ \kappa_{salt}}{c}$$
   As we know, salt is sparingly soluble; the solution is treated as the solution at infinite dilution. Then, $\Lambda_m = \Lambda_0$. Λ_0 is limiting equivalent conductance and its value is available in literature. Therefore, the concentration of the sparingly soluble salt is measured as:
   $$c = \frac{1000 \ \kappa_{salt}}{\Lambda_m}$$
3. Conductometry can be used for determination of purity and ionic product of water.
4. This technique is also very useful in quantitative analysis of number of compounds.

Advantages/Disadvantages   Conductometry is highly selective and an appropriate analytical technique for dilute solutions. For calculating the results of the titrations, there is no need of either specific conductivity of the solution or any indicator. In conductometric titrations, the end point can be easily determined by plotting the graph. However, it is less accurate and satisfactory when compared to other methods because solutions with high concentrations cannot be analysed using conductometric titrations.

12.11.1 THERMOGRAVIMETRIC ANALYSIS

Principle: In this technique of thermal analysis, mass of a sample is monitored which may change with change in the temperature due to the various processes like decomposition, sublimation, vaporization, absorption or desorption. For that reason, TGA is used for the analysis of polymers, composites, films, fibers, coatings, and paints.

Instrumentation: The instrument used for TGA analysis is known as Thermobalance (Fig. 12.28). Thermobalance consists of a furnace, temperature programmer, recorder, and an electronic microbalance.

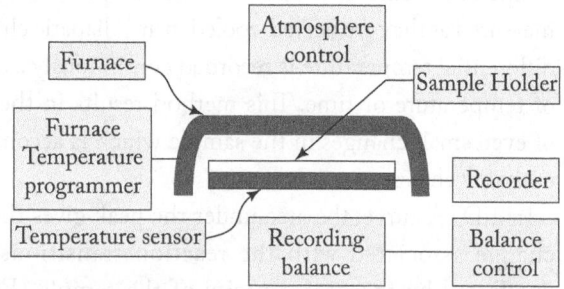

Fig. 12.28 Block diagram of a thermobalance

TGA curve is the plot of percentage change in mass versus temperature or time as shown in Fig. 12.29.

In Fig. 12.29, T_i represents the temperature at which the mass change starts and T_f (final decomposition temperature) is the lowest temperature at which it gets completed.

TGA curves may have different shapes depending on one step or multi-step decomposition. A S-shaped curve shows single stage decomposition whereas a stair-case shaped curve shows multi stage decomposition. If no mass change is observed over the entire range of the temperature, it suggests that the material is stable in that range of temperature. Evaporation of volatile products during desorption/drying or polymerization reactions results in TGA curve showing a large mass loss followed by a plateau.

Check Your Progress

60. Define conductivity. What are the factors upon which conductance of an electrolytic solution depends?
61. What is the effect of dilution on specific conductance?
62. What will be the molar conductivity of a 0.5M solution of an electrolyte whose resistivity is 50 ohm. cm? (cell constant is 1)

Usually the reactions show weight loss in TGA experiments, but there are some reactions which shows weight gain like adsorption or absorption or surface oxidation reactions in the presence of an interacting atmosphere. There are some reactions which show both weight gain as well as weight loss depending upon the reaction conditions, for example, solid–gas reactions and magnetic transitions.

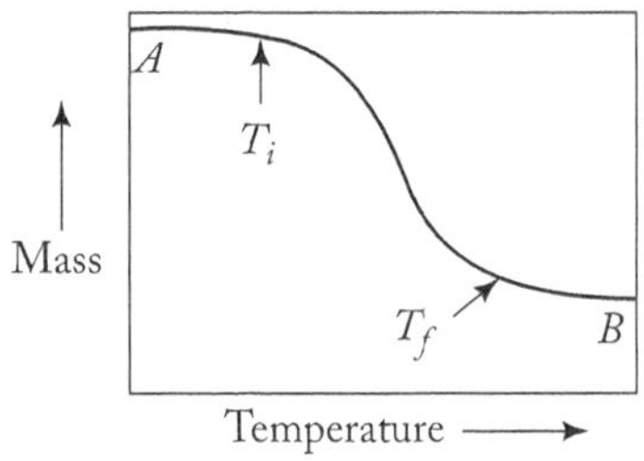

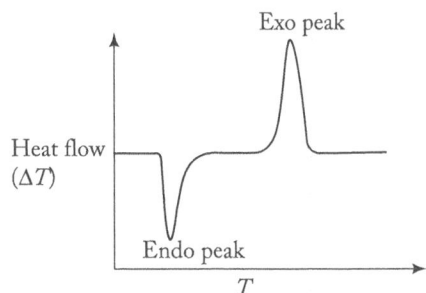

Fig. 12.29 The plot of mass change with temperature

Applications TGA possess many applications.

(i) It is used to compare the thermal stability of related materials which help in elucidating the decomposition mechanisms.

(ii) TGA is also used for materials characterization as it can be used to fingerprint materials for identification or quality control.

(iii) Compositional analysis can be done using TGA curves.

(iv) TGA curves also help in studying kinetic aspects of many reactions.

(v) Corrosion studies can be done using TGA, as it provides a means of studying oxidation or some reactions with other reactive gases or vapours.

12.11.2 Differential Thermal Analysis

Differential thermal analysis (or DTA) is another analytical technique for thermal analysis. In this thermal technique, temperature difference between the sample and reference is measured as they are heated/cooled in an adiabatic chamber. The differential temperature is recorded continuously as a function of temperature or time. This method results in the detection of even small changes in the sample which is accompanied by enthalpy change.

Fig. 12.30 A DTA curve showing exothermic and endothermic transition peaks

In a DTA curve, the area under the peak gives the enthalpy change associated with the reaction transitions and it is unaffected by the heat capacity of the sample (Fig. 12.30). This method is highly responsive to endothermic and exothermic processes such as phase transitions, dehydration, and decomposition, redox, or solid-state reactions. Therefore, the shape and size of the peak provide information about the nature of the test sample; for example, a sharp endothermic peak implies phase changes transition such as melting, fusion, etc., while a broad endothermic peak is obtained as a result of dehydration reactions.

In a typical DTA curve, an endothermic curve is usually obtained as a result of physical changes, whereas chemical reactions like oxidative reactions result in exothermic peaks.

Instrumentation A typical DTA instrument (Fig. 12.31) consists of the following components:

- A thermocouple assembly with a sample and reference holder
- Furnace to heat the sample
- Temperature programmer to control the furnace temperature
- Atmosphere control system to maintain the proper atmosphere in the furnace and the sample holder
- A DC amplifier
- Recorder

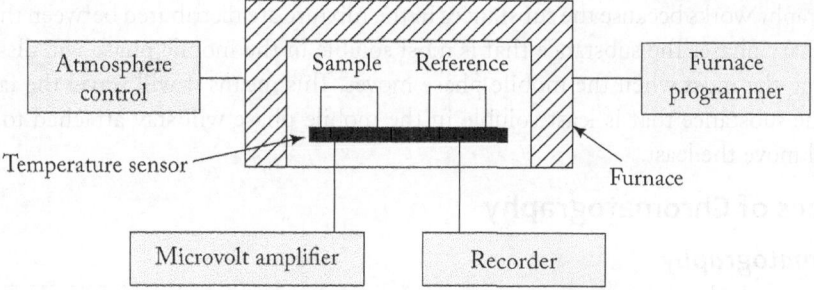

Fig. 12.31 Set-up of DTA instrument

Factors affecting the DTA curve A DTA curve can be affected by the amount and particle size of the sample. Also some of the instrumental factors affect the DTA curve; for example, size and shape of the sample holder. Even the material of the sample holder and the furnace can alter the result if it is corrosive in nature. A DTA curve can be modified by varying the furnace heating rate.

Applications DTA has a wide range of applications.
 (i) It can be used as a fingerprint for identification of samples as these curves are not identical for any two substances.
 (ii) It can be used to study the characteristics of polymeric materials.
 (iii) It is also used in the pharmaceutical and food industries for testing the purity of drug samples.
 (iv) It is also used to test the quality control of number of substances like cement, soil, and glass.
 (v) DTA curves are used for the determination of heat of reaction, specific heat, and energy change occurring during melting, etc.
 (vi) With the help of DTA, thermal stability of ligands can be studied and thus, the trend in ligand stability gives information about the ligands in the coordination sphere. DTA curves may also be used to date bone remains or to study archaeological materials.

12.12 CHROMATOGRAPHY

Chromatography is an important method of analysis that enables the separation, identification, and purification of substances of a mixture for qualitative and quantitative analyses.

12.12.1 Principle

There are different types of chromatography, but the principle on which they perform is the same. There are two important parts of chromatography, *a stationary phase* and a mobile phase. Stationary phase is composed of a solid or a liquid which is supported on a solid, while a mobile phase is composed of either a liquid or a gas. The sample mixture is dissolved in the mobile phase, which carries it through a structure holding the stationary phase. The speed of the different constituents of the sample mixture with which they travel is an important factor which cause them to separate.

Check Your Progress

63. In thermogravimetric analysis, what are the factors upon which the initial and final temperatures depend?
64. For a hydrated sample, state the schematic DTA sequence having reversible and irreversible changes.
65. Define rapid TGA method.
66. State the use of DTA.

Chromatography works because the substances in the mixture are distributed between the mobile phase and the stationary phase. The substance that is most soluble in the mobile phase will dissolve more and be carried along the most when the mobile phase moves. This means it will travel the farthest. On the other hand, the substance that is least soluble in the mobile phase will stay attached to the stationary phase and will move the least.

12.12.2 Types of Chromatography

Paper Chromatography

In paper chromatography, the stationary phase is a piece of chromatography paper. A spot of sample is put onto the strip of chromatography paper. The paper is placed in a container with a solvent.

The solvent rises slowly upwards, separating the substances in the spot. The distances travelled by the sample and the solvent are recorded to calculate the R_f values (Fig. 12.32).

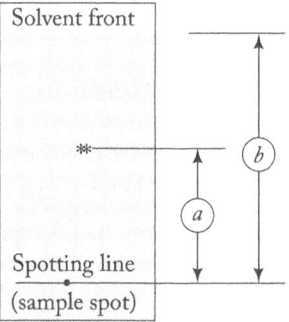

Retention factor = a/b

Fig. 12.32 Calculation of retention factor (R_f) in a paper chromatogram

$$R_f \text{ value} = \frac{\text{Distance travelled by sample}}{\text{Distance travelled by solvent}}$$

R_f values can be used to identify the substances in a mixture by comparing the R_f values of the unknown substances with the R_f values for known reference substances. R_f values are always the same for a particular substance as long as the same solvent is used at the same temperature. The observed values may slightly vary from the reference values depending upon the solvent used and type of chromatographic paper.

R_f values vary from 0 to 1. The value of 0 signifies that the substance is not attracted to the mobile phase, whereas value of 1 shows that the substance is not attracted to the stationary phase.

Paper chromatography is often used for coloured dissolved substances, such as inks, food colourings, dyes and plant pigments.

Thin Layer Chromatography (TLC)

In thin-layer chromatography (TLC), the stationary phase is a plate of plastic or glass coated with a thin layer of solid (e.g. silica or alumina). Multiple samples can be separated in one go, making TLC a very versatile technique. It is extensively used for screening of drugs. TLC has advantages over paper chromatography since it runs faster, a much better separation is achieved leading to better quantitative analysis, and different adsorbents can be chosen depending on the sample. **High-performance TLC** can be used for better resolution and faster separation.

Gas Chromatography

Gas chromatography (GC), also sometimes known as gas–liquid chromatography, (GLC) works in the same way as paper chromatography, but uses a different stationary and mobile phase. It works for mixtures that turn into a gas easily when heated. The stationary phase is a solid packed into a long column. The column is coiled up so that it takes up less space. The substances are moved through the column by a gas. This is called the 'carrier gas' and is the mobile phase. The gas used must be inert so that it does not react with the substances being tested. Nitrogen or argon are often used. The column is surrounded by an oven. This heats up and controls the temperature of the column.

When the sample consisting of a mixture of substances moved through the column, some substances are attracted to the solid in the tube more than the others, therefore each substance is carried along by the carrier gas at different speeds. The substance spread out in the column, the ones that travel fastest comes out at the end of the column first. The gas chromatogram gives a print out to show the time taken for each substance to travel through the column. This can be used to identify each substance.

Gas chromatography is a simple and highly sensitive technique which can be applied for the separation of very minute molecules.

Liquid Chromatography

Liquid chromatography (LC) is a chromatographic technique which uses liquid as a mobile phase. An improved form of liquid chromatography, high-performance liquid chromatography (HPLC), utilizes very small packing particles and works at relatively high pressure.

Ion Exchange Chromatography

Ion exchange chromatography uses a charged stationary phase to separate components of a mixture based on their respective charges. In these methods, an ion exchange resin with charged functional groups is utilized as stationary phase. These charged groups interact with oppositely charged groups present in the sample compound and keep them attached while letting others to pass. Salts are used to elute these attached ions. Based on charge on the exchangeable ions, ion exchange chromatography can be of two types: Cation-Exchange and Anion-Exchange. The exchangeable ions are cation and anion in the cation-exchange chromatography and anion-exchange chromatography, respectively. Ion exchange chromatography is commonly used to purify proteins

Size-exclusion Chromatography

Size-exclusion chromatography (SEC) is also known as gel permeation chromatography (GPC) and separates molecules according to their size. In SEC, the stationary phase is formed by the inert molecules with small pores.

Larger molecules in the sample move across the spaces between porous particles, therefore, move quickly through the column. However, molecules that are smaller than the average pore size of the packing are diffused into the pores and hence take proportionally longer time to leave the column. The average residence time in the pores depends upon the effective size of the analyte molecules.

SEC is used to determine the molecular weight of polymer and biomolecular samples.

Affinity Chromatography

Affinity chromatography is a separation technique employed for purification of mixtures. This technique is based on the specific interaction of a component of mixture with a ligand. The ligand is put onto a solid matrix to create a stationary phase. The target molecules from the mobile phase will get attached to the stationary while the rest of the component in the sample pass through the column unaffected. Fig. 12.33 shows the schematic representation of the different steps involved in affinity chromatography.

An appropriate selection of the ligand and purification procedure will always help to the increase the interaction between and ligand and its target molecules, and thereby increase the efficiency of the affinity purification technique.

Affinity chromatography has a wide range of applications. It is used in nucleic acid purification,

Step 1: Binding   Step 2: Washing   Step 3: Eluting

Different components of complex protein mixture

Target protein

Affinity resin with as ligand

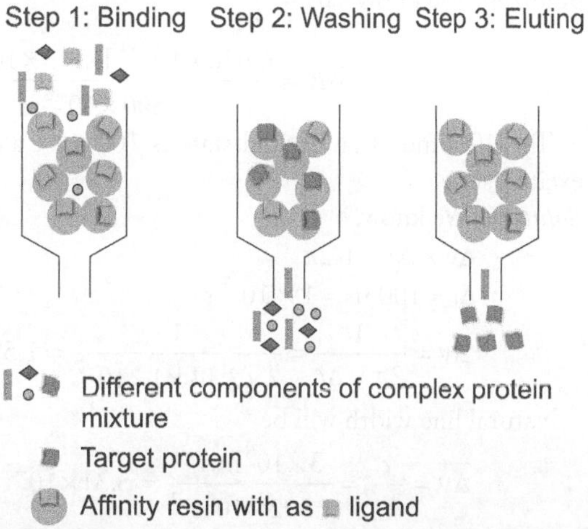

Fig. 12.33 Affinity chromatography for protein purification

protein purification from cell free extracts, and purification from blood. It can also be used to study biological interactions.

The different types of chromatography and their applications are tabulated in Table 12.16.

Table 12.16 Types of chromatography

Technique	Stationary Phase	Mobile Phase
Paper Chromatography	Chromatographic paper	solvent
Thin Layer Chromatography (TLC)	a plate of plastic or glass coated with a thin layer of solid (e.g. silica or alumina)	solvent
Gas Chromatography	a solid packed into a long column	carrier gases like nitrogen or argon
Liquid Chromatography	Resin of different kinds	a liquid containing sample
Ion-exchange Chromatography	resin with charged functional groups	Salt solutions
Size-exclusion Chromatography	inert molecules with small pores	Solvent with appropriate co-solvents
Affinity Chromatography	Ligand immobilized to a solid matrix	Sample in appropriate solvent.

Check Your Progress

67. What is the principle of chromatography?
68. What is the difference between retention time and dead time?
69. What is the principle of TLC?
70. Define retention factor. What is the retention factor for a sample that travels 3.5 cm up to the paper, while solvent travels 6 cm?
71. Hydrogen is not suitable to use as a carrier gas in gas chromatography. Why?
72. What are the factors that enhance the efficiency of liquid chromatography?

SOLVED EXAMPLES

1. What is the energy difference between energy states 1 and 2 if the wavelength of transition from 1 to 2 is 500 nm?

Solution: We know $\Delta E = \dfrac{hc}{\lambda}$

$$\Delta E = \frac{6.626 \times 10^{-34}\,\text{J.s} \times 3 \times 10^8\,\text{m.s}^{-1}}{500 \times 10^{-9}\,\text{m}} = 3.976 \times 10^{-19}\,\text{J}$$

2. The life time of an excited state is 100 ns. Calculate the natural line width of transition from ground to excited state.

Solution: We know,

$\Delta v \times \Delta t = 1/2\pi$

$\Delta t = 100 \text{ ns} = 1 \times 10^{-7}$ s

$$\Delta v = \frac{1}{2\pi \times \Delta t} = \frac{1}{2 \times 3.14 \times 1 \times 10^{-7}} = 1.592 \times 10^6\,\text{s}^{-1}$$

Natural line width will be

$$\overline{\Delta v} = \frac{c}{\Delta v} = \frac{3 \times 10^8\,\text{m.s}^{-1}}{1.592 \times 10^6\,\text{s}^{-1}} = 5.31 \times 10^{-3}\,\text{m}^{-1}$$

3. (a) What is the Doppler-shifted wavelength of a red (680 nm) traffic light approached at 60 km h^{-1}?
(b) Estimate the life time of a state that gives rise to a line of width 0.2 cm^{-1}.
Solution: (a) Doppler-shifted wavelength,

$$\Delta\lambda = \frac{\lambda_o v}{c} = \frac{680 \text{ nm} \times 16.67 \text{ m.s}^{-1}}{3 \times 10^8 \text{ m.s}^{-1}} = 3.77 \times 10^{-5} \text{ nm}$$

(b) Life-time of a state is given as:

$$\Delta v \times \Delta t = 1/2\pi$$

$$\Delta v = \frac{c}{\lambda} = c\bar{v} = 3 \times 10^8 \text{ m.s}^{-1} \times 0.23 \times 10^2 \text{ m}^{-1} = 0.6 \times 10^{10} \text{ s}^{-1}$$

$$\Delta t = \frac{1}{2\pi \times \Delta v} = \frac{1}{2 \times 3.14 \times 0.6 \times 10^{10}} = 2.65 \times 10^{-11} \text{ s} = 26.5 \text{ ps}$$

4. The rotational constant of HBr is 10 cm^{-1}. What will be the rotational constant for DBr?
Solution: Rotational constant for HBr, B_{HBr} = 10 cm^{-1}

$$B \propto 1/\mu \quad B_{DBr} = \frac{B_{HBr}}{2} = \frac{10}{2} = 5 \text{ cm}^{-1}$$

5. The separation between successive rotational lines in CO molecule is 3.82 cm^{-1}. What is the equilibrium bond length of the given molecule?

Solution: Gap between two successive lines will be $\dfrac{1}{\lambda'} - \dfrac{1}{\lambda} = 2B(J+2) - 2B(J+1) = 2B$

where, $B = \dfrac{h}{8\pi^2 \mu r_o^2 c} = 1.91 \text{ cm}^{-1} = 191 \text{ m}^{-1}$

Reduced mass for CO, $\mu = \dfrac{m_C m_O}{m_C + m_O} \times 1.66 \times 10^{-24} \text{ kg} = 1.138 \times 10^{-23} \text{ kg}$

The equilibrium bond length will be

$$r_o = \sqrt{\frac{h}{8\pi^2 \mu Bc}} = \sqrt{\frac{6.626 \times 10^{-34}}{8 \times (3.14)^2 \times 1.138 \times 10^{-23} \times 191 \times 3 \times 10^8}} = 3.589 \times 10^{-12} \text{ m}$$

r_o = 3.598 pm

6. The fundamental vibrational frequency of HBr is 2000 cm^{-1}. What will be the fundamental vibrational frequency of DBr?
Solution: Fundamental vibrational frequency of HBr, $\bar{v}_{HBr}$ = 2000 cm^{-1}

$$\bar{v} \propto 2000/\sqrt{\mu} \quad \bar{v}_{DBr} = \frac{\bar{v}_{HBr}}{\sqrt{2}} = \frac{2000}{\sqrt{2}} = 1410 \text{ cm}^{-1}$$

7. How will you distinguish between polythene and polystyrene using Infra-Red spectroscopy?
Solution: IR spectrum of polystyrene have a band above 3000 cm^{-1} which is due to aromatic benzene ring. This band is absent in the IR spectrum of polythene. Also, a strong band around 1600 cm^{-1} is observed in spectrum of polystyrene for the bending of C=C, again which is not observed in case of polythene.

8. A fluorescent species (X) with life time (τ_0) = 5 ns is quenched by a quencher (Q) with k_q = 2.5 × 10^{10} dm^3 mol^{-1} s^{-1}. What will be the concentration of Q required to decrease the fluorescent intensity of X to 50 % of the unquenched value?
Solution: According to Stern–Volmer equation:

$$\frac{F_0}{F} = 1 + k_q \tau_o [Q]$$

We have, $k_q = 2.5 \times 10^{10}$ dm^3 . mol^{-1} .s^{-1}, $\tau_o = 5$ ns
the amount of quencher required to decrease the fluorescent intensity to 50 % of unquenched value will be given as

$$[Q] = \frac{\left(\dfrac{F_0}{F} - 1\right)}{k_q \tau_o} = \frac{\left(\dfrac{F_0}{0.5 F_0} - 1\right)}{k_q \tau_o} = \frac{1}{k_q \tau_o}$$

$$[Q] = \frac{1}{2.5 \times 10^{10}\, \text{dm}^3 .\text{mol}^{-1} .\text{s}^{-1} \times 5 \times 10^{-9}\, \text{s}} = \frac{1}{125\ \text{dm}^3 .\text{mol}^{-1}} = 8\ \text{mM}$$

9. In a 200-MHz NMR spectrometer, a molecule shows two lines separated by 4 ppm. What will be the separation between these two signals in a 400-MHz spectrometer?
Solution: 8 ppm

10. What is the chemical shift position of proton in (a) CH_3COCH_3 (b) CH_3CH_2CHO (c) CH_3CH_2OH (d) CH_3COOCH_3?

Solution: (a) $\underset{a}{C H_3} \underset{b}{COC H_3}$

Both protons a and b are equivalent. Therefore, only one signal around 2.2 ppm will be visible.

(b) $\underset{a}{CH_3} \underset{b}{CH_2} \underset{c}{CHO}$

Three peaks will be observed for propanal. The chemical shift for proton a will be around 1.1, for proton b it is ~2.5, and for proton c the chemical shift will be ~9.7 ppm.

(c) $\underset{a}{CH_3} \underset{b}{CH_2} \underset{c}{OH}$

In case of ethanol, two peaks will be observed. The chemical shift for protons a and b will be ~1.2 and ~3.6 ppm, respectively. The proton of hydroxyl group is exchangeable with solvent, therefore it is not visible. However, if spectra is taken in restricted conditions, then a sharp peak for proton of hydroxyl group can also be observed with chemical shift ~4.9.

(d) $\underset{a}{CH_3} \underset{b}{COOCH_3}$

In methyl acetate, two peaks are observed with chemical shift ~2 for proton a and ~3.7 ppm for proton b.

11. Identify the number of signals that the aldehyde $(CH_3)_3CCH_2CHO$ have in ^{1}H NMR spectra?
Solution: There are three signals observed in ^{1}H NMR spectra. One corresponds to the proton of *tert*-butyl group, one for methylene group, and one for aldehydic group.

12. Deduce the structure of the organic compound with molecular formula $C_4H_9NO_2$. ^{1}H NMR of the compound shows chemical shifts: 5.30 (broad, IH), 4.10(q, 2H), 2.80 (d, 3H), 1.20 (t, 3H) ppm.
Solution: $CH_3NHCOOCH_2CH_3$

13. What is the multiplicity of NMR signal in proton spectra of (i) $CH_3CH_2CH_3$ (ii) $HOCH_2CH_2CH_2OH$ (iii) $CH_3COOCH_2CH_3$?

Solution: (a) $\underset{a}{CH_3} \underset{b}{CH_2} \underset{c}{CH_3}$

Multiplicity for proton a: triplet, for proton b: septet and for proton c: triplet. However, proton a and proton c are equivalent, therefore only one signal is observed for both protons a and c.

(b) $\underset{a}{HO} \underset{b}{CH_2} \underset{c}{CH_2} \underset{d}{CH_2} \underset{e}{OH}$

Multiplicity for proton c: sextet

In this case also, protons b and d are equivalent, therefore only one signal is observed and the multiplicity for that signal will be triplet. The protons of hydroxyl group are exchangeable with solvent; therefore these are not available for coupling. However, in restricted conditions, they will also show coupling; in that case the signal for protons b and d will be doublet of triplet.

The proton of hydroxyl group is exchangeable with solvent, therefore it is not visible. However, if spectrum is taken in restricted conditions, then a triplet will be observed for protons a and e.

(c) $\underset{a}{\underbrace{CH_3}}\ COO\underset{b}{\underbrace{CH_2}}\ \underset{c}{\underbrace{CH_3}}$

Multiplicity for proton a: singlet; proton b: quartet; proton c: triplet

14. If the relative intensity of the M+1 to M peak is 7.80, calculate the number of carbon atoms in the molecule.
Solution:
Answer: 7.8/1.1 is around 7. Thus the compound will have 7 carbon atoms.

15. An organic molecule has molecular ion peak at $m/z = 107$. If the relative intensity of the M + 1 peak is 7.8 with respect to molecular ion peak, what is the molecular formula of the molecule?
Solution: M^+ at $m/z = 107$ implies an odd number of nitrogens
M^+ peak It is the base peak, hence recalculation is not required.
M + 1 peak 8.00/1.1 = 7.3; 7 carbons
Number of hydrogen atoms: $107 - (1)(14) - (7)(12) = 9$
Answer: C_7H_9N

16. What is the use of solvent in paper chromatography?
Solution: The solvent is called the mobile phase while the paper contains the stationary phase.

17. What are the possible numbers of spot that a pure substance can produce during chromatography?
Solution: A pure substance will produce only one spot. Two or more spots show a mixture of two substances.

18. Calculate the R_f value of a substance that travels 2 cm up the paper, compared with the solvent which travels 7 cm.
Solution: R_f = (distance travelled by substance) ÷ (distance travelled by solvent) = $2 \div 7 = 0.29$

SUMMARY

- Spectroscopic methods constitute a set of methods in which interaction of electromagnetic radiation with matter/chemical molecules is observed and analysed to obtain information about their structure (e.g. bond lengths, bond angles, planarity, presence of functional groups) and other fundamental properties such as the energy levels of a molecule.

- Time-domain spectra need to be converted into frequency-domain spectra through Fourier transformation.

- The uncertainty in wavelength (line width) will depend on life time (Δt) of the excited state.

- Rotational spectroscopy is based on the transition between the quantized rotational levels. Transition takes place on absorption in the microwave region.

- The gap between the successive rotational energy levels is inversely proportional to I.

- The separation between successive rotational lines in CO molecule can be used to calculate equilibrium bond length of the given molecule.

- There are many molecules whose energies are less than kT and many of these rotational levels will be appreciably occupied at room temperature.

- Frequency of transition in vibrational spectroscopy is directly proportional to the strength of the bonding between the two atoms ($v \propto k$) and inversely proportional to the reduced mass of the two atoms ($v \propto 1/\mu$).

- The peak values in IR spectra help us to know the functional groups present in an unknown molecule.

- No two unique molecules have the same IR spectrum. Thus, the IR spectrum can be used to identify a particular molecule.

- The most significant electronic transition is $\pi \rightarrow \pi^*$ transition in a conjugated molecule. The value typically lies between 200 to 700 nm. The wavelength of such transition can be calculated by using the particle in a box.

- Absorption by transition metal complexes leads to d–d transition.

- Absorbance at absorbance maximum is related to concentration and hence can be used in the determination of concentration of molecule.

- Molecule can lose energy from excited state singlet to ground state singlet by emitting light. This phenomenon is called fluorescence.

- Time scale of absorption is in femtoseconds to picoseconds. Time scale of absorption is much shorter in comparison to the time scale of molecular motion that might take place during the absorption process.

- NMR deals with interaction of electromagnetic radiation in radiofrequency range with nuclear spin of certain nuclei.

- The nuclei with I spin has $(2I + 1)$ degenerate levels. The degeneracy of the energy levels is lifted by the interaction of nuclear dipole moment with an intense external magnetic field (B_0).

- Excitation of transitions between energy levels is stimulated using radio-frequency region of electromagnetic radiation.

- The frequency of the absorption will be different for same spin at NMR instruments with different magnetic strength.

- To keep the position of peaks same on two NMR machines, the position in NMR peak is not expressed in terms of frequency/wavelength/wave number, but in terms of chemical shift.

- Every nucleus is associated with circulating electrons that generate a local magnetic field (B_e) opposite to the direction of external magnetic field. Thus, a nucleus experiences a smaller field than an external magnetic field. In others, spins are shielded from the external magnetic field.

- Shielding is proportional to the external magnetic field.

- The interaction between one spin with neighbouring spin results in splitting of lines in the NMR spectrum into two or more components.

- Coupling constant is independent of the applied field.

- The multiplicity of the signal for a set of equivalent protons is dependent on the number of protons of the adjacent atoms. The multiplicity is given by $n + 1$ where n is the number of protons attached to adjacent atom.

- In a simple mass spectrometer, the molecules (M) are bombarded with high-energy electrons. The molecules lose an electron forming a radical cation, a species with positive charge and one unpaired electron (M^+).

- The cations formed are separated by magnetic deflection. The amount of deflection is proportional to the mass to charge ratio (m/z).

- The mass spectrum shows the mass of each cation vs its relative abundance.

- Base peak is the most intense peak in the spectrum.

- The presence of heavier isotopes leads to peaks that have masses that are higher than the parent ion peak.

- Flame emission spectroscopy is based on the measurements of emitted photons produced when a sample is introduced into the flame.

- The wavelength of the emitted light helps in the identification of the element present in the sample while the intensity of the flame quantifies the amount of the element present in the sample.

- de Broglie wavelength of the electron is much smaller than that of light (photons) and thus electron microscopy gives better resolution than optical microscopy.

- Transmission electron microscopy (TEM) is an electron microscopy technique in which wavelike property of moving electron and its interaction with a sample is utilized. Transmitted electron after passing through a sample forms an image. The image is magnified and focused on an imaging device.

- Scanning Transmission Microscopy (STM) is a type of scanning probe microscope which uses tunnelling property of electron to image the sample. Electrons tunnel to forbidden region due to wave-like nature of electron.

- X-ray diffraction is one of the best analytical techniques for determination of the crystal structure.

- Conductometric measurements are used for determination of the end points in various titrations.

- Chromatography is an important method of analysis that enables the separation, identification, and purification of substances for qualitative and quantitative analysis.

GLOSSARY

Spectrum: A display of wavelengths/frequencies absorbed or emitted by an object/atom/molecule.

Spectral resolution: A measure of the ability to separate nearby features in the wavelength space.

UV/visible spectroscopy: Deals with light absorption in the UV/visible part of the spectrum (210 – 900 nm).

Quantum yield: The ratio of the number of photons emitted to the number absorbed.

Coupling constant: The spacing between lines obtained due to spin–spin coupling .

Molecular ion peak: Also called parent ion peak, it is peak of radical cation corresponding to the mass of the original molecule.

Microscopy: A technique which is used to see the objects of much smaller size than 0.1 mm.

Scanning electron microscopy: An electron microscopy technique in which an electron beam scans rapidly over the surface of the sample specimen.

Atomic force microscopy: A type of scanning probe microscope which makes use of a cantilever with a sharp probe to scan the sample.

Conductometry: An electrochemical technique of analysis which is used to measure the electrical conductance of an electrolyte solution.

EXERCISES

Multiple Choice Questions

1. The wavenumber of a given transition is 2000 cm^{-1}. The region of electromagnetic spectrum corresponding to this transition is
   - (a) infrared
   - (b) UV-visible
   - (c) radiowave
   - (d) microwave

2. The frequency (ν) corresponding to transition at 500 nm is
   - (a) 3×10^{14} Hz
   - (b) 6×10^{14} Hz
   - (c) 3×10^{15} Hz
   - (d) 6×10^{15} Hz

3. The frequency of the transition due to absorption of 2.0×10^{-23} J energy is
   - (a) 3.1×10^{10} Hz
   - (b) 3.1×10^{10} s
   - (c) 2.4×10^{8} Hz
   - (d) 2.4×10^{8} s

4. The concentration of a solute in an aqueous solution which has an absorbance of 0.8 is

   (take path length as 1 cm and the molar absorptivity of the solute in aqueous solution is 8000 mol^{-1} L cm^{-1})
   - (a) 10 μM
   - (b) 100 μM
   - (c) 10 mM
   - (d) 100 mM

5. A solution of dichromate ions absorbs light with wavelength of 505 nm. The dichromate ions absorbs in the
   - (a) visible region
   - (b) IR region
   - (c) UV region
   - (d) microwave region

6. Among the following spectra, sharpest signal is observed in
   - (a) UV–Vis
   - (b) rotational
   - (c) vibrational
   - (d) NMR

7. Among the following spectra, the broadest signal is observed in
   - (a) UV–Vis
   - (b) rotational
   - (c) vibrational
   - (d) NMR

8. The number of translational, rotational and vibrational degree of freedom in NO_2, are respectively
   - (a) 3,2, 4
   - (b) 2,3, 5
   - (c) 3, 3, 4
   - (d) 3, 3, 3

9. Spacing between successive lines in a spectrum of rigid rotor is equal to
   - (a) B
   - (b) 2B
   - (c) 4B
   - (d) B/2

10. The rotational constant (in cm^{-1}) is given by
    - (a) $\dfrac{h}{8\pi^2 I}$
    - (b) $\dfrac{h^2}{4Ic}$
    - (c) $\dfrac{h^2}{8\pi^2 Ic}$
    - (d) $\dfrac{h}{8\pi^2 Ic}$

11. The molecule showing a pure rotational microwave spectrum is
    - (a) H_2
    - (b) HCl
    - (c) Br_2
    - (d) CO_2

12. The rotational constant and the fundamental vibrational frequency of *HBr* are, respectively, 10 cm^{-1} and 2000 cm^{-1}. The corresponding values for *DBr* approximately are

    (a) 20 cm^{-1} and 2000 cm^{-1}
    (b) 10 cm^{-1} and 1410 cm^{-1}
    (c) 5 cm^{-1} and 2000 cm^{-1}
    (d) 5 cm^{-1} and 1410 cm^{-1}

13. The molecule having IR active vibrations is

    (a) CF_4 (b) D_2
    (c) NO   (d) H_2

14. The vibration which is not a bending molecular vibration is

    (a) wagging   (b) twisting
    (c) stretching   (d) rocking

15. Among the following, the weakest transition in terms of intensity is

    (a) $n \rightarrow n^*$ (b) $\sigma \rightarrow \sigma^*$
    (c) $\pi \rightarrow \pi^*$ (d) $n \rightarrow \sigma^*$

16. Decrease in the intensity of the spectroscopic signal is called

    (a) bathochromic shift
    (b) hypsochromic shift
    (c) hyperchromic shift
    (d) hypochromic shift

17. Which of the following region in spectrum is used for NMR study?

    (a) Microwave   (b) Infrared
    (c) Ultra-violet   (d) Radio waves

18. What will be the maximum number of orientations that a magnetic nuclei can have?

    (a) I (b) $2I + 1$
    (c) $2I$ (d) $2I-1$

19. ^1H NMR spectrum of an organic compound recorded on a 500 MHz spectrometer showed a quartet with line positions at 1759, 1753, 1747, 1741 Hz. Chemical shift and coupling constant (Hz) of the quartet are: -

    (a) 3.5 ppm, 6Hz   (b) 3.5 ppm, 12Hz
    (c) 2.5 ppm, 6Hz   (d) 2.5 ppm, 12Hz

20. The incorrect statement in the context of NMR spectroscopy is

    (a) Static magnetic field is used to induce transition between the spin states.
    (b) Magnetization vector is perpendicular to the applied static magnetic field.

    (c) Static magnetic field is used to create population difference between the spin states.
    (d) Static magnetic field induces spin-spin coupling

21. In a 200 MHz NMR spectrometer, a molecule shows two doublets separated by 2 ppm. The observed coupling constant is 10 Hz. The separation between these two signals and the coupling constant in a 400 MHz spectrometer will be, respectively

    (a) 400 Hz and 10 Hz
    (b) 800 Hz and 10 Hz
    (c) 400 Hz and 20 Hz
    (d) 800 Hz and 20 Hz

22. The number of signals observed in acetone is

    (a) 1   (b) 2
    (c) 3   (d) 4

23. In an NMR spectrum, the number of signals shown by 1-propanol and 2-propanol is respectively,

    (a) 2 and 4   (b) 2 and 3
    (c) 4 and 2   (d) 4 and 3

24. Resistivity is defined as

    (a) reciprocal of resistance of a substance
    (b) reciprocal of conductivity of a substance
    (c) cell constant of a substance
    (d) Resistance/cell constant

25. On dilution, conductivity

    (a) increases
    (b) decreases
    (c) remains unaffected
    (d) first increases and then decreases

26. Find out the correct order of increasing molar conductivity among the following ions:

    (a) $Li^+ < Na^+ < K^+ < Rb^+$
    (b) $Rb^+ < K^+ < Na^+ < Li^+$
    (c) $Li^+ < Na^+ < Rb^+ < K^+$
    (d) $Li^+ < K^+ < Na^+ < Rb^+$

27. Which of the following bridges is not used in the measurement of conductance?

    (a) Wheatstone bridge
    (b) Kohlrausch bridge
    (c) Salt bridge
    (d) Both (a) and (b)

28. In a conductometric titration between HCl and NaOH, the decrease in the conductance of acid on addition of NaOH is due to
    (a) the change in pH
    (b) the replacement of fast-moving H^+ ions by the slow-moving Na^+ ions
    (c) the replacement of slow-moving H^+ ions by the fast-moving Na^+ ions
    (d) the consumption of hydrogen by hydroxide ions

29. A conductometric titration is carried out between $BaCl_2$ and Na_2SO_4 by adding Na_2SO_4 to $BaCl_2$. Which of the following will show the conductions variation of $BaCl_2$?

    (a)

    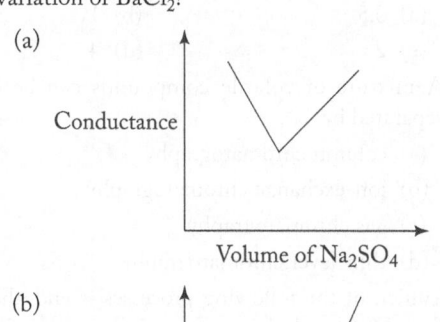

    (b)

    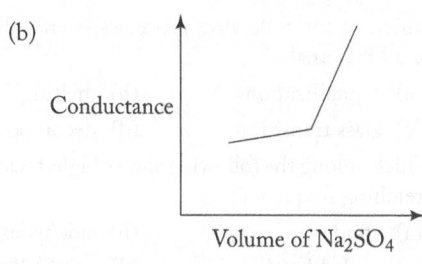

    (c)

    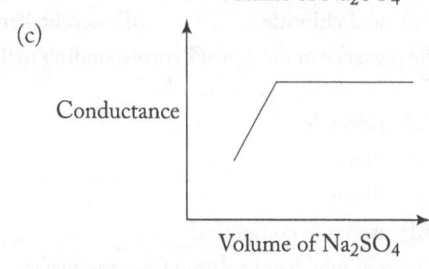

    (d)

    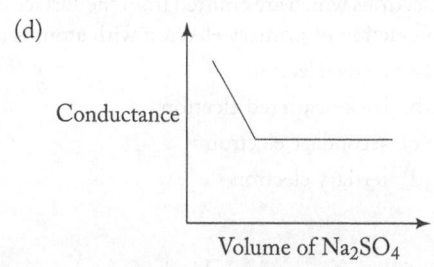

30. Which of the following statements is true for XRD?
    (a) The wavelengths used in XRD are comparable to the inter-atomic distances.
    (b) X-rays have very high energy which enables them to penetrate through solids.
    (c) These are electromagnetic radiations, therefore results in no interaction with crystals.
    (d) High frequency of these radiations leads to fast analysis.

31. For the first order reflection, the smallest interplanar spacing from which diffraction can be obtained is
    (a) λ
    (b) $\lambda/2$
    (c) 2λ
    (d) $\lambda/3$

32. What will be the distance between the diffracted planes if the diffraction with X-rays ($\lambda = 2.45 \ \lambda$) gives first order reflection at 28°?
    (a) 2.45 Å
    (b) 2.61 Å
    (c) 5.90 Å
    (d) 2.95 Å

33. What will be the angle at which first order reflection from a crystal takes place with X-rays ($\lambda = 0.154$ nm)? The inter layer distance is 0.3145 nm.
    (a) 14.2°
    (b) 20.2°
    (c) 17.9°
    (d) 15.0°

34. If X-rays are diffracted from a NaCl crystal (d = 253 pm) at an angle of 32°, assuming first order reflection, calculate the wavelength of the X-rays.
    (a) 208.1 nm
    (b) 0.208 nm
    (c) 0.208 pm
    (d) 0.208 Å

35. Thermogravimetric analysis is the type of thermal analysis which measures:
    (a) change in concentration as a function of temperature
    (b) change in mass/weight as a function of temperature
    (c) change in position of the crystal point as a function of temperature
    (d) change in all physical properties of the material as a function of temperature

36. Which of the following statements is correct?
    (a) DTA stands for direct thermal analysis.
    (b) Area under a DTA curve represents the enthalpy of the change.
    (c) In a typical DTA technique, the reference and sample both undergo a thermal change.
    (d) None of the above.

37. Which of the following processes shows weight gain in a TGA experiment?
    (a) Adsorption
    (b) Absorption
    (c) Both (a) and (b)
    (d) Decomposition

38. Which of the following type of reactions show both weight gain as well as weight loss in TGA study?
    (a) Surface oxidation reactions
    (b) Magnetic transition
    (c) Adsorption
    (d) Dehydration

39. In a TGA experiment, a multi- stage decomposition (stair-case curve) can be converted into single stage decomposition (S-shaped curve) by
    (a) increasing heating rate
    (b) increasing surface area of the sample
    (c) involvement of intermediates
    (d) increasing pressure

40. Which of the following factors does not affect a DTA curve?
    (a) Temperature of the sample
    (b) Concentration of the sample
    (c) Heat capacity of the sample
    (d) None of these

41. In gas chromatography, the difference in which of the following properties is used for separation of components?
    (a) Partition coefficient
    (b) Molecular weight
    (c) Molarity
    (d) Conductivity

42. Ion exchange chromatography is a technique in which separation is based on
    (a) electrostatic attraction
    (b) adsorption
    (c) electrical mobility of ionic speices
    (d) conductivity

43. In a TLC, if the solubility of one amino acid in mobile phase is lower than other, then
    (a) it must have higher molecular mass.
    (b) it will have low R_f value.
    (c) it will move with a speed of the solvent.
    (d) all of the above.

44. Immobilized biochemical is used as stationary phase in
    (a) ion-exchange chromatography
    (b) exclusion chromatography
    (c) gel permeation chromatography
    (d) affinity chromatography

45. If sample A, sample B, and solvent travel 2 cm, 4 cm and 8 cm up the paper, respectively; the ratio of the retention factor of A to B is
    (a) 0.5   (b) 1
    (c) 2   (d) 4

46. A mixture of volatile compounds can be easily separated by
    (a) column chromatography
    (b) ion-exchange chromatography
    (c) gas chromatography
    (d) thin-layer chromatography

47. Which of the following processes is endothermic for a DSC analysis?
    (a) crystallization   (b) melting
    (c) glass transition   (d) decomposition

48. Which among the following shows highest carbonyl stretching frequency?
    (a) amide   (b) aldehydes
    (c) acid chloride   (d) acyclic ketone

49. The presence of mass peaks corresponding to 91 and 77 indicates the presence of
    (a) aldehyde
    (b) alkane
    (c) alkene
    (d) aromatic compound

50. Electrons which are emitted from the surface on the interaction of primary electron with atom is called
    (a) auger-electron
    (b) back-scattered electron
    (c) secondary electron
    (d) tertiary electron

Review Questions

1. Give an example where scattering affects the colour of an object.

2. Emission from which kind of species gives line spectrum and continuous spectrum?

3. What are the characteristics of black body radiation?

4. Write about two characteristics of Fourier transformation.

5. What is natural line width? Give reason for it.

6. How does velocity of the source relative to observer affect the broadening of peaks?

7. What is the transition probability and what is its value for allowed transition?

8. How is the intensity of transition affected by population of the states?

9. How can information of spacing between two rotational lines of a molecule obtained from rotational spectrum be used to calculate bond length?

10. How do you distinguish the following pairs of isomers based on their IR and mass spectral data?

    (a) $PhCOCH_2CH_3$ (b) $PhCH_2COCH_3$

11. Is it possible to distinguish 2-methylcyclopentan-1,3-dione and 2,2-dimethylcyclopentan-1,3-dione based on IR spectroscopy? Rationalize your answer.

12. Two compounds having the same molecular formula $C_4H_6O_2$ have the following IR spectral data: (a) 1723 cm^{-1} (b) 1775 cm^{-1} and 1630 cm^{-1}. Identify the compounds.

13. How do you distinguish fumaric acid from maleic acid by IR spectroscopy by comparing their C=C stretching frequencies?

14. 4-Methyl-2,6-di-*tert*-butylphenol shows a strong and sharp IR absorption frequency at 3470 cm^{-1}. What is this frequency due to? This absorption frequency remains the same irrespective of the concentration of the sample. What conclusion could you arrive based on this information? Explain.

15. Food is often packed in clear foil, which may be either cellulose acetate (cellophane) or polypropylene. How could you use IR spectroscopy to distinguish them?

16. Which of the following two compounds, *m*-chlorobenzoic acid or *p*-chlorobenzoic acid, will have a higher C=O stretching frequency? Why?

17. Discuss the electronic transitions observed in transition metal complexes.

18. Discuss the effect of solvents on the absorption spectra.

19. What is Stoke's shift? Explain the reason for this shift.

20. Explain Kasha's Rule.

21. What is the difference between collisional and static quenching?

22. It is well known that the NMR signals of ^2H (deuterium) and ^{13}C will not overlap ^1H spectra (i.e. their signals will not appear in the ^1H-NMR spectrum). Verify this statement by showing that the frequency (v) will be different for ^1H, ^2H and ^{13}C at a field strength (B_0) of 4.7 Tesla. The gyromagnetic ratios for ^1H, ^2H and ^{13}C are 2.7 $\times 10^8$, 0.41×10^8 and 0.67×10^8, respectively.

23. Are the protons of the methyl groups and olefinic protons of *cis*-2-butene and *trans*-2-butene chemically and magnetically equivalent or not?

24. Three compounds of the same molecular formula C_9H_{12} displayed the following NMR spectral data. Identify the structure of these compounds.

    Compound **A**: δ (ppm) 2.31 (s, 3H), 2.35 (s, 6H), 7.10 (d, 2H, J = 6.8 Hz), 71.8 (t, 1H, J = 6.8 Hz)

    Compound **B**: δ (ppm) 2.37 (s, 9H), 7.12 (s, 3H

    Compound **C**: δ (ppm) 2.30 (s, 3H), 2.31 (s, 3H), 2.34 (s, 3H), 7.09 (d, 1H, J = 2.9 Hz), 7.12 (dd, 1H, J = 7.0, 2.9 Hz), 7.16 (d, 1H, J = 7.0 Hz)

25. Draw (clearly) the splitting pattern of a proton that has coupling constants of 12.0, 9.0 and 6.0 Hz with three neighbouring protons. From the splitting pattern (that you have drawn) explain how will you determine the various three coupling constants?

26. A compound of molecular formula C_8H_8O shows an IR absorption band at 1695 cm^{-1}. Its mass spectral value are given below. M/z 120(29), 105(100), 78(10), 77(88), 51(40), 50(21), 43(17). Identify the compound. Explain the fragmentation leading to the formation of an ion at m/z 78.

27. Identify the structures of the compounds from their mass spectral data given below:

    Compound **A**: m/z 86 (M$^+$, 15), 71 (32), 57 (100), 43(80), 29 (63)

    Compound **B**: m/z 156 (M$^+$, 100), 127 (35), 29 (75)

28. Compound **C**: m/z 102 (10), 60 (22), 45 (28), 43 (100). Account for the formation of the fragment at m/z 45 by writing suitable mechanism for its formation.

29. Compound **D**: m/z 75 (M$^+$, 100), 77 (M+2, 33), 48 (92), 50 (30), 40. Account for the formation of the fragment at m/z 40 by writing suitable mechanism for its formation.

30. *o*-Nitrophenol shows a daughter ion peak at m/z 122 in its EI mass spectrum, while *p*-nitrophenol does not. Explain.

31. Compound **E** and **F** are both isomeric amino alcohols of molecular weight 89. Identify the isomers

based on their mass spectral data and explain their fragmentation pattern.

Compound **E**: m/z 89, 58 and 31

Compound **F**: m/z 89, 59 and 30

32. What are the applications of flame photometry?

33. Why does electron microscopy have higher resolution than optical microscopy?

34. Explain the difference between secondary and Auger electrons.

35. What is the principle of scanning tunnelling microscopy?

36. Conductometry can be used to get end-points in acid–base titration. Explain.

37. What is the difference between TGA and DTA?

38. What is the principle of chromatography?

39. What is the difference between size-exclusion chromatography and affinity chromatography?

40. Explain thin layer chromatography.

NUMERICAL PROBLEMS

1. If transition between two electronic states requires a radiation of wavelength 400 nm, calculate the energy gap between the two states. **(Ans: 299.3 kJ mol^{-1})**

2. If the life time of an excited state is 10 ns, what will be the natural line-width of the transition? **(Ans: 15.92 MHz)**

3. Energy gap between ground state and first excited state of a molecule is 4×10^{-23} J per molecule. Calculate the number of molecules present in excited state at 500 K if the number of molecules in ground state is 5000? **(Ans: 4971)**

4. A man driving a hypothetical car saw red light (660 nm) as green light (520 nm). What is the speed with which the man was driving this hypothetical car? **(Ans: 6.4×10^7 m s^{-1})**

5. For a gaseous HCl molecule, the internuclear distance is found to be 143 pm. If the rotational spectrum of HCl comprises equally distance lines, calculate the spacing between the lines. (Given: atomic masses of ^{1}H and ^{35}Cl are 1.673×10^{-27} kg and 58.06×10^{-27} kg, respectively.) **(Ans: 16.80 cm^{-1})**

6. Where will the first and second line fall in microwave spectrum of a CN$^+$ molecule, if the bond length of CN$^+$ is 0.129 nm? **(Ans: 3.146 cm^{-1}, 6.292 cm^{-1})**

7. For a diatomic molecule AB, calculate the moment of inertia and bond distance. Given B = 1.92 cm^{-1}, atomic mass of A = 12.041 amu and atomic mass of B = 15.995 amu. **(Ans: 1.46×10^{-46} kg m^2, 1.13 A^0)**

8. In a pure rotational spectrum of a NO molecule, if the bond length is found to be 113.9 pm, then calculate the frequency of the transition from J = 2 to J = 3. (Atomic mass of N = 14.003 amu and Atomic mass of O = 15.995) **(Ans: 3.13×10^{11} s^{-1})**

9. What will be the separation between the lines in pure rotational spectrum of a molecule of NH$_3$? (Given: I = 2.82×10^{-47} kg.m^2) **(Ans: 19.8 cm^{-1})**

10. Identify the allowed vibrational transition for a diatomic molecule: (a) v = 1 to v = 3; (b) v = 3 to v = 4 and (c) v = 5 to v = 4.

11. What will be the spacing (in eV) between the vibrational energy levels of a CO molecule, if the force constant is 1748 Nm^{-1}. **(Ans: 0.258 eV)**

12. What will be the fundamental frequency and zero point energy of CO molecule, if the force constant is 1748 N.m^{-1}. **(Ans: 6.24×10^{13} Hz, 2.06×10^{-20} J)**

13. The force constant of ^{12}CO is 1748 Nm^{-1}. Assuming the force constant to be same for ^{13}CO, calculate the difference between zero point energy of ^{12}CO and ^{13}CO. **(Ans: 4.6×10^{-22} J)**

14. Calculate the molar absorption coefficient of a 0.5 mM aqueous solution of an unknown compound which showed 23% transmittance at 250 nm. (assuming cell length 1 cm) **(Ans: 1.276×10^3 M^{-1}cm^{-1})**

15. What will be the required magnetic field for ^{19}F to observe a signal at 60 MHz? (Given, g$_N$ for ^{19}F = 5.257 and μ_N = 5.05 × 10-27 JT^{-1}). **(Ans: 1.497 T)**

16. What will be the gyromagnetic ratio for ^{19}F, if its NMR signal is observed at 35.5 MHz in a spectrometer operating at 1 T? **(Ans: 2.22×10^{38} M^{-1}cm^{-1})**

17. On a 60 MHz NMR spectrometer, if a proton resonates at 120 Hz downfield from TMS, what will be the value of chemical shift (δ) in ppm for the proton? **(Ans: 2 ppm)**

18. If an NMR active entity shows a shift from TMS at 300 Hz, calculate its chemical shift in ppm on a 500 MHz NMR spectrometer. **(Ans: 0.6 ppm)**

19. Calculate the magnetic field (in Tesla) required by a ^{1}H nucleus to resonate in an NMR spectrometer operating at 500 MHz. (Given: $\gamma = 4.29 \times 10^8$ T^{-1} s^{-1}) **(Ans: 7.32 T)**

20. The chemical shift of a proton recorded in an NMR spectrometer with a magnetic field strength of 2.33 Tesla is 4.2 ppm. What will be the resonance frequency of the same proton (with respect to TMS) when the spectrum is recorded in a 300 MHz NMR instrument? **(Ans: 1260 Hz)**

21. The precisional frequency of an unknown nucleus is 30 MHz if the spectrum is recorded in a NMR spectrum having a magnetic field strength of 7 Tesla. Given the gyromagnetic ratio of proton as 2.7×10^8 rad T^{-1} s^{-1}, calculate the gyromagnetic ratio of the unknown nucleus. **(Ans: 2.69×10^7 rad T^{-1} s^{-1})**

22. Three compounds of the same molecular formula C_9H_{12} displayed the following NMR spectral data. Identify the structure of these compounds.

    Compound **A**: δ (ppm) 2.31 (s, 3H), 2.35 (s, 6H), 7.10 (d, 2H, J = 6.8 Hz), 71.8 (t, 1H, J = 6.8 Hz)

    Compound **B**: δ (ppm) 2.37 (s, 9H), 7.12 (s, 3H)

    Compound **C**: δ (ppm) 2.30 (s, 3H), 2.31 (s, 3H), 2.34 (s, 3H), 7.09 (d, 1H, J = 2.9 Hz), 7.12 (dd, 1H, J = 7.0, 2.9 Hz), 7.16 (d, 1H, J = 7.0 Hz) **(Ans: A:1,2,3- trimethyl benzene, B:1,3,5, trimethyl benzene, C: 1,3,4 trimethyl benzene.)**

23. A proton resonated as a triplet at δ 3.4 higher than TMS in a 200 MHz NMR spectrometer with a coupling constant of 7.0 Hz. Calculate its frequency with respect to TMS. What will be the frequency of the proton if the spectrum is recorded in a 270 MHz NMR instrument. Write the frequency of all the three signals. **(Ans: 918, 911, 925 Hz)**

FURTHER READING

1. Banwell, C. N. and E.M. McCash, *Fundamentals of Molecular Spectroscopy*, Tata McGraw-Hill Publishing Company Ltd.

2. Dogra, S.K. and H.S. Randhawa, *Atomic and Molecular Spectroscopy*, Pearson.

3. Hollas J.M., *Modern Spectroscopy*, Wiley.

4. Keeler, James, *Understanding NMR Spectroscopy*, Wiley.

5. Khopkar, S.M., *Basic Concepts of Analytical Chemistry*, New Age international Publishers.

6. Skoog, D. A. D.M. West, and H. Hollar, *Fundamentals of Analytical Chemistry*, Cengage Learning

7. Mendham, J., et.al. *Vogel's Textbook of Quantitative Chemical Analysis*, Pearson.

ANSWERS

Check Your Progress

1. When electromagnetic radiation is irradiated upon matter, this radiation can be absorbed by the matter or the absorbed radiation can be emitted and transmitted. Other phenomena such as e.g. scattering, reflection, refraction, etc., can also take place

2. Absorption can be explained by particle nature of electromagnetic radiation while reflection can be explained by considering EMR as a wave.

3. A hypothetical object that absorbs all the radiation falling on it, at all wavelengths and emits all of the absorbed radiation.

4. According to Rayleigh–Jeans law, the energy density is proportional to the square of frequency v; thus spectral brightness is expected to increase with increase in v. Prediction by this law matches with experimental observations at low frequencies (v), but fails at high frequencies which is known as 'ultraviolet catastrophe'.

5. Upon absorption of IR light, transition takes place between vibrational levels in a molecule.

6. Transition between various rotational levels of molecule occurs on absorption of microwave radiation.

7. Ultra violet–visible radiation is required for electronic transitions.

8. In time-domain spectroscopy, amplitude of electromagnetic radiation is measured as a function of time while in frequency domain spectroscopy absorption is measured as the function of frequency.

9. Fourier transformation of time-domain data gives frequency-domain data.

10. $\Delta E = 1000$ kJ mol^{-1}; $h = 6.626 \times 10^{-34}$ J s; $c = 3 \times 10^8$ m s^{-1}; $N_A = 6.022 \times 10^{23}$ mol^{-1}

$$\Delta E = N_A\, hc/\lambda$$

$$\lambda = N_A\, hc/\Delta E = 119.7 \text{ nm}$$

11. $\dfrac{N}{N_o} = \exp\left(\dfrac{\Delta E}{kT}\right) = \exp(-0.462) = 0.62$

12. Life time $\tau = 10$ ns

Natural line width of transition $\Delta\upsilon_L$,

$$\Delta\upsilon_L = \dfrac{1}{2\pi\tau} = 15.92 \text{ MHz}$$

13. $\lambda_0 = 680$ nm; $\upsilon = 100$ km h^{-1}; $c = 3 \times 10^8$ m s^{-1}

Doppler-shifted wavelength $\Delta\lambda$,

$$\dfrac{\Delta\lambda}{\lambda o} = \dfrac{\upsilon}{c}$$

$$\Delta\lambda = 6.2963 \times 10^{-5} \text{ nm}$$

14. From uncertainty principle,

$$\Delta E\, \Delta t \geq h/2\pi$$

Δt can be replaced by life-time of a state 'τ' and $\Delta E = 0.5$ cm^{-1}; 1 cm$^{-1} = 1.9863 \times 10^{-23}$ J

$$\Delta = h/2\pi\Delta E = 10.6237 \text{ ps}$$

15. Intensity of spectroscopic signal depends on many factors, e.g., concentration of the sample, transition probability between the two states, population difference between two states, degeneracy, etc.

16. HCl and CH_3Cl are rotationally active since only these two molecules possess non-zero dipole moment.

17. $I = \mu.r^2 = \dfrac{(1.673 \times 10^{-27}) \times (5.68 \times 10^{-26})}{(1.673 \times 10^{-27}) + (5.68 \times 10^{-26})} \times (1.353 \times 10^{-10})^2 = 2.97 \times 10^{-47} \text{ kg.m}^2$

18. Rotational constant,

$$B = \dfrac{h}{8\pi^2 \mu r^2 c}$$

$$= \dfrac{6.626 \times 10^{-34} \text{ Js}}{8 \times (3.14)^2 \times (1.459 \times 10^{-26}) \text{kg} \times (1.392 \times 10^{-10})^2 \text{m}^2 \times (3 \times 10^{10}) \text{ cm.s}^{-1}} = 0.99 \text{ cm}^{-1}$$

19. Equilibrium bond length,

$$r = \sqrt{\dfrac{h}{8\pi^2 \mu Bc}} = \sqrt{\dfrac{6.626 \times 10^{-34} \text{ Js}}{8 \times (3.14)^2 \times (4.585 \times 10^{-26}) \text{kg} \times 0.142 \text{ cm}^{-1} \times (3 \times 10^{10}) \text{cm.s}^{-1}}} = 2.07 \times 10^{-10} \text{m} = 2.07 \text{ Å}$$

20. If the intensity of rotational line of molecule A*X (A* is isotope of A) is x time that of AX, then natural abundance of A*X (y) is given by

$$y = \dfrac{x}{(1+x)} \times 100 \%$$

$$x = 1/124$$

$$y = \dfrac{1/124}{(1+1/124)} \times 100 \%$$

$$y = \dfrac{1}{(125)} \times 100 \%$$

$$y = \frac{100}{125}$$

21. (a) H_2O: 3 and (b) SO_2: 4

22. (a) C–Cl and (b) C≡N

23. C≡C: 2112 cm^{-1} and terminal C-H: 3309 cm^{-1}.

24. $\bar{v} = \dfrac{1}{2\pi c}\sqrt{\dfrac{k}{\mu}} = \dfrac{1}{2\times 3.14 \times 3\times 10^8\, m.s^{-1}}\sqrt{\dfrac{450\ N.m^{-1}}{1.58\times 10^{-26}\ kg}} = 8.96\times 10^4\, m^{-1}$

25. Benzoic acid, C_6H_5COOH

26. Hyperchromic shift refers to the increase in absorption intensity whereas hypochromic shift occurs when there is decrease in absorption intensity of compound.

27. $\varepsilon = \dfrac{A}{Cl} = \dfrac{1.25}{1\times 10^{-4}\, M \times 1\ cm} = 1.25\times 10^4$

28. Methanol is less polar than water. Therefore, it exhibit lesser amount of hydrogen bonding which decrease the energy gap and thereby increase the wavelength associated with the n → π* transition.

29. (b), there are two ketonic groups adjacent to each other which facilitates conjugation. Due to conjugation, there is a decrease in energy gap and thereby increase in wavelength of the n → π* transition.

30. $A = \varepsilon Cl = 1000 \times 0.2 \times 10^{-3} \times 1 = 0.2$

    as, $A = \log(I_o/I)$

    the intensity ratio will be, $I_o/I = 1.58$

31. Sensitivity, since fluorescence intensity does not depend upon the concentration of the analyte unlike UV–Vis signal. Fluorescence intensity depends on the power of source so we can increase the signal intensity by taking high-power light source and can detect very low concentration of analytes. In fluorescence, life time of excited state is high and thus can be used to look at the event at ns scale.

32. As a fluorophore possesses larger dipole moment in excited state, it is more stabilized by a highly polar solvent and therefore, the emission shifts to the lower energy.

33. Quantum yield of a fluorophore is the measure of the efficiency of photon emission through fluorescence. Quantum yield 'Φ' is represented as,

    $$\Phi = \frac{\text{Number of photon emitted through fluorescence}}{\text{Number of photon absorbed}}$$

    Typical life time of fluorescence is around 10 ns.

34. Stoke's shift = 345–275 = 70 nm

35. $\dfrac{F_o}{F} = 1 + k_q \tau_o [Q]$

    $\tau_o = \left(\dfrac{F_o}{F} - 1\right)\Big/ k_q[Q] = \dfrac{1}{3.58\times 10^{11} \times 3\times 10^{-3}} = 9.31\times 10^{-10}\, s = 93.1\ ns$

36. All compounds give one single peak because all have only one type of hydrogen.

37. A peak of a proton nuclei splits when it comes in contact with the neighbouring proton nuclei. The distance between these peaks is known as coupling constant.

    The multiplicity of the signal of H_a is doublet and the relative intensity of the lines is 1:1.

38. Larmor frequency = $\gamma . B_o$ = 9.68 × 3.1 = 30.0 MHz

39. No. of observable spin states of the ^1H with spin $I = 2I + 1$

    Here, $I = ½$, therefore

    No. of observable spin state = 2

40. $v_o = \dfrac{g_N \mu_N}{h} B_o = \dfrac{5.585 \times 5.05 \times 10^{-27}}{6.63 \times 10^{-34}} \times 1.5 = 63.8\ MHz$

41. $C^aH_3N^bHCOOC^cH_2C^dH_3$

^{a}H: 2.73 (doublet because of neighboring 1H attached to N), ^{b}H: 5.2 (broad because H is attached to N), ^{c}H: 4.16 (quatret because of neighboring 3H) and ^{d}H: 1.28 (triplet because of neighboring 2H)

42. In mass spectroscopy, the given sample is first fragmented into its multiple ions which are then separated according to their mass-to-charge ratio (m/z). Using the relative abundance of each ion type, the molecular formula of the sample is determined.

43. It resolves the ions of the given sample into their characteristic mass components based on their mass-to-charge ratio.

44. M^+ at m/z = 181:
    implies an odd number of N
    $M^+ + 1$: 8.06/1.1 = 7.3: 7C
    $M^+ + 2$: 1:1
    Br is present
    Molecular Formula: C_7H_4BrN

45. McLafferty rearrangement is very famous in mass spectroscopy, organic compounds containing keto group undergoes γ cleavage by abstracting γ H atom forming radical cation. It was first described in 1959 by the Americal chemist Fred McLafferty.

46. A molecular ion peak is the peak in the mass spectrum that has resulted from the original molecule after gaining charge but retaining its mass. The base peak is the strongest signal that corresponds to the most stable ion but it is not necessarily the molecular ion peak, it can result from a fragment also.

47. (i) Molecular ion peak is fairly noticeable due to stabilization from aromatic ring.

    (ii) In aromatic rings substituted by alkyl group a prominent peak at m/z 91 is observed due to the formation of tropylium ion.

    (iii) A peak at m/z 65 can also be observed frequently resulting from the elimination of neutral acetylene molecule from tropylium ion.

    (iv) When alkyl group is longer than two carbon, a peak at m/z 92 is observed.

48. In AAS, atoms in gaseous state absorb energy and are excited to the higher state. Being unstable in higher state, they return to ground state with emission of radiation.

    Yes, the AAS is also called flame photometry because here the sample is sprayed into the flame.

49. The flame is used to reduce the sample into its atomic state.

50. The atomizer breaks the large mass of liquid into small drops. It also introduces the liquid sample into the flame at a steady rate.

51. Acetylene, propane, hydrogen are some of the fuels used in flame photometry.

52. Flame photometry is a very simple, economical, selective, and sensitive method for both qualitative and quantitative analysis. This method can be used to estimate the elements which are rarely analysed.

    However we cannot measure the accurate concentration of metal ion in solution by this method. It cannot directly detect and determine the presence of inert gases.

53. In a light microscope, light waves and glass lenses are used while in an electron microscope electron beams and magnetic field are used to produce the image. This gives much higher resolution with extremely short wavelength of the electron beam.

54. In SEM, the surface of the sample is irradiated by a very narrow beam of electrons. This irradiation leads to ejection of some low-energy electrons from the sample also known as secondary electrons. These secondary electrons are then collected on anode, a positively-charged plate thereby generating an electric signal.

55. The intensity of the secondary electrons depends upon the shape and the chemical composition of the irradiated sample. It also depends upon the number of electrons ejected and reabsorbed by the surroundings.

56. In SEM, the secondary electrons are collected on anode plate and results in contrast while in TEM the contrast results from the differential scattering of the electrons by the specimen. The degree of scattering depends upon the number and mass of the atoms lying in the electron path.

57 (i) Radiation of the source (should be monochromatic); (ii) Sample (should be either single crystal, powdered or a solid piece); and (iii) Detector (should be either radiation counter or photographic film).

58. Na^+ shows higher scattering than Li^+ because the amount of scattering is directly proportional to the amount of electrons present around the atom.

59. $n\lambda = 2d \sin \theta$

$$d = \frac{n\lambda}{2 \sin \theta} = \frac{2 \times 225}{2 \times \sin(25.3)} = 526.5 \text{ pm}$$

60. Conductivity is the ability to carry current. It is measured by the flow of electrons and charges through the conductor.

Conductance of an electrolytic solution depends upon temperature, pressure, number of charge carriers and dielectric constant of the solvent.

61. Dilution of the solution decreases the concentration of the solution and thereby decrease the specific conductance of the electrolyte.

62. $\Delta_m = \dfrac{1000 \times \kappa}{M} = \dfrac{1000}{M \times \rho} = \dfrac{1000}{0.5 \times 50} = 40 \text{ S.cm}^{-1}.\text{mol}^{-1}$

63. The initial and final temperatures of the study depend upon the heating rate, nature of the sample and the atmosphere above the sample.

64. In a DTA study of a hydrated sample, first of all dehydration occurs on heating and observed as an endotherm. After that, upon cooling the melt crystallizes exothermally with some polymeric changes but there is no rehydration takes place upon cooling.

65. A rapid TGA method is used to study the decomposition reactions isothermally. In such study, the TGA furnace is preset at high temperature and the sample is directly introduced at that temperature. This process may be repeated at set of different temperatures. Such analysis is used in determination of the reaction mechanism.

66. Differential thermal analysis is used as a powerful tool for determination of the phase diagram. Conjugation with X-ray diffraction increases the efficiency of DTA for the identification of crystalline phases present in the sample.

67. The principle of chromatography is based on the different rate of movement of solute in the column.

68. Retention time is the time taken by the analyte to reach the detector after sample injection. Dead time is the time taken by the molecules of the mobile phase to pass through the column.

69. In thin-layer chromatography (TLC), the stationary phase is a plate of plastic or glass coated with a thin layer of solid (e.g. silica or alumina). Multiple samples can be separated in one go, making TLC a very versatile technique.

70. It is defined as the ration of the moles of solute in stationary phase to that in mobile phase.

$R_f = 3.5/6 = 0.58$

71. Because of high thermal conductivity and lower density, hydrogen is dangerous to use as a carrier gas.

72. The efficiency of the liquid chromatography can be increased by reducing of the sample size and the diameter of the column.

Multiple Choice Questions

1. (a)	2. (b)	3. (a)	4. (b)	5. (a)	6. (d)	7. (a)
8. (d)	9. (b)	10. (d)	11. (b)	12. ()	13. (c)	14. (c)
15. (a)	16. (d)	17. (d)	18. (b)	19. (a)	20. (b)	21. (a)
22. (a)	23. (d)	24. (b)	25. (b)	26. (a)	27. (c)	28. (b)
29. (a)	30. (a)	31. (b)	32. (b)	33. (a)	34. (b)	35. (b)
36. (b)	37. (c)	38. (b)	39. (a)	40. (c)	41. (a)	42. (a)
43. (b)	44. (d)	45. (a)	46. (c)	47. (b)	48. (b)	49. (d)
50. (a)						

Part II: Applied Chemistry

Water Chemistry

LEARNING OBJECTIVES

After reading this chapter, you will be able to:

- classify the different forms of impurities and hardness in water.
- describe the various water softening methods.
- explain the various treatment methods for drinking water purposes.
- describe the methods of chemical analysis of water for chloride, fluoride, and sulphate in water.
- solve numerical problems based on water hardness, EDTA method, lime–soda process, zeolite, dissolved oxygen, and ion-exchange processes.

13.1 INTRODUCTION

Water is a critical and essential factor for the sustenance of all life forms on the earth. It is pertinent for today's engineers to have an in-depth understanding of the characteristics of renewable resource 'water' and its applications in varied fields, especially chemical, civil, mechanical, and particularly environmental engineering. However, it is not limited to these fields. Water is required in industrial processes such as steam generation, crystallization, scrubbing, extraction, and domestic uses include washing, cooking, bathing, and drinking. It is an important engineering material used for construction.

The earth's surface has about 1.386×10^9 km^3 of water, two-thirds of which (constituting about 97 %) exist as ice and polar caps. Of the 3 % freshwater available, only 1 % is usable water in the form of surface and ground waters. Chemically, water is a universal solvent and is known to dissolve almost everything. Water is a colourless, odourless, non-toxic, and renewable resource. Table 13.1 shows some of the characteristic properties of water.

According to the World Bank, nearly 77 million people in India lack access to safe drinking water, with over 21% human population inflicted with water-borne diseases. UNICEF (2013) reports India as a country under 'water stress'. The ever increasing population and overall shortage of natural water reservoirs have become a major concern today. Various initiatives have been taken to conserve and recycle water. In India, National Water Policy (1987), National Mission for Clean Ganga (2009), and National Water Mission (2011) are some of the initiatives that propagate water conservation, improve water quality, and its recycling methods. The UN annually releases World Water Development reports under World Water Assessment Programme (WWAP) highlighting water issues of different regions and measures to deal with water scarcity.

Table 13.1 Characteristic properties of water

Molar mass	18.09153 g/mol
Boiling and melting points (°C)	100 and 0 respectively
Triple point (°C)	0.01
Density of liquid at 25°C (g/cm^3)	0.9970
Dipole moment (Debye)	1.84
Surface tension at 20°C (mN/m)	72.8
H—O—H angle	104°
Bond length, H—O (pm)	96
Viscosity at 20°C (poise)	0.01

13.2 SOURCES AND IMPURITIES

There are various sources of water (Fig. 13.1) as movement of water is a dynamic process taking place on the earth, called the hydrological (or water) cycle which includes various processes such as evaporation, run-offs, and precipitation.

In spite of the variety of water sources, potable water is a rare commodity. All sources of water are contaminated and need conventional processes to treat them. Potable water is an improved form of drinking water deem fit for human consumption. Water is treated by municipalities to obtain potable water meeting drinking water standards.

Many dissolved impurities like cations (Ca^{2+}, Mg^{2+}, Na^+, K^+, Al^{3+}, traces of Zn^{2+} and Cu^{2+}), anions (Cl^-, SO_4^{2-}, NO_3^-, F^-), gases (CO_2, O_2, NH_3, H_2S), and organic salts are generally present in water. Dissolved gases cause acidity of water

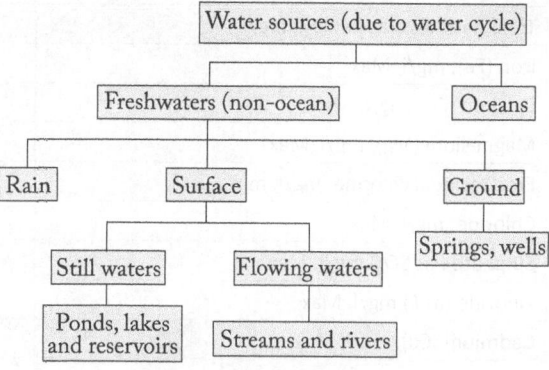

Fig. 13.1 Sources of water

while dissolved substances like $NaOH$, KOH, $Na(HCO)_3$, and Na_2CO_3, add alkalinity to water. Suspended impurities such as clay, sand, oils, sludge, particulates, and vegetable and animal matter are common types of impurities found in natural water bodies.

Physical impurities affect the quality of water such as its colour, odour, and taste. Metallic salts of iron, chromium, and manganese from industrial effluents impart colour to water. Any variation in water colour makes it unfit for drinking purpose. Colloidal, fine suspended particles of clay, organic/inorganic mater, and microbes tend to impart turbidity to water. The presence of impurities like decaying microorganisms, fungi, planktons, and industrial effluents also results in disagreeable odour that is also linked to the taste of water. Chemical impurities, which can be either organic or inorganic, dissolve in water and adversely affect its overall quality. Organic and inorganic compounds in industrial effluents are released in the form of dyes, paints, drugs, insecticides, detergents, etc., into water bodies resulting in pH changes, hardness, and higher dissolved oxygen demand of water.

Biological contamination of water is caused due to the presence of algae, bacteria, and or viruses. The disease-causing microorganisms are called *pathogens* that cause typhoid, dysentery, and gastroenteritis. *Giardia* and *Cryptosporidium* are commonly found in rivers contaminated with animal faeces and sewage. Radiological impurities emit radioactivity when they undergo decomposition. The sources of radioactive material are either geological rocks on which water moves through or industrial wastes. Table 13.2 presents BIS specifications (Bureau of Indian Standards) (ISO: 2012) for potable water and the acceptable (ideal) and permissible (allowed) limits for major elements are enlisted.

Table 13.2 Significant Indian Standard drinking water specifications (BIS: 10500:2012)

Parameters	Requirement (Acceptable or desirable limits)	Permissible limit (in absence of alternate source)
Colour, Hazen units, Max	5	15
Odour	Agreeable	Agreeable
Taste	Agreeable	Agreeable
Turbidity, NTU, Max	1	5
pH value	6.5–8.5	No relaxation
TDS (total dissolved solids), mg/l	500	2000

(Contd)

Table 13.2 (Contd)

Parameters	Requirement (Acceptable or desirable limits)	Permissible limit (in absence of alternate source)
Total alkalinity (as $CaCO_3$), mg/l	200	600
Total hardness (as $CaCO_3$), mg/l	200	600
Iron (Fe), mg/l, Max	0.3	No relaxation
Calcium (Ca), mg/l, Max	75	200
Magnesium (Mg), mg/l, Max	30	100
Free residual chlorine, mg/l, min	0.2	1
Chloride, mg/l, Max	250	1000
Sulphate (as SO_4) mg/l, Max	200	400
Fluoride (as F) mg/l, Max	1.0	1.5
Cadmium (Cd), mg/l, Max	0.003	No relaxation
Total chromium (Cr), mg/l, Max	0.05	No relaxation
Mercury (Hg), mg/l, Max	0.001	No relaxation
Cyanide (CN), mg/l, Max	0.05	No relaxation
Lead (Pb), mg/l, Max	0.01	No relaxation
Selenium (Se), mg/l, Max	0.01	No relaxation
Polynuclear aromatic hydrocarbons (PAH), mg/l, Max	0	No relaxation

13.3 HARDNESS OF WATER

Water containing dissolved salts that prevent lathering of soap is called *hard water*. The presence of large amounts of dissolved salts of calcium, magnesium, and other heavy metals imparts hardness to water. Water that lathers easily with soap is called *soft water* as it contains no dissolved calcium and magnesium salts. Soft water reacts with soap to give lather in the following way:

$$C_{17}H_{35}COONa + H_2O \rightarrow C_{17}H_{35}COOH + OH^- + Na^+$$

$$C_{17}H_{35}COONa + C_{17}H_{35}COOH \rightarrow Lather$$

The primary function of soap is to produce lather and act as a cleansing agent, but water hardness is a limiting factor. Calcium and magnesium ions tend to react with soaps (salts of fatty acids like stearic or palmitic acid) forming insoluble precipitates of calcium or magnesium stearate or palmitate.

$$2\,C_{17}H_{35}COONa + CaCl_2 \rightarrow (C_{17}H_{35}COO)_2\,Ca \downarrow (scum) + 2\,NaCl$$

$$2\,C_{17}H_{35}COONa + MgSO_4 \rightarrow (C_{17}H_{35}COO)_2\,Mg \downarrow (scum) + Na_2SO_4$$

The property of water to prevent lathering of soap is called *hardness of water*. In the above reactions, sodium stearate reacts with calcium forming an insoluble precipitate of calcium stearate. Similarly, magnesium present as a salt of sulphate reacts with sodium stearate forming magnesium stearate. Other metal ions such as Fe^{2+}, Mn^{2+}, and Al^{3+} also react with soap in a similar manner, thereby contributing to the hardness of water. However, for practical purposes only Ca^{2+} and Mg^{2+} are considered for deciding water hardness.

13.3.1 Types of Hardness

Based on the type of dissolved salts present in water, hardness is of two types, temporary and permanent hardness.

Temporary or Alkaline or Carbonate Hardness

This is due to the presence of dissolved bicarbonates of calcium and magnesium in water. As the name suggests, hardness of water can be easily removed by boiling. On boiling, bicarbonates get converted to insoluble carbonates and hydroxides which are removed by filtration process.

$$Ca(HCO_3)_2 \xrightarrow{\Delta} CaCO_3\downarrow + H_2O + CO_2$$

Calcium bicarbonate (soluble in water) → Calcium carbonate (insoluble ppt)

$$Mg(HCO_3)_2 \xrightarrow{\Delta} MgCO_3\downarrow + H_2O + CO_2$$

Magnesium bicarbonate (soluble in water) → Magnesium carbonate (insoluble ppt)

$$MgCO_3 + H_2O \xrightarrow[-CO_2]{\Delta} Mg(OH)_2\downarrow$$

Magnesium hydroxide (sparingly soluble)

Permanent or Non-alkaline or Non-carbonate Hardness

This is imparted by dissolved sulphates, chlorides, nitrates, phosphates of calcium, and magnesium in water. As the name suggests, permanent hardness cannot be removed easily on boiling, but requires special chemical treatment.

As per the definition, it is clear that the amount of dissolved salts in water is responsible for hardness. Further, temporary and permanent hardness together gives the total hardness of water. The water hardness limits are shown in Table 13.3.

When a water sample is sent for analysis in a laboratory, it is assumed that the sample contains different types of salts in varying concentrations. A reference standard is chosen so as to compare the hardness of water. Calcium carbonate is completely insoluble in

Table 13.3 Water classification as per Ca and Mg salt concentrations

Type of water	Dissolved Ca and Mg (mg/l)
Soft	0–50
Moderately soft	50–100
Slightly hard	100–150
Moderately hard	150–200
Hard	200–300
Very hard	> 300

water and hence, easy to form precipitates and quantified. Also, with a molecular weight of 100 (Ca – 40, C – 12, and O – 16) and equivalent weight of 50, it is easy to equate the masses of different impurities with calcium carbonate. Calcium carbonate ($CaCO_3$) is the reference standard and the total hardness of water is generally expressed in terms of equivalents of $CaCO_3$. The amount of dissolved impurities is always converted in terms of $CaCO_3$ equivalents as follows.

$$\text{Equivalent of } CaCO_3 = \frac{\text{Equivalent weight of } CaCO_3}{\text{Equivalent weight of hardness causing salt}} \times \text{Mass of hardness causing salt}$$

(13.1)

On substituting, we get,

$$\text{Equivalent of } CaCO_3 = \frac{50 \times W_{salt}}{E_{salt}}$$

(13.2)

where W_{salt} is the mass of hardness causing salt (mg/l) and E_{salt} is its equivalent weight.

One can also calculate hardness w.r.t molecular weight, hence, Eq. (13.2) can be written as,

$$\text{Equivalent of } CaCO_3 = \frac{100 \times W_{salt}}{M_{salt}}$$

(13.3)

where W_{salt} is the mass of hardness causing salt (mg/l) and M_{salt} is its molecular weight.

In this chapter, let us consider Eq. (13.2) as it is easier and when one considers the equivalents of $CaCO_3$, then one must calculate in terms of equivalent weights. Also, it will be easier to follow the same method for calculating hardness of complex Al and Fe salts.

13.3.2 Units of Hardness

Hardness of water is expressed in terms of calcium carbonate equivalents in various ways.

Parts per million (ppm)   This expresses the concentration of hardness causing salts as the number of parts of solute by weight in million parts (10^6) by weight of solvent (water). Ideally, 1 ppm hardness means one part of $CaCO_3$ equivalent of hardness present in one million parts of water.

$$1 \text{ ppm} \equiv 1 \text{ mg/litre}$$

Milligrams per litre (or mg/l)   This is the number of milligrams of $CaCO_3$ equivalent of hardness present in one litre of water.

$$1 \text{ mg/l} \equiv 1 \text{ mg } CaCO_3 \text{ equivalents in 1 litre of water}$$

We know, 1 litre = 1 kg = 1000 g (as density of water is 1 g/cm^3) = 1000×1000 mg = 10^6 mg

Thus, 1 mg/l = 1 mg of $CaCO_3$ equivalents of hardness in 10^6 mg of water or 1 ppm.

Grains per imperial gallon (gpg)   This is an expression where hardness units are reported in grains,

$$1 \text{ grain} = \frac{1}{7000} \text{ lb per gallon (10 lbs)}$$

Degree Clark (°Cl)   This is the number of grains of $CaCO_3$ equivalent of hardness per gallon of water. It is also expressed as parts of $CaCO_3$ equivalent of hardness per 70,000 parts of water.

$$1°\text{Clark} = 1 \text{ grain of } CaCO_3 \text{ equivalent hardness in one gallon of water}$$

or,   1 °Clark = 1 part present per 70,000 parts of water

French Unit (°Fr)   This is expressed as the number of parts of $CaCO_3$ equivalent hardness present in per 10^5 parts of water.

$$1°\text{Fr} \equiv 1 \text{ part of } CaCO_3 \text{ equivalent of hardness present per } 10^5 \text{ parts of water}$$

The various units of hardness are interrelated and by using the following information, hardness in one unit can be expressed in another way as follows:

$$1 \text{ ppm} \equiv 1 \text{mg/l} \equiv 0.1 \text{ °Fr} \equiv 0.07 \text{ °Cl}$$
$$1 \text{ mg/l} \equiv 1 \text{ ppm} \equiv 0.1 \text{ °Fr} \equiv 0.07 \text{ °Cl}$$
$$1 \text{ °Cl} \equiv 14.3 \text{ ppm} \equiv 14.3 \text{ mg/l} \equiv 1.43 \text{ °Fr}$$
$$1 \text{ °Fr} \equiv 10 \text{ ppm} \equiv 10 \text{ mg/l} \equiv 0.7 \text{ °Cl}$$

Table 13.4 Conversion factors for dissolved salts in terms of $CaCO_3$

Dissolved impurity/salt/ compound/ion	Molar mass	Chemical equivalent	Conversion factor for $CaCO_3$ equivalents (CF)
$CaCO_3$	100	50	50/50
$Ca(HCO_3)_2$	162	81	50/81
$Mg(HCO_3)_2$	146	73	50/73
$CaCl_2$	111	55.5	50/55.5
$MgCl_2$	95	47.5	50/47.5
$CaSO_4$	136	68	50/68
$MgSO_4$	120	60	50/60
$FeSO_4.7H_2O$	278	139	50/139
$Al_2(SO_4)_3$	342	57	50/57
CO_2	44	22	50/22
$Ca(NO_3)_2$	164	82	50/82
$Mg(NO_3)_2$	148	74	50/74

(Contd)

Table 13.4 (Contd)

Dissolved impurity/salt/compound/ion	Molar mass	Chemical equivalent	Conversion factor for CaCO₃ equivalents (CF)
H_2SO_4	98	49	50/49
HCl	36.5	36.5	50/36.5
Ca^{2+}	40	20	50/20
Mg^{2+}	24	12	50/12
OH^-	17	17	50/17
CO_3^{2-}	60	30	50/30
$NaAlO_2$	82	82	50/82
H^+	1	1	50/1

13.3.3 Disadvantages of Hard Water

There are innumerable disadvantages of using hard water for domestic and industrial purposes. Some of them are listed below.

(a) Washing: No lather formation due to dissolved salts leading to wastage of soap and the primary function of cleansing by soap is adversely affected.

(b) Drinking: Though hard water does not immediately causes health disorders, on continued drinking, can affect the metabolism. Calcium oxalate stones may develop in urinary tracts, leading to severe infections.

(c) Bathing: Hard water with soap forming insoluble white precipitate tends to stick to the body and cleansing action does not take place.

(d) Cooking: Due to dissolved salts, elevated boiling point is observed that increases the overall cooking time and wastage of fuel and energy resources.

(e) Textiles and paper industry: The dissolved salts of iron, manganese, and chromium react with dyes and result in the formation of undesirable colour on fabric and paper substrates.

(f) Sugar: The presence of chloride, sulphate, and nitrate salts of calcium and magnesium adversely affect crystallization of sugar, making it deliquescent.

(g) Boilers: Boilers are large containers to boil water and produce supersaturated steam. The dissolved calcium and magnesium salts tend to form scales and sludges in the boiler pipelines, which if unattended may result in corrosion of boilers and steamers. Besides these we will discuss some of the problems like priming and foaming in boilers in the following sections.

(h) Pharmaceutical industry: If hard water is used in drug, injection and ointments preparations, it results in undesirable products.

(i) Concrete: The sulphates and chlorides of calcium and magnesium tend to affect hydration and strength of concrete.

Check Your Progress

1. List the various sources of water.
2. Define hardness of water.
3. Why does hard water prevent lathering of soap?
4. State the relationship between ppm, mg/l, degree Clarke, and degree French.
5. Why is total hardness of water expressed in terms of $CaCO_3$ equivalents?
6. Name some salts responsible for causing temporary and permanent hardness in water.

13.4 BOILER TROUBLES / PROBLEMS

Boilers are designed to generate large volumes of steam and hot water continuously for days/weeks together. Water is utilized to generate steam in industries and power plants. Such water is called *boiler feed water*. The boiler feed water must be free from dissolved calcium and magnesium salts. If boiler feed water contains impurities beyond prescribed levels, they cause serious boiler problems such as: scales and sludges in boilers; boiler corrosion, and priming and foaming.

13.4.1 Scales and Sludges in Boilers

In boilers, water evaporates continuously and concentration of dissolved salts increases progressively. When the concentration reaches saturation point, they are thrown out of water in the form of precipitate in the inner walls of the boiler. If the precipitate is loose, slimy, and soft, it is called *sludge*. If the nature of precipitate is hard, crusty, and adherent, it is called *scale* (Fig. 13.2). Sludges tend to form due to $MgCO_3$, $CaCO_3$, and $CaCl_2$ having greater solubility in water in hot conditions. Sludges are easy to remove by simple scrapping process; however, scales are difficult to remove.

Scale formation in boilers is generally due to:

(a) The deposition of calcium bicarbonate as, $Ca(HCO_3)_2 \rightarrow CaCO_3 + H_2O + CO_2$.

(b) $CaSO_4$ deposition is yet another problem as its solubility decreases with increasing temperatures.

(c) Hydrolysis of magnesium salts such as $MgCl_2 + 2\,H_2O \rightarrow Mg(OH)_2 + 2\,HCl$, yields magnesium hydroxide, a type of scale in the boilers.

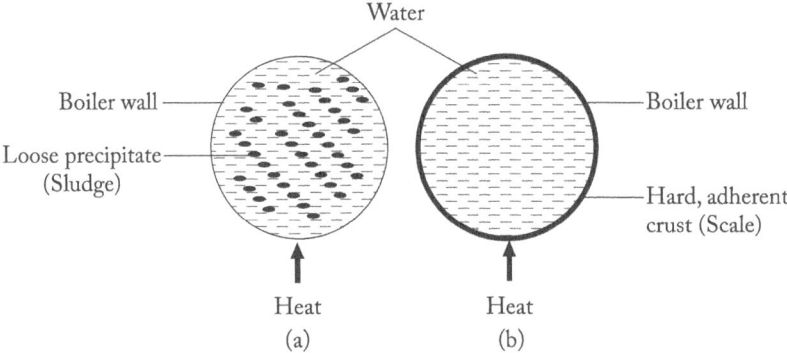

Fig. 13.2 (a) Sludge (b) Scale inside the boiler walls

(d) Trace amounts of silica in water results in the formation of $CaSiO_3$ and $MgSiO_3$ which are deposited as hard, crusty scales.

The major limitation is that such precipitates (scales and sludges) are poor conductors of heat, thereby adversely affecting boiler functioning. Also, such precipitate formation leads to loss of fuel and may also lead to corrosion of the inner boiler walls.

Prevention of Scale Formation

The following are some common methods adopted to remove or inhibit scale formation in boilers.

Colloidal conditioning   Scale formation can be prevented by adding organic substances such as agar and tannins. These compounds adsorb on the scale and form precipitates, thereby leading to non-sticky, loose deposits that can be easily removed.

Phosphate conditioning   In this case, sodium phosphate is added that reacts with scales forming non-adherent precipitates that are easily removed as sludge of calcium phosphate.

$$3\,CaCl_2 + 2\,Na_3PO_4 \rightarrow Ca_3(PO_4)_2 + 6\,NaCl$$

The demerit of this method is continuous blow-down boiler operation to remove calcium phosphate.

Carbonate conditioning   Calcium sulphate can be removed as a loose precipitate of calcium carbonate by adding sodium carbonate to the boiler water.

$$CaSO_4 + Na_2CO_3 \rightarrow CaCO_3 \text{ (loose)} + Na_2SO_4$$

Calgon conditioning   Chemically known as sodium hexametaphosphate, Calgon ($Na_2[Na_4(PO_3)_6]$) is commonly employed to eliminate scales and sludges. Calgon avoids sludge–scale formation by forming a soluble complex with $CaSO_4$.

$$Na_2[Na_4(PO_3)_6] \rightarrow 2\,Na + [Na_4P_6O_{18}]^{2-}$$
$$2\,CaSO_4 + [Na_4P_6O_{18}]^{2-} \rightarrow [Ca_2P_6O_{18}]^{2-} + 2\,Na_2SO_4$$

Sodium aluminosilicate   On hydrolysis, sodium aluminosilicate forms a gelatinous precipitate of aluminum hydroxide and sodium hydroxide. The resulting sodium hydroxide formed reacts with magnesium salts forming a soft sludge.

$$NaAlO_2 + 2\,H_2O \rightarrow NaOH + Al(OH)_3$$
$$MgCl_2 + 2\,NaOH \rightarrow Mg(OH)_2 + 2\,NaCl$$

Both, $Mg(OH)_2$ and $Al(OH)_3$ precipitates tend to trap particulate matter and their loose precipitates can be easily removed by blow-down operations.

Electrical conditioning   This is carried out by using a sealed glass bulb filled with mercury and connected to a battery and allowed to float in the boiler. When water boils, mercury boils emitting electrical discharges that inhibit the precipitates to adhere to inner boiler walls.

Radioactive conditioning   This involves using small tablets with radioactive salts placed in boiler water. When water boils, the tablets release radiations and inhibit scale–sludge formation.

13.4.2 Boiler Corrosion

The degradation or decay of the boiler body material due to various chemical or electrochemical reactions with its environment is called *boiler corrosion*. Boilers are primarily used in industries to generate supersaturated steam. Water fed in the boiler to generate steam is called *boiler feed water*. The boiler feed water should be free from the following impurities before being sent for steam generation.

 (i) Hardness causing salts such as Ca and Mg salts as these impurities causes scale formation.
 (ii) Oils, non-scaling dissolved impurities, and turbidity so as to avoid foaming in boiler water.
(iii) Mineral acids, caustic alkalies, dissolved gases such as oxygen and carbon dioxide so as to protect boiler material from corrosion or caustic embrittlement.

The mineral acids formed by hydrolysis of dissolved salts and dissolved gases such as oxygen and carbon dioxide are major factors contributing to boiler corrosion. As seen in earlier section, hydrolysis of magnesium salts forms hydrochloric acid that will react with iron present in boiler material leading to corrosive attack.

$$Fe + 2\,HCl \rightarrow FeCl_2 + H_2(g) \uparrow$$
$$FeCl_2 + 2\,H_2O \rightarrow Fe(OH)_2 \downarrow + 2\,HCl$$

The above reactions can be prevented by neutralizing the acids by adding alkali solutions thereby protecting the boiler from corrosion. The bicarbonate salts on decomposition produce dissolved carbon dioxide that can be eliminated by mechanical de-aeration or by adding calculated amounts of ammonium hydroxide.

$$2\,NH_4OH + CO_2 \rightarrow (NH_4)_2CO_3 + H_2O$$

Generally, about 8ppm of dissolved oxygen is present in water. However, the high temperatures in boilers cause oxygen to corrode the boiler material and if unchecked, may result in the formation of rust.

$$2\,Fe + 2\,H_2O + O_2 \rightarrow 2\,Fe(OH)_2\downarrow \text{ and, } 4\,Fe(OH)_2 + O_2 \rightarrow (Fe_2O_3\,.2H_2O)\,(rust)$$

Hydrazine, sodium sulphite, or sodium sulphide are used as oxygen scavengers.

$$N_2H_4\,(hydrazine) + O_2 \rightarrow N_2 + 2\,H_2O$$

$$2\,Na_2SO_3\,(sodium\ sulphite) + O_2 \rightarrow 2\,Na_2SO_4$$

$$Na_2S\,(sodium\ sulphide) + O_2 \rightarrow Na_2SO_4$$

Of the above, hydrazine is ideally suited as a reducing agent to remove oxygen since it produces non-toxic nitrogen as against other reagents that tend to form sulphate salts in water thereby causing increasing salt content in boiler water.

13.4.3 Priming and Foaming

While steam generation occurs in the boiler, rapid boiling causes some particles of liquid water to get carried along with the steam. The steam containing water droplets is called *wet steam* and the process of formation of wet steam in boilers is called *priming*. Dissolved solids, sudden boiling, higher water levels in boiler, high steam velocity are some of the factors that cause priming. Priming can be prevented by softening the feed water, fitting steam purifiers, and maintaining low water levels in the boiler.

Foaming is the formation of stable bubbles over the surface of boiler water. These foams do not break off easily and are carried over by steam leading to excessive priming. Priming and foaming generally occur together and tend to carry along the dissolved salts that get deposited on the turbine blades and other boiler machinery parts, thereby lowering the machine efficiency. Using anti-foaming agents such as synthetic polyamides and castor oil can prevent foaming.

13.4.4 Langelier Index

In many industries, a large amount of water is utilized for cooling purposes. The water used for cooling purposes should necessarily have the following characteristics:

(a) The water should be non-scaling, non-corrosive, and non-staining to the metal pipes and containers used as conveyors of cooling water.

(b) The cooling water should be free from colloidal impurities, algae, slime, and other organic matter.

Thus, cooling waters need to be treated for preventing scale and slime formation, corrosion of heat exchangers, condensers, cooling towers, and metal pipes conveying the water. The corrosive behaviour of water to be used as coolants is determined by Langelier index (L.I.) or $CaCO_3$ saturation index. An algebraic difference between the actual pH and the saturation pH is defined as *Langelier index* which is expressed as, L.I. = pH − (pH)$_s$; where (pH)$_s$ is the saturation pH.

Usually, carbonate hardness is removed by lime treatment. Sodium hexametaphosphate is used to prevent scaling of metal pipes. The suspended impurities are coagulated by settling and filtration methods. Chlorine, chloramine, and copper sulphate are added to prevent algae and slime growth in the cooling pipes. Corrosion of metals in cooling towers are prevented by: (a) pH control of water; (b) removal of oxygen from water by deaeration or mechanical agitation; and (c) using corrosion inhibitors such as sodium nitrite, tannins, and Calgon. Chapter 14 on corrosion discusses corrosion inhibitors in greater more detail.

The saturation pH can be expressed as,

$$(pH)_s = \log\frac{K_{sp}}{K_d} - \log\left[Ca^{2+}\right] - \log\left[HCO_3^-\right] \tag{13.4}$$

where, K_{sp}, the solubility product of calcium carbonate ($CaCO_3$) is expressed as,

$$K_{sp} = \left[Ca^{2+} \right]\left[CO_3^{2-} \right]$$

and, K_d is the dissociation constant for the reaction, $HCO_3^- \rightleftharpoons H^+ + CO_3^{2-}$.

$[Ca^{2+}]$ denotes calcium hardness expressed as $CaCO_3$ equivalents in mg/l and $\left[HCO_3^- \right]$ is called bicarbonate alkalinity. Hence, saturation pH can be expressed as:

$$(pH)_s = \left(pK_d - pK_{d2} \right) + p\left[Ca^{2+} \right] + p\left[alk \right] \tag{13.5}$$

Here, p represents reciprocal log, that is, $px = \log_{10} x^{-1} = -\log_{10} x$. K_d and K_{d2} are the dissociation constants of bicarbonate ions and $CaCO_3$ salt, respectively. The term $[alk]$ represents the titratable equivalents of alkali per 1000 ml. The above expression also depicts that one can alter the L.I. by adjusting water pH, alkalinity, and calcium hardness.

If L.I. is zero, water is considered as stable and fit for use as cooling water. A positive L.I. indicates scale formation, while negative L.I. implies chances of corrosion. For general industrial cooling purposes, it is difficult to maintain L.I. of water as zero (practically not possible due to environmental and storage conditions). Hence, L.I. of water used for cooling purposes is always maintained within the range of 0.4 – 1.0.

13.5 CHEMICAL ANALYSIS OF WATER

There are various parameters of water that can be quantitatively measured such as: (a) total hardness, (b) alkalinity; (c) chlorides; (d) sulphates, and (e) fluorides.

13.5.1 Determination of Hardness of Water by Complexometry

Any chemical process that eliminates or reduces hardness of water is called *softening of water*. Before, subjecting the sample to water softening process, one needs to determine its hardness. Complexometric titrations are used to estimate the total amount of Ca^{2+} and Mg^{2+} ions in water. The procedure involves titration of the water sample with standard EDTA solution using an organic dye indicator such as EBT (Eriochrome Black T).

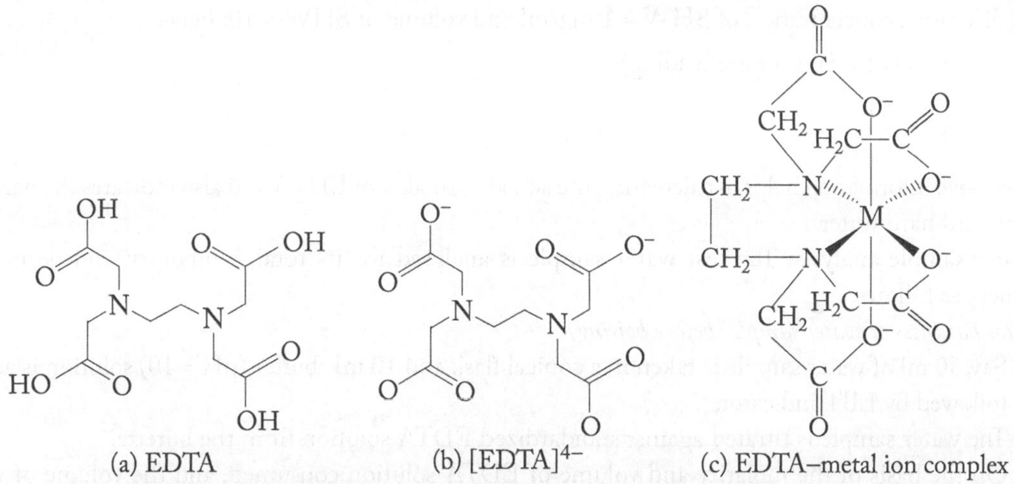

(a) EDTA   (b) [EDTA]$^{4-}$ (c) EDTA–metal ion complex

Fig. 13.3 Structures of EDTA; tetracarboxylate ion, formed via EDTA dissociation in buffered solution, and its coordination complex

In an aqueous solution buffered at pH 10, EDTA dissociates to form the maximum amount of tetracarboxylate ion, $[EDTA]^{4-}$. This ion is electron-rich as it has six bonding sites, four carboxylate groups and two nitrogen atoms, (Fig. 13.3) and each site has electron pair available for bonding with the metal ion. The $[EDTA]^{4-}$ anion forms a complex with Ca^{2+} or Mg^{2+} ions by sharing its six electron pairs with them. The reactions occurring during EDTA titration are:

Metal ion + EBT (blue) → [Metal ion ---EBT]

[Metal ion ---EBT] (wine red unstable complex) → [Metal ion —EDTA] (stable) + EBT (blue)

Experimental Procedure of EDTA Titration

Preparation of standard hard water (1 mg/ml concentration): A standard hard water (SHW) is a reference solution that is used to compare hardness of water sample. As already discussed, calcium carbonate is chosen to prepare standard hard water. One gm of pure and dry calcium carbonate is dissolved in a minimum quantity of dilute hydrochloric acid. The resulting solution is evaporated to dryness on a water bath until a white residue is formed. The residual salt is dissolved in double-distilled water and the solution is diluted to one litre. The final hardness of the solution should be 1 mg of $CaCO_3$ equivalent/ml.

EDTA solution is generally available as its disodium salt; hence it can be prepared in water. The concentration should be known while performing the experiment. EDTA solution will act as the complexing agent. Eriochrome Black T (denoted as EBT) indicator is prepared in ethanol. As complexes are formed at pH of 10, a range of ammonia-based buffer solutions can be used; for example, NH_4Cl + conc.NH_3, NH_4Cl + NH_4OH, liquid ammonia, etc., are some commonly employed buffer systems.

Standardization of EDTA solution: In this case, 50 ml of standard hard water is taken in a conical flask and 10 mL of ammonia-based buffer is added to it. The pH is maintained between 8.5 and 10. After adding the buffer, about 4 – 5 drops of EBT indicator is added. The solution is titrated against a known strength of EDTA solution from the burette until a colour change is observed. The distinct colour changing from wine red to deep sea blue is easily observed and recorded.

Stoichiometrically, EDTA ≡ SHW

$$C_1V_1 \equiv C_2V_2$$

As discussed, concentration of SHW = 1 mg/ml and volume of SHW = 10; hence

$$x \times V_1 \equiv 1 \times 50 \text{ (burette reading)}$$

Thus, $x = 1 \times \dfrac{50}{V_1}$

The above information helps in calculating the actual normality of EDTA and also indicates the hardness of standard hard water.

Water sample analysis: The test water sample is analysed for its total, temporary, and permanent hardness as follows:

Total hardness of water sample (before boiling)

(i) Say, 50 ml of water sample is taken in a conical flask and 10 mL buffer (pH = 10) solution is added, followed by EBT indicator.

(ii) The water sample is titrated against standardized EDTA solution from the burette.

(iii) On the basis of the molarity and volume of EDTA solution consumed, and the volume of water titrated, one can calculate the total Ca^{2+} and Mg^{2+} ion concentration in water sample in terms of $CaCO_3$ equivalents of hardness.

Permanent hardness of water sample (after boiling)

(i) Finally, 50 mL of water sample is boiled off and the salts causing temporary hardness are removed by filtration.

(ii) The filtrate is diluted with double distilled water and analysed for permanent hardness by titrating it against standard EDTA solution.

(iii) Once, total and permanent hardness of water are analysed, one can calculate temporary hardness by subtracting the values.

13.5.2 Determination of Chloride by Argentometry

Chloride ions are present as $NaCl$, $CaCl_2$, $MgCl_2$, etc. in natural waters. The acceptable limit of chloride in water is 250 ppm (BIS 2012) and can be determined by argentometric method. Argentometry is a precipitation method that involves the use of silver nitrate as the precipitating agent. Potassium chromate (K_2CrO_4) is used as an indicator to detect the end point of precipitation reaction.

$$AgNO_3(aq) + Cl^-(aq) \rightarrow AgCl(s) \downarrow (white\ ppt) + NO_3^-(aq)$$

$$2\ AgNO_3(aq) + CrO_4^{2-}(aq) \rightarrow AgCrO_4(s) \downarrow (reddish-brown\ ppt) + 2\ NO_3^-(aq)$$

After all the chloride ions have precipitated as white silver chloride ($K_{sp} = 1.82 \times 10^{-10}$ at 25 °C), the added extra silver ions react with chromate ions that results in a reddish-brown precipitate of silver chromate at the end of the reaction.

Procedure

(i) 10 ml of standard NaCl solution is taken in a conical flask. About 2 ml of potassium chromate is added as indicator. The solution turns yellow.

(ii) The solution is titrated with $AgNO_3$ solution from the burette until a persistent reddish-brown colour is observed. Record the burette reading and calculate the molarity of $AgNO_3$ solution.

(iii) About 100 ml of the water sample is pipetted out into a clean conical flask and 3–4 drops of K_2CrO_4 are added to the water sample.

(iv) The water sample is titrated against standard $AgNO_3$ solution (from step(i)). The end point of the titration is observed as the colour changes from yellow to reddish-brown.

The chloride content in water is determined as follows:

$$1000\ ml\ of\ 1M\ AgNO_3 \equiv 35.5\ g\ chloride$$

$$\text{Concentration of chloride ions} = \frac{V \times \text{Molarity of } AgNO_3 \times 35.5}{1000} \text{ g chloride per 100 ml water.}$$

$$= V \times \text{molarity of } AgNO_3 \times 35.5 \times 10\ mg\ chloride/100\ ml\ water\ sample$$

Check Your Progress

7. What is boiler feed water?
8. Why is boiler feed water softened before use?
9. What are scales and sludges in boilers?
10. What is Calgon conditioning? Write the chemical reactions.
11. What is colloidal conditioning?
12. Why is Calgon conditioning better than phosphate conditioning for boiler treatment?
13. What is EDTA? State its application.
14. Why is $NH_4OH–NH_4Cl$ buffer added while determining water hardness by EDTA method?
15. Name the indicator used in EDTA method. Mention the end point observed in the titration.

13.5.3 Determination of Sulphate by Gravimetry

Sulphate occurs naturally in drinking waters. Most of the sulphate compounds found in nature are due to oxidation of sulphite ores, shales, and rain water. The presence of higher concentrations of sulphate causes digestive disorders such as flatulence and diarrhoea. According to BIS 2012, the allowed limit is about 200 ppm of sulphates in drinking water. Gravimetric analysis of sulphate involves the precipitation of sulphates using a suitable reagent like barium chloride. Barium chloride is preferred since the resulting precipitate, $(BaSO_4)$ is of low solubility (K_{sp} = 1.1 $\times 10^{-10}$ at 25°C that allows for accurate quantitative analysis).

One can determine sulphate concentration in water by gravimetric methods described as follows

Procedure
  (i) Prepare a standard $BaCl_2$ solution. Dissolve 10 g of barium chloride dihydrate ($BaCl_2.2H_2O$) in water and dilute it with deionized water; make up to 100 ml in a volumetric flask.
 (ii) Pipette out 200 ml of the water sample into a 500 mL clean beaker. Add 3 ml of bromine solution (or saturated bromine water) and 1 ml of concentration HCl.
(iii) Heat the mixture to boiling until all the bromine leaves off (Careful: perform this step in fume hood). Now, make up the volume to 200 ml with deionized water.
 (iv) Using a graduated pipette, add 10 ml of barium chloride dropwise into the solution with continuous stirring.
  (v) Keep on boiling the solution for 4–5 min. Place the beaker on a water bath and stir the contents intermittently to obtain the precipitate.

$$BaCl_2 + SO_4^{2-} \rightarrow BaSO_4 \downarrow + Cl^-$$

 (vi) Allow the precipitate to settle completely at the bottom of the beaker.
(vii) Filter the $BaSO_4$ precipitate using Whatmann No. 42 by vacuum suction and quantitatively transfer all the precipitate on the filter paper.
(viii) Wash the precipitate with mother liquor and transfer it in to a clean silica crucible.
 (ix) Incinerate the filter paper with residual precipitate in a silica crucible and heat the residue in an electric oven at 800 °C for 30 min.
  (x) Cool the residue in the crucible and determine the mass of the precipitate using an electronic balance.

The concentration of water-soluble sulphates is calculated as follows:

Consider, 233.4 g of $BaSO_4$ contains 96.06 g of sulphate.

Hence, W g of $BaSO_4$ = $\dfrac{96.06 \times W}{233.4}$ g of sulphate

As we had taken 200 ml water sample, it will contain = $\dfrac{96.06 \times W}{233.4}$ g of sulphate

Hence, for 1000 ml water sample = $\dfrac{96.06 \times W \times 1000}{233.4 \times 200}$ g of sulphate

The gravimetric analysis of sulphate is accurate and reported in terms of ppm or mg/l.

13.5.4 Determination of Fluoride by Colorimetry

About 0.3 % of the earth's crust consists of fluoride in the form of compounds such as sodium fluoride and hydrogen fluoride that are present in minerals such as fluorspar, fluorapatite, cryolite, and topaz. Thus, fluoride occurs in water supplies naturally. Higher concentration of fluorides in water can cause dental fluorosis, whereas its deficiency results in dental decay. The acceptable level of fluoride concentration in drinking water should be less than 1.5 ppm. The major sources of fluoride are various industries that manufacture glass, pesticides, steel, and ceramics. Fluoride in water is usually calculated by SPADNS method.

SPADNS Method of Fluoride Determination (Colorimetric Analysis)

Principle   Chemically, SPADNS is trisodium salt of 4,5-dihydroxy-3-(4-sulphophenylazo)-2,7-naphthalene disulfonic acid. For determining fluoride levels in water, zirconyl acid-SPADNS is employed as a reagent. SPADNS–$ZrOCl_2$ is a red coloured complex that changes colour when it reacts with fluoride. The change in concentration of SPADNS–$ZrOCl_2$ causes a change in the transmitted light that can be detected by the colorimeter. SPADNS reacts with zirconyl chloride to give a wine-red complex which further reacts with fluoride to give a new complex as shown in Fig. 13.4. This method is rapid and accurate for determining fluoride levels in water samples.

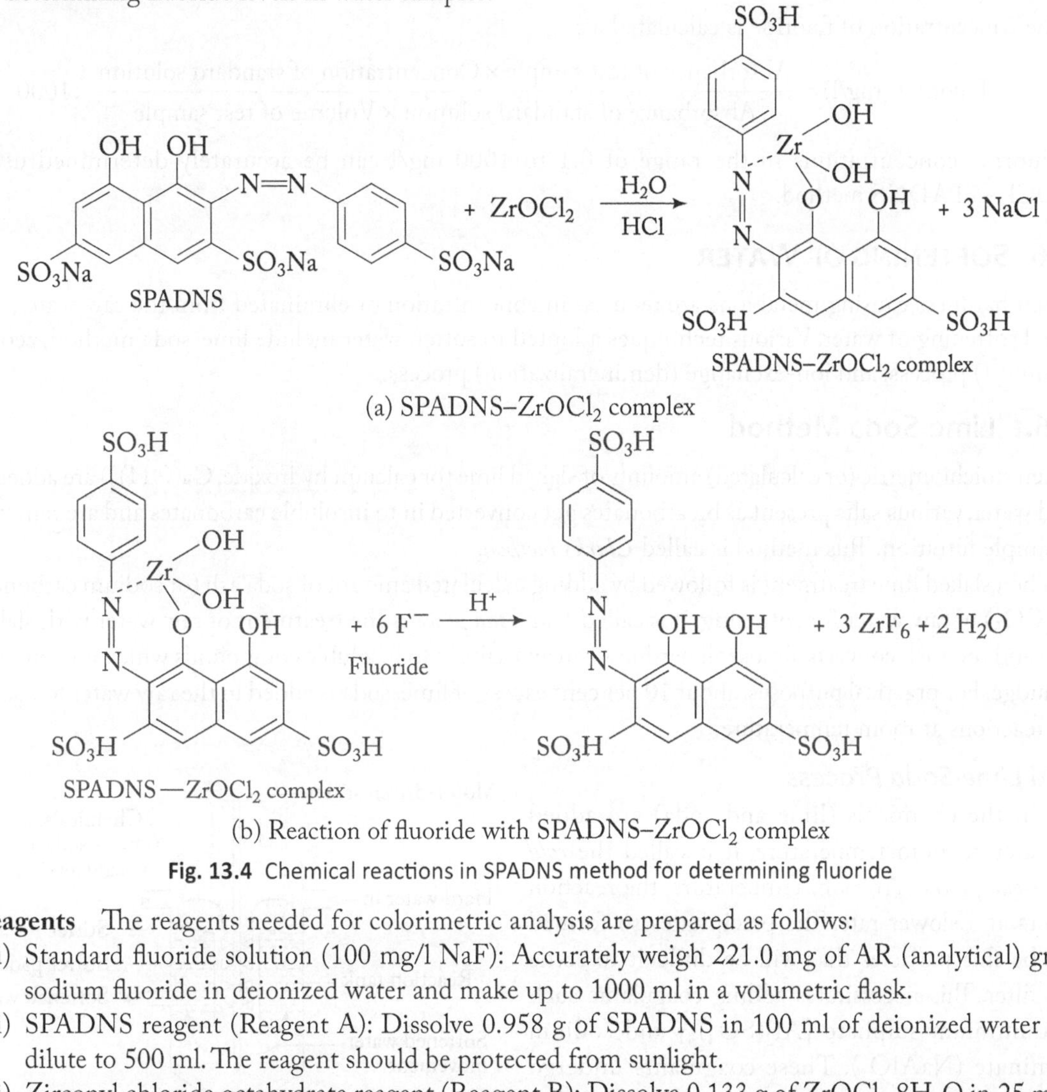

(a) SPADNS–$ZrOCl_2$ complex

(b) Reaction of fluoride with SPADNS–$ZrOCl_2$ complex

Fig. 13.4 Chemical reactions in SPADNS method for determining fluoride

Reagents   The reagents needed for colorimetric analysis are prepared as follows:
 (i) Standard fluoride solution (100 mg/l NaF): Accurately weigh 221.0 mg of AR (analytical) grade sodium fluoride in deionized water and make up to 1000 ml in a volumetric flask.
 (ii) SPADNS reagent (Reagent A): Dissolve 0.958 g of SPADNS in 100 ml of deionized water and dilute to 500 ml. The reagent should be protected from sunlight.
 (iii) Zirconyl chloride octahydrate reagent (Reagent B): Dissolve 0.133 g of $ZrOCl_2.8H_2O$ in 25 ml of deionized water. Add 350 ml of conc. HCl (350 ml) to it and make up the volume to 500 ml with deionized water.
 (iv) Zirconyl acid–SPADNS reagent: Mix Reagents A and B in equal volumes and protect it from sunlight.
 (v) A reference solution (blank reagent) is prepared by mixing 20 ml of distilled water with 5 ml of $ZrOCl_2$–SPADNS reagent. Store in a 50 ml volumetric flask in a dark place.

Procedure

(i) Prepare a series of solutions of sodium fluoride in the range of 0.005 – 0.150 mg/l.

(ii) Measure the absorbance of the blank reagent using a colorimeter set at 570 nm. Next, measure the absorbances of all the solutions of sodium fluoride (from step (i)) and plot a calibration plot.

(iii) Take 20 ml of the test water sample in a test tube and add 4 ml of $ZrOCl_2$–SPADNS reagent along with the blank reagent. The reaction mixture is allowed to continue for 30 min until complete development of colour. The absorbance of the reaction mixture is recorded using a colorimeter set at 570 nm.

The concentration of fluoride is calculated as:

$$\text{Fluoride (mg/l)} = \frac{\text{Absorbance of test sample} \times \text{Concentration of standard solution}}{\text{Absorbance of standard solution} \times \text{Volume of test sample}} \times 1000$$

Fluoride concentrations in the range of 0.1 to 1000 mg/l can be accurately determined using $ZrOCl_2$–SPADNS method.

13.6 SOFTENING OF WATER

When hardness causing metal ions are reduced in concentration or eliminated from the raw water, it is called softening of water. Various techniques adopted to soften water include lime-soda method, zeolite (permutit) process, and ion-exchange (demineralization) process.

13.6.1 Lime-Soda Method

When stoichiometric (or calculated) amounts of slaked lime (or calcium hydroxide, $Ca(OH)_2$) are added to hard water, various salts present as bicarbonates get converted in to insoluble carbonates and are removed by simple filtration. This method is called *Clark's method*.

When slaked lime treatment is followed by adding calculated amount of soda ash (or sodium carbonate, Na_2CO_3) in raw water for softening, it is called *lime-soda process*. The treatment of raw water with slaked lime and soda ash converts almost all hardness causing salts in to insoluble compounds which are removed as sludge. For practical purposes, about 10 per cent excess of lime-soda is added in the raw water to hasten the reactions at room temperature.

Cold Lime-Soda Process

When the chemicals (lime and soda) are added to water at room temperature, it is called the *cold lime-soda process*. At room temperature, the reaction occurs at a slower rate. The precipitates so formed will be fine particles that will be difficult to settle and filter. This necessitates adding coagulants such as aluminium sulphate $[Al_2(SO_4)_3]$ and sodium aluminate $(NaAlO_2)$. These coagulants undergo hydrolysis forming gelatinous precipitates that trap the particulate matter and fine particles and eventually allow faster sedimentation.

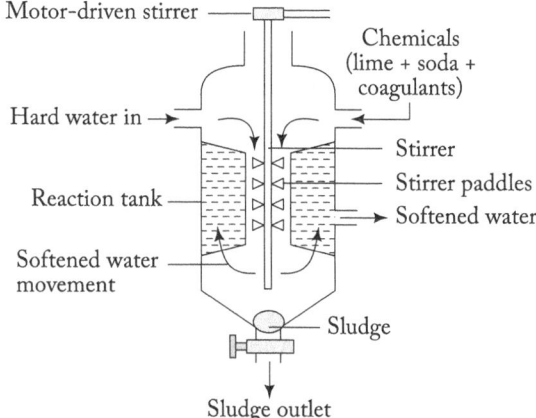

Fig. 13.5 Cold lime-soda softener

$NaAlO_2 + 2H_2O \rightarrow NaOH + Al(OH)_3$ (ppt)

$Al_2(SO_4)_3 + 3\ Ca(OH)_2 \rightarrow 2\ Al(OH)_3$ (ppt) $+ 3\ CaSO_4 + 6\ CO_2$

In Fig. 13.5, cold lime-soda softener is shown, that is, a reaction tank equipped with stirrer and paddles. Water to be softened is sent in the reaction tank through the raw water feed inlet and the chemicals inlet allows addition of lime, soda, and the coagulants. The quantities of chemicals are decided on the basis of the volume of water to be treated. Generally, 10–12 % excess chemicals are added to hasten up the reactions and the overall water softening.

The motor-driven stirrer enables vigorous agitation for the thorough mixing of chemicals and water. During softening, soft water tends to rise upwards due to reduced viscosity and sludge settles down which is removed and sent for disposal. The softened water is filtered and collected in suitable vessels. Cold lime-soda process results in treated water with residual hardness in the range of 50–60 ppm.

Hot Lime-soda Process

Hot lime-soda method is similar to cold lime-soda process except that the reactions are carried out at higher temperatures maintained at about 80–110 °C and there is no need for coagulants. So, it is obvious that as against cold method, there are several advantages of hot lime-soda method.

A typical hot lime-soda softener (Fig. 13.6) has a reaction tank and a sand filter system. Raw water is sent in to the reaction tank. Superheated steam is passed through the tank at 110 °C along with the chemicals. The softened water, so obtained, is sent to the sand-filtration system. The filter system consists of fine, coarse, and gravel sand that removes particulate matter and other impurities from the water.

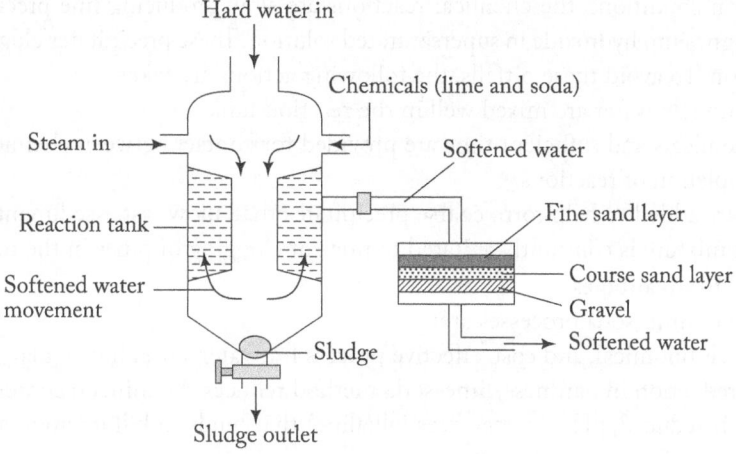

Fig. 13.6 Hot lime-soda softener

As the reaction takes place at high temperature, the following advantages are noted:
(a) The reaction will be faster, as lime and soda will quickly react with hardness causing metal ions with quicker formation of precipitates.
(b) As precipitates are coarse in nature, the resulting sludge settles down rapidly without coagulants.
(c) Dissolved gases such as oxygen and carbon dioxide are easily removed.
(d) Due to reduced viscosity of soft water, it can be easily filtered.
(e) The overall residual hardness is much lower, about 15–30 ppm as compared to the cold method.
The soft water obtained from lime-soda process is used for feeding boilers.

Reactions with lime   The following reactions take place in the lime-soda process:
(i) with free acids (HCl, H_2SO_4, etc.)

$$2\ HCl + Ca(OH)_2 \rightarrow CaCl_2 + 2\ H_2O$$

$$H_2SO_4 + Ca(OH)_2 \rightarrow CaSO_4 + 2\ H_2O$$

(ii) with aluminum and iron salts

$$Al_2(SO_4)_3 + 3\ Ca(OH)_2 \rightarrow 2\ Al(OH)_3 + 3\ CaSO_4$$
$$FeSO_4 + Ca(OH)_2 \rightarrow Fe\ (OH)_2 + CaSO_4$$

(iii) with dissolved CO_2 gas

$$CO_2 + Ca(OH)_2 \rightarrow CaCO_3 + H_2O$$

(iv) with calcium and magnesium salts (temporary Ca, Mg hardness, and permanent Mg salts)

$$Ca(HCO_3)_2 + Ca(OH)_2 \rightarrow 2\ CaCO_3 + 2\ H_2O$$
$$Mg(HCO_3)_2 + 2\ Ca(OH)_2 \rightarrow Mg(OH)_2 + CaCO_3 + 2\ H_2O$$
$$MgCl_2 + Ca(OH)_2 \rightarrow Mg(OH)_2 + CaCl_2$$

(v) with HCO_3^- ions (eg., $NaHCO_3$, $KHCO_3$)

$$2NaHCO_3 + Ca(OH)_2 \rightarrow CaCO_3 + 2H_2O + Na_2CO_3$$

Reactions with soda Soda removes all the soluble permanent hardness due to calcium salts (present originally and also those introduced during the removal of Mg^{2+}, Fe^{2+}, Al^{3+}, HCl, H_2SO_4, CO_2, etc., by lime.)

$$CaCl_2 + Na_2CO_3 \rightarrow CaCO_3 + 2\ NaCl$$
$$CaSO_4 + Na_2CO_3 \rightarrow CaCO_3 + Na_2SO_4$$

In actual reaction conditions, the chemical reactions are slow producing fine precipitates of calcium carbonate and magnesium hydroxide in supersaturated solution. These precipitates clog the reaction tanks and cause corrosion. To avoid these pitfalls, the following actions are taken.

(a) Chemicals and raw water are mixed well in the reaction tank.
(b) Excess of chemicals and sufficient time are provided for contact between chemicals and raw water to allow completion of reactions.
(c) Coagulants are added to help form coarse precipitates that allow faster sedimentation of sludge.
(d) The reaction mixture is constantly agitated to prevent clogging of pipes in the reaction tank.

Advantages and Disadvantages

Some **advantages** of lime-soda processes are:

(i) It is a simple, economical, and cost-effective process for water softening at a large scale.
(ii) Along with reduction in hardness, lime-soda method reduces the mineral content in water.
(iii) As hardness is reduced, pH of water rises (alkaline) that tends to kill microorganisms to a certain extent.

Some **disadvantages** include:

(i) The large amount of sludge sent for disposal makes this method to show adverse environmental impact.
(ii) As there is residual hardness in soft water, it is not employed in high-pressure boilers.

Points to remember while solving numerical problems

(i) NaCl, KCl, KNO_3, SiO_2, Fe_2O_3 do not react with lime and soda; hence, ignore their values.
(ii) One equivalent of $Mg(HCO_3)_2$ consumes two equivalents of lime.
(iii) Impurities such as $CaCO_3$ and $MgCO_3$ are considered as bicarbonate hardness of calcium and magnesium respectively.
(iv) Extra bicarbonate ions consume one equivalent of lime and also produce one equivalent of CO_3^{2-}. So, their value should be added with lime and subtracted from soda.

$$2\ HCO_3^- + Ca(OH)_2 \rightarrow CaCO_3 + CO_3^{2-} + 2\ H_2O$$

(v) Sodium aluminate is added as a coagulant. It forms one equivalent of OH⁻ (hydroxyl) ions that is equivalent to one equivalent of lime; hence, this amount is subtracted from lime.

$$NaAlO_2 + 2\,H_2O \rightarrow NaOH + Al(OH)_3$$

(vi) If purity factors of lime and soda are not mentioned, consider them as 100 % pure.

The general formula for lime requirement is as follows,

$$= \frac{74}{100}\left[\begin{array}{l}\text{Temporary Ca.hardness} + 2 \times \text{Temp. Mg hardness} + \text{Perm. (Mg + Fe + Al)} \\ \text{hardness} + (HCl + H_2SO_4 + HCO_3^- + CO_2 - NaAlO_2)\ CaCO_3\ \text{equivalents}\end{array}\right] \times \frac{V_{water}}{10^6}\ 100/\%\ \text{purity}$$

here, 74 is mol. wt of lime and 100 is mol. wt of $CaCO_3$.

The general formula for soda requirement is as follows,

$$= \frac{106}{100}\left[\begin{array}{l}\text{Perm.Ca hardness} + \text{Perm.(Mg + Fe + Al)hardness} + \\ (HCl + H_2SO_4 - HCO_3^-)\ CaCO_3\ \text{equivalents}\end{array}\right] \times \frac{V_{water}}{10^6}\ 100/\%\ \text{purity}$$

where, 106 is the molecular weight of soda and 100 is the molecular weight of calcium carbonate.

13.6.2 Zeolite (Permutit) Process

Coming under the class of sodium aluminosilicates, zeolites (or permutit) are commercial adsorbents for softening water. The hydrated form of zeolite has the general formula, $Na_2O.\,Al_2O_3.\,xSiO_2.\,yH_2O$, where x is $2 - 10$ and y is $2 - 6$. Zeolite is also denoted as Na_2Ze, where Ze refers to zeolite.

A long glass column is packed with sodium zeolite called zeolite bed (Fig. 13.7). Hard water is passed through the zeolite bed at a constant rate. During softening, zeolite reversibly exchanges hardness-causing ions present in the water sample with sodium ions. The softened water is collected at the bottom of the column and is withdrawn after each cycle.

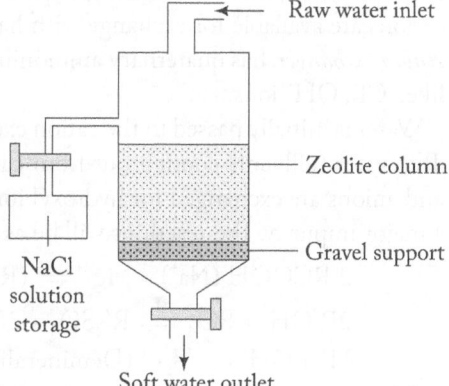

Fig. 13.7 Zeolite water softening process

When raw water flows through sodium zeolite, Ca^{2+} and Mg^{2+} ions are adsorbed by the zeolite pores and simultaneously release equivalent Na^+ ions in exchange for them. If calcium chloride is the major impurity to be removed from water, the following reaction is carried out along the zeolite column.

$$CaCl_2 + Na_2Ze \rightarrow CaZe + 2\,NaCl$$

In the above reaction sodium zeolite gets converted to calcium zeolite. If the softening cycles are continued, zeolite column gets completely exhausted and requires regeneration. The zeolite bed is treated with 10 % brine solution. So, during regeneration of zeolite column, the following reaction will occur.

$$CaZe + 2\,NaCl \rightarrow Na_2Ze + CaCl_2$$

So, it is clear that sodium zeolite is obtained again and fresh cycles of water softening can be resumed.

Some of the **advantages** of zeolite softening are:
(i) Softened water of about 10 ppm residual hardness is produced after the process.
(ii) The zeolite pores automatically adjust themselves for varying hardness of incoming water.
(iii) The overall process requires less skill in maintenance, preparation, and operation.

Some **disadvantages** are as follows.
(i) Zeolite replaces only Ca^{2+} and Mg^{2+} ions in exchange of Na^+ ions, but leaves out ions such as HCO_3^- and CO_3^{2-} in the treated water.
(ii) The treated soft water containing $NaHCO_3$, Na_2CO_3 tends to decompose and releases dissolved CO_2 thereby causing corrosion of steel walls of the containers.

13.6.3 Demineralization Using Ion-exchange Method

When water is free from all cations and anions, it is called demineralized or deionized water. Ion-exchange method is an advanced sophisticated analytical technique that can achieve demineralization of water. Ion-exchange resins are insoluble, cross-linked, long-chain organic polymers with porous structure. During the process, there is a reversible exchange of ions of similar charge between the mobile liquid phase (water) and the insoluble solid stationary phase (functional groups attached to the long-chain polymers). Generally, there are two types of exchangers, cation exchanger and anion exchanger.

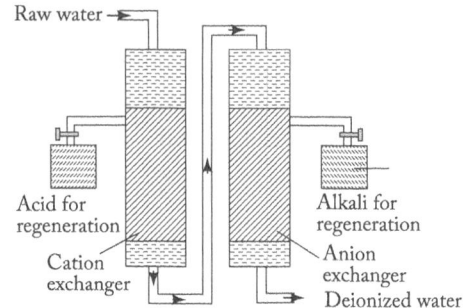

Fig. 13.8 Ion-exchange process for demineralization

Cation exchanger, as the name suggests have functional groups such as sulphonic ($-SO_3H$), carboxylic (-COOH), or phenolic (–OH) groups as part of the resin and an equivalent amount of cations. These cations are available for exchange with hardness-causing ions present in water.

Anion exchanger, has quaternary ammonium groups ($-N^+R_2$) containing an equivalent amount of anions like, Cl^-, OH^- ions.

Water is initially passed to the cation exchanger in acid form. All cations will be exchanged for H^+ ions. The water (effluent) coming out from cation exchanger is passed in to anion exchanger in basic form and anions are exchanged for hydroxyl ions. Let us consider that water contains magnesium sulphate as a major impurity. The reactions will be as follows:

$$2\ RCOOH\ (Na^+) + Mg^{2+} \rightleftharpoons (RCOO)_2Mg + 2H^+ + 2Na^+\ (Cation\ exchange)$$

$$2R'OH + SO_4^{2-} \rightleftharpoons R'_2SO_4 + 2OH^-\ (Anion\ exchange)$$

$$H^+ + OH^- \rightleftharpoons H_2O\ (Demineralized\ water)$$

After some time, the resin gets exhausted and requires regeneration. Cation exchangers are regenerated using strong acids, whereas anion exchangers are treated with strong bases.

$$(RCOO)_2Mg + H_2SO_4 \rightarrow 2RCOOH + Mg^{2+}$$

$$R'_2SO_4 + NaOH \rightarrow R'OH + SO_4^{2-}$$

Check Your Progress

16. What is softening of water?
17. Name two methods of water softening.
18. State the role of coagulants in cold lime-soda method of water softening.
19. What is zeolite? Write its general formula.
20. Justify the statement, 'Hard water is initially passed through a cation resin followed by anion resin during ion -exchange demineralization process.'
21. Justify, 'Hot lime-soda method is superior to cold lime-soda method.'
22. Differentiate between lime-soda and zeolite water softening methods.
23. Differentiate between cold and hot lime-soda water softening methods.
24. Name the types of ion-exchanges used in ion-exchange methods. State the functional groups in each of the ion-exchangers.
25. Name the method used to determine fluoride in water.
26. State the characteristics desired in water used for cooling purposes.
27. What is Langelier index? Give its expression.

Some of its **advantages** are:
(i) The method can be employed to soften highly acidic and alkaline water samples.
(ii) The final soft water has a residual hardness of about 1.5–2 ppm making it ideal for industrial purposes.

Some **disadvantages** are:
(i) The equipment is costly and its operation requires constant check for dealing with effluent waters.
(ii) The incoming water turbidity should be as low as 10 ppm, otherwise it may block the ion-exchanger.

13.7 DRINKING WATER PURIFICATION

Municipalities are regulatory bodies to supply potable water fit for human consumption to city households and industry. Generally water is pumped from surface or ground natural water sources and sent for screening to remove the suspended matter. The water is then allowed to stand undisturbed in a large sedimentation tank for about 6 hours that results in the sedimentation of suspended particles or impurities. After sedimentation, the supernatant liquid is drawn for the next treatment cycle via pumps. Clay and colloidal matter are eliminated using coagulants. The coagulants form flocculant precipitates with colloidal impurities which are removed as sludge. Alum, sodium aluminate, ferrous sulphate are commonly employed as coagulants while treating water. All the colloidal impurities and microorganisms are further passed through a sand filtration system.

After subjecting to sedimentation, coagulation, and filtration, water still contains a host of pathogenic bacteria. The process of destroying disease-producing bacteria and microorganisms from water and making it fit for drinking purposes is called *sterilization* or *disinfection*. The chemicals added (in calculated amounts) to water to kill microbial population in water are known as *disinfectants*.

13.7.1 Chlorination

Chlorine is the most widely used disinfectant. Chlorine (gas or concentrated solution) produces hypochlorous acid (HOCl) that acts as an effective germicide.

$$Cl_2 + H_2O \rightarrow HOCl + HCl$$
$$Bacteria + HOCl \rightarrow Bacteria\ killed$$
$$HOCl \rightleftharpoons H^+ + OCl^-$$

It is observed that hypochlorite ion (OCl⁻) causes inactivation of enzymes present in microorganisms that are essential for their metabolism. Due to this, growth and multiplication of microorganisms is arrested, leading to their death.

Ideally, municipalities carry out chlorination in a chlorinator equipped with a number of baffle plates (Fig. 13.9). Next, water and a calculated quantity of chlorine solution (0.3–0.5 ppm) are introduced at its top. During their passage through the tower, they get mixed and treated water is withdrawn out from the bottom.

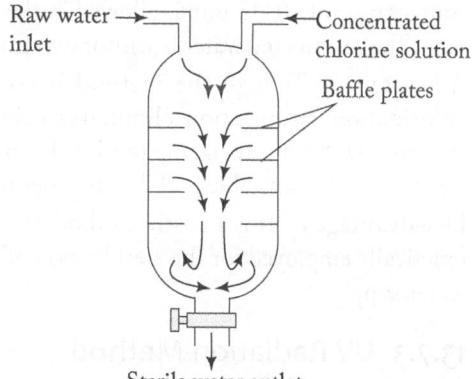

Raw water inlet → ← Concentrated chlorine solution

Baffle plates

Sterile water outlet

Fig. 13.9 Chlorinator

Advantages of chlorine as a disinfectant:
(i) Chlorine is cheap, economic, and easy to store.
(ii) It is stable and does not degrade at general room temperature conditions.

Disadvantages of chlorine as a disinfectant:
(i) One needs to strictly add only calculated amounts of chlorine. If added in excess, it may produce undesirable and unpleasant taste and odour. It can act as an irritant to the mucous membranes.
(ii) Ideally, the concentration of chlorine needs to be maintained up to 0.2 ppm, depending upon the volume of water that needs to be treated.

Alternative to chlorine, bleaching powder (calcium hypochlorite $Ca(ClO)_2$) is also used for disinfecting drinking water and swimming pool waters. 1 kg of $Ca(ClO)_2$ per 1000 kilo litre of water is mixed and allowed to stand for a contact time of several hours. The chemical action produces hypochlorous acid, a powerful germicide. The disinfecting action of bleaching powder is due to chlorine made available as follows.

$$CaOCl_2 + H_2O \rightarrow Ca(OH)_2 + Cl_2$$
$$Cl_2 + H_2O \rightarrow HCl + HOCl$$

Microorganisms + HOCl → Microorganisms killed

Calcium hypochlorite tends to introduce calcium in the treated water, thereby making it unfit for domestic use. Also, bleaching powder gets degraded on storage for long periods of time. Similar to chlorine, bleaching powder also imparts unpleasant taste and odour making it unfit.

If chlorine and ammonia are mixed in the ratio of 2:1, chloramine (NH_2Cl) is formed. When water is treated with chloramine, hypochlorous acid is produced that eventually kills the bacterial population.

$$Cl_2 + NH_3 \rightarrow NH_2Cl + HCl$$
$$NH_2Cl + H_2O \rightarrow HOCl + NH_3$$

Microorganisms + HOCl → Microorganisms killed

If chloramine is added in excess, it does not cause bad taste and odour to the treated waters. Also, it is known to have a longer effect as against chlorine and bleaching powder.

13.7.2 Ozonization

Ozonolysis involves sterilizing water with ozone, which is a proven harmless and effective disinfectant. Ozone can be produced by passing an electric discharge through oxygen.

$$3O_2 \rightarrow 2O_3$$

As it is known, ozone is highly unstable and breaks down liberating nascent oxygen.

$$O_3 \rightarrow O_2 + [O]$$

Nascent oxygen is a powerful oxidizing agent and kills all microorganisms along with the oxidation of organic matter content in water. As shown in Fig. 13.10, ozone is injected (2–3ppm) in water and contact time of 10–15 min is allowed in the sterilizing tank. The disinfected water is withdrawn from the top.

Advantages Though the method is costlier than chlorination, ozonization eliminates colour, odour, and taste without leaving any residue. Its excess is not harmful, as it is unstable and decomposes in oxygen.

Disadvantages It is a costly method and cannot be practically employed for the sterilization of municipal water supply.

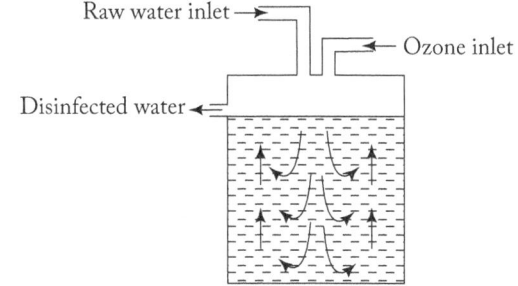

Fig. 13.10 Ozone sterilizer

13.7.3 UV Radiation Method

Disinfecting water with UV rays is a simple, effective, and environmentally safe method. UV systems are known to destroy 99.9 per cent of harmful microorganisms, without adding chemicals or changing the colour, taste, or odour of water.

Ultraviolet radiations kill microbes by arresting their DNA and reproduction. The effectiveness of this method depends on the exposure time and UV lamp intensity. According to the US Department of Health and Human Services, a minimum exposure of 16,000 μwatt-s/cm^2 is the allowed UV dosage for sterilizing water.

13.8 DESALINATION OF BRACKISH WATER

A separation process used to reduce the concentration of dissolved salts in saline waters is called desalination. Reverse osmosis and electrodialysis are the commonly adopted methods for desalinating brackish waters.

13.8.1 Reverse Osmosis

As shown in Fig. 13.11, two solutions of unequal concentrations are placed in a container separated by a semipermeable membrane. Under normal equilibrium conditions, the flow of the solution is from lower concentration to higher concentration due to osmosis. If a hydrostatic pressure is applied greater than the osmotic pressure at the concentrated (seawater) side, the solvent is forced to move from higher to lower concentration side. As the normal equilibrium is reversed, it is called reverse osmosis (or hyperfiltration).

The semipermeable membrane, also called RO membrane, is made from cross-linked polyamide and cellulose acetate with operating pressures of 15–30 kg/cm^2. RO membrane comprises a thin layer of 0.1–0.2 μm polymeric material and a fabric support that are stable over a wide range of pH and temperature.

Reverse osmosis is the best membrane separation technique that can reduce the concentration of dissolved metal ions, suspended particles, and some organic compounds. Ease of operation, continuous supply of pure water, removal of organic contaminants are some of the **advantages** of this method. The requirement of membranes and their replacement is a **limitation** of employing reverse osmosis method on a large-scale setting.

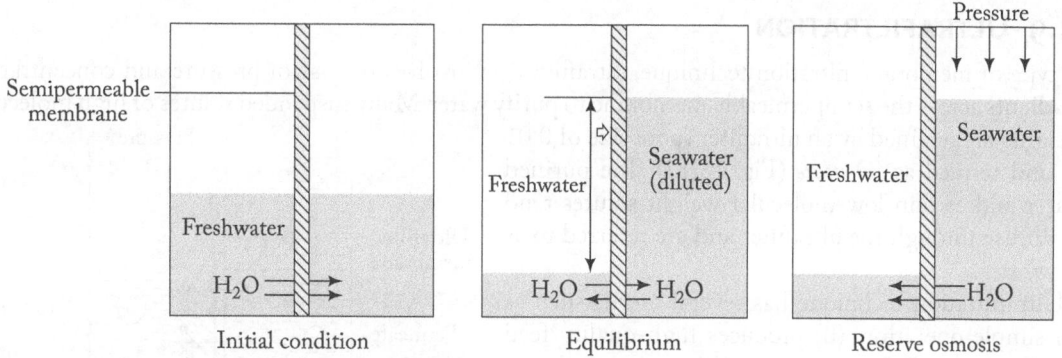

Fig. 13.11 Reverse osmosis set-up

13.8.2 Electrodialysis

A specialized membrane separation technology, electrodialysis, is commonly employed for water desalination, particularly for brackish waters with salinity of about 500 to 30,000 ppm. The process involves separating the dissolved salts from saline water in the form of ions in the presence of direct current (DC) using special-type of membranes called ion-selective membranes. As the name suggests, an ion-selective membrane is a specialized membrane that is only permeable for one type of ions with specific charge. If one employs a cation-selective membrane, it will only allow cations to pass through it and not anions. Similarly, anion-selective membrane

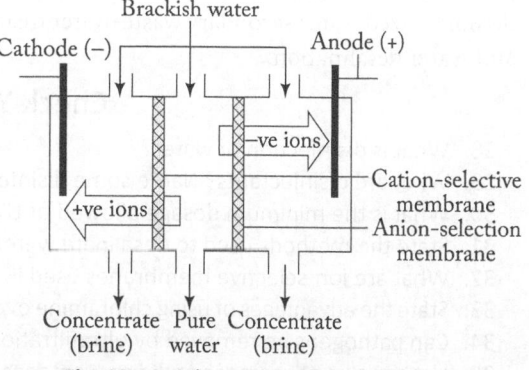

Fig. 13.12 Schematic representation of electrodialysis

will strictly allow only anions to permeate. Depending upon the nature of separation desired, membranes are selected for the purpose.

The process of electrodialysis is carried out in an electrodialyzer that is equipped with numerous pairs of ion-selective membranes, also called *membrane stack*. Saline or brackish water to be treated is sent in to the unit at a pressure of about 5–6 kg/m². During operation (Fig. 13.12), a DC is passed through saline water in the electrodialyzer equipped with ion-selective membranes. An electric field is applied perpendicular to the direction of water flow. Due to the electric current, ions will be formed. Soon, the ions will start migrating towards oppositely charged electrodes via the membrane stack. Cations will move towards the cathode and anions will move towards the anode.

All the ion permeation occurs through the series of ion-selective membranes. The process is continued to attain reduction in the concentration of ions present in saline water.

Some of the **merits** of this method are:
(i) Equipment is compact taking less of laboratory space.
(ii) Installation cost is minimal.
(iii) All the processes can be applied at ambient (room) temperatures.

Some **demerits** of electrodialysis are:
(i) Utilization of electricity to drive the process continuously makes it less economical, if one accounts for energy factors.
(ii) It can eliminate only ionic impurities such as sodium, potassium, chloride ions, whereas colloidal impurities and non-ionic impurities cannot be separated.

13.9 ULTRAFILTRATION

A type of membrane filtration technique, ultrafiltration involves the use of pressure and concentration gradients across the semipermeable membrane to purify water. Many suspended solutes of high molecular weights are retained by an ultrafilter (pore size of 0.01 μ) and termed as *retentate* (Fig. 13.13). The purified water and certain low-molecular weight solutes tend to diffuse through the ultrafilter and are referred to as *permeate*.

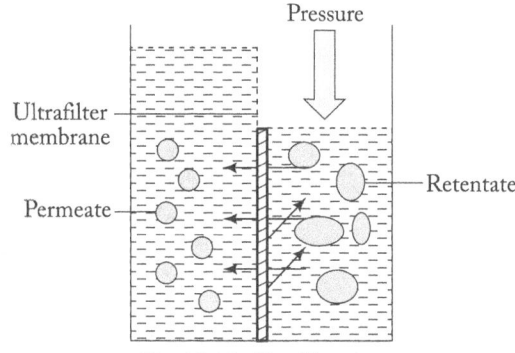

Ultrafiltration technique has several merits such as (a) simple operation, (b) produces high-quality feed waters to the RO systems, (c) lower chemical use, and (d) removal of pathogens. Various applications of ultrafiltration technique include drinking water purification, water used in beverages, pre-treatment of demineralized water, secondary waste-water treatment, and water desalination.

Fig. 13.13 Ultrafiltration

Check Your Progress

28. What is disinfection of water?
29. What are disinfectants? Name some disinfectants used to treat drinking water.
30. What is the minimum dosage allowed of UV radiations for sterilizing water?
31. State the methods used to desalinate water.
32. What are ion-selective membranes used in electrodialyzer?
33. State the advantages of using chloramine over other chlorine-based substances as a water disinfectant.
34. Can pathogens be removed by ultrafiltration?
35. List any two advantages and disadvantages of chlorination.
36. State the advantages and disadvantages of ozonolysis.

13.10 DISSOLVED OXYGEN

It is known that oxygen gas is slightly soluble in freshwater. *The number of moles of molecular oxygen that is dissolved in one litre of water at a given temperature is* called dissolved oxygen. Ideally, sources of dissolved oxygen in water are either the atmosphere or the product of photosynthesis of aquatic planktons. Dissolved oxygen is a crucial factor for supporting aquatic life and various biochemical reactions carried out by organisms in the natural waters.

The amount of dissolved oxygen is a frequently measured parameter and an ideal indicator of the quality of water. While determining dissolved oxygen (DO), two requirements must be met:

(a) As oxygen is present in trace quantity, the method adopted to measure it must be highly accurate.

(b) The method should be easy to carry out in any field operations.

Dissolved oxygen levels in water range from $1-18$ mg O_2 per litre and may fluctuate based on the time of the day, biochemical activity of aquatic organisms, temperature, and other weather-related conditions. The extent of water pollution can be clearly indicated by the oxygen levels in water. Higher the DO, better is the quality of water. Today, highly advanced oxygen-selective electrodes, polarographic-oxygen sensors, fibre-optic micro-sensory probes are designed to measure even trace levels of oxygen in water (up to 1 ppm).

The amount of free oxygen (in mg/l) needed to oxidize organic matter by bacteria and other microorganisms at 20°C for a period of 5 days is called biological oxygen demand (BOD_5). As oxygen consumed by various inorganic matter is also measured, it is also called *biochemical oxygen demand*.

Eutrophication is a natural process in freshwater that has a rich supply of nutrients. It is also a part of ageing process, as nutrients tend to accumulate over a period of time. It tends to become abnormally high due to fertilizers, animal excreta, sewage, and other toxic chemicals ending up in the water bodies. This causes excessive growth called algal blooms of microorganisms and aquatic vegetation. Under abundant sunlight, algal blooms contribute oxygen to water through photosynthesis. However, after a period of time, they start decomposing due to bacterial attack and this process consumes dissolved oxygen from the natural waters robbing the aquatic organisms from the supply. If continued, it leads to anaerobic conditions that result in death of fishes and aquatic organisms. Hence, there exists a close relationship between organic content and dissolved oxygen in water.

13.10.1 Biological Oxygen Demand (BOD)

Biological Oxygen Demand indicates the extent of pollution of water bodies and even sewage waters, as it also detects decomposing organic matter. So, if higher BOD is obtained, greater is the level of pollution in the water body. Rapidly depleting levels of dissolved oxygen, higher will be the BOD which has a detrimental effect on the aquatic life and humans.

Dissolved oxygen is determined by iodometric titration method developed by Winkler (1888). When manganese hydroxide is precipitated in the water sample

(a) it is oxidized to form basic manganese oxide by dissolved oxygen present in the water sample

(b) acidification of the hydroxide floc with sulphuric acid to pH 1 results in floc dissolution and immediately liberating nascent oxygen;

(c) the liberated nascent oxygen oxidizes the previously added iodide ions to iodine

(d) I_2 reacts with excess iodide ions to form tri-iodide (I_3^-) ion

(e) iodine is titrated with standard thiosulphate solution which reduces the iodine to iodide and itself gets oxidized to tetrathionate ions.

$$2\,NaOH + MnSO_4 \rightarrow Mn(OH)_2 + K_2SO_4$$

$$2\,Mn(OH)_2 + O_2 \rightarrow 2\,MnO(OH)_2$$

$$MnO(OH)_2 + H_2SO_4 \rightarrow MnSO_4 + 2\,H_2O + [O]$$

$$2\,KI + H_2SO_4 + [O] \rightarrow K_2SO_4 + H_2O + I_2$$

$$I_2 + I^- \rightarrow I_3^-$$

$$2\,S_2O_3^{2-} + I_3^- \rightarrow 3I^- + S_4O_6^{2-}$$

BOD estimation involves water sampling in two different bottles (BOD bottles). Similarly, incubated sample is also estimated for its DO level using Winkler's method. The known volume of water sample is diluted with known volume of distilled water. The diluted water sample is filled in two different BOD bottles. One of the bottles (DO_1) is incubated at 20 °C for a 5-days period and estimated as per Winkler method. The other bottle (DO_2) containing water sample is estimated for its DO level using Winkler's method as explained above immediately after sampling.

BOD can be estimated as, BOD (mg/l) = $(DO_1 - DO_2)/x$

where x is the fraction of sample = Vol. of sample/Total vol. to which it was diluted.

BOD = $(DO_1 - DO_2)$ Vol. of sample after dilution/Vol. of sample before dilution, where DO_1 and DO_2 are dissolved oxygen estimated in bottles 1 and 2, respectively.

13.10.2 Chemical Oxygen Demand (COD)

Chemical oxygen demand is also a measure of oxygen in water required to oxidize all organic matter into carbon dioxide and water. COD values are always found greater than BOD values, but COD measurements can be completed in a few hours, while BOD measurements take up to five days. It is considered as a quick basis of analysing dissolved oxygen in sewage and industrial effluent waters. In COD estimation, a known quantity of water sample is refluxed in the presence of a known excess of potassium dichromate (oxidant) in 50 % sulphuric acid and traces of silver sulphate catalyst. Mercuric sulphate is added to remove chloride interferences during the experiment. Under reflux conditions, organic matter present in the water sample gets oxidized to carbon dioxide, water, and ammonia. The excess dichromate that is not consumed by organic content is titrated against a known strength of ferrous ammonium sulphate (FAS). COD can be calculated as follows (in mg/l),

$$COD = \frac{(V_1 - V_2) \times N \times 8 \times 1000}{x}$$

where V_1 and V_2 are volumes of FAS (of known normality N) in blank and main experiments respectively and x is the volume of water sample under test.

Check Your Progress

37. What is eutrophication?
38. Define BOD.
39. State the importance of BOD analysis.
40. What is COD?

> ## Activity-based Questions
>
> ### A. ZLD Technology in Industries
>
> Zero-liquid discharge (ZLD) is a water treatment process adopted by industries in which wastewater effluent is purified, recycled, and at times, reused. A concerted treatment method that includes ultrafiltration, reverse osmosis, filtration, evaporation or crystallization, and electro-deionization is known to achieve zero-to-minimal discharges from polluting industries such as textile, paper, dyes, metal-finishing, tanning, pharmaceuticals, and others. Installing zero-to-minimal discharge technology is beneficial for the plant's water management; encouraging monitoring of water usage, avoiding wastage and promotes recycling by simpler, cost-effective methods.
>
> Present an overview of ZLD technology adopted in industry and state its processes, costs, challenges, and advantages.
> Reference: http://www.gewater.com/zero-liquid-discharge-zld.html
>
> ### B. Cavitation and Plasma-based Water Disinfection
>
> Risks associated with current disinfection techniques (chlorination) including the formation of disinfection by-products (DBPs) and multidrug resistant bacterial species have prompted the search for advanced disinfection techniques in major water plants.
>
> Explain cavitation-acoustic-hydrodynamic (ultrasound) and plasma methods of water disinfection. Cite references of the ongoing work on these techniques on a pilot scale with a brief account on potentials and challenges foreseen.
>
> ### C. Ultrafiltration in Dialysis
>
> Dialysis is required for patients with failed kidney functioning. It works on the principle of ultrafiltration of fluids across semipermeable membrane. Describe the construction and working of a hemodialysis machine and how it helps to remove the blood impurities.

SOLVED EXAMPLES

1. A water sample has a hardness of 747 mg/l of $CaCO_3$ equivalent. Express its hardness in terms of degree Clark and degree French.
Solution: 0.07 degree Clark = 1 ppm
Hence, 747 ppm = 52.29 °Cl
0.1 degree French = 1 ppm
Hence, 747 ppm = 74.7 °Fr

2. Classify the following impurities into temporary, permanent, and non-hardness causing impurities in water: $Ca(HCO_3)_2$, $MgSO_4$, $CaCl_2$, CO_2, HCl, $Mg(HCO_3)_2$, H_2SO_4, $Ca(NO_3)_2$, NaCl, KCl, $FeSO_4$, Na_2SO_4, KNO_3, $Al_2(SO_4)_3$.

Impurities in water	Type of hardness
$Ca(HCO_3)_2$, $Mg(HCO_3)_2$,	Temporary hardness
$MgSO_4$, $CaCl_2$, $Ca(NO_3)_2$, $FeSO_4$, $Al_2(SO_4)_3$	Permanent hardness
CO_2, HCl, H_2SO_4, NaCl, KCl, Na_2SO_4, KNO_3	Non- hardness

3. Hard water contains 272 mg of magnesium sulphate in one litre. Express the hardness of water in terms of $CaCO_3$ equivalents.
Solution: $MgSO_4 \equiv CaCO_3$

As per Eq. (13.2), Equivalent of $CaCO_3 = \dfrac{50 \times W_{salt}}{E_{salt}} = \dfrac{50 \times 272}{60} = 226.66\ CaCO_3$ equivalents

Result: Hardness of water is 226.66 $CaCO_3$ equivalents.

4. A water sample on laboratory analysis gave the following data:

$$MgCl_2 = 0.182°Fr, MgSO_4 = 0.662°Fr, Ca(NO_3)_2 = 0.311°Fr, \text{ and } Ca(HCO_3)_2 = 2.216°Fr.$$

Calculate the total hardness of water in terms of ppm.

Solution: Since, $0.1°Fr = 1$ ppm,

$0.1832°Fr$ of $MgCl_2 = 1.832$ ppm $= \dfrac{1.832 \times 50}{47.5}$ ppm $CaCO_3$ equivalent $= 1.928$ ppm

$0.662°Fr$ of $MgSO_4 = 6.62$ ppm $= \dfrac{6.62 \times 50}{60}$ ppm $CaCO_3$ equivalent $= 5.51$ ppm

$0.311°Fr$ of $Ca(NO_3)_2 = 3.11$ ppm $= \dfrac{3.11 \times 50}{82}$ ppm $CaCO_3$ equivalent $= 1.896$ ppm

$2.216°Fr$ of $Ca(HCO_3)_2 = 22.16$ ppm $= \dfrac{22.16 \times 50}{81}$ ppm $CaCO_3$ equivalent $= 13.679$ ppm

Result: Total hardness = 1.928 + 5.51 + 1.896 + 13.679 = 23.013 ppm
Total hardness of the given water sample is 23.013 ppm.

5. Calculate temporary, permanent and total hardness of a water sample containing $Ca(HCO_3)_2 = 16.2$ mg/l, $Mg(HCO_3)_2 = 7.3$ mg/l, $CaSO_4 = 13.6$ mg/l, $MgCl_2 = 9.5$ mg/l, and $NaCl = 20.2$ mg/l.

Solution:

Salts	Quantity (W_{salt})(mg/l)	Conversion factor	$CaCO_3$ equivalent (W_{salt} x CF) (mg/l)*
$Ca(HCO_3)_2$	16.2	50/81	16.2 x 50/81 = 10
$Mg(HCO_3)_2$	7.3	50/73	7.3 x 50/73 = 5
$CaSO_4$	13.6	50/68	13.6 x 50/68 = 10
$MgCl_2$	9.5	50/47.5	9.5 x 50/47.5 = 10
NaCl	20.2	-	Does not contribute hardness to water (Ignored).

* refer to Eq. (13.1) and Table 13.4.
Temporary hardness is due to salts of $Ca(HCO_3)_2$ and $Mg(HCO_3)_2$.
Hence, their respective $CaCO_3$ equivalents are added
Temporary hardness = 10 + 5 = 15 mg/l.
Permanent hardness is due to $CaSO_4$ and $MgCl_2$. Hence, their respective $CaCO_3$ equivalents are added.
Permanent hardness = 10 + 10 = 20 mg/l.
Total hardness is due to $Ca(HCO_3)_2$, $Mg(HCO_3)_2$, $CaSO_4$, and $MgCl_2$.
= Temporary hardness + Permanent hardness = 15 + 20 = 35 mg/l
Result: The temporary, permanent, and total hardness of water sample are 15, 20, and 35 (mg/l) respectively.

6. Accurately weighed 0.5 g of $CaCO_3$ was dissolved in dilute HCl and diluted with 500 ml of distilled water to prepare standard hard water (SHW). 20 ml of SHW consumed 18 ml of EDTA solution, while 20 ml of water sample consumed 15 ml of EDTA solution. After boiling, cooling, and filtering, the same volume of hard water sample consumed 10 ml of EDTA solution. Calculate all types of hardness present in the water sample.
Solution: Concentration of SHW = 0.5g / 500 ml, or 1 mg/ml
  (a) Standardization of EDTA (burette) solution:
    20 ml of SHW ≡ 18 mL EDTA
    20 ml of SHW = (20 × 1) mg $CaCO_3$

    ∴ 18 ml EDTA = $20 \times 1/18$ mg $CaCO_3$
    ∴ 1 ml EDTA = 1.111 mg $CaCO_3$
  (b) Water sample analysis:
    Total hardness of water sample (before boiling)
    20 ml of water sample (W.S.) ≡ 15 ml EDTA = 15 × 1.111 mg $CaCO_3$

$$\therefore 1000 \text{ ml water sample} = \frac{1000 \times 15 \times 1.111}{20}$$

$\therefore$ Total hardness of water sample = 833.25 ppm

Permanent hardness (after boiling)

20 ml of water sample (W.S) ≡ 10 ml EDTA = 10 × 1.111 mg $CaCO_3$

$$\therefore 1000 \text{ ml water sample} = \frac{1000 \times 10 \times 1.111}{20}$$

$\therefore$ Permanent hardness = 555.5 ppm

$\therefore$ Temporary hardness = (Total − Permanent) hardness

$$= (833.25 - 555.5) = 277.75 \text{ ppm}$$

Result: Therefore, total, temporary, and permanent hardness of water sample are 833.25, 277.75, and 555.5 (ppm), respectively.

7. When 50 ml of water sample was titrated against 0.01 N EDTA solution, burette reading was recorded as 14 ml. The same water sample, on boiling and filtering consumed 10 ml EDTA solution. Calculate the total, temporary, and permanent hardness of the given water sample.

Solution:

(a) Water sample analysis:

Total hardness of water sample (before boiling)

50 ml of water sample ≡ 14 ml EDTA

$$\therefore 1000 \text{ ml water sample} = \frac{1000 \times 14}{50} \text{ mg } CaCO_3$$

$\therefore$ Total hardness of water sample = 280 ppm

Permanent hardness of water sample (after boiling)

50 ml of water sample ≡ 10 ml EDTA

$$\therefore 1000 \text{ ml water sample} = \frac{1000 \times 10}{50} \text{ mg } CaCO_3$$

$\therefore$ Permanent hardness = 200 ppm

$\therefore$ Temporary hardness = (Total − Permanent) hardness

$$= (280 - 200) = 80 \text{ ppm}$$

Result: Therefore, total, temporary, and permanent hardness of water sample are 280, 80, and 200 (ppm), respectively.

8. A standard hard water sample is prepared by dissolving 0.51 g $CaCO_3$ in 400 ml distilled water. 50 ml of SHW consumed 25 ml EDTA solution. 50 ml of water sample consumed 20 ml EDTA solution. After boiling, filtering, and cooling, 50 ml of the same sample consumed 15 ml EDTA solution. Calculate all types of hardness present in the water sample.

Solution: Concentration of SHW = 0.51 g per 400 ml

i.e., in 1000 ml = $0.51 \times \dfrac{1000}{400} = 1.275$ mg/ml

(a) Standardization of EDTA (burette) solution:

50 ml of SHW ≡ 25 ml EDTA

and, 50 ml of SHW = (50 × 1.275) mg $CaCO_3$

$\therefore$ 25 ml EDTA = 50 × 1.275 mg $CaCO_3$

$\therefore$ 1 ml EDTA = $\dfrac{50}{25} \times 1.275$ mg $CaCO_3$ = 2.55 mg $CaCO_3$

(b) Water sample analysis:

Total hardness of water sample (before boiling)

50 ml of water sample ≡ 20 ml EDTA

$$\therefore \ 1000 \text{ ml} = \frac{1000 \times 20 \times 2.55}{50} \text{ mg CaCO}_3$$

$\therefore$ Total hardness of water sample = 1020 ppm

Permanent hardness of water sample (after boiling)

50 ml of water sample $\equiv$ 15 ml EDTA

$$\therefore \ 1000 \text{ ml water sample} = \frac{1000 \times 15 \times 2.55}{50}$$

$\therefore$ Permanent hardness = 765 ppm

$\therefore$ Temporary hardness = (Total – Permanent) hardness

$$= (1020 - 765) = 255 \text{ ppm}$$

Result: Therefore, total, temporary, and permanent hardness of water sample are 1020, 255, and 765 (ppm), respectively.

9. Calculate the amounts of lime and soda required for softening 7500 litres of hard water containing 60 ppm of $MgSO_4$.

Solution: Calculation of $CaCO_3$ equivalents

Impurity	W_{salt} (ppm)	Conversion factor (CF)	CaCO$_3$ equivalent (W_{salt} x CF)	Requirement
MgSO$_4$	60	50/60	50	L + S

$$\text{Lime required} = \frac{74}{100} \times 50 \ \times \frac{7500}{10^6} \times \frac{100}{100} = 0.277 \text{ kg}$$

$$\text{Soda required} = \frac{106}{100} \times 50 \ \times \frac{7500}{10^6} \times \frac{100}{100} = 0.398 \text{ kg}$$

Result: Therefore, 0.277 kg lime and 0.398 kg soda will be required to soften 7500 litres of water with 60 ppm $MgSO_4$.

10. Water having the following composition (in ppm) has to be softened by lime-soda process: $Ca(HCO_3)_2$ = 220, $Mg(HCO_3)_2$ =76, $MgCl_2$ = 150, $CaSO_4$ = 100, CO_2 =44, $MgSO_4$ = 60. Calculate the amount of lime and soda required to soften 10^6 litres of water. (Given: purities of lime and soda = 100%).

Solution: Calculation of $CaCO_3$ equivalent of impurities

Impurities in water	W_{salt} (ppm)	CF	CaCO$_3$ equivalent (W_{salt} x CF)	Requirement L/S
Ca(HCO$_3$)$_2$	220	50/81	135.80	L
Mg(HCO$_3$)$_2$	76	50/73	52.05	2 L
MgCl$_2$	150	50/47.5	157.9	L+S
CaSO$_4$	100	50/68	73.52	S
CO$_2$	44	50/22	100	L
MgSO$_4$	60	50/60	50	L+S

$$\text{Lime required} = \frac{74}{100} \times \left[135.80 + 2(52.05) + 157.9 + 50 + 100 \right] \ \times \frac{10^6}{10^6} \times \frac{100}{100} = 405.32 \text{ kg}$$

$$\text{Soda required} = \frac{106}{100} \times \left[157.9 + 50 + 73.5 \right] \ \times \frac{10^6}{10^6} \times \frac{100}{100} = 298.28 \text{ kg}$$

Result: Therefore, lime and soda required to soften 10^6 litres of water are 405.37 kg and 298.28 kg respectively.

11. Calculate the amounts of lime and soda needed for softening 50,000 l of hard water whose chemical analyses are as follows (in mg/l):

$Ca(HCO_3)_2$ = 40.5, $Mg(HCO_3)_2$ = 73.0, $MgSO_4$ = 60.0, $CaCl_2$ = 27.5, $CaSO_4$ = 34.0, KCl = 20.0. (Given: % purity of lime = 80 and % purity of soda = 90).

Solution: Calculation of $CaCO_3$ equivalent of impurities.

Impurities in water	W_{salt} (ppm)	CF	$CaCO_3$ equivalent (W_{salt} x CF)	Requirement L/S
$Ca(HCO_3)_2$	40.5	50/81	25	L
$Mg(HCO_3)_2$	73.0	50/73	50	2 L
$MgSO_4$	60	50/60	50	L+S
$CaCl_2$	27.5	50/55.5	24.77	S
$CaSO_4$	34.0	50/68	25	S
KCl	20	—	—	Does not react

$$\text{Lime required} = \frac{74}{100} \times \left[25 + 2(50) + 50\right] \times \frac{50000}{10^6} \times \frac{100}{80} = 8.093 \text{ kg}$$

$$\text{Soda required} = \frac{106}{100} \times \left[50 + 24.77 + 25\right] \times \frac{50000}{10^6} \times \frac{100}{90} = 5.874 \text{ kg}$$

Result: Therefore, lime and soda required to soften 50,000 l of water are 8.093 kg and 5.874 kg respectively.

12. Calculate the amount of lime and soda required to soften 10^6 l water which contains the following impurities (in mg/l) viz, HCl = 9.3, $Al_2(SO_4)_3$ = 57, $CaCl_2$ = 55.5 , NaCl =39.3. Purity factor of lime is 90 % and that of soda is 98 %. If 5 % of chemicals are added in excess in above water sample to complete the reaction, determine the amounts of lime and soda for the same.

Solution: Calculation of $CaCO_3$ equivalent of impurities

Impurity	W_{salt} (ppm)	CF	$CaCO_3$ equivalent (W_{salt} x CF)	Requirement L/S
HCl	9.3	50/36.5	12.73	L+S
$Al_2(SO_4)_3$	57	50/57	50	L+S
$CaCl_2$	55.5	50/55.5	50	S

$$\text{Lime required} = \frac{74}{100} \times \left[12.73 + 50\right] \times \frac{10^6}{10^6} \times \frac{100}{90} = 51.57 \text{ kg}$$

$$\text{Soda required} = \frac{106}{100} \times \left[12.73 + 50 + 50\right] \times \frac{10^6}{10^6} \times \frac{100}{98} = 121.932 \text{ kg}$$

If 5 % excess chemicals are added,

Total lime required (using 5 % excess) = 51.57 × 105/100 = 54.14 kg.

Total soda required (using 5 % excess) = 121.932 × 105/100 = 128 kg.

Result: Actual amounts of lime and soda required are 51.57 and 121.932 (kg) respectively. After adding 5 % excess chemicals, lime and soda required are 54.14 and 128 (kg) respectively.

13. Calculate the quantities of lime (90% pure) and soda (95% pure) required to soften 10^6 litres of water with impurities: $CaCO_3$ = 15.0 ppm, $MgCO_3$ = 9.0 ppm, $CaCl_2$ = 20.0 ppm, $MgCl_2$ = 8.0 ppm, CO_2 = 30.0 ppm, and HCl = 9.2 ppm.

Solution: Conversion of $CaCO_3$ equivalents of impurities:

Impurities	W_{salt} (ppm)	CF	CaCO₃ equivalent (W_{salt} x CF)	Requirements (L/S)
$CaCO_3$	15.0	$\dfrac{50}{50}$	$\dfrac{50}{50} \times 15.0 = 15.0$	L
$MgCO_3$	9.0	$\dfrac{50}{42}$	$\dfrac{50}{42} \times 9.0 = 10.7$	2L
$CaCl_2$	20.0	$\dfrac{50}{55.5}$	$\dfrac{50}{55.5} \times 20.0 = 18.0$	S
$MgCl_2$	8.0	$\dfrac{50}{47.5}$	$\dfrac{50}{47.5} \times 8.0 = 8.42$	L + S
CO_2	30.0	$\dfrac{50}{22}$	$\dfrac{100}{44} \times 30.0 = 68.18$	L
HCl	9.2	$\dfrac{50}{36.5}$	$\dfrac{50}{36.5} \times 9.2 = 12.6$	L + S

$$\text{Lime required} = \frac{74}{100} \times \left[15 + 2(10.7) + 8.42 + 68.18 + 12.6 \right] \times \frac{10^6}{10^6} \times \frac{100}{90} = 103.26 \text{ kg}$$

$$\text{Soda required} = \frac{106}{100} \times \left[18 + 8.42 + 12.6 \right] \times \frac{10^6}{10^6} \times \frac{100}{95} = 43.53 \text{ kg}$$

Therefore, lime and soda required to soften 10^6 litres of water are 103.26 kg and 43.53 kg respectively.

14. On analysis, a water sample gave the following data (in ppm): Ca^{2+} = 30, Mg^{2+} = 24, CO_2 = 24, HCl = 50 and K^+ = 5. Calculate the quantities of lime and soda required for softening 10^6 litres of the water, if purities of lime and soda are 90 % and 94 % respectively.

Solution: Calculation of CaCO₃ equivalent of impurities

Impurity	W_{salt} (ppm)	CF	CaCO₃ equivalent (W_{salt} x CF)	Requirement L/S
Ca^{2+}	30	50/20	75	S
Mg^{2+}	24	50/12	100	L + S
CO_2	24	50/22	54.5	L
HCl	50	50/36.5	68.5	L + S

$$\text{Lime required} = \frac{74}{100} \times \left[Mg + CO_2 + HCl \text{ as } CaCO_3 \text{ equivalents} \right] \times \frac{10^6}{10^6} \times \frac{100}{90}$$

$$= \frac{74}{100} \times \left[100 + 54.5 + 68.5 \right] \times \frac{10^6}{10^6} \times \frac{100}{90} = 183.4 \text{ kg}$$

$$\text{Soda required} = \frac{106}{100} \times \left[Ca + Mg + HCl \text{ as } CaCO_3 \text{ equivalents} \right] \times \frac{10^6}{10^6} \times \frac{100}{94}$$

$$= \frac{106}{100} \times \left[75 + 100 + 68.5 \right] \times \frac{10^6}{10^6} \times \frac{100}{94} = 274.5 \text{ kg}$$

Result: The amounts of lime and soda required to soften 10^6 litres of the sample water are 183.4 kg and 274.5 kg respectively.

15. 1 litre of hard water containing 5.6 g $CaCl_2$ is allowed to pass through a zeolite softener. Calculate the quantity of NaCl produced in the soft water.

Solution: As per the reaction, $CaCl_2 + Na_2Ze$ (sodium zeolite) $\rightarrow$ $CaZe + 2NaCl$

Hence, 111g $CaCl_2$ leaves 2×58.5 g NaCl in soft water.

$\because$ 111 g $CaCl_2 \equiv 117$ g NaCl

$$\therefore 5.6 \text{ g } CaCl_2 \equiv 5.6 \times \frac{117}{111} = 5.902 \text{ g}$$

Result: The quantity of NaCl produced in soft water is 5.902 g.

16. 60,000 litres of hard water was softened using zeolite softener. The zeolite softener, on regeneration required 300 litres of NaCl solution containing 130 g/l of NaCl. Calculate the hardness of the incoming hard water.

Solution: As per the regeneration reaction, $CaZe + 2NaCl \rightarrow Na_2Ze + CaCl_2$

As 1 litre of NaCl $\equiv 130$ g NaCl

Hence, 300 litres of NaCl will be $300 \times 130 = 39000$ g NaCl needed for regeneration.

$$\text{Hardness in terms of } CaCO_3 \text{ equivalents} = 39000 \times \frac{50}{58.5} = 33333.33 \text{ g } CaCO_3 \text{ equivalents}$$

$$\text{Hardness in 60,000 l of hard water} = \frac{33333.33}{60000} = 0.555 \text{ g } CaCO_3 \text{ equivalents}$$

Hence, hardness of incoming water = 555.55 mg $CaCO_3$ equivalents.

Result: Hardness of the incoming water sample is 555.55 ppm.

17. A water sample with hardness of 500 ppm was softened using zeolite. The exhausted zeolite was regenerated by 300 litres of NaCl solution having concentration of 70 g/l NaCl. Calculate the volume of water softened using zeolite.

Solution: Let the quantity of water be 'x' litres.

Hardness of water = 500 ppm.

As 1 litre of NaCl $\equiv 70$ g NaCl

Hence, 300 litres of NaCl = $70 \times 300 = 21000$ g NaCl needed for regeneration

$$\text{Hardness in terms of } CaCO_3 \text{ equivalents} = 21000 \times \frac{50}{58.5} = 17948.71 \text{ g } CaCO_3 \text{ equivalents}$$

$\because$ 1 litre water $\equiv 500$ ppm

$$\therefore \quad x \text{ litre of water} = \frac{17948.71}{500} \times 1000 = 35897.43 \text{ litres of water}$$

Result: The volume of water softened by zeolite softener is 35897.43 litres

18. 1,00,000 litres of water is treated using ion-exchanger. The cation-exchanger required 200 litres of 0.1 N HCl and anion-exchanger required 200 litres of 0.1 N NaOH solution. Determine the total hardness of water sample.

Solution: The cation-exchanger removes all the cations, whereas the anion-exchanger removes all the anions from the water. Hence, the amount of acid required for regeneration of cation resin refers to its hardness and similarly for anionic resin.

Hardness of 1,00,000 litres of water = 200 litres of 0.1 N HCl = 200 l of 0.1 N $CaCO_3$ eq.

$= 200 \times 0.1 = 20$ litres of 1 N $CaCO_3$ eq.

$= 20 \times 50$ g $CaCO_3$ eq = 1000 g $CaCO_3$ eq

$$\therefore \quad \text{Hardness in 1 litre water} = \frac{1000}{1,00,000} = 0.01 \text{ g } CaCO_3 \text{ eq}$$

$= 0.01 \times 1000$ mg = 10 mg $CaCO_3$ eq

Result: The hardness of water sample is 10 mg/l.

19. 20 ml of waste water sample is diluted to 200 ml with distilled water and equal volumes are filled in two BOD bottles. In the blank titration, 100 ml of diluted water sample on titration consumed 6.2 ml of 0.02 N sodium thiosulphate. 100 ml of incubated sample after 5 days consumed 3.8 ml of 0.02 N sodium thiosulphate. Determine the BOD of waste water.

Solution:

Data: Volume of waste water (V_1) = 20 ml

Volume of water after dilution (V) = 200 ml

Volume of dilute water taken for titration (V_2) = 100 ml

Blank titration reading = 6.2 ml

Titration reading after 5 days = 3.8 ml

Dissolved oxygen at the start of the experiment is equivalent to 6.2 ml thiosulphate and dissolved oxygen after 5 days is equivalent to 3.8 ml thiosulphate.

Hence, Actual DO consumed by waste water in 5 days = A − B ml = 6.2 − 3.8 = 2.4 ml of 0.02 N thiosulphate.

1 ml of 1 N $Na_2S_2O_3$ ≡ 8 mg oxygen

Hence, 2.5 ml of 0.02 N $Na_2S_2O_3$ = 2.4 × 0.02 × 8 = 0.384 mg oxygen

So, 100 ml of diluted water = 0.384 mg oxygen

BOD of 1000 ml waste water is given as,

$$BOD = \frac{(A-B) \times N \times 8 \times 0.02 \times 200 \times 1000}{100 \times 25}$$

$$BOD = \frac{(6.2-3.8) \times 0.02 \times 8 \times 200 \times 1000}{100 \times 20} = 38.4 \text{ mg/l}$$

Therefore, BOD of waste water is 38.4 mg/l.

20. 30 ml of sewage water was reacted with 30 ml potassium dichromate solution, acidified and refluxed for 3 hours. After refluxing, the unreacted dichromate consumed 10 ml of 0.1 N FAS. In the blank titration, 30 ml dichromate solution consumed 17 ml of the same 0.1 N FAS solution.

Difference in volumes of FAS consumed in blank and sample waters = 17 − 10 = 7 ml

Now, 1000 ml 1 N FAS ≡ 8 g oxygen

or, 7 ml of 0.1 N FAS ≡ $\dfrac{8 \times 7 \times 0.1}{1000}$

∴ 30 ml water sample = $\dfrac{8 \times 7 \times 0.1}{1000}$ g of oxygen

So, 1000 ml water sample = $\dfrac{8 \times 7 \times 0.1}{1000} \times \dfrac{1000}{30}$ g of oxygen

Hence, COD of waste water = 0.186 g or 186 mg/l of oxygen.

Thus, COD of waste water was found to be 186 mg/l.

21. Determine the COD of effluent water sample when 30 mL of the sample consumed 8.5 mL of 0.001 M potassium dichromate solution. [Given: Molar mass of $K_2Cr_2O_7$ = 294].

Solution: 6 N of $K_2Cr_2O_7$ ≡ 1 M $K_2Cr_2O_7$

1000 mL of 0.001 M $K_2Cr_2O_7$ = 6 × 8 × 0.001 g oxygen

$$= \frac{6 \times 8.5 \times 0.001 \times 8000}{30} \text{ mg oxygen}$$

Hence, COD of effluent = 13.6 ml

SUMMARY

○ Water is an essential limited resource that needs to be conserved. The main sources of freshwater are rains, surface and ground waters.

○ Various impurities namely physical, chemical, biological, and radiological are present in natural waters.

○ Hardness of water is the tendency to prevent lather formation and categorized as temporary and permanent hardness. A pertinent problem of water hardness is found in boilers as scales and sludge formation. Hardness of water is expressed in terms calcium carbonate equivalents.

○ EDTA method (or complexometry) is employed to estimate hardness of water sample that involves complexation reaction of metal ions with EDTA in the presence of metal-ion indicator.

○ Some of the water softening methods include lime-soda, zeolite, and ion-exchange. Chlorination, ozonization, and UV-ray treatments are some common water disinfection methods. Desalination of saline waters is done by reverse osmosis, electrodialysis, and ultrafiltration.

○ Chemical analysis of water allows quantitative determination of various ions in water. Chlorides are determined routinely by argentometric method, sulphates and fluorides are estimated by gravimetric and colorimetric methods, respectively.

○ BOD and COD are measures of the extent of pollution in natural waterbodies such as lakes and rivers.

GLOSSARY

Coagulant: A chemical compound that facilitates coalescence of finely suspended particles effecting quick sedimentation of sludge.

Complexometry: A volumetric analysis involving measurement of complexes.

Desalination: The removal of dissolved salts from water.

Disinfection: The killing of microorganisms and pathogens present in water.

Foam: The stable bubbles formed over the boiler water that do not break off easily.

Hard water: Water containing dissolved salts that prevent lathering of soap.

Ion-selective membrane: A special polymeric membrane that allows preferential permeability of one type of ion in the solution.

Permanent hardness: The presence of dissolved sulpates, chlorides, nitrates, phosphates, calcium, and magnesium in water.

Potable water: Water that is fit for drinking purpose.

Priming: Wet steam formation in boiler waters caused due to dissolved salts.

Scales: The deposition of hard, adherent, crusty precipitate in the inner boiler walls.

Semi-permeable membrane: A polymer-based membrane that allows movement of solvent molecules through it and retains solute molecules.

Sludge: The dissolved salt precipitate depositing as a loose, slimy, soft layer in the boiler walls.

Sterilization: Complete destruction of living organisms present in water.

Temporary hardness: The presence of dissolved bicarbonates of calcium and magnesium in water which can be easily removed by boiling.

Total hardness: The presence of all dissolved salts in water that prevents lathering of soap.

KEY FORMULAE

• Equivalent of $CaCO_3 = \dfrac{\text{Equivalent weight of } CaCO_3}{\text{Equivalent weight of hardness causing salt}} \times \text{Mass of hardness causing salt}$

• Lime requirement

$$= \frac{74}{100}\left[\begin{array}{l}\text{TempCa.hardness} + 2 \times \text{Temp. Mg hardness} + \text{Perm. (Mg + Fe + Al) hardness} \\ + HCl + H_2SO_4 + HCO_3^- + CO_2 - NaAlO_2 \text{ all in terms of } CaCO_3 \text{ equivalents}\end{array}\right] \times \frac{V_{water}}{10^6} 100 \, / \, \%\text{purity}$$

- Soda requirement,

$$= \frac{106}{100} \left[\frac{\text{Perm.Ca hardness} + \text{Perm.(Mg + Fe + Al) hardness}}{\text{HCl} + \text{H}_2\text{SO}_4 - \text{HCO}_3^- \text{ all in terms of CaCO}_3 \text{ equivalents}} \right] \times \frac{V_{\text{water}}}{10^6} 100 / \% \text{purity}$$

- BOD (mg/l) = $(DO)_1 - (DO)_2/x$

- $\text{COD} = \dfrac{(V_1 - V_2) \times N \times 8 \times 1000}{X} \text{ mg / l}$

EXERCISES

Multiple Choice Questions

1. A water sample contains 220 mg of Ca^{2+} ions per litre; the hardness in terms of $CaCO_3$ equivalents is
   (a) 500 ppm   (b) 550 ppm
   (c) 220 ppm   (d) 275 ppm

2. Total hardness in water sample is determined by
   (a) precipitation titration
   (b) acid–base titration
   (c) redox titration
   (d) complexometric titration

3. Desalination of saline water is done by
   (a) EDTA titration   (b) boiling
   (c) reverse osmosis   (d) sedimentation

4. An exhausted zeolite column bed is regenerated by
   (a) conc. HCl   (b) conc. NaOH
   (c) conc. NH_3 (d) 10% brine

5. Ion-exchange resins are
   (a) polymeric   (b) aluminosilicate
   (c) separating agents (d) softening agents

6. The chemical formula of alum is
   (a) $KNO_3.Al_2(SO_4)_3.24H_2O$
   (b) $K_2SO_4.Al_2(SO_4)_3.20H_2O$
   (c) $K_2SO_4.Al_2(SO_4)_3$
   (d) $K_2SO_4.Al_2(SO_4)_3.24H_2O$

7. If BOD is higher in the water, it means
   (a) pollution levels is higher
   (b) pollution levels is lower
   (c) no pollution detected
   (d) none of these.

8. As per BIS, the permissible limit of total hardness content in water is
   (a) 400 ppm   (b) 500 ppm
   (c) 550 ppm   (d) 600 ppm

9. UV water sterilization dosage is
   (a) 16,000 µwatt-s/cm^2
   (b) 16 µwatt-s/cm^2
   (c) 1000 µwatt-s/cm^2
   (d) 10,600 watt-s/cm^2

10. Lime-soda method tends to leave behind a residual hardness of
    (a) 10 ppm   (b) 50–60 ppm
    (c) 12–15 ppm   (d) 2 ppm

11. Ion-selective membranes are used in
    (a) lime-soda method
    (b) EDTA titration
    (c) electrodialysis
    (d) filtration

12. Ultrafiltration tends to eliminate
    (a) water hardness
    (b) pathogens from water
    (c) water alkalinity
    (d) water acidity

13. Bleaching powder is a
    (a) chlorinating agent
    (b) sterilant
    (c) insecticide
    (d) water impurity

14. If water is boiled and filtered, the type of hardness eliminated is
    (a) total hardness
    (b) permanent hardness
    (c) temporary hardness
    (d) none of these

15. The major impurity responsible for scale formation in boilers is
    (a) Na_2SO_4 (b) $CaSO_4$
    (c) $MgCO_3$ (d) $CaCO_3$

16. Fluoride is performed/determined using the reagent
    (a) $ZrCl_2$–SPADNS
    (b) $ZrOCl_2$–SPADNS
    (c) $ZrCl_2$
    (d) none of these

17. The chemical formula of bleaching powder is
    (a) $CaOCl_2$ (b) Cl_2
    (c) $HOCl$ (d) $CaCl_2$

18. Soft, loose, slimy precipitate formed in the boiler walls is called
    (a) scale   (b) coagulant
    (c) flocculant   (d) sludge

19. By ion-exchange process, residual hardness of water is
    (a) 100 ppm   (b) 10 ppm
    (c) 0 ppm   (d) 15 ppm

20. Priming and foaming in boiler feed water are due to
    (a) wet steam
    (b) air bubbles and wet steam
    (c) sludge formation
    (d) scale formation

Review Questions

1. Define temporary and permanent hardness of water. Write the chemical reactions of water with soap.

2. Explain hardness of water. List the disadvantages of using hard water.

3. State the various mathematical units for expressing hardness of water.

4. Differentiate between scales and sludges.

5. Explain boiler corrosion and state the various conditioning techniques employed to prevent scaling and sludge formation in boilers.

6. Write a short informative note on 'priming and foaming.'

7. State the principle of EDTA titration. Describe the procedure for estimating hardness of water.

8. Explain the various conditioning methods employed for boiler water.

9. Write the chemical reactions involved in lime-soda method. Compare hot and cold lime-soda methods.

10. With a suitable diagram, explain cold-lime soda method and state its advantages and disadvantages.

11. Explain priming and foaming in boilers. How can these processes be minimized in boilers?

12. What is Calgon? State its application in the treatment of water.

13. Explain the role of coagulants in lime-soda method.

14. Explain carbonate and electrical conditioning of boiler water.

15. Write a short note on Langelier index.

16. Describe the chemical analysis of water for determining chloride by argentometry.

17. Explain the method for determining fluorides in water using SPADNS method.

18. Explain the method of chemical analysis of sulphates in water by gravimetry.

19. What is electrodialysis? Mention the principle and add a note on ion-selective membranes.

20. Explain zeolite softening process with a suitable diagram w.r.t the following: (a) softening method, (b) reactions involved, (c) regeneration, (d) advantages and disadvantages.

21. Discuss ion-exchange process with a suitable diagram w.r.t the following: (a) Ion-exchangers, (b) deionization process, (c) regeneration, (d) advantages and disadvantages.

22. What is disinfection of water? Explain chlorination of water.

23. Explain ozonization of water. Write the chemical reactions involved in the process.

24. Explain disinfection of water by chlorination and UV treatment.

25. How does one perform demineralization using ion-exchange method? Explain with chemical reactions.

26. Explain ion-exchange process for demineralization of water. Justify the statement, 'During ion-exchange, water is first sent to the cation-exchanger and then in to the anion-exchanger.'

27. Distinguish between zeolite and lime-soda methods.

28. What is dissolved oxygen? Explain the steps involved in determining BOD of water.

29. State the principle of BOD and COD. Explain the method to determine COD of water.

30. Explain ultrafiltration of water.

31. How does one perform desalination of brackish waters using reverse osmosis?

32. Write a short note on 'reverse osmosis and electrodialysis.'

NUMERICAL PROBLEMS

1. Classify the following impurities as permanent, temporary, and non-hardness causing salts: $Ca(NO_3)_2$, KNO_3, Na_2SO_4, $CaSO_4$, $MgCO_3$, H_2S, SO_2, H_3PO_4, $MgCl_2$.   (**Ans:** Permanent hardness salts: $Ca(NO_3)_2$, $CaSO_4$, $MgCl_2$; Temporary hardness salts: $MgCO_3$; Non-hardness salts: KNO_3, Na_2SO_4, H_2S, SO_2, H_3PO_4)

2. SHW contains 0.5 g/500 ml of $CaCO_3$. 50 ml of this solution consumed 48 ml EDTA solution. 50 ml of water sample required 15 ml EDTA solution. After boiling, the same water sample required 10 ml EDTA solution. Calculate all types of hardness present in the water sample.
   (**Ans:** Total hardness = 312.3 ppm, Permanent hardness = 208.2 ppm, Temporary hardness = 104.1 ppm)

3. Water sample contains impurities (in ppm) viz: $Ca(NO_3)_2$ = 25.3, KCl = 3.2, $MgSO_4$ = 12.6, HCl = 10.2. Calculate the total, temporary, and permanent hardness present in water sample.
   (**Ans:** Total hardness = 25.92 ppm; Temporary hardness = 0 ppm; perm. hardness = 25.92 ppm)

4. Calculate the quantities of lime and soda for softening 100000 l of water with impurities (mg/l) viz, $Ca(HCO_3)_2$ = 30.2, $Mg(HCO_3)_2$ = 20.8, $CaCl_2$ = 28.1, $MgCl_2$ = 8.7, $CaSO_4$ = 3.4, $MgSO_4$ = 6.7. lime purity = 70% and soda purity = 85%.   (**Ans:** Lime = 6.535 kg, Soda = 5.301 kg)

5. 10 ml water sample consumed 5 ml EDTA solution before boiling and 1.5 ml EDTA solution after boiling. Calculate all types of hardness present in the water sample.
   (**Ans:** Total hardness = 500 ppm, permanent hardness = 150 ppm, temporary hardness = 350 ppm)

6. If 1 litre of water contains 4.3 g $CaCl_2$ and is softened using permutit softener, calculate the quantity of NaCl produced in soft water.   (**Ans:** 4.532 g)

7. 55,000 litres of raw water was softened using zeolite. Zeolite, on regeneration required 200 litres of NaCl solution with strength of 120 g/l. Calculate the total hardness of water sample.   (**Ans:** 372.76 ppm)

8. A water sample contains impurities (mg/l), $Mg(HCO_3)_2$ = 14.6, $Ca(HCO_3)_2$ = 8.1, $MgCl_2$ = 19. From this data calculate (a) Quantities of lime and soda needed to soften 10,000 litres of water; (b) total hardness of water.   (**Ans:** Lime = 0.333 kg, Soda = 0.212 kg)

9. 20 ml of waste water sample is diluted with 200 ml distilled water and equal volumes are filled in two BOD bottles. In the blank titration, 100 ml of diluted water sample on titration consumed 9.2 ml of 0.01 N sodium thiosulphate. 100 ml of the incubated sample after 5 days consumed 2.7 ml of 0.01 N sodium thiosulphate. Determine BOD of the waste water sample.   (**Ans:** 52 mg/l)

10. Estimate COD of effluent water sample when 20 ml of sample consumed 8.3 ml of 0.001 M potassium dichromate solution.   (**Ans:** 19.92 mg/l)

FURTHER READING

1. Brezonik.P, and W. Arnold. *Water Chemistry*, Oxford University Press, NY, 2011.
2. Central Ground Water Board (CGWB), Government of India, BIS (2012) http://cgwb.gov.in
3. Dore, M. (ed) *Chemistry of Oxidants and Treatment of Water*. VCH, New York, 1996.
4. Manahan, S.E. *Water chemistry. Green Science and Technology of Nature's Most Renewable Resource.* CRC Press. Taylor and Francis Group, 2010.
5. Salvato. J.A. *Environmental Engineering*, John Wiley & Sons, New Jersey, 2003.
6. Water Quality Association. www.wqa.org/

ANSWERS

Check Your Progress

1. Rain water, sea water, ground water, and surface water are various sources of water.
2. The property of water that prevent of lathering of soap due to the presence of dissolved salts of calcium and magnesium is called hardness of water.

3. The Ca^{2+} and Mg^{2+} ions present in hard water tend to react with soaps forming insoluble precipitates of calcium or magnesium stearate or palmitate, thereby preventing lathering of soap.

4. 1 ppm ≡ 1mg/l ≡ 0.1 °Fr ≡ 0.07 °Cl

5. As calcium carbonate is completely insoluble in water and easy to form precipitates and quantified, it is the reference standard and hence the total hardness of water is generally expressed in terms of equivalents of $CaCO_3$.

6. Temporary hardness causing salts: $Ca(HCO_3)_2$ and $Mg(HCO_3)_2$

   Permanent hardness causing salts: $CaCl_2$, $MgCl_2$, $CaSO_4$, $MgSO_4$, $FeSO_4$

7. Water used in boilers to generate steam is called boiler feed water.

8. Boiler feed water should be softened to prevent various boiler problems such as scale–sludge formation, priming and foaming, boiler corrosion.

9. When the concentration of dissolved salts in boiler feed water reaches the saturation point, they are thrown out of water in the form of precipitate in the inner walls of the boiler. If the precipitate is loose, slimy, and soft, it is called *sludge*. If the nature of precipitate is hard, crusty, and adherent, it is called *scale*.

10. Refer to Section 13.5.1

11. Refer to Section 13.5.1

12. Phosphate conditioning method involves continuous blow-down boiler operation to remove calcium phosphate; whereas in Calgon conditioning, a soluble complex of $[Ca_2P_6O_{18}]^{2-}$ is formed on reacting with calcium salts in water which is easy to remove from boilers as sludge.

13. EDTA is ethylenediaminetetraacetic acid. It is used as complexing agent to determine the hardness of water.

14. The indicator used in EDTA method is Eriochrome Black T, which works in the pH range of 8.5–10; hence alkaline buffer of NH4OH–NH4Cl is added.

15. Eriochrome Black T is the indicator used in EDTA method. The end point in EDTA titration method is colour change from wine red to deep blue.

16. When hardness causing metal ions are reduced in concentration or eliminated from the raw water, it is called softening of water.

17. Lime-soda method and zeolite softening are two methods of water softening.

18. Sodium aluminate and aluminium sulphate are used as coagulants in cold lime-soda method to hasten the reaction. These coagulants undergo hydrolysis thereby forming gelatinous precipitates that trap particulate matter and fine particles and eventually allow faster sedimentation.

    $NaAlO_2 + 2 H_2O \rightarrow NaOH + Al(OH)_3$ (ppt)

    $Al_2(SO_4)_3 + 3 Ca(OH)_2 \rightarrow 2 Al(OH)_3$ (ppt) $+ 3 CaSO_4 + 6 CO_2$

19. Zeolites (or permutit) are hydrated sodium aluminosilicates used as commercial adsorbents for softening water. The hydrated form of zeolite has the general formula, $Na_2O.Al_2O_3.xSiO_2.yH_2O$, where x is 2 – 10 and y is 2 – 6.

20. Cation exchangers are easily attacked by alkalies due to acidic functional groups in their resin structure. However, all types of ion-exchangers are resistant to acids. When water is passed through cation resin, salts present in the water are converted into the corresponding acids, which on passing through anion exchanger do not harm it. If the reverse sequence is followed, then alkalies produced on passing the water through the anion exchanger will harm the cation exchanger in the subsequent step thereby resulting in column bleeding.

21. Refer to p. 16

22. Lime-soda and zeolite methods

Lime soda method	Zeolite method
1. Produces water having residual hardness of 15–60 ppm	1. The soft water so produced is of residual hardness of almost 0 ppm.
2. Use of coagulants in cold lime-soda method	2. No use of coagulants in the process.
3. Temporary hardness-causing salts are converted to $NaHCO_3$ which makes the water not suitable for use in boilers.	3. Temporary hardness-causing salts are completely removed by zeolites.
4. Chemicals such as lime, soda, coagulants are to be fed each time for from cycle.	4. Exhausted zeolites can be regenerated with brine.

23. Cold and hot lime-soda methods

Cold lime-soda method	Hot lime-soda method
1. The process is very slow and takes several hours for completion.	Requires less time to complete the softening process.
2. Reaction is hastened by adding coagulants like sodium aluminate and alum.	Coagulants are not required, precipitate settles down rapidly due to high temperatures.
3. Dissolved gases like oxygen, CO_2 are not removed.	Dissolved gases like oxygen, CO_2 are removed.
4. Filtration takes place very slowly.	Filtration takes place rapidly as viscosity of water decreases.
5. Cold lime-soda method produces soft water with residual hardness of 50–60 ppm.	Hot lime-soda method produces soft water with residual hardness of 15–30 ppm.

24. Refer to Section 13.7.3
25. $ZrOCl_2$-SPADNS colorimetric method is used to determine fluoride in water.
26. Water used for cooling purposes must be non-scaling, non-corrosive, and non-staining.
27. An algebraic difference between the actual pH and the saturation pH is defined as Langelier index (L.I.) which is expressed as, $L.I = PH - (pH)_S$ where $(pH)_s$ denotes saturation pH.
28. The process of destroying disease-producing bacteria and microorganisms from water and making it fit for drinking purposes is called sterilization or disinfection.
29. The chemicals added (in calculated amounts) to water to kill microbial population in water are known as disinfectants. Chloramine, chorine, ozone, bleaching powder are some disinfectants.
30. A minimum exposure of 16,000 μwatt-s/cm^2 is the allowed UV dosage for sterilizing water.
31. Reverse osmosis and electrodialysis are the commonly adopted methods for desalinating water.
32. Ion selective membrane is a specialized membrane that is only permeable for one type of ions with specific charge.
33. If chloramine is added in excess, it does not cause bad taste and odour to the treated waters. Also, it is known to have a longer effect as against chlorine and bleaching powder.
34. Yes, ultrafiltration process can eliminate pathogens from water.
35. Refer to Section 13.8.1
36. Refer to Section 13.8.2
37. A natural process in freshwater that has a rich supply of nutrients. It is also a part of ageing process, as nutrients tend to accumulate over the period of time
38. The amount of free oxygen (in mg/l) needed to oxidize organic matter by bacteria and other microorganisms at 20°C for a period of 5 days is called biological oxygen demand (or BOD_5).
39. BOD analysis is important as it indicates the extent of pollution of water bodies and sewage waters. Higher the BOD, greater is the level of pollution of the water body.
40. COD is the measure of oxygen in water required to oxidize all organic matter into carbon dioxide and water.

Multiple Choice Questions

1. (b)	2. (d)	3. (c)	4. (d)	5. (a)	6. (d)	7. (a)
8. (d)	9. (a)	10. (b)	11. (c)	12. (b)	13. (a)	14. (c)
15. (b)	16. (b)	17. (a)	18. (d)	19. (c)	20. (b)	

Corrosion

14.1 INTRODUCTION

We have all seen rusty iron nails, metal fences, window grills, and perforated water pipelines around us. Corrosion is everywhere around us and is difficult to inhibit or repair. In an industrial scenario, if corrosion remains unchecked, there can be major failures of engineered components and machinery. Annually, India loses 6 lakh crores (4–5% GDP) to corrosion failures.

Coming under the branch of material sciences, corrosion engineers are challenged to find ways to prevent corrosion. Metallurgical engineering, corrosion engineering, and coating science are the major fields that deal with materials and their degradation so as to solve real-life corrosion problems.

Corrosion (Latin: *corrodere* means to gnaw to pieces) is the degradation of metals due to their reactions with the surrounding environment. The environment surrounding the metal could be dry gases or an electrolyte solution. Stainless steels in sea water (contains chlorine), wine in lead bottles (acetic acid in wine corrodes lead), red powder of iron oxide (Fe_3O_4) formed during rusting of iron, green layer of basic carbonate $[CuCO_3+Cu(OH)_2]$ on copper metal when exposed to air containing moisture, and dissolved CO_2 are some of the classic examples of corrosion.

Corrosion is a natural spontaneous process as it is the return of metals to their more natural state as minerals (oxides). Let us assume that a given metal occurs as an oxide in an ore. Various metallurgical operations are carried out to extract the pure metal from the ores. All metallurgical processes are reduction reactions that give metals in their pure excited state. The pure metal will have a natural tendency to return to the ground state as its oxide form (oxidation).

Hence, corrosion is a natural oxidative process observed in metals and is considered as metallurgy in reverse (Fig. 14.1).

14.2 MECHANISM OF CORROSION

Corrosion can be broadly classified into two groups: dry or atmospheric corrosion and wet or electrochemical corrosion. (Fig. 14.2).

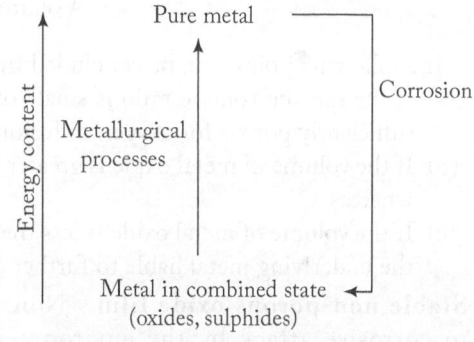

Fig. 14.1 Corrosion as a natural process

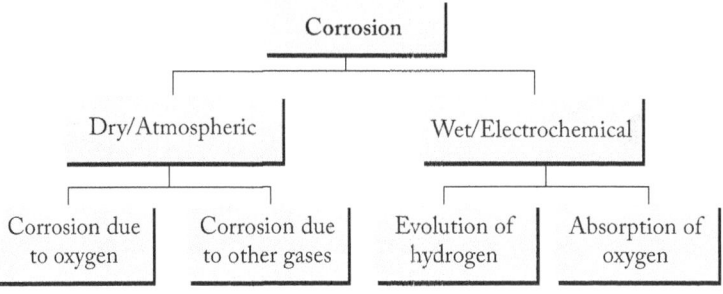

Fig. 14.2 Types of corrosion

14.2.1 Direct Chemical Corrosion or Dry Corrosion

Dry corrosion occurs when atmospheric gases such as oxygen, sulphur dioxide, hydrogen sulphide react with metals to form solid films of corrosion products. The extent of dry corrosion depends on chemical affinity of the gases and metals along with the nature of corrosion products.

Corrosion Due to Oxygen

Oxygen gas does not attack most of the metals at ambient temperatures. However, alkali metals such as sodium and potassium, and alkaline earth metals such as magnesium and calcium, are readily oxidized at room temperature. At higher temperatures, all metals except the noble metals (silver, gold, platinum) get oxidized and corroded. When metal surface forms an oxide layer, it acts as a barrier between the metal and the environment. If oxygen is continuously supplied, the rate of corrosion of the underlying metal will depend on the nature of the oxide film formed on the metal surface (Fig. 14.3).

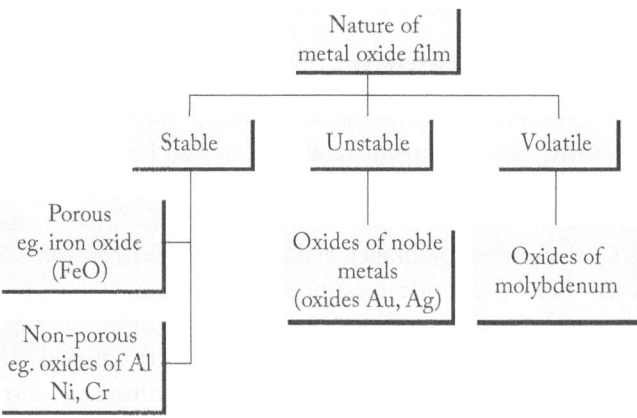

Fig. 14.3 Nature of metal oxide film

Pilling–Bedworth rule explains porosity of metal oxide layers formed during corrosion by calculating the specific volume ratio.

$$\text{Specific volume ratio} = \frac{\text{Volume of metal oxide}}{\text{Volume of metal}}$$

The following points can be concluded from the above equation.

(a) If the specific volume ratio is small, oxidation corrosion will take place as the metal oxide film is sufficiently porous for oxygen diffusion.

(b) If the volume of metal oxide is greater than the volume of metal, the oxide film will be non-porous, whereas

(c) If the volume of metal oxide is less than the volume of metal, the oxide film will be porous making the underlying metal liable to further corrosion.

Stable non-porous oxide film Non-porous adherent layers on metals will allow access to corrosive attack by the environment. When aluminium is exposed to oxygen it forms non-porous oxide film (Al_2O_3), thereby protecting the underlying metal from further corrosive attack.

Porous non-protective layers If the oxide layer formed is porous, oxygen tends to diffuse through the pores and oxidation continues until the metal decays completely (Fig. 14.4). When iron metal is exposed to oxygen, its oxides (e.g., Fe_2O_3) are highly corrosive in nature.

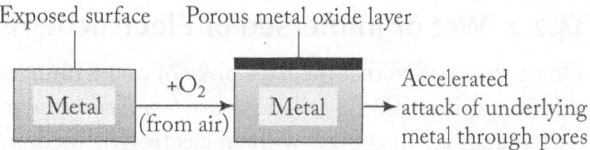

Fig. 14.4 Formation of porous metal oxide layer

Unstable non-protective layer Gold, silver, and platinum form oxides, which when exposed to the atmosphere tend to decompose back into the metal and oxygen (Fig. 14.5). Such noble metals do not oxidize in air or oxygen at any given temperature and do not exhibit corrosion.

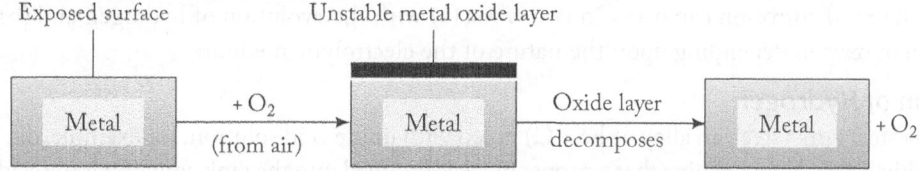

Fig. 14.5 Formation of unstable metal oxide layer

Volatile non-protective layers Molybdenum oxide layer (MoO_2), if formed on the metal quickly volatilizes and exposes the underlying fresh metal to further corrosive attack (Fig. 14.6).

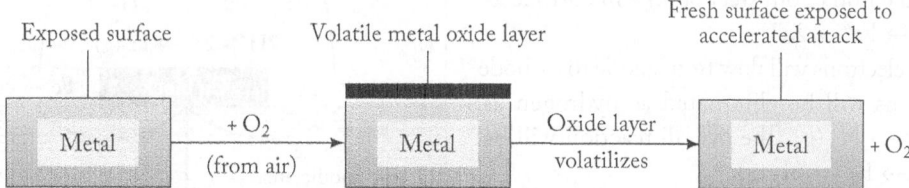

Fig. 14.6 Formation of volatile metal oxide layer

Corrosion by Other Gases

Gases like hydrogen, carbon dioxide, sulphur dioxide, chlorine, hydrogen sulphide react with metals forming the corresponding corrosion products; the extent of corrosion depends on the chemical affinity between the metal and gas in the environment. It also depends on the formation of protective and non-protective layers on the metal surface. When the corrosion film formed is adherent and non-porous, the extent of corrosion is reduced as the film acts like a barrier between the metal and the corrosive substance.

The corrosive action of chlorine on silver metal forming silver chloride film is one such example. When the film formed is porous or non-protective, the metal surface gets degraded. The action of dry chlorine gas on tin forms a volatile layer of stannic chloride ($SnCl_4$), which makes the underlying metal vulnerable to further corrosive attack.

Hydrogen gas attacks metals at room temperature, leading to *hydrogen embrittlement*. If iron is exposed to hydrogen sulphide gas, ferrous sulphide is formed on the metal surface along with nascent hydrogen atoms that diffuse into the iron metal, where they combine to form molecular hydrogen.

$$Fe + H_2S \rightarrow FeS + 2[H]$$
$$[H] + [H] \rightarrow H_2 (g)$$

The molecular hydrogen, thus formed, escapes the metal in the form of gas leaving a cavity in the metal. When there is continuous evolution of hydrogen molecules, the metal loses its tensile strength, ductility, and malleability and becomes weak.

14.2.2 Wet or Immersed or Electrochemical Corrosion

On the basis of Nernst theory, a piece of metal immersed in a liquid medium has the tendency to go into solution as ions. This is called *solvation of metal ions* and electrochemical corrosion sets in when:
  (a) metals are in contact with an electrolytic medium, and
  (b) two dissimilar metals are immersed in an electrolytic solution.

The metal with higher electrode potential (active metal) will act as the anode and the one with lower electrode potential will be the cathode. Anode undergoes solvation and gets corroded, whereas cathode accepts ions and forms a protective coating, thereby resisting corrosion. This process of corrosion that proceeds due to ionic reactions in the presence of a conducting medium, is called *electrochemical corrosion*.

Electrochemical corrosion can occur in two ways: (a) with the evolution of hydrogen and (b) with the absorption of oxygen depending upon the nature of the electrolytic medium.

Evolution of Hydrogen

Consider a steel tank (steel, an alloy of Fe + C) filled with dilute acid solution, for example, dil. HCl acid (pH < 7) (Fig. 14.7). Now, imagine that a copper block is dropped into the tank containing the acid solution.

Due to differing potential of the metals, the junction at which copper is in contact with the steel tank starts getting corroded. Hence, a small cathodic area is formed within the larger anodic area (steel tank). The metallic iron (from steel tank) will corrode as,

$$Fe \rightarrow Fe^{2+} + 2e^-$$

Further, electrons will flow from anode to cathode and H^+ ions will be eliminated as hydrogen as, $2H^+ + 2e^- \rightarrow H_2$ (g). The overall reaction will be, $Fe + 2H^+ \rightarrow Fe^{2+} + H_2$ (g).

Thus, all metals above hydrogen in the electrochemical series will have a tendency to get dissolved in the acidic media with simultaneous evolution of hydrogen.

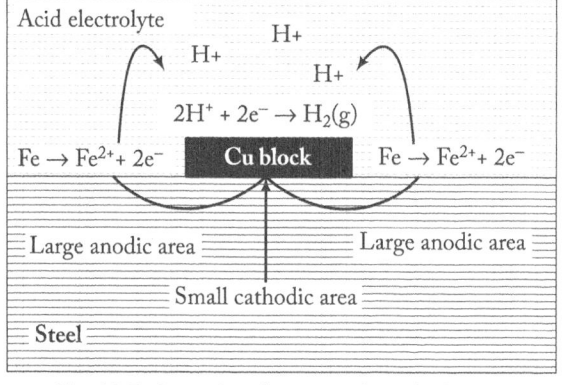

Fig. 14.7 Corrosion due to evolving hydrogen

Absorption of Oxygen

Let us consider that a steel plate is coated with an oxide film and placed in an aqueous electrolyte. Imagine that a crack develops at one point in the oxide film (Fig. 14.8). Iron (present in steel) gets exposed to the electrolyte and turns anodic with respect to the large cathodic (steel) area.

At anode: $Fe \rightarrow Fe^{2+} + 2e^-$ (Oxidn)

At cathod: $1/2\ O_2 + H_2O + 2e^- \rightarrow 2OH^-$ (Redn)

In the cathodic reaction, electrons flow from the cracked site (anodic) to the cathodic area (steel) and are consumed by dissolved oxygen present in the electrolyte. The ferrous ions and hydroxide ions diffuse and form ferric hydroxide, $[Fe(OH)_2]$, which further gets oxidized to ferric oxide or rust, Fe_2O_3, and gets deposited around the cracked site (anode).

$$4\ Fe(OH)_2 + 2H_2O + O_2 \rightarrow 2\ (Fe_2O_3.3H_2O)$$

The process of rust formation will continue until the dissolved oxygen present in the electrolyte is exhausted.

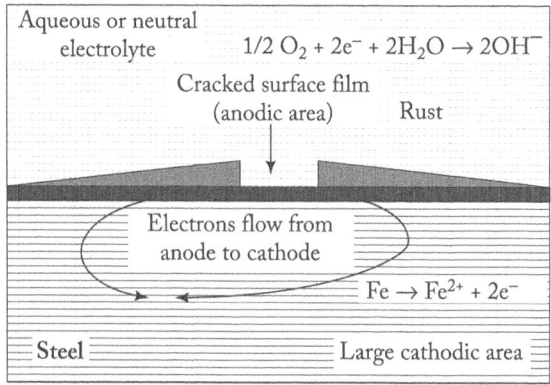

Fig. 14.8 Corrosion due to oxygen absorption

14.3 DIFFERENTIAL AERATION

Differential aeration (differential oxygenation) is an oxygen concentration cell-type of corrosion. This type of electrochemical corrosion sets in when a metal surface is exposed to varying concentrations of electrolyte, thereby with varying concentrations of dissolved oxygen in the electrolyte. Imagine a drop of water (electrolyte) on a steel surface (Fig. 14.9). The area of steel coming under the water droplet experiences lesser access to dissolved oxygen. Oxygen from the atmosphere can only diffuse from the edges of the water droplet. The other parts of the steel surface will have more access to oxygen. Due to differing oxygen concentrations on the steel surface, the parts below the water droplet will become anodic and undergo corrosion, whereas the remaining surface will be cathodic.

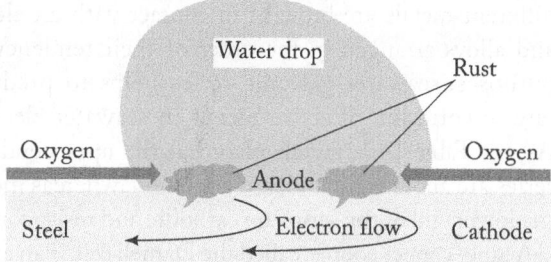

Fig. 14.9 Illustration of droplet corrosion (oxygen concentration cell)

Some classic instances of differential aeration are:

(a) Waterline corrosion of metals can be seen on ships and ocean liners. Due to the presence of minerals, sea water is a highly conducting electrolyte. When ships enter the sea water, the metal part just below the waterline will have lesser access to oxygen making it less oxygenated (anodic) than the metal part above the waterline. Due to differential oxygen supply, corrosion sets in and the corrosion product gets deposited on the bottom of the ships.

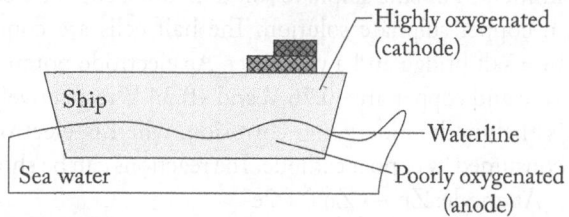

Fig. 14.10 Ship body showing corrosion at the waterline

(b) If an iron rod is partially immersed in water (Fig. 14.11), the metal parts below the waterline will be poorly oxygenated and hence undergo corrosion.

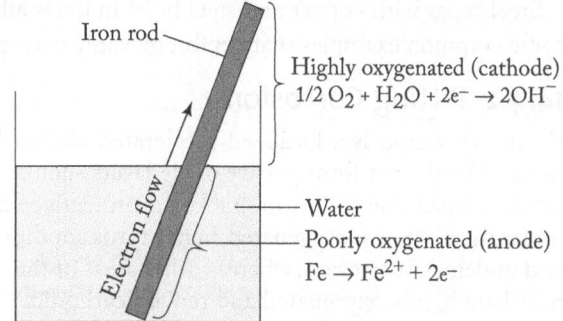

Fig. 14.11 Immersed portion of iron rod showing corrosion

(c) An iron water pipeline is placed in soil of differing porosities. Corrosion tends to occur forming pits in those portions of the pipeline buried under the soil.

Check Your Progress

1. What is corrosion?
2. What is dry corrosion?
3. Does the nature of metal oxide film help to predict corrosion of metals? Which type of metal oxide film causes rapid corrosion?
4. State the two conditions for electrochemical corrosion of metals.
5. When designing ocean-liners or ships, one should use only one type of metal for all parts including nuts and bolts. Why?
6. What is electrochemical corrosion?
7. Imagine partially dipping a metal piece in 30 % saline solution. After three hours, a layer of rust is seen just at the waterline. What type of corrosion is observed?

14.4 Types of Corrosion

14.4.1 Galvanic Corrosion

Galvanic corrosion or differential metal corrosion is also called *bimetallic corrosion* because it sets in when different metals are brought in contact with an electrolyte. Seawater galvanic series comprises metals and alloys arranged in the order of their tendency to corrode in marine environments. For practical purposes, seawater galvanic series helps to predict the rate of corrosion of active metals in seawater electrolyte. As per Table 14.1, metals placed at the upper end of the series are anodic and undergo corrosion, whereas metals in the lower end of the series are cathodic and resist corrosion.

A zinc–copper couple, called the Daniell cell, is an example of galvanic corrosion. The cell consists of a zinc electrode immersed in zinc sulphate solution and a copper electrode in copper sulphate solution. The half cells are connected by a salt bridge and a voltmeter. As electrode potentials of zinc and copper are –0.76 V and +0.34 V respectively, zinc is the anode undergoing corrosion, whereas electrons are consumed by copper cathode. The reactions can be shown as:

At anode: $Zn \rightarrow Zn^{2+} + 2e^-$

At cathode: $Cu + 2e \rightarrow 2e^- + Cu(s)$

Steel pipes with copper and steel bolts in brass alloy are some common examples that exhibit galvanic corrosion.

Table 14.1 Galvanic series

Magnesium	(Highly anodic)
Magnesium alloys	
Zinc	
Aluminum	
Aluminum alloys	
Low carbon steels	
Cast iron	
Stainless steel (active)	
Lead–tin alloys	
Lead	
Brass	
Copper	
Bronze	
Copper–nickel alloys	
Silver	
Stainless steel (Passive)	
Monel	
Titanium	
Gold	
Platinum	(Highly cathodic)

14.4.2 Pitting Corrosion

Pitting corrosion is a localized accelerated electrochemical attack that produces 'holes' or 'pits' in the metal. Metals that form passive oxide layers such as aluminium and steel are susceptible to pitting. Pits are developed due to dirt or cracks on protective oxide layers on metal surface. In Fig. 14.12, the area under the dirt is less oxygenated, hence turns anodic and undergoes corrosion, whereas other areas of the metal are highly oxygenated and remain cathodic.

Corrosion by pitting is difficult to detect and leads to equipment failures with low weight losses. Pitting corrosion can be prevented by removal of dust/dirt from the metal surfaces or by using corrosion inhibitors such as chromates and phosphates.

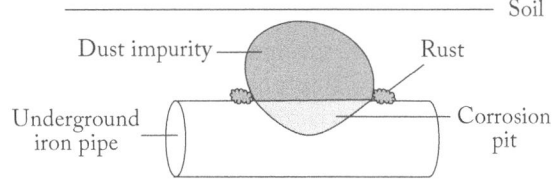

Fig. 14.12 Pitting corrosion on underground pipe

14.4.3 Intergranular Corrosion

If one observes the microstructure of metals and alloys, it appears as grains separated by grain boundaries. It is observed that corrosion occurs along the grain boundaries than the grain centres. Intergranular attack (Fig. 14.13) occurs due to the sensitization of the material due to 'inadequate' heat treatment during welding. A classic example showing intergranular corrosion is austenitic steel, where chromium reacts with carbon during heat treatment at 550°–800°C to produce chromium carbide as per the reaction, $23Cr + 6C \rightarrow Cr_{23}C_6$.

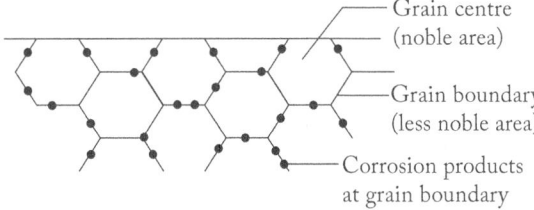

Fig. 14.13 Chromium depletion due to intergranular corrosion

During weld decay, chromium carbide gets deposited at the grain boundaries. The formation of chromium carbide indicates the depletion of chromium content, which provides passivity to austenitic steel. Such depletion of chromium makes austenitic steel susceptible to corrosion. It is suggested that applying adequate heat treatment followed by rapid quenching process could mitigate the chances of such type of corrosion.

14.4.4 Stress Corrosion

Similar to intergranular corrosion, stress corrosion is also a localized electrochemical attack. It occurs due to the combined effect of static tensile stresses and corrosive environment (like strong alkali, chlorides) on a metal. The presence of stress produces strains due to displacement of atoms making it anodic. Stressed metallic areas possess higher electrode potentials and hence become chemically active. Such metals are highly susceptible to corrosive environments. Pure metals are comparatively immune to stress corrosion. Welding, bending, riveting, and quenching causes stress in metal and alloys. Figure 14.14 shows a metal strip bent due to stress corrosion. At the stress portion, the atoms in the stressed region get displaced and become anodic, thereby liable to corrosion. Stress corrosion results in the formation of cracks that tend to grow and propagate in perpendicular direction to the operating tensile stress within the metal bulk structure until failure occurs.

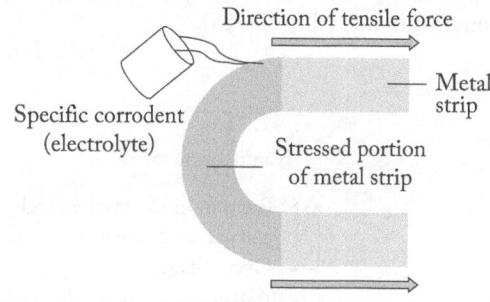

Fig. 14.14 Stressed metal strip under tensile force and alkaline corrodent

The various forms of stress corrosion include (a) season cracking and (b) caustic embrittlement.

Season cracking   It is observed in brass articles due to the presence of moisture. In its pure form, copper is quite resistant to corrosion, but when alloyed with zinc, it exhibits stresses at the grain boundaries and gets corroded in moist or alkaline environments.

Caustic embrittlement   This type of stress corrosion is observed at joints, screws, bolts, rivets, nuts, and bent portions of steamers and boilers made of mild steel that are under high pressures and exposed to alkaline waters. Boiler water is generally softened by adding trace amounts of sodium carbonate making it alkaline at higher pressures.

$$Na_2CO_3 + H_2O \rightarrow 2NaOH + CO_2$$

The NaOH so formed flows in to the holes and cracks by capillary action. The concentration of NaOH keeps increasing as the boiler operates at higher temperatures and pressures. Further, the iron in steel reacts with NaOH forming sodium ferrate, leading to weakening of the boiler points, which if unchecked may result in explosion.

$$2Na_2FeO_2 + 4H_2O \rightarrow 6NaOH + Fe_3O_4 + H_2$$

Iron (in mild steel) exposed to higher concentration of NaOH behaves as an anode and the rest of the areas in contact with dilute NaOH becomes cathodic, thereby setting up a concentration cell where anodic areas undergo corrosion. Sodium ferrate, on decomposition leads to NaOH regeneration that can further attack the mild steel and rust will be deposited leading to boiler bursting and failure.

14.4.5 Soil and Microbial Corrosion

Soil is considered a highly conducting medium due to the presence of minerals, moisture, organic matter, dirt, impurities, and even microbes. If soil is highly acidic, corrosion occurs due to evolving hydrogen, whereas in non-acidic soils, corrosion sets in due to differential aeration. The presence of minerals, moisture, and organic matter in the soil results in the formation of soluble metal complexes, thereby resulting in accelerated corrosion. Underground pipelines are susceptible to corrosion if buried under compact soil rather than porous-aerated soil as per differential aeration. Certain bacteria influence the rate of corrosion processes. If oxygen-consuming bacteria are present in the soil, they tend to decrease the oxygen content in the medium. When a metal structure is partially exposed to such an electrolytic medium with depleted oxygen, the rate of corrosive attack is higher. Iron bacteria or iron-oxidizing bacteria cause pitting corrosion of underground pipelines. Other corrosive bacteria are *Desulfovibrio, Clostridia*, mucoids, and *Pseudomonas*, which are known to cause accelerated attack on metals.

14.5 FACTORS INFLUENCING CORROSION

The rate of corrosion majorly depends on (a) nature of the metal and (b) nature of the electrolyte (or environment) (Fig. 14.15).

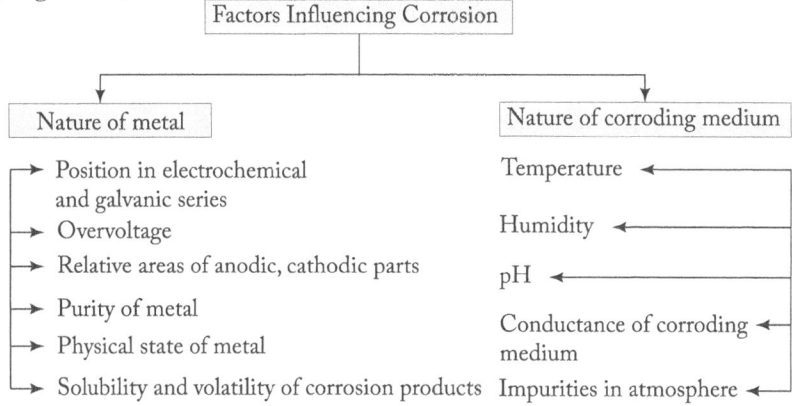

Fig. 14.15 Factors influencing corrosion

14.5.1 Nature of Metal

Position of metal in electrochemical and galvanic series A metal placed at the *higher end of the series is an active anodic metal and thereby undergoes corrosion*. If two or more metals are farther from each other in the series, greater will be their oxidation potential and they exhibit higher rates of corrosion.

Overvoltage When a metal is immersed in an electrolytic solution, the solution acquires a potential. If the metal begins to corrode in the electrolyte, then its potential also alters in value. The difference of this potential is called overvoltage. Hydrogen overvoltage is the potential difference between an electrode and a reversible hydrogen electrode in the same electrolyte solution. If zinc metal is placed in 1 N sulphuric acid, it undergoes corrosion along with the evolution of hydrogen from the metal surface. Initially, the reaction is gradual due to higher overvoltage of zinc (0.72V) that reduces the overall electrode potential for corrosion to an infinitesimally small factor. Now, imagine adding a few drops of 1 N copper sulphate in the electrolyte with immersed zinc metal. Copper will start depositing on

Table 14.2 Standard electrode potentials of various metals at 25°C

Metal electrode	Potential (Volts)	
Li/Li^+	−3.04	Reactive (Anodic)
K/K^+	−2.925	
Ca/Ca^{++}	−2.866	
Na/Na^+	−2.714	
Mg/Mg^{++}	−2.363	
Al/Al^{+++}	−1.662	
Zn/Zn^{++}	−0.763	
Cr/Cr^{+++}	−0.744	
Fe/Fe^{++}	−0.441	
Ni/Ni^{++}	−0.23	
Sn/Sn^{++}	−0.136	
Fe/Fe^{+++}	−0.045	
$H_2/2H^+$	0.000	(Reference)
Cu/Cu^{++}	0.337	
Ag/Ag^+	0.80	
Hg/Hg^{++}	0.854	
Pd/Pd^{++}	0.987	
Pt/Pt^+	1.2	
Au/Au^{+++}	1.42	Noble (Cathodic)

zinc metal forming small cathodic areas and zinc will corrode faster resulting in hydrogen overvoltage drop to 0.33 V. Hence, *higher the overvoltage, lower the rate and extent of corrosion.*

Relative cathodic and anodic parts   When two dissimilar metals are in contact, the rate of corrosion *is directly dependent on the cathodic area available for attack.*

$$\text{Corrosion rate} \propto \frac{\text{Cathodic area}}{\text{Anodic area}}$$

So, if a small copper pipe is fitted in a large steel vessel, the entire vessel is available for corrosive attack.

Purity of metals   The presence of impurities in the electrolyte or in the metal bulk tends to set up a galvanic cell and promotes corrosion. The iron pillar, located in the Qutub complex at Delhi is known for its exemplary rust-resistant wrought iron composition. The protective layer of iron phosphate is claimed to protect the ancient structure from corrosion.

> Iron Pillar is a classic example representing metallurgical wonder for scientists around the world. In 2000, the journal *Corrosion Science* reported the presence of a protective layer on iron pillar. The rust characterization revealed the presence of crystalline iron hydrogen phosphate hydrate ($FePO_4 \cdot H_3PO_4 \cdot 4H_2O$), α-, γ-, δ-FeOOH and magnetite. This protective layer is known to prevent further attack on the underlying metal. (see Further Reading)

Physical state of the metal   As discussed in Section 14.3.5, metal areas under stress tend to be anodic and undergo corrosion in suitable corrosive environments. Iron wire mesh exhibits stress corrosion at the mesh joints. When metals or their alloys exhibit much higher resistance to corrosion than their respective positions in the electrochemical series, they are said to exhibit passivity. Chromium, tin, nickel, aluminium, and their alloys are known to exhibit passivity. Some metals show a decreasing order of passivity as:

$$Ti \rightarrow Al \rightarrow Cr \rightarrow Be \rightarrow Mo \rightarrow Mg \rightarrow Ni \rightarrow Co \rightarrow Fe \rightarrow Mn \rightarrow Zn \rightarrow Cd \rightarrow Pb \rightarrow Cu$$

Stainless steel exhibits passivity due to oxide of chromium (Cr_2O_3) that forms a protective layer on steel.

Solubility and volatility of corrosion products   If the corrosion layer is soluble in an electrolytic medium, then corrosion of the metal proceeds at a faster rate. However, an insoluble corrosion layer forms an adherent barrier between the metal and the electrolyte, thereby inhibiting corrosion. Lead–acid battery consists of lead electrodes in sulphuric acid. The lead sulphate formed during battery operations acts as an insoluble barrier, thereby controlling lead corrosion. Also, if corrosion products are volatile, they will vaporize from the metal surface, thereby making the underlying metal available for further attack.

14.5.2 Nature of Electrolyte (Environment)

Temperature   If temperature of the environment is high, diffusion rate of ionic species increases thereby enhancing the corrosion rate.

Humidity   It is known that corrosion rate increases in a humid environment than in dry air. Humid air tends to dissolve corrosive gases, such as carbon dioxide, oxygen, sulphur dioxide, and acid vapours, thereby setting up a corrosion cell on the metal surface.

Influence of pH (hydrogen ion concentration)   Another important parameter is influence of hydrogen ion concentration on the rate of corrosion. It is observed that acidic media are more corrosive than alkaline and neutral media. In acidic pH (< 7), the rate of corrosion is higher as electrochemical corrosion proceeds with the evolution of hydrogen gas at the cathode. In alkaline pH (>7), electrochemical corrosion by absorption of oxygen takes place, forming a metal oxide film as cathodic product. The oxide film gets adhered to metal surface and hence the rate of corrosion depends on the nature of corrosion product.

Conductance of the medium   It is of prime importance in case of underground pipelines. Conductance of dry sandy soils is lower than those of clay soils. Stray currents will damage the metal buried under clay or mineral-rich soil than dry soil.

Presence of impurities   When gases like CO_2, SO_2, H_2S, and acid fumes of HCl and H_2SO_4 are present in electrolytes, they result in extreme corrosion of metals. The presence of these gases increases the acidity of the medium surrounding the metal, thereby increasing the rate of corrosion. Further, suspended particles such as NaCl and $(NH_4)_2SO_4$ that are hygroscopic in nature (i.e., absorb moisture), act as strong electrolytes, thereby further degrading the metal.

14.6 METHODS OF CORROSION CONTROL

Corrosion is a natural interaction between a metal and its environment. To inhibit or control corrosion, one needs to employ various methods either by treating the metal or the environment (or a combination of both). The various methods for controlling corrosion are discussed as follows.

14.6.1 Proper Selection and Design of Materials

Selection of the right type of material is a crucial factor for inhibiting corrosion. The choice of the metal should be made not only on the cost and structure, but also on its chemical properties and the environment. Noble metals are immune to corrosion, but are expensive to use. As far as possible, purer metals should be utilized to design metal work-pieces, as even traces of impurities can result in severe corrosion. If two dissimilar metals have to be in contact, they should be so selected that their oxidation potentials are as close as possible in the galvanic series. Impingement attack can be reduced by careful filtration of suspended solids from the liquid stream and by preventing turbulent flow.

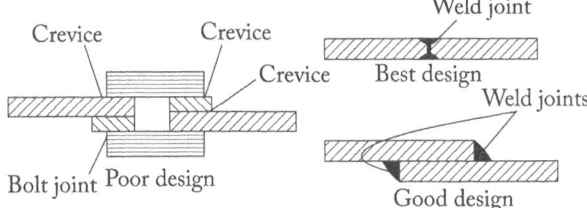

Fig. 14.16 (a) Joint welds – poor and good designs

When a structure consists of two dissimilar metals, it is beneficial to use a third, more active metal in contact with that structure so as to inhibit corrosion. Crevices should be avoided between adjacent parts of the structure to avoid formation of concentration cells. (Fig. 14.16(a)) Bolts and rivets should be replaced by proper welding. Bolts and nuts should be made of the same metal; otherwise crevices present around the metal joints cause corrosion.

A good design should not allow accumulation of corrosion products, dust, and water. This is especially true while working with metallic angles. One must avoid sharp corners of metal joints. It is necessary that a good design involves smooth bends at the corners of metallic work-pieces (Fig. 14.16(b)).

During metal work-piece design, welding is performed at higher temperatures, followed by

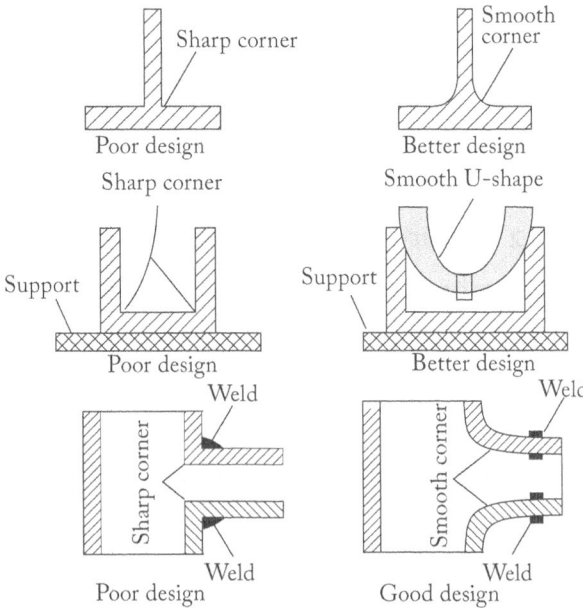

Fig. 14.16 (b) Metal work-pieces showing sharp and smooth corners

rapid cooling. These processes cause stresses in the metal structure and can be avoided by reducing the welding times.

14.6.2 Alloy Formation

Ferrous alloys with silicon, nickel, and chromium provide good resistance towards corrosion. The formation of non-porous, adherent oxide film resists further attack on the metal. The chromium content in stainless steel on interacting with air forms a self-healing oxide film that protects the underlying steel from corrosion.

14.6.3 Purification of Metals

The impurities present in the metal act as an anode and results in corrosion. A purer metal offers more corrosion resistance. Further, removal of stray currents from the base metal and prevention of bacterial growth in the corroding medium are some of the measures used to reduce the rate of corrosion.

14.6.4 Cathodic Protection

Cathodic protection is used to reverse the flow of current between two dissimilar metals in an electrolytic medium, thereby reversing the action of metals in contact. The principle of cathodic protection involves connecting an external anode to the metal structure to be protected and passing electric current so that all areas of the metal surface become cathodic; thus it becomes passivated and resists corrosion. As corrosion involves the oxidation of metal, it can be understood that cathodic polarization, which discourages oxidation and favours reduction at the metal surface, should tend to provide protection. It is achieved by two methods, namely by the use of sacrificial anodes or by 'impressed' current.

14.6.5 Anodic Protection

Metals like Al, Cr, Ti, and Ni exhibit passivity and hence cathodic protection cannot be employed for such metals. In anodic protection method, the metallic structure to be protected is made more anodic by applying an external impressed direct current to it. The method is applicable only to metals exhibiting

Check Your Progress

8. If a piece of impure zinc and pure zinc are placed in a salt solution, which one will corrode faster?
9. Determine the amount of rust that will be formed if 200 g of iron is allowed to undergo complete rusting.
10. What is passivity?
11. Arrange metals in the order of their decreasing tendency to undergo oxidation in a given solution: Au, Ag, Fe, Mg, Zn.
12. What will happen if a copper rod is partially immersed in a beaker containing 30% NaCl solution? Write the chemical reactions.
13. Bolts and nuts should be made from same metal. Why?
14. Justify, 'Wire mesh corrodes at the joints rapidly.'
15. Under identical conditions, impure metal corrodes faster than pure metal. Justify.
16. List the various factors that influence rate of corrosion.
17. Name the types of stress corrosion of metals.
18. What is hydrogen embrittlement?
19. What is intergranular corrosion?
20. What is microbial corrosion?
21. What is the effect of pH of electrolyte on the rate of corrosion?

active–passive behaviour in extremely corrosive environments (eg., stainless steel tanks in high saline regions). Due to application of external impressed anodic current, a thin oxide layer is formed that protects the base metal from further corrosive attack. A typical anodic protection cell is illustrated in Fig. 14.17.

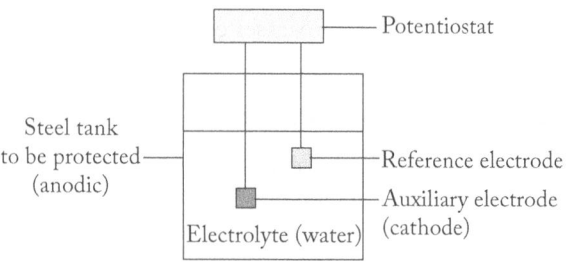

Fig. 14.17 Anodic protection system for tanks

The steel tank structure that requires protection should be made anodic. The steel tank is connected to a potentiostat (device used to maintain constant potential) and a reference electrode. An auxiliary electrode is provided that is corrosion-resistant (passive) metal acting as a cathode. Platinum-clad materials or corrosion-resistant alloys are generally used as cathode materials. The potentiostat is adjusted for a specific potential and current, such that the tank is passivated and resists corrosion. Further, an optimum potential is maintained between the tank and the reference electrode that is calculated using electrochemical measurements. Some of the **advantages** of anodic protection method are low current requirements and ability to protect complex structures in extreme corrosive environments. However, the major **limitations** include high installation costs, higher currents needed to induce passivity, and ability to precisely maintain constant potential.

Sacrificial Anodic Protection (SAP)

In this method, an active metal like zinc, magnesium, or aluminum is connected to the structure to be protected (Fig. 14.18). The more active metal behaves anodic thereby undergoing destruction.

These anodes are called sacrificial anodes and they can be replaced when corroded. This method is used for protecting cables, pipelines, and other buried structures from soil corrosion. It is also applicable for mitigating corrosion in ship hulls, cables, submarines, etc., from sea water. Magnesium rods or pieces are inserted into domestic water boilers or tanks for preventing corrosion.

Impressed Current Cathodic Protection (ICCP)

In this method, an external emf from a source like battery or rectifier is applied to the corroding system in such a manner that the article is forced to behave like cathode and thus gets protected (Fig. 14.19).

The anode is either an inert material like graphite or one which undergoes destruction and needs to be replaced. Other anode materials in use are high-silica iron, carbon, stainless steel, and platinum. The anodes are buried in backfill such as coke, breeze, or gypsum so as to increase the electrical contact with the surrounding soil. This method is useful in protecting buried items such as tanks, pipelines, and transmission-line tower condensers.

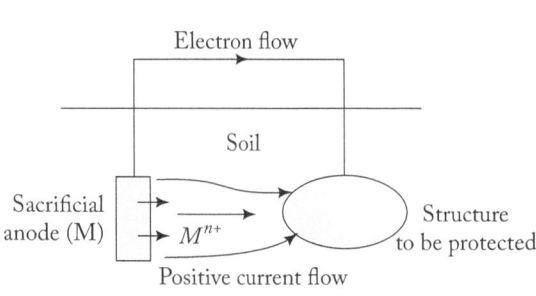

Fig. 14.18 Sacrificial anodic protection

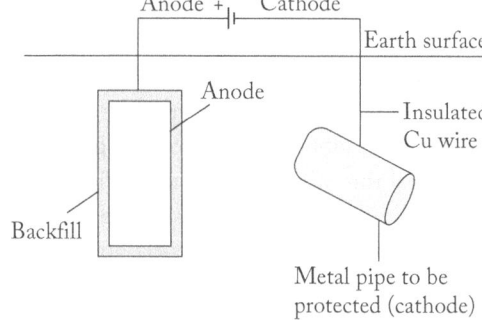

Fig. 14.19 Impressed current cathodic protection

Drawbacks

(a) This method requires high current which increases maintenance costs.

(b) The current may not be uniform over the entire surface of the metal.

(c) Sometimes hydrogen gas is liberated at the cathode and this may result in hydrogen embrittlement.

14.6.6 Metallic Coatings

Metal coatings that are anodic to the base metal are termed as anodic metal coatings (Fig. 14.20). In order to protect a metal surface from corrosion, any metal placed higher up in the galvanic series such as Zn, Mg, or Sn are coated on the metal. These active metals are anodic with respect to the base metal. Zinc coating on the base metal iron acts as the anodic area, while iron is cathodic.

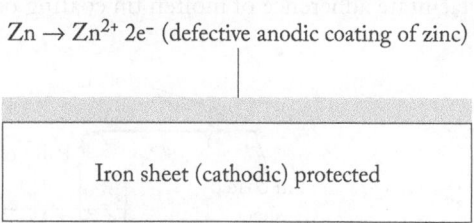

$Zn \rightarrow Zn^{2+} 2e^-$ (defective anodic coating of zinc)

Iron sheet (cathodic) protected

Fig. 14.20 Anodic coating

A hot dipping method involving coating of zinc on iron or steel surface (base metal) is called *galvanizing*. In this process, the base metal is cleaned with an organic solvent followed by treatment with dilute sulphuric acid solution for 15–20 minutes at 60–90°C. This results in the removal of impurities from the base metal surface. The material is then washed, dried, and dipped in molten zinc bath maintained at 450°C. As the melting point of zinc is 419.5°C, its molten bath is kept at a much higher temperature and covered with a flux of ammonium chloride to prevent oxide formation (Fig. 14.21).

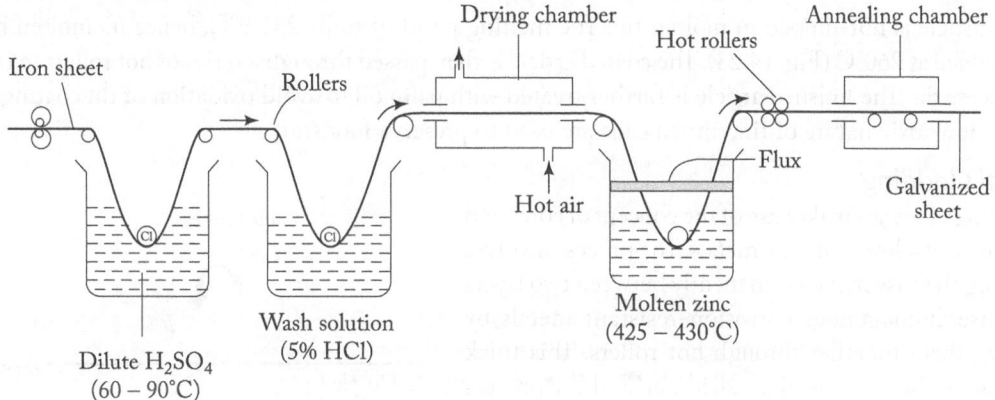

Fig. 14.21 Set-up illustrating the process of galvanization

Once the base metal is coated with molten zinc, it is passed through a series of hot rollers, so as to form a uniform, homogeneous coat. The coating is annealed at 650°C and cooled completely. Galvanized iron is used to make water pipelines, roof-sheets, nails, screws, nuts, and bolts. If the coating is broken, corrosion rate does not increase, since iron (base metal) acts as a cathode and zinc

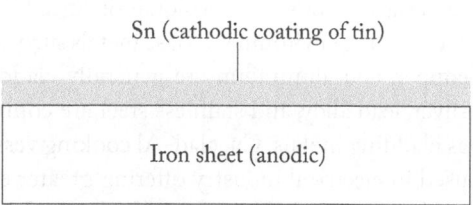

Sn (cathodic coating of tin)

Iron sheet (anodic)

Fig. 14.22 Cathodic coating

is the anode. Hence galvanizing provides better protection to iron than tin. Galvanized containers cannot be used to preserve food and beverages, since zinc metal coat tends to dissolve in acidic media (i.e., food) and results in the formation of toxic zinc compounds, detrimental to human health.

When a noble metal is used as the coating material on base metal, it is termed as *cathodic coating* (Fig. 14.22). It protects the base metal due to its high corrosion resistance than the base metal. A thin film coating of Sn, Cr or Ni is deposited on the iron metal by electroplating process. If cathodic coating is continuous, surface coating provides effective protection to the base metal. If the coating is irregular, it leads to serious corrosion damages to the base metal.

Another hot dipping method involving application of a thin coat of tin metal on the base metal surface is called *tinning*. It involves treating the base metal with organic solvents followed by cleaning with dilute sulphuric acid at 80°C to remove any impurities. The base metal is dipped in a flux of $ZnCl_2$ solution to facilitate adherence of molten tin coating on the surface.

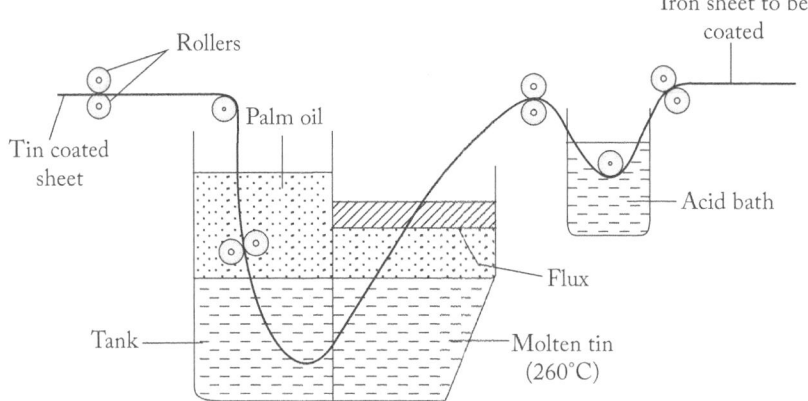

Fig. 14.23 Set-up showing the process of tinning

The article is hot-dipped in molten tin. The melting point of tin is 231.9°C, hence its molten bath is maintained at 260°C (Fig. 14.23). The coated article is then passed through a series of hot rollers to remove the excess tin. The finished article is further treated with palm oil to avoid oxidation of tin coating. Due to the non-toxic nature of tin, tinned cans are used to preserve foodstuff.

Metal Cladding

Cladding of metals makes use of the concept of corrosion resistance of alloys and pure metals. The process involves bonding the base metal permanently between two layers of dense, homogenous corrosion-resistant metals by passing them together through hot rollers. This thick coating on the base metal is called *clad* and the process is known as *cladding* (Fig. 14.24). The choice of cladding material depends on corrosion resistance of the metal in the working environment. Base metals such as mild steel, copper, and aluminium are generally cladded. Nickel, silver, lead alloy, and stainless steel are commonly used

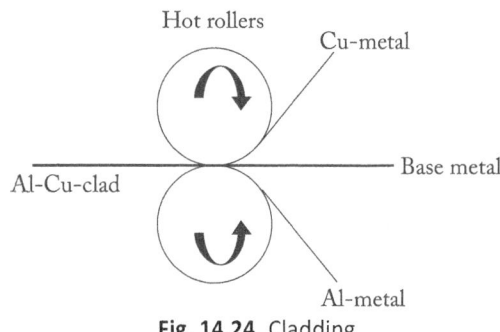

Fig. 14.24 Cladding

as cladding metals. Cu-clad-Al cooking vessels have shown increased heating efficiency. Cu-clad-steel is used in electrical industry offering greater electrical conductivity and high strength.

Aluminium-clad is a classic example of protection by metal cladding used in aircraft industry. Al-clad, as the name suggests, is a plate of duralumin (alloy of Al; composition: 95% Al, 4% Cu, 0.5% Mn, & 0.5% Mg) sandwiched between two layers of 99.5% pure aluminium using hot rollers.

Metal Spraying

As the name suggests, the molten metal to be deposited is sprayed as a thin, continuous, uniform coat on the base metal surface followed by drying. A versatile technique, it can be easily employed on any finished metal article and also in huge civil structures such as metal roofs, rail bridges, and so on.

Cementation or Diffusion Coatings

It is very commonly employed to coat uneven, small, and rough surfaces of varying shapes. The base metal is heated in a powder of metal whose coating is desired. The temperature is kept high enough to allow diffusion of metal powder in to the base metal. This results in the formation of an adherent alloy layer on the base metal surface; hence the process is called cementation. Al, Zn, Si, Cr are used to form protective alloy layers on metal surfaces thereby inhibiting corrosion.

(a) When zinc is coated on ferrous base metals it is called *sherardizing*. Generally, sherardizing is performed on small metal articles such as screws, nuts, bolts, and so on. In this process, the base metal is heated to 350 – 450°C in the presence of zinc dust–zinc oxide powder in a revolving closed drum in an inert atmosphere for about 12 hours. Beyond 300°C, zinc evaporates and diffuses into the steel base metal, thereby forming thick bonded Zn–Fe alloy layer.

(b) When chromium is coated on ferrous base metal, it is called *chromizing*. Articles such as turbines and blades are chromized so as to inhibit erosion of metals. The base metal is placed in a revolving closed drum containing 55–60 % chromium powder with the inert filler, alumina (40–45%). The processing temperature is maintained at 1300 – 1400°C for 4 hours. The alumina filler prevents coalescence of chromium powder on the base metal. A halide salt is added that changes to vapour phase and acts as carrier gas to bring chromium metal powder to the metal surface. Chromized articles exhibit higher resistance towards chemical and thermal shocks.

(c) If aluminium is used as a diffusion coating metal, the process is called *calorizing*. Aluminium diffusion coatings are primarily performed on furnaces, chemical retorts, condensers, and safety valves. The metal article is heated in aluminium powder–alumina mixture with traces of ammonium chloride flux in a hydrogen atmosphere at 1000°C for 6 hours. This treatment gives higher resistance to corrosive, toxic gases.

14.6.7 Electroplating

The process of electro-deposition of metals on metals, non-metals, and alloys is called electroplating. It involves depositing a thin film of a noble metal or an alloy over an active base metal by passing direct current through an electrolytic solution containing the soluble salt of the coating metal. Nickel, chromium, gold, and tin plating methods are some common examples of electroplating. The metal object to be coated is thoroughly cleaned and made cathodic. Further, the coating metal block acts as the anode. These electrodes are suitably suspended in an electrolytic bath containing a solution of salt of the metal to be coated. The metal salt in aqueous solution undergoes ionization and a potential difference is applied to the salt solution through the electrodes. Due to the applied potential difference, metal ions migrate towards the cathode and get deposited.

The factors that govern electroplating processes are as follows.

(a) If anode is of the same metal of which the salt is in solution, the salt is re-formed by anode material passing into the solution in the form of ions.

(b) The conducting power, solubility, and metal content of the electrolytic solution should be high. It should not undergo hydrolysis, oxidation, or other chemical or environmental changes. It should possess sufficient covering power and the deposit obtained should be compact and adherent, thereby giving a uniform deposit on the base metal.

Applications of electroplating include the following.

(a) It improves the appearance of the base metal, thereby also making it resistant to corrosion.

(b) Engineering coatings (called *functional coatings*) are used for enhancing specific properties of the surface, such as solderability, wear resistance, reflectivity, and conductivity. Metals for engineering purposes include gold (Au) silver (Ag), and lead (Pb). Decorative protective coatings are primarily used for adding an attractive appearance to some protective qualities. Metals in this category include copper (Cu), nickel (Ni), chromium (Cr), zinc (Zn), and tin (Sn).

14.6.8 Electroless Plating

In electroless plating the base metal object is dipped in an electrolyte bath of a noble metal salt. The noble metal layer gets deposited on the base metal object by displacement of the base metal by the noble metal. Hence, it is also termed as *displacement plating* or *immersion plating*. If nickel coating is desired, the base metal is dipped in nickel sulphate bath and sodium hypophosphite at 100^{0}C maintained at pH 4.5–5.0. Nickel ion from electrolytic solution reduces to nickel and nickel phosphide adherent film is formed on the base metal object.

14.6.9 Corrosion Inhibitors

Substances or mixtures added in trace amounts to aqueous corrosive environment to prevent corrosion of metals are called corrosion inhibitors. Corrosion inhibitors are chemicals that could be synthetic or natural in origin. Inorganic and organic chemical compounds can act as effective inhibitors. Generally, two or more compounds are used as corrosion inhibitors and hence are called *synergistic* in nature.

Anodic Inhibitors or Passivators (Barrier-type Inhibitors)

Substances that inhibit anodic reactions (solvation of metals) are called anodic inhibitors. Inorganic salts that contain anions form sparingly soluble compounds with metal ions and are capable of suppressing anodic reactions. Ions of transition elements with high content of oxygen, such as chromates, tungstates, phosphates, nitrates, molybdates, hydroxides, and silicates are generally used. These compounds react with metallic ions produced on the anode, forming insoluble precipitates which are deposited on the metal surface as insoluble film, impermeable to metallic ions, thereby causing passivation and eventually inhibiting corrosion.

Cathodic Inhibitors (Barrier-type Inhibitors)

Compounds that suppress cathodic reactions are termed as cathodic inhibitors. They can be used in both acidic and neutral solution. In an acidic medium, the metal corrodes at the anode and the cathodic reaction proceeds due to evolving hydrogen, $2H + 2e^{-} \rightarrow H_2$ (gas).

Arsenic compounds get adsorbed on cathode surface that restricts the liberation of hydrogen. In neutral solutions, dissolved oxygen in the electrolyte gets reduced and forms hydroxide ions at the cathodic area. Under such conditions, the rate of corrosion can be reduced by restricting diffusion of oxygen to cathodic areas. $1/2O_2 + H_2O + 2e^{-} \rightarrow 2OH^{-}$

Salts of zinc, magnesium, nickel, etc., react with hydroxide ions forming insoluble metal hydroxides that deposit as a protective layer on the metal, hence inhibiting corrosion.

Chromates, tungstates, phosphates, nitrates, molybdates, hydroxides, and silicates are common inorganic anodic inhibitors. The efficiency of organic corrosion inhibitors is related to the presence of polar functional groups with S, O, or N atoms in the molecule. Heterocyclic compounds and pi electrons generally have hydrophilic or hydrophobic parts that are ionizable. The polar function is usually regarded as the reaction centre for the establishment of the adsorption process. Organic inhibitors that contain oxygen, nitrogen, and/or sulphur are adsorbed on the metallic surface, blocking the active corrosion sites. Amines, urea, mercaptans, benzotriazole, aldehydes, heterocyclic nitrogen compounds, tryptamine, caffeine, and other natural product extracts are examples of organic inhibitors.

Organic adsorption inhibitors such as sulphides and mercaptans form an oily layer on the metal surface that prevents the adsorption of evolving hydrogen. It also prevents solvation of the metal ion. Inorganic adsorption inhibitors such as bicarbonates and phosphates form tough adherent layers on the metal and prevent corrosion. Dicyclohexylamine nitrite (DCHN) is a common vapour phase inhibitor added to the electrolytic medium. As it has a low vapour pressure (3×10^{-7} atm. at 25°C), it quickly evaporates and forms a thin barrier film on the metal surface and inhibits corrosion.

14.6.10 Organic Coatings

Paints, varnishes, enamels, and lacquers are common organic surface coatings. Such surface coatings are inert organic barriers added on the surface of base metal for providing corrosion resistance and for decoration purposes. The protective value of such coatings depends on various factors such as

  (a) their chemical inertness to the surrounding environment,
  (b) adequate surface adhesion,
  (c) impermeability to water, salts, gas, and electrolytes, and
  (d) suitable application method.

Paints

Paint is a uniform dispersion mixture of pigments, driers, and fillers in a liquid called 'vehicle' or medium. Various constituents of paints include pigment, vehicle or drying oil, thinners, driers, fillers or extenders, plasticizers, and anti-skinning agents.

Pigment This is an important constituent of paint, present in about 70–80% in vehicle medium. Apart from providing strength and aesthetic appeal to paints, it increases the life of paint film by preventing harmful UV rays that degrade the oil film. The pigment should be opaque, chemically inert, non-toxic, and miscible in liquid vehicle. Pigments with higher refractive indices than the vehicle (also called *prime pigments*) provide good opacity and hiding power. Some of the commonly used pigments and their respective colour are listed in Table 14.3.

Table 14.3 Pigments and their colour

Name of pigments	Colour
White lead, $2PbCO_3.Pb(OH)_2$	White
Titanium dioxide, TiO_2	White
Red chrome, $PbCrO_4.Pb(OH)_2$	Red
Prussian blue, $Fe_4[Fe(CN)_6]_3$	Blue
Carbon black	Black
Chromium oxide, Cr_2O_3	Green

Vehicle or drying oil It is the liquid medium that binds the pigments to the surface. Oils such as linseed oil, dehydrated castor oil, tung oil, or a mixture of drying and semi-drying oils are commonly used as vehicle. Chemically, these oils are esters of unsaturated organic higher fatty acids. When paint is applied onto a surface, unsaturated fatty acids present in the oil undergo oxidation forming oxides or peroxides at the double bond. Further, it undergoes polymerization to form a tough, adherent, uniform, and impermeable layer of polymer on the surface.

Thinners It is a volatile solvent added to the paint so as to adjust the consistency of the paint. Turpentine and petroleum fractions such as benzene, spirit, and naphtha are commonly used as thinners. They also increase the penetrating power of the vehicle and elasticity of the paint film on the applied surface.

Driers Borates, resonates, and linoleates of metals such as Zn, Co, Ni, and Mn are used as driers that are added to reduce the drying time of paint coatings. Driers act as oxygen-carrier catalysts that accelerate the drying of oil film by oxidation, polymerization, and condensation.

Fillers or extenders Inorganic substances such as aluminium silicate, barium sulphate, asbestos, chalk, gypsum, and clay are used as fillers to reduce the cost of paint. They tend to fill the voids in the paint film and reduce the cracking of paint coat.

Check Your Progress

22. What is chromizing?
23. What is displacement plating?
24. Justify the statement, 'Food and beverage containers are always made using tin-coats.'
25. What are corrosion inhibitors? State the types.
26. Which of the compounds can act as effective corrosion inhibitor/s when added in an electrolytic medium—dicyclohexylamine nitrite, sodium chloride, hydrazine, sodium sulphite?

Plasticizers Tricresyl phosphate and triphenyl phosphate are commonly used plasticizers in paints. They enhance the elasticity of paint film and also the overall durability of the coat.

Anti-skinning agents Compounds such as polyhydric phenols prevent the formation of insoluble skins on a painted surface due to polymerization or oxidation. They inhibit cracking or peeling of the paint coat from the surfaces.

Varnish

Varnish is a colloidal dispersion of natural or synthetic resins in a vehicle. Various constituents of varnish include resin, vehicle or drying oil, thinners, driers, and anti-skinning agents.

Resins Natural resins such as copal, rosin, and shellac and synthetic resins such as phenol formaldehyde, urea formaldehyde, and vinyl resins are commonly used. Varnishes do not contain pigments. The primary role of resins is to provide resistance to weathering and chemical attack. It also causes hardening of the dried varnish coat.

Vehicle or drying oils Generally varnishes are oil or spirit based, that is, the resin is either dissolved in oils (linseed oil, fish oil, soy oil, etc.) or in volatile solvents (alcohols, spirit, etc). They reduce the drying time of varnish film by oxidation and polymerization.

Thinners Turpentine, naphtha, kerosene, spirit, alcohols, etc., are generally used as thinners so as to adjust the consistency of the varnish.

Driers These include linoleates or naphthenates of lead, manganese, or cobalt and are added to enhance the quality of the oil coat and further accelerate the drying of varnish coat by polymerization, oxidation, and condensation.

Anti-skinning agents Tertiary amyl alcohols and phenols are used as anti-skinning agents that provide good adherence of varnish coat on the applied surface.

Enamel

Enamel is a pigmented varnish that contains pigments, oil or resin vehicle, and thinners and driers. Enamels provide a combined form of paint and varnish, imparting a glossy, lustrous appearance to the surface.

Resins Phenolic resins (such as phenol–formaldheyde), aldehyde resins, etc., are commonly used in enamels.

Drying oils Oils like linseed oil, soya bean oil, fish oil, dehydrated castor oil are used.

Driers Naphthenates, resonates, and linoleates of lead, manganese, cobalt, and zinc are used for quick drying of enamel coat.

Thinners As in paints and varnish, turpentine, xylol, and acetone are used so as to adjust the consistency of enamel coat.

Lacquers

Lacquer is a colloidal dispersion of cellulose derivatives, resin, and plasticizer in a volatile solvent. When lacquer is applied on a surface, it forms a transparent, adherent, and waterproof coating. The various constituents of a lacquer are cellulose derivate, volatile solvent, resin, plasticizer, and diluents.

Cellulosic derivatives These are a major constituent that provides hardness and durability to the film. Cellulose acetate, cellulose nitrate, ethyl cellulose, etc., are commonly employed in lacquers.

Resins Various formaldehyde resins impart thickness, gloss, adhesion, and lustre to the lacquer coat.

Solvents Ethyl acetate, acetone, dioxane, etc., are commonly used to dissolve cellulose derivatives and resins.

Plasticizers Dibutyl phthalate, dibutyl phosphate, tricresyl phosphate, etc., impart plasticity to the lacquer coat.

Diluents These are low-cost solvents that decrease the viscosity of the medium (similar to thinners). They also provide tough, smooth, and glossy coating to the surface. Petroleum naphtha, toluol, etc., are commonly used as diluents

Check Your Progress

27. What are paints and varnishes?
28. List the ingredients of paints.
29. What is lacquer?
30. What is enamel? List its ingredients.
31. State the role of plasticizers in paints.
32. What is the role of anti-skinning agents in paints and varnishes?

Activity-based Questions

A. Superhydrophobic surfaces or coatings

Extremely water-repellent surfaces, on which water droplets adhere in a spherical shape with contact angles > 150°, are called superhydrophobic surfaces. Extreme water repellency is desired in several applications, especially in corrosion resistance or inhibition. Superhydrophobicity can potentially be exploited in non-wetting, non-fogging, non-icing, and self-cleaning surfaces, or in micro- and macrofluidic devices. SiO_2 and TiO_2 nanoparticles are known to exhibit superhydrophobicity.

Present a brief review on superhydrophobic SiO_2 and TiO_2 coatings (preparation, applications, current trends, challenges).

B. Greener corrosion inhibitors

Chemical corrosion inhibitors are organic or inorganic substances or mixtures added in trace amounts to the electrolytic medium to inhibit corrosion. The harmful effects of chemical inhibitors have prompted the search for innocuous and greener corrosion inhibitors. Various plant extracts and natural materials have shown significant anti-corrosive activity with minimal environmental impact.

Present a study on green corrosion inhibitors with their classification, mechanistic details, and industrial applications listing eco-friendly substances with supporting examples.

SUMMARY

- Understanding of corrosion process is vital in engineering fields, such as metallurgy and material sciences.
- Corrosion of metals is essentially a spontaneous chemical or an electrochemical phenomenon. Various corrosive gases such as hydrogen, chlorine, sulphides, and even aqueous medium are known to enhance the rate of corrosion.
- Based on the nature of the surrounding environment of metals, there are two broad categories of corrosion mechanism—atmospheric and electrochemical corrosion.
- Differential aeration explains the electrochemical type of corrosion of metals due to varying concentrations of oxygen in the electrolyte. Water line corrosion in ships is a classic example elucidating differential aeration.

- The other types of corrosion include galvanic, pitting, intergranular, stress, soil, and microbial corrosion.
- The nature of the metal (such as its position in the galvanic series, overvoltage, physical state, passivity, etc.) and the nature of electrolyte (such as its pH, temperature, forms of corrosion products, and so on) are major factors that influence the rate of corrosion.
- Proper design, material selection, alloy formation, cathodic–anodic protection, use of corrosion inhibitors and coatings (metallic and organic) are various methods to control corrosion of metals.
- Metallic and organic coatings are known to control corrosion and also provide aesthetic appeal to the metal object. Organic coatings include paints, varnishes, enamels, and lacquers.

GLOSSARY

Base metal: The metal that needs protection from corrosion.

Cladding: The process of sandwiching an alloy between two pure metals, eg., Al-clad.

Chromizing: The process of applying chromium coating on a base metal by cementation.

Corrosion: The degradation of metals due to their reactions with the surrounding environment.

Corrosion inhibitors: Substances or their mixtures added in trace amounts to an electrolyte to inhibit corrosion of metal.

Dry corrosion: A direct form of corrosion that takes place between the metal and the environmental gases like oxygen and hydrogen.

Electrochemical corrosion: The corrosion occurring due to ionic reactions between anodic and cathodic areas of the metal in an aqueous medium.

Electroless plating: The deposition of a layer of noble metal on the base metal object by displacement of the base metal by the noble metal.

Electroplating: The process of electro-deposition of metals on metals, non-metals, and alloys to enhance aesthetic appeal of the metal object along with corrosion protection.

Enamel: A type of organic coating that is a pigmented varnish.

Galvanizing: The process of applying a zinc coat on iron base metal by hot dipping method.

Lacquer: An organic coating of a colloidal dispersion of cellulose derivatives, resin, and plasticizer in a volatile solvent.

Tinning: The process of applying a thin tin coat on iron base metal by hot dipping method.

Paint: An organic coating that is a dispersion of pigments in a fluid vehicle.

Sherardizing: The process of coating zinc on ferrous base metals by diffusion coating process.

Varnish: An organic coating that is a resin in a vehicle of drying oils, thinners, and driers.

EXERCISES

Multiple Choice Questions

1. During electrochemical corrosion in acidic environment,
   - (a) O_2 evolves
   - (b) H_2 evolves
   - (c) O_2 is absorbed
   - (d) H_2 is absorbed

2. Galvanizing is a process of coating iron with
   - (a) Zn
   - (b) Sn
   - (c) Al
   - (d) Mg

3. Ships and ocean-liners exhibit _____ corrosion.
   - (a) stress
   - (b) waterline
   - (c) galvanic
   - (d) none of these

4. Which of the following factors affect rusting of iron?
   - (a) Moisture
   - (b) Air
   - (c) Impurities
   - (d) All of these

5. As per Pilling–Bedworth rule, greater the specific volume ratio
   - (a) higher is the oxidation corrosion.
   - (b) lower is the oxidation corrosion.
   - (c) higher is the reduction corrosion.
   - (d) none of these.

6. If a metal rod exhibits holes and crevices on its surface, the type of corrosion is,
   - (a) waterline
   - (b) galvanic
   - (c) pitting
   - (d) stress corrosion

7. Season cracking and caustic embrittlement are special cases of _____ corrosion.
   - (a) chemical
   - (b) stress
   - (c) electrochemical
   - (d) galvanic

8. Hydrazine added to corrosive environment
   - (a) inhibits anodic reaction
   - (b) enhances hydrogen overvoltage
   - (c) controls cathodic reactions by consuming dissolved oxygen
   - (d) all of these

9. Corrosion is maximum when the pH of the corroding medium is
   - (a) at 7.0
   - (b) above 7.0
   - (c) at 1.0
   - (d) below 7.0

10. Iron pillar of Delhi exhibits corrosion resistance due to
    (a) iron hydrogen phosphate layer
    (b) more impurities in metal
    (c) use of corrosion inhibitors
    (d) least metal stresses

11. In impressed current cathodic protection, anode is provided with a gypsum backfill because
    (a) it enhances the rate of reaction.
    (b) it enhances electrical contact with surrounding soil.
    (c) it decreases metal–metal contact.
    (d) none of these.

12. The role of anti-skinning agent in paints is to
    (a) prevent chipping off of paint coat
    (b) allow removal of paint coat
    (c) allow quick drying of paint coat
    (d) provide viscosity to the paint film

13. Al-clad is
    (a) duralumin cladded between pure Al metals
    (b) pure Al metal cladded between duralumin
    (c) pure duralumin cladded with Fe
    (d) example of Zn alloy

14. If anodic area is small, corrosion rate is
    (a) lower   (b) higher
    (c) not observed   (d) inhibited

15. Corrosion inhibitors are
    (a) added in trace amounts in the medium to decrease corrosion rate
    (b) used in synergism
    (c) of both anodic and cathodic types
    (d) all of the above

16. During electroless plating process, base metal object is dipped in
    (a) highly active metal salt
    (b) more noble metal salt
    (c) a mixture of active–passive metal salts
    (d) none of these

17. The process where an iron pipe is buried in soil and connected to a Mg metal block is called
    (a) galvanic protection
    (b) impressed current cathodic protection
    (c) sacrificial anodic protection
    (d) passivation

18. Which of the following is not a type of organic coating?
    (a) Lacquer   (b) Enamel
    (c) Varnish   (d) Cladding

19. A metal that is protected by its own oxide layer is
    (a) Al   (b) Fe
    (c) Cu   (d) Mo

20. During electroplating process, the base metal (to be protected) is made
    (a) cathode   (b) anode
    (c) passive   (d) none of these

Review Questions

1. What is corrosion? Explain the term giving suitable examples.

2. Justify the following statements with suitable examples:
   (a) Corrosion is metallurgy in reverse.
   (b) Ships exhibit corrosion just at the waterline.
   (c) Wire mesh exhibits corrosion at the bent joints.
   (d) Impure metal corrodes faster than a purer metal.
   (e) Silver, gold, and platinum do not undergo oxidation corrosion.
   (f) Galvanized containers are not used to store foodstuff.

3. How do metals undergo corrosion? Explain dry corrosion.

4. Discuss the mechanism of electrochemical corrosion.

5. Explain the concept of corrosion. Give a brief account of electrochemical corrosion taking place in the case of metals of different electrode potentials.

6. Explain electroplating and state the various factors affecting the deposition of metals.

7. Explain the various factors that influence the rate of corrosion of metals.

8. How is the rate of corrosion influenced by the following factors: (a) Design of the material, (b) pH, (c) purity of metal, (d) physical state, and (e) nature of corrosion products?

9. Describe differential aeration. Provide suitable examples to illustrate differential aeration corrosion.

10. What is corrosion? Discuss the mechanism of wet corrosion.

11. Discuss the different methods used for controlling the corrosion of metals.

12. How can corrosion be controlled by (a) control of the environment (b) cathodic protection?

13. Discuss the method of controlling corrosion by proper design and material selection.

14. Mention the various methods used for prevention of corrosion and discuss the use of sacrificial anode for preventing corrosion.

15. Write a short note on paints and varnishes.

16. Discuss stress corrosion of metals. How can it be prevented?

17. Explain the cathodic protection method of corrosion control.

18. Write a short informative note on 'Corrosion control by proper design and metal selection.'

19. Explain the principle of differential aeration with a supporting example.

20. What are the factors that control the increasing sensitivity of an alloy to intergranular corrosion? Discuss chromium depletion causing intergranular corrosion.

21. What is meant by cathodic protection? Explain the types of cathodic protection and their applications.

22. What are the factors influencing corrosion? Mention the methods to overcome the same.

23. Define corrosion. Explain the basic reason of metallic corrosion.

24. Explain the mechanism of wet corrosion in neutral medium with a schematic diagram.

25. Explain concentration cell corrosion with the help of a suitable example.

26. What is cathodic protection? Describe the impressed current method of corrosion control.

27. Discuss the effect of the following factors on the rate of corrosion: (a) nature of corrosion product, (b) overvoltage, (c) relative areas of anode and cathode, (d) passivity, (e) temperature and pH of medium, (f) solubility of corrosion products, and (g) nature of ions present.

28. What are the necessary conditions for electrochemical corrosion? Explain the mechanism of electrochemical corrosion in acidic medium.

29. Discuss the corrosion due to combination of metals of different electrode potentials.

30. What is cathodic protection? Discuss sacrificial anode method of corrosion control.

31. Explain impressed current cathodic protection method with a suitable diagram.

32. Explain concentration cell corrosion with the help of suitable example.

33. Explain how various factors influence the rate of corrosion.

34. Explain galvanic corrosion with a suitable example.

35. Define corrosion. Explain stress corrosion with an appropriate diagram and examples.

36. What is metal cladding? State the composition and uses of Al-clad.

37. What are metallic coatings? Explain metal spraying to protect surfaces of base metals.

38. Write a note on 'cementation.'

39. What are corrosion inhibitors? State the role of cathodic and anodic inhibitors in controlling corrosion. Give examples of organic and inorganic compounds used as corrosion inhibitors.

40. Compare and contrast 'galvanizing and tinning of metals.'

41. Write a short note on organic coatings.

42. What are varnishes? List the constituents in varnish and functions of each of them with examples.

43. Define lacquer. State the constituents of lacquers and list their functions.

44. Compare and contrast 'paints and varnishes.'

45. Define enamel. List the constituents in an enamel and their specific functions.

Further Reading

1. Asadi, Z.S. and R.E. Melchers., 'Long-term External Pitting and Corrosion of Buried Cast Iron Water Pipes', *Corrosion Engineering, Science & Technology*, 53, pp.1–9, Taylor Francis, 2017.

2. Balasubramaniam. R., 'On the Corrosion Resistance of the Delhi Iron Pillar', *Corrosion Science*, 42 (12), pp. 2103–2129, December 2000.

3. Davis, J.R., *Corrosion: Understanding the Basics*, ASM International, Ohio, 2000.

4. Forsgren, A., *Corrosion Control Through Organic Coatings*, CRC Press (imprint), Taylor & Francis, Florida, 2006.

5. Finsgar, M. & J. Jackson, 'Application of Corrosion Inhibitors for Steels in Acidic Media for the Oil and Gas Industry', *Corrosion Science*, (86), pp. 17–41, Elsevier, 2014.

6. McCafferty, E. *Introduction to Corrosion Science*, Springer, New York, 2010.

7. National Association of Corrosion Engineers. http://www.nace.org/index.asp

8. National Metallurgical Laboratory (CSIR), http://www.nmlindia.org/corrosion.html

9. Revie, R.W., *Uhlig's Corrosion Handbook*, Wiley, New Jersey, 2011.

10. Sastri, V.S., *Corrosion Inhibitors: Principles and Applications*, John Wiley & Sons, New York, 1998.

11. Song, J. and W.A. Curtin, 'Atomic Mechanism and Prediction of Hydrogen Embrittlement in Iron', *Nature Materials*, (12), pp.145–151, 2013.

ANSWERS

Check Your Progress

1. Corrosion is the degradation of metals due to chemical or electrochemical reactions with the surrounding environment.

2. When gases such as O_2, CO_2, SO_2 react with metal surfaces, it is called dry corrosion.

3. Yes, the nature of metal oxide film helps to predict corrosion of metals. If the metal oxide is porous and volatile in nature, it causes rapid and continued corrosion.

4. (i) Metals are in contact with an electrolytic medium, and (ii) two dissimilar metals are immersed in an electrolytic solution.

5. Ocean-liners and ships should be made of only one type of metal with the same type of bolts and nuts to avoid galvanic cell corrosion.

6. When a metal is in contact with a wet medium (electrolyte), there will be flow of electrons from anode to cathode resulting in corrosion. This type of corrosion that proceeds due to ionic reactions in the presence of an electrolyte is called electrochemical corrosion.

7. This is an example is of waterline corrosion, a classic case of differential aeration principle. The corrosion product will be formed just at the waterline.

8. Impure zinc will corrode faster as impurities tend to promote the formation of small galvanic cells across the metal surface.

9. Solution: The formula of rust is $Fe_2O_3.3H_2O$ and its molecular weight is 214 g.

   So, the amount of iron present in 214 g of rust = 112 g.

   Hence, 200 g iron will form $\dfrac{214 \times 200}{112} = 382.14$ g of rust.

10. Passivity is a phenomenon of metals and alloys exhibiting higher corrosion resistance than predicted on the basis of the galvanic series.

11. As per the electrochemical series, the decreasing tendency of metals to undergo oxidation in a solution will be as follows:

    Mg ($E° = -2.38$V) > Zn (-0.76V) > Fe (-0.441V) > Ag ($+0.8$V) > Au ($+ 1.68$V noble)

12. As copper rod is partially immersed in NaCl solution, corrosion will occur due to differential aeration. The chemical reactions will be:

    $$Cu(s) \rightarrow Cu^{2+} + 2e^- \text{ (oxidation)}$$

    $$1/2\ O_2 + H_2O + 2e^- \rightarrow 2OH^- \text{ (reduction)}$$

    The presence of NaCl will merely accelerate the oxidation of copper as it raises the conductance of water. Thus, corrosion of copper rod will occur with rust formation just at the waterline.

13. Using different metals as bolts and nuts will result in galvanic corrosion. Hence, nuts and bolts are always made of the same metal.

14. It is an example of stress corrosion. The joints in a wire mesh are stressed due to welding and hence turn anodic. Oxidation of the metal occurs at the joints of wire mesh causing corrosion at a rapid rate.

15. The impurities in a metal cause heterogeneity and forms tiny electrochemical cells on the metal surface (anodic). Hence, impure metals corrode faster than pure metals.

16. The factors that affect rate of corrosion are:

    Nature of metal: The position of metal in electrochemical series, overvoltage, purity of metals, passivity, relative anodic–cathodic areas, type of corrosion products.

    Nature of electrolyte: Temperature, humidity, pH, conductance, impurities in electrolyte.

17. Season cracking and caustic embrittlement are types of stress corrosion in metals.

18. Refer to P. (Corrosion by other gases)

19. Refer to Section 14.4.3

20. Refer to Section 14.4.5

21. Lower the pH of the electrolyte, higher will be the rate of corrosion.

22. When a mixture of chromium powder, alumina, and steel is heated at $1300 - 1400°C$ for 4 hours, chromium gets coated on a steel article. The process is called chromizing.

23. When a noble metal layer gets deposited on the base metal object by displacement of the base metal by the noble metal, it is termed as *displacement plating* or *immersion plating*.

24. Galvanized containers have zinc that tends to get dissolved in acidic components of food (such as juice, pulp, etc.) forming toxic compounds. Tin-coated containers are known to be passive and non-toxic and hence used to store foodstuff and beverages.

25. Corrosion inhibitors are substances or mixtures added in trace amounts to aqueous corrosive environment to prevent corrosion of metals. They are of two types, anodic and cathodic.

26. Dicyclohexylamine nitrite, hydrazine, and sodium sulphite are commonly used as corrosion inhibitors. NaCl enhances conductance of the electrolyte thereby accelerating the rate of corrosion. Hence, NaCl cannot be used as a corrosion inhibitor.

27. Paint is a uniform dispersion mixture of pigments, driers, and fillers in a liquid called vehicle or medium. Varnish is a colloidal dispersion of natural or synthetic resins in a vehicle.

28. Pigment, drying oil, thinners, driers, fillers, plasticizers, and anti-skinning agents form the ingredients of paints.

29. Lacquer is a colloidal dispersion of cellulose derivatives, resin, and plasticizer in a volatile solvent.

30. Enamel is a pigmented varnish that contains pigments, oil or resin vehicle, and thinners and driers.

31. Plasticizers enhance the elasticity of paint film and also the overall durability of the coat.

32. Anti-skinning agents avoid cracking of applied paint coat. They also provide good adherence of varnish coat on the applied surface.

Multiple Choice Questions

1. (b)	2. (a)	3. (b)	4. (d)	5. (b)	6. (c)	7. (b)
8. (c)	9. (d)	10. (a)	11. (b)	12. (a)	13. (a)	14. (b)
15. (d)	16. (b)	17. (c)	18. (d)	19. (a)	20. (a)	

CHAPTER 15

Metals and Alloys

LEARNING OBJECTIVES

After reading this chapter, you will be able to:

- understand the concepts and thermodynamics of metallurgy.
- understand the applications of powder metallurgy.
- list the composition of various important alloys of copper, aluminium, and lead.
- explain the preparation of metal ceramic powders.
- understand the behaviour and importance of shape memory alloys.

15.1 INTRODUCTION

In Chapter 4, we learnt about phase diagrams and phase rule. Primarily, studying phase diagrams is fundamental to understand various alloy systems. The branch of metallurgical science and technology deals with obtaining metals from their ores found in the earth's crust. Apart from this, alloy preparation is another area which is of prime importance in mechanical engineering, civil engineering, and material sciences. An ore is a starting material which refers to a complex mixture of metallic matter (elements or compounds) and other undesired materials called *gangue* of lower significance.

Some important physical properties of metals are as follows:

Hardness is an inherent physical property of a metal that enables it to resist penetration, abrasion, or scratching by other materials. When a metal possesses the capability to scratch a comparatively softer material, it is termed as hardness. *Vickers hardness test* is done to determine the ability of a metal to resist plastic deformation from a standard source (diamond indenter). Steels possess good hardness.

Malleability is the property of a metal that enables it to form sheets when hammered, pressed, or rolled in to thin sheets without breaking.

Ductility is the physical property of a metal that allows it to be drawn in to wires when stretched without undergoing stress or breaking.

Stiffness refers to a metal's capability to resist deflection by an external applied force.

Specific heat is the amount of heat in calories needed to raise the temperature of 1 g of metal by 1°C.

Elasticity refers to the property of a metal that enables it to regain its original shape once the load that caused the deformation is removed. The atomic lattices of metallic solids tend to change shape and size when a load is applied. On load removal, they regain their original dimensions.

Plasticity refers to the inability of a metal to regain its original shape when the load causing deformation is removed.

Fusibility or fluidity refers to the ease with which a metal can be melted on heating and flows into a mould.

Machinability refers to the ease with which a metal can be cut (or machined) using cutting tools so as to obtain satisfactory finish. Steel has good machinability and thus is used in many industrial designs.

Magnetism is a metal's capability to be attracted towards a magnet. Such metals are called magnetic. Iron, cobalt, chromium, nickel are magnetic metals. Metals that get feebly attracted to a magnet are called diamagnetic; examples are zinc, copper, lead.

Tensile strength refers to load bearing capability of a metal that can withstand tensile stresses before failure. Steel alloys exhibit higher tensile strength.

Electrical and thermal conductivities refer to the ability of metals to conduct electricity. Metals that do not conduct electricity are called insulators.

15.2 METALLURGY

Metallurgy is the process of extracting metals from their ores usually by a series of reduction reactions. The process usually involves four steps—*mining of ores*, *concentration of ores* and *extraction of metals* from the ore followed by its *purification*.

Concentration of ore The impurities and undesired materials present in the mined ore are removed by *concentration* or *dressing* or *benefication*. Concentration of ore can be accomplished by several ways, viz., hydraulic washing, magnetic separation, froth floatation, and leaching.

Magnetic separation This method is based on the differences in magnetic properties of the ore components. The powdered ore is kept on a conveyer belt passing over magnetic rollers ((Fig. 15.1); iron ores such as haemetite, limonite, and siderite will fall close to the rollers, whereas non-magnetic materials will fall away as a heap.

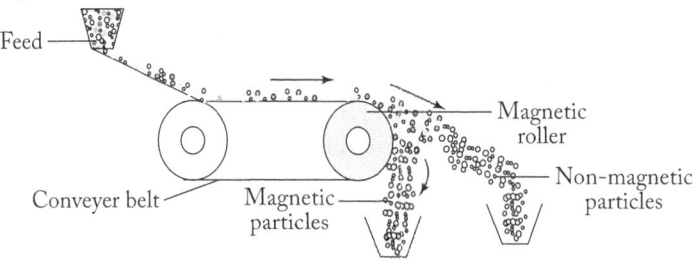

Fig. 15.1 Magnetic separator

Extraction of metals Before reduction, the concentrated ores are usually converted into their oxides as it is easier to reduce them and extract the crude metals. Calcination and roasting are examples of oxide conversion processes.

Reduction Metal oxides are heated with reducing agents such as carbon, carbon monoxide, or another metal. The reducing agent (e.g., carbon) combines with the oxygen present in the metal oxide to form the crude metal.

$$M_xO_y + yC \rightarrow xM + yCO$$

Purification of metals Some of the commonly employed purification methods are: distillation, liquation, electrolysis, zone refining, vapour phase refining, and chromatographic methods. Of these, electrolytic, zone refining, and chromatography are the preferred methods to obtain pure metals.

Zone refining This method is based on the principle that impurities are more soluble in the melt than in the solid state of metal. An impure metal rod placed in an inert quartz tube is heated using a movable external heating coil (Fig. 15.2). As the heating continues, pure metal separates out from impurities and crystallizes out of the melt.

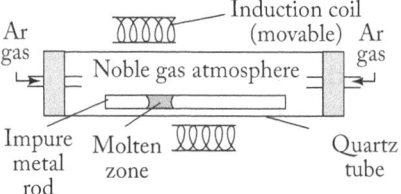

Fig. 15.2 Zone refining

The process is repeated several times until all impurities are concentrated at one end of the tube and removed. The high-purity metals so obtained such as germanium and silicon are used in semiconductor making.

In our intermediate classes we have learnt about metals and their extraction processes. In this chapter, we will discuss the thermodynamics of metallurgy and various important alloys for industrial purposes.

15.2.1 Use of Free Energy Considerations in Metallurgy Using Ellingham Diagram

We have learnt that Gibbs equation helps to predict the spontaneity of a chemical reaction on the basis of enthalpy and entropy values. Gibbs free energy (ΔG) refers to the thermodynamic driving force for a process to occur spontaneously. In 1944, H.G.T. Ellingham proposed a diagram to predict the spontaneity of reduction of various metal oxides occurring during metallurgical operations. Ellingham diagram is a plot of the relation between temperature and stability of compounds. Figure 15.3 is an Ellingham diagram that relates change in Gibbs free energy with respect to temperature. The plots of $\Delta G°$ versus temperature help in predicting the feasibility of thermal reduction of various ores and also choosing the reducing agents.

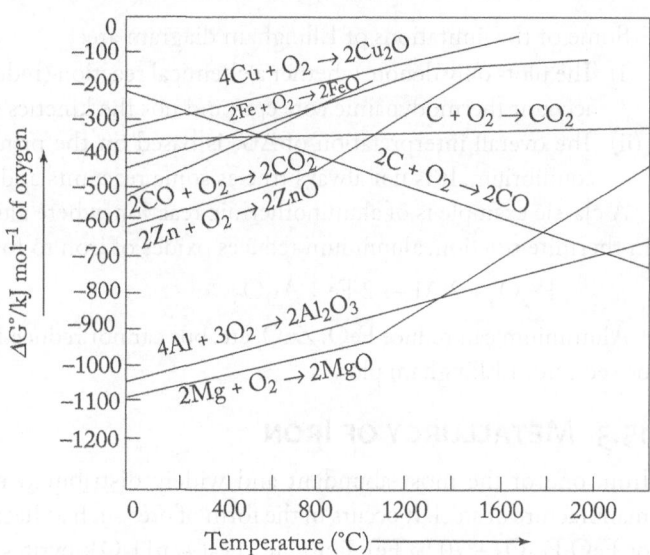

Fig. 15.3 Plots of $\Delta G°$ versus temperature for oxide formation of some elements (Ellingham diagram)

For a spontaneous chemical reaction, the free energy change (ΔG) is negative as per the equation,

$$\Delta G = \Delta H - T\Delta S \tag{15.1}$$

where, ΔH is enthalpy change of the reaction, T is temperature, and ΔS is entropy change of the reaction. The reaction of metal oxide formation can be written as,

$$M \text{ (metal)} + O_2 \rightarrow MO \text{ (metal oxide)}$$

We know that gases have high entropy values due to their disorderly nature. Since oxygen gas is consumed during metal oxide formation, the entropy decreases, that is, ΔS is negative. If temperature is increased, the $T\Delta S$ factor will become more negative. Hence as per Eq. (15.1), ΔG will be negative which implies that on increasing the temperature of the chemical reaction, change in free energy decreases.

The following are the salient features of Ellingham diagram.

1. The change in free energy varies linearly with temperature. When temperature increases, free energy change decreases which means that stability of metal oxide decreases with increase in temperature.
2. Plots of almost all elements seem like a straight line except for those exhibiting phase changes, such as solid → liquid or liquid → gas. The temperatures at which such phase changes occur are depicted as positive increase in the slope.

For example, in the curve of ZnO formation, an abrupt change is noticed that indicates melting process. Metal oxide (M_xO) is stable at the point in a curve below which ΔG is negative. Above this point, metal oxides are generally unstable and undergo decomposition on their own.

3. As Ellingham diagram depicts $\Delta G°$ for oxide formation of elements at varying temperatures, it allows easier interpretation of reduction reactions. Such diagrams are also constructed for halides and sulphides.

4. If an oxide of a metal is present above another metal in Elligham plot, it means that it can be reduced by the latter. For example, coke can reduce FeO and itself get oxidized to CO (see Fig. 15.3).

Some of the limitations of Ellingham diagram are:

(i) The plots only denote whether a chemical reaction (reduction) is possible or not as it only takes into account thermodynamic concepts and not the kinetics of reduction reactions.

(ii) The overall interpretation of $\Delta G°$ is based on the principle that all reactants and products are in equilibrium. It is not always true as some reactants and products may be solid.

A classic example is of aluminothermic reaction, where aluminium acts as the reducing agent. As seen in thermite reaction, aluminium reduces oxides of iron to form pure iron as,

$$Fe_2O_3 + 2\,Al \rightarrow 2\,Fe + Al_2O_3$$

Aluminium can reduce FeO, ZnO, etc. but cannot reduce MgO at temperatures below 1500 °C as can be seen from Ellingham plot.

15.3 METALLURGY OF IRON

Iron, one of the most abundant and widely distributed metals, is commercially important for the manufacture of steel. It occurs in the form of ores such as haematite ($Fe_2O_3 - 72\,\%$ Fe) magnetite (Fe_3O_4 or $FeO.Fe_2O_3 - 70\,\%$ Fe); limonite ($Fe_2O_3.nH_2O$); pyrites (FeS_2) and chalcopyrite ($CuFeS_2$); siderite ($FeCO_3$) – 48 % iron.

Haematite ore is commonly used for extraction of iron because it has the highest percentage of iron.

15.3.1 Extraction of Iron

Blast furnace   It is a tall cylindrical steel structure of about 16 to 65 m high, lined inside with fire bricks that can withstand high temperatures.

The wider middle portion called *bosh is* the hottest region in the furnace. The narrower base of the blast furnace is called *hearth*, from where molten iron covered with slag is collected. There are two separate tap holes to remove molten iron and slag. The top portion of furnace from where it starts to widen is called *stack* or *shaft* of the furnace. As the blast furnace is a tall tower-like structure, the temperature is not same throughout but goes on decreasing from bottom to top.

Charge of the furnace   The charge added to the blast furnace contains calcined iron ore (8 parts), good quality coke (4 parts) and limestone ($CaCO_3$) or silica (SiO_2) (flux 1 part). A blast of hot and dry

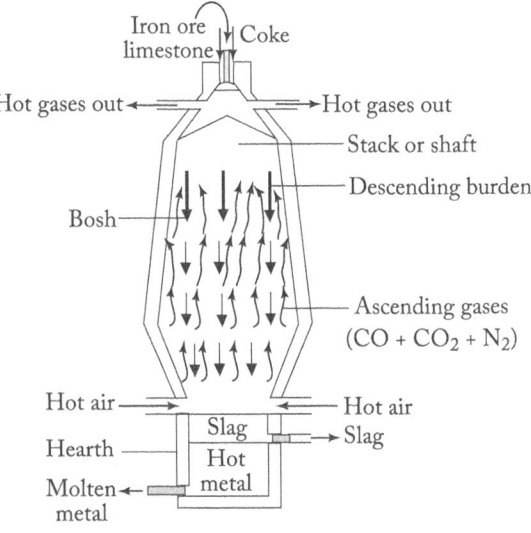

Fig. 15.4 Blast furnace

air at 800 °C under pressure is blown at high speed from the bottom of the furnace. This allows quick burning of coke that forms carbon monoxide used as a reducing agent in the operation. Flux is added to the charge to remove impurities from the ore forming slag.

15.3.2 Working of the Furnace

A blast of hot dry air is blown from the bottom of the furnace by burning coke in the lower portion providing a temperature of 1900 °C. On burning the coke, CO and heat move towards the upper portion of the furnace. In the upper part, the temperature is about 500 °C and the iron oxides (Fe_2O_3 and Fe_3O_4) coming from the top are reduced to FeO. The chemical reactions occurring in a blast furnace are explained below.

Reactions during reduction   Coke burns in hot dry air producing CO_2 which is a highly exothermic process. This reaction occurs near the hearth of the furnace. A temperature of 1300 °C is attained and carbon dioxide so formed quickly reacts with coke forming carbon monoxide. This reaction occurs at the middle portion of the furnace and is called *zone of absorption* as heat is absorbed during the process.

$$C + O_2 \rightarrow CO_2 \text{ (exothermic = –393 kJ)}$$
$$CO_2 + C \rightleftharpoons 2\,CO \text{ (endothermic = +171 kJ)}$$

Much of the iron oxide gets reduced in the upper portion of the furnace which is called the *zone of reduction*. Here, the temperature is 500–800 °C where Fe_2O_3 is first reduced to Fe_3O_4 and then to FeO.

$$3Fe_2O_3 + CO \rightarrow 2Fe_3O_4 + CO_2 \;\; (-20\text{kJ})$$
$$Fe_3O_4 + 4\,CO \rightarrow 3Fe + 4\,CO_2 \;\; (-25 \text{ kJ})$$
$$Fe_2O_3 + CO \rightarrow 2FeO + CO_2 \;(-33.4 \text{ kJ})$$

As the charge moves down towards the hotter region, any Fe_2O_3 if present gets reduced by carbon to iron at about 1000 °C.

$$Fe_2O_3 + 3C \rightarrow 2Fe + 3CO \;(+452 \text{ kJ})$$

Reactions during slag formation   Near the hearth of the furnace, both metal and slag are obtained as per the following chemical reactions and this region of the furnace is called the *zone of fusion*.

Limestone is also decomposed to CaO above 800 °C which removes silicate impurity of the ore as slag. The slag is in molten state and separates out from iron. At 1300 °C, carbon dioxide quickly gets reduced to carbon monoxide by coke. At 1000 °C, CO so formed reduces FeO to iron. Lime (CaO) fuses with silica present in the charge forming silicate slag.

$$CaCO_3 \rightarrow CaO + CO_2$$
$$C + CO_2 \rightarrow 2\,CO$$
$$FeO + CO \rightarrow Fe + CO_2$$
$$CaO + SiO_2 \rightarrow CaSiO_3$$

The liquid slag is lighter and immiscible in molten iron and thus forms an upper layer in the hearth of the furnace and prevents oxidation of iron by the hot blast of air.

Usually, the extraction process takes up to 6–7 hours and pig iron (92–95 % iron, 2–4.5 % carbon, 0.7–3 % silicon, 0.5–1 % phosphorus, and 0.1–0.2 % manganese) is obtained. The molten iron is removed and freed from the slag by passing through rollers. Pig iron can be used for production of steel, cast iron, and wrought iron. When pig iron is treated to remove impurities in a cupola furnace it is called *cast iron*. Using a puddling furnace, *wrought iron* is obtained from pig iron.

The reduction processes occurring in the blast furnace can be explained using Ellingham diagram.

$$FeO(s) \rightarrow Fe(s) + 1/2\ O_2(g),\ (\Delta G^\circ_{FeO,\ Fe})$$

$$C(s) + 1/2\ O_2(g) \rightarrow CO\ (g)\ (\Delta G^\circ_{C,\ CO})$$

Net reaction: $FeO(s) + C(s) \rightarrow Fe(s) + CO\ (g)$

Therefore the net Gibbs free energy will be,

$$\Delta G^\circ_{C,\ CO} + \Delta G^\circ_{FeO,\ Fe} = \Delta G^\circ_r$$

On plotting ΔG°_r versus temperature (where r is reaction) the change of Fe $\rightarrow$ FeO in Fig. 15.3 is upwards and the change of C $\rightarrow$ CO goes downward. These two changes are represented as lines that cross each other at about 1073 K. At temperatures above 1073 K, it is denoted that the C, CO line is below the Fe, FeO line. Hence, at temperatures above 1073 K (in the temperature range 900–1500 K), coke will reduce FeO and will itself be oxidized to CO. As per Fig. 15.3, for overall reaction, the value be –53 kJ mol^{-1}; thus, the reduction reaction is feasible and spontaneous.

15.4 PRODUCTION OF STEEL

Steel is an important engineering material belonging to iron–carbon system (see Chapter 4). Approximately 2500 different grades of steel are produced that are of structural importance ranging from daily life to aerospace engineering.

Globally, Nippon Steel (Japan), Tata Steel and SAIL (India) are the major producers of steel. According to World Steel Association Report 2016, India produces 89 million tonnes of steel annually.

Steelmaking involves manufacturing steel from iron ore and steel scrap usually by the following two main processes.

15.4.1 Basic Oxygen Steelmaking

Figure 15.5 shows the basic oxygen steelmaking furnace or converter, which is a pear-shaped vessel charged with molten iron and steel scrap from a ladle. The inner lining of the converter is made of refractory bricks. Pure oxygen is blown for 15 – 20 minutes from the top through a water-cooled lance lowered into the converter.

This method is called *top blown steelmaking* during which all the impurities are converted to their corresponding oxides. The chemical reactions during the process are:

$$Fe + O_2 \rightarrow FeO$$
$$2\ C + O_2 \rightarrow 2\ CO$$
$$Si + O_2 \rightarrow SiO_2$$
$$2\ Mn + O_2 \rightarrow 2\ MnO$$
$$4\ P + 5\ O_2 \rightarrow 2\ P_2O_5$$

An important feature of the converter process is that no heat is externally provided; rather the heat from the chemical reactions drives the reactions.

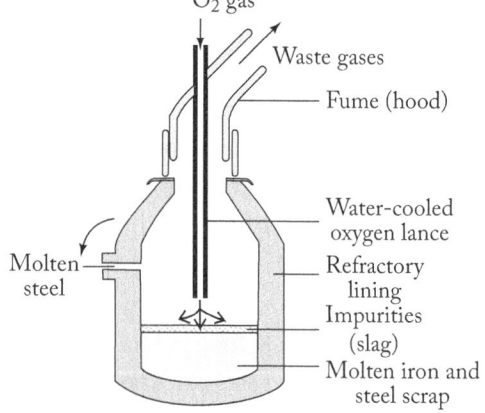

Fig. 15.5 Basic oxygen steelmaking

Except CO, other products react with lime added during the oxygen blow forming slag. The above reactions are all exothermic and controlled quantities of scrap are added as coolant to maintain the desired temperature. Steel so obtained from this process contains about 0.04 % carbon.

15.4.2 Electric Arc Furnace

In electric arc furnace steelmaking, scrap, hot iron metal, and reduced iron are used to produce plain carbon steel.

As shown in Fig. 15.6, the furnace is a circular bath with a capacity to hold over 100 tonnes of liquid steel equipped with a movable roof, supporting three graphite electrodes that can be raised or lowered.

The furnace is charged with steel scrap and the huge amount of heat generated melts the steel scrap.

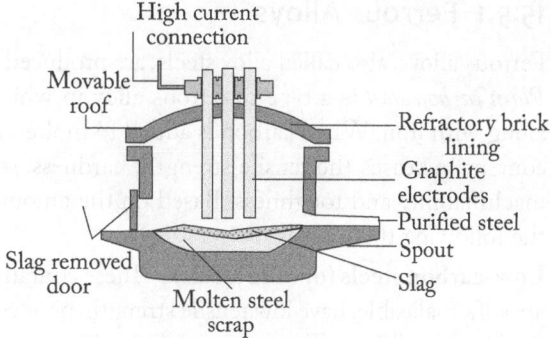

Fig. 15.6 Electric arc furnace

Lime (as calcium oxide or calcium carbonate), fluorspar (helps to keep the slag as a fluid) and iron ore are added and these combine with impurities to form slag. Once steel has reached the correct composition, slag is poured off and the steel tapped from the furnace for alloying or other commercial purposes.

Casting of steel is done in the form of a slab or pipes and then sent for hot rolling process. After the production, scrap steel present as charge in the furnace is recycled in the process.

15.5 ALLOYS

An alloy is a solid engineering material obtained by the solidification of a metallic solution of two or more elements, of which one is essentially a metal. Almost all engineering materials are made of alloys rather than pure metals. Ideally, alloying is done to enhance the inherent properties of a pure metal such as ductility, malleability, machinability (welding, forging, etc.) tensile strength, hardness, and particularly to impart corrosion and chemical resistance to the base metal. On the basis of the presence of iron, alloys are broadly classified into two major groups, ferrous and non-ferrous alloys. Figure 15.7 depicts the general classification of alloys.

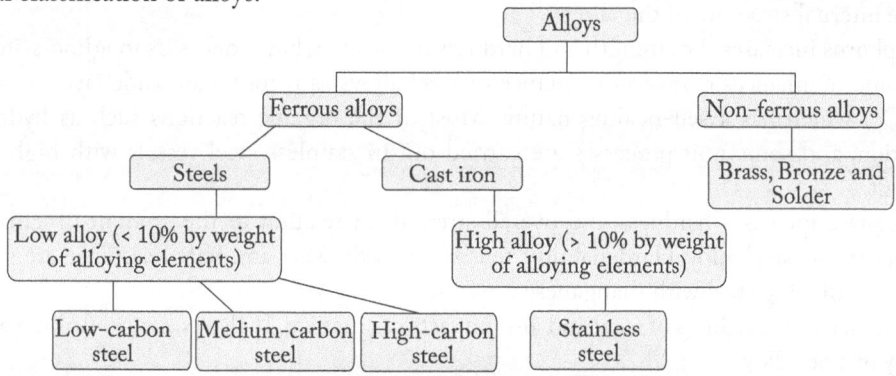

Fig. 15.7 Classification of alloys

Check Your Progress

1. State any five properties of metals.
2. What is metallurgy?
3. List the various types of iron ores.
4. What is steelmaking?
5. Name the zones observed in blast furnace during the extraction of iron.
6. Name the various types of iron.
7. Name two methods of producing steel.
8. What is an aluminothermic reaction? Give an example.

15.5.1 Ferrous Alloys

Ferrous alloys, also called alloy steels, are produced in large quantities with iron as the major constituent. *Plain carbon steel* is a type of ferrous alloy in which carbon is the principal hardening element present along with iron. When carbon is added to molten iron, iron carbides are formed in steel. Higher carbon content increases the tensile strength, hardness, wear-abrasion resistance of the alloy, but decreases its machinability and toughness. Based on the amount of carbon (% by weight), plain carbon steels are of the following three types.

Low-carbon steels (or mild steels) These contain less than 0.25 %C. As the name suggests, these alloys are soft, malleable, have low tensile strength, poor corrosion resistance, and good machinability under cold working conditions. They are used for making automotive panels, rivets, bolts, screws, and boiler tubes.

Medium-carbon steels These contain 0.25–0.6 %C making them slightly tougher than mild steel alloys. Due to their good tensile strength and fair corrosion resistance, they are employed in fabricating machine parts, such as turbines, automotive parts, and railway axles.

High-carbon steels These contain 0.6–1.4 %C, making them tougher and harder material than mild steels. As the carbon content is high, machinability is reduced to some extent. High-carbon steels can be heat treated and adding elements like Ni, Cr, Co will further enhance their alloying characteristics. These are used in railway wheels, gears, etc.

Sometimes, other elements are also added to Fe along with carbon to impart special properties to steel. Such steels are called *special steels*.

Effects of alloying elements on steel The specific effects of various alloying elements on steel are enumerated as follows.

(a) Nickel improves the tensile strength, toughness, elasticity, and ductility of steel. Corrosion and heat resistance are also improved along with overall increase in impact strength of steel by reducing cracks in the internal structure of the alloy.
(b) Phosphorus increases the strength and hardness of the alloy, but reduces its toughness and ductility.
(c) Chromium enhances corrosion resistance of steel alloys as it forms an oxide layer on the surface (Cr_2O_3) which has a self-healing nature. Most of the organic reactions such as hydrogenation, nitration, and some unit processes are carried out in stainless steel vessels with high chromium content.
(d) Manganese increases hardness and overall strength in relation to the amount of carbon already present in the steel alloy. Hardenability, tensile strength, wear and heat resistance are some of the effects of alloying steel with manganese.
(e) Cobalt increases hardness of steel and also imparts magnetism. It also enhances drilling and cutting ability of steel alloy.
(f) Molybdenum tends to impart specific effects only in combination with other elements. Overall hardness, creep resistance, tensile strength, grain structure, and machinability are greatly enhanced, when alloyed with steel.
(g) Tungsten is an excellent carbide former which is hard in nature. It imparts good toughness, wear resistance, and overall strength to the alloy.
(h) Vanadium also forms carbides with steel thereby imparting toughness, hardness, and shock resistance.

Heat resisting steels These are special steels that can be employed at high working temperature conditions. Such steels are usually alloyed with 3.5 % Mo, 12 % Cr, or other elements. Nichrome (Ni=80%, Cr= 20%) is a type of heat resisting steel that can bear temperatures up to 1100 °C. Nichrome alloy possesses ductility, resistance to oxidation, heat resistance, resistance to electron flow, and has low cost of manufacture. Hence, it is commonly used as electric heating materials like coils in hair dryers and electric iron. It is used as wire loops in salt analysis during flame tests in place of platinum.

Stainless steels   These are alloy steels that exhibit heat resistance, corrosion resistance, and remain unaffected by chemicals and atmosphere. The stainless nature of steel comes from the presence of chromium (> 16% wt) that forms Cr_2O_3 on contact with air. The oxide of chromium formed on the steel surface protects the underlying metal from corrosive attack. Generally, there are two major types of stainless steels, namely heat treatable stainless steels and non-heat treatable stainless steels.

Heat treatable stainless steel (< 18 % Cr, ≈ 1.2 % C) is tough, corrosion resistant, magnetic, and can withstand temperatures up to 800°C. It is especially used to fabricate surgical instruments.

Non-heat treatable stainless steel can be workable in cold conditions and exhibit corrosion resistance. Magnetic non-heat treatable stainless steel (≈ 12–22 % Cr, < 0.35 % C) is generally rolled, cold-drawn, forged, and used in making chemical equipment. Non-magnetic non-heat treatable stainless steel (≈ 18–26 % Cr, 8–21 % Ni, 0.15 % C) is highly corrosion resistant and used in dental and surgical equipment.

15.5.2 Non-ferrous Alloys

Non-ferrous alloys do not possess iron as the major constituent. Duralumin and magnalumin are important alloys of aluminium. Brasses and bronzes are two classical types of copper alloys. Copper and zinc in varying composition is called brass, whereas varying compositions of copper and tin form bronze. Alloys of lead are known as solders as they possess good flow properties and hence, used as solders. Table 15.1 lists the properties and applications of important alloys of aluminium, copper, and lead.

Table 15.1 Alloys of Al, Cu, Pb

Alloy	Elemental composition (%)	Properties	Applications
Duralumin	Al: 95, Cu: 4 Mn: 0.5, Mg: 0.5	Light, tough, ductile, good conductor of heat and electricity, corrosion-resistant with high machinability.	Aircraft and locomotive parts, surgical and scientific instruments.
Magnalumin	Al: 70–95 Mg: 5–30	Hard yet lighter than base metal Al, possess mechanical properties similar to that of brass.	Fabricating electric balances, aircraft parts and scientific instruments.
Brasses			
Commercial Brass or French gold	Cu = 90 Zn = 10	Golden colour, harder than pure copper.	Forging, rivets, screws, ornaments.
German Silver	Cu = 60 Zn = 15 Ni = 25	Silver white colour, strong, hard, resistance towards corrosion to salt water, high ductility and malleability.	Utensils, bolts, screws, cutlery, coins.
Bronzes			
Gun Metal	Cu = 85, Zn = 4 Sn = 8, Pb = 3	Hard, tough, resist explosion, high tensile strength	Hydraulic fittings, marine pumps, water fittings, load bearings.
High-phosphorus bronze	Cu = 93–96 Sn = 10–13 P = 0.4–1	Hard, brittle, and abrasion resistant	Bearings, gears, springs, turbine blades, fibres for moving coil galvanometers.
Wood's metal	Bi = 50, Pb = 25 Sn = 12.5, Cd = 12.5	Easily fusible alloy with melting point of 70°C.	Making fire alarms, safety plugs for cookers, castings for dental works.
Tinman's solder	Sn = 66, Pb = 34	Fusible alloy with melting point of 183°C.	Soldering, welding, and tinning.

15.6 POWER METALLURGY

Commercially, alloys are supplied in the form of pipes, slabs, blocks, and powders. Powder metallurgy is a technique of manufacturing commercial grade metals, alloy articles, ceramics, and composite materials. When metals and alloys cannot be shaped through smelting, powder metallurgy is performed, in which metal powders are produced and compressed to desired shapes and sizes for commercial uses.

The processes involved in powder metallurgy are: (a) metal powder preparation, (b) mixing and blending, (c) compaction, (d) pre-sintering and sintering, and (e) finishing operations.

Metal Powder Production

Pulverization   Metal or alloy powders are powdered in large mechanical pulverizers having rotating hammers and chisels. These mechanical forces disintegrate the metal particles to form a fine powder (0.1 to 1000 µ). Atomization, chemical reduction, and electrolysis are other methods to obtain metal powders.

Atomization   It involves forcing the molten metal through a nozzle into a stream of compressed air or water. The metal stream is broken by jets of inert gas, air, or water. The inert gas inhibits oxidation of metals. Figure 15.8 depicts the atomization process to obtain high-purity fine metal powder.

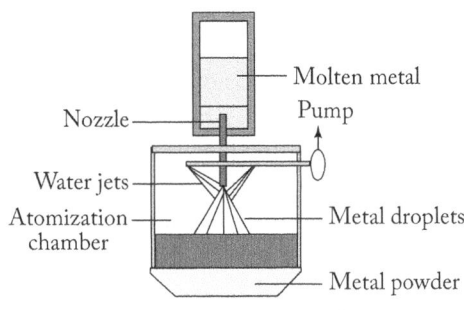

Fig. 15.8 Atomization

Chemical reduction   It involves heating metal oxides either in a current of hydrogen or carbon resulting in spongy metal powder that can be sent for compaction. Generally, metals like Cu, Ni, Mo are pulverized by chemical reduction. Reduction is carried out at temperatures well below the melting point of metals. Cuprous oxide is reduced in a current of hydrogen gas to give spongy copper; magnetic ores like Fe_3O_4 on reacting with carbon forms spongy iron; for metals such as Ti and Zr, chemical reduction using reactive metals is generally preferred.

$$Cu_2O + H_2 \rightarrow 2\,Cu + H_2O$$
$$Fe_3O_4 + 4\,CO \rightarrow 3\,Fe + 4\,CO_2$$
$$TiCl_4 + 2\,Mg \rightarrow Ti + 2\,MgCl_2$$

Electrolysis   The method involves simple electroplating process in which the electrolyte is a metal salt solution and anode is of the same metal whose powder is to be obtained. The positive ions get deposited on the cathode, which can be scrapped, washed, dried, and pulverized. The powder obtained in the above process is resistant to oxidation.

Check Your Progress

9. What are alloys? Name any two alloys of aluminium.
10. State three advantages of preparing alloys.
11. What is the purpose of Ellingham diagram?
12. What is plain carbon steel?
13. Why is stainless steel resistant to corrosion?
14. What are brasses and bronzes?
15. State the differences between brass and bronze.

Thermal decomposition   Metal carbonyls are allowed to undergo thermal decomposition yielding fine, pure, spherical powder of the respective metal. Iron and nickel powders are generally prepared through this method: $Fe(CO)_5 \rightarrow Fe + 5CO$ and $Ni(CO)_4 \rightarrow Ni + 4\,CO$. The precipitated metals are obtained and sent for compaction and sintering.

Mixing and Blending

Generally mixing and blending of metallic powders is carried out in centrifugal mixers by adding volatilizing agents and lubricants to facilitate homogeneous dispersion of metal particles of same shape, size, and density. This process ensures that the metal powders on compacting in machines do not stick to the mould.

Compaction or Briquetting

This is done after blending in which a known quantity of metal powder is placed in a die and compressed at a pressure of about 100–1000 MN/m^2 to obtain green compact. Some of the important compacting techniques are cold compaction or cold pressing, powder injection moulding, cold powder extrusion, and roll and hot compaction.

Cold compaction   It is the most important method in compaction in powder metallurgy. The starting material is the bulk powder with or without a lubricant or binder. Cold compaction can be done by *axial die pressing* or *isostatic pressing* methods.

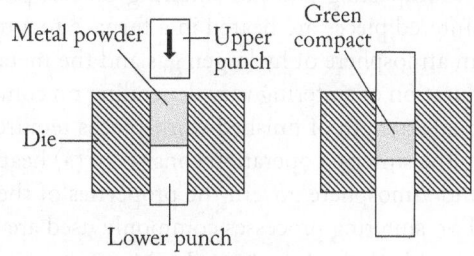

Fig. 15.9 Axial die pressing

Axial die pressing   In this case, the metal powder is compacted in die moulds (Fig. 15.9) by axially loaded punches. The metal powder is compacted between punch faces and dies at a mechanical pressure of about 600 MPa.

Cold isostatic pressing   It is performed either by wet-bag or dry-bag methods (Fig. 15.10). The metal powder is sealed in an elastic mould and exerted to a hydrostatic pressure of a liquid pressure medium. In wet bag method, mould is removed and refilled after each pressure cycle. This method is suitable for compaction of large and complicated parts. However, in dry bag method, mould is an integral part of the vessel and is used for compaction of simpler and smaller parts.

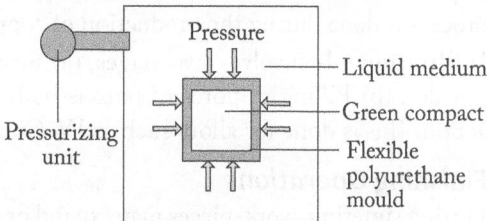

Fig. 15.10 Cold isostatic pressing

Powder injection moulding   abbreviated as PIM (or metal injection moulding), is synonymous to plastic moulding process (Fig. 15.11). The metallic powder is transformed into a mouldable state by adding binders called *feedstock*. The blended powder is cast in a mould and heated. The plasticity of metals allows desired shaping of metal compact, which is then cooled and sent for sintering.

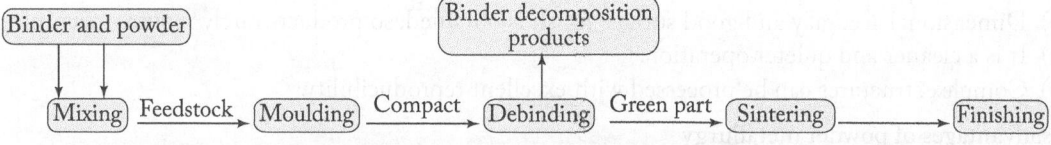

Fig. 15.11 Stages in powder injection moulding

Cold powder extrusion   It is similar to injection moulding of plastics. In this method, metal powder is first plasticized with organic binders and forced by a position or screw unit in to a die mould. The ejector pins allow for removal of compacted material. This technique is used in producing rod shaped tools for metal drilling purposes.

Roll compaction It involves passing metal powder in between two rollers to obtain strips of metal or metal sheets (Fig. 15.12). The metallic strips are used for making batteries, thermostats, and other bimetallic applications.

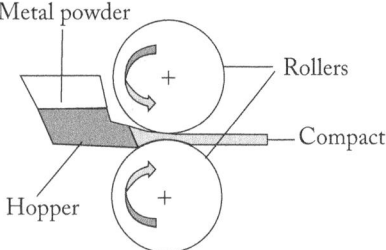

In *hot compaction* high temperatures (>2500 °C) and pressures are employed to activate the metal powder followed by compaction in a die. It is used in the fabrication of non-oxide ceramics and metal-bound diamond tools, beryllium pieces, blocks, and cemented carbide parts.

Fig. 15.12 Roll compaction

Pre-sintering and Sintering The objective of this operation is to remove lubricants and binders added while blending. Pre-sintering involves heating 'green compact' to temperatures below its sintering temperature, thereby enhancing overall strength of the compact.

Compacting and pre-sintering do not provide enough strength and cohesiveness. Compacted, pre-sintered pieces are heated in a furnace to a temperature close to the melting point of the basic metal in an atmosphere of hydrogen gas and the metal particles get sintered. The sintering temperature and time duration of sintering vary depending on compression load used during compacting, nature of metal, and final strength of finished work-pieces required.

The sintering operation consists of (a) heating; (b) soaking; and (c) cooling cycle. Sintering conditions and atmosphere govern the properties of the final product such as its toughness, hardness, and density, Two sintering processes commonly used are described here.

Liquid-phase sintering In this process, melting of low melting constituent takes place and the liquid so produced and other component solid particles (of higher melting point) get arranged uniformly. This process is done during the production of copper–tungsten (W–Cu) and silver–tungsten (W–Ag) alloys.

Infiltration It involves two stages, (a) formation of porous body of high melting constituent metal powder; (b) Filling of pores of porous body by molten metal of lower melting point due to capillary action. This is done for alloys such as W–Cu and Mb–Cu.

Finishing Operations
During sintering, work-pieces may expand or contract depending on the nature of the metal. The finished operations are the final steps for getting the perfect dimensions of the work-pieces suitable for commercial purposes.

Uses of powder metallurgy Various products are manufactured such as electric filaments, refractory metal composites, bearings for automobiles, gas turbine parts, and diamond cutting tools. Thermal appliances such as nozzles, burners, stoves are prepared using powder metallurgy techniques.

Advantages of powder metallurgy
  (a) The rate of production is high and products obtained through the technique possess uniform structures.
  (b) Dimensional accuracy and good surface finish is obtained, so products rarely require machining.
  (c) It is a cleaner and quieter operation.
  (d) Complex structures can be processed with excellent reproducibility.

Disadvantages of powder metallurgy
  (a) It involves the use of expensive precision dies.
  (b) Complex structures produced by casting cannot be done by powder metallurgy.
  (c) The articles produced by powder metallurgy technique possess poor ductility.
  (d) The overall process involves application of high pressures.

15.7 METAL CERAMIC POWDERS

Cermets are composite materials prepared by mixing ceramic and metallic constituents by powder metallurgy technique. In cermets, even a small proportion of metal is significant as the metal acts as a binder for high refractory ceramic particles. In cermets, 80 % ceramic and 20 % metal composition (as binder) is generally maintained. Composite materials are multiphase systems in which the matrix phase can be metal or ceramic. Thus, the two types of cermets, are MCC (metal ceramic composites) and CMC (ceramic metal composites). Metals such as Fe, Cr, Co, Al, Cu, Ca, and ceramics such as carbides of Zr, W, Mb, Ti, are commonly employed. Cermets possess greater refractoriness, thermal conductivity, electric and thermal shock resistance. They are usually applied in the fabrication of high refractory materials like ovens, furnaces, engine parts (brakes, clutches), turbine blades, and also as bullet-proof vests.

Today, advanced ceramic materials (or functional ceramics) are utilized extensively as refractory materials. Conventional ceramics generally comprised clay and silica, whereas advanced ceramics are oxide or non-oxide engineering materials with a simple lattice structure and chemical composition. These functional ceramics possess higher mechanical strength, refractoriness, corrosion resistance, and magnetic and optical properties.

Manufacture of Important Ceramic Powders

Silicon carbide (or carborundum) is one of the most commonly used refractory and abrasive material. It is a non-oxide ceramic having a hexagonal and cubic lattice structure. It is generally prepared by the reaction of silica (sand) with coal in an electric resistance furnace.

$$SiO_2 + 3C \rightarrow SiC \text{ (silicon carbide)} + 2CO$$

Saw dust is added in trace amounts during the preparation to decrease the packing density and 1–3 % of NaCl for purification. This is heated with graphite electrodes at 2000 to 2300 °C for 30 hours.

Silicon nitride is prepared by reacting silica metal with nitrogen gas (solid–gas reaction).

$$3\ Si + 2\ N_2 \rightarrow Si_3N_4 \text{ (silicon nitride)}$$

Boron carbide B_4C is similarly prepared in an electric furnace by reacting boron oxide with coke.

$$2\ B_2O_3 + 7\ C \rightarrow B_4C + 6\ CO$$

Sometimes, boron nitride is admixed with silicon carbide to form high-quality abrasives and refractory materials.

15.8 SHAPE MEMORY ALLOYS (SMAS)

As the name suggests, an alloy that can remember or recall its original shape before undergoing deformation is called shape memory alloy. Such alloys have the ability to recover their shape when the temperature is increased. An increase in temperature can result in shape recovery even under high applied loads. Shape memory alloys find applications in robotics, automotives, aerospace, and biomedical industries.

Generally, there are three main types of shape memory alloys, namely copper–zinc–aluminium–nickel, copper–aluminium–nickel, and nickel–titanium (Ni–Ti) alloys. The most widely studied are the Ni–Ti alloys (Nitinols) which change from austenite to martensite upon cooling; M_f is the temperature at which the transition to martensite is finished during cooling. Austenite is metallic, non-magnetic allotrope of iron and in plain carbon steel, it exists above the critical eutectoid temperature of 727 °C in iron–carbon system. Martensite is a hard constituent of steel formed by rapid cooling of austenite and is a metastable phase in iron–carbon phase diagram.

As shown in Fig. 15.13, during heating A_s and A_f are the temperatures at which the transformation from martensite to austenite starts and finishes. The repeated use of shape memory effect may lead to a shift of the characteristic transformation temperatures. This effect is known as *functional fatigue*, as it is closely related to a change of microstructural and functional properties of the material.

When an SMA is in its cold state (below *As*), the metal can be bent or stretched and will hold this shape until heated above the transition temperature. On heating, the shape changes to its original shape. Later, when the metal cools again, it will remain in the hot shape until deformed again. In this case, cooling from high temperature does not cause macroscopic shape change. This is called *one-way shape memory effect*.

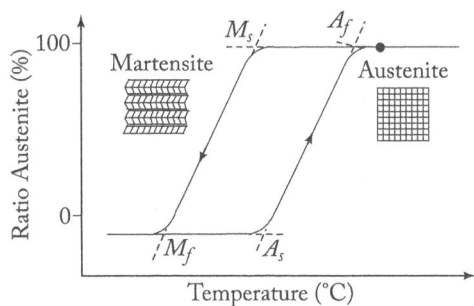

Fig. 15.13 Shape memory effect

Contrary to one-way shape memory effect, if the metal remembers two shapes, one at higher temperature and the other at lower temperature, it is called *two-way shape memory effect*. Such metals show shape memory effect during both cooling and heating cycles. The material can be trained to leave some reminders of the deformed low-temperature condition in the high-temperature phases. Above a certain temperature, the metal loses this memory effect and is termed as *amnesia effect*.

Some important applications of SMAs are in fabricating actuators, sensory devices, dental braces, robotics, eye-glass frames, and autofocus actuator in smartphones.

Check Your Progress

16. What is powder metallurgy?
17. List the steps involved during powder metallurgy.
18. What are solders? Name any two with elemental composition.
19. What are shape memory alloys? Cite one example.
20. List the applications of shape memory alloys.
21. What is a cermet? State its composition.
22. Write the preparation of (a) silicon carbide, (b) silicon nitride and (c) boron nitride.
23. List the stages in powder injection moulding.
24. What are the two types of cermets?

Activity-based Questions

A. Mesophase transition behavioural studies

Liquid crystals are distinct mesophasic engineering materials that are widely used in electronic devices. Various phases of liquid crystals are nematic, smectic, and cholesteric. It is also known that mixtures of liquid crystals of different types show solidification as those of isotropic liquids.

p-Azoxy anisole, a liquid crystal (say 1 g) is placed in a test tube under vacuum and sealed. Applying the concept of phase rule, present a critical review on the phase transitions of the compound. Infer predictions about the melting and transition temperatures of the compound. If *p*-azoxy anisole forms an eutectic mixture with *p*-azoxy phenetole, draw the two-component phase diagram and explain the various phases involved therein.

B. Superalloys in aerospace

Superalloys (also called high performance alloys) are smart engineering materials that exhibit excellent mechanical strength, resistance to creep deformation, surface stability, and corrosion resistance. Some examples of high performing alloy systems are Hastelloy, Inconel, Rene alloys, and TMS alloys. Today, superalloys have found varied applications in designing and fabricating design parts in aerospace, steam power plants, space vehicles, nuclear control rods, and in automotive engines.

Perform a critical review on superalloys in aerospace industry in terms of its preparation methods, characteristic properties, recent trends in aerospace design, and other newer applications. Cite suitable examples of various high performing alloys currently utilized in commercial setting.

SUMMARY

- Metallurgy is the process of extracting metals from their ores and usually involves mining and concentration of ores, followed by extraction and purification of metals.
- Ellingham diagram helps in predicting the feasibility of thermal reduction of various metals.
- Haematite ore has the highest percentage of iron and hence preferred for the extraction of iron; blast furnace, which can withstand very high temperatures is used for the purpose.
- Steelmaking involves manufacturing steel from iron ore and steel scrap usually by two main processes— basic oxygen steelmaking and electric arc furnace steelmaking.

- Alloys are classified into ferrous and non-ferrous alloys based on the presence or absence of iron in them. Ferrous alloys are also called alloy steels. Duralumin, gunmetal, solder, brasses, and bronzes are non-ferrous alloys.
- The steps of powder metallurgy are pulverizing, blending, compacting, sintering, and finishing.
- Cermets are composite materials composed of usually 80 % ceramic and 20 % metallic constituents; examples include silicon carbide, boron carbide, and silicon nitride.
- Shape memory alloys are newer smart engineering materials that recall their original shape on heating after deformation. SMAs are widely employed in telecommunication and robotics.

GLOSSARY

Actuator: A device causing machine movements.

Austenite: Metallic, non-magnetic allotrope of iron and in plain carbon steel, it exists above the critical eutectoid temperature of 727 °C of iron–carbon system.

Benefication: The process of concentration of ore.

Cermet: A composite material prepared by mixing ceramic and metallic constituents by powder metallurgy technique.

Ellingham diagram: Plot of $\Delta G°$ against temperature for depicting a class of related compounds such as metal oxides and sulphides which help in deducing thermal stabilities and reduction tendencies of such compounds.

Ductility: The ability of a material to be bent or stretched without rupturing.

Hardenability: The capability of ferrous alloys to form martensite on quenching from a temperature above its critical point.

Machinability: It indicates an alloy material's hardness, chemical composition, and qualities such as microstructure, propensity to work, hardness in cold and hot conditions.

Martensite: The hard constituent of steel formed by rapid cooling of austenite and is a metastable phase in iron–carbon phase diagram.

Powder metallurgy: Processes in which metal particles are fused under various combinations of heat and pressure to prepare solid commercial metals.

Quenching: The rapid cooling of metal work-piece with air or a liquid medium.

Shape memory alloy: An alloy that has the ability to recover its shape when the temperature is increased.

Sintering: It involves bonding of adjacent surfaces in metal powder by molecular or atomic attraction on heating below melting temperature of the material.

Tensile strength: The capacity of an alloy to withstand loads.

EXERCISES

Multiple Choice Questions

1. Percentage of aluminium in duralumin alloy is
   (a) 95 %
   (b) 93.5 %
   (c) 99.9 %
   (d) 98.01 %

2. Alloys of copper are generally called
   (a) brasses
   (b) bronzes
   (c) solders
   (d) both (a) and (b)

3. Which of the following methods is used to obtain metal powders?
   (a) Sintering
   (b) Compaction
   (c) Atomization
   (d) Moulding

4. Combining ceramic and metal components yields _____.
   (a) alloys
   (b) composites
   (c) cermets
   (d) high resisting alloys

5. The composition of German silver is
   (a) 60% Cu, 15% Zn and 25% Ni
   (b) 60% Cu, 25% Zn and 15% Ni
   (c) 60% Zn, 25% Cu and 15% Ni
   (d) 60% Zn, 25% Ni and 15% Cu

6. Which of the following is a refractory material?
   (a) Low steels
   (b) Carbon nanotubes
   (c) Silicon carbide
   (d) None of these

7. The most studied shape memory alloy is
   (a) Nitinol
   (b) Cu–Al–Ni
   (c) Nickel steel
   (d) Martensite

8. Ellingham diagram helps to predict
   (a) Free energy changes
   (b) Metal oxide reduction during metallurgy
   (c) Entropy changes
   (d) Enthalpy changes

9. Tinman's solder comprises
   (a) 60% Sn and 40% Pb
   (b) 70% Sn and 30% Pb
   (c) 66% Pb and 34% Sn
   (d) 66% tin and 34% lead

10. While alloying steels, chromium imparts
    (a) toughness
    (b) corrosion resistance
    (c) softness
    (d) tensile stress

11. The steelmaking process is carried out in
    (a) blast furnace
    (b) electric arc furnace
    (c) converter-type furnace
    (d) Both (b) and (c)

12. The percentage of iron in an impure pig iron is
    (a) 92–95 %
    (b) 98 %
    (c) 50 %
    (d) 25–30 %

13. The role of carbon monoxide in blast furnace is of
    (a) reducing agent
    (b) oxidizing agent
    (c) flux
    (d) charging

14. Which of the following is not an ore of iron?
    (a) Haemetite
    (b) Pyrite
    (c) Limonite
    (d) Brass

15. Silicon carbide is also called
    (a) carborundum
    (b) nitinol
    (c) austenite
    (d) martensite

Review Questions

1. Explain the construction and working of the blast furnace for extracting iron with a suitable diagram. Write the chemical reactions taking place in the blast furnace.

2. With the help of Ellingham diagram, explain the thermodynamics of metallurgy.

3. 'Metallurgy is a technological process of obtaining metals'. Comment.

4. Define ore and list the stages of metallurgy to obtain metals.

5. What is an alloy? State the purpose of making alloys. How are alloys classified?

6. What is plain carbon steel? Write a note on the types of plain carbon steels.

7. Write a short informative note on stainless steels.

8. Using Ellingham plot, explain reduction reaction occurring during iron extraction in a blast furnace.

9. Explain steel production using basic oxygen and electric arc furnace methods.

10. State the composition, properties, and applications of: (a) Gun metal (b) German silver (c) Tinman's solder (d) Wood's metal (e) Magnalumin (f) Duralumin (g) French gold (h) High phosphor bronze.

11. Explain the effect of alloying metals such as Cr, W, Mo, Co on the properties of steel.

12. Write a short note on different types of steel.

13. What is powder metallurgy? Discuss the various steps involved in the technique.

14. State the principle involved in powder metallurgy. Describe atomization and sintering processes involved in the technique.

15. What are the applications of powder metallurgy?

16. What are metal ceramic powders? Explain the methods of preparing silicon carbide and silicon nitride.

17. Explain the various compaction methods in powder metallurgy.

18. Explain cold powder extrusion method.
19. Explain powder injection moulding method.
20. Explain roll and hot compaction techniques in powder metallurgy.
21. State the specific effects of alloying on the following elements on steel: Ni, Mo, P, Cr, Co, Mn.
22. What are special steels? Explain the various types of heat treatable steels.
23. Citing suitable examples, justify the statement, 'alloying of metals provide corrosion resistance.'
24. Explain the uses, advantages, and disadvantages of powder metallurgy.
25. Explain shape memory alloys with suitable examples.

Further Reading

1. Benvenuto, M.A. *Metals and Alloys – Industrial Applications*, Walter de Gruyter Publishing, 2016.
2. https://www.worldsteel.org/
6. Jones W. D., *Fundamental Principles of Powder Metallurgy*. London: Edward Arnold Ltd, 1960.
3. Lavakumar, A., *Concise Concepts in Physical Metallurgy*, USA: Morgan & Claypool Publishers, 2017.
4. Lecce, L., A. Concilio, *Shape Memory Alloy Engineering*, Elsevier, 2013.
5. Tsukerman S. A., *Powder Metallurgy*, Pergamon Press, 2013.

ANSWERS

Check Your Progress

1. Refer to Section 15.1
2. The process of extracting metals from their ores usually by a series of reduction reactions.
3. Haematite (Fe_2O_3), Magnetite (Fe_3O_4 or $FeO.Fe_2O_3$), Limonite ($Fe_2O_3.nH_2O$), Pyrite, and Siderite ($FeCO_3$).
4. The process of producing steel from iron ore and steel scrap.
5. Zone of reduction, zone of absorption, and zone of fusion.
6. Pig iron, wrought iron, cast iron.
7. Basic oxygen steelmaking and electric arc furnace methods.
8. When aluminium acts as the reducing agent it is called aluminothermic reaction. In thermite reaction, aluminium reduces oxides of iron to form pure iron as, $Fe_2O_3 + 2 Al \rightarrow 2 Fe + Al_2O_3$.
9. Alloys are engineering materials obtained by the solidification of a metallic solution of two or more elements, of which one is essentially a metal. Duralumin and Magnalumin are two alloys of aluminium.
10. Improves hardness of metal, enhances corrosion resistance, and allows easy machinability.
11. The main purpose of Ellingham diagram is to predict spontaneity of reduction of various metal oxides occurring during metallurgical operations.
12. Plain carbon steel is a type of ferrous alloy in which carbon is the principal hardening element present along with iron.
13. The stainless nature of steel comes from the presence of chromium (> 16% wt) that forms Cr_2O_3 on contact with air. The oxide of chromium formed on the steel surface protects the underlying metal from corrosive attack.
14. Brasses and bronzes are two classical types of copper alloys. Copper and zinc in varying composition is called brass, whereas varying composition of copper and tin is called bronze.
15. Brass is an alloy of copper and zinc whereas bronze is an alloy of copper and tin.
16. A technique of manufacturing commercial grade metals, alloy articles, ceramics, and composite materials.
17. Metal powder production, mixing and blending, compaction, sintering, and finishing are the steps involved during powder metallurgy.
18. Solders are low-melting alloys made of lead and tin. Wood's metal (Bi = 505, Pb = 25%, Sn and Cd = 12.5% each) and Tinman's solder (Sn = 66%, Pb = 34&) are examples of solder alloys.
19. An alloy that can remember or recall its original shape before undergoing deformation is called shape memory alloy. Nickel-titanium (NiTi) alloy is an example.

20. Shape memory alloys are used in preparing actuators, sensory devices, dental braces, robotics, eye-glass frames, and autofocus actuator in smartphones.

21. Composite materials prepared by mixing ceramic and metallic constituents by powder metallurgy technique. In cermets, 80% ceramic and 20% metal composition (as binder) is generally maintained.

22. (a) Silicon carbide is prepared by reacting silica (sand) with coal in an electric resistance furnace. The reaction is, $SiO_2 + 3C \rightarrow SiC$ (silicon carbide) $+ 2CO$.

    (b) Silicon nitride is prepared by reacting silica metal with nitrogen gas (solid-gas reaction) as,

    $3Si + 2N_2 \rightarrow Si_3N_4$ (silicon nitride)

    (c) Boron carbide B_4C is prepared in an electric furnace by reacting boron oxide with coke

    $2B_2O_3 + 7C \rightarrow B_4C + 6CO$

23. Refer to Fig. 15.11

24. MCC (Metal Ceramic Composites) and CMC (Ceramic Metal Composites) are two types of cermets.

Multiple Choice Questions

1. (a)	2. (d)	3. (c)	4. (c)	5. (a)	6. (c)	7. (a)	8. (b)
9. (d)	10. (b)	11. (d)	12. (a)	13. (a)	14. (d)	15. (a)	

Polymers

16.1 INTRODUCTION

From natural proteins and DNA in human body to plastic buckets, toys, office chairs, cars, clothing, and surface coatings like paints, polymers are omnipresent. They are the most versatile engineering materials that form the backbone of major industries such as plastics, rubbers (elastomers), fibres, and surface coatings. Polymers (Greek word: *poly* means many and *mer* means part) are defined as high molecular weight molecules with mass of 10^3–10^7 amu. When repeat structural units obtained from simple and reactive molecules called *monomers* are linked by covalent bonds, we get *polymers*. The process of forming long-chain polymers is called *polymerization*. The formation of polythene from ethene is depicted here, where the double bond in ethene is the functionality.

$$n\, CH_2 = CH_2 \xrightarrow{\text{Polymerization}} n\!\left[CH_2-CH_2\right] \longrightarrow \left[CH_2-CH_2\right]_n$$

Ethene   Repeating unit   Poly (ethene) polymer

The degree of polymerization (denoted as n) is defined as the average number of repeat units in a polymeric chain. Let us take an example of polythene PE, $(C_2H_4)_n$. The molecular weight of PE is given as $M = n\, Mo$, where Mo, is the molecular weight of monomer and M is the molecular weight of polymer.

By controlling the length of the polymer chain and also its molecular weight, it is possible to vary the physical properties of polymers.

16.2 CLASSIFICATION OF POLYMERS

On the basis of *sources*, polymers are classified into the following three categories.

Natural polymers These are obtained from plants and animals; examples include silk and wool fibres, proteins, cellulose, starch, resins, and rubber.

Synthetic polymers These are man-made polymers such as plastics, synthetic fibres, polyamides, teflon, and polyvinyl chloride (PVC). These polymers have great versatility and are extensively used in industries and other products of everyday use.

Semi-synthetic polymers These are generally cellulosic derivatives such as rayon (cellulose acetate) and cellulose nitrate.

On the basis of *structural complexity*, polymers are studied under the following three categories.

Linear polymers As the name suggests, these are long and straight chains polymers as shown here. Polyethene (PE) and polyvinyl chloride (PVC) are common linear polymers.

Branched chain polymers These are linear chains of polymers with some level of branching. Low density polyethene (LDPE), glycogen, amylopectin are some examples of branched chain polymers.

Cross-linked or network polymers These are formed due to combining of two or more functional groups in the repeat unit, thereby forming strong covalent bonds between linear polymeric chains.

These polymers have high rigidity and toughness compared to the above two categories. Rubber, Bakelite, urea–formaldehyde resins are examples of cross-linked polymers. Figure 16.1 shows (a) Linear polymer (b) branched chain polymers, and (c) cross-linked polymers

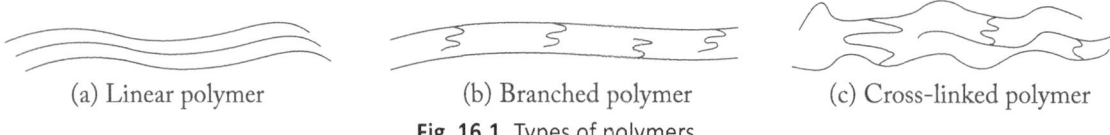

(a) Linear polymer   (b) Branched polymer   (c) Cross-linked polymer

Fig. 16.1 Types of polymers

On the basis of *polymerization methods*, polymers are categorized as follows.

Addition polymers These are obtained by the simple joining of monomers, usually containing one or more double bonds. When only one type of monomeric units tends to form a polymer, it is called a *homopolymer*. Poly ethene (PE) is an example of homopolymer as it is formed by adding repeat units of only ethene molecule, without loss of any small molecule in the process.

Polymers formed of two different monomers is called a *copolymer*, for example, Buna-S elastomer.

Copolymers may be linear, branched or cross-linked. In a copolymer, monomers are arranged in a polymer chain either regularly or irregularly. The different types of copolymers are as follows:

Alternating copolymer Here, monomers are distributed in a regular alternate fashion.

- A-B-A-B-A-B-A-B-A- (where, A and B represent two different monomers)

Random copolymer Here, monomeric units are distributed randomly or sometimes unevenly in the polymer chain.

- A-B-B-A-A-A-B-A-A-B-B-B-A-B-A-A-B-A-

Block copolymer Here, a long sequence or block of one monomer is joined to a block of a second monomer.

-A-A-A-A-A-B-B-B-B-B-B-B-A-A-A-A-A-A-A-B-B-B-

Graft copolymer Here, the side chains of a monomer are attached to the main chain of another monomer.

Condensation polymers When monomers with two or more functional groups react together with the loss of a small molecule, it is called condensation polymerization. Nylon – 6, 6 is an example of a condensation polymer obtained from adipic acid and hexamethylene diamine.

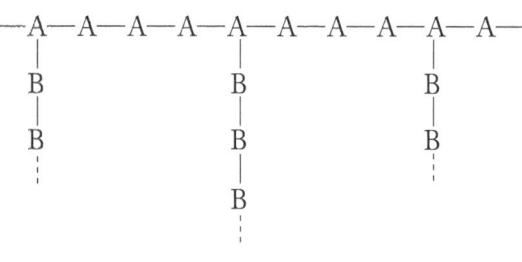

On the basis of the *intermolecular forces* present between polymer chains, they can be classified as follows.

Elastomers (or rubbers) These have the weakest intermolecular forces of attraction making them highly elastic. Buna-S and Buna-N are examples of elastomers.

Fibres These are thread-like polymers possessing higher tensile strength. They tend to have strong intermolecular forces like hydrogen bonding. Examples include Nylon-6,6 and terylene.

Thermoplastic polymers   These have linear or branched polymeric chains with intermolecular forces as seen in rubbers and fibres. Such polymers tend to become soft on heating. They can be moulded into any shape which retains on cooling. PVC, PE, and nylon sealing wax are some examples.

Thermosetting polymers   These have cross-linked branched polymeric chains with strong intermolecular forces. On heating, such polymers undergo cross-linking and become infusible that cannot be reshaped on cooling. Formaldehyde resin is a classic example.

16.3 Types of Polymerization Reactions

The various polymerization reactions are discussed here.

16.3.1 Addition or Chain Growth Polymerization

In addition polymerization, monomers of same or different types tend to add together to form long linear polymer chains. Alkenes, alkadienes, and their analogues are the monomers forming addition polymers by free radical mechanism as shown below.

The free radical mechanism is induced by heat or sunlight and includes three steps; viz., initiation, propagation, and termination.

Chain initiation   In this step free radicals are generated by homolytic dissociation of initiators such as acetyl peroxide and benzoyl peroxide.

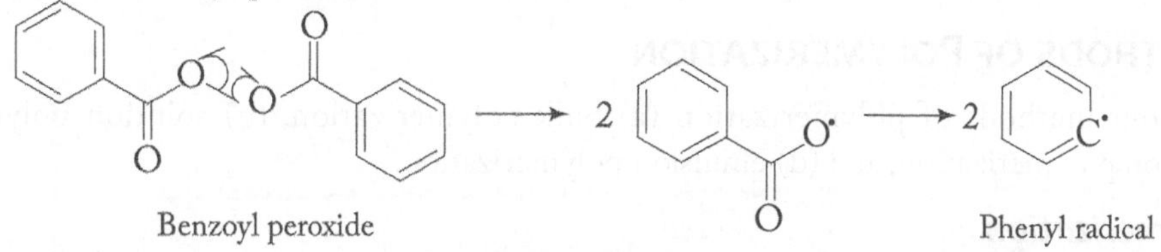

Benzoyl peroxide   Phenyl radical

The reactive free radical opens the carbon–carbon double bond in ethene by joining to one side of the monomer as shown here. Now, the resulting species reacts with another monomer (ethene) as shown below:

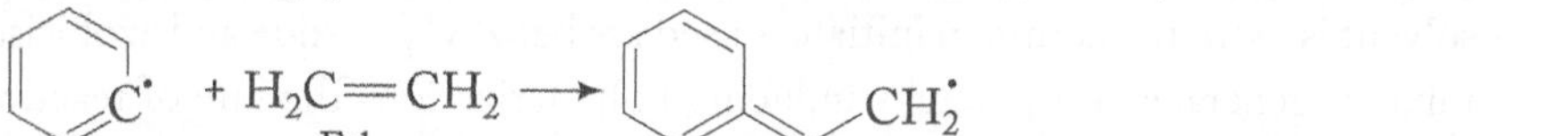

Chain propagation   This involves continued addition of successive monomeric units.

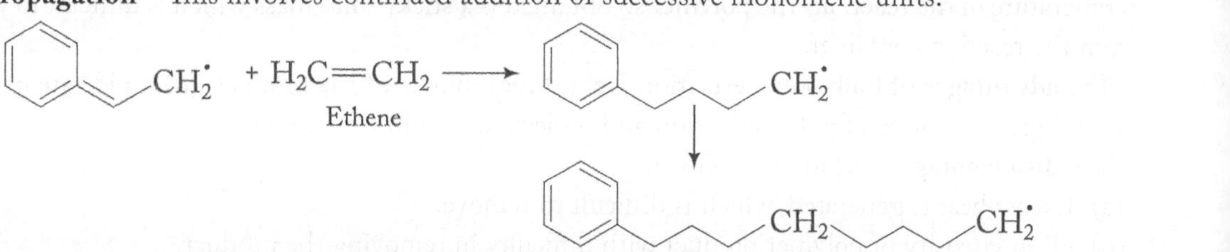

Chain termination   This occurs once the required length of polymer is obtained. Free radicals combine together to form the final polymeric chain.

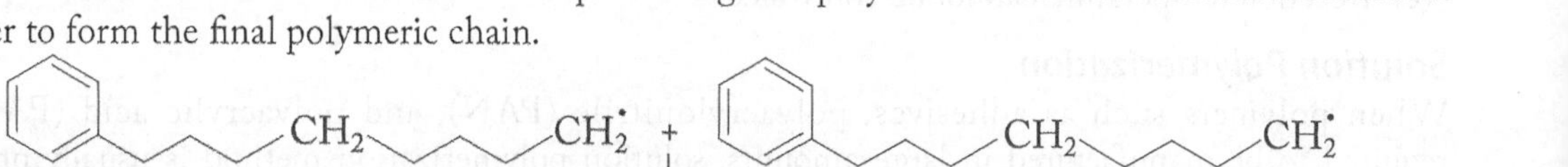

$$C_6H_5-(CH_2-CH_2-)_n-CH_2-CH_2-CH_2-(CH_2-CH_2-)_n-C_6H_5$$

16.3.2 Condensation Polymerization

Bifunctional groups in a monomer react with the loss of a small molecule such as water or ammonia, leading to the formation of condensation polymers. It is also called *step growth polymerization*.

Nylon-6,6 is an example of condensation polymer.

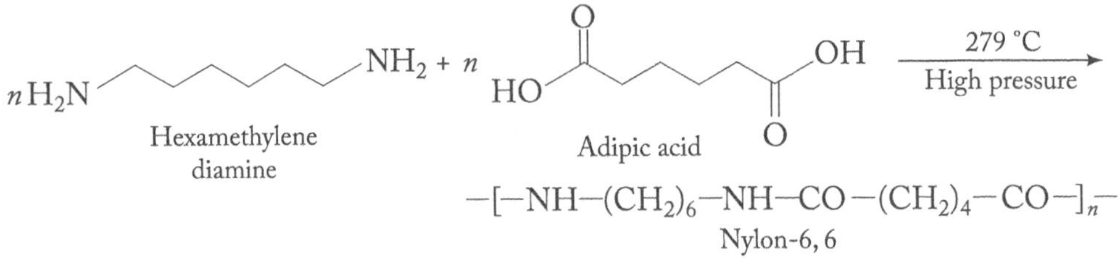

$$-[-NH-(CH_2)_6-NH-CO-(CH_2)_4-CO-]_n-$$
Nylon-6, 6

16.3.3 Copolymerization

Similar to addition polymerization, copolymerization involves the joining of two monomers of different type. Upon copolymerization, 1,3-butadiene and styrene forms Buna-S rubber.

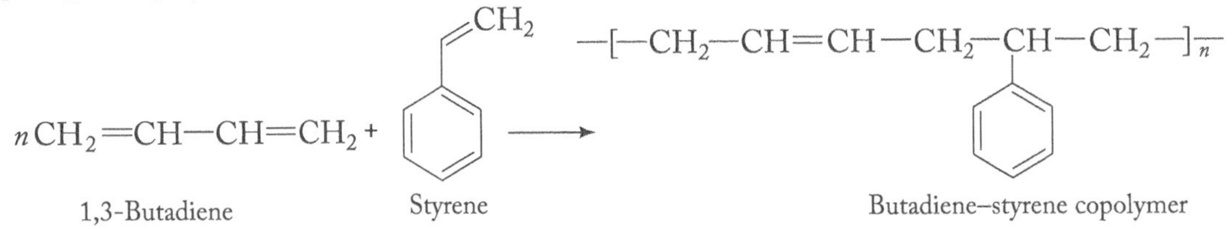

$$-[-CH_2-CH=CH-CH_2-CH-CH_2-]_n-$$

Butadiene–styrene copolymer

$$n\,CH_2=CH-CH=CH_2 +$$

1,3-Butadiene   Styrene

16.4 METHODS OF POLYMERIZATION

There are four methods of polymerization (a) bulk polymerization, (b) solution polymerization, (c) suspension polymerization, and (d) emulsion polymerization.

Bulk Polymerization

Bulk polymerization is used to fabricate polystyrene and organic glasses. It involves free radical polymerization on a vinyl group monomer in the presence of an initiator. In bulk polymerization, no solvent is used. The common initiators used are benzoyl peroxide and azobisisobutyronitrile. This reaction mixture generates heat, thereby inducing polymerization. The rate of reaction is quite rapid and heat is evolved during polymerization. As the reaction proceeds with selective heating, it is difficult to adjust the temperature of the reaction. The polymer so obtained is a sticky solid mass which is difficult to withdraw from the reaction container.

The **advantages** of bulk polymerization are: (a) high purity of the product; (b) rapid rate of reaction, and (c) greater degree of polymerization and molecular weight of the polymer.

The **disadvantages** include the following.
(a) Excess heat is generated, which is difficult to remove.
(b) High viscosity of polymer product with difficulty in removing the product.
(c) Reaction temperature cannot be controlled.

Solution Polymerization

When polymers such as adhesives, polyacrylonitrile (PAN), and polyacrylic acid (PAA), are required to be manufactured in large amounts, solution polymerization method is usually preferred. Solvents such as toluene, methyl acetate, and ethylacetate that are non-reactive are added to the monomer. It is important to select a solvent in which the polymer product is soluble and unlike bulk polymerization, also allows removal of heat produced during the reaction. Monomer, initiator, and solvent are added in the reactor and continuously stirred to obtain the polymer. The excess solvent is filtered and solid, pure polymer product is obtained.

The **advantages** of solution polymerization are:

(a) The solvent acts a diluent and allows heat removal during polymerization process. Hence thermal control is far better than bulk polymerization.

(b) The solvent helps to adjust the viscosity of polymer product that allows easier product withdrawal from the reactor.

The **disadvantages** of solution polymerization are:

(a) Difficulty in eliminating solvents from final polymer product causing degradation.

(b) Solvent release may result in environmental pollution.

Suspension Polymerization

Polymers such as polymethyl methacrylate (PMMA) and polystyrene (PS) are prepared using suspension technique. In this method, insoluble monomer, initiator, and an emulsifier are added to a continuously stirred reactor. As the monomer is insoluble, it is dispersed and suspended in the liquid phase such as water along with continuous agitation. Benzoyl peroxide initiator is added to the dispersed monomer. As both monomers and polymer products are insoluble in water phase, a dispersing agent such as methyl cellulose is added as a dissolved suspending agent. The suspending agent prevents coalescence of droplets in the reaction mixture. The reaction proceeds due to heat and is later water-cooled. As the suspending agent transforms monomeric droplets in to spherical solid polymer beads, suspension polymerization is also called *bead (or pearl) polymerization*. The polymer beads obtained by this method are of particle size in the range of 10–500 µm. The bead size is dependent on agitation speed and suspension stabilizers. The swelling ratio of polymeric bead depends on the degree of cross-linking and solvent.

The **advantages** of suspension polymerization are:

(a) The polymer product obtained is of moderate viscosity and are in the form of beads.

(b) Heat generated during the process can be easily removed due to the high heat capacity of water.

(c) Finely divided, stable beads of polymers obtained can be directly used in paints, ion-exchange resins, and adhesives.

The **disadvantages** of suspension polymerization are:

(a) The reaction cannot be controlled and isolating the polymer is difficult.

(b) Polymer product obtained in this method is not highly pure due to use of suspending agents. Hence, polymer needs isolation and purification.

(c) This method cannot be employed for polymers whose glass transition temperature is less than the polymerization temperature; otherwise polymer bead aggregation will occur.

Emulsion Polymerization

Emulsion polymerization is used to prepare paints, adhesives, polymethyl methacrylate, polyvinyl chloride, and polystyrene. Quite contrary to bulk and suspension polymerization, here the emulsifier such as polyvinyl alcohol is dissolved in water. Next, water-insoluble (or poorly soluble) monomer is added in a continuously stirred reactor along with benzoyl peroxide as the initiator. The emulsifier gets associated with water and forms a micelle. When a monomer is added to the solution containing the micelles, water gets adsorbed in the micelles and thus water will now exist as a droplet. The polymerization is initiated with initiator within the miceller as free radicals are formed. Once the monomer present in the micelle is consumed in the reaction, another monomer droplet is reinforced from solution, and the polymerization continues. The heat released during polymerization is absorbed by water present around the micelles.

The **advantages** of emulsion polymerization are:

(a) Moderate viscosity polymers are obtained with higher molecular weights.

(b) Higher rates of polymerization with thermal control, which is better than bulk and suspension polymerization.

Some **disadvantages** of emulsion polymerization are:
(a) Difficulty in removal of surfactants after the polymerization process.
(b) Water removal is an energy-intensive process.
(c) It is designed to operate for higher conversion of monomers causing significant chain transfer to polymer product.

16.5 MOLECULAR WEIGHT OF POLYMER

The determination of molecular weight of a polymer is one of the most important measurements as most of the mechanical properties of polymers depend on their molecular weights. Molecular weight of a polymer depends on the monomers joined together by polymerization, that is, degree of polymerization. Since, a polymeric chain can be broken at different stages (chain termination), the final chain generally contains monomers of differing weights. When a low molecular weight compound ethylene ($CH_2=CH_2$), is polymerized to poly(ethene), it possesses an indefinite chemical structure, $-(-CH_2-CH_2-)_n-$, where n is a variable value for different polyethene molecules in the same polymer sample. This is due to the fact that during polymerization, numerous polymeric chains start growing at any instant, but all of them do not terminate after growing to the same size and length. Hence, each polymer molecule will have different numbers of monomer units and hence different molecular weights. So, a polymer sample of the same type will be a mixture of molecules of the same chemical type, but of different molecular weights. Therefore, average molecular weight is considered such as *number average* and *weight average* molecular weights.

Number average molecular weight   Consider a polymer sample with $n_1, n_2, n_3 \ldots$ number of molecules with molecular weights $M_1, M_2, M_3\ldots$ respectively. Hence, number average molecular weight ($\overline{M_n}$) can be expressed as,

$$\overline{M_n} = \frac{n_1 M_1 + n_2 M_2 + n_3 M_3 + \ldots}{n_1 + n_2 + n_3 + \ldots} \text{ a.m.u}$$

$$\overline{M_n} = \frac{\sum n_i M_i}{\sum n_i} = \sum x_i M_i$$

where n_i is the number of molecules of the i th type with molecular weight M_i and x_i is the mole fraction of polymer of molecular weight M_i.

Weight average molecular weight   If $m_1, m_2, m_3 \ldots$ are the masses of species with molecular weights $M_1, M_2, M_3\ldots$ respectively ($m_1 = n_1 M_1$ where n_1 is the number of molecules with molecular weight M_1), then weight average molecular weight ($\overline{M_w}$) can be expressed as,

$$\overline{M_w} = \frac{m_1 M_1 + m_2 M_2 + m_3 M_3 + \ldots}{m_1 + m_2 + m_3 + \ldots} \text{ a.m.u}$$

or, $$\overline{M_w} = \frac{\sum m_i M_i}{\sum m_i}$$

Check Your Progress

1. What are monomers and polymers?
2. What is polymerization? Name various types of polymerization.
3. What is Bakelite? State its uses.
4. Classify the polymers as addition and condensation polymers: Polyethene, Bakelite, polyvinyl chloride, Nylon– 6,6.
5. Differentiate between homopolymer and copolymer.
6. Consider a polymer ABABABABABA. Is it a graft copolymer, homopolymer, or copolymer?
7. Name the various polymerization methods.

But, $m_i = n_i M_i$, $\therefore$ $\overline{M_w} = \dfrac{\sum n_i M_i^2}{\sum n_i M_i} = \dfrac{\sum x_i M_i^2}{\sum x_i M_i}$

For all synthetic polymers, $\overline{M_w}$ is greater than $\overline{M_n}$. If they are equal, the polymer sample is considered to be homogeneous, that is, each molecule has the same molecular weight, but is never true in the practical sense. The ratio of $\dfrac{\overline{M_w}}{\overline{M_n}}$ is called *polydispersity index* which provides information on molecular weight distribution. Polydispersity index is always greater than 1, but if found equal to unity, it denotes that the polymer is monodisperse with all polymer chains of equal lengths.

16.6 Tacticity of Polymers (Configuration)

The orientation or configuration of monomers in a polymer is either in an orderly or disorderly manner with respect to the main polymeric chain. *The stereochemistry or arrangement of chiral centres of molecules in a polymer is called tacticity.* The position of monomeric units across the polymer chain dictates the physical properties of the polymer such as its solubility in solvents, optical activity, melting temperatures, and so on. Various types of tacticity can be depicted as three-dimensional structures by drawing the zigzag polymer chain of a vinyl polymer and the respective positions of its substituents as follows.

Isotactic polymer   The functional groups are on the same side of the chain.

H_3C H H_3C H H_3C H H_3C H H_3C H

Natural rubber is an example of isotactic polymer discussed later in the chapter.

Syndiotactic polymer   The arrangement of the functional groups is in alternating fashion with respect to polymer chain.

H_3C H H CH_3 H_3C H H CH_3 H_3C H

Gutta-percha is an example of syndiotactic polymer.

Atactic polymer   The arrangement of the functional groups is random around the main polymeric chain.

H_3C H H_3C H H CH_3 H_3C H H CH_3

Polypropylene is an example of atactic polymer.

The tacticity of polymers refers to the stereoregularity of monomers in its chain. Hence, atactic polypropylene will be a gummy solid, whereas its isotactic form will be a crystalline and tough material.

16.7 Crystallinity of Polymers

No polymer can be 100 % crystalline in structure. A polymer consists of both crystalline and amorphous regions. *Degree of crystallinity (f_c^m) is defined as the ratio of the mass of crystalline regions to the total mass of the polymer sample.*

$$f_c^m = \dfrac{\text{Mass of crystalline regions}}{\text{Total mass of polymer sample}}$$

From density measurements, one can determine the degree of crystallinity using the expression:

$$\% C = \dfrac{\rho_c (\rho_s - \rho_a)}{\rho_s (\rho_c - \rho_a)}$$

where ρ_c is the density of perfect crystalline polymer, ρ_s is the density of polymer specimen, and ρ_a is the density of amorphous polymer.

Degree of crystallinity tells us the volume fraction of polymer that is crystalline in a semi-crystalline polymer. A completely amorphous polymer can be imagined as random arrangement of entangled polymer chains, whereas a semi-crystalline polymer will have somewhat ordered arrangement of polymer chains.

Factors affecting polymer crystallinity The degree of crystallinity depends on the polymer structure and its configuration. Polymer crystallinity depends on the following factors.

Length of polymer chain Long polymer chains represent high degree of polymerization. Thus, such polymers are less likely to crystallize as they get entangled resulting in amorphous regions.

Branching Branching in a polymer chain results in the formation of amorphous regions. The branched polymers are difficult to fold and may come in the way of crystalline packing of the polymer.

Tacticity Stereoregular polymers which exhibit isotactic and syndiotactic configuration are mostly crystalline due to their even arrangement of monomeric units. Atactic polymers are completely amorphous due to random arrangement of monomers.

Plasticizers These are low molecular weight compounds added to polymers to prevent crystallization as they prevent polymer chains to come in contact with each other.

Copolymers Amongst all types of copolymers, only alternating copolymers have an inherent periodicity within the chain itself, and thus may exhibit crystallinity. Random, block, and graft copolymers will not exhibit crystallinity.

The total fraction of amorphous region in a polymer can be controlled by copolymerization that lowers structural symmetry, thus decreasing the tendency to crystallise; for example, vinylidene chloride (85–90 %) is copolymerized with vinyl chloride (10–15 %) to obtain pure polyvinylidene polymer. Polyvinylidene has greater degree of flexibility and on adding plasticizers can be made even more mouldable and used in piping products, paints, and in food packaging.

16.8 GLASS AND MELTING TRANSITION TEMPERATURES

The behaviour of polymers with respect to their flow properties is temperature sensitive. Ideally, amorphous polymers do not possess melting points, but softening point. At low temperatures, polymers exist as glassy materials. In this state, a solid tends to shatter if it is hit, since the molecular chains cannot move easily.

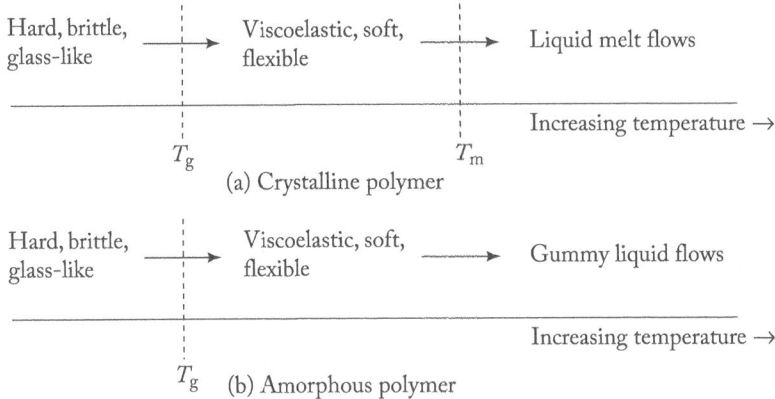

Fig. 16.2 Effect of temperature on polymers

The lowest temperature beyond which a polymer is hard, brittle, and glass-like is called *glass transition temperature* denoted as T_g. The temperature above which it turns out to be flexible, elastic, and rubbery is called *melting transition temperature* denoted as T_m.

Rubber band (very flexible) is placed in a container of liquid nitrogen. When removed rubber band is solid and inflexible (glass state) and can be shattered. Upon standing at room temperature the rubber band will again become flexible and rubbery. Both T_g and T_m are significant parameters which give an indication of the range of temperatures at which a polymer transforms from a rigid solid to a soft viscous material. They also help in choosing the right processing temperature at which materials can be converted into finished products.

Factors influencing T_g

(i) T_g is directly proportional to the molecular weight of the polymer.
(ii) Greater the degree of cross-linking, higher the T_g. This is due to the fact that crystalline polymer chains are arranged in a regular manner and each chain is bound by strong forces like hydrogen bonding. Polymers with strong intermolecular forces of attraction have high T_g values.
(iii) Side groups, especially benzene and aromatic groups, hinder free rotation when attached to the main chain thereby increasing T_g.
(iv) Stereoregularity of polymers increases T_g. Thus T_g of an isotactic polymer is higher than that of a syndiotactic polymer whose T_g is in turn higher than that of an atactic polymer.

Significance of T_g   The value of T_g is a measure of flexibility of a polymer and also gives an idea of thermal expansion, heat capacity, as well as its electrical and mechanical properties. Thus the workability and usefulness of a polymer over a range of temperature can be assessed from its T_g.

16.9 VISCOELASTICITY

Viscoelasticity is defined as the characteristic property of materials that exhibit both viscous and elastic behaviour on application of stress and deformation. Studying the viscoelasticity of a polymer helps in understanding its mechanical strength over varying temperatures. When strain is applied on a polymer, it elongates up to a particular length. When the strain is removed, the elongated polymer soon retracts at the same rate as during elongation. This means that when load strain is applied on a polymer for a long period of time, it remains deformed. *The tendency to behave as an elastic solid and also a viscous liquid is called viscoelasticity and materials exhibiting this behaviour are called viscoelastic materials.*

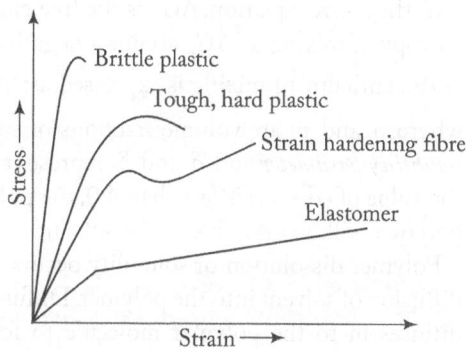

Fig. 16.3 Stress–strain curves of various polymers

The response of a polymer to stress depends on temperature in relation to glass transition temperature (T_g) and also on the time period of deformation. Silly Putty is one such classic example that exhibits viscoelasticity. Chemically, Silly Putty is polydimethyl siloxane (PDMS) which tends to be a bouncy elastic solid and also a viscous flowy liquid, thus exhibiting viscoelasticity. Figure 16.3 depicts the stress–strain curves of various polymers under load.

The graph depicts that at the end of line, polymers tend to rupture or break. At lower temperatures below the glass transition temperature, brittle failure is observed as a break at low strain rate at maximum stress. At higher temperatures, polymers undergo change from brittle to ductile solid in deformation and fracture. This temperature is called the *brittle–ductile transition temperature* (denoted as, T_β). For brittle plastics, there is a negligible stress on increasing the strain. Tough, hard plastics tend to show flexibility up to certain levels of strain, after which they undergo complete deformation. Polymers like fibres, on strong stretching, break at a much higher stress. This phenomenon is called *strain hardening*. In elastomers, a slight increase in stress results in greater strain as they are easy to elongate and retract. The behaviour of polymers before breakpoint will depend upon their cross-linking; low cross-linked materials will exhibit higher elastic deformation, whereas non cross-linked polymers will show viscoelasticity.

16.10 SOLUBILITY OF POLYMERS

Solubility of polymers is an important property as it finds application in membrane science and plastics recycling. It is known that polymers do not dissolve instantaneously. However, it is an essential requirement in many industrial processes. From a practical standpoint, only linear polymers can undergo dissolution, whereas cross-linked polymers may merely swell up due to solvent diffusion. Though cross-linked polymers may get solvated, the cross-linking bonds do not allow solvent molecules to interact with the whole polymer molecule and thus inhibits the polymer to enter the solution phase. In polymers, solvent diffusion followed by disentanglement of polymer chains result in dissolution.

There are many water-soluble polymers that are used commercially. The presence of polar functional groups causes them to be water soluble. Polyvinyl alcohol and polyacrylic acid are classic examples of water-soluble polymers. Even though these polymers are water soluble, they possess polar functional groups along with hydrophobic polymeric backbone. Hence, they behave as hydrocarbons, resisting complete dissolution in a solvent.

To put it simply, solubility works on the principle 'like dissolves like' and polymers too behave in the same way. Polar polymers will dissolve readily in polar solvents and aromatic polymers will dissolve in organic solvents. This can be expressed thermodynamically using Gibbs equation, $G_m = \Delta H_m - T\Delta S_m$.

In the above equation, ΔG_m is the free energy of mixing, ΔH_m is the enthalpy of mixing, and ΔS_m is the entropy of mixing. If ΔG_m attains a negative value, the polymer will undergo dissolution.

The enthalpy of mixing is expressed as: $\Delta H_m = v_s v_p (\delta_s - \delta_p)^2$

where v_s and v_p are volume fractions of solvent and polymer, respectively. The quantity δ is termed as *solubility parameter* and δ_s and δ_p represent solubility parameters of solvent and polymer, respectively. If the value of $(\delta_s - \delta_p)$ is less than 4.0, the polymer will dissolve in a given solvent and if it is above 4.0, the polymer will be insoluble in the solvent.

Polymer dissolution or solubility occurs when there is disentanglement of the polymer chains due to diffusion of solvent into the polymer. Diffusion occurs in two stages; in the first stage, the solvent gradually diffuses in to the polymer molecule to form a gel and in the second stage the solvent disintegrates the gel resulting in the formation of polymer solution. However, dissolution of polymer in a solvent may take several days or weeks. The dissolution of a polymer depends on the differences in the size of solvent and polymer molecules, viscosity of solution, and molecular weight of the polymer. Cross-linked polymers lack solubility in most solvents; rather they swell in the solvent.

Table 16.1 Solubility parameters of selected solvents and polymers

Solvent	(δ_s) in J/cm³	Polymer	(δ_p) in J/cm³
n-Hexane	14.8	Polyethene	16.2
Toluene	18.3	Polystyrene	17.6
Methanol	29.3	Polyvinyl chloride	19.4
Water	47.9	Nylon 6,6	27.8

Polyethene has a δ_p value of 16.2 J/cm³. If we consider hexane as a solvent with δ_s value of 14.8 J/cm³, then the parameter $(\delta_s - \delta_p)$ for such a polymer–solvent system will be – 1.4 J/cm³ which is much less than 4.0. Thus, it indicates that polyethene is soluble in hexane. Now, if we consider water as the solvent, polyethene will be insoluble in it as $(\delta_s - \delta_p)$ value is found to be 31.7 J/cm³. All these expressions only provide details on the energetics of polymer solubility and no information is obtained on the kinetics. The above accounts for only non-polar systems but can be modified for polar systems by taking hydrogen bonding and polarities into consideration.

16.11 PLASTICS

Plastics are organic materials with high molecular weights. They are prepared by polymerization and possess the ability to be shaped on application of heat and pressure in the presence of catalysts. Almost all commercial packaging, furniture, pipes, machines, toys, etc., are made from plastics, making them one of the most versatile materials used by mankind in every walk of life.

Some of the characteristic properties of plastics are: (a) lightweight; (b) excellent plasticity and flexibility; (c) easy mouldability; (d) corrosion resistant and chemically inert (e) good shock-absorber and (f) possess good transparency.

All these properties make plastics a lucrative engineering material. However, there are many demerits, some of which include: (a) plastics are combustible, (b) non-biodegradable, (c) have high cost, (d) poor ductility and (e) deform under heavy load. The versatility of plastics is reflected by the applications which are listed below.

(a) Plastics are commonly used as commercial packaging materials in food and beverage industries.

(b) They are used to make furnitures, door knobs, floorings, and wall-linings.

(c) Plastics are used in making everyday items such as toys, combs, toothbrush bristles, plastic bags, syringes, spoons, and so on.

(d) They are also used to prepare fibres such as nylon and terylene.

(e) As they possess good insulating property, plugs, switches, cables, and refrigerator parts are made from plastics.

(f) They are also used in paints, adhesives, and water-softening resins.

(g) They are used as storage tanks, water-proof vessels, safety glasses, light fixtures, and pipes.

16.11.1 Compounding of Plastics

The majority of the commercial plastic products are not made of only high and pure polymers. Before fabricating commercial plastic products, they are generally mixed with various additives to enhance their mechanical and aesthetic properties. Such process is called compounding. The following are the major types of compounding ingredients or additives.

Resins   They act as binding agents. Natural or synthetic resins are added to the polymer to hold all the additives together.

Plasticizers   These additives increase the plasticity and flexibility of the polymer. Phthalate esters, oils, camphor, tricresyl phosphate, tributyl phosphate are commonly used plasticizers. Plasticizer molecules penetrate into the polymer and reduce the intermolecular forces of attraction between the polymer chains. This increases the mobility of polymer chains as they slide over each other. Thus, plasticizers act as an internal lubricant.

Stabilizers   Most polymers undergo chemical, thermal, or photolytic degradation during processing of plastics. Stabilizers are additives that tend to chemically stabilize the polymer from degradation.

Check Your Progress

8. What are the stages for polymer dissolution?

9. What is a criterion for an organic molecule to act as a monomer? Give example.

10. Polymers do not have exact molecular weights; rather average molecular weights are calculated. Explain.

11. What is polydispersity index?

Organometallic salts, lead chromate, alkaline earth oxides (CaO, BaO), and epoxy compounds are used as stabilizing agents. It is noted that stabilizers negatively impact the environment due to use of heavy metals. Today, lead stabilizers are replaced with calcium and zinc salts.

Fillers or extenders These are inert materials added to polymers to enhance their mechanical strength and also to reduce the cost of the final plastic product. Particulate fillers such as asbestos powder, sawdust, mica, talc, clay, and fibrous fillers such as cotton, pulp, fibreglass, are used.

Catalysts Various antioxidants such as peroxides, ammonia, and certain transition metals are added to accelerate the formation of cross-links in thermoset polymers.

Lubricants These provide glossy finish to the final commercial plastic product thereby enhancing the aesthetic appeal. Lubricant additives such as waxes and oils also prevent plastics from sticking to the moulding equipment.

Colouring materials Organic dyes and pigments impart the desired colour, thereby providing aesthetic appeal to the commercial plastic product. Carbon black (black), ultramarine (blue), zinc oxide (white), ferric oxide (red) are some commonly employed colouring pigments.

16.11.2 Fabrication Methods of Plastics

Moulding is commonly employed to fabricate plastic products. Compression, injection, transfer and extrusion moulding methods are explained here.

Compression Moulding

This is one of the simplest methods of fabricating thermoset plastics and thermoplastics. In this method, a compression die mould is used (Fig. 16.4) in which the required amount of plastic ingredients is taken. During moulding, a pressure of about 100–500 kg/cm^2 is applied and the temperature is maintained at 100–200 °C. In such conditions, plastic becomes fluidic and fills the mould cavity which is set in to the desired shape. After curing by heating (thermoset) and cooling (thermoplastics), the moulded plastic product is de-moulded.

Injection Moulding

Injection moulding is commonly employed for making thermoplastics. In this method the compounded mixture is fed in to the hopper as a granular powder (Fig. 16.5). The mixture is injected at a controlled rate (1758 kg/cm^2, 90–250 °C) in a mould using a piston-type plunger. The plasticized material is pushed through the nozzle by the feed screw equipped with heater. The molten plastic enters the mould cavity where the desired shape is obtained. The final plastic product is cooled and ejected from the mould.

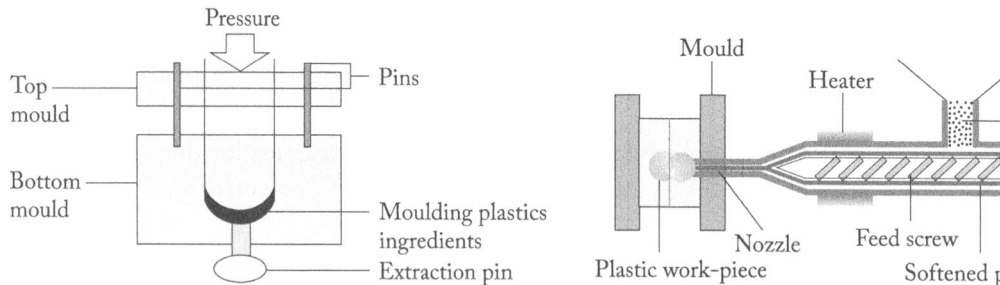

Fig. 16.4 Compression moulding unit   **Fig. 16.5** Injection moulding unit

Transfer Moulding

A combined compression–injection process for thermosets, here the compounded material is plasticized by heat and pressure (Fig. 16.6) in a separate chamber. The resulting plasticized material is injected through the orifice in to the mould by a plunger at a pressure of 1758 kg/cm^2.

Due to friction developed at the orifice, the temperature of material rises at the time of injection from the orifice opening. This causes the plasticized material to become a fluid. The fluidic plastic quickly flows in to the mould that is finally heated up to curing temperature for obtaining the final plastic product.

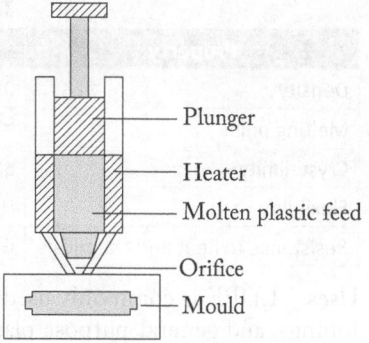

Fig. 16.6 Transfer moulding unit

Extrusion Moulding

A type of injection moulding used to fabricate thermoplastics, extrusion process is used to make plastic materials of uniform cross-section such as rods, tubes, and cables. The compounded plastic powder is forced through the screw conveyor into a heated chamber (Fig. 16.7).

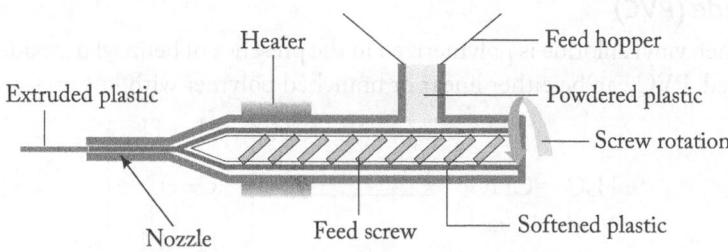

Fig. 16.7 Extrusion moulding unit

When the screw is rotated, molten material in the form of continuous uniform shaped plastic articles is pushed out of the orifice into cold water, resulting in solidification.

16.11.3 Preparation, Properties, and Applications of Commercial Plastics

In this section, we will discuss the preparation of a variety of commercial plastics. The properties and applications of commercial plastics are also listed.

Polyethene (PE)

Polyethene is one of the simplest addition homopolymers and is generally of two types: low density polyethene (LDPE) and high density polyethene (HDPE).

Preparation   LDPE is obtained by addition polymerization (chain reaction) of ethylene gas at 350 °C at 1000–2500 atmospheric pressures in the presence of a catalyst. HDPE is prepared by polymerizing ethylene in a hydrocarbon solvent employing Ziegler–Natta catalyst at 50–60 °C.

$$n\,CH_2{=}CH_2 \xrightarrow[\text{O_2 or peroxide catalyst}]{\Delta\ 350\ °C,\ 1000-2500\ \text{Atm}} \left[CH_2\text{---}CH_2\text{---}CH_2 \right]_n$$

Ethene   Poly(ethene) polymer

$$TiCl_4 + Al(C_2H_5)_3$$
Ziegler-Natta catalyst
Δ 50–60 °C

Structure   LDPE has a linear structure with extensive branching whereas HDPE is a linear chain with minimum branching.

Properties   PE is a common plastic which is non-biodegradable. It possesses flexibility, chemical resistance, and good tensile strength. Table 16.2 lists the various properties of LDPE and HDPE.

Table 16.2 Comparison of LDPE and HDPE

Property	LDPE	HDPE
Density	0.91– 0.94 g/cm^3	0.95–0.97 g/cm^3
Melting point	~115 °C	~135 °C
Crystallinity	50–60 % (low)	> 90 % (high)
Flexibility	High	Rigid
Resistance to heat and chemicals	Good heat and chemical resistance	Useful above 100 °C, chemically inert

Uses LDPE is commonly used to make plastic bags, dispensing bottles, ink-refills in ball-point pens, tubings, and general-purpose plastic containers. HDPE is used to make bottle caps, plastic containers, pipes, and milk and beverage bottles.

Polyvinyl Chloride (PVC)

Preparation When vinyl chloride is polymerized in the presence of benzoyl peroxide catalyst, polyvinyl chloride is produced. PVC can be either linear or branched polymer with low crystallinity.

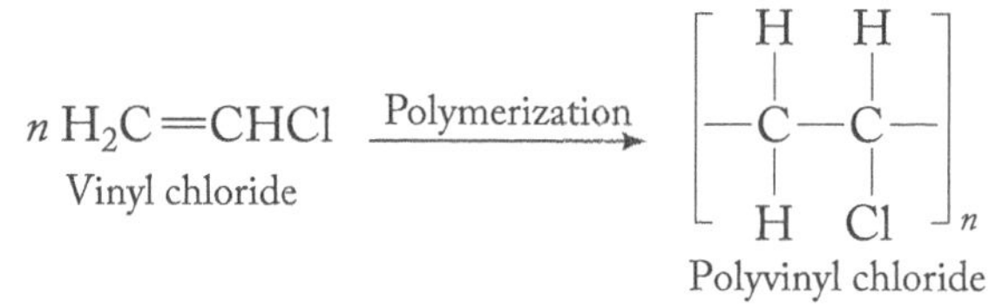

$$n\, H_2C{=}CHCl \xrightarrow{\text{Polymerization}} \left[\begin{array}{c} H \quad H \\ | \quad\ | \\ -C-C- \\ | \quad\ | \\ H \quad Cl \end{array}\right]_n$$

Vinyl chloride

Polyvinyl chloride

PVC is difficult to process due to its brittle nature and usually compounded with plasticizers. Plasticizers provide flexibility to polymer which makes it easier to process. PVC is insoluble in alcohol, water, and alkalies. It is commonly used to prepare pipes, cable insulations, laminates, adhesive coatings, and fibres.

Polytetrafluoroethylene (PTFE or Teflon)

PFTE has a structure similar to that of polyethene, with only difference that hydrogen atoms are replaced by fluorine atoms.

Preparation Chloroform on treatment with hydrofluoric acid gives difluorochloro methane which quickly forms tetrafluoroethylene monomer. On emulsion polymerization of tetrafluoroethylene in the presence of peroxide initiator, PTFE is obtained.

$$CHCl_3 + 2\, HF \longrightarrow CHClF_2 + 2\, HCl$$

Chloroform   Difluorochloro methane

$$2\, CHClF_2 \longrightarrow CF_2{=}CF_2 + 2\, HCl$$

Difluorochloro methane   Tetrafluoroethylene

$$n\left[CF_2{=}CF_2\right] \xrightarrow[\text{Initiator}]{\text{Peroxide}} \left(CF_2{-}CF_2\right)_n$$

Tetrafluoroethylene   PTFE or Teflon

Fig. 16.8 Preparation of PTFE

Properties PTFE is a linear polymer with about 95 % of crystallinity. It is practically insoluble in almost all solvents. It cannot be easily wetted with oil or water. It is chemically inert and cannot be attacked by acids, alkalies, and oxidizing and reducing agents. It is thermally stable and has good insulating properties.

Uses Due to its chemical inertness and non-wetting tendency, Teflon is commercially used to form non-stick coatings on frying pans. It can also be used as an insulator in motors, generators, transformers, and other electrical equipment. PTFE is also used as a lubricant (see Chapter 18 Lubricants).

Polymethyl Methacrylate (PMMA)

PMMA is an addition polymer obtained from methyl methacrylate and is a good substitute for glass, hence also called *plexiglass*.

Acetone and hydrogen cyanide undergo addition to form cyanohydrin which on hydrolysis gives methyl lactic acid. When methyl lactic acid is treated with concentrated sulphuric acid at 125 °C, methyl acrylic acid is formed. On esterification with acidified methanol, the monomer, methyl methacrylate is obtained. Methyl methacrylate, on addition polymerization in the presence of acetyl peroxide, gives polymethyl methacrylate.

Properties and Uses PMMA is a transparent, thermoplastic material. Due to its optical clarity and resistance to sunlight and gases, it is generally used as a substitute to glass. Due to excellent transparency, PMMA is used as aquarium windows. It is used to fabricate contact lenses, dentures, bone splints, and in cosmetic surgery. It is also used in acrylic paints and optical fibres.

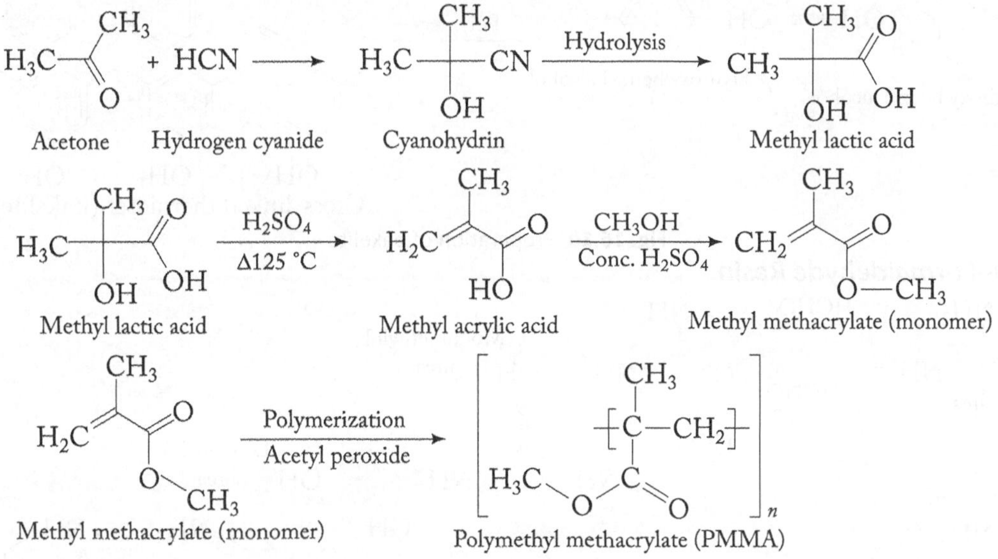

Fig. 16.9 Preparation of PMMA

Phenol–Formaldehyde Resin

One of the first synthetic polymers, it is the most versatile material for preparing various plastic products. It involves condensation polymerization of phenol with formaldehyde in the presence of an acid or base catalyst. With three sites on phenol and two sites on formaldehyde, the nature of products formed depends on the ratio of both reactants. With mole ratio of phenol–formaldehyde of 1:1, *o*- and *p*-isomers of hydroxyl benzyl alcohol are formed. If the mole ratio of phenol : formaldehyde is 1: 3, (2-hydroxybenzene-1,3,5 triyl)trimethanol is formed.

o-Hydroxy benzyl alcohol, on condensation, polymerizes to form linear, thermoplastic Novolac resin. If *o*- and *p*-hydroxyl benzyl alcohol are polymerized, a cross-linked, thermoset bakelite is formed.

Phenolic resins are good adhesive materials. They are readily soluble in organic solvents and exhibit lability towards alkalies due to free hydroxyl groups in the structure. They act as good adhesive materials. Bakelite is used to fabricate switches, toys, phones, etc.

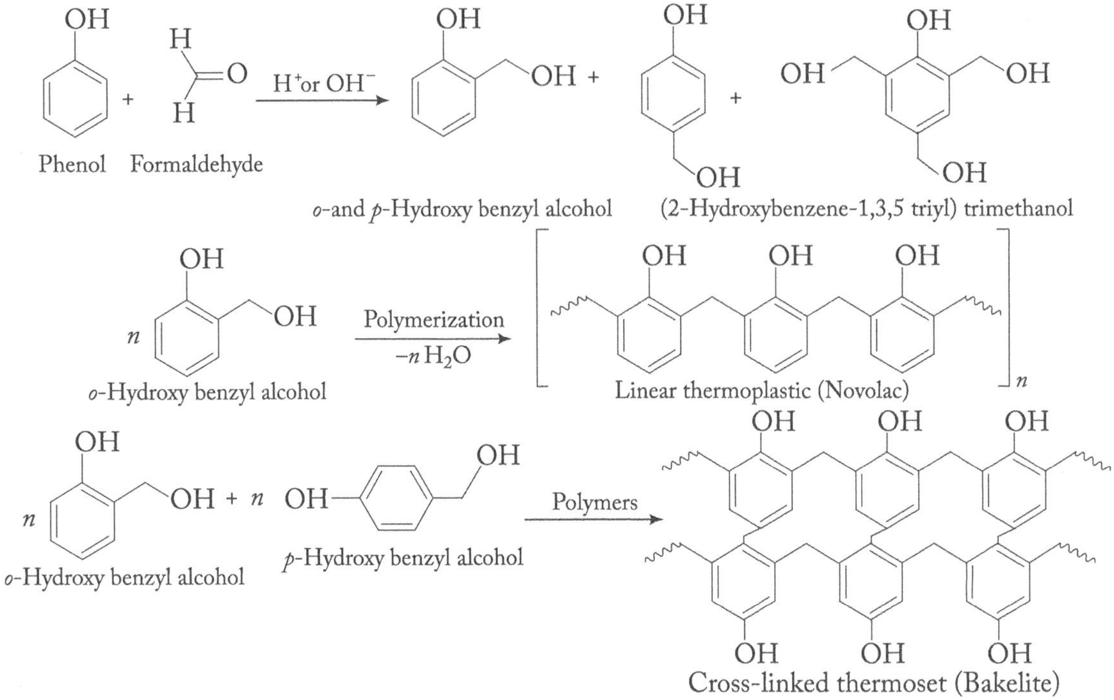

Fig. 16.10 Preparation of bakelite

Urea–Formaldehyde Resin

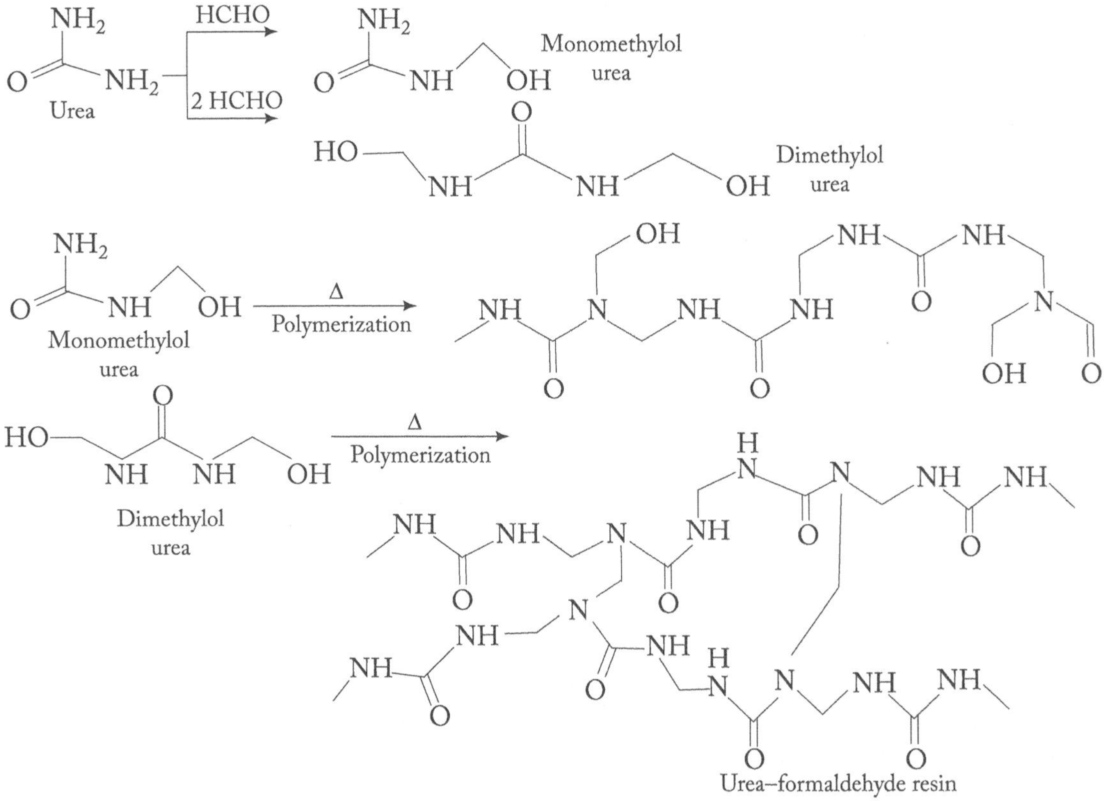

Fig. 16.11 Preparation of urea–formaldehyde resin

Urea on reaction with formaldehyde forms monomethylol and dimethylol urea. Monomethylol urea undergoes condensation polymerization to form a linear polymer as shown in Fig. 16.11. Dimethylol urea forms a hard, tough cross-linked thermosetting urea–formaldehyde resin. Clear-white plastic products are obtained from urea–formaldehyde resins. They exhibit good tensile strength, good chemical and abrasion resistance with inherent surface hardness. UF polymers are employed as cation exchanger resin, foam insulations, adhesives, decorative articles, and crockery items such as glasses and plates.

Melamine–Formaldehyde Resin

When melamine and formaldehyde undergo condensation, melamine–formaldehyde resin is formed.

Melamine–formaldehyde resins are water and chemical resistant. They possess good hardness and tensile strength. These resins are used as surface adhesives and lacquers, and in the manufacture of unbreakable crockery and durable cups.

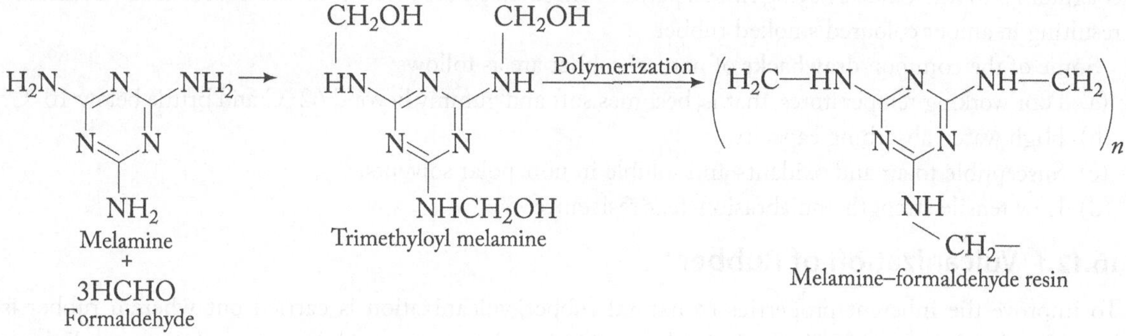

Melamine + 3HCHO Formaldehyde → Trimethyloyl melamine → Polymerization → Melamine–formaldehyde resin

16.12 ELASTOMERS

Rubber is a naturally occurring polymer that possesses remarkable elastic properties. It is also termed as 'elastomer,' as it tends to elongate at least thrice its original length under stress and regains the original shape when the stress is released. Natural rubber is obtained from the latex of rubber tree *Hevea brasiliensis* as a colloidal dispersion. Chemically, natural or raw rubber is a linear polymer of *cis*-isoprene units (or 2-methyl-1,3-butadiene). *cis*-Polyisoprene, as polymer chains, are held together by weak van der Waals forces that result in coiled elastic structure similar to a spring.

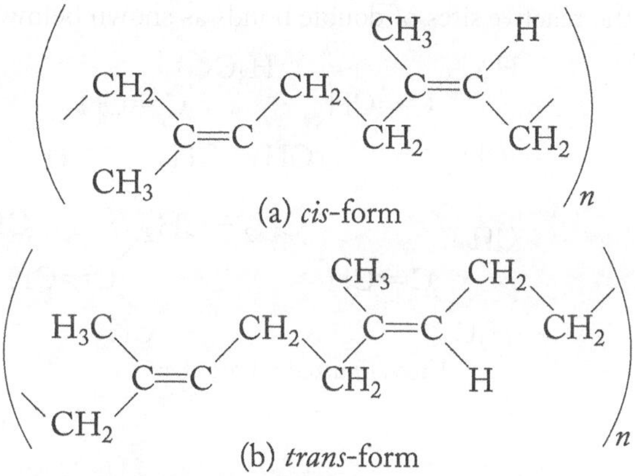

(a) *cis*-form

(b) *trans*-form

Fig. 16.12 Natural rubber-Gutta percha

Check Your Progress

12. Why is plasticizer added before moulding of plastics?
13. Distinguish between thermoplastics and thermoset polymers.
14. Write the chemical structures of thiokol, neoprene, Nylon 6,6, silicone rubber.
15. Write the preparation of Kevlar.
16. Justify the statement, 'PVC is soft and flexible whereas bakelite is a hard and brittle polymer.'
17. List the various plastic moulding techniques.
18. List the characteristic properties of plastics.
19. List the applications of plastics.
20. Name the various ingredients added during compounding of plastics.

Processing of natural rubber The latex contains 25–45 % rubber along with water, proteins, and other resinous material collected as a milky white colloidal emulsion. Small incisions are made on the bark of the rubber tree to collect the latex. After some time, the flow of latex decreases and to prolong the latex flow, slanting cuts are again made. This is called *tapping*. The collected latex is diluted to make 15 to 20 % rubber and filtered to eliminate dirt particles. Further, it is coagulated by adding acetic acid (about 1 kg for every 200 kg rubber). Coagulated rubber is washed, dried, and placed in smoke houses to prevent microbial attack.

Crepe rubber is obtained by adding trace quantities of sodium sulphite for bleaching the rubber, which is then rolled in to thin sheets of about 1 mm thickness. Thin sheets have even surfaces and look similar to a crepe paper that is dried at 50 °C in air.

Smoked rubber is obtained by removing the bleaching agent using sodium sulphite and rolling the coagulum into thick sheets having ribbed pattern. These sheets are dried in smoke houses at 50 °C, thereby resulting in amber coloured smoked rubber.

Some of the common **drawbacks** of natural rubber are as follows:
(a) Poor working temperatures, that is, becomes soft and gummy beyond 62 °C and brittle below 10 °C.
(b) High water-absorbing capacity.
(c) Susceptible to air and oxidants and soluble in non-polar solvents.
(d) Low tensile strength and abrasion-tear resistance.

16.12.1 Vulcanization of Rubber

To improve the inherent properties of natural rubber, vulcanization is carried out wherein rubber is heated with sulphur at 135 °C for 1 to 4 hours. During the reaction, sulphur atoms form cross-links at the reactive sites of double bonds as shown below.

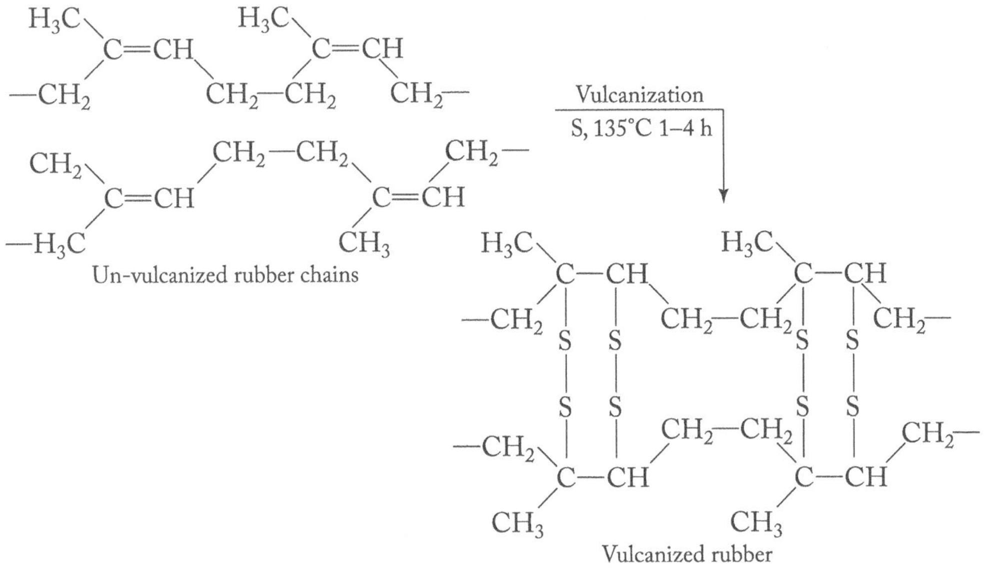

Un-vulcanized rubber chains

Vulcanization
S, 135°C 1–4 h

Vulcanized rubber

This process enhances the tensile strength and chemical resistance. About 5 % sulphur along with carbon filler is added to prepare car tyres. The toughest rubber called *ebonite* contains about 30 % sulphur and is used in laboratories to demonstrate static electricity. Ebonite is used to make electric plugs, fountain pen nibs, anti-corrosive lining of storage vessels, and musical instruments such as saxophones and clarinets.

Table 16.3 lists the properties of raw rubber and vulcanized rubber.

Table 16.3 Raw rubber vs vulcanized rubber

Property	Raw rubber	Vulcanized rubber
Elasticity	Very high	Low (depend on % S)
Tensile strength	200 kg/cm^2	2000 kg/cm^2
Resilience	Good	Very good
Working temperature	10 to 60 °C	−40 to 100 °C
Chemical resistance	Poor	High
Quality	Tacky, inherent	Controlled by vulcanization
Durability	Low	High

16.12.2 Compounding of Rubber

Before compounding, rubber is masticated by means of milling process. Natural rubber is a high molecular weight polymer thereby having high viscosity. It is practically impossible to mix any powder or liquid ingredients in to it. Mastication shortens the molecular chains of rubber and tends to impart plasticity. During milling process a mechanical equipment is used to shred raw rubber in to smaller mouldable units followed by compounding. The masticated rubber is mixed with the following compounding ingredients.

Vulcanizers   Sulphur, sulphur monochloride, hydrogen sulphide, and benzoyl chloride are some vulcanizing agents that are added to rubber to enhance tensile strength, etc.

Accelerators   These are positive catalysts employed in vulcanization so as to reduce the reaction time. Lime, magnesia, white lead (inorganics), thiocarbamates, benzothiazole, 2-mercaptol (organics) are added (0.5 – 1 % w.r.t rubber weight) while compounding.

Antioxidants   Raw rubber tends to perish due to oxidation by air and light. Phenolics, phosphites, phenyl naphthyl amine are added (about 1%) while compounding rubber to retard oxidation.

Reinforcing agents   Carbon black, ZnO, $CaCO_3$, $MgCO_3$, $BaSO_4$, etc. provide strength, rigidity, and toughness to rubber. Generally, 35 % of reinforcers are added to rubber, especially in automotive tyres.

Plasticizers   Waxes, vegetable oils, stearic acid, etc., are used as plasticizers that provide tenacity and good adhesion properties to rubber.

Fillers   Sawdust, chalk, and asbestos are commonly added fillers that lower the cost of the final finished product.

Colouring agents   Aesthetic appeal of the final rubber product is enhanced by adding colouring matter such as titanium dioxide (white), ultramarine (blue), ferric oxide (red), lead chromate (yellow), and so on.

16.12.3 Artificial Rubber

Styrene–Butadiene Rubber (SBR or GR-S or Buna-S or Cold Rubber)

$$n[CH_2\!\!=\!\!CH\!-\!CH\!\!=\!\!CH_2]_x + n \text{(Styrene)} \xrightarrow{\text{Na catalyst}} -\{-[CH_2\!-\!CH\!\!=\!\!CH\!-\!CH_2]_x\!-\!CH_2\!-\!CH\!-\}_n$$

1, 3-Butadiene   Styrene   Styrene–butadiene rubber (SBR)

When 1, 3-butadiene (75 % by weight) is copolymerized with styrene (25 % by weight), SBR is obtained in the presence of peroxides. It is also called cold rubber as the reaction is carried out at 0–5 °C. SBR has high abrasion resistance and good load-bearing capacity. It can be easily attacked by oxidizing agents such as oxygen and ozone. Adhesives, tank-liners, motor tyres, shoe soles, floor tiles, and wire and cable insulations are fabricated from SBR.

Buna-N Rubber (GR-N)

It is a copolymer of 1, 3-butadiene and acrylonitrile (ratio of 75 : 25 parts respectively).

$$x CH_2=CH-CH=CH_2 + y \ CH=CH_2 \xrightarrow{\text{Peroxides}} -(CH_2-CH=CH-CH_2)_x-(CH_2-\underset{\underset{CN}{|}}{CH})_y$$

1, 3-Butadiene $\underset{\underset{CH}{|}}{}$ Acrylonitrile   Buna-N

Properties Due to the presence of cyano (-CN) groups and hydrocarbon-like nature, Buna-N is highly resistant to water, oils, acids, and salts. However, they are easily attacked by alkalies. Buna-N possesses good tensile strength. It exhibits good resistance to abrasion, heat, and sunlight.

Uses Buna-N rubber is used as conveyor belts, aircraft components, tank-linings, hoses, gaskets, and automotive parts. It can be used as latex in textiles, paper, and leather. Buna-N can be used as an adhesive.

Neoprene

On polymerization of chloroprene in the presence of peroxides, polychloroprene, also called neoprene is obtained.

$$x \ CH_2=\underset{\underset{Cl}{|}}{C}-CH=CH_2 \xrightarrow{\text{Polymerization}} -(CH_2=\underset{\underset{Cl}{|}}{C}-CH=CH_2)_x-$$

Chloroprene   Neoprene

Properties Neoprene is highly resistant to vegetable and mineral oils. However, it is easily soluble in organic solvents.

Uses Neoprene is used to prepare laptop sleeves, orthopaedic braces, electrical insulation, conveyor belts, tubes carrying corrosive liquids, and adhesives.

Thiokol

It is also called polysulphide rubber (or GR-P) which is prepared by condensing ethylene dichloride and sodium polysulphide (Na_2S_4).

$$n Cl-CH_2-CH_2-Cl + n Na-\overset{\overset{S}{||}}{S}-\overset{\overset{S}{||}}{S}-Na \xrightarrow[-NaCl]{\Delta} \left[CH_2-CH_2-\overset{\overset{S}{||}}{S}-\overset{\overset{S}{||}}{S} \right]_n$$

Ethylene dichloride   Sodium polysulphide   Thiokol

Properties Thiokol rubber has excellent resistance towards organic solvents, oils, fuels, oxygen, ozone, and sunlight. It shows low permeability towards gases. These rubbers cannot be vulcanized and thus do not form hard rubbers.

Uses Thiokol rubber is used in rafts, balloons, life jackets which are inflated by carbon dioxide. It is also used to prepare conveyor belts, linings, hoses for conveying gasoline, gaskets, diaphragms, and printing rolls.

Polyurethane Rubber (Isocyanate Rubber)

$$[OH(CH_2)_2-OH + n O=C-N-(CH_2)_2-N-C=O \xrightarrow{\text{Polymerize}} -[O-(CH_2)_2-O-CO-NH-(CH_2)_2]-NH-CO]_n$$

Ethylene glycol   Ethylene disocyanate   Isocyanate rubber

When ethylene glycol reacts with ethylene di-isocyanate, polyurethane rubber is obtained. As the structure is highly saturated, it is resistant to oxidation and organic solvents. It is used to make foams, fibres, thermal insulations, surface coatings.

Silicone Rubbers (Silicones)

Silicones have alternate silicon–oxygen bonds and an organic radical attached to silicon atom. When silicon or silicon halide is treated with an alkyl halide or Grignard reagent respectively we get,

$$2\ CH_3 - Cl + Si \xrightarrow{Cu} CH_3SiCl_3 + (CH_3)_2SiCl_2$$

$$SiCl_4 + CH_3MgCl \rightarrow (CH_3)_3SiCl + (CH_3)_4Si \ \text{(organo silicon chlorides)}$$

The organo silicon chlorides mixture is separated by fractional distillation. Further, it is polymerized as follows.

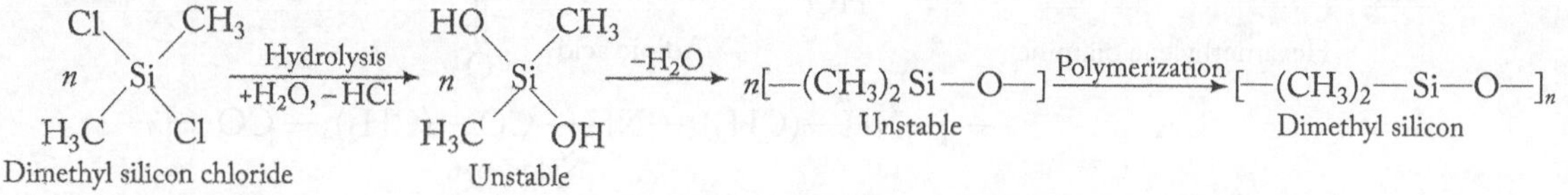

As dimethyl silicon chloride is bifunctional, long-chain silicone rubbers are obtained. If the starting material is trimethyl silicon chloride (mono functional), we get a shorter polymer chain as shown below.

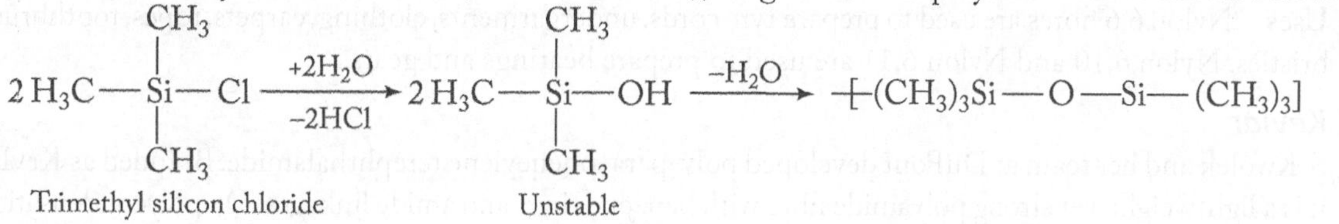

Monomethyl silicon chloride is trifunctional that yields a cross-linked polymer.

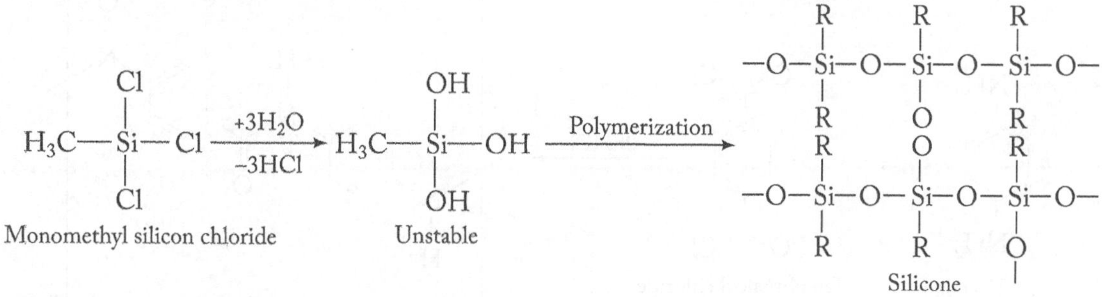

Properties   Silicone rubbers possess excellent resistance to acids, alkalies, oils, weathering, and sunlight. They exhibit greater flexibility within the temperature range of 90 – 250 °C.

Uses   Silicones are used in various products as automotive parts, sportswear, sealants, insulating wires, as adhesives, in fabricating artificial hearts, lungs, and implants.

16.12.4 Polyamides

Polyamides are polymers which are made of repeat units of amide (-NHCO-) linkages. In this section we will discuss the preparation, properties, and uses of nylon and Kevlar.

Nylon

Nylon is composed of synthetic polyamide fibres characterized by repeat units of amide linkages. When a dicarboxylic acid and diamine undergo condensation reaction, amide linkage is formed. Nylon 6, 6, Nylon 6, 10, Nylon 6, 11 are some examples of nylon fibres.

Check Your Progress

21. What is an elastomer?
22. What is vulcanization of rubber? Write the chemical reaction.
23. What is natural rubber? Write the structure.
24. State the repeat unit of natural rubber.
25. Name some vulcanizing agents added while compounding rubber.

Preparation   Nylon 6, 6 is prepared by condensing hexamethylene diamine and adipic acid under high pressure at temperature of about 279 °C.

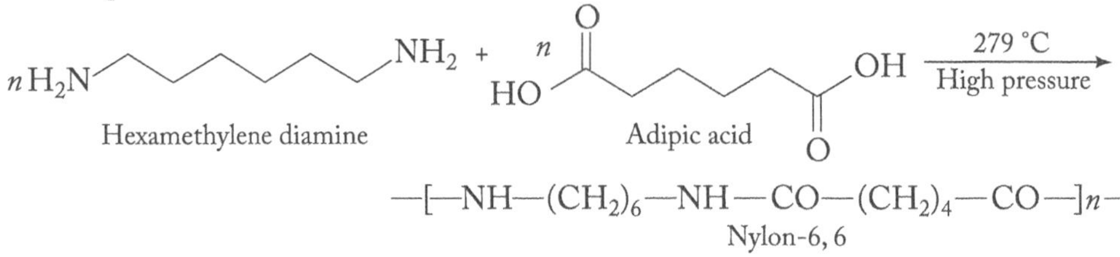

Hexamethylene diamine   Adipic acid

$$-[-NH-(CH_2)_6-NH-CO-(CH_2)_4-CO-]n-$$

Nylon-6, 6

Properties   Nylon fibres are translucent, flexible, white, high melting polymers with good abrasion resistance, insoluble in organic solvents, but soluble in formic acid.

Uses   Nylon 6,6 fibres are used to prepare tyre cords, undergarments, clothing, carpets, ropes, toothbrush bristles. Nylon 6,10 and Nylon 6,11 are used to prepare bearings and gears.

Kevlar

S. Kwolek and her team at DuPont developed poly-para-pheneylene terephthalamide. Branded as Kevlar, it is a lightweight, yet strong polyamide fibre with benzene rings and amide linkages. On polycondensation of terephthaloyl chloride (acid chloride of terephthalic acid) and *p*-phenylene diamine, Kevlar is obtained.

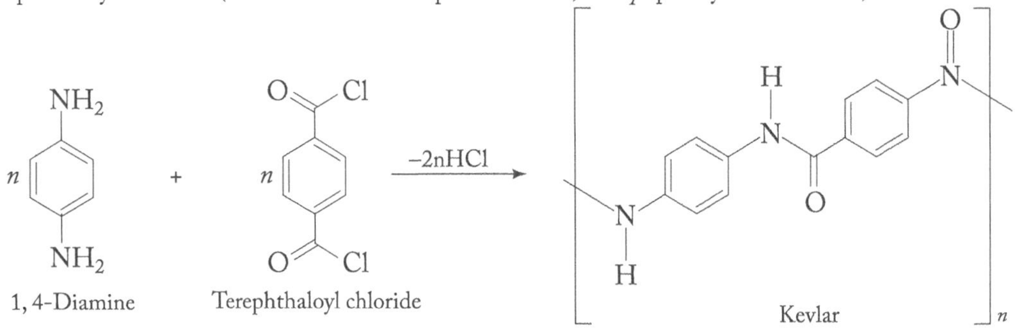

1, 4-Diamine   Terephthaloyl chloride   Kevlar

Properties   Kevlar is a lightweight, strong, thermally stable fibre. It is resistant to abrasion and has strong resilience even at – 196 °C. Due to absence of aliphatic units in the main chain, Kevlar possesses high thermo-oxidative stability. It is highly crystalline and forms fibres whose strength and modulus are higher than that of steel on an equal-weight basis.

Uses   Due to its low thermal conductivity, it is employed in cryogenics. Kevlar is widely used in making personal armour (face masks, bullet-proof vests, etc.), gloves, and also used as an alternative to Teflon in non-stick cooking pans.

16.12.5 Conducting Polymers

It is known that polymers are inherently insulating materials. The challenge to make it conducting on par to those of metals was achieved by Shirakawa et. al. (1978). The modifications in the organic polymer framework were first demonstrated by doping semi-conducting *trans* polyacetylene with iodine vapour. The uptake of iodine by the polymer exhibited marked conductivity of 10^3 S/cm. Later, a series of conducting polymers such as polypyrrole (PPy), polyaniline (Pan), polythiophene (PTh) were studied. To make conducting polymers, the following conditions need to be met.

   (i) The polymer must possess conjugated double bonds.

   (ii) Its structure can be disrupted by either removing electrons or by adding (doping) them.

   Conducting polymers generally possess band structure similar to semiconductors (Si) as depicted in Fig. 16.13. Such polymers, on doping exhibit dramatic conductivities and are hence called *zero-band gap conducting materials*.

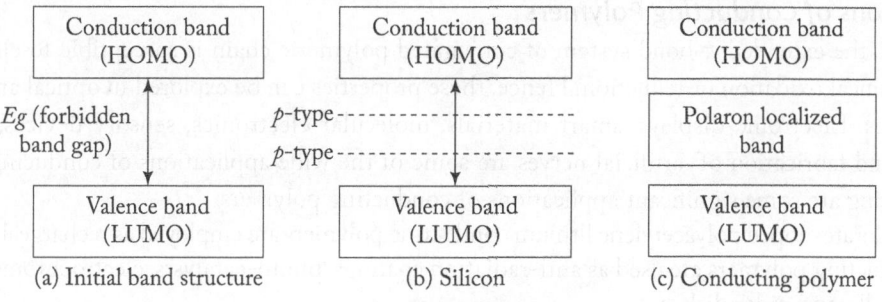

Conduction band (HOMO)	Conduction band (HOMO)	Conduction band (HOMO)
E_g (forbidden band gap)	p-type - - - - - - - - -	Polaron localized band
	p-type - - - - - - - - -	
Valence band (LUMO)	Valence band (LUMO)	Valence band (LUMO)
(a) Initial band structure	(b) Silicon	(c) Conducting polymer

Fig. 16.13 Band gaps

Polymers are made conducting in various ways and are of broadly three types, namely intrinsic, extrinsic, and coordination conducting polymers (Fig. 16.14).

Intrinsic conducting polymers ICPs have conjugated π electrons that are extended in the polymer chain. When an electric field is applied, π electrons get excited and move through the polymer chain thereby making it conducting. They are also called conjugated intrinsic conducting polymers.

Doped conducting polymers These are obtained by disturbing the polymeric structure by either adding or removing electrons.

(a) *p-Doping* can be best illustrated by treating polyacetylene with iodine (Fig. 16.15). When iodine molecule abstracts an electron from polyacetylene chain and becomes I^{3-}, the resulting polymer becomes positively charged as a radical cation. The lone electron removed from the double bond can now move freely across the polymer chain making it conducting solitons.

(b) *n-Doping* involves treating the conjugated polymer with a Lewis base (reduction). If polyacetylene is doped with sodium naphthalide it results in an n-doped conducting polymer.

Extrinsic conducting polymers As the name suggests, these are polymers whose conductivity is due to the addition of external agents.

(a) *Conducting element-filled polymer* is obtained by binding polymers with conducting materials like metal oxides, carbon, and metallic fibres.

(b) *Blended conducting polymer* is made by blending normal polymeric materials with a conducting polymer.

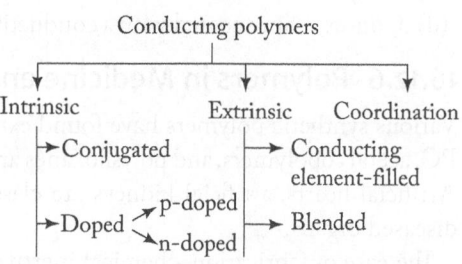

Fig. 16.14 Types of conducting polymers

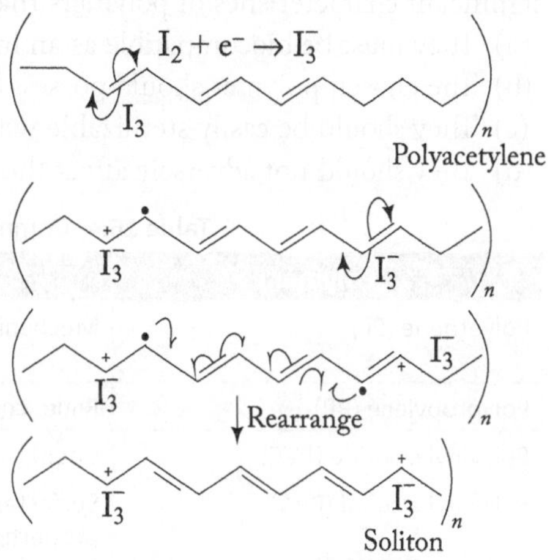

(a) *p*-Doping of polyacetylene

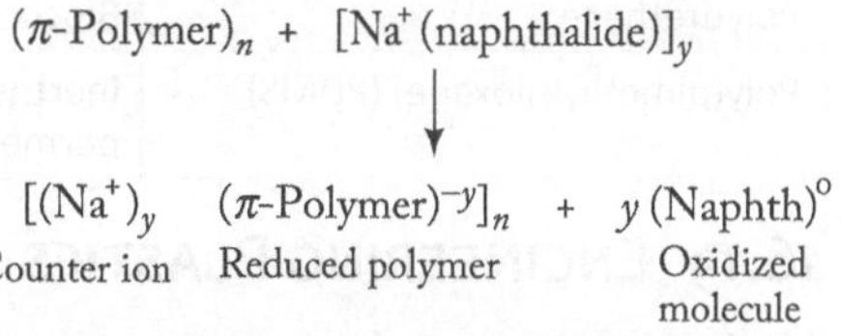

$(\pi\text{-Polymer})_n$ + $[Na^+(\text{naphthalide})]_y$

$\downarrow$

$[(Na^+)_y \quad (\pi\text{-Polymer})^{-y}]_n$ + $y(\text{Naphth})^0$

Counter ion   Reduced polymer   Oxidized molecule

(b) *n*-Doping of conjugated polymer

Fig. 16.15

Coordination conducting polymers These are generally charge-transfer complexes bound with a polymer. The presence of a coordination bond between the metal in the complex and the polymer allows movement of electrons, thereby making it a conducting material.

Applications of Conducting Polymers

As studied, the extended π-bond system of conjugated polymeric chain is susceptible to chemical and electrochemical oxidation or reduction. Hence, these properties can be explored in optical and electrical applications. Electronic displays, smart materials, molecular electronics, sensory devices, anti-static clothing, and fabrication of artificial nerves are some of the wide applications of conducing polymers. The following are some significant applications of conducting polymers.

(a) Perchlorate-doped polyacetylene lithium conducting polymers are employed as rechargeable batteries.
(b) Conducting polymers are used as anti-radiation coatings, photo-catalysts, electrochromic windows, fuel cells, and radar dishes.
(c) They are used in the manufacture of photovoltaic devices, for example, Al/conducting polymer/Au.
(d) Contex, a fibre coated with a conductive polymer, namely polypyrrole is used as anti-static in fabrics.

16.12.6 Polymers in Medicine and Surgery

Various synthetic polymers have found extensive applications in the biomedical field. PVC, PE, PP, PS, PC, acetal copolymers, and polysiloxanes are conventional polymers employed in biomaterial preparation. Artificial hearts, artificial kidneys are classic examples of artificial organs that are designed to replace diseased organs.

The ease of fabrication, chemical inertness, and lower toxicity make high polymers an ideal choice to fabricate surgical prosthetics. Some of the common polymeric biomaterials are listed in Table 16.4. The significant characteristics of polymers that make them suitable as biomaterial are as follows.

(a) They must be biocompatible as an implant in the human body.
(b) The chosen polymer should possess high purity and reproducibility.
(c) They should be easily sterilizable with no changes in their inherent properties.
(d) They should not adversely affect the bodily fluids such as blood, enzymes, and electrolytes.

Table 16.4 Common polymers in biomedical applications

Polymer	Key properties	Applications
Polyethene (PE)	Mechanical strength with lubricity	Disposable syringes, catheters, orthopaedic implants
Polypropylene (PP)	Rigid, chemical inertness	Heart walls, blood filters, sutures
Polyvinyl chloride (PVC)	Chemical inertness	Syringes, ampoules
Polyvinyl alcohol (PVA)	Surfactant and gel-forming properties	As emulsifiers for drug delivery
Polymethyl methacrylate (PMMA)	Excellent optical transparency	Contact lens, cosmetic surgery
Polyurethane	Biocompatible	Artificial heart, surgical gauze
Poly(dimethylsiloxane) (PDMS)	Inertness to body fluids, excellent oxygen permeability and optical transparency	Ocular lens, biomembranes

16.13 ENGINEERING PLASTICS

High-performance plastics or engineering plastics are a class of advanced polymer materials that have found wide applications in challenging areas such as defense, aerospace, automotives, and so on. They are generally fabricated from polymeric resins and possess advanced properties such as ease of fabrication, high mechanical strength, very high working temperatures in extreme conditions, dimensional stability, excellent abrasion, and chemical, gas, and degradation resistance. Today, high-performance plastics are replacing metals and alloys to fabricate various crucial structures. The civil structures made from these specialized plastics are low-cost, lightweight with easy fabrication process. Some commonly employed high-performance plastics are as follows:

Polyamides (eg., Kevlar, nylons) are easy to mould and recast, exhibit abrasion and chemical resistance, high service temperatures. They are applied in fabricating gears, automotive tyres, race cars, bullet-proof vests, aircraft panels, etc.

Polyurethanes (Perlon-U) are versatile polymers that are tough, flexible and show abrasion and chemical resistance. They are utilized in defense, oceanographic sensory studies, and other marine applications.

Polycarbonates offer high-impact resistance and durability. They are used in CDs, DVDs, sound walls, bullet-proof glass, capacitors, etc.

Polytetrafluoroethylene (PTFE or Teflon) is a thermoplastic, hydrophobic, non-reactive, heat-resistant material commonly used as non-stick coatings in cookware. Other uses include sports gear, air-filters, grease additive, and bullet coatings.

Polyacetals (polyoxymethylene or POM) are engineering thermoplastic materials that possess high rigidity and dimensional stability, but are labile towards aerial oxidation and acids. Various applications include appliance parts, electronic gears, bearings, plumbing fixtures.

Polysiloxanes (silicone) are inert, synthetic rubber-like, heat-resistant material commonly used as adhesives, gaskets, insulators, and surgical implants.

Hyperbranched conducting polymers (HBPs) are specialty performance polymers that have conducting properties and are employed in making photovoltaic devices.

16.14 SELF-HEALING POLYMERS

Self-healing polymers are smart materials that have the ability to repair themselves when they are damaged without the need for detection or repair by manual intervention. The concept of 'self-healing' is inspired from wound healing process in human beings, called biomimetic approach to extend the life span of engineering materials. In polymers, failure occurs through crack formation and propagation. The emerging self-healing technologies are designed to give polymeric materials the capability to arrest crack propagation at an early stage, thereby preventing catastrophic failures.

The healing process is accomplished by adding a microencapsulated healing agent like monomers and a catalyst or an initiator within the polymeric matrix. Endo-dicyclopentadiene (or endo-DCPD), as self-healing agent microencapsulated with urea–formaldehyde resin shell is the most common microcapsule model. Let us imagine a propagating crack through the polymer, as shown in Fig. 16.16, which will rupture the microcapsule and release the healing agent in to the crack plane by capillary action. This will initiate polymerization as soon as it comes in contact with the embedded catalyst or initiator. The bonding results in repairing the cracks and restoring the structural continuity. Sometimes, an external stimulus such as heat or UV radiations is required for complete healing to occur.

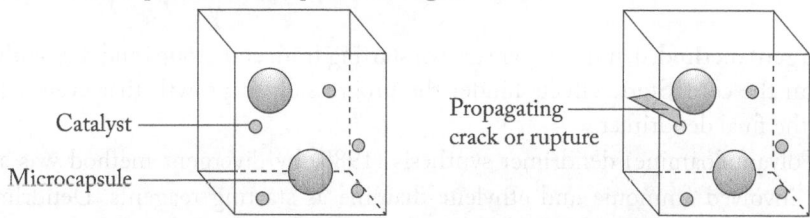

Catalyst

Microcapsule

Propagating crack or rupture

Fig. 16.16 Self-healing polymers by microencapsulation

Terminator polymer (T-100) is a classic example of self-healing polymer. Chemically, it is poly (urea-urethane) elastomeric matrix, a network of complex molecular interactions that will spontaneously cross-link to 'heal' almost any break. In this context, the word 'spontaneous' means that the material needs no outside intervention (catalyst or additional reactant) to begin the healing process. Experimentally, a sample cut in half with a razor blade at room temperature healed the cut with 97 % efficiency in two hours.

Applications

  (i) Nissan Motor Co. Ltd has commercialized world's first self-healing clear coat for car surfaces. The trade name of this product is 'Scratch Guard Coat.'
 (ii) In construction industry, self-healing concretes may soon become a reality.
(iii) Self-healing corrosion resistant coatings can be beneficial for structural metallic components.
 (iv) Self-healing materials are now used as composite materials in aircraft.

16.15 DENDRIMERS

Dendrimer (Greek: dendron = tree) is a monodisperse, hyperbranched, highly symmetric, spherical macromolecule. Fritz (1978) reported the first dendrimer synthesis that was followed by D. Tomalia and his group in 1983 and finally Newkome, in 1985 reported the preparation of high dendrimeric molecules called *arborols* (Latin: arborol = tree). A dendrimer consists of three parts, namely interior core, inner shell, and an outer shell of functional groups. One can prepare various dendrimers by chemically altering the functionalities in each part. Due to the presence of many chain-ends, dendrimers exhibit higher solubility or miscibility with marked chemical reactivity. Due to its spherical shape and presence of inner cavities, a dendrimer can encapsulate different molecules within its structure. Some of the characteristic properties of dendrimers are: nanoscopic size, globular or spheroidal shape, hyperbranched, hydrophobic core with hydrophilic exterior surfaces, non-crystalline, lower T_g, lower viscosity, higher solubilities, and low compressibility.

Dendrimers can be prepared by two methods, viz., divergent and convergent methods. In the **divergent method,** dendrimers are grown from core to the periphery. The core molecules react with monomer molecules with two dormant and one reactive group (Michael reaction). Every step of the reaction must proceed towards the outside, otherwise, trailing generations of dendrimers will be formed, that is, some branches will be shorter than the others. Such dendrimers will possess some imperfections and are difficult to purify.

In the **convergent method,** dendrimers are grown starting from end groups and the synthesis progresses inwards to form the core. Steric effects hinder the progression of growth that eventually dictates the overall size of the final dendrimer.

PAMAM (Polyamidoamine) dendrimer synthesis (1985) by divergent method was a pathbreaking discovery that involved ammonia and ethylene diamine as starting reagents. Dendrimers find wide applicability in medicine and drug development. They are used as MRI contrast agents, drug delivery agents, in anti-tumour therapy, and also as gene transfer reagents. Cadmium sulphide/polypropylenimine tetrahexacontamine dendrimer composite is used as fluorescent signal quencher. PEG (polyethylene glycol)-dendrimers are known to heal ocular injuries.

Dendrimers

16.16 BIODEGRADABLE POLYMERS

Polymers that get decomposed under aerobic or anaerobic conditions due to the action of microorganisms or enzymes are called biodegradable polymers. The higher chemical stability of polymers which seemed to be an advantage has turned out to be a major disposal problem. Plastic products, especially those used as packaging material, are generally recycled that can curb the problem to a certain extent. However, polymer-based materials get trashed after use and pose a huge threat to the environment. Estimates reveal that of the 300 million tonnes of plastics produced annually in the world, only 3 % is recycled. Since the 1980s, biodegradable polymers are being developed that disintegrate spontaneously after their intended use.

Biodegradation generally takes place through the action of enzymes or chemicals associated with living organisms. The first step involves fragmentation of polymers in to low molecular weight species either by oxidation, photolysis, hydrolysis, or enzymatic action of microorganisms.

For degradation of polymers, C–C bonds need to be broken. As these bonds are inert to enzymes, certain functional groups are added such as ester group (-COOR) to make them biodegradable.

Poly-hydroxybutyrate-co-β-hydroxy valerate (PHBV) is a biodegradable polymer widely used as packaging materials, orthopaedic devices, etc. It is prepared by copolymerizing 3-hydroxy butanoic acid and 3-hydroxy pentanoic acid.

$$CH_3CH(OH)CH_2COOH + CH_3CH_2CH(OH)CH_2COOH \longrightarrow$$

3-Hydroxy butanoic acid   3-Hydroxy pentanoic acid

Weak linkage

$$\left(\!\!\! \begin{array}{c} O - CH - CH_2 - C - O \\ | \\ R \qquad\qquad\; \| \\ \qquad\qquad\; O \end{array} \!\!\!\right)_{\!n}$$

(PHBV)

Nylon 2-nylon 6 is a polyamide prepared by the co-polymerization of glycine and amino caproic acid.

$$nH_2NCH_2COOH + nH_2N(CH_2)_5COOH \longrightarrow$$

Glycine   Amino caproic acid

$$\left(\!\!\! \begin{array}{c} \qquad\qquad O \qquad\qquad\qquad\quad O \\ \qquad\qquad \| \qquad\qquad\qquad\quad \| \\ NH - CH_2 - C - NH - (CH_2)_5 - C \end{array} \!\!\!\right)_{\!n}$$

(Nylon 2-nylon 6)

The ester linkage present in PHBV is attacked by enzymes and hence easily hydrolysed and degraded. Polyglycolic acid (PGA), polylactic acid (PLA), and polyhydroxy butyrate (PHB) are commonly employed as biodegradable polymers.

Since 2017, various countries have reported the biodegradability of certain polymers, especially oxo-biodegradable plastics resulting in their ban across various countries. The European Commission is assessing the impact of oxo-degradable plastics on the environment. It is found that oxo-fragmentable plastics with additives do not biodegrade spontaneously, but merely fragment into small pieces that end up in the environment. Due to high cost and other concerns, biodegradable plastics are yet to find large-scale applications.

Check Your Progress

26. How can one obtain a cross-linked silicone?
27. What are dendrimers?
28. If the glass transition temperature of polystyrene and nylon-6,6 are 100 °C and 45 °C respectively, then what will happen if these polymers are hit with a hammer at (a) 0 °C (b) 120 °C?
29. Name any one biodegradable polymer and write its preparation.
30. Why are polymers considered versatile materials?

Activity-based Questions

1. Electroactive Polymers

Electroactive polymers (EAPs) tend to exhibit changes in size or shape when stimulated by an electric field. These materials are commonly applied in sensors and actuators. Mechatronic devices and systems based on so-called 'electroactive polymers' represent a rapid-growing and promising scientific field of research and development.

Present the various types of EAPs in the field of electronics and explain their applications in biomedical science and tactile display technologies. State the technology trends and their challenges.

2. Fuels from plastic waste materials

The majority of the plastic products end up as non-degradable wastes polluting the environment. The basic premise is to melt the plastic wastes and convert them (thermally or catalytically) in to cleaner fuels. Various fuels obtained from the method can be utilized in furnaces, stoves, boilers, etc. If this method has the ability to scale up from grams to tonnes along with reduced costs and cleaner fuel, the current plastic-to-fuel approach is expected to reach a 1.9 billion dollar sector by 2024. Enumerate the various recycling methods for converting plastics in to fuels.

Highlight the various efforts taken so far. List the various merits and demerits of the method and comment on the challenges foreseen in the trend. (Refer to science direct in Further Reading)

SOLVED EXAMPLES

1. If the molecular weight of polythene is 49,000 and the molecular weight of its repeat unit is 49, determine the degree of polymerization of polymer.

Solution:

Given: Mol.wt (M) of polymer = 49,000 and mol. wt of repeat unit (monomer) is 49.

As we know, M = nMo, Hence, $n = \dfrac{M}{Mo} = \dfrac{49000}{49} = 1000$

Result: Degree of polymerization of the polymer is 1000.

2. A polymer sample contains 10, 20, 30, 40 molecules of polymers with molecular weights 10,000, 12,000, 14,000 and 16,000, respectively. Calculate the mole fraction of each type of polymer molecule and determine the number average and weight average molecular weights of the polymer sample.

Solution: Mole fraction, $x_i = \dfrac{\text{Number of molecules of type } i}{\text{Total number of molecules}}$

It is given that total number of polymer molecules = 100.

Hence, mole fractions (x_i) are 10/100 = 0.1; 20/100 = 0.2; 30/100 = 0.3, and 40/100 = 0.4.

$$\overline{M_n} = \sum x_i M_i = 0.1 \times 10000 + 0.2 \times 12000 + 0.3 \times 14000 + 0.4 \times 16000 = 14,000$$

$$\overline{M_w} = \frac{\sum x_i M_i^2}{\sum x_i M_i} = \frac{0.1(10000)^2 + 0.2(12000)^2 + 0.3(14000)^2 + 0.4(16000)^2}{14000}$$

$$= \frac{10^7 + 28.8 \times 10^6 + 58.8 \times 10^6 + 102.4 \times 10^6}{14000} = \frac{200 \times 10^6}{14000} = 14,285$$

Result: Hence number average and weight average molecular weights are 14,000 and 14,285 amu respectively.

3. What will be the maximum percentage of sulphur that can be added in vulcanized rubber?

Solution: Rubber has isoprene units and 2 isoprene units require two sulphur atoms for cross-linking. Mol. wt of isoprene is 68.

Thus, 2×68 g isoprene will need $= 2 \times 32$ g sulphur

We can say, 68 g isoprene needs = 32 g sulphur

$(68 + 32)$g of vulcanized rubber = 32 g of sulphur

Result: Maximum % sulphur in vulcanized rubber $= 32/100 \times 100 \% = 32 \%$.

4. A polymer sample contains:

Degree of polymerization	500	600	700	800	900
% S	15	25	15	25	20

Calculate the average degree of polymerization.

Solution:

$$(DP) \text{ average} = \frac{15 \times 500 + 25 \times 600 + 15 \times 700 + 25 \times 800 + 20 \times 900}{15 + 25 + 15 + 25 + 20} = 710$$

Result: The average degree of polymerization = 710.

5. If 216 g of 1, 3-butadiene is copolymerized with 104 g of styrene, predict the molecular formula of the copolymer.

Solution: 216 g of 1, 3-butadiene = 216/54 = 4 mol and 104 g of styrene = 104/104 = 1 mol.

Thus, molecular formula of the copolymer will be:

$$\left[CH_2 - CH = CH - CH_2 \right]_{4n} \left[\begin{array}{c} CH_2 - CH \\ | \\ C_6H_5 \end{array} \right]_n$$

6. If number average molecular weight and weight average molecular weight of a polymer sample A are 50,000 and 80,000 respectively, calculate polydispersity index of the polymer. Based on the value, predict the nature of polymer.

Solution: Polydispersity index $= \dfrac{\overline{M_w}}{\overline{M_n}} = \dfrac{80000}{50000} = 1.6$

Result: As the polydispersity index is greater than 1, it may have monomer units arranged in chains of different lengths.

SUMMARY

○ Polymers are high molecular weight macromolecules consisting of repeat monomeric units. They are of natural or synthetic origin and can be classified in numerous ways.

○ Polymerization is the fundamental process to obtain these giant macromolecules. Addition, condensation, and co-polymerization are different types of polymerization.

- Various methods of polymerization include bulk, solution, suspension, and emulsion techniques.
- Viscoelasticity, molecular weight, tacticity, T_g-T_m are characteristic properties of polymers.
- Elastomers or natural rubber is a linear polymer of *cis*-isoprene units and has many applications. Vulcanization is the process of heating rubber with sulphur to improve its inherent properties.
- Various agents such as vulcanizers, accelarators, antioxidants, fillers, and colouring agents are added to rubber during compounding.
- Examples of artificial rubber include Buna-N rubber, neoprene, Thiokol, and silicones.
- Polyamides are polymers made of repeated units linked by amide linkages; examples include Kevlar and nylon.
- Polymers are made conducting by adding dopants; they find application in the manufacture of photovoltaic and sensory devices, anti-static clothing, and fabrication of artificial nerves.

KEY FORMULAE

- Degree of crystallinity

$$f_c^m = \frac{\text{Mass of crystalline regions}}{\text{Total mass of polymer sample}}$$

$$\% C = \frac{\rho_c(\rho_s - \rho_a)}{\rho_s(\rho_c - \rho_a)}$$

- Degree of polymerization: $M = nM_0$
- Number average molecular weight: $\overline{M_n} = \sum x_i M_i$
- Weight average molecular weight: $\overline{M_w} = \dfrac{\sum x_i M_i^2}{\sum x_i M_i}$
- Polymer solubility: $\Delta H_m = v_s v_p (\delta_s - \delta_p)^2$

GLOSSARY

Bakelite: A condensation thermoset polymer made from phenol and formaldehyde.

Biomaterial: A natural or synthetic substance engineered to interact with biological systems in order to direct medical treatment.

Biocompatible: Compatible with human body; does not evoke harmful respnses.

Dendrimer: A monodisperse, hyperbranched, highly symmetric, spherical macromolecule.

Degree of polymerization: The average number of repeat units in a polymeric chain.

Ebonite: The hardest vulcanized rubber with 32 % sulphur.

Elastomer: A high molecular weight polymer that elongates on stretching and reverts back to its original shape.

Kevlar: A strong polyamide fibre with benzene rings and amide linkages.

Latex: The milky white colloidal emulsion obtained from the sap (inside tree bark) of rubber trees.

Plexiglass: A common term for polymethyl methacrylate.

Polydispersity index: The measure of molecular weight distribution, i.e., ratio of weight average molecular weight to number average molecular weight.

Polymer dissolution: The disentanglement of polymer chains due to diffusion of solvent into the polymer.

Rubber mastication: The softening of raw rubber by milling process before adding the compounding ingredients.

Self-healing polymers: Polymers that repair themselves after damage without manual intervention.

Tacticity: The orientation of monomer units with respect to main polymeric chain (stereochemistry).

Thermoplastics: Polymers that soften on heating and become rigid on cooling.

Thermosetting polymers: On heating, polymers form cross-links that are set and do not regain original shape on cooling.

Viscoelasticity: A material that behaves as both elastic solid and viscous liquid.

Vulcanization: Compounding rubber with chemicals like sulphur, to enhance the inherent properties.

EXERCISES

Multiple Choice Questions

1. When two different monomers undergo polymerization, it is called
   (a) co-polymerization
   (b) addition polymerization
   (c) condensation polymerization
   (d) none of the above

2. Novolac resin is prepared by condensation of
   (a) benzene and formaldehyde
   (b) benzene and acetaldehyde
   (c) phenol and formaldehyde
   (d) phenol and acrylic acid

3. The building block of polymers is called
   (a) thermo units   (b) monomers
   (c) functionality   (d) proteins

4. Buna-S rubber is a polymer of
   (a) styrene and butadiene
   (b) styrene and ethylene
   (c) only butadiene
   (d) only styrene

5. Which of the following is an example of synthetic polymer?
   (a) PET   (b) Protein
   (c) Cellulose   (d) Starch

6. Which of the following has the largest molecular mass?
   (a) Oligomer   (b) Monomer
   (c) Polymer   (d) None of these

7. The monomer in a natural rubber is
   (a) isoprene   (b) butadiene
   (c) vinyl alcohol   (d) phenols

8. Terminator polymer is an example of
   (a) conducting polymer
   (b) self-healing polymer
   (c) thermoplastic
   (d) thermoset polymer

9. What will happen if polystyrene which has T_g of 101 °C is hit on a hard surface at 25 °C?
   (a) Melt   (b) Expand
   (c) Shrink   (d) Break

10. Kevlar is prepared by condensing_____.
    (a) only *p*-amino aniline
    (b) terephthalic acid and aniline
    (c) only aniline
    (d) terephthalic acid chloride & *p*-amino aniline

11. Which of the following is the **wrong** statement?
    (a) Natural rubber has *trans*-configuration across all its double bonds.
    (b) Buna-S is a copolymer of butadiene and styrene.
    (c) Natural rubber is a 1, 4-polymer of isoprene.
    (d) Sulphur bridges in vulcanized rubber makes it harder and stronger.

12. Phenol–formaldehyde resin is commercially called
    (a) PE   (b) Elastomer
    (c) Nylon   (d) Bakelite

13. Monomers with double bonds undergo
    (a) condensation polymerization
    (b) copolymerization
    (c) addition polymerization
    (d) none of these

14. The polymer that exhibits non-wetting property towards oil and water is
    (a) nylon   (b) Teflon
    (c) polyethene   (d) PVC

15. An example of dendrimer is
    (a) PE   (b) PAMAM
    (c) PVC   (d) elastomer

16. PGA, PLA, and PHB are examples of
    (a) conducting polymers
    (b) dendrimers
    (c) self-healing polymers
    (d) biodegradable plastics

17. Which of the following is NOT a viscoelastic material?
    (a) Nylon   (b) Kevlar
    (c) Ethene   (d) Buna N rubber

18. Polymer that can exhibit conducting behaviour is
    (a) nylon   (b) polyaniline
    (c) protein   (d) elastomer

19. On condensing sodium polysulphide and ethylene dichloride, we get
    (a) neoprene   (b) Buna-N rubber
    (c) nylon   (d) thiokol

20. Buna-N is a copolymer of
    (a) butadiene and acrylonitrile
    (b) acyl nitrile and propene
    (c) chloroprene and butadiene
    (d) none of these

21. Polymer solubility depends on
    (a) molecular weight
    (b) crystallinity
    (c) size of polymer
    (d) all of these

22. Kevlar is the commercial name of
    (a) glass fibres
    (b) carbon fibres
    (c) aramid fibres
    (d) cermets

23. A unique property of polymeric materials is
    (a) elasticity
    (b) viscoelasticity
    (c) plasticity
    (d) tensile strength

24. The commercial polymer that contains sulphur is
    (a) thiokol
    (b) PMMA
    (c) PVC
    (d) polyurethane

25. The resins prepared on condensing formaldehyde, urea, and melamine are collectively called
    (a) epoxy resins
    (b) amino resins
    (c) phenolic resins
    (d) alkyl resins

26. Unbreakable crockery is made from
    (a) melamine
    (b) PMMA
    (c) PVC
    (d) silicone

27. Mastication of rubber means
    (a) its hardening
    (b) its softening
    (c) providing antioxidation
    (d) moulding

28. Which of the following is **NOT** a thermoplastic?
    (a) Urea-formaldehyde
    (b) PVC
    (c) PE
    (d) Polystyrene

29. Polymerized product of C_2F_4 is popularly called
    (a) neoprene
    (b) silicone rubber
    (c) Teflon
    (d) Polyurethane

30. Monomer of neoprene is
    (a) chloroprene
    (b) acetylene
    (c) isoprene
    (d) urea

31. The polymer used to make contact lenses is
    (a) PVC
    (b) PMMA
    (c) SBR
    (d) LDPE

32. Adhesives, polyacrylonitrile (PAN), and polyacrylic acid (PAA) are prepared by
    (a) bulk polymerization
    (b) suspension polymerization
    (c) solution polymerization
    (d) none of these

33. Of the following, the unique property of plastics is
    (a) chemical reactivity
    (b) rigidity
    (c) brittleness
    (d) flexibility

34. Artificial heart can be made from
    (a) polyurethane
    (b) polyacetal
    (c) polyethene
    (d) silicone rubber

35. Dendrimers were first reported by
    (a) Tomalia
    (b) Newkome
    (c) Einstein
    (d) Carothers

Review Questions

1. Explain the terms: (a) monomer (b) polymer (c) polymerization (d) degree of polymerization

2. How are polymers classified? Give examples.

3. Justify the statement, 'Polymers are versatile engineering materials.'

4. Distinguish between thermoplastic and thermosetting polymers.

5. What are natural and synthetic polymers? State two examples of each type.

6. What is functionality of a monomer? Give an example.

7. Distinguish between addition and condensation polymerization.

8. Explain 'copolymerization' with a suitable example.

9. Explain free radical polymerization with a suitable example.

10. List the advantages and limitations of emulsion polymerization.

11. Discuss bulk and solution polymerization methods. Illustrate with suitable diagrams.

12. List the various methods of polymerization. Explain suspension polymerization.

13. Why does one calculate the average molecular weight of a polymer? State various formulae to determine their molecular weights.

14. 'Weight average molecular weight is always higher than number average molecular weight.' Justify with suitable explanation.

15. Explain viscoelasticity of a polymer.

16. What is degree of crystallinity of a polymer? State the various factors that affect them.

17. Define degree of crystallinity. How is it calculated?

18. What are T_g and T_m of a polymer? State the various factors that affect T_g of a polymer.

19. Explain glass transition temperature. State its significance.

20. What is tacticity of a polymer? Explain various types of tacticity observed in polymers with suitable examples.

21. What is compounding of plastics? Name the various ingredients employed during compounding and state the role of each.

22. Describe compression moulding for fabrication of plastics.

23. Explain extrusion and injection moulding of plastics with suitable diagrams.

24. Explain the preparation, properties, and applications of: (a) Bakelite (b) PMMA (c) Kevlar (d) Silicone rubber (e) polyurethane (f) Buna-S (g) PVC (h) Melamine-formaldehyde (i) Urea-formaldehyde (j) Teflon. (k) PE.

25. Distinguish between LDPE and HDPE.

26. What is natural rubber? How is it obtained from latex?

27. What are the drawbacks of natural rubber? Explain vulcanization of rubber.

28. How can one make polymers conducting? Explain it with a suitable example.

29. Explain healing process in self-healing polymers.

30. List the various applications of self-healing polymers.

31. What are terminator polymers? State their applications.

32. Write short informative notes on:
    (a) High performance plastics
    (b) Surgical polymers
    (c) Dendrimers

33. Write a short note on 'solubility of polymers.'

34. Compare and contrast plastics and elastomers.

35. Define 'polymer dissolution' and state the various equations to determine solubility parameters of polymer.

36. Write a note on 'silicone rubber preparation and applications.'

37. Explain crystallinity of polymers.

38. What is latex? How does one obtain crepe rubber and smoking rubber?

39. Explain the preparation of Nylon-6,6 and give its properties and uses.

40. Explain the energetics of polymer solubility. Write the thermodynamic equations.

41. Discuss the preparation, properties, and uses of the following commercial rubbers:
    (a) Thiokol (b) polyurethane (c) Buna-N (d) Neoprene (e) Buna-S

42. Explain conducting polymers with a detailed account on doping processes.

43. List the various applications of conducting polymers.

44. Discuss the structure of dendrimer and its preparation methods.

45. How are dendrimers prepared by divergent and convergent methods?

46. Explain vulcanization of rubber. How is raw rubber different from vulcanized rubber?

47. Describe the preparation, properties, and uses of melamine–formaldehyde resin.

48. Discuss high-performance plastics? List the applications of polyacetals, polyurethane, and polysiloxanes as high-performing plastics.

49. What are biodegradable polymers? How can one degrade polymers? Cite an example.

50. Explain biopolymers with suitable examples.

NUMERICAL PROBLEMS

1. If the molecular weight of a polymer is 50,000 and the molecular weight of its repeat unit is 50, determine its degree of polymerization. (**Ans: 1000**)

2. A polymer sample contains 10, 20, 30, 50 molecules of polymers with molecular weights 10,000, 12,000, 14,000 and 16,000 respectively. Calculate the mole fraction of each type of polymer molecule and determine the number average and weight average molecular weights of the polymer sample.
$$\text{(Ans: } \overline{M_n} = 15600 \text{ amu; } \overline{M_w} = 14461 \text{ amu).}$$

3. What will be the maximum percentage of sulphur that can be added in vulcanized rubber? (**Ans: 32%**).

4. A polymer sample contains:

Degree of polymerization	500	600	700	800
% S	15	25	15	25

Calculate the average degree of polymerization. (**Ans: 662.5 amu**)

5. If the number average molecular weight and weight average molecular weight of a polymer sample are 30,000 and 60,000 respectively, calculate the polydispersity index of the polymer. (**Ans: 2**)

FURTHER READING

1. Kasapis, S., I.T. Norton, and J.B. Ubbink. *Modern Biopolymer Science*, Elsevier Inc, 2009.

2. Mohammad, F., *Specialty Polymers–Materials and Applications*, I.K. International Publishing House Pvt. Ltd, 2007.

3. Young, R.J. and P.A. Lovell, *Introduction to Polymers*, CRC Press, 2011.

4. Mascia, L., *Polymers in Industry from A-Z*, Wiley VCH, 2012.

5. Binder, W.H., *Self-Healing Polymers-From Principles to Applications*, Wiley VCH, 2013.

6. Scrosati, B., *Applications of Electroactive Polymers*, Springer, 1993.

7. Ueberreiter, K. 'The Solution Process'. *Diffusion in Polymers*, (ed: J. Crank and G.S. Park), Academic Press, New York 1968. p. 219–57.

8. http://ndeaa.jpl.nasa.gov/nasa-nde/lommas/aa-hp.htm

ANSWERS

Check Your Progress

1. Small molecules which combine to form polymers are called monomers. Polymers are high molecular weight substances consisting of numerous monomeric units.

2. Polymerization is the process of forming long-chain polymers. Addition, condensation, and copolymerization are the different types of polymerization.

3. Bakelite is a cross-linked thermosetting polymer made by condensing phenol and formaldehyde.

4. Addition polymers: Polythene, Polyvinyl chloride.
   Condensation polymers: Bakelite, Nylon – 6, 6

5. Homopolymer consists of the same monomers, whereas copolymers are made of different chemical structures.

6. The polymer **ABABABABABA** is a copolymer.

7. Bulk, suspension, solution, and emulsion polymerization methods.

8. Solvent enters the polymer molecule causing it to swell up as a gel and it further disintegrates the polymer so as to obtain it in solution phase.

9. A monomer is a simple molecule which possesses certain functionalities. Ethylene is bifunctional and thus can form polyethene. Organic molecules such as C_2H_6 and C_2H_5Cl are monofunctional and thus do not polymerize.

10. Refer to Section 16.5

11. Polydispersity index is the measure of molecular weight distribution; ratio of weight average molecular weight to number average molecular weight.

12. Plasticizer reduces the intermolecular forces of attraction between polymer chains. This increases the mobility of polymer chains as they slide over each other making it easier to process and mould.

13.

Thermoplastic polymer	Thermosetting polymer
Prepared by addition polymerization	Prepared by condensation polymerization.
Softens on heating and on cooling, it can regain original shape.	Softens on heating and once moulded cannot regain original shape.
Low molecular weight polymers, are usually linear in shape.	High molecular weight polymers with cross-links
Soft, weaker, less brittle	Hard, strong and quite brittle
Eg., PVC, PE, Polystyrene	Eg., Urea–formaldehyde resins, melamine–formaldehyde resin

14. Refer to Section 16.12

15. Refer to Section 16.12.4

16. PVC is a linear thermoplastic polymer and its polymer chains are held together by van der Waals forces of attraction. Thus, they are mouldable, soft, and flexible polymers. On the other hand, bakelite is a cross-linked thermoset polymer held by strong covalent bonds. Thus, it is hard and brittle in nature.

17. Various plastic moulding techniques include injection, compression, extrusion, and transfer moulding techniques.

18. Lightweight, chemical inertness, corrosion-resistance, easy to mould are some of the characteristic properties of plastics.

19. Plastics are used in preparing insulations, safety glasses, water-proof vessels, pipes, household materials like floors, wall-linings, water-exhange resins, adhesives, paints.

20. Resins, plasticizers, stabilizers, catalysts, colouring agents, lubricants, and fillers are compounding agents.

21. A polymeric substance that can be stretched at least thrice its length yet returns to original shape on releasing the stress. Rubber is a common example of elastomer.

22. Refer to Section 16.12.1

23. Refer to Section 16.12

24. Isoprene is the repeat unit of natural rubber.

25. Sulphur, sulphur monochloride, hydrogen sulphide, and benzoyl chloride are examples of vulcanizing agents.

26. Cross-linked silicones can be obtained by adding calculated quantities of trifunctional methyl silicon trichloride to condensing organosilicon chloride.

27. Monodisperse, hyperbranched, highly symmetric, spherical macromolecules are termed as dendrimers. It consists of three parts, interior core, inner shell, and an outer shell of functional groups.

28. (a) At 0°C, both polymers will shatter (below T_g); (b) Both polymers will not break (above their T_g).

29. Poly-hydroxybutyrate-Co-β-hydroxy valerate (PHBV) is widely used as a biodegradable polymer prepared by copolymerizing 3-hydroxy butanoic acid and 3-hydroxy pentanoic acid.

30. Due to flexibiltity, easy workability, chemical and corrosion resistance, impermeability towards water makes polymers a versatile material.

Multiple Choice Questions

1. (a)	2. (c)	3. (b)	4. (a)	5. (a)	6. (c)	7. (a)
8. (b)	9. (d)	10. (d)	11. (a)	12. (d)	13. (c)	14. (b)
15. (b)	16. (d)	17. (c)	18. (b)	19. (d)	20. (a)	21. (d)
22. (c)	23. (b)	24. (a)	25. (b)	26. (a)	27. (b)	28. (a)
29. (c)	30. (a)	31. (b)	32. (c)	33. (d)	34. (a)	35. (a)

Important Engineering Materials

After reading this chapter, you will be able to:

- list the various types of engineering materials such as cement, concrete, refractories, abrasives, adhesives, ceramics, glass, nanomaterials, liquid crystals, and composites.
- appreciate the expanse of engineering materials and their current trends in technological aspects.
- understand the characteristic properties of engineering materials.
- describe the manufacture of cement, concrete, refractory bricks, glass, nanomaterials, and composites.
- list the applications of important engineering materials.

17.1 INTRODUCTION

Irrespective of the engineering specialization, studying materials and understanding their chemistry is of much significance. Material sciences and engineering thrives on the concepts of chemistry of metals and non-metals. In the fields of civil, mechanical, and metallurgical engineering, inorganic and building materials such as cement, concrete, metals, semiconductors, refractories, abrasives, adhesives are of prime importance. The criterion for material selection has its genesis in the understanding of their chemical composition and properties. Computer engineers need a thorough understanding of nanomaterials and liquid crystal displays.

In this chapter, we will discuss the conventional engineering materials such as cement, concrete, adhesives, refractories, abrasives, glass and ceramics, and composite materials, as well contemporary materials such as nanomaterials and liquid crystals that have revolutionized the world of engineering.

17.2 CEMENT AND CONCRETE

An important engineering material, generally employed in construction, cement is known to have adhesive and cohesive properties that can bond materials such as bricks and stones. The principal constituents in cement are calcium (calcareous) and aluminium–silica (argillaceous) that react with water and result in setting and hardening and are referred to as *hydraulic cements.*

Cements are of four types, viz., natural, Puzzolana, Slag, and Portland. The various features of each type of cement are as follows.

Natural cement is prepared by calcination and crushing of naturally-occuring limestone at high temperatures. During calcination, calcium silicates and aluminates are produced. At times, sand is mixed with natural cement which can be commercially used.

Puzzolana cement is an age-old cement used to construct dams and other civil structures. Natural pozzolana and slake lime are mixed and ground together to obtain Puzzolana cement.

Slag cement is a mixture of lime and slag (obtained from blast furnace) and consists of calcium and aluminium silicates which are granulated and poured over cold-water streams. The mixture is dried, powdered, and used for various concrete-based constructions.

Portland cement is made by heating limestone and clay and grinding it in to fine powder and usually made in to a slurry in water to obtain hard rock-like material. It is used in the construction of various civil structures.

> The name 'Portland cement' originated from its similarity to Portland Stone that was found quarried on Isle of Portland in England. Joseph Aspdin obtained a patent for Portland cement in 1824.

17.3 PORTLAND CEMENT

Portland cement is a completely pulverized finely ground powder obtained by calcining a known proportion of a mixture of argillaceous and calcareous raw materials together at 1500 °C.

The following are the **raw materials** required for the manufacture of Portland cement.

(a) **Calcareous materials** such as lime, limestone, chalk, etc. having calcium oxide

(b) **Argillaceous materials** such as clay, shale, and slate, having alumina and silicon dioxide

(c) **Pulverized** coal and gypsum ($CaSO_4.2H_2O$)

17.3.1 Manufacture of Portland Cement

The manufacture of Portland cement involves the following three processes—mixing, burning, and grinding.

Mixing of Raw Materials

Mixing of raw materials is performed either by either dry or wet process.

Dry mixing process   In this process, limestone (or chalk) and clay (or shale) are crushed separately in gyratory crushers to obtain pieces of 2–5 cm size. The pulverized powder is further ground to fine powder in ball mills and the powdered limestone and clay are placed in separate hoppers. These pulverized materials are mixed in requisite proportions to obtain a 'dry-mix,' which is stored in silos and kept ready to be sent to rotary kilns. The proportion of raw materials used in the dry process is as follows:

(a) **Lime (CaO)** It is the principal constituent of cement (60–69 %) and its proportion is regulated because if there is any excess or lack of lime, it reduces the strength of cement causing disintegration.

(b) **Silica (SiO_2)** It is an important constituent (17–25 %) that imparts strength to cement.

(c) **Alumina (Al_2O_3)** It constitutes 3–8 % of cement and imparts strength to allow quick-setting. An excess of alumina can weaken the cement bonding.

(d) **Iron oxide (Fe_2O_3)** It constitutes about 2–4 % and imparts strength, colour, and hardness to the cement.

(e) **Magnesium oxide (MgO)** It constitutes about 1–5 % and any excess of it can cause disintegration of cement.

(f) **Sulphur trioxide (SO_3)** 1 – 3 % Sulphur trioxide, in trace amounts, is desirable as it imparts soundness of cement.

(g) **Alkali oxides ($Na_2O + K_2O$)** constitute only 0.3–1.5 %. An excess of alkalis results in effervescence of cement and may cause disintegration.

Wet mixing process   During the wet process, calcareous raw materials are powdered and stored in silos. Clay is thoroughly mixed with water to eliminate any organic matter and stored separately. Both these raw materials are then mixed in calculated proportions and sent to grinding mills, where about 40 % water is added to form the slurry. The corrected slurry is then sent to the rotary kiln.

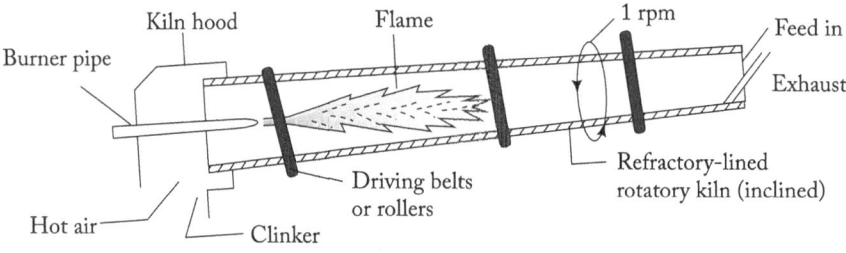

Fig. 17.1 Rotary cement kiln

Burning

The rotary kiln is a long steel tube of about 90–120 m and about 2.5–3 m in diameter. The inner lining of the kiln is made of refractory bricks. Rotary kiln is kept in an inclined position of 1 in 25 to 1 in 30 which is allowed to rest on roller bearings. The cement kiln is allowed to rotate at 1 rpm about its longitudinal axis. Coal and air are allowed to pass from the lower end of the kiln for obtaining long flame that heats up the inner kiln lining to about 1750 °C.

During the manufacture of cement in rotary kiln, the corrected slurry is passed through the upper part of the heated kiln called the *drying zone*. Due to the high temperatures (above 400 °C) most of the water in the slurry evaporates. The dry slurry glides down towards the hotter central zone (1000 °C) called the *calcination zone*, where it decomposes forming quicklime (CaO) and carbon dioxide. The CO_2 escapes leaving behind a lumpy mass, which reaches the lowest portion of the kiln, *clinkering zone*, maintained at 1750 °C where clay and lime undergo fusion resulting in the formation of calcium aluminates and silicates, that is, cement.

The chemical reactions are represented as:

$$2\,CaO + SiO_2 \rightarrow Ca_2SiO_4 \text{ (Dicalcium silicate, } C_2S)$$
$$3\,CaO + SiO_2 \rightarrow Ca_3SiO_5 \text{ (Tricalcium silicate, } C_3S)$$
$$3\,CaO + Al_2O_3 \rightarrow Ca_3Al_2O_6 \text{ (Tricalcium aluminate, } C_3A)$$
$$4\,CaO + Al_2O_3 + Fe_2O_3 \rightarrow Ca_4Al_2Fe_2O_{10} \text{ (Tetracalcium aluminoferrite, } C_4AF)$$

These aluminates and silicates fuse together forming small clinkers which are of about 1 cm diameter stones and are hard, grey in colour.

Grinding

The cooled clinkers are ground in ball mills with about 2–3 % powdered gypsum ($CaSO_4$), to prevent early setting of cement. Gypsum retards the dissolution of tricalcium aluminate by forming tricalcium sulphoaluminate, which is practically insoluble and prevents early setting and hardening of cement.

$$Ca_3Al_2O_6 + x\,CaSO_4.7H_2O \rightarrow 3\,CaO.Al_2O_3.x\,CaSO_4.7H_2O \text{ (Tricalcium sulphoaluminate)}$$

17.3.2 Chemical Composition of Portland Cement

The overall chemical composition of cement depicts quality of the material. If chemical constituents are in right proportions, setting and hardening of cement will be efficient and strong, leading to strengthening of the construction. Table 17.1 shows the average chemical composition of Portland cement. Of all the chemical constituents, tricalcium silicate is mainly responsible for rendering strength to cement. Its rate of hydration is moderate with heat of hydration 880 kJ/kg.

Table 17.1 Average chemical composition of Portland cement

Compound	Molecular formula	Average percentage (%)	Setting time
Tricalcium silicate	$3CaO.SiO_2$	45	7 days
Tricalcium aluminate	$3CaO.Al_2O_3$	1	1 day
Dicalcium silicate	$2CaO.SiO_2$	25	28 days
Tetracalcium aluminoferrite	$4CaO.Al_2O_3.Fe_2O_3$	9	1 day
Calcium sulphate	$CaSO_4$	5	–
Magnesium oxide	MgO	4	–
Calcium oxide	CaO	2	–

17.3.3 Setting and Hardening of Cement

When water is added to cement, it first forms a paste which then hardens; the process is referred to as setting and hardening of cement. Setting involves initial stiffening of cement paste due to gel formation, whereas hardening refers to strength development due to crystallization. Thus, the extent of gelation followed by crystallization dictates the strength of cement. The following chemical reactions occur during setting–hardening processes in cement:

Hydration of tricalcium aluminate and tetracalcium aluminoferrite gives hydrated crystalline products along with gel formation.

$$3\,CaO.Al_2O_3 + 6\,H_2O \rightarrow 3\,CaO.Al_2O_3.6\,H_2O \text{ (hydrated tricalcium aluminate)} + 880 \text{ kJ/kg}$$

$$4\,CaO.Al_2O_3.Fe_2O_3 + 7\,H_2O \rightarrow 3\,CaO.Al_2O_3 + 6\,H_2O \text{ (crystal)} + CaO.Fe_2O_3.\,H_2O \text{ (gel)} + 420 \text{ kJ/kg}$$

$$2\,[2CaO.SiO_2] + 4\,H_2O \rightarrow 3\,CaO.2\,SiO_2.6\,H_2O \text{ (Tobermonite gel)} + Ca(OH)_2 + 250 \text{ kJ/kg}$$

$$2\,[3CaO.SiO_2] + 6\,H_2O \rightarrow 3\,CaO.2\,SiO_2.3\,H_2O \text{ (Tobermonite gel)} + 3\,Ca(OH)_2 + 550 \text{ kJ/kg}$$

From the above exothermic reactions, it is evident that final setting and hardening of cement occurs due to the formation of tobermonite gel and crystallization of calcium hydroxide and hydrated tricalcium aluminate.

After initial hydration of tricalcium aluminate and tetracalcium aluminoferrite, hydration of tricalcium silicate is carried out within 24 hours and completed in 7 days. Further, the aluminate gel crystallizes out simultaneously and dicalcium silicate begins to hydrate that completes in 28 days. In a nutshell, we can say that strength enhancing takes place between 7 to 28 days due to hydration of dicalcium and tricalcium silicates.

17.3.4 Concrete and Reinforced Concrete

Concrete is an important building and structural construction material obtained by mixing binding materials (cement or lime), inert materials (sand, stones gravel, slag, bricks), and water in calculated proportions. Such materials can be molded in to any desired shape and set in a compact, rigid structure that is strong and durable. Concrete is composed of cement, sand, and coarse aggregate materials in the ratio of $1:1^{1/2}:3$ or 1:2:4, or 1:3:6, respectively.

Reinforced concrete (RCC) is concrete reinforced with steel rods and at times with heavy wire mesh. It is known that concrete has higher compression strength and lower tensile strength that tends to bear certain load, beyond which it may get crushed. Reinforcement with steel, and metal wires enhances the overall tensile strength of concrete. To prepare RCC, concrete is poured over steel or metal rods. On setting, concrete stiffens and hardens around the metal rods leading to reinforcements capable of withstanding compressive stresses and heavy loads.

Some of the advantages of RCC over ordinary concrete are listed as follows.

(a) RCC is simple to manufacture and can be cast into many desired shapes that can withstand various types of stresses and loads.

(b) It has greater rigidity, tensile strength, and fire resistance; and its maintenance is quite negligible.

(c) It can evenly distribute compressive and shrinkage stresses, thus cannot result in cracks and structure failure.

17.3.5 Decay of Concrete

Concrete is susceptible to chemical attack particularly by water, carbon dioxide, chlorides, and sulphur dioxide. Let us have a look at the susceptibility of concrete in the presence of various chemicals.

Water The pH of water plays a key role in determining the rate of decay of concrete. If water is acidic (pH < 7) the lime present in concrete will dissolve, thereby making it weak and less durable. Thus, while preparing slurry, it is essential to maintain a neutral or slightly alkaline pH of water.

Carbon dioxide The quality of concrete depends on carbon dioxide that usually enters through carbonation reaction. Carbonation reaction occurs between carbon dioxide and hydrated cement thereby causing shrinkage. Usually concrete pores contain alkaline water (pH~12.5) and when exposed to carbon dioxide (acidic oxide), neutralization reaction sets in between water and carbon dioxide within the concrete pores. This causes shrinkage and cracks develop affecting the structure. One can check the extent of carbonation of concrete using phenolphthalein indicator. Upon addition of a few drops to concrete, pink colour formation indicates that concrete water is alkaline and thus carbonation has not occurred (uncarbonated concrete is alkaline). Once neutralization occurs due to carbonation, pink colour will not develop denoting the presence of carbon dioxide. Thus, concrete porosity, cement content, water : cement ratio, degree of compactness dictate the strength of concrete and its subsequent deterioration.

Sulphur dioxide A serious form of deterioration of concrete occurs due to sulphur dioxide and its corresponding acids. Sulphur dioxide present in air combines with rain water to form sulphurous acid, which eventually comes in contact with concrete and causes corrosion of the material. On atmospheric oxidation of sulphurous acid, a more corrosive sulphuric acid is obtained, which deteriorates concrete, which if unchecked, may result in structural failure. Sulphates in water react with tricalcium aluminates present in concrete forming sulphoaluminates thereby resulting in the expansion of concrete. Due to this, cracks develop that reduce the life of concrete. Thus, to avoid decay of concrete, tricalcium aluminate is replaced by tetracalcium aluminoferrite in the cement.

Chlorides Chlorides react with calcium aluminate and calcium aluminoferrite present in the concrete forming insoluble chloroaluminates and chloroferrites, respectively, thereby resulting in its decay. However, if the reaction remains incomplete, traces of chloride in equilibrium with aqueous phase of concrete may result in corrosion of steel in reinforced concrete. Pitting and generalized form of corrosion are observed in reinforced concrete due to corrosive chloride content.

Catastrophic failures in construction and civil works result due to such corrosion problems. To prevent decay of concrete the following can be adopted:

(i) The choice of water is critical while preparing cement and concrete. Water should be devoid of chlorides so as to prevent decay of concrete.

(ii) One can protect concrete by applying coal coating such that there is little contact between water and concrete.

(iii) If one coats the surface of concrete with silicon fluoride in soluble form along with oxides of magnesium, zinc or aluminium, precipitate of calcium fluoride will be formed in the pores of concrete that seal them against chemical attacks and prevent lime dissolution.

17.3.6 Plaster of Paris

Plaster of Paris (POP) is a white, fine, odourless powder obtained on heating gypsum ($CaSO_4.2H_2O$) at 150 °C. Chemically, it is hemihydrate of calcium sulphate represented as $CaSO_4.1/2H_2O$. When gypsum, a naturally occurring mineral is heated, the following cementing materials are obtained as follows:

$$CaSO_4.2H_2O \xrightarrow[\Delta]{150°C} CaSO_4.\frac{1}{2}H_2O \xrightarrow{600°C} CaSO_4 \xrightarrow{800°C} CaO + SO_3$$

$$\text{Gypsum} \qquad\qquad \text{Plastics of paris} \qquad \text{Anhydrite}$$

Plaster of Paris on heating at 600 °C yields anhydrite, which further decomposes at 800 °C to form calcium oxide and sulphur trioxide.

Preparation   When Plaster of Paris is mixed with water, a plastic-type paste is obtained which quickly sets into a hard mass accompanied with slight expansion. This hardening of plaster occurs due to hydration of POP to gypsum. When POP is mixed with sufficient amount of water, a super saturated solution is obtained causing needle-like crystals of gypsum to form and interlock into a hard mass. The rate of hardening of plaster depends on the solubility of dehydrated products. If the more soluble portion of hemihydrate hardens and sets rapidly, then it cannot be used as a plastering material as one requires some time to shape the desired material. Hence, retarders such as glue, rosin, calcium stearate, boric acid are added in to the POP-water mixture.

Uses   Plaster of Paris is used as surgical bandages, plastering of building walls, false ceilings, art objects such as ceramic objects, pottery, and so on.

17.4 ADHESIVES

An adhesive is a non-metallic substance that sticks or bonds two dissimilar bodies together based on the molecular forces present in the area of contact. Glue, cement, mucilage, resins, beeswax are common examples of adhesive substances. The adhesion forces that bind the two surfaces together are due to surface adhesion and cohesion forces imparted by the inner stability of adhesive layer. It is primarily utilized for surface attachments. The dissimilar bodies which are bonded together using an adhesive are called *adherends*. The final assembly of two adherends along with adhesive is called *joint*. Figure 17.2 shows the assembly of two surfaces bonded using adhesives.

The quality of adhesive is determined by its tackiness (or stickiness), quick bonding, bonding strength post-drying, and overall durability.

Check Your Progress

1. Name the different types of cement.
2. Compare dry and wet processes of raw material mixing in cement manufacture.
3. What is Plaster of Paris?
4. What are the main chemical constituents in Portland cement?
5. State the functions of (a) lime (b) iron oxide (c) gypsum (d) alumina in cement.
6. Illustrate the setting and hardening of cement.
7. What is the main cause of final setting and hardening of cement?
8. What is concrete and reinforced concrete?
9. Name some chemicals that can decay concrete.
10. State any two measures to prevent concrete decay.
11. What is setting and hardening of cement?

The boundary layer refers to the zone between adherend surface and adhesive layer where strength of the bonding is highly effective. The space between the two surfaces of the adherend filled with an adhesive substance refers to the glueline. Apart from other joining technologies such as welding, brazing, and riveting, use of adhesives tends to bind materials together by distributing the stresses evenly across the bonded joint, resulting in an aesthetic material design. From an engineering perspective, adhesives and sealants are usually considered synonymous, because they both adhere and seal and also possess resistance towards operating conditions.

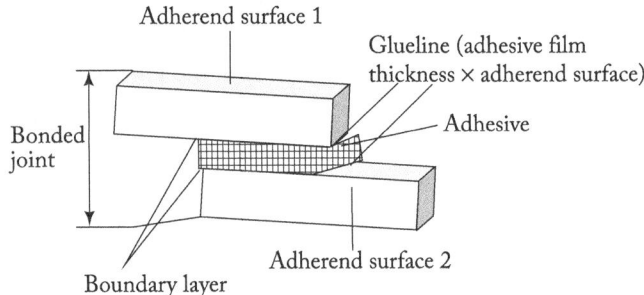

Fig. 17.2 A typical adhesion process assembly

17.4.1 Classification of Adhesives

Adhesives can be broadly classified in to two groups, namely structural and non-structural adhesives. **Structural adhesive** is a collective term for substances that possess high shear strength (1000 psi) and excellent environmental resistance, thereby rendering longer durability to the assembly on which it is applied. Various polymeric materials such as thermoset acrylics, epoxy resins, urethane-based polymers are examples of structural adhesives.

Non-structural adhesives have much lower strength and poor adhesion permanence. These adhesives are usually utilized for temporary binding of the surfaces and are found to be pressure sensitive. Wood glue, sealants, elastomers are examples of non-structural adhesives. Table 17.2 lists the various advantages and disadvantages of adhesive bonding.

Table 17.2 Advantage and disadvantages of adhesive bonding

Advantages	Disadvantages
Provides a larger stress-bearing area for bonding.	Surfaces must be completely clean before application of adhesive films.
Imparts good fatigue strength.	Longer curing time is required
Dampens vibration and shock absorbent.	Cannot be applied for higher operating temperatures.
Minimizes galvanic corrosion of dissimilar materials.	Higher pressures and heat treatment is required for adhesive action.
Smooth contour design is retained.	Fixtures are required.
Can seal joints thereby reducing stresses and corrosion.	Operating temperatures, environment, health, safety are some limiting factors.
Provides adequate strength-to-weight ratio.	Inspection of adhesive failure is difficult and requires training.

17.4.2 Mechanism of Adhesion

Several theories describe the mechanism of adhesion. Due to the complexity of the adhesive action, one cannot state a single theory of adhesive mechanism. In a nutshell, attachment of surfaces by adhesives is proposed to occur primarily due to intermolecular forces of attraction called *specific adhesion*. Adhesive substance fills the voids of joining adherend surfaces and holds the surfaces by interlocking, usually termed as *mechanical adhesion*. At times, material surfaces partially dissolve in the adhesive formulation or its solvent and bonding of the surface takes place which is termed as *fusion adhesion*. Some commonly accepted adhesive mechanisms are adsorption, mechanical interlocking, diffusion-electrostatic interaction, and weak-boundary layers.

Adsorption It is proposed that adhesion occurs due to the adsorption of adhesive molecules onto the surface and the resulting van der Waals forces of attraction. These intermolecular forces can be developed between the adherend surfaces only when adhesive makes an intimate molecular contact with them and is not separated for more than 5 Å distance. This is referred to as the 'wetting' process.

Wetting is determined by contact angle measurements governed by Young's equation, which is given as, $\gamma_{LV} \cos \theta = \gamma_{SV} - \gamma_{SL}$ where θ is the equilibrium contact angle, γ_{LV} denotes the surface tension of the fluid material in equilibrium with its vapour, γ_{SV} is interfacial tension of the solid material in equilibrium with fluid vapour, and γ_{SL} is the surface tension of the fluid material in equilibrium with its vapour. It can be summarized that for complete spontaneous wetting $\cos \theta > 1.0$.

Mechanical interlocking mechanism We know that even smooth material surfaces have irregularities in the form of peaks and asperities. Roughening of surfaces is known to increase the total surface area over which the adhesion forces can be developed. The roughness of adherend surface enhances the adhesive bonding by interlocking effect. The irregularities of the surface can interlock or even use of abrasives can generate greater roughness of joining adherends.

Electrostatic forces and diffusion Electrostatic forces can form at the adhesive–adherend interface and result in resistance and separation of joining surfaces. It was observed that electrical discharges were generated when the adhesive gets peeled out from the adherend, particularly in thin films of metal sputtered onto polymeric surfaces. Adhesion can also arise due to inter-diffusion of molecules in the adhesive and adherend surfaces.

Weak-boundary-layer mechanism Did you ever notice the way plasticized binder covers tend to stick together on your bookshelf? This can be explained by weak-boundary-layer concept proposed by Bikerman that describes failures of adhesive action. According to Bikerman, weak boundary layers occur on the adhesive or adherend if an impurity tends to concentrate near the bonding surface and form a poor attachment to the substrate. When failure occurs, it is the weak boundary layer that fails and not the adhesive–adherend interface.

17.4.3 Steps to Apply Adhesives

An adhesive joining technology deals with all aspects of material selection, joint design, and its production. The requirements of a successful adhesive application are enumerated as:
  (a) Substrate surface cleaning
  (b) Wetting of substrate surface (intimate contact of adhesive on the adherend)
  (c) Solidification of adhesive
  (d) Forming a joint structure in a given operating stresses
  (e) Designing of the joint
  (f) Selection and control of materials and manufacturing processes

Adherend preparation As the success of adhesive action relies on surface contact, adherend surfaces must be cleaned properly before application. Even a single molecular layer of contaminant can adversely affect the wetting of the surfaces. Metallic surfaces are machined so as to provide plain surfaces for proper joining using thin adhesive films. In some cases, even chemical treatment is performed to eliminate grease, dirt, or debris which improves specific adhesion due to removal of metallic oxides from the adherend surfaces. Plastic-type surfaces are sand-blasted so as to allow fresh surfaces to bond with adhesive substances.

Adhesive formulation An adhesive formulation usually comprises an adhesive base or binder—a major component that holds the materials together. For instance, epoxy adhesive has epoxy resin as the binder. A hardener is a substance added to an adhesive to promote the curing reactions. In some cases, a suitable catalyst is incorporated in adhesive formulation to hasten the reaction between the base and the hardener. To optimize spreading consistency of the formulation, organic solvents are added so as to disperse the adhesive. Diluents are added to an adhesive to reduce the concentration of the base material. Fillers are non-adhesive materials added in the formulation so as to improve the strength, permanence, thermal expansion, conductivity, shrinkage, viscosity, and resistance.

Adhesive film application   A uniform layer of adhesive coating is essential for the smooth and proper contour design of the adherend assembly. Liquid adhesives are usually coated by simple brushing technique. Dry solid adhesives can be coated as a thin layer at the interface by applying adequate pressures. Thermosetting and polymeric resins are commonly utilized as adhesive films on a variety of materials. Spray guns are also employed to apply adhesives in many cases for quick fixing of the material assembly.

Joining and setting of adhesive   The molten, cross-linked thermosetting adhesives dry quickly and impart excellent fatigue strength to the material assembly. After application of cross-linked polymeric resin in organic solvents, they are allowed to interact with air so as to volatilize the solvent. Air-drying results in setting and joining of the adherend surfaces without any void formation in adhesive film. When metal adhesives are to be coated, they are heated in an oven either before assembling the material surfaces or at the end of the process. This process is called as *pre-curing*. After setting, pressure is also applied for complete bonding of adhesives and is especially true for slow-setting adhesives such as thermoset polymeric resins followed by cooling and annealing.

Conditioning post-adhesive bonding   When adhesive layers are allowed to interact with adherend surfaces under high pressure and temperature, it needs to be cooled before bonding pressures are released, otherwise a weaker contact will result in adhesive failure.

Since the use of thermoplastic adhesives involves heating under pressure, adhesive film produced in this process must be cooled below a certain temperature, before bonding pressure is released, or else a weak bond will be formed. Cooling and proper conditioning inhibit formation of internal stresses of adhesive bond and avoids skinning or peeling of adhesive layers.

17.5 ABRASIVES

Abrasive is a hard inorganic substance used to grind away another substance. Abrasives are generally used for cutting, machining, grinding, or finishing a work-piece. Machining involves various processes where the work-piece is cut in desired shape and size by controlled material removal process such as turning, milling, or drilling processes. In *turning process*, the work-piece is rotated in a machine tool, thereby moving it against the abrasive. In *milling process*, the abrasive rotates so as to bring the cutting edges to bear against the work-piece, whereas in *drilling process* holes are produced by rotating cutter with cutting edges allowed to make contact with the work-piece. *Finishing process* is the final step of machining process that conditions, cleans, and smoothes out the entire work-piece giving final aesthetic appeal and design.

Check Your Progress

12. What is an adhesive?
13. What are structural and non-structural adhesives?
14. Define (a) adherend, (b) bonding, and (c) bond.
15. All glues are adhesives, but all adhesives are not glues. Justify.
16. Name the various theories of adhesive mechanism.

A material to act as an ideal abrasive must possess the following characteristics.

(a) The abrasive substance should be very hard.

(b) It must be tough and resistant to mechanical shock.

(c) It should possess high refractoriness and remain unaffected by heat treatment.

The capacity of an abrasive substance to rub other objects is termed as its *abrasive power*. Abrasive substances that can scratch or leave cutting marks on another object is termed as *hardness of abrasives*. There is a lack of an exact scale by which one can estimate the hardness of a given abrasive. However, using a reference substance, one can adjudge abrasive hardness. Mohs proposed a hardness scale for a series of substances that came to be known as 'Mohs scale of hardness.' According to Mohs scale defined by using 10 indicators, substances are arranged in the order of increasing hardness as shown in Table 17.3; a mineral with higher index scratches those below it. Mohs scale ranges from talc (phyllosilicate), one of the softest substances to diamond (carbon allotrope), which is the hardest natural substance. Hence, from the table we can assume that topaz will scratch calcite more easily than fluorite.

Table 17.3 Mohs scale of hardness

Abrasive substance	Chemical formula	Mohs number
Talc	$3MgO.4SiO_2.H_2O$	1
Gypsum	$CaSO_4.2H_2O$	2
Calcite	$CaCO_3$	3
Fluorite	CaF_2	4
Apatite	$CaF_2.3Ca_3(PO_4)_2$	5
Feldspar	$K_2O.Al_2O_3.6H_2O$	6
Quartz	SiO_2	7
Topaz	$AlF_3.SiO_2$	8
Corundum	Al_2O_3	9
Diamond	C (allotrope)	10

> Friedrich Mohs was a mineralogist who developed the Mohs scale of hardness based on the ability of one mineral to scratch another. If a mineral cannot be scratched, it can be considered hard such as the diamond.

17.5.1 Classification of Abrasives

Abrasives are broadly classified on the basis of their origin in two types, viz., natural and synthetic abrasives. Figure 17.3 depicts the general classification of abrasives.

Natural Abrasives

These are obtained from the earth's crust and can be siliceous or non-siliceous in their chemical composition. Diamond, graphite, corundum, quartz, emery, garnet are major natural abrasives.

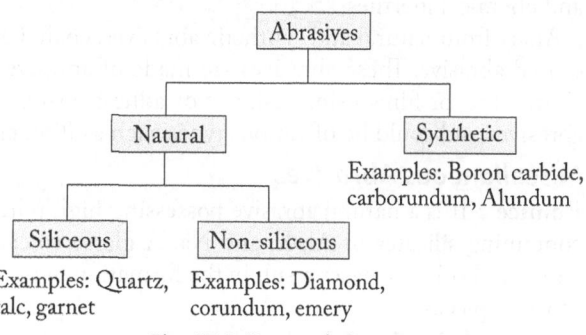

Fig. 17.3 Types of abrasives

Diamond It is a crystalline form of carbon and known to be the hardest material. Diamond does not get attacked by acids and alkalies. Carbonado (or black diamond) is a commonly used as an abrasive material since it is not used as jewellery. Diamond has a high scratching power and used in cutting tools, polishing gems, grinding wheels, etc.

Graphite An allotrope of carbon, graphite is used to manufacture crucibles, lead pencils, and lubricants.

Corundum It is a crystalline form of alumina and used in metal industry, grinding gems, glasses, etc.

Garnet It is a mixture of trisilicates of alumina, magnesia, and ferrous oxide with a Mohs hardness of 6.0 – 7.5. Garnet can be used for manufacturing abrasive paper, grinding glass, and polishing metals.

Quartz It is pure crystalline silica (SiO_2) with hardness of 7 on Mohs scale and employed in grinding pigments present in paints, grinding ores, sharpening stones, etc.

Emery   It is a fine-grained, black, opaque coloured abrasive substance composed of 55 – 75 % alumina, 20 – 40 % magnesite, and 12 % other minerals. With a hardness of 8 on Mohs scale it is used as tips of drilling and cutting tools and used to make emery cloth.

Synthetic Abrasives

These include abrasive materials that can be manufactured using various chemical processes. Silicon carbon (carborundum, SiC), boron carbide, boron nitride, alumina, and synthetic graphite are common examples of synthetic abrasives. In Chapter 15 Metals and Alloys, we have discussed silicon and boron carbide. Here, we will discuss the preparation of calcium carbide, alundum, and boron nitride.

Calcium carbide (CaC_2)   Coke (or anthracite) and quicklime (CaO) are mixed in an electric furnace and heated to about 2200 °C to obtain calcium carbide.

$$CaO + 3\,C \xrightarrow{\;2200\,°C\;} \underset{\text{Calcium carbide}}{CaC_2} + CO$$

Care is taken so that the ash content in coal is lower than 3%, otherwise carbide so obtained will be viscous, making it difficult to extract from the furnace.

Calcium carbide is obtained in the molten form and cooled, crushed, screened, and sieved to obtain a very fine powder for commercial purposes. Calcium carbide is black to reddish-black in colour and used as an abrasive material, reducing agent in metallurgical reactions, and drying agent in organic synthesis.

Alundum (Al_2O_3)   It is prepared by heating a mixture of calcined bauxite, coke, and iron at 4000 °C in an electric furnace, $4\,Al + 3\,O_2 \rightarrow 2\,Al_2O_3$.

Alundum is stable at higher operating temperatures and is brittle, tough, and hard (9 on Mohs scale). It is also found to resist acid attack. Alundum is used in grinding of steels and abrasive wheels.

Boron nitride (Borazon, BN)   When boron nitride (hexagonal close packed structure) is heated to about 1650 °C at pressure of 1 million psi, borazon is obtained (cubic crystal structure). Borazon is the second hardest material next to diamond possessing high thermal conductivity, excellent wear resistance, and chemical inertness.

Apart from natural and synthetic abrasives, coated abrasives are also prepared. Sandpaper is a form of coated abrasive. These abrasives are made of abrasive grits bonded to various substrates such as paper, cloth, fibre, or films using resinous or adhesive substances. Sandpapers are usually considered as sheet abrasives and could be of various types such as flint, emery, and garnet.

Miscellaneous Abrasives

Pumice   It is a natural abrasive possessing high porosity and is relatively soft. It is pale grey in colour containing silicates of Al, K, and Na. A glassy volcanic lava having a cellular structure caused due to steam and volcanic gases result in the formation of pumice. It is used in cleaning, polishing, and finishing of work-pieces.

Steel wool abrasive   As the name suggests, steel wool is used to smoothen wood and is quite similar to sandpaper.

Pumacite   It is a volcanic dust and used admixed with sand and clay for various abrasive processes such as a cleaning, polishing, and finishing

Pulpstone   While preparing paper, pulpstone is used for grinding wood and is a cheap abrasive substance.

Check Your Progress

17. What is an abrasive?
18. List the characterisitcs of a good abrasive material.
19. What is abrasive power?
20. How is hardness of an abrasive material determined?
21. Name some siliceous and non-siliceaous natural abrasives.
22. Name two synthetic abrasives.

17.6 REFRACTORIES

An engineering material utilized in metallurgical and industrial construction such as furnaces and ovens is called refractory material. Refractory materials are known to withstand extreme temperatures without any deformation, abrasion, and corrosion. Various linings of furnaces, kilns, crucibles, ovens, tanks, etc. are made of refractory materials as they exhibit excellent resistance from attack of molten matter, slag, sludges, and corrosive gases under high temperatures. Figure 17.4 shows the general classification of refractories.

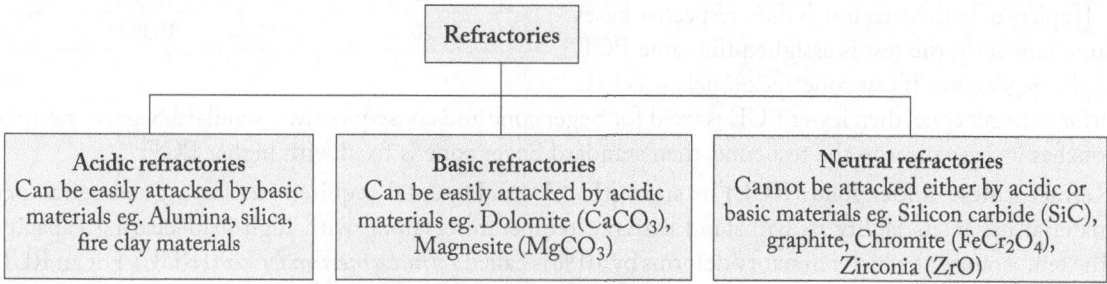

Fig. 17.4 Classification of refractories

The **characteristics** of an ideal refractory material are as follows.
(a) It should not be infusible or melt at any operating temperature.
(b) It should possess excellent chemical resistance towards all corrosive gases, chemicals, and molten matter.
(c) There should not be any deformation of refractory material under any operating load or temperature.
(d) During fluctuating operation conditions (temperatures), the expansion and contraction of refractory material should be uniform without any crack formation.

Applications of refractories The choice of a refractory material depends primarily on the reaction conditions. Zirconia, silicon carbide, graphite are excellent refractory materials usually used in the construction of tanks, furnaces, ovens, kilns, and crucibles. Refractory materials are also employed in the manufacture of glass, ceramics, and cement.

17.6.1 Properties of Refractory Materials

The following is a comprehensive account of various properties of refractory materials.

Refractoriness The ability of an engineering material to withstand heat without considerable deformation or softening is called refractoriness. As the primary constituents in a refractory are a mixture of various metallic oxides, they do not exhibit sharp fusion (melting) points. Hence, for practical purposes softening temperatures are considered to determine the refractoriness of a material. In an ideal refractory, softening temperature of a given material should be higher than the operating temperature of furnace in which it is used. Hence, the inner lining of furnace is usually maintained at a much higher temperature than the outer lining. Pyrometric cones (or Seger cone) test is performed to determine refractoriness of the material.

Pyrometric cone equivalent (PCE) is a number that represents softening temperature of refractory specimen of standard dimension (38 mm and 19 mm with a triangular base) and composition. Seger cone test is performed to determine the fusion

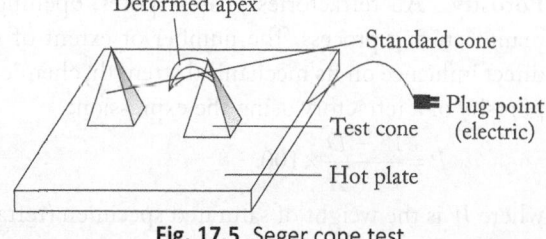

Fig. 17.5 Seger cone test

temperatures of refractory specimens, classify them, and also estimate their purity. While performing Seger cone test, Seger cones with pyramidal shape of uniform dimensions are placed on a hot plate and uniformally heated at the rate of 20 °C per hour. When the temperature rises, one has to observe the apex of the cone touching its base.

Table 17.4 Seger cone number (PCE) and fusion temperatures

Seger cone (PCE)	Fusion temperature (°C)
1	1110
5	1180
10	1300
15	1435
20	1530

If apices of both cones touch their respective bases simultaneously, the test is assigned the same PCE as the Seger cone. If test cone apex touches its base prior to Seger cone, then lesser PCE is fixed for Seger cone and consequently, if standard Seger cone apex touches its base prior to the test cone, then standard Seger cone is fixed with higher PCE.

Refractoriness-under-load (RUL) or strength A fundamental requirement for a good refractory material lies in its ability to withstand extreme temperatures along with high load-bearing capacity. The temperature at which refractory deforms by 10 % is called *refractoriness under load* (RUL). For an RUL test, refractory specimen (75 cm height and 5 cm^2 base) is applied with a standard load of 1.75 or 3.5 kg/cm^2. Refractory specimen is heated in a furnace at the rate of 10 °C per minute. Depending on the per cent deformation observed, refractory specimens can be classified into:

High heat duty refractories —do not deform more than 10 % at 1350 °C.

Intermediate heat duty refractories — do not deform more than 10 % at 1300 °C.

Low heat duty refractories —do not deform more than 10 % at 1110 °C.

Dimensional stability It is essential for a good refractory material to maintain dimensional stability during metallurgical operations; otherwise the unwanted reactions may occur resulting in collapse of the operation. The dimensional changes can be permanent contraction and permanent expansion.

(a) *Permanent contraction* occurs if a refractory material transforms from one crystalline form to another of lower density when exposed to high temperatures. For instance, silica exhibits such crystalline changes as follows:

Quartzite		Tridymite		Cristobalite
Crystalline form of silica	⟷	Orthorhombic (α) form of silica	⟷	Cubic (β) form of silica
(specific gravity = 2.65)		(specific gravity) = 2.26)		(specific gravity = 2.32)

(b) *Permanent expansion* occurs when the refractory is exposed to high operating temperatures for a prolonged period of time. Such expansion in volume of refractory material affects the transformation of crystalline forms in to one another of higher densities. For instance, magnesium oxide may transform to a denser crystalline form as follows.

MgO
Amorphous form
(specific gravity = 3.05)

Exposed to high temperatures
for prolonged time
⟶

Pericase
Crystalline form
(specific gravity = 3.54)

It also affects fusion of refractory material and causes shrinkage and vitrification.

Porosity All refractories possess pores, openings, or minute channels that are formed during the manufacturing process. The number or extent of refractory pores is an important parameter as it has direct influence on its mechanical strength, chemical stability, and abrasion resistance. One can calculate porosity of a refractory using the expression:

$$P = \frac{W - D}{W - A} \times 100$$

where W is the weight of saturated specimen (refractory), D is the weight of dry specimen, and A is the weight of saturated refractory specimen submerged in water.

If refractory material has greater porosity, gases and sludge may enter into the crevices and adversely affect the life of the material. Thus, a refractory material with low porosity is an ideal situation that can resist corrosion and shock.

Chemical inertness   A good refractory material must necessarily possess chemical inertness and should not form fusible compounds with ash content, furnace gases, sludges, and slag materials.

Thermal expansion   An ideal furnace design allows for thermal expansion of refractory as all solids expand on heating and contract on cooling. A good refractory material must possess least thermal expansion, since it adversely affects furnace design, and repeated expansion and contraction may cause wear of the refractory lining and overall furnace structure, making it prone to structure failure.

Thermal spalling   Cracking, peeling, or fracturing of a refractory material under high operating temperature is called thermal spalling. A good refractory must possess excellent resistance towards spalling. Spalling can be reduced by avoiding sudden furnace temperature changes and protecting the refractory from undue stresses.

17.6.2 Manufacture of Refractories

Refractory materials are ceramic-based chemically inert materials that must withstand very high temperatures, low thermal conductivities, low thermal expansion, and resist shock, abrasion, and corrosion. Thus, during the manufacture of such engineering materials, various steps are involved that allow all these properties. Figure 17.6 explains the various steps in the manufacture of tough and high-duty refractory materials.

17.6.3 Types of Refractory Brick Materials

Silica bricks   These contain 90–95% silica and is prepared by crushing siliceous rocks with about 2 % lime and water. The thick paste so obtained is made in to bricks and air dried and burnt in kiln followed by cooling. Silica bricks are yellowish coloured material with brown specks on them and have a specific gravity of 2.2–2.6. They possess a homogenous texture and has low porosity and good mechanical strength of about 3700 lbs/in^2. Silica bricks can withstand operating loads at about 1680 °C beyond which they may disintegrate. Silica bricks are used for buliding arches, side-walls, converter linings, open-hearth arch in furnaces due to their good mechanical strength.

Dolomite bricks   These are prepared by mixing dolomite (CaO + MgO) in definite proportion with silicate binder fired at about 1500 °C for a day and moulded and sent for cooling. Dolomite bricks are quite porous and soften easily. Usually, these bricks withstand high temperatures up to 1650 °C under a load of 3.5 kg/cm^2. Dolomite bricks are susceptible to corrosion by silica. Dolomite refractories are employed as a repairing aid rather than direct refractory due to their high porosity, shrinkage, and

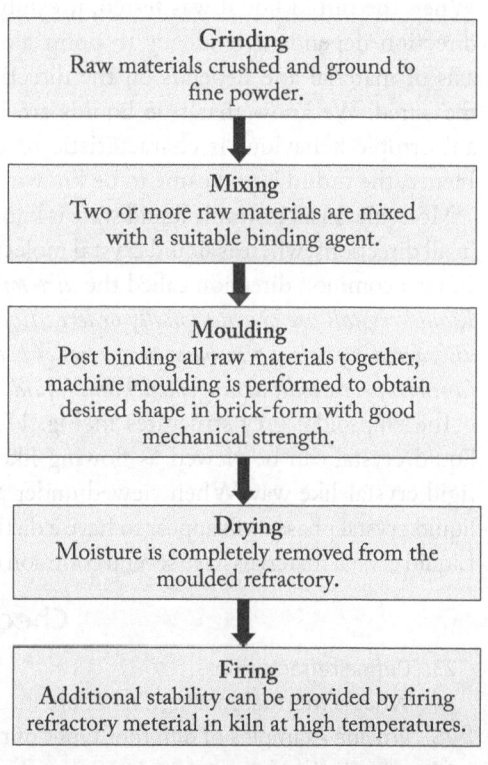

Fig. 17.6 Schematic representation of manufacture of refractory material

softness. Stabilized dolomite material could be used in electric arc furnaces, Bessemer convertors, etc., as cheaper alternative to magnesia bricks.

Silicon carbide bricks   These are hard covalently bonded refractory materials prepared in an electric furnace by heating silica sand and coke packed around electrodes at 2200 °C along with traces of sawdust filler and salt. Silicon carbide, so formed, as a greyish coloured fine powder is moulded in to bricks. It exhibits excellent resistance towards chemicals, thermal spalling, and possess high electrical conductivity. SiC bricks are routinely used as furnace rods, partition walls in cement kilns, coke ovens, and muffle furnaces.

17.7 LIQUID CRYSTALS

Liquid crystals are substances whose properties lie between those of solids and liquids. F. Reinitzer (1888) studied that cholesteryl benzoate had two distinct melting points. It was observed that on increasing the temperature, the solid crystal changes into a hazy-cloudy liquid at 145 °C and on further heating changed into a clear liquid at 178 °C. On cooling, these transformations were reversed and can be depicted as:

$$\underset{\text{Solid}}{p\text{-Cholesteryl benzoate}} \overset{145\,°\text{C}}{\rightleftharpoons} \underset{\substack{\text{Liquid crystal}\\ \text{(Mesomorphic state)}}}{p\text{-Cholesteryl benzoate}} \overset{178°\text{C}}{\rightleftharpoons} \underset{\text{Liquid}}{p\text{-Cholesteryl benzoate}}$$

The temperature at which p-cholesteryl benzoate transforms in to a liquid crystal is termed as *phase transition temperature* (or transformation temperature). The temperature at which it forms a true isotropic liquid is called *melting temperature*. When the turbid liquid was tested, it exhibited anisotropy, a direction-dependent tendency to point along one common axis of material and depends on the direction in which it is measured. We know that true liquids are isotropic whereas anisotropic behaviour is characteristic of crystalline solids. Hence, the turbid liquid came to be known as liquid crystals.

Molecules in an isotropic liquid state is known to move freely in all directions, whereas liquid crystal molecules tend to point along a common direction called the *director n* (see Fig. 17.7). *Liquid crystals are orientationally ordered liquids or positionally disordered crystals which have properties of both crystal structure (anisotropy, periodic arrangement) and liquid (molecular fluidity, coalescence of droplets) states.*

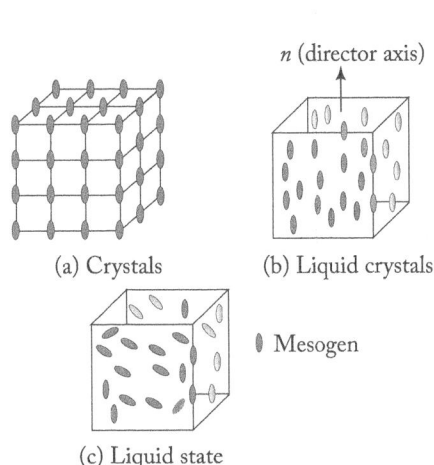

(a) Crystals   (b) Liquid crystals

(c) Liquid state

Fig. 17.7 Schematic representation of molecular packing

The ellipsoidal grey structures in Fig. 17.7 depict *mesogen*, a fundamental unit of liquid crystal. A liquid crystal can be viewed as flowing like a liquid, but have the molecules oriented in a somewhat rigid crystal-like way. When viewed under a microscope using a plane-polarized light source, different liquid crystal phases will appear to have a distinct texture which is exploited in various display technologies. Liquid crystal materials have several common characteristics such as possessing rod-like molecular structure;

Check Your Progress

23. Define refractoriness.
24. What is thermal spalling? What are its causes?
25. Provide examples of different types of refractories.
26. What is the test employed to determine refractoriness of a refractory material?

rigidity of director axis; strong dipoles; and ease of polarizability. Temperature is a measure of randomness of molecules and if one increases the temperature of liquid crystal environment, it will exhibit greater mobility or lesser order and if continued, there will be transition from solid to liquid form through the intermediate mesomorphic (or liquid crystal) state.

The orientation of molecules in an electric field and birefringence of molecules are interesting phenomena observed in liquid crystals. Liquid crystals are known to exhibit dielectric anisotropy and when placed in an electric field, they align in the direction of the field. Such a phenomenon can be utilized for preparing switchable liquid crystalline medium (i.e., switching between opaque and transparent states) by application of voltages across a liquid crystal cell. Liquid crystals have birefringence, that is, they have two indices of refraction due to their inherent anisotropic nature. When a medium possesses a refraction index which is direction-dependent, it is called *birefringence medium*. This phenomenon is of great importance as it dictates the propagation of polarized light across liquid crystal and consequently in liquid crystal displays.

17.7.1 Classification of Liquid Crystals

Figure 17.8 shows the broad classification of liquid crystalline substances based on their chemical properties.

Thermotropic liquid crystals These are long-chain organic solids that undergo sharp transformation to liquid crystals at a particular temperature. If higher temperature is applied, thermal motion disturbs the highly ordered solid crystal structure to an isotropic liquid. *p*-Cholesteryl benzoate is a thermotropic liquid crystal which at 145 °C transforms to a liquid crystal. Thermotropic liquid crystals find applications as temperature sensory devices and electro-optic displays.

```
                    Liquid crystals
                         |
            +------------+------------+
            |                         |
       Thermotropic              Lyotropic
            |
         → Nematic
         → Smectic
         → Cholesteric
```

Fig. 17.8 Classification of liquid crystals

Lyotropic liquid crystals These can be imagined as the phospholipid cell membranes. A typical lipid structure comprises a flexible hydrophobic chain (tail) and a polar hydrophilic head (ionic or non-ionic). Such molecules are called *amphiphilic molecules* due to their varying solvent interactions. When such long-chain organic molecules are dissolved in an isotropic solvent, lyotropic liquid crystals are obtained. The formation of such liquid crystals is temperature and concentration dependent, that is, it depends on the number of long chain molecules (mesogens) in a given solvent at a particular temperature.

A crucial condition for the stability of lyotropic liquid crystals lies in the solute–solvent interactions, that is, organic molecules in a polar solvent (usually water). Soapy water, phospholipid membranes, DNA, synthetic polypeptides are classic examples of lyotropic liquid crystals.

Smectic liquid crystals These have a layered structure and can possess varied molecular

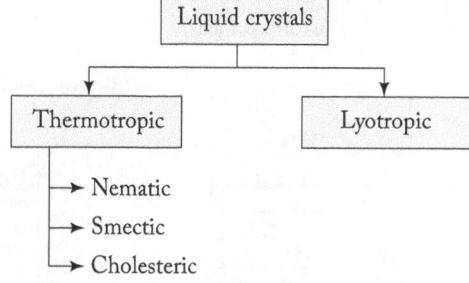

Hydrophobic tail

Hydrophilic head

The tail portion is usually an alkyl chain with 6-20 methylene (–CH_2–) groups

(a) Ionic

(b) Non-ionic

(c) Zwitterionic

where R = alkyl group

Fig. 17.9 Mesogens of lyotropic liquid crystals

arrangements within each layer. The inter layer attractions are weak as compared to the lateral forces of attraction between molecules. When a stress is applied on smectic liquid crystals, their layers slide past one another just like soap, while retaining their parallelism. Ethyl *p*-azoxy benzoate is one of the examples of smectic liquid crystals. A large number of different smectic phases are known which are characterized by varying orientational–positional orders.

As shown in Fig. 17.10, if a mesogen is oriented along the normal smectic layer, it is referred to as smectic A phase, whereas if it seems to be tilted away from the layer, it is called smectic C phase.

Examples of smectic liquid crystals: Ethyl *p*-ethoxybenzal cinnamate (77–118 °C), p'-n-octadecyloxy 3'-nitro diphenyl carboxylic acid (159 – 195 °C)

Nematic liquid crystals These are parallelly arranged thread-like mesogens lying along the director axis. These crystals have fluidity similar to true liquid; however, they are known to align themselves in the presence of electric or magnetic fields.

Compared to smectic phase, nematic liquid crystals exhibit greater fluidity and dielectric anisotropy. In smectic phase, molecules move past one another in layers whereas in nematic phase, mesogens seem to move parallel to each other. Some examples of nematic liquid crystals are *p*-azoxy anisole (117–137 °C), p-n-hexyl-p'-cyano biphenyl (14–28 °C).

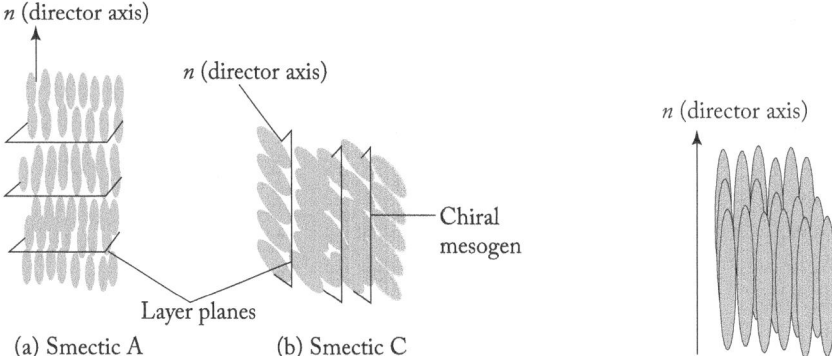

Fig. 17.10 Typical scheme of smectic phases (a) Smectic A (b) Smectic C

Fig. 17.11 Typical representation of nematic liquid crystal phase

Cholesteric liquid crystals These are observed in cholesterol derivatives which are optically active showing strong diverse colour effects in the presence of polarized light. The chiral molecules which do not possess inversion symmetry can result in cholesteric phases. In other words, cholesteric liquid crystals are chiral variants of nematic phases.

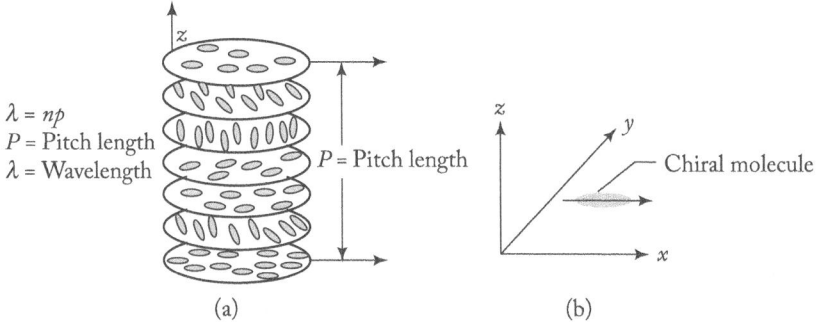

Fig. 17.12 (a) Helical structure of the cholesteric phase and, (b) director lies in the *xy* plane, perpendicular to the direction of helix (z), and rotates in a plane that defines helical structure

Cholesteric liquid crystals can be viewed as a molecule that is helically twisted around a director axis and the twist may be either left-handed or right-handed depending on molecular conformation. Due to the presence of helical twists, these phases exhibit unique optical properties. If plane-polarized light interacts with such chiral cholesteric phases, its plane of polarization is rotated along the direction of the helix. When pitch of the helix corresponds to a wavelength in visible region (~400–800 nm) of electromagnetic spectrum, these phases show textures and colours.

17.7.2 Liquid Crystalline Behaviour and Chemical Structure

The molecular structure plays an important role in determining the phases, phase changes, transition temperatures, and optical and electronic properties of liquid crystals. They usually possess weaker intermolecular forces of attraction, which on application of electric field shows various colours, textures, and patterns. The general structure of a liquid crystal is

$$R\text{---}\langle A \rangle\text{---}Z\text{---}\langle B \rangle\text{---}X$$

where R is a side chain group, A and B are the main aromatic rings of the compound, Z is the linking group, and X is the terminal group. The side chains (R) can be alkyl, alkoxy, or alkenyl groups. The length and flexibility of side chain affect the phase transition temperature and the type of liquid crystal phase. If the number of carbon atoms is about 3–7, nematic phase occurs. Smectic phase appears when the number of carbon atoms in R is more than 7. The aromatic rings A and B may be same or different and if there are any polar substituents such as – Cl, – F, – CN a drastic change is observed in the dielectric properties of liquid crystals. The linking group (Z) influences the phase transition temperature and other physical properties of a liquid crystal. Esters (–COO–), ethylene (–CH$_2$–CH$_2$–), azo (–N=N–), etc., act as linking groups which help in delocalization across the chemical structure and electronic transitions take place at longer wavelength.

Table 17.5 Common liquid crystals with their phase transition temperatures

Compound	Transition temperature (°C)	Melting temperature (°C)
p-Azoxy anisole	116	135
p-Azoxy phenetole	137	167
p-Methoxy cinnamic acid	170	186
p-Cholesteryl benzoate	145	178
Anisaldizine	165	180
Dibenzal benzidine	234	260

The terminal group (X) contributes to dielectric anisotropy and could be –CN, –OR, –R, –CNO, –CF$_3$, –Cl, also referred to as the electron-withdrawing groups. This results in overall increase in dielectric anisotropic properties of a liquid crystal.

17.7.3 Applications of Liquid Crystals

Display technology Due to their fluidic nature, liquid crystals can be prepared as thin films along with retaining their optical properties. Further, the orientation of the molecules in liquid crystal films can be modulated on a relatively short time scale using a low electric field. The orientation of nematic liquid crystal can be changed by electric field or pressure changes causing different orientations and varying colours. When an electric field is applied on a thin LC film with the help of electrodes, the patterns of molecules become visible. This principle is used in LCD (liquid crystal display) in today's television sets, calculators, optical displays, computers, and mobile screens.

Temperature sensors Cholesteric liquid crystals can reflect light having a wavelength proportional to the magnitude of pitch. Since pitch is temperature-dependent, the corresponding colours formed are also dependent on temperature. Thus, cholesteric liquid crystals are used as fever strips, for detecting tumours in human body and in disposable thermometers. Liquid crystal colour transitions are used on many aquarium and pool thermometers.

Other applications   Liquid crystals are used as solvents for spectroscopic study of anisotropic solids. Liquid crystalline solvents are employed to vary the reaction rates of molecular thermal and photochemical reactions. Liquid crystals tend to change colour on being stretched or stressed, thus can be used to locate hotspots, map heat flow, and measure stress distribution patterns. Cholesteric LCDs are utilized during chiral recognition studies. Lyotropic LCDs are exploited for application in commercial detergents, cosmetics, and in simulation studies of biomembranes. These specialized engineering materials have found application as a solvent medium for controlled drug release in pharmaceutics.

17.8 NANOMATERIALS

When Richard Feynman (1959) gave his popular lecture titled 'There's Plenty of Room at the Bottom,' at the annual meeting of the American Physical Society, it opened up a whole new field of science, known as 'nanotechnology'. He detailed about manipulating and controlling things on a small scale. In this section, we will discuss various nanomaterials that have been conceptualized till date and their far-reaching applications in medicine, surgery, environment, catalysis, and other fields of science and engineering. Nanotechnology is the fabrication of nanostructures, functional materials, devices, and components through control of matter on the nanometre scale. Coming under the branch of material sciences, nanotechnology is the study of nanomaterials. *Nanomaterials are defined as substances or materials whose at least one dimension is less than 100 nm.* We know that 1 nanometre is equal to 10^{-9} metre, that is, one-billionth of a metre which can be imagined as about 10,000 times smaller than the diameter of human hair. Usually, one can picture nanomaterials as composed of grains which may or may not be visible to the naked eye. Thin films, nanoparticles, nanotubes, nanorods, nanocones, nanodots, nanocomposites, and nanoelectronics are examples of nanomaterials which are currently studied in the field of nanotechnology. All these materials, being nanomaterials, satisfy the criterion that they have structured components with dimensions in the range of 10–100 nm. Nanoscience involves the manipulation of materials on atomic, molecular, or macromolecular scales. Nanotechnology is used to manufacture, design, and also alter the properties of these materials on the nanoscale.

These sophisticated materials are of particular interest because at this scale unique optical, magnetic, electrical, and other properties emerge.

Check Your Progress

27. What are liquid crystals? Give a broad classification of liquid crystals.
28. How do liquid crystals differ from isotropic liquids?
29. State the importance of birefringence in liquid crystals.
30. What is the major difference between nematic and cholesteric liquid crystal phases?
31. What is the fundamental unit of a liquid crystal?
32. Which type of liquid crystal shows helical twists?
33. What are thermotropic liquid crystals?
34. How are lyotropic liquid crystals obtained?
35. Give some examples of (a) lyotropic , (b) smectic and (c) nematic liquid crystals.
36. List various applications of liquid crystals.
37. Name a liquid crystal that possesses a chiral centre.

17.8.1 Classification of Nanomaterials

According to Siegel, there are four types of nanomaterials according to the number of dimensions.

Zero-dimensional All dimensions (x, y, z) are at nanoscale, examples: metallic, semiconducting and ceramic nanoparticles (spheres and slusters).

1D nanomaterials Two dimensions (x, y) are at nanoscale and the other dimension is outside nanoscale; for example, nanowires, nanotubes, and nanorods.

2D nanomaterials One dimension (x) is at nanoscale and the other two are outside nanoscale, examples: Thin films (plates and networks).

3D nanomaterials These nanomaterials are not confined to nanoscale in any dimension (x, y, z) and are characterized arbitrarily having all three dimensions above 100 nm, that is they are composed of particles in macroscale. Examples include bundles of nanowires and multilayered nanotubes.

Structural Features of Nanomaterials

Nanomaterials possess structural features in between those of atoms and bulk materials. The properties of nanomaterials are significantly different or rather improved from those of atoms and bulk materials.

Due to nanometre size, these engineering materials exhibit large fraction of surface atoms, high surface energy, and spatial confinement with reduced imperfections. Table 17.6 lists the nanoscale properties that are exploited for various applications.

Table 17.6 Nanoscale properties and their applications

Properties	Applications
Large surface-to-volume ratio	Battery technology, adsorption, catalysis, solar cells
Increased hardness along with decreasing grain size	Protective layers or hard coatings, eg., superhydrophobic surfaces
Narrow band gap, spatial confinement, and reduced imperfections	Faster electronic gadgets, optoelectronics, quantum computers
Lightweight with immense mechanical strength	Sports goods

17.8.2 Nanofabrication

Nanofabrication primarily relies on two different approaches, namely top-down and bottom-up approaches.

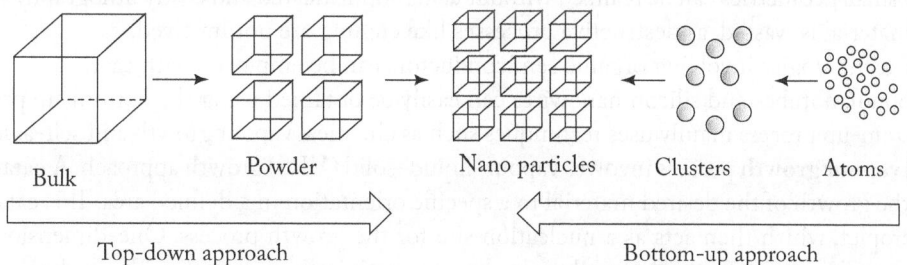

Fig. 17.13 Top-down and bottom-up approaches of nanofabrication

Top-Down Approach

A typical top-down approach is shown in Fig. 17.13. In this approach, nanomaterials are obtained by breaking down the bulk materials into smaller pieces using mechanical, chemical, or other forms of energy. This approach utilizes photolithography to pattern a layer of material over a given substrate. This is then followed by patterning the bulk substrate by different etching techniques. Silicon is the most commonly used substrate.

In the process, first a suitable substrate is realized. In the example illustrated, it consists of a silicon wafer with a top silicon dioxide layer [Fig. 17.14(a)]. A positive photoresist (a positive resist is a more popular choice than a negative resist) is then coated on the silicon dioxide layer and suitably treated. A mask with the requisite pattern is then used to expose the photoresist layer to UV rays in a selective manner, as shown in Fig. 17.14(b). The exposed portion of the resist is then developed using suitable solvents [Fig. 17.14(c)]. The windows opened in the resist are then used to etch the silicon dioxide to realize the desired pattern, as shown in Fig. 17.14(d).

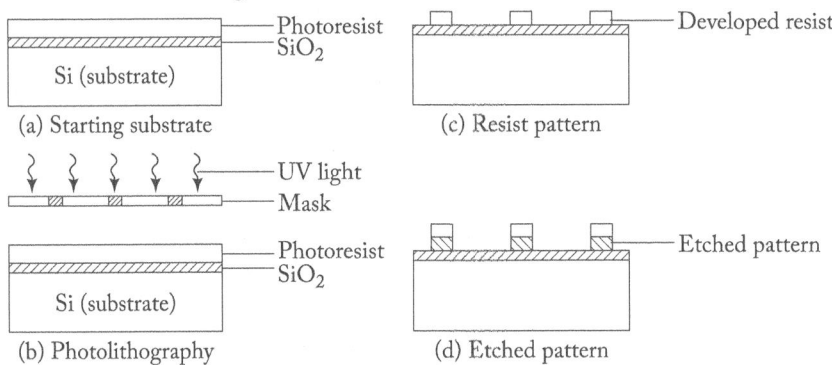

Fig. 17.14 Schematic representation of top-down approach

In addition to the basic processes of photolithography and etching, top-down approach also utilizes some additive processes such as deposition, growth, and ion implantation. The deposition process utilizes consumption of energy to deposit a layer of the desired material. This can be achieved using physical vapour deposition techniques such as evaporation and sputtering, or chemical vapour deposition (CVD) techniques.

Bottom-Up Approach

As the name suggests, a bottom-up approach involves fabrication of nanomaterials from atomic or molecular species using self-assembly or chemical reactions. This approach employs controlled addition of desirable atoms to create nanostructures. The bottom-up approach has the following distinct advantages over top-down approach of realizing nanostructures:

(a) Very small geometries can be realized without using sophisticated and costly lithography techniques.
(b) No material is wasted, as destructive processes like etching are not involved.
(c) New technologies involving organic semiconductors can be employed with ease.
(d) Carbon nanotubes and silicon nanowires can easily be obtained using the bottom-up process.

The bottom-up process mainly uses techniques such as chemical vapour growth and self-assembly. The **chemical vapour growth** process involves vapour–liquid–solid (VLS) growth approach. A catalyst is used to direct the growth of the desired material to a specific orientation in a defined area. This catalyst forms a liquid droplet, which then acts as a nucleation site for the growth process. One-dimensional growth results due to the saturation of the catalyst, leading to precipitation of a solid. This technique has been used to obtain silicon nanowires.

The **self-assembly** process utilizes colloidal chemistry. Nanoparticles or molecules aggregate via chemical and physical interactions to result in desirable nanoscale structures. The process does not require photolithography, and therefore, involves low cost. Gold and silver nanoparticles have been realized using this approach.

Sol–gel method The sol–gel process uses colloidal suspension (sol) and gelatin to form a network in a continuous liquid phase (gel). The precursor for synthesizing these colloids consists of ions of metal alkoxides and alkoxysilanes. The most commonly used materials are tetramethoxysilane (TMOS) and tetraethoxysilanes (TEOS), which form silica gels. Alkoxides are immiscible in water. They are organometallic precursors for silica, aluminium, titanium, zirconium, and many others. A solvent such as alcohol is used. The sol–gel process initially uses a homogeneous solution of one or more selected alkoxides. A catalyst is used to start the reaction and control the pH. The formation of sol–gel involves the following stages; hydrolysis, condensation, growth of particles, and agglomeration of particles; a detailed discussion on these is beyond the scope of this book.

17.9 STRUCTURAL FEATURES AND PROPERTIES OF NANOMATERIALS

Let us have a detailed account on various nanomaterials such as graphene, carbon nanotubes, fullerenes, fullerols, nanowires, nanocones, and haeckelites.

17.9.1 Graphene

Geim, Novoselov, and co-workers (2004) delicately cleaved a sample of graphite with sticky tape to obtain graphene. Graphene is a single sheet of crystalline carbon just one-atom thick having a single-layered, two-dimensional honeycomb lattice structure and is also referred to as 'mother' of all carbon-based nanomaterial systems.

Graphene is one of the strongest known material from which many nanomaterials can be obtained such as carbon nanotubes and fullerenes. A single-layer graphene (SLG) is one-atom thick, optically transparent and has high electrical conductivity. The characteristic electronic properties of graphene are directly related to its thickness.

Honeycomb lattice structure of graphene causes electrons moving in the material to behave as if they have no mass. Electrons in graphene move at an effective speed 300 times less than the speed of light in vaccum. The electrons in graphene can travel large distances without being scattered, making it a promising material for fast electronic components. The π bonds in graphene also contribute to the electron movements, electrical conduction, and weak interactions between graphene layers.

Some of the peculiar properties of graphene are as follows.
(i) Graphene is the lightest, yet strongest known material. It weighs only 0.77 mg/m^3.
(ii) It has a high tensile strength (Young modulus = 1.1 TPa terapascal). It is about 200 times stronger than steel.
(iii) The thermal conductivity of graphene at 27 °C is about 5000 W.m$^-$.K^{-1}.
(iv) Graphene exhibits quantum Hall effect (i.e., production of a voltage difference) at room temperature, tunable band gap, ballistic conduction of charge carriers, and is a highly transparent material.
(v) Though graphene is a flat layered nanomaterial, it exhibits ripples due to thermal conductive behaviour.

17.9.2 Carbon Nanotubes (CNTs)

Imagine graphene sheet being rolled-up longitudinally along its axis which is referred to as carbon nanotubes. They are among the stiffest and strongest known fibres, and have remarkable electronic properties. CNTs are allotropes of carbon with a nanostructure that can have a length-to-diameter ratio of up to 28,000,000:1, which is significantly larger than that of any other material. These cylindrical carbon molecular tubes have novel properties that make them potentially useful in many applications in nanotechnology, electronics, optics, and other fields of material sciences.

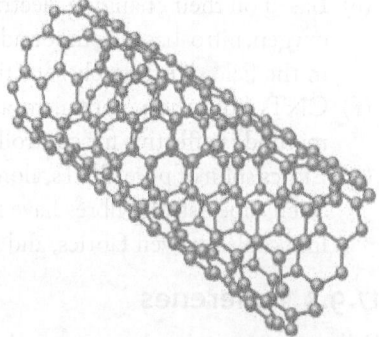

Fig. 17.15 Carbon nanotube

Each carbon in a carbon nanotube is sp^2 hybridized and each atom is joined to three neighbours, in the same way as in graphite. CNTs can be single walled or multiwalled.

Single-walled carbon nanotubes (SWCNTs) are long wrapped graphene sheets with length-to-diameter ratio of 1000. It can be obtained by wrapping one atomic layer of graphene into a seamless cylinder.

Multi-walled carbon nanotubes (MWCNTs), on the other hand, consist of concentric SWCNTs with different diameters with an interlayer spacing of 3.4 Å. MWCNTs have more than one surface within it.

Properties of CNTs

(a) CNTs are one of the strongest and stiffest materials known, in terms of tensile strength and elastic modulus, respectively.

(b) Young's modulus is in the range 1–5 TPa for single-walled CNTs, and the tensile strength is about 150 GPa for multiwalled CNTs. These values are comparable to high-carbon steel. They possess extraordinary mechanical strength and can act as satellite tethers and space elevators.

(c) CNTs can be either metallic or semiconducting, depending upon the chirality. For example, the armchair form is metallic, whereas the zigzag form is found to be semiconducting. The energy gap of semiconducting CNTs is inversely proportional to the diameter of the tube. The energy band gap can also be affected by localized defects.

(d) CNTs are known to exhibit superconductivity below 20 °C. They are highly flexible nanomaterials and do not break during mechanical deformation.

(e) CNTs will be able to achieve conductivities up to 20 times more than that of copper, a metal well known for its conductivity. The temperature stability of CNTs is estimated to be up to 2800 °C in vacuum and about 750 °C in air.

Applications of CNTs

CNTs have wide-ranging applications. The strength and flexibility of CNTs make them potentially useful in controlling other nanoscale structures. This suggests an important role for CNTs in nanotechnology engineering.

(a) Due to their superior mechanical properties, many CNT-based products have been proposed, ranging from everyday items such as clothes and sports gear to combat jackets and space elevators.

(b) CNTs are employed as electrodes in capacitors and batteries due to their higher electrical conductivity, high surface area, and linear geometry making them easily accessible to electrolytes in batteries.

(c) They can act as unique catalyst supports in various industrial chemical reactions.

(d) They make ideal components in electrical circuits. They are used in terahertz sources (switching instruments) and sensors. CNT-based transistors are capable of digital switching with a single electron.

(e) Based on their changing electrical properties, CNTs can detect very small concentrations (ppm) of oxygen, nitro-based compounds, ammonia, etc. This opens up the possibilities of several applications in the field of extremely sensitive gas sensors.

(f) CNTs hold the potential to allow drug dosage to be lowered by localizing its distribution. This method is effective for controlled delivery and distribution of drugs inside the body.

(g) Fibres spun of pure CNTs, along with CNT composite fibres, have exceptional mechanical strength. Such super-strong fibres have many applications, such as in body and vehicle armour, transmission line cables, woven fabrics, and stain-resistant textiles.

17.9.3 Fullerenes

Fullerenes are molecules composed of carbon atoms arranged in the form of a hollow sphere, ellipsoid, or tube. The spherical form which has a geodesic dome-type structure is called *buckyball*. It was given the name buckminsterfullerene in honour of Buckminster Fuller, who did pioneering work on geodesic domes. C_{60} is an allotrope of carbon with a structure shown schematically in Fig. 17.16 (a) One can have C_{60} and C_{70} as fullerene molecules. Today, the method to prepare fullerene molecule employs helium gas in an evacuated chamber to produce a black soot-like material on application of a suitable voltage between electrodes placed within the chamber. The black soot is then scraped to obtain a high yield of C_{60}.

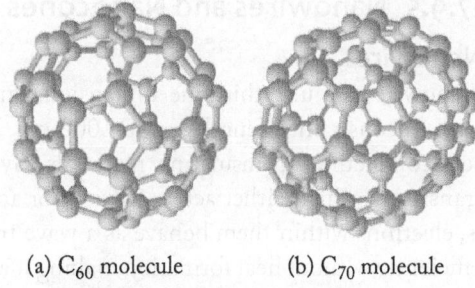

(a) C_{60} molecule   (b) C_{70} molecule

Fig. 17.16 3D view of fullerene

C_{60} molecule consists of perfect hollow spherical cages of 60 carbon atoms arranged in interlocking 20 hexagons and 12 pentagons. The number of hexagonal faces can vary and each carbon is bonded to other carbons in the hexagon in a pseudo-spherical arrangement consisting of alternating pentagonal and hexagonal rings similar to a soccer ball. Many structural variations of fullerenes have since been discovered. These include, buckyball clusters, mega tubes, and fullerene rings, among others.

Properties

(i) Fullerenes are extremely strong and are known to resist high pressures. These fascinating molecules can bounce back to their original shape after subjected to extreme pressures as high as 3000 Atm.

(ii) They are employed as catalysts for various chemical transformations, such as in the conversion of ethyl benzene to styrene, an important raw material in rubber industry.

(iii) Fullerenes aggregate together through van der Waals forces.

(iv) These molecules are ferromagnetic and exhibit superconductivity.

(v) Their cage-like unreactive structure is exploited in intercalation reactions, where alkali metal atoms can be stored within them causing fullerenes to exhibit metallic behaviour. They can be also be used to store radioactive materials.

(vi) Fullerenes are also employed as rocket fuel.

(vii) They are aromatic molecules and are usually stable with low reactivity to most solvents and are known to exhibit marvelous colours in them.

(viii) Fullerenes are employed in light emitting diodes, data storage devices in organic solar cells, and photodetectors for X-rays.

17.9.4 Fullerols

Figure 17.17 shows the typical structure of a fullerol. In spite of the vast utility of fullerene, its application at times becomes limited due to its insolubility in water. Today, many synthetic routes have been developed to allow introducing various hydrophilic functional groups on C_{60} cage. Fullerenol, commonly referred to as fullerol, is an extensively hydroxylated fullerene derivative written as [nano-$C_{60}(OH)_n$] and is a water-soluble nanomaterial. S. Manickam, et. al. recently reported a greener approach of preparation of fullerol that employed acoustic cavitation induced by ultrasound at ambient temperature in the presence of dilute hydrogen peroxide solution that completed the conversion within one hour.

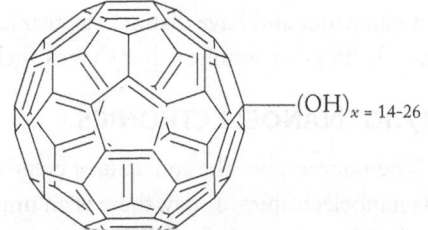

$(OH)_{x = 14-26}$

Fig. 17.17 Schematic structure of fullerol

Properties and Applications

Recently, fullerols have shown promise as drug delivery agents that can even bypass ocular barriers but are known to be phototoxic to human lens and retinal tissues.

17.9.5 Nanowires and Nanocones

Nanowires

Nanowires are ultrathin one-dimensional nanomaterials with diameter of the order 10^{-9} nm or less with unconstrained lengths up to 1000 nm. Nanowires exhibit properties encompassing from metallic, semiconducting, to insulating materials. By proper configuration of nanowires, engineers can create transistors, which either act as a switch or an amplifier. Certain nanowires are ballistic conductors, that is, electrons within them behave as a wave travelling without any collisions, thus conducting electricity efficiently without heat formation making it an energy-efficient choice for fabricating advanced materials and devices.

Nanocones

Nanocones, as the name suggests, are conical nanostructures made from carbon with one dimension in the order of 1μm or less. As shown in Fig. 17.18, nanocone structure consists of a hexagonal plane with varying number of pentagonal defects ranging from 1 to 5. Each pentagonal disclination is of angle $2\pi/6$ and these nanocones are held together by weaker van der Waals forces. Nanocones can be closed or open and can be engineered in various ways.

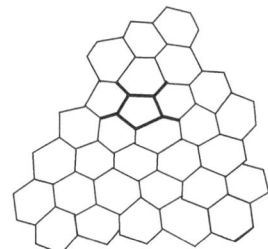

Fig. 17.18 Representation of top portion of nanocone. The centre pentagon (dark shaded) denotes the top portion of nanocone

Solitary graphitic nanocones are obtained by carbon condensation on a graphite substrate and by pyrolysis of heavy oils. Nanocones act as excellent electron field emitters and are employed as probe tips in scanning tunnelling microscope and atomic force microscope and even as 'nano-sponges' for hydrogen in fuel cells.

17.9.6 Haeckelites

Ernst Haeckel hypothesized that introducing intensive defects like pentagons and heptagons in a fullerene molecule modifies its electronic properties appreciably resulting in faster electronics. If one imagines a graphene sheet with extra pentagons, hexagons, and heptagons such that total number of pentagons and heptagons is in the same order to compromise for positive curvature of pentagons and negative curvature of heptagons. Such arrangements are called Haeckelites which resemble the micro-skeleton of zooplankton (Fig. 17.19).

The haeckelite tube surface appears rippled due to protruding pentagonal and hexagonal faces in opposite directions. These nanomaterials are metallic in behaviour and have distinct electronic properties. Usually tubular in shape,

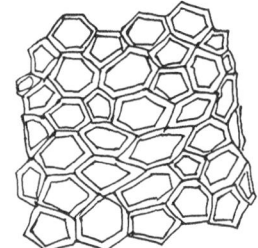

Fig. 17.19 A typical haeckelite structure

heackelites can assume other shapes such as coiled, double-screw, sheet, and cylindrical types.

17.10 NANOELECTRONICS

When one applies the concepts of creating electronic components using nanotechnology, it is referred to as nanoelectronics. Today, there is an unprecedented need for downsizing electronic gadgets, particularly reducing the sizes of transistors to such an extent along with uninterrupted speed. It was predicted by Gordon Moore that on reducing sizes of transistors and adding them onto integrated circuits doubled every two year (or accurately 18 months) since 1965, the speed can only increase resulting in faster electronics. This is also referred to as the Moore's law.

Nanoelectronics includes all the electronic devices and materials with small sizes that have physical effects due to nanoscale dimensions imparting different interatomic interactions and quantum mechanical properties to the materials. Tunnelling and atomic disorder are the characteristic features of nanoscale devices. Today, the tiniest working transistor is about 7 nm long, that is, it measures 1.4 million times smaller than the original transistor that was first built which was about 1 cm long. Let us enumerate the different nanoelectronics devices which have varied applications.

Spintronics   Also referred to as magneto-electronics, spintronics exploit the dual property of electrons, namely their charge and spin. The differences in their electron spin directions cause a weak change in the magnetic energy state and are identified as either 'spin up' or 'spin down' states along with magnetic moments. Such materials allow data storage and information will be carried both by charge and spin states resulting in greater functionality of the devices. Today, majority of the hard disc drives employs spin effects called the *giant magneto-resistive effect* which aids in massive data storage.

Optoelectronics   These are devices that can detect and control light. Light emitting nanofibres are employed as optoelectronic textiles. Analogue electric devices that were once used in communication technology, are being replaced by optoelectronic devices as they provide enormous bandwidth and capacity. Photonic crystals and quantum dots are classic examples of optoelectronic materials. Photonic crystals are materials whose refractive indices vary periodically and whose lattice constant is half the wavelength of the light used, quite similar to semiconductors with particular band gap. Such photonic crystals find applications in LCDs, optical fibres, nanoscopic lasers, radio frequency antennas, and photonic integrated circuits.

Organic LEDs, electronic paper, field emission displays are some advances in display technology that utilizes various nanomaterials.

Wearable electronics   Due to the growing demand of continuous health monitoring, researchers are working towards portable, wearable, and flexible devices which can revolutionize health care. Fitness bands are a common feature today when you want to keep track of various physiological parameters during physical activities. These wearable-flexible electronics use nanomaterials such as carbon nanotubes to sense, act, store, emit, and move resulting in high-advanced sensory devices. Nanomaterial-enabled wearable and flexible sensors that can detect temperature, electrophysiological, stress, strain, tactile, electrochemical, and environmental sensors are being investigated. Carbonized silk nanofibre (monitoring wrist pulse, respiration), graphene-nanocellulose nanopaper-PDMS (data gloves), CNT/PDMS (detecting knee bending and breathing), CNT/silicone/CNT (robotics) are some examples of wearable nanoelectronics utilizing nanotubes along with advanced polymeric materials.

17.11 APPLICATIONS OF NANOMATERIALS

Catalysis   Nanomaterials are particularly heterogeneous in catalytic action. Due to their large surface area-to-volume ratio and nanoscale sizes (10–80 nm), these materials are extremely efficient than conventional transition catalysts. Nano-TiO_2 is an efficient catalyst in photocatalysis due to the presence of large number of defects in their structure and catalytic activity takes place at defects on the surfaces. Aluminium dodeca-tungsto-phosphate nanotubes are employed during biodiesel production.

Medicine   In biomedicine, nanoparticles with sizes smaller than 10 μm can move freely in tissues and bind to biological membranes. Today, nanorobots are created to deliver drugs to the target within the human body to alleviate disease or infection only in the affected area resulting in negligible side effects. These nanorobots are also utilized to remove fat deposits during heart and cosmetic surgeries. Quantum dots are known to glow brightly in UV light and thus can be used to detect tumours in the body. Magnetic nanoparticles (MNPs) are used in biological imaging techniques and have been successfully used as magnetic biosensors and in cancer therapy. Ferromagnetic nanoparticles are developed and optimized for targeted delivery of therapeutic drugs, genetic material, and even radionuclides.

Environmental analysis   Oil spillage, gaseous pollutants, and odoriferous contaminants are adsorbed by nanofoams. Reusable nanocomposites made from polymeric foams and gold nanoparticles are utilized to eliminate organic pollutants from contaminated waters. Nanoporous membranes made of advanced polymer membranes with adjustable pore size are used as microfilters for removal of particulate matter from noxious air. Nanoporous aluminosilicates called nano-zeolites are used to remove radionuclides and heavy transition metals from wastewater. Nano-zinc oxide is popularly used to degrade chlorinated phenols from wastewater. Today, carbon nanotubes are used to fabricate sensory devices that can detect low quantities of pollutants with high precision and are used to analyse gas mixtures.

Electronics and display technology   Today, miniaturized organic light emitting devices (OLED), thin film transistors (TFT), and thin film organic photovoltaics made of advanced nanomaterials are made available that provide enhanced efficiencies. Nanoelectronics such as spintronics, wearable nanoelectronics devices, quantum computers have revolutionized the electronics field. Today, there is a huge market demand for compact, lightweight, flat-panel displays especially in televisions and classroom smartboards. Nanocrystalline zinc selenide, zinc sulphide, cadmium sulphide, and lead telluride are ideal candidates for next generation light emitting phosphors. With the advancement in portable electronics such as smartphones, laptops, and remote sensors, there is a great demand for high-performance batteries with long lives. Nickel–metal hydride battery made of nanocrystalline nickel metal and metal hydrides is one classic example of nano-based batteries.

Nanomechanics   We know that molecules can undergo conformational changes due to pH, temperature, spin states, and electron transfer. Today, nano cars driven thermally using fullerene-based wheel-type motion are fabricated. Composite textiles consist of carbon nanotubes and polymeric membranes prepared from nano-tungsten carbide, tantalum carbide, and titanium carbide are employed to manufacture bullet-proof clothes, protective fabrics, and helmets.

17.12 COMPOSITE MATERIALS

Composite materials are multiphase material systems comprising a mixture of two or more macro-constituents which are essentially insoluble and differ in form and composition leading to the formation of different phases, namely matrix phase and reinforcing phase. When one combines two or more distinct materials, a new engineering material is obtained referred to as 'composite,' having the desired properties achieved by the principle of combined constitution action. Such engineering materials exhibit a higher degree of stiffness (ability to resist elastic deformation on loading) without any brittleness and hence find

Check Your Progress

38. Define nanotechnology.
39. What are nanomaterials? Name any two important nanomaterials.
40. How are nanomaterials different from conventional bulk materials?
41. What is top-down approach of nanofabrication?
42. What is bottom-up approach of nanofabrication?
43. What are CNTs? Mention their types.
44. What are nanoelectronics? Give some examples.
45. What are fullerenes and fullerols?
46. Differentiate between nanowires and nanocones.
47. Name the nanomaterial that resembles the micro-skeleton of zooplankton.
48. List any four applications of nanomaterials.

applications in the fabrication of various structural materials. A classic example of incorporating composite material is the fuselage of Boeing 787 aircraft that was made from carbon fibre-reinforced plastic in an epoxy matrix. Carbon fibres impart strength and stiffness to the material and the epoxy matrix binds the fibres together. High specific strength, low specific gravity, toughness, shock and impact resistance, corrosion resistance are some characteristic properties of composite materials. Wood is a classic example of a composite material made of cellulose fibres and lignin. Bone, which is composed of collagen, proteins, and apatite, is also a composite material.

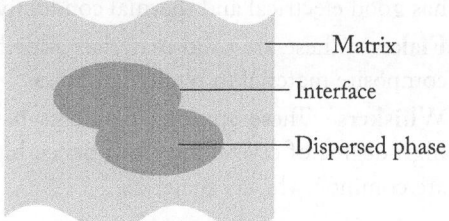

Fig. 17.20 A typical representation of a composite material

Composite materials have two phases (Fig. 17.20), *a matrix* that is usually the main bulk that surrounds the other phase called *dispersed phase*. The chemically dissimilar phases are separated by a distinct interface.

Matrix Phase

Matrix phase is the major constituent that encompasses the composite providing the bulk and could be metal, ceramic, or polymer substances. Based on the type of matrix, composite materials can be of the following three types.

Metal matrix composites (MMCs) are mixtures of ceramics and metals. Cemented carbides and cermet are examples of metal matrix composites.

Ceramic matrix composites (CMCs) are alumina- or silicon carbide-embedded with fibres to enhance the properties of composites especially when used in high-temperature conditions.

Polymer matrix composites (PMCs) are thermosetting resins such as epoxy and polyester with fibre reinforcements.

The main function of matrix phase is to bind the dispersed phases together. It can distribute the stresses evenly around the composite material when load is applied on them and hence protects the dispersed phases from crack, rupture, or any failure. Hence, it is essential that the matrix phase itself should be of higher ductility, greater strength, toughness, and resistant to corrosive attack. Aluminium, copper as metallic phases and polymers are commonly employed as matrix phases.

Dispersed Phase

It is the structural constituent that directly influences the internal structure of composite material. The phase that can act as a dispersed phase should be thin, possess large length-to-diameter ratio, and high tensile strength. Some of the important dispersed phases are as follows.

Fibres   These are long thin filaments of polymer, metal, or ceramic possessing a high length-to-diameter ratio, high stiffness, and tensile strength. Fibres could be of the following three types.

Glass fibres are obtained by forcing glass melt through numerous small orifices of a spinneret, with rapid pulling, followed by cooling. The emerging continuous filaments of about 10 μm in diameter are referred to as *monofilaments*. Glass fibres are high-performance materials which have low cost, are readily available, and possess chemical inertness.

Carbon fibres are obtained as long filaments by pyrolysis of organic material, usually cellulose or polyamylonitrile in an inert atmosphere. The monofilaments so formed are about 8 μm in diameter. Carbon fibres are used as reinforcing phase with polyester resins and are high-performance materials found to be affected by moisture, acids, and various organic solvents.

Aramid fibres are obtained from aramid polymer, Kevlar, which has great strength and if used along with polymer matrix phase, the composite material will be of high tensile strength, resistant to creep, and fatigue failure with excellent toughness.

Particulates or particles These are small coarse metallic or non-metallic powders (carbon black, carbides, silica, mica) mixed with polymeric matrix to produce a composite material. Due to the addition of these particulate materials, the surface is rendered good hardness even at higher operating temperatures and has good electrical and thermal conductivities.

Flakes These are solid materials added in to the composite materials. Mica flakes are added to the composite material to render hardness.

Whiskers These are long filaments having large length-to-diameter ratio. These materials possess high degree of crystallinity and hence high strength. Graphite, alumina, silicon carbide, silicon nitride are common whisker materials.

Interface
The interface is the locus of composite material and allows for the even distribution of stresses from matrix phase to dispersed phase.

17.12.1 Classification of Composite Materials
Composite materials are broadly classified into three major classes based on reinforcements as particle-reinforced, fibre-reinforced, and structural composites (Fig. 17.21).

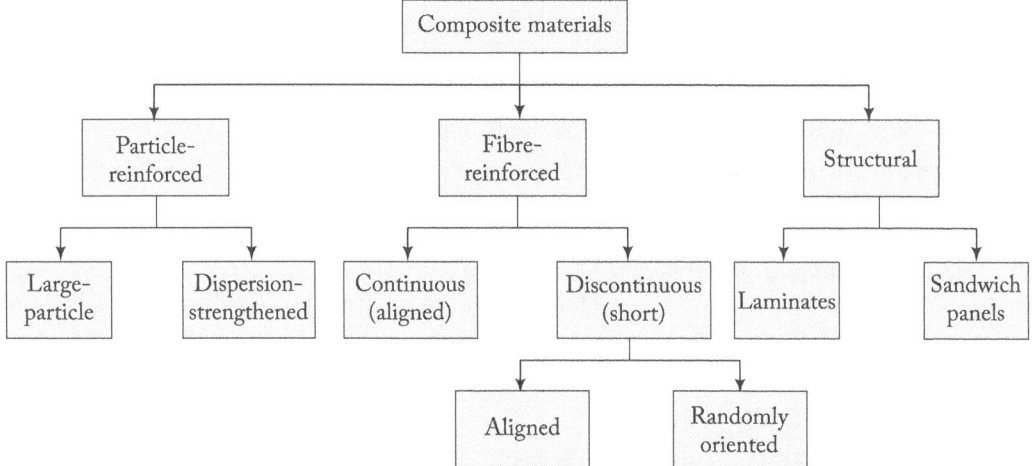

Fig. 17.21 Classification of composite materials based on reinforcements

Fibre-reinforced Composites
These are composed of fibres embedded in matrix phase. The fibres are made of three components: filament, a polymer matrix, and a bonding agent. Glass and metallic fibres are usually employed as the dispersed phase. When small fibres are pushed axially, they bend easily within the composite and provide high tensile strength. Fibres could be of two types, continuous and discontinuous fibres. Many types of organic and inorganic fibres can be employed as reinforced composite materials. Usually, organic fibres possess good flexibility, elasticity, and low density; whereas inorganic fibres exhibit good thermal stability and possess greater rigidity as compared to organic fibres.

Fibres possess higher aspect ratio, that is, their length could be several times more than their diameters. If the fibres are aligned in longitudinal direction, the material gains more strength and is relatively less prone to load failures. The reinforcement efficiency of a discontinuous fibre is quite low compared to that of continuous aligned fibres. Glass fibres, Kevlar, and carbon are some fibres commonly used as discontinuous fibres.

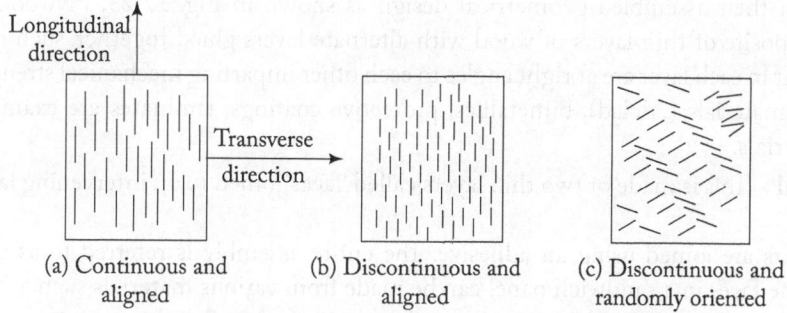

(a) Continuous and aligned   (b) Discontinuous and aligned   (c) Discontinuous and randomly oriented

Fig. 17.22 Schematic representation of types of fibre-reinforced composites

Some of the important types of fibre-reinforced polymer composites are as follows.

Glass fibre-reinforced polymer composites are the earliest known materials that employ glass fibres as a dispersed phase added in the polymeric matrix phase. These materials possess high tensile strength and good resistance towards chemicals, shock, and corrosion. Automotive parts and chemical plant pipelines are usually made from these composites.

Carbon fibre-reinforced polymer composites have carbon fibres in the polymeric matrix. Such composite materials provide good corrosion resistance and are employed in fabricating various aeroplane parts, sports goods, building materials, etc.

Aramid fibre-reinforced polymer composites contain short discontinuous or long continuous aramid fibres in a polymeric matrix. These composite materials possess good surface area, large aspect ratio and are employed as structural, civil, and advanced engineering materials in aircraft and automotive parts.

Particle-reinforced Composites

These are prepared by adding particles of varying sizes and shapes in to the matrix phase, usually metals or ceramic. There are two types of particle-reinforced composites which are discussed below.

Large-particle composites In these, the particles are harder than the matrix phase and help in the restricted movement of the matrix. The particle–matrix interface influences the degree of reinforcement and the mechanical properties of composite materials. Cement admixed with particulate sand gravel that forms concrete is a classic example of such composite materials, where gravels form the particulate phase and cement is the matrix phase. Cermets and other refractory materials are also examples of composite materials.

Dispersion strengthened composites In these, particles of size about 10–100 nm are dispersed in the matrix phase. Various alloy systems such as Cu–Sn, Cu–Be, Al–Cu, and ferrous alloys are usually hardened and produced in the form of composites with ceramics.

Structural Composite Materials

Laminar and sandwich panel are two broad categories of structural composite materials.

Laminar composite materials When several layers of two or more metal materials are placed alternately in a determined order, they are called layered or laminar composites. Such laminar composites are prepared using powder metallurgy techniques such as hot pressing, diffusion-bonding, brazing of different alloy sheets, foils, and powders or sprayed materials. Their properties depend on every constituent

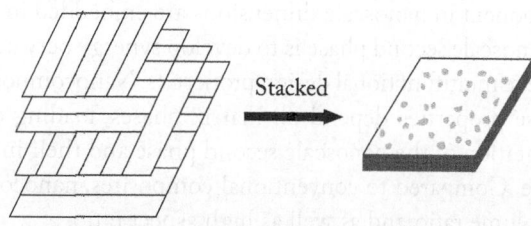

(a) Oriented stacked layers   (b) Fabricated laminar composite

Fig. 17.23 Laminar composites

material and on their assembled geometrical design as shown in Fig. 17.23. Plywood is an example of laminar composite of thin layers of wood with alternate layers glued together, such that constituent materials present in each layer are at right angles to each other imparting mechanical strength and rigidity. Metal cladded materials (Alclad), bimetallics, protective coatings, laminates are examples of laminar composite materials.

Sandwich panel This is made of two thin layers called 'faces' joined to an intervening layer called 'core' between them.

All these layers are joined using an adhesive. The entire assembly is referred to as 'sandwich panel' (Fig. 17.24). The faces in a sandwich panel can be made from various materials such as fibre-reinforced plastics, aluminium alloys, titanium, steel, plywood. Faces of a sandwich panel are known to bear most in-plane loading and transverse bending stresses. The core materials could be cement, rubber, balsa wood, formed polymers, etc. Core material separates the faces of the sandwich panel assembly and tends to resist deformations perpendicular to the face plane. Core also provides rigidity along the planes perpendicular to the faces and usually assumes a honeycomb-like structure.

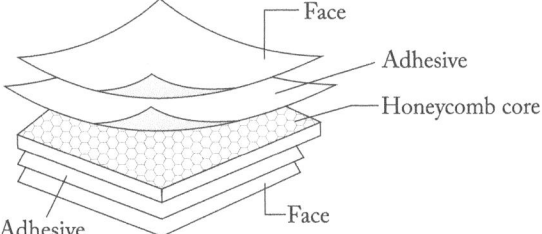

Fig. 17.24 A typical sandwich panel-type structural composite material

17.12.2 Applications of Composite Materials

Structural materials In construction, various structural gratings and claddings to full structural systems for industrial support materials, buildings, roof structures, tanks, bridge systems employ high-performance structural composites.

Aircraft Commercial aircraft applications rely on strength and toughness of composites. Fibre-reinforced composites that impart strength, durability, and corrosion and fatigue resistance are utilized during their manufacture.

Transportation Reinforced plastics are commonly employed in various surface transportation materials. Polyester resin along with fillers and reinforcements are employed as composites in road transportation.

Miscellaneous applications Carbon fibre-reinforced polymer matrix is used as a composite limb (prosthetics). Other miscellaneous applications of composite materials include fabricating bearing materials, pressure vessels, abrasives, electrical machinery, cutting tools, and crankshaft in automotive engines.

17.12.3 Nanocomposites

Nanocomposites are a class of engineering materials in which one or more phases with at least one component in nanoscale dimensions are embedded in a metal, ceramic, or polymer matrix. The addition of nanoscale second phase is to develop synergy between various constituents in composite materials and achieve multifunctional design properties. Nanocomposites are considered as high-performance materials whose properties depend on matrix phases, loading conditions, degree of dispersion, size, shape, and orientation of the nanoscale second phase and their interactions between the matrix and the reinforcing phase. Compared to conventional composites, nanocomposites possess exceptionally high surface area-to-volume ratio and as well as high aspect ratio.

Classification of Nanocomposite Materials

Broadly, nanocomposites are of two types, polymer-based and non-polymer-based nanocomposites. Figure 17.25 elucidates the classification of nanocomposites.

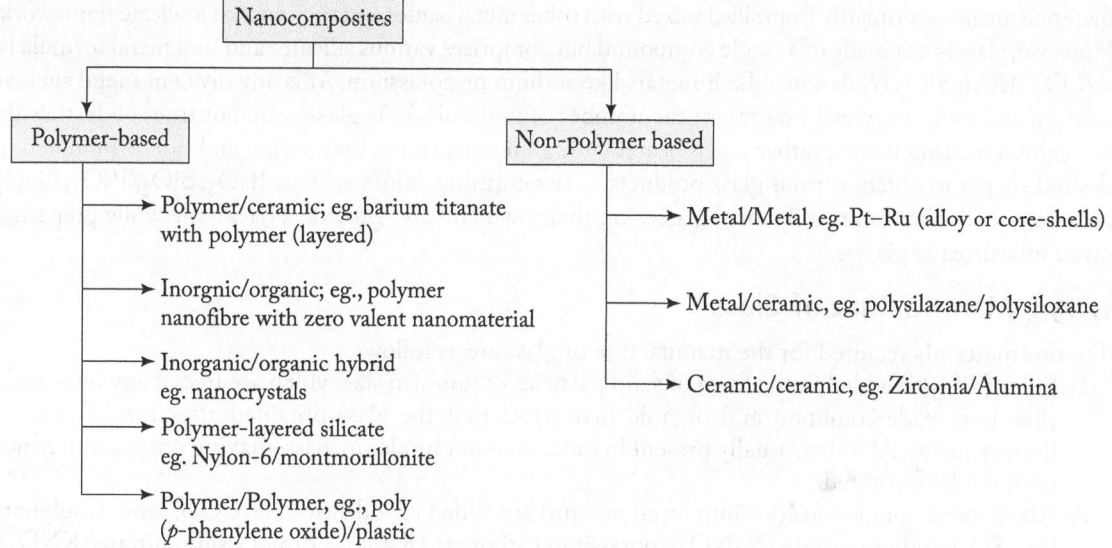

Fig. 17.25 General classification of nanocomposites

Applications of nanocomposite materials

(i) Silicon–carbon nanocomposite materials are employed as anodes in lithium-ion batteries which are in close contact with lithium electrolyte.

(ii) Carbon nanotubes are fabricated as nanocomposite fibres and employed in various structural design materials.

(iii) Hybrid-based nanocomposites made of poly(dimethyl siloxane) rubber and nano-silica are specifically designed to shape various sports goods such as golf balls, badminton racquets, etc.

(iv) Polymer nanocomposites act as good barriers for various gases like carbon dioxide, oxygen, nitrogen, and solvents such as nitric acid, hydrochloric acid. Hence, many chemical protection gears are made from advanced polymeric nanocomposites in order to protect against chemical attack or even warfare.

(v) Clay-based polymer nanocomposites are employed in plastic bottle manufacturing industries for improving mechanical properties and shelf-life of the product.

(vi) Nanocomposite membranes made of silylated-sulfonated poly(ether ether ketone), poly(benzimidazole)/sulfonated silica nanoparticle nanocomposite, fluorinated polybenzimidazole/silica nanoparticle composites are successfully employed as proton conducting membranes in fuel cells with enhanced conductivities compared to pure polymeric membranes.

(vii) Polyurethane/clay-based nanocomposites exhibit high flame retardancy and are commercialized in the manufacture of automobile seats (foams).

Check Your Progress

49. What are composite materials? Name any two natural composites.
50. Name a layered composite material.
51. List the characteristics of composite materials.
52. Name three fibres used to prepare composite materials.
53. What are nanocomposites?
54. State a few applications of composite materials.

17.13 GLASS

Glass is a supercooled material, usually obtained by cooling a liquid below its freezing point resulting in decrease of viscosity and after a particular temperature below freezing point, viscosity becomes zero. Many glass materials are made primarily from silica mixed with other metal oxides and thus possess a silicate framework. However, glass is not made of a single compound but comprises various silicates and its general formula is, $x\,R_2O.y\,MO.6\,SiO_2$ (R denotes alkali metals like sodium or potassium, M is any divalent metal such as calcium and lead, and x and y represent the number of molecules). As glasses are not true solids, they do not exhibit melting points; rather when heated at high temperatures, they soften and can be moulded in desired shapes to obtain various glass products. Glass-forming oxides such as B_2O_3, SiO_2, P_2O_5, GeO_2 are compounds that can readily form glasses on their own and also provide a backbone while preparing other mixed-oxide glasses.

17.13.1 Manufacture of Glass

The **raw materials** required for the manufacture of glass are as follows.
1. Silica (SiO_2) is added in the form of sand particles of uniform size which are free of any impurities (like iron oxide-colourant and organic matter) so that the glass obtained after manufacture is homogeneous. Alumina, usually present in sand, does not hinder in glass-making process and hence need not be separated.
2. Alkali metal compounds (sodium or potassium) are added as soda ash (Na_2CO_3), sodium sulphate (Na_2SO_4), sodium nitrate ($NaNO_3$), potassium carbonate (K_2CO_3), or potassium nitrate (KNO_3).
3. Alkaline earth metal compounds (calcium or barium) such as lime (CaO), limestone ($CaCO_3$), barium carbonate ($BaCO_3$) are added and particularly barium-based compounds are added to obtain glasses with high refractive indices.
4. Heavy metal oxides such as lead oxide (PbO) or red lead (Pb_3O_4) are also employed as raw material. Today, heat-resisting glass includes zinc oxide along with borax in place of lime which allows decrease in coefficient of expansion of the glass material.
5. Calcium phosphates along with arsenic and antimony oxides are added during the preparation of opalescent glasses.
6. Colouring matter is added while preparing coloured glasses. Certain metallic oxides like iron oxide, chromium oxide and cupric oxide, are added to the molten mass of fused silicates so as to impart aesthetic colours to the final glass product.

Glass Manufacturing Process

Glass manufacture has four stages, viz., melting, shaping, annealing, and finishing.

Melting   Typically two types of furnaces are employed during glass manufacture, which are:

Pot furnace: It is a large dome-shaped clay crucible and numerous such crucibles are placed in a circular manner in a pot furnace. Raw materials required for glass are charged and melted in the furnace. The furnace is fired and the resulting flue gases are allowed to pass through fire-brick refractory arrangement which gets heated up. Thus, heat of burnt flue gases is itself utilized for heating purposes.

Tank furnace: It consists of a huge rectangular clay tank where melting of glass takes place. Just like in pot furnace, heat is generated to prepare molten mass of glassy material. The glass formed at the end is withdrawn from the opposite end of the tank. Over a period of continuous operation, tank walls wear out and needs replacement. Generally, furnace design is regenerative or recuperative, that is, in both types heat is obtained from exhaust gases to preheat air utilized for fuel combustion. Once the furnace is heated, a temperature of about 1205 °C is maintained and huge amount of heat is lost as radiation which is necessary so as to reduce corrosive action of molten glass on the furnace wall. At times, cooling pipes are installed so as to provide cooling of furnace walls.

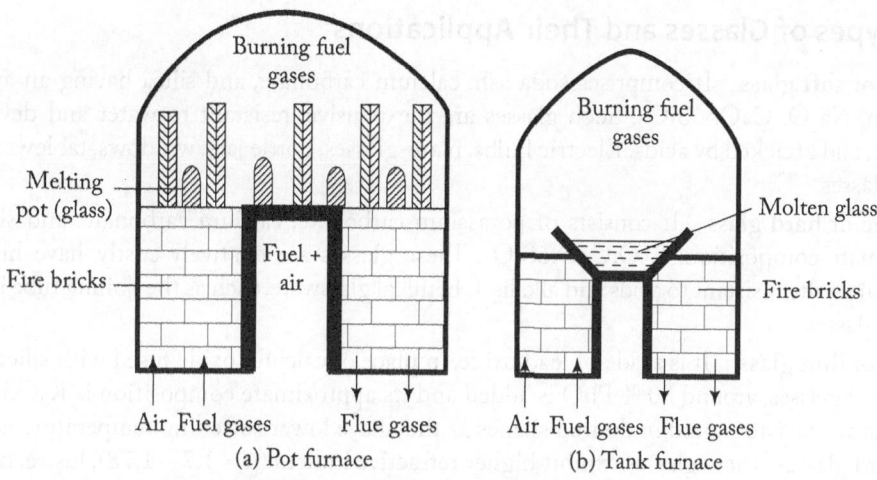

Fig. 17.26 Furnaces for glass manufacture

Chemical reactions occurring within the furnace Ordinary glass can be represented as $Na_2O.CaO.6 SiO_2$ and is obtained as per the following reactions.

Sand (SiO_2) fuses with soda ash forming a glassy mass called *water glass* (Na_2SiO_3) which is soluble in water.

$$Na_2CO_3 + SiO_2 \rightarrow Na_2 SiO_3 + CO_2$$

Calcium carbonate reacts with sand forming calcium silicate that is insoluble in water, but soluble in acids.

$$CaCO_3 + SiO_2 \rightarrow CaSiO_3 + CO_2$$

However, if both soda ash and limestone are present, sodium calcium silicate is obtained that is insoluble in water and acids.

$$Na_2CO_3 + CaCO_3 + SiO_2 \rightarrow Na_2O.CaO.6SiO_2 + 2CO_2$$

The proportions of soda ash, limestone, and sand are 35:15:50 and one can replace soda ash with a mixture of Na_2SO_4 and coal and in such cases, the chemical reactions will be,

$$2 Na_2SO_4 + SiO_2 + C \rightarrow 2 Na_2SiO_3 + CO_2 + 2 SO_2$$
$$Na_2SiO_3 + CaCO_3 + 5 SiO_2 \rightarrow Na_2O.CaO.6SiO_2 + CO_2$$

During melting process, raw materials are thoroughly mixed and charged so as to obtain a homogeneous mixture referred to as *batch*. The batch is transferred to the furnace and mixed with some broken pieces of glass called *cullet*. Cullet melts at lower temperature and allows melting of the entire mixture. The hot molten mass is fused with hot gases taking care because a lot of frothing takes place due to the presence of carbon dioxide which is driven out so as to obtain clear fused viscous melt referred to as *plain*. Various impurities, unreacted sulphates, chlorides of calcium, and alkali metals rise to the top as scum called *glass ball* which is skimmed out easily.

Shaping or forming The molten mass is used sent for obtaining various glass products by the process of blowing or moulding in to desired shapes and sizes.

Annealing If glass obtained after shaping is rapidly cooled, its exterior portion will be cooler than the inner portions leading to stresses and strain that may result in cracks due to unequal expansion and complete failure of the material. Thus, annealing is performed where the glass product is cooled slowly at room temperature by passing them through a long-tunnel furnace which is hot at one end but cold at the exit end.

Finishing operations Commonly employed finishing operations, post annealing, include cleaning, grinding, cutting, enameling, polishing, and cutting.

17.13.2 Types of Glasses and Their Applications

Soda-lime or soft glass   It comprises soda ash, calcium carbonate, and silica having an approximate composition, $Na_2O. CaO. 6SiO_2$. Such glasses are inexpensive, resistant to water and devitrification, easy to melt, and attacked by acids. Electric bulbs, plate-glasses, bottle jars, windows, tableware are made from soft glasses.

Potash-lime or hard glass   It consists of potassium carbonate, calcium carbonate, and silica having an approximate composition: $K_2O.CaO.6SiO_2$. These glasses are relatively costly, have high melting points, and are quite resistant to acids and alkalis. Chemical glassware, such as the boiling tube is fabricated using hard glasses.

Lead glass or flint glass   It is made of lead oxide, in place of calcium oxide fused with silica. To obtain denser optical glasses, around 80 % PbO is added and its approximate composition is $K_2O.PbO.6SiO_2$. Flint glasses are easy to mould in desired shapes as they have lower softening temperature compared to soft and hard glasses. These glasses exhibit higher refractive indices ($\mu = 1.7 - 1.78$), lustre, high specific gravity (3 – 3.3), and excellent electrical properties. Flint glasses are used as lenses, prisms, neon sign tubing, electric insulators, cathode-ray and X-ray tubes employed in medical devices.

Borosilicate glass or Pyrex   It is a mixture of sodium aluminoborosilicate and contains boron trioxide (rich in silica, but poor in alumina) and its general composition is, $7Na_2O.4CaO.3ZnO.36SiO_2$ (80–83 % SiO_2 + 10–13% B_2O_3). As their coefficient of expansion is quite low, such glasses can withstand high operating temperatures. Higher softening points, good shock and chemical resistance are the properties of such glasses and are employed in laboratory glassware and kitchenware.

Special Glasses

Vycor glass   It is a high-grade silica glass (96% silica glass) which is quite difficult to prepare as pure silica is hard to melt and mould into glassy state. Its general composition is 96 % silica, 3 % B_2O_3, and the rest is alumina. These glasses have low coefficient of expansion, high softening temperatures, good chemical resistance to most acids, except for HF and H_3PO_4. Laboratory crucibles and electric insulators are usually fabricated using Vycor glasses.

Vitreosil glass   It is prepared by heating pure silica (99.5% silica glass) to about 1750 °C (melting temperature) and the glass so obtained is difficult to mould due to dense viscosity. Such glasses can be moulded below 1650 °C (softening point) to obtain opaque glassy state. If this glass is heated above its melting point, a clear silica glass is obtained. These glasses are used in chemical plant pipelines which are employed to carry concentrated acids.

Safety glass or laminated glass   It is fabricated by adding a thin vinyl plastic layer between two to three glass sheets and heat and pressure are employed to obtain tough glassy sandwich composite material. When the material is cooled, a clear yet tough glass is obtained. Such glasses are shatter-proof (on breaking, it does not fly in pieces) and shock-resistant. These find application in automobile wind panes, bullet-proof glasses, locking windows, washing machines shields, windshield, and airplane windows and panes.

Coloured glass   It consists of oxides of transition metals of iron, chromium, copper, cadmium, titanium, etc., that absorbs certain light frequencies giving aesthetic colours when added to hot molten glass. Colloidal metal oxide particles are infused in hot molten glass to obtain coloured glass and employed in traffic lights, signal systems, architecture designs, etc.

Glass wool or fibre glass   It is a wool-type fibrous material made of entangled fine glass threads. Each glass thread diameter is as small as 0.005 mm which is obtained by forcing glass through tiny orifices in a continuous process and thrown rapidly so as to get glass wool. Glass wool is fire-proof, possess poor thermal conductivity, high tensile strength, and resistant to chemicals. It is used as an insulator, as plugs in salt bridges of electrochemical cells, textiles, etc.

17.14 CERAMICS

Ceramics are engineering materials made from clays (hydrous aluminosilicates) such as sand, Kaolinite, feldspar, and montmorillonite. Broadly, ceramics are of three types: (a) natural ceramics (clay-based) such as limestone ($CaCO_3$), granite (aluminosilicate), silica (SiO_2); (b) structural ceramics such as alumina, zirconia, silicon carbide, etc., and (c) functional ceramics such as piezoelectrics and superconductors such as zirconia (ZrO), alumina (Al_2O_3) which are used as catalytic converters in automotive engines. Most classes of engineering materials are considered as ceramic-based materials such as cement, glasses, refractories, abrasives, pottery, and porcelain. In all these materials, the common feature is the presence of silicate framework.

Ceramics are inorganic non-metallic solids that are prepared by firing or heating in a mixture of appropriate composition of clays, sand, and feldspar, followed by cooling. Pottery and porcelain are clay-fired products which are generally glazed or enamelled.

17.14.1 Properties of Ceramics

The physical characteristics of ceramics depend on the chemical composition and their crystal structures. Ceramics could possess either ionic or covalent bonds that are much stronger than metallic bonds making them brittle than metals. Ceramics possess very high melting points and high elastic modulus. These materials are resistant to various acids and organic solvents. Ceramic materials also possess excellent creep resistance, wear resistance, good hardness, and high compressive strength. The microstructure of ceramics directly influences their mechanical, optical, electrical, thermal, and magnetic properties.

On the basis of suitable microstructure, various ceramic materials exhibit mechanical properties such as ductility, malleability, strength, fracture toughness, and elasticity. Si_3N_4 is employed in the fabrication of radial rotor in turbine engines. Silicon carbide is utilized to prepare disc brakes.

Most ceramics are semi-conducting in their behaviour. Zinc oxide is used in blue light emitting diodes. Lead zirconate titanate and barium titanate are used as piezoelectric materials. An interesting feature of ceramics is that when certain gases pass through them, their electrical resistance changes. Thus, such ceramic materials can be employed as gas sensors. At extreme low temperatures, some ceramic materials exhibit superconductivity, though its mechanism is unknown. Many transparent and glassy ceramics are used to fabricate optical fibres, switches, lenses, and laser amplifiers.

17.14.2 Classification of Ceramics

Depending on the glaze, ceramics are of two types, viz., glazed and unglazed ceramics.

Glazed tiles comprise a glaze layer (a vitreous coating) which is fired to fuse on to the ceramic object so as to impart strength, colour, and water-proof properties to them. Usually, glazing materials consist of silica and other metal oxides and calcium is added as a fluxing agent to reduce the overall melting point of the mixture. Further, alumina is also added to impart stiffness to the ceramic product. Various transition metal oxides are added to impart colour to the final product. Glazing process allows to enhance the aesthetic appeal of the finished ceramic product. Tiles are usually coated with glaze so as to obtain a lustrous finish and such coatings are carried out at about 1100 °C. These coatings can provide different colours and designs to the finished product. Glazed materials are easy to clean with water and can also be made non-slippery by etching grout lines on them. Bathroom tiles and kitchenware usually employ glazed materials.

Unglazed tiles do not contain any glazed coating on their surfaces. These hard and dense materials possess homogeneous colours and are known to withstand wear and tear. Compared to glazed materials, these are non-slippery, easy to clean, and safe to use.

17.14.3 Some Common Ceramics

Kaolin (China clay) It is a pure white clay-burnt engineering material possessing low plasticity and usually obtained due to weathering reaction of feldspar.

$$K_2O.\,Al_2O_3.\,6SiO_2 + CO_2 + 2H_2O \rightarrow K_2CO_3 + Al_2O_3.\,2SiO_2.\,2H_2O + 4SiO_2$$

$$\text{Feldspar} \qquad\qquad\qquad\qquad \text{Kaolinite}$$

As the composition of kaolinite is $Al_2O_3.\,2SiO_2.\,2H_2O$, it has good amount of silica and when it undergoes fusion at about 1550 °C, kaolinite is obtained. Kaolinite, so formed is employed in the manufacture of electric insulators, pottery, tableware, bricks, tiles, sanitary goods, etc.

Porcelain The formation of porcelain begins from china clay heated at about 600 °C followed by silicate formation around 1000 °C. Porcelain is used to prepare dinner sets, crucibles, insulators, floor tiles, sanitary goods, etc. The chemical reactions can be depicted as:

$$3Al_2O_3.\,2SiO_2.\,2H_2O \xrightarrow{\;600°C\;} 6H_2O\uparrow + 3Al_2O_3 + 6SiO_2$$

$$\text{China clay} \qquad\qquad\qquad\qquad \text{Amorphous}$$

$$\Big\downarrow 900°C$$

$$\underset{\text{Crystal}}{SiO_2} \xleftarrow{\;1400°C\;} \underset{\text{Mullite}}{4SiO_2 + 3Al_2O_3\cdot + 2SiO_2} \xleftarrow[\text{formation}]{\underset{\text{Silicate}}{1000°C}} \underset{\text{(Crystalline)}}{\gamma\text{-}3Al_2O_3} + \underset{\text{(Amorphous)}}{6SiO_2}$$

17.14.4 Applications of Ceramic Materials

From simple earthenware, stoneware to construction materials, chemical plants, and architecture designs, ceramics find applications in almost all fields of engineering. Ceramic materials find wide use as refractory materials in the manufacture of furnaces and ovens utilized in several metallurgical operations. Floor tiles, sanitary products, kitchenware, bricks, gypsum, glass, and concrete come under the class of ceramics and find major use in construction.

Alumina It possesses good hardness, durability, chemical resistance, smoother surfaces, and good electrical insulating properties. Hence, it is employed as cutting tools, abrasives, medical components, thermocouple protection tubes, grinding wheels, heat exchanger balls, glass tank fittings, orthodontic brackets, and in sodium vapour lamps.

Silicon carbide It exhibits high thermal conductivities and can be employed at very high operating temperatures. Thus, silicon carbide is used as refractory, in preparing abrasives, bearings in semiconducting industry, laser mirrors, igniters, heating elements, and as an additive for metal reinforcement composites.

Silicon nitride It possesses good hardness, moderate thermal conductivities, high elastic modulus, fracture toughness, and hence is employed in making high-temperature-resistant components, turbine parts, abrasives, and ball bearings.

Zirconia (ZrO_2) It has high strength and is resistant to crack propagation. Thus, it finds use in medical prosthetics such as hip implants and body armour.

Boron nitride It is known for its chemical and thermal stability and thus used commonly as an abrasive substance. It is also used as lubricant and to prepare cutting tools.

Yttrium barium copper oxide ($YBa_2Cu_3O_{7-x}$) and **magnesium diboride** (MgB_2) are employed as superconducting materials. **Zinc oxide** is used as semiconductor whereas **magnesium silicate** is an electrical insulator. **Titanium carbide** is used to prepare capacitors, scratch-proof watches, piezoelectric materials, cutting tools, catalyst supports, and even as superconductors.

Check Your Progress

55. What is glass?
56. State the purpose of annealing glass.
57. Name some varieties of glasses.
58. List the stages of glass manufacture.
59. State the composition of (a) Pyrex (b) flint glass (c) potash glass, (d) and soda-lime glass.
60. What are ceramics?
61. Classify ceramics. Name any two ceramic materials.
62. Bring out the comparisons between ceramics and metals.
63. What is glaze? Name any two glazes.

Activity-based Questions

A. Advanced nanoelectronics

With advances in nanoelectronics, various advanced devices have sprung up in our daily lives. Electronic sensing is one such development witnessed in the recent times in the field of nanotechnology. Electronic nose is particularly designed to sense and detect odours, scents, and certain classes of pollutants. Many industries need odour assessment tests to be performed on a variety of products.

Perform a comprehensive review on the functioning of electronic nose and skin, their features, utility, working, and industrial applications citing suitable examples.

B. Non-invasive medical sensing

Today, health care finds solutions for delivering therapeutic drugs in a non-invasive manner in patients. Syringes and catheters cause pain and inconvenience to ailing patients and non-invasive drug delivery and sensing is gaining attention. Wearable nanoelectronics is already making headlines with alternative diagnostics and monitoring devices. Numerous critical challenges such as selectivity, recognition of target body parts, sample handling, and patient comfort are yet to be overcome successfully.

Present a comprehensive review on different types of wearable-non-invasive sensors, nanomaterials used, biomimetics, and their applications.

C. Concrete 3D printing

Since late 2010, 3D printing concretes in construction industry have become popular. It is a novel way to deal with concrete material inside of an automated robot. There have been developments in the manufacturing process using robotic arms to deposit material so as to obtain desired shapes and design. 3D printed houses, bridges, etc. will have exceptional strength and will be developed in a lesser time with lesser human resource.

Discuss 3D printing technology in construction industry, types of 3D designs, manufacturing processes, advantages and limitations of the process.

SUMMARY

- Important engineering materials are of various types which are either building materials or advanced materials. Building materials include cement, concrete, composites, refractories, abrasives, adhesives, glass, and ceramics. Advanced engineering materials include nanomaterials and liquid crystals, among others.

- The manufacture of Portland cement uses calcareous, argillaceous materials in cement kilns. After mixing of raw materials, they are burnt in kilns where different chemical reactions take place such as drying, calcination, and clinkering. The final setting and hardening of cement occur due to formation of tobermonite gel and crystallization of calcium hydroxide and hydrated tricalcium aluminate.

- Concrete and reinforced concrete are important building materials; however these are known to undergo decay in the presence of carbon dioxide, sulphur dioxide, water, and chloride content.

○ Adhesives are important non-metallic materials utilized in joining dissimilar surfaces or bodies employed in construction. They could be structural and non-structural adhesives.

○ Abrasives are cutting tools that find greater applications in construction industries. The abrasive power and hardness are measured by Mohs' hardness scale. Natural and synthetic abrasives are two common classes; examples of abrasives include silicon carbide, boron carbide, calcium carbide, alundum, and boron nitride.

○ Refractory materials are used to manufacture inner linings of ovens and furnaces that are employed in metallurgical operations as they can withstand extreme temperatures without any deformation or corrosion. Refractoriness, RUL, dimensional stability, porosity, thermal spalling are characteristic properties of refractories.

○ Composite materials are used in fabricating strong parts of structures such as aircraft and jet planes.

○ Liquid crystals and nanomaterials are advanced sophisticated materials used in a variety of electronic devices. Liquid crystals are of two types, namely thermotropic and lyotropic. Nematic, smectic, and cholesteric are types of thermotropic liquid crystals. All these crystals find applications in electronic displays and medicine.

○ Nanomaterials include carbon nanotubes, fullerenes, fullerols, nanocones, nanowires, and haeckelites. They find applications in the fields of catalysis, medicine, environment, nanomechanics, and electronics.

○ Glass and ceramics are silicate-framework materials which find wide use in a variety of tableware and as piezoelectric substances.

GLOSSARY

Abrasives: A hard inorganic substance used to grind away another substance generally used for cutting, machining, grinding, or finishing a work-piece.

Adhesive: A non-metallic substance that sticks or bonds two dissimilar bodies together based on the molecular forces present in area of contact.

Ceramic: Includes engineering materials made from clays (hydrous aluminosilicates) such as sand, Kaolinite, feldspar, montmorillonite.

Clinker: The greyish lumps or nodules that remain after sintering process during Portland cement manufacture.

Composite: The multiphase material system comprising a mixture of two or more macro-constituents that are insoluble and differ in form and composition leading to the formation of matrix phase and reinforcing phase.

Concrete: Construction material obtained by mixing cement, sand, stones gravel, slag, bricks, and water in calculated proportions.

Cullet: The broken pieces of glass mixed with raw materials during glass manufacture.

Devitrification: The crystal-like growths within glass causing opacity of the material.

Glass: A fusion product of inorganic materials which has been supercooled to a rigid condition without crystallization.

Kiln: A type of oven that produces temperatures sufficient to complete physical processes such as hardening, drying, or chemical changes.

Liquid crystal: Orientationally ordered liquids or positionally disordered crystals which have properties of both solid and liquid states.

Nanomaterials: Substances or materials whose at least one dimension is less than 100 nm.

Portland cement: The pulverized finely ground powder that is obtained by calcining known proportions of argillaceous and calcareous raw materials together at 1500 °C.

Pyrometric cone equivalent: A number denoting softening temperature of refractory specimen of standard dimension and composition.

Refractoriness: The resistance of abrasive material to breaking of its grain and softening with rising temperatures.

Reinforced concrete: Concrete reinforced with steel rods and at times with heavy wire mesh.

Thermal spalling: Cracking, peeling, or fracturing of a refractory material under high operating temperature.

EXERCISES

Multiple Choice Questions

1. The mixture of slag and cement is called
   (a) Portland cement
   (b) Puzzolana cement
   (c) slag cement
   (d) natural cement

2. Setting and hardening of Portland cement occurs due to
   (a) formation of tetraclcium aluminoferrite
   (b) formation of tobermonite gel
   (c) $Ca(OH)_2$ and hydrated tricalcium aluminate crystallization
   (d) Both (b) and (c)

3. The cause of thermal spalling in a refractory is
   (a) rapid changes in furnace temperature
   (b) porosity
   (c) low furnace temperatures
   (d) molten slag

4. Higher the porosity of a refractory
   (a) lesser is its electrical conductivity
   (b) higher is its thermal conductivity
   (c) higher is its refractoriness
   (d) low expansion

5. In a blast furnace, refractory should have
   (a) high thermal conductivity
   (b) low tensile strength
   (c) low thermal conductivity
   (d) moderate thermal conductivity

6. Chemical reactions occurring in upper, central, and lower zones of cement rotary kiln are:
   (a) drying, calcination, and clinkering
   (b) clinkering, calcination, and drying
   (c) drying, clinkering, and calcination
   (d) calcination, drying, and grinding

7. A type of coated abrasive is
   (a) pumice
   (b) sandpaper
   (c) pulpstone
   (d) diamond

8. The chemical formula of kaolinite is
   (a) $Al_2O3.4SiO_2.3H_2O$
   (b) $Al_2O_3.2C.4H_2O$
   (c) $Al_2O_3.2SiO_2.2H_2O$
   (d) $Al_2O_3.2SiO_2$

9. The number of carbons in Buckminsterfullerene is
   (a) 270
   (b) 60
   (c) 75
   (d) 220

10. 'There's plenty of room at the bottom.' This was stated by _____.
    (a) Eric Drexler
    (b) Richard Feynmann
    (c) Harold Croto
    (d) Richard Smalley

11. Which of these consumer products is already being made using nanotechnology methods?
    (a) Golf clubs
    (b) Sunscreen lotion
    (c) LCDs
    (d) All of these

12. LCDs are widely used in electronic systems. Which of the following describes construction of such a display?
    (a) Two sheets of conducting material separated by a layer of insulating dielectric
    (b) A *pn* junction diode made of gallium arsenide or gallium phosphide
    (c) Two sheets of polarized glass with a thin layer of compound between them
    (d) Two polarizers with insulating material

13. Of the following, the one that is NOT an example of dispersed phase in a composite material is:
    (a) fibre
    (b) polymer
    (c) flake
    (d) whisker

14. The approximate composition of flint glass is
    (a) $K_2O.3PbO.SiO_2$
    (b) $K. PbO.5SiO_2$
    (c) $K_2O.6SiO_2$
    (d) $K_2O.PbO.6SiO_2$

15. Plywood is an example of
    (a) refractory
    (b) ceramic
    (c) composite
    (d) glass

16. Of the following, which one is NOT an example of nanoelectronic devices?
    (a) Spintronics
    (b) Optoelectronics
    (c) Smartphones
    (d) Silica

17. An example of a lyotropic liquid crystal is
    (a) spintronics
    (b) smartphones
    (c) phospholipid
    (d) glass

18. Liquid crystals are
    (a) true liquids
    (b) isotropic
    (c) pure solids
    (d) anisotropic

19. A liquid crystal that can assume a helical structure is
    (a) smectic
    (b) cholesteric
    (c) nematic
    (d) None of these

20. The space between the two surfaces filled with an adhesive substance is called
    - (a) glueline
    - (b) adherend
    - (c) assembly
    - (d) adherend–adhesive contact

21. The term *nano* means
    - (a) one billionth of a metre
    - (b) one billionth of a millimetre
    - (c) one billionth of an inch
    - (d) one billionth of a kilometre

22. Chemically, tobermonite gel is
    - (a) hydrated tricalcium aluminate
    - (b) hydrated dicalcium ferrite
    - (c) hydrated tricalcium silicate
    - (d) calcium sulphate

23. Portland cement does not contain
    - (a) $Ca_3Al_2O_6$
    - (b) Ca_3SiO_3
    - (c) Ca_2SiO_4
    - (d) $Ca_3(PO_4)_2$

24. An example of an acid refractory is
    - (a) silica
    - (b) dolomite
    - (c) chromite
    - (d) magnesite

25. Silica brick refractory material belongs to
    - (a) basic
    - (b) neutral
    - (c) acidic
    - (d) none of these

26. Magnesite brick refractory materials are
    - (a) acidic
    - (b) basic
    - (c) neutral
    - (d) none of these

27. The chemical formula of Plaster of Paris is
    - (a) $CaSO_4.1/2H_2O$
    - (b) $CaSO_4.7H_2O$
    - (c) $CaSO_4$
    - (d) $CaSO_4.2H_2O$

28. An example of adhesive is/are
    - (a) glue
    - (b) mucilage
    - (c) wax
    - (d) all of them

29. The ripple effect is shown by
    - (a) graphite
    - (b) graphene
    - (c) fullerene
    - (d) haeckelite

30. Structural composite materials are
    - (a) laminar
    - (b) sandwich panel
    - (c) both (a) and (b)
    - (d) none of the above

31. Which of the following is a refratory material?
    - (a) Low steels
    - (b) Carbon nanotubes
    - (c) Silicon carbide
    - (d) None of these

Review Questions

1. With suitable chemical reactions, explain the manufacture of Portland cement.
2. List the ingredients of cement and explain their functions.
3. Describe the chemical reactions of water with constituents of cement during setting and hardening processes.
4. Explain 'dry and wet mixing process' of various ingredients of cement.
5. What is decay of concrete? Mention the effects of carbon dioxide and sulphides on concrete decay.
6. Draw the cement rotary kiln and explain the manufacture of Portland cement.
7. Write a short note on 'Plaster of Paris.'
8. Discuss the various factors that influence adhesive action.
9. What are adhesives? Explain the various mechanisms of adhesion.
10. Classify adhesives and explain each class in detail.
11. What is Mohs scale of abrasive hardness?
12. How are abrasives classified? Cite suitable examples.
13. State the various applications of adhesives.
14. Define abrasive power and hardness of an abrasive.
15. List the applications of abrasives.
16. What are refractories? Name three main forms of refractory materials.
17. List the characteristic properties of refractories.
18. Explain the manufacture of refractories. Write a note on silica bricks.
19. Define the terms:(a) refractoriness (b) thermal spalling (c) porosity of refractory (d) refractoriness-under-load (e) dimensional stability.
20. Discuss the procedure of Seger cone and RUL tests performed on refractory materials.
21. Write a short informative note on dolomite and silicon carbide bricks.
22. What are liquid crystals? State the features of nematic liquid crystals.
23. What are smectic liquid crystals? State their features.
24. State the various properties of liquid crystal. How is liquid crystalline behaviour related to chemical structure?
25. Explain the various applications of liquid crystals.
26. Compare and contrast smectic, nematic, and cholesteric liquid crystals.

27. What are thermotropic and lyotropic liquid crystals? Cite suitable examples for each.

28. What are composite materials? Describe fibre-reinforced composite materials.

29. Explain particle-reinforced composites.

30. Classify composite materials. State the features of each class of composite materials.

31. What are laminates and sandwich panel? Draw and explain each of these materials.

32. Discuss the applications of composite materials.

33. Explain nanocomposites with suitable examples.

34. Describe the preparation, properties, and applications of fullerenes.

35. What are nanomaterials? How are nanomaterials different from conventional bulk materials?

36. Explain the structure, properties, and uses of (a) carbon nanotubes, (b) fullerenes, (c) nanocones, (d) nanowires, (e) haeckelites.

37. Explain the applications of nanomaterials in (a) medicine (b) environment (c) catalysis, (d) electronics and (e) nanomechanics

38. What are nanocomposites? Classify them and discuss their applications.

39. What is glass? List the raw materials employed during glass manufacture.

40. Classify glasses and write the composition of each type.

41. Discuss melting process of glass in pot and tank-type furnaces.

42. Explain the manufacturing process of glass.

43. Write the composition and applications of the following glasses: (a) Safety glass, (b) high grade silica glass, (c) glass wool, (d) Pyrex, (e) hard and soft glasses, (f) Vycor and (g) Vitreosil glasses.

44. What are ceramics? State their characteristic properties.

45. Classify ceramic materials. Discuss the different applications of ceramics.

FURTHER READING

1. Afreen, S., K. Kokubob, K. Muthoosamy, and S. Manickam, 'Hydration or Hydroxylation: Direct Synthesis of Fullerenol from Pristine Fullerene [C60] via Acoustic Cavitation in the Presence of Hydrogen Peroxide', RSC Adv., 2017, 7, 31930-31939 (for preparation of fullerenol using greener approach).
   https://pubs.rsc.org/en/content/articlehtml/2017/ra/c7ra03799f

2. Bye, G.C., *Portland cement*. Pergamon Press, New York, 1986.

3. Cognard P., *Handbook of Adhesives and Sealants*, vol 2. Elsevier, Amsterdam/San Diego, 2006.

4. Dierking, Ingo. *Textures of Liquid Crystals*. Wiley-VCH, 2003.

5. Karmarkar, B., *Functional Glasses and Glass-Ceramics*, Butterworth-Heinemann, UK, 2017.

6. Lubin, G. (ed), *Handbook of Composites*. Van Nostrand Reinhold, New York, 2013.

7. Ozin, G. A. and A. C. Arsenault, *Nanochemistry*, RSC Publishing, Cambridge, 2006.

8. R. Feynman Lecture 'There's Plenty of Room at the Bottom,' Full text:
   http://www.zyvex.com/nanotech/feynman.html.

9. Yanagida, H., *The Chemistry of Ceramics*. Wiley, New York, 1996.

ANSWERS

Check Your Progress

1. Cements are of four types viz, natural, Puzzolana, slag, and Portland.

2.

Dry process	Wet process
It is carried out when raw materials are harder in physical state.	This process can be performed on any type of raw materials.
Process of mixing is quite slow with lower fuel consumption.	Process of mixing is relatively quick with higher fuel usage.
Cement obtained by such mixing process is of inferior grade (granular powder).	Cement obtained using wet process is of superior grade (slurry).
Dry process is quite expensive in terms of production.	Wet process is comparatively a cheaper and easy method.

3. Refer to Section 17.2.3

4. Refer to Section 17.2.1

5. Refer to Section 17.2.1

6. Refer to Section 17.2.1

7. The formation of tobermonite gel and crystallization of calcium hydroxide and hydrated tricalcium aluminate results in the final setting and hardening of cement.

8. Concrete is an important building and structural construction material obtained by mixing binding materials (cement or lime), inert materials (sand, stones gravel, slag, bricks), and water in calculated proportions. Reinforced concrete (RCC) is concrete reinforced with steel rods and at times with heavy wire mesh.

9. Water, carbon dioxide, sulphur dioxide, chlorides can decay concrete.

10. To prevent concrete decay, chloride-free water should be used in the preparation. Also coal coating can be given to avoid little contact with water and concrete.

11. Setting involves initial stiffening of cement paste due to gel formation, whereas hardening refers to strength development due to crystallization.

12. An adhesive is a non-metallic substance that sticks or binds two dissimilar bodies together based on the molecular forces present in the area of contact.

13. Refer to Section 17.3.1

14. When dissimilar bodies are held together with adhesive, they are called adherends. (b) The process of holding adherends together by means of adhesive is called bonding; (c) and the overall assembly is termed as bond.

15. Glues represent one of the many classes of adhesives. Thermoset polymers, resins, starch are adhesives. Hence every adhesive need not be glue but all glues are adhesives.

16. Adsorption, mechanical interlocking, electrostatic diffusion, weak-boundary layer are the various theories of adhesive mechanism.

17. An abrasive is a hard inorganic substance used to grind away another substance.

18. A good abrasive should be hard, tough, and resistant to mechanical shock. It should possess high refractoriness and remain unaffected by heat treatment.

19. The capacity of an abrasive substance to grind away other objects is called its abrasive power.

20. Hardness is determined using Mohs scale.

21. Siliceous: Talc, quartz, garnets; Non-siliceaous: diamond, emery, corundum

22. Carborundum and boron nitride are examples of synthetic abrasives.

23. Refractoriness is the ability of a material to withstand heat without undergoing any appreciable deformation in shape under operating conditions.

24. Thermal spalling refers to peeling, cracking, or fracture of a refractory material due to high temperatures. Rapid changes in heating of furnaces and seepage of molten metal into the pores of refractory material cause thermal spalling.

25. Alumina and silica are acidic refractories; dolomite and magnesite are basic refractories; and silicon carbide and graphite are neutral refractories.

26. Seger cone test is used to determine refractoriness of a refractory material.

27. Refer to Section 17.6.

28. Molecules in an isotropic liquid state are known to move freely in all directions, whereas liquid crystal molecules tend to point along a common direction called the director.

29. Refer to Section 17.6.

30. Cholesteric liquid crystal phase is a nematic liquid crystal, but is optically active.

31. Mesogen is the fundamental unit of liquid crystal.

32. Cholesteric liquid crystals show helical twists.

33. Thermotropic liquid crystals are long-chain organic solids that undergo sharp transformation to a liquid crystal at a particular temperature.

34. Lyotropic liquid crystals are obtained when amphiphilic molecules are dissolved in an isotropic solvent.

35. (a) Soapy water, phospholipid biological membranes, DNA, and synthetic polypeptides are examples of lyotropic liquid crystals. (b) Ethyl p-ethoxybenzal cinnamate and p'-n-octadecyloxy 3'-nitro diphenyl carboxylic acid are examples of smectic liquid crystals. (c) *p*-Azoxy anisole and p-n-hexyl p'-cyano biphenyl are examples of nematic liquid crystals.

36. Liquid crystals find applications in display technology (LCDs), thermal imaging, thermometers, locating hotspots, detecting tumour cells, and in drug delivery.

37. Cholesteric liquid crystal possesses a chiral centre.

38. Nanotechnology is the fabrication of nanostructures, functional materials, devices, and components through control of matter on the nanometre scale.

39. Nanomaterials are substances whose at least one dimension is less than 100 nm. Carbon nanotubes and fullerenes are examples.

40. Nanomaterials exhibit large surface-to-volume ratio, higher surface energy, and spatial confinement with reduced imperfections.

41. In top-down approach nanomaterials are obtained by breaking the bulk materials to smaller pieces using mechanical, chemical, or other forms of energy.

42. The fabrication of nanomaterials from atomic or molecular species using self-assembly or chemical reactions is called bottom-up approach.

43. Rolled up (longitudinally) sheets of graphene are called CNTs. SWCNTs and MWCNTs are two types of CNTs.

44. When one applies the concepts of creating electronic components using nanotechnology, it is referred to as nanoelectronics. Examples: wearable electronics, spintronics, optoelectronics, smart displays.

45. A fullerene is a molecule composed of carbon atoms arranged in the form of a hollow sphere, ellipsoid, or tube. Fullerol is an extensively hydroxylated fullerene derivative written as [nano-$C_{60}(OH)_n$] and is a water-soluble nanomaterial.

46. Nanowires are ultrathin 1D nanomaterial having diameter of the order 10^{-9} nm or less with unconstrained lengths up to 1000 nm. Nanocones are conical nanostructures made from carbon with one dimension in the order of 1μm or less.

47. Haeckelite resembles the micro-skeleton of zooplankton.

48. Catalysis, nanoelectronics, drug delivery, nanomechanics are some applications of nanomaterials.

49. Two or more microconstituents essentially insoluble in each other and differ in composition due to formation of different phases are called composite materials. Wood and bone are examples of natural composites.

50. Laminar composite (eg. Plywood) is a layered composite material.

51. High strength and stiffness at high operating temperatures, resistant to corrosion, good coefficient of thermal expansion, higher aspect ratio of reinforced phase are some characteristics of composite materials.

52. Glass fibres, carbon fibres, and aramid fibres are employed to prepare composite materials.

53. Refer to Section 17.12.3

54. Composite materials find applications in marine, aeronautics, automobile, safety, and communication equipment.

55. Glass is a fusion product of inorganic materials which are supercooled to a rigid condition without crystallization.

56. Refer to Section 17.13.1

57. Pyrex, soft and hard glasses, safety glass, optical glass are some types of glass.

58. Melting, shaping, annealing, and finishing are the various stages of glass manufacture.

59. Refer to Section 17.13.2

60. Ceramics are engineering materials made from clays (hydrous aluminosilicates) such as sand, kaolinite, feldspar, montmorillonite.

61. Ceramics are of three types: natural, structural, and functional. They can be either glazed or unglazed ceramics. Kaolinite, alumina, zirconia, pottery and porcelain are examples of ceramics.

62. Metals have metallic bonds, are good conductors of heat and electricity, possess opacity, have uniform grain sizes,

non-porous, high tensile strength, and are ductile and malleable. Ceramics contain ionic or covalent bonds, are bad conductors of heat and electricity, possess transparency, grain microstructures are different, highly porous, have low tensile strength with low density.

63. Glaze refers to glossy finish or coating on the ceramic material so as to impart water resistance, making it non-porous. Lead silicate, feldspar along with traces of boric oxide are some glazing materials.

Multiple Choice Questions

1. (c)	2. (d)	3. (a)	4. (b)	5. (c)	6. (a)	7. (b)
8. (c)	9. (b)	10. (b)	11. (d)	12. (c)	13. (b)	14. (d)
15. (c)	16. (d)	17. (c)	18. (d)	19. (b)	20. (a)	21. (a)
22. (c)	23. (d)	24. (a)	25. (c)	26. (b)	27. (a)	28. (d)
29. (b)	30. (c)	31. (c)				

Lubricants

18.1 INTRODUCTION

The important property of a lubricant is to reduce friction and wear of metals, particularly between machine parts. A substance introduced between two moving machine parts to reduce frictional resistance between them is known as *lubricant*. The process of reducing the friction between moving machine parts by introducing lubricants between them is called *lubrication*. Lubrication and lubricants are of particular importance in the field of mechanical engineering, especially under the branch of tribology. Tribology is the science of friction, wear, and lubrication.

A tribological system can be depicted as two materials of different hardness sliding over one another. The resulting sliding friction causes erosion of surface metal causing wear and tear. Friction of moving surfaces is accompanied by frictional heat. The presence of frictional heat at rubbing surfaces results in higher local temperatures forming welded junctions.

As shown in Fig. 18.1, a typical tribological system involves two moving surfaces separated by a lubricant. The machine is considered with a moderate load in a system boundary or working conditions. To keep apart the moving parts of a machine and to reduce frictional resistance, lubrication is provided. The implementation of tribological systems and its application in selecting the lubricant for particular machinery provides economic benefits by reducing energy loss due to friction and breakdowns.

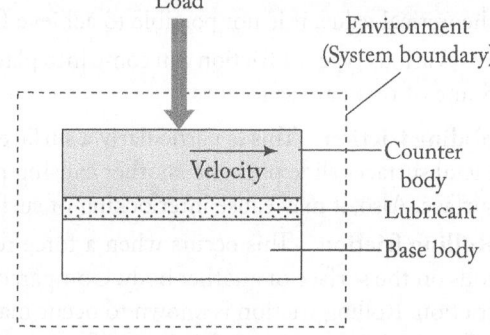

Fig. 18.1 Tribological system

Lubricants are classified into industrial and automotive lubricants.

Of all the types of lubricants, the most common examples are mineral oils and greases. Undoubtedly, mineral oils and greases are general purpose lubricating machinery oil.

As per world rankings (2005), in India, Indian Oil Corporation (IOC) holds the major share of manufacturing lubricating oils, whereas in the US, Exxon Mobil Company is the market leader for industrial oils.

The functions of a lubricant are listed as follows:

(a) **Reduces surface wear, tear, and friction** as direct contact between moving machine parts is avoided.
(b) **Reduces frictional heat** thereby controlling expansion of metals by maintaining the shape of metal parts of the moving machine components.
(c) **Acts as sealant**; for example, the lubricant used between piston and cylinder wall of an internal combustion (IC) engine acts as a seal, thereby preventing leakage of gases under high pressure.
(d) **Controls corrosion** of moving machine parts as it prevents attack due to moisture.
(e) **Facilitates smooth operation** of the equipment by removing and suspending potentially harmful products, such as carbon, dirt, and wear debris.

To achieve the above desired properties in lubricating oils, they are primarily composed of 93 % base lubricant, 7 % chemical additives, and other components.

18.2 FRICTION

When viewed at a molecular scale, even smooth metallic surfaces show irregularities generally referred to as peaks (or asperities) and valleys. If such surfaces are brought together, they form welded junctions at the contact of peaks and the real contact between two metallic surfaces is reduced. Hence, applying load on such joints may cause the welded junctions to bear all of it, causing unequal distribution of load and consequently resulting in friction and wear of the metal. *Friction is broadly defined as the force at the surface of contact of two layers which resists their sliding on one another.* The friction force, F is the force required to maintain motion. Let us say, if W is the applied load, its coefficient of friction μ is expressed as,

$$\mu = F/W$$

Table 18.1 Coefficient of friction of various metals

Metal	Coefficient of friction (μ)
Iron	1.2
Aluminium	1.5
Copper	1.5
Gold	2.5
Platinum	3

Coefficient of friction can assume values from 0 to 1, but one needs to bear in mind that zero is a mere theoretical value; it is not possible to achieve frictionless surfaces.

Various types of friction can come into play when two bodies are brought in contact with one another. Some of them are as follows.

Sliding friction This is particularly a surface phenomenon and occurs when two unlubricated moving metal surfaces slide past one another causing peaks of the softer metal to be broken by the harder metal surface. Also, it may cause interlocking of surface valleys resulting in greater friction between them.

Rolling friction This occurs when a force resists the motion of a sphere or when a cylindrical object rolls on the surface of another body. Compared to sliding friction, rolling friction has lower coefficient of friction. Rolling friction is known to occur mainly due to plastic deformation of materials and adhesion effects in the contact area. It is interesting to note that lubricants cannot reduce deformation caused during rolling; hence lubrication will rarely have any effect on such frictional bodies. Hence, in this chapter we will particularly focus on sliding friction.

18.3 MECHANISM OF LUBRICATION

To understand the selection of suitable lubricating oil for specific machinery, it is essential to have proper knowledge of the mechanism of lubrication. The operating conditions of the machinery such as load, operating temperatures, and pressure dictate the success of lubrication. The following is a discussion on the various types of mechanisms of lubrication, namely hydrodynamic lubrication, boundary lubrication, and extreme pressure lubrication.

18.3.1 Hydrodynamic Lubrication

In hydrodynamic lubrication or *fluid film* or *thick film lubrication*, the moving or sliding surfaces are separated from each other by a thick lubricant film of approximately 1000Å thickness. The lubricant layer prevents direct surface-to-surface contact so that the small peaks and valleys present on the moving surfaces do not interlock. It reduces the coefficient of friction (μ) to about 0.001 to 0.03. Hydrodynamic lubrication is required in scientific instruments such as microscopes and journal bearings. If the value of μ is about zero, it indicates that there is minimal-to-no friction occurring between the two moving surfaces.

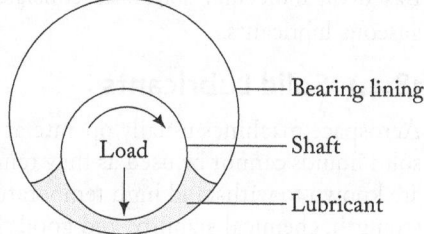

Fig. 18.2 Fluid film in bearings

The mechanism of hydrodynamic lubrication can be understood from the working of a journal bearing. A journal bearing consists of a shaft rotating at a moderate speed and a load along the axle. The lubricant is applied in the annular space between the lining of the bearing and rotating shaft as shown in Fig. 18.2.

When the bearing is in stationary position, the two surfaces remain in contact. A thick lubricating film is applied within the annular space between the lining of the bearing and the shaft. When it is set in motion, the shaft begins to rotate and the lubricant film rotates between the metallic surfaces. As the shaft gains speed, the fluid film flows at a rapid rate and fills up all the irregularities of the two metal surfaces. Eventually pressure is developed that keeps the two surfaces separated, thereby reducing the wear. Hydrocarbon oils and mineral oils are considered to be satisfactory lubricants for fluid film lubrication. Mineral oils are often blended with polymer additives to prevent their degradation due to climatic variations.

18.3.2 Boundary Lubrication

A thick lubricant film is not suitable when machines are operated at a low speed or under heavy load with a lubricant of low viscosity. Under such cases, a thin film of lube oil is applied in the moving machine parts; this is called boundary lubrication (Fig. 18.3). Generally, the boundary lubrication mechanism is applied in machines like gears, tractors, rollers, and rail axle boxes.

A thin layer of lubricant is adsorbed by physical or chemical forces onto the metallic surfaces thereby avoiding direct metal-to-metal contact. In this case, the coefficient of friction (μ) is about 0.05 to 0.15.

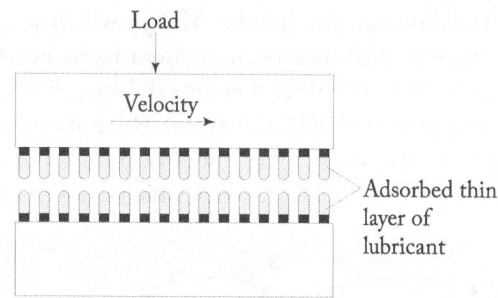

Fig. 18.3 Boundary film lubrication

The effectiveness of boundary lubrication depends on the chemical properties of the lubricating oil. For boundary lubrication, the oil must have good oiliness and possess long hydrocarbon chains and polar groups. Soaps of vegetable or animal oils, mineral oils blended with fatty acids, solid lubricants such as graphite, MoS_2, and greases are generally employed for boundary lubrication.

18.3.3 Extreme Pressure Lubrication

Under heavy-load and high-speed operating conditions, machines attain very high temperatures. Due to this high frictional heat, liquid lubricants do not adhere to the surfaces and eventually decompose or vaporize. In such cases, extreme pressure lubricants are employed. Additives such as phosphorus, sulphur or chloro-based compounds are added to the base mineral oils. Such additives that improve the specific characteristics of lubricating oil are called *extreme pressure additives*.

These additives react with metal surfaces and form the corresponding durable metallic films of phosphides, sulphides, and chlorides. These metallic films have high melting points and act as good lubricants under extreme pressures and temperatures. As chemical reaction occurs, certain amount of wear is expected. Solid lubricants like graphite, molybdenum disulphide act as extreme pressure lubricants.

18.4 CLASSIFICATION OF LUBRICANTS

Based on molecular state and consistency, lubricants are classified into solid, semi-solid, liquid, and gaseous lubricants.

18.4.1 Solid Lubricants

Aerospace machines usually operate at higher temperatures and are under heavy load; liquid and semi-solid liquids cannot be used as they tend to decompose. In such cases, solid lubricants are employed that are known to withstand high temperatures and heavy load. An ideal solid lubricant must have low shear strength, chemical stability, and good thermal conductivity. Solid lubricants act as dry film lubricant to minimize the frictional resistance of moving machine parts. Due to their layered lattice structure, the layers of atoms glide over one another resulting in the lubricating action.

Graphite, molybdenum disulphide, tungsten disulphide, cadmium dichloride, borax, lead iodide, talc, mica, and vermiculite are some of the commonly used solid lubricants. The solid lubricant is applied either in powdered form or mixed with greases.

Graphite

Graphite has a layered or lamellar structure (Fig. 18.4); when added as a dry powder or colloidal dispersion it fills up the annular spaces of machine parts, thereby reducing friction and wear. Graphite acts as a lubricant up to a working temperature of about 790 °C.

When graphite is dispersed in water, it is called *aquadag*. If dispersed in oil, it is called *oildag*.

Molybdenum disulphide

Molybdenum disulphide (MoS_2), which is a metal–solid lubricant, also has a sandwich-like layered structure with molybdenum atom layers between two layers of sulphur atoms (Fig. 18.5). These layers glide over each other thereby exhibiting lubricating properties. MoS_2 can act as a lubricant only up to a temperature of 400 °C, beyond which it undergoes oxidation and decomposition. While choosing solid lubricants for machines under vacuum conditions, MoS_2 is preferred over graphite. On decomposition, MoS_2 forms molybdic oxide which is also a good high-temperature lubricating oil.

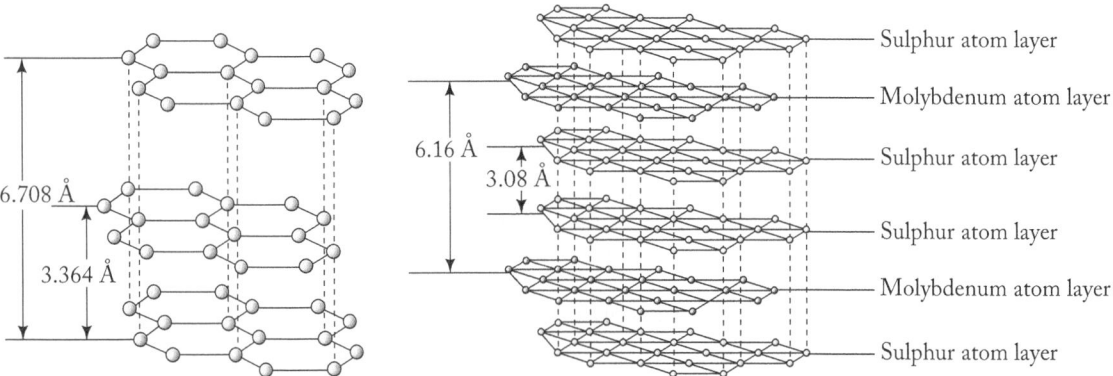

6.708 Å
3.364 Å

6.16 Å
3.08 Å

Sulphur atom layer
Molybdenum atom layer
Sulphur atom layer
Sulphur atom layer
Molybdenum atom layer
Sulphur atom layer

Fig. 18.4 Lamellar structure of graphite

Fig. 18.5 Lamellar structure of molybdenum disulphide

PTFE

PTFE (Polytetrafluoroethylene), a polymer of tetrafluoroethylene is another commonly preferred solid lubricant. In PTFE the fluorine atoms replace all hydrogen atoms of tetrafluoroethylene monomers and hence exhibit good lubricating properties. Due to smooth orientation of polymeric chains in PTFE structure, it can easily slide and slip. This property reduces friction and lowers the coefficient of friction to about 0.04. PTFE (or teflon) is known to exhibit higher chemical stability making it an ideal choice for cryogenic machines (up to $-260°C$). The limitations of using polymeric lubricants such as PTFE are: (a) they decompose at high temperatures and (b) they have higher wear rate.

18.4.2 Semi-solid Lubricants

Semi-solid lubricants are used for machines operating at low speeds and high-pressure conditions. They are employed in cases where spilling of lubricant is undesirable such as food and paper automated machines. Grease is a classic example of semi-solid lubricant. A greyish or yellow sticky mass, grease is a combination of base oil (75–95 %), soap (5–20 %), and additives (0–5 %). It is considered as a multipurpose oil, as it has a lower coefficient of friction (0.04–0.3), good adhesion to surfaces, water-resistance, and also enables the use of solid lubricants. They do not spill under squeezing conditions; hence greases are ideal for bearings. They also act as excellent sealants under heavy load; has water-resistance and good adhesion even when the machine experiences shock loading and reversing operation conditions.

Grease is prepared by dispersing a gelling agent (soaps of Ca, Al, Li) in petroleum oil (lubricating oil). The presence of soap in grease acts as a thickening agent and enables it to cling or adhere to the machine parts firmly. The nature of soap, that is, consistency, water-resistance, and oxidation, determines the temperature at which greases can be employed. The following are the different types of greases that find application in many types of machines.

Lime of calcium base greases (cup greases)   These are prepared from fats, lube oils, and slaked lime. They have good water resistance and are applicable at wider consistency ranges.

Aluminium soap greases   These have excellent oiliness, good adhesion to surfaces, have lower soap content, and high water-resistance. However, they undergo thermal decomposition above 90 °C and hence cannot be used at higher temperatures.

Lithium soap greases   These have excellent water-resistance and are used as all-purpose lubricant oil.

Barium soap greases   These also exhibit water-resistance, good clinginess, and adhesion to surfaces that require lubrication.

Some disadvantages of grease are poor heat dissipation and difficult contaminant removal.

18.4.3 Liquid Lubricants

Liquid lubricants or lubricating oils are versatile oils that assist in lubrication and also provide cooling effects, thereby reducing frictional heat. Mineral oils, vegetable oils, and blended and synthetic oils are some of the most commonly employed liquid lubricants.

Vegetable Oils

Vegetable oils are prone to aerial oxidation, but exhibit good oiliness. Animal and vegetable oils possess good oiliness. Due to chemical instability, particularly of vegetable oils (oxidizes easily), they are usually blended with mineral oils. Animal fats or fixed oils do not decompose or volatilize rapidly as vegetable oils, but their limited availability is a demerit.

Mineral Oils

Mineral oils, namely paraffinic, naphthenic, and aromatic hydrocarbons, are multipurpose lube oils employed in mostly all machines. These oils are obtained by continuous distillation of crude petroleum.

Crude petroleum contains impurities such as asphalt, wax, and coke, which are completely removed before the distillation process. If wax is present, the lubricant will exhibit higher pour points, making it unfit for its usage at lower temperatures. Asphalt, if present in the oil, may undergo decomposition at higher temperatures resulting in carbon residue and sludge. Waxy impurities are removed from crude oil using solvents, such as propane and tricholoro ethylene and allowed to refrigerate so that the wax solidifies. It is then centrifuged to remove the waxy impurities. The de-waxed oil is further treated with acids such as sulphuric acid to remove asphaltic impurities. *Paraffinic oils* are preferred over naphthenic and aromatic oils as they exhibit good thermal stability, higher viscosity indices (90–115), and higher flash–fire points.

Synthetic Lubricants

Any lubricant which cannot be classified as petroleum oils are grouped under synthetic lubricants. Silicone oils, polyglycols, polyphenyl ethers, fluoro compounds, chlorofluorocarbon polymers, and phosphate esters are examples of synthetic lubricants. The major advantage of these fluids is that they are tailor-made for a specific function and usually require few additives.

Of all the oils, today synthetic lubricants have major utility in machinery that work under harsh conditions such as those in jet engines, cryogenics, and turbines. Synthetic oils are far better than natural oils when considered for lubrication since they are non-inflammable, have high viscosity indices, possess higher thermal stability at higher temperatures, and are chemically stable.

Synthetic oils are commercially prepared to address specific conditions such as higher temperatures, heavy load, and speed. These oils do not undergo viscosity changes on varying temperatures and are resistant to oxidation and chemicals. Silicone oils (Viscosity index VI = 300), polyglycols (VI = 200), and perfluoropolyether or PFPE (high oxidation stability at 320 °C, high thermal stability at 370 °C, chemically inert) are some commonly employed synthetic liquid lubricants.

The following are the properties of some of the synthetic lubricants.

Silicone oils   These are usually employed for high-temperature operations and can be polymerized to any desired viscosity. Silicones have higher viscosity indices, good chemical stability, anti-foaming property, and low volatility. They are ideally suited in torque converters and ball bearings. However, they tend to decompose forming solid silicates and silica, which damage the lubricated joints. Silicones possess lower surface tension and thus cannot be employed in thin film lubrication.

Polyglycols   These are water insoluble. On treating with ethylene oxide, polyglycols are made water soluble by forming ethylene oxide–glycol polymer. Polyglycols possess high viscosity index (about 160), fire resistance, and lower volatility. The degraded products of these oils are soluble in fluids or are usually volatile, resulting in low sludge formation.

Fluorolubes   These consist of fluorinated hydrocarbons and polyethers known to function effectively from −90 °C to 250 °C. They have greater thermal stability with lower volatility and are non-explosive. Fluorolubes are used in machines with higher corrosive environment practically lasting for longer times.

18.4.4 Gaseous Lubricants

Gaseous lubricants are generally employed in machines like air compressors, turbines, and light bearings. Gases like air, helium, nitrogen, and argon are employed in the annular spaces of machinery. Gaseous lubricants find application in a variety of machines as their viscosities do not change with increasing temperatures. Moreover, their non-drip property adds to the ease of using them in spacecraft and cryogenic equipment.

Gas lubrication is best suited to continuous operation in radiation fields which would normally degrade regular lubricating oils. Gaseous lubricants are non-contaminating with coefficient of friction known to reach near-zero values. However, they also suffer limitations, such as limited load capacity and instability and require precise machining.

18.5 PROPERTIES OF LUBRICANTS

For the proper selection of a lubricant, mechanical engineers must be adept with the important properties of lubricating oil.

18.5.1 Viscosity and Viscosity Index

The most important property of an oil to act as a lubricant is its viscosity. *Viscosity is defined as the property by virtue of which a liquid or fluid (oil) offers resistance to its own flow.* The formation of lubricant film between two moving machine parts is dictated by viscosity of the oil. Viscosity is inversely dependent on temperature; hence on increasing the temperature viscosity decreases and oils undergo thinning. If oils undergo thinning, lubricant film cannot persist between the surfaces leading to eventual failure of operation. Thus, to achieve good lubrication, oils must possess adequate viscosity.

Generally, a lubricant is selected whose viscosity does not vary rapidly with temperature. This variation is determined by plotting viscosity–temperature curves or viscosity indices. *The rate at which viscosity of oil changes with temperature is measured by an empirical number, known as viscosity index* (VI).

Viscosity index compares the kinematic viscosity of test oil to the viscosities of two reference oil standards that have considerable temperature sensitivity difference. The reference oils are so selected that the first oil has a VI of zero and the second oil has a VI of 100. As per ASTM (American Society for Testing and Materials), Pennsylvania oils and Gulf Coast oils have same viscosity at 210 °C, with Pennsylvania oil having VI as 100 and Gulf oil has VI of zero. Viscosity index can be calculated by the formula.

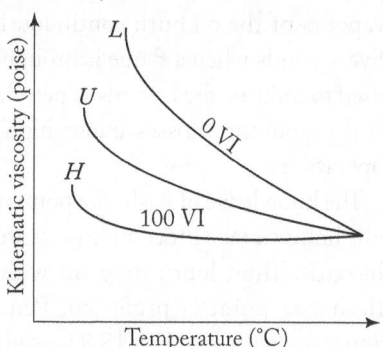

Fig. 18.6 Viscosity index curves from viscosity–temperature plot

$$VI = \frac{V_L - V_T}{V_L - V_H} \times 100$$

where, V_T is the viscosity of the test oil.

V_L is the viscosity of Gulf oil reference with same viscosity as the test oil at 210 °C.

V_H is the viscosity of Pennsylvania oil reference at 100 °C with same viscosity as the test oil at 210 °C.

Higher the VI of the oil, lower is the decrease in the viscosity of the oil with increasing temperature (Fig. 18.6).

Redwood Viscometer

Redwood viscometer is used to determine the viscosity and viscosity index of lubricating oils.

Redwood viscometer (Fig. 18.7) consists of a metallic oil cup placed in a water bath. A thin capillary tube is present in the apparatus through which oil flows. The oil cup is equipped with a ball valve to restrict the flow of the oil through the capillary tube. A heating coil is placed in the water bath along with a stirrer for uniform heating. During operation, oil is heated in the oil cup at a definite temperature. When the ball valve is opened, the oil starts to flow from the oil cup

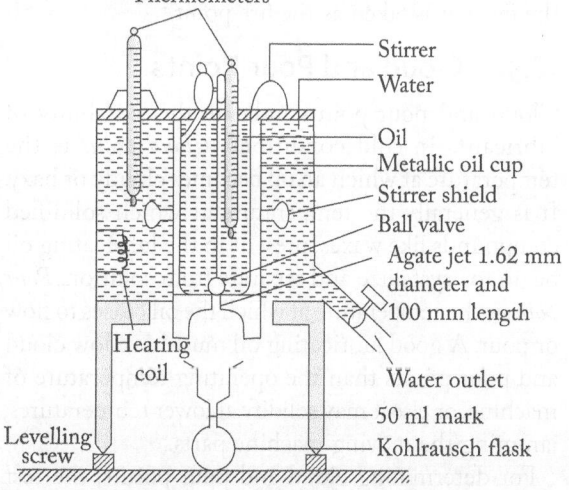

Fig. 18.7 Redwood viscometer

through the capillary tube and gets collected in the graduated Kohlrausch flask (50 ml) placed below the capillary opening agate jet having a diameter of 1.62 mm and an internal length of 10 mm. Simultaneously, a stopwatch is started to record the time taken for 50 ml of the test oil to be collected in the flask. This time is reported in Redwood seconds. Commercially, Redwood viscometers are available as brand no.1 for liquids having Redwood flow of 20 to 2000 s and brand no. 2, for liquids whose flow time exceeds 2000 s.

18.5.2 Flash Point and Fire Point

Flash point and fire point are measures of flammability. It indicates the chances of fire hazard while storing various lubricating oils. *Flash point* is the minimum temperature at which a liquid gives off vapours that will ignite for a moment (but do not burn) when a small flame is brought near it. Fire point is the minimum temperature at which the vapours of the oil burn continuously for at least five seconds when a flame is brought near it. It is used to indicate fire hazards of petroleum products and evaporation losses under high temperature operations.

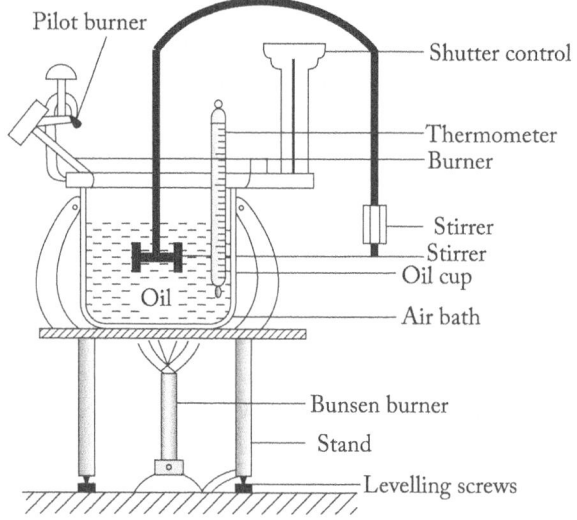

Fig. 18.8 Pensky–Martens flash point apparatus

The knowledge of flash–fire points in lubricating oils helps to take preventive measures against fire hazards. Thus, lubricating oil with a moderate flash-fire point is preferred. Pensky–Martens apparatus shown in Fig. 18.8 is used to determine the flash–fire points of lubricating oils.

Determination of Flash–Fire Points Using Pensky–Martens Apparatus

In this method, oil cup is filled with the test oil. The stirrer and thermometer are placed inside the oil cup. The apparatus is equipped with burner arrangement. During operation, oil is heated and slowly the shutter of the burner is opened and a small flame is lit near the burner nozzle. The flash sound is recorded as the flash point of the test oil. If the flame continues to burn for more than 5 seconds after the flash, it is taken as the fire point.

18.5.3 Cloud and Pour Points

Cloud and pour points indicate the suitability of lubricants in cold conditions. *Cloud point* is the temperature at which an oil becomes cloudy or hazy. It is generally the temperature at which solidified compounds like waxes, present in the lubricating oil begin to crystallize and separate from solution. *Pour point* is the temperature at which the oil ceases to flow or pour. A good lubricating oil must have low cloud and pour points than the operating temperature of machine, or else it may solidify at lower temperatures, jamming the moving machine parts.

For determining cloud and pour points, the test oil is placed in a freezing mixture as shown in Fig. 18.9. In intervals of every 5 seconds, the oil is

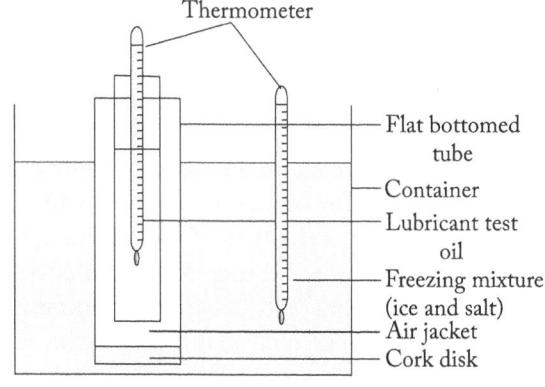

Fig. 18.9 Cloud–pour point apparatus

checked for cloudy appearance. The temperature at which the oil appears cloudy is recorded as its cloud point. Cooling is continued and the oil is tested after every 30 °C fall in temperature. The test tube is tilted horizontally to check the oil's pouring ability. The temperature at which test oil ceases to flow is recorded as its pour point.

18.5.4 Other Significant Properties

Oiliness

Oiliness is the ability of lube oil to cling or adhere to the machine parts under heavy load or pressure. Lubricating oils with poor oiliness will be easily squeezed out when machines are operated under heavy load. Petroleum oil has poor oiliness, but vegetable oils have good oiliness. Petroleum oils are mixed with animal/vegetable oils or fatty acids like oleic acid and stearic acid.

Emulsification and De-emulsification

Oils form intimate mixtures with water called emulsions and this is referred to as emulsification. Such emulsions have a tendency to trap debris, dirt, particulate matter formed during machinery working and results in clogging of machine parts. An ideal lubricating oil should form an emulsion with water that breaks off easily which is denoted as de-emulsification number. De-emulsification number is determined by recording the time required (in seconds) for a given volume of oil to separate as distinct layer from an equal volume of condensed steam under standard conditions (ASTM test) called the *steam emulsion number*. A good lubricant should possess a low steam emulsion number.

Aniline Point

Aniline point, as the terms suggests, is the property of oils to completely mix with aniline. It indicates the aromatic content in a lubricating oil. Lubrication oils with high aromatic hydrocarbon content are known to degrade polymeric materials, like rubber sealants and tubings. *Aniline point* is defined as the lowest temperature at which an oil is completely miscible with equal volumes of aniline (pure).

It is determined by taking equal volumes of aniline and oil in a glass tube as shown in Fig. 18.10. The mixture is stirred vigorously in a glass tube equipped with aniline point thermometer (appropriate range). The oil–aniline mixture is heated until the two forms a homogeneous phase and finally cooled slowly. The temperature at which the two phases separate out is recorded as the aniline point. A lubricant with higher value of aniline point (generally 65.5 °C or higher) indicates low aromatic content and is desirable, especially when using rubber sealants in machines.

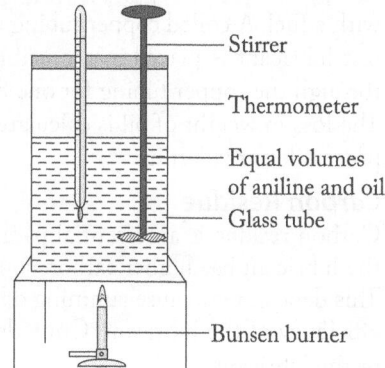

Fig. 18.10 Aniline point estimation apparatus

Saponification Value

Saponification value is the number of milligrams of potassium hydroxide required to saponify or neutralize one gram of oil. It is a characteristic property of vegetable/animal oil; mineral/synthetic oils do not undergo saponification. The reaction involves alkaline hydrolysis of pure oil that yields soap and glycerol. Saponification value helps to estimate the stability of lubricating oil in aqueous or alkaline medium. It also provides an indication of the drying property of the lubricating oil which is harmful to the machine parts.

For determining the saponification value, a known quantity of the oil (W g) is treated with a known excess volume of alcoholic KOH. It is then stirred vigorously and refluxed for two hours in a water bath using a water condenser. During the reaction, fatty acids form potassium salts (soaps) and glycerol is obtained as a by-product.

$$\text{Lubricant} + \text{KOH} \longrightarrow \text{Soap} + \text{Glycerol} + \text{Unreacted KOH}$$

The unreacted potassium hydroxide is back titrated against standard hydrochloric acid. The volume of unreacted KOH is known from the volume of HCl consumed; thus the quantity of KOH consumed can be calculated.

$$\text{Saponification value} = \frac{\text{Amount of KOH consumed (blank} - \text{back)} \times \text{Normality of KOH} \times 56}{\text{Weight of oil (g)}} \dots \text{mg of KOH}$$

Acid Value

Acid value is the number of milligrams of potassium hydroxide needed to neutralize free fatty acids in per gram of oil. It is used to determine the content of free acids in oil; their presence can harm machines during lubrication. On prolonged exposure to oxygen, mineral oils undergo oxidation forming carboxylic acids which makes oil unsuitable for lubrication. Acid value should be less than 0.1. Higher the acid value, more will be the corrosion of machine parts with eventual higher maintenance cost.

Add 50 ml of ethanol to 5g of the test lubricant oil and heat over a water bath for 30 minutes. Alcoholic hydrolysis takes place resulting in the formation of free acids. The release of free acids is tested by a simple acid–base titration. A few drops of phenolphthalein indicator is added to the hot solution and titrated against standard KOH until a pale permanent pink colour appears at the end point.

$$\text{Acid value} = \frac{\text{Amount of KOH consumed} \times \text{Normality of KOH} \times 56}{\text{Weight of oil (g)}} \dots \text{mg of KOH}$$

Volatility

An ideal lubricating oil should not volatilize off when applied under extreme operating machinery conditions. On volatilization, the oil leaves behind a residual oil which may possess properties (increased viscosity) quite different from the original oil.

Vaporimeter is an apparatus used to determine the volatility of a lubricant. It consists of a furnace heated with a fuel. A coiled copper tubing and a thermometer are placed in the furnace. A known weight of the test lubricant is placed in a crucible (Fig. 18.11). Dry air is allowed to pass at the rate of 2 litre/min through the copper tubing for one hour. The crucible is withdrawn from the furnace and allowed to cool. The loss in weight of oil is calculated as the percent weight of oil taken initially. A good lubricant must possess lower volatility.

Carbon Residue

Carbon residue is an essential parameter in a lubricant if employed in internal combustion engines. If the lubricant has higher carbon content, the residual carbon deposits in the lube oil during combustion. This deposit may cause jamming of internal combustion engines. Lower the carbon residue, better is the efficiency of the lubricant. Conradson's apparatus (Fig. 18.12) is used to determine the carbon residue of the lubricant.

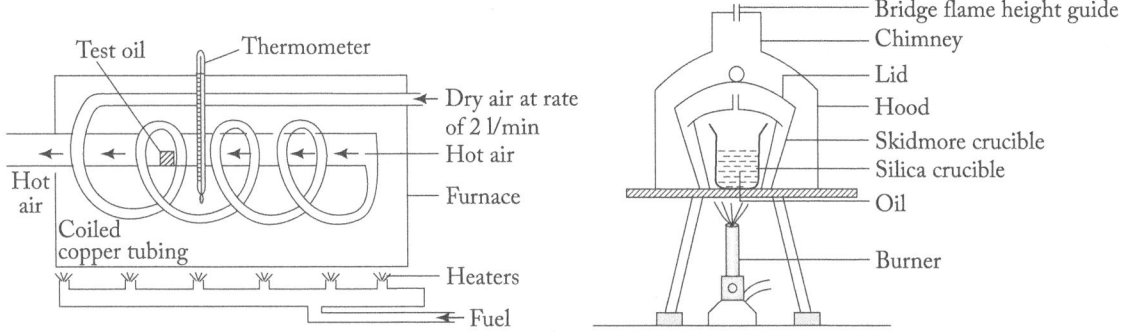

Fig. 18.11 Vaporimeter   **Fig. 18.12** Conradson's apparatus

A crucible is filled with known quantity of oil. A skidmore crucible is placed which is provided with a lid and a small opening that allows venting of volatile matter. Both the crucibles are placed in a wrought iron crucible covered with an iron hood, which acts as a chimney to allow escape of combustion gases. The oil is slowly heated for 5–10 minutes followed by stronger red-hot heating for 15 minutes until all the volatile matter is completely burnt. After cooling, the residual weight of carbon matter is expressed as percent of the initial weight of oil.

Drop (or Dropping) Point

As grease is a mixture of lubricating oil and soap thickeners, it does not exhibit melting point, hence it is imperative to know the temperature at which greases turn so hot that it loses its inherent semi-solid consistency. The temperature at which grease passes in to the liquid state is called drop point under standard test conditions. It refers to the upper temperature limit of applicability of greases. The drop point test apparatus (Fig. 18.13) consists of a metallic sample cup placed in a glass case provided with a lid. The sample cup is provided with an opening at the bottom.

A thermometer is placed in the sample cup such that bulb of the thermometer is just above the grease sample. The entire assembly is placed in a glass beaker filled with water and provided with a stirrer and a thermometer. Table 18.2 shows a list of common industrial greases and their drop points.

The water is heated at the rate of 1°C/minute with continuous stirring until the grease sample passes in to fluid state (up to 17 °C below estimated drop point) and the temperature at which its first drop falls from the opening in the glass case is recorded as the drop point.

Hence, if grease is heated above its drop point and further allowed to cool, it may not regain its semi-solid consistency, and its performance will be adversely affected. Thus, during operation the drop point of greases should not be exceeded.

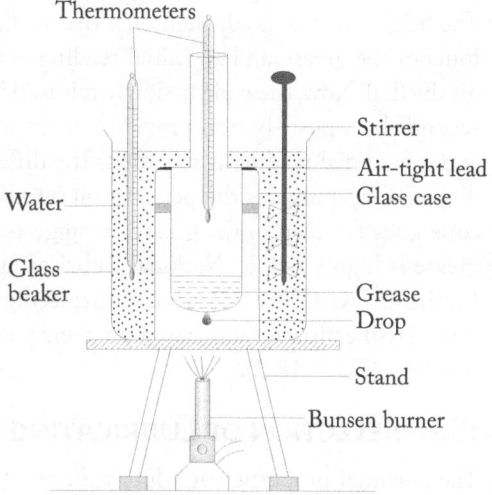

Fig. 18.13 Drop point apparatus

Table 18.2 Drop points of common greases

Grease	Drop point (°C)
Lithium greases	190–220
Calcium greases	135– 145
Sodium greases	175
Aluminium greases	110–115

Penetration Number

The consistency of grease is usually expressed in terms of penetration number. Grease penetrometer (Fig. 18.14) is an apparatus used to determine the consistency of grease. Penetration number is the distance in tenth of a millimetre that the standard cone (of penetrometer) penetrates vertically in the grease sample under standard conditions of load (150 g), temperature (25°C) and time (5 s).

A penetrometer comprises a base made of cast-iron alloy equipped with levelling screws, spirit level, and a table with a box containing the grease sample. A vertical stand made of iron rod is fitted to the base. A standard penetration cone is held with the vertical stand.

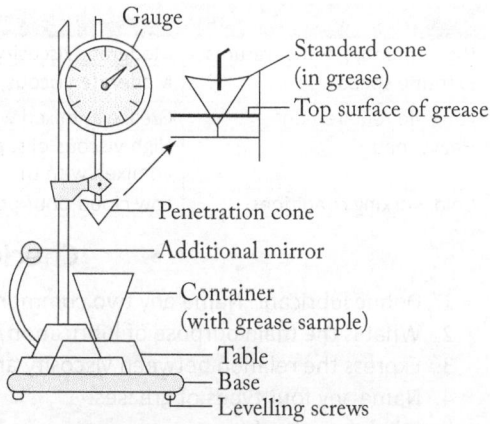

Fig. 18.14 Grease penetrometer

The stand is equipped with holder and screw to move the holder vertically during the experiment. The holder also carries a dial gauge which provides the depth of penetration of grease in millimetres. An additional mirror is provided for positioning the standard cone in the grease sample and eliminating any parallax errors.

During working, the apparatus assembly is properly levelled on a flat working surface. Grease sample is placed in a box and placed below the standard cone. The height of cone is adjusted such that its tip merely touches the grease and an initial reading is recorded on the dial. Now, the cone is slowly released just for 5 seconds by a push-button provided in the apparatus and a final dial reading is recorded. The difference in dial readings provides the penetration number. If the cone goes far deep in the grease, it suggests that the grease is highly fluidic. National Lubricating Grease Institute (NLGI) categorizes lubricating greases into classification groups based on their penetration numbers (Table 18.3).

Table 18.3 Grease categories as per NLGI with penetration number ranges

NLGI Class	Penetration number range (mm/10)	Consistency of grease (at 25°C)
000	445–475	Very liquid
00	400 – 430	Liquid
0	355 – 385	Semi-liquid
1	310 – 340	Very soft
2	265 – 295	Soft
3	220 – 250	Semi-solid
4	175 – 205	Solid
5	130 – 160	Very solid
6	85 – 115	Extremely solid

18.6 SELECTION OF LUBRICATING OIL

The essential properties of a lubricant are: (a) adequate lubricant film strength and (b) chemical–thermal stability under machinery operating conditions. As given in Fig. 18.15, a ready-reference is generated to select the lubricating oil for the machinery under test. Load and stress are major factors that affect selection of lube oils for the given machinery type.

Most lubricants are chosen primarily to reduce friction, and wear and tear (Table 18.4). However, the selected lubricant should be able to resolve corrosion problems. Ideally one must select a low-cost oil and check for minimum maintenance of the lubricant film during machine operation.

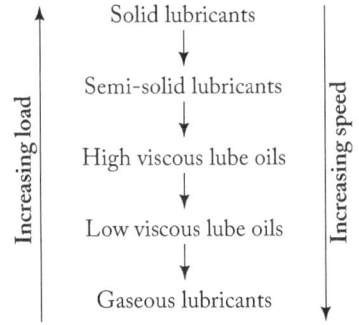

Fig. 18.15 Schematic representation of selecting lubricants

Table 18.4 Choice of general lubricants based on the type of machinery

Type of machinery	Lubricant
High operating temperatures	Moderate viscosity index oil, antioxidant additives, solid lubricants
Extreme speed	Moderate viscous oil with greater flow rate, gas lubricants such as N_2, Ar, and air
Long life requirement	Grease admixed with anti-wear additives
Heavy load	High viscous oils, greases, solid lubricants like graphite, extreme pressure additives admixed with thermal stabilizing additives, coolants
Cold working conditions	Low cloud-pour points, naphthenic base oils

Check Your Progress

1. Define lubricant. Name any two commonly used lubricants.
2. What is the main purpose of lubrication?
3. Express the relation between viscosity and viscosity index.
4. Name any four types of greases.
5. State the role of soap in greases.

To get better performance, additives are generally blended with the lubricating oils. Various additives are employed as follows:

(a) Detergent and dispersant additives such as calcium and barium salts remove corrosion products of the oil from the surfaces.

(b) Corrosion inhibitors prevent corrosion due to combustion and oxidation products. Refrigeration system lubricants work at very low temperatures, and hence they need to have low pour points.

(c) Anti-wear additives prevent metal–metal contact and are employed under moderate load conditions. Zinc dithiophosphate and stearic acid are typical anti-wear, anti-corrosive additives.

(d) Antioxidant additives like phenols, amines, organic phosphates retard the oxidation of oil that cause corrosion of intrinsic machine parts by minimizing sludge formation.

(e) Anti-foam additives like silicone and glycols prevent the formation of stable foams thereby reducing chances of corrosion.

(f) Polymeric additives are admixed with low pour point lube oils (cold conditions). Waxy crystals are formed at lower temperature, therefore polymeric additives like methacrylates and polyalkylphenol esters are added.

(g) Extreme pressure additives such as sulphurized fats, chlorinated hydrocarbons, metallic soaps (lead naphthenate) tend to adsorb on the metal surface, either physically or chemically, and form tough adherent films inhibiting corrosive attack of underlying metal.

The following is a comprehensive discussion on lubricant selection for specific machineries.

Refrigeration systems As refrigerators operate at low temperatures, oils with lower cloud and pour points are preferred. A lubricant oil with a pour point of about –25 to 4.4 °C is employed for the purpose; for example, naphthalene-based oils.

IC engines The lubricant should be able to withstand high temperature operating conditions. Higher viscosity indices along with high thermal stability are preferred for engine lubricants. The lubricant should not emulsify in the presence of water as it causes corrosion of engine parts (cylinder and piston). Examples include petroleum-based mineral oils mixed with antioxidant additives.

Steam turbines High viscous oils are preferred in steam turbines. Mineral oils are used for superheated steam whereas blended oils are employed in wet saturated steaming conditions. It is essential to add de-emulsifier additives in blended oils so that stable oil–water emulsion (that may hinder steam turbine operations) is not formed. The lubricants must possess chemical stability and antioxidant and anti-foaming properties.

Transformers In transformers, lubricants are used to insulate the windings and transfer away the heat generated during loading operations. An ideal lubricant must possess heat transfer efficiency, good dielectric properties, chemical stability and low viscosity as it is circulated in windings.

Spindles Light-load spindles work at high speeds and thin oils with slightly lower viscosities are employed. Anti-corrosion and rust inhibitors are added to the lubricant.

Gears The gear lubricants are usually subjected to extreme pressure conditions; thus they must possess adequate oiliness to withstand high centrifugal force applications. They should have higher oxidation resistance and heavy load-bearing capacity. Extreme pressure additives such as sulphur- or phosphorus-based compounds are added in mineral oils that act as gear lubricants.

Aircraft engines There are stringent requirements for aircraft lubricants such as:

(a) cooling and cleaning properties at all altitudes and atmospheric conditions;

(b) low viscosity for easier flow within engine parts;

(c) low pour points so that it can readily flow under extreme lower temperatures; and

(d) low carbon residue; practically the lubricant must be contaminant-free.

Cutting Fluids and Lubricant Emulsions

Cutting fluids (or machining fluids) reduce friction through lubrication and cooling when applied to machines with operations that produce huge frictional heat. These fluids are required for machine tools required for cutting, sawing, drilling, polishing, etc. Cutting fluids can also hinder corrosion of machine tools and remove machined chips (wear-out material). An ideal cutting fluid should be non-toxic, odourless, non-corrosive, possess good lubricating property, chemical stability, and have higher thermal conductivity, and low viscosity. All these properties of cutting fluids help in increasing the life span of machines. Oil–water emulsion and water–oil emulsion are commonly employed as cutting fluids.

The composition of a cutting fluid includes a base oil, emulsifiers, corrosion prevention additives, pH regulator, wear protection agents, biocides, and anti-foaming agents. The base oils could be mineral oils followed by emulsifiers such as surfactants, particularly sulphonate soaps. Certain amines, amides, and fatty acids act as pH regulators. Wear-protecting agents are sulphur- or phosphorus-based compounds that minimize wear of machine parts. Biocides such as formaldehyde are used to arrest microbial attack in the cutting fluids. Anti-foaming agents such as silicone oils, wax, and emulsions are employed to prevent foam generation (which reduces visibility) during the machining operation.

Table 18.5 Differences between lubricant emulsion and microemulsion

Property	Emulsion	Microemulsion
Appearance	Cloudy or turbid	Transparent
Droplet size	> 500 nm	20 – 200 nm
Stability	Quite unstable	High stability
Viscosity	High	Lower

We know that oil and water do not mix intimately due to high interfacial surface tension between the two phases. On adding an emulsifier (or surfactants), these phases can be mixed intimately as they reduce the surface tension at the interphase of the liquids. A surfactant has a polar head and a non-polar hydrocarbon tail. The polar end of surfactant will get attracted towards the water phase and the tail end will be attracted towards oil. This results in greater miscibility of oil–water phases. It should be noted that interphase still exists and one cannot intimately prepare a uniform oil–water solution. Apart from reducing the surface tension of liquids, emulsifier also provides lubrication, anti-foaming property, and good heat stability to the base oil. Today, multiple emulsions are also employed such as oil-in-water-in-oil (O/W/O) type and water-in-oil-in-water (W/O/W) type that provide greater surface area and consequently, better lubricating property.

18.7 BIODEGRADABLE LUBRICANTS

Either during operation or after use, lubricants are known to pollute the environment. It can also occur as a result of spillage, emissions, or leaks from the machinery. Worldwide, about 40 million tonnes of lubricants are consumed annually, of which the major share is of petroleum-based oils. Though an easier option, petroleum oils are known to have adverse environmental impacts. Today, biodegradable lubricants are being considered as an alternative and viable option for long-term sustainable lubrication. *Biodegradability refers to the breakdown of organic matter by microorganisms to innocuous (non-toxic) products with little or no impact on environment (air, water, or land).*

Check Your Progress

6. Express the equation of viscosity index.
7. List the various functions of a lubricant.
8. How does Teflon reduce friction of machine parts?
9. Name some solid lubricants. In what form can one employ solid lubricants?
10. What is a cutting fluid?

Disposal of used lubricant oils is a serious environmental concern. According to Karnataka State Pollution Board, over 25,000 barrels of waste and used oil are generated in the State which requires reprocessing. Hence, biodegradable oils that require lesser processing after use are being considered as viable lubricants.

The term 'biodegradable lubricants,' covers bio-based oils that are either inherently degradable or readily degradable. According to OECD (Organization for Economic Co-operation and Development), lubricants that exhibit at least 60% degradability are called 'bio-based lubricants.' Biodegradable lubricants are generally prepared from vegetable oils (e.g., soy bean, rapeseed, and castor oil) or animal oils and often mixed with polymers such as starch or chitosan. Vegetable oils are composed of triglycerides (tri-esters) with long chain fatty acid chain with unsaturation across the chain. The presence of polar groups makes it ideal for hydrodynamic and boundary lubrication. However, to act as a biodegradable lubricant, vegetable oils are modified by several chemical reactions such as esterification, epoxidation, additive treatment, and others to enhance their stability under the machinery operating conditions.

Advantages   Some of the advantages of biodegradable lubricants are as follows:
1. As they are biodegradable, no residue is left behind in the machinery after the operation.
2. Low toxicity towards the environment (environmentally compatible).
3. They have high flash–fire points making them suitable to employ at higher temperatures.
4. Vegetable oils possess excellent oiliness with adequate viscosity making them suitable for high pressure conditions.

Disadvantages   The disadvantages of biodegradable lubricants are as follows:
1. Bio-based oils are susceptible to oxidation leading to undesirable structural changes in them.
2. They have high pour points making them unsuitable for employing in machinery working at lower temperatures.
3. When additives are mixed with bio-oils, they themselves do not possess biodegradability as most of them are polymeric in nature.
4. Presence of waxes in vegetable oils tends to clog the internal machine parts, hence continuous maintenance is required.

Mobil Corporation is a pioneer in biodegradable lubricants that introduced 'Environmental Awareness Lubricants (EAL) for hydraulic fluids. Lubrizol Corporation is yet another pioneer in sunflower-based lubricating oils and eco-friendly additives. Today, the recycle and reuse of biodegradable lubricants is being researched and it is not long before when most mineral oils will be successfully replaced with eco-friendly oils.

Check Your Progress

11. Why should a good lubricant possess low steam emulsion number?
12. List the various additives employed in lubricating oils.
13. Justify the statement, 'a good lubricant should have high viscosity index.'
14. What are synthetic lubricants? Name any two synthetic lubricants.
15. What type of lubricant is preferred for refrigeration systems?
16. What type of lubrication is employed in delicate instruments like watches, scientific instruments?

SOLVED EXAMPLES

1. 5 g of vegetable oil was saponified using an excess of 0.5N alcoholic KOH. The mixture required 16 ml of 0.5N HCl. Blank titration reading was 45 ml. Calculate saponification value of the oil.

Given: Weight of oil = 5 g; Blank titration reading = 45 ml; Back titration reading = 16 ml

Solution: Since,

Volume of 0.5N KOH required by the oil for saponification in terms of 0.5N HCl
$$= (45 - 16) \text{ ml } = 29 \text{ ml}$$

$$\therefore \text{Saponification value} = \frac{\text{Amount of KOH consumed (blank - back)} \times \text{Normality of KOH} \times 56}{\text{Weight of oil (g)}}$$

$$= \frac{(45 - 16) \times 0.5 \times 56}{5} = 162.4$$

Result: Saponification value of the oil was found to be 162.4 mg of KOH.

2. 1.53 g of oil was saponified with 23 ml of N/2 alcoholic KOH. After refluxing the mixture, it required 15 ml of N/2 HCl solution. Calculate the saponification value of oil.

Solution: Given: Weight of oil = 1.53 g; Blank titration reading = 23 ml; Back titration reading = 15 ml;

Normality of KOH and HCl = N/2 = 0.5N

Volume of 0.5N KOH required by the oil for saponification in terms of 0.5N HCl = (23–15) ml = 8 ml

$$\text{Saponification value} = \frac{\text{Amount of KOH consumed (blank - back)} \times \text{Normality of KOH} \times 56}{\text{Weight of oil (g)}}$$

$$= \frac{8 \times 0.5 \times 56}{1.53} = 146.40$$

Result: Saponification value of the oil was found to be 146.40 mg of KOH.

3. 1g of lubricating oil of saponification value 148 mg of KOH was saponified using 0.4 N alcoholic KOH. The blank titration reading was 45 ml of 0.4 N HCl solution. Find the quantity of alcoholic KOH required for the saponification reaction.

Solution: Given: Saponification value of oil = 148 mg of KOH; Blank titration reading = 45 ml

Weight of oil = 1g; Normality of KOH= 0.4N

$$\text{Saponification value} = \frac{\text{Amount of KOH consumed (blank - back)} \times \text{Normality of KOH} \times 56}{\text{Weight of oil (g)}}$$

Consider the back titration reading as x ml; hence,

$148 = (\text{Blank} - \text{Back}) \times 0.4 \times 56$; $148 = (45 - x) \ 0.4 \times 56$

Hence, $x = 1008 - 148/22.4 = 38.39$

Result: Quantity of alcoholic KOH required by oil for saponification was found to be 38.39 ml.

4. 7 g of oil was saponified with 25 ml of 0.5 N alcoholic KOH. The reaction mixture then required 15 ml of 0.5N HCl. Calculate the saponification value of the oil.

Solution: Given: Weight of oil = 7 g; Normality of KOH and HCl = 0.5N

Volume of 0.5N KOH required by oil for saponification in terms of 0.5N HCl = (25–15) ml = 10 ml

$$\text{Saponification value} = \frac{\text{Amount of KOH consumed (blank - back)} \times \text{Normality of KOH} \times 56}{\text{Weight of oil (g)}}$$

$$= \frac{(25 - 15) \times 0.5 \times 56}{7} = 40$$

Result: Saponification value of the oil was found to be 40 mg of KOH.

5. A vegetable oil was tested for its acid value. 10 g of the oil required 2.2 ml of 0.02 N KOH. Calculate the acid value and determine whether the oil is suitable for lubrication.

Solution: Data: Weight of oil = 10 g; Volume of KOH = 2.2 ml; Normality of KOH = 0.02 N

$$\text{Acid value} = \frac{\text{Amount of KOH consumed} \times \text{Normality of KOH} \times 56}{\text{Weight of oil (g)}}$$

$$= \frac{2.2 \times 0.02 \times 56}{10} = 0.246$$

Result: Acid value of vegetable oil was found to be 0.246 mg of KOH per gram of oil. As the acid value of vegetable oil is greater than 0.1 mg of KOH, the oil is not suitable for lubrication.

6. 9.0 ml of oil taken from a machine was found to require 1.5 ml of 0.04 N KOH. Calculate its acid value, if density of the oil is 0.81 g/cm^3.

Solution: Given: Volume of oil = 9 ml; Volume of KOH = 1.5 ml; Density of oil = 0.81 g/cm^3;

  Normality of KOH = 0.04 N

  Mass of oil = Volume × Density = 9 × 0.81 = 7.29 g

$$\text{Acid value} = \frac{\text{Amount of KOH consumed} \times \text{Normality of KOH} \times 56}{\text{Weight of oil (g)}}$$

$$= \frac{1.5 \times 0.04 \times 56}{7.29} = 0.461$$

Result: Acid value of the oil was found to be 0.461 mg of KOH per gram of oil. As the acid value is greater than 0.1 mg of KOH, the oil is not suitable for lubrication purpose.

7. 1.3 g of oil required 0.8 ml of 0.001N KOH for neutralization. Calculate the acid value and mention whether the oil is suitable to be used as a lubricant.

Solution: Given: Weight of oil = 1.3 g; Volume of KOH = 0.8; Normality of KOH = 0.001N

$$\text{Acid value} = \frac{\text{Amount of KOH consumed} \times \text{Normality of KOH} \times 56}{\text{Weight of oil (g)}}$$

$$= \frac{0.8 \times 0.001 \times 56}{1.3} = 0.034$$

Result: Acid value of the oil was found to be 0.034 mg of KOH. As the acid value is less than 0.1 mg of KOH, the oil is suitable to be used as a lubricant.

Activity-based Questions

A. Aviation Lubricants

With the introduction of gas turbines for aircraft propulsion, it is not possible to employ mineral oil based lubricants. The critical nature of the aircraft machinery necessitates critical requirement for specified details on the performance characteristics of the lubricants. Various properties like cloud-pour points, viscosities and their indices are just some of the crucial factors in aviation lubricants.

Present a technical study on any one type of aircraft engine and explain the various lubricant technologies adopted in India. Include the examples of aircraft failures (any one) citing reasons.

B. Failure Analysis of Lubricants

Lubrication-related failures are generally due to a variety of reasons; viz., (a) poor lubricant selection, (b) poor application, (c) lubricant contamination or (d) lubricant degradation during operation. Precision lubrication is important to prevent failures while operating machines. The common types of failures are catastrophic and functional. A third type of failure is the lubricant-related failures which are largely ignored.

Discuss the above observations and give a suitable real-time example of lubrication failure that led to a catastrophic event.

SUMMARY

- ○ Lubricants and the concept of lubrication, friction and wear are studied under tribological systems.
- ○ The primary functions of lubricants are to reduce friction between two moving machine parts and to also provide cooling and sealing effects.
- ○ Hydrodynamic, boundary, and extreme pressure are the three major mechanisms of lubrication systems.
- ○ The important properties of lubricating oils are viscosity, flash–fire points, cloud–pour points, oiliness, emulsification, aniline point, and saponification and acid values.
- ○ Solid, liquid, and gaseous lubricants are the most commonly studied lubricants.
- ○ Grease, mineral oils, and synthetic oils are considered versatile and multipurpose lubricating oils.

- ○ While selecting the lubricating oil, one must choose oils with adequate viscosity providing good adherence to applied surfaces. The type of load and stresses on the machine dictates the choice of lubricating oil.
- ○ Various additives such as extreme pressure, antioxidants, anti-corrosive, and anti-wear additives are employed so as to attain desired lubrication.
- ○ The speed, load, and operating parameters like temperature and pressure of a machine are taken into account while selecting a lubricant.
- ○ Biodegradable lubricants have been found to be environmentally compatible and will soon replace mineral oils.

GLOSSARY

Acid value: The number of milligrams of KOH required to neutralize free fatty acids in one gram of lubricating oil.

Aniline point: The lowest temperature at which an oil is completely miscible with equal volumes of aniline.

Biodegradable lubricants: Include bio-based oils that are either inherently degradable or readily degradable (up to 60% as per OECD).

Cloud point: The temperature at which an oil becomes cloudy or hazy in appearance.

Drop point: The temperature at which grease passes in to liquid state under standard test conditions.

Emulsification: The property of oils to get intimately mixed with water forming an emulsion.

Fire point: The minimum temperature at which the vapours of the oil burn continuously for at least five seconds when a flame is brought near it.

Flash point: The minimum temperature at which a lubricant/oil gives off vapours that will ignite for a moment when a small flame is brought near it.

Kinematic viscosity: The measure of relative flow of a lubricant under the influence of gravity.

Lubricant: A substance introduced between two moving or sliding surfaces (moving machine parts) to reduce frictional resistance.

Lubrication: The process of reducing friction between moving surfaces by introduction of lubricants.

Oiliness: The ability of lube oil to cling or adhere to the machine parts under heavy load or pressure.

Penetration number: The distance in one-tenth of a millimetre that the standard cone (of penetrometer) penetrates vertically in the grease sample under standard conditions of load (150 g; 25 °C; 5 s).

Pour point: The temperature at which an oil ceases to flow or pour.

Saponification value: The number of milligrams of potassium hydroxide required to saponify one gram of oil.

Steam emulsion number: The time taken for a given volume of oil to separate as distinct layer from an equal volume of condensed steam under standard conditions.

Viscosity: The property by virtue of which a liquid or fluid (oil) offers resistance to its own flow.

Viscosity index: An empirical number, defined as the rate at which viscosity of oil changes with temperature is measured.

KEY FORMULAE

- Coefficient of friction, $\mu = F/W$

- Viscosity index (VI) is $VI = \dfrac{V_L - V_T}{V_L - V_H} \times 100$

- Saponification value = $\dfrac{\text{Vol. of KOH consumed (blank - back)} \times \text{Normality of KOH} \times 56}{\text{Weight of oil (g)}}$

- Acid value = $\dfrac{\text{Vol. of KOH consumed} \times \text{Normality of KOH} \times 56}{\text{Weight oil (g)}}$

EXERCISES

Multiple Choice Questions

1. Of the following, _____ is a semi-solid lubricant.
   (a) grease
   (b) graphite
   (c) WS_2
   (d) MoS_2

2. Calcium soap base grease is also called
   (a) soap grease
   (b) axle grease
   (c) cup grease
   (d) oily grease

3. The reference oil(s) in determining viscosity index is/are
   (a) blended oils
   (b) Gulf oils
   (c) Pennsylvanian oils
   (d) both (b) and (c)

4. Acid value of the lubricant should be
   (a) > 1
   (b) > 0.1
   (c) < 0.1
   (d) 0

5. Coefficient of friction for a fluid film is
   (a) 0.01 – 0.03
   (b) 0.001 – 0.03
   (c) 0 – 0.1
   (d) 0 – 0.01

6. Coefficient of friction for a boundary film is
   (a) 0.05 – 0.15
   (b) 0.5 – 15
   (c) 0.005 – 0.15
   (d) 0.5 – 10

7. The apparatus used to determine viscosity of a lubricant is
   (a) Redwood viscometer
   (b) Ostwald viscometer
   (c) colorimeter
   (d) Bomb calorimeter

8. The apparatus used to determine flash–fire points of a lubricant is
   (a) spectrometer
   (b) polarimeter
   (c) Pensky–Martens' apparatus
   (d) Redwood viscometer

9. PTFE exhibits
   (a) high thermal stability
   (b) chemical inertness
   (c) non-toxicity
   (d) all of these

10. The thickness of fluid film is
    (a) 1 Å
    (b) 10 Å
    (c) 1000 Å
    (d) 100 Å

11. Of the following, the most versatile class of lubricants is
    (a) solid lubricants
    (b) liquid lubricants
    (c) gaseous lubricants
    (d) none of these

12. Oildag and aquadag refer to
    (a) dispersion of graphite in oil and water
    (b) dispersion of grease in oil and water
    (c) dispersion of mineral oil in grease
    (d) dispersion of mica in oil and water

13. Carbon residue of a lubricant is measured by
    (a) penetrometer
    (b) Conradson apparatus
    (c) vaporimeter
    (d) none of these

14. The aromatic content of a lubricant is found by
    (a) oiliness
    (b) saponification value
    (c) acid value
    (d) aniline point

15. Unit to express viscosity is
    (a) calorie
    (b) poise
    (c) ml/min
    (d) mg KOH

16. Of the following, the oil that is a synthetic lubricant is
    (a) grease
    (b) mineral oil
    (c) Fluorolube
    (d) MoS_2

17. In a Redwood viscometer, volume of oil collected is
    (a) 50 ml
    (b) 600 ml
    (c) 100 ml
    (d) 450 ml

18. Environment-friendly lubricant oil is
    (a) graphite
    (b) bio-oil
    (c) grease
    (d) all of them

19. Lubricant volatility is determined using
    (a) penetrometer
    (b) Conradson apparatus
    (c) vaporimeter
    (d) none of these

20. An ideal lubricant must possess
    (a) adequate viscosity
    (b) low carbon residue
    (c) low flash-fire points
    (d) all of these

Review Questions

1. What are Lubricants? How are they classified?
2. Enlist some common lubricating oils and state their properties.
3. What is lubrication? Explain hydrodynamic lubrication with a suitable example.
4. Explain the properties of the following lubricants and add a note on their significance: (a) Viscosity and viscosity index, (b) Flash and fire points, (c) Saponification value (d) Acid value, (e) Oiliness, (f) Emulsification, (g) Aniline point (h) Drop point (i) Steam emulsion number
5. Discuss the conditions under which solid lubricants cab be used. Explain the use of graphite as a lubricant.
6. List the various functions of lubricating oil.
7. State the properties of MoS_2 and graphite as solid lubricants.
8. What are greases? State the role of soap in greases.
9. Distinguish between hydrodynamic, boundary, and extreme pressure lubrication.
10. How will you determine the viscosity of lubricating oil by using Redwood viscometer?
11. Write short notes on: (a) semi-solid lubricants (b) solid lubricants (c) liquid lubricants.
12. Explain the procedure to determine viscosity of a lubricant using Redwood viscometer.
13. Explain the method to determine cloud and pour points. State the significance of these points in lubricating oils.
14. Discuss the desirable properties of lubricants to be employed for aircraft, steam turbines, refrigerators, and IC engines.
15. Write a short note on cutting fluids and lubricant emulsions.
16. Explain the principle and procedure of extreme pressure lubrication.
17. How does one select a lubricating oil for (a) aircraft, (b) gears, (c) IC engine (d) spindles, (e) refrigerators, (f) transformers.
18. Justify the statement—'graphite is used as a solid lubricant'.
19. Explain the statement—'grease and mineral oils are considered multipurpose oils'.
20. Explain the laboratory procedures to determine saponification and acid values.
21. What are chemical additives? List the various additives employed in lubricants and add a note on the functions of each.
22. What are gaseous lubricants? State their properties with examples.
23. Compare and contrast graphite and MoS_2 solid lubricants.
24. Write a short informative note on 'selection of lubricant oils'.
25. What are biodegradable lubricants? State their advantages and disadvantages.
26. Justify—'MoS_2 is a better lubricant than graphite.'
27. What are the different types of greases? State their applications.
28. Explain various compounds used as additives in lubricants.
29. Explain boundary film lubrication with a suitable example.
30. What are cutting fluids? Name some of the common cutting fluids.

Numerical Problems

1. 1.2 g of cottonseed oil was treated with 26 ml of 0.5N of alcoholic KOH in a flask. After the reaction, the back titre reading was recorded as 12.5 ml of 0.5 N HCl. The blank titre reading was found to be 26 ml. Calculate the saponification value of the cottonseed oil.   (**Ans:** 315 mg of KOH)
2. 5 g of vegetable oil was refluxed with 50 ml of alcohol and 50 ml of 0.5 N KOH solution in a flask. To achieve equivalence point, 20 ml of 0.5 N HCl was consumed in the reaction. The blank titre reading was 52 ml (no sample). Based on the information, calculate the saponification value of vegetable oil.   (**Ans:** 179.2 mg of KOH)

3. 2.5 g of oil was saponified with 0.5 N alcoholic KOH solution. The blank titre reading with 0.5 N HCl was found to be 40 ml, while the back titre reading was 21ml of 0.5N HCl. Based on the information provided, calculate the SV of oil. **(Ans: 212.8 mg of KOH)**

4. 12 g of oil was titrated against 0.02 N standard KOH solution and the titre reading was recorded as 1.6 ml. Comment on the quality of the lube oil.
(Ans: 0.149 acid value i.e, greater than 0.1, hence not suitable for lubrication)

5. 10 ml of lubricant consumed 0.2 ml of 0.02 N of alcoholic KOH solution. If the density of the oil is 0.83g/ cm^3, determine the suitability of the oil for lubrication purposes.
(Ans: 0.026 acid value, oil is suitable for lubrication)

6. 1 g of oil sample with a saponification value 172 was saponified using 0.2 N alcoholic KOH. The blank titration reading was 32 ml of 0.2 N HCl solution. Determine the quantity of alcoholic KOH required for the saponification reaction. **(Ans: 16.6 ml)**

FURTHER READING

1. Ghosh M.K, B.C. Majumdar, and M. Sarangi. *Theory of Lubrication*. Tata McGraw Hill Pvt. Ltd, New Delhi, 2013.

2. Halling, J. *Introduction to Tribology*. Wykeham Publications Ltd., London, 1976.

3. Kogelsky, I.V. and V.V. Allison (eds). *Tribology, Lubrication, Friction & Wear*. Wiley, New York, 2005.

4. Mortier. R.M. and S.T. Orszulik (eds.). *Chemistry & Technology of Lubricants*. Springer Science & Business Media, New York, 1992.

5. Mang. T, and W. Drese (eds). *Lubricants and Lubrication* – Vol.1. Wiley VCH, Germany, 2017.

6. Wills, J.G. *Lubrication Fundamentals*. Marcel Dekker Inc., New York, 1980.

ANSWERS

Check Your Progress

1. A substance introduced between two moving machine parts to reduce frictional resistance between them is called lubricant. Grease and mineral oil are commonly used lubricants.

2. Lubricants reduce friction of sliding surfaces.

3. With increasing temperature, viscosity quickly falls to a lower value, it will have a low viscosity index. On slight decrease in viscosity on increasing the temperature, it will have a higher viscosity index.

4. Greases can be lithium-based, calcium-based, barium-based, axle-based.

5. Soap in grease acts as a thickening agent and enables it to cling or adhere to the machine parts firmly.

6. Refer to Section 18.5.1.

7. The general functions of a lubricant are: (a) reduces surface wear, tear, and friction (b) reduces frictional heat, (c) acts as a sealent and (d) controls corrosion of machine parts.

8. Refer to Section 18.4.1

9. Graphite, molybdenum disulphide, tungsten disulphide, cadmium dichloride, borax, lead iodide, talc, mica, and vermiculite are some of the commonly used solid lubricants. Solid lubricants can be used as a powder or mixed with grease.

10. A substance that can act as a lubricant as well as a coolant is called a cutting fluid.

11. Some oils form emulsions easily. These collect dirt, debris, etc., which may cause abrasion of the lubricated machine parts. Thus, a good lubricant should form an emulsion that breaks off easily, i.e., steam emulsion number should be low.

12. Anti-wear, antioxidants, anti-foam, corrosion inhibitors, polymeric and extreme pressure additives are used in lubricating oils.

13. Higher the viscosity index of the oil, lower is the decrease in the viscosity of the oil with increasing temperature.
14. Non-petroleum based lubricating oils are synthetic lubricants. Silicone oils and polyglycol are examples of synthetic lubricants.
15. Lubricants with lower cloud and pour points with lower viscosity is preferred for referigerators.
16. Fluid-film or hydrodynamic lubrication in employed in watches and scientific construments.

Multiple Choice Questions

1. (a)	2. (c)	3. (d)	4. (c)	5. (b)	6. (a)	7. (a)
8. (c)	9. (d)	10. (c)	11. (b)	12. (a)	13. (b)	14. (d)
15. (b)	16. (c)	17. (a)	18. (b)	19. (c)	20. (d)	

Energy Resources

After reading this chapter, you will be able to:

- distinguish between renewable and non-renewable sources of energy.
- understand the implications of energy usage on the environment.
- appreciate the importance of solar, tidal, wind, ocean, geothermal, hydro, and nuclear energies as non-conventional energy resources.
- explain the concept of ocean thermal energy conversion (OTEC).
- appreciate the importance of nuclear energy as an alternative fuel resource and learn about the mechanism of nuclear fission and working of reactors.
- understand the use of biomass as energy source.

19.1 INTRODUCTION

Energy is defined as the ability to perform work and it can be harnessed from various sources. Energy generation and its consumption are directly correlated to the standard of living of a nation. The energy requirements of the world have steadily increased over the centuries, and is expected to grow further. According to World Bank, nearly 300 million people do not have access to electricity. The conventional fossil fuels such as coal, petroleum, oil, and natural gas are utilized to meet the major energy demands globally. However, due to increasing energy demands, fossil fuels are expected to become depleted in the near future. Further, using fossil fuels have caused an increase in the concentration of carbon dioxide leading to global warming. It is estimated that CO_2 concentration has increased from its normal range of 180–300 ppm to about 278–390.5 ppm. Today, various non-conventional energy resources such as solar, wind, biomass, tidal, geothermal, and nuclear have gained significant momentum. These energy resources are abundant in nature, pollution-free, and eco-friendly. India has estimated that about 175 GW of renewable power will be made available by 2022 comprising 100 solar, 60 wind, 10 hydro, and 5 biomass-based plants.

19.2 RENEWABLE AND NON-RENEWABLE ENERGY SOURCES

Depending on their usability, energy resources can be categorized in to two types, namely renewable and non-renewable sources.

A renewable energy source is defined as the energy source that can be used for harnessing energy using resources that can be continually replenished. Solar energy, hydropower, tidal, biomass, geothermal energy are examples of renewable resources.

Non-renewable energy source is defined as a resource that once consumed cannot be replenished. Coal, petroleum, and natural gas are examples of non-renewable energy resources.

With ever increasing population, conventional energy sources such as firewood, coal, petroleum are not sufficient to meet the energy requirements. Hence, there is an urgent need to find non-conventional or alternative energy resources to meet this ever-increasing demand. Solar, tidal, hydro, biomass, nuclear energy sources are showing promise as energy providers and its full potential is yet underway.

19.3 SOLAR ENERGY

A major source of enormous amounts of energy is the sun which produces about 4×10^{23} kJ/s of radiant energy, out of which only 5×10^{21} kJ/year reaches the earth's surface. The sun is known to contain about 80 % hydrogen, 20 % helium, and 1 % carbon, oxygen, and nitrogen. Solar energy is formed during the thermonuclear fusion reactions hydrogen to helium which can occur in either of the below two mechanisms.

Bethe mechanism (1939) (at higher temperatures):	Salpeter mechanism (1953)
$^{12}C + {}^1H \rightarrow {}^{13}N + \gamma\,(Q = 1.94 \text{ MeV})$	$^{1}H + {}^1H \rightarrow {}^2H + e^+ + v\,(Q = 0.42 \text{ MeV})$
$^{13}N \rightarrow {}^{13}C + c^+ + v\,(Q = 1.2 \text{ MeV})$	$^{2}H + {}^1H \rightarrow {}^3He + v\,(Q = 5.49 \text{ MeV})$
$^{13}C + {}^1H \rightarrow {}^{14}N + \gamma\,(Q = 7.55 \text{ MeV})$	$^{3}He + {}^3He \rightarrow {}^4He + 2\,{}^1H\,(Q = 12.86 \text{ MeV})$
$^{14}N + {}^1H \rightarrow {}^{15}O + \gamma\,(Q = 7.29 \text{ MeV})$	$^{4}H \rightarrow {}^4He + 2e^+ + 2v\,(Q = 24.68 \text{ MeV})$
$^{15}O \rightarrow {}^{15}N + c^+ + v\,(Q = 1.74 \text{ MeV})$	(Q refers to liberated energy)
$^{15}N + {}^1H \rightarrow {}^{12}C + {}^4He\,(Q = 4.96 \text{ MeV})$	
$4\,{}^1H \rightarrow {}^4He + 2e+ + 2v\,(Q = 24.68 \text{ MeV})$	
$2e+ + 2e- \rightarrow 2\,\gamma\,(Q = 2.04)$	

Under the Ministry of New and Renewable Energy (MNRE), India has set up the National Institute of Solar Energy (NISE, Faridabad) which is committed to work towards solar renewable resource. Solar energy can be categorized as thermal, biomass, photovoltaics, and photogalvanics. Let us discuss the various applications of solar energy.

19.3.1 Solar Water Heating

Solar water heating involves the use of flat plate thermal collectors for heating water. A flat plate thermal collector (Fig. 19.1) works on the principle of black body. It consists of a large heat absorber plate made either of copper or aluminium that is blackened for allowing maximum absorption of solar radiation. The blackened heat absorbing surface has several parallel copper tubes called *risers* that are placed lengthwise containing a heat transfer fluid, typically water. The pipes and absorber plate

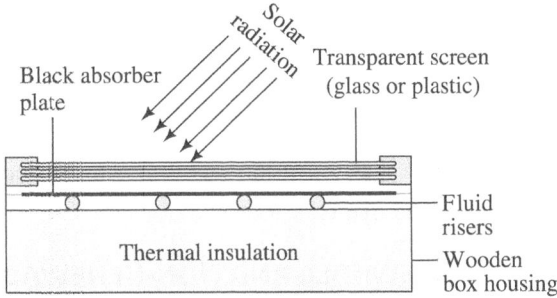

Fig. 19.1 Solar flat plate thermal collector

are enclosed in an insulated wooden box covered with transparent screen (glass or plastic case) so as to protect the enclosed absorber plate and create an insulating air space. The collector is allowed to face the sun and the black bottom becomes hot as it absorbs the sunlight. Further, the water in the circulating tubes also gets heated and is conveyed to the tank where it is stored.

Solar water heating can be active or passive. An *active system* involves pumping water from a storage tank and passing it through a flat thermal collector for heating (Fig. 19.2). Once the water is hot, it leaves

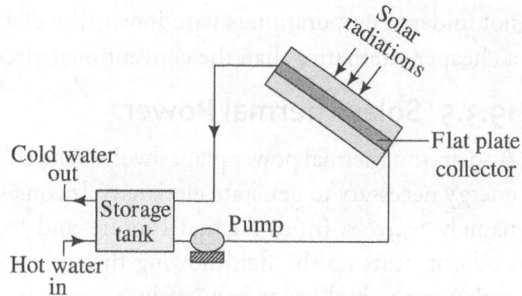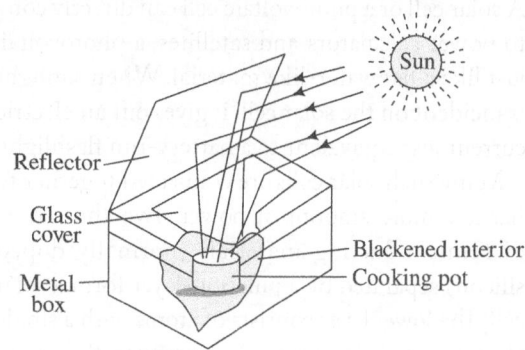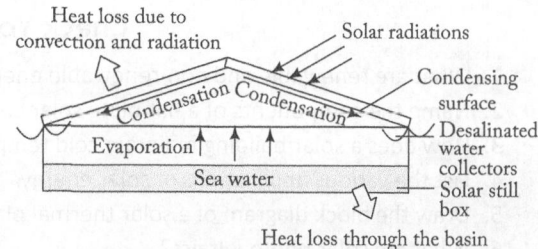

the collector and returns to the storage tank, thereby flowing in a continuous loop. From the storage tank, water is pumped back whenever hot water is required. In the *passive system*, a solar flat plate collector is combined with a horizontally mounted storage tank located immediately above the flat plate collector.

Due to convection, the hot water rises through the solar collector pipes and enters the storage tank. As soon as the hot water enters the storage tank, the cooler water flows down to the collectors by gravity for heating.

Fig. 19.2 Water heating system

19.3.2 Solar Heating of Buildings

The building is so designed that it acts as the flat plate collector with windows generally oriented to face the solar radiations. The windows, walls, and floors are made to collect, store, and distribute solar energy in the form of heat during the winters and reject solar heat during the summer. In the summer, as temperatures rises, a solar building utilizes its thermal mass to keep the building cool. To achieve this, the sun's radiations are kept away from reaching the thermal mass of the building by placing a shade or by covering the window panels.

Indira Paryavaran Bhavan (New Delhi) is the first of its kind solar building in India which houses its own solar-run power plant, sewage treatment facility, and heat exchanger system.

19.3.3 Solar Cooking

A solar cooker is a well-insulated rectangular metal box (see Fig. 19.3) in which the inner sides are etched black and fitted with a flat glass cover. The raw food is placed in the metal box and the cooker is allowed to face the sunlight.

The solar radiations penetrate the glass cover and get completely absorbed by the inner blackened surface of the cooker. Due to solar heat absorption, the vessel gets heated and the food is cooked quickly. Solar cookers are of great use in areas with limited access to water and also provides a pollution-free alternative as no fuels are utilized for cooking.

Fig. 19.3 A solar cooker

19.3.4 Solar Desalination

Figure 19.4 shows a solar still that is utilized for desalinating water using solar energy. A solar still comprises an airtight insulated blackened basin covered with a tilted glass sheet. Solar radiation passes through the transparent glass cover and is absorbed by sea water in the basin. The sea water is heated and causes evaporation.

The water vapour condenses at the inner side of the glazing as purified water and flows due to gravity into a trough or storage tank. The excess brine that does

Fig. 19.4 A typical solar still

not undergo evaporation is withdrawn. The distilled water obtained from the solar still is comparatively a cheaper alternative than the conventional electric-run distillation plants.

19.3.5 Solar Thermal Power

A solar-run thermal power plant involves the collection and concentration of sunlight to produce thermal energy necessary to generate electricity. It consists of solar energy collectors with two main components, namely *reflectors* (mirrors) that capture and focus sunlight on to the *receiver tubes*. The concentrated sunlight heats up the fluid flowing through the receiver tubes. The heated fluid is then sent to a heat exchanger to boil water and produce steam in a conventional steam-turbine generator. In the turbine, steam is converted into mechanical energy that powers the generator to produce electricity (see Fig. 19.5).

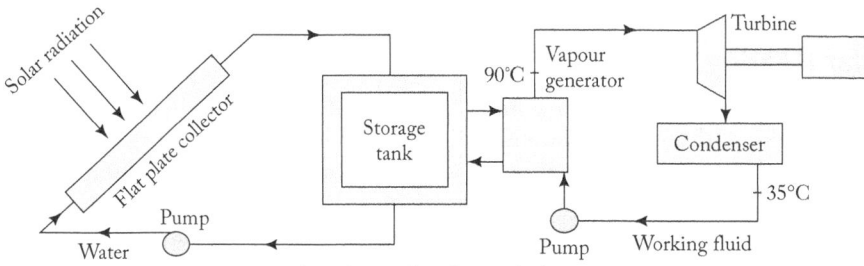

Fig. 19.5 Solar thermal plant

19.3.6 Solar Photovoltaics

A solar cell or a photovoltaic cell can directly convert solar radiations into electric current. Generally used to power calculators and satellites, a photovoltaic cell is made of a semiconducting material and looks just like a flat wafer-like material. When sunlight is incident on the solar cell, it gives out an electric current just equivalent to a battery-run flashlight.

Many such solar cells are connected together to harness more amount of power. Two thin layers of semiconducting materials (normally doped silicon) separated by a junction-layer form a solar cell. The lower layer consists of atoms with a single electron in their outer orbit, whereas the upper layer consists of atoms lacking electrons in the outer orbit. When sunlight falls upon the solar cell material, the kinetic energy of the photons dislodges electrons from the lower layer; this generates a current that flows back to the upper layer.

Fig. 19.6 Solar power plant with solar panels arranged together. (Courtesy: Pixabay permission from sarangib: https://pixabay.com/en/solar-panels-renewable-energy-3507947/)

Check Your Progress

1. What are renewable and non-renewable energy sources?
2. Name the components of a flat plate solar collector.
3. How does a solar building maintain cold temperature during summer?
4. List the various applications of solar energy.
5. Draw the block diagram of a solar thermal plant.
6. What are solar photovoltaics?

19.4 Wind Energy

Wind energy is a converted form of solar energy; also the most widely distributed renewable energy resource. It is considered the second largest and fastest growing source of electricity. In India, it is estimated to have a potential of about 302 GW power at 100 metres. The major wind power states in India include Tamil Nadu, Gujarat, Karnataka, Maharashtra, and Rajasthan.

Fig. 19.7 Horizontal wind turbines (Courtesy: Pixabay permission from sarangib: https://pixabay.com/en/wind-mill-energy-alternative-2251810/)

Generally, wind results from air movements due to atmospheric pressure gradients. Wind flows from regions of higher pressure to those of lower pressure. Larger the atmospheric pressure gradient, higher the speed of wind and hence greater will be the wind power captured by wind turbines. Mathematically, the power available from a windmill can be expressed as, $P_{max} = 0.5 \, \rho a V^3$; where ρ is the density of air, a is the cross-sectional area of windmill disc, and V is the velocity of air.

Some of the **advantages** of wind energy are as follows.

(a) It is a pollution-free renewable source available in large amounts in the atmosphere.

(b) It does not generate greenhouse gases, toxins, or any radioactive waste.

(c) No fuel is utilized to harness wind energy.

(d) Wind energy can further be stored in storage cells or batteries so as to utilize in the absence of wind.

The following are the **limitations** of wind energy.

(a) Windmill farms can be established only where wind blows for the greater part of a year with a minimum wind speed of 15 km/h to keep the turbines moving.

(b) Wind energy farms require large areas of land with higher capital costs.

(c) As the wind tower and blades are exposed to the environment, it requires constant maintenance.

19.5 Geothermal Energy

When natural heat present within the earth's crust is tapped for various energy needs, it is called geothermal energy. The earth's core temperature is about 1000°C and if drilled till the depths of 10 km, one can obtain thermal energy to produce electricity. The primary sources of geothermal energy are hot springs, hot rocks, and molten rocks (called magma) present beneath the earth's surface. It is known that heat is continuously produced within the earth's crust due to decay of natural radioactive elements such as uranium and the frictional dissipation of heat due to movement of tectonic plates. Further, the heat gets transmitted to the sub-surface water and transforms it in to steam which forces out of the surface as hot springs. Normally, geothermal energy is captured from hotspots such as volcanic zones, mineral deposits, and hot springs or *geysers*.

Norris geyser basin at Yellowstone National Park (USA) erupts in every 3 to 4 hours. Chumathang hot spring in Puga Valley, Ladakh is estimated to generate 5000 MW of geothermal energy.

The **advantages** of geothermal energy are as follows.

(a) It is a clean, inexpensive, pollution-free, sustainable alternate energy resource.

(b) It can produce electricity 24 hours a day.

(c) Geothermal power plants usually require less space and hence do not have much effect on the nearby environment.

(d) As no fuel is utilized to generate the power from geothermal heat, the running costs or geothermal power plants are quite lower as compared to the oil extraction or nuclear power plants.

The **limitations** of geothermal energy are the following.

(a) Improper harnessing of geothermal energy may produce toxic pollutants, hazardous minerals, and gases.

(b) Power generation through geothermal energy is still at a nascent stage and requires skill and the support of government.

Various international organizations are involved in exploiting geothermal energy resource to meet the household and industrial power requirements. Several Indian organizations such as National Geophysical Research Institute (Hyderabad), Geological Survey of India (Ministry of Mines, Delhi), Regional Research Laboratory (Jammu), and many others are also devoted towards harnessing the geothermal energy. In 2016, India proposed to harness 10,000 MW (10 GW) of geothermal energy by 2030 with the aid of international collaboration.

19.6 HYDROPOWER

Energy harnessed from falling or moving water is called hydropower (or water power). Hydroelectricity is generated by capturing the kinetic energy of moving water which provides mechanical energy to drive turbines. Turbines can efficiently convert mechanical energy in to electrical energy using generators. The *International Journal on Hydropower and Dams* 2005 and *World Atlas and Industry Guide* (IJHD, 2005) provide an exhaustive inventory of current installed capacities, annual electricity generation, and hydropower potential. As shown in Fig. 19.8, a hydropower station consists of three parts, namely

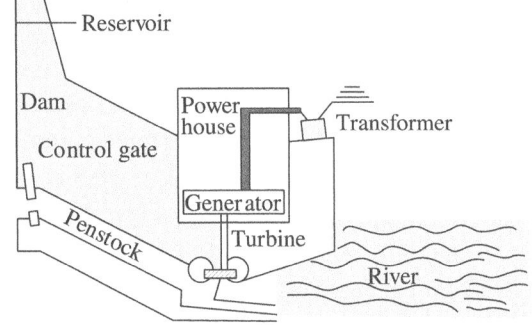

Fig. 19.8 Hydropower station

(a) A water reservoir (site for hydropower generation)

(b) A dam that comprises a flow control system

(c) An electric plant, where the electricity is produced

The working of a hydropower plant involves water flowing from the reservoir through the dam and striking against the turbine blades with a force. The force of water is so maintained that the turbine blades move at a particular speed. The turbine then spins a generator and electricity is generated.

The **advantages** of hydropower are as follows.

(a) It is a robust renewable source of energy.

(b) Smaller hydropower plants are of low cost and eco-friendly in nature.

(c) It can power small localities, where electric grid system is uneconomical.

The **disadvantages** of hydropower are the following.

(a) Construction of dams raises social and environmental concerns.

(b) As hydropower plant is site-specific, there are limited water reservoirs which can be utilized to full potential.

19.7 TIDAL ENERGY

Tidal energy involves harnessing power from tidal waves. The tidal waves are generated due to the gravitational pull on ocean water by the sun and the moon. Tidal flow consists of vertical movement of the rise and fall of ocean water level and horizontal motion in the tidal currents towards and away from the shore. The large amount of energy present in oceanic and tidal waves can be harnessed and stored both when the tidal waves rise and also during its fall. With 7517 km coastline, India has the potential to develop more than 8000 MW of electricity using tidal energy. A dam equipped with turbines allows generation of power by tidal waves in all directions. Tidal steam generators, tidal barrages, tidal lagoons are some of the common tidal power generating methods.

The **advantages** of tidal energy are as follows.
(a) It is a renewable energy resource.
(b) Tidal energy is considered as a carbon neutral method of producing electricity, that is, no carbonaceous residue is generated as obtained on burning coal, wood, and other fossil fuels.

The following are some of the **limitations** of tidal energy.
(a) High capital cost.
(b) Use of tidal turbines may adversely affect the marine life.
(c) As it is dependent on the movement of tidal waves, there will be inconsistent power supply.
(d) Land-locked areas cannot utilize this form of energy.
(e) In spite of the huge potential, due to the lack of policies, tidal energy has not been harnessed to its full extent.

19.8 Ocean Thermal Energy Conversion (OTEC)

Over 70 % of the earth's surface is covered by oceans which are continually absorbing solar radiations. With a solar penetration to only 3 % at 100 m ocean level, a temperature gradient is naturally present in oceans. In 1881, French physicist Jacques D'Arsonval was the first person to suggest extracting heat energy from the oceans. The temperature differences at the surface and the deeper ocean waters can be harnessed for electrical energy. It works on the principle of Carnot thermodynamic efficiency (second law of thermodynamics) which states that a heat engine can perform work while operating between two temperatures. Greater the temperature difference, higher will be the efficiency by which heat is converted in to work. At least a temperature difference of about 20°C is essential to enable working of ocean thermal energy conversion at the power plants. Today, it is envisioned to employ OTEC plants that pump large quantities of deep ocean waters (cold) and surface water to generate electricity.

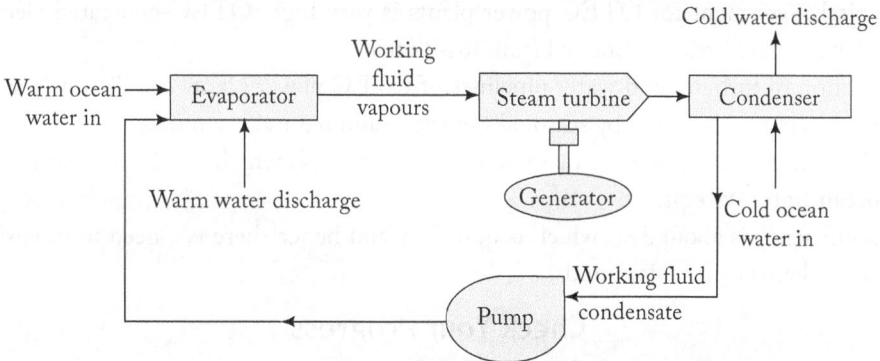

Fig. 19.9 Schematic diagram of closed cycle OTEC system

There are two types of OTEC plants, namely closed cycle and open cycle OTEC plants.

Closed cycle OTEC system It is also referred to as the Rankine or vapour power cycle (see Fig. 19.9). The working fluid (usually, ammonia or propane as it is a low boiling liquid) is vaporized by heat transfer from the warm ocean water in the evaporator. The vapour, so formed, undergoes expansion through the steam turbine generator and is condensed by heat transfer to cold ocean water in the condenser. Closed-cycle OTEC power systems operate at elevated pressures and thus require smaller turbines than open cycle systems.

Open cycle OTEC system Here, the warm ocean water is used directly as the working fluid that is flash evaporated in the evaporator (see Fig. 19.10). The vapour, so formed, undergoes expansion through the turbine and is condensed with cold ocean water in the condenser. Open cycle OTEC power systems are operated at lower operating pressures and this require larger turbines to allow greater flow rates of steam. The open OTEC system allows both desalination of ocean waters and electricity generation.

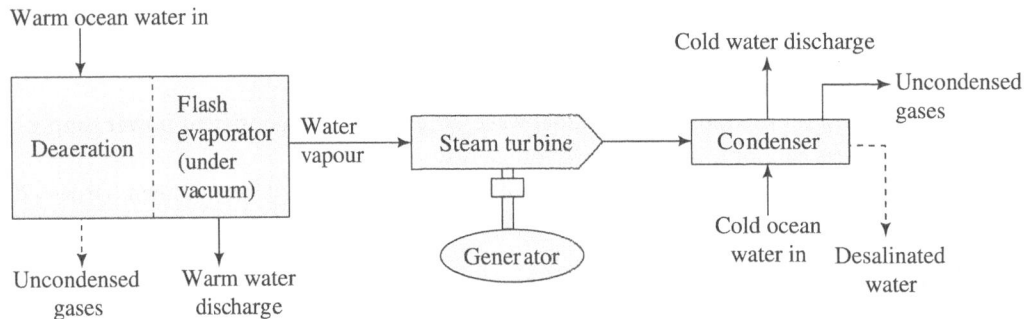

Fig. 19.10 Schematic diagram of open cycle OTEC system

India is geographically ideal to generate energy from ocean water as the solar thermal gradient is always about 20°C throughout the year. According to the 2001 report of NIOT (National Institute of Ocean Technology), OTEC has a potential of about 1,80,000 MW electricity generation. However, issues such as installation and poor electricity generation have made it difficult to achieve any success.

Some of the **advantages** of ocean thermal energy are the following.
(a) It is clean, renewable, and pollution-free using natural resources.
(b) OTEC open cycle power plant can be used for desalination of water as well as to generate electricity.

The **limitations** of ocean thermal energy are as follows.
(a) The capital investment for OTEC power plants is very high. OTEC-generated electricity costs much more than electricity obtained from fossil fuels.
(b) Construction of turbines and water pipelines of OTEC power plants can disrupt sea life. OTEC power plants may be affected by seasonal variations and natural calamities.
(c) Solar thermal gradient has to be maintained at about 20 °C which is difficult to attain every time in all ocean water systems.
(d) Carnot efficiency is about 3 %, which is quite low and hence there is a need to intensify the solar thermal gradients using solar ponds.

Check Your Progress

7. Provide an equation to determine power availability from windmills.
8. What is geothermal energy?
9. What is tidal energy? List its advantages and limitations.
10. What is the underlying principle for ocean thermal energy conversion?
11. What are the components of a hydropower station?
12. Name the types of OTEC power plants.

19.9 NUCLEAR ENERGY

In spite of the known hazards of nuclear energy, it has gained importance as an essential alternative fuel source. Nuclear energy is considered as a clean, emission-free, and competitive energy source compared to fossil fuels. Nuclear fuels are materials that can generate enormous amounts of energy in less than a millionth of a second. It is estimated that U^{235} has 2.5 million times more energy per kilogram than coal. The energy harnessed from nuclear fuel can be utilized to produce electricity, hydrogen, and steam. Until 2016, India has 22 operative nuclear reactors in eight nuclear power plants with a total installed capacity of 6,780 MW producing more than 35,000 GWh of electricity.

19.9.1 Mass Defect and Binding Energy

It is known that the mass of an atomic nucleus is always lesser than the sum of the masses of its constituent nucleons (i.e, protons, neutrons, and electrons). This is called *mass defect*, denoted as Δm. The loss of mass is in accordance with Einstein's mass–energy relationship, where the emitted energy is called *binding energy* of the nucleus. Consider an isotope (atomic number Z and mass number A) and let m_p, m_e, and m_n be the masses of proton, electron, and neutron, respectively.

Hence, mass of the isotope will be,

$$M' = Zm_p + Zm_e + (A - Z)m_n = Zm_n + (A - Z)m_n \text{ (since, } m_p + m_e = m_H)$$

If m is the actual mass of an element, then

$$\Delta m = Zm_H + (A - Z)m_n - M$$

where Δm is mass defect, that is, the loss of mass during nuclear formation from its constituents. The minimum energy needed to disassemble the nucleus from its constituent particles (protons and neutrons) is called *binding energy* and can be expressed as,

$$\text{B.E. per nucleon} = 1/A\ Zm_n + (A - Z)m_n - M]c^2 \text{ (as per } E = mc^2)$$

Greater the binding energy per nucleon, higher is the nuclear stability.

Q-value of Nuclear Reactions

Nuclear reactions involve a change in the composition of the nucleus. Rearrangement of atoms along with energy changes is a common feature of nuclear reaction. The changes in energy during nuclear reactions are very high, in the order of million electron volts. The total energy liberated or taken up in a nuclear reaction is called *nuclear energy* or *Q-value of the nuclear reaction*. The Q-value is obtained from the energies and masses of the nuclei involved in the nuclear process.

Let us consider the following nuclear process: $a + M \rightarrow N + b$, where a is the projectile, M is the target nucleus, N is recoil nucleus, and b is the ejected particle. While calculating the Q-value of a nuclear reaction, the target nucleus is assumed to be at rest; thus, its kinetic energy is zero. As the total energy is conserved in a nuclear reaction and the energy of a particle is equal to the sum of its rest energy and kinetic energy (from $E = mc^2$), it follows that for the above reaction,

$$(E_a + m_ac^2) + m_Mc^2 = (E_n + m_Nc^2) + E_b + m_bc^2)$$ (19.1)

where E depicts the respective kinetic energies and m denotes the corresponding masses. Hence, if Q is the difference between the kinetic energy of the products and reactants,

$$Q = E_N + E_b - E_a$$ (19.2)

On combining Eqs (19.1) and (19.2), we get

$$Q = E_N + E_b - E_a = (m_M + m_a - m_N - m_b)\ c^2$$ (19.3)

Considering the masses (in amu), then the value of Q (in MeV) will be

$$Q = \{(m_M + m_a) - (m_N + m_b)\} \times 931 \text{ MeV}$$ (19.4)

As per Eq. (19.4), the Q-value can be determined from the differences in energies of species involved in the nuclear reaction. Q-value can be either positive or negative; if it is negative, energy is taken up in the process and is termed as *endoergic process;* whereas a positive Q-value means energy is liberated in the process, and is called *exoergic process.* In an endoergic process additional energy called *threshold energy* (i_{th}) needs to be supplied to break the compound nucleus, additional energy called *threshold energy* (E_{th}) needs to be supplied given by the expression

$$E_{th} = Q\left(1 + \frac{m_a}{m_M}\right) \tag{19.5}$$

19.9.2 Energy Changes in Nuclear Reactions

Nuclear energy follows Einstein's mass–energy relationship; that is, $E = mc^2$. Nuclear energy is generally expressed in MeV (million electron volts); hence, energy in MeV corresponding to mass of 1 amu (atomic mass unit) is calculated as:

1 amu = 1.658×10^{-27} kg and c = 3×10^8 m/s and 1 MeV = 1.602×10^{-13} J

If the above values are substituted in the equation, $E = mc^2$, then

$$E = 1 \times 1.658 \times 10^{-27} \times (3 \times 10^8)^2 = 1.492 \times 10^{-10} \text{ J} = \frac{1.492 \times 10^{-10}}{1.603 \times 10^{-13}} = 931 \text{ MeV}$$

Hence, 1 amu = 931 MeV

Consider the nuclear reaction: $Li^7 + H^1 \rightarrow 2He^4$ (mass of Li = 7.0182 amu; mass of He = 4.0038 amu)
Thus, mass change = mass of products – mass of reactants

= (4.0038 × 2) – (7.0182 + 1.0081) = – 0.0187 amu

Thus, E = 0.0187 × 931 = 17.41 MeV. Hence, in a nuclear reaction, 0.0187 amu of mass is lost and 17.41 MeV of energy is released.

Generally nuclear reactions are of two types—nuclear fission and nuclear fusion.

19.10 NUCLEAR FISSION

O. Hahn and F. Strassman (1938) discovered that on bombarding uranium nucleus with neutrons, it undergoes fission. *The phenomenon of splitting of a nucleus into two appropriately equal fragments is called nuclear fission.* U^{235} gets bombarded with neutrons to form the following fission products:

$$^{235}_{92}U + ^{1}_{0}n \rightarrow ^{236}_{92}U \rightarrow ^{141}_{56}Ba + ^{92}_{36}kr + 3^{1}_{0}n$$

The neutrons formed again bombard another uranium atom for another fission reaction and this phenomenon occurs as a chain reaction within a span of 10^{-8} seconds, along with the release of large amounts of energy. The atom bomb works on the principle of nuclear fission.

19.10.1 Liquid Drop Model of Nucleus

George Gamow in 1935 proposed the liquid drop model to explain the characteristics of nuclear fission reaction that was further developed by Neils Bohr and John Wheeler (1939). The model considers the nucleus as a homogeneous entity with a specific number of nucleons similar to a liquid drop. The postulated similarities between a liquid drop and the nucleus of an atom are as follows:

(a) A liquid drop consists of a large number of molecules, as also an atomic nucleus comprises a large number of nucleons per unit volume.

(b) Both are incompressible. The physical properties such as density and charge of both the liquid drop and the nucleus are also the same. Both are homogeneous and exhibit surface effects.

(c) The force between the nucleons is same and this denotes that nuclear force is independent of the charge and spin of the nucleons. This behaviour is similar to that of an ideal solution, where solute particles and the solvent interact without distinction.

(d) The nucleons in an atomic nucleus interact only with their immediate neighbours, similar to molecules in a liquid drop.

(e) When a small drop of liquid is added to a liquid drop, it grows in size. In the same way, when a particle impinges on an atomic nucleus, it gets captured resulting in the formation of a large-size compound nucleus. The excess energy of the captured particle is shared among the nucleons.

(f) Smaller droplets of a liquid can fuse together to form a bigger drop. Quite similarly, smaller nuclei can be made to undergo nuclear fusion.

19.10.2 Mechanism of Nuclear Fission

As depicted in Fig. 19.11, suppose an atomic nucleus (say A) is like a spherical droplet of liquid. When a neutron is made to impinge the nucleus, it supplies additional energy to the nucleus. Due to the excess energy, nucleus (A) tends to deform from its spherical shape and becomes ellipsoidal in shape (B). The presence of surface forces on the nucleus makes it to return to its original spherical shape, while the additional energy tends to deform it still further. If the energy supplied is sufficiently large, the nucleus assumes a dumbbell shape (C). The repulsive forces will push the two ends of the dumbbell further apart until it breaks in to two spherical droplets or stable nuclei (D).

19.11 NUCLEAR REACTORS

A nuclear reactor is a device in which nuclear energy due to fission is generated in a controlled manner. The isotope U^{235} undergoes fission by neutrons of all energies and those neutrons that possess energy in excess of 1 MeV only can cause fission of U^{238} isotope. Hence, nuclides that can cause fission by neutrons of all energies are called *fissionable nuclides*. Pu^{239} and U^{233} are also fissionable nuclides. When U^{238} absorbs neutrons, Pu^{239} is formed as follows.

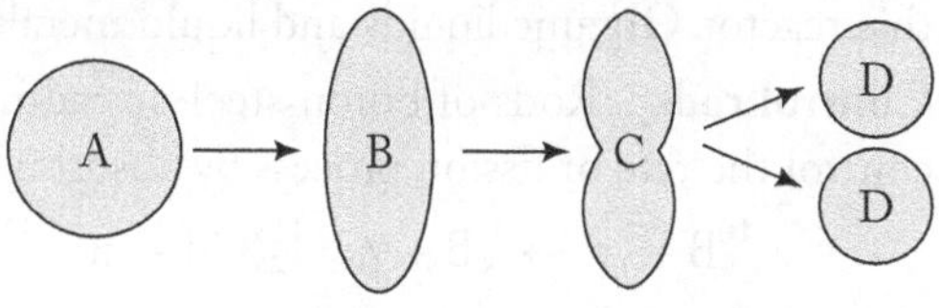

Fig. 19.11 Elucidation of nuclear fission by Bohr–Wheeler liquid drop model

$$_{92}U^{238} + {}_0n^1 \rightarrow {}_{92}U^{239} + \gamma$$

$$_{92}U^{239} \rightarrow {}_{93}Np^{239} + {}_{-1}\beta^0$$

$$_{93}Np^{239} \rightarrow {}_{94}Pu^{239} + {}_{-1}\beta^0$$

Pu^{239} undergoes fission in the similar manner as U^{235} and this concept is applied for converting abundant U^{235} isotope in to Pu^{239} in a convertor or breeder reactor. Generally, the following four competing reactions take place in a breeder reactor:

(a) Fission of uranium nuclei followed by emission of two or more neutrons per fission

(b) Non-fission neutron capture by uranium

(c) Non-fission neutron capture by other materials

(c) Escape of neutrons and complete loss of neutrons from the reactor assembly

There are various types of nuclear reactors, but certain features remain common in all of them. A nuclear reactor (Fig. 19.12) consists of a core containing fissionable material like U^{235} and U^{238}. The fissionable material captures neutrons and undergoes fission in a controlled manner. The neutrons are slowed down by a moderator such as light or heavy water and graphite rods.

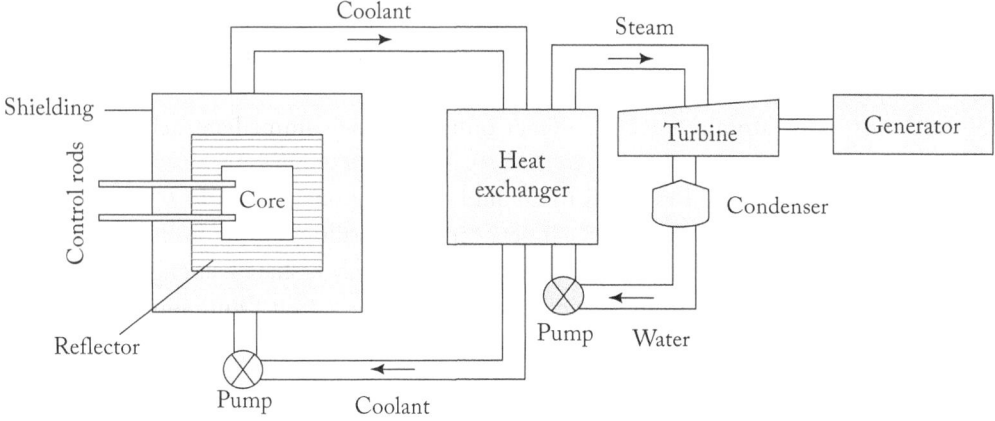

Fig. 19.12 A typical nuclear reactor

19.11.1 Thermal Reactors

When ordinary water is used as a moderator (or coolant) to slow down the neutrons, the reactor is called light water reactor (LWR). The reactor design can be explained as follows:

Core and fuel   The fissionable material, called nuclear fuel (U^{235}), is placed in the core of the reactor, where the nuclear reactions take place. The nuclear fuel is generally in the form of a solid rod or pellets that are shielded by placing stainless steel tubes in the core.

Moderator   For efficient nuclear fission, slow neutrons are required. Moderators (or coolants) slow down the neutrons obtained during fission and do not absorb them. Light water is commonly used in this reactor. Organic liquids and liquid metals are also used as coolants.

Control rods   Rods of boron steel or cadmium are suspended between fuel rods in the core so as to control the rate of fission process by absorbing the neutrons. The reactions can be shown as,

$$^{10}_{5}B + ^{1}_{0}n \rightarrow ^{11}_{5}B + \gamma; \quad ^{113}_{48}Cd + ^{1}_{0}n \quad ^{114}_{48}Cd + \gamma$$

Shielding   The reactor core is enclosed in a stainless steel vessel which is housed in a concrete building. The personnel who operate the reactor uses proper protection while working near the reactor.

Cooling system   Light water is pumped through the core of the reactor and it absorbs the heat generated during the fission. The coolant is further sent to a heat exchanger, where it transfers its heat to water and produces steam. The coolant is recycled back to the reactor core. The steam, so produced, is sent to a turbine that drives the generator and produces electricity.

> **Nuclear Power Plants in India**
> Tarapur Atomic Power Station (Maharashtra), Narora Atomic Power Station (Uttar Pradesh), Kakrapar Atomic Power Station (Gujarat), Kudankulam Nuclear Power Plant (Tamil Nadu), Kaiga Atomic Power Station (Karnataka) are operational nuclear reactors in our country.

19.11.2 Breeder Reactors

A breeder reactor is a nuclear fission-type reactor that produces large amounts of fissionable materials than what it consumes in the operation. Breeding is a term used to describe a process in which for every fissionable nucleus, say Pu^{239} undergoing fission, more than one neutron is captured by U^{238} to produce more Pu^{239} than is used in the fission. This process increases the stockpile of nuclear fuel. U^{238} and Th^{232} can be easily converted in to fissionable material and hence are called *fertile nuclides*, whereas

Pu^{239} and U^{235} are fissionable and termed as *fissile nuclides*. Breeder reactors involve the use of fast-moving neutrons and hence, no moderator is required.

$$^{238}_{92}U + {}^{1}_{0}n \rightarrow {}^{239}_{92}U \rightarrow {}^{239}_{93}Np + \beta \rightarrow {}^{239}_{93}Pu + \beta$$

As seen above, Pu^{239} is obtained by neutron bombardment and two successive decays. Similarly, U^{233} is obtained by bombarding Th^{232} with neutrons as shown below.

$$^{232}_{90}Th + {}^{1}_{0}n \rightarrow {}^{233}_{90}U \rightarrow {}^{239}_{93}Np + \beta \rightarrow {}^{239}_{93}Pu + \beta$$

19.11.3 Nuclear Fusion

A process in which lighter nuclei combine to form heavier nuclei is called nuclear fusion. Nuclear fusion reactions are known to occur in stellar bodies. Fusion reactions account for the huge amounts of energy found in stars and other stellar objects occurring at temperatures close to 10 million degree on Celsius scale. Hence, nuclear fusion reactions are also called *thermonuclear reactions*. Hydrogen bomb is an example of thermonuclear reaction. The mechanism of nuclear fusion at higher temperatures occur via the carbon cycle shown by the following reactions.

$$_{1}H^{1} + {}_{6}C^{12} \rightarrow {}_{7}N^{14} + energy$$
$$_{7}N^{14} \rightarrow {}_{6}C^{13} + {}_{+1}e^{0}$$
$$_{1}H^{1} + {}_{6}C^{13} \rightarrow {}_{7}N^{14} + energy$$
$$_{1}H^{1} + {}_{7}N^{14} \rightarrow {}_{8}O^{15} + energy$$
$$_{8}O^{15} \rightarrow {}_{7}N^{15} + {}_{+1}e^{0}$$
$$_{1}H^{1} + {}_{7}N^{15} \rightarrow {}_{2}He^{4} + {}_{6}C^{12}$$

Hence, the net reaction will be: $4_{1}H^{1} \rightarrow {}_{2}He^{4} + 2_{+1}e^{0} + energy$

The conversion of four protons in to a helium nucleus and two positrons release about 30 MeV or 7.1 × 10^{8} kJ/g of energy. At lower stellar temperatures, the reactions occur as:

$$_{1}H^{1} + {}_{1}H^{1} \rightarrow {}_{1}H^{2} + {}_{+1}e^{0} + energy (\times 2)$$
$$_{1}H^{2} + {}_{1}H^{1} \rightarrow {}_{2}He^{3} + energy$$
$$_{2}He^{3} + {}_{2}He^{1} \rightarrow {}_{2}He^{4} + 2_{1}H^{1} + energy$$

The net reaction will be: $4_{1}H^{1} \rightarrow {}_{2}He^{4} + 2_{+1}e^{0} + energy$

One can study nuclear fusion using particle accelerators, stellar reactions, or by studying fusion bombs. Due to attainment of high temperatures during nuclear fusion, it is difficult to harness energy on a large scale. As the containers holding the reactions itself undergo vaporization at high temperatures, no successful attempts have been made till date to design nuclear fusion reactors.

Check Your Progress

13. Distinguish between nuclear fission and nuclear fusion.
14. Define mass defect.
15. What is binding energy of a nucleon?
16. Define Q-value of a nuclear reaction.
17. Name the isotope used in a nuclear reactor.
18. Provide examples of moderators used in a nuclear reactor.
19. What is a breeder reactor?
20. What is a light water reactor?

19.12 Nuclear Waste Management

At least four types of wastes are generated from a nuclear power plant that can prove to be hazardous to the environment; these are:
 (a) The spent nuclear fuel at the site of the reactor
 (b) Waste rock materials at uranium mines
 (c) Used gloves, uniforms, etc., of persons handling nuclear reactors
 (d) Small release of radioactive isotopes during nuclear fission, such as I^{131} and Cs^{137} which are toxic to the environment

Nuclear waste materials can be of solid, liquid, and gaseous types and each of these have their own characteristic method of disposal. Some of the safety disposal methods for managing nuclear waste are as follows:
 (a) Radioactive solid waste is generally buried at suitable geologic sites in near-surface or deep-geologic sites such as trenches or tunnels.
 (b) In case of a liquid nuclear waste, it is segregated, filtered, and adequately diluted such that its radioacive discharge is minimized and safely recycled.
 (c) Radioactive gaseous products can be volatile and non-volatile fission products, particulate dispersions of nuclear fuel material. As these fission products are known to be electrically charged, they easily adhere to dust particles present in air. I^{131}, with a half-life of about eight days, if inhaled can cause respiratory problems and on prolonged exposure may adversely affect thyroid function. Hence, such air needs to be filtered using evaporative gas strippers. After stripping, the gases are compressed and allowed to undergo natural decay for a shorter time period. After complete decay, gases will be suitably recycled before releasing into the environment.

19.12.1 Safety Measures of Nuclear Reactors

The operating persons, the public, and the environment need safety from nuclear reactors. Generally, there are three levels of safety in a nuclear facility.

In the first level, the nuclear reactor is so designed to minimize leakage and accidents and maximize surveillance and security. In the second level, operators handling the nuclear reactor are made to strictly follow the stipulated safety protocols. Finally, in the third level, comprehensive measures are put in place to contain/prevent damages in case of a nuclear accident.

The safety measures to be adopted while designing and working with a nuclear reactor are as follows:
 (a) As the nuclear reactions occurring in the core of the reactor are exothermic in nature, the complete reactor set-up should be contained inside a well-cemented building.
 (b) During operation and accidents, the nuclear reactor core should be completely shielded from the outside environment. As fission occurs in the nuclear core, proper care should be ensured for the control rods and cooling systems.
 (c) The operators working in a nuclear facility must be under complete surveillance to monitor radiation exposure levels. Both the working environment and personnel should be checked for radiation doses.

19.13 Biomass

Any organic matter obtained from plant and animal remains is called biomass and is a well-known alternative source of energy. Wood, solid waste, animal excreta, sewage, agricultural waste, landfill gas, etc., are materials from which biomass is obtained. Biomass energy can be utilized either directly (by burning) or indirectly (by converting into biofuels). As plants store energy from the sun by photosynthesis, when biomass material is burned, the stored chemical energy is released as heat. However, huge amounts of pollutants such as carbon dioxide, nitrous oxide, sulphur dioxide, and particulate matter are also released that harm the environment.

Biogas consists of a mixture of different gases obtained due to the anaerobic digestion of biomass matter. Methane, carbon dioxide, hydrogen sulphide, carbon monoxide gases are found in biogas which can undergo combustion to produce energy.

Biofuels are obtained by the anaerobic digestion of organic waste material. Biodiesel obtained from petro crops such as Jatropha, is a classic example of biofuel energy (see Chapter on Fuels and Combustion for more details). Today, algal fuels are also utilized to obtain biomass energy and are considered as an alternative to conventional liquid fuels. During photosynthesis, algae traps carbon dioxide and with the help of sunlight converts it in to oxygen and biomass, which is utilized to obtain heat and algal oils.

Check Your Progress

21. List the hazardous materials generated from a nuclear power plant.
22. List any two safety measures to be adopted in a nuclear facility.
23. What is biomass? Give examples.
24. How is biogas obtained from biomass?
25. How can one obtain energy from algal matter?

Activity-based Questions

A. Hybrid Energy Power Generation

In recent years, power crisis is a daunting global problem. Generally, power is generated using fossil-based materials like coal and petroleum. Even renewable sources such as the solar radiation, wind, nuclear energy are considered for power generation. Rather than employing simply one renewable energy source, hybrid energy systems are installed for power generation. A hybrid energy system comprises one or more renewable sources working in parallel with a standby secondary non-renewable module and storage units. Combining solar energy with photovoltaics for power generation is a classic example of a hybrid energy system. Many solar and wind hybrid energy systems are being considered for optimizing energy supply in remote areas.

Perform a comprehensive review on different hybrid energy systems and hybrid design models, and also highlight the power output efficiencies, simulation studies, costs, challenges, improvement, and future scope.

B. Natural Gas Hydrates as Energy Source

Natural gas hydrates (NGHs) gained attention post-1960 and has been a subject of great interest while searching for alternative fuels. NGHs are ice-like gas resources that contain various hydrocarbons found in oceans and permafrost regions.

Perform a critical review of the chemical structures of gas hydrates, hotspots, properties, carbon capture using NGHs, challenges, and prospects of using NGHs.

SUMMARY

- As fossil fuels are fast depleting, it is imperative to search for alternative energy resources that are eco-friendly, renewable, cheaper, and sustainable.
- Solar, wind, geothermal, tidal, biomass, and nuclear energy resources are considered as safer alternatives to fossil fuels.
- Solar energy, hydropower, tidal, biomass, geothermal energy are examples of renewable resources, whereas coal, petroleum, and natural gas are examples of non-renewable energy resources.

○ Water heating, desalination, cooking, power generation, photovoltaics are some of the applications that utilize solar energy.

○ Wind energy is the most widely available renewable energy resource. It does not generate greenhouse gases, toxins, or any radioactive waste, and no fuel is required to harness it.

○ The natural heat present within the earth's crust which is tapped for various energy needs is called geothermal energy. The primary sources of geothermal energy are hot springs, hot rocks, and molten rocks present beneath the earth's surface.

○ Tidal energy involves harnessing power from tidal waves.

○ The temperature differences prevalent between the surface and deeper ocean waters can be harnessed for electrical energy, forming the basis of Ocean Thermal Energy Conversion plants. There are two types of OTEC plants, namely closed cycle and open cycle OTEC plants.

○ An essential alternative fuel source is nuclear energy, which is considered as clean, emission-free, and generates large amounts of energy when compared to the fossil fuels.

○ A nuclear reactor is a device in which nuclear energy due to fission is generated in a controlled manner.

○ Due to the hazardous nature of the radioactive material, extensive safety measures are adopted during operations and also during the disposal of nuclear waste.

○ Wood, solid waste, animal excreta, sewage, etc are materials from which biogas is obtained.

○ Biofuels such as biodiesel and algal fuels show promise as renewable sources of energy.

GLOSSARY

Binding energy: The minimum energy needed to disassemble the nucleus from its constituent particles.

Biogas: A mixture of gases obtained due to the anaerobic digestion of biomass matter.

Biomass: A known alternative energy resource, it is any organic matter obtained from plant and animal remains.

Breeder reactor: A nuclear fission-type reactor producing huge amount of fissionable materials than what it consumes in the operation.

Geothermal energy: The natural heat present within the earth's crust tapped for various energy needs.

Nuclear fission: The phenomenon of splitting of a nucleus into two appropriate equal fragments.

Nuclear fusion: Lighter nuclei combine to form heavier nuclei, a natural phenomenon in stellar bodies.

Nuclear reactor: A device in which nuclear energy due to fission is released in a controlled manner.

Photovoltaic cell: A device that directly converts solar radiations into electric current.

Q-value: Total energy liberated or taken up in a nuclear reaction.

Radioisotope: A natural or man-made isotope that exhibits radioactivity.

Renewable energy: Energy derived from renewable resources that get naturally replenished and utilized repeatedly.

EXERCISES

Multiple Choice Questions

1. Which of the following energy resources do not harm the environment?
   - (a) Wind energy
   - (b) Petroleum
   - (c) Coal
   - (d) Nuclear

2. Geothermal energy is harnessed from
   - (a) atmosphere
   - (b) heat from earth's interior
   - (c) biomass
   - (d) organic matter

3. The generation of power from sea tides is called
   - (a) solar energy
   - (b) nuclear energy
   - (c) tidal energy
   - (d) fossil fuel energy

4. Control rods in nuclear reactors are made of
   - (a) beryllium
   - (b) strontium
   - (c) boron
   - (d) iron

5. The major constituent of biogas is
   - (a) oxides of sulphur
   - (b) methane
   - (c) carbon
   - (d) fuel
6. Hydrogen bomb is an example of
   - (a) nuclear fusion
   - (b) nuclear fission
   - (c) chemical reaction
   - (d) none of these
7. The mechanism of nuclear fission is explained by
   - (a) chemical reactions
   - (b) atomic structure model
   - (c) liquid drop model
   - (d) chemical bonding
8. An endoergic nuclear reaction denotes that
   - (a) Q-value is negative.
   - (b) Q-value is positive.
   - (c) Q-value is zero.
   - (d) Q could assume values between 0–1.
9. Breeder reactor involves
   - (a) slow moving neutrons
   - (b) fast moving neutrons
   - (c) slow moving positrons
   - (d) fast moving positrons
10. The hotspot(s) for obtaining geothermal energy are
    - (a) volcanoes
    - (b) hotsprings
    - (c) mineral deposits
    - (d) All of them
11. The moderator used in the nuclear reactor is
    - (a) water
    - (b) heavy water
    - (c) graphene
    - (d) sodium
12. Indira Paryavaran Bhavan is an example of
    - (a) solar building
    - (b) solar thermal plant
    - (c) solar still
    - (d) none of these

13. India has the potential of generating tidal electricity of about
    - (a) 7000 MW
    - (b) 7800 MW
    - (c) 8000 MW
    - (d) 8800 MW
14. Liquid drop model was first proposed by
    - (a) Bohr
    - (b) George Gamow
    - (c) Wheeler
    - (d) Hahn
15. Energy harnessed from moving water is called
    - (a) tidal power
    - (b) wind energy
    - (c) solar energy
    - (d) hydropower
16. Energy resources that do not get depleted are called
    - (a) renewable
    - (b) non-renewable
    - (c) original
    - (d) none of these
17. Which of the below helps in harnessing solar energy?
    - (a) Solar heaters
    - (b) Solar cells
    - (c) Solar cookers
    - (d) All of these
18. Pu^{239} and U^{235} are
    - (a) non-fissionable
    - (b) fissionable
    - (c) non-radioactive
    - (d) neutrons
19. Liquid drop model was improvized by
    - (a) Bohr
    - (b) Wheeler
    - (c) Both (a) and (b)
    - (d) George Gamow
20. Nuclear energy follows the principle of
    - (a) mass–energy relation
    - (b) photoelectric effect
    - (c) blackbody radiation
    - (d) Bohr atomic model
21. OTEC working is based on
    - (a) mass–energy relation
    - (b) blackbody radiation
    - (c) Carnot cycle
    - (d) None of these

Review Questions

1. What are solar flat plate collectors? State any three uses of flat plate collector in harnessing energy.
2. Explain the working of a photovoltaic cell.
3. List the applications of solar energy.
4. What is wind energy? State the advantages and limitations of wind energy.
5. How can geothermal energy be utilized for power generation?
6. What is geothermal energy? State the advantages and limitations of geothermal energy.
7. Write a short informative note on nuclear fusion.
8. Discuss tidal energy as an alternative resource. List its advantages and limitations.
9. List the advantages and limitations of wind energy.
10. Discuss generation of electricity using hydropower plant.
11. Discuss closed and open cycle ocean thermal energy conversion for producing electricity.
12. Write a short informative note on OTEC systems.

13. Define mass defect and nuclear binding energy. How are they related?

14. Define Q-value of a nuclear reaction. What are exoergic and endoergic reactions?

15. Explain nuclear fission on the basis of liquid drop model of nucleus.

16. What is a nuclear reactor? With a suitable diagram, describe the components of a light water nuclear power plant.

17. Distinguish between nuclear fission and fusion reactions.

18. Explain the safety measures to be adopted while working with nuclear reactors.

19. What is nuclear waste? State the steps to manage nuclear waste.

20. Discuss the working of a thermal nuclear reactor.

21. Write a note on breeder reactors.

22. What is biomass? Discuss in detail the various biofuels.

23. Write a short note on biomass as non-conventional energy source.

FURTHER READING

1. Glassley, W.A., *Renewable Energy and the Environment*, CRC Press, 2015.
2. National Institute of Solar Energy (NISE) https://nise.res.in/
3. National Institute of Wind Energy (NIWE) http://niwe.res.in/
4. National Institute of Ocean Technology (NIOT) https://www.niot.res.in/
5. Twidell, J. and T. Weir, *Renewable Energy Resources*, Routledge, 2015.
6. Tiwari, G.N., R.K. Mishra, *Advanced Renewable Energy Sources*, RSC Publishing, 2011.
7. World Energy Resource- *International Energy Annual*, http://www.eia.doe.gov/ (interesting and relevant articles on energy resources are published in this journal)

ANSWERS

Check Your Progress

1. The resources than can produce energy repeatedly and can be utilized naturally. Solar, tidal, geothermal, are some renewable sources of energy. Non-renewable energy sources get exhausted after use and cannot be replaced. Coal, oils, etc. are some examples of non-renewable energy resources.

2. A flat plate solar collector comprises a blackened absorber plate made of copper or aluminium and risers which are copper tubes containing heat transfer fluid, typically water.

3. In the summer, as the temperature rises, a solar building utilizes its thermal mass to keep the building cool. To achieve this, the sun's radiation is kept away from reaching the thermal mass of the building by placing a shade or covering the window panels.

4. Solar heating, solar building, solar cooking, solar desalination, solar electricity, and batteries are various applications of solar energy.

5. Refer to Section 20.3.5

6. Also called solar cell, solar photovoltaics directly convert solar radiations into electric current.

7. Power available from a windmill can be expressed as, $P_{max} = 0.5 \ \rho a V^3$; where ρ is the density of air, a is the cross-sectional area of windmill disc, and V is the velocity of air.

8. Geothermal energy is the natural heat present within the earth's crust, which is tapped for various energy needs.

9. Refer to section 20.7.

10. Ocean thermal energy conversion is based on Carnot cycle (second law of thermodynamics).

11. The components of a hydropower station include a water reservoir (site for hydropower generation), a dam that comprises a flow control system, and an electric plant where electricity is generated.

12. Closed cycle OTEC and open cycle OTEC power plants

13.

Nuclear Fission	Nuclear Fusion
1. Heavier nucleus is fragmented into smaller nuclei.	1. Lighter nuclei combine to form a heavier nucleus.
2. Initiated by neutrons of suitable energy.	2. Occurs only at higher temperatures close to 10 million °C.
3. A chain reaction; can be controlled using a nuclear reactor.	3. Occurs as a carbon or proton cycle (stellar); cannot be controlled.
4. Generation of nuclear waste.	4. No generation of nuclear waste.

14. The mass of an atomic nucleus is always lesser than the sum of the masses of its constituent nucleons (i.e, protons, neutrons, and electrons). This is called *mass defect* denoted as (Δm).

15. The minimum energy needed to disassemble the nucleus of its constituent particles (protons and neutrons) is called *binding energy.*

16. The total energy liberated or taken up in a nuclear reaction is called *nuclear energy* or *Q-value of nuclear reaction.*

17. U^{235} (Uranium-235) is as isotope used in nuclear reactors.

18. Heavy water and graphite rods are examples of moderators.

19. A nuclear fission-type reactor that produces large amounts of fissionable materials than what it consumes in the operation is a breeder reactor.

20. When ordinary light water is used as a moderator (or coolant) to slow down the neutrons, the reactor is called light water reactor.

21. Spent fuel, used coolants, uranium rock waste, used gloves, and uniforms are some of the waste generated from a nuclear plant.

22. (a) As the nuclear reactions occurring in the core of the reactor are exothermic in nature, the complete reactor set-up should be contained in a well-cemented building.

    (b) During operations, the nuclear reactor core should be completely shielded from the outside environment.

23. Any organic matter obtained from plant and animal remains is called biomass. Wood, solid waste, animal excreta, sewage, agricultural waste, landfill gas are types of biomass.

24. Biogas consists of a mixture of different gases obtained due to the anaerobic digestion of biomass matter.

25. During photosynthesis, algae traps carbon dioxide and with the help of sunlight converts it in to oxygen and biomass which is utilized to obtain heat and algal oils.

Multiple Choice Questions

1. (a)	2. (b)	3. (c)	4. (c)	5. (b)	6. (a)	7. (c)
8. (a)	9. (b)	10. (d)	11. (b)	12. (a)	13. (c)	14. (b)
15. (d)	16. (a)	17. (d)	18. (b)	19. (c)	20. (a)	21. (c)

Fuels and Combustion

After reading this chapter, you will be able to:

- classify the various types of fuels.
- explain calorific values of coal and their elemental analyses.
- discuss crude oil mining, cracking, reforming, refining processes along with commercially viable fuels such as coal, gasoline, diesel, CNG, LPG, biodiesel, and power alcohol.
- solve numerical problems on calorific values, combustion of fuels, elemental analysis of solid fuels, and flue gas analysis.
- understand the working of explosives and propellants.

20.1 INTRODUCTION

The concept of fuels and their combustion characteristics finds relevance in civil and mechanical engineering fields. Combustion engineering is the science of combustion and deals with harnessing energy from fuels for commercial purposes. Generally, combustion engineers plan and design novel energy systems with low carbon footprint. Fuel is a combustible substance that produces a large amount of heat. The heat produced can be harnessed for various domestic and industrial purposes. Wood, charcoal, kerosene, petrol, diesel, producer gas, etc., are some commonly employed fuel materials. Generally, fuels contain carbon as the main constituent. The combustion process involves oxidation of carbon, hydrogen, and other elements in the fuel to carbon dioxide, water, and other products of combustion. The difference in energy of reactants and products is liberated as a large amount of heat energy which is commercially utilized.

$$\text{Fuel} + \text{Oxidizer} \rightarrow \text{Products of combustion} + \text{Heat}$$

Explosives and propellants are highly energetic materials that produce huge amounts of energy that can be utilized in warfare and even for peaceful purposes. They are high melting crystalline solids or liquids at room temperature that undergo decomposition at rapid rates forming various combustion products. In this chapter we will discuss various combustion reactions and the manufacture, analysis, and types of fuels. We will also look at the characteristics of explosives and propellants.

20.2 CLASSIFICATION OF FUELS

Fuels are broadly classified on the basis of their occurrence, viz., as primary and secondary fuels (Fig. 20.1). Further, based on their physical state, fuels are classified into solid, liquid, and gaseous fuels. Primary fuels occur naturally whereas secondary fuels are derived or manufactured from primary fuels. Coal, petroleum, and natural gas are examples of primary fuels. Coke, petrol, diesel, LPG are examples of secondary fuels.

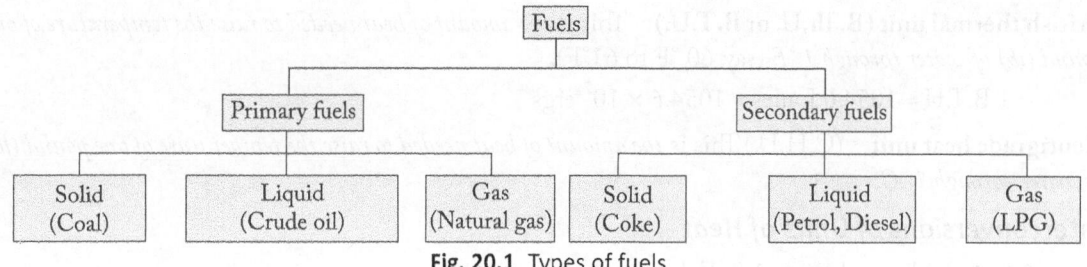

Fig. 20.1 Types of fuels

20.3 CALORIFIC VALUE

The performance criterion of any given fuel is adjudged by its calorific value (CV). *The total quantity of heat liberated by burning a unit mass (for solid fuels) or volume (for liquid or gaseous) of fuel completely is called calorific value.*

20.3.1 Gross Calorific Value

The gross calorific value or higher calorific value, denoted as GCV or HCV, *is the total amount of heat produced when a unit mass or volume of the fuel is burnt completely and the combustion products are cooled to room temperature* (i.e., 15 °C or 60 °F).

Upon heating coal, hydrogen present in it will undergo combustion generating steam as the combustion product, which on cooling will condense to form water. The latent heat evolved is called the *latent heat of condensation of steam,* which is included in gross calorific value equation as follows.

$$GCV = \frac{1}{100}\left[8080 \times C + 34500\left(H - \frac{O}{8}\right) + 2240 \text{ S kcal/kg}\right] \tag{20.1}$$

where C, H, O, and S are the percentages of carbon, hydrogen, oxygen, and sulphur in the fuel.

The numerical factors in the above equation, that is, 8080, 34500, and 2240 are the calorific values (in kcal/kg) of carbon, hydrogen, and sulphur, respectively. Equation (20.1) is called the Dulong–Petit's formula that allows theoretical determination of calorific values.

20.3.2 Net Calorific Value

The net calorific value or low calorific value denoted as NCV or LCV, *is the net heat produced when a unit mass or volume of fuel is burnt completely and the products are allowed to escape,* and is depicted using the following equation.

NCV = GCV – Mass of hydrogen (%) × 9/100 × Latent heat of water vapour $\qquad$ (20.2)

The factor '9' in Eq. (20.2) denotes that one part by weight of hydrogen produces 9 parts (1 + 8) by mass of water.

20.3.3 Units of Calorific Value

Various units are used to express calorific value of fuels. The following are some commonly used expressions for calorific values.

Calorie (cal) or gram calorie (gcal)   This is the most practical unit for the expression of calorific value. One calorie is defined as *the amount of heat needed to raise the temperature of 1 gram of water through 1 °C* (say, 15°C to 16°C).

$\qquad$ 1 Calorie = 4.185 Joules = 4.185 × 10^7 ergs

Kilocalorie or kilogram calorie (kcal or kg cal)   This can be considered equal to 1000 calories and is *the amount of heat needed to raise the temperature of 1 kilogram of water through 1 °C* (say, 15 °C to 16 °C).

$\qquad$ 1 kcal = 1000 cal

British thermal unit (B.Th.U. or B.T.U.) This is *the amount of heat needed to raise the temperature of one pound (lb) of water through 1 °F* (say, 60 °F to 61°F).

$$1 \text{ B.T.U} = 1054.6 \text{ Joules} = 1054.6 \times 10^7 \text{ ergs}$$

Centigrade heat unit (C.H.U.) This is *the amount of heat needed to raise the temperature of one pound (lb) of water through 1 °C.*

Interconversions of Units of Heat

Since, 1 kg = 2.2 lbs and 1°C = 1.8 °F; hence,

1 kcal = 1000 cal = 3.968 B.T.U. = 2.2 C.H.U.

Also, 1 B.T.U. = 252 cal and 100,000 B.T.U.

Generally, calorific values of solid and liquid fuels are expressed as calories per gram (cal/g) or kilocalories per kilogram (kcal/kg) or even British Thermal Unit per pound (B.T.U./lb). For gaseous fuels, calorific value is expressed as kilocalories per cubic metre (kcal/cu.m or kcal/m^3) or B.T.U./ft^3 or centigrade heat unit per pound or centigrade heat unit per cubic foot (C.H.U./lb or C.H.U./ft^3).

So, 1 cal/g = 1 kcal/kg = 1.8 B.T.U./lb

1 kcal/m^3 = 0.1077 B.T.U./ft^3

1 B.T.U./ft^3 = 9.3 kcal/m^3

20.4 CHARACTERISTICS OF A GOOD FUEL

The following criteria need to be fulfilled while selecting a fuel to be employed for a given purpose.

(a) A good fuel must possess high calorific value. The overall amount of heat evolved depends on the calorific value, making it a performance criterion while selecting fuels.

(b) Ignition temperature of the fuel should be moderate as it is the lowest temperature at which the fuel needs to be preheated to start burning smoothly. A low-ignition temperature fuel will be unsafe to store and handle, whereas a higher value will make it difficult to ignite.

(c) An ideal fuel should have low moisture content as the presence of moisture directly affects its heating value, that is, calorific value.

(d) Ash content is whitish or brownish coloured non-combustible remains of the fuel left behind after complete combustion. Lower the ash content, better is the performance of the fuel.

(e) An ideal fuel on combustion should not form harmful products and gases, such as carbon monoxide, sulphur dioxide, hydrogen sulphide, and phosphine, which tend to result in environmental pollution.

(f) An ideal fuel should be the one whose combustion can be controlled easily, is low on cost, easy to transport, and burns without smoke formation.

20.5 DETERMINATION OF CALORIFIC VALUE USING BOMB CALORIMETER

Principle Bomb calorimeter is used to determine the calorific value of solid and liquid fuels. A known mass of solid or liquid fuel (test fuel) sample is completely burnt in excess of oxygen in a bomb calorimeter. The heat liberated is absorbed by the surrounding water and copper calorimeter. The overall increase in temperature of the calorimeter and water is noted by a precision-based sensitive thermometer. The calorific value of the test fuel is calculated using the principle: *heat liberated by a fuel on combustion is equal to the heat absorbed by water and the calorimeter.*

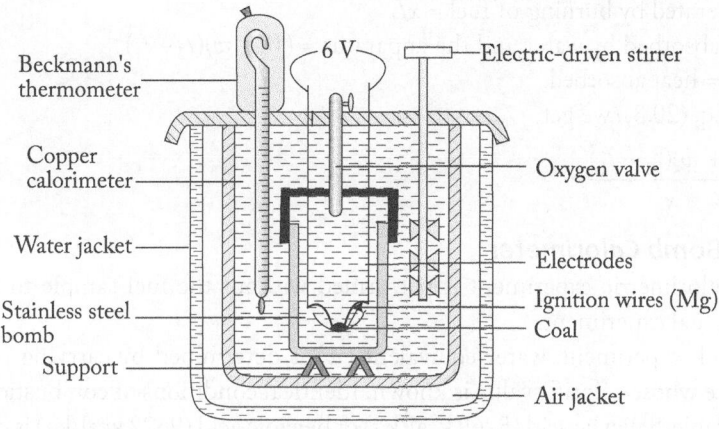

Fig. 20.2 Bomb calorimeter

Construction of Bomb Calorimeter

(i) It is a strong cylindrical stainless steel bomb equipped with a lid screw. It is gas-tight sealed from all sides. The bomb lid is provided with an inlet valve for oxygen and two stainless steel electrodes.

(ii) The electrodes are connected externally to a 6 volt battery. Of the two electrodes, one is connected to a stainless steel crucible and the other with a magnesium fuse wire for ignition of fuel.

(iii) The bomb is placed in a copper calorimeter and further surrounded by air-jacket and water-jacket to prevent heat losses due to radiation.

(iv) The copper calorimeter is provided with electrically operated stirrer and Beckmann's thermometer that accurately reads temperature difference up to 1/100 th of a degree (Fig. 20.2).

Working of Bomb Calorimeter

(i) A weighed mass of about 0.5–1.0 g of air-dried and powdered fuel sample is taken in a clean and dry crucible.

(ii) A clean magnesium wire connected to one of the electrodes ignites the test fuel. The bomb lid is tightly screwed and filled with oxygen maintaining a pressure of about 25 atm.

(iii) The steel bomb is placed in the copper calorimeter containing a known mass of water. The electrically operated stirrer ensures uniform temperature and the initial temperature of water is noted as t_1.

(iv) The electrodes connected to 6 V battery completes the circuit. The fuel sample burns and heat is evolved.

(v) The stirring of water is continued and the maximum temperature attained is recorded as t_2.

Calculations

Let x be the mass in g of fuel taken in the crucible.

W is the mass of water in calorimeter.

w is the water equivalent in g of calorimeter, stirrer, thermometer, bomb, etc.

As per the principle, the total amount of heat liberated by the fuel is absorbed strictly by water present in the copper calorimeter. However, it is possible that some fraction of heat evolved may be absorbed by the wall of the bomb container, electric stirrer, calorimeter, and even the thermometer. This heat gain is called *water equivalent of the bomb calorimeter* denoted as 'w' in Eq. (20.3). Hence,

$$xL = (W + w)(t_2 - t_1) \qquad (20.3)$$

where, t_1 and t_2 are the initial and final temperatures of water in copper calorimeter.

L is the gross calorific value of fuel in cal/g or kcal/kg.

Then, the heat liberated by burning of fuel = xL.

Further, the heat absorbed by water and the apparatus = $(W + w)(t_2 - t_1)$

But heat evolved = heat absorbed

On rearranging Eq. (20.3), we get,

$$L = \frac{(W + w)(t_2 - t_1)}{x} \qquad \qquad ...(20.4)$$

Corrections in Bomb Calorimeter

Ideally, a blank calorimetric experiment is performed without the fuel sample to ascertain the values obtained in the actual experiment.

During the blank experiment, water equivalent (w) is determined by carrying out combustion of a standard substance whose calorific value is known. Identical conditions of combustion are maintained as that of the fuel sample. Salicylic acid (5269 kcal/kg) or benzoic acid (9622 kcal/kg) is commonly employed as the standard substance to determine the water equivalent parameter.

It is essential to consider the following corrections, to reach the exact practical calorific value of the given fuel sample.

Fuse wire correction (t_f) A metal wire such as magnesium is taken to initiate ignition of the fuel. However, once ignited, it is possible for the heat liberated by the fuel to be partially absorbed by the metal wire itself. Hence, it should be subtracted from the final rise in temperature.

To determine t_f, magnesium wire of the same dimensions is taken and allowed to ignite in the absence of fuel sample (blank calorimetric reading).

Acid correction (t_a) Most of the fuels contain nitrogen and sulphur and on combustion they are oxidized to form the corresponding acids.

$$S + 2O_2 + 2H \rightarrow H_2SO_4 + \text{heat}$$

$$N_2 + 3O_2 + 2H \rightarrow 2HNO_3 + \text{heat}$$

Thus, the formation of acids is an exothermic reaction (heat energy is released) and hence the value needs to be subtracted from the final rise in temperature. The acid correction is determined from amounts of nitrogen and sulphur present in the fuel. The correction values are 1.43 cal/ml of 0.1 N nitric acid and 2.25 cal/mg of sulphur.

Cooling correction (t_c) As the temperature is recorded by using Beckmann's thermometer placed in the calorimeter, some time elapses before the heat liberated by the fuel is sensed. During the elapsed time, there can be a slight decrease in temperature in the water in the calorimeter. Hence, cooling correction is to be added to the final rise in temperature. Cooling correction is calculated by taking the product of $t \times dt$, that is, the time taken to cool water in the calorimeter from the maximum temperature to room temperature (in minutes) and dt is the rate of cooling (dt°/min).

By considering fuse wire correction, acid correction, and cooling corrections, the gross calorific value can be calculated using Eq. (20.5).

$$L = \frac{[\{(W + w)(t_2 - t_1 + tc)\} - \{t_a + t_f\}]}{x} \qquad \qquad (20.5)$$

If we consider % of hydrogen in the given fuel sample, mass of water from 1 g of fuel will be 0.09 H g. Hence, heat absorbed by water to form steam will be 0.09 H × 587 cal

Thus, NCV = GCV – 0.09 H × 587 cal/g ...(same as Eq. (20.2))

20.6 DETERMINATION OF CALORIFIC VALUE BY JUNKERS CALORIMETER

Junkers gas calorimeter is employed to determine calorific values of gaseous and volatile liquid fuels. As shown in Fig. 20.3, it consists of the following components.

Combustion chamber or **gas calorimeter** is a vertical cylinder surrounded by annular space for heating water and interchange coils. It is further surrounded by an outer jacket to avoid heat and radiation losses.

Bunsen burner is a special burner system clamped at the bottom of the combustion chamber.

Pressure governor controls the supply of volume of gas at a given pressure.

Gasometer is used to measure the volume of gas undergoing combustion per unit time. Further, the gasometer is attached to a manometer equipped with a thermometer to read pressure and temperature of gas before burning.

The apparatus is installed on a rigid and flat platform near an uninterrupted water supply source and a drain pipe. The gas source is connected in series to a pressure regulator, gasometer, and burner respectively. Thermometers are placed so as to measure water inlet-outlet temperatures and exhaust gas temperatures.

Water is allowed to enter the gas calorimeter at a constant flow rate and allowed to drain through overflow. The gaseous fuel whose calorific value is to be determined is sent through a pressure governor and gasometer and burns in Bunsen burner. The products

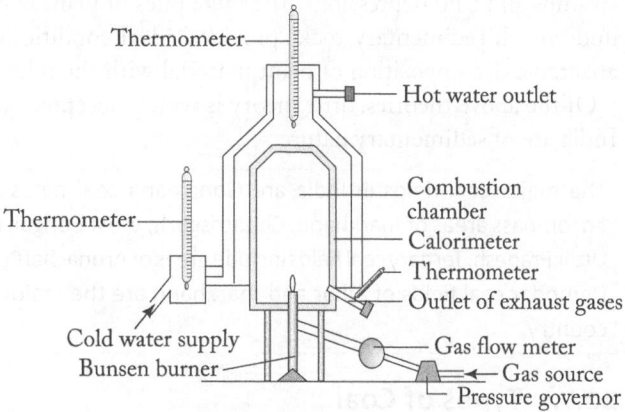

Fig. 20.3 Junkers calorimeter

of combustion move upwards in the chamber and then downwards and finally escape through the exit and the temperature is recorded. The heat generated by combustion of gaseous fuel is absorbed by the cold water circulating in the annular space outside the calorimeter chamber. After steady conditions are established, the following readings are recorded, viz,

(a) Volume of gaseous fuel (V) at given pressure and temperature at a particular period of time (t).

(b) Quantity of water (W kg) passing through the annular spaces at the same interval of time.

(c) Rise in temperature ($T_2 - T_1$).

(d) Mass of water (m) condensed in the outlet water (kg).

(e) Gross calorific value = L

(f) Specific heat of water = S

(g) Heat absorbed by circulating water = $W(T_2 - T_1) \times S$

(h) Heat produced by gaseous fuel combustion = VL

Hence, $VL = W(T_2 - T_1) \times S$

$$\therefore \quad L = W(T_2 - T_1) \times S/V \text{ kcal/m}^3$$

$$NCV = L - \frac{m \times 587}{V} \text{ kcal/m}^3$$

20.7 COAL

Coal, a solid fuel, is a major source to generate heat and electricity. It occurs in strata called seams that are bound by upper and lower layers of sedimentary rocks. It comprises mainly carbon and other elements such as hydrogen, oxygen, nitrogen, and sulphur. Based on the carbon and hydrogen ratio, ranking of coal (in increasing order) is as follows.

Peat → Lignite → Semi-bituminous coal → Bituminous coal → Anthracite coal

The above order also depicts the transformation of peat to coal. Coal is a black coloured naturally occurring organic rock formed due to accumulation and eventual physical and chemical changes of plant debris over long periods of time within the earth's crust. There are two popular theories that propose the accumulation of plant debris and its transformation to coal.

In situ theory   It is envisioned that over a period of time, land started to sink and vegetation went underneath the water bodies but did not undergo decomposition. However, land continued to sink resulting in shrinking and submergence of plants with eventual decomposition. This cycle continued resulting in the formation of coal seams and strata.

Drift theory   Plant materials were uprooted and transported by rivers and lakes and got deposited in swamps and land depressions. The huge piles of plant wood got buried in sedimentary rocks. This burial underneath sedimentary rocks provided ideal conditions such as high temperatures and pressures for anaerobic decomposition of plant material with the release of carbon dioxide and methane gases.

Of the above theories, drift theory is widely accepted since evidence suggests that coal seams found in India are of sedimentary nature.

> The major coal fields in India are Gondwana coal fields and Tertiary coal fields. Gondwana coal fields encompass areas of Jharkhand, Chhattisgarh, West Bengal, Madhya Pradesh, , Telengana, Maharashtra, and Uttar Pradesh. Tertiary coal fields include areas of Arunachal Pradesh, Assam, Meghalaya, Rajasthan, Gujarat, etc. Damodar coal fields of Bihar and Jharkhand are the major suppliers of coal to iron and steel plants in the country.

20.7.1 Types of Coal

Peat   This is the brownish mass obtained after coalification of wood and is dug out manually from coal mines. As it contains 80–90 % moisture, peat is a low-quality and an uneconomical fuel source that can be ignited quickly giving long flames. The average % composition (by weight) of an air-dried peat sample contains: C = 57, H = 6, O = 35, and ash = 2.5 – 6. Peat has a calorific value of 5400 kcal/kg and is used as domestic fuel. Peat briquettes are used in steamers, boilers, power plant stations, and gas producers.

Lignite (brown coal)   This is the lowest ranking coal. It is soft, woody, brownish black coloured amorphous material. If exposed to atmosphere, lignite tends to disintegrate as it dries out completely. The average % composition (by weight) of an air-dried lignite sample contains: C = 60 – 70 and O = 30 – 50. Lignite has a calorific value of about 6500 – 7100 kcal/kg. It is used as domestic fuel for boilers and in the manufacture of producer gas.

Bituminous coal (common coal)   This is a dark grey-to-black mass that is the most commonly employed fossil fuel. The following are the different types of bituminous coal.

Sub-bituminous coal is a black coloured material with a smooth homogeneous appearance. It is denser and harder than lignite with a calorific value of about 6800 – 7600 kcal/kg. Its air dried % composition (by weight): C = 70 – 78, H = 4.5 – 5.5, O = 20. It ignites easily and is used in the manufacture of gaseous fuels.

Bituminous coal has a band-type appearance (in Hindi it is called *koela*) and is a black coal material with a calorific value of about 8000 – 8500 kcal/kg. Bituminous coal is commercially used in domestic ovens, furnaces, boilers, railway locomotives, and thermal power stations.

Semi-bituminous coal includes all those varieties of bituminous coal with high carbon content of about 95 % and low volatile matter. They are employed in the manufacture of coke.

Anthracite   This is the highest ranking coal with carbon content of about 98 % and lowest amount of volatile matter and moisture. It is the hardest solid fuel with sub-metallic lustre or a graphite like appearance. Its calorific value is about 8400–8700 kcal/kg. Anthracite is mainly used in domestic stoves, boilers, metallurgical furnaces, production of carbon electrodes, and so on.

20.7.2 Selection of Coal

Various factors are considered while selecting coal for different purposes which are listed here.

Calorific value This is an important parameter that determines the quality of coal. If calorific value of coal is high, even smaller quantities of coal are sufficient to harness huge amount of energy.

Calorific intensity (theoretical flame temperature) This is the maximum temperature reached when coal is burnt completely in theoretical amount of air. It depends on the nature, quantity, and specific heat of combustion products. The calorific intensity is calculated by dividing the heat of combustion of fuel by the heat capacity of the products of combustion given by the following expression:

$$\text{Calorific intensity} = \frac{\text{Heat of combustion} + \text{Sensible heat of air}}{\sum(\text{Combustion products} \times \text{Specific heat})}$$

Calorific intensity is an important characteristic while judging the maximum possible efficiency of an internal combustion engine (Carnot cycle efficiency), while judging the rate of heating coal. Ideally, it is difficult to attain flame temperature, but coal must possess high calorific intensity.

Size of coal It must be uniform to allow even heating and easy handling of material during combustion operations. Generally, pulverized coal is employed for various domestic and industrial operations.

Moisture The presence of moisture reduces the heating value and thus should be as low as possible. Moisture content in coal could be either inherent or extraneous. Inherent moisture can be removed by heating coal above 100 °C and extraneous moisture is removed by air-drying of coal.

Ash content It should be very low as it is residual non-combustible white matter after burning coal.

Sulphur and phosphorus These should be as low as possible as they are polluting factors and corrode the combustion equipment.

20.7.3 Pulverized Coal

Usually, the combustion rate of solid fuels is slow as it is difficult to permit complete contact between the solid fuel and oxygen. One can increase the combustion rate either by:

(a) *Increasing the rate of oxygen supply;* but this leads to wastage of a large amount of heat carried by air itself. Thus, it is essential to reduce the excess air required for complete combustion of coal.

(b) *Finely dividing the coal* referred to as pulverization of coal. When coal is pulverized to a fine powder, its surface area increases and thus can easily allow greater contact with air during combustion.

Pulverized coal particles are small (< 100 μm) that allow higher heating rates (10^3 to 10^5 K/s) and are commonly employed in various power plants and industrial operations.

Some of the advantages and disadvantages of pulverized coal are as follows:

Advantages

(a) Pulverized coal is easy to transport by conveyors or by forcing stream of air.

(b) It is easy to control the rate of combustion by controlling the amount of coal powder (free flow).

(c) The pulverized coal allows even heating, thus fuel wastage is avoided. Hence, thermal efficiency is increased and higher temperatures are obtained.

Disadvantages

(a) There is an extra cost for pulverizing coal before employing it in various domestic and industrial operations.

(b) Post-combustion of pulverized coal leads to formation of finely divided ash causing pollution. Cottrell's precipitator is used in such a case (Sec. 21.5.2).

20.8 ANALYSIS OF SOLID FUELS

For a chemist, it is essential to determine the elemental composition of coal sample so as to employ it as a source of carbon. Engineers need relevant information on combustion value, moisture content, ash residue, volatile matter, etc., of the coal to utilize in power generation plants. Coal analysis is broadly of two types, namely proximate and ultimate analysis (Fig. 20.4).

ASTM (American Society for Testing and Materials) has provided detailed definitions (D121-85) and tests (D388-88) for coal rank classification. All these methods are as per ASTM and hence reproducible for coal samples employed in domestic and industrial uses.

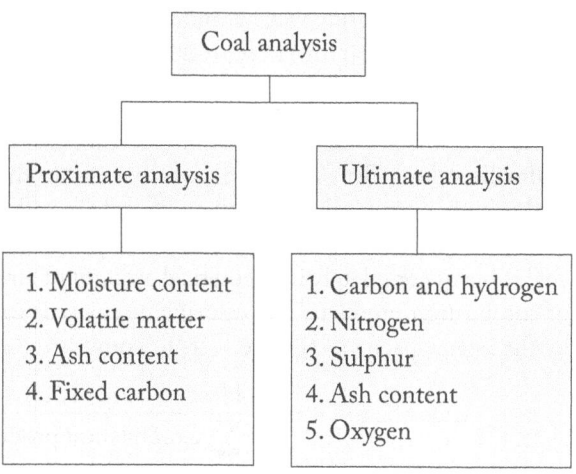

Fig. 20.4 Steps involved in coal analysis

20.8.1 Proximate Analysis of Coal

Proximate analysis is an easy, rapid method that provides a fair idea about the quality of coal. Let us discuss the various steps (ASTM D3172) involved in this process.

Step 1. Determination of moisture (ASTM D3173) The coal sample is pulverized so as to pass through a 250 μm (60 mesh) sieve and is air dried completely. It is then taken in a silica crucible and heated in a hot air oven at temperatures maintained between 105–110 °C for one hour. At the end of one hour, the contents of the crucible are allowed to cool in a desiccator and the loss in moisture is determined.

$$\% \text{ Moisture} = \frac{\text{Loss in weight due to removal of moisture}}{\text{Weight of coal (g)}} \times 100$$

Higher percentage of moisture is undesirable as it reduces the calorific value of coal and quenches the fire in the furnace. Lesser the moisture content, better is the quality of coal as a fuel.

Step 2. Determination of volatile matter (ASTM D3175) Moisture-free coal from Step 1 is taken in a silica crucible and partially covered with a lid. It is then heated in a muffle furnace at 925 +/– 20 °C (inert atmosphere) for 7 minutes. After flash heating, the crucible (red hot condition) is taken out, cooled in a desiccator, and weighed again. The loss in weight as volatile matter on percentage basis is determined.

$$\% \text{ Volatile matter} = \frac{\text{Loss in weight due to removal of volatile matter}}{\text{Weight of coal (g)}} \times 100$$

Volatile matter is present in the form of combustible gases such as hydrogen, methane, and lower hydrocarbons. It results in smoke formation during combustion of coal and hence it is a demerit for the given fuel sample and also requires a huge combustion chamber. Hence, lower the volatile matter, better is the quality of coal as a fuel.

Step 3. Determination of ash content (ASTM D3174) The residual coal from Step 2 is taken in a silica crucible and heated in a muffle furnace at 700–750 °C for 30 minutes. After heating, the contents in the crucible are cooled in air and desiccated. The entire process is repeated until a white non-combustible ash is obtained and a constant weight is recorded.

$$\% \text{ Ash} = \frac{\text{Weight of residue left}}{\text{Weight of coal (g)}} \times 100$$

Ash—the non-combustible component of coal is carried off with the combustion products as very fine particulates called fly ash. Today, fly ash is used as a low-cost adsorbent to eliminate various textile dyes from water.

The presence of ash reduces the calorific value of coal. Ash content increases the handling, storage, and transportation costs. Hence, lower the ash content, better is the quality of coal.

Step 4. Determination of fixed carbon (ASTM D3172) The material remaining after determining the moisture content, volatile matter, and ash content, is considered as the fixed carbon. It is this fixed carbon that burns in solid fuel to give heat of commercial value. It is determined as follows.

%Fixed carbon = 100 – % of (Moisture + Volatile matter + Ash)

Higher the percentage of fixed carbon, greater will be the calorific value of coal and better will be its overall quality as a fuel.

20.8.2 Ultimate Analysis of Coal

After performing proximate analysis, ultimate analysis is essential for calculating heat balances in any process in which coal is employed as a fuel. It involves the elemental analysis of coal sample as weight percentage of C, H, N, S, and O.

Determination of Carbon and Hydrogen (ASTM D3178 Liebig's Method)

Carbon and hydrogen constitute about 70–95 % and 2–6 % respectively and are the major elements in coal. About 1–2 g of powdered (60 mesh) coal sample in a porcelain dish is placed in a combustion tube. A stream of pure and dry oxygen at 25 atm is allowed to pass through the combustion tube with furnace temperature maintained at 850–900 °C (Fig. 20.5). Complete combustion of coal sample takes place with the release of carbon dioxide and steam through heated copper oxide pellets. Carbon dioxide and steam are absorbed by previously weighed tubes containing KOH solution and anhydrous calcium chloride tubes, respectively.

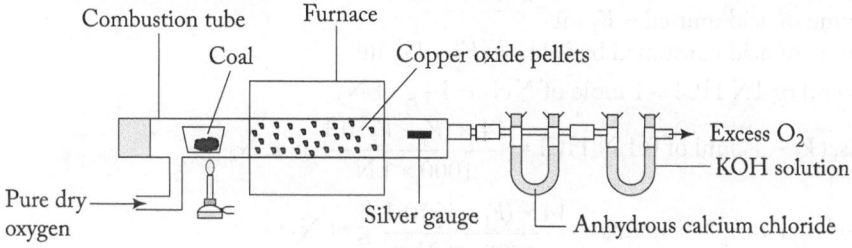

Fig. 20.5 Carbon and hydrogen combustion

The increase in weight of KOH (potash tube) represents the weight of CO_2 and increase in weight of $CaCl_2$ represents the weight of H_2O.

$$C + O_2 \rightarrow CO_2 \quad \text{and} \quad 2\,KOH + CO_2 \rightarrow K_2CO_3 + H_2O$$
$$12 \qquad\qquad 44 \qquad\qquad\quad 112\,g \qquad\qquad\quad 138\,g$$
$$H_2 + 1/2\,O_2 \rightarrow H_2O \quad \text{and} \quad CaCl_2 + 7\,H_2O \rightarrow CaCl_2.7H_2O$$
$$2 \qquad\qquad 18 \qquad\qquad 111\,g \qquad\qquad\qquad 237\,g$$

$$\% \, C = \frac{\text{Increase in weight of KOH tube} \times 12 \times 100}{\text{Weight of coal sample} \times 44}$$

$$\% \, H = \frac{\text{Increase in weight of CaCl}_2 \text{ tube} \times 2 \times 100}{\text{Weight of coal sample} \times 18}$$

Higher percentage of carbon and hydrogen increases the calorific value of coal and hence better is the quality of coal sample as a fuel.

Determination of Nitrogen (ASTM D3179, Kjeldahl's Method)

In this method, a known amount (say, 1 g) of powdered coal is heated with Conc. H_2SO_4 and K_2SO_4 (catalyst) in a Kjeldahl's flask (Fig. 20.6). The elemental nitrogen present in coal is converted to ammonium sulphate. The clear solution is treated with an excess of 50 % sodium hydroxide causing the liberation of ammonia gas, which is distilled over a known excess of standard 0.1 N hydrochloric acid solution. Some amount of the standard acid will remain unused which can be determined by titrating it against standard 0.1 N sodium hydroxide solution.

$$N \xrightarrow[\substack{K_2SO_4, \text{catalyst}}]{\substack{\text{c. } H_2SO_4, \Delta}} (NH_4)_2SO_4 \xrightarrow[\Delta]{\text{Alkali}} NH_3 \longrightarrow \text{Distilled over HCl}$$

In coal   Ammonium sulphate   Ammonia
(clear solution)   gas

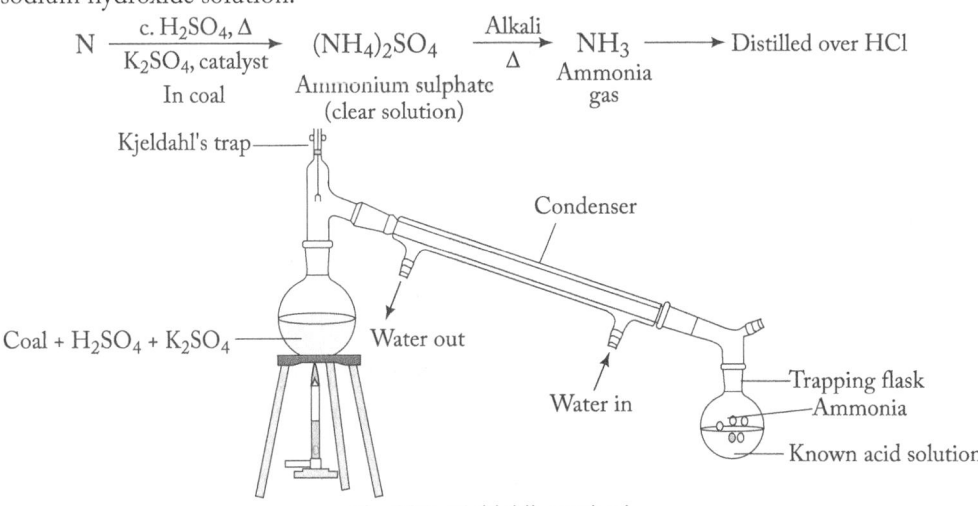

Fig. 20.6 Kjeldahl's method

From the volume of acid used by the liberated ammonia, the percentage of nitrogen can be estimated.

Mass of coal = x gm
Volume of acid in which NH_3 is passed = V_1 ml
Volume of acid unused = V_2 ml
Volume of acid consumed by NH_3 = $(V_1 - V_2)$ ml
1000 ml of 1N HCl ≡ 1 mole of NH_3 ≡ 14 g of N_2

Thus, $(V_1 - V_2)$ ml of 0.1 N HCl = $\dfrac{14 \times (V_1 - V_2)0.1}{1000 \times 1\,N}$ g of N_2

x g of coal sample contains = $\dfrac{14 \times (V_1 - V_2)0.1}{1000 \times 1\,N}$ g of N_2

$\% \, N = \dfrac{14 \times (V_1 - V_2) \, \text{Normality of HCl}}{1000 \times \text{Weight of coal sample } (x\,g)} \times 100$

Determination of Sulphur (ASTM D2015 and ASTM D3286)

A known amount of coal is burnt in a current of oxygen in a bomb calorimeter. After complete combustion of coal, washings from the inside of the bomb is done quantitatively. All the washings are collected and the sulphates are precipitated with barium chloride as barium sulphate. The precipitate is filtered, washed, dried, and weighed until a constant weight is recorded. The percentage of sulphur as weight percent is calculated.

$\% \, S = \dfrac{\text{Weight of } BaSO_4 \times 32}{\text{Weight of coal (g)} \times 233} \times 100$

Sulphur, though contributes to calorific value (see Eq. 20.1), is undesirable due to its polluting properties as it forms SO_2 on combustion.

Determination of Ash

Ash is determined as per proximate analysis of coal (ASTM D3172)

Determination of Oxygen (ASTM D 3176)

The sum percentages of C, H, N, S, and ash subtracted from 100 gives the oxygen content.

$$\% \text{ Oxygen} = 100 - \text{percentage of } (C + H + N + S + \text{ash})$$

Higher the percentage of oxygen, lower is the calorific value and lower is the coking (combustion) power. Also, oxygen, when combined with hydrogen in coal becomes unavailable for combustion.

20.9 LIQUID FUELS

Crude oil or petroleum (Latin: *petra* = rock, *oleum* = oil) is a dark-brown viscous liquid fuel found in deep earth's crust. Due to decomposition, vegetative matter buried for millions of years is subjected to intense heat and pressure, thereby forming petroleum. The various hydrocarbons in petroleum are formed due to radioactivity or anaerobic bacterial decomposition.

> Digboi (Assam) is the oldest oil producing region of India. The major oil fields in India are in Ankaleshwar, Dholka, Mehsana, Lunej, Kathana, Sananda, Kalol (Gujarat), and Mumbai High in Arabian Sea (Mumbai).

20.9.1 Classification of Petroleum

Crude oil is composed of mainly carbon (83–87 %), hydrogen (12–14 %), and a mixture of hydrocarbons of differing molecular weights. It also contains organic compounds of oxygen, nitrogen, and sulphur along with metals such as nickel, iron, and copper. Generally, the composition of crude oil depends largely on the location of the oil reserves. On the basis of major hydrocarbons, there are three main types of petroleum.

Paraffinic hydrocarbons or **alkanes** These having the general molecular formula C_nH_{2n+2} (where $n = 1$–20) are present as waxes in crude oil. The alkanes could be straight-chained or branched. Methane, ethane, propane, butane, isobutene, etc., are some of the examples of paraffins found in crude oil.

Naphthenes or **cycloalkanes** These hydrocarbons have the general formula C_nH_{2n} (where, $n = 1$–20) and ring-like structures.

Aromatic compounds These have the general formula $C_6H_5 - A$ (where A could be a straight chain molecule connected to aromatic ring); examples include benzene and naphthalene.

Let us have a look at the mining of crude oil and processes involved to obtain valuable fractions such as petrol, diesel, lubricants, and other important fuels of commercial value.

20.9.2 Mining of Petroleum

The most important criterion before mining is selecting the oil reservoir site on the basis of geological evidences. In the oil reservoir, crude oil floats over brine and natural gas surrounds the porous oil-bearing rocks. Mining is done by drilling holes up to the oil-bearing surface and digging pipes till the oil-bearing rocks.

Due to hydraulic pressure of natural gas, crude oil gushes out through the pipelines. At times, two coaxial tubes are dug in to the oil reservoir and

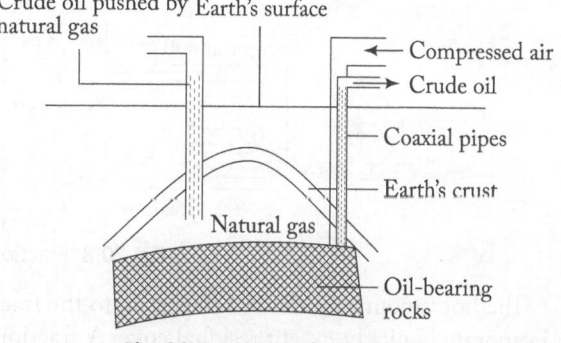

Fig. 20.7 Mining of petroleum

compressed air is forced through the outer pipe, while oil gushes out from the inner pipe and is sent to the refinery for further processing.

20.9.3 Refining of Crude Oil

The major problem while processing petroleum is that it comprises a complex mixture of hydrocarbons. It is known that different hydrocarbon chain lengths tend to have progressively higher boiling points. The process of separating hydrocarbon fractions based on their differences in boiling points is called *fractional distillation* and is carried out in an oil refinery. Before fractionating them, water, dirt, and other contaminants are removed from the crude oil by the following processes.

> The first recorded commercial use of petroleum products was done in 1840 by Dr. Abraham Gesner who distilled a kerosene fraction to use as an illuminant.

Step 1. Separation of Water (Cottrell's Process) The crude oil is a stable emulsion of oil and salt water. The process of freeing oil from brine consists of flowing them between two highly charged electrodes and applying weak electric fields. This process termed as *electric demulsification* was proposed by Cottrell. On application of electric field, colloidal water droplets rapidly coalesce to form larger drops, thereby separating from the oil.

Step 2. Removal of harmful sulphur compounds It is essential to eliminate sulphur compounds as they act as a polluting factor. After removal of water, oil is treated with copper oxide leading to the formation of copper sulphide precipitate, which is removed by filtration. Crude oil can also be treated with sodium plumbite that converts thiols in to disulphides.

$$R\text{—SH} + Na_2PbO_2 \xrightarrow{\text{Sulphur (in NaOH)}} R{\diagdown}S{-}S{\diagup}R + PbS + 2NaOH$$

Thiol   Sodium plumbite   Alkyl disulphide

The above process is referred to as *sweetening of petrol.*

Step 3. Fractional distillation Figure 20.8 depicts the set-up of the fractionating tower. The crude oil is heated to about 400 °C in an iron retort.

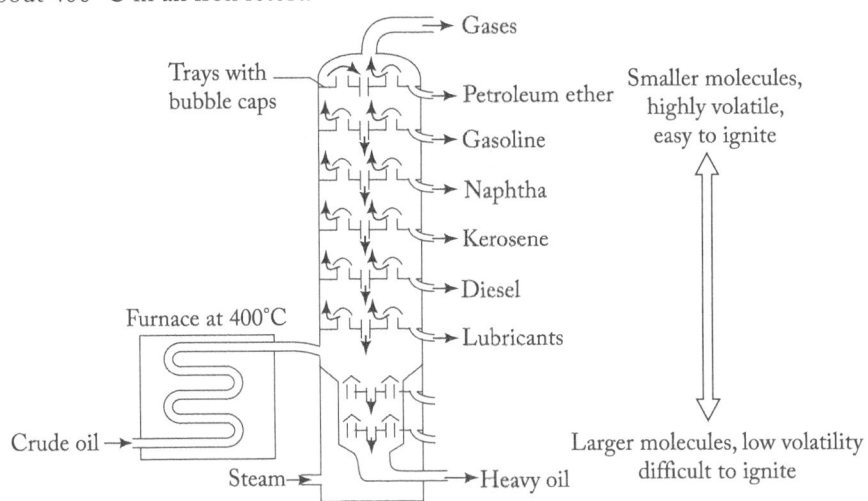

Fig. 20.8 Fractional distillation unit

The hot vapours are allowed to pass in to the fractionating tower. Most of the volatile components get evaporated quickly, except residual coke. A fractional column or tower has bubble caps in trays so as to allow maximum contact time between the liquid and vapour phases. Imagine the fractionating column with varying temperature differences with hottest bottom zone and colder top zone. So, when a mixture of crude oil is heated and sent in to the column, the most volatile component will reach the top zone of the column and undergoes condensation forming a liquid. The liquid, so formed, will move downwards and at its boiling point will be collected through a pipeline. A dynamic equilibrium exists between ascending vapours and descending liquids across the trays in the column. Overall, the components with the lowest boiling point will condense at the topmost zone in the column followed by components with higher boiling points condensing at the lower zone in the column.

Table 20.1 Fractions obtained by distillation of crude oil

Fractions	Hydrocarbon composition	Boiling ranges (°C)	Applications
Uncondensed gases	*C1 – C4	< 30	Domestic fuel (LPG)
Petroleum ether	C5 – C7	30 – 70	Solvent
Gasoline (Petrol)	C5 – C9	40 – 120	Motor fuel
Naphtha	C9 – C10	120 – 180	Solvent
Kerosene	C10 – C16	180 – 250	Engine oil
Diesel	C15 – C18	250 – 320	Engine oil
Heavy oil	C17 – C30	320 – 400	Sent for cracking to obtain gasoline
Lubricating oils	-	-	Lubricants for various machines (see Chapter 18)
Petroleum jelly	-	-	Cosmetics
Residual asphalt	> C30	> 400	Water-proofing material
Residual coke	-	-	Fuel

*C indicates carbon chain length of the hydrocarbon.

20.9.4 Chemical Processing of Crude Oil

Most of the components separated by fractional distillation column are of commercial value and ready for the market (Table 20.1). However, a few of them need to be chemically processed to obtain fuel-based products. Continued process of fractional distillation is expensive and an energy-intensive process. It is a better proposition for oil companies to chemically process valuable fractions obtained from the fractional tower to prepare gasoline. Chemical processing increases the overall yield of gasoline from each barrel of crude oil. One can change one fraction into another by one of the three following methods.

(i) Breaking large hydrocarbons into smaller pieces, called *cracking*
(ii) Combining smaller hydrocarbons to make larger valuable components, called *unification*
(iii) Rearranging various hydrocarbon components to obtain desired hydrocarbons, called *alteration*

Cracking of Petrol

Cracking of petrol is considered as the process of thermal or catalytic decomposition of high-molecular weight hydrocarbons present in crude oil. The higher boiling fractions dissociate in to valuable lower boiling fractions suitable for spark-ignition engines of automobiles. The primary objective of the cracking process is to obtain a higher yield of pure gasoline. Also, it is found that cracked gasoline provides better engine performance by producing lesser knocking of IC engines. The higher saturated hydrocarbons are converted to simpler molecules like paraffinic and olefinic hydrocarbons; for example, octane undergoes cracking to form hexane and ethane; it can also form pentane and propene.

$$C_8H_{18} \rightarrow C_6H_{14} + C_2H_4; \quad C_8H_{18} \rightarrow C_5H_{12} + C_3H_6$$
$$\text{Hexane} \quad \text{Ethene} \qquad \text{Pentane} \quad \text{Propene}$$

Ethene and propene are commercially used to make polymers such as polyethene and polypropylene respectively.

Cracking is an endothermic reaction; hence external heat is supplied by maintaining suitable temperatures. The quality and yield of gasoline depend directly on the cracking temperature, catalyst involved, and duration of the process. Chemical cracking is of two types, namely thermal and catalytic cracking.

Thermal cracking   The heavy oil is subjected to high temperatures and pressure that causes the high boiling fractions to break in to lower hydrocarbons of paraffinic and olefinic series. Generally, the overall yield is 7–30 % of cracked gasoline. Thermal cracking processes can be carried out in various phases.

Liquid phase cracking   The oil is cracked at 475–530 °C under a pressure of 100 kg/cm^2. The final cracked hydrocarbon products are separated in a fractionating column.

Vapour phase cracking   The oil is vapourized and cracked at 600–650 °C under a pressure of 10–20 kg/cm^2. As the reaction takes place in vapour phase, completion of the reaction is rapid, but product stability is lower than those obtained in the liquid phase. The octane value of the gasoline obtained by vapour phase is higher than that obtained by liquid phase cracking method.

Catalytic cracking   This is a widely employed refinery process for converting heavy oil in to gasoline of high octane value. The catalyst does not take part in the cracking reaction, but enhances the overall reaction kinetics at much lower temperatures, that is, in the range of 300 – 450 °C and at 1 – 5 kg/cm^2 pressure, respectively.

Fixed bed catalytic cracking   In this process, the heavy oil charge is passed through a preheater maintained at a cracking temperature of 425–450 °C. The oil vaporizes and enters the catalytic chamber, which is packed with clay and zirconium oxide. The temperature at the preheater and the catalytic chamber is maintained constant at the cracking temperature and pressure of 1.5 kg/cm^2 is maintained throughout the plant design.

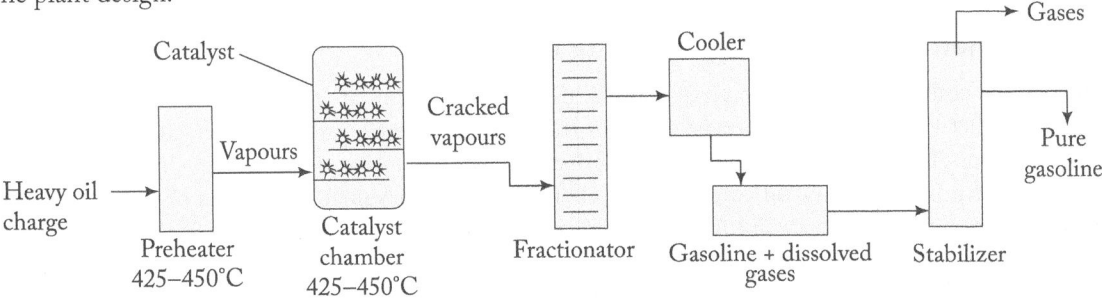

Fig. 20.9 Fixed bed catalytic cracking

The actual cracking takes place in the catalytic reactor. The hot cracked vapours are allowed to enter the fractionator. In the fractionator, gasoline vapours and gaseous products are recovered from the head of the column, whereas heavy oil fraction gets condensed at the bottom. The remaining uncondensed gases are allowed to move further and escape as flue gases. The condensate is now sent to the stabilizer to eliminate any dissolved gases to obtain pure gasoline. After 8–10 hours of continued cracking operations, the catalyst ceases to function and requires regeneration, which is done by blowing a stream of hot air in the catalyst bed to remove carbon deposits. Figure 20.9 shows the schematic of the set-up of the process.

Fluid bed catalytic cracking   In fluid bed cracking, finely powdered catalyst is kept agitated by vapour streams from heavy oil feedstock coming from the fractionator (Fig. 20.10). Thus, the catalyst can be handled like a fluid system and pumped just as a liquid phase. This permits close contact between the catalyst and the reactants, resulting in an efficient cracking process. Heavy oil is fed into the reactor and the fluid catalyst is introduced from the regenerator. The cracking takes place at the catalytic bed that circulates with oil vapours at a temperature of 530 °C and pressure between 3–5 kg/cm^2. The volatile gasoline moves to the head of the reactor and enters the fractionator. Further, the cracked gasoline is obtained from the top of the fractionating column. Gasoline vapours are cooled and the condensate is sent into the stabilizer to eliminate any dissolved gases. The overall process results in a faster reaction and purer gasoline with higher yields.

The **advantages of catalytic cracking** are listed as follows.

(a) The yield and quality of petrol obtained from catalytic cracking are very high.

(b) The cracking reaction can be controlled and carried out at low pressures.

(c) The products obtained from catalytic cracking contain higher amounts of aromatics and hence possess better anti-knock characteristics. As already known, catalysts are selective in their action and result in cracking of higher boiling point hydrocarbons into valuable lower boiling point fractions.

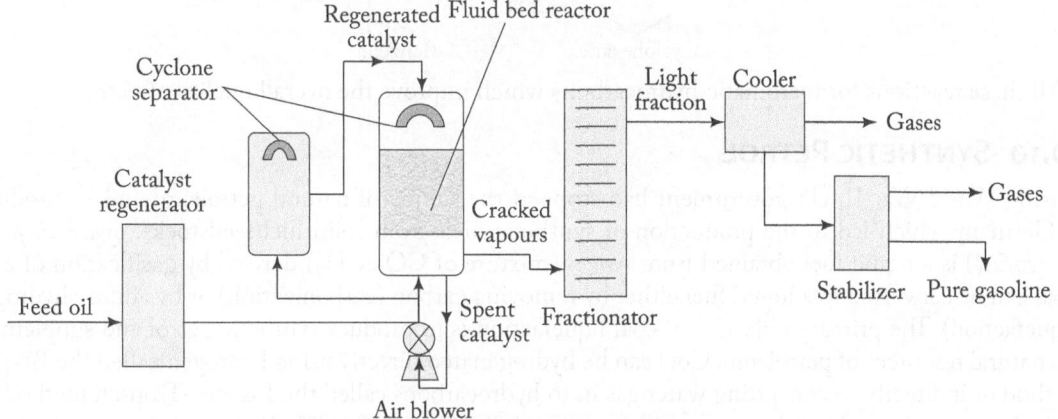

Fig. 20.10 Fluid bed catalytic cracking

20.9.5 Reforming of Petrol

Reforming is a process of bringing structural modifications in straight-run gasoline thereby improving its anti-knock characteristics. It involves the chemical conversion of open chain aliphatic hydrocarbons and cycloalkanes in the presence of a catalyst in to aromatic hydrocarbons containing the same number of carbon atoms. It is carried out thermally at 500–600 °C under 85 atm or by employing platinum (0.75%) catalyst supported over alumina at 460–530 °C maintained at pressure of 30–35 atm. Isomerization, alkylation, polymerization, and dehydrogenation are examples of reforming reactions.

In **isomerization**, straight-chain hydrocarbons undergo rearrangement to form branched chain hydrocarbons. n-Hexane forms branched hydrocarbons 2-methyl pentane and 2, 3-dimethyl butane.

$$H_3C \diagup\diagdown\diagup\diagdown CH_3 \longrightarrow H_3C \diagup\diagup CH_3 \quad \text{2-Methyl pentane}$$

n-Hexane

2,3-Dimethyl butane

n-Alkanes can be converted into cyclic compounds and then dehydrogenated to yield aromatic compounds. At times, n-alkanes are reacted with hydrogen in the presence of a catalyst to yield lighter hydrocarbons; for example, n-hexane on heating with hydrogen in the presence of platinum catalyst (on alumina) yields the following reformed products, namely ethane, propane, and butane. Such reactions are called *hydrocracking*.

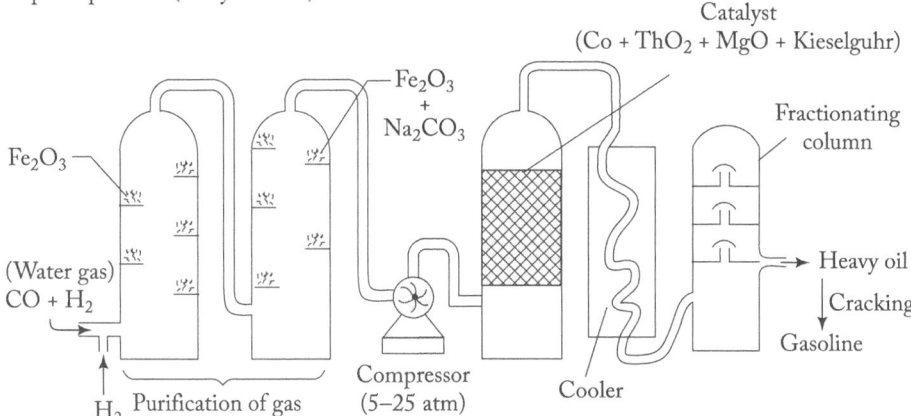

Dehydrogenation involves converting cycloalkanes to cycloalkenes and other aromatic compounds; for example, cyclohexane is dehydrogenated to form benzene.

All these reactions form aromatic hydrocarbons which improve the overall quality of petrol.

20.10 SYNTHETIC PETROL

During World War II, US government had stopped the supply of natural petroleum and its products to Germany which led to the production of synthetic fuels from solid fuel feedstocks. *Synthetic petrol* (or *synfuel*) is a liquid fuel obtained from *syngas* (mixture of CO & H_2) derived by gasification of coal. Coal can be converted to a liquid fuel either by removing carbon (carbonization) or by adding hydrogen (liquefaction). The primary objective of coal liquefaction is to produce synthetic petrol and supplement the natural resources of petroleum. Coal can be hydrogenated directly using hydrogen called the Bergius method or indirectly by converting water gas in to hydrocarbons called the Fischer–Tropsch method. In both the methods, coal is first converted into gas and finally into liquid fuel.

20.10.1 Fischer–Tropsch Method

The process was developed by Franz Fischer and Hans Tropsch in 1920 and came to be known as Fischer–Tropsch process (or synthesis).

Fig. 20.11 Fischer–Tropsch method

The schematic diagram of Fischer-Tropsch method is shown in Fig. 20.11. In this method, coal is first gasified to make syngas. Next, the Fischer–Tropsch catalysts convert syngas in to lighter hydrocarbons which are further processed to obtain gasoline.

When oven-heated coke is mixed with hydrogen and steam is allowed to pass through it, water gas is obtained. Water gas is purified by passing it through ferrous oxide (Fe_2O_3) to eliminate hydrogen sulphide. Further, water gas is sent to a chamber containing ferrous oxide and sodium carbonate to eliminate organic sulphur compounds. The purified gas is compressed at 5–25 atm and sent to catalyst chamber containing cobalt, magnesia, thoria, and Kieselguhr (100 : 8 : 5 : 200 parts respectively) that is maintained at 200–300 °C. The following exothermic reactions take place.

$$n\,CO + 2n\,H_2 \rightarrow C_nH_{2n} \text{ (unsaturated hydrocarbons)} + n\,H_2O$$
$$n\,CO + (2n + 1)\,H_2 \rightarrow C_n\,H_{2n+2} + n\,H_2O$$

Hot gases obtained after hydrogenation are sent to the cooler, where liquid fuel is obtained that is passed through the fractionating column. After fractionation, petrol (or gasoline) and heavy oil are obtained. On cracking heavy oil, gasoline is obtained.

20.10.2 Bergius Method

A schematic diagram of Bergius method is shown in Fig. 20.12. In 1913, Friedrick Bergius developed a direct method to obtain liquid fuel from coal. In this method, a paste of finely powdered coal, heavy oil, and tin or nickel oleate catalyst is pumped into converter along with hydrogen and heated at 450 °C and about 200–250 atm for 2 hours. This process is called hydrogenation as hydrogen reacts with coal and gives saturated hydrocarbons.

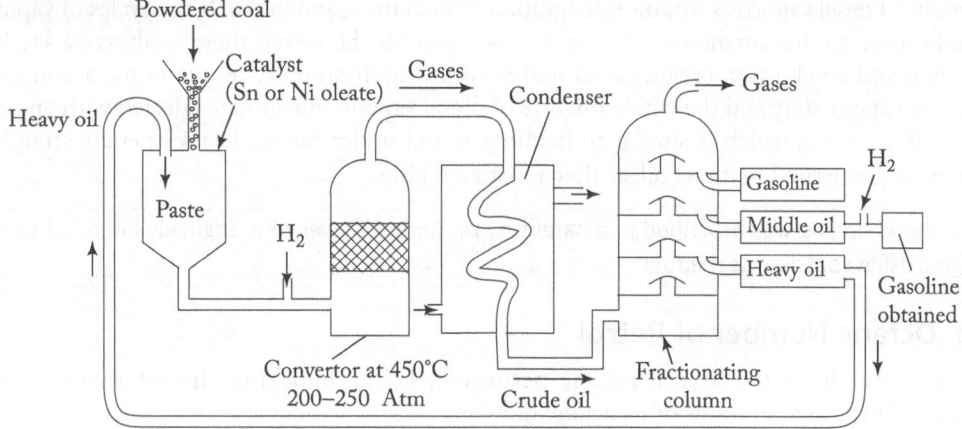

Fig. 20.12 Bergius method

$$n\,C + (n - x + 1)\,H_2 \rightarrow C_nH_{2n-2x+2} \text{ (where } x \text{ is the degree of unsaturation)}$$

The hydrocarbons are sent to the condenser to obtain crude oil. The crude oil is fractionated to obtain gasoline, middle oils, and heavy oils. The middle oil is further hydrogenated in the presence of a solid catalyst to obtain gasoline. Heavy oil is recycled back in the process.

20.11 Knocking in IC Engines

Knocking is a phenomenon seen in spark-ignition IC engines that run on petrol. As discussed earlier, petrol is a mixture of different hydrocarbons. Diesel oil is difficult to vaporize than petrol and hence is not interchangeable in car engines. A car engine is a four-stroke device that operates on a four-cycle path called the 'Otto cycle.' In IC engines, a mixture of gasoline vapours and air is compressed and ignited by an electric spark. The four strokes are intake, compression, combustion, and exhaust. In the simplest sense, an engine comprises a cylinder and a piston that moves up and down within the cylinder. As the piston moves to the bottom of the cylinder, a mixture of fuel and air flows in. Next, as the piston moves upward towards the top of the cylinder, it causes compression of the air–fuel mixture. As soon as the piston reaches the top of the cylinder, spark plug is ignited. This spark creates a small, controlled explosion that causes the piston to move to bottom of the cylinder. The fuel–air mixture undergoes combustion and finally, in the last cycle, the piston moves upwards to expel the combustion (exhaust) gas out of the cylinder. Once the exhaust gas has been pushed out, the entire cycle repeats again. If all the cycle paths are complete, the engine will run smoothly. However, at times, the movement of the piston itself adds pressure that causes the air and fuel mixture to ignite prematurely in the compression cycle. This is called *pre-ignition* and results in a small explosion that is heard as a pinging sound under the hood of the car. The sharp, metallic, pinging sound resulting due to premature ignition of air–fuel mixture in an IC engine is called *knocking*. If this remains unchecked, knocking may affect the engine performance and lead to damage of the cylinder–piston assembly. Knocking depends on various factors, such as the structure of hydrocarbons in fuel (petrol), sudden burning of hydrocarbons, engine design such as shape of head, location of plug, and operating conditions. Thus, the tendency of gasoline to knock is in the following order:

Straight chain paraffins > branched chain paraffins > olefins > cycloparaffins > aromatics

Cyclic, aromatic, and branched chain hydrocarbons in a petrol sample are considered desirable as these compounds resist knocking.

However, it follows the reverse order with diesel oils. Diesel engines are compression ignition (CI) engines that do not have spark plug. The fuel oil is straight chain hydrocarbon mixture within the boiling range of $100 - 360\,°C$. In the suction stroke, only air is drawn in the cylinder. Next, air is compressed till its temperature is about $500\,°C$, called compression stroke, followed by injecting the fuel (diesel) as a spray in the engine. Diesel vaporizes, attains self-ignition temperature, and burns. If the cycles of vaporization and combustion are instantaneous, the fuel burns smoothly. However, there is always a lag between vaporization and combustion cycles, called *ignition delay*. Ignition delay results in the accumulation of diesel in the vapour state and the whole mixture of diesel vapour and air gets injected with an explosion termed as *diesel knock*, which is similar to hammer sound under the vehicle. Generally straight chain hydrocarbons are desired in diesel oil, as they resist knocking.

> In 1892, diesel engine was described in a patent by Dr. Rudolf Diesel who originally intended to operate the engine using coal dust as the fuel.

20.11.1 Octane Number of Petrol

Engine knock reduces the overall vehicle performance. The knocking characteristics of a fuel is measured and expressed in terms of its octane number. Octane number (performance of petrol engines) is the measure of a fuel's ability to resist knocking. Octane rating of a fuel is also an indication of the tendency of the fuel to auto-ignite. Hence, lower the octane number, higher are the chances of auto-ignition of the fuel. An ideal fuel must exhibit higher octane numbers. Iso-octane (C_8H_{18}) has good combustion characteristics and exhibit little tendency to detonate when mixed with air and ignited at high temperatures. Hence, its octane number is generally taken as 100. n-Heptane (C_7H_{16}) has a higher tendency to auto-ignite, hence its octane number is considered as zero.

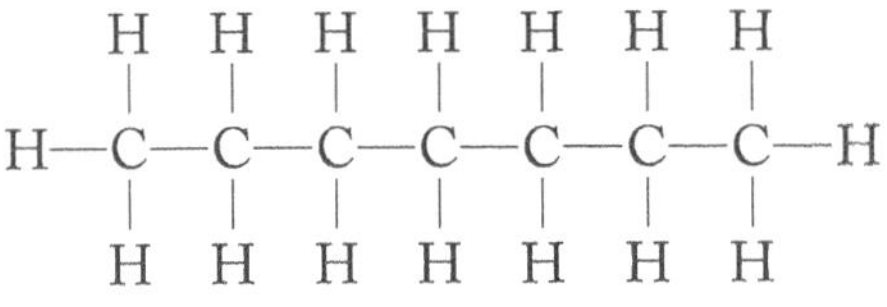

n-Heptane

$CH_3-(CH_2)_5-CH_3$

Octane number = 0
Knocking = very high

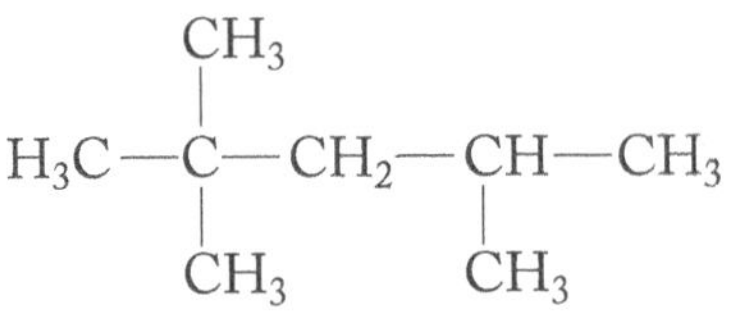

2,2,4-Trimethyl pentane
(Iso-octane)

C_8H_{18}

Octane number = 100
Knocking = minimum

Octane number is defined as the percentage of iso-octane in the mixture of iso-octane and n–heptane that just matches with the knocking characteristics of the petrol sample under consideration. Thus, if a sample of petrol gives as much knocking as a mixture of 85 parts of iso-octane and 15 parts of n-heptane, then its octane value is taken as 85. In practice, octane rating of petrol obtained in this manner does not necessarily correspond to the fuel behaviour, when used in a car under standard conditions. Thus, a fuel with an octane number of 85 may not be always superior to the one with an octane number of 78.

Anti-knocking agents   Anti-knocking agents are compounds that help to increase the octane number of fuel and eventually decrease the knocking. It is observed that by adding tetraethyl lead (TEL), $Pb(C_2H_5)_4$, knocking is reduced with an increase in the octane number. The process of admixing anti-knocking compounds is called *doping*. TEL is a colourless liquid with a sweet odour having a specific gravity of 1.62. It is highly poisonous and special care has to be taken so that it does not enter the body through wounds. Gasoline containing TEL is coloured to indicate its poisonous nature and to detect engine leak. The use of TEL also tends to form PbO_2 or free lead in the cylinder that deposits in the cylinder walls causing engine damage. To avoid this, a trace amount of ethylene dibromide (scavenger) is added that reacts with free lead or PbO_2 at 200–300 °C to form $PbBr_2$. $PbBr_2$ is highly volatile and can be easily expelled from IC engines along with other exhaust gases. However, all these products result in air pollution. Due to the poisonous nature of lead, TEL has been banned in many countries and replaced by other octane enhancers. Leaded petrol cannot be used in automobiles with catalytic converters, as lead present in exhaust gas poisons the catalyst (rhodium catalyst), thereby destroying the active sites of the catalyst.

Unleaded petrol   A petrol sample whose octane number is increased without the addition of lead compounds is called *unleaded petrol*. An alternative method of enhancing octane number of gasoline is to add octane enhancer compounds such as isopentane, isooctane, isopropyl benzene, or methyl tertiary butyl ether (MTBE). MTBE is the most preferred anti-knocking agent, as it contains oxygen in the form of an ether group and this additional oxygen assists in the combustion of petrol in IC engines, along with reducing the peroxide formation. Also, it tends to increase the molecules with branched and aromatic rings in the fuel structure.

20.11.2 Cetane Number of Diesel

The knocking characteristic of diesel is usually expressed in terms of cetane number. *Cetane number is the measure of ease with which diesel will ignite under compression.* Cetane ($C_{16}H_{34}$) is a straight-chain saturated hydrocarbon with short ignition lag as against any other diesel fuel. Hence, its cetane number is considered as 100. It is seen that straight-chain compounds undergo easy compression ignition and n-cetane is chosen as the upper limit of cetane number 100. It is observed that 1-methyl naphthalene does not undergo compression ignition easily and its cetane number is taken as zero.

Cetane number (performance of diesel engines) is defined as the percentage by volume of cetane in a mixture of cetane and 1-methyl naphthalene which exactly matches in its ignition delay characteristics with the diesel under test. These values can be improved by adding trace amounts of doping agents (~ 2 %) such as isoamyl nitrate, ethyl nitrite, ethyl nitrate, and acetone peroxide.

As discussed, petrol knock is due to the sudden combustion of fuel sample, whereas diesel knock is due to delay in spontaneous ignition of fuel. Hence, a fuel with high octane number will be the one with lower cetane number, and vice versa. So, greater the aliphaticity (or straight chain) of hydrocarbons, lower will be the ignition delay and overall higher cetane value of diesel oils.

1-Methyl naphthalene

$C_{11}H_{10}$
Cetane number = 0
ignition delay = maximum

CH_3—$(CH_2)_{14}$—CH_3
n-Hexadecane

$C_{16}H_{34}$
Cetane number = 100
ignition delay = minimum

20.12 POWER ALCOHOL

Ethanol is an alternative non-fossil fuel that can be prepared from a variety of renewable sources such as corn, straw, grass, and sugarcane molasses. The fermentation of sugar and starch from various agricultural sources yields ethanol. It is called power alcohol, when employed as 5–25 % blended with gasoline to power IC engines. Ethanol–gasoline blends (22 % ethanol) replace lead as antiknock additive and octane enhancer. Calorific value of ethanol is quite low (~7000 cal/g). The admixture of alcohol with fuels is called *gasohol*. Ethanol is also employed in diesel engines with 4.5 % ignition improvers like isoamyl nitrate and ethylene glycol dinitrate in oil. At times, 0.5–1 % castor oil is added to the ethanol to lubricate engine parts.

For manufacturing power alcohol, molasses is diluted with distilled water until the concentration of sugar is about 10–12 %. Nutrients such as ammonium salts (sulphates or phosphates) and dilute sulphuric acid are added to maintain a pH of about 5. A known quantity of yeast (enzymatic catalysis) is added at room temperature. The presence of invertase in yeast converts sucrose in to glucose and fructose.

$$\underset{\text{Sucrose}}{C_{12}H_{22}O_{11}} \xrightarrow[\text{30 °C}]{\text{Invertase}} \underset{\text{Glucose}}{C_6H_{12}O_6} + \underset{\text{Fructose}}{C_6H_{12}O_6}$$

$$C_6H_{12}O_6 \xrightarrow[\text{30 °C}]{\text{Zymase}} 2C_2H_5OH + 2CO_2$$

The enzyme zymase (of yeast) converts glucose and fructose in to ethanol and carbon dioxide. Frothing occurs during the reaction due to the release of carbon dioxide. Overall, fermentation takes about two days to yield the product. The fermented liquid comprises 18–20 % alcohol and on distillation, it is separated as rectified spirit (90–95 % alcohol). When rectified spirit is digested with lime for 48 hours and distilled, absolute alcohol is obtained.

Advantages   Addition of alcohol in gasoline enhances the octane value of the fuel along with better anti-knock characteristics. It is capable of removing traces of moisture from the fuel and also assists in the combustion process.

Disadvantages   It lowers the calorific value of fuel, since alcohols have only two-third calorific value as that of petrol. Alcohol has considerable surface tension making it difficult to atomize and ignite, hence a special arrangement is necessary for the carburetor. During combustion, alcohols tend to oxidize in to acids that can lead to corrosion of the internal parts of the engine.

Check Your Progress

1. What are fuels?
2. What is calorific value of a fuel?
3. Name the different varieties of coal.
4. Justify the statement, 'An ideal fuel should have moderate ignition temperature.'
5. What are the various criteria for selecting a fuel?
6. List the steps involved in the proximate analysis of coal sample.
7. What is ultimate analysis of coal? State its significance.
8. What is the objective of performing proximate and ultimate analysis of coal?
9. Differentiate between octane and cetane numbers.
10. What is a flue gas? State its composition.
11. What is knocking? Is it a good feature of IC engine?
12. Why does an IC engine undergo knocking? How can it be prevented?
13. Justify the statement, 'During catalytic cracking, catalyst requires regeneration.'
14. What is sweetening of petrol?
15. State the parameters to be considered while selecting coal.
16. List the advantages and disadvantages of pulverized coal.

20.13 BIODIESEL

The limited fossil fuel reserves, increased CO_2 emissions, and the need for energy security are the major drivers to seek alternative fuels without impacting the engine performance. According to US Energy Information Administration (2011), fuel consumption may increase from 86 million barrels to 112 million barrels (per day) by 2035. Amongst the alternative fuels, biodiesel is widely investigated due to its biodegradability, comparable properties to diesel oil, no poisonous emission products (sulphur, CO) on usage, and low-cost production methods.

Biodiesel is a non-fossil fuel which is obtained from vegetable or animal fat-based oil feedstock. It consists of long-chain alkyl esters. Vegetable oils have 90–98 % triglyceride content with trace amounts of mono-diglycerides, free fatty acids, phospholipids, carotenes, tocopherols, phosphatides, sulphur, and moisture. Triglycerides are esters of long-chain stearic, palmitic, linoleic, or oleic acids. The transesterification of triglycerides in the presence of excess methanol and an alkaline catalyst yields biodiesel along with soap and glycerol.

where, R and R_1 alkyl chain lengths are of different lengths or saturation.

The triglycerides in vegetable oil are treated with excess methanol in the presence of an alkaline catalyst like potassium hydroxide. The reaction mixture is heated and kept standing for 2 hours to obtain biodiesel (fatty acid alkyl esters) with excess of glycerol as a co-product. The reaction is reversible, the equilibrium is shifted towards biodiesel formation by adding an excess of one of the substrates (generally methanol) or continuously removing the water formed during the reaction.

Biodiesel is an easy to use, biodegradable, eco-friendly, and non-toxic fuel that can be employed as an alternative to diesel. *Jatropha*, a known petrocrop, has been used to obtain biodiesel. The latex of *Jatropha* plant consists of hydrocarbons similar to petroleum that can be extracted by organic solvents. The extracted latex is called *biocrude* which is cracked with hydrogen to obtain biodiesel. Biocrude is further purified and converted in to important petroleum products.

Production of Biodiesel

The following methods are used to produce biodiesel.
  (i) Batch process/conventional method
 (ii) Supercritical extraction
(iii) Ultrasonication and microwaves

In the **batch process**, vegetable oil is heated with excess methanol and alcoholic potassium hydroxide at 60–70 °C in a batch reactor. After 2 hours, the reaction mixture is allowed to settle and the lower glycerol layer is withdrawn. The upper layer of methyl esters is separated, washed, and purified to obtain biodiesel. The unreacted methanol is recycled for further use.

In **supercritical extraction method**, methanol is used in its supercritical phase and mixed with oil. The reaction is carried out in such a way that the reactants are in the same phase at higher temperatures and pressures than the batch process. Hence, the reaction rate is enhanced. Compared to the batch process, the supercritical extraction and purification of biodiesel is faster (takes less than 2 hours). In this method, catalyst is not employed, thereby reducing the efforts to regenerate the catalyst and obtain the product.

In **ultrasonication**, cavitation (bubbles) formed in the reaction mixture allows both heating and mixing of the reactants to form biodiesel. The reaction time is reduced to just 20 – 30 minutes. Similarly, **microwave irradiation** is used to carry out esterification. Oil is preheated to 65 °C using microwave irradiation. Methanol–KOH catalyst is fed in to the heated oil through a condenser. The reaction mixture is irradiated under agitation of 400 rpm and refluxed to obtain biodiesel.

Applications
(a) Pure form of biodiesel (B100) or 20 % blended with petroleum diesel can improve the overall engine performance.
(b) Biodiesel is used as a heating fuel in domestic and commercial boilers.
(c) Backup biodiesel-fuelled generators can be employed during powercuts in industries which is a pollution-free option.

When ethanol is blended with diesel, it is called *diesohol*. There are commercially available systems in diesel engines, which allow ethanol to be added to an air intake of the engine. It cools the diesel engine, but also on mixing with diesel increases its power output.

20.14 Gaseous Fuels

Amongst all fuels, gaseous fuels are considered as the most convenient fuel. It requires the simplest burning arrangement, but poses the risk of leak and fire hazard. It can be obtained either naturally or by burning solid or liquid fuels. The performance criteria of heating by a gaseous fuel are dictated by its calorific value and specific gravity. Hydrogen, acetylene, natural gas, liquefied natural gas, producer gas, water gas, and coal gas are some of the classic examples of gaseous fuels.

20.14.1 LPG (Liquified Petroleum Gas)

LPG is obtained as a by-product during the cracking of heavy oils or from natural gas. It is a mixture of n-butane, isobutane, butylene, and propane with traces of organic sulphides, such as thiols (R-SH) so as to identify a leak. Generally, it is bottled gas supplied under pressure in cylinders used as domestic and industrial fuel. The average composition of LPG (% by volume of air) is: C_3H_8 = 24.7, C_4H_{10} = 38.5, isobutane = 36.7 LPG has a calorific value of 27,800 kcal/m^3. An odourless, colourless, volatile hydrocarbon gas with a faint smell is admixed with ethyl mercaptan (stenching agent) to allow early detection in case of a gas leak. As per IS-4576, the amount of stenching agent should be sufficient to allow detection in atmosphere at odour level of only 2.

LPG provides higher heating efficiency as its calorific value is three times that of natural gas and seven times that of coal gas. It is a cleaner fuel with easy handling, storage, and use. It is well designed as a domestic fuel with easy control on combustion. The explosive range of LPG is about 1.8 to 9.5 (% by volume) in air which is comparatively narrower in range as compared to other gaseous fuels and hence is a hazardous gas. The auto-ignition temperature of LPG is around 410–580 °C and hence, it will not ignite on its own at normal room temperature.

20.14.2 CNG (Compressed Natural Gas)

The principal constituents of natural gas is (% by volume in air) CH_4 = 88.5, C_2H_6 = 5.5, C_3H_8 = 3.7, C_4H_{10} = 1.8, C_5H_{12} = 0.5 with a calorific value of 3000 kcal/m^3. As the name suggests, CNG is obtained by compressing natural gas up to 1000 atm to less than 1 % of the volume it occupies at standard atmospheric pressure. It is normally stored and distributed in cylinders at a pressure of 2900–3600 psi. If a cylinder contains about 15 kg of CNG, it amounts to 2×10^4 litres of natural gas at 1 atmospheric pressure. During combustion, no unregulated pollutants such as smoke, SO_2, SO_3, benzene, HCHO are formed and hence, CNG is considered as an alternative fuel for automotive vehicles.

CNG is preferred over other gaseous fuels, because of the following reasons.

(i) It is comparatively a safer alternative fuel, as it ignites at a higher temperature than petrol and diesel.

(ii) Gasoline operated automobiles are easy to be interconverted as a CNG-run vehicle.

(iii) The operating cost of CNG fuel is lower compared to gasoline operations.

(iv) Combustion of CNG leads to lesser greenhouse gases as against other fuels.

(v) In case of a leak, CNG disperses better with air than LPG.

20.14.3 Producer Gas

Producer gas or suction gas is a mixture of carbon monoxide and nitrogen with traces of hydrogen. The calorific value of producer gas is about 1300 kcal/m³. The % by volume composition of producer gas is: CO: 30; N_2: 51 – 56; H_2: 10–15; and the rest is CO_2 + CH_4.

Manufacture of Producer Gas

The manufacture of producer gas is carried out in a reactor called gas producer (Fig. 20.13). The gas producer consists of a stainless steel tank lined with refractory bricks. At the top of the tank is a cup-cone feeder. Coke bed is present within the reactor which is maintained at 1100 °C. An outlet is provided for collecting producer gas. There are two inlets provided at the bottom of the tank for air and steam supply in the reactor.

During production, the air–steam mixture is passed over red hot coke maintained at 1100 °C. Producer gas is obtained which can be explained as chemical reactions that occur in four zones of the reactor.

Ash zone It is the lowermost zone of the reactor which consists of ash. The incoming air–steam mixture is heated in this zone.

Oxidation (or combustion) zone The zone above ash zone, here the actual combustion takes place. Coke is oxidized to carbon monoxide and carbon dioxide. The chemical reactions in this zone are:

$$C + 1/2\,O_2 \rightarrow CO \quad \text{and} \quad C + O_2 \rightarrow CO_2$$

As these reactions are exothermic, coke bed temperature rises above 1100 °C.

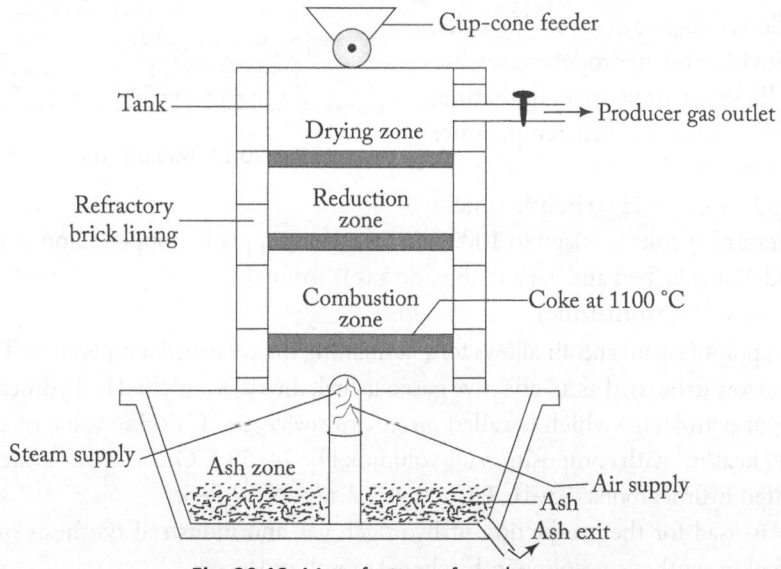

Fig. 20.13 Manufacture of producer gas

Reduction zone It is the middle zone where both carbon dioxide and steam are produced and the chemical reactions are:

$$C + O_2 \rightarrow CO_2 \text{ (exothermic)} \quad \text{and} \quad C + H_2O \rightarrow CO + H_2 \text{ (endothermic)}$$

Due to endothermic reaction during the formation of CO and hydrogen, coke bed temperature falls to about 1000 °C.

Distillation (or drying) zone The topmost zone of the reactor, here the coke bed is heated at 400–800 °C by outgoing combustion gases.

Uses:
(a) Producer gas is used as a reducing agent in various metallurgical operations.
(b) It is used as the heating source for muffle furnaces and open-hearth furnaces.

20.14.4 Water Gas

Water gas or blue gas is a mixture of carbon monoxide and hydrogen with trace amounts of nitrogen. The calorific value of water gas is about 2800 kcal/m^3. The % by volume composition of water gas is: CO : 41; H$_2$: 51; N$_2$: 4, and the rest is CO$_2$ and CH$_4$.

Due to high carbon monoxide content, water gas gives a characteristic blue flame and is also known as blue gas.

Manufacture of Water Gas

The apparatus consists of a tall stainless steel tank lined with refractory bricks. At the top of the tank is a cup-cone feeder. Similar to gas producer tank, here also there are two inlet pipes provided in the tank for air and steam supply. An outlet is present in the tank for collecting water gas for commercial use (Fig. 20.14).

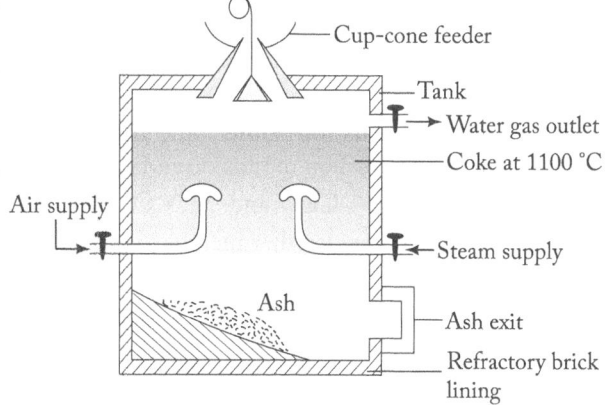

Air and steam are passed alternately on a bed of red-hot coke maintained at 1000 °C in the tank. The reactions occur in the following two stages.

(i) Steam is allowed to pass over red-hot coke. Carbon dioxide and hydrogen gases are produced. These reactions are endothermic in nature and thus coke-bed temperature falls slightly below 1000 °C.

$$C + H_2O \rightarrow CO + H_2 \text{ (endothermic)}$$

Fig. 20.14 Manufacture of water gas

(ii) The temperature of coke is raised to 1000 °C and steam supply is stopped temporarily. Air is blown over the red-hot coke bed and carbon dioxide gas is formed.

$$C + O_2 \rightarrow CO_2 \text{ (exothermic)}$$

The alternate supply of steam and air allows for maintaining the coke-bed temperature. The calorific value of water gas is too low to be used as an effective gaseous fuel, thus it is enriched by hydrocarbons (obtained during cracking of petroleum) which is called *carbureted water gas*. Calorific value of carbureted water gas is about 4500 kcal/m^3 with composition (% volume): H$_2$: 34–38, CO: 23–28, saturated hydrocarbons: 17–21, unsaturated hydrocarbons: 13–16, CO$_2$: 0.2–2.2, and N$_2$: 2.5–5.

Uses: Water gas is used for the production of hydrogen gas and industrial synthesis of ammonia. It is employed as a fuel in synthetic gasoline in Fischer–Tropsch process.

20.14.5 Coal Gas

Coal gas is also called *town gas* as it was used as an illuminant in townships and villages before the advent of electricity. It is a colourless gas with a characteristic odour having a calorific value of about 4900 kcal/m^3. Coal gas is lighter than air and burns with non-smoky type flame. The composition of coal gas is shown in Table 20.2.

Table 20.2 % by volume composition of coal gas

Composition	% by volume
H_2	40
CH_4	32
CO	7
N_2	4
C_2H_4	3
C_2H_2	2
CO_2	1

Manufacture of Coal Gas

When coal is heated in the absence of air at about 1350 °C in a silica retort, coke and coal gas are obtained.

$$\text{Coal} \xrightarrow{\ 1350°C\ } \text{Coke} + \text{Coal gas} \uparrow$$

Coal gas, so obtained is contaminated with impurities, such as tarry vapours, ammonia, nitrogen, and sulphur containing compounds. It is passed in hydraulic main for scrubbing to eliminate any tar-like material. The gas is then sent to condenser for cooling and further scrubbed with water to eliminate ammonia and traces of tar. It is further treated with creosote oil that dissolves naphthalene and benzol. The final purification involves removal of hydrogen sulphide by passing coal gas in a purifier packed with moist ferric oxide.

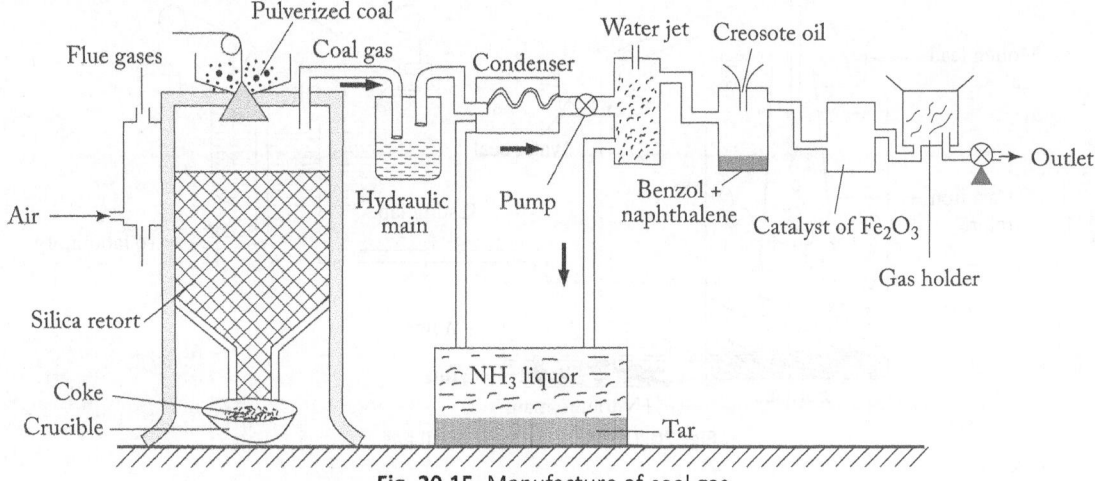

Fig. 20.15 Manufacture of coal gas

The following chemical reactions occur in the process:

$$NH_3(g) + H_2O \rightarrow NH_4OH$$
$$CO_2(g) + H_2O \rightarrow H_2CO_3$$
$$6\ HCN + Fe_2O_3\ (s) \rightarrow 2Fe(CN)_3 + 3\ H_2O$$
$$3\ H_2S\ (g) + Fe_2O_3(s) \rightarrow Fe_2S_3\ (s)\ + 3\ H_2O$$

The ferric oxide purifier on continued operation tends to get exhausted due to the formation of excess ferrous sulphide. Hence, it is essential to activate the catalyst surface by blowing dry air. The reaction can be depicted as:

$$2\ Fe_2S_3\ (s) + 3\ O_2\ (g) \rightarrow 2\ Fe_2O_3(s) + 6\ H_2S\ (g)$$

Pure coal gas is stored over water in a gas holder.

20.14.6 Oil Gas

Oil gas burns with a smoky flame and has a characteristic colour. It has a calorific value of about 4500 – 5400 kcal/m^3 and has a composition as shown in Table 20.3.

Oil gas is obtained by cracking kerosene oil as shown in Fig. 20.16. Kerosene oil is taken in a cast iron retort which is enclosed in a coal-blast furnace. At the mouth of the iron retort, a bonnet is fitted with a molten lead seal. Hydraulic main is connected to the bonnet and a gas holder is attached to collect the oil gas.

During oil gas manufacture, iron retort is made red hot and a continuous stream of kerosene oil is allowed to fall on the red-hot retort. Upon contact of oil with the red-hot furnace cracking occurs thereby forming various lower hydrocarbons. The gases rise and enter the hydraulic main through the bonnet. In the hydraulic main, tar gets condensed and pure gas fills up the gas holder. The cracking reaction that takes place can be shown as:

Table 20.3 % by volume composition of oil gas

Composition	% by volume
CH_4	25–30
CO	10–15
H_2	50–55
CO_2	3

$$C_{12}H_{26} \rightarrow CH_4 + C_2H_6 + C_2H_4 + C_3H_6 + \text{tar}$$

Kerosene   Methane   Ethane   Ethane   Propene

Uses: Oil gas is commonly employed as a laboratory gas.

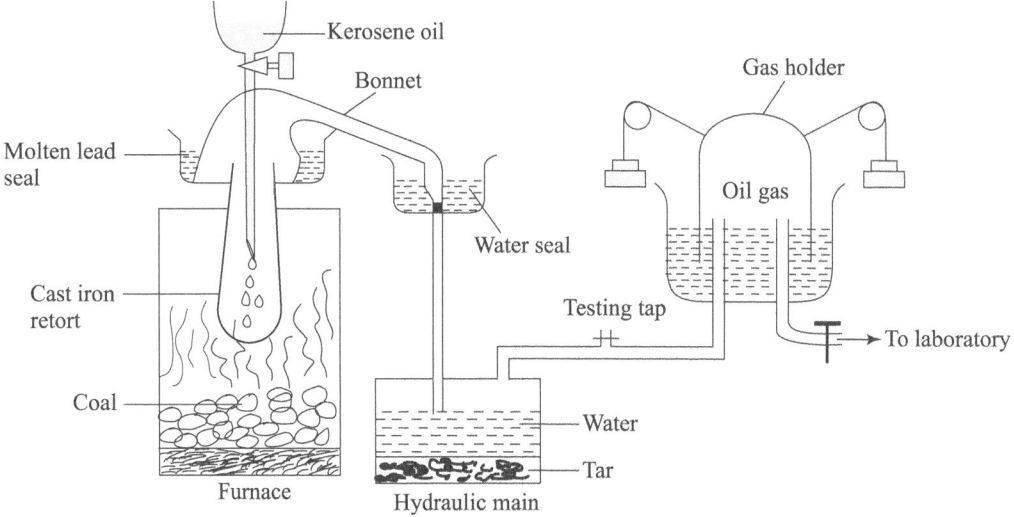

Fig. 20.16 Manufacture of oil gas

20.15 FLUE GAS ANALYSIS BY ORSAT APPARATUS

Have you seen industries releasing smoke from long pipes? Power plants release flue gases which are combustion products comprising CO_2, CO, and O_2 gases. Flue gases are released from ovens through long tubes called flue pipes. One can determine whether combustion of fuel is complete on analysing flue gases. If flue gases consist of huge amounts of carbon monoxide, it indicates incomplete combustion and low supply of oxygen. Incomplete combustion also leads to wastage of fuel. If flue gases have considerable amounts of oxygen, it indicates the presence of an excess supply of oxygen that was not consumed during combustion.

Check Your Progress

17. State the composition of producer gas and water gas.
18. What is power alcohol? State its uses.
19. Which organic reactions are generally carried out for reforming petrol?
20. What is synthetic petrol? Name the methods to prepare them.

Construction   Orsat apparatus shown in Fig. 20.17 consists of the following parts; (a) three horizontal absorption tubes, (b) three-way stopcocks, and (c) a graduated burette.

The apparatus is equipped with a U-tube packed with fused calcium chloride and glass wool from where flue gases are introduced. Next, absorption tubes filled with potassium hydroxide solution, alkaline pyrogallic acid, and ammonical cuprous chloride respectively are connected in series. These tubes are connected to a three-way stopcock. The burette is connected to absorption tubes in series through a separate stopcock. The burette is surrounded by a water jacket and further connected to a water reservoir. The water level in the burette is monitored by raising or lowering the reservoir.

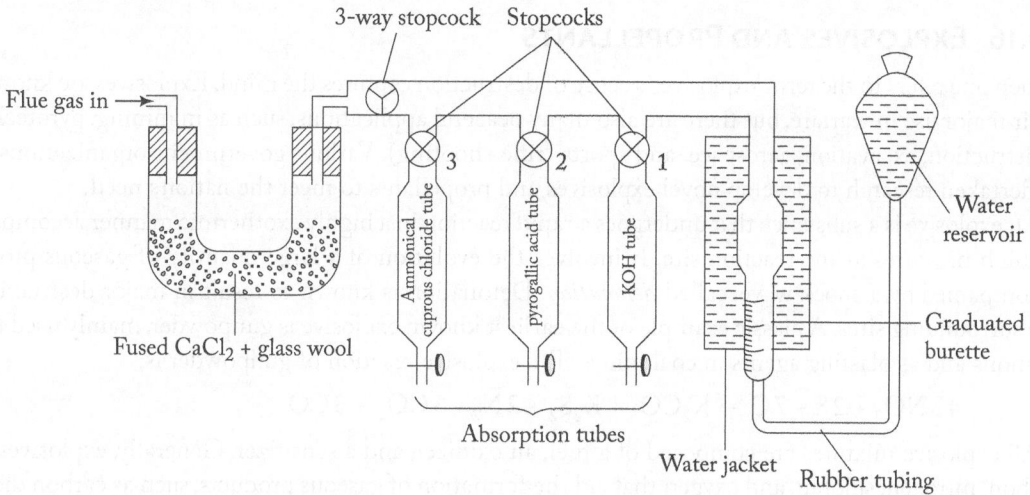

Fig. 20.17 Orsat apparatus for flue gas analysis

Working

(a) The apparatus is cleaned with acetone, stopcocks are greased, and absorption tubes are packed suitably with potassium hydroxide solution, alkaline pyrogallic acid, and ammoniacal cuprous chloride respectively and labelled as 1, 2, and 3 respectively.

(b) Water jacket and water reservoir are completely filled with water and air is purged out from the apparatus. For excluding excess air from the apparatus, stopcocks are opened and water reservoir is lowered so that no air bubbles are trapped.

(c) 100 ml of flue gas is sent to the burette and forced through the first bulb containing potassium hydroxide which absorbs CO_2 from the flue gas. The procedure is repeated until all the CO_2 gases are absorbed by KOH tube.

(d) For proper analysis, flue gas is first passed through KOH tube that absorbs CO_2. Pyrogallic acid can absorb both oxygen and carbon dioxide, but now when flue gas is passed through it, only oxygen will be absorbed as CO_2 is already absorbed by KOH tube.

(e) Ammonical cuprous chloride can absorb O_2, CO_2, and CO gases, but as O_2 and CO_2 are removed, it will only absorb CO gases.

(f) The remaining gases are sent back to the burette and the stopcock of first absorption tube is closed. Water levels in the reservoir and burette are normalized and the decrease of volume of gas is noted which denotes the volume of carbon dioxide in flue gas.

(g) In the same way, volumes of oxygen and carbon monoxide are determined by passing them through absorption tubes containing pyrogallic acid and ammonical cuprous chloride, respectively.

The calculations for Orsat flue gas analysis can be depicted as follows:

Let the volume of flue gas be 100 ml. Thus, the volume of gas after passing through KOH tube can be denoted as A ml. The volume of gas after passing through pyrogallic acid tube is denoted as B ml and the volume of gas after passing through ammonium cuprous chloride will be C ml.

Thus, % CO_2 = $100 - A$ = _____ ml, % CO = $A - B$ = _____ ml

and % O_2 = $B - C$ = _____ ml

Non-combustible nitrogen is also found in flue gases which can be estimated as,

% Nitrogen = Volume of flue gas − (% CO_2 + % CO + % O_2)

20.16 EXPLOSIVES AND PROPELLANTS

When one refers to the term 'explosive,' a sense of destruction captures the mind. Explosives are known for their major use in warfare, but there are also many peaceful applications, such as in mining, pyrotechnics, construction, excavation, aerospace, and sports (rifle shooting). Various government organizations have undertaken research to develop novel explosives and propellants to meet the nation's need.

An explosive is a substance that undergoes a rapid reaction in a highly exothermic manner accompanied by high pressures at the reaction site. It involves the evolution of a large quantity of gaseous products accompanied by a shock wave, called *detonation*. Detonation is known to result in major destruction to the surrounding sites. A classic example of the earliest known explosive is gunpowder, mainly used to fire cannons and as blasting agents in coal mines. The explosive reaction of gunpowder is,

$$4\,KNO_3 + 2\,S + 7\,C \rightarrow K_2CO_3 + K_2S_2 + 2\,N_2 + 3\,CO_2 + 3\,CO$$

All explosive mixtures are composed of a fuel, an oxidizer, and a sensitizer. Generally, explosives have carbon, nitrogen, sulphur, and oxygen that aid the formation of gaseous products, such as carbon dioxide, carbon monoxide, oxygen, nitrogen, water vapour, and large amounts of heat. The presence of C, H, S, N, etc. provides the necessary fuel for the oxygen in the oxidizer. Detonation is primarily an oxidation reaction. The use of sensitizer in the mixture enhances the reaction of explosive with an initiator.

20.16.1 Characteristics of Explosives

The following are the general characteristics required for a compound to be considered as an explosive material.

(a) The decomposition rate of an explosive should be rapid and a large volume of hot gases should be liberated.

(b) The explosive molecule should have low energy of dissociation.

Check Your Progress

21. In an Orsat apparatus, which gases are absorbed by which type of compounds or solutions?
22. Water gas gives a characteristic flame – why?
23. Why are TEL and MTBE used as anti-knock agents?
24. Name the calorimeters employed to determine calorific values of fuels.
25. Name some petroleum products.
26. Is it necessary to eliminate sulphur compounds from fuels? If so, give the reason.
27. Why is it necessary to analyse flue gases?
28. What is reforming of petrol?
29. What is coal liquefaction? Name any one method of coal liquefaction.
30. In catalytic cracking, catalyst requires regeneration, Justify.

(c) Oxygen balance of an explosive is a measure of oxygen present in the molecule.
  • As detonation is primarily an oxidation reaction, all the C form CO_2, all H form H_2O, and all N forms N_2.
  • On this basis, an explosive with the composition $C_xH_yO_zN_D$ will have an oxygen balance,
    OB = $z - 2x - y/2 \times 1600$/molecular weight of explosive.   (20.6)
  • Ideally, oxygen balance must be zero or a positive value. A positive value of OB indicates the additional oxygen is present within the explosive, whereas a negative OB indicates the oxygen needs to be supplied externally to result in successful reaction.
  • The detonation of a mixture of explosive is not only dependent on OB, but also on the combustion reactions.
(d) It should be sensitive towards specific impact and explode immediately after the application of the stimuli like heat, spark, friction, or mechanical force.
(e) Explosives should be economical, non-volatile, non-hygroscopic, and should not decompose during storage.

Table 20.4 highlights the characteristic features of a few primary explosives.

Table 20.4 Selected primary explosives and their characteristic properties

Primary	Ignition temp. (°C)	Activation energy (kJ/g)	Detonation (mm/μs)	Velocity (km/s) at density (g/cm^3)	Sensitivity to impact (kgm)
Mercury Fulminate, $Hg(ONC)_2$	215	30	4.5	3.3	0.18
Lead Azide, $Pb(N_3)_2$	330	170	4.5	3.8	0.41
Tetrazene, $C_2H_8N_{10}O$	130	--	--	--	5
Diazodinitrophenol, $C_6H_2N_4O_5$	170	230	6.9	1.6	16

Blasting Fuses

Fuses or blasting fuses are required to ignite the explosive material. A blasting fuse is a thin water-proof canvas length of tube packed with gunpowder (or TNT) so arranged to burn at a given speed for setting off charges of an explosive (Fig. 20.18). Generally, it is in the form of a small cotton rope with a core continuous thread containing gunpowder. When explosion is initiated at one end of the detonator, an explosive wave travels along the fuse length at a high velocity and on contact with the explosive material causes explosion. The main use of a blasting fuse is to allow the person firing the explosive to immediately reach safety before explosion takes place.

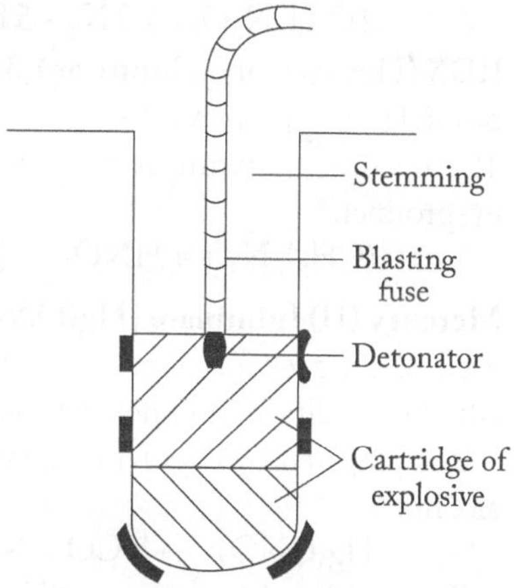

Generally there are two types of blasting fuses.

Safety fuse It is used in initiating caps where electrical firing is not used. It consists of a black powder wrapped in a water-proof fabric and a burning speed of 30–40 s/ft is maintained. On firing, a sufficient length is maintained so that appropriate time is allowed for the person firing the explosive to reach safety before explosion takes place.

Fig. 20.18 Blasting fuse

Detonating fuse (French: *cordeau detonant*) allows a firing speed of over 6000 m/s and consists of powerful explosives such as TNT placed in a small bent tube and thus allows explosion to be conducted in deep holes within the earth's surface.

20.16.2 Manufacture of Important Explosives

Let us have a look at the preparation of some important explosive materials.

A word of caution: Extreme care has to be taken while working with explosive and propellant materials. Students and teachers are instructed not to perform any chemical reactions in the laboratory given in the chapter.

Lead azide, $Pb(N_3)_2$ It is prepared by reacting sodamide (formed by reacting ammonia with sodium) with nitrous oxide. Next, the resulting sodium azide is treated with lead acetate. Lead azide is obtained as shown in the below reaction.

$$6\,NH_3 + 2\,Na \rightarrow 2\,NaNH_2 + 3\,H_2\,;\ NaNH_2 + Pb(OAc) \rightarrow Pb(N_3)_2 + 2\,AcONa$$

The high thermal stability of $Pb(N_3)_2$ makes it suitable for long-term storage. Lead azide, on decomposition, forms lead and releases nitrogen gas as, $Pb(N_3)_2 \rightarrow Pb + 3\,N_2$.

Nitroglycerin (1,3-Dinitrooxypropan-2-yl nitrate) It is prepared by adding glycerol to a cold mixture of concentrated sulphuric acid (60 %) and concentrated nitric acid (40 %) at 10 °C with constant stirring and agitation.

$$\begin{array}{l} CH_2{-}OH \\ | \\ CH_2{-}OH \\ | \\ CH_2{-}OH \end{array} + 3HNO_3 \longrightarrow \begin{array}{l} CH_2{-}ONO_2 \\ | \\ CH_2{-}ONO_2 \\ | \\ CH_2{-}ONO_2 \end{array} + 3H_2O$$

Trinitrotoluene (TNT) It is readily made by reacting concentrated HNO_3 and H_2SO_4 with toluene as a solvent medium. The crude TNT obtained, is treated with aqueous sodium sulphite solution to remove the less-stable isomers from the reaction media. It is further recrystallized using absolute alcohol.

The explosive TNT has extensive applications as bombs and grenades in warfare. On decomposition, TNT forms steam, nitrogen, carbon monoxide, and carbon.

2, 4, 6 - Trinitrotoluene

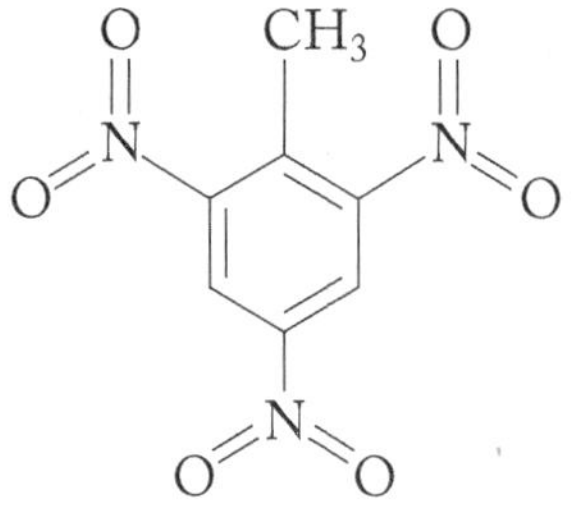

$$2\,C_7H_5N_3O_6 \rightarrow 3\,N_2 + 5\,H_2O + 7\,CO + 7\,C$$

RDX (Hexogen or cyclonite or 1,3,5-Trinitroperhydro-1,3,5-triazine) It is mostly used as a smokeless propellant, is prepared by reacting ammonia with formaldehyde to form hexamethylenetetramine. The resulting mixture is nitrated forming RDX with tetramethylene tetranitramine (HMX) as a by-product.

$$(CH_2)_6N_4 + 4\,HNO_3 \rightarrow (CH_2NNO_2)_3 + NH_4NO_3 + 3\,HCHO$$

Mercury (II) fulminate [$Hg(ONC)_2$] It is a primary explosive that is highly sensitive to shock. It is used as a trigger for other explosives in blasting caps. It is relatively weak and does not store well under adverse conditions and hence is not used much at present.

It is prepared by dissolving mercury in excess nitric acid and the resulting solution is poured in ethyl alcohol.

$$Hg(CNO)_2 \rightarrow 2\,CO + N_2 + Hg + 117\text{ kcal}$$

On decomposition, mercury fulminate forms carbon monoxide, nitrogen, and mercury with the release of huge amounts of energy.

20.17 PROPELLANTS

Rockets are long-tube missiles used in aerospace studies to launch satellites. They are usually fuelled with a propellant. A propellant is a high oxygen-containing mixture of fuel along with an oxidant present in the rocket combustion chamber. In the rocket, the propellant is burned in a definite and controlled manner. Due to reaction of the fuel and oxygen, combustion takes place with large evolution of gases. The large volume of gases is formed at a temperature of 3000 °C and a pressure of 300 kg/cm^2 accompanied with a supersonic velocity of 3 kg/s. All combustion gases exit through a small nozzle of the combustion chamber in the rocket. As per Newton's third law of motion, this act of pushing combustion gases downwards results in an equal and opposite reactions, propelling the rocket upwards into space.

The rocket moves forward due to the opposed reaction from the exhaust blast, not the exhaust itself. Greater the speed of the blast, higher the reaction is against the rocket, which in turn produces a greater overall velocity. Exhaust produces a force against air, thus propelling it forward. Since exhaust blast does not require anything to push against, rockets can operate with greater efficiency in space. Common propellants in use today are O_2/H_2, O_2/C_2H_5OH, peroxides, and hydrazine. Nitrocellulose, hydrazine, RDX, TNT, etc. are manufactured under strict guidelines.

20.17.1 Characteristics of a Good Propellant

A good propellant should have the following general characteristics.
  (i) High specific impulse, which is the thrust delivered divided by the rate of propellant burnt.
 (ii) Produce low-molecular weight products (H_2, CO, CO_2, N_2) during combustion.
(iii) Burn at slow and steady rate.
 (iv) Low ignition delay, that is, time taken by propellant to catch fire in the presence of an oxidizing agent especially in milliseconds. Also, on ignition, the propellant should not leave behind solid residues.
  (v) Stable over a wide range of temperatures.
 (vi) Safe to handle and store under ordinary conditions, that is, it should not detonate under shock, heat, or impact. Also, it should be non-corrosive, non-hydroscopic, and produce non-toxic combustion products.

20.17.2 Classification of Propellants

Propellants are primarily classified on the basis of their physical state into solid and liquid propellants.

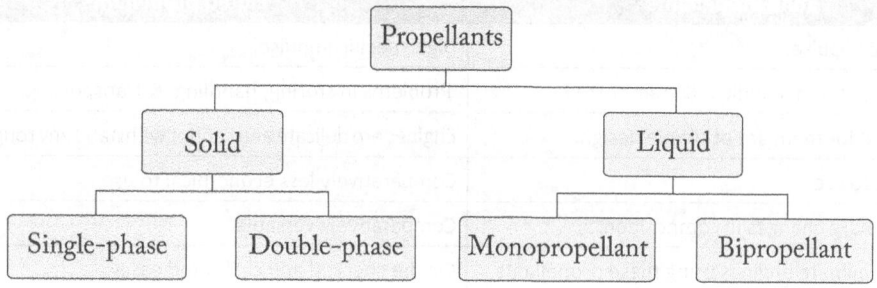

Fig. 20.19 Types of propellants

Solid Propellants

These can be homogeneous and composite materials.

Homogeneous propellant   This is a solid propellant or a mixture of propellants thoroughly mixed in a colloidal state. Further, homogeneous propellants can be single or double-phase systems.

Check Your Progress

31. What are explosives? Name any two explosive compounds.
32. What is a rocket propellant?
33. Define oxygen balance of an explosive.
34. What is blasting fuse? State its use.
35. State the Newtons law applicable to rocket propulsion.

Single-phase propellant is a type of homogeneous solid propellant wherein single propellant is employed; for example, nitrocellulose ($C_6H_7O_2(NO_3)_3$/gun cotton/smokeless powder).

Double-phase propellant contains two materials, that is, an oxidizing agent and plasticizers dispersed in the propellant solid mass. Examples include Ballistile (Nitrocellulose and nitroglycerine mixture), Cordite (65 % nitrocellulose, 30 % nitroglycerine, 5% petroleum jelly plasticizer).

Composite propellants These are heterogeneous solid propellants in which the solid fuel is admixed with a polymeric matrix and an oxidizing agent. Gunpowder, a composite propellant (75 % KNO_3 + 15 % charcoal + 10 % S) is a known explosive. It gives flame temperature at 800–1500 °C and volume of gases is $\approx$ 400 times the volume of the original content. While selecting an oxidizer, it should be non-hygroscopic, stable in contact with fuel mass, and should not form any corrosive products.

Liquid Propellants

These are versatile compounds as against solid propellants and engines using them can be easily checked and calibrated. Liquid propellants are delicate and cannot withstand rough handling. Some classic types of liquid propellants are as follows.

Monopropellants These have a fuel and an oxidizer in the same molecule or in a solution containing both these materials. Hydrogen peroxide, nitromethane, ethylene oxide, hydrazine, propyl nitrate, and mixture of 21.4 % CH_3OH are common monopropellants.

They must be easy to store and should burn smoothly. However, H_2O_2 is not easy to store or handle due to its reactivity and also metal oxides catalyse its decomposition. So, storage tanks are made of special materials.

Table 20.5 compares the properties of solid and liquid propellents.

Table 20.5 Characteristic properties of solid and liquid propellants

Solid propellants	Liquid propellants
Low specific impulse.	High specific impulse.
Can be easily stored, handled, & transported.	Problems in storing, handling, & transporting.
Engines used for them are of simple design.	Engines are delicate and cannot withstand any rough handling.
Economical to use.	Comparatively less economical to use.
Difficult to make changes in composition.	Comparatively versatile.
Difficult to calibrate engines using these propellants.	Can be checked and calibrated easily.

Bipropellants In this case, liquid fuel and oxidizer, kept separately, are injected into combustion chamber separately (Fig. 20.20). Commonly used fuels are liquid hydrogen, hydrazine, ethanol, aniline, kerosene oil, etc.

The most common oxidizers employed are liquid oxygen, ozone, fuming nitric acid, liquid fluorine. Liquid oxygen is non-toxic, safe, good oxidizing agent, but it has to be stored under pressure in insulated containers. Ozone is a powerful oxidizing agent. It is toxic and can explode at high concentration. Liquid fluorine is volatile, toxic, corrosive, and reactive, but a good oxidizing agent. It is difficult to store and handle.

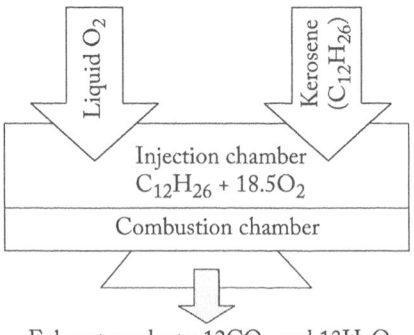

Fig. 20.20 Elucidation of bipropellant in a rocket

Activity-based Questions

A. Can fuel cells replace conventional IC-run vehicles?

Road transport is considered to be a major polluting factor globally. It is essential to find alternative lower emitting fuel engines. In the present scenario, fuel cells are considered as an alternative fuel engine for vehicles. An intensively studied fuel cell is an electrochemical device that converts oxygen and hydrogen in to heat, water, and electricity that is known to power batteries. Fuel cells have already proven its mettle in batteries and space vehicles to provide water to astronauts. Conceptualizing this idea, the Black cab was designed as a zero- emissions vehicle (ZEV) on the roads in London. The companies Intelligent Energy, Lotus Engineering, TRW Conekt, and London Taxis International (LTI) came forward with a proposal to the British government with support of Technology Strategy board to support their research idea. In 2008, they won the funding for the innovative idea and developed a taxi that runs on fuel cell with overall zero emission.

Present a case study on 'zero emission hybrid fuel cell taxis,' showing project details, comparative costs, and merits–demerits. Also, present a counterview about the replacement of IC vehicles with fuel cell-hybrid vehicles.

B. CNG-run vehicles

In 2002, Supreme Court of India directed the Delhi government to replace diesel engine-run public transport by CNG-run vehicles to combat air pollution in the city. After a long dispute against private transporters, the government succeeded in implementing the order. The results were surprising and visible on the national highways.

Do you think that replacement of diesel fuel with CNG led to this change? Compare CNG, LPG, diesel fuels with respect to their calorific value, handling, combustion products, emissions, efficiency, mileage, running cost, environmental problems, etc., to find answer to this question.

C. Safer fuels as propellants

Currently, most satellite thrusters are powered by hydrazine, a toxic and corrosive fuel, which is difficult to store. Its difficult handling necessitated replacing hydrazine with an eco-friendly fuel. NASA is testing space missiles propelled by green propellants that can provide better performance than hydrazine. Ionic liquid-based blends are being utilized, namely hydroxylammonium nitrate-based propellant and borohydride-rich ionic liquids.

Through the Green Propellant Infusion Mission (to be launched in 2017), NASA is developing a green alternative to conventional chemical propulsion systems for next-generation launch vehicles and spacecraft. Can greener propellants replace hydrazine as a fuel? Describe the propellant technology (chemical reactions), comparison of hydrazine-based propellants with greener propellant fuels. State the comparative costs, merits and demerits, and future trends.

SOLVED EXAMPLES

Based on calorific values

1. Calculate the gross and net calorific values of coal with composition: C = 85 %, H = 8 %, S = 1 %, N = 2 %, ash = 4% (Given: latent heat of steam = 580 cal/g).

Solution: According to Dulong–Petit's formula,

$$GCV = \frac{1}{100}\left[8080 \times C + 34500\left(H - \frac{O}{8}\right) + 2240\ S\right] \text{ kcal/kg}$$

As % oxygen is not given, consider it as 0.

On substituting the values in the above formula,

$$GCV = \frac{1}{100}\left[8080 \times 85 + 34500\left(8 - \frac{O}{8}\right) + 2240 \times 1\right]$$

$$= \frac{1}{100}\left[686800 + 276000 + 2240\right] = \frac{1}{100}\left[965040\right] = 9650.4 \text{ kcal/kg}$$

$$\text{NCV} = \text{GCV} - \text{Mass of hydrogen (\%)} \times \frac{9}{100} \times \text{Latent heat of water vapour}$$

$$= 9650.4 - 8 \times \frac{9}{100} \times 580 = 9232.8 \text{ kcal/kg}$$

Result: Gross and net calorific values are found to be 9650.4 kcal/kg and 9232.8 kcal/kg respectively.

2. Calculate the gross and net calorific values of a coal sample having the following composition: C = 83 %, H = 6 %, O = 3 %, S = 3.7 %, N = 2.5 %, ash = 1.8 %, (latent heat of steam is 587 cal/g).

Solution: According to Dulong-Petit's formula,

$$\text{GCV} = \frac{1}{100} \left[8080 \times C + 34500 \left(H - \frac{0}{8} \right) + 2240\,S \right] \frac{\text{kcal}}{\text{kg}}$$

On substituting the values in the above formula,

$$\text{GCV} = \frac{1}{100} \left[8080 \times 83 + 34500 \left(6 - \frac{3}{8} \right) + 2240 \times 3.7 \right]$$

$$= \frac{1}{100} \left[670640 + 194062.5 + 8288 \right] = 8729.905 \text{ kcal/kg}$$

$$\text{NCV} = \text{GCV} - \text{Mass of hydrogen (\%)} \times \frac{9}{100} \times \text{Latent heat of water vapour}$$

$$= 8729.905 - 6 \times \frac{9}{100} \times 587 = 8729.905 - 316.98 = 8412.925 \text{ kcal/kg}$$

Result: Gross and net calorific values are found to be 8729.905 kcal/kg and 8412.925 kcal/kg respectively.

3. A coal sample has the following % composition by weight: C = 92, O = 2, S = 1, N = 1, and ash = 4. If the net calorific value of coal is 8590.3 kcal/kg, determine the % hydrogen and gross calorific value of the coal sample. (latent heat of steam = 590 cal/g)

Solution: GCV = NCV + 0.09 × % H × 590 kcal/kg

$$= 8590.3 + 0.09 \times \text{\% H} \times 590 \text{ kcal/kg}$$

$$= 8590.3 + 53.1 \text{ H kcal/kg} \qquad \qquad \qquad \ldots\text{(a)}$$

Now, $$\text{GCV} = \frac{1}{100} \left[8080 \times C + 34500 \left(H - \frac{0}{8} \right) + 2240\,S \text{ kcal/kg} \right]$$

$$= \frac{1}{100} \left[8080 \times 92 + 34500 \left(H - \frac{2}{8} \right) + 2240 \times 1 \text{ kcal/kg} \right]$$

$$= 7433.6 + 345 \times H - 86.25 + 22.4 \text{ kcal/kg}$$

$$= 7369.75 + 345 \text{ H} \qquad \qquad \qquad \ldots\text{(b)}$$

Now equating (a) and (b), we get,

$$8590.3 + 53.1 \text{ H} = 7369.75 + 345 \text{ H}$$

$$291.9 \text{ H} = 1220.55$$

Thus % H = 1220.55/291.9 = 4.18

Thus, GCV = 8590.3 + 53.1 × 4.18 = 8812.25 kcal/kg

Result: % hydrogen = 4.18 % and GCV = 8812.25 kcal/kg.

4. When 0.973 g of a fuel sample undergoes complete combustion in excess of oxygen, the rise in temperature of water in the calorimeter containing 1325 g of water was 2.4 °C. Calculate the gross calorific value of the fuel sample, if water equivalent of calorimeter is 135 g. If the fuel contains 5.5 % hydrogen, calculate the net calorific value of the fuel sample. (Given: latent heat of steam = 580 cal/g).

Solution: Given: Mass of fuel sample x = 0.973 g, Weight of water in bomb calorimeter W = 1325 g, Weight of water in the calorimeter w = 135 g, Raise in temperature = $t_2 - t_1$ = 2.4 °C.

As per Eq. (20.4),

$$\text{GCV of the fuel sample } (L) = \frac{(W + w)(t_2 - t_1)}{x} = \frac{(1325 + 135)(2.4)}{0.973} = 3601.2 \text{ cal/g}$$

As per Eq. (20.2), NCV = GCV – Mass of hydrogen (%) × 9/100 × Latent heat of water vapour

$$= 3601.2 - 5.5 \times 9/100 \times 580 = 3314.1 \text{ cal/g}$$

Result: GCV and NCV of the fuel were found to be 3601.2 cal/g and 3314.1 cal/g, respectively.

5. The following data was obtained in a bomb experiment when solid fuel (% C = 93 % and % H = 7) undergoes complete combustion in pure oxygen. (Given: latent heat of steam = 590 cal/g).

Weight of the crucible = 3.349 g, weight of the crucible + fuel = 4.278 g, water equivalent of calorimeter = 550 g, water taken in the calorimeter = 2300 g, rise in temperature = 2.3°C, cooling correction = 0.04°C, acid correction = 52.8 calories, fuse wire correction = 3.8 calories.

Based on the above data, calculate the GCV and NCV of the solid fuel. Express the calorific values in kJ/kg.

Solution: Weight of the fuel *x* = (Weight of the crucible + fuel) – Weight of the crucible

$$= (4.278 - 3.349) \text{ g } = 0.929 \text{ g}$$

As per Eq. (20.5),

$$L = \frac{[\{W + w\}(t_2 - t_1 + tc)\} - \{ta + tf\}]}{x}$$

$$L = \frac{[\{2300 + 550\}\,(2.3 + 0.04)\} - \{52.8 + 3.8\}]}{0.929} = \frac{[\{2850\}\,(2.34)\} - \{56.6\}]}{0.929} = 7117.76 \text{ cal/g}$$

NCV = GCV – 0.09 (% H) × Latent heat of condensation of steam

$$= 7117.76 - 0.09 \times 7 \times 590 = 6746.06 \text{ cal/g}$$

Express the above values in kJ/kg, we get,

GCV = 7117.76 × 4.2 = 29894.6 kJ/kg and NCV = 6746.06 × 4.2 = 28333.45 kJ/kg

Result: GCV and NCV of the fuel are found to be 7117.76 cal/g (or 29894.6 kJ/kg) and 6746.06 cal/g (or 28333.45 kJ/kg), respectively.

6. In a bomb calorimeter experiment, 0.680 g of fuel (% C = 93 and % H = 6, % ash = 1) was burnt in excess of oxygen. The rise in temperature due to heat generated was 3.61°C for 1080 g of water in the calorimeter. If the water equivalent of the calorimeter is 150 g, calculate the gross calorific value of the fuel under test.

Solution: Given: Mass of the fuel under test *x* = 0.681g; Water equivalent of calorimeter *W* = 150 g

Mass of water in calorimeter *w* = 1080 g; Rise in temperature = $t_2 - t_1$ = 3.61°C

On substituting the given values in Eq. (20.4), we get,

$$\text{GCV of the fuel sample } (L) = \frac{(W + w)(t_2 - t_1)}{x}$$

$$= \frac{(1080 + 150)(3.61)}{0.680} = 6529.85 \text{ cal/g (before corrections)}$$

When the above experiment was repeated for acid, cooling, and fuse wire corrections, data obtained was: acid correction = 50.0 cal; fuse wire correction = 5.0 cal; cooling correction = 0.05 °C. Calculate the gross and net calorific values for the fuel under test. (Given: latent heat of steam = 590 cal/g.)

Solution: As per Eq. (20.5), we get,

$$\text{GCV} = \frac{[\{W + w\}(t_2 - t_1 + tc)\} - \{ta + tf\}]}{x}$$

$$= \frac{[\{1080 + 150\}(3.61 + 0.05)\} - \{50 + 5\}]}{0.680} = 6539.41 \text{ cal/g(after corrections)}$$

$$\text{NCV} = \text{GCV} - 0.09 \, (\%H) \times \text{Latent heat of condensation of steam}$$
$$= 6539.41 - 0.09 \times 6 \times 590 = 6220.81 \text{ cal/g}$$

Result: GCV (before corrections) = 6529.85 cal/g, GCV (after corrections) = 6539.41 cal/g
NCV (after corrections) = 6220.81 cal/g

7. 0.1 m^3 of a gaseous fuel on combustion in a Junkers calorimeter gave the following observations:
Weight of water heated = 25 kg; Inlet water temperature = 20°C; Outlet water temperature = 33°C; Mass of water condensed = 0.025 kg.
Calculate GCV and NCV per cu.m at NTP. (Given: Latent heat of steam = 580 kcal/kg)
Solution: GCV = $W(T_2 - T_1)/V$ kcal/m^3

$$\text{GCV} = 25 \frac{(33-20)}{0.1} = 3250 \text{ kcal/m}^3$$

Further, $\text{NCV} = \text{GCV} - \dfrac{m \times 580}{V}$ kcal/m^3

So, on substituting, we get, $= 3250 - \dfrac{0.025 \times 580}{0.1} = 3105 \text{ kcal/m}^3$

Result: The GCV and NCV per cu.m at NTP are found to be 3250 and 3105 kcal/m^3.

Based on analysis of coal
8. 1.5 g of air-dried coal sample was weighed in a silica crucible. After heating for 1 hour at 110°C, the dry coal residue weighed 0.985 g. The crucible was partially covered with a lid and heated strongly for 7 minutes at 975°C. The residue left behind weighed 0.813 g. The crucible was further heated strongly in air until a constant weight was obtained. The weight of the final white residue was found to be 0.13 g. Perform proximate analysis of the coal sample.
Solution: Loss in weight due to moisture of coal sample = 1.5 – 0.985 = 0.515 g
  Loss in weight due to volatile matter = 0.985 – 0.813 = 0.172 g
  Weight of white residual ash = 0.13 g

(a) % Moisture = $\dfrac{\text{Loss in weight due to removal of moisture}}{\text{Weight of coal (g)}} \times 100 = \dfrac{0.515}{1.5} \times 100 = 34.3\%$

(b) % Volatile matter = $\dfrac{\text{Loss in weight due to removal of volatile matter}}{\text{Weight of coal (g)}} \times 100 = \dfrac{0.172}{1.5} \times 100 = 11.4\%$

(c) % Ash = $\dfrac{\text{Weight of residue}}{\text{Weight of coal(g)}} \times 100 = \dfrac{0.13}{1.5} \times 100 = 8.6\%$

(d) Determination of % fixed carbon = 100 – % [moisture + volatile matter + ash]
$$= 100 - [34.3 + 11.4 + 8.6] = 45.7\%$$

Result: % Moisture = 34.3; % volatile matter = 11.4; % ash = 8.6, % fixed carbon = 45.7.

9. Determine C, H, and N elements (as %) from the following observations in experiments of analysis of coal.
(a) 0.25 g coal on burning in a combustion tube and passing the gases through tubes containing anhydrous CaCl$_2$ and KOH increases their weight by 0.09 g and 0.8 g respectively.
(b) In Kjeldahl's method, ammonia evolved by 0.42 g coal was absorbed in 49.5 ml of 0.12 N HCl solution. After absorption, the excess acid required 36.5 ml of 0.12 N NaOH for neutralization.
Solution: For C and H determination,

$$\% \, C = \frac{\text{Increase in weight of KOH tube} \times 12 \times 100}{\text{Weight of coal sample} \times 44} = \frac{0.8 \times 12 \times 100}{0.25 \times 44} = 87.3\%$$

$$\% \, H = \frac{\text{Increase in weight of CaCl}_2 \text{ tube} \times 2 \times 100}{\text{Weight of coal sample} \times 18} = \frac{0.09 \times 2 \times 100}{0.25 \times 18} = 4.0\%$$

$$\% \text{ N} = \frac{14 \times (V_1 - V_2) \times \text{Normality of HCl}}{1000 \times \text{Weight of coal sample}(x \text{ g})} \times 100 = \frac{14 \times (49.5 - 36.5)0.12}{1000 \times 0.42} \times 100 = 5.2\,\%$$

Result: % C : % H : % N = 87.3 : 4.0 : 5.2

10. In Kjeldahl's method, 1.5 g of coal sample was analysed. The ammonia gas evolved was absorbed in 50.0 ml of 0.1 N H_2SO_4 solution. After absorption, the excess H_2SO_4 required 35.0 ml of 0.1 N NaOH for neutralization. 2.7 g of coal sample in ultimate analysis gave 0.175 g of barium sulphate. Calculate the % nitrogen and % sulphur in the coal sample.

Solution: $\% \text{ N} = \dfrac{14(V_1 - V_2)\text{Normality of HCl}}{1000 \times \text{Weight of coal sample}(x \text{ g})} \times 100 = \dfrac{15 \times 0.1 \times 1.4}{1.5} = 1.4\,\%$

$$\% \text{ S} = \frac{\text{Weight of BaSO}_4 \times 32}{\text{Weight of coal}(g) \times 233} \times 100 = \frac{0.175 \times 32}{2.7 \times 233} \times 100 = 0.890\,\%$$

Result: On ultimate analysis, % nitrogen and % sulphur of coal sample are 1.4 % and 0.890 % respectively.

11. A coal sample was subjected to ultimate analysis and the following results were recorded.
   (a) 0.3 g of sample on combustion gave 0.79 g of CO_2 and 0.022 g of H_2O while determining carbon and hydrogen elements by Liebig's method.
   (b) In Kjeldahl's method, the ammonia liberated by 1.3 g of coal was absorbed in 50.0 ml of 0.1 N H_2SO_4. The resultant solution consumed 10 ml of 0.1 N NaOH for complete neutralization. The same coal sample gave 0.33 g of $BaSO_4$ in bomb combustion method.

Determine the percentages of C, H, N, and S in the given coal sample.

Solution: $\% \text{ C} = \dfrac{\text{Increase in weight of KOH tube} \times 12 \times 100}{\text{Weight of coal sample} \times 44} = \dfrac{0.79 \times 12 \times 100}{0.3 \times 44} = 71.8\,\%$

$$\% \text{ H} = \frac{\text{Increase in weight of CaCl}_2 \text{ tube} \times 2 \times 100}{\text{Weight of coal sample} \times 18} = \frac{0.022 \times 2 \times 100}{0.3 \times 18} = 0.81\,\%$$

$$\% \text{ N} = \frac{14 \times (V_1 - V_2)\text{Normality of HCl}}{1000 \times \text{Weight of coal sample}(x \text{ g})} = \frac{14 \times (50 - 10)0.1}{1000 \times 1.3} \times 100 = 4.3\%$$

$$\% \text{ S} = \frac{\text{Weight of BaSO}_4 \times 32}{\text{Weight of coal}(g) \times 233} \times 100 = \frac{0.33 \times 32}{1.3 \times 233} \times 100 = 3.48\,\%$$

Result: On ultimate analysis, % C, H, N, and S are found to be 71.8, 0.81, 4.3 and 3.48 (%), respectively.

12. An air-dried sample of coal weighing 3.0 g was taken for volatile matter determination. After losing the volatile matter, the coal sample weighed 2.93 g. If it contains 2.5 % moisture, find the % volatile matter in the coal sample.

Solution: Given: Weight of coal = 3.0 g; After heating weight of coal = 2.93 g; % Moisture = 2.5 %

Let us calculate the weight of coal after losing moisture.

Let us assume that the weight after losing moisture be '*x*' g.

$$\% \text{Moisture} = \frac{\text{Weight of coal} - x}{\text{Weight of coal}} \times 100$$

$$2.5 = \frac{3.0 - x}{3.0} \times 100 \quad \text{or} \quad \text{i.e.,} \quad \frac{2.5 \times 3.0}{100} = 2.9 - x$$

Thus, $x = 2.825$ g.

$$\% \text{Volatile matter} = \frac{\text{Loss in weight due to removal of volatile matter}}{\text{Weight of coal}(g)} \times 100$$

$$= \frac{2.825 - 2.63}{3.0} \times 100 = 6.5 \%$$

Result: Percent volatile matter of air-dried coal sample is found to be 6.5 %.

Based on combustion of fuels

Combustion is a process in which oxygen from the air reacts with the elements or compounds to give heat. As the elements or compounds combine in indefinite proportions with oxygen, we need to calculate the minimum quantity oxygen or air required for the complete combustion of compounds. From the equations, the quantity of oxygen and air for complete combustion of fuels can be calculated.

1. $C + O_2 \rightarrow CO_2$
2. $2 H_2 + O_2 \rightarrow 2 H_2O$
3. $S + O_2 \rightarrow SO_2$
4. $2 CO + O_2 \rightarrow 2 CO_2$
5. $CH_4 + 2 O_2 \rightarrow CO_2 + 2 H_2O$
6. $2 C_2H_6 + 7 O_2 \rightarrow 4 CO_2 + 6 H_2O$
7. $C_2H_4 + 3 O_2 \rightarrow 2 CO_2 + 2 H_2O$
8. $2 C_2H_2 + 5 O_2 \rightarrow 4 CO_2 + 2 H_2O$
9. $2 C_3H_6 + 9 O_2 \rightarrow 6 CO_2 + 6 H_2O$
10. $C_3H_8 + 5 O_2 \rightarrow 3 CO_2 + 4 H_2O$
11. $C_4H_8 + 6 O_2 \rightarrow 4 CO_2 + 4 H_2O$
12. $2 C_4H_{10} + 13 O_2 \rightarrow 8 CO_2 + 10 H_2O$

Carbon dioxide, moisture, nitrogen, ash do not take part in combustion; these factors are ignored during calculations. The composition of air is taken as 23 % oxygen and 77 % nitrogen (by weight), whereas by volume it is 21 % of oxygen and 79 % of nitrogen. The mean molecular weight of air is 28.94 considering that air consists of 21 moles of oxygen, 78 moles of nitrogen, and 0.94 moles of argon. As argon is an inert gas, it is taken along with nitrogen for combustion calculations. Hence, % nitrogen in air by volume is 79%. Mean molecular weight of air is calculated as follows,

$$\text{Mean molecular weight} = \frac{[(21 \times 32) + (78 \times 28) + (0.94 \times 39.94)]}{100} = 28.94$$

Weight of air required for complete combustion of fuel is given as,

$$W_{\text{air}} = 100/23 \, [C \times 32/12 + H \times 32/4 + S \times 32/32 - O]$$

Volume of air required for complete combustion of fuel is given as,

$$28.94 \text{ g of air} \equiv 22.4 \text{ litres of volume of air at NTP}$$

13. Calculate the minimum weight of air required for the complete combustion of 1 kg of coal sample having the following percentage composition by weight: C = 90 %, H = 3.5 %, S = 0.5%, O_2 = 3 %, N_2 = 0.5 % and ash = 1.5 %. Calculate the percentage composition by weight of dry products of combustion.

Solution: First, calculate the per kg constituents and the amount of oxygen necessary to combine with each element. This can be tabulated as below.

Elements	Reactions	%	per kg constituents	Wt of O_2 required (kg)
C	$C + O_2 \rightarrow CO_2$	90	0.9	$0.9 \times 32/12 = 2.4$
H	$2 H_2 + O_2 2 \rightarrow H_2O$	3.5	0.035	$0.035 \times 32/4 = 0.28$
O_2	No reaction	3	0.03	In fuel
N_2	No reaction	0.5	0.005	-
S	$S + O_2 \rightarrow SO_2$	0.5	0.005	$0.005 \times 32/32 = 0.005$
Ash	Non-combustible	1.5	-	-
			Total O_2 required	2.685
			O_2 in fuel	0.03
			Net (Actual) O_2 required	$2.685 - 0.03 = 2.655$

In the above problem, 1 kg of coal sample contains 0.9 kg C, 0.035 kg H, 0.005 kg S, 0.03 kg of O_2, and 0.005 kg of N_2. First, determine the weight of oxygen necessary for complete combustion of coal. Stoichiometrically, 12 kg of carbon combines with 32 kg of O_2.

So, 0.9 kg carbon will require = $0.9 \times 32/12$ = 2.4 kg of O_2 (as shown above).

32 kg O_2 combines with 4 kg H_2.

Hence, 0.035 kg of H_2 will require = $0.035 \times 32/4$ = 0.28 kg of O_2.

32 kg of S combines with 32 kg of O_2.

Hence, 0.005 kg S in coal will require, $0.005 \times 32/32$ = 0.005 kg of O_2.

Nitrogen and ash are ignored as they do not take part in combustion reaction. So, theoretical oxygen required for the complete combustion of 1 kg coal will be

2.4 + 0.28 + 0.005 = 2.685 kg oxygen (theoretical oxygen by weight).

0.03 kg of O_2 is already present in the coal sample; hence this oxygen will assist in combustion.

Practical (net) amount of O_2 required = Theoretical O_2 required – O_2 present in fuel
$$= 2.685 - 0.03 = 2.655 \text{ kg (practical oxygen by weight)}$$

As air contains 23 % O_2 by weight,

Weight of air necessary for complete combustion of 1 kg coal = $(100/23) \times 2.655$ = 11.54 kg.

Volume of air necessary for complete combustion of 1 kg coal will be,

28.94 g (molar mass of air) $\equiv$ 22.4 litres of volume of air at NTP

Hence, 11540 g air will be equal to $11540 \times \dfrac{22.4}{28.94}$ = 8932.13 litres of air

We are calculating air, which is ideally present in the gaseous state. One uses cubic metre as the unit to express the values. Rather than using litre as the unit, we use cubic metre as, 1 litre = 10^{-3} m^3

Hence, 8932.13 litres can be expressed as 8.932 m^3.

If one has to calculate dry products of combustion, let us assume that composition of dry flue gases is $CO_2 + SO_2 + N_2$ (in fuel + air) + O_2 (in fuel + air).

Stoichiometrically,
  (a) $CO_2 = 44/12 \times C$ (kg); $44/12 \times 0.9$ = 3.3 kg
  (b) $SO_2 = 64/32 \times S$ (kg); $64/32 \times 0.005$ = 0.01 kg
  (c) N_2 = 77 % by weight of air + in fuel = $(77/100) \times 11.54 + 0.005$ = 8.889 kg

Thus, the total weight of dry flue gases of combustion = Weight of $CO_2 + SO_2 + N_2$
$$= 3.3 + 0.01 + 8.889 = 12.2 \text{ kg}$$

In terms of % composition,
  (a) % $CO_2 = (3.3/12.2) \times 100$ = 27.0 %
  (b) % $SO_2 = (0.01/12.2) \times 100$ = 0.08 %
  (c) % $N_2 = (8.889/12.2) \times 100$ = 72.9 %

Result: Weight of air required for complete combustion of 1 kg coal = 11.54 kg.

Volume of air required for complete combustion of 1 kg coal = 8.932 m^3

Minimum weight of air for complete combustion of 1 kg coal is 11.54 kg.

14. Calculate the weight and volume of air required for complete combustion of 5 kg coal having the elemental composition, viz., C = 85 %, H =10 %, O_2 = 5 %.

Solution: (a) Conversion of % composition into per kg constituents

Elements	Reactions	%	Per kg	Wt of O_2 needed (kg)
C	$C + O_2 \rightarrow CO_2$	85	0.85	$0.85 \times 32/12$ = 2.266
H	$2 H_2 + O_2 \rightarrow H_2O$	10	0.1	$0.1 \times 32/4$ = 0.8
O	In fuel	5	0.05	–
			Total O_2 needed	3.066
			Net O_2 needed	3.066 – 0.05 = 3.016

Thus, 3.016 kg of oxygen is needed by 1 kg coal sample for complete combustion. Now, we can extrapolate it for 5 kg of coal sample.

(b) Weight of air needed by 1 kg coal sample for complete combustion = $100/23 \times 3.016 = 13.11$ kg
   For 5 kg of coal sample, weight of air needed = 13.11×5 kg = 65.55 kg of air.

(c) Volume of air needed for complete combustion of 1 kg coal,
   Stoichiometrically, 28.94 g air is equivalent to 22.4 litres of volume of air at NTP.

Thus, volume of air will be $65550 \times \dfrac{22.4}{28.94} = 50736.69$ litres at NTP

But, 1 litre = 10^{-3} cu.m; Thus, 50736.69 litre = 50.73 m^3

Result: Weight of air needed for complete combustion of 5 kg of coal = 65.55 kg air.

Volume of air needed for complete combustion of 5 kg of coal = 50.73 m^3 at NTP.

15. A coal sample on ultimate analysis has composition (% weight) C = 66.2, H = 4.2, S = 2.9, N_2 = 1.4, O_2 = 6.1, and the rest is ash. Determine the percentage of dry exhaust gases, if 1 kg of coal is burnt with 25 % excess of air.

Solution: 1 kg coal contains 0.662 kg C, 0.042 kg of H_2, 0.029 kg of S, 0.061 kg of O_2, and 0.014 kg of N_2.

12 kg of C requires 32 kg of O_2

0.662 kg of C requires = $32/12 \times 0.662 = 1.765$ kg of O_2

4 kg of H_2 requires 32 kg of O_2.

0.042 kg H_2 requires = $32/4 \times 0.042 = 0.336$ kg of O_2

32 kg of S requires 32 kg of O_2

0.029 kg of S requires = $32/32 \times 0.029 = 0.029$ kg of O_2

N_2 does not burn. Hence, it does not consume any O_2.

Theoretical O_2 required for 1 kg of the fuel = $1.765 + 0.336 + 0.029 = 2.13$ kg of O_2

Net amount of O_2 required for 1 kg of fuel = Theoretical O_2 required – O_2 present in fuel

$$= 2.13 - 0.061 = 2.069 \text{ kg } O_2$$

23 kg of O_2 is supplied by 100 kg of air.

Hence, 2.069 kg of O_2 is supplied by = $100/23 \times 2.069 = 9.0$ kg

If 25 % excess air is used, the amount of air required for the complete combustion of 1 kg coal

$$= 9.0/100 \times 125 = 11.25 \text{ kg}$$

Masses of the dry products of combustion

12 kg of C gives 44 kg of CO_2.

Hence, 0.6662 kg of C gives = $44/12 \times 0.662 = 2.4$ kg of CO_2

32 kg of S gives 64 kg of SO_2.

Hence, 0.029 kg of S gives = $64/32 \times 0.029 = 0.058$ kg of SO_2

Mass of N_2 = N_2 present in fuel + N_2 present in air = $0.014 + 77/100 \times 11.25 = 8.674$ kg of N_2

Mass of O_2 = $(11.25 - 9.0) \times 23/100 = 0.5$ kg

Total mass of combustion products = $2.4 + 0.058 + 8.674 + 0.5 = 11.632$ kg

Percentage composition of the dry products of combustion

CO_2 = $2.4/11.632 \times 100 = 20.6$ %

SO_2 = $0.058/11.632 \times 100 = 0.5$ %

O_2 = $0.5/11.632 \times 100 = 4.3$ %

N_2 = $8.674/11.632 \times 100 = 74.6$ %

Result: Dry products of combustion (exhaust gases) were found to be % CO_2 : % SO_2 : % N_2 : % O_2 as 20.6 % : 0.5 % : 74.6 % : 4.3 % respectively.

16. A gas has composition (% volume) viz, $H_2 = 20$, $CO = 22$, $CH_4 = 6$, $CO_2 = 4$, $O_2 = 8$, $N_2 = 40$. Calculate the volume of air actually supplied per m^3 of the gas.

Solution: (A) Conversion of % into per m^3 constituents

Constituents	Reaction	%	Vol (m^3)	Vol. of O_2 required (m^3)
H_2	$2 H_2 + O_2 \rightarrow 2H_2O$	20	0.20	0.1
CH_4	$CH_4 + 2O_2 \rightarrow CO_2 + 2 H_2O$	6	0.06	0.12
CO	$2 CO + O_2 \rightarrow 2CO_2$	22	0.22	0.11
CO_2	No reaction	4	-	-
O_2	In fuel	8	0.08	In fuel
N_2	No reaction	40	-	-
			Vol. of O_2 required	0.33 m^3
			Net O_2 (Actual O_2 required)	$0.33 - 0.08 = 0.25$ m^3

According to stoichiometry, 100 m^3 of air ≈ 21 m^3 O_2
Thus, 0.25 m^3 $O_2 \times 100/21 = 1.190$ m^3 of air
Result: Volume of air actually supplied is 1.190 m^3.

17. A gaseous fuel has the following composition by volume: $CH_4 = 35$ %, $CO = 10$ %, $H_2 = 6$ %, $N_2 = 2$ % and $C_3H_8 = 12$ %. Calculate the volume of air required for complete combustion of 3.3 m^3 of gas.

Solution: (a) Calculation of oxygen required by each constituent in the fuel

Constituents	Reaction	%	Vol (m^3)	Vol. of O_2 required (m^3)
CH_4	$CH_4 + 2 O_2 \rightarrow CO_2 + 2 H_2O$	35	0.35	0.7
CO	$2 CO + O_2 \rightarrow 2CO_2$	10	0.1	0.05
H_2	$2 H_2 + O_2 \rightarrow 2H_2O$	6	0.06	0.03
N_2	No reaction	–	–	–
C_3H_8	$C_3H_8 + 5 O_2 \rightarrow 3CO_2 + 4 H_2O$	12	0.12	0.6
			Vol. of O_2 required	1.38 m^3

According to stoichiometry, 100 m^3 of air $= 21$ m^3 O_2.
Thus, 1.38 m^3 $O_2 \times 100/21 = 6.571$ m^3 of air.
For 3.3 m^3 of gas, volume of air will be $6.751 \times 3.3 = 21.684$ m^3.
Result: Volume of air required for complete combustion of gas is 21.684 m^3.

18. Petrol contains 85 % carbon and 15 % hydrogen. Calculate the volumes of oxygen and air necessary for the complete combustion of 1 kg petrol (by weight) under standard conditions.

Solution: Given: Weight of petrol $= 1$ kg $= 1000$ g,

Weight of C $= \dfrac{85}{100} \times 1 = 0.85$ kg

Weight of hydrogen $= \dfrac{15}{100} \times 1 = 0.15$ kg

32 g oxygen at standard conditions will occupy 22.4 litres of air.

So, 3.466×1000 g of O_2 in standard conditions occupies

Reactions	Weight of O_2 required (in kg)
$C + O_2 \rightarrow CO_2$	$= \dfrac{32}{12} \times 0.85 = 2.266$
$2 H_2 + O_2 \rightarrow 2H_2O$	$= \dfrac{32}{4} \times 0.15 = 1.2$
	Total oxygen $= 3.466$ kg

$$= 3.466 \times 1000 \times \dfrac{22.4}{32} = 2426.2 \text{ litres}$$

$\therefore$ Vol. of air $=$ Vol. of oxygen $\times 100/21$

Thus, $2426.2 \times 100/21 = 11553.33$ litres of air

Result: The volumes of oxygen and air are 2426.2 litres and 11553.33 litres respectively.

19. Composition of coal (% wt): C = 81, H = 4, N = 1.5, S = 1.2, O = 3, ash = 9.3. The coal undergoes complete combustion on passing 20 % excess air. From the data calculate the following:
   (a) Theoretical and practical weights of oxygen and air for 250 g of above coal sample.
   (b) Volume of air required for complete combustion of 250 g coal sample.

Solution: First, calculate the volume of oxygen required for per kg coal sample. 1 kg coal will consists of 0.81 kg C; 0.04 kg H_2, and 0.012 kg S.

Theoretical weight of air $2.492 \times 100/23 = 10.83$ kg air

Practical weight of air required = $2.462 \times 100/23 = 10.70$ kg air

As 20 % excess air is required for the complete combustion, the actual air can be calculated as,

$$10.70 \times \frac{120}{100} = 12.845 \text{ kg air}$$

As 1 kg coal (1000 g) coal will require 12.845 kg air.

$$\therefore \quad 250 \text{ g coal requires} = \frac{250}{1000} \times 12.845 = 3.211 \text{ kg air}$$

We know that 28.94 g air = 22.4 litres air at NTP.

So for 3211.25 g air will be $3211.25 \times \dfrac{22.4}{28.94} = 2485.55$ litres air

As 1 litre ≡ 0.001 m^3, hence volume of air = 2.485 m^3

Reactions	Weight of O_2 required (in kg)
$C + O_2 \rightarrow CO_2$	$= \dfrac{32}{12} \times 0.81 = 2.16$
$2H_2 + O_2 \rightarrow 2H_2O$	$= \dfrac{32}{4} \times 0.04 = 0.32$
$S + O_2 \rightarrow SO_2$	$= \dfrac{32}{32} \times 0.012 = 0.012$

Total (theoretical) oxygen = 2.492 kg
Net (practical) oxygen = 2.492 − 0.03
= 2.462 kg

Result: (a) Theoretical and practical oxygen = 2.492 and 2.462 (kg) respectively. Theoretical and practical air are 10.83 and 10.70 (kg) respectively; (b) volume of air = 2.485 m^3.

20. A gaseous fuel has % composition (volume basis): CH_4 = 40, C_2H_4 = 6, C_2H_6 = 24, C_3H_8 = 16, CO = 5, O_2 = 3, N_2 = 6. If 40 % excess air is supplied for complete combustion, calculate the air : fuel ratio and % composition of dry combustion products for 1 m^3 of fuel.

Solution: (a) Calculation of oxygen and combustion gases

Constituents	Reactions	Vol. of O_2 (m^3)	Vol. of CO_2 (m^3)
CH_4	$CH_4 + 2 O_2 \rightarrow CO_2 + 2 H_2O$	$0.4 \times 2 = 0.8$	$0.4 \times 1 = 0.4$
C_2H_4	$C_2H_4 + 3 O_2 \rightarrow 2 CO_2 + 2 H_2O$	$0.06 \times 3 = 0.18$	$0.06 \times 2 = 0.12$
C_2H_6	$2 C_2H_6 + 7 O_2 \rightarrow 4 CO_2 + 6 H_2O$	$0.24 \times 7/2 = 0.84$	$0.24 \times 4/2 = 0.48$
C_3H_8	$C_3H_8 + 5 O_2 \rightarrow 3CO_2 + 4 H_2O$	$0.16 \times 5 = 0.8$	$0.16 \times 3 = 0.48$
CO	$2 CO + O_2 \rightarrow CO_2$	$0.05 \times 1/2 = 0.025$	0.05
N_2	No reaction		0.06
		Total O_2 = 2.645 Net O_2 = 2.645 − 0.03 = 2.615	Total CO_2 = 1.53 N_2 = 0.06

(b) Calculation of air : fuel ratio.

If 40 % excess air is supplied, then actual O_2 supplied $= 2.615 \times \dfrac{140}{100} = 3.661$ m^3

Volume of air $= 3.661 \times 100/21 = 17.433$ m^3

For 1 m^3 of gas, 17.433 m^3 air needs to be supplied for complete combustion.

∴ Air : fuel ratio = 17.433 : 1

(c) % Composition of dry combustion products.

From the above table, total CO_2 = 1.53 m^3.

Total N_2 = N_2 present in fuel + N_2 in air $= 0.06 + \left[17.433 \times \dfrac{79}{100} \right] = 13.832$ cu.m

Excess air is supplied; hence $O_2 = 17.433 \times \dfrac{40}{100} \times \dfrac{21}{100} = 1.464$ cu.m

Total amount of combustion products = 1.53 + 13.832 + 1.464 = 16.826 m^3

$$\therefore \qquad \% \, CO_2 = \frac{1.53}{16.826} \times 100 = 9.09 \, \%$$

$$\% \, N_2 = \frac{13.832}{16.826} \times 100 = 82.2 \, \%$$

$$\% \, O_2 = \frac{1.464}{16.826} \times 100 = 8.7 \, \%$$

Result: Air : fuel ratio = 17.433 m^3 and % CO_2 : % N_2 : % O_2 = 9.09 : 82.2 : 8.7 (%) for 1 m^3 of gaseous fuel.

21. Calculate the total quantity of exhaust gases generated from the combustion of gaseous fuel (% volume) CH_4 = 35, CO = 10, H_2 = 6, N_2 = 1, C_3H_8 =10.

Solution: (a) Calculation of oxygen and combustion gases

Constituents	Reactions	Vol. of O_2 (m^3)	Vol. of CO_2 (m^3)	Vol. of H_2O (m^3)
CH_4	$CH_4 + 2O_2 \rightarrow CO_2 + 2H_2O$	$0.35 \times 2 = 0.7$	$0.35 \times 1 = 0.35$	$0.35 \times 2 = 0.7$
CO	$2CO + O_2 \rightarrow 2CO_2$	$0.1 \times 1/2 = 0.05$	$0.1 \times 1 = 0.1$	-
H_2	$2H_2 + O_2 \rightarrow 2H_2O$	$0.06 \times 1/2 = 0.03$	-	0.06
C_3H_8	$C_3H_8 + 5O_2 \rightarrow 3CO_2 + 4H_2O$	$0.1 \times 5 = 0.5$	0.3	0.4
N_2	-	Total O_2 = 1.28 m^3		-

As per stoichiometry, 100 m^3 of air is equivalent to 21 m^3 of O_2.

Hence, volume of air $= \dfrac{1.28 \times 100}{21} = 6.09$ m^3 air

As per above table, total combustion products will be CO_2 : H_2O : N_2

For N_2 present in air + in fuel = $0.01 + \dfrac{79}{100} \times 6.09 = 4.8211$ m^3

Hence total combustion products = 0.35 + 0.1 + 0.3 + 0.7 + 0.06 + 0.04 + 4.8211 = 6.731 m^3.

Result: Total quantity of exhaust gases for the above fuel is 6.731 m^3.

22. On ultimate analysis, a coal sample was found to have the following (%) composition: C = 54, H = 6, O = 3, N = 2, moisture = 18.2, and the rest is ash. When 1 kg of this coal was burnt in excess air, it released 20.3 kg of dry exhaust gases. Determine the % of excess air that was required for complete combustion of the coal sample.

Solution: Given: 1 kg coal contains C = 0.54, H = 0.06, O = 0.03, N = 0.018 (in kg). Minimum weight of air

needed for combustion $= \dfrac{100}{23} \left[0.54 \times \dfrac{32}{12} \times 0.06 \times \dfrac{32}{4} - 0.03 \right] = 8.217$ kg

Weights of dry combustion products

$$CO_2 = 0.54 \times \frac{44}{12} = 1.98 \text{ kg}$$

$$N_2 = 0.02 + 8.217 \times \frac{77}{100} = 6.347 \text{ kg}$$

Total dry combustion products = 1.98 + 6.347 (kg) = 8.327 kg

Actual weight of flue gases = 20.3 kg
However, the actual weight is higher than theoretical value. This means, the balance air may have come from the excess air supplied for combustion,

∴ Weight of flue gases = 20.3 − 8.327 = 11.973 kg

$$\% \text{ Excess air} = \frac{11.973}{8.217} \times 100 = 145.71\%$$

Result: Excess air for complete combustion of coal sample is 145.71 %.

Based on explosives

23. Calculate the oxygen balance for RDX $(CH_2NNO_2)_3$ and state its significance.

Solution: As given in Eq. (20.6), $z − 2x − y/2 \times 1600/$Molecular weight of the explosive

Oxygen balance (OB) = $\dfrac{6 − (2 \times 3) − (6/2) \times 1600}{222.1163}$ = −21.61 (% by weight)

As the OB for RDX is negative, additional oxygen needs to be supplied for detonation.

24. Calculate oxygen balance for nitroglycerine $C_3H_5N_3O_9$ with a comment on the same.

Solution: As given in Eq. (20.6), $z − 2x − y/2 \times 1600/$Molecular weight of the explosive

Oxygen balance (OB) = $9 − 2 \times 3 − 5/2 \times 1600 \div 227.0865$ = +3.52 (% by weight)

As the OB for nitroglycerine is positive, surplus oxygen is present in the explosive for allowing complete detonation.

SUMMARY

○ Combustion engineering involves the study of fuels and their combustion characteristics along with harnessing energy for various domestic and industrial purposes.

○ Calorific value is the main performance parameter for any given fuel (solid, liquid, or gas). Gross and net calorific values are two types of calorific values. Dulong-Petit's formula gives the theoretical calorific values, whereas bomb calorimeter gives practical values.

○ Proximate and ultimate analyses of coal signify the quality of coal sample. The characteristics of a coal sample are found out by analysing the content of moisture, volatile matter, ash, carbon, hydrogen, sulphur, nitrogen, and oxygen.

○ Petroleum (or crude oil) is just second in line to coal and is an important fossil fuel. Mining of petroleum and its subsequent fractional distillation yields a variety of important fuels such as gasoline (petrol), diesel, LPG, kerosene, and even lube oils.

○ Continuous fractionation is an energy-consuming process; hence companies perform cracking of major hydrocarbon fractions. Thermal and catalytic cracking yield pure gasoline that has anti-knocking characteristics.

○ Knocking of engines can be improved by adding agents such as TEL, MTBE, and isoamyl nitrates. Both petrol and diesel engines differ in their modes of functioning. Petrol-run engines are four-stroke combustion engines (Otto cycle), whereas diesel-run are compression-ignition engines.

○ Gaseous fuels such as producer gas, water gas, coal gas, and oil gas are commonly used for various industrial purposes.

○ Octane and cetane numbers express the performance of petrol and diesel fuels respectively.

○ Analysis of flue gases is indicative of combustion characteristic of a fuel and carried out using Orsat apparatus.

○ On detonation, explosives and propellants release huge amounts of energy that find applications in warfare, aerospace, pyrotechnics, and ballistics.

GLOSSARY

Anthracite: A type of coal that has the highest carbon content.

Anti-knocking agent: A compound that helps to increase the octane number of fuel and eventually decrease the knocking.

Biodiesel: A non-fossil fuel obtained from vegetable or animal fat-based oil feedstock.

Calorific value: The total quantity of heat liberated by burning a unit mass (for solid fuels) or volume (for liquid or gaseous) of fuel completely.

Carburettor: A device that blends air and fuel for an IC engine in the proper ratio for combustion.

Cetane number: The percentage by volume of cetane in a mixture of cetane and 1-methyl naphthalene, which exactly matches in its ignition delay characteristics with the diesel under test.

Combustion engineering: The science of combustion dealing with harnessing energy from fuels for commercial purposes.

Cracking: The process of thermal or catalytic decomposition of high-molecular weight hydrocarbons present in crude oil.

Octane number: The percentage of iso-octane in the mixture of iso-octane and n-heptane, which just matches the knocking characteristics of petrol sample under consideration.

Coal: A high carbonaceous matter obtained by compression of vegetable matter in earth's crust.

Coke: The whitish matter left behind after destructive distillation of bituminous coal.

Flue gas: The uncondensed gases leaving the combustion chamber.

Gross calorific value: The total amount of heat produced when a unit mass or volume of the fuel is burnt completely and combustion products are cooled to room temperature (15 °C or 60 °F).

Knocking: The sharp, metallic, pinging sound due to premature ignition of air–fuel mixture in an IC engine.

Net calorific value: The net heat produced when a unit mass or volume of fuel is burnt completely and the products are allowed to escape.

Petroleum: The crude oil found in the earth's crust due to ages' of compression of decaying organic matter.

Power alcohol: An alternate fuel, such as ethanol employed as a fuel in IC engines.

Proximate analysis: The coal analysis for moisture, volatile matter, ash, and carbon content.

Reforming: The structural modifications in the components of straight-run gasoline to improve its anti-knock characteristics.

Stenching agent: A sulphurous compound added to gaseous fuel to add odour for identifying leaks.

Straight-run gasoline: Gasoline obtained directly after fractional distillation and used as a fuel.

Transesterification: The process by which triglycerides in excess methanol and alkaline catalyst forms biodiesel and glycerol.

Ultimate analysis: The elemental analyses of coal for estimating percentage of C, H, O, N, S, and ash.

Detonation: An explosion with a resulting shock wave noise.

Explosive: A substance that undergoes a rapid reaction in a highly exothermic manner accompanied by high pressures and detonation.

Propellant: A low explosive, that is, high oxygen-containing mixture of fuel along with an oxidant used to propel rockets.

KEY FORMULAE

- $GCV = \dfrac{1}{100}\left[8080 \times C + 34500\left(H - \dfrac{O}{8}\right) + 2240\,S\right] \text{ kcal/kg}$

- $NCV = GCV - \text{Mass of hydrogen (\%)} \times \dfrac{9}{100} \times \text{Latent heat of water vapour}$

- For bomb calorimetry:

  $L = \dfrac{(W + w)(t_2 - t_1)}{x}$ ...(before corrections)

  $L = \dfrac{[\{W + w\}(t_2 - t_1 + tc)\} - \{ta + tf\}}{x}$ ...(after corrections)

- For Junker's calorimeter: $GCV = W(T_2 - T_1) \times S/V$ kcal/m^3; $NCV = GCV - \dfrac{m \times 587}{V}$ kcal/m^3

- $\%C = \dfrac{\text{Increase in weight of KOH tube} \times 12 \times 100}{\text{Weight of coal sample} \times 44}$

- $\%H = \dfrac{\text{Increase in weight of CaCl}_2 \text{ tube} \times 2 \times 100}{\text{Weight of coal sample} \times 18}$

- Oxygen balance $OB = z - 2x - y/2 \times 1600/\text{molecular weight of explosive}$

- $\%N = \dfrac{14 \times (V_1 - V_2)\,\text{Normality of HCl}}{1000 \times \text{Weight of coal sample } (Xg)} \times 100$

- $\%S = \dfrac{\text{Weight of BaSO}_4 \times 32}{\text{Weight of coal } (g) \times 233} \times 10$

- $\% \text{ Moisture} = \dfrac{\text{Loss in weight due to removal of moisture}}{\text{Weight of coal}(g)} \times 100$

- $\% \text{ Volatile matter} = \dfrac{\text{Loss in weight due to removal of volatile matter}}{\text{Weight of coal}(g)} \times 100$

- $\% \text{ Ash} = \dfrac{\text{Weight of residue left}}{\text{Weight of coal}(g)} \times 100$

- % Fixed carbon = 100 – % [moisture + volatile matter + ash]

EXERCISES

Multiple Choice Questions

1. An ideal fuel must have
   - (a) low calorific value
   - (b) high moisture content
   - (c) moderate ignition temperature
   - (d) high ash content

2. In bomb calorimeter, calorific value is determined for
   - (a) liquid fuels
   - (b) solid fuels
   - (c) gaseous fuels
   - (d) both (a) and (b)

3. Proximate analysis of coal is used to determine
   - (a) carbon, moisture, volatile matter, ash
   - (b) only moisture
   - (c) moisture, CO, C, and ash
   - (d) calorific value

4. Ultimate analysis of coal is used to determine
   - (a) C, H, N, O, S
   - (b) only C, H, N
   - (c) fixed carbon
   - (d) ignition temperature

5. The lowest ranking coal is
   - (a) bituminous
   - (b) anthracite
   - (c) peat
   - (d) lignite

6. The mean molecular weight of air is
   - (a) 28.94
   - (b) 77
   - (c) 29.84
   - (d) 34500

7. The main constituent of natural gas is
   - (a) propane
   - (b) methane
   - (c) butane
   - (d) ethane

8. The process of breaking higher hydrocarbons in to lower hydrocarbons of value is called
   - (a) splitting
   - (b) reforming
   - (c) refining
   - (d) cracking

9. Iso-octane is chosen to determine the octane number of petrol because
   - (a) it has better combustion characteristics.
   - (b) it does not cause smoke formation
   - (c) it has lower calorific value.
   - (d) it ignites rapidly.

10. Which among the following possesses zero cetane number?
    - (a) Diesel
    - (b) Petrol
    - (c) 1-Methyl naphthalene
    - (d) Hexadecane

11. Better petrol characteristics is due to the presence of
    - (a) straight-chain hydrocarbons
    - (b) branched hydrocarbons
    - (c) side chain hydrocarbons
    - (d) aromatic hydrocarbons

12. In a diesel-run engine, there is no
    - (a) spark plug
    - (b) combustion
    - (c) ignition
    - (d) formation of exhaust gases

13. The performance criterion for selecting a good fuel is its
    - (a) combustion properties
    - (b) smoke forming ability
    - (c) calorific value
    - (d) both (a) and (b)

14. Octane rating of n-heptane is
    - (a) 0
    - (b) 100
    - (c) 50
    - (d) none of these

15. LPG mainly comprises
    - (a) methane
    - (b) ethane
    - (c) propane
    - (d) butane

16. When alcohol is blended with gasoline, it is called
    - (a) petrol
    - (b) power alcohol
    - (c) diesel
    - (d) biodiesel

17. The highest-ranking coal is
    (a) bituminous   (b) anthracite
    (c) peat   (d) lignite
18. The chemical reaction to prepare biodiesel is
    (a) trans esterification   (b) combustion
    (c) pyrolysis   (d) cracking
19. Which of the following statement(s) is true about knocking in IC engines?
    (a) It lowers performance.
    (b) It affects engine parts adversely
    (c) It improves octane and cetane values.
    (d) Both (a) and (b).
20. Dehydrocyclization and isomerization are carried out to achieve
    (a) combustion   (b) mining
    (c) reforming   (d) refining
21. The performance characteristics of diesel is measured by its
    (a) cetane number
    (b) octane number
    (c) percentage of hydrogen and carbon
    (d) % ash
22. Four-stroke engine is called
    (a) four-stroke combustion
    (b) Otto cycle
    (c) Pattinsons cycle
    (d) Carnot cycle
23. The catalyst employed in cracking process is
    (a) clay   (b) zeolite
    (c) clay & zirconium oxide (d) none of these
24. While removing sulphur from crude oil, it is treated with
    (a) zeolites   (b) clay
    (c) air   (d) sodium plumbite
25. The Dulong's formula is used to determine
    (a) weight of air needed for combustion
    (b) calorific values
    (c) W_{air}, V_{air} of coal
    (d) none of the above
26. The presence of which of the following elements is considered objectionable in a fuel?
    (a) Sulphur   (b) Carbon
    (c) Hydrogen   (d) Oxygen

27. Biodiesel and power alcohol are considered to be
    (a) fossil fuels   (b) catalyst
    (c) octane enhancers (d) alternative fuels
28. TEL and MTBE are used as
    (a) catalysts
    (b) octane enhancers
    (c) cetane enhancers
    (d) alternative fuels
29. Iso amyl nitrite is a/an
    (a) octane enhancer   (b) cetane enhancer
    (c) catalyst for cracking (d) fossil fuel
30. Flue gas composition is determined by
    (a) calorimeter
    (b) colorimeter
    (c) Orsat apparatus
    (d) Redwood viscometer
31. Rockets propel on the basis of
    (a) Newton's 1st law of motion
    (b) Newton's 3rd law of motion
    (c) Pauli's principle
    (d) law of gravity
32. Which one of the following is used as an explosive?
    (a) PCl_3 (b) Mercuric oxide
    (c) Graphene   (d) Nitroglycerin
33. Water gas gives characteristic blue flame because of
    (a) high CO content
    (b) high oxygen content
    (c) low steam content
    (d) high CO_2 content
34. Orsat apparatus is used to determine
    (a) specific heat of components
    (b) volumetric analysis of flue gases
    (c) gravimetric analysis of flue gases
    (d) combustion rate
35. The gas with least calorific value is
    (a) water gas   (b) coal gas
    (c) producer gas   (d) natural gas
36. Blasting fuse helps to
    (a) fire explosive safely
    (b) diffuse explosives
    (c) create massive explosion
    (d) None of these

Review Questions

1. What are fuels? How does one classify them? State suitable examples for each class.
2. Define GCV and NCV. Write Dulong's formula. Write the characteristics of a good fuel.
3. What is the principle of bomb calorimetry? With a neat labelled diagram, discuss the construction, working, and corrections involved in determining calorific value by bomb calorimetry.

4. State the construction and working of Junkers calorimeter.

5. Justify the statement, 'Calorific value is the performance parameter for fuels.'

6. Discuss proximate and ultimate analysis of coal and mention its significance.

7. What is coal? State the properties of different types of coal.

8. Describe the various steps involved in proximate analysis of coal. State the significance of each step with formulae.

9. Describe the various steps involved in ultimate analysis of coal. State the significance of each step with formulae.

10. How is mining of petroleum done? State the types of petroleum.

11. Discuss 'pulverized coal.'

12. State the various factors that have to be considered while selecting coal.

13. What are the fractions obtained on distilling crude oil? State any six valuable fractions and their properties and uses.

14. Justify the statement, 'Crude oil is a versatile liquid fuel for harnessing energy.'

15. What are octane and cetane numbers? Mention their significance.

16. Differentiate between octane and cetane numbers.

17. Explain knocking in IC engines. State the various cycles in the working of a petrol-run engine.

18. Discuss knocking in IC engines. How is the fuel structure related to knocking characteristics?

19. What is diesel knock? Explain in detail the engines working on diesel.

20. What are anti-knocking agents? State the role of TEL and MTBE as anti-knock agents.

21. Explain fixed bed catalytic cracking with a neat diagram elucidating the temperature, pressure, catalyst used, and quality of gasoline obtained by this method.

22. What is cracking? Explain fluid bed catalytic cracking with a neat diagram elucidating the temperature, pressure, catalyst used, and quality of gasoline obtained by this method.

23. What is vapour phase cracking? How is it different from catalytic cracking?

24. Distinguish between catalytic and thermal cracking processes.

25. State the advantages of catalytic cracking for gasoline.

26. What is unleaded petrol? Explain reforming of petrol.

27. How is refining of petrol carried out? Write a short note on power alcohol.

28. What is biodiesel? Describe the method to obtain biodiesel from vegetable oil. State the advantages of biodiesel as an alternative fuel.

29. What are LPG and CNG? State their composition, properties, and applications.

30. What is water gas? State its composition and properties. How is it prepared on a large scale? State the uses of water gas.

31. State the composition of producer gas. Explain the manufacture of producer gas.

32. What is coal gas? State its composition, manufacture, and uses.

33. Discuss the manufacture of oil gas with a note on its properties and uses.

34. Define flue gas. Explain the determination of flue gases using Orsat apparatus.

35. What is synthetic petrol? Explain the preparation of synthetic petrol by Fischer–Tropsch method.

36. List the characteristics of a good explosive. Mention the significance of oxygen balance of an explosive.

37. Discuss the preparation of explosives: (a) RDX; (b) Mercury fulminate; (c) Lead azide; (d) Nitroglycerin.

38. Justify the statement, 'detonation is oxidation reaction,' with supporting example.

39. What is the use of blasting fuse? Explain its types.

40. What are propellants? Classify them and provide examples.

41. List the characteristics of a good propellant.

42. What are bio-propellants? Elucidate their use in rockets.

NUMERICAL PROBLEMS

1. A coal sample has composition (in % by weight), C = 80, H = 10, O = 1, N = 3, S = 2, ash = 4. Calculate gross and net calorific values of the coal sample using Dulong's formula.

(**Ans:** GCV = 9915.6 kcal/kg, NCV = 9387.3 kcal/kg)

2. A coal sample has the following % composition by weight: C = 92, O = 2, S = 1, N = 1, and ash = 4. If NCV of coal is 8590.3 kcal/kg, determine % hydrogen and gross calorific value of coal sample. (latent heat of steam = 590 cal/g). **(Ans: 4.18% H and 8812.25 kcal/kg)**

3. A coal sample (C = 93%, H = 6%) was tested in a bomb calorimeter and the following observations were made: Weight of coal = 0.92 g, Water equivalent of calorimeter = 2200 g, Water taken in calorimeter = 550 g, Rise in temperature = 2.42°C, fuse wire correction = 10 cal, acid correction = 50 cal. Calculate GCV and NCV for the above coal sample. Latent heat of steam = 580 cal/g **(Ans: GCV = 7233.7 cal/g and NCV = 6920.4 cal/g)**

4. A gaseous fuel has composition by volume viz., CH_4 = 35%, CO = 10%, H_2 = 6%, N_2= 2 % and C_3H_8 = 12%. Calculate the volume of air required for complete combustion of 3.8 m^3 of gas. **(Ans: 24.969 m^3)**

5. 2.499 g air-dried coal was heated in a crucible at 105°C for 1 hour. The residue weighed 2.414 g. Calculate the moisture content (% basis) present in coal sample. **(Ans: 3.40 %)**

6. 1 g of coal sample was weighed in a silica crucible and heated at 105°C for 1 hour. The residue weighed 0.985 g. The crucible was partially covered with a lid and heated for 7 min at 955°C and the residue weighed 0.8 g. On strong heating, the final residue weight was found to be 0.1 g. Determine % moisture, % volatile matter, ash content, and % fixed carbon.

   (Ans: % moisture = 1.5, % volatile matter = 18.5, % ash = 10, % fixed carbon = 70)

7. 3.2 g of coal is heated in a Kjeldahl's method resulting in the formation of ammonia. The ammonia gas is absorbed in 35 ml of 0.1 N H_2SO_4 solution. After absorption, excess acid required 20 ml of 0.1 N NaOH for complete neutralization. 2.7 g coal on ultimate analysis is allowed to undergo combustion in a bomb calorimeter. After combustion, 0.65 g of barium sulphate is formed. Calculate the percentage of nitrogen and sulphur in coal.

   (Ans: %N = 0.656 and %S = 2.789)

8. Coal has composition (% weight) as: C = 54, H = 6.5, O = 3, N, 1.8, ash = rest, moisture = 17.3. On burning with excess air, it gave 21.5 kg dry exhaust gases per kg of coal. Calculate % excess air used for combustion.

   (Ans: 154.31 %)

9. During ultimate analysis, 1.58 g coal was allowed to undergo combustion in pure oxygen. Combustion gases evolved were absorbed in anhydrous calcium chloride and potash tubes causing increased weights recorded as 1.27 g and 5.12 g, respectively. Calculate %C : %H present in the coal sample.

   (Ans: %C = 88.37 and %H = 8.93)

10. Calculate the total quantity of exhaust gases obtained after combustion of gaseous fuel (% volume) CH_4 = 35, CO = 10, H_2 = 6, N_2 = 1, C_3H_8 =10. **(Ans: 6.731 m^3)**

11. The following data was obtained in a bomb experiment when solid fuel (%C = 93% and %H = 7) undergoes complete combustion in pure oxygen, (Given: latent heat of steam = 590 cal/g).

    Weight of the crucible = 3.5 g, weight of the crucible + fuel = 4.5 g, water equivalent of calorimeter = 500 g, water taken in the calorimeter = 2000 g, rise in temperature = 2 °C, cooling correction = 0.1 °C, acid correction = 50 calories, fuse wire correction = 10 calories.

    Based on the above observations, calculate the GCV and NCV of the solid fuel.

    (Ans: GCV = 5190 cal/g, NCV = 4818.3 cal/g)

12. A petrol sample composition is (% weight basis), C = 80 and H = 20. Calculate the volumes of oxygen and air required for the complete combustion of 1 kg petrol sample under standard conditions.

    (Ans: 2613.1 L, 12443.33 L)

13. A coal sample required 20 % excess air for complete combustion. Calculate the weight and volume of air, if 300 g of coal has elemental composition: C = 81 %, H = 4 %, N = 1.5 %, S = 1.2 %, O = 3 % and ash = rest.

    (Ans: weight and volume of air = 3.852 kg and 2.981 m^3 respectively)

14. Producer gas has % composition by volume as: CO = 30, H_2 = 12, CO_2 = 4, CH_4 = 2, N_2 = 52. If 50% excess air was used to perform combustion of 1 m^3 of gas, then determine the composition of dry flue gases.

    (Ans: %CO_2 : %N_2 : %O_2 = 11.63 : 78.5 : 9.85)

15. On ultimate analysis, a coal sample has (%) composition: C = 54, H = 6, O = 1, N = 3, moisture = 18.1, and ash = rest. If 1 kg of this coal was burnt in excess air, it released 20.3 kg of dry exhaust gases. Determine the percentage of excess air required for complete combustion of the coal sample. **(Ans: 145.7%)**

FURTHER READING

1. Dawe, Richard A. (ed), *Modern Petroleum Technology*, Vol 1, Upstream, Wiley–Blackwell 2000.
2. ISRO Propulsion Center, http://www.isro.gov.in/about-isro/isro-propulsion-complex-iprc
3. Keating, Eugene L. *Applied Combustion*, CRC Press (Taylor and Francis Group), NY, 2007.
4. NASA, Green Propellant Infusion Mission https://www.nasa.gov/mission_pages/tdm/green/overview.html
5. Rao, B.K.B., *Modern Petroleum Refining Processes*, Oxford & IBH Publishing Co. Ltd. 2002.
6. Sarkar, S. *Fuels and Combustion*, University Press, India. 2009.

ANSWERS

Check Your Progress

1. Combustible substances that produce a large amount of heat which can be harnessed for various domestic and industrial purposes.

2. Total quantity of heat liberated by burning a unit mass (for solid fuels) or volume (for liquid or gaseous) of fuel completely is called calorific value.

3. Peat, lignite, bituminous and anthracite are the different varieties of coal.

4. If ignition temperature is low, the fuel can cause fire hazards and thus dangerous to store and transport. If fuel has higher ignition temperature, it will be difficult to ignite and start the combustion of fuel.

5. High calorific values, moderate ignition temperature, low ash content and moisture content, low cost, and easy to handle during transportation, are some of the criteria for selecting a fuel.

6. Determination of moisture, volatile matter, ash content, and fixed carbon are the steps in the proximate analysis of coal sample.

7. The determination of the elemental composition of coal is termed as ultimate analysis. Elements such as carbon, hydrogen, nitrogen, sulphur, oxygen are determined. Higher percentages of carbon and hydrogen enhance the calorific value of coal. Higher percentage of nitrogen and sulphur is a limiting factor as it contributes to environment pollution.

8. Proximate and ultimate analysis of coal gives an idea on the practical utility of coal sample. It lays a basic premise for grading coal. Ultimate analysis gives the elemental composition of coal.

9.

Octane number	Cetane number
1. It is defined as the percentage of iso-octane in a mixture of iso-octane and n-heptane that matches with knocking of petrol sample under consideration.	1. It is defined as the percentage by volume of cetane in a mixture of cetane and 1-methyl naphthalene that matches in its ignition delay of diesel under test.
2. It signifies knocking characteristics of petrol.	2. It depicts ignition quality of diesel.
3. Greater the aromaticity of hydrocarbons in petrol, higher the octane number.	3. Greater the aliphaticity of hydrocarbons in diesel, higher the cetane number.
4. Examples of octane enhancers include tetra ethyl lead and methyl tertiary butyl ether.	4. Examples of cetane enhancers are isoamyl nitrate, ethyl nitrite, acetone peroxide, and ethyl nitrate.

10. Flue gas (or exhaust gas) is the gas released into the atmosphere through a flue pipe after combustion of fuel. The gases such as CO_2, CO, N_2, SO_2 are some of the exhaust gases present in a flue gas.

11. The sharp, metallic, pinging sound resulting due to premature ignition of air–fuel mixture in an IC engine is called *knocking*. Knocking affects engine performance and lead to damage of engines.

12. Knocking in IC engines occurs due to the premature ignition of fuel. Using TEL or MTBE as an anti-knocking agent in petrol can lower the chances of knocking in engines.

13. In a catalytic cracking process, carbon gets adsorbed on the catalyst surface that tends to deactivate its activity. Hence, dry air is periodically blown over the catalyst surface so as to eliminate any particulate matter. This is called catalyst regeneration.

14. Crude oil is treated with sodium plumbite that converts thiols in to disulphides. This process helps to eliminate sulphur from crude oil.

$$R-SH + Na_2PbO_2 \xrightarrow{\text{Sulphur (in NaOH)}} R-S-S-R + PbS + 2NaOH$$

Thiol Sodium plumbite Alkyl disulphide

15. Higher calorific value, even particle size, high flame temperature, low moisture, ash, sulphur and phosphorus content are the parameters to be considered while selecting coal.

16. See Section 20.7.3

17. The composition of producer gas (% by volume): $CO = 30$, $N_2 = 51 – 56$, $H_2 = 10 – 15$, $CO_2 + CH_4 = $ rest. Water gas composition (% by volume): $CO = 41$, $H_2 = 51$, $N_2 = 4$, $CO_2 + CH_4 = $ rest.

18. Power alcohol is an alternative non-fossil fuel in which 5–25% ethanol is blended with gasoline to power engines.

19. Isomerization and dehydrogenation are the organic reactions performed for reforming petrol.

20. Liquid fuel from coal through artificial method is called synthetic petrol. Fischer–Tropsch and Bergius methods are used to prepare synthetic petrol.

21. KOH solution: CO_2; pyrogallic acid: O_2 and CO_2; ammonical cuprous chloride: CO, O_2, and CO_2.

22. Water gas has high carbon monoxide content and thus exhibits characteristic blue flame.

22. TEL improves the anti-knocking characteristics of IC engines but is considered a polluting factor. MTBE contains oxygen in the form of an ether group and this additional oxygen assists in the combustions of petrol in IC engines, along with reducing the peroxide formation. Also, it tends to increase the molecules with branched and aromatic rings in the fuel structure.

24. Bomb calorimeter is used for solid fuels and Junkers calorimeter for gaseous and liquid fuels.

25. LPG, CNG, gasoline, kerosene, diesel are examples of petroleum products.

26. Sulphur compounds pollute the environment and if present in fuels may undergo combustion to form undesirable toxic gases.

27. Flue gases are a mixture of CO, CO_2, and O_2 and thus signify whether combustion process of a given fuel is complete or not.

28. Reforming of petrol involves structural modifications in straight-run gasoline to improve its anti-knock characteristics.

29. Coal can be converted to a liquid fuel by adding hydrogen which is referred to as coal liquefaction. Bergius and Fischer–Tropsch methods produce synthetic petrol by liquefaction of coal.

30. During cracking process, the catalyst gets exhausted due to carbon deposits thereby losing its catalytic activity. Hot air is blown on the catalyst to remove carbon deposits and thus regenerated for further process.

31. An explosive is a substance that undergoes a rapid reaction in a highly exothermic manner accompanied by high pressures at the reaction site; e.g., TNT and RDX.

32. Rocket propellant is a high oxygen-containing mixture of fuel along with an oxidant present in the rocket combustion chamber.

33. Oxygen balance is a measure of oxygen present in an explosive molecule.

34. Fuses or blasting fuses are required to ignite the explosive material. It allows the person firing the explosive to immediately reach safety before explosion takes place.

35. As per Newton's third law of motion, the act of pushing combustion gases downwards results in an equal and opposite reaction propelling the rocket upwards into space.

Multiple Choice Questions

1. (c)	2. (d)	3. (a)	4. (a)	5. (c)	6. (a)	7. (b)
8. (d)	9. (a)	10. (c)	11. (d)	12. (a)	13. (c)	14. (a)
15. (d)	16. (b)	17. (b)	18. (a)	19. (d)	20. (c)	21. (a)
22. (b)	23. (c)	24. (d)	25. (b)	26. (a)	27. (d)	28. (b)
29. (b)	30. (c)	31. (b)	32. (d)	33. (a)	34. (b)	35. (c)
36. (a)						

Pollution and Its Control

After reading this chapter, you will be able to:

- list the various types of pollution such as air, water, soil, noise, and radiation.
- account for the release of various pollutants and their optimal levels in the environment.
- enumerate the adverse effects of pollutants on human health and environment.
- explain global warming, smog, acid rains, ozone depletion, and their impact on the environment.
- identify efficient ways to manage solid wastes such as biomedical waste and e-waste so as to have the least environmental impact.
- discuss the various control measures of pollution.

21.1 INTRODUCTION

Pollution refers to the contamination of the environment with materials that interfere with human health, quality of life, or the natural functioning of the ecosystems. The undesirable foreign matter which is released into the environment that contaminates air, water, and land is referred to as a *pollutant*. Today, we live in a world where environment is threatened. Overpopulation, rapid industrialization, deforestation, global warming, climate change, and calamities are major causes of environmental pollution.

The origin of the term 'environment' comes from the French word 'environ' which means to encircle or surround. The surroundings in which an organization operates, including air, water, land, nature resources, flora, fauna, humans is called environment (International Organization for Standardization, ISO 14001). All these natural resources are adversely affected by pollution. The major forms of pollution include water pollution, air pollution, noise pollution, soil, and radiation pollution.

Environmental engineering focuses on the protection of environment by reducing waste and pollution. Environmental engineers strive relentlessly to improve the environmental conditions through design of technologies and processes that control pollutant release and clean up the existing contamination, whether in air, water, or land.

In this chapter, we will discuss the different types of pollution, effects on health and environment, and measures to deal with them.

21.2 AIR POLLUTION

An excessive release of any chemical, physical, or biological matter in air that causes harm to living organisms and deteriorates the natural atmosphere is called air pollution. An invisible killer, air pollution is, by far, a serious form of pollution, which if unchecked, may have severe impact on human health and environment. The Air (Prevention and Control of Pollution) Act (1987) provides information on the prevention, control, and abatement of air pollution in India. According to the WHO Global Ambient

Air Quality Database (2018), nearly 80 % population residing in urban areas is exposed to air quality levels that exceed the WHO limits. In fact, it should not be thought of that air pollution is prevalent just outside our households; in 2014, WHO cited that indoor air is highly polluted due to dampness, cooking fuel emissions, microorganisms, mosquito repellants, and radiation. Table 21.1 lists some of the Indian legislations related to environment taken up over the decades.

21.2.1 Classification of Air Pollutants

The undesirable foreign matter present in the air causing contamination and changes in the air composition is called air pollutant. Air pollutants are categorized as per origin into two types: primary and secondary. *Primary pollutants* are those which are directly released in the atmosphere. Table 21.2 lists the global optimum levels for each pollutant laid down by WHO.

Dust particles, carbon monoxide, sulphur dioxide, nitrogen oxides, carbon dioxide, particulate matter (PM), volatile organic compounds (VOCs), and chlorofluoro carbons (CFCs) are examples of primary air pollutants. *Secondary pollutants* are produced in the atmosphere by physical and chemical reactions occurring among the natural constituents in air. Ozone is one such example which is produced due to hydrocarbons and nitrogen oxides in the presence of sunlight. Other examples of secondary pollutants are nitric acid (from nitric oxide) and sulphuric acid (obtained by SO_2 oxidation).

According to chemical composition, air pollutants can be organic or inorganic. *Inorganic pollutants* include carbon monoxide, carbon dioxide, oxides of sulphur, nitrogen oxides, peroxides, silica, asbestos, fly ash, and ozone, whereas hydrocarbons, amines, alcohols, aldehydes, and ketones are *organic pollutants*.

Based on the state of matter, air pollutants can be gaseous or particulate matter. *Gaseous pollutants* easily disperse in air and spread across leading to respiratory problems, besides hindering visibility. Particulate matter is one of the worst forms of air pollutant, as finely divided particles of dust, fly ash, silica, asbestos can enter the lungs and cause serious pulmonary problems. Generally, particulate matter exists in colloidal state as *aerosols*.

Various sources of air pollution include vehicle exhaust, power generation plants, industrial processes, fertilizer manufacture, construction, mining, solid waste disposal, and fires. Let us now focus on major air pollutants in terms of their sources of release, ill-effects, chemical reactions, and suitable remedies.

Table 21.1 India's environmental legislation

Legislation[*]
Wildlife Protection Act, 1972
Water (Prevention and Control of Pollution) 1974
Air (Prevention and Control of Pollution) Act, 1981
Environment (Protection) Act, 1986
Hazardous Wastes (Management and Handling) Rules, 1989
Public Liability Insurance Act, 1991
National Environment Tribunal Act, 1995
Biomedical Waste (Management and Handling) Rules, 1998
Noise Pollution (Regulation and Control) Rules, 2000
Ozone Depleting Substances (Regulation and Control) Rules, 2000
National Green Tribunal Act, 2010

(*For more information on legislation, Ministry of Environment, Forest and Climate change visit http://www.moef.nic.in/)

Table 21.2 Global optimum levels as per WHO, 2005

Global optimum levels ($\mu g/m^3$)		
$PM_{2.5}$	1 year	10
	24 h (99th percentile)	25
PM_{10}	1 year	20
	24 h (99th percentile)	50
Ozone, O_3	8h, daily maximum	100
Nitrogen dioxide, NO_2	1 year	40
	1 h	200
Sulphur dioxide	24 h	20
	10 minutes	500

21.2.2 Oxides of Sulphur

Sulphur is present in the atmosphere in trace amounts due to natural phenomena as SO_2 (volcanic eruptions), H_2S (decaying plants, animals), and as an aerosol, sulphate. However, vast amounts of its oxides are formed due to various fuel combustion processes. The major sources of oxides of sulphur emission are the following.

Thermal power stations   Probably one of the major sources of SO_2 release, combustion of fuel releases a huge amount of sulphur in to the atmosphere.

Sulphuric acid industry   In the contact process of H_2SO_4 manufacture, 96–98% SO_2 gets converted to SO_3 along with the formation of acid. Huge amounts of SO_2 release are from these processes.

Roasting of sulphide ores   This is a step in metallurgy where sulphide ores are converted into oxides. The chemical reaction involved is as follows:

$$2\,FeS + 3\,O_2 \rightarrow 2\,FeO + 2\,SO_2$$

As evident from the reaction, large amounts of SO_2 are released during the process of ore roasting.

Sulphur dioxide does not remain in the gaseous form for prolonged periods in the atmosphere. It quickly reacts with atmospheric moisture in the presence of sunlight to form harmful acids.

In the presence of sunlight, sulphur dioxide is converted into SO_3 and further into sulphuric acid. However, photochemical dissociation of sulphur dioxide is not feasible in the atmosphere as it requires 566 kJ/mol of energy that cannot be provided by absorbing the solar radiations; the reaction can be depicted as, $SO_{2(g)} + h\nu \rightarrow {}^3SO_{2(g)} + {}^1SO_{2(g)}$.

During photochemical reaction, SO_2 produces excited triplet and singlet states occurring at 385 nm and 294 nm respectively.

Molecular oxygen or ozone can further oxidize sulphur dioxide

$$SO_2 + 1/2\,O_2 \rightarrow SO_3 \text{ and } SO_2 + O_3 \rightarrow SO_3 + O_2$$

The oxidation of sulphurous acid (H_2SO_3) that is formed due to SO_2 dissolution in rainwater is quick in the presence of metallic oxides found in aerosols.

$$SO_2 + H_2O \rightarrow H_2SO_3; \quad H_2SO_3 + 1/2\,O_2 \rightarrow H_2SO_4$$

These reactions depict the various ways in which sulphur compounds can be formed in the atmosphere.

Control of SO_2 Pollution

The combustion gases containing large amounts of oxides of sulphur are allowed to pass through a slurry of limestone. Sulphur dioxide gets absorbed in limestone and air is blown through the slurry to complete the conversion of calcium sulphite to calcium sulphate.

$$CaCO_3 + SO_2 \rightarrow CaSO_3 + CO_2; \quad CaSO_3 + 1/2\,O_2 \rightarrow CaSO_4$$

All these chemical reactions are carried out in a fluidized combustion bed, where pulverized coal and limestone are fed in the chamber and fluidized by blowing pure air at 1000 °C.

21.2.3 Oxides of Nitrogen

Nitrogen can form various oxides of which, nitrous oxide (N_2O), nitric oxide (NO), and nitrogen dioxide (NO_2) are found in appreciable concentrations in unpolluted air. One can use NO_x as a representative of various oxides of nitrogen present in air.

The major sources of oxides of nitrogen are vehicles and fuel combustion processes and the main source of N_2O is the soil. Microorganisms present in the soil can degrade nitrogenous matter to form nitrogen gas and nitrous oxide, and thus, N_2O cannot be considered as an air pollutant. Nitrous oxide can undergo photodissociation in the atmosphere forming the major air pollutant nitric oxide.

$$N_2O + h\nu \rightarrow NO + N \ (\lambda = 250 \text{ nm})$$
$$N_2O + h\nu \rightarrow N_2 + O \ (\lambda = 337 \text{ nm})$$

The lightning in the atmosphere combines nitrogen and oxygen to form nitric oxide, which further oxidizes to form NO_2. In IC engines, nitrogen and oxygen from the air enter the cylinder and form NO and NO_2 which are released during the exhaust cycle, thereby causing air pollution.

21.2.4 Oxides of Carbon

The major oxides of carbon are carbon monoxide and carbon dioxide. Of these two, the more harmful air pollutant is carbon monoxide. Naturally, methane (from decomposed organic matter), on oxidation in the presence of oxygen produces carbon monoxide and water.

$$CH_4 + 1/2 \ O_2 \rightarrow CO + 2 \ H_2O$$

Anthropogenic sources of carbon monoxide are incomplete combustion of fossil fuels like coal and coke. Internal combustion engines release huge amounts of carbon monoxide due to incomplete combustion taking place in the cylinder. Carbon dioxide on reacting with carbon forms carbon monoxide.

$$CO_2 + C \rightarrow 2 \ CO$$

Adverse Effects

One of the adverse effects of high levels of atmospheric CO is carbon monoxide poisoning affecting the human respiratory system. The inhaled carbon monoxide combines with haemoglobin (abbreviated as Hb; an oxygen-transporter) in the blood stream, forming a harmful metal–protein complex, also referred to as carboxyhaemoglobin (COHb).

$$HbO_2 \ (\text{oxyhemoglobin}) + CO \rightarrow COHb + O_2$$

Due to the formation of COHb, the oxygen-carrying capacity of Hb is considerably reduced, resulting in oxygen deficiency in tissues. It has been observed that cigarette smokers have higher levels of COHb in their bloodstream with carbon monoxide levels as high as 400 ppm which causes pulmonary system collapse, breathlessness, and even myocardial (heart) attacks.

Control of CO pollution Automobile exhaust is a major source of carbon monoxide release in the atmosphere. Carbon monoxide adsorbs on activated charcoal or one can employ catalytic converters in the exhaust, where carbon monoxide is oxidized to carbon dioxide that helps in controlling CO concentrations.

21.3 GREENHOUSE EFFECT AND GLOBAL WARMING

Global warming refers to *an increase of average global temperature as a result of greenhouse effect.* J. Fourier (1822) coined the term 'greenhouse effect'. Today, we have recognized that global warming is a significant environmental issue that is addressed in almost all government policies and awareness campaigns. The

Check Your Progress

1. Define environment.
2. What is pollution? Define a pollutant.
3. Define air pollution.
4. What do you mean by primary and secondary pollutants? Give an example of each.
5. What are organic and inorganic pollutants?
6. Name the sources of oxides of sulphur.
7. What is the major source of carbon monoxide in air?

warming of the earth occurs due to greenhouse gases such as carbon dioxide, nitrous oxide, water vapour, and methane that trap heat and light coming from the sunlight in the atmosphere. With ever-increasing levels of greenhouse gases, climate change is currently a pressing challenge for the future development and sustenance of environment. The temperature details are not the only indicators of a changing climate. There are many other indicators such as receding mountain glaciers, decreased snow cover in the Northern hemisphere, rising sea levels, and thinning of the Arctic ice. The

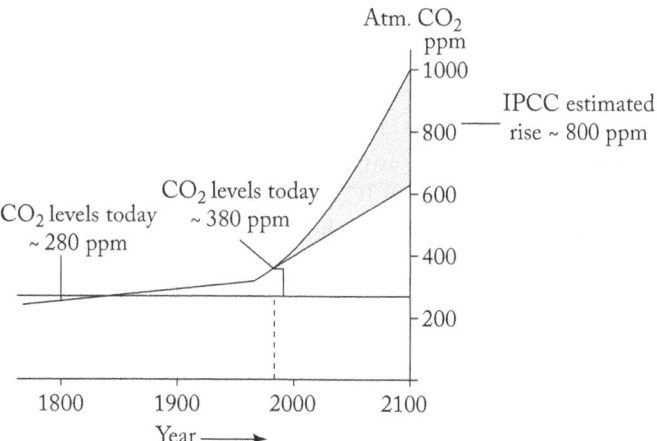

Fig. 21.1 Projected atmospheric carbon dioxide concentrations

combined data proves beyond reasonable doubt that climate is changing globally, which will have serious implications for all forms of life, if no stringent plan is put in action.

The earth absorbs most of the sun's incident radiation (visible, UV, and IR) and re-emits a part of it back, thereby maintaining equilibrium and constant temperatures conducive for life. The radiations reflected back from the earth has the wavelength range of about 4000–5000 nm and falls in the thermal IR region. Greenhouse gases absorb these thermal infrared radiations and re-emit them to the earth's surface in the same way as glass in a greenhouse making it warm (just like your closed car on a hot day!). This is called greenhouse effect or global warming. Due to this effect, earth's temperature is at least 15 °C above the optimum temperature which worsens with increasing pollution levels. Witnessing numerous emission scenarios on the rise, IPCC (Intergovernmental Panel on Climate Change) has concluded that global temperatures may rise between 1.4 and 5.8 °C by the year 2100 (Fig. 21.1).

Carbon dioxide is the main culprit when it comes to global warming and climate changes. Large amounts of carbon dioxide are released during fuel combustion, vehicle exhausts, thermal power plants, etc. Deforestation and urbanization have together resulted in the decrease in carbon dioxide sinks (which remove CO_2 from the air and act as carbon reservoir).

Control Measures

There is a need to establish smoke-free zones with stringent legislation. People should be aware of using safer fuels in cars, kitchen, aeroplanes, power stations, etc. Replacing fossil-fuels with electric or hybrid cars are underway and shows promise that may revolutionize the automotive industry. Reforestation and increasing the number of carbon sinks will help reduce higher levels of greenhouse gases. Today, carbon capture and utilization techniques are used to exploit carbon dioxide in a number of chemical reactions.

21.4 OZONE DEPLETION

Ozone is present as two layers in the atmosphere, of which stratosphere has natural ozone layer (about 10 ppm) extending from about 6 to 30 miles protecting the earth from harmful UV rays. Troposphere, a layer closest to the earth contains 'bad' ozone and categorized as an air pollutant. Ozone is formed in the stratosphere by photochemical reactions.

Molecular oxygen absorbs UV radiations and dissociates to atomic oxygen, which again combines with oxygen forming ozone. Ozone, so formed, absorbs short wavelength UV and decomposes rapidly as oxygen and nascent oxygen.

$$O_2 \xrightarrow[\text{UV at 240 nm}]{h\nu} 2(O); \quad O_2 + (O) \rightarrow O_3$$
$$O_3 \rightarrow O_2 + (O)$$

> *The thickness of ozone layer is measured in Dobson Units (DU), where 1 DU is the number of ozone molecules required to create ozone layer of 0.01 mm thick at 0 °C and 760 mmHg pressure.*

Thus, a dynamic equilibrium between oxygen and ozone is attained in the stratosphere. A thin umbrella of ozone exists in the stratosphere and protects the earth from the harmful UV rays that are known to cause DNA damage, mutation, and serious forms of cancer (skin and lungs).

Over the recent years, man-made chemicals, particularly chlorofluorocarbons (CFCs), hydro-chlorofluorocarbons (HCFCs), $CHCl_3$, and CCl_4 are released that deplete the natural ozone layer. Such chemicals are collectively called ozone-depleting substances (ODS) and the phenomenon is called *ozone depletion*.

> *According to World Meteorological Organization (Geneva), the ozone layer is depleting at around 4.3 % every decade in the northern hemisphere since 2000 and is known to deplete further.*

CFCs, HCFCs, CCl_4, and methyl chloroform are still used as coolants, refrigerants, foaming agents, pesticide sprays, propellants, fire extinguishers, and blowing agents. These ODS degrade very slowly in the environment and hence lurk for longer duration causing chain reactions with ozone thereby depleting it. If these substances dissociate, chlorine radicals will form that will destroy the natural ozone layer. The main culprit in ozone depletion is CFCs as they are stable, inert, non-toxic when released in the atmosphere. Due to its inertness, they do not have natural sinks where it can safely be absorbed, rather they increase in concentration in the air.

$$CCl_2F_2 \rightarrow CClF_2^{\bullet} + Cl^{\bullet}$$

The chlorine radicals catalyse ozone dissociation as,

$$Cl^{\bullet} + O_3 \rightarrow ClO^{\bullet} + O_2 \text{ and } ClO^{\bullet} + O \rightarrow Cl^{\bullet} + O_2$$

The regenerated chlorine radicals continue the chain reactions that have resulted in an ozone hole in the stratosphere. Replacing CFCs and other ODS are underway and will soon be phased out.

21.4.1 Acid Rain

Acid rain refers to the presence of excessive amounts of acids in rainwater. The sulphur dioxide and nitrogen oxide gases present in the atmosphere cause acid rains. Rainwater is slightly acidic due to the presence of dissolved CO_2 from air having a pH of about 5.6.

$$H_2O + CO_2 \rightarrow H_2CO_3 \text{ (Carbonic acid)}$$

However, pH of rainwater comes down to about 4.5 or in severe cases to about pH 2.3 due to the presence of oxides of sulphur and nitrogen. Acid rains are formed due to the oxidation of these gases in the atmosphere and return back to the ground as dissolved water drops.

> *Robert Angus Smith (1852), a Scottish Chemist was the first to coin the term 'acid rain,' and is known for his detailed analysis on acid rains documented in a classic book in 1872, 'Air and Rain: the Beginning of a Chemical Climatology,' in Glasgow Medical Journal. For more info visit:* https://www.ncbi.nlm.nih.gov/pmc/articles/PMC5878585/)

The formation of sulphuric acid from sulphur dioxide and the resulting rain is referred to as acid rain.

$$SO_2 + H_2O \rightarrow H_2SO_3 \text{ (Sulphurous acid)}$$
$$H_2SO_3 + 1/2\, O_2 \rightarrow H_2SO_4$$

Sulphurous acid forms sulphuric acid in the atmosphere in the presence of soot-like metallic oxide catalyst. Further, nitric oxide is released through vehicle exhausts and on combining with oxygen from air, forms nitrogen dioxide which on dissolution in water form toxic acids such as nitrous and nitric acids.

$$NO + 1/2\, O_2 \rightarrow NO_2; \quad 2\, NO_2 + H_2O \rightarrow HNO_2 + HNO_3$$

Adverse effects

(a) Higher concentrations of sulphur dioxide have adverse effects on the ciliary movements of the respiratory tract in human beings. These ciliary movements are vital in clearing microorganisms and toxins from the respiratory tract. A concentration of about 20 ppm SO_2 may cause irritation of the epithelial lining of eyes, digestive tract, and if it exceeds beyond 400 ppm, it may prove fatal.

(b) Even as low as 1 ppm SO_2 can cause wilting and yellowing of plant leaves which also acts as a pollution indicator.

(c) If the atmospheric sulphuric acid comes down with the rainwater may destroy vegetation, architectural structures, and human health too.

(d) Acid rains damage crops by washing away the nutrients from soils.

(e) Acid rains destroy forests and other biodiverse areas.

(f) They cause leaching and yellowing of buildings; for example, the yellow tinge on the Taj Mahal at Agra is a striking example of the adverse impacts of acid rain.

21.5 SMOG

Smog is a hazardous combination of fog and smoke in the air that reduces atmospheric visibility by forming a grey haze. London smog and photochemical smog are classic environmental hazards witnessed till date.

London smog (1952) prevailed for five days with a reported casualty of about 5000 people. The higher concentrations of unburnt carbon soot and sulphur dioxide in the atmosphere caused the smog.

$$S + O_2 \rightarrow SO_2; \quad SO_2 + 1/2\, O_2 \rightarrow SO_3$$

The above reactions occur in the presence of sun's radiations and particulate matter. Sulphur dioxide so formed, on combining with fog and moisture gave sulphurous acid.

$$SO_2 + H_2O \rightarrow H_2SO_3$$

$$H_2SO_3 + 1/2\, O_2 \rightarrow H_2SO_4 \text{ (in the presence of metallic salts in atmosphere)}$$

$$H_2SO_4 + 2\, NH_3 \rightarrow (NH_4)_2SO_4$$

It could also follow the chemical reaction: $SO_2 + 2\, NH_3 + H_2O + 1/2\, O_2 \rightarrow (NH_4)_2SO_4$

Photochemical smog refers to an atmospheric reaction that involves nitric oxide and nitrogen dioxide that usually occurs in warm sunny weather. Smog can be characterized as brown and hazy fumes in the air that causes haziness, eye irritation, and vegetation decay on prolonged exposure. Nitrogen dioxide is a major reactant in the formation of smog.

The chemical reaction during photochemical smog formation can be represented as shown in Fig. 21.2.

Reactive hydrocarbons react with oxygen to form hydrocarbon free radical ($RCH_2^\bullet$) which reacts with oxygen forming ($RCH_2O_2^\bullet$). $RCH_2O_2^\bullet$ radical reacts with nitric oxide forming nitrogen dioxide and $RCH_2O^\bullet$, yet another free radical. On further oxidation, aldehyde (RCHO) and hydroperoxyl ($HO_2^\bullet$) radical are formed. Hydroperoxyl radical combines with another molecule of nitric oxide forming nitrogen dioxide and hydroxyl ($^\bullet OH$) radical, which rapidly reacts with stable hydrocarbon in the atmosphere forming regenerated $RCH_2^\bullet$ radical, thus completing the reaction cycle.

PAN (peroxyacyl nitrate) is a secondary pollutant in the smog that is formed through several oxidation reactions as shown in Fig. 21.2.

Adverse effects Eye irritation, dry throat, respiratory problems, and even lung cancer are some serious complications due to chronic exposure of nitrogen oxides on the human body. PAN causes eye irritation (lachrymator) and hinders respiratory tract. There are no reports on any adverse effects on plants on NOx absorption. Acid rains and photochemical smog are other exemplary impacts of oxides of nitrogen.

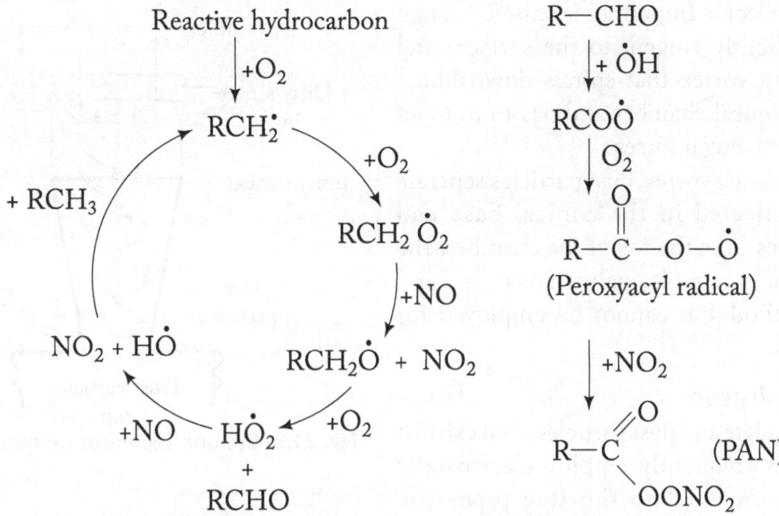

Fig. 21.2 Photochemical smog formation

Control of NOx emissions By employing a catalytic converter with Pt/Rh catalyst in IC engines, oxides of nitrogen can be decomposed to nitrogen and oxygen.

21.5.1 Particulate Matter

According to Environment Protection Agency (EPA), particulate matter is a complex mixture of air-borne particles and liquid droplets composed of acids (such as nitrates and sulphates), ammonium salts, water, carbon, organic chemicals, metals, and soil crust materials. 'Coarse particles' (PM$_{10\text{-}2.5}$) are found to be released from construction work and industries, and range in diameter from 2.5 to 10 mm (or µ, microns). The 'coarse' particle standard (known as PM$_{10}$) includes all particles less than 10 µ in size. 'Fine particles' (or PM$_{2.5}$) are released during smoking and smog haze that have diameters less than 2.5 µ. Particles in this size range constitute a large proportion of dust that can be drawn deep into the lungs. The larger particles get trapped in the nose, mouth, or throat causing cough, asthma, and other respiratory irritations. EPA released a global mandate through the National Ambient Air Quality Standard for PM$_{2.5}$ and PM$_{10}$ to be within optimum levels of 12 µg/m^3 and 50 µg/m^3 respectively.

Sources The various sources of particulate matter are:

Natural sources include dust storms, cyclones, volcanic eruptions, forest fires, and so on.

Anthropogenic sources are soot formation by combustion of fossil fuels, construction dust, vehicular soot, smoking, pesticide sprays, mosquito repellants, fly-ash from thermal power plants, cement, pigments, volatile organic compounds (VOCs), etc that are known to be released as aerosols in the atmosphere.

Adverse effects Respiratory problems arise due to inhalation of air contaminated with particulate matter. Silica dust is known to cause asbestosis of lungs. Kidney malfunctions and pulmonary disorders are other health problems due to dust and aerosol in the air. As per the air quality mandate of CPCB (Central Pollution Control Board), non-communicable diseases are reportedly on the rise in India due to exceedingly high PM levels in the atmosphere.

21.5.2 Methods to Reduce Particulate Matter

Various methods are employed to reduce particulate matter in air. We will highlight two main methods, viz., cyclone separator (centrifugal separator) and Cottrell precipitation (electrostatic precipitator).

Cyclone Separator

The removal of particulate matter is based on the principle of using centrifugal forces on the gas stream. Cyclone separator is made up of a cylindrical chamber as shown in Fig. 21.3, an inner tube, and a conical base with dust collector. Impure air is passed through the chamber (slightly tangent to the surface) and made to create a vortex that spirals down like a cyclone in the conical chamber and passes in to an inner tube by centrifugal forces.

Due to the spiraling vortex, dust particles separate out and are collected in the conical base and purified gas leaves from the top of the chamber. The cyclone separator is a simple, maintenance-free, and economical method, but cannot be employed for smaller particles.

Cottrell Precipitator

To remove particulate and dust particles from exhaust gases, industries frequently employ electrostatic precipitators connected to the flue pipes (or chimneys) before releasing the gases into the air. Exhaust gases are passed through the precipitator (see Fig. 21.4) chamber equipped with discharge electrodes that produce high current intensities and corona in which gas molecules are ionized releasing electrons from them.

These electrons collide with suspended dust particles present in impure gas and thus particles acquire negative charge. These negatively charged particles are attracted by collector electrodes and get collected in a conical base collector. The purified gases leave

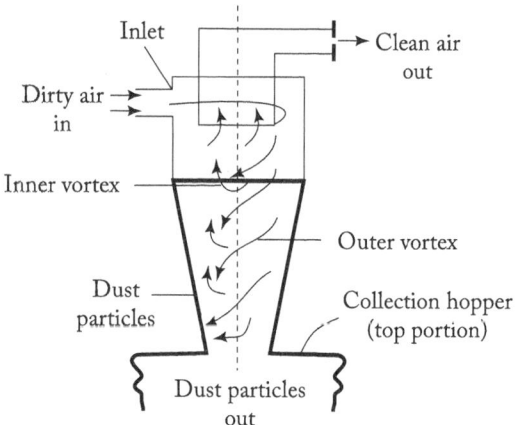

Fig. 21.3 Cyclone separator to remove dust from air

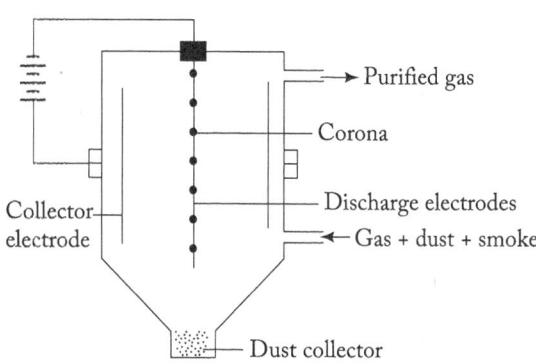

Fig. 21.4 Cottrell precipitator

from the top of the precipitator and are released safely in the environment. Cottrell precipitator has better dust collection efficiency and requires low maintenance. Only particulate matter and dust particles can be removed by this method, not other gaseous impurities.

21.6 SOLID WASTE DISPOSAL

Another nuisance witnessed in the environment is rampant solid waste generation and its disposal. Due to increasing population and improper handling, solid waste has landed up as piles. Lack of public awareness and absence of stringent legislation have led to this problem, which is growing by each passing day.

Sources   The various sources of solid waste are largely classified as follows.

(a) *Garbage dump:* Municipalities from various localities collect the garbage which consists of domestic, kitchen, and human wastes, and decomposed matter. The major problem is non-decomposable materials like plastics that land up with these materials in the dump. Many states like Maharashtra and Karnataka have implemented complete ban on the use of plastics, yet the problem remains persistent in India.

(b) *Debris:* Various construction debris, used metal parts, ashes, etc. also add up as solid waste.

Check Your Progress

8. What is smog? Name any two smog formations.
9. Define particulate matter.
10. Name the major air pollutant in photochemical smog.
11. State the methods employed to reduce particulate matter in air.

(c) *Electronics:* Obsolete phones, discharged batteries, wires, computers, switches, etc., all come under electronic waste (e-waste), which is way more challenging to recycle, thereby compromising the environment.

(d) *Industrial waste:* Chemicals, detergents, surfactants, paints, dyes, and sand, add up as solid waste in dumping sites. Slag, fly ash, coal refuse, sludge are common solid waste from the mining industries.

(e) *Medical centres:* Biomedical waste such as used ampoules, blood-stained cotton swabs, disposable syringes, and injections once used are disposed off as hospital waste which are major sources of diseases, if not treated properly.

(f) *Agricultural waste:* Decomposed plant leaves, tree barks left behind after natural calamities, destroyed crops, grass, fruit peels, etc., comprise agricultural waste which is quite simpler to handle, but if mixed with non-decomposed matter, become difficult to manage.

21.6.1 Solid Waste Management and Recycling

For an effective solid waste management, the following steps needs to be performed.
 (a) Identifying the type, source, and quantity of solid waste
 (b) Potential hazards of solid waste
 (c) Safe collection and transportation to treatment sites and safe disposal

Landfill and incineration are commonly employed methods to treat solid waste but looking at the complexity of waste matter, one needs to find safer techniques to manage solid waste. Composting is an eco-friendly way of dealing with solid waste. In this method, decomposable organic matter is placed in the soil which enhances plant growth. Complete monitoring needs to be performed while composting to measure moisture and chemical levels in the soil. The only disadvantage of composting is the requirement of trained personnel while handling large-scale composting.

Recycling is yet another safer method to manage solid waste. Glass bottles, plastic containers, aluminium or tin cans, paper are some materials that can be recycled to have lower carbon footprint. One of the limitations of recycling involves expense factors. Reuse of electronic waste is a far better alternative than landfilling or burning.

21.7 BIOMEDICAL WASTE

The solid waste generated from hospitals, dispensaries, clinics, veterinary houses, and pathology laboratories are termed as *biomedical waste*. Biomedical waste contains pathogenic microorganisms and can lead to serious infections if they are disposed directly in the environment.

Types   Human and animal waste, used syringes (or scalpels, forcep, scissors), used bandages, blood-soaked cotton swabs, body fluids, discarded glassware, biotechnological wastes (tissue cultures, engineered organisms, etc.) are some common biomedical wastes.

Characteristics of biomedical waste   Biomedical waste or BMW can be categorized as non-infectious, infectious, hazardous, or cytotoxic.

Non-infectious waste includes solid waste materials that are similar to kitchen or household waste such as edibles, used wrappers, food items, etc.

Infectious waste includes all pathological, surgical waste like used syringes, intravenous sets, blood-soaked cotton swabs, used plasters, bandages, ampules, etc. that pose dangers of spreading infection.

Cytotoxic or hazardous includes chemical waste (disinfectants, insecticides), outdated medicines, microbial cultures, biotechnological waste.

Disposal methods   BMW are usually segregated, collected, stored, disinfected, and disposed. According to Biomedical Waste (Management and Handling) Rules, 1998, various categories of biomedical waste and their safe disposal techniques have been drafted. Some of the methods of BMW disposal are as follows:
 (a) People handling and transporting biomedical waste should wear protective gears such as face masks, gloves, aprons.

(b) Segregation involves bar-coded separation of different types of biomedical waste. Human waste, animal waste, and microbial waste are placed in yellow-coded bins. Soiled dressings, microbial cultures, used bandages, blood-soaked cotton swabs, etc. are disinfected and placed in red-coded bins.

(c) Used plastics, ampules, and sharp objects such as used blades, catheters, forceps, and scissors are disinfected and placed in blue-coded bins. Discarded medicines, chemical waste, incinerated ash (obtained from incinerating solid waste) are placed in black-coded bins.

(d) The bins are properly sealed before transporting to disposal sites. The vehicles carrying such hazardous waste should symbolize them as 'Biohazard' or 'Cytotoxic.'

(e) Autoclave and microwave treatment are carried out for some types of solid waste which are placed in black-coded bins. Deep burial is done for most of the biomedical waste after disinfecting the solid waste in rural areas, though this method causes soil pollution.

21.8 Electronic Waste Management

Electronic waste or e-waste is a category of surplus, obsolete, broken, or discarded electrical or electronic devices. It includes all secondary computers, entertainment device electronics, mobile phones, and other items such as television sets and refrigerators, whether sold, donated, or discarded by their original owners destined for reuse, resale, recycling, or disposal.

Most electronic gadgets after their useful life are harder to upgrade, easy to break, and impractical to repair. E-waste has resulted in global toxic emergency. Silicon Valley is the most poisoned community in the US. An IBM report stated that their workers involved in the manufacture of silicon chips had 40 % chances of miscarriages and serious forms of cancer*.

The processing of electronic waste in developing countries causes serious health and pollution problems due to the fact that electronic equipment contains serious contaminants such as lead, beryllium, cadmium, and brominated flame retardants. Even in developed countries recycling and disposal of e-waste involves significant risk. Most of the electronic gadgets are 'designed for the dump.'

Electronic gadgets/products have become obsolete due to the following reasons.

(a) Latest advancement in existing technologies; for example, mobiles phones replaced pagers within a year or two.

(b) Changes in design, style, and status; for example, advanced versions of cell phones are regularly and rapidly replacing existing handsets.

(c) The gadgets have exhausted their useful life.

21.8.1 Classification and Sources of E-waste

E-waste encompasses an ever-growing range of obsolete electronic devices such as the following.

(a) Household appliances: Washing machines, dryers, refrigerators, air-conditioners, vacuum cleaners, coffee machines, irons, toasters

(b) Office, information and communication equipment: PCs, laptops, mobiles, telephones, fax machines, copiers, printers

(c) Entertainment and consumer electronics: Televisions, VCR/DVD/CD players, Hi-Fi sets, radio, DVDs, CDs, floppies, tapes

(d) Lighting equipment: Fluorescent tubes, sodium lamps; (except: bulbs and halogen bulbs)

(e) Electric and electronic tools: Drills, electric saws, sewing machines, lawn mowers

(f) Toys, leisure, sports and recreational equipment: Electric train sets, coin slot machines, treadmills

---

* Clapp, Richard, 'Mortality among US employees of a large computer manufacturing company: 1969–2001'. *Environmental Health*, Springer, 2006: https://ehjournal.biomedcentral.com/articles/10.1186/1476-069X-5-30

Table 21.3 Hazardous substances found in e-waste and their health effects

Substances in e-waste	Source	Hazards
Arsenic	LEDs	Chronic exposure can lead to skin diseases, lung cancer
Barium	Cathode ray tubes (CRTs)	Short-term exposure can lead to brain swelling, muscle weakness, heart damage
Cadmium	Ni–Cd batteries, printer inks	Lung cancer, kidney damage, bone disorder (osteoporosis)
Lead	CRT screens, batteries, printer boards	Appetite loss, constipation, fatigue, persistent headaches
Mercury	Alkaline batteries	Brain and liver damage, if ingested or inhaled
Chromium VI	CDs, data tapes	DNA damage, permanent eye injury
Tetrabromobisphenol A (TBBA)	Fire retardants in plastics	Severe hormonal disorder
PVC	Cable insulations	Skin disorders
Americium (radioactive)	Medical devices, smoke detectors	Poisoning, acute respiratory distress leading to death
Dioxins	Burning of e-waste, plastics	Hormonal disorders
CFCs	Electrical equipment	Skin cancer, genetic damage
Beryllium	Batteries, cell phones	Classified human carcinogen-chronic beryllium disease affecting the lungs
Selenium	Batteries, lighting equipment	Selenosis - neurological abnormalities

21.8.2 Hazardous E-waste Disposal Techniques

Incineration, open fire burning, and landfills are commonly employed as processing techniques in large as well as small scale. But these techniques pose the problem of environmental pollution in surrounding areas. It simultaneously affects human beings, animals, and the natural vegetation cover. The following is a discussion on the health and environmental problems arising due to the application of these techniques.

Incineration   It is a hazardous disposal technique wherein onsite workers burn the bulk of e-waste in a closed incinerator at temperatures of 800 °C–1000 °C.

Incineration is associated with the major risk of releasing contaminants and toxic substances in the environment because of the hazardous substances found in e-waste (Table 21.3). This is especially true for incineration without prior treatment. Much of the e-waste is metal–polymer composite with PVC as a common polymer. It is known that if polyvinyl chloride is burned, it forms toxic HCl fumes that are detrimental, particularly to the human respiratory tract. Incineration also leads to the loss of valuable trace metals which could have been recovered and reused, if they had been sorted and processed separately.

Open burning   It is a common and the cheapest method wherein on-field workers burn the surplus e-waste without any control on temperature. Such on-site workers, on chronic exposure to the resulting fumes face serious health conditions like respiratory tract infections, hormone disruption, and even serious forms of cancer. Generally, on burning e-waste, severe air pollution is observed due to release of hydrogen chloride, carbon monoxide, CO_2, SO_2, along with particulate matter (aerosols). Toxic gases and particulate matter give rise to serious health concerns to people residing nearby dump ground sites.

Land-filling   It involves breaking the e-waste into smaller parts and dumping in landfills or dumping grounds. Over a period of time the toxic metals (mercury and other transition metals) begin to leach and pollute the soil environment, along with the ground water. For example, when circuit boards are destroyed, mercury leaches into the soil, and can potentially bioaccumulate in living beings.

21.8.3 Eco-friendly Methods to Manage E-Waste

It is estimated that about 75 % of electronic waste is stored due to uncertainty of how to manage it. These electronic junks lie unattended in houses, offices, warehouses, etc., and are finally disposed at landfills, usually mixed with household wastes. This necessitates implementing state-of-the-art recycling techniques.

Detoxication

(a) Electronic waste processing usually first involves dismantling the equipment into various parts (metal frames, power supplies, circuit boards, plastics).
(b) In this process, critical components are removed from the e-waste in order to avoid dilution and (or) contamination of these materials with toxic substances during the downstream processes.
(c) Critical components include, for example lead glass from CRT screens, CFC gases from refrigerators, light bulbs, and batteries.

The **advantage** of this process is the human being's ability to recognize and save working and repairable parts, including chips, transistors, RAM, and so on.

The **disadvantage** of the process is that labour is cheapest in countries with the lowest health and safety standards.

Shredding

Mechanical processing or shredding is the next step in e-waste treatment.

(a) It is normally an industrial large-scale operation involving sophisticated mechanical separator, with screening and granulating machines to separate constituent metal and plastic fractions. These are then sold to smelters or plastics recyclers.
(b) Shredding is done to obtain concentrates of recyclable materials and also to further separate hazardous materials.
(c) Typical components of a mechanical processing plant are: (i) crushing units and (ii) shredders.
(d) Magnets and eddy currents are employed to separate glass, plastic, and ferrous and non-ferrous metals, which can then be further separated at a smelter.
(e) Hazardous smoke and gases are captured, contained, and treated to mitigate environmental threat.

Refining

Refining of resources in e-waste is possible with technologies to get back the raw material with minimal environmental impact. Most of the fractions are refined or conditioned in order to be sold as secondary raw materials or to be disposed of in a final disposal site.

For example, leaded glass from CRTs is reused in car batteries, ammunition, and lead wheel weights, or sold to foundries as a fluxing agent in processing raw lead ore. Copper, gold, palladium, silver, and tin are valuable metals sold to smelters for recycling.

An ideal electronic waste recycling plant combines dismantling for component recovery with increased cost-effective processing of bulk electronic waste.

A growing trend in electronic waste management is *reuse*. Reuse is preferable to recycling because it extends the lifespan of a device. Devices still need eventual recycling, but by allowing others to purchase used electronics, recycling can be postponed and value can be gained from device use. Making electronic company manufacturers to deal with e-waste is called 'extended producer responsibility' or 'product takeback'. There is a need for turning the company strategy from 'designed for the dump' to 'designed to last.'

Check Your Progress

12. What is e-waste?
13. Name some hazardous substances found in e-waste and mention their health effects.
14. Name some of the methods to deal with hazardous solid waste.
15. Name some eco-friendly ways to deal with e-waste.
16. Name two important sources of biomedical waste.
17. List the various categories of biomedical waste.

21.9 WATER POLLUTION

Water pollution is defined as alteration in the physical, chemical, and biological characteristics of waste due to discharge of foreign matter making the water unsuitable of its intended purpose. These foreign materials that are discharged in to the natural water bodies (lakes, rivers, streams, oceans) are called effluents. Water pollution is primarily caused due to the following.

Natural processes The decomposed plant and animal remains, if discharged directly in to the water, contaminate it with organic matter which adversely affects the water characteristics.

Anthropogenic processes Various activities such as agriculture, industrial, urban, domestic, mining, thermal, and radioactivity results in serious forms of water pollution.

Sewage Domestic sewage comprises kitchen waste, soaps, detergents, excreta, which if released in the water through drainage system will have detrimental impact on human health.

Fig. 21.5 Polluted bed of Ganges river (Courtesy: Sayanta Mukherjee: https://www.publicdomain pictures.net/en/view-image.php?image=27815& picture=pollution)

21.9.1 Sewage and Its Treatment

Sewage is liquid waste from the community which is of foul nature that comprises latrine discharges, sullage, industrial waste, hospital waste, and even sewage left behind after natural calamities. Waste water is a combination of liquid and solid wastes from residences, commercial units, industries, etc. Such waters if released directly in the environment will cause adverse health effects both to humans and animal life. Due to the complex nature of waste water, just one treatment method will not suffice. The waste water treatment usually involves physico-chemical and biological methods to reduce the pollutants in waste water.

Some of the major objectives of sewage treatment are:

(a) Preventing adverse effects on the environment by making sewage water less offensive.

(b) Avoiding dangers of contamination of ground water by preventing disposal as landfills.

Waste water treatment is accomplished in the following three stages.

Primary or Mechanical Treatment

In this method, large, suspended, and coarse solid materials are screened, coagulated, and allowed to sediment for easy removal. Sewage is allowed to sediment at the bottom that comprises 60 % of suspended matter. During primary treatment about 35 % BOD–COD, 60 % TSS (Total Suspended Solids) and about 30 % of nitrogenous matter are reduced.

Secondary Treatment

Trickling filters and activated sludge processes are employed to reduce the biological contaminants in waste water.

Trickling filters These are either circular or rectangular tanks about 1–3 m deep filled with coarse rocks, coal, broken bricks, gravel, sand, slag, etc. (40–150 mm size) that act as the filtration medium. The sewage water is allowed to flow over the filtration medium using a distributer. The waste water trickles downwards and allows microbial slime (biofilm) to grow covering the bed of filtration media. Air is allowed to pass through to maintain aerobic conditions. The filtered effluent is then sent to a clarifier where sludge is separated from the treated water. If the treated effluent (water) still has objectionable suspended matter, it is recycled using mechanical pumps to the filtration medium tank.

Activated sludge method In this process, primary effluent is allowed to enter the aeration tank where it comes in contact with sludge. The sludge is added as an inoculum that comprises various microorganisms that aerobically degrade the organic matter into nitrates, phosphates, carbon dioxide, and new microbial cells. This microbial flora grows and remains as a floc that is called 'activated sludge'. After the process the treated water is separated from the sludge in a sedimentation tank and sent for digestion. A small part of the sludge is recycled and used as an inoculum for fresh treatment of influent waste water. About 95 % BOD is removed using activated sludge process.

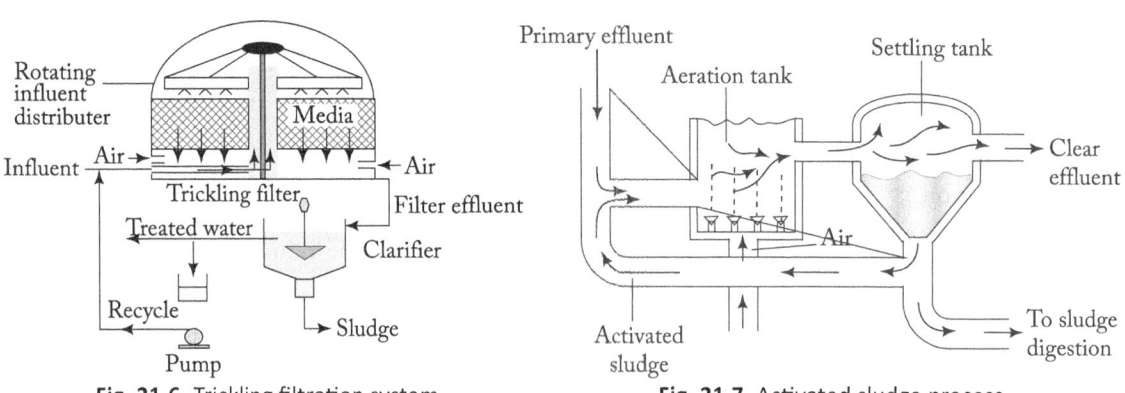

Fig. 21.6 Trickling filtration system   **Fig. 21.7** Activated sludge process

Tertiary Treatment

After biological treatment, the effluents free from sludge are sent for disinfection by chlorination before safe discharge in to the environment. Often, reverse osmosis coupled with ultrafiltration is utilized to further eliminate any microscopic or nanoscopic suspended matter from the effluent.

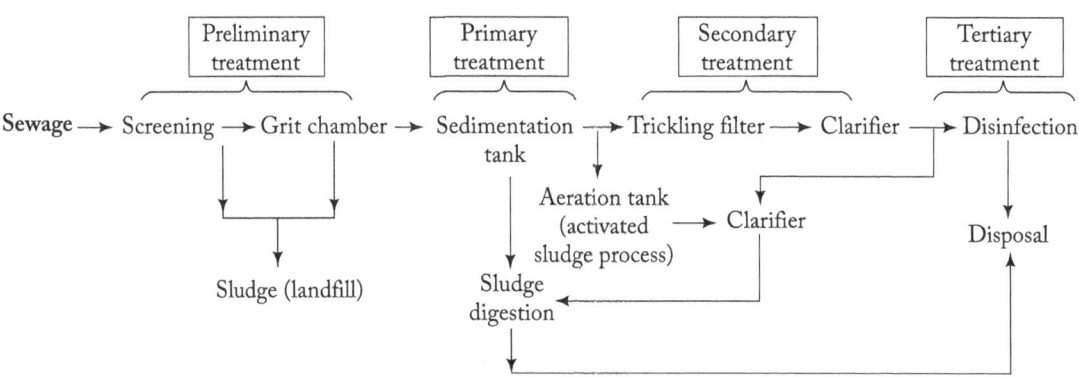

Fig. 21.8 Schematic showing municipal sewage treatment process

Check Your Progress

18. Define water pollution.
19. What is sewage and waste water?
20. Name the methods employed to reduce biological contaminants in waste water.

21.10 NOISE POLLUTION

The mere mention of the term 'noise' brings an image in our minds about the unwanted or disrupting sounds in crowded cities, religious ceremonies, vehicular traffic, electronic devices, and even the classroom where we study! When there is a regular exposure to elevated levels of noise that lead to adverse effects on the health of human beings and other living organisms, it is referred to as *noise pollution*.

Sound is an acoustical energy that is released in the atmosphere by moving or vibrating bodies. Usually, molecules vibrate in an oscillatory mode and energy travels through a medium in the form of vibrations. If oscillations in the medium range from 20 Hz to 20 KHz it is audible to human ears (sound), frequencies below 20 Hz are called sub-sonic, whereas those above 20 KHz are ultrasonic inaudible sounds. But, noise is an unwanted, unpleasant, disagreeable sound product.

Sources Transportation is a major contributor of noise pollution especially in cities and towns. Just second-in-line are noises generated during construction and related operations. Sadly, noise pollution is visualized as only a 'nuisance' rather than a serious 'environmental hazard.' Continued exposure to undesirable and loud noises can have severe psychological effects, such as mental stress, fatigue, and even cardiac complications.

Noise pollution primarily arises due to (a) vehicular noise, (b) industrial noise, and (c) noise from the neighbourhood.

Sound intensity is measured in decibels (dB) which is a logarithmic scale and if an increase in sound is 3 dB, it means doubling of sound volume. Decibel is the tenth part of the longest unit 'Bel' named after A. Graham Bell. To put in simpler context, one dB is equal to the faintest sound audible to the human ear.

The maximum level of noise is below 80 dB beyond which it is considered as a pollutant. According to WHO, 45 dB is the optimum level of noise in a busy city. Table 21.3 lists some typical sound levels of various sources.

Table 21.3 Sound levels of various sources

Source	Decibel (dB)
Normal hearing threshold	0
Empty auditorium, breathing	10
Library	30
Polite conversations	50
Office	55
Vehicular traffic	70
Printing press	80
Industry or factory	90

Adverse effects The question arises that how do continued exposures to noise manifest in health complications. We can categorize the adverse effects of noise as: (a) auditory effects and (b) non-auditory effects.

Auditory effects refer to whistling, buzzing, or ringing-type sounds in the ear usually occuring around 90 dB which results in auditory fatigue. Chronic exposure to high pitched sounds beyond 90 dB may result in temporary (40–60 dB) or permanent (100 dB and beyond) deafness. *Non-auditory effects* refer to adverse health effects to other than human ear, such as high blood pressure, sleep disorder, emotional stress, and hormonal imbalances (increasing adrenaline due to noise related to stress). The effects of noise adversely affect animal and wildlife too. Animals tend to feel threatened or get frightened in the presence of noise and may affect their mode of communication.

Blue whales are known to use sonar for communication, mating, and hunting which are severely hindered due to increasing noise levels.

Control It is not possible to completely eliminate noise from our surroundings; however several measures can be adopted to alleviate the adverse effects of noise. Some of the salient control measures are as follows.

(a) Use of silencing devices at homes, offices, malls, engines, automobiles, machinery, kitchen appliances may assist in controlling unwanted noise.

(b) Use of ear-aids, noise-cancelling headsets, etc., may also help in reducing undesirable health effects, though for shorter periods of time.

(c) Transmission control: Various sound absorbing materials such as acoustic tiles, doors, windows help in reducing noise to reach our ears. In automobiles, muffler acts as an acoustic material present in the exhaust system of the car by attenuating (or dampen) the noise output from the engine.

(d) Laws and regulations: In 2017, Supreme Court of India had ordered a complete ban on using loudspeakers during festivals; however, this law needs to be strengthened across the country. Educating the public through campaigns, lectures, and streetshows is the need of the hour to create awareness on the issues of noise pollution.

21.11 RADIATION POLLUTION

Radioactive pollution refers to an increase in the natural radiation levels due to human activities that can lead to serious environmental problems. Post the discovery of artificial radioactivity, nuclear weapons, installed nuclear reactors, radiation pollution (or nuclear pollution) is a concern today as it poses a high threat to human health and environment. We know nuclear energy is a cleaner source of energy; yet it faces criticism for its use in power generation due to associated hazards and risks. In 2011, the world witnessed Fukushima nuclear reactor leak during a tsunami in Japan thereby raising concerns on installations of nuclear reactors. Japan, by far is the most seismologically studied nations and is known for nuclear hazards faced by their population. The optimum threshold of radiation dose is 1 Gray (or 1 Sievert) above which there is potential hazard to the environment and human health.

Types of radiation There are many natural radiations such as the sun's rays, cosmic rays (galaxy), radon-222 in soil (traces), C-40, U235, and K-40 (in lithosphere). These radiations are not harmful as they are present in trace amounts and omnipresent referred to as background radiation. Alpha, beta, and gamma ionizing radiations are commonly an outcome of radioactivity. Alpha rays are blocked by paper and skin, β rays penetrate through skin but can be blocked by glass or metal, whereas γ rays can completely penetrate human cells and cause damage which can only be blocked by massive concrete pieces.

Sources The sources of radiation pollution, apart from natural sources, are anthropogenic such as:
  (a) Use of X-rays, gamma-rays in diagnostic medical applications
  (b) Nuclear tests and controlled explosions on land sites
  (c) Nuclear reactors and power plants
  (d) Disposal of nuclear waste
  (e) Mining of isotopes such as uranium, thorium, and so on

Adverse effects Chronic exposure to ionizing radiations may have genetic or non-genetic effects on the human body. Genes and chromosomes may undergo mutations which may manifest in foetus and result in deformations in offsprings. Leukaemia, tumours, cancer, miscarriages, fertility problems are serious non-genetic effects observed in human beings.

Control measures The following measures can be adopted to reduce the effects due to nuclear radiations:
  (a) Nuclear tests or explosions should not be carried out in open atmosphere.
  (b) Closed cycle coolant system should be used in nuclear reactors, so that radiation leakage does not occur through coolant.
  (c) Radioactive wastes generated by nuclear reactors or from nuclear weapons must be disposed off safely that allows natural decay without leaving any traces in the environment. Nuclear wastes should always be sealed in double-walled tanks to avoid any leakage.

(d) Nuclear fission reactions should be minimized as far as possible.

(e) People working in nuclear reactors should undergo medical diagnosis every three or six months to check the radiation levels in their body.

(f) Since 2003, Awaaz Foundation, an Indian-based NGO is working towards this serious form of pollution by propagating information through campaigns and legal advocacies. National Green Tribunal (2015) issued directives to ensure stringent adherence on noise pollution but these reforms need to be strengthened throughout the country.

21.12 SOIL (LAND) POLLUTION

Soil is a complex mixture comprising organic matter, inorganic matter, microorganisms, moisture, and air. When man-made activities degrade the land and its soil, it is referred to as soil pollution. The substances that can change the productivity of soil is called soil pollutants.

Sources   The various sources that lead to soil pollution are as follows.

(a) Poor agricultural practices such as improper tillage, harsh fertilizers and chemicals, etc.

(b) Digging and disposal of solid waste on land

(c) Mining activities

(d) Demolition of buildings and other solid structures

(e) Erosion of soil due to overgrazing, deforestation, etc.

(f) Heavy metals such as lead, copper, cadmium, iron, and zinc released from various industries

The pollutants reach the soil from the above sources in more than one way, some of which are listed here.

(a) Pollutants may get absorbed by soil and seep within the soil.

(b) The soil pollutants could be taken up by plants through the soil and enters the food chain.

(c) Pollutants may undergo chemical reactions with the components of the soil.

(d) Pollutants present in the soil may get broken down into various chemical species by microorganisms.

Adverse effects   Since pollutants enter soil in more than one way, there are many adverse effects.

(a) Soil erosion is the washing away of the soil that contains essential nutrients required for plants.

(b) The use of harmful pesticides such as polychlorinated biphenyls can cause lung disorders, nervous disorders, and even cancer.

(c) If soil pollutant enters the food chain, it can cause bioaccumulation of harmful chemicals in living organisms.

(d) The presence of heavy metals such as lead, mercury, cadmium in the soil can cause damage to brain development of growing children.

(e) Soil pollutants, if vaporized may cause air pollution by releasing volatile components in the atmosphere.

(f) Soils can get severely affected due to acid rains and pH imbalance of soil can harm the vegetation.

(g) If radioactive waste is dumped within the soil, it may contaminate the land resources, since the radionuclides remain active for prolonged periods of time.

Control measures   Some common methods to control soil pollution are as follows.

(a) Solid wastes such as garbage and debris, should not be disposed off on land. Proper segregation, collection, recycle, and reuse of solid waste are essential to control soil pollution. Solid waste such as plastics, metals, glasses, oils, and industrial effluents should be suitably treated before disposing in the environment.

(b) Improvements in agricultural practices such as proper tillage and using safer fertilizers.

(c) Nuclear explosions and improper dumping of radioactive waste should be banned.

(d) Informal and awareness campaigns to general public and strengthening laws to reduce soil pollution.

Check Your Progress

21. Define noise pollution.
22. What is noise? State the unit of measurement.
23. Define radiation pollution.
24. List the sources of radiation pollution.
25. List the adverse effects of radiation on human beings.
26. What is soil pollution?
27. List the adverse effects of soil pollution.

Activity-based Questions

A. Delhi Smog

Smog is fog mixed with smoke and other harmful pollutants rendering poor quality of air that also hinders visibility. On 8 November 2017, Delhi recorded the most severe pollution of 999 on Air Quality Index. The Air Quality Guideline released by WHO in September 2011 recommended PM_{10} as 20 µg/m3. However, it was found that Delhi's air quality was about ten times the WHO recommended value. Vehicular pollution is one of the worst situations that has hampered the air quality in present times.

Present a comprehensive case study on the impact of Delhi smog, its formation, chemical reactions, current situation, legal and public awareness regimes.

B. Is Banning Plastics a Solution?

Over the years, many states in India have imposed a ban on manufacture, usage, sale, transport of plastic bags and single-use plastics. It is already known that plastics do pose an environmental risk, the ban was imposed leading to penalizing the offenders. The ban particularly covers various plastic materials such as plastic bags and disposable plastic products (cups, spoons, plates). However, there are many plastic-based materials that are exempted from the ban, such as medicine packets, plastics used in manufacturing, plastics used for large-scale solid waste disposal, agricultural storage bags, and bags used during transportation.

Present a comprehensive review on the recent plastic ban, its consequences, effects, and reflect your views on the issue.

C. Initiatives in e-waste management—An Indian scenario

E-waste has been a subject of concern globally. Some notable initiatives are adopted in India and globally for recycling/reusing e-waste. E-Parisara, Saahas, Plug-in to eCycling, etc., are some initiatives taken up to deal with e-waste.

As per the gazette notification by Ministry of Environment and Forests (2011), e-waste management and handling rules were drafted that made it mandatory for e-waste recyclers to acquire authorization for safe disposal of surplus scrap. In Bengaluru, e-bins to ensure safe disposal of e-waste generated at government offices are installed. Saahas, a Bengaluru-based NGO involved in this pioneering effort, plans to hold campaigns in government offices to create awareness about e-waste and the need to dispose it safely.

Review various initiatives taken or launched in your locality and their outreach in the society in dealing with e-waste recycle. Perform a field-survey on e-waste recyclers in your locality and waste management methods adopted. Suggest remedial measures to recycle or reuse of electronic waste.

SUMMARY

○ Environmental pollution can manifest in many forms like air, water, land, thermal, nuclear, and others.

○ Air pollution is among the most serious and worst forms of pollution that involves contamination of air by various impurities such as aerosols, particulate matter, oxides of sulphur, nitrogen, and carbon, hydrocarbons, to name a few.

- Pollutants are foreign substances that contaminate the environment and cause many ill-effects on human health and environment. Carbon monoxide, nitrogen oxides, carbon dioxide, methane and sulphur dioxide are the major pollutants of air.
- Greenhouse effect and global warming are pressing challenges today. Acid rains, ozone depletion, smog formation are various manifestations of air pollution.
- Water pollution is the contamination of water by harmful substances which if exceeds optimum levels may cause adverse effects on living organisms. Municipal sewage is treated using trickling filters and activated sludge methods before reuse.
- Solid waste can be generated as garbage dump, debris, e-waste, biomedical waste, industrial, and agricultural waste.
- Open burning, incineration, and landfills are hazardous methods to dispose solid waste. Dismantling,

segregation, detoxifying, shredding, composting, recycling are some of the environmentally friendly methods to manage solid waste.
- Noise pollution is any unwanted or undesirable sound that on chronic exposure may hinder hearing ability of man besides other complications.
- Radiation can be both beneficial as well as hazardous to mankind. If atmosphere gets contaminated with radiations, there will be hazardous effects on the environment. Fertility problems, genetic mutations, cancer are a few serious health effects radiation has on mankind.
- Managing e-waste in a greener way is essential to achieve the least environment impact. Hazardous substances in e-waste are known to cause serious health effects depending on the period of exposure, ranging from short-term to chronic exposures.

GLOSSARY

Acid rain: The pH of rainwater lowers to about 4.5 causing damage to buildings and marine life.

Air pollution: The introduction of chemicals, particulate matter (PM), or biological matter that causes harm to human health, plants, and animals and deteriorates the natural atmosphere.

Biohazard: Biological substances that pose health threats to living organisms, particularly human beings.

Effluent: The liquid discharge from various sources like industries and households that carries harmful pollutants.

Electronic waste: The term used to refer to surplus, obsolete, broken, or discarded end-of-the-lifespan electrical or electronic devices.

Environment: The surroundings in which an organization operates, including air, water, land, nature resources, flora, fauna, and humans.

Electronic waste: The term used to refer to surplus, obsolete, broken, or discarded end-of-the-lifespan electrical or electronic devices.

Environmental engineering: The branch which focuses on environment protection by reducing waste and pollution.

Inoculum: A small portion of pure culture of microorganisms added in a suitable medium to grow a new microbial culture.

Lachrymator: A substance that causes eye irritation and tear secretion.

Lithosphere: The outermost part of the earth comprising the crust and mantle.

Ozone depleting substances: Chlorofluorocarbons (CFCs), hydro-chlorofluorocarbons (HCFCs), $CHCl_3$, CCl_4 released in the atmosphere that depletes the natural ozone layer.

$PM_{2.5}$: Particulate matter released in atmosphere having diameters less than 2.5 μ in size.

PM_{10}: Particulate matter that have diameters less than 10 μ in size.

Pollution: The contamination of the environment with materials that interfere with human health, quality of life, or the natural functioning of the ecosystems.

Pollutant: The undesirable foreign matter which is released in the environment that contaminates air, water, and land.

Sewage: The liquid waste from the community which is of foul nature and comprises household wastes, sullage, industrial discharge, and hospital waste.

Smog: The atmospheric product of photochemical processes induced by pollutants in air that reduces visibility and hampers air quality.

Stratosphere: The layer of earth's atmosphere lying between the troposphere and mesosphere.

EXERCISES

Multiple Choice Questions

1. Which of the following is not a source of air pollution?
   (a) Automobile exhaust
   (b) Windmill
   (c) Burning of firewood
   (d) Power plant

2. The type of pollution which has affected Taj Mahal in Agra to a greater extent is
   (a) air pollution   (b) soil pollution
   (c) water pollution   (d) noise pollution

3. Which of the following is a major greenhouse gas?
   (a) nitrogen gas   (b) methane gas
   (c) water vapour   (d) carbon dioxide

4. Formation of ozone is a/an
   (a) oxidation reaction
   (b) reduction reaction
   (c) photochemical reaction
   (d) none of these

5. The thickness of ozone layer is measured as a unit of
   (a) Debye   (b) Dobson
   (c) ppm   (d) milligram

6. Noise is measured using sound meter and the unit is
   (a) Hertz   (b) Decibel
   (c) Joule   (d) Sound

7. Carbon monoxide is a serious pollutant because
   (a) it reacts with oxygen and hydrocarbons.
   (b) it inhibits glycolysis.
   (c) it causes nervous system disorders.
   (d) it reacts with hemoglobin.

8. The main component/s of photochemical smog is
   (a) water vapour   (b) sulphur dioxide
   (c) oxides of nitrogen   (d) all of these

9. Sound becomes hazardous noise pollution at decibels
   (a) above 30   (b) above 80
   (c) above 100   (d) above 120

10. Ozone layer absorbs
    (a) X-rays   (b) α-rays
    (c) UV rays   (d) IR rays

11. Which one of the following is classified as solid waste?
    (a) e-waste
    (b) methyl chloroform
    (c) carbon dioxide
    (d) hydrocarbons

12. Fukushima disaster is an example of
    (a) air pollution
    (b) radiation pollution
    (c) water pollution
    (d) noise pollution

13. An environment-friendly way of managing solid waste is
    (a) composting   (b) open burning
    (c) incineration   (d) landfill

14. The optimum radiation threshold is
    (a) 2 Sievert   (b) 12 Sievert
    (c) 1 Sievert   (d) 0.53 Sievert

15. Trickling filters is employed for
    (a) air treatment   (b) sludge treatment
    (c) noise abetment   (d) radiation control

16. Which of the following substances found in e-waste have no reported health hazard?
    (a) Cd   (b) Dioxins
    (c) Be   (d) Li

17. On burning PVC, _____ is formed.
    (a) HCl   (b) vinyl compounds
    (c) chlorine   (d) CO

18. The best method to deal with e-waste is
    (a) recycle and reuse
    (b) incinerate and burn
    (c) disposal in land
    (d) none of the above.

19. Photochemical smog may cause
    (a) reduced visibility
    (b) eye irritation
    (c) respiratory problems
    (d) all of the above.

20. The best method to safely dispose biomedical waste is
    (a) landfill   (b) burning
    (c) autoclave   (d) roasting

21. The scientist who coined the term 'greenhouse effect' is
    - (a) J. Fourier
    - (b) L. Fourier
    - (c) Robert A. Smith
    - (d) Dobson
22. COHb stands for
    - (a) carboxymonoglobin
    - (b) carboxyhaemoglobin
    - (c) carbon monoxide
    - (d) haemoglobin
23. Soil pollutant may enter living organisms through
    - (a) food chain
    - (b) bioaccumulation
    - (c) through air
    - (d) all of the above
24. Of the following, the one that is NOT a particulate matter is
    - (a) soot
    - (b) fly ash
    - (c) plastic
    - (d) dust
25. Alteration in the physical, chemical, and biological characteristics of waste due to discharge of foreign matter is called
    - (a) water pollution
    - (b) air pollution
    - (c) soil pollution
    - (d) radiation pollution
26. The approximate % of BOD removal in an activated sludge method is
    - (a) 96
    - (b) 95
    - (c) 90
    - (d) 89
27. Of the following, the secondary pollutant is
    - (a) PAN
    - (b) sulphur dioxide
    - (c) carbon monoxide
    - (d) None of these
28. Primary pollutant is
    - (a) ozone
    - (b) PAN
    - (c) SO_2
    - (d) dust
29. The one that is NOT a source of solid waste is
    - (a) garbage
    - (b) electronic items
    - (c) waste water
    - (d) hospital waste
30. Of the following, the one that is NOT a source of radioactive pollution
    - (a) nuclear reactors
    - (b) nuclear waste
    - (c) sewage
    - (d) mining of isotopes

Review Questions

1. Define environmental pollution and list the various causes of pollution.
2. Define air pollution and state the various sources of air pollutants.
3. Classify various air pollutants and mention their health effects.
4. Write a short note on oxides of sulphur, nitrogen, and carbon.
5. Give an informative account on 'acid rains.'
6. Explain greenhouse effect and write a note on green house gases.
7. Explain measures to control carbon monoxide in air.
8. Explain, 'Smoking of cigarettes have detrimental effects on human health.
9. What are the reactions of nitrogen oxides in the atmosphere? State the various methods to control them.
10. Comment on the serious forms of pollution and suggest various control measures.
11. Justify the statement, 'Ozone depletion occurs due to chloroflurocarbons.'
12. What is smog? Explain London and Photochemical smog formation with chemical reactions.
13. What is photochemical smog? Explain the mechanism of smog formation and its health effects.
14. Explain 'acid rain' and its adverse effects on the environment.
15. What is acid rain? State the adverse effects of acid rain with suitable chemical reactions.
16. 'Particulate matter is an indicator of air quality,' Justify.
17. What is particulate matter? Why is it harmful? Describe the Cottrell electrostatic precipitator for removing them.
18. Describe removal of particulate matter using Cyclone separator.
19. What is sewage water? With help of a schematic diagram, explain the various steps involved in sewage water treatment.
20. How do trickling filters and activated sludge methods help in sewage water treatment?
21. Why is it challenging to manage biomedical waste? Cite relevant examples.
22. 'Recycle and Reuse are environmental-friendly methods for managing solid waste'- Justify.
23. Explain the categories of biomedical waste and methods to dispose them.

24. Suggest various eco-friendly methods to manage solid waste.

25. Define e-waste. Describe the hazardous e-waste disposal methods with a note on their demerits.

26. How is e-waste classified? List the various hazardous substances found in e-waste and their health effects.

27. Explain the various harmful methods adopted to deal with e-waste. Mention their disadvantages with suitable examples.

28. Detail an account of general e-waste handling methods. Suggest greener methods to deal with e-waste citing suitable examples.

29. Describe the eco-friendly methods to manage e-waste effectively. State their advantages.

30. Discuss noise pollution. Add a note on its adverse effects and control measures.

31. What is radioactive pollution? How can it be prevented?

32. Write a note on 'control of radiation pollution.'

33. Comment on the control measures for dealing with nuclear waste.

34. Explain the sources, adverse effects and control measures of soil pollution.

35. Discuss soil pollution and mention their health effects and methods to control them.

36. Write a short note on soil pollution.

FURTHER READING

1. Bharucha, Erach, *Textbook of Environmental Studies for Undergraduate Courses for UGC*, Universities Press, 2005.

2. Clapp, Richard, 'Mortality among US employees of a large computer manufacturing company: 1969–2001'. *Environmental Health*, Springer, 2006: https://ehjournal.biomedcentral.com/articles/10.1186/1476-069X-5-30

3. Duarte, Armando C., Anabela Cachada, et al, *Soil Pollution: From Monitoring to Remediation*, Academic Press, 2018.

4. Fourier, J. (1822). Analytical Theory of Heat, Paris: Cambridge University Press, 2010. (This iconic paper was translated in 1878 and finally republished as a book in 2010. https://www.cambridge.org/core/books/analytical-theory-of-heat/F6D4802336FABD1116DDA4AA3FE6EFAA

5. Guangyin, Z. and Z. Youcai, *Pollution Control and Resource Recovery–Sewage Sludge*, BH-Elsevier, US, 2017.

6. Joshi, P., 'Carbon Dioxide Utilization: A Comprehensive Review', *International Journal of Chemical Sciences*, 12(4), 2014, pp. 1208–1220.

7. Sharma, B.K, *Environmental Chemistry*, Goel Publishing House, 2007.

ANSWERS

Check Your Progress

1. The surroundings in which an organization operates, including air, water, land, nature resources, flora, fauna, and humans is called environment.

2. Refer to Sec. 21.1

3. Refer to Sec. 21.2

4. Refer to Sec. 21.2.1

5. Refer to Sec. 21.2.1

6. Thermal power stations, roasting sulphide ores, sulphuric acid industries are various sources of oxides of sulphur.

7. Incomplete combustion of fossil fuels is the major source of carbon monoxide in air.

8. Refer to Sec. 21.5

9. Refer to Sec. 21.5.1

10. PAN (peroxyacyl nitrate) is the major air pollutant in photochemical smog.

11. Cyclone separator and Cottrell precipitator are two methods employed to reduce particulate matter in air.

12. The category of surplus, obsolete, broken, or discarded electrical or electronic devices constitute e-waste.

13. Refer to Table 21.3

14. Incineration, open fire burning, and landfills are some of the methods to deal with hazardous solid waste.
15. Detoxication, shredding, refining, recycling, and reuse are some eco-friendly ways to deal with e-waste.
16. Hospitals and veterinary houses are important sources of biomedical waste.
17. Biomedical wastes can be non-infectious, infectious, hazardous, or cytotoxic.
18. Refer to Sec. 21.9
19. Refer to Sec. 21.9.1
20. Trickling filters and activated sludge processes are employed to reduce the biological contaminants.
21. Refer to Sec. 21.10
22. Noise is any undesirable and unwanted sound. Sound intensity is measured in decibels (dB).
23. Refer to Sec. 21.11
24. Refer to Sec. 21.11
25. Leukaemia, tumours, cancer, miscarriages, fertility problems are serious non-genetic effects observed in human beings.
26. When man-made activities degrade the land and its soil, it is referred to as soil pollution.
27. Refer to Sec. 21.12

Multiple Choice Questions

1. (b)	2. (a)	3. (d)	4. (c)	5. (b)	6. (b)	7. (d)
8. (c)	9. (b)	10. (c)	11. (a)	12. (b)	13. (a)	14. (c)
15. (b)	16. (d)	17. (a)	18. (a)	19. (d)	20. (c)	21. (a)
22. (b)	23. (d)	24. (c)	25. (a)	26. (b)	27. (a)	28. (d)
29. (c)	30. (c)					

Green Chemistry

After reading this chapter, you will be able to:

- appreciate the economic and environmental significance of green chemistry.
- understand the twelve principles of green chemistry.
- elucidate the chemical synthesis of adipic acid, indigo, ibuprofen, carbaryl, and acrylamide.
- apply greener approaches to bring out the correlation of the topic with engineering field through case studies provided at the end of the chapter.

22.1 INTRODUCTION

The term 'Green Chemistry' was coined by Paul T. Anastas (1991) in a special programme launched by US Environmental Protection Agency (USEPA). Today, this term is universally accepted to describe the environmentally friendly chemical processes and products.

The application of greener and eco-friendly methods pertains mainly to the chemical engineering field. Chemical engineers are primarily involved in understanding the technological facets of a chemical reaction. Mass and heat transfer, kinetics of chemical reactions, production of chemical products, and design of chemical processes are some of the common areas dealt with in chemical engineering.

Green chemistry can be defined as the design of chemical products and processes that are environmentally benign and reduce the negative impact to human health and environment. It incorporates a new approach to the synthesis, processing, and application of chemical substances in such a manner so as to reduce threats to human health and the environment. This new approach is also known as *environmentally benign chemistry* or *clean chemistry* or *sustainable chemistry*.

Most of the organic chemical reactions involve benzene or its derivatives as raw materials. The major by-product of these reactions is found to be hydrochloric acid. Benzene and its analogues are known to be carcinogenic compounds. Hence, it is essential to redesign such reactions and also find methods in which hydrochloric acid can be reused as a solvent. Today, chemists are encouraged to adopt processes that lead to environmentally benign synthesis of new products and develop greener methods of synthesis of existing chemicals by replacing hazardous organic solvents. Chemical reactions and processes described in most textbooks do not give information on mass–energy inputs and outputs for a given organic reaction. Chemists generally rely upon past experiences and trial and error methods to set up viable and economic chemical processes and technologies.

Conceptually, green chemistry is a sustainable science. It costs less in strictly economic terms with minimal waste discharge and zero environmental impact. By efficiently using low-cost and non-toxic compounds in lesser amounts, green chemistry is sustainable with respect to starting raw materials. Moreover, by reducing the volume of effluent generation, green chemistry is sustainable with respect to waste.

Thus, the main objective of green chemistry is to replace toxic, existing organic reactions with greener, energy-efficient reactions. With the ability to effect changes in the already existing reactions, first comes the responsibility to ensure that new materials, processes, and reaction conditions have the least environmental impact, reducing risk and hazard to the chemist, better utilization of resources, lower costs on waste management, and no compromise on quality and yield of the products.

22.2 Need and Significance of Green Chemistry

Due to stringent laws regarding chemical effluent release and rising pollution concerns, much emphasis is laid on reframing the chemical processes so as to have minimum environmental impact. In view of this, industries are adopting methods to formulate greener chemical processes with minimal discharge of effluents. Further, steps need to be taken to introduce green chemistry as an integral part of the curriculum in schools, colleges, and universities. Incorporation of low-cost, innocuous raw materials, minimal solvents, and energy efficient processes are some of the major significant factors of green chemistry. The major outcome of greener processes is that overall costs incurred to manage effluent treatment will be reduced.

The significance of green chemistry can be depicted by R4M4. Theory that implies Reduce, Reuse, Recycle, Redesign, Multipurpose, Multidimensional, Multifaceting, and Multitracking for overall chemical process. Apart from redesigning chemical reactions, there have been developments even in apparatus design. Conventional burettes used for volumetric analysis can be replaced by semi-micro econoburette that has an in-built pipette and a conical flask. Econoburette is an economical, greener alternative apparatus used to analyse microlitre volume of chemicals. Similarly, survismeter is a one-for-all apparatus to determine the primary parameters of a liquid, such as its surface tension and viscosity. Survismeter can easily replace redundant devices like stalagmometers and viscometers. Both econoburette and survismeter greatly reduce the laboratory space, electricity, chemicals, and even water to a great extent.

To achieve the overall objective of green chemistry, life cycle assessment (LCA) is done to adjudge the environmental impact of the green chemical reaction right from the initial production stages till the final disposal. This is referred to as the *cradle-to-grave approach*.

22.3 Principles of Green Chemistry

Green chemistry is presented as a set of twelve principles proposed by Paul T. Anastas and John Warner. The principles comprise instructions for professional chemists to implement new chemical compounds, newer syntheses, and new technological processes. In a nutshell, the first principle describes the general idea of waste management. The remaining principles are focussed on issues like redesigning chemical reactions, atom efficiency, toxicity, solvent and other media, consumption of energy, application of raw materials from renewable sources, and degradation of chemical products to simple, non-toxic substances that are friendly for the environment.

1. Prevention of Waste

It is better to prevent waste than to treat or clean up waste after it has been created.

Industries are known to release a huge amount of effluents into the environment. The treatment methods and disposal of effluent waste add to the overall production cost. Hence, it is plausible to carry out the chemical synthesis by designing organic reaction pathway in such a way that formation of waste or by-products is minimized. It is ideal to say in this case that 'prevention is better than cure.' This is also a viable option from an economic point of view.

For example, zero liquid discharge or minimal liquid discharge is an emerging trend in industrial waste water management. It deals with the concept of treating industrial effluents, and essentially involves the recycle and reuse of waste water. As per the second law of thermodynamics, conversion of thermal energy to useful work occurs along with the release of some amount of energy. Hence, it is not possible to have zero discharge, but minimal discharge of effluent serves the purpose of this green chemistry principle.

2. Atom Economy

Synthetic methods should be designed to maximize the incorporation of all materials used in the process into the final product.

In most of the organic reactions, along with the desired products, a number of by-products are formed which have limited utility. Any by-product of a chemical reaction for which there is no profitable use is a 'waste'. Green chemistry requires that new processes should be devised such that maximum amount of the starting material is converted into the desired product.

The efficiency of a conventional organic reaction can be determined in terms of product yield. Organic reactions that have more than 90 % practical product yield are regarded as successful methods. The percentage yield is calculated as follows,

$$\text{Percentage yield} = \frac{\text{Actual yield of the product}}{\text{Theoretical yield of the product}} \times 100$$

If one mole of a starting material produces one mole of the product, the yield is 100 %.

If one obtains a 100 % yield, it does not imply that the reaction is a green method. For example, in Grignard and Wittig reactions, the yield is generally 100%, but large amounts of by-products are formed. Further, use of Grignard reagent is detrimental since a large amount of heavy metals is left behind after the disposal.

A reaction or a synthesis is considered to be green if maximum amount of the starting materials or reagents is converted into final product. The percentage atom utilization (atom economy/AE) can be determined by the following equation:

$$\% \text{ AE} = \frac{\text{Formula weight of atoms in desired product}}{\text{Formula weight of all reactants}} \times 100$$

Consider the following reaction,

$$\text{Theoretical yield} = \text{Stoichiometric ratio} \times \frac{\text{Mwt of } C_7H_8}{\text{Mwt of } C_6H_6} \times \text{Wt of } C_6H_6 \text{ (1 g)}$$

$$= \frac{1}{1} \times \frac{92}{78} \times 1.0 = 1.18 \text{ g}$$

If the practical yield of the reaction was found to be 1.16 g, then

$$\text{Percentage yield} = \frac{1.16}{1.18} \times 100 = 98.3 \text{ \%}$$

As per green chemistry, a higher yield is not sufficient. We are looking at actual mass of reactants incorporated in products. Atom economy is the measure of how efficiently the atoms of the reactants in any reaction are incorporated into the desired product.

$$\% \text{AE} = \frac{\text{Formula weight of atoms in desired product}}{\text{Formula weight of all reactants}} \times 100$$

$$= \frac{\text{Formula weight of } C_7H_8}{\text{Formula weight of } C_6H_6 + CH_3Cl} \times 100 = \frac{92}{78 + 50.5} \times 100 = 71.6 \text{ \%}$$

Atom economy shows the actual number of atoms incorporated into products. Around 20% is waste that needs treatment in the given reaction.

Rearrangement and addition reactions in organic chemistry are considered as atom efficient methods as there is no loss of small molecules as by-products. However, substitution and elimination reactions are highly atom inefficient reactions as there is always a release of by-product(s) at the end of the reaction. The formation of salts is also considered as an atom-efficient method; for example, the preparation of potash alum $(K_2SO_4.Al_2(SO_4)_3.24H_2O)$ has 100 % atom economy.

Some classic examples of atom economy calculations are as follows:

1. Oxidation of benzene in the presence of vanadium pentoxide catalyst yields maleic anhydride.

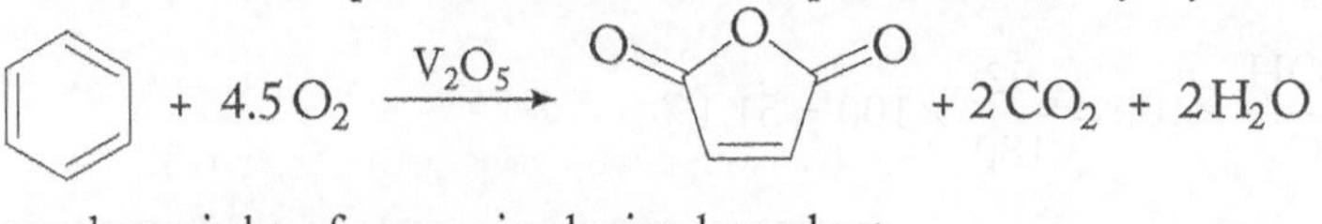

$$\% \text{ AE} = \frac{\text{Formula weight of atoms in desired product}}{\text{Formula weight of all reactants}} \times 100$$

$$= \frac{C_4H_2O_3}{C_6H_6 + 4.5O_2} \times 100 = \frac{98}{78 + 144} \times 100 = 44.1\%$$

2. $CH_3-CH=CH_2 + Cl_2 \rightarrow Cl-CH_2-CH=CH_2 + HCl$

$$\% \text{ AE} = \frac{C_3H_5Cl}{C_3H_6 + Cl_2} \times 100 = \frac{76.5}{42 + 71} \times 100 = 67.7\%$$

3. $C_2H_6 + Br_2 \rightarrow C_2H_5Br + HBr$ (At.wt of Br = 80)

$$\% \text{ AE} = \frac{C_2H_5Br}{C_2H_6 + Br_2} \times 100 = \frac{109}{30 + 160} \times 100 = 57.3\%$$

4. $C_4H_8 + 3O_2 \rightarrow C_4H_2O_3 + 3H_2O$

$$\% \text{ AE} = \frac{C_4H_2O_3}{C_4H_8 + 3O_2} \times 100 = \frac{98}{56 + 96} \times 100 = 64.47\%$$

5. $C_6H_6 + CH_3COCl \rightarrow C_6H_5COCH_3 + HCl$

$$\% \text{ AE} = \frac{C_8H_8O}{C_6H_6 + CH_3COCl} \times 100 = \frac{120}{78 + 78.5} \times 100 = 76.6\%$$

6. $C_6H_6 + HNO_3 \rightarrow C_6H_5NO_2 + H_2O$

$$\% \text{ AE} = \frac{C_6H_5NO_2}{C_6H_6 + HNO_3} \times 100 = \frac{123}{78 + 63} \times 100 = 87.2\%$$

7. $CH_3-NH_2 + COCl_2 \rightarrow CH_3-N=C=O$ (methyl isocyanate) + 2HCl

$$\% \text{ AE} = \frac{C_2H_3NO}{CH_5N + COCl_2} \times 100 = \frac{57}{31 + 99} \times 100 = 43.8\%$$

8. $C_6H_5-CHO + CH_3-CHO \rightarrow C_6H_5-CH=CH-CHO + H_2O$

$$\% \text{ AE} = \frac{C_9H_8O}{C_7H_6O + C_2H_4O} \times 100 = \frac{132}{106 + 44} \times 100 = 88\%$$

9. Benzene + $Cl_2 \rightarrow$ Chlorobenzene + HCl

$$\% \text{ AE} = \frac{C_6H_5Cl}{C_6H_6 + Cl_2} \times 100 = \frac{112.5}{78 + 71} \times 100 = 75.5\%$$

10. $CH_3-(CH_2)_4-CH_2-OH + SOCl_2 \rightarrow CH_3-(CH_2)_4-CH_2-Cl + SO_2 + HCl$

$$\% \ AE = \frac{C_6H_{13}Cl}{C_6H_{14}O + SOCl_2} \times 100 = \frac{120.5}{102 + 119} \times 100 = 54.5\%$$

11. $2NH_3 + NaOCl \rightarrow N_2H_4 \text{ (hydrazine)} + NaCl + H_2O$

$$\% \ AE = \frac{N_2H_4}{2NH_3 + NaOCl} \times 100 = \frac{32}{34 + 74.5} \times 100 = 29.5 \ \%$$

12. $C_6H_{12}O_6 \rightarrow 2C_2H_5OH + 2CO_2$

$$\% \ AE = \frac{2C_2H_5OH}{C_6H_{12}O_6} \times 100 = \frac{92}{180} \times 100 = 51.1\%$$

3. Less Hazardous Chemical Synthesis

Wherever possible, synthetic methods should be designed to use and generate substances that possess little or no toxicity to human health and the environment.

The starting material should be selected such that it is the least toxic. The reactions of benzene and heterocyclic moieties such as aromatic substituted benzene and naphthalene compounds led to the generation of intermediates, final products, and by-products, most of which are carcinogenic. Hence, synthetic routes starting with such compounds should be avoided. An alternative route of chemical reactions must be approached in terms of 'benign by design' at the design stage that will ensure sustainability of new products and processes. While redesigning chemical reactions, one must remember to not compromise on the yield of the final product.

Let us have a look at the synthesis of indigo, a classic dye mainly used to dye denim jeans.

Conventional Method

As shown, aniline is the starting compound which is toxic in nature. It is treated with chloroacetic acid which is a corrosive solvent. The resulting intermediate (I) is fused with sodamide forming an indole derivative (II). On aerial oxidation of derivative (II), indigo is obtained.

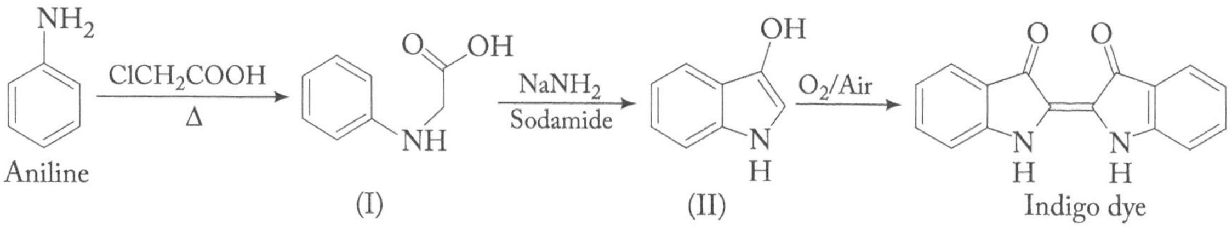

Fig. 22.1 Preparation of indigo using conventional method

Greener Method

Enzyme catalysis is considered as a greener alternative in most of the organic reactions. Overall, the greener method is a two-step reaction catalysed by enzymes.

In this case, the starting material is L-tryptophan, an amino acid obtained from protein-based natural sources. Using the enzyme, tryptophanase, cyclization is carried out with the removal of the amino acid side chain, giving indole. On dihydroxylation, using naphthalene dioxygenase, and aerial oxidation, indigo is obtained.

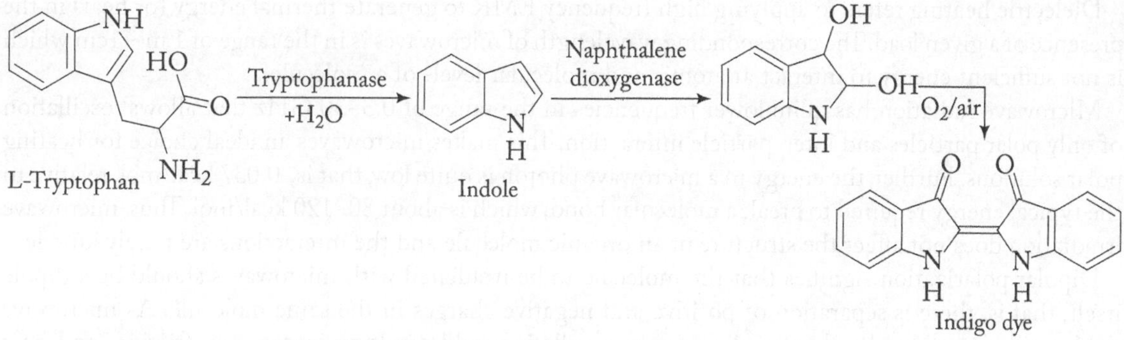

Fig. 23.2 Preparation of indigo using greener method

4. Designing Safer Chemicals

Chemical products should be designed to effect their desired function while minimizing toxicity.

After a drug is synthesized, clinical trials are carried out. If the medicine causes serious side effects in human population, its use is suspended. It is then modified keeping the basic chemical structure same so as to maintain its desired function, while reducing its toxicity.

For example, thalidomide (1957) was a therapeutic drug used for treating morning sickness in pregnant women. It was found that women who were administered thalidomide drug had babies with foetal abnormalities (malformed limbs). Hence, thalidomide was recalled from the market and banned.

5. Safer Solvents and Auxiliaries

The use of auxiliary substances (e.g., solvents, separation agents, etc.) should be made unnecessary wherever possible and innocuous when used.

Chloroform, carbon tetrachloride, benzene, and other aromatic hydrocarbons are used in a majority of organic reactions as versatile solvent media. Solvents end up as more waste than that generated during the reaction. Ether, acetone, benzene are highly inflammable as they cause fire hazards. All these solvents and their analogues carry health risks to the analyst using them.

Using safer solvents like water and liquid carbon dioxide, wherever possible is the recommended greener approach. For example, greener solvents like supercritical fluids such as $SF-CO_2$, $SF-H_2O$ have efficiently replaced toxic chemical reagents. A detailed account of greener solvents is given later in the chapter.

6. Design for Energy Efficiency

Energy requirements of chemical processes should be recognized for their environmental and economic impacts and should be minimized. If possible, synthetic methods should be conducted at ambient temperature and pressure.

Catalysts accelerate the kinetics of the main reaction and decelerate the by-product formation. Further, optimum temperatures and pressures are employed while using catalysts in the chemical processes, making the overall process an energy efficient method.

The use of fuel-consuming equipment like burners in the laboratory needs to be avoided as they cause pollution. *Microwave irradiation* and *sonication* are alternative methods to bring in energy efficiency in laboratories. Irradiation using microwaves coupled with sonication allows the reaction to proceed at a rapid rate due to uniform, homogeneous heat transfer to the organic media. There is no compromise in the overall yield and purity of the final product(s). Such reactions are termed as *solvent-less processes*.

Microwave Irradiation

Microwave radiation lies in a lower energy region between radiowaves and infrared region in the electromagnetic spectrum at frequencies of 0.3 GHz – 300 GHz. Microwave irradiation works on the principle of dipolar dielectric polarization and conduction allowing cleaner and rapid synthesis.

Dielectric heating refers to applying high frequency EMR to generate thermal energy (or heat) in the presence of a given load. The corresponding wavelength of microwaves is in the range of 1 m – 1cm which is not sufficient energy to interact at atomic and molecular levels of a molecule.

Microwave radiation has quite lower frequencies in the range of 0.3–30 GHz that allows oscillation of only polar particles and inter-particle interaction. This makes microwaves an ideal choice for heating polar solutions. Further, the energy in a microwave photon is quite low, that is, 0.037 kcal/mol, relative to the typical energy required to break a molecular bond, which is about 80–120 kcal/mol. Thus, microwave irradiation does not affect the structure of an organic molecule and the interactions are purely kinetic.

Dipolar polarization signifies that the molecule to be irradiated with microwaves should be a dipole itself, that is, there is separation of positive and negative charges in the same molecule. As microwave field oscillates, molecular dipoles align to the oscillating field resulting in rotation, friction, and heat energy. Thus, microwave heating depends on the dielectric properties of organic molecules and higher the dielectric constant, greater will be the heating efficiency.

Microwave irradiation is energy-efficient than traditional heating methods as there is direct coupling of microwaves with molecular dipoles in the reaction mixture. This direct interaction of molecules (solvents, compounds, reagents, catalysts) with microwaves is not possible in traditional refluxing on water bath using burners. Due to 'in-core' heating (no initial heating of the vessel surface as observed in traditional heating), microwave irradiation results in energy-efficient organic reactions. Further, microwaves allow rapid heating at a lower temperature without formation of many by-products. This is an added advantage while using microwaves for carrying out organic reactions as it allows higher product yields.

Some examples where microwave irradiation are used are:

(a) Oxidation of toluene with potassium permanganate by traditional refluxing takes about 12 hours as against microwave conditions, which takes fewer minutes with good yield (about 98 %). (R. N. Gedye)

(b) Hydrolysis of benzamide takes about one hour for obtaining benzoic acid. Under microwave conditions, hydrolysis is completed within 7 minutes and 99 % yield of benzoic acid has been reported. (R. N. Gedye et al.)

Sonication

When ultrasound wave is utilized to carry out organic reactions, it is termed as sonication. A branch of chemical research dealing with the chemical effects and applications of ultrasonic waves which are mechanical waves beyond the audible frequency range of 20 Hz – 16 kHz. Ultrasound waves allow reactions to occur due to *acoustic cavitation* phenomenon (Latin: *cavus* meaning cavity). The ultrasonic waves propagate through the medium as alternate compressions and rarefactions (or expansions). If rarefaction cycle exceeds the intermolecular forces of attraction in a molecule, cavitation bubble is formed, which grows after a few cycles by taking in gas from the medium to a size that matches the frequency of bubble resonance to that of applied ultrasound frequency. These compression and rarefaction cycles are more rapid than thermal transport and hence allows for rapid reaction with practically no by-products.

The rapid bubble formation, growth of bubble, and subsequent collapse of bubbles is referred to as *cavitation*. Sonic waves result in rapid reactions since each microbubble behaves similar to a micro-reactor that produces different reactive species and heat during its collapse.

Some examples where ultrasound radiation are used are as follows.

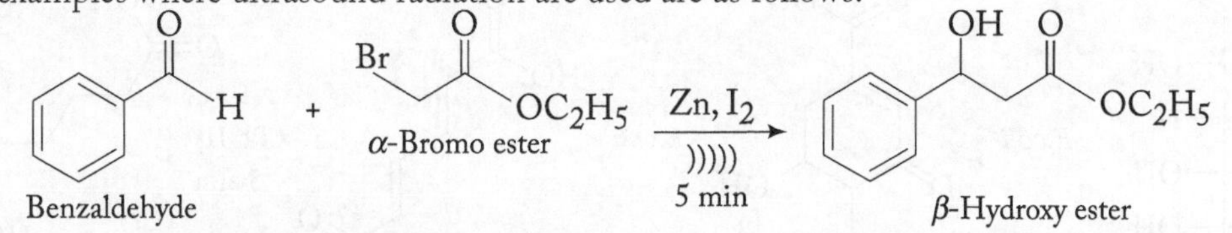

(a) The Reformatsky reaction of benzaldehyde with α-bromo ester in the presence of zinc dust and iodine using ultrasound radiation reaches completion in about 5 minutes.

(b) 3-Vinyl indole is synthesized via Pd/C–Cu catalysed coupling between 3-iodo-1-methyl-1H-indole and terminal alkenes under ultrasound irradiation.

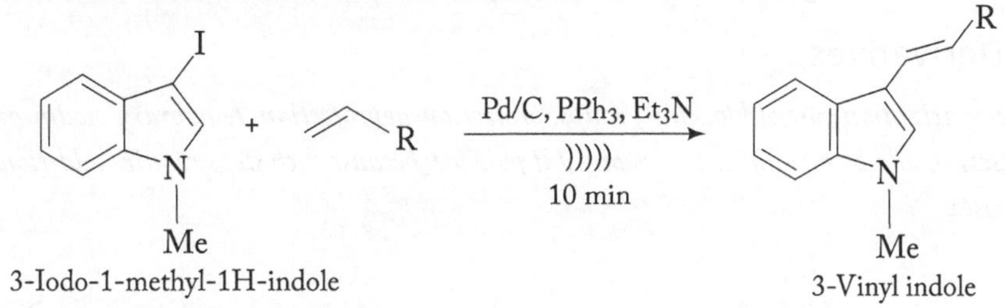

7. Use of Renewable Feedstocks

A raw material or feedstock should be renewable rather than depleting whenever technically and economically practicable.

Renewable feedstock materials are efficient to replace toxic organic compounds as starting raw materials. Let us take the example of adipic acid [$HOOC(CH_2)_4COOH$] which is the starting material in the production of nylon-6,6.

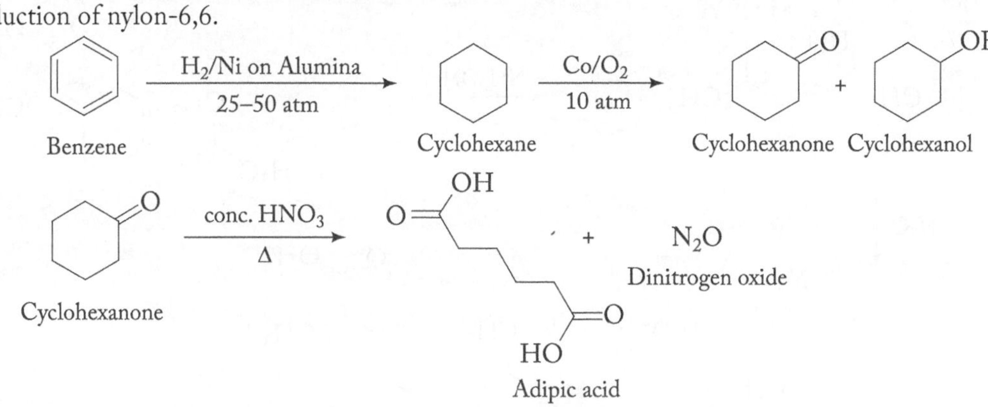

Fig. 22.3 Preparation of adipic acid by conventional method

In the conventional route (Fig. 22.3) benzene, a known human class I carcinogen, is used as a substrate for preparing adipic acid. On hydrogenation using nickel catalyst supported on alumina, cyclohexane is obtained. The oxidation of cyclohexane in the presence of cobalt catalyst is a highly explosive reaction forming cyclohexanone and the by-product, cyclohexanol.

Nickel and cobalt catalysts are difficult to regenerate after the process is complete making them less-efficient transition catalysts. The disposal of heavy transition metals in the environment is another demerit of the process. The treatment of cyclohexanone with concentrated nitric acid results in the formation of adipic acid and a greenhouse gas, nitrous oxide (N_2O) as the by-product.

In the greener pathway, (Fig. 22.4) glucose, which is a renewable starting material, is used. Glucose is converted into adipic acid by *Escherichia coli* that catalyses two steps of the reaction. This reduces the use of chemical reagents with significant toxicity.

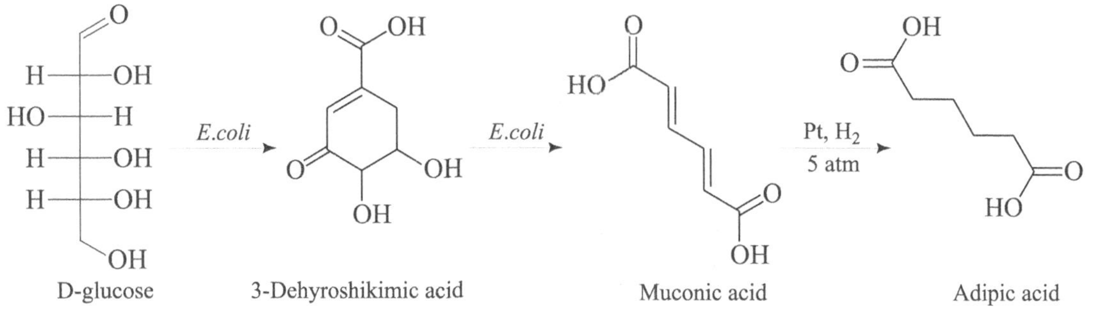

Fig. 22.4 Preparation of adipic acid by greener method

8. Reduce Derivatives

Unnecessary derivatization (use of blocking groups, protection/deprotection, temporary modification of physical/ chemical processes) should be minimized or avoided if possible, because such steps require additional reagents and can generate waste.

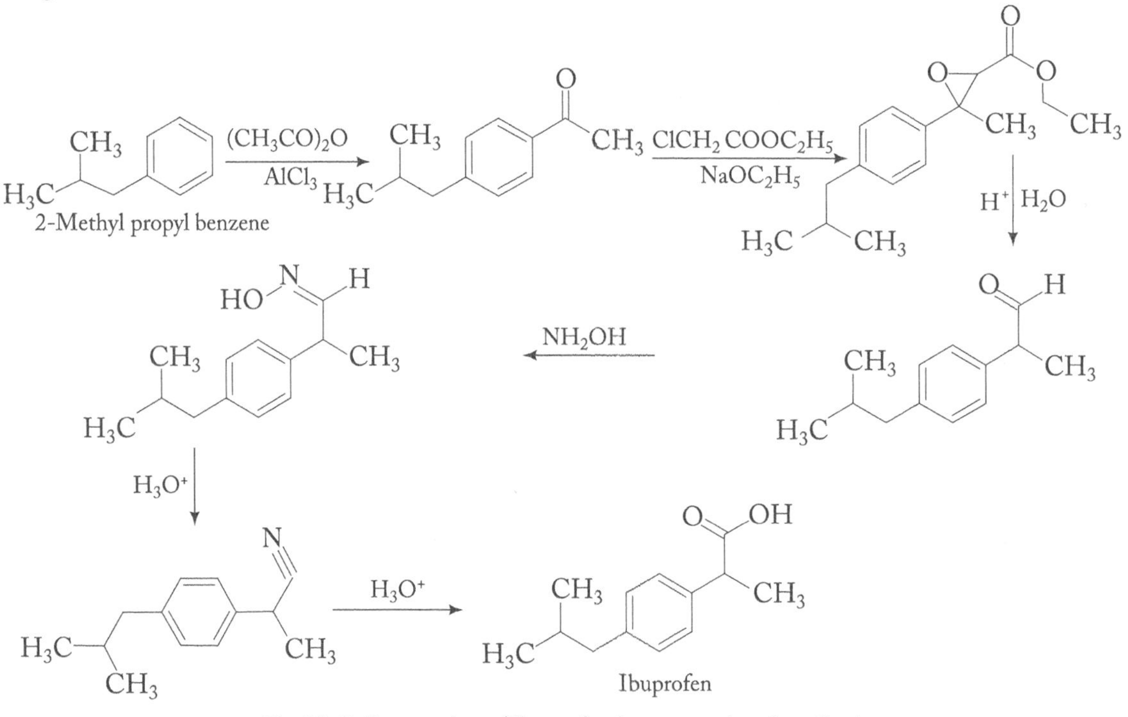

Fig. 22. 5 Preparation of ibuprofen by conventional method

While redesigning organic reactions, it is necessary to avoid the formation of too many derivatives as it consumes a lot of solvents and stoichiometric reagents.

Let us take the example of ibuprofen (2-(4-Isobutylphenyl)propanoic acid) a much-sought-after analgesic and compare the conventional and modified methods.

The conventional process is an energy-intensive, multi-step process with each step releasing various by-products.

A derivative, 2-methylpropyl benzene, obtained from crude oil acts as the starting material. It is treated with acetic anhydride in the presence of Lewis acid, AlCl$_3$. The resulting keto derivative is converted to an epoxide using a corrosive reagent, namely chloroacetic acid in ethoxide solution. Further, derivatization is continued to obtain, ibuprofen.

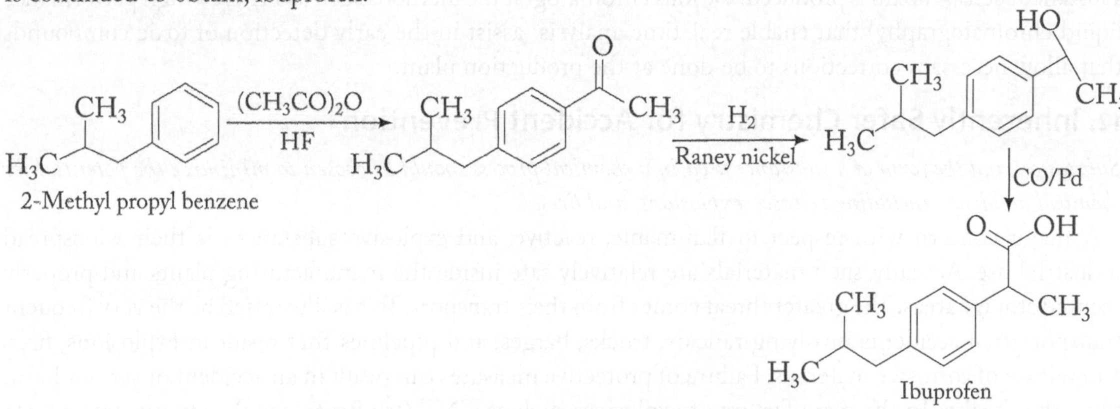

Fig. 22.6 Preparation of Ibuprofen by greener method

The starting material and reagent are the same as in the conventional method, but the Lewis acid $AlCl_3$ is replaced with hydrofluoric acid. HF is a true catalyst and generates minimal waste after completion of the reaction. The resulting derivative is hydrogenated using Raney nickel, followed by CO/Pd catalysis to obtain ibuprofen.

The above process is not a green method per se the definition; however, it is considered as a better alternative in comparison to the conventional route of synthesizing ibuprofen.

9. Catalysis

Catalytic reagents (as selective as possible) are superior to stoichiometric reagents.

Replacing organic reagents with catalysts, in certain cases prove to be a beneficial means to achieve the green chemistry objective. Catalysts are known to lower the activation energy of the overall chemical reaction, thereby hastening the kinetics of the reaction. The classic example is cracking of petrol using zirconium oxide supported on clay catalyst (for more details, see chapter Fuels and Combustion).

10. Design for Degradation

Chemical products should be designed so that at the end of their function they break down into innocuous degradation products and do not persist in the environment.

Materials that are produced must be benign (harmless) or readily biodegradable (easily digestible by microorganisms to produce simpler and safer products). If they are not easily degraded, they accumulate resulting in environmental issues.

Some pertinent examples are:

(a) DDT residues remain in the soil, animal and plant tissues for a long time, causing bioaccumulation.
(b) Biological waste particularly, hospital waste poses a serious health concern to human population. Hydrogen peroxide, in dilute concentration, is used as a sterilant for biomedical syringes, ampules and needles before sending for disposal.

11. Real-time Analysis for Pollution Prevention

Analytical methodologies need to be further developed to allow for real-time, in-process monitoring and control prior to the formation of hazardous substances.

Processes should be continuously monitored and all the products formed should be analysed. The prevention and minimization of the generation of hazardous substances in chemical processes may be controlled by using sensors, monitors, and analytical techniques. If any toxic material is detected, then the conditions should be modified to prevent its formation.

A pertinent case is monitoring the presence of ethylene glycol, since at higher temperatures, a toxic substance such as dioxin is produced. Various chromatographic methods such as HPLC (high performance liquid chromatography) that enable real-time analysis assist in the early detection of toxic compounds that allow necessary corrections to be done at the production plant.

12. Inherently Safer Chemistry for Accident Prevention

Substances and the form of a substance used in a chemical process should be chosen to minimize the potential for chemical accidents, including releases, explosions, and fires.

A major concern with respect to flammable, reactive, and explosive substances is their widespread industrial use. Actually, such materials are relatively safe inside the manufacturing plants and properly secured storage areas. The greater threat comes from their transport. This is illustrated by the very frequent transportation accidents involving railcars, trucks, barges, and pipelines that result in explosions, fires, and release of corrosive materials. Failure of protective measures can result in an accident or serious harm to worker health. In the manufacture of explosives such as TNT (trinitrotoluene), extreme precautions should be taken.

One is familiar with the Bhopal gas tragedy, wherein accidental leak of methyl isocyanate (abbreviated as MIC) resulted in the loss of thousands of lives. Hence, occurrences of accidents in chemical industries must be avoided by using safer chemicals, in optimum temperature, pressure, and other working conditions. The escape of solvents into the atmosphere should also be prevented.

Take the case of Carbaryl (1-naphthyl methylcarbamate) synthesis by comparing conventional and greener route of synthesis.

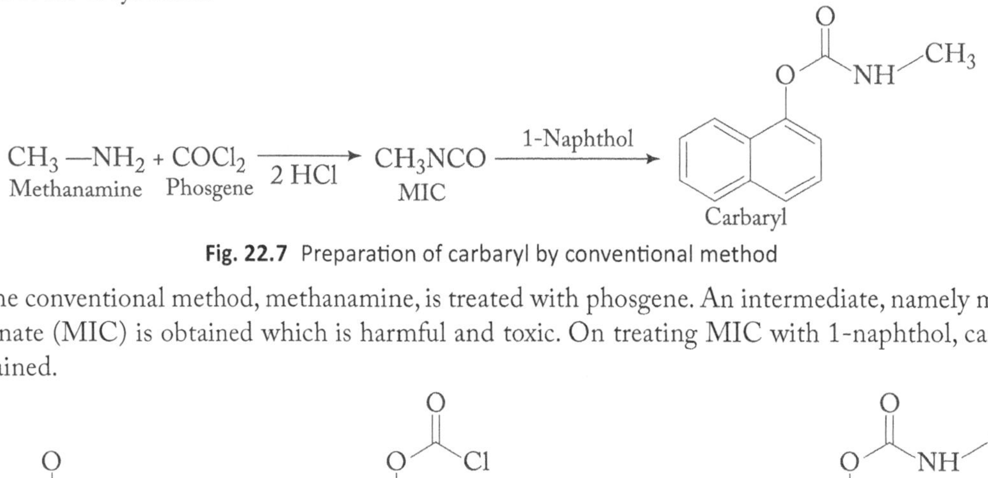

$$CH_3-NH_2 + COCl_2 \xrightarrow{\ \ \ \ } CH_3NCO \xrightarrow{\text{1-Naphthol}}$$

Methanamine   Phosgene   2 HCl   MIC

Fig. 22.7 Preparation of carbaryl by conventional method

In the conventional method, methanamine, is treated with phosgene. An intermediate, namely methyl isocyanate (MIC) is obtained which is harmful and toxic. On treating MIC with 1-naphthol, carbaryl is obtained.

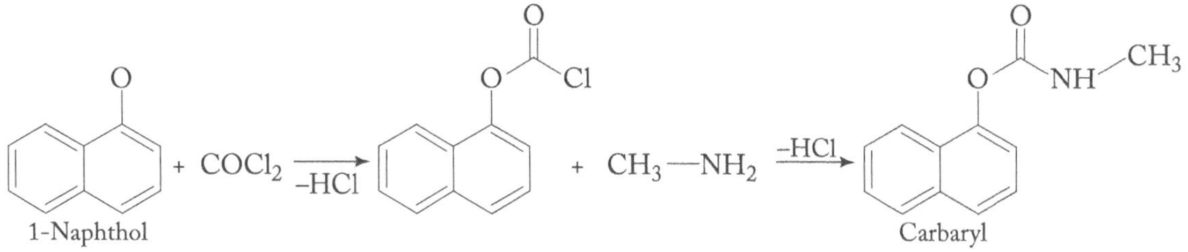

1-Naphthol $+ COCl_2 \xrightarrow{-HCl}$ $+ CH_3-NH_2 \xrightarrow{-HCl}$ Carbaryl

Fig. 22.8 Preparation of carbaryl by greener method

The alternative method to obtain carbaryl begins with 1-naphthol, which on treatment with phosgene gives carbaryl. Hydrochloric acid is obtained as the by-product. If one looks at the greener route, it is still a question whether to employ as an alternative method. The presence of phosgene and removal of hydrochloric acid seems controversial to be considered as a greener alternative; though these two products are less hazardous than MIC.

22.4 INDUSTRIAL APPLICATIONS OF GREEN CHEMISTRY

Green chemistry finds wide applications in chemical processes such as extraction and manufacture of fats, oils, pigments, drug molecules, polymers, and in biotechnology. The following are some of the recent trends that reflect the growth of green chemistry as a viable sector in industries.

22.4.1 Green Solvents

Supercritical fluids (SFCs) and ionic liquids (ILs) are fast emerging as green solvents for organic solvents in various chemical reactions. SFCs are often used to analyse low concentrations of compounds and high molecular weight molecules. They are routinely used in the analyses of drug products, food additives, explosives, petroleum, polymers, and propellants.

Supercritical CO_2   Carbon dioxide, as a supercritical fluid, is most frequently used as a solvent medium for reactions due to the following reasons:

(a) It is easily available (from natural sources and power engineering) and is cheap.

(b) Its application gives considerable energy savings, because critical point (T = 31.1 °C and P = 72.9 atm) is easy to reach due to low evaporation heat of CO_2.

(c) It dissolves a variety of solutes accompanied with diffusivity as that of a gas. All these properties make the transport phenomenon rapid, thereby resulting in an energy efficient process.

SF-CO_2 has been successfully employed as an extracting solvent for caffeine from tea and coffee.

Routine separations of chiral compounds, herbal products, and essential oil extractions and their purification can be accomplished using supercritical fluids.

Ionic liquids   Ionic liquids (ionic solvents) are used as green solvents with the advantage of lower vapour pressure. Room-temperature ILs, organic salts that are liquid below 100 °C, are nonflammable, non-volatile, and recyclable. Due to their remarkable properties, like high solvation power, thermal stability, and tunable properties by suitable choices of cations and anions, they are considered ideal for various chemical syntheses. The first synthesized IL was an ammonium-based compound (ethanol ammonium nitrate, EOAN) used widely as an electrolyte in high-energy electrochemical devices owing to its good electrochemical cathodic stabilities, low melting point, and low viscosities.

22.4.2 Products from Natural Sources

Obtaining commercial products from plant-based materials is considered as a viable green chemistry option. Utilization of enzymes and agro-based materials as catalysts is one such example.

Use of natural products to manufacture various commercial goods is taken up by many industries. The oleochemical section of Cognis (US), delivers a whole range of chemical products obtained from natural materials. Using vegetable oils as starting materials, the company produces a variety of products such as fatty acids, fatty methyl esters, glycerol, and long chain alcohols. These products can be used as starting material for various items used in body care, pharmaceutical, and food industries.

Plastic products are known as non-biodegradable materials. Cargill-Dow Polymers (USA) produces eco-plastics from corn waste materials. Eco-plastics are then utilized as packaging materials, foil, and plastic wraps that do not affect the environment on disposal.

Enzyme catalysis is yet another example of greener method. As per green chemistry principles 6 and 9, catalysis is proven to be an energy efficient route in a chemical reaction. Today, most of the catalysts include transition metals which are difficult to regenerate after completion of the reaction. Replacing metal-based catalysts with enzymes is a green alternative. Like metal catalysts, enzymes also reduce the energy of activation, thereby hastening the reaction kinetics. After completion of the reaction, the enzyme is not consumed and hence can be reused.

Let us have a look at the synthesis of acrylamide, which is the starting compound to prepare polyacrylamide, used as a common supporting medium for electrophoresis.

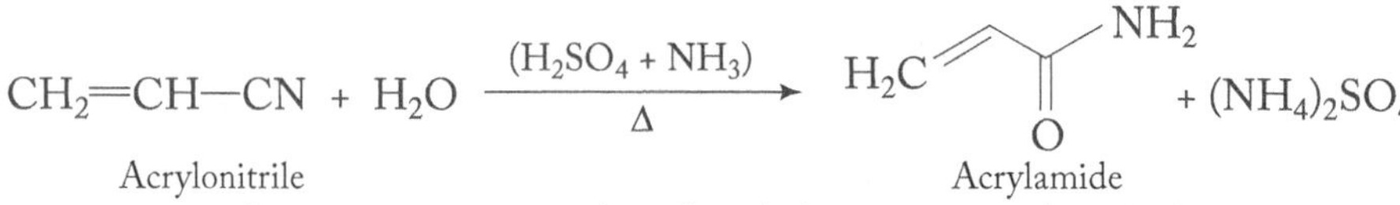

$$CH_2\!=\!CH\!-\!CN + H_2O \xrightarrow[\Delta]{(H_2SO_4 + NH_3)} H_2C \diagdown\diagup \stackrel{NH_2}{\underset{O}{|}} + (NH_4)_2SO_4$$

Acrylonitrile   Acrylamide

Fig. 22.9 Preparation of acrylamide by conventional method

In the above method, acrylonitrile undergoes acid hydrolysis in the presence of strong toxic solvents such as sulphuric acid and ammonia to produce acrylamide with ammonium sulphate as the by-product.

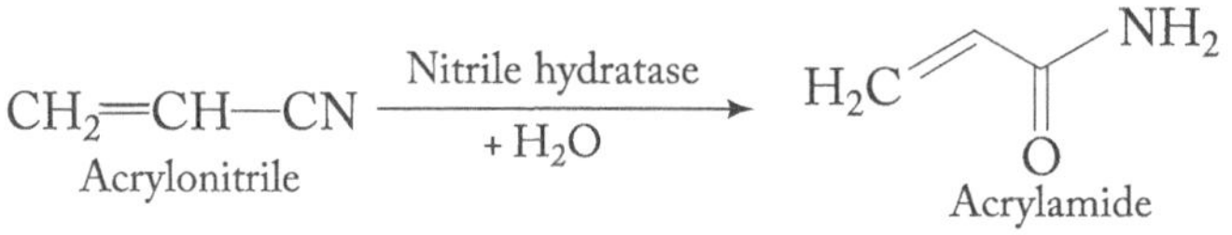

$$CH_2\!=\!CH\!-\!CN \xrightarrow[+\,H_2O]{\text{Nitrile hydratase}} H_2C \diagdown\diagup \stackrel{NH_2}{\underset{O}{|}}$$

Acrylonitrile   Acrylamide

Fig. 22.10 Preparation of acrylamide by greener method

In the above green method, enzymatic hydrolysis is carried out using nitrile hydratase obtained from *Rhodococcus* species. As compared to the conventional method, there is no by-product formation in the greener route.

Check Your Progress

1. Define green chemistry.
2. What is atom economy? Express its formula.
3. Name any two green solvents.
4. Name any one reaction which is an atom-inefficient method in chemistry.
5. Can one carry out solventless synthesis? If yes, state the method used for solventless synthesis.
6. Compare microwave heating with traditional heating processes of organic reactions.

Activity-based Questions

A. Green Cloud Computing in India has taken IT by storm with various initiatives taken up by Google Inc.

Present the various energy saving initiatives taken in green cloud computing and review the challenges and future vistas of the technology in India.

B. Deep Eutectic solvents (DESs) are employed as eco-friendly solvents in metal processing, organic reactions, nanoscale materials media, etc.

Present a comprehensive literature review on DESs, types, recyclability, mechanism (if any), advantages, limitations, applications (two to three).

SUMMARY

○ Green Chemistry involves the design of chemical processes and products so as to achieve least environmental impact.

○ Rather than yield of the organic reaction, achieving atom efficiency is a prime requisite for success of a green chemical reaction.

○ The 12 principles of green chemistry deal with waste reduction, redesigning chemical reactions, atom efficiency, energy efficiency, applying renewable materials, catalysis, and preparing products that result in zero environmental impact.

○ Greener methods of preparation of indigo dye, ibuprofen, adipic acid, acrylamide, and carbaryl are preferred over conventional methods.

○ Ionic liquids and supercritical fluids are commonly employed as green solvents. Supercritical carbon dioxide and water are used in a variety of organic reactions.

GLOSSARY

Atom economy: The measure of how efficiently the atoms of the reactants in any reaction are incorporated into the desired product.

Cavitation: The rapid formation, growth and collapse of bubbles.

Energy of activation: The minimum energy required for the reactants to result in a chemical reaction.

Enzyme catalysis: It involves an increase in the rate of a chemical reaction by the action of enzyme catalyst.

Green chemistry: Design of chemical products and processes that are environmentally benign and reduce negative impact on human health and environment.

Oleochemical: A compound obtained from agro-based vegetable oils.

KEY FORMULAE

- $\% \text{AE} = \dfrac{\text{Formula weight of atoms in desired product}}{\text{Formula weight of all reactants}} \times 100$

- $\text{Theoretical yield} = \text{Stoichiometric ratio} \times \dfrac{\text{M Wt of products}}{\text{M Wt of reactants}} \times \text{Wt of reactants}$

- $\text{Percentage yield} = \dfrac{\text{Actual yield of the product}}{\text{Theoretical yield of the product}} \times 100$

EXERCISES

Multiple Choice Questions

1. The term 'green chemistry' was coined by _____
   - (a) Barry Trost
   - (b) Paul Anastas
   - (c) J.C. Warner
   - (d) R.A. Sheldon

2. Green chemistry is also called _____
   - (a) sustainable chemistry
   - (b) waste management chemistry
   - (c) industrial chemistry
   - (d) solvent chemistry

3. Ionic liquids and supercritical fluids are examples of
   - (a) natural products
   - (b) green solvents
   - (c) renewable materials
   - (d) salts

4. The starting material to prepare indigo dye in greener method is
   - (a) β-naphthol
   - (b) benzene
   - (c) aniline
   - (d) tryptophan

5. The number of step(s) involved in the conventional preparation of ibuprofen is/are
   - (a) 2
   - (b) 4
   - (c) 6
   - (d) 10

6. Bhopal gas tragedy was due to release of
   - (a) methyl iodide
   - (b) methyl isocyanate
   - (c) ethylene
   - (d) benzene

7. Which of the following represents a green method?
   - (a) Minimal effluent discharge
   - (b) Using toxic solvents
   - (c) combustion of fuels
   - (d) using burners for heating

8. Which of the following classes of organic reactions is/are considered as atom-efficient?
   - (a) Rearrangement
   - (b) Substitution
   - (c) Elimination
   - (d) Redox

9. Microwaves and sonication results in
   - (a) energy-efficient heating
   - (b) energy-inefficient heating
   - (c) poor yields
   - (d) toxins in environment

10. The atom economy for the following reaction is:
    $2NH_3 + NaOCl \rightarrow N_2H_4$ (hydrazine) $+ NaCl + H_2O$
    - (a) 29.5 %
    - (b) 30.2 %
    - (c) 35 %
    - (d) 15 %

11. Benzene is an important industrial solvent which is classified as
    - (a) non-toxic
    - (b) non-flammable
    - (c) biodegradable
    - (d) carcinogenic

12. The rapid bubble formation, its growth and subsequent collapse is called
    - (a) cavitation
    - (b) sonication
    - (c) microwave heating
    - (d) dielectric heating

13. Atom economy of reaction; Benzene + $Cl_2 \rightarrow$ Chlorobenzene + HCl is
    - (a) 70 %
    - (b) 76.3 %
    - (c) 75.5 %
    - (d) 84.3 %

14. When a chemical reaction is studied from initial production stages till the final disposal, the approach is called
    - (a) R4M4
    - (b) cradle-to-grave
    - (c) green chemistry
    - (d) None of these

15. The starting material to prepare Ibuprofen is
    - (a) 2-methylpropyl benzene
    - (b) 2-methyl benzene
    - (c) 1-naphthol
    - (d) benzene

16. Carbon dioxide is commonly used as supercritical fluid because of
    - (a) easy availability
    - (b) dissolves all solutes
    - (c) low evaporation heat
    - (d) all of these

Review Questions

1. What is green chemistry? List the 12 principles of green chemistry.

2. Define green chemistry. State *any six* principles of green chemistry and explain with suitable examples.

3. With a suitable example, explain the application of green chemistry in waste utilization.

4. Explain synthesis of adipic acid and highlight the green chemistry principle addressed in this case.

5. Elucidate the conventional and green routes of indigo dye preparation and describe the relevant green chemistry principle in this case.

6. Explain 'reducing derivatives,' in chemical process with the example of ibuprofen synthesis.

7. Explain green chemistry and mention its significance.

8. Discuss 'microwave irradiation' and 'sonication' as energy efficient chemical methods.

9. Write the chemical synthesis (conventional and greener routes) of Carbaryl and acrylamide .

10. Justify the statement w.r.t green chemistry principle, 'Prevention is better than cure.'

11. Discuss ionic liquids and supercritical fluids.

12. What are green solvents? Name some commonly used green solvents in chemical reactions and state their properties.

13. What is enzyme catalysis? State various enzymes employed to catalyse reactions.

14. Write a short informative note on green solvents.

15. How is atom economy different from practical yield of an organic reaction? Cite suitable examples to support your answer.

NUMERICAL PROBLEMS

1. Calculate % AE for the reaction: $CH_3 - CH = CH_2 + Br_2 \rightarrow Br - CH_2 - CH = CH_2 + HBr$
   (Ans: 59.9 %)

2. Benzene is treated with methyl chloride in the presence of aluminium trichloride catalyst. Toluene is obtained as the product along with hydrochloric acid. Write the chemical reaction and calculate its atom economy.
   (Ans:

   Benzene   Methyl chloride   Toluene   Hydrochloric acid

   Atom economy: 92/78 + 50.5 = 71.6 %)

FURTHER READING

1. Anastas, P. T. and J. C. Warner, *Green Chemistry: Theory and Practice*, Oxford University Press, New York, 1998.
2. Adewuyi Y. G. (2001) *Sonochemistry: Environmental Science and Engineering Applications*, Ind Eng Chem Res 40:4681–4715.
3. C. O. Kappe, A. Stadler, D. Dallinger, *Microwaves in Organic and Medicinal Chemistry*, 2012, Wiley-VCH, Weinheim.
4. Dwivedy M. and R.K. Mittal, 'Estimation of Future Outflows of e-Waste in India', *Waste Management*, 30 (2010) 483–491.
5. Gedye, M. N., F. C. Smith, and K. C. Westaway, *Can. Journal of Chemistry*, 69(700), 1991.
6. op. cit. 66(17) 1988.
7. Holmes. I., *Dumping, Burning and Landfill*, R.E Hester and R. M. Harrison (ed.) *Electronic Waste Management: Design, Analysis and Applications*, Royal Society of Chemistry, UK, 2009.
8. Kerton F.M. and R. Marriott, *Alternative Solvents for Green Chemistry*, Royal Society of Chemistry, UK, 2013.
9. Vallero, D. and Braiser Chris *Sustainable Design*: The Science of Sustainability and Green Engineering*, John Wiley & Sons Inc, US. (* use of recycled paper was main feature of the book.), 2008.

ANSWERS

Check Your Progress

1. Green chemistry is the design of chemical products and processes that are environmentally benign and reduce the negative impact to human health and environment.
2. The measure of how efficiently the atoms of the reactants in any reaction are incorporated into the desired product is called atom economy.

$$\% \text{ AE} = \frac{\text{Formula weight of atoms in desired product}}{\text{Formula weight of all reactants}} \times 100$$

3. Supercritical fluids and ionic liquids are examples of green solvents.
4. Substitution reactions are atom-inefficient as the by-products are formed after the reaction is complete.
5. Yes, solventless synthesis can be carried out using microwave irradiation.

6.

Traditional heating	Microwave irradiation
Heating of reaction mixture proceeds from a surface, usually the insides of reaction vessels.	Heating proceeds directly within the reaction mixture.
Thermal or electric source for heating is essential.	Electromagnetic wave heating takes place; electric source is required.
Heating mechanism involves conduction.	Heating mechanism occurs due to dielectric polarization and conduction.
The highest temperature that can be achieved is limited by the boiling point of reaction mixture.	Superheating (with uniformity) can be attained.

Multiple Choice Questions

1. (a)   2. (a)   3. (b)   4. (d)   5. (c)   6. (b)   7. (a)
8. (d)   9. (a)   10. (a)   11. (d)   12. (a)   13. (c)   14. (b)
15. (a)   16. (d)

Laboratory Experiments

EXPERIMENT 1. DETERMINATION OF SURFACE TENSION USING STALAGMOMETER

Aim   To determine the surface tension of a given liquid by drop weight method using a stalagmometer.

Principle   Stalagmometer is a pipette-like glassware consisting of a thin bulbed capillary tube. Surface tension of a given liquid can be measured by drop weight method using a stalagmometer. When a liquid is allowed to flow through a capillary tube, a drop is formed at its lower end. It increases to a certain size and falls off. The size of the drop depends on the radius of the capillary and surface tension of the liquid. The measurement of surface tension of a liquid is based on the fact that drop of the liquid at lower end of capillary falls down when the weight of drop becomes equal to surface tension. The same numbers of drops are counted for water and the given liquid. Surface tension of the given liquid is then determined relative to water at room temperature. (Surface tension of water = 72.8 mN/m at 20 °C at R.T.)

Requirements   Stalagmometer, beakers, electronic balance, thermometer, water, lubricants, rubber bulbs, filter papers.

Procedure

1. Weigh a 50 ml beaker on an electronic balance and record the weight as W_1 g.
2. Fill the stalagmometer (Fig. A.1) with water. Allow the water to drop and count 20 drops.
3. Check the weight of the beaker with 20 drops of water and record as, W_2 g.
4. Calculate the weight of 20 drops as, $W_3 = W_2 - W_1$ g.
5. Calculate the weight of the single drop$= \dfrac{W_3}{20} = M_1$ for water
6. In the same manner, determine the drop weight for the given liquid and denote it as M_2. Obtain the surface tension from the given formula.

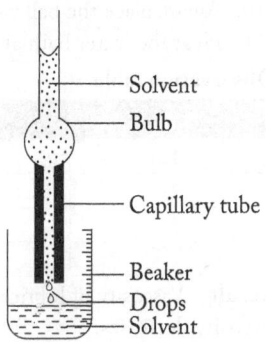

Solvent
Bulb

Capillary tube

Beaker
Drops
Solvent

Fig. A.1 Stalagmometer

Observation table

Liquid	Weight of 20 drops of liquid (W_3)	Weight of single drop
Water		(M_1)
Lubricant		(M_2)

Calculations

γ_1 = Surface tension of water = 72.8 mN/m

$\gamma_2 = M_2/M_1 \times \gamma_1$

Result   Relative surface tension of given lubricant w.r.t water at room temperature is, _____

(**Note:** The number of drops can be varied depending on type of liquid under experimentation)

2. DETERMINATION OF EFFECT OF TEMPERATURE ON VISCOSITY OF LUBRICANTS

Aim To determine the effect of temperature on viscosity of lubricating oil using Redwood viscometer.

Construction Redwood viscometer is an efflux-type viscometer. The apparatus consists of an oil cup equipped with a lid along with provision for thermometer and a ball valve. The oil cup is surrounded by a water bath equipped with an electric stirrer. At the base of the oil cup, a capillary is present. A ball valve placed within the oil cup controls the opening and closing of the capillary. A 50 ml capacity Kohlrausch flask is also provided for collecting oil (Fig. A.2).

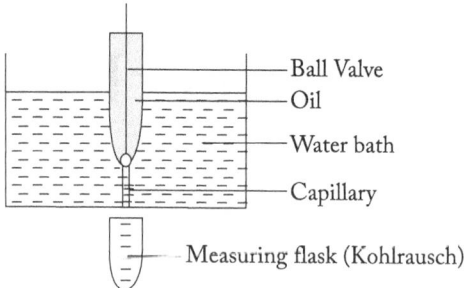

Ball Valve
Oil
Water bath
Capillary
Measuring flask (Kohlrausch)

Fig. A.2 Typical representation of Redwood viscometer

Procedure

1. Clean the oil cup properly with the help of a suitable solvent (like CCl_4) and dry it completely to remove any traces of solvent.
2. Level the viscometer with the help of levelling screws on the working bench-top.
3. Fill the outer bath surrounding the oil cup with water for determining the viscosity of oil at room temperature.
4. Insert a clean thermometer in the cup and cover it with a lid. Electrically operated stirrer placed in the water batch is started and room temperature is recorded.
5. Place a ball valve on the jet of the oil cup to close. Pour the test oil into the oil cup.
6. Secure the ball valve of the oil cup and place a 50 ml capacity Kohlrausch flask below the capillary.
7. Record the room temperature and lift the ball valve and start the stop watch.
8. Once the ball valve is opened, oil starts flowing through the capillary and gets collected in Kohlrausch flask.
9. Record the time taken for 50 ml of oil to collect in the flask at room temperature.
10. Again, place the ball valve on the jet and pour the test oil into the cup up to the tip of indicator.
11. Heat the water bath at $60\,°C$, $70\,°C$, and $80\,°C$ and record the flow times and report as Redwood seconds.

Observation table

S.no	Temperature of oil (°C)	Flow time (in Redwood seconds)
1.	Room (____)	
2.	60	
3.	70	
	80	

Result Viscosity of lubricating oil is inversely proportional to the temperature, i.e., with increase of temperature, viscosity decreases.

3. DETERMINATION OF VISCOSITY OF CASTOR OIL AND GROUNDNUT OIL BY USING OSTWALD'S VISCOMETER

Aim To determine viscosities of given oils using Ostwald's viscometer.

Principle Viscometer is an apparatus to measure relative viscosities of liquids. The most common viscometer employed in laboratories is Ostwald's viscometer. It is a U-tube gravity-type glassware with two reservoir bulbs as shown in Fig. A.3 and is etched with various markings on it. The fundamental premise of viscosity measurements is based on Poiseuille's equation that relates viscosity of a flowing liquid in a capillary tube and volume of liquid, pressure of liquid radii, length of the tube, and time required for liquid flow.

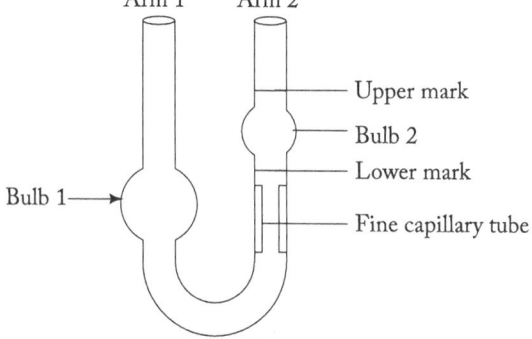

Arm 1   Arm 2
Upper mark
Bulb 2
Lower mark
Bulb 1
Fine capillary tube

Fig. A.3 Ostwald's viscometer

$$\eta = \frac{\pi P r^4 t}{8lv}$$

Here, v is volume of liquid flowing through capillary in time t, P is pressure, r is radius of capillary, and l is length of tube through which liquid flows. Generally, viscosity of a liquid is always measured relative to water at room temperature and the equation is,

$$\frac{\eta_l}{\eta_{H_2O}} = \frac{d_l t_l}{d_{H_2O} \cdot t_{H_2O}}$$

where, t_l and t_{H_2O} are obtained from experimental results and once d_l, d_{H_2O} and η_{H_2O} are known, η_l i.e., viscosity of the liquid can be determined.

Requirements Ostwald's viscometer, castor oil, ground nut oil, distilled water, beakers, wash bottles, rubber bulbs

Procedure

1. Ostwald's viscometer is rinsed with chromic acid solution followed by distilled water. Clamp the apparatus securely in an upright position as in Fig. A.3. Attach a rubber bulb for suction of liquid.
2. A definite volume of distilled water is filled in the Ostwald viscometer through Arm 1 using a pipette. Attach a rubber bulb at Arm 2 and suck the water to reach slightly above the upper mark present on bulb 2.
3. The water is now allowed to flow back and immediately the stopwatch is started to record the time taken for its flow from mark upper mark to lower mark. Repeat the procedure to check for reproducibility.
4. The viscometer is dried and in the same way experimental liquids, castor oil, and groundnut oil are tested for their flow relative to water.

Calculations Calculate viscosities as per equation,

$$\frac{\eta_l}{\eta_{H_2O}} = \frac{d_l t_l}{d_{H_2O} \cdot t_{H_2O}}$$

If the densities of liquids are similar, they are ignored and hence equation becomes,

$$\frac{\eta_l}{\eta_{H_2O}} = \frac{t_l}{t_{H_2O}}$$

Result The viscosities of castor oil and groundnut oils are _____ & _____ respectively.

4. ANALYSIS OF THIN LAYER CHROMATOGRAPHY OF ASPIRIN

Aim To analyse the TLC of aspirin.

Principle Chromatography is a technique for separation of two or more compounds based on their distribution between moving phase and stationary phase. Thin-layer chromatography or TLC, is a solid–liquid form of chromatography, where the stationary phase is a polar absorbent and mobile phase is generally a single solvent or combination of solvents (also called solvent system). A quick, inexpensive and a micro-scale technique TLC is routinely employed in the following cases.

- Identifying given substance or compound.
- Determine the number of components present in a mixture
- Monitor the progress of a reaction
- Analyse the fractions obtained from column chromatography

Requirements TLC chamber, TLC plates, aspirin tablets, mortar–pestle, forceps, capillary tubes, glass vials, ethanol, water, iodine, cotton.

Procedure

1. Prepare aspirin sample by crushing ¼ of the tablet and add the powder in a small glass vial along with a few drops of ethanol (labelled 1).

2. Prepare a suspension of the aspirin tablet such that only some part of it remains undissolved. Due to the presence of binders, additives, starch or silica the tablet will not completely dissolve in ethanol.
3. A standard solution of pure aspirin is also prepared in ethanol and kept in another glass vial (labelled 2)
4. Take 2 clean capillary tubes and fill one of it with the standard aspirin and the other with tablet suspension.

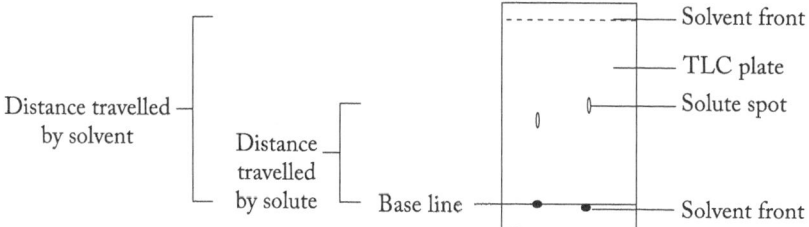

Fig. A.4 Developed TLC plate

5. Spot the sample and standard on the TLC plate such that the spot is not larger than 1 – 2 mm in diameter. (*Remember:* Smaller the spot on the TLC plate, better is the chromatographic analyses.)
6. After ethanol solvent evaporates, place the spotted TLC plates in the development chamber containing 50 : 50 ethyl acetate : hexane solvent system. Allow the solvent to run across the spotted TLC plate until it reaches at least 1cm below the top of the plate.
7. Using a clean forcep, carefully remove the plate from the chamber and allow it to dry.
8. Place the TLC plate in iodine chamber and observe the brown spots. One can alternatively examine the spots in a UV chamber. Outline the spots of the sample along with standard aspirin and calculate R_f value for the spots and report the results.

Calculations Determine the solute and solvent front as given in the diagram below:
Calculate the R_f value as given by the formula,

$$R_f = \frac{\text{Distance from baseline travelled by solute (solute front)}}{\text{Distance from baseline travelled by solvent (solvent front)}}$$

Result On performing the TLC of aspirin, the following results are obtained:
(a) Solute front _____, (b) Solvent front _____ and (c) R_f = _____

5. ESTIMATION OF HARDNESS OF WATER USING ION EXCHANGE COLUMN

Aim To remove hardness causing ions from the given water sample using ion-exchange resin column.

Principle In laboratory, chemical method used for determining hardness in water is complexometric titration. In this method, water sample is titrated against EDTA using Eriochrome Black T (EBT) as an indicator. An initial unstable complex between calcium and magnesium ions and EBT gives a wine red colour due to the formation of an unstable metal ion-indicator complex (1). Towards the end point of the titration, a highly stable metal ion-EDTA complex (2) is formed and the colour changes from wine red to deep blue color.

$$\text{EBT} + Mg^{2+}/Ca^{2+} \text{ (aq)} \longrightarrow [\text{Ca-EBT}]/[\text{Mg-EBT}] \qquad (1)$$
$$\text{(Indicator)} \qquad\qquad\qquad \text{Wine red-metal ion-indicator complex}$$

$$[\text{Ca-EBT}]/[\text{Mg-EBT}] + \text{EDTA} \longrightarrow [\text{Ca-EDTA}]/[\text{Mg-EDTA}] + \text{EBT} \qquad (2)$$
$$\text{Metal-EDTA complex-more stable} \quad \text{Deep blue colour}$$

Ion exchange technique is used to deionize or demineralize the hard water sample. In this experiment, incoming water sample hardness is determined using complexometric titration. The diluted water sample is allowed to undergo exchange in a cation-exchange column and the effluent is titrated against EDTA to check for hardness of water post-ion exchange process.

Requirements   Standard hard water (1 ppm $CaCO_3$), water sample, EDTA solution, NH_4Cl–NH_3 buffer solution, Eriochrome Black T (EBT) indicator, 50 ml burette, 25 ml pipettes, conical flasks, 10 ml measuring cylinders, glass droppers, rubber bulbs.

Amberlite (cation exchanger), deionized water, 1M HCl, hard water sample (1 ppm $CaCO_3$), 0.01M EDTA, NH_4Cl–NH_3 buffer, Eriochrome Black T indicator, glassware used in titrimetry.

Procedure
1. Prepare standard hard water (SHW) keeping its concentration to about 1 ppm calcium carbonate.
2. Pipette 10 ml of 0.01M SHW into a conical flask and add 2 ml of NH_4Cl–NH_3 buffer (pH = 10) and 2 drops of Eriochrome Black T indicator.
3. Titrate the above solution against EDTA solution from the burette and observe the color change from wine red to deep sea-blue colour. Repeat the procedure to obtain concordant readings and report the strength of EDTA solution.
4. Measure 100 ml of the water sample in a conical flask and add 2 ml of NH_4Cl–NH_3 buffer (maintain pH =10) and 5–6 drops of EBT indicator. Titrate the solution against standard EDTA solution until deep blue colour appears.
5. Prepare the ion-exchange column by soaking Amberlite resin in 1M HCl before starting the experiment. Ideally, resins are activated a day before the experiment.
6. On the day of experiment, thoroughly wash the resin with deionized water completely to remove traces of acid.
7. Fill the ion-exchange column with the cation-exchange resin taking care that no air bubbles are formed in the column. Always keep some amount of HCl or deionized water at the top of the resin column bed throughout the experiment. Drying of the resin column leads to failure of the experiment.
8. Determine the total hardness of hard water sample before the ion-exchange process.
9. For removal of hardness, pipette 25 ml of the hard water sample and add it from top of the resin column.
10. Arrange a 250 ml volumetric flask below the ion-exchange column for collecting the effluent.
11. Perform the elution process using deionized water and collect all the effluent in the volumetric flask at the rate of 2 ml per minute.
12. Dilute the contents of the volumetric flask to 100 ml with deionized water and pipette out 50 ml of diluted effluent into a conical flask. Determine total hardness of treated effluent.

Calculations
1 ml of standard hard water = 1 mg of $CaCO_3$ (given)

10 ml of standard hard water = 10 mg of $CaCO_3$

V_1 ml of EDTA is required for 10 mg of $CaCO_3$

As, 100 ml of water = V_2 ml of EDTA, where V_2 is total hardness of given water sample

Thus, 1000 ml of water = 10 × V_2 ml of EDTA

$$= 10 \times V_2 \times \frac{10}{V_1} \text{ mg } CaCO_3 = 100 \times \frac{V_2}{V_1} \text{ mg } CaCO_3 \text{ per litre of water.}$$

The above calculations give the total hardness of water sample before ion-exchange treatment. Post-ion exchange treatment, the following calculations are done.

25 ml of hard water requires = x ml of 0.01 M EDTA

50 ml of treated water requires = y ml of 0.01 M EDTA

Thus, 250 ml of treated water required = $y \times 5$ ml of 0.01 M EDTA = z

% Efficiency of the operation can be determined as,

$$\% \text{ Efficiency} = \frac{z}{x} \times 100$$

If the value of y is zero, then efficiency of the operation is 100 %.

$x - 5y = z$ ml of 0.01M EDTA

Compare the total hardness before treatment with the hardness value obtained after ion-exchange process. If the total hardness of the given water sample is found to be less than 3 times the actual hardness, the process is deemed to be efficient.

Result Total hardness of given sample post ion-exchange is _____.

6. Determination of Chloride Content of Water

Aim To estimate chloride content in the given water sample by Mohr's method.

Principle Chloride ions are present in water in ppm levels as $CaCl_2$, $MgCl_2$, NaCl. This is a precipitation titration method that uses $AgNO_3$ as the precipitating agent; hence is also called 'argentometric titration'. Potassium chromate is used as the indicator to detect the end point of the argentometric titration. When all the chloride ions have precipitated as white silver chloride, the added extra silver ions react with the chromate ions to form a reddish brown precipitate of silver chromate. Thus, end point of the titration is the formation of a reddish-brown precipitate as per the following reactions.

$$AgNO_3 \text{ (aq)} + Cl^- \text{ (aq)} \rightarrow AgCl \text{ (s)} \downarrow \text{ white ppt} + NO_3^- \text{(aq)}$$

$$2\,AgNO_3 \text{ (aq)} + CrO_4^{2-} \text{ (aq)} \rightarrow Ag_2CrO_4 \text{ (s)} \downarrow \text{ reddish-brown ppt} + 2\,NO_3^- \text{ (aq)}$$

Reagents and equipment Silver nitrate solution, standard NaCl solution, potassium chromate (K_2CrO_4) indicator, chloride sample solution, burette, pipette, conical flasks, measuring cylinders.

Procedure

Standardization of $AgNO_3$ solution

a. Pipette out 10 ml of standard NaCl solution in a conical flask and add 2 ml of K_2CrO_4 indicator.
b. Titrate the above solution against $AgNO_3$ solution (in burette) until the precipitate formed acquires a persistent reddish brown colour.
c. Note the burette readings and use the constant burette reading (CBR) for calculating the molarity of silver nitrate solution.

Estimation of chloride in water sample

a. Using a measuring cylinder, add 100 ml of water sample in a conical flask and 5 ml of K_2CrO_4 indicator.
b. Titrate the above solution against $AgNO_3$ solution until the precipitate formed acquires a persistent reddish brown colour.
c. Note the burette readings and calculate molarity of silver nitrate solution.

Calculations For estimating molarity of $AgNO_3$ solution,

$$M_1V_1 \equiv M_2V_2$$

$$M_1 \times V_1 = 0.1 \times 10$$

$$\therefore M_1 = \frac{1}{V_1} \text{ (molarity of silver nitrate solution)} = \underline{\quad}$$

We know, 1000 ml of 1 M $AgNO_3$ = 35.5 g chloride

$$V_2 \text{ (from step 2) of } AgNO_3 = \frac{V_2 \times M_1 \times 35.5}{1000} \text{ g chloride per 100 ml water sample.}$$

$$= V_2 \times M_1 \times 35.5 \text{ mg chloride per 100 ml water sample}$$

$$= V_2 \times M_1 \times 35.5 \times 10 \text{ mg chloride per 100 ml water sample}$$

Result The given water sample has _____ ppm of chloride ions.

7. Colligative Properties using Freezing Point Depression

Aim To determine freezing point depression of a given solvent by Beckmann method.

Principle Usually, in a solution, solvent properties like vapour pressure, boiling point, and freezing point, are modified by added solute. Colligative properties refer to the modifications in solvent properties that depend on the relative number of solute and solvent molecules, but not on the chemical nature of the added solute. Freezing point of a solvent is the temperature at which vapour pressure of given solvent becomes equal to that of the solid in which solvent freezes under 1 Atm pressure. If a non-electrolyte is dissolved in a given solvent, vapour pressure of solvent is lowered and consequently the solution (solute and solvent) freezes at a temperatre below freezing point of pure solvent.

Requirements   Beckmann freezing point apparatus, urea, deionized water

Procedure

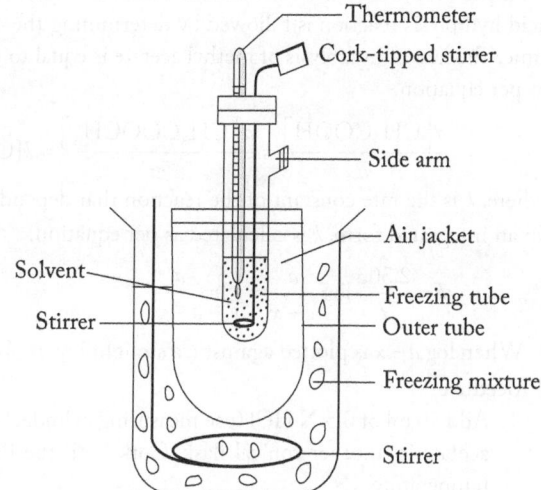

1. Beckmann freezing point apparatus is shown in Fig. A.5. The apparatus consists of a freezing tube with a side-arm. The provision of side-arm is for introducing solute in the solvent kept in the freezing tube. Another tube encloses the freezing tube and provides an air-jacket that allows uniform rate of cooling. A jar containing freezing mixture (ice+ NaCl) is used to contain the freezing tube. Two stirrers are provided; one in the freezing tube and another in the jar containing freezing mixture.

2. 15 g of deionized water is placed in a freezing tube which is equipped with a stirrer and a Beckmann thermometer (can record temperature changes up to 0.001 °C precisely).

3. The freezing tube with water (solvent) is placed in another tube keeping air space so as to prevent freezing of the solvent at the walls of the tube. The entire assembly is further placed in a container consisting of a freezing mixture.

Fig. A.5 Beckmann freezing point apparatus

4. With the help of a stirrer the freezing mixture is agitated and temperature of the solvent is allowed to fall to about 0.5 °C below the freezing point of solvent. This allows supercooling of the solvent.

5. Crystallization of solvent begins on supercooling and stirring is continued. During this time, temperature rises to the freezing point and remains constant which is denoted as T_0 i.e, freezing point of solvent.

6. Detach the freezing tube and add 0.2 g urea to the solvent and freezing point of solution (T) is determined in the same manner as discussed above.

Calculations   Report T_0 and T and calculate depression in freezing point (ΔT_f) of solvent as per the equation: $\Delta T_f = T_0 - T$

Molal depression constant (K_f) is defined as the depression in freezing point of 1 kg solvent when one mole of known solute is added to it.

$$K_f = \frac{RT_0^2}{1000 l_f}$$

Here, l_f is heat of fusion per gram of solvent and the equation can be rewritten as,

$\Delta T_f = K_f \times m$; where, m is molality of the solution.

For water, K_f = 1.86 K kg/mol, thus the freezing point depression and molecular weight of solute (urea) can be calculated using the expression,

$$M_2 = \frac{1000 \, K_f W_{solute}}{W_{solvent} \, \Delta T_f}$$

where, M_2 is molecular weight of the solute.

Result ΔT_f = _____ and molecular weight of urea = _____.

8. DETERMINATION OF RATE CONSTANT OF ACID-CATALYSED HYDROLYSIS OF METHYL ACETATE

Aim To determine the rate constant of acid-catalysed hydrolysis of methyl acetate.

Principle The hydrolysis of methyl acetate catalysed by acid is a convenient reaction to perform kinetic investigation at room temperature. It follows first order kinetics shown as follows:

On hydrolysis, methyl acetate gives, $CH_3COOCH_3 + H_2O \rightarrow CH_3COOH + CH_3OH$

As per the reaction, for each mole of ester hydrolysed, one mole of acetic acid is produced. Hence, the progress of acid hydrolysis reaction is followed by determining the change in concentration of methyl acetate as a function of time. The rate of hydrolysis of methyl acetate is equal to the rate of formation of acetic acid which can be expressed as per equation.

$$\frac{d[CH_3COOH]}{dt} = \frac{d[CH_3COOCH_3]}{dt} = k[CH_3COOCH_3]$$

where, k is the rate constant of the reaction that depends on temperature and solvent employed during hydrolysis. In an integrated form, k is calculated as per equation,

$$k = \frac{2.303}{t}\log\frac{a}{a-x} = \frac{1}{t}\ln\frac{a}{a-x}$$

When $\log a - x$ is plotted against t, a straight line is obtained and the rate constant 'k' is calculated from the slope.

Procedure

1. Add 50 ml of 0.5 N HCl (use measuring cylinder) in a clean 100 ml conical flask. Pipette out 10 ml of methyl acetate in another conical flask. Cork both the flasks and place them in a thermostat maintained at room temperature.
2. Arrange 5 conical flasks (50 ml capacity) and add 25 ml ice cold water in each of them.
3. After maintaining the acid and ester at room temperature, measure 2 ml methyl acetate and dispense it into the conical flask containing 50 ml HCl solution.
4. Once added, immediately start the stopwatch and shake the mixture at once. Immediately pipette 2 ml of reaction mixture and transfer quantitatively in 25 ml cold water to arrest the reaction.
5. Titrate it quickly against 0.05 N NaOH from the burette using phenolphthalein indicator till it turns pale pink.
6. In the same way, perform four more trials at successive intervals of 10, 20, 30, 40, 50 and 60 minutes and record the volume of alkali consumed in each titration.
7. To determine infinite time when the hydrolysis is complete, the following procedure is done:

Pipette 10 ml reaction mixture in a conical flask. Cork the flask and place in a water bath at 50° C for 1 hour. After the reaction is complete, cool the contents of the flask and pipette out 2 ml of the reaction mixture. Titrate the reaction mixture against 0.05 N NaOH using phenolphthalein indicator and note it as V_∞, which represents the concentration of total acid after hydrolysis is complete.

Observations Room temperature _____ °C. V_∞ = _____ ml

Time, t (minutes)	Titre value (ml)	$(a-x)$ ml	$\log (a-x)$	k
0	$V_0 =$			
10	$V_{10} =$	$V_\infty - V_{10}$		
20	$V_{20} =$	$V_\infty - V_{20}$		
30	$V_{30} =$	$V_\infty - V_{30}$		
40	$V_{40} =$	$V_\infty - V_{40}$		
50	$V_{50} =$	$V_\infty - V_{50}$		
60	$V_{60} =$	$V_\infty - V_{60}$		

Calculations If V_0 is proportional to the amount of HCl present in 2 ml reaction mixture at zero time when no acetic acid is formed, V_∞ is proportional to amount of acid present in 2 ml reaction mixture when reaction is complete, i.e., initial HCl + acetic acid formed.

Thus, $(V_\infty - V_0)$ is the amount of acetic acid formed when hydrolysis is complete and is proportional to the amount of methyl acetate hydrolysed in 2 ml reaction mixture. Let it be equal to 'a'. V_t is the amount of acid in 2 ml reaction mixture at time, t.

$(V_t - V_0)$ is the amount of acetic acid formed up to time t or the amount of methyl acetate hydrolysed up to time t and let us consider it as 'x'. Hence, $(a - x) \propto (V_\infty - V_0) - (V_t - V_0)$ or $(V_\infty - V_t)$

The first order reaction equation can be written as,

$$k = \frac{2.303}{t} \log \frac{(V_\infty - V_0)}{(V_\infty - V_t)}$$

Plot the graphs of time versus $\log (V_\infty - V_t)$ and time versus $\log \dfrac{(V_\infty - V_0)}{(V_\infty - V_t)}$ which should be straight lines and

obtain k values from the slope. Intercepts will be obtained from the time axes.

Result The rate constant for methyl acetate hydrolysis is _____.

9. Determination of Cell Constants and Conductance of Solutions

Aim To estimate the amount of HCl (strong acid) and acetic acid (weak acid) in a given solution by conductometry.

Principle Prior to titration, initial conductance is generally high due to complete dissociation of strong acid, as H^+ ions have high conductance value. The addition of strong base (NaOH) in the acid solution gives the following reaction,

$$H^+ + Cl^- + Na^+ + OH^- \rightarrow Na^+ + Cl^- + H_2O$$

As NaOH is added, H^+ ions combine with hydroxyl ions forming undissociated water molecules and H^+ ions get replaced by Na^+ ions that possess much lower conductance value. Hence, the conductance of the solution goes on decreasing and reaches a minimum at equivalence point. After the equivalence point, NaOH added remains unreacted, but ionizes completely producing hydroxyl ions, hence the conductance of the solution increases again. A 'V' shaped curve is obtained and the point of intersection of two lines in the graph represents end point of the titration.

Requirements Conductometer and conductivity cell, burette, beaker, pipette, conductivity water, 0.1 N NaOH solution, glass rod, wash bottle.

Procedure

1. Dilute the given acid solution with conductivity water to 100 ml in a volumetric flask.
2. Rinse and fill the burette with 0.1N NaOH solution.
3. Pipette out 10 ml of dilute acid solution into a clean 100 ml beaker.
4. Place conductivity cell in the beaker containing acid solution and add sufficient quantity of conductivity water so as to immerse the electrodes completely.
5. Stir the solution well using a glass rod and keep it immersed in acid solution until the titration is complete.
6. Connect the cell to a conductometer and measure initial conductance of the solution by selecting a proper range.
7. Add a small volume of 0.1 N NaOH solution at a time from the burette. Stir the solution well and note the conductance value.
8. Continue adding NaOH until the conductance values start increasing. Perform five trials so as to get concordant readings.
9. Plot a graph of conductance against volume of NaOH added as shown in Fig._ which is a 'V' shaped curve.

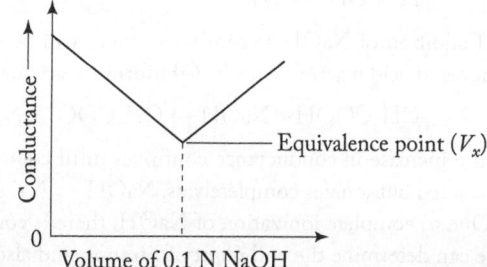

Fig. A.6 Conductometric titration of NaOH vs HCl

10. Volume corresponding to the intersection of two limes (V_x) gives end point of the titration. Note the volume as V_x. Calculate normality and strength of acid solution as given in the calculations.

Observations

  a. Temperature = _____ °C (or _____K)

  b. Normality of NaOH solution = 0.1N

  c. Volume of diluted acid pipette out = 10ml

  d. Equivalent weight of acid (HCl) = 36.5

Observation table

S. No.	Volume of 0.1N NaOH added (in ml)	Conductance (1/R) mhos (S)
1.		
2.		
3.		

Calculations

Volume of NaOH required for end point (V_x) = _____ ml (from graph)

$$\text{Normality of HCl} = \frac{N_{\text{NaOH}} \times V_x}{V_{\text{HCl}}} = \frac{0.1 \times V_x}{10}$$

10 ml of diluted solution requires V_x ml of 0.1 N NaOH solution.

Thus, 100 ml of diluted solution will require $10 \times V_x$ ml of 0.1 N NaOH solution.

Stoichiometrically, 1000 ml of 1N NaOH = 36.5 g HCl

$$\therefore 10\,V_x \text{ of } 0.1\,\text{N NaOH} = \frac{36.5 \times 10\,V_x \times 0.1}{1000} = \underline{\quad} \text{ g HCl}$$

Result

  a. Volume of NaOH required for the end point (V_x) = _____ ml

  b. Normality of HCl = _____ N

  c. Amount of HCl present in the solution = _____ g

10. ESTIMATION OF ACETIC ACID BY CONDUCTOMETRIC TITRATION

Aim To estimate amount of acetic acid using conductometry.

Principle In this experiment, conductometric titration is carried out between acetic acid and sodium hydroxide. Initially the conductance will be due to partial ionization of acetic acid as per reaction.

$$CH_3COOH \rightleftharpoons CH_3COO^- + H^+$$

When NaOH is added, the initial conductance of the solution decreases due to free H^+ ions formed during ionization which combine with hydroxyl ions to form water.

$$H^+ + OH^- \rightarrow H_2O$$

If addition of NaOH is continued, there will be a gradual increase in the conductance of the solution due to unionized acid reacting with NaOH forming acetate and sodium ions that have considerable conductance.

$$CH_3COOH + NaOH \rightarrow CH_3COO^- + Na^+ + H_2O$$

The increase in conductance continues until equivalence point is reached and after that excess NaOH remains unreacted but ionizes completely as, $NaOH \rightarrow Na^+ + OH^-$.

Due to complete ionization of NaOH, there is continued rise in conductance and on intersecting the two lines, one can determine the end point of titration and also the concentration of acid present in the given solution.

Requirements Conductometer and conductivity cell, burette, beaker, pipette, conductivity water, 0.1N NaOH solution, glass rod, wash bottle.

Procedure

1. Fill the burette with 0.1 N NaOH. Dilute the given acid solution (acetic acid) with conductivity water in a 100 ml volumetric flask.

2. Pipette out 10 ml of diluted acid solution in a beaker and add sufficient quantity of conductivity water so that electrodes can be immersed completely.

3. Place a glass rod or magnetic stirrer for agitating the solution during titration. Record the initial conductance before titration.

4. Start the titration by adding small quantities of NaOH from the burette and keep stirring with the glass rod. Note the conductance and continue the titration so as to collect around 5 – 6 conductance readings.

5. Plot a graph of conductance against volume of strong base added and determine the end point of titration. Further, calculate the strength of acid solution as provided here in the calculations and report the results.

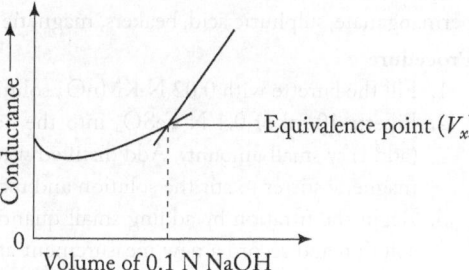

Fig. A.7 Conductometric titration of NaOH vs acetic acid

Observation table

S. No.	Volume of 0.1N NaOH added (in ml)	Conductance (1/R) mhos (S)
1.		
2.		
3.		

Calculations

Volume of NaOH required for end point (V_x) = _____ ml (from graph)

$$\text{Normality of HCl} = \frac{N_{NaOH} \times V_x}{V_{CH_3COOH}} = \frac{0.1 \times V_x}{10}$$

10 ml of diluted solution requires V_x ml of 0.1 N NaOH.

Thus, 100 ml of diluted solution will require $10 \times V_x$ ml of 0.1 N NaOH.

Stoichiometrically, 1000 ml of 1N NaOH = 60 g CH_3COOH

$$\therefore 10\, V_x \text{ of } 0.1\, \text{N NaOH} = \frac{60 \times 10\, V_x \times 0.1}{1000} = \underline{\quad\quad} \text{ g } CH_3COOH$$

Result

(a) Volume of NaOH required for the end point (V_x) = _____ ml.

(b) Normality of CH_3COOH = _____ N.

(c) Amount of CH_3COOH present in the solution = _____ g.

11. POTENTIOMETRY–DETERMINATION OF REDOX POTENTIALS AND EMF

Aim To estimate Fe^{2+} by potentiometric titration using $KmnO_4$

Principle Iron(II) is determined potentiometrically through titration with potassium permanganate and such titrations are called manganometric titration. The titration of ferrous ions and potassium permanganate is a redox titration that proceeds without using any additional indicator as permanganate itself acts as self-indicator. The reaction studied in this case is,

$$MnO_4^- + 5Fe^{2+} + 8H^+ \rightleftharpoons Mn^{2+} + 5Fe^{3+} + 4H_2O$$

Here, MnO_4^- ions reduce to Mn^{2+} ions (colour change from violet to colourless) and a measurable change in the voltage is observed from which equivalence point is determined using a potentiometer.

Requirements Potentiometer, redox electrode (or glass electrode), burette, pipette, ferrous sulphate, potassium permanganate, sulphuric acid, beakers, magnetic stirrer, glass rods,

Procedure
1. Fill the burette with 0.02 N KMnO$_4$ solution. Prepare 0.1 N ferrous sulphate solution using distilled water.
2. Pipette 10 ml of 0.1 N FeSO$_4$ into the beaker and acidify with a known volume of diluted sulphuric acid (add very small amount). Add distilled water to immerse the redox electrode in to the solution. Turn on the magnetic stirrer to stir the solution and record the initial potential.
3. Begin the titration by adding small quantities of potassium permanganate solution at a time to the FeSO$_4$ solution and record a new measurement after each addition using a potentiometer. Keep stirring during the entire titration.
4. Continue adding until there is a steep increase in the measurement curve followed by a plateau. Plot the graph to obtain the equivalence point.

Observations At the beginning of the experiment, a clear FeSO$_4$ solution is present in the beaker. As potassium permanganate solution is added, a gradual discolouration of the liquid is observed initially that disappear again instantaneously. As titration proceeds, the solution in the beaker acquires a reddish-violet colour. After reaching the end point, solution will have the strong colour of potassium permanganate.

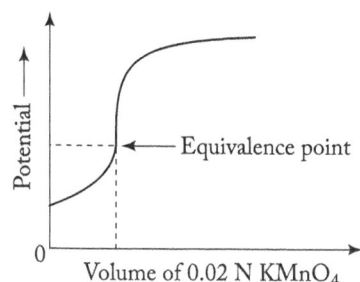

Fig. A.8 Measurement curve for manganometric titration of iron(II) ions

Calculations Determine the end point of the titration using the graph as follows:

From the curve, determine the concentration of ferrous ions as,

$$C\left(FeSO_4\right) = \frac{5.\,C\left(KMnO_4\right) \times V\left(KMnO_4\right)}{V\left(FeSO_4\right)}$$

Result The end point of the titration is _____ ml and the concentration of ferrous ions in solution is _____ mol/l.

12. SYNTHESIS OF A DRUG

Aim To prepare aspirin by esterification.

Reaction On esterification of salicylic acid, acetyl derivative is obtained as per:

| Salicylic acid | Acetic anhydride | conc. H$_2$SO$_4$ / 50°C – 60°C, 20 min | Acetyl salicylic acid (Aspirin) | Acetic acid |

Requirements Salicylic acid, acetic anhydride, concentrated sulphuric acid, beakers, glass rod, funnels, filter papers, hot-plate, watch glass, ice-cold water, test tubes, absolute ethanol, electronic balance, butter paper, spatula.

Procedure
1. In a 100 ml beaker, add salicylic acid (2 g) and acetic anhydride (3 ml).
2. Carefully, add 1 drop of concentrated sulphuric acid and stir the mixture.
3. Heat the reaction mixture for 15–20 minutes while stirring continually with a glass rod.

4. Add 25–30 ml of water, swirl the mixture and carry out a filtration. Weigh the mass of crude product (aspirin) obtained.
5. The crude aspirin is purified by recrystallization. It is dissolved in minimum quantity of hot ethanol. (Caution: Use a hot water bath to obtain ethanol).
6. The resulting solution is poured into 20 ml of ice-cold water. After recrystallization, the solid is filtered and dried. Weigh the mass of pure product (aspirin) obtained.
7. Determine the melting point of the pure product and report your results.

Result   (a)   Weight of the crude product = _____ g.
          (b)   Melting point of the product = _____ °C.

13. Acid Value of Oil

Aim   To estimate acid value of given lubricating oil sample.

Principle   Acid value is the number of milligrams of potassium hydroxide required to neutralize the free acids in per gram of given oil sample. Any increase in acid value is an indication of oxidation of the oil that may lead to gum and sludge formation and eventual corrosion of machine parts where such lubricating oils are employed. In this experiment, a known weight of oil sample is dissolved in a suitable solvent and titrated with an alcoholic solution.

Requirements   Conical flasks, water bath, burettes, pipettes, oil sample, potassium hydroxide, hydrochloric acid, ethanol, phenolphthalein indicator.

Procedure
1. Pipette 25 ml of 0.03 N alcoholic KOH into a conical flask and add 2 drops of phenolphthalein indicator. Titrate the solution against 0.03 N HCl from the burette till the pink colour disappears. Report the reading as 'x' ml.
2. Weigh an empty conical flask on an electronic balance and add 5 ml of oil and weigh it again. The difference in weights will provide the actual weight of oil.
3. Add 50 ml of neutralized alcohol to the oil and heat the mixture over a hot water bath for about 20 minutes.
4. Cool the reaction mixture to room temperature and add 3 drops of phenolphthalein indicator to the mixture.
5. Titrate the mixture against standard 0.03 N KOH solution till a pale pink colour appears at the end point. Perform two more trials to obtain concordant readings and report it as 'y' ml.

Calculations

Normality of KOH solution; $N_{KOH} = \dfrac{0.03 \times x}{25}$

Volume of 0.035 N KOH required in estimation = y ml
Volume of 0.035 KOH used in the test $(x - y)$ml = V ml
Weight of oil taken (W) = _____ g.
On substituting above values we get,

$$\text{Acid value} = \frac{V \times 56 \times N_{KOH}}{W(\text{in g})}$$

Result   Acid value of oil sample is _____.

(Note: The acid value should be less than 0.1 so as to be utilized for lubrication purposes.)

14. Lattice structures and packing of Spheres

Aim   To demonstrate lattice structure and close packing of spheres.

Principle   It is already studied that metals exhibit crystalline structure though they do not possess well-developed crystal faces. Most metals we use are made of several crystals having different orientations separated by irregular boundaries. The properties of a metal are directly influenced by its crystal structure and orientation. Solidification, shaping and treatment steps for transforming metals to various materials can be understood only by demonstrating crystal lattice structures and their orientations which form the fundamental premise of metallurgical science and engineering. The most common crystal structures of metals are body centred cubic (BCC), face centrred cubic (FCC), and hexagonal close packed (HCP). In the case of the first two each point represents one metal atom, whereas in the case of hcp structure each point represents a pair of points.

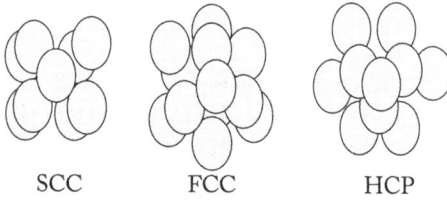

SCC   FCC   HCP

Fig. A.9 BCC, FCC, and HCP crystal structures in metals

Procedure

1. Draw the lattice structures of BCC, FCC, and HCP crystals.
2. One can demonstrate the above crystal structures and close packing using a plastic box and adding various spheres in them.
3. Spheres will first form a single layer and later keep on packing as closely as possible arranging them such that their centres are at the corners of an equilateral triangle and each sphere is surrounded by a hexagon of six other spheres (see Fig. A. 10).
4. Hence, second layer can be demonstrated by stacking spheres in the depressions on the top of the first layer.

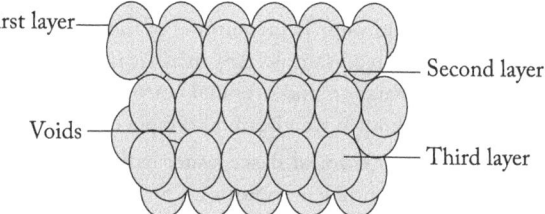

Fig. A.10 Close packing of spheres

5. When adding the third layer, spheres may now transpose on the first layer making first and third layer identical. If there are depressions even after adding the second layer, add the third layer of spheres to cover the depressions.

During close packing, two types of voids are created; tetrahedral voids or holes that lie immediately above each sphere of the lower layer and immediately below each sphere of the upper layer. Octahedral voids are unfilled hollows of the first layer.

15. CHEMICAL OSCILLATIONS OF IODINE CLOCK REACTION

Aim To study chemical oscillating reactions between H_2O_2 and KI.

Principle A hydrogen peroxide-potassium iodide 'clock' reaction involves hydrogen peroxide mixed with potassium iodide, starch, and sodium thiosulphate. After a few seconds, the colourless mixture turns dark blue. This is one of a number of reactions loosely called the *iodine clock*. It is generally used as an introduction to experiments on reaction kinetics.

The kinetics of the reaction between H_2O_2 and HI is investigated as follows:
$$H_2O_2 + 2HI \rightarrow I_2 + 2H_2O$$
The overall reaction consists of the following two steps:
(i) $H_2O_2 + I^- \rightleftharpoons H_2O + IO^-$ (slow)
(ii) $IO^- + 2H^+ + I^- \rightleftharpoons H_2O + I_2$ (rapid)

The first step is the rate determining step. In the presence of a constant excess of I^- the reaction follows pseudo-first order with respect to H_2O_2. This condition is achieved by continuously adding trace amounts of $Na_2S_2O_3$ to convert iodine to iodide ions. The progress of the reaction can be denoted by recording the time of appearance of iodine, indicated by the appearance of blue colour after the addition of a small known volume of thio solution. The amount of H_2O_2 reacted during the time lapse for the appearance of blue colour corresponds to the $Na_2S_2O_3$ added. Since the reaction rate depends on the concentration of H_2O_2 present in the reaction mixture, the time for the reappearance of blue colour will increase with the progress of the reaction.

Procedure

1. Take 20 ml H_2SO_4 (1:2 acid : water mixture) in a conical flask and dilute it to 100 ml with distilled water.
2. To the above mixture add 2 g of solid KI and 10 ml of H_2O_2 solution. Mix well and warm to 30 °C. Titrate the liberated iodine with 0.1 M $Na_2S_2O_3$ solution using starch solution as an indicator.

3. Place 250 ml of KI (4g/l) solution in a 500 ml conical flask and add 15 ml of dilute sulphuric acid and 10 ml of starch solution.
4. Suspend the flask in a thermostatic bath at 25 °C. Add 10 ml of H_2O_2 (2 Vol) from a stock solution previously maintained at 25 °C.
5. Start the stop watch when the pipette is half discharged into the flask containing potassium iodide, etc.
6. Shake vigorously and slowly start adding 1 ml of 0.1M $Na_2S_2O_3$ solution from a burette.
7. Record the time at which blue colour appears. Again add immediately 1 ml of $Na_2S_2O_3$ and note the time of appearance of blue colour after each addition until the time of appearance of blue colour becomes 5–6 times the initial time. Keep on shaking the reaction mixture throughout the addition of $Na_2S_2O_3$ solution from the burette.
8. Activation energy of the reaction may also be determined by repeating the reaction at different temperatures, e.g. 10, 15, 30, and 35 °C keeping the other reaction parameters constant.

Calculations   Volume of $Na_2S_2O_3$ solution added will correspond to the amount of H_2O_2 reacted with I^- during the time of appearance of blue colour of starch. As one mole of iodine is liberated for every mole of hydrogen peroxide, the amount of peroxide remaining and amounts decomposed (x) at any time can be determined.

V ml of $Na_2S_2O_3$ solution with 10 ml H_2O_2 is proportional to the initial concentration (a) of hydrogen peroxide. V_t is the volume of $Na_2S_2O_3$ at time t corresponding to the appearance of blue colour. Hence, x is proportional to V_t and $(a - x)$ will be $(V - V_t)$. On applying first-order rate equation,

$$k_{obs} = \frac{2.303}{t} \log \frac{V}{V - V_t}$$

On plotting $\log (V - V_t)$ against time t is a straight line depicting first order and the slope gives values for $\frac{k_{obs}}{2.303}$. A plot of $\log k$ against $1/T$ gives a straight line with slope of $-\frac{E}{2.303R}$ and hence E, i.e., activation energy is determined.

Further, for determining the rate order, $k_{obs} = k [HI]n$, where, n is order of the reaction w.r.t [HI].
On taking log on both the sides, we get, $\log k_{obs} = \log k + n \log [HI]$

$$\therefore n = \frac{\log(k_{obs})_2 - \log(k_{obs})_1}{\log[HI]_2 - \log[HI]_1}$$

Result   The order of reaction and activation energy of the reaction are _____ and _____ respectively.

16. DETERMINATION OF PARTITION COEFFICIENT OF A SUBSTANCE BETWEEN TWO IMMISCIBLE LIQUIDS

Aim   To determine the distribution coefficient of iodine between CCl_4 and water at room temperature.

Principle   Nernst distribution law states that, 'At constant temperature, a given solute will distribute itself between two immiscible solvents in a particular ratio,' expressed as:

$$\frac{C_1}{C_2} = K_D$$

where C_1 and C_2 are molar concentrations of a given solute in two immiscible solvents at a given temperature. K_D denotes the distribution coefficient or partition coefficient.

In this experiment, iodine distributes itself between aqueous and organic CCl_4 layers in the presence of potassium iodide in the aqueous layer. Equilibrium between iodine and potassium iodide is established in the aqueous layer shown as:

$$KI + I_2 \rightleftharpoons KI_3$$

The equilibrium constant can be expressed as:

$$\frac{[KI_3]}{[KI][I_2]} = K_{eq}$$

Procedure

1. Take three separating funnels and prepare the following mixtures:

Separating funnel No.	Contents
1.	20 ml of iodine in CCl_4 + 40 ml distilled water
2.	20 ml of iodine in CCl_4 + 40 ml known KI
3.	20 ml of iodine in CCl_4 + 40 ml unknown KI

2. Stopper the funnels securely and shake the contents for 15 – 20 minutes vigorously. Keep on removing the glass lid one in a time to release the pressure build-up in the funnel.
3. After shaking, allow the contents to rest for 20 minutes at equilibrium in a water tray at ambient conditions.
4. After equilibrium is attained, pipette 2 ml of organic CCl_4 layer from the funnel into a conical flask. Add 10 ml of 10 % KI and 25 ml distilled water.
5. The solution is titrated against standard 0.1 N sodium thiosulphate solution using starch indicator (standardize thiosulphate solution with potassium dichromate). The disappearance of blue colour denotes the end point.
6. Pipette out 10 ml of aqueous layer from each funnel and titrate it with 0.1N sodium thiosulphate solution using starch indicator.
7. Repeat the procedure for funnel nos. 2 and 3. Perform the titration to obtain consistent burette readings.

Observations

Separating funnel nos.	Volume of CCl_4 layer (ml)	Burette readings		Volume of $Na_2S_2O_3$ (ml)	Concordant value (ml)
		Initial (ml)	Final (ml)		
	2.0				
	2.0				
	2.0				
	2.0				
	2.0				
	2.0				

Separating funnel nos.	Volume of aqueous layer (ml)	Burette readings		Volume of $Na_2S_2O_3$ (ml)	Concordant value (ml)
		Initial (ml)	Final (ml)		
	10.0				
	10.0				
	10.0				
	10.0				
	10.0				
	10.0				

Separating funnel Nos.	Solvent layers	Volume of $Na_2S_2O_3$ (ml)	Strength of iodine	$\dfrac{C1}{C2} = K_D$
	CCl_4			
	Aqueous			
	CCl_4			
	Aqueous			
	CCl_4			
	Aqueous			

Result The distribution coefficient of iodine between CCl_4 and water is _____.

17. Determination of Partition Coefficient of Acetic Acid between N-Butyl Alcohol and Water

Aim   To determine partition coefficient of acetic acid between butanol and water.

Principle   Nernst distribution law states that, 'At a constant temperature, a given solute will distribute itself between two immiscible solvents in a particular ratio,' expressed as, $C_1/C_2 = K_D$, where C_1 and C_2 are molar concentrations of a given solute in two immiscible solvents at a given temperature. K_D denotes the distribution coefficient or partition coefficient.

Chemicals/Reagents   n-Butanol, acetic acid, deionized water, NaOH solution.

Apparatus   Burette, pipette, 100 ml conical flask, 250 ml glass stoppered bottles, wash bottle.

Procedure

1. A clean 250 ml stoppered bottle is taken and rinsed with deionized water. Now, fill it with 20 ml n-butanol, 25 ml distilled water, and 5 ml of acetic acid. Stopper the bottle and place it on a mechanical shaker for about an hour.
2. After shaking, keep the contents in the bottle aside for about 5–10 minutes. The two layers will separate out distinctly. Record the temperature of the mixture using a thermometer.
3. Pipette out 5ml of aqueous layer (lower layer- Be careful while siphoning the layer in the pipette!) in a 100 ml conical flask and add 2–3 drops of phenolphthalein indicator. Titrate the contents against standard NaOH solution until color changes to pink.
4. In the same way, upper organic (alcohol) layer is estimated. Pipette out 5ml of organic layer is pipette out in a 100 ml conical flask and add 2–3 drops of phenolphthalein indicator. Titrate the contents against standard NaOH solution until colour changes to pink.
5. Repeat the titrations so as to get concordant values.

Observation table   Room temperature = _____ °C.

Layers	Volume taken (ml)	Burette reading (ml)			Volume of NaOH consumed (ml)	$\dfrac{C1}{C2} = K_D$
		Initial	Final	Difference		
Aqueous layer						
Organic layer						

Calculations   For aqueous layer (C_1)

Volume of aqueous layer $\times C_1$ = Volume of NaOH required $\times$ Strength of NaOH

Hence, $C_1 = \dfrac{\text{Volume of NaOH} \times \text{Normality of NaOH}}{5\ ml(\text{volume pipetted})}$

For organic layer (C_2)

Volume of organic layer $\times C_2$ = Volume of NaOH required $\times$ Strength of NaOH

Hence, $C_2 = \dfrac{\text{Volume of NaOH} \times \text{Normality of NaOH}}{5\ ml\ (\text{volume pipetted})}$

Further, $\dfrac{C_1}{C_2} = K_D$

Result   Partition coefficient (K_D) of acetic acid between n-butyl alcohol and water is _____.

18. Adsorption of Acetic Acid on Charcoal (Verification of Freundlich Adsorption Isotherm)

Aim   To study adsorption of acetic acid on charcoal and verify Freundlich adsorption isotherm.

Principle   Adsorption can be defined as 'mass transfer process that involves accumulation of substance at the interface of two phases, such as, liquid–liquid, liquid–solid, gas–liquid, or gas–solid.' The substance accumulated or collected on to the surface is called adsorbate, whereas the substance that adsorbs or collects is called adsorbent. Usually, adsorption processes are studied through graphs called adsorption isotherms. Freundlich adsorption isotherm is an empirical relation between the amount of gas adsorbed per unit mass of solid adsorbent and pressure at a given temperature. It can be expressed as,

$$\frac{x}{m} = k.p^{1/n} \text{ where, } n > 1 \tag{A.1}$$

In the above expression, x is the mass of adsorbed gas on m gram of adsorbent, p is the pressure of gas in the space above the adsorbent, and k and n are constants that depend on the type of adsorbent and gas at a particular temperature. For the purpose of calculations, taking logarithms on both sides of the Eq 1, gives,

$$\log \frac{x}{m} = \log k + \frac{1}{n} \log p \tag{A.2}$$

Requirements   10 glass-stoppered 125 ml capacity conical flasks, glacial acetic acid, dry charcoal powder (20 g), phenolphthalein indicator, 50 ml capacity burette, filter paper, NaOH.

Procedure

1. Accurately weigh 1.5 g of charcoal on an electronic balance and add it to each of the 10 glass stoppered flasks.
2. Prepare a series of acetic acid solutions of varying dilutions as given below. Add 100 ml of each solution to each charcoal sample.

Sample no.	0.4 M acetic acid (in ml)	Aliquot for analysis (in ml)
1.	100	10
2.	75	10
3.	50	10
4.	25	25
5.	10	25
6.	5	50

3. Swirl the contents of the flasks vigorously or place it on an agitator for 15–20 minutes.
4. Let the flasks rest overnight. Filter the solutions and titrate a known volume of the solution against 0.1 M NaOH solution using phenolphthalein indicator.

Report   Calculate the concentration of each solution (mol/l); the initial concentration. Also, calculate the concentration of each filtrate using titration data (mol/l); the final concentration. The difference between initial and final concentration is the amount adsorbed by charcoal.

Convert the molar quantity to x/m by multiplying it with the volume of initial solution and the molecular weight of the adsorbed material as well as dividing by the mass of adsorbent.

Plot values of y against x/m and draw the adsorption isotherm. Plot $\log y$ against $\log x/m$ and ascertain the values of k and n from the curve obtained.

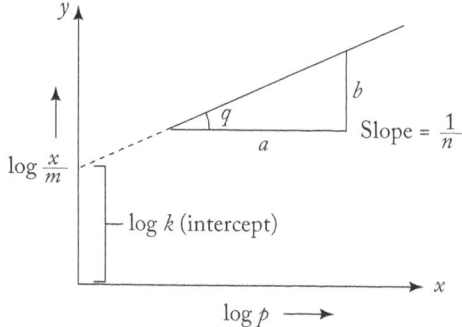

Fig. A.11 Freundlich isotherm plot

Index

About the Authors

Payal B. Joshi is Assistant Professor at Department of Basic Science & Humanities, SVKM's NMIMS, Mukesh Patel School of Technology Management and Engineering, Mumbai. She obtained M.Sc. degree (Organic Chemistry) from University of Mumbai in 2005. While pursuing M.Sc. degree (Organic Chemistry) at University of Mumbai, she was conferred with National Scholarship by the Ministry of Human Resource Development (HRD), Government of India. She obtained her Ph.D. (Chemical Science) from Sunandan Divatia School of Science, SVKM's NMIMS (Mumbai) in 2011.

Dr Joshi has more than a decade's experience in teaching and research. She has taught undergraduate courses ranging from Organic Chemistry, Biotechnology, Environmental Sciences, Applied and Engineering Chemistry. Her research areas of interest include green chemistry, catalysis, and environmental sciences. She serves as a life member for IPA (Indian Pharmaceutical Association) and CASSS (California Separation Science Society).

Shashank Deep is Professor, Department of Chemistry, Indian Institute of Technology Delhi. After obtaining his Ph.D. from Indian Institute of Technology Delhi, he took up a post-doctoral assignment at Prof. Hinck Laboratory at Department of Biochemistry, University of Texas Health Science Center at San Antonio, Texas, USA. This was followed by a second postdoctoral work with Prof. Erik Zuiderweg at Department of Biophysics, University of Michigan, Ann Arbor, MI, USA.

Dr Deep's research involves application of thermodynamics, chemical kinetics, and spectroscopy in understanding the reasons responsible for various diseases and formulating strategies to counter them. He has 45 publications in reputed journals. Dr. Deep is a member of American Chemical Society, Protein Society, and Indian Biophysical Society. Prof. Deep is involved in the development of online courses at various levels. He has delivered lectures at numerous teacher training workshops and conferences.

RELATED TITLES

9780199498741

G.K. Suraishkumar Professor, Dept. of Biotechnology, IIT Madras, Chennai.

Biology for Engineers is an interdisciplinary textbook designed for students of various engineering streams and covers the AICTE syllabus.

- Adopts a narrative presentation to ensure ease of learning for students from varied engineering streams
- Provides reflection points interspersed in every chapter to kindle further interest in the students' mind
- Includes additional information at the end of each chapter to encourage deeper and wider learning

9780199455461

Subrata Sen Gupta, Former Reader in Chemistry, R.P.M. College, West Bengal

Problems and Solutions in Organic Chemistry has been written with an emphasis on learning the subject through questions and answers. It will also be of immense help to students preparing for competitive examinations such as Joint Admission Test for Masters (JAM), National Eligibility Test (CSIR-UGC NET), and Graduate Aptitude Test in Engineering (GATE), where advanced knowledge in organic chemistry is essential.

- Comprises more than 1500 solved problems for complete understanding and practice of the concepts in organic chemistry
- Includes 700 chapter-end exercise problems to aid self-evaluation
- Contains problems relating to stereochemistry, spectroscopy, and pericyclic reactions

9780198814740

Peter Atkins, Julio de Paula & James Keeler

Atkins' Physical Chemistry is widely acknowledged by students and lecturers around the globe to be the textbook of choice for studying physical chemistry. This International Edition has been carefully developed to provide the information, explanation, and guidance you need to master the subject.

- Provides a comprehensive overview of physical chemistry to meet the needs of current students
- Exceptional mathematical support enables students to master the maths which underlies physical chemistry
- The development of problem solving and analytical skills is actively encouraged by worked examples, discussion questions, and exercises

9780199480173

Reema Thareja Asst. Professor, Dept. of Computer Science, Shyama Prasad Mukherji College for Women, University of Delhi.

Python Programming is designed as a textbook to fulfil the requirements of the first-level course in Python programming. It is suited for undergraduate degree students of computer science engineering, information technology as well as computer applications. This book will enable students to apply the Python programming concepts in solving real-world problems.

- Numerous programming examples along with their outputs to help students master the art of writing efficient Python programs.
- Notes and programming tips to highlight the important concepts and help readers avoid common programming errors.
- Case studies on creating calculator, calendar, hash files, compressing strings and files, tower of Hanoi, image processing, shuffling a deck of cards, and mail merge demonstrate the application of various concepts.

Other Related Titles

9780195686173 Loudon: *Organic Chemistry, 4/e*
9780199456819 Sen Gupta: *Reaction Mechanisms in Organic Chemistry, 1/e*
9780195699258 Stevens: *Polymer Chemistry, 3/e*
9780199599608 Atkins et al.: *Shriver & Atkin's Inorganic Chemistry, 5/e*
9780199658503 Li et al.: *Problems in Structural Inorganic Chemistry*
9780199228867 vanLoon & Duffy: *Environmental Chemistry, 3/e*

Visit us at *www.oup.co.in* and *www.oupinheonline.com*